Lecture Notes in Computer Science 3991

Commenced Publication in 1973
Founding and Former Series Editors:
Gerhard Goos, Juris Hartmanis, and Jan van Leeuwen

Editorial Board

Vassil N. Alexandrov
Geert Dick van Albada   Peter M.A. Sloot
Jack Dongarra (Eds.)

Computational Science – ICCS 2006

6th International Conference
Reading, UK, May 28-31, 2006
Proceedings, Part I

 Springer

Volume Editors

Vassil N. Alexandrov
University of Reading
Centre for Advanced Computing and Emerging Technologies
Reading RG6 6AY, UK
E-mail: v.n.alexandrov@rdg.ac.uk

Geert Dick van Albada
Peter M.A. Sloot
University of Amsterdam
Department of Mathematics and Computer Science
Kruislaan 403, 1098 SJ Amsterdam, The Netherlands
E-mail: {dick,sloot}@science.uva.nl

Jack Dongarra
University of Tennessee
Computer Science Department
1122 Volunteer Blvd., Knoxville, TN 37996-3450, USA
E-mail: dongarra@cs.utk.edu

Library of Congress Control Number: 2006926429

CR Subject Classification (1998): F, D, G, H, I, J, C.2-3

LNCS Sublibrary: SL 1 – Theoretical Computer Science and General Issues

ISSN   0302-9743
ISBN-10   3-540-34379-2 Springer Berlin Heidelberg New York
ISBN-13   978-3-540-34379-0 Springer Berlin Heidelberg New York

Springer is a part of Springer Science+Business Media

springer.com

© Springer-Verlag Berlin Heidelberg 2006
Printed in Germany

Typesetting: Camera-ready by author, data conversion by Scientific Publishing Services, Chennai, India
Printed on acid-free paper   SPIN: 11758501   06/3142   5 4 3 2 1 0

Preface

The Sixth International Conference on Computational Science (ICCS 2006) was held in Reading, United Kingdom, May 28-31 and continued the traditions of previous conferences in the series: ICCS 2005 in Atlanta, Georgia, USA; ICCS 2004 in Krakow, Poland; ICCS 2003 held simultaneously at two locations in, Melbourne, Australia and St. Petersburg, Russia; ICCS 2002 in Amsterdam, The Netherlands; and ICCS 2001 in San Francisco, California, USA.

Since the first conference in San Francisco, rapid developments in Computational Science as a mainstream area facilitating multi-disciplinary research essential for the advancement of science have been observed. The theme of ICCS 2006 was "Advancing Science through Computation", marking several decades of progress in Computational Science theory and practice, leading to greatly improved applications science. The conference focused on the following major themes: tackling Grand Challenges Problems; modelling and simulations of complex systems; scalable algorithms and tools and environments for Computational Science. Of particular interest were the following major recent developments in novel methods and modelling of complex systems for diverse areas of science, scalable scientific algorithms, advanced software tools, computational grids, advanced numerical methods, and novel application areas where the above novel models, algorithms and tools can be efficiently applied such as physical systems, computational and systems biology, environmental systems, finance, and others.

Keynote lectures were delivered by Mateo Valero (Director, Barcelona Supercomputing Centre) - "Tackling Grand Challenges Problems"; Chris Johnson (Distinguished Professor, University of Utah) - "Visualizing the Future"; José Moreira (IBM, Chief Architect, Commercial Scale Out) - "Achieving Breakthrough Science with the Blue Gene/L Supercomputer"; Martin Curley (INTEL, Global Director of Innovation and IT Research) - "IT Innovation: A New Era"; Vaidy Sunderam (Samuel Candler Dobbs Professor of Computer Science, Emory University, USA) - "Metacomputing Revisited: Alternative Paradigms for Distributed Resource Sharing"; and Ron Bell (AWE plc.) - "The AWE HPC Benchmark".

In addition, two special sessions were held - one by industry and one by the funding bodies. Three tutorials preceded the main technical program of the conference: "Tools for Program Analysis in Computational Science" by Dieter Kranzlmüller; "P-GRADE Portal" by P. Kascuk, T. Kiss and G. Sipos; and "Scientific Computing on Graphics Hardware" by Dominik Göddeke. We would like to thank all the keynote, the invited, and the tutorial speakers for their inspiring talks.

Apart from the plenary sessions and tutorials the conference included twelve parallel oral sessions and two poster sessions. Since the first ICCS in San

Francisco the conference has grown steadily attracting increasing numbers of researchers in the field of Computational Science. For ICCS 2006 we received over 1,400 submissions, around 300 for the main track and over 1,100 for the originally proposed workshops. Of these submissions, 98 were accepted as a full papers and 29 as posters for the main track; and 500 were accepted as full papers, short papers or posters for the 32 workshops. This selection was possible due to the tremendous work done by the Program Committee and the 720 reviewers. The author index contains over 1,000 names and over 600 participants from all the major continents. The papers cover a wide variety of topics in Computational Science, ranging from Grand Challenges problems and modelling of complex systems in various areas to advanced numerical algorithms and new scalable algorithms in diverse application areas and software environments for Computational Science. The ICCS 2006 Proceedings consist of four volumes, 3991 to 3994, where the first volume contains the papers from the main track and all the posters; the remaining three volumes contain the papers from the workshops. ICCS this year is primary published on a CD and we would like to thank Springer for their cooperation and partnership. We hope that the ICCS 2006 Proceedings will be a major intellectual resource for many computational scientists and researchers for years ahead. During the conference the best papers from the main track and workshops as well as the best posters were nominated and commended on ICCS 2006 website. A number of selected papers will also be published in special issues of relevant mainstream journals.

We would like to thank all workshop organisers and the program committee for the excellent work, which further enhanced the conference's standing and led to very high quality event with excellent papers. We would like to express our gratitude to Advanced Computing and Emerging Technologies Centre staff, postgraduates and students for their wholehearted support of ICCS 2006. We would like to thank the School of Systems Engineering, Conference Office, Finance Department and various units at the University of Reading for different aspects of the organization and for their constant support in making ICCS 2006 a success. We would like to thank the Local Organizing Committee for their persistent and enthusiastic work towards the success of ICCS 2006. We owe special thanks to our sponsors: Intel, IBM, SGI, Microsoft Research, EPSRC and Springer; and to ACET Centre and the University of Reading for their generous support. We would like to thank SIAM, IMACS, and UK e-Science programme for endorsing ICCS 2006.

ICCS 2006 was organized by the Advanced Computing and Emerging Technologies Centre, University of Reading, with support from the Section Computational Science at the Universiteit van Amsterdam and Innovative Computing Laboratory at the University of Tennessee, in cooperation with the Society for Industrial and Applied Mathematics (SIAM), the International Association for Mathematics and Computers in Simulation (IMACS), and the UK Engineering and Physical Sciences Research Council (EPSRC). We invite you to visit the ICCS 2006 website (http://www.iccs-meeting.org/iccs2006/) and ACET Centre website (http://www.acet.reading.ac.uk/) to recount the events leading up to the conference, to view the technical programme, and to recall memories of three and a half days of engagement in the interest of fostering and advancing Computational Science.

June 2006

Vassil N. Alexandrov
G. Dick van Albada
Peter M.A. Sloot
Jack J. Dongarra

Organisation

ICCS 2006 was organised by the Centre for Advanced Computing and Emerging Technologies (ACET), University of Reading, UK, in cooperation with the University of Reading (UK), the Universiteit van Amsterdam (The Netherlands), the University of Tennessee (USA), Society for Industrial and Applied Mathematics (SIAM), International Association for Mathematics and Computers in Simulation (IMACS) and Engineering and Physical Sciences Research Council (EPSRC). The conference took place on the Whiteknights Campus of the University of Reading.

Conference Chairs

Scientific Chair - Vassil N. Alexandrov (ACET, University of Reading, UK)
Workshops Chair - G. Dick van Albada (Universiteit van Amsterdam,
    The Netherlands)
ICCS Series Overall Chair - Peter M.A. Sloot (Universiteit van Amsterdam,
    The Netherlands)
ICCS Series Overall Co-Chair - Jack J. Dongarra (University of Tennessee, USA)

Local Organising Committee

Vassil N. Alexandrov
Linda Mogort-Valls
Nia Alexandrov
Ashish Thandavan
Christian Weihrauch
Simon Branford
Adrian Haffegee
David Monk
Janki Dodiya
Priscilla Ramsamy
Ronan Jamieson
Ali Al-Khalifah
David Johnson
Eve-Marie Larsen
Gareth Lewis
Ismail Bhana
S. Mehmood Hasan
Sokratis Antoniou

Sponsoring Institutions

Intel Corporation
IBM
SGI
Microsoft Research
EPSRC
Springer
ACET Centre
University of Reading

Endorsed by

SIAM
IMACS
UK e-Science Programme

Program Committee

D. Abramson - Monash University, Australia
V. Alexandrov - University of Reading, UK
D.A. Bader - Georgia Tech, USA
M. Baker - University of Portsmouth, UK
S. Belkasim - Georgia State University, USA
A. Benoit - Ecole Normale Superieure de Lyon, France
I. Bhana - University of Reading, UK
R. Blais - University of Calgary, Canada
A. Bogdanov - Institute for High Performance Computing and Information
     Systems, Russia
G. Bosilca - University of Tennessee, USA
S. Branford - University of Reading, UK
M. Bubak - Institute of Computer Science and ACC Cyfronet - AGH, Poland
R. Buyya - University of Melbourne, Australia
F. Cappello - Laboratoire de Recherche en Informatique, Paris Sud, France
T. Cortes - Universitat Politecnica de Catalunya, Spain
J.C. Cunha - New University of Lisbon, Portugal
F. Desprez - INRIA, France
T. Dhaene - University of Antwerp, Belgium
I.T. Dimov - University of Reading, UK
J. Dongarra - University of Tennessee, USA
C. Douglas - University of Kentucky, USA
G.E. Fagg, University of Tennessee, USA
M. Gerndt - Technical University of Munich, Germany

Y. Gorbachev - Institute for High Performance Computing and Information
    Systems, Russia
A. Goscinski - Deakin University, Australia
A. Haffegee - University of Reading, UK
L. Hluchy - Slovak Academy of Science, Slovakia
A. Hoekstra - Universiteit van Amsterdam, The Netherlands
A. Iglesias - University of Cantabria, Spain
R. Jamieson - University of Reading, UK
D. Johnson - University of Reading, UK
J. Kitowski - AGH University of Science and Technology, Poland
D. Kranzlmüller - Johannes Kepler University Linz, Austria
A. Lagana - Universita di Perugia, Italy
G. Lewis - University of Reading, UK
E. Luque - University Autonoma of Barcelona, Spain
M. Malawski - Institute of Computer Science AGH, Poland
M. Mascagni - Florida State University, USA
E. Moreno - Euripides Foundation of Marilia, Brazil
J. Ni The - University of Iowa, Iowa City, IA, USA
G. Norman - Russian Academy of Sciences, Russia
S. Orlando - University of Venice, Italy
B. Ó Nulláin - UUniversiteit van Amsterdam, The Netherlands
M. Paprzycki - Computer Science Institute, SWSP, Warsaw, Poland
R. Perrott - Queen's University of Belfast, UK
R. Renaut - Arizona State University, USA
A. Rendell - Australian National University, Australia
D. Rodriguez-García - University of Reading, UK
P. Roe Queensland - University of Technology, Australia
S.L. Scott - Oak Ridge National Laboratory, USA
D. Shires - U.S. Army Research Laboratory, USA
P.M.A. Sloot - Universiteit van Amsterdam, The Netherlands
G. Stuer - University of Antwerp, Belgium
R. Tadeusiewicz - AGH University of Science and Technology, Poland
A. Thandavan - University of Reading, UK
P. Tvrdik - Czech Technical University, Czech Republic
P. Uthayopas - Kasetsart University, Thailand
G.D. van Albada - Universiteit van Amsterdam, The Netherlands
J. Vigo-Aguiar - University of Salamanca, Spain
J.A. Vrugt - Los Alamos National Laboratory, USA
J. Wasniewski - Technical University of Denmark, Denmark
G. Watson - Los Alamos National Laboratory, USA
C. Weihrauch - University of Reading, UK
Y. Xue - Chinese Academy of Sciences, China
E. Zudilova-Seinstra - Universiteit van Amsterdam, The Netherlands

Reviewers

A. Adamatzky
A. Arenas
A. Belloum
A. Benoit
A. Bielecki
A. Bode
A. Cepulkauskas
A. Chkrebtii
A. Drummond
A. Erzan
A. Fedaravicius
A. Galvez
A. Gerbessiotis
A. Goscinski
A. Griewank
A. Grösslinger
A. Grzech
A. Haffegee
A. Hoekstra
A. Iglesias
A. Jakulin
A. Janicki
A. Javor
A. Karpfen
A. Kertész
A. Knuepfer
A. Koukam
A. Lagana
A. Lawniczak
A. Lewis
A. Li
A. Ligeza
A. Mamat
A. Martin del Rey
A. McGough
A. Menezes
A. Motter
A. Nasri
A. Neumann
A. Noel
A. Obuchowicz
A. Papini
A. Paventhan

A. Pieczynska
A. Rackauskas
A. Rendell
A. Sánchez
A. Sánchez-Campos
A. Sayyed-Ahmad
A. Shafarenko
A. Skowron
A. Sosnov
A. Sourin
A. Stuempel
A. Thandavan
A. Tiskin
A. Turan
A. Walther
A. Wei
A. Wibisono
A. Wong
A. Yacizi
A. Zelikovsky
A. Zhmakin
A. Zhou
A.N. Karaivanova
A.S. Rodinov
A.S. Tosun
A.V. Bogdanov
B. Ó Nualláin
B. Autin
B. Balis
B. Boghosian
B. Chopard
B. Christianson
B. Cogan
B. Dasgupta
B. Di Martino
B. Gabrys
B. Javadi
B. Kahng
B. Kovalerchuk
B. Lesyng
B. Paternoster
B. Payne
B. Saunders

B. Shan
B. Sniezynski
B. Song
B. Strug
B. Tadic
B. Xiao
B.M. Rode
B.S. Shin
C. Anthes
C. Bannert
C. Biely
C. Bischof
C. Cotta
C. Douglas
C. Faure
C. Glasner
C. Grelck
C. Herrmann
C. Imielinska
C. Lursinsap
C. Mastroianni
C. Miyaji
C. Nelson
C. Otero
C. Rodriguez Leon
C. Schaubschläger
C. Wang
C. Weihrauch
C. Woolley
C. Wu
C. Xu
C. Yang
C.-H. Huang
C.-S. Jeong
C.G.H. Diks
C.H. Goya
C.H. Kim
C.H. Wu
C.K. Chen
C.N. Lee
C.R. Kleijn
C.S. Hong
D. Abramson

H.Y. Lee
I. Bhana
I. Boada
I. Kolingerova
I. Lee
I. Mandoiu
I. Moret
I. Navas-Delgado
I. Podolak
I. Schagaev
I. Suehiro
I. Tabakow
I. Taylor
I.T. Dimov
J. Abawajy
J. Aroba
J. Blower
J. Cabero
J. Cai
J. Cao
J. Chen
J. Cho
J. Choi
J. Davila
J. Dolado
J. Dongarra
J. Guo
J. Gutierrez
J. Han
J. He
J. Heo
J. Hong
J. Humble
J. Hwang
J. Jeong
J. Jurek
J. Kalcher
J. Kang
J. Kim
J. King
J. Kitowski
J. Koller
J. Kommineni
J. Koo
J. Kozlak

J. Kroc
J. Krueger
J. Laws
J. Lee
J. Li
J. Liu
J. Michopoulos
J. Nabrzyski
J. Nenortaite
J. Ni
J. Owen
J. Owens
J. Pang
J. Pjesivac-Grbovic
J. Quinqueton
J. Sanchez-Reyes
J. Shin
J. Stefanowski
J. Stoye
J. Tao
J. Utke
J. Vigo-Aguiar
J. Volkert
J. Wang
J. Wasniewski
J. Weidendorfer
J. Wu
J. Yu
J. Zara
J. Zhang
J. Zhao
J. Zivkovic
J.-H. Nam
J.-L. Koning
J.-W. Lee
J.A. Vrugt
J.C. Cunha
J.C. Liu
J.C. Teixeira
J.C.S. Lui
J.F. San Juan
J.H. Hrusak
J.H. Lee
J.J. Alvarez
J.J. Cuadrado

J.J. Korczak
J.J. Zhang
J.K. Choi
J.L. Leszczynski
J.M. Bradshaw
J.M. Gilp
J.P. Crutchfield
J.P. Suarez Rivero
J.V. Alvarez
J.Y. Chen
K. Akkaya
K. Anjyo
K. Banas
K. Bolton
K. Boryczko
K. Chae
K. Ebihara
K. Ellrott
K. Fisher
K. Fuerlinger
K. Gaaloul
K. Han
K. Hsu
K. Jinsuk
K. Juszczyszyn
K. Kubota
K. Li
K. Meridg
K. Najarian
K. Ouazzane
K. Sarac
K. Sycara
K. Tai-hoon Kim
K. Trojahner
K. Tuncay
K. Westbrooks
K. Xu
K. Yang
K. Zhang
K.-J. Jeong
K.B. Lipkowitz
K.D. Nguyen
K.V. Mikkelsen
K.X.S. Souza
K.Y. Huang

L. Borzemski
L. Brugnano
L. Cai
L. Czekierda
L. Fernandez
L. Gao
L. Gonzalez-Vega
L. Hascoet
L. Hluchy
L. Jia
L. Kotulski
L. Liu
L. Lopez
L. Marchal
L. Neumann
L. Parida
L. Taher
L. Xiao
L. Xin
L. Yang
L. Yu
L. Zheng
L. Zhigilei
L.H. Figueiredo
L.J. Song
L.T. Yang
M. Aldinucci
M. Baker
M. Bamha
M. Baumgartner
M. Bhuruth
M. Borodovsky
M. Bubak
M. Caliari
M. Chover
M. Classen
M. Comin
M. Deris
M. Drew
M. Fagan
M. Fras
M. Fujimoto
M. Gerndt
M. Guo
M. Hardman

M. Hobbs
M. Houston
M. Iwami
M. Jankowski
M. Khater
M. Kim
M. Kirby
M. Kisiel-Dorochinicki
M. Li
M. Malawski
M. Mascagni
M. Morshed
M. Mou
M. Omar
M. Pérez-Hernández
M. Palakal
M. Paprzycki
M. Paszynski
M. Polak
M. Rajkovic
M. Ronsse
M. Rosvall
M. Ruiz
M. Sarfraz
M. Sbert
M. Smolka
M. Suvakov
M. Tomassini
M. Verleysen
M. Vianello
M. Zhang
M.A. Sicilia
M.H. Zhu
M.J. Brunger
M.J. Harris
M.Y. Chung
N. Bauernfeind
N. Hu
N. Ishizawa
N. Jayaram
N. Masayuki
N. Murray
N. Navarro
N. Navet
N. Sastry

N. Sundaraganesan
N.T. Nguyen
O. Beckmann
O. Belmonte
O. Habala
O. Maruyama
O. Otto
O. Yasar
P. Alper
P. Amodio
P. Balbuena
P. Bekaert
P. Berman
P. Blowers
P. Bonizzoni
P. Buendia
P. Czarnul
P. Damaschke
P. Diaz Gutierrez
P. Dyshlovenko
P. Geerlings
P. Gruer
P. Heimbach
P. Heinzlreiter
P. Herrero
P. Hovland
P. Kacsuk
P. Li
P. Lingras
P. Martineau
P. Pan
P. Praxmarer
P. Rice
P. Roe
P. Sloot
P. Tvrdik
P. Uthayopas
P. van Hooft
P. Venuvanalingam
P. Whitlock
P. Wolschann
P.H. Lin
P.K. Chattaraj
P.R. Ramasami
Q. Deng

W. Miller	Y. Cotronis	Y.J. Ye
W. Rachowicz	Y. Cui	Y.Q. Xiong
W. Yan	Y. Dai	Y.S. Choi
W. Yin	Y. Li	Y.Y. Cho
W. Zhang	Y. Liu	Y.Z. Cho
W. Zheng	Y. Mun	Z. Cai
W.K. Tai	Y. Pan	Z. Hu
X. Huang	Y. Peng	Z. Huang
X. Liao	Y. Shi	Z. Liu
X. Wan	Y. Song	Z. Pan
X. Wang	Y. Xia	Z. Toroczkai
X. Zhang	Y. Xue	Z. Wu
X.J. Chen	Y. Young Jin	Z. Xin
X.Z. Cheng	Y.-C. Bang	Z. Zhao
Y. Aumann	Y.-C. Shim	Z. Zlatev
Y. Byun	Y.B. Kim	Z.G. Sun
Y. Cai	Y.E. Gorbachev	Z.M. Zhou

Workshop Organisers

Third International Workshop on Simulation of Multiphysics Multiscale Systems

V.V. Krzhizhanovskaya - Universiteit van Amsterdam, The Netherlands and
    St. Petersburg State Polytechnical University, Russia
Y.E. Gorbachev - St. Petersburg State Polytechnic University, Russia
B. Chopard - University of Geneva, Switzerland

Innovations in Computational Science Education

D. Donnelly - Department of Physics, Siena College, USA

Fifth International Workshop on Computer Graphics and Geometric Modeling (CGGM 2006)

A. Iglesias - University of Cantabria, Spain

Fourth International Workshop on Computer Algebra Systems and Applications (CASA 2006)

A. Iglesias - University of Cantabria, Spain
A. Galvez - University of Cantabria, Spain

Tools for Program Development and Analysis in Computational Science

D. Kranzlmüller - GUP, Joh. Kepler University, Linz, Austria
R. Wismüller - University of Siegen, Germany
A. Bode - Technische Universität München, Germany
J. Volkert - GUP, Joh. Kepler University, Linz, Austria

Collaborative and Cooperative Environments

C. Anthes - GUP, Joh. Kepler University, Linz, Austria
V.N. Alexandrov - ACET, University of Reading, UK
D.J. Roberts - NICVE, University of Salford, UK
J. Volkert - GUP, Joh. Kepler University, Linz, Austria
D. Kranzlmüller - GUP, Joh. Kepler University, Linz, Austria

Second International Workshop on Bioinformatics Research and Applications (IWBRA'06)

A. Zelikovsky - Georgia State University, USA
Y. Pan - Georgia State University, USA
I.I. Mandoiu - University of Connecticut, USA

Third International Workshop on Practical Aspects of High-Level Parallel Programming (PAPP 2006)

A. Benoît - Laboratoire d'Informatique du Parallélisme, Ecole Normale
    Supérieure de Lyon, France
F. Loulergue - LIFO, Université d'Orléans, France

Wireless and Mobile Systems

H. Choo - Networking Laboratory, Sungkyunkwan University, Suwon, KOREA

GeoComputation

Y. Xue - Department of Computing, Communications Technology and
    Mathematics, London Metropolitan University, UK

Computational Chemistry and Its Applications

P. Ramasami - Department of Chemistry, University of Mauritius

Knowledge and Information Management in Computer Communication Systems (KIMCCS 2006)

N.T. Nguyen - Institute of Control and Systems Engineering,
    Wroclaw University of Technology, Poland

A. Grzech - Institute of Information Science and Engineering,
   Wroclaw University of Technology, Poland
R. Katarzyniak - Institute of Information Science and Engineering,
   Wroclaw University of Technology, Poland

Modelling of Complex Systems by Cellular Automata (MCSCA 2006)

J. Kroc - University of West Bohemia, Czech Republic
T. Suzudo - Japan Atomic Energy Agency, Japan
S. Bandini - University of Milano - Bicocca, Italy

Dynamic Data Driven Application Systems (DDDAS 2006)

F. Darema - National Science Foundation, USA

Parallel Monte Carlo Algorithms for Diverse Applications in a Distributed Setting

I.T. Dimov - ACET, University of Reading, UK
V.N. Alexandrov - ACET, University of Reading, UK

International Workshop on Intelligent Storage Technology (IST06)

J. Shu - Department of Computer Science and Technology, Tsinghua University,
   Beijing, P.R. China

Intelligent Agents in Computing Systems

R. Schaefer - Department of Computer Science, Stanislaw Staszic University
   of Science and Technology in Kraków
K. Cetnarowicz - Department of Computer Science, Stanislaw Staszic University
of Science and Technology in Kraków

First International Workshop on Workflow Systems in e-Science (WSES06)

Z. Zhao - Informatics Institute, University of Amsterdam, The Netherlands
A. Belloum - University of Amsterdam, The Netherlands

Networks: Structure and Dynamics

B. Tadic - Theoretical Physics Department, J. Stefan Institute, Ljubljana,
   Slovenia
S. Thurner - Complex Systems Research Group, Medical University Vienna,
   Austria

Evolution Toward Next Generation Internet (ENGI)

Y. Cui - Tsinghua University, P.R. China
T. Korkmaz - University of Texas at San Antonio, USA

General Purpose Computation on Graphics Hardware (GPGPU): Methods, Algorithms and Applications

D. Göddeke - Universität Dortmund, Institut für Angewandte Mathematik
    und Numerik, Germany
S. Turek - Universität Dortmund, Institut für Angewandte Mathematik
    und Numerik, Germany

Intelligent and Collaborative System Integration Technology (ICSIT)

J.-W. Lee - Center for Advanced e-System Integration Technology,
    Konkuk University, Seoul, Korea

Computational Methods for Financial Markets

R. Simutis - Department of Informatics, Kaunas Faculty, Vilnius University,
    Lithuania
V. Sakalauskas - Department of Informatics, Kaunas Faculty, Vilnius University,
    Lithuania
D. Kriksciuniene - Department of Informatics, Kaunas Faculty,
    Vilnius University, Lithuania

2006 International Workshop on P2P for High Performance Computational Sciences (P2P-HPCS06)

H. Jin - School of Computer Science and Technology, Huazhong University of
Science and Technology, Wuhan, China
X. Liao - Huazhong University of Science and Technology, Wuhan, China

Computational Finance and Business Intelligence

Y. Shi - Graduate School of the Chinese Academy of Sciences, Beijing, China

Third International Workshop on Automatic Differentiation Tools and Applications

C. Bischof - Inst. for Scientific Computing, RWTH Aachen University, Germany
S.A. Forth - Engineering Systems Department, Cranfield University,
    RMCS Shrivenham, UK
U. Naumann - Software and Tools for Computational Engineering,
    RWTH Aachen University, Germany
J. Utke - Mathematics and Computer Science Division, Argonne National
    Laboratory, IL, USA

2006 Workshop on Scientific Computing in Electronics Engineering

Y. Li - National Chiao Tung Univeristy, Hsinchu City, Taiwan

New Trends in the Numerical Solution of Structured Systems with Applications

T. Politi - Dipartimento di Matematica, Politecnico di Bari, Itali
L. Lopez - Dipartimento di Matematica, Universita' di Bari, Itali

Workshop on Computational Science in Software Engineering (CSSE'06)

D. Rodríguez García - University of Reading, UK
J.J. Cuadrado - University of Alcalá, Spain
M.A. Sicilia - University of Alcalá, Spain
M. Ruiz - University of Cádiz, Spain

Digital Human Modeling (DHM-06)

Y. Cai - Carnegie Mellon University, USA
C. Imielinska - Columbia University

Real Time Systems and Adaptive Applications (RTSAA 06)

T. Kuo - National Taiwan University, Taiwan
J. Hong - School of Computer Science and Engineering, Kwangwoon University,
    Seoul, Korea
G. Jeon - Korea Polytechnic University, Korea

International Workshop on Grid Computing Security and Resource Management (GSRM'06)

J.H. Abawajy - School of Information Technology, Deakin University, Geelong,
    Australia

Fourth International Workshop on Autonomic Distributed Data and Storage Systems Management Workshop (ADSM 2006)

J.H. Abawajy - School of Information Technology, Deakin University, Geelong,
    Australia

Table of Contents – Part I

Keynote Abstracts

Modelling and Simulation in Economics and Finance

Modelling and Simulation of Complex Systems and in the Natural Sciences

Modelling and Simulation of Complex Systems

Advanced Numerical Algorithms

Data Driven Computing

Advanced Numerical Algorithms and New Algorithmic Approaches to Computational Kernels and Applications

Modelling and Simulation in the Natural Sciences

Modelling and Simulation in the Natural Sciences

Advanced Numerical Algorithms

Applications of Computing as a Scientific Paradigm

Applications of Computing as a Scientific Paradigm

Modelling and Simulation in Engineering

Modelling and Simulation in Engineering

Parallel and Distributed Algorithms

Other Aspects of Computational Science

Computational Science Aspects of Data Mining and Information Retrieval

Hybrid Computational Methods and New Algorithmic Approaches to Computational Kernels and Applications

Simulations and Systems

Advances in Parameter Estimation in Computational-Science: Strategies, Concepts, and Applications

Efficient Fault Tolerance Techniques for Large Scale Systems

Poster Session I

Poster Session II

Table of Contents – Part II

Third International Workshop on Simulation of Multiphysics Multiscale Systems

Innovations in Computational Science Education

Fifth International Workshop on Computer Graphics and Geometric Modeling (CGGM 2006)

Fourth International Workshop on Computer Algebra Systems and Applications (CASA 2006)

Tools for Program Development and Analysis in Computational Science

Collaborative and Cooperative Environments

Second International Workshop on Bioinformatics Research and Applications (IWBRA06)

Third International Workshop on Practical Aspects of High-Level Parallel Programming (PAPP 2006)

Wireless and Mobile Systems

Table of Contents – Part III

GeoComputation

Computational Chemistry and Its Applications

Knowledge and Information Management in Computer Communication Systems (KIMCCS 2006)

Modelling of Complex Systems by Cellular Automata (MCSCA 2006)

Dynamic Data Driven Application Systems (DDDAS 2006)

Parallel Monte Carlo Algorithms for Diverse Applications in a Distributed Setting

International Workshop on Intelligent Storage Technology (IST06)

Intelligent Agents in Computing Systems

First International Workshop on Workflow Systems in e-Science (WSES06)

Networks: Structure and Dynamics

Table of Contents – Part IV

Evolution Toward Next Generation Internet (ENGI)

General Purpose Computation on Graphics Hardware (GPGPU): Methods, Algorithms and Applications

Intelligent and Collaborative System Integration Technology (ICSIT)

Computational Methods for Financial Markets

2006 International Workshop on P2P for High Performance Computational Sciences (P2P-HPCS06)

Computational Finance and Business Intelligence

Third International Workshop on Automatic Differentiation Tools and Applications

2006 Workshop on Scientific Computing in Electronics Engineering

New Trends in the Numerical Solution of Structured Systems with Applications

Workshop on Computational Science in Software Engineering (CSSE'06)

Digital Human Modeling (DHM-06)

Real Time Systems and Adaptive Applications (RTSAA 06)

International Workshop on Grid Computing Security and Resource Management (GSRM'06)

Fourth International Workshop on Autonomic Distributed Data and Storage Systems Management Workshop (ADSM 2006)

Metacomputing Revisited: Alternative Paradigms for Distributed Resource Sharing

Vaidy Sunderam

Department of Math & Computer Science
Emory University, Atlanta, GA 30322, USA
vss@emory.edu
http://www.mathcs.emory.edu/dcl/

Abstract. Conventional distributed computing paradigms such as PVM and MPI(CH) have had mixed success when translated to computing environments that span multiple administrative and ownership domains. We analyze fundamental issues in distributed resource sharing particularly from the viewpoint of different forms of heterogeneity – i.e. not just in the computing platforms, but also in storage, network characteristics, availability, access protocols, robustness, and dynamicity. We argue that effective multidomain resource sharing in the face of such variability is critically dependent upon minimizing global state between providers, and between providers and clients. The H2O framework has made one step in this direction, by decoupling provider concerns from client requirements, and enabling clients to (re)configure resources as required. H2O is based on a "pluggable" software architecture to enable flexible and reconfigurable distributed computing. A key feature is the provisioning of customization capabilities that permit clients to tailor provider resources as appropriate to the given application, without compromising control or security. Through the use of uploadable "pluglets", users can exploit specialized features of the underlying resource, application libraries, or optimized message passing subsystems on demand. The next generation of this framework takes the second step, to virtualize and homogenize resource aggregates at the client side, thereby further reducing the degree of coupling between providers, and between providers and clients. The system also supports dynamic environment preconditioning to automate many of the tasks required in multidomain resource sharing. The architecture and design philosophies of these software frameworks, their implementation, recent experiences, and planned enhancements are described, in the context of new paradigms and directions for metacomputing.

V.N. Alexandrov et al. (Eds.): ICCS 2006, Part I, LNCS 3991, p. 1, 2006.
© Springer-Verlag Berlin Heidelberg 2006

Achieving Breakthrough Science with the Blue Gene/L Supercomputer

José E. Moreira

IBM Thomas J. Watson Research Center
Yorktown Heights
NY, USA
jmoreira@us.ibm.com
http://www.research.ibm.com/people/m/moreira

Abstract. The Blue Gene project started in the final months of 1999. Five years later, during the final months of 2004, the first Blue Gene/L machines were being installed at customers. By then, Blue Gene/L had already established itself as the fastest computer in the planet, topping the TOP500 list with the breathtaking speed of over 70 Teraflops. Since the beginning of 2005, many other systems have been installed at customers, the flagship machine at Lawrence Livermore National Laboratory has greatly increased in size, and Blue Gene/L has established itself as a machine capable of breakthrough science. We here examine why Blue Gene/L is such an effective science machine, and report on some of the more significant research being performed by scientists using Blue Gene/L systems today. We also explain why this is just the beginning, and why there is more excitement ahead of us than behind us in the Blue Gene project.

V.N. Alexandrov et al. (Eds.): ICCS 2006, Part I, LNCS 3991, p. 2, 2006.

Visualizing the Future

Chris Johnson

Scientific Computing and Imaging Institute
Distinguished Professor
School of Computing
University of Utah
`crj@sci.utah.edu`
`www.sci.utah.edu`

Abstract. Computers are now extensively used throughout science, engineering, and medicine. Advances in computational geometric modeling, imaging, and simulation allow researchers to build and test models of increasingly complex phenomena and thus to generate unprecedented amounts of data. These advances have created the need to make corresponding progress in our ability to understand large amounts of data and information arising from multiple sources. In fact, to effectively understand and make use of the vast amounts of information being produced is one of the greatest scientific challenges of the 21st Century.

Visual computing, which relies on and takes advantage of, the interplay among techniques of visualization, computer graphics, virtual reality, and imaging and vision, is fundamental to understanding models of complex phenomena, which are often multi-disciplinary in nature. In this talk, I will first provide several examples of ongoing visual computing research at the Scientific Computing and Imaging (SCI) Institute as applied to problems in computational science, engineering, and medicine, then discuss future research opportunities and challenges.

V.N. Alexandrov et al. (Eds.): ICCS 2006, Part I, LNCS 3991, p. 3, 2006.
© Springer-Verlag Berlin Heidelberg 2006

IT Innovation: A New Era

Martin Curley

Senior Principal Engineer
Global Director, Innovation and IT Research
Intel Corporation
martin.g.curley@intel.com

Abstract. This presentation will discuss the emerging discipline of IT Innovation, a discipline where unique value can be created at the intersection of Information Technology and Innovation. The presentation also shares the structure of an emerging Innovation Capability Maturity Framework and Model while discussing how computing trends such as multicore technology and virtualization will re-ignite Moore's law and provide a platform for Innovation that is unparalleled in history.

IT Innovation is emerging as a new discipline, one which exists at the intersection of two relatively immature disciplines, that of Information Technology and Innovation. Whilst IT is increasingly being recognized as a discipline and the profession is reasonably well developed, it is just in the last couple of years that business schools have started to recognize that Innovation is not just something that happens by luck but perhaps is a process that can be mastered. Driven by the power of Moore's Law IT is a unique innovation resource, it's transformative power and it's ever improving performance/price dynamic making it a very attractive transformation resource. However in parallel IT is also a great resource for helping automate and manage the process of Innovation itself.

For any IT Innovation to be successful, at least six vectors need to be managed successively in parallel

- Vision (Opportunity/Problem)
- IT enabled Solution
- Business Case
- Business Process Change
- Organizational Change
- Customer or Societal Change

To start with somebody has to have a vision of how a particular opportunity can be exploited or how a particular problem can be solved through using information technology or some combination thereof. Typically the vision will be evolved over time. While IT Professionals often think the hardest part of any IT innovation is delivering the IT solution, this often can be the easier part of achieving a new innovation. Since the dotcom crash it is increasingly important and necessary to have a business case associated with any new innovation. The dotcom crash meant that there was a swing from irrational exuberance to irrational pessimism with respect to IT investment. Thankfully there is a

V.N. Alexandrov et al. (Eds.): ICCS 2006, Part I, LNCS 3991, pp. 4–6, 2006.

modicum of rationality returning to IT investment but increasingly a solid business case is a requirement for an innovation to proceed.

The next three vectors are where soft issues come into play and often can be the most problematic part of delivering an IT innovation. Typically an IT Innovation will deliver some form of business process change - in the past typically we saw business process automation but now more often that not we seen business process transformation. This is often accompanies by organizational change, either in the form of new behaviors, new roles or indeed new organization forms. Resource fluidity and organizational and ecosystem dynamics will have a significant influence on the organizational vector. Finally and perhaps most importantly an Innovation is only successful if it is adopted by the customer. Many of today's IT enabled Innovations have societal impact and society's willingness and ability to accept an Innovation is a crucial modulator in the success of an Innovation.

Looking at Innovation as a process there is increasing awareness that Innovation perhaps could be managed as a process. Typically there are three processes involved with an Innovation - cognitive, organizational/logistical and economic. In collaboration with the National University of Ireland, Intel is working to develop an IT Innovation Capability Maturity Framework (ITI CMF) and Model (ITI CMM) to provide organizations with a roadmap and assessment tool to help deliver more value through IT innovation. This framework and model is based on the concept of systemic innovation where synchronous improvements are made to organizational strategy, culture, tools and metrics to help create a virtuous circle of Innovation.

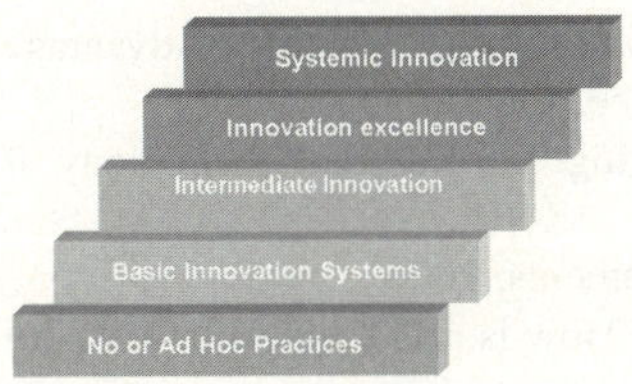

Fig. 1. IT Innovation Capability Maturity Framework (IT CMF). Source: IVI Intel/National University of Ireland.

At the bottom level of the maturity curve the organization has no or ad hoc practices for innovation but as the organization traverses through the curve it steadily increases the maturity and quality of its Innovation Strategy, Culture, Tools and Metrics and ultimately the innovation results. Level 5 maturity, i.e. systemic innovation is when Innovation becomes a way of life and this typically means sustained new profitable growth for a commercial organization.

In parallel with the maturing of IT Innovation as a discipline, IT as a technology continues to evolve. Whilst some were concerned that Moore's Law would peter out as the laws of Physics became barriers for ever increasing frequency , the adoption of a Multicore approach will revitalize and give Moore's law another lease of life. There is a fundamental shift happening in the PC industry as companies move away from increasing the frequency of the CPU to increase performance to the adoption of parallelism as the key parameter for delivering better performance. This "right-hand" turn will likely deliver an acceleration of Moore's' Law. The years 2000 to 2004 saw an approximately 3X improvement in CPU performance but the advent of multicore is projected to deliver a 10X improvement in just the next four years. Multicore is essentially the configuration of multiple logical microprocessors on a single physical die.

Multicore will not just bring better performance, it will also bring better performance per watt, larger throughput and better scalability. As a company Intel is on its third generation of parallelism. Intel's 1st generation of parallel computing was with introduction of purpose-built multi-way servers (DP, MP and SMP). Intel's 2nd generation of parallelism was with Hyper-threading or 2 virtual threads per core, this typically yield up to 30% performance gains on threaded code and enabled outstanding multi-tasking usage benefits. Intel is now on its 3rd generation of threading with dual core and beyond that combines multicore core with hyper threading for increased parallelism, throughput and versatility. In addition as Virtualization technology diffuses into the marketplace the opportunity expands. Also multicore designs will deliver better performance/watt as easing concerns on the growing power consumption for more and more powerful machines.

As multicore technology develops and as the process of IT Innovation is more managed and ultimately mastered, the future does look bright as a new era of IT Innovation is ushered in.

References

1. Curley, M. 2004. "IT Innovation for Competitive Advantage and Adaptiveness", Kluwer, Proceedings IFIP 8.6 Conference. June.
2. Curley, M. 2004. "Managing Information Technology for Business Value", Intel Press. January.
3. Intel internal technical documents, 2006
4. Swanson, Burton E, 2004. "How is an IT Innovation assimilated", Kluwer, Proceedings IFIP 8.6 Conference. June.
5. Pickering, Cindy. 2004. "Using IT Concept Cars to accelerate Innovation", Kluwer, Proceedings IFIP 8.6 Conference. June.

The AWE HPC Benchmark

Ron Bell

AWE plc
Aldermaston
Berkshire
RG7 4PR
`ron.bell@awe.co.uk`

Abstract. Near the end of 2005, AWE, Aldermaston placed an order for a Cray XT3 system with 3936 dual-core nodes (over 40 Tflops peak) to replace its existing HPC system. This paper describes the design of the benchmark used during the preceding competitive procurement including separate capacity and capability requirements analysis. Details are presented of the evaluation process and the relative results of the six HPC systems evaluated. These include the more than 2-times speedup obtained by Cray by tuning the source code of the most important application.

V.N. Alexandrov et al. (Eds.): ICCS 2006, Part I, LNCS 3991, p. 7, 2006.

Newton's Method for the Ellipsoidal l_p Norm Facility Location Problem

Yu Xia[*]

The Institute of Statistical Mathematics,
4-6-7 Minami-Azabu, Minato-ku,
Tokyo 106-8569, Japan
`yuxia@ism.ac.jp`

Abstract. We give the minisum ellipsoidal l_p norm facility location model. Our model includes both $p \geq 2$ and $p < 2$ cases. We derive the optimality conditions for our model and transform the optimality conditions into a system of equations. We then solve the system by perturbed Newton's method. Some numerical examples are presented.

1 Introduction

We want to locate a new facility. The objective is to minimize the weighted sum of distances from it to some existing facilities. In addition, the new facility is required to be within some region. The distances from the new facility to some resources, such as transportation centers, should not exceed some upperbounds.

Because it usually takes a long time and a large amount of capital to build a facility, after its establishment, the facility is supposed to operate for a long period. Since traffic between the existing locations and the new one occurs on a regular basis during the facility's life circle, it is important to calculate the location of the new facility as accurate as possible. The approximation of road distances by weighted l_p distance measure has been studied in a series of papers ([3,4], etc.), where it is argued with empirical study that l_p distances weighted by an inflation factor tailored to given regions can better describe the irregularity in the transportation networks such as hills, bends, and are therefore superior to the weighted rectangular and Euclidean norms. Denote the l_p norm (Minkowski distance of order p, $p > 1$) of a d-dimensional Euclidean space $\mathbb{R}^d$ as

$$l_p(\mathbf{z}) \overset{\text{def}}{=} \left[\sum_{i=1}^{d} |z_i|^p \right]^{1/p} .$$

In this paper, we measure distances by the ellipsoidal l_p norm distance:

$$l_{pM}(\mathbf{z}) \overset{\text{def}}{=} l_p(M\mathbf{z}) ,$$

---

[*] Research supported by a JSPS postdoctoral fellowship for foreign researchers. I thank suggestions and comments of an anonymous referee.

V.N. Alexandrov et al. (Eds.): ICCS 2006, Part I, LNCS 3991, pp. 8–15, 2006.

where M is a linear transformation. It is not hard to see that l_{pM} is a norm when M is nonsingular. Note that the l_{pM} distance is the l_p norm when M is the identity. The Euclidean distance (2-norm distance), the Manhattan distance (l_1 norm distance), and the Chebyshev distance (l_∞ norm distance) are special cases of l_{pM}. The l_{pM} distance also includes the l_{pb} norm (see, for instance [2]):

$$l_{pb}(\mathbf{z}) \overset{\text{def}}{=} \left[\sum_{i=1}^{n} b_i |x_i|^p \right]^{1/p}, \quad b_i > 0 \, (i = 1, \ldots, n) \ .$$

The l_{pM} distance measure can better describe the actual transportation networks than the weighted l_p distance or the l_{pb} distance can. In real word, the road connecting two points is not likely a straight line, but a curve, due to existing landmarks or geographic conditions. The curvature of the road can be better described via the rotation of the axes of the coordinates for the facilities and a proper choice of p. In addition, the transportation costs on different segments of a road may be different due to the physical conditions and traffic flows of different segments. For instance, some segments may consist of bridges or uphills. The difference can be modeled by the scaling of the axes. The model is more accurate with different p's and different transformation matrices for different locations instead of one p and one M for all locations.

The rest of the paper is organized as follows. In §2, we give the ellipsoidal l_p norm facility location model. In §3, we derive optimality conditions for the model and reformulate them into equations. We then give our perturbed Newton's method for the equations. In §4, we present some numerical examples.

2   The Mathematical Model

Let vectors $\mathbf{f}_1, \ldots, \mathbf{f}_n$ represent the n existing facilities, or demand points. Let vectors $\mathbf{f}_j \, (j = n+1, \ldots n+s)$ represent the resources. It is possible that some resources $\mathbf{f}_j \, (n < j \leq n+s)$ are just some existing facilities $\mathbf{f}_i \, (1 \leq i \leq n)$. Let vector $\mathbf{x}$ denote the new facility to be located. Below is our model:

$$\min_{\mathbf{x}} \quad \sum_{i=1}^{n} w_i \, \| M_i(\mathbf{x} - \mathbf{f}_i) \|_{p_i} \tag{1a}$$

$$\text{s.t.} \quad A\mathbf{x} \leq \mathbf{b} \tag{1b}$$

$$\| M_{n+j}(\mathbf{x} - \mathbf{f}_{n+j}) \|_{p_{n+j}} \leq r_j \quad (j = 1, \ldots, s) \ . \tag{1c}$$

In the above model, the weight $w_i > 0 \, (i = 1, \ldots, n)$ is a combination of demands and inflation factor which counts for the condition of transportation equipments, cost of manpower, etc., for the existing facility $\mathbf{f}_i$. And $M_i \, (i = 1, \ldots, n+s)$ are d_i-by-d matrices, which are not necessarily nonsingular. A is a given m-by-d matrix. And $\mathbf{b}$ is a given vector of dimension m. r_i are constants. The constraints (1b) give the region in which the new facility can be located. The constraints (1c) specify that the new facility should be within certain distances from some resources.

Model (1) is reduced to the classic l_p norm location problem if all M_i ($i = 1, \ldots, n+s$) are the identity matrix and the constraints (1b) and (1c) are empty.

3 The Algorithm

In this section, we derive optimality conditions for (1) and reformulate them into a system of equations. Then we present our global algorithm for the equations.

We first show that (1) is a convex program in order to give necessary and sufficient conditions for optimality.

For any $0 \leq \lambda \leq 1$, and $\mathbf{x}_1, \mathbf{x}_2 \in \mathbb{R}^d$, by Minkowski's inequality, we have

$$\sum_{i=1}^{n} w_i \left\| M_i \left[\lambda \mathbf{x}_1 + (1-\lambda)\mathbf{x}_2 - \mathbf{f}_i \right] \right\|_{p_i} = \sum_{i=1}^{n} w_i \left\| \lambda M_i(\mathbf{x}_1 - \mathbf{f}_i) \right.$$

$$\left. +(1-\lambda)M_i(\mathbf{x}_2 - \mathbf{f}_i) \right\|_{p_i} \leq \lambda \sum_{i=1}^{n} w_i \left\| M_i(\mathbf{x}_1 - \mathbf{f}_i) \right\|_{p_i} + (1-\lambda) \sum_{i=1}^{n} w_i \left\| M_i(\mathbf{x}_2 - \mathbf{f}_i) \right\|_{p_i}.$$

Therefore, the objective function (1a) is convex in $\mathbf{x}$.

For $i = 1, \ldots, n+s$, define q_i as the scalar satisfying $\frac{1}{q_i} + \frac{1}{p_i} = 1$.

Next, we consider the Lagrangian dual to (1).

We use z_{il} to denote the lth element of the vector $\mathbf{z}_i$, and M_{il} to denote the lth row of the matrix M_i.

By Hölder's inequality, when $M_i(\mathbf{x} - \mathbf{f}_i) \neq \mathbf{0}$,

$$w_i \left\| M_i(\mathbf{x} - \mathbf{f}_i) \right\|_{p_i} \geq_{(\|\mathbf{z}_i\|_{q_i} \leq w_i)} \mathbf{z}_i^T M_i(\mathbf{x} - \mathbf{f}_i) ,$$

where the equality holds iff

$$\|\mathbf{z}_i\|_{q_i} = w_i, \quad \|M_i(\mathbf{x}-\mathbf{f}_i)\|_{p_i}^{p_i}|z_{il}|^{q_i} = w_i^{q_i}|M_{il}(\mathbf{x}-\mathbf{f}_i)|^{p_i}, \quad \mathrm{sign}(\mathbf{z}_{il}) = \mathrm{sign}\left[M_{il}(\mathbf{x}-\mathbf{f}_i)\right].$$

The subdifferential of $\|M_i(\mathbf{x} - \mathbf{f}_i)\|$ at any point where $M_i(\mathbf{x} - \mathbf{f}_i) = \mathbf{0}$ is $\{M_i^T \mathbf{z}_i : \|\mathbf{z}_i\|_{q_i} \leq 1\}$.

Let $\boldsymbol{\eta} \overset{\mathrm{def}}{=} (\eta_1, \ldots, \eta_s)^T$, $\boldsymbol{\lambda} \overset{\mathrm{def}}{=} (\lambda_1, \ldots, \lambda_m)^T$ denote the dual variables. Then the dual to (1) is

$$\max_{\boldsymbol{\lambda} \geq \mathbf{0}, \boldsymbol{\eta} \geq \mathbf{0}} \min_{\mathbf{x}} \sum_{i=1}^{n} w_i \left\| M_i(\mathbf{x}-\mathbf{f}_i) \right\|_{p_i} + \boldsymbol{\lambda}^T (A\mathbf{x} - \mathbf{b}) + \sum_{j=1}^{s} \eta_j \left(\left\| M_{n+j}(\mathbf{x}-\mathbf{f}_{n+j}) \right\|_{p_{n+j}} \right.$$

$$\left. - r_j \right) = \max_{\boldsymbol{\lambda} \geq \mathbf{0}, \boldsymbol{\eta} \geq \mathbf{0}} \min_{\mathbf{x}} \max_{\substack{\|\mathbf{z}_i\|_{q_i} \leq w_i \\ (i=1,\ldots,n) \\ \|\mathbf{z}_j\|_{q_j} \leq \eta_j \\ (j=n+1,\ldots,n+s)}} \left[\sum_{i=1}^{n} \mathbf{z}_i^T M_i(\mathbf{x} - \mathbf{f}_i) + \boldsymbol{\lambda}^T (A\mathbf{x} - \mathbf{b}) \right.$$

$$\left. + \sum_{j=1}^{s} \mathbf{z}_{j+n}^T M_{j+n}(\mathbf{x} - \mathbf{f}_{j+n}) - \sum_{j=1}^{s} \eta_j r_j \right] = \max_{\boldsymbol{\lambda} \geq \mathbf{0}, \boldsymbol{\eta} \geq \mathbf{0}} \min_{\mathbf{x}} \max_{\substack{\|\mathbf{z}_i\|_{q_i} \leq w_i \ (i=1,\ldots,n) \\ \|\mathbf{z}_j\|_{q_j} \leq \eta_j \ (j=n+1,\ldots,n+s)}}$$

$$\left[\left(\sum_{i=1}^{n+s} \mathbf{z}_i^T M_i + \boldsymbol{\lambda}^T A \right) \mathbf{x} - \sum_{i=1}^{n+s} \mathbf{z}_i^T M_i \mathbf{f}_i - \boldsymbol{\lambda}^T \mathbf{b} - \sum_{j=1}^{s} \eta_j r_j \right].$$

Let $\mathbf{z}(\mathbf{x})$ denote the optimal solution of $\mathbf{z}$ to the inner maximization problem for a fixed $\mathbf{x}$. The first term in the above expression implies $\sum_{i=1}^{n+s} \mathbf{z}(\mathbf{x})_i^T M_i + \boldsymbol{\lambda}^T A = \mathbf{0}$; otherwise, the value of the second minimization is unbounded. Therefore, the Lagrangian dual to (1) is:

$$
\begin{aligned}
\min_{\boldsymbol{\lambda},\boldsymbol{\eta},\mathbf{z}_i} \quad & \sum_{i=1}^{n+s} \mathbf{z}_i^T M_i \mathbf{f}_i + \boldsymbol{\lambda}^T \mathbf{b} + \sum_{j=1}^{s} \eta_j r_j \\
\text{s.t.} \quad & \sum_{i=1}^{n+s} M_i^T \mathbf{z}_i + A^T \boldsymbol{\lambda} = \mathbf{0} \\
& \|\mathbf{z}_i\|_{q_i} \le w_i \quad (i = 1,\ldots,n) \\
& \|\mathbf{z}_j\|_{q_j} \le \eta_j \quad (j = n+1,\ldots,n+s) \\
& \boldsymbol{\lambda} \ge \mathbf{0} \\
& \boldsymbol{\eta} \ge \mathbf{0} \ .
\end{aligned}
\qquad (2)
$$

The dual of the linearly constrained l_p norm facility location problem has been studied in [5]; however, our model includes the resource constraints (1c).

Assume there is a strict interior solution, i.e., $\exists\, \tilde{\mathbf{x}} \in \mathbb{R}^d$ such that $A\tilde{\mathbf{x}} \le \mathbf{b}$ and $\|M_j(\mathbf{x} - \mathbf{f}_j)\|_{p_j} < r_j\ (j = n+1,\ldots,n+s)$. Then there is no duality gap between (2) and (1) (see for instance [6]), since the objective and the constraints are convex. In the course of deriving (2), we also obtain that at an optimum, the variables $\mathbf{x}, \mathbf{z}, \boldsymbol{\lambda}, \boldsymbol{\eta}$ satisfy the following conditions:

$$
\sum_{i=1}^{n+s} M_i^T \mathbf{z}_i + A^T \boldsymbol{\lambda} = \mathbf{0} \ ,
$$

$$
\eta_{j-n}\left(r_{j-n}^{p_j} - \|M_j(\mathbf{x} - \mathbf{f}_j)\|_{p_j}^{p_j} \right) = 0 \quad (j = n+1,\ldots,n+s) \ ,
$$

$$
\eta_j \ge 0 \quad (j = 1,\ldots,s) \ ,
$$

$$
\|M_{n+j}(\mathbf{x} - \mathbf{f}_{n+j})\|_{p_{n+j}} \le r_j \quad (j = 1,\ldots,s) \ ,
$$

$$
\lambda_i(b_i - A_i\mathbf{x}) = 0 \quad (i = 1,\ldots,m) \ ,
$$

$$
\lambda_i \ge 0 \quad (i = 1,\ldots,m) \ ,
$$

$$
A_i\mathbf{x} \le b_i \quad (i = 1,\ldots,m) \ ,
$$

$$
\alpha_i |z_{il}|^{q_i} = |M_{il}(\mathbf{x} - \mathbf{f}_i)|^{p_i} \quad (i = 1,\ldots,n;\ l = 1,\ldots,d_i) \ ,
$$

$$
\alpha_j |z_{jl}|^{q_j} = |M_{jl}(\mathbf{x} - \mathbf{f}_j)|^{p_j} \quad (j = 1+n,\ldots,n+s;\ l = 1,\ldots,d_j) \ ,
$$

$$
\mathrm{sign}(z_{il}) = \mathrm{sign}\left[M_{il}(\mathbf{x} - \mathbf{f}_i) \right] \quad (i = 1,\ldots,n+s;\ l = 1,\ldots,d_i) \ ,
$$

$$
\alpha_i(w_i - \|\mathbf{z}_{il}\|_{p_i}) = 0 \quad (i = 1,\ldots,n) \ ,
$$

$$
\|\mathbf{z}_i\|_{q_i} \le w_i \quad (i = 1,\ldots,n) \ ,
$$

$$
\alpha_{n+j}(\eta_j - \|\mathbf{z}_{n+j}\|_{p_{n+j}}) = 0 \quad (j = 1,\ldots,s) \ ,
$$

$$
\|\mathbf{z}_{n+j}\|_{q_{n+j}} \le \eta_j \quad (j = 1,\ldots,s) \ ,
$$

where we define both $\mathrm{sign}(0) = 1$ and $\mathrm{sign}(0) = -1$ to be true and use A_i to denote the ith row of A.

Usually the new facility is not located on the existing facility or resource sites. Therefore, we assume the linear constraints $A\mathbf{x} \le \mathbf{b}$ excluding $\mathbf{f}_i\ (i = 1,\ldots,n+s)$. Then, we can omit the variables $\alpha_i\ (i = 1,\ldots,n+s)$, since

$$
\|z_i\|_{q_i} = w_i\ (i = 1,\ldots,n) \ , \qquad \|z_j\|_{q_j} = \eta_j\ (j = n+1,\ldots,n+s) \ .
$$

The optimality conditions are reduced to the following:

$$\sum_{i=1}^{n+s} M_i^T \mathbf{z}_i + A^T \boldsymbol{\lambda} = \mathbf{0} \, ,$$

$$\eta_j \left(r_j^{p_{n+j}} - \|M_{n+j}(\mathbf{x} - \mathbf{f}_{n+j})\|_{p_{n+j}}^{p_{n+j}} \right) = 0 \quad (j = 1, \ldots, s) \, ,$$

$$\eta_j \geq 0 \quad (j = 1, \ldots, s) \, ,$$

$$\|M_{n+j}(\mathbf{x} - \mathbf{f}_{n+j})\|_{p_{n+j}} \leq r_j \quad (j = 1, \ldots, s) \, ,$$

$$\lambda_i(b_i - A_i\mathbf{x}) = 0 \quad (i = 1, \ldots, m) \, ,$$

$$\lambda_i \geq 0 \quad (i = 1, \ldots, m) \, ,$$

$$A\mathbf{x} \leq \mathbf{b} \, ,$$

$$\left(\sum_{l=1}^{d_i} |M_{il}(\mathbf{x} - \mathbf{f}_i)|^{p_i} \right)^{\frac{1}{q_i}} |z_{il}| = w_i \, |M_{il}(\mathbf{x} - \mathbf{f}_i)|^{\frac{p_i}{q_i}} \quad (i = 1, \ldots, n; \ l = 1, \ldots, d_i) \, ,$$

$$\left(\sum_{l=1}^{d_j} |M_{jl}(\mathbf{x} - \mathbf{f}_j)|^{p_j} \right)^{\frac{1}{q_j}} |z_{jl}| = \eta_{j-n} \, |M_{jl}(\mathbf{x} - \mathbf{f}_j)|^{\frac{p_j}{q_j}}$$

$$(j = n + 1, \ldots, n + s; \ l = 1, \ldots, d_j) \, ,$$

$$\text{sign}(z_{il}) = \text{sign}\left[M_{il}(\mathbf{x} - \mathbf{f}_i) \right] \quad (i = 1, \ldots, n + s; \ l = 1, \ldots, d_i) \, .$$

However, the above system is not easy to solve, because it includes both equalities and inequalities. Next, we use the nonlinear complementarity function min to reformulate the complementarity conditions in order to transform the above system into a system of equations (3), which can be solved by Newton's method. Note that $\min(a, b) = 0$ is equivalent to $a \geq 0$, $b \geq 0$, and at least one of a, b is 0. To avoid some nonsmooth points, we distinguish between $p_i \geq 2$ and $p_i < 2$ in our formulation. Notice that $p_i \geq 2 \Rightarrow q_i \leq 2$ and $p_i \leq 2 \Rightarrow q_i \geq 2$.

From $\frac{1}{p_i} + \frac{1}{q_i} = 1$, $(i = 1, \ldots, n + s)$, we have

$$\frac{p_i}{q_i} = p_i - 1 = \frac{1}{q_i - 1}, \qquad \frac{q_i}{p_i} = q_i - 1 = \frac{1}{p_i - 1} \, .$$

The system of equations (3) is given on page 13.

Let F represent the left-hand-side of (3). Let $\Psi \overset{\text{def}}{=} \frac{F'F}{2}$. Any global optimization method that locates a global minimal solution to Ψ solves (1). We use the gradient decent method with perturbed nonmonotone line search for almost everywhere differentiable minimization [7] to find a global minimum of Ψ.

The Algorithm

Denote $\mathbf{u} \overset{\text{def}}{=} (\mathbf{x}; \boldsymbol{\lambda}; \boldsymbol{\eta}; \mathbf{z})$. Let $\Delta \mathbf{u}$ represent the Newton's direction to (3).

Initialization. Set constants $s > 0$, $0 < \sigma < 1$, $\beta \in (0, 1)$, $\gamma \in (\beta, 1)$, $nml \geq 1$. For each $k \geq 0$, assume Ψ is differentiable at $\mathbf{u}^k$. Set $k = 0$.

Do while. $\|F\|_\infty \geq opt$, $\|\mathbf{u}^{k+1} - \mathbf{u}^k\|_\infty \geq steptol$, and $k \leq itlimit$.
1. Find the Newton's direction for (3): $\Delta\mathbf{u}^k$.
2. (a) Set $\alpha^{k,0} = s$, i $= 0$.
   (b) Find the smallest nonnegative integer l for which

$$\Psi(\mathbf{u}^k) - \Psi\left(\mathbf{u}^k + \beta^l \alpha^{k,i} \Delta\mathbf{u}^k\right) \geq_{0 \leq j \leq m(k)} -\sigma\beta^l\alpha^{k,i}\nabla\Psi(\mathbf{u}^k)^T\Delta\mathbf{u}^k.$$

   where $m(0) = 0$ and $0 \leq m(k) \leq \min[m(k-1)+1, nml]$.
   (c) If Ψ is nondifferentiable at $\left(\mathbf{u}^k + \beta^l\alpha^{k,i}\Delta\mathbf{u}^k\right)$, find $t \in [\gamma, 1)$ so that
   Ψ is differentiable at $\left(\mathbf{u}^k + t\beta^l\alpha^{k,i}\Delta\mathbf{u}^k\right)$, set $\alpha^{k,i+1} = t\beta^l\alpha^{k,i}$, $i +$
   $1 \to i$, go to step 2b.
   Otherwise, set $\alpha^k = \beta^l\alpha^{k,i}$, $\mathbf{u}^{k+1} = \mathbf{u}^k + \alpha^k\Delta\mathbf{u}$, $k+1 \to k$.

The Equations $F(\mathbf{u}) = 0$

$$\sum_{i=1}^{n+s} M_i^T \mathbf{z}_i + A^T\boldsymbol{\lambda} = 0 \tag{3a}$$

$$\min\left[\eta_{j-n}, r_{j-n}^{p_j} - \|M_j(\mathbf{x} - \mathbf{f}_j)\|_{p_j}^{p_j}\right] = 0 \quad (j = n+1, \ldots, n+s), \tag{3b}$$

$$\min\left(\lambda_i, b_i - A_i\mathbf{x}\right) = 0 \quad (i = 1, \ldots, m), \tag{3c}$$

$$\left(\sum_{l=1}^{d_i} |M_{il}(\mathbf{x} - \mathbf{f}_i)|^{p_i}\right)^{\frac{1}{q_i}} z_{il} - w_i |M_{il}(\mathbf{x} - \mathbf{f}_i)|^{\frac{p_i}{q_i}} \operatorname{sign}\left(M_{il}(\mathbf{x} - \mathbf{f}_i)\right) = 0 \tag{3d}$$

$$(i \in \{1, \ldots, n\}, p_i \geq 2; \; l = 1, \ldots, d_i)$$

$$\left(\sum_{l=1}^{d_i} |M_{il}(\mathbf{x} - \mathbf{f}_i)|^{p_i}\right)^{\frac{1}{p_i}} |z_{il}|^{\frac{q_i}{p_i}} \operatorname{sign}(z_{il}) - w_i^{\frac{q_i}{p_i}} M_{il}(\mathbf{x} - \mathbf{f}_i) = 0 \tag{3e}$$

$$(i \in \{1, \ldots, n\}, p_i < 2; \; l = 1, \ldots, d_i)$$

$$\left(\sum_{l=1}^{d_j} |M_{jl}(\mathbf{x} - \mathbf{f}_j)|^{p_j}\right)^{\frac{1}{q_j}} z_{jl} - \eta_{j-n} |M_{jl}(\mathbf{x} - \mathbf{f}_j)|^{\frac{p_j}{q_j}} \operatorname{sign}\left(M_{jl}(\mathbf{x} - \mathbf{f}_j)\right) = 0 \tag{3f}$$

$$(j \in \{n+1, \ldots, n+s\}, p_j \geq 2; \; l = 1, \ldots, d_j)$$

$$\left(\sum_{l=1}^{d_j} |M_{jl}(\mathbf{x} - \mathbf{f}_j)|^{p_j}\right)^{\frac{1}{p_j}} |z_{jl}|^{\frac{q_j}{p_j}} \operatorname{sign}(z_{jl}) - \eta_{j-n}^{\frac{q_j}{p_j}} M_{jl}(\mathbf{x} - \mathbf{f}_j) = 0 \tag{3g}$$

$$(j \in \{n+1, \ldots, n+s\}, p_j < 2; \; l = 1, \ldots, d_j).$$

4   Numerical Experiments

We have implemented the above algorithm in MATLAB. We adopt the suggested parameters in [1]. The machine accuracy of the computer running the code is $\epsilon = 2.2204e - 16$. Our computer program stops either

$\|F\|_\infty < opt = \epsilon^{1/3} = 6.0555e - 5$, or the infinity norm of the Newton's direction is less than $steptol = \epsilon^{2/3}$; or the number of iterations exceeds $itlimit = 100$. We set $s = 1$, $\beta = 0.5$, $\sigma = 1.0e - 4$, $nml = 10$.

Below is a numerical example with 10 existing facilities and 2 resources:

$f_1 = (0.8175669, 0.9090309, 0.2491902)$, $f_2 = (0.3332802, 0.4928804, 0.0171481)$,
$f_3 = (0.5420494, 0.8212501, 0.3767346)$, $f_4 = (0.5911344, 0.9217661, 0.6447813)$,
$f_5 = (0.2692597, 0.2433543, 0.6320366)$, $f_6 = (0.9949809, 0.8636484, 0.3300618)$,
$f_7 = (0.5093652, 0.9573201, 0.1044826)$, $f_8 = (0.2043801, 0.8580418, 0.2739731)$,
$f_9 = (0.1839796, 0.2708408, 0.7940208)$, $f_{10} = (0.4267598, 0.0453742, 0.1054062)$,
$f_{11} = (0.2745187, 0.5993611, 0.1221155)$, $f_{12} = (0.2181376, 0.9671702, 0.5442553)$.

The weights for $\mathbf{f}_1, \ldots, \mathbf{f}_{10}$ are: $\mathbf{w} = (0.7781196, 0.2842939, 0.0304183, 0.4431445, 0.2179517, 0.5138524, 0.3852747, 0.4875440, 0.4408099, 0.6238188)$.

The p for $\mathbf{f}_1, \ldots, \mathbf{f}_{12}$ are $\mathbf{p} = (1.7, 1.5, 2.5, 2.1, 2.5, 2.1, 2.3, 2.5, 2.5, 2.1, 2.9, 1.9)$.

The transformation matrices for the distance measure between the new facility and the existing facilities and the resources are:

$$M_1 = \begin{bmatrix} 0.4043783 & 0.2878919 & 0.5358692 \\ 0.4152851 & 0.5412418 & 0.2719560 \\ 0.3998512 & 0.3457063 & 0.3102344 \end{bmatrix} \quad M_2 = \begin{bmatrix} 0.6493642 & 0.4302950 & 0.8370728 \\ 0.9980712 & 0.2561977 & 0.7632986 \\ 0.2469849 & 0.1924696 & 0.8870779 \end{bmatrix}$$

$$M_3 = \begin{bmatrix} 0.3580290 & 0.1878973 & 0.5333254 \\ 0.5734145 & 0.0221892 & 0.1189078 \\ 0.8408229 & 0.0575776 & 0.8342093 \end{bmatrix} \quad M_4 = \begin{bmatrix} 0.3855365 & 0.5720651 & 0.8868140 \\ 0.0013690 & 0.1720985 & 0.6703435 \\ 0.5097000 & 0.4759156 & 0.7431721 \end{bmatrix}$$

$$M_5 = \begin{bmatrix} 0.1805674 & 0.2649339 & 0.6593017 \\ 0.9350963 & 0.9854972 & 0.6605832 \\ 0.9342599 & 0.8032064 & 0.4336484 \end{bmatrix} \quad M_6 = \begin{bmatrix} 0.6954798 & 0.2586548 & 0.0355097 \\ 0.7623554 & 0.3843390 & 0.4729695 \\ 0.4053601 & 0.0080771 & 0.2474898 \end{bmatrix}$$

$$M_7 = \begin{bmatrix} 0.0132654 & 0.9779720 & 0.0447804 \\ 0.6306576 & 0.7188327 & 0.8326453 \\ 0.5479869 & 0.5564485 & 0.2328634 \end{bmatrix} \quad M_8 = \begin{bmatrix} 0.1321284 & 0.6354911 & 0.6550676 \\ 0.4827887 & 0.5106806 & 0.1679761 \\ 0.5706509 & 0.5329994 & 0.1684392 \end{bmatrix}$$

$$M_9 = \begin{bmatrix} 0.5942561 & 0.9369910 & 0.3738004 \\ 0.8647487 & 0.7414568 & 0.2842745 \\ 0.1066821 & 0.7481810 & 0.2772064 \end{bmatrix} \quad M_{10} = \begin{bmatrix} 0.7579091 & 0.6877772 & 0.4022239 \\ 0.1689570 & 0.2506596 & 0.7048818 \\ 0.1025312 & 0.9542088 & 0.6352403 \end{bmatrix}$$

$$M_{11} = \begin{bmatrix} 0.0136439 & 0.2677678 & 0.4533213 \\ 0.8115766 & 0.2669812 & 0.5141040 \\ 0.2874519 & 0.7550024 & 0.1576553 \end{bmatrix} \quad M_{12} = \begin{bmatrix} 0.8747555 & 0.7809548 & 0.8933810 \\ 0.8366846 & 0.5110550 & 0.1887794 \\ 0.8529809 & 0.5310465 & 0.3032273 \end{bmatrix}$$

The data for the feasible region are

$$A = \begin{pmatrix} -0.3725763 & -0.6852143 & -0.2041526 \\ -0.9844071 & -0.2049625 & -0.8364794 \\ 0.4568025 & 0.3087572 & 0.0137238 \end{pmatrix},$$

$$\mathbf{b} = (-0.8759974, -1.2869180, 0.5790744)^T.$$

The upper bounds for the distances are $\mathbf{r} = (0.6361848, 0.5080279)^T$.

We start from $\mathbf{x} = \mathbf{0}$, $\mathbf{z} = \mathbf{0}$, $\boldsymbol{\lambda} = \mathbf{0}$, $\boldsymbol{\eta} = \mathbf{0}$. Iterations are depicted in the figure "Example 1". We then perturb each element of $\mathbf{b}$, A, $\mathbf{f}$, randomly by a real number in $(-0.5, 0.5)$ respectively. The starting point for each perturbed problem is the solution to "Example 1". The iterations are described in the figures: "Perturbation of A", "Perturbation of b", and "Perturbation of f".

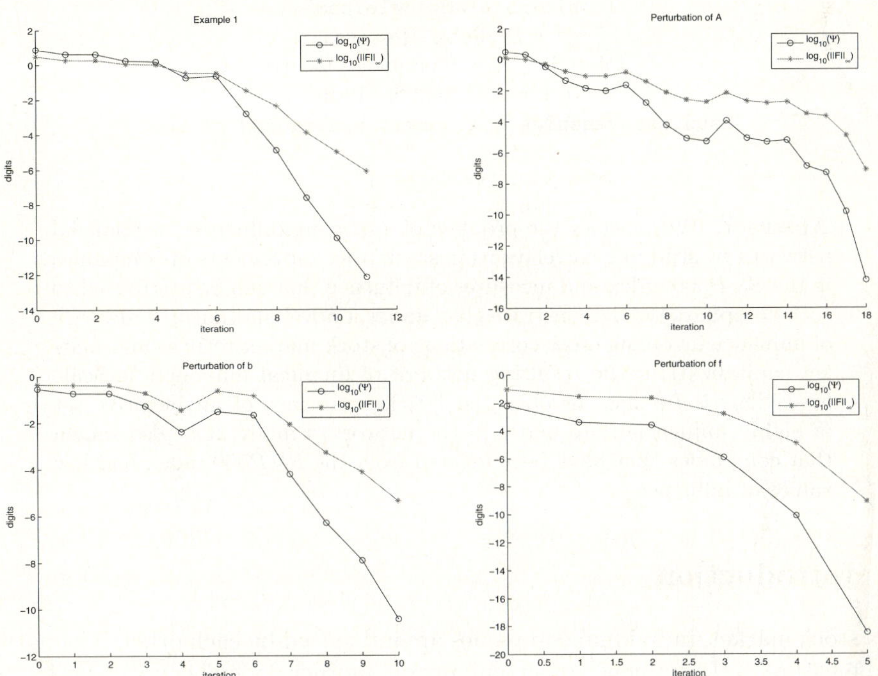

References

1. John E. Dennis, Jr. and Robert B. Schnabel. *Numerical methods for unconstrained optimization and nonlinear equations.* Prentice Hall Series in Computational Mathematics. Prentice Hall Inc., Englewood Cliffs, NJ, 1983.
2. J. Fernández, P. Fernández, and B. Pelegrin. Estimating actual distances by norm functions: a comparison between the $l_{k,p,\theta}$-norm and the $l_{b_1,b_2,\theta}$-norm and a study about the selection of the data set. *Comput. Oper. Res.*, 29(6):609–623, 2002.
3. R.F. Love and JG Morris. Modelling inter-city road distances by mathematical functions. *Operational Research Quarterly*, 23:61–71, 1972.
4. R.F. Love and JG Morris. Mathematical models of road travel distances. *Management Science*, 25:130–139, 1979.
5. Robert F. Love. The dual of a hyperbolic approximation to the generalized constrained multi-facility location problem with l_p distances. *Management Sci.*, 21(1):22–33, 1974/75.
6. R. Tyrrell Rockafellar. *Convex analysis.* Princeton University Press, Princeton, N.J., 1970. Princeton Mathematical Series, No. 28.
7. Yu Xia. An algorithm for perturbed second-order cone programs. Technical Report AdvOl-Report No. 2004/17, McMaster University, 2004.

Financial Influences and Scale-Free Networks

Nitin Arora[1], Babu Narayanan[2], and Samit Paul[2]

[1] Google Inc., Bangalore 560001, India
nitinarora@google.com*
[2] GE Global Research,
John F. Welch Technology Centre,
Bangalore 560066, India
babu.narayanan@ge.com, samit.paul@geind.ge.com

Abstract. We consider the problem of analyzing influences in financial networks by studying correlations in stock price movements of companies in the $S\&P500$ index and measures of influence that can be attributed to each company. We demonstrate that under a novel and natural measure of influence involving cross-correlations of stock market returns and market capitalization, the resulting network of financial influences is Scale Free. This is further corroborated by the existence of an intuitive set of highly influential hub nodes in the network. Finally, it is also shown that companies that have been deleted from the $S\&P500$ index had low values of influence.

1   Introduction

In a stock market, individual companies are influenced by each other. The nature and measure of the influence of a company on another depends on various factors including their sectors of operation, direct buyer-seller relation, acquisition, etc. A quantity that measures the direct influence among companies is the relationship between their stock price changes. This can be measured using the pairwise correlations of the two stock prices averaged over a time window. Analysis of such correlations, usually using matrices, is a well studied topic in computational finance. In this paper, following Kim et al [1], we pursue a graph theoretic approach to obtain interesting structural properties in this correlation matrix. A weighted discrete graph is constructed with the nodes representing the companies and the weight of the edge between two companies equals the correlation. The analysis is carried out on this graph.

Scale free graphs and networks have generated a lot of interest lately. A graph is said to be scale free if the degree sequence of the graph follows a power law - the probability that a vertex has k neighbors is proportional to k^γ. Scale free networks show up in a large number of real world scenarios, especially large complex networks. For example, it is shown in [2] that the scale free property applies to the world wide web graph consisting of documents as vertices, two

---

* The work of the first author was carried out while he was visiting the John F. Welch Technology Centre, Bangalore 560066, India.

V.N. Alexandrov et al. (Eds.): ICCS 2006, Part I, LNCS 3991, pp. 16–23, 2006.

documents being connected if a hyperlink points one document to the other. Many social and business networks are also scale free [3]. In nature, one finds that cellular metabolic networks and protein interaction networks are scale free [4, 5]. Scale free networks have some interesting properties. There are a few nodes with very high degree of connectivity - called hubs. Most nodes on the other hand have very few neighbors. A computer network that is scale free would be resistant to random security attacks, but would be vulnerable to planned attacks on the hubs, see [6]. A popular theoretical model for the evolution of scale free networks uses the notion of preferential attachment and growth - a new vertex that is added attaches itself, with higher probability, to vertices that are already highly connected. See [7, 8, 9].

In this paper, we study the network of companies in the $S\&P500$ index. [1] used cross correlations as the metric and defined the influence of a company as the absolute sum of the crosscorrelations involving the company and conjectured that the resulting influence graph would be scale free. Here, note that the graph is a weighted fully connected graph and scale free refers to the existence of a power law governing the set of influences. Our analysis of similar data shows that under the measure in [1] the companies with the largest influence tended to all come out of the same sector, namely, semiconductor sector and included many relatively smaller companies. We present a novel alternate measure involving cross correlations and market capitalization under which the $S\&P500$ influence network is scale free. Also, the list of companies with large influence is intuitive. We also examine the companies that have been deleted from the index in 2004-2005 and observe that the deleted ones indeed had low influence over a period of time.

The rest of the paper is organized as follows. Section 2 provides the details of the data used. Section 3 discusses the different measures of influence and analyzes the results obtained. Finally, section 4 summarizes the results and provides direction for further work.

2   Data

The daily stock price data, adjusted for splits and dividends, for $S\&P500$ companies from 1994 to 2005 was downloaded from the website www.finance.yahoo.com. Only companies that remained in the index from 1994 to 2004 were considered for the main part of the analysis. The final experiments were done for 416 companies that remained after data cleaning. The market capitalization values for these companies were downloaded for a particular day in July 2005 as a representative value. We calculated the cross correlations and influence values over the period 1994 to 2005 for various window sizes ranging from 3 months to 4 years. In this paper, we present some representative results.

3   Analysis and Results

Let the stock price of company $i(i = 1, ..., N)$ at time t be $Y_i(t)$. The return of the stock price after a time-interval Δt is given by

$$S_i(t) = \ln Y_i(t + \Delta t) - \ln Y_i(t), \tag{1}$$

meaning the geometrical change of $Y_i(t)$ during the interval Δt. We take $\Delta t = 1$ day for the following analysis. The cross-correlations between individual stocks are considered in terms of a matrix C, whose elements are given as

$$C_{i,j} = \frac{<S_i S_j> - <S_i><S_j>}{\sqrt{(<S_i^2> - <S_i>^2)(<S_j^2> - <S_j>^2)}}, \tag{2}$$

where the brackets mean a temporal average over the period we studied. Then $C_{i,j}$ can vary between $[-1, 1]$, where $C_{i,j} = 1(-1)$ means that two companies i and j are completely positively(negatively) correlated, while $C_{i,j} = 0$ means that they are uncorrelated. It is known that the distribution of the coefficients $\{C_{i,j}\}$ is a bellshaped curve, and the mean value of the distribution is slowly time-dependent, while the standard deviation is almost constant. The time-dependence of the mean value might be caused by external economic environments such as bank interest, inflation index, exchange rate, etc, which fluctuates from time to time. To extract intrinsic properties of the correlations in stock price changes, we look at the following quantity as in [1],

$$G_i(t) = S_i(t) - \frac{1}{N} \sum_i S_i(t), \tag{3}$$

where $G_i(t)$ indicates the relative return of a company i to its mean value over the entire set of N companies at time t. The cross-correlation coefficients are redefined in terms of G_i,

$$w_{i,j} = \frac{<G_i G_j> - <G_i><G_j>}{\sqrt{(<G_i^2> - <G_i>^2)(<G_j^2> - <G_j>^2)}}. \tag{4}$$

To check that the distribution $P(w)$ is time-independent, [1] looked at $P(w)$ by taking temporal average in equation (4) over each year from 1993 to 1997. The distribution of $\{w_{i,j}\}$ for the period $1994 - 1997$ are shown in Figure 1 and match the distribution obtained by [1].

Let us now look at the weighted graph defined as follows - the vertices of the graph are individual companies, each pair (i, j) of vertices has an edge between them with an associated weight $e_{i,j}$. The value $e_{i,j}$ represents the influence that the companies exert on each other. The total influence strength s_i of company i can then be defined as the sum of the weights on the edges incident upon the vertex i, that is,

$$s_i = \sum_j e_{i,j}, \tag{5}$$

What will be a good way to calculate $e_{i,j}$? Does it depend only on the cross-correlation coefficients? Should the influence of company i on j be the same as that of j on i? [1] took $e_{i,j} = w_{i,j}$ in the above definition of influence strength (note that it is symmetric). For them,

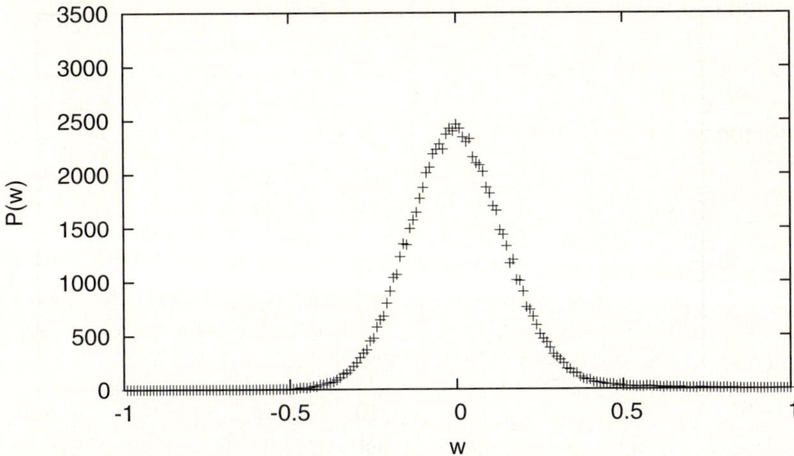

Fig. 1. Plot of the distribution of the cross correlation coefficients over the period 1994 to 1997. $P(w)$ denotes the number of pairs of companies with cross correlation coefficient $w \pm 0.005$.

$$s_i = \sum_{j \neq i} w_{i,j}. \tag{6}$$

We obtained the values $w_{i,j}$ using stock prices of $S\&P500$ companies over a period of 4 years(1994-1997). It was observed that the definition in (6) assigns the highest values of influence strengths to companies such as AMAT, PMCS, KLAC, XLNX, LSI - surprisingly all of them in the semiconductor sector, while the companies such as GE, JNJ, PG, XOM and MSFT all landed in lower influence region (the symbols used for companies can be looked up for company name at www.finance.yahoo.com). Upon analysis, it was inferred that this was caused by a strong correlation between semiconductor companies, which in some cases was as high as 0.6, a very rare value otherwise. The typical graph of the distribution $P(|s|)$ of the influence strength $|s|$ looked like in Figure 2.

Above observations suggest using the values of market capitalizations of individual companies while calculating the influence strengths. We first attempted the following formula to calculate s_i,

$$s_i = \sum_{j} w_{i,j} \times M_j, \tag{7}$$

where M_j is the market capitalization of company j. That is, the influence of two companies A and B on the market should not be the same value if A exerts high influence on bigger companies but B exerts high influence on smaller companies. Also, for the case $j = i(w_{i,j} = 1)$, we are adding the market capitalization of company i to s_i which makes intuitive sense. This definition of influence strength moved many of the bigger companies to the top. The most influential companies given by this formula are shown in Table 1. The influence strength was calculated using a window of size two years(2003-04).

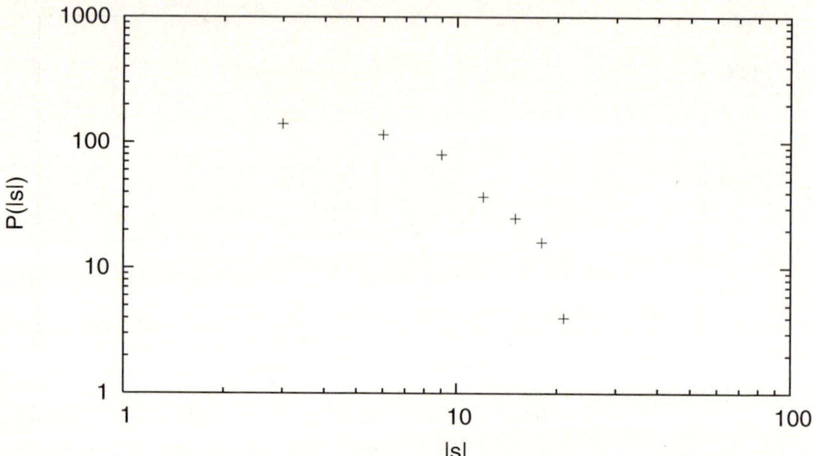

Fig. 2. Plot of the influence strength distribution $P(|s|)$ versus the absolute value of the influence strength $|s|$ as defined in [1]. (in double-logarithmic scales) $P(|s|)$ denotes the number of companies with influence $s \pm 1.5$.

Table 1. Top influential companies under equation (7)

Symbol	Mkt Cap(B$)	Infl Str(B$)
WFC	102.66	1069.33
PG	130.96	1047.74
XOM	376.28	1032.79
JNJ	192.66	980.272
BUD	35.67	964.955
BAC	181.41	938.318
PMCS	1.73	913.625
GE	368.24	875.932
C	241.69	842.594
PEP	88.05	832.869

Table 2. Top influential companies under equation (8)

Symbol	Mkt Cap(B$)	Infl Str(B$)
XOM	376.28	720.45
PG	130.96	590.82
JNJ	192.66	589.969
BAC	181.41	555.194
GE	368.24	552.366
WFC	102.66	539.617
C	241.69	509.567
WMT	206.51	471.956
PEP	88.05	404.739
PFE	199.04	392.209

3.1   Scale Free Network

We now propose our final measure. To motivate, let us calculate the influence of company i on the market in two steps. The symmetric measure of cross correlation $w_{i,j}$ between i and j can not be interpreted as influence of i on j as well as the influence of j on i. This is too simplistic. Suppose a company i with market capitalization $M_i(\sim 1B)$ is supplying a raw material to a manufacturer j with market capitalization $M_j(\sim 300B)$. Suppose further that the stock price of i is correlated to that of j with $w_{i,j}(\sim 0.2)$. We propose that the influence exerted by i on j, denoted $Infl(i \rightarrow j)$, is $w_{i,j} \times \frac{M_i}{M_i+M_j}$. Similarly define the influence by j on i. Now to calculate the influence of i on the market we will sum

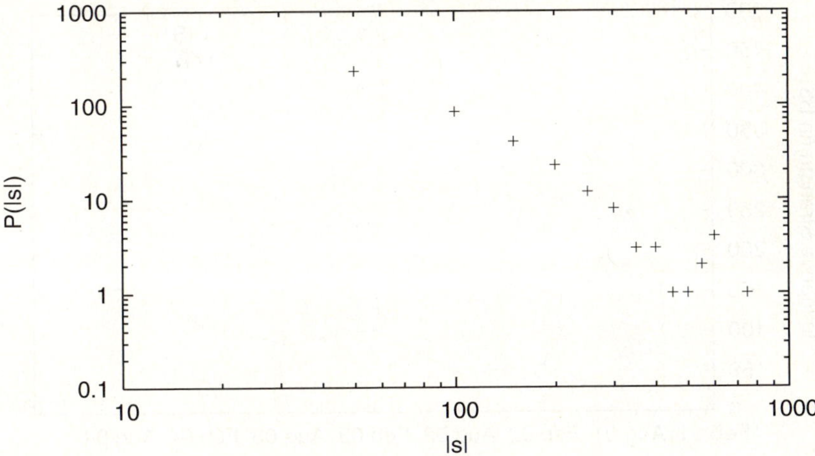

Fig. 3. Plot of the influence strength distribution $P(|s|)$ versus the absolute value of the influence strength $|s|$ as given by equation (8)(in double-logarithmic scales). We looked at the stock returns for two years 2003-2004. $P(|s|)$ denotes the number of companies with influence $s \pm 25$

over all companies j the value of influence of i on j multiplied by the market capitalization of j. This gives

$$s_i = \sum_j Infl(i \to j) \times M_j = \sum_j w_{i,j} \times \frac{M_i M_j}{M_i + M_j}. \tag{8}$$

This definition of influence strength gives a much more satisfactory ranking to companies. The top 10 companies are shown in Table 2.

Moreover, the plot of influence strength distribution $P(|s|)$ as a function of $|s|$ follows a power-law, $P(|s|) \sim |s|^{-\delta}$, with $\delta \approx 2.2$. See Figure 3. Thus the network of influences among these companies is indeed scale free. The outliers towards the high influence region may be attributed to extremely dominant companies in separate sectors. In particular, note that $P(|s|)$ in these cases is small.

Comparison of the Walt Disney Co and Texas Instruments Inc(both with market capitalization $\sim 50B$) calculated over various 6 month windows between 2001 and 2004 shows that our measure does not give undue importance to market capitalization (See Figure 4).

3.2   Deletion

Finally, we looked at some companies that were deleted from the $S\&P500$ recently, namely, American Greetings(Apr '04), Thomas and Belts(Aug '04), Worthington Industries(Dec '04) and Deluxe Corp.(Dec '04). Figure 5 shows the influence strength of these companies for the eight half-yearly periods between Feb 2001 and Feb 2005. A downward trend in their influence strengths just be-

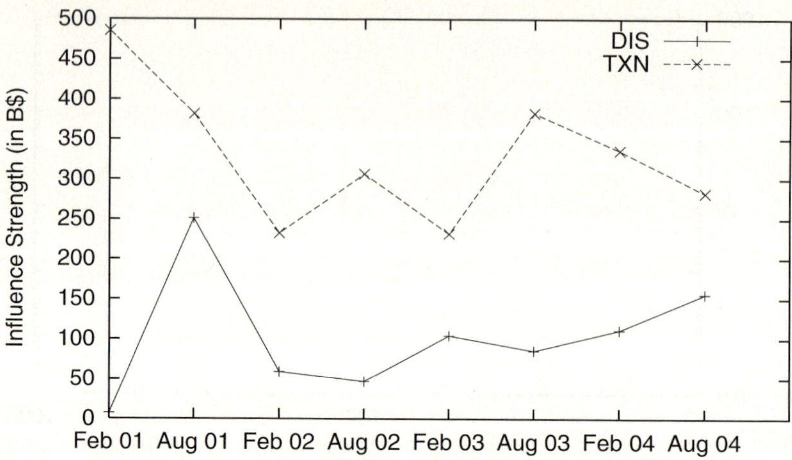

Fig. 4. Comparison of the influence strengths of DIS and TXN between 2001 and 2004

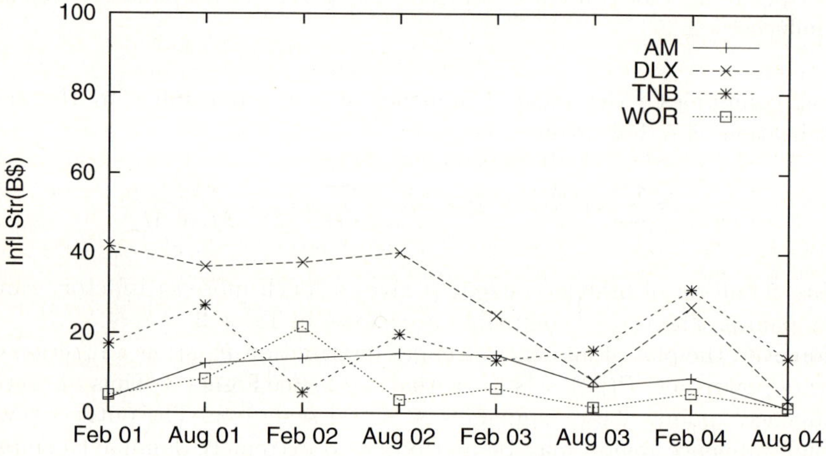

Fig. 5. Influence strengths of AM, DLX, TNB and WOR between Feb 2001 and Feb 2005. These were removed from the list of S&P500 companies in 2004. Note the low values relative to the companies in Table 2 having influence > 350.

fore being removed from the $S\&P500$ index is seen. Also these influence strength values are substantially less than most of the other companies.

4   Conclusion

We demonstrate a weighted scale free network of influences that exists among $S\&P500$ companies. We have come up with a measure for the influence of a company on another as well as on the whole market by utilizing cross-correlations on movements of stock prices as well as market capitalization. Our definition for the influence strength is observed to intuitively give a satisfactory ranking to the $S\&P500$ companies and can also explain the deletion of companies from this index in the period from 2004 to 2005. Some open questions include analysis of other financial networks, analysis of the theoretical models for scale free networks in this context. An important problem is to derive methods of risk analysis for finance, supply chain and other domains incorporating concepts from the theory of scale free networks.

References

1. Kim, H.J., Lee, Y., Kim, I.M., Kahng, B.: Weighted scale-free network in financial correlations. Physical Society of Japan **71** (2002) 2133–2136
2. Albert, R., Jeong, H., Barabási, A.L.: Scale-free characteristics of random networks: The topology of the world-wide web (2000)
3. Barabási, A.L., Bonabeau: Scale-free networks. SCIAM: Scientific American **288** (2003)
4. Hu, X.: Mining and analysing scale-free protein–protein interaction network. Int. J. of Bioinformatics Research and Applications **1** (2005) 81–101
5. Jeong, H., Tomber, B., Albert, R., Oltvai, Z.N., Barabási, A.L.: The large-scale organization of metabolic networks. Nature **407** (2000) 651–654
6. Albert, R., Jeong, H., Barabási, A.L.: Error and attack tolerance of complex networks. NATURE: Nature **406** (2000)
7. Bollobás, B., Riordan: Mathematical results on scale-free random graphs. In Bornholdt, Schuster, eds.: Handbook of Graphs and Networks: From the Genome to the Internet. (2003)
8. Bollobás, B., Riordan, Spencer, Tusnady: The degree sequence of a scale-free random graph process. RSA: Random Structures & Algorithms **18** (2001)
9. Bollobás, B., Borgs, C., Chayes, J.T., Riordan, O.: Directed scale-free graphs. In: SODA. (2003) 132–139

Comparison of Simulation and Optimization Possibilities for Languages: DYNAMO and COSMIC & COSMOS – on a Base of the Chosen Models

Elżbieta Kasperska, Elwira Mateja-Losa, and Damian Słota

Institute of Mathematics,
Silesian University of Technology,
Kaszubska 23, 44-100 Gliwice, Poland
{e.kasperska, e.mateja, d.slota}@polsl.pl

Abstract. On the base of the chosen models, the comparison of simulation and optimization possibilities for languages DYNAMO and COSMIC & COSMOS, is presented. The computational techniques, for example: integration facilities, optimization opportunities, are the main point of interest for this comparison.

1 Introduction

The problem of modelling, simulation and optimization of complex, nonlinear, dynamical and multilevel systems, authors already have undertaken in many papers (see [3, 4, 5, 6, 7, 8, 9, 10, 11]). In mentioned works the main attention on the problem of modelling, structure of object, results of experiments, were paid. Now we are concentrating on some technical aspects of realization of these experiments using languages: Professional DYNAMO and COSMIC & COSMOS [1,13]. The computational techniques, for example: integration facilities, optimization opportunities, are the main point of interest. Specially, two different philosophies of embedding simulation and optimization, will be undertaken. The locally or globally solutions can be confronted with the decision-makers preferences and objectives, giving interesting issues for building, for example, decision support systems.

2 Some Technical Aspects of Computational Opportunities for Languages: DYNAMO and COSMIC & COSMOS

Realization of simulation and optimization experiments on models type System Dynamics [2, 3, 4, 5, 6, 7, 8, 9, 10, 11] requires special languages and computational techniques.

V.N. Alexandrov et al. (Eds.): ICCS 2006, Part I, LNCS 3991, pp. 24–29, 2006.

Two main problems have occured. The one is connected with integration in simulation computation in models of complex, nonlinear, dynamic and multi-level systems. The second problem is related with two philosophies of, so called, embedding simulation in optimization and vice versa on models type System Dynamics.

The Professional DYNAMO [13] provides two types of integration:

- Euler method,
- Runge-Kutta method.

If the Euler's method is used, attention should be paid to a proper choice of the simulation step.If the Runge-Kutta method is used, the simulation step is automatically partitioned to obtain a solution for a given exactness. The example of using both methods is described in paper [7].

In COSMIC (Computer Oriented System Modelling – Integrated Concept) the main attention is paid on simulation step and delays (different orders). To ensure that instability does not occur from delays, the value of DT should be one quarter of the smallest first order delay or one twelfth of the smallest third order delay duration specified in the model (see [1, pp. 34–35]).

The second technical aspect of computational opportunities of mentioned languages is optimization facilities. COSMOS (Computer – Oriented System Modelling Optimization Software) is a software tool which automatically links a dynamic simulation model to an optimization package. This facility makes it possible to apply powerful optimization techniques to:

- the fine tuning of policies in the model (Direct Optimization),
- sensitivity analysis of the model (Base Vector Analysis),
- simplification of the structure of the model (Simplification),
- exploring the effects of forecasting and forward planning in the model (Planning Horizon).

All of these types of optimization are, so called, simulation embedded in optimization. Different philosophy is the optimization embedded in simulation model type System Dynamics. Generally speaking, locally in structure of the model is embedded the optimized decision rule, which dynamically, during horizon of simulation, is changing the structure of model, giving solution (or pseudosolution [12]) which optimized the chosen objective function. Such embedding, authors already have applied in some models [6,9].

3   Comparison of Simulation and Optimization Possibilities

The scope of the paper not allows to undertaken all interesting aspects of mentioned comparison. Lets concentrate of one of them. The problem of embedding optimization in simulation of System Dynamics models is connected with evolution of structure in these models, and applying "hybrid" ideas. Below the idea of such evolution, in the scheme of the main structure of the computational

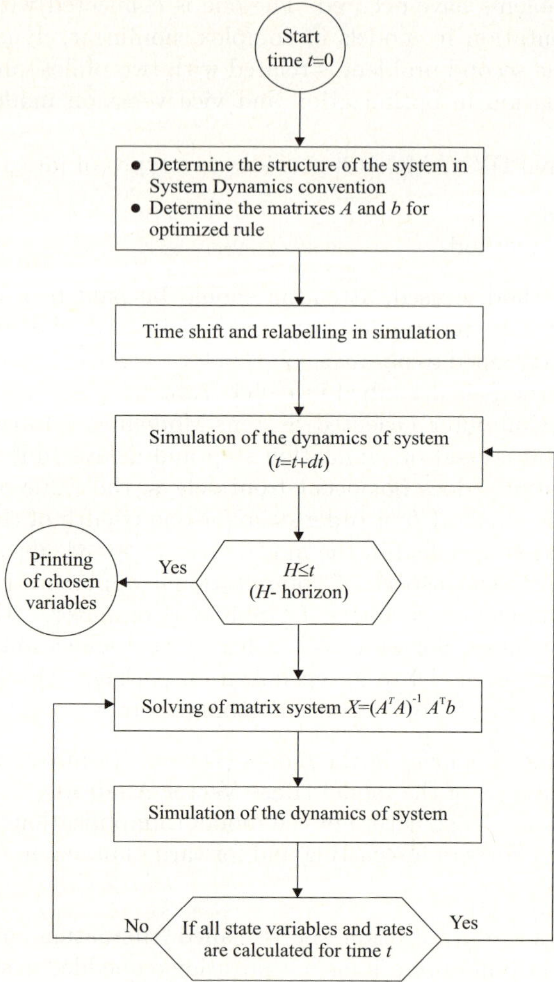

Fig. 1. Block diagram of optimization embedded in simulation on System Dynamics model – on a base of the example of model DYNBALANCE(2-2) [9]

program is presented on Figure 1. For comparison, the block diagram of simulation embedded in optimization on models type System Dynamics is presented on Figure 2.

Comparing both diagrams, we can see that in first case the achieved solution has locally meaning. Contrary, in second case we have obtained the globally optimized solution. Reminding, that objective function measures interesting aspects of dynamic behaviour of systems (we put attention, our preferences on them), we obtain solution optimal in whole horizon of simulation. Such are the technical possibilities of both computational programs in appropriate languages: DYNAMO and COSMIC & COASMOS.

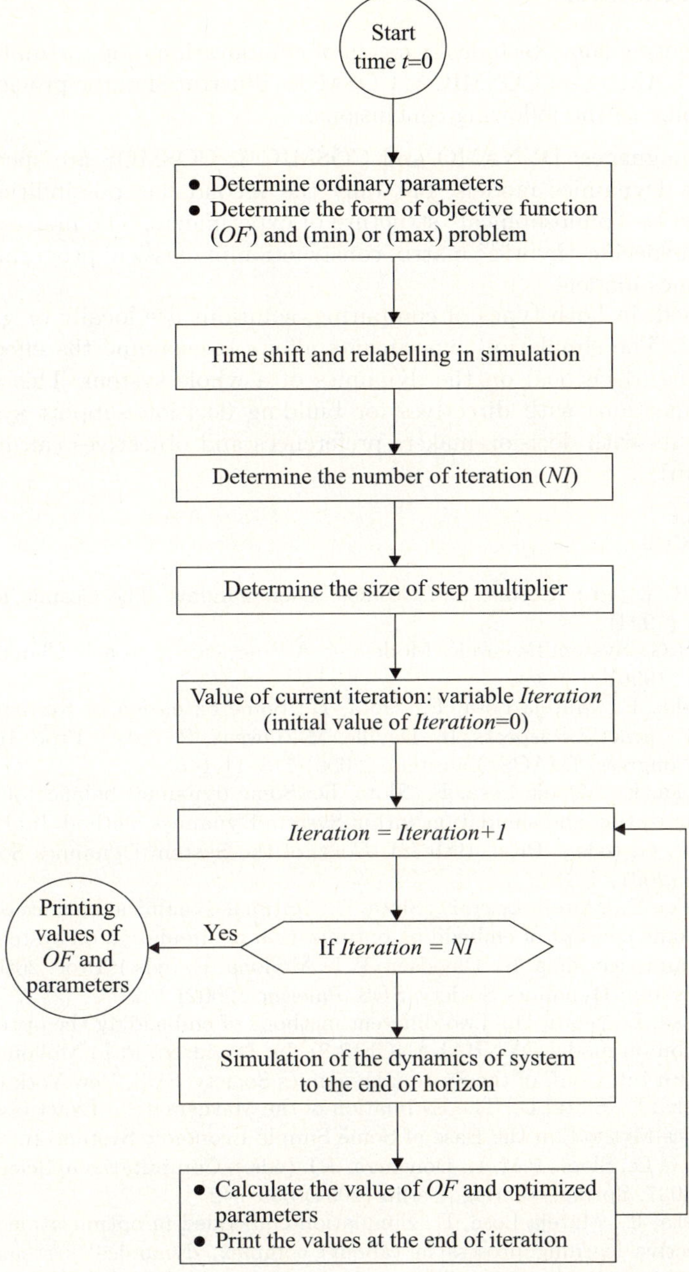

Fig. 2. Block diagram of simulation embedded in optimization on System Dynamics model – on a base of the example of model DYNBALANCE(3-1-III) [8]

4 Conclusions

After presenting some technical aspects of computational opportunities for languages DYNAMO and COSMIC & COSMOS, illustrated in the previous section, we have come to the following conclusions:

- Both languages: DYNAMO and COSMIC & COSMOS are specialized for System Dynamics models, but only the second has possibilities that are build in its "philosophical" structure of experiments. The first required embedded specific "hybrid" matrix construction in classical program of System Dynamics models.
- Obtained, in both types of computing, solutions are locally or globally optimized. The simulation experiments allows to examine the effects of such solutions (decisions) on the dynamics of a whole system. This aspect has its connections with directives for building decision support systems (experiments with decision-makers preferences and objectives can help in this problem).

References

1. Coyle, R.G. (ed.): Cosmic and Cosmos. User manuals. The Cosmic Holding Co, London (1994)
2. Coyle, R.G.: System Dynamics Modelling. A Practical Approach. Chapman & Hall, London (1996)
3. Kasperska, E., Mateja-Losa, E., Słota, D.: Some extension of System Dynamics method – practical aspects. In: Deville, M., Owens, R. (eds.): Proc. 16th IMACS World Congress. IMACS, Lausanne (2000) 718–11 1–6
4. Kasperska, E., Mateja-Losa, E., Słota, D.: Some dynamics balance of production via optimization and simulation within System Dynamics method. In: Hines, J. H., Diker, V. G. (eds.): Proc. 19th Int. Conf. of the System Dynamics Society. SDS, Atlanta (2001) 1–18
5. Kasperska, E., Mateja-Losa, E., Słota, D.: Optimal dynamical balance of raw materials – some concept of embedding optimization in simulation on system dynamics models and vice versa. In: Davidsen, P. I., Mollona, E. (eds.): Proc. 20th Int. Conf. of the System Dynamics Society. SDS, Palermo (2002) 1–23
6. Kasperska, E., Słota, D.: Two different methods of embedding the optimization in simulation on model DYNBALANCE(2-2). In: Davidsen, P. I., Mollona, E. (eds.): Proc. 20th Int. Conf. of the System Dynamics Society. SDS, New York (2003) 1–27
7. Kasperska, E., Słota, D.: The Estimation of the Mathematical Exactness of System Dynamics Method on the Base of Some Simple Economic System. In: Bubak, M., Albada, G.D., Sloot, P.M.A., Dongarra, J.J. (eds.): Computational Science, Part II. LNCS 3037, Springer-Verlag, Berlin (2004) 639–642
8. Kasperska, E., Mateja-Losa, E.: Simulation embedded in optimization – a key for the effective learning prosess in (about) complex, dynamical systems. In: Sunderam, V.S., Albada, G.D., Sloot, P.M.A., Dongarra, J.J. (eds.): Computational Science, Part III. LNCS 3516, Springer-Verlag, Berlin (2005) 1040–1043
9. Kasperska, E., Słota, D.: Optimization embedded in simulation on models type System Dynamics – some case study. In: Sunderam, V.S., Albada, G.D., Sloot, P.M.A., Dongarra, J.J. (eds.): Computational Science, Part I. LNCS 3514, Springer-Verlag, Berlin (2005) 837–842

10. Kasperska, E.: Some remarks about chosen structural aspect of System Dynamics method. In: 6 éme Congrés Européen de Science des Systémes, AFSCET, Paris (2005) 1–5
11. Kasperska, E., Mateja-Losa, E., Słota, D.: Multicriterion choices in System Dynamics – some way of modelling and simulation. In: 6 éme Congrés Européen de Science des Systémes, AFSCET, Paris (2005) 1–6
12. Legras, J.: Methodes et Techniques De'Analyse Numerique. Dunod, Paris (1971)
13. Pugh, A.L. (ed.): Professional Dynamo 4.0 for Windows. Tutorial Guide. Pugh-Roberts Associates, Cambridge (1994)

Bond Pricing with Jumps and Monte Carlo Simulation

Kisoeb Park[1], Moonseong Kim[2], and Seki Kim[1],[*]

[1] Department of Mathematics, Sungkyunkwan University,
440-746, Suwon, Korea
Tel.: +82-31-290-7030, 7034
{kisoeb, skim}@skku.edu
[2] School of Information and Communication Engineering,
Sungkyunkwan University,
440-746, Suwon, Korea
Tel.: +82-31-290-7226
moonseong@ece.skku.ac.kr

Abstract. We derive a general form of the term structure of interest rates with jump. One-state models of Vasicek, CIR(Cox, Ingersol, and Ross), and the extended model of the Hull and White are introduced and the jump-diffusion models of the Ahn & Thompson and the Baz & Das as developed models are also investigated by using the Monte Carlo simulation which is one of the best methods in financial engineering to evaluate financial derivatives. We perform the Monte Carlo simulation with several scenarios even though it takes a long time to achieve highly precise estimates with the brute force method in terms of mean standard error which is one measure of the sharpness of the point estimates.

1 Introduction

We introduce one-state variable model of Vasicek[9], Cox, Ingersoll, and Ross (CIR)[3], the extended model of the Hull and White[6], and the development of the models which are the jump-diffusion model of the Ahn and Thompson[1] and the Baz and Das[2]. Conventionally, financial variables such as stock prices, foreign exchange rates, and interest rates are assumed to follow a diffusion processes with continuous paths when pricing financial assets.

In pricing and hedging with financial derivatives, jump-diffusion models are particularly important, since ignoring jumps in financial prices will cause pricing and hedging rates. For interest rates, jump-diffusion processes are particularly meaningful since the interest rate is an important economic variables which is, to some extent, controlled by the government as an instrument. Term structure model solutions under jump-diffusions are justified because movements in interest rates display both continuous and discontinuous behavior. These jumps are caused by several market phenomena money market interventions by the Fed, news surprise, and shocks in the foreign exchange markets, and so on.

---

[*] Corresponding author.

V.N. Alexandrov et al. (Eds.): ICCS 2006, Part I, LNCS 3991, pp. 30–37, 2006.

Ahn and Thompson[1] extended the CIR model by adding a jump component to the square root interest rate process. Using linearization technique, they obtained closed-form approximations for discount bond prices. Also, Baz, and Das[2] extended the Vasicek model by adding a jump component to the Ornstein-Uhlenbeck(O-U) interest rate process, and obtained closed form approximate solutions for bond price by the same linearization technique.

All of the models mentioned above take special functional forms for the coefficients of dt, dW, and $d\pi$ in the stochastic differential equation for r. We derive a general form of the term structure of interest rate with jump and study a solution of the bond pricing for the above models. As above present a model which allows the short term interest rate, the spot rate, the follow a random walk. This leads to a parabolic partial differential equation for the prices of bonds and to models for bonds and many other interest rate derivative products. Above in result, we look into as the Vasicek, the CIR, the Hull and White, and the jump-diffusion models.

In addition, we introduce the Monte Carlo simulation. One of the many uses of Monte Carlo simulation by financial engineers is to place a value on financial derivatives. Interest in use of Monte Carlo simulation for bond pricing is increasing because of the flexibility of the methods in handling complex financial instruments. One measure of the sharpness of the point estimate of the mean is Mean Standard Error(MSE). Numerical methods that are known as Monte Carlo methods can be loosely described as statistical simulation methods, where statistical simulation is defined in quite general terms to be any method that utilizes sequences of random numbers to perform the simulation.

The structure of the remainder of this paper is as follows. In Section 2, the basic of bond prices with jump are introduced. In Section 3, the term structure models with jump are presented. In Section 4, we calculate numerical solutions using Monte Carlo simulation for the term structure models with jump. In Section 5, we investigate bond prices given for the eight models using the Vasicek and the CIR models. This paper is finally concluded in Section 6.

2   Bond Pricing Equation with Jump

In view of our uncertainty about the future course of the interest rate, it is natural to model it as a random variable. To be technically correct we should specify that r is the interest rate received by the shortest possible deposit. The interest rate for the shortest possible deposit is commonly called the **spot rate**. In the same way that a model for the asset price is proposed as a lognormal random walk, let us suppose that the interest rate r is governed by a **stochastic differential equation(SDE)** of the form

$$dr = u(r, t)dt + \omega(r, t)dW + Jd\pi \tag{1}$$

The functional forms of $\omega(r, t)$, $u(r, t)$(the instantaneous volatility and the instantaneous drift, respectively), and jump size J is normal variable with mean μ

and standard deviation γ determine the behavior of the spot rate r. We consider a one-dimensional jump-diffusion process $r(t)$ is satisfying

$$r(t) = r(0) + \int_0^t u(r, s)ds + \int_0^t \omega(r, s)dW(s) + \sum_{i=1}^{\pi(t)} J_i, \qquad (2)$$

where $\pi(t)$ represents the number of jumps happening during the period between time 0 and t. When interest rates follow the **SDE**(1), a bond has a price of the form $V(r, t)$; the dependence on T will only be made explicit when necessary. We set up a riskless portfolio and the jump-diffusion version of Ito's lemma to functions of r and t. And then, we derive the partial differential bond pricing equation.

Theorem 1. *If r satisfy Stochastic differential equation $dr = u(r, t)dt + \omega(r, t) dW + Jd\pi$ then the zero-coupon bond pricing equation in jumps is*

$$\frac{\partial V}{\partial t} + \frac{1}{2}\omega^2\frac{\partial^2 V}{\partial r^2} + (u - \lambda\omega)\frac{\partial V}{\partial r} - rV + hE[V(r + J, t) - V(r, t)] = 0, \qquad (3)$$

where $\lambda(r, t)$ is the market price of risk. The final condition corresponds to the payoff on maturity and so $V(r, T, T) = 1$. Boundary conditions depend on the form of $u(r, t)$ and $\omega(r, t)$.

3   Term Structure Models with Jump

We denote by $V(r, r, T)$ the price at time t of a **discount bond**. It follows immediately that $V(r, T, T) = 1$. In our framework, the yield curve is the same as **term structure of interest rate**, as we only work with zero-coupon bonds. Now consider a quite different type of random environment. Suppose $\pi(t)$ represents the total number of extreme shocks that occur in a financial market until time t. The time dependence can arise from the cyclical nature of the economy, expectations concerning the future impact of monetary policies, and expected trends in other macroeconomic variables. In this study, we extend the jump-diffusion version of equilibrium single factor model to reflect this time dependence. This leads to the following model for r:

$$dr(t) = [\theta(t) - a(t)r(t)]dt + \sigma(t)r(t)^\beta dW(t) + Jd\pi(t), \qquad (4)$$

where $\theta(t)$ is a time-dependent drift; $\sigma(t)$ is the volatility factor; $a(t)$ is the reversion rate. We investigate the $\beta = 0$ case is an extension of Vasicek's jump-diffusion model; the $\beta = 0.5$ case is an extension of CIR jump-diffusion model.

3.1   Jump-Diffusion Version of Extended Vasicek's Model

We proposed the mean reverting process for interest rate r is given by the equation(4) with $\beta = 0$:

$$dr(t) = [\theta(t) - a(t)r(t)]dt + \sigma(t)dW(t) + Jd\pi(t) \qquad (5)$$

We will assume that the market price of interest rate diffusion risk is a function of time, $\lambda(t)$. Let us assume that jump risk is diversifiable. From equation (5) with the drift coefficient $u(r,t) = \theta(t) - a(t)r(t)$ and the volatility coefficient $\omega(r,t) = \sigma(t)$, we get the partial differential difference bond pricing equation:

$$[\theta(t) - a(t)r(t) - \lambda(t)\sigma(t)]V_r + V_t + \frac{1}{2}\sigma(t)^2 V_{rr} - rV$$
$$+ hV[-\mu A(t,T) + \frac{1}{2}(\gamma^2 + \mu^2)A(t,T)^2] = 0. \tag{6}$$

The price of a discount bond that pays off \$ 1 at time T is the solution to (6) that satisfies the boundary condition $V(r,T,T) = 1$. A solution of the form:

$$V(r,t,T) = \exp[-A(t,T)r + B(t,T)] \tag{7}$$

can be guessed. Bond price derivatives can be calculated from (7). We omit the details, but the substitution of this derivatives into (6) and equating powers of r yields the following equations for A and B.

Theorem 2

$$-\frac{\partial A}{\partial t} + a(t)A - 1 = 0 \tag{8}$$

and

$$\frac{\partial B}{\partial t} - \phi(t)A + \frac{1}{2}\sigma(t)^2 A^2 + h[-\mu A + \frac{1}{2}(\gamma^2 + \mu^2)A^2] = 0, \tag{9}$$

where, $\phi(t) = \theta(t) - \lambda(t)\sigma(t)$ and all coefficients is constants. In order to satisfy the final data that $V(r,T,T) = 1$ we must have $A(T,T) = 0$ and $B(T,T) = 0$.

3.2   Jump-Diffusion Version of Extended CIR Model

We propose the mean reverting process for interest rate r is given by the equation(4) with $\beta = 0.5$:

$$dr(t) = [\theta(t) - a(t)r(t)]dt + \sigma(t)\sqrt{r(t)}dW(t) + Jd\pi(t) \tag{10}$$

We will assume that the market price of interest rate diffusion risk is a function of time, $\lambda(t)\sqrt{r(t)}$. Let us assume that jump risk is diversifiable.

In jump-diffusion version of extended Vasicek's model the short-term interest rate, r, to be negative. If Jump-diffusion version of extended CIR model is proposed, then rates are always non-negative. This has the same mean-reverting drift as jump-diffusion version of extended Vasicek's model, but the standard deviation is proportional to $\sqrt{r(t)}$. This means that its standard deviation increases when the short-term interest rate increases. From equation(3) with the drift coefficient $u(r,t) = \theta(t) - a(t)r(t)$ and the volatility coefficient $\omega(r,t) = \sigma(t)\sqrt{r(t)}$, we get the partial differential bond pricing equation:

$$[\theta(t) - a(t)r(t) - \lambda(t)\sigma(t)r(t)]V_r + V_t + \frac{1}{2}\sigma(t)^2 r(t)V_{rr} - rV$$
$$+ hV[-\mu A(t,T) + \frac{1}{2}(\gamma^2 + \mu^2)A(t,T)^2] = 0. \tag{11}$$

Bond price partial derivatives can be calculated from (11). We omit the details, but the substitution of this derivatives into (7) and equating powers of r yields the following equations for A and B.

Theorem 3

$$-\frac{\partial A}{\partial t} + \psi(t)A + \frac{1}{2}\sigma(t)^2 A^2 - 1 = 0 \tag{12}$$

and

$$\frac{\partial B}{\partial t} - (\theta(t) + h\mu)A + \frac{1}{2}h[(\gamma^2 + \mu^2)A^2] = 0, \tag{13}$$

where, $\psi(t) = a(t) + \lambda(t)\sigma(t)$ and all coefficients is constants. In order to satisfy the final data that $V(r,T,T) = 1$ we must have $A(T,T) = 0$ and $B(T,T) = 0$.

Proof). In equations (12) and (13), by using the solution of this Ricatti's equation formula we have

$$A(t,T) = \frac{2(e^{\omega(t)(T-t)} - 1)}{(\omega(t) + \psi(t))(e^{\omega(t)(T-t)} - 1) + 2\omega(t)} \tag{14}$$

with $\omega(t) = \sqrt{\psi(t)^2 + 2\sigma(t)}$. Similarly way, we have

$$B(t,T) = \int_t^T \left\{ -(\theta(t) + h\mu)A + \frac{1}{2}h(\gamma^2 + \mu^2)A^2 \right\} dt'. \tag{15}$$

These equation yields the exact bond prices in the problem at hand. Equation (15) can be solved numerically for B. Since (14) gives the value for A, bond prices immediately follow from equation (7).

4   Monte Carlo Simulation of the Term Structure Models with Jump

Recent methods of bond pricing do not necessarily exploit partial differential equations(PDEs) implied by risk-neutral portfolios. They rest on converting prices of such assets into martingales. This is done through transforming the underlying probability distribution using the tools provided by the Girsanov's theorem. We now move on to discuss Monte Carlo simulation. A Monte Carlo simulation of a stochastic process is a procedure for sampling random outcomes for the process. This uses the risk-neutral valuation result. The bond price can be expressed as:

$$V(r_t, t, T) = E_t^Q\left[e^{-\int_t^T r_s ds}|r(t)\right] \tag{16}$$

where E^Q is the expectations operator with respect to the equivalent risk-neutral measure. To execute the Monte Carlo simulation, we discretize the equations (5) and (12). we divide the time interval $[t, T]$ into m equal time steps of length Δt each. For small time steps, we are entitled to use the discretized version of the risk-adjusted stochastic differential equations (5) and (12):

$$r_j = r_{j-1} + [(\theta \cdot t) - (a \cdot t)r_{j-1} \cdot t - (\lambda \cdot t)(\sigma \cdot t)]\Delta t \\ + (\sigma \cdot t)\varepsilon_j \sqrt{\Delta t} + J_j N_{\Delta t} \tag{17}$$

and

$$r_j = r_{j-1} + [(\theta \cdot t) - (a \cdot t)r_{j-1} - (\lambda \cdot t)(\sigma \cdot t)\sqrt{r_{j-1} \cdot t}]\Delta t \\ + (\sigma \cdot t)\sqrt{r_{j-1} \cdot t}\,\varepsilon_j \sqrt{\Delta t} + J_j N_{\Delta t}, \tag{18}$$

where $j = 1, 2, \cdots, m, \varepsilon_j$ is standard normal variable with $\varepsilon_j \sim N(0, 1)$, and $N_{\Delta t}$ is a Poisson random variable with parameter $h\Delta t$. We can investigate the value of the bond by sampling n short rate paths under the discrete process approximation of the risk-adjusted processes of the equations (17) and (18). The bond price estimate is given by:

$$V(r_t, t, T) = \frac{1}{n} \sum_{i=1}^{n} \exp\left(-\sum_{j=0}^{m-1} r_{ij}\Delta t\right), \tag{19}$$

where r_{ij} is the value of the short rate under the discrete risk-adjusted process within sample path i at time $t + \Delta t$. Numerical methods that are known as Monte Carlo methods can be loosely described as statistical simulation methods, where statistical simulation is defined in quite general terms to be any method that utilizes sequences of random numbers to perform the simulation. The Monte Carlo simulation is clearly less efficient computationally than the numerical method. The precision of the mean as a point estimate is often defined as the half-width of a 95% confidence interval, which is calculated as

$$Precision = 1.96 \times MSE, \tag{20}$$

where MSE$= \nu/\sqrt{n}$ and ν^2 is the estimate of the variance of bond prices as obtained from n sample paths of the short rate:

$$\nu^2 = \frac{\sum_{i=1}^{n} \left[\exp\left(-\sum_{j=0}^{m-1} f_{ij}\Delta t\right) - \nu\right]}{n - 1} \tag{21}$$

Lower values of Precision in Equation(20) correspond to sharper estimates. Increasing the number of n is a brute force method of obtaining sharper estimates. This reduces the MSE by increasing the value of n. However, highly precise estimates with the brute force method can take a long time to achieve. For the purpose of simulation, we conduct three runs of 1,000 trials each and divide the year into 365 time steps.

5   Experiments

In this section, we investigate the Vasicek, the CIR, the Hull and White, and jump diffusion version of three models. Experiments are consist of the numerical method and Monte Carlo simulation. Experiment 1, 2 plot estimated term structure using the four models. In Experiment 1 and 2, the parameter values are assumed to be $r = 0.05$, $a = 0.5$, $b = 0.05$, $\theta = 0.025$, $\sigma = 0.08$, $\lambda = -0.5$, $\gamma = 0.01$, $\mu = 0$, $h = 10$, $t = 0.05$, and $T = 20$.

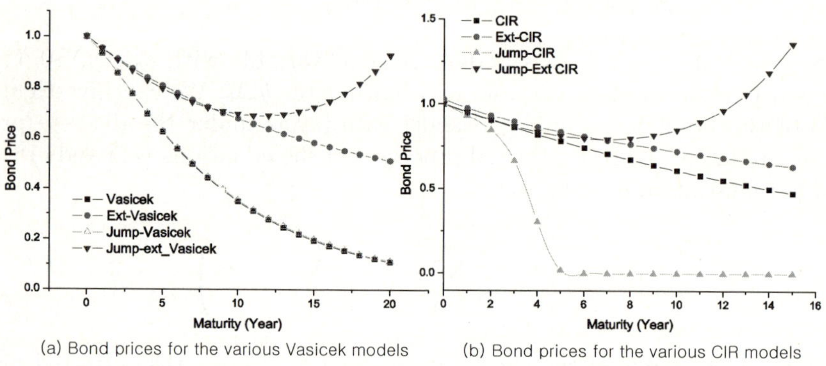

(a) Bond prices for the various Vasicek models   (b) Bond prices for the various CIR models

Fig. 1. *Experiment 1 and 2:* The various bond prices

Table 1. *Experiment 3:* Bond prices estimated by the Monte Carlo simulation for the Vasicek, the extended Vasicek, and the jump diffusion version of the Vasicek and the extended Vasicek, the CIR, the extended CIR, and the jump diffusion version of the CIR and the extended CIR models

	CFS	MCS	CFS − MCS	Variance	Precision
Vasicek	0.93585	0.93492	0.0013423	0.001159	0.00182868
Ext_Vasicek	0.95408	0.95122	5.0085E-06	7.0771E-05	0.0056056
jump-Vasicek	0.93596	0.93391	0.001508	0.001228	0.004018
jump-Ext_Vasice	0.95370	0.95031	0.000287	0.000536	0.00665224
CIR	0.95127	0.94747	7.2915E-05	0.00027	0.0074284
Ext_CIR	0.97604	0.95123	2.42154E-07	1.5561E-05	0.0486472
jump-CIR	0.942005	0.947482	0.005478	0.000286321	0.0010488
jump-Ext_CIR	0.95348	0.95169	0.0017904	0.000306335	0.00108482

Experiment 3 examines bond prices using by Monte Carlo simulation. In experiment 3, the parameter values are assumed to be $r = 0.05$, $a = 0.5$, $b = 0.05$, $\theta = 0.025$, $\sigma = 0.08$, $\lambda = -0.5$, $\Delta t = (T - t)/m$, $m = 365$, $n = 1000$, $\gamma = 0.01$, $\mu = 0$, $h = 10$, $t = 0.05$, and $T = 20$.

6 Conclusions

The Monte Carlo simulation is both harder and conceptually more difficult to implement than the other numerical methods. Interest in use of Monte Carlo simulation for bond pricing is getting stronger because of its flexibility in evaluating and handling complicated financial instruments. However, it takes a long time to achieve highly precise estimates with the brute force method. In this paper we investigate bond pricing models and their Monte Carlo simulations with several scenarios. The bond price is generally a decreasing function of the maturity, but we found the fact that the bond price is humped in the jump versions of the extended Vasicek and CIR models. In Monte Carlo simulation, we know that the bond prices of the arbitrage models are larger than those of the equilibrium models. Also lower values of precision in the equilibrium models correspond to sharper estimates.

References

1. C. Ahn and H. Thompson, "Jump-Diffusion Processes and the Term Structure of Interest Rates," Journal of Finance, vol. 43, pp. 155-174, 1998.
2. J. Baz and S. R. Das, "Analytical Approximations of the Term Structure for Jump-Diffusion Processes : A Numerical Analysis," Journal of Fixed Income, vol. 6(1), pp. 78-86, 1996.
3. J. C. Cox, J. Ingersoll, and S. Ross, "A Theory of the Term Structure of Interest Rate," Econometrica, vol. 53, pp. 385-407, 1985.
4. D. Health, R. Jarrow, and A. Morton, "Bond Pricing and the Term Structure of Interest Rates," Econometrica, vol. 60. no. 1, pp. 77-105, 1992.
5. F. Jamshidian, "An Exact Bond Option Formula," Journal of Finance, vol. 44, 1989.
6. J. Hull and A. White, "Pricing Interest Rate Derivative Securities," Review of Financial Studies, vol. 3, pp. 573-92, 1990.
7. J. Hull and A. White, "Options, Futures, and Derivatives," Fourth Edition, 2000.
8. M. J. Brennan and E. S. Schwartz, "A Continuous Time Approach to the Pricing of Bonds," Journal of Banking and Finance, vol. 3, pp. 133-155, 1979.
9. O. A. Vasicek, "An Equilibrium Characterization of the Term Structure," Journal of Financial Economics, vol. 5, pp. 177-88, 1977.

On Monte Carlo Simulation for the HJM Model Based on Jump

Kisoeb Park[1], Moonseong Kim[2], and Seki Kim[1],[*]

[1] Department of Mathematics, Sungkyunkwan University
440-746, Suwon, Korea
Tel.: +82-31-290-7030, 7034
{kisoeb, skim}@skku.edu
[2] School of Information and Communication Engineering
Sungkyunkwan University
440-746, Suwon, Korea
Tel.: +82-31-290-7226
moonseong@ece.skku.ac.kr

Abstract. We derive a form of the HJM model based on jump. Heath, Jarrow, and Morton(HJM) model is widely accepted as the most general methodology for term structure of interest rate models. We represent the HJM model with jump and give the analytic proof for the HJM model with jump. We perform the Monte Carlo simulation with several scenarios to achieve highly precise estimates with the brute force method in terms of mean standard error which is one measure of the sharpness of the point estimates. We have shown that bond prices in HJM jump-diffusion version models of the extended Vasicek and CIR models obtained by Monte Carlo simulation correspond with the closed form values.

1 Introduction

Approaches to modeling the term structure of interest rates in continuous time may be broadly described in terms of either the equilibrium approach or the no-arbitrage approach even though some early models include concepts from both approaches. The no-arbitrage approach starts with assumptions about the stochastic evolution of one or more underlying factors, usually interest rate. Bond prices are assumed to be functions of the these driving stochastic processes.

Heath, Jarrow and Morton (HJM)[4] is widely accepted as the most general methodology for term structure of interest rate models. The major contribution of the HJM model [4], as it allows the model to be no-arbitrage, a major improvement over the Ho and Lee[5] and other similar models. We will represent the HJM model with jump. In pricing and hedging with financial derivatives, jump models are particularly important, since ignoring jumps in financial prices will cause pricing and hedging rates. Term structure model solutions under HJM model with jump is justified because movements in forward rates display both continuous and discontinuous behavior. These jumps are caused by several market

---

[*] Corresponding author.

V.N. Alexandrov et al. (Eds.): ICCS 2006, Part I, LNCS 3991, pp. 38–45, 2006.
© Springer-Verlag Berlin Heidelberg 2006

phenomena money market interventions by the Fed, news surprise, and shocks in the foreign exchange markets, and so on. The HJM model with jump uses as the driving stochastic dynamic variable forward rates whose evolution is dependent on a specified volatility function. The most models of forward rates evolution in the HJM framework result in non-Markovian models of the short term interest rate evolution. This model depend on the entire history of forward rates. Therefore, this model is difficult of the actual proof analysis of the HJM model with jump. In this study, we got achieved to make the actual proof analysis of the HJM model with jump easy. The HJM model with volatility function was studied by Hull and White, Carverhill, Ritchken and Sankarasubramanian (RS)[9], Inui and Kijima, and Bhar and Chiarella in their attempt to obtain Markovian transformation of the HJM model. We examines the one-factor HJM model with jump which we use restrictive condition of RS.

We investigate the restrictive condition of RS. In addition, we introduce the Monte Carlo simulation. One of the many uses of Monte Carlo simulation by financial engineers is to place a value on financial derivatives. Interest in use of Monte Carlo simulation for bond pricing is increasing because of the flexibility of the methods in handling complex financial institutions. One measure of the sharpness of the point estimate of the mean is Mean Standard Error. Numerical methods that are known as Monte Carlo methods can be loosely described as statistical simulation methods, where statistical simulation is defined in quite general terms to be any method that utilizes sequences of random numbers to perform the simulation.

The structure of the remainder of this paper is as follows. In the section 2, the HJM model with jump are introduced. In the section 3, we calculate numerical solutions using Monte Carlo simulation for the HJM model with jump. In the section 4, we investigate the HJM model with the jump version of the extended Vasicek and CIR models. This paper is finally concluded in section 5.

2   Heath-Jarrow-Merton(HJM) Model with Jump

The HJM consider forward rates rather than bond prices as their basic building blocks. Although their model is not explicitly derived in an equilibrium model, the HJM model is a model that explains the whole term structure dynamics in a no-arbitrage model in the spirit of Harrison and Kreps[6], and it is fully compatible with an equilibrium model. If there is one jump during the period $[t, t + dt]$ then $d\pi(t) = 1$, and $d\pi(t) = 0$ represents no jump during that period. We will ignore taxes and transaction costs. We denote by $V(r, r, T)$ the price at time t of a **discount bond**. It follows immediately that $V(r, T, T) = 1$. We consider the multi-factor HJM model with jump of term structure of interest rate is the stochastic differential equation(SDE) for forward rate

$$df(t,T) = \mu_f(t,T)dt + \sum_{i=1}^{n} \sigma_{f_i}(t,T)dW_i(t) + \sum_{i=1}^{n} J_i d\pi_i(t) \qquad (1)$$

where, $\mu_f(t,T)$ represents drift function; $\sigma^2{}_{f_i}(t,T)$ is volatility coefficients; J_i is the magnitude of a jump with $J_i \sim N(\theta,\delta^2)$; in this stochastic process n independent Wiener processes and Poisson processes determine the stochastic fluctuation of the entire forward rate curve starting from a fixed initial curve.

The main contribution of the HJM model is the parameters $\mu_f(t,T)$ and $\sigma_{f_i}(t,T)$ cannot be freely specified; drift of forward rates under the risk-neutral probability are entirely determined by their volatility and by the market price of risk. We introduce the no-arbitrage condition as follows:

$$\mu_f(t,T) = -\sum_{i=1}^{n} \sigma_{f_i}(t,T)\left(\lambda_i(t) - \int_t^T \sigma_{f_i}(t,s)ds\right) \tag{2}$$

where, $\lambda_i(t)$ represents the instantaneous market price of risk and that is independent of the maturity T. Furthermore, by an application of Girsanov's theorem the dependence on the market price of interest rate risk can be absorbed into an equivalent martingale measure. Thus, the Wiener processes is

$$dW_i^Q(t) = dW_i(t) + \lambda_i(t)ds$$

We consider the one-factor HJM model with jump of the term structure of interest rate(that is, $n = 1$). Substituting the above the equation into no-arbitrage condition(3), we represent the stochastic integral equation the following:

$$f(t,T) - f(0,T) = \int_0^t \sigma_f(u,T)\int_t^T \sigma_f(t,s)dsdu$$
$$+ \int_0^t \sigma_f(s,T)dW^Q(s) + \sum_{j=1}^{\pi(t)} J_j \tag{3}$$

where, dW_i^Q is the Wiener process generated by an equivalent martingale measure Q. The spot rate $r(t) = f(t,t)$ is obtained by setting $T = t$ in the equation (5), so that

$$r(t) = f(0,t) + \int_0^t \mu_f(s,T)ds + \int_0^t \sigma_f(s,T)dW^Q(s) + \sum_{j=1}^{\pi(t)} J_j \tag{4}$$

where, $\mu_f(t,T) = \sigma_f(t,T)\int_t^T \sigma_f(t,s)ds$, and $dW^Q(t)$ is a standard Wiener process generated by the risk-neutral measure Q. Under the corresponding risk-neutral measure Q, the explicit dependence on the market price of risk can be suppressed, and we obtain the differential form of (3) is given by

$$df(t,T) = \mu_f(t,T)dt + \sigma_f(t,T)dW^Q(t) + Jd\pi. \tag{5}$$

We know that the **zero coupon bond prices** are contained in the forward rate informations, as bond prices can be written down by integrating over the forward rate between t and T in terms of the risk-neutral process

$$V(t,T) = \exp\left(-\int_t^T f(t,s)ds\right). \tag{6}$$

From the equation (3), we derive the zero coupon bond prices as follow:

$$V(t,T) = e^{-\int_t^T f(t,s)ds}$$
$$= \frac{V(0,T)}{V(0,t)} e^{-(\int_0^t \int_t^T \mu_f(u,s)dsdu + \int_0^t \int_t^T \sigma_f(u,s)dsdW^Q(u) + \sum_{j=1}^{\pi(t)} J_j \int_t^T ds)} \quad (7)$$

where, we define as $V(0,t) = e^{-\int_0^t f(0,s)ds}$, $V(0,T) = e^{-\int_0^T f(0,s)ds}$, and $\mu_f(t,T) = \sigma_f(t,T) \int_t^T \sigma_f(t,s)ds$.

The most models of forward rates evolution in the HJM framework result in non-Markovian models of the short term interest rate evolution. As above the equation (7), these integral terms depend on the entire history of the process up to time t. But, numerical methods for Markovian models are usually more efficient than those necessary for non-Markovian models.

We examines the one-factor HJM model with jump which we use restrictive condition of RS[9]. RS have extended Carverhill results showing that if the volatilities of forward rates were differential with respect to maturity date, for any given initial term structure, if and only if for the prices of all interest rate contingent claims to be completely determined by a two-state Markov process is that the volatility of forward rate is of the form

$$\sigma_f(t,T) = \sigma_r(t)\exp\left(-\int_t^T a(s)ds\right) \quad (8)$$

where, σ_r and a are deterministic functions. For the volatility of forward rate is of the form (8), the following formula for the discount bond price $V(t,T)$ was obtained in restrictive condition of RS.

Theorem 1. *Let $\sigma_f(t,T)$ be as given in (8), then discount bond price $V(t,T)$ is given by the formula*

$$V(t,T) = \frac{V(0,T)}{V(0,t)}\exp\left\{-\frac{1}{2}\varphi^2(t,T)\phi(t) + \varphi(t,T)\xi(t)]\right\} \quad (9)$$

$$where, \quad \begin{cases} \varphi(t,T) = \int_t^T \exp\left(-\int_u^t a(s)ds\right)du \\ \phi(t) \quad = \int_0^t \sigma_f{}^2(s,t)ds \\ \xi(t) \quad = [f(0,t) - r(t)] \end{cases}$$

As we mentioned already, a given model in the HJM model with jump will result in a particular behavior for the short term interest rate. We introduce relation between the short rate process and the forward rate process as follows. In this study, we jump-diffusion version of Hull and White model to reflect this restriction condition. We know the following model for the interest rate r;

$$dr(t) = a(t)[\theta(t)/a(t) - r(t)]dt + \sigma_r(t)r(t)^\beta dW^Q(t) + Jd\pi(t), \quad (10)$$

where, $\theta(t)$ is a time-dependent drift; $\sigma_r(t)$ is the volatility factor; $a(t)$ is the reversion rate. We will investigate the $\beta = 0$ case is an extension of Vasicek's jump diffusion model; the $\beta = 0.5$ case is an extension of CIR jump diffusion model.

Theorem 2. *Let be the jump-diffusion process in short rate $r(t)$ is the equation (10). Let be the volatility form is*

$$\sigma_f(t,T) = \sigma_r(t)(\sqrt{r(t)})^\beta \eta(t,T) \tag{11}$$

with $\eta(t,T) = \exp\left(-\int_t^T a(s)ds\right)$ is deterministic functions. We know the jump-diffusion process in short rate model and the "corresponding" compatible HJM model with jump

$$df(t,T) = \mu_f(t,T)dt + \sigma_f(t,T)dW^Q(t) + Jd\pi(t) \tag{12}$$

where $\mu_f(t,T) = \sigma_f(t,T)\int_t^T \sigma_f(t,s)ds$. Then we obtain the equivalent model is

$$f(0,T) = r(0)\eta(0,T) + \int_0^T \theta(t)\eta(s,T)ds$$
$$- \int_0^T \sigma_r^2(s)(r(s)^2)^\beta \eta(s,T)\int_s^T (\eta(s,u)du)ds \tag{13}$$

that is, all forward rates are normally distributed. Note that we know that $\beta = 0$ case is an extension of Vasicek's jump diffusion model; the $\beta = 0.5$ case is an extension of CIR jump diffusion model.

Note that the forward rates are normally distributed, which means that the bond prices are log-normally distributed. Both the short term rate and the forward rates can become negative. As above, we obtain the bond price from the theorem 1. By the theorem 2, we drive the relation between the short rate and forward rate.

Corollary 1. *Let be the HJM model with jump of the term structure of interest rate is the stochastic differential equation for forward rate $f(t,T)$ is given by*

$$df(t,T) = \sigma_f(t,T)\int_t^T \sigma_f(t,s)dsdt + \sigma_f(t,T)dW^Q(t) + Jd\pi(t) \tag{14}$$

where, dW_i^Q is the Wiener process generated by an equivalent martingale measure Q and $\sigma_f(t,T) = \sigma_r(t)(\sqrt{r(t)})^\beta \exp\left(-\int_t^T a(s)ds\right)$.
Then the discount bond price $V(t,T)$ for the forward rate is given by the formula

$$V(t,T) = \frac{V(0,T)}{V(0,t)}\exp\{-\frac{1}{2}\left(\frac{\int_t^T \sigma_f(t,s)ds}{\sigma_f(t,T)}\right)^2 \int_0^t \sigma_f^2(s,t)ds$$
$$- \frac{\int_t^T \sigma_f(t,s)ds}{\sigma_f(t,T)}[f(0,t) - r(t)]\}$$

with the equation (13).

Note that we know that $\beta = 0$ case is an extension of Vasicek's jump diffusion model; the $\beta = 0.5$ case is an extension of CIR jump diffusion model.

3   Monte Carlo Simulation of the HJM Model with Jump

Recent methods of bond pricing do not necessarily exploit partial differential equations(PDEs) implied by risk-neutral portfolios. They rest on converting prices of such assets into martingales. This is done through transforming the underlying probability distribution using the tools provided by the Girsanov's theorem. A **risk-neutral measure** is any probability measure, equivalent to the market measure P, which makes all discounted bond prices martingales.

We now move on to discuss Monte Carlo simulation. A Monte Carlo simulation of a stochastic process is a procedure for sampling random outcomes for the process. This uses the risk-neutral valuation result. The bond price can be expressed as:

$$V(t,T) = E_t^Q \left[e^{-\int_t^T f(t,s)ds} \right] \tag{15}$$

where, E_t^Q is the expectations operator with respect to the equivalent risk-neutral measure. Under the equivalent risk-neutral measure, the local expectation hypothesis holds(that is, $E_t^Q \left[\frac{dV}{V} \right]$). According to the local expectation hypothesis, the term structure is driven by the investor's expectations on future short rates. To execute the Monte Carlo simulation, we discretized the equation (15). We divide the time interval $[t, T]$ into m equal time steps of length Δt each(that is, $\Delta t = \frac{T-t}{m}$). For small time steps, we are entitled to use the discretized version of the risk-adjusted stochastic differential equation (14):

$$f_j = f_{j-1} + \left[\sigma_f(t,T) \int_t^T \sigma_f(t,s)dsdt \right] \Delta t + \sigma_f(t,T)\varepsilon_j \sqrt{\Delta t} + J_j N_{\Delta t} \tag{16}$$

where, $\sigma_f(t,T) = \sigma_r(t)(\sqrt{r(t)})^\beta \exp\left(-\int_t^T a(s)ds\right)$, $j = 1, 2, \cdots, m, \varepsilon_j$ is standard normal variable with $\varepsilon_j \sim N(0,1)$, and $N_{\Delta t}$ is a Poisson random variable with parameter $h\Delta t$. Note that we know that $\beta = 0$ case is an extension of Vasicek's jump diffusion model; the $\beta = 0.5$ case is an extension of CIR jump diffusion model. We can investigate the value of the bond by sampling n spot rate paths under the discrete process approximation of the risk-adjusted processes of the equation (16). The bond price estimate is given by:

$$V(t,T) = \frac{1}{n} \sum_{i=1}^{n} \exp\left(-\sum_{j=0}^{m-1} f_{ij}\Delta t\right), \tag{17}$$

where f_{ij} is the value of the forward rate under the discrete risk-adjusted process within sample path i at time $t + \Delta t$. Numerical methods that are known as Monte Carlo methods can be loosely described as statistical simulation methods, where statistical simulation is defined in quite general terms to be any method that utilizes sequences of random numbers to perform the simulation. The Monte Carlo simulation is clearly less efficient computationally than the numerical method.

The precision of the mean as a point estimate is often defined as the half-width of a 95% confidence interval, which is calculated as

$$Precision = 1.96 \times MSE. \tag{18}$$

where, $MSE = \nu/\sqrt{n}$ and ν^2 is the estimate of the variance of bond prices as obtained from n sample paths of the short rate:

$$\nu^2 = \frac{\sum_{i=1}^{n}\left[\exp\left(-\sum_{j=0}^{m-1} f_{ij}\Delta t\right) - \nu\right]}{n - 1}. \tag{19}$$

Lower values of Precision in Equation(18) correspond to sharper estimates. Increasing the number of n is a brute force method of obtaining sharper estimates. This reduces the MSE by increasing the value of n. However, highly precise estimates with the brute force method can take a long time to achieve. For the purpose of simulation, we conduct three runs of 1,000 trials each and divide the year into 365 time steps.

4   Experiments

In this section, we investigate the HJM model with the jump version of the extended Vasicek and CIR models. In experiment 1, the parameter values are assumed to be $r = 0.05$, $a = 0.5$, $\theta = 0.025$, $\sigma_r = 0.08$, $\lambda = -0.5$, $t = 0.05$, $\beta = 0$, and $T = 20$.

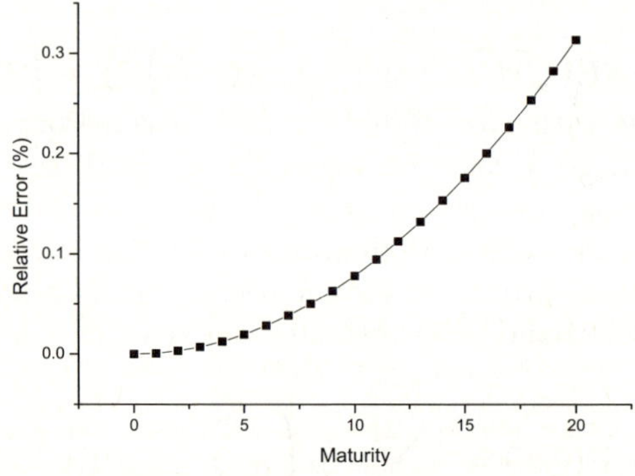

Fig. 1. *Experiment 1*: The relative error between the HJM model with the jump version of the extended Vasicek and CIR models

Experiment 2, contrasts bond prices by Monte Carlo simulation. In experiment 2, the parameter values are assumed to be $r[0] = 0.05$, $f[0, t] = 0.049875878$, $a = 0.5$, $\theta = 0.025$, $\sigma_r = 0.08$, $\lambda = -0.5$, $\beta = 0$, $\Delta t = (T - t)/m$, $m = 365$, $n = 1000$, $t = 0.05$, and $T = 20$.

Table 1. *Experiment 2*: Bond price estimated by the Monte Carlo simulation for the HJM model with the extended Vasicek model, CIR model, the jump diffusion version of the extended Vasicek model and CIR model.

	HJME_V	HJME_CIR	Jump−HJME_V	Jump−HJME_CIR
CFS	0.954902	0.95491	0.954902	0.95491
MCS	0.951451	0.951456	0.951722	0.950465
CFS−MCS	5.03495E-06	1.27659E-06	0.000319048	0.000289694
Variance	7.09574E-05	0.000112986	0.00178619	0.00170204
Precision	0.00676342	0.00676933	0.00623192	0.00871051

5 Conclusion

In this paper, we derive and perform the evaluation of the bond prices of the HJM-Extended Vasicek and the HJM-CIR models with forward interest rates instead of short rates using numerical methods. The results show that the values obtained are very similar. Even though it is hard to achieve the value of bond prices to term structure models when forward rates follow jump diffusions, we have shown that bond prices in HJM jump-diffusion version models of the extended Vasicek and CIR models obtained by Monte Carlo simulation correspond with the closed form solution. Lower values of precision in the HJM model with jump of the extended Vasicek model correspond to sharper estimates.

References

1. C. Ahn and H. Thompson, "Jump-Diffusion Processes and the Term Structure of Interest Rates," Journal of Finance, vol. 43, pp. 155-174, 1998.
2. J. Baz and S. R. Das, "Analytical Approximations of the Term Structure for Jump-Diffusion Processes: A Numerical Analysis," Journal of Fixed Income, vol. 6(1), pp. 78-86, 1996.
3. J. C. Cox, J. Ingersoll, and S. Ross, "A Theory of the Term Structure of Interest Rate," Econometrica, vol. 53, pp. 385-407, 1985.
4. D. Health, R. Jarrow, and A. Morton, "Bond Pricing and the Term Structure of Interest Rates," Econometrica, vol. 60, no.1, pp. 77-105, 1992.
5. T. S. Ho and S. Lee, "Term Structure Movements and Pricing Interest Rate Contingent Claims," Journal of Finance, vol. 41, pp. 1011-1028, 1986.
6. M. J. Harrison and D. M. Kreps, "Martingales and arbitrage in multiperiod securities markets," Journal of Economic Theory, vol. 20. pp. 381-408, 1979.
7. J. Hull and A. White, "Pricing Interest Rate Derivative Securities," Review of Financial Studies, vol. 3, pp. 573-92, 1990.
8. M. J. Brennan and E. S. Schwartz, "A Continuous Time Approach to the Pricing of Bonds," Journal of Banking and Finance, vol. 3, pp. 133-155, 1979.
9. P. Ritchken and L. Sankarasubramanian, "Volatility Structures of Forward Rates and the Dynamics of the Term Structure," Mathematical Finance, vol. 5, pp. 55-72, 1995.
10. O. A. Vasicek, "An Equilibrium Characterization of the Term Structure," Journal of Financial Economics, vol. 5, pp. 177-88, 1977.

Scalable Execution of Legacy Scientific Codes

Joy Mukherjee, Srinidhi Varadarajan, and Naren Ramakrishnan

660 McBryde Hall, Dept of Computer Science,
Virginia Tech, Blacksburg, VA 24061
{jmukherj, srinidhi, naren}@cs.vt.edu

Abstract. This paper presents Weaves, a language neutral framework for scalable execution of legacy parallel scientific codes. Weaves supports scalable threads of control and multiple namespaces with selective sharing of state within a single address space. We resort to two examples for illustration of different aspects of the framework and to stress the diversity of its application domains. The more expressive collaborating partial differential equation (PDE) solvers are used to exemplify developmental aspects, while freely available Sweep3D is used for performance results. We outline the framework in the context of shared memory systems, where its benefits are apparent. We also contrast Weaves against existing programming paradigms, present use cases, and outline its implementation. Preliminary performance tests show significant scalability over process-based implementations of Sweep3D.

1   Introduction

The past decade has witnessed increasing commoditization of scientific computing codes, leading to the prevailing practice of compositional software development. The ability to combine representations for different aspects of a scientific computation to create a representation for the computation as a whole is now considered central to high-level problem solving environments (PSEs)[1]. Common solutions for compositional scientific software [2],[3],[4] are primarily targeted at the higher end of distribution and decoupling among processors – clusters, distributed memory supercomputers, and grids. As a result, they do not require expressive mechanisms for realizing certain facets of parallelism. For instance, most of them follow the process model along with message passing for data exchanges. Nevertheless, processes can lead to problems of scalability when resources are limited.

This paper concentrates on shared memory multiprocessor (SMP) machines. It introduces Weaves—a language neutral framework for scalable execution of scientific codes—for such platforms. Weaves exploits typical shared memory properties to enrich scalability of unmodified scientific applications. It facilitates creation of multiple namespaces within a single process (address) space with arbitrary sharing of components (and their state). The framework supports lightweight threads of control through each namespace thus enabling scalable exchange of state, code, and control information. We resort to two examples

V.N. Alexandrov et al. (Eds.): ICCS 2006, Part I, LNCS 3991, pp. 46–53, 2006.

for illustration of different aspects of the framework and to stress the diversity of its application domains. The more expressive collaborating partial differential equation (PDE) solvers are used to exemplify developmental aspects, while freely available Sweep3D is used for performance results. We discuss related work, outline Weaves' design and implementation, and present use-cases.

2   Collaborating PDE Solvers

Our first example application involves collaborating partial differential equation (PDE) solvers [4], an approach for solving heterogeneous multi-physics problems using interface relaxation [5], [6] (see Fig. 1). Mathematical modeling of the multi-physics problem distinguishes between solvers and mediators. A parallel PDE solver is instantiated for each of the simpler problems and a parallel mediator is instantiated for every interface, to facilitate collaboration between the solvers. Fig. 1 illustrates typical solver and mediator codes. Among other pa-

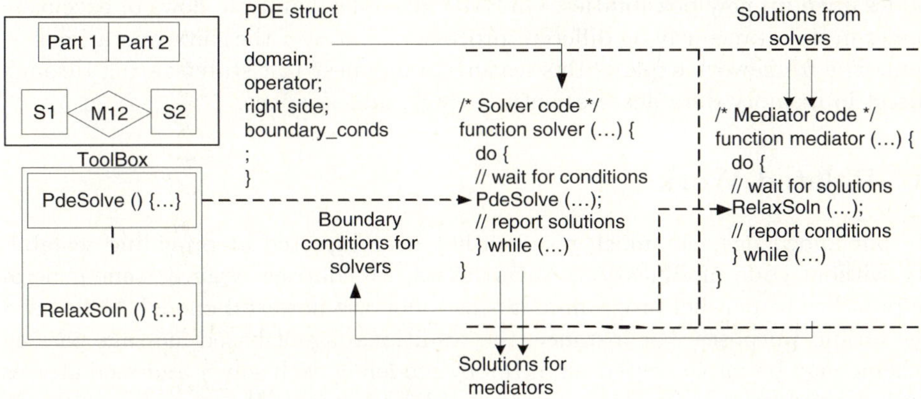

Fig. 1. *Inset (top-left)*: Simple composite multy-physics problem with two subdomains or parts. Each part is modeled by a PDE solver (S). The mediator (M) is responsible for agreement along the interface. Realistic scenarios can involve several solvers and mediators with complex graph-like connections. *Center*: Typical solver and mediator codes are shown. *PdeSolve* and *RelaxSoln* routines are chosen from a PSE toolbox.

rameters, a solver takes boundary conditions as inputs to compute solutions. The *PdeSolve* routine is chosen from a problem solving environment (PSE) toolbox depending on the PDE problem characteristics. Different *PdeSolve* routines may implement different algorithms, but could use identical names and signatures. Further, a composite problem might use the same solver on all subdomains or adopt different solvers. After computing solutions, a solver passes the results to mediators and waits till the mediators report back fresh boundary conditions. Upon the receipt of new conditions, it may recompute the solutions and repeat the whole process till a satisfactory state is reached. A mediator relaxes the solutions from solvers and returns improved values [6]. The *RelaxSoln* routines are also chosen from the PSE toolbox depending on the problem instance. Once again, multiple *RelaxSoln* routines may expose identical names and signatures and there is a choice of using the same or different mediator algorithms.

The PDE solver problem exemplifies three requirements from a parallel programming perspective: (i) Arbitrary state sharing: A part of solver state (corresponding to a boundary) should be accessible to a mediator. Additionally, different segments of solver state should be accessible to different mediators. (ii) Transparency: PDE solution and relaxation routines are mostly legacy procedural codes validated over decades of research. Modification to their sources should be minimized. (3) Scalability: Complex problem instances such as modeling of turbines and heat engines may involve thousands of solvers and mediators. Solution approaches should, therefore, be scalable. Traditionally, the collaborative PDE solvers problem has been approached using agent technology [4] in distributed environments such as clusters. Agent-based solutions use message passing as an indirect representation of procedural invocations for arbitrary state sharing. Nevertheless, continuing rise of low-cost SMPs and increases in 'arity' of nodes open up new possibilities. On SMP machines, multiple flows of execution may run simultaneously on different processors, but over the same operating system. The framework exploits this feature to manifest fast state sharing through direct in memory data accesses witin a single addess space.

3   Related Work

To our knowledge, not much research has been directed at providing scalability without code modification. Nevertheless, we contrast against some general approaches to parallel programming since they are powerful enough to be used for various purposes. For instance, the traditional agent-based message passing scheme may be implemented on SMPs by modeling each solver and mediator as an independent process. This approach aids code-reuse. However, inter-process communication and process switching overheads hamper scalability [1]. Scalability issues with multiple independent processes emphasize the use of lightweight intra-process threads for parallel flows of control. Techniques for concurrent [3], [7], [8], compositional [9], [10], and object-oriented [2] programming can enable scalable state sharing over lightweight threads through the use of in-memory data structures. However, all of these mechanisms resort to varying degrees of source-level constructs to enforce encapsulation (separation of namespace) and therefore do not meet our transparency requirement. (Recall that legacy PDE solver and relaxation codes may use identical names or symbols for functions and data).

4   Weaves

From the previous section we deduce that on SMP machines, the transparency and scalability requirements of the collaborating PDE solver example reduce to (1) use of traditional procedural codes and programming techniques, and (2) use of lightweight intra-process threads for modeling parallel flows of control. The need is selective separation of state and namespace of intra-process threads without resorting to source-level programming. Scalability and transparency debar OS level solutions. These observations lead to the first step towards Weaves.

4.1   Design and Definitions

The Weaves framework creates encapsulated intra-process components called modules from source written in any language. A module is an encapsulated runtime image of a compiled binary object file. Each module defines its own namespace. Multiple identical modules have independent namespaces and global data within the address space of a single process. A module may make references to external definitions. Weaves offers application control over reference-definition bindings. While encapsulation of modules enforces separation, individual references may be explicitly redirected to achieve fine-grain selective sharing. References from different modules pointing to a particular definition result in sharing of the definition among the modules. The minimal definition of the module allows Weaves to flexibly work with high-level frameworks, models, and languages.

At the core of the Weaves framework is the definition of a weave. A weave[1] is a collection of one or more modules composed into an intra-process subprogram. From the viewpoint of procedural programs, weave composition from modules is similar to linking object files for executable generation. However, there are important differences: (1) Weave composition is an intra-process runtime activity (recall that modules are intra-process runtime entities). (2) Weave composition does not necessitate resolution of all references within constituent modules. (3) Reference redirections may be used later to fulfill completeness or to transcend weave boundaries. A weave unifies namespaces of constituent modules. Hence, identical modules cannot be included within a single weave. However, different weaves may comprise similar, but independent, modules. This facility of weaves helps create multiple independent copies of identical programs within a process space. Going a step further, the Weaves framework allows a single module to be part of multiple weaves. These weaves, therefore, share the contents of the common module. This lays the foundation for selective sharing and separation of state within the Weaves framework. Individual reference redirections extend such sharing among weaves in arbitrary ways.

A string is the fundamental unit of execution under Weaves. It is a lightweight intra-process thread. A string executes within a weave. Multiple strings may simultaneously run through the same or different weaves. The Weaves framework emulates the process model when one string is initiated through each one of multiple independent weaves. It emulates the traditional threads model when multiple strings are initiated through a single weave. Apart from these extremes, the framework also provides for arbitrary compositions of strings (realized through

---

[1] We use 'weave' to indicate the unit of composition and 'Weaves' to refer to the overall framework.

the composition of associated weaves). A runtime issue can arise when a module is part of more than one weave – external references from a shared module may have to resolve to different definitions depending on the weave. We use late binding mechanisms to resolve this on-the-fly. Whenever a string accesses such an external reference, the framework's runtime environment queries the string for its weave and resolves to a definition accordingly. Weaves requires a minimal bootstrapping module (which runs on the main process thread of every application) to set up the target application – load modules, compose weaves, and start strings. The main process thread may later be used to monitor and externally/asynchronously modify the application constitution at runtime [1]. The entire application—including all modules, weaves, strings, and the monitor—runs within a single OS process. The framework provides a meta-language for

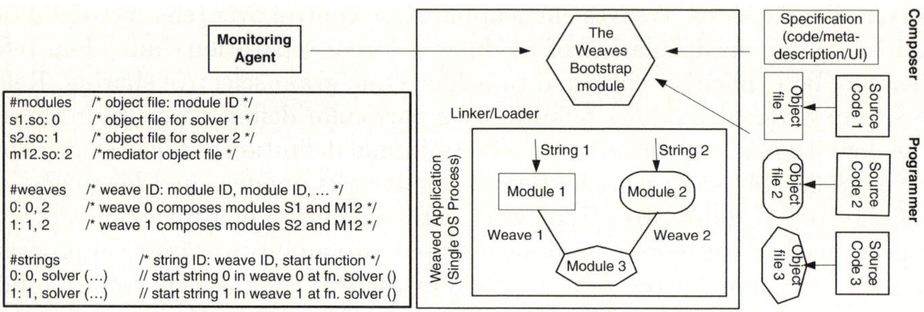

Fig. 2. *Inset (bottom-left)*: A simple configuration file for a Weave-based approach to realize the Fig 1. (inset) scenario. Here, codes for S1, S2, and M12 are compiled into objects S1.so, S2.so, and M12.so respectively. Each solver module is composed into a distinct weave with the single mediator module. *Center*: Diagrammatic illustration of the development process of a general application using Weaves.

specification of application configuration in a file and a script that automatically generates a bootstrap module, builds it, and initiates a live application from such a meta-description. A simple configuration file for a Weaves-based approach to realize the Fig. 1 (inset) scenario is shown in Fig. 2 (inset). One direction of current research on Weaves aims at an integrated GUI for tapestry specification and automatic execution. Fig. 2 diagrammatically illustrates the complete development process of a general application using Weaves.

4.2   Implementation

Weaves' current prototype implementation works on x86 (32 and 64 bit) and ia64 architectures running GNU/Linux. Being a binary-based framework, it relies on the Executable and Linkable File Format (ELF) [11] used by most UNIX systems for native objects. It recognizes shared object (.so) files as loadable modules. Shared objects define encapsulated/independent namespaces and are easily created from most relocatable objects (.o) compiled with position independent options (-fPIC for gcc). The implementation of a weave follows from the basic design. A string can be customized to use either POSIX threads (pthreads) or GNU's user-level threads, Pth.

Weaves's runtime environment requires extensive binary loading and linking capabilities to load and compose modules, and randomly manipulate reference-definition bindings. However, traditional binary loader services are not sufficient to support the demands of Weaves. For instance, typical loaders do not provide an explicit interface to connect a reference to an arbitrary definition. Hence, Weaves provides its own tool—Load and Let Link (LLL)—for dynamic loading and linking of modules [13]. The LLL loader maps given object files on disk to corresponding modules in memory. It can load multiple identical, but independent, modules from the same shared object file. LLL does not try to resolve external references at load-time, since any attempt to resolve them at this time would result in avoidable overhead. All cross-binding actions involving multiple modules are delegated to the linker, which dynamically composes a weave given an ordered set of modules. Additionally, the linker is invoked for runtime relocations of external references from a shared module. Finally, it provides an interface for explicitly binding a reference in a module to a definition in another.

5   Weaving PDE Solvers

Weaves opens up various possibilities for implementing collaborating PDE solvers. However, due to space limitations, we discuss only the most radical approach that reuses unmodified solver and mediator codes from traditional agent-based implementations. Here, unmodified solver and mediator agent codes are compiled into object components. Additionally, a communication component is programmed to emulate dependable and efficient messages transfers between intra-process threads through in-memory data structures. The component's interfaces are identical to those of the distributed communication library used in the agent-based solution. At runtime, all solvers and mediators are loaded as distinct modules. The communication component is loaded as a single module. Every solver module is composed with the communication module into a solver weave. Every mediator module is composed with the communication module into a mediator weave. Parallel strings are then fired off at the main functions of solver and mediator modules. The solvers and mediators run as independent vir-

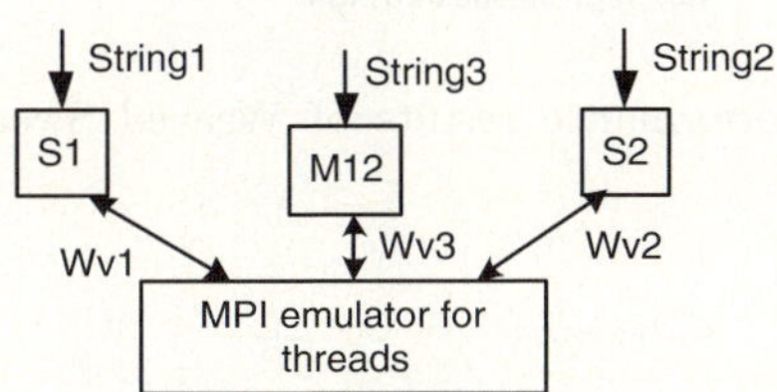

Fig. 3. Weaving unmodified agent-based codes (Wv implies a weave)

tual machine abstractions unaware of Weaves. Fig. 3 diagrammatically illustrates a corresponding tapestry for a simple case assuming MPI for communication.

6   Performance Results

For preliminary performance results, we ran a Weaved version of Sweep3D [14] (an application for 3 dimensional discrete ordinates neutron transport) on an 8-way x86_64 SMP and compared results with traditional MPI-based implementations. The Weaved-setup was similar to Fig. 3. We developed a simple MPI emulator for in-memory data-exchange among threads. Multiple Sweep3D modules were composed with a single MPI module for communication. We used a 150-cube input file with a 2x3 split (6 processes/strings) as a start point and increased the split to 2x4, 4x6, 6x9, and so on upto 10x15 (150 processes/strings). The performance of the Weaved implementation matched that of LAM [12] and MPICH [15] as long as the number of processes/strings was lesser than the number of processors. Beyond that, the Weaved implementation performed much better thereby clearly demonstrating scalability (Fig.4). Both the MPI implementations (compiled and run with shared memory flags) crashed beyond 24 processes (4x6 split). The sharp performance degradation of LAM and MPICH are primarily due to systems-level shared memory schemes, which do not scale beyond the number of processors. The use of systems-level shared-memory is a direct consequence of reliance on the process-model. Weaves works around this problem by emulating processes within a single address-space.

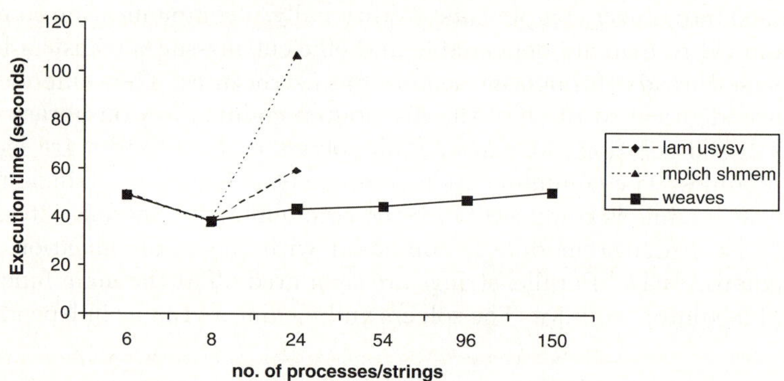

Fig. 4. Comparison of performance results of Weaved Sweep3D against LAM and MPICH based ones.

7   Discussions

Weaved implementations of collaborating PDE solvers exploit lightweight intra-process threads and direct in-memory state sharing for scalability. Furthermore, they reuse legacy procedural codes for *PdeSolve* and *RelaxSoln* routines for transparency. Lastly, the bootstrap module requires minimal information about the internals of solver and mediator codes. Thus, Weaves can be used to flexibly compose a wide range of solver-mediator networks as well as applications from other domains. Preliminary performance tests show significant scalability over process-based implementations of Sweep3D. A prototype of the Weaves framework is available for download from http://blandings.cs.vt.edu/joy.

References

1. Varadarajan, S, Ramakrishnan, N.: Novel Runtime Systems Support for Adaptive Compositional Modeling in PSEs. Future Generation Computing Systems (Special Issue), 21(6) (June 2005), 878-895.
2. Chandy, K. M., Kesselman, C.: Compositional C++: Compositional Parallel Programming. Technical Report CaltechCSTR:1992.cs-tr-92-13, California Institute of Technology CA USA (2001).
3. Foster, I.: Compositional Parallel Programming Languages. ACM Transactions on Prog. Lang. and Sys., 18(4) (July 1996), 454-476.
4. Drashansky, T. T., Houstis, E. N., Ramakrishnan, N., Rice J. R.: Networked Agents for Scientific Computing. Communications of the ACM, 42(3) (March 1999), 48-54.
5. McFaddin, H. S., Rice, J. R.: Collaborating PDE Solvers. Applied Numerical Mathematics, 10 (1992), 279-295.
6. Rice, J. R.: An Agent-based Architecture for Solving Partial Differential Equations. SIAM News, 31(6) (August 1998).
7. Carriero, N., Gelernter, D.: Linda in Context. Communications of the ACM, 32(4). (April 1989) 444-458.
8. Sato, M.: OpenMP: Parallel Programming API for Shared Memory Multiprocessors and On-Chip Multiprocessors. In Proceedings of the 15th International Symposium on System Synthesis (ISSS ?02), Kyoto Japan (October 2-4 2002).
9. Common Component Architecture: http://www.cca-forum.org/
10. Mahmood, N., Deng, G., Browne, J. C.: Compositional Development of Parallel Programs. In Proceedings of the 16th Workshop on Langs. and Compilers for Parallel Computing (LCPC?03), College Station TX (2003).
11. Tools Interface Standards Committee: Executable and Linkable Format (ELF) Specification, (May 1995).
12. LAM MPI: http://www.lam-mpi.org/
13. Mukherjee, J., Varadarajan, S.: Weaves: a framework for reconfigurable programming. International Journal forParallel Programming, 33(2) (June 2005) 279-305.
14. Koch, K. R., Baker, R. S., Alcouffe, R. E.: Solution of the First-Order Form of the 3D Discrete Ordinates Equation on a Massively Parallel Processor. Transactions of the American Nuclear Society, 65(198) (1992).
15. Mpich: http://www-unix.mcs.anl.gov/mpi/mpich/

Parallel Solvers for Flexible Approximation Schemes in Multiparticle Simulation*

Masha Sosonkina[1] and Igor Tsukerman[2]

[1] Ames Laboratory/DOE, Iowa State University,
Ames, IA 50011, USA
masha@scl.ameslab.gov
[2] Department of Electrical and Computer Engineering,
The University of Akron,
Akron, OH 44325-3904, USA
igor@uakron.edu

Abstract. New finite difference schemes with flexible local approximation are applied to screened electrostatic interactions of spherical colloidal particles governed by the Poisson-Boltzmann equation. Local analytical approximations of the solution are incorporated directly into the scheme and yield high approximation accuracy even on simple and relatively coarse Cartesian grids. Several parallel iterative solution techniques have been tested with an emphasis on suitable parallel preconditioning for the nonsymmetric system matrix. In particular, flexible GMRES preconditioned with the distributed Schur Complement exhibits good solution time and scales well when the number of particles, grid nodes or processors increases.

1   Introduction

The study described in this paper has three key ingredients: (1) A new class of finite difference (FD) schemes; (2) Efficient parallel solvers; (3) Applications to the Poisson-Boltzmann equation (PBE) for electrostatics in solvents. These three components are intertwined: the new schemes are particularly suitable for electrostatic fields of dielectric particles (for other applications, see [1, 2]), and their practical use is facilitated greatly by suitable parallel solvers.

Under the mean field theory approximation, the electrostatic potential of multiple charged particles in a solvent is governed by the Poisson-Boltzmann equation (Sect. 2). Several routes are available for the numerical simulation. First, if particle sizes are neglected and the PBE is linearized, the solution is simply the sum of the Yukawa potentials of all particles. If the characteristic length of the exponential field decay (the Debye length) is small, the electrostatic interactions are effectively short-range and therefore relatively inexpensive to compute. For

* The work of M.S. was supported in part by the U.S. Department of Energy under Contract W-7405-ENG-82, NERSC, by NSF under grant NSF/ACI-0305120, and by University of Minnesota Duluth. The work of I.T. was supported in part by the NSF awards 0304453, 9812895 and 9702364.

V.N. Alexandrov et al. (Eds.): ICCS 2006, Part I, LNCS 3991, pp. 54–62, 2006.

weak ionic screening (long Debye lengths) Ewald-type methods [3] or the Fast Multipole Method [4] can be used. However, our goal is to develop algorithms that would be applicable to finite-size particles and extendable to nonlinear problems. Ewald-type methods and FMM are not effective in such cases.

An alternative approach is the Finite Element Method (FEM) and the Generalized Finite Element Method (GFEM) [5]. FEM requires very complex meshing and re-meshing even for a modest number of moving particles and quickly becomes impractical when the number of particle grows. In addition, re-meshing is known to introduce a spurious numerical component in force calculation (e.g., [6]). GFEM relaxes the restrictions of geometric conformity in FEM and allows suitable non-polynomial approximating functions to be included in the approximating set. This has been extensively discussed in the literature [7], including our own publications [8, 5, 9]. Unfortunately, GFEM has a substantial overhead due to numerical quadratures and a higher number of degrees of freedom in generalized finite elements around the particles.

A two-grid approach from computational fluid mechanics has been adapted to colloidal simulation: a spherical mesh around each particle and a common Cartesian background grid [10, 11]. The potential has to be interpolated back and forth between the local mesh of each particle and the common Cartesian grid; the numerical loss of accuracy in this process is unavoidable. In contrast, the Flexible Local Approximation MEthod (FLAME) (Sect. 2, [1, 2]) has only one global Cartesian grid but incorporates *local* approximations of the potential near each particle into the difference scheme. The Cartesian grid can remain relatively coarse – on the order of the particle radius or even coarser. This is in stark contrast with classical FD schemes, where the grid size has to be much smaller than the particle radius to avoid the spurious 'staircase' effects.

2   Formulation of the Problem

The electrostatic field of charged colloidal particles ('macroions') in the solvent is screened by the surrounding microions of opposite charge. If microion correlations are ignored (a good approximation for monovalent ions under normal conditions), the electrostatic potential is known to be governed by the PBE

$$\epsilon_s \nabla^2 u = -\sum_\alpha n_\alpha q_\alpha \exp\left(-\frac{q_\alpha u}{k_B T}\right), \tag{1}$$

where summation is over all species of ions present in the solvent, n_α is volume concentration of species α in the bulk, q_α is the charge of species α; ϵ_s is the (absolute) dielectric permittivity of the solvent; k_B is the Boltzmann constant, and T is the absolute temperature. The right hand side of (1) reflects the Boltzmann redistribution of microions in the mean field with potential u. Note that inside the colloidal particles the potential simply satisfies the Laplace equation.

If the electrostatic energy $q_\alpha u$ is smaller than thermal energy $k_B T$, PBE can be approximately linearized to yield (after taking electroneutrality into account)

$$\nabla^2 u \; - \; \kappa^2 u \; = \; 0, \quad \text{with} \;\; \kappa \; = \; (\epsilon_{\mathrm{s}} k_B T)^{-\frac{1}{2}} \left(\sum_{\alpha} n_\alpha q_\alpha^2 \right)^{\frac{1}{2}}. \qquad (2)$$

(This value of the *Debye-Hückel parameter* κ corresponds to linearization about $u = 0$.)[1] Typically, the sources of the electrostatic field are surface charges on the colloidal particles. Standard boundary conditions apply on the particle-solvent interface: the potential is continuous, while the normal component of the displacement vector $\mathbf{D} = -\epsilon \nabla u$ has a jump equal to the surface charge density. The Dirichlet condition at infinity is zero (in practice, the domain boundary is taken sufficiently far away from the particles). Alternative boundary conditions (e.g. periodic) are possible but will not be considered here.

The remainder of the paper deals only with the linearized equation and with the relevant FD schemes and parallel solvers for it. However, the numerical procedure can be extended to the nonlinear PBE by applying the Newton-Raphson-Kantorovich method to the *continuous* problem [1, 2].

Flexible Local Approximation MEthods (FLAME). The accuracy of classical Taylor-based FD schemes deteriorates substantially if the solution is not smooth enough (e.g. at material interfaces where one or more components of the field are discontinuous). In FLAME schemes [1, 2], the Taylor expansions of classical FD are replaced with local basis functions that in many important cases provide very accurate and physically meaningful approximation. In the 'Trefftz' version of FLAME, these basis functions are chosen as *local* solutions of the underlying differential equation. Various examples ranging from the Schrödinger equation to PBE to electromagnetic scattering are given in [1, 2].

Let r_α ($\alpha = 1, 2, \ldots, M$) be the node positions of an M-point grid stencil; usually a regular Cartesian grid is used. Further, let ψ_β ($\beta = 1, 2, \ldots, m$) be a set of *local* approximating functions which (in Trefftz-FLAME) satisfy the underlying differential equation; the number m of these functions is typically equal to $M - 1$. First, let the underlying linear differential equation be homogeneous (no sources) in the vicinity of the stencil. Then the FLAME scheme is a coefficient vector $s \in R^M$ such that $s \in \mathrm{Null}(N^T)$, where matrix N comprises the nodal values of all basis functions on the stencil: $N_{\alpha\beta} = \psi_\beta(r_\alpha)$, $1 \leq \alpha \leq M$, $1 \leq \beta \leq m$. As shown in [2], consistency of FLAME schemes follows directly from the approximation properties of the basis set.

For the electrostatic problem governed by the linearized PBE (2), a FLAME scheme on a global Cartesian grid is obtained in the following way. First, one chooses a suitable grid stencil (standard seven-point grid stencils are used in the numerical experiments reported in Sect. 4; higher-order schemes on expanded stencils will be considered elsewhere. In the vicinity of a given particle, Trefftz-FLAME basis functions – local solutions of the electrostatic problem – can be generated by matching harmonic expansions inside and outside the particle:

$$\psi_{nm}(r, \theta, \phi) \; = \; \begin{cases} r^n Y_{nm}(\theta, \phi), & r \leq r_p \\ (a_{nm} j_n(\mathrm{i}\kappa r) + b_{nm} n_n(\mathrm{i}\kappa r)) Y_{nm}(\theta, \phi), & r \geq r_p \end{cases}, \qquad (3)$$

---

[1] For a systematic account of "optimal" linearization procedures, see [12, 13].

where Y_{nm} are the spherical harmonics, r_p is the radius of the particle, and the coefficients a_{nm}, b_{nm} can be determined from the boundary conditions. The spherical Bessel functions $j_n(i\kappa r)$ and $n_n(i\kappa r)$ in (3) are expressible in terms of hyperbolic sines / cosines and hence relatively easy to work with. The actual expressions for the coefficients a_{nm}, b_{nm} are given in [1]. For the seven-point stencil, one gets a valid FLAME scheme by adopting six basis functions: one 'monopole' term ($n = 0$), three dipole terms ($n = 1$) and any two quadrupole harmonics ($n = 2$). Away from the particles, the classical seven-point scheme for the Helmholtz equation is used for simplicity. For inhomogeneous equations, the FLAME scheme is constructed by splitting the potential up into a particular solution $u_f^{(i)}$ of the inhomogeneous equation and the remainder $u_0^{(i)}$ satisfying the homogeneous one [1, 2]. For the linearized PBE, the inhomogeneous part can be taken as the Yukawa potential that satisfies the PBE in the solvent, the Laplace equation (in a trivial way) inside the particle, and the boundary conditions:

$$u_{Yukawa} = \begin{cases} q\,[4\pi\epsilon_s r_p(\kappa r_p + 1)]^{-1}, & r \leq r_p \\ q\,\exp(-\kappa(r - r_p))\,[4\pi\epsilon_s r(\kappa r_p + 1)]^{-1}, & r \geq r_p \end{cases}. \tag{4}$$

To summarize, the FLAME scheme in the vicinity of charged particles is constructed in the following way: (i) compute the FLAME coefficients for the homogeneous equation; for each grid stencil, this gives the nonzero entries of the corresponding row of the global system matrix; (ii) apply the scheme to the Yukawa potential to get the entry in the right hand side.

3   Parallel Solution Methods

Realistic multiparticle simulations require three-dimensional grids with millions of grid points, which renders direct solvers infeasible. Iterative methods provide a suitable alternative. FLAME matrices have a regular sparsity structure (7-diagonal if the standard 7-point stencil is used) but are not symmetric and not in general diagonally dominant. This increases the importance of parallel preconditioning in making the iterative solution scalable with respect to problem size and the number of processors.

In parallel iterative methods, each processor holds a set of equation and the associated unknowns. The resulting distributed sparse linear system may be solved using techniques similar to those of Domain Decomposition [14]. Three types of variables are distinguished: (1) Interior variables coupled only with local variables; (2) Inter-domain interface variables coupled with external as well as the local ones; and (3) External interface variables that belong to neighboring processors. These external interface variables must be first received from neighboring processor(s) before a distributed matrix-vector product can be completed. Each local vector of unknowns x_i ($i = 1, \ldots, p$) is also split into two parts: the sub-vector u_i of interior variables followed by the sub-vector y_i of inter-domain interface variables. The right-hand side b_i is conformally split into sub-vectors f_i and g_i. The local matrix A_i residing in processor i is block-partitioned according to this splitting. The equations assigned to processor i can be written as follows:

$$\begin{pmatrix} B_i & F_i \\ E_i & C_i \end{pmatrix} \begin{pmatrix} u_i \\ y_i \end{pmatrix} + \begin{pmatrix} 0 \\ \sum_{j \in N_i} E_{ij} y_j \end{pmatrix} = \begin{pmatrix} f_i \\ g_i \end{pmatrix}. \tag{5}$$

The term $E_{ij} y_j$ is the contribution to the local equations from the neighboring sub-domain number j, and N_i is the set of sub-domains that are neighbors to sub-domain i.

Additive Schwarz and Schur Complement Preconditioning. Additive Schwarz (AS) procedures are the simplest parallel preconditioners available. They are easily parallelized and incur communications only in the exchange of interface variables, which may be the same as communications during a distributed matrix vector product and must precede the update of the local residual vector. This vector is used as the right-hand side for the local system to find the local update. There are several options for solving the local system: a (sparse) direct solver, a standard preconditioned Krylov solver, or a forward-backward solution associated with an accurate ILU preconditioner [14, 15].

Schur complement (SC) methods iterate on the inter-domain interface unknowns only, implicitly using interior unknowns as intermediate variables. SC systems are derived by eliminating the variables u_i from (5):

$$S_i y_i + \sum_{j \in N_i} E_{ij} y_j = g_i - E_i B_i^{-1} f_i \equiv g_i', \tag{6}$$

where S_i is the "local" SC, $S_i = C_i - E_i B_i^{-1} F_i$. Equations (6) for all sub-domains i ($i = 1, \ldots, p$) constitute a global system of equations $Sy = g'$ involving only the inter-domain interface unknown vectors y_i. Once the global SC system is (approximately) solved, *each* processor computes the u-part of the solution vector by solving the system $B_i u_i = f_i - E_i y_i$ obtained by substitution from (5) [16].

4   Numerical Results

We have used parallel iterative methods as implemented in the pARMS package [17, 18]. The pARMS package contains several state-of-the-art algebraic parallel preconditioning techniques, such as several distributed SC algorithms. For the subdomain solution, pARMS has a wide range of options including the Algebraic Recursive Multilevel Solver (ARMS) [19]. The following is a subset of pARMS preconditioners used here for the numerical experiments: `add_ilu(k)` denotes an AS procedure described in Sect. 3. The local system is solved approximately by applying incomplete LU factorization with level k fill-in (called ILU(k)) as solver. `add_arms` is similar to `add_ilu(k)` but ARMS is used as an approximate solver for local systems. `sch_ilu(k)` consists of solving the (global) SC system, associated with the inter-domain interface variables, with a few iterations of GMRES preconditioned by Block-Jacobi, in which ILU(k) is used locally on inter-domain interface variables. `sch_arms` is the same as `sch_ilu(k)` but ARMS acts as local solver for the Block-Jacobi preconditioner. In the pARMS package, typical input parameters, such as the maximum number of iterations, convergence tolerance,

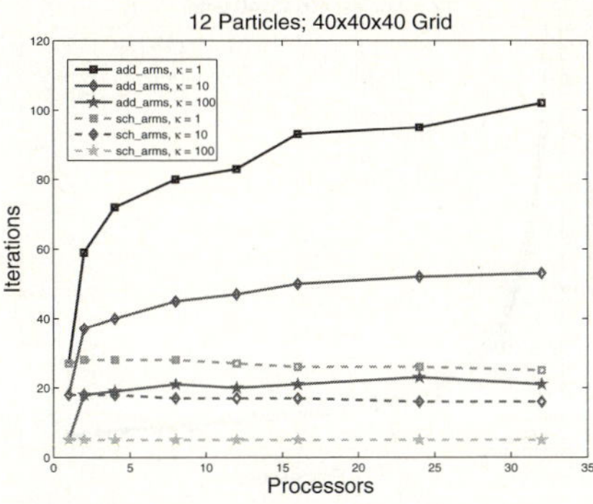

Fig. 1. Iterations for `add_arms` and `sch_arms` and various κ values

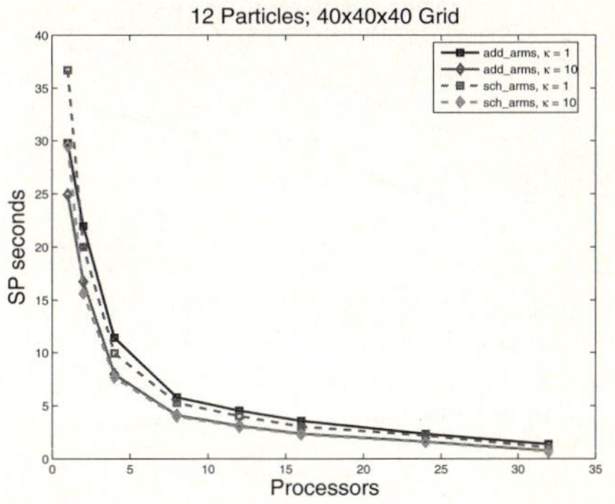

Fig. 2. Timings for `add_arms` and `sch_arms`

and the restart value (for GMRES), are augmented with the parameters that fine-tune the preconditioners. To estimate these parameters, consider that the FLAME matrices for PBE are non-symmetric and the weight of the main diagonal, defined as the *weak diagonal dominance*, depends on the Debye-Hückel parameter κ in (2). For example, for the $60 \times 60 \times 60$ grid the relative weak diagonal dominance is 99% when $\kappa = 10$ but decreases by 10% when $\kappa = 1$. We have investigated the preconditioners described earlier with a different amount of fill governed by the input parameter `lfil`. It means either the level of fill, for the (local) ILU(k) preconditioner or the number of entries to be retained in the L and U factors of ARMS. or the latter, the drop tolerance has been defined as 10^{-3}. The problems have been solved using flexible GMRES (FGMRES) [15] with the restart value 20 and tolerance 10^{-9}. Flexible GMRES allows preconditioners containing inner iterations and thus changing for each outer iteration. In particular, the `sch_arms` preconditioner is set to perform five inner iterations, which is a rather small number taken to accelerate the convergence without large increases of the execution time. The experiments have been performed on the IBM SP RS/6000 supercomputer at NERSC, which has 6,080 processors on 16-way SMP nodes interconnected by the IBM proprietary switching network. Each processor has the peak performance of 1.5 GFlops and shares between 16 and 64 GBytes of main memory.

For a 3D 12-particle problem with 40 grid nodes in each direction, Fig. 1 shows outer iterations to convergence for `add_arms` and `sch_arms` when different κ values are used and `lfil=25`. Although the iteration numbers grow as κ decreases for both `add_arms` and `sch_arms`, the latter exhibits a more scalable behavior as the number of processors increases. Since `sch_arms` performs inner iterations and solves the global system at each step, it is somewhat more costly than `add_arms` (Fig. 2) for a small number of processors. To study the effect of the amount of fill, we have varied the `lfil` parameter for the same twelve-particle problem. Figures 3 and 4 indicate that, even for small κ, decreasing fill-in yields acceptable convergence and may reduce the execution time. Thus for large

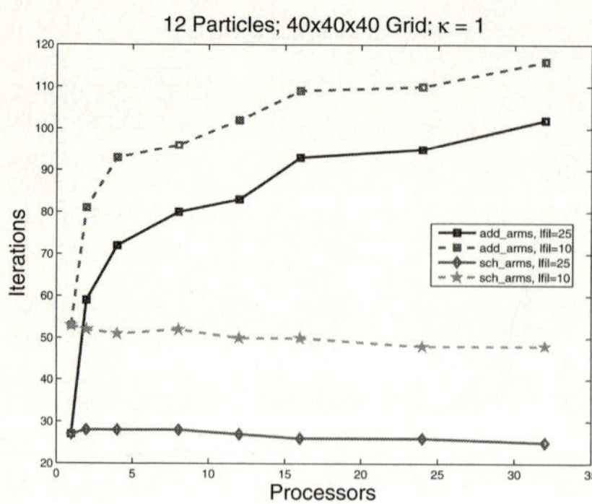

Fig. 3. Iteration numbers of `add_arms` and `sch_arms` for two fill-in amounts

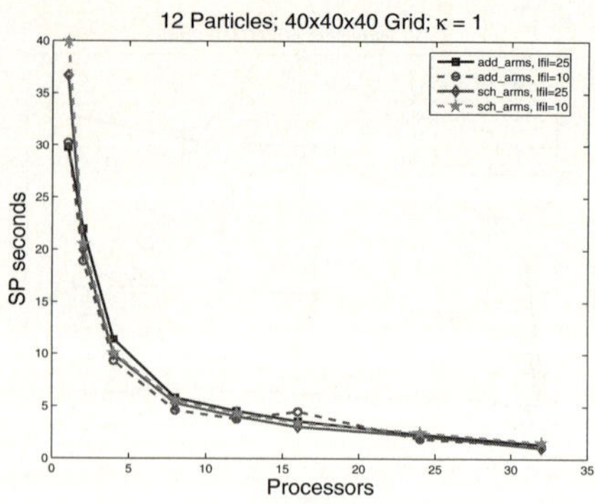

Fig. 4. Timings of `add_arms` and `sch_arms` for two fill-in amounts

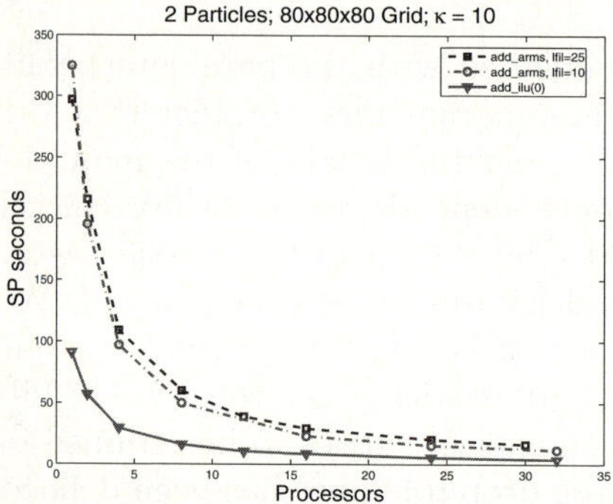

Fig. 5. Timings for a large problem with different fill-in values in `add_arms`

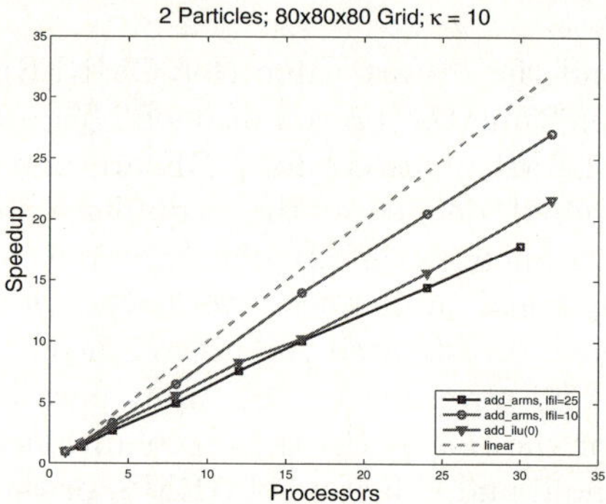

Fig. 6. Speedup of `add_arms` for different fill-in amounts on a large problem

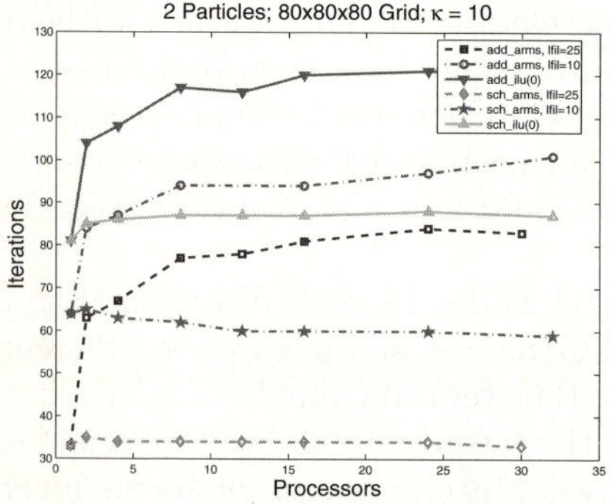

Fig. 7. Iteration numbers for a two-particle large problem with different fill

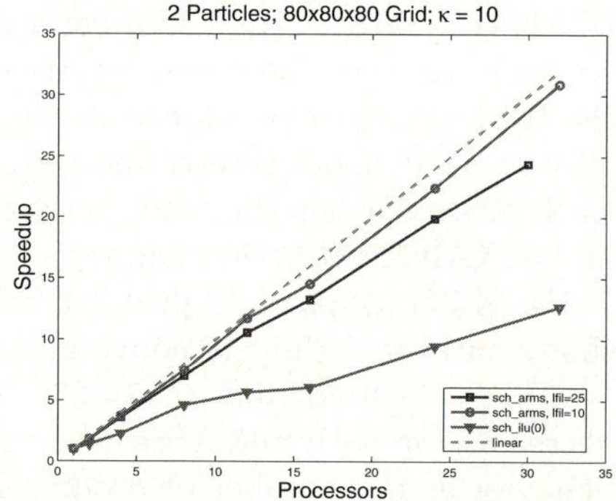

Fig. 8. Speedup of `sch_arms` for different fill-in amounts on a large problem

problem sizes, it may be beneficial to very small fill-in, as in `arms_ilu(0)`, on smaller numbers of processors. Figure 5 presents a comparison of the execution times for a two-particle problem on an 80^3 grid. For AS with different amounts of fill, the speedups are shown in (Fig. 6). Note that the speedup is defined as the ratio of the execution time of the corresponding sequential algorithm to the parallel execution time. Due to a rather drastic increase, by 38% on average, in the number of iterations from sequential to parallel execution for AS (Fig. 7), the best speedup is only 84% of the ideal (linear) case. On the other hand, the speedups for `sch_arms` are almost linear for a sufficient amount of fill (Fig. 8).

5   Conclusion

Several parallel iterative solvers have been applied to linear FLAME systems for PBE, with the goal of finding the best methods and preconditioning techniques in parallel environments for varying problem difficulty and size. It has been observed that Schur Complement preconditioning with a small amount of fill and a few inner iterations scales well and becomes competitive with Additive Schwarz for large number of processors, while attaining almost linear speedup. The number of iterations and the computational time depends only mildly on the Debye parameter. Overall, the parallel distributed Schur Complement solver is promising for the simulation of colloidal suspensions, as well as polymer and protein molecules in solvents.

References

1. Tsukerman, I. IEEE Trans. Magn. **41** (2005) 2206–2225
2. Tsukerman, I. J. Comput. Phys. **211** (2006) 659–699
3. Salin, G., Caillol, J.M. J. Chem. Phys. **113** (2000) 10459–10463
4. Greengard, L.F., Huang, J. J. Comput. Phys. **180** (2002) 642–658
5. Plaks, A., Tsukerman, I., Friedman, G., Yellen, B. IEEE Trans. Magn. **39** (2003) 1436–1439
6. Tsukerman, I. IEEE Trans. Magn. **31** (1995) 1472–1475
7. Strouboulis, T., Babuška, I., Copps, K. Comput. Meth. Appl. Mech. Eng. **181** (2000) 43–69
8. Proekt, L., Tsukerman, I. IEEE Trans. Magn. **38** (2002) 741–744
9. Basermann, A., Tsukerman, I. Volume LNCS 3402., Springer (2005) 325–339
10. Fushiki, M. J. Chem. Phys. **97** (1992) 6700–6713
11. Dobnikar, J., Haložan, D., M.Brumen, von Grünberg, H.H., Rzehak, R. Comput. Phys. Comm. **159** (2004) 73–92
12. Deserno, M., von Grünberg, H.H. Phys. Review E **66** (2002) 15 pages
13. Bathe, M., Grodzinsky, A.J., Tidor, B., Rutledge, G.C. J. of Chem. Phys. **121** (2004) 7557–7561
14. Smith, B., Bjørstad, P., Gropp, W.: Domain Decomposition: Parallel Multilevel Methods for Elliptic Partial Differential Equations. Cambridge University Press, New York (1996)
15. Saad, Y.: Iterative Methods for Sparse Linear Systems. SIAM (2003)

16. Saad, Y., Sosonkina, M. SIAM J. Sci. Comput. **21** (1999) 1337–1356
17. Li, Z., Saad, Y., Sosonkina, M. Numer. Lin. Alg. Appl. **10** (2003) 485–509
18. Sosonkina, M., Saad, Y., Cai, X. Future Gen. Comput. Systems **20** (2004) 489–500
19. Saad, Y., Suchomel, B. Numer. Lin. Alg. Appl. **9** (2002) 359–378

Alternate Learning Algorithm on Multilayer Perceptrons

Bumghi Choi[1], Ju-Hong Lee[2,*], and Tae-Su Park[3]

[1] Dept. of Computer Science & Information Eng., Inha University, Korea
neural@inha.ac.kr
[2] School of Computer Science & Eng., Inha University, Korea
juhong@inha.ac.kr,
[3] taesu@datamining.inha.ac.kr

Abstract. Multilayer perceptrons have been applied successfully to solve some difficult and diverse problems with the backpropagation learning algorithm. However, the algorithm is known to have slow and false convergence aroused from flat surface and local minima on the cost function. Many algorithms announced so far to accelerate convergence speed and avoid local minima appear to pay some trade-off for convergence speed and stability of convergence. Here, a new algorithm is proposed, which gives a novel learning strategy for avoiding local minima as well as providing relatively stable and fast convergence with low storage requirement. This is the alternate learning algorithm in which the upper connections, hidden-to-output, and the lower connections, input-to-hidden, alternately trained. This algorithm requires less computational time for learning than the backpropagation with momentum and is shown in a parity check problem to be relatively reliable on the overall performance.

1   Introduction

Backpropagation (BP) algorithm was developed by Rumelhart, Hinton, and Williams [1] as a learning algorithm for multilayered perceptrons. Though the BP algorithm has not yet been able to learn an arbitrary computational task in a network, it can solve many problems such as XOR, which the simple one-layer perceptrons can not solve. BP algorithm is described as follows in a two-layered network.

Our usual error measure or cost function is described as

$$E[w] = \frac{1}{2} \sum_{\mu i} \left[\zeta_i^{\mu} - g\left(\sum_j w_{ij}^2 g\left(\sum_k w_{jk}^1 \xi_k^{\mu} \right) \right) \right]^2 \tag{1}$$

where ζ_i^{μ} is the ideal value of output unit i at pattern μ, g is an activation function, w_{ij}^2 is the weight from hidden unit j to output unit i, w_{jk}^1 is the weight from input unit k to hidden unit j.

For hidden-to-output connections the gradient-descent rule gives

$$\Delta w_{ij}^2 = -\eta \frac{\partial E}{\partial w_{ij}^2} = \eta \sum_{\mu} \left[\zeta_i^{\mu} - z(2)_i^{\mu} \right] g'\left(h(2)_i^{\mu} \right) z(1)_i^{\mu} \tag{2}$$

---

* Corresponding author.

V.N. Alexandrov et al. (Eds.): ICCS 2006, Part I, LNCS 3991, pp. 63 – 67, 2006.

From input-to-hidden connections,

$$\Delta w^1_{jk} = -\eta \, \frac{\partial E}{\partial w^1_{jk}} = \eta \sum_{\mu i} \left[\zeta^\mu_i - z(2)^\mu_i \right] g'\!\left(h(2)^\mu_i\right) w^2_{ij} g'\!\left(h(1)^\mu_j\right) \xi^\mu_k \tag{3}$$

BP is known to have convergence problems such as local minima or the plateau. The plateau causes the problem of very slow convergence. In the local minima, the gradient equals to zero. If the training process falls into a local minimum, the process of updating weight vectors stops.

Dynamic change of the learning rate and momentum[2,3,4] or the selection of a better function for activation or error evaluation followed by a new weight-updating rule have been proposed to avoid the problems. Quickpro[7], the resilient propagation(RPROP)[8], the genetic algorithm[5], the conjugate gradient[6], and the second-order method such as Newton's method[9,10] appear to pay some trade-off between the convergence speed and the stability of convergence avoiding the traps in wide range of parameters, or between the overall performance and storage requirement.

2 Alternate Learning Algorithm

In this section, a new learning strategy will be introduced, providing a conjecture why this can avoid local minima and plateaus. It just adopts the alternate learning strategy that combines the solvability of two-layered networks and the training power of the simple perceptrons. Based on that strategy, we develop a new learning algorithm. The new algorithm will be called *Alternate learning with the target values of hidden units*.

2.1 Analogy to Detour

In the proposed algorithm the learning process is separated into two components. First the upper connections are trained with the lower connections fixed. Whenever the training slows due to a local minima or plateau, the training is forcibly stopped. Then, the lower connections are trained with the upper connections fixed until the process meets the slow training time. This process is repeated until the error function reaches to the global minima.

At this point an analogy to traffic flow clarifies the philosophy of the learning technique. At a glance, BP algorithm and its variations seem to give the shortest way to the global minimum. But there are too many traffic jams like local minima or plateaus on the road of BP. When faced with a traffic jam the simple solution is to make a detour. The alternate learning method is based on this simple rule.

2.2 Approximation of the Target Values of Hidden Units

We need the target values of hidden units for the training of the lower connection. Here a target value of a hidden unit is intended to mean the value of a hidden unit which makes a selected output approximate to its ideal value as close as possible.

Though the exact values may not be figured out directly, the Newton-like approximation of the inverse function can be possible. We can draw the errors of hidden units from a selected output error by using the inverse function. To explain this process in terms of mathematical formulae, consider the expected error of output $z(2)_i$ from a point, $z(1)$, at the hidden units space

$$= \frac{\left(\zeta_i^{\mu} - z(2)_i^{\mu}\right)}{\left|\nabla z(2)_i^{\mu}\left(z(1)^{\mu}\right)\right|} \tag{4}$$

The component in the direction of $z(1)_j$ of $\nabla z(2)_i$

$$= \frac{\dfrac{\partial z(2)_i^{\mu}}{\partial z(1)_j^{\mu}}}{\left|\nabla z(2)_i^{\mu}\left(z(1)^{\mu}\right)\right|} = \frac{w_{ij}^2 g'\left(h(2)_i^{\mu}\right)}{\left|\nabla z(2)_i^{\mu}\left(z(1)^{\mu}\right)\right|} \tag{5}$$

By multiplying two factors, the expected error is

$$\gamma_j^{\mu} - z(1)_j^{\mu} = \frac{\left(\zeta_i^{\mu} - z(2)_i^{\mu}\right)w_{ij}^2}{g'\left(h(2)_i^{\mu}\right)\left|w_i^2\right|} \tag{6}$$

where γ_j is the target value, $z(1)_j$ is the current value of j'th unit of the hidden layer, $w_i^w = \left(w_{i1}^2, w_{i2}^2, ..., w_{in}^2\right)$, n is the dimension of hidden space.

The algorithm is summarized as follows

1. Train upper-connections with the ordinary gradient-descent rule until the process converges or meets a slow training point.
2. If the process converges then stop the program, otherwise propagate the error and target value of each hidden unit using formula (6)
3. Train the lower-connections from input units to the hidden units with produced target values of the hidden units until the training is slow or converges.
4. Go to 1.

3 Test and Evaluation

3.1 Test Environment

In order to verify the effectiveness of the alternate learning, 4-2-1 parity problem is used. For comparison, we used the BP with momentum 0.1 and 0.9.

3.2 Test Results

The convergence rate is defined as the inverse of the averaged convergence epochs. Fig. 1. shows the comparison of the stability between the alternate learning and the BP with momentum.

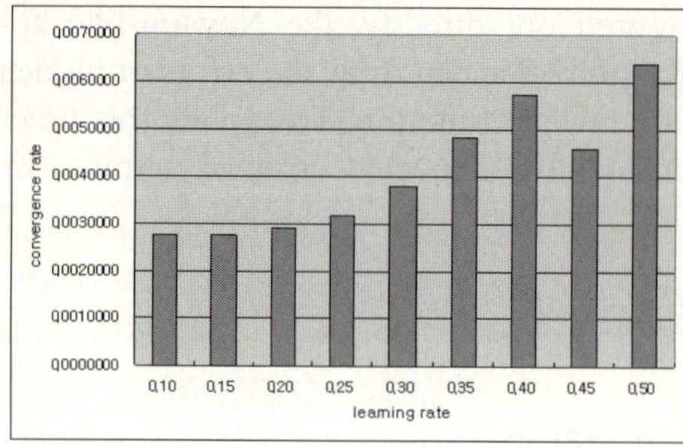
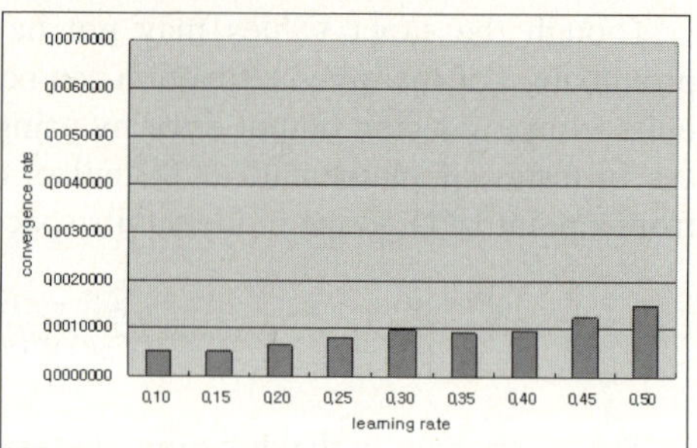

Fig. 1. The comparison of the alternate learning, the left, and BP with momentum, the right

Table 1. shows the experimental results of the three methods based on 100 runs randomly chosen from learning rate [0.1, 0.5], and initial weight [-1, 1] at 4-2-1 parity problem. The case of epochs over 1000 is treated as failed.

Table 1. Experimental results from 4-2-1 parity problem

Method	Average epochs	Minimum epochs	Success rate
BP with momentum 0.1	155	3	72%
Bp with momentum 0.9	150	2	54%
The alternate learning	78	3	100%

4 Conclusion

Thus far, the alternate learning algorithm on multilayer perceptrons have been derived, tested and compared with BP. From logical conjecture and the experimental results, the alternate learning algorithm is more stable in convergence than BP. Furthermore, since it is a kind of learning strategy, combined with existing methods announced as substitutes for BP, the additional effectiveness can be achieved.

Acknowledgments. This research was supported by the Ministry of Information and Communication, Korea, under the Information Technology Research Center support program supervised by the Institute of Information Technology Assessment.

References

1. Rumelhart, D.E., G.E. Hinton, and R.J. Williams.: Learning Internal Representations by Error propagation. In Parallel Distributed Processing, vol. 1, chap8 (1986)
2. Jacobs, R. A.: Increased Rates of Convergence Through Learning Rate Adaptation. Neural Networks 1 (1988) 293-280
3. Vogl, T. P., J.K. Magis, A.K. Rigler, W.T. Zink, and D.L. Alkon.: Accelerating the Convergence of the Back-Propagation Method. Biological Cybernetics 59 (1988) 257-263
4. Allred, L. G., Kelly, G. E.: Supervised learning techniques for backpropagation networks. In Proc. of IJCNN, vol. 1 (1990) 702-709

5. Montana D. J., Davis L.: Training feedforward neural networks using genetic algorithms, in Proc. Int. Joint Conf. Artificial Intelligence, Detroit (1989) 762-767
6. Moller, M. S.: A scaled conjugate gradient algorithm for fast supervised learning, Neural Networks, vol. 6 (1993) 525-534
7. Fahlman, S. E.: Fast learning variations on backpropagation: An empirical study, in Proc. Connectionist Models Summer School (1989)
8. Riedmiller, M. and Braun, H.: A direct adaptive method for faster backpropagation learning: The RPROP algorithm, in Pro. Int. Conf. Neural Networks, vol. 1 (1993) 586-591
9. Ricoti, L. P., Ragazzini, S. and Martinelli, G.: Learning of word stress in a suboptimal second order back-propagation neural networks, in Proc. 1st Int. Conf. Neural Networks, vol. I (1988) 355-361
10. Watrous, R. L.: Learning algorithms for connectionist network: applied gradient methods of nonlinear optimization, in Proc. 1st. Int . Conf. Neural Networks, vol. II (1987) 619-628

A Transformation Tool for ODE Based Models

Ciro B. Barbosa, Rodrigo W. dos Santos, Ronan M. Amorim,
Leandro N. Ciuffo, Fairus Manfroi, Rafael S. Oliveira, and Fernando O. Campos

FISIOCOMP, Laboratory of Computational Physiology
Department of Computer Science - Universidade Federal de Juiz de Fora (UFJF)
PO Box 15.064 - 91.501-970 - Juiz de Fora - MG - Brasil
{ciro, rodrigo}@dcc.ufjf.br,
ronanrmo@ig.com.br, leandro@areaweb.com.br, fayrus@gmail.com,
rsachetto@gmail.com, fernando.ocampos@terra.com.br

Abstract. This paper presents a tool for prototyping ODE (Ordinary Differential Equations) based systems in the area of computational modeling. The models, tailored during the project step of the system development, are recorded in MathML, a markup language built upon XML. This design choice improves interoperability with other tools used for mathematical modeling, mainly considering that it is based on Web architecture. The resulting work is a Web portal that transforms an ODE model documented in MathML to a C++ API that offers numerical solutions for that model.

1 Introduction

This work is within the scope of Computational Modeling of Electrophysiology [1]. Under this area of research, biological cell models are often based on large non-linear systems of Ordinary Differential Equations (ODEs). Nowadays, modern cardiac cell models comprises of ODE systems with tens to near hundred of free variables and hundreds of parameters. Recently, the computational biology community has come out with a XML based standard for the description of cellular models [2]. The CellML standard provides the community with both a human- and computer-readable representation of mathematical relationships of biological components.

In this work we extend the CellML goals with a transformation tool that automatically generates C++ code that allows one to manipulate and numerically solve CellML based models.

The transformation tool described here alleviates several problems inherent to the development, implementation, debugging and use of cellular biophysical models.

The implementation of the mathematical models is a time consuming and error prone process, due mainly to the ever rising size and complexity of the models. Even the setup process of the simulations, where all initial values and parameters are to be set, is time consuming and error prone. In addition, ODE numerical resolution typically demands high performance computing environments and the programming expertise adds more complexity to this multidisciplinary area of research.

V.N. Alexandrov et al. (Eds.): ICCS 2006, Part I, LNCS 3991, pp. 68–75, 2006.
© Springer-Verlag Berlin Heidelberg 2006

To minimize the above mentioned problems, we have built a systematic transformation process that automatically turns mathematical models into corresponding executable code. The tool is an API (Application Program Interface) generator for ODE Solution (AGOS) [1]. AGOS is an on-line tool that automatically builds up an object-oriented C++ class library that allows its users to manipulate and numerically solve Initial Value Problems based on ODE systems described by the CellML or MathML standard. Manipulation here means to set initial values, parameters and some features of the embedded model and of the numerical solver.

Finally, although the AGOS tool was initially tailored to support models described by the CellML standard, currently it works for any initial value problem based on non-linear system of first-order ODEs documented in the MathML standard. Therefore, AGOS is a powerful and useful transformation tool that aims to support the development of many scientific problems in the most diverse areas of research. Biological, ecological, neural and cardiac prototype models are available at the AGOS web page [1] as examples.

In this paper we present the systematization of the transformation process, showing a compromise with implementation correctness. Some other tools described in the Internet [3][4] pursue similar goals. However, the lack of scientific documentation does not allow a proper evaluation and comparison to the AGOS tool.

The next sections present the transformation process, the tool architecture and its components, and some concluding remarks.

2 Transformation Process

The input data for this process is a CellML [2] or a Content MathML [5] file, i.e., XML-based languages. MathML is a W3C standard for describing mathematical notation. CellML is an open-source mark-up language used for defining mathematical and electrophysiological models of cellular functions. A CellML file includes Content MathML to provide both a human- and a computer-readable representation of mathematical relationships of biological components. Therefore, the AGOS tool allows the submission of a complete CellML file or just its MathML subset.

Once submitted, the XML file is translated to an API. The AGOS application was implemented in C++ and makes use of basic computer data structures and algorithms in order to capture the variables, parameters and equations that are embedded in a MathML file and to translate these to executable C++ code, i.e. the AGOS API.

The transformation process consists of identifying and extracting the ODE elements documented in the XML file and generating the corresponding API classes. The conceptual elements in the ODE are: independent variable, dependent variables, auxiliary variables, equation parameters, differential equations and algebraic equations.

The structural elements in the API are methods that can be classified as private or public. The public ones include methods that: set and get the values of the dependent variables (Set/GetVar), set the number of iteration cycles and the discretion interval

(Setup), set the equation parameters (ParSet), calculate the numerical solution via the Explicit Euler scheme (SolveODE).

In addition, the API offers public reflexive functions used, for example, to restore the number of variables and their names. These reflexive functions allow the automatic creation of model-specific interfaces. This automatic generated interface enables one to set any model initial condition or parameter, displaying their actual names, as documented in the CellML or MathML input file.

The algebraic equation solver (SolveAE) is an example of a AGOS private method that is used by the numerical solution method (SolveODE) to obtain the values of auxiliary variables.

Figure 1 synthesizes the relations between the conceptual elements of the ODEs and the basic methods of the API. ODE elements are presented with circles and API methods with rectangles. Arrow directions define the relationship dependency. For instance, algebraic equations depend on parameters, dependent, auxiliary and independent variables; the SolveAE method depends on the algebraic equations; and in turn it influences the auxiliary variables.

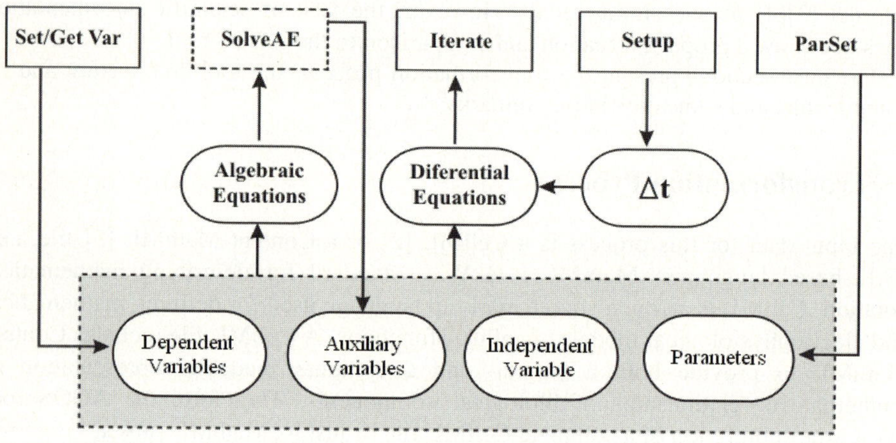

Fig. 1. ODE to API Mapping

The next example better illustrates the transformation process and the relationship between ODE elements and API methods. Consider the following ODE, known as the bistable equation [6]:

$$dVm/dt = - (I\_ion) / Cm, \tag{1}$$

$$I\_ion = a (Vm - b) (c - Vm) Vm. \tag{2}$$

AGOS identifies the ODE elements: Eq. 1 is a differential equation and Eq. 2 is an algebraic equation; Vm, I_ion, and t are dependent, auxiliary and independent variables, respectively; and the ODE parameters are Cm, a, b and c. Using the Forward Euler method a numerical implementation of the above ODE can be written as:

$$Vm^i = -\Delta t \, a \, (Vm^{i-1} - b) \, (c - Vm^{i-1}) \, Vm^{i-1} / \, Cm + Vm^{i-1}, \tag{3}$$

where Δt is the time step and Vmi is the discretization of $Vm(i\,\Delta t)$, for $i \geq 0$.

Based on the extracted ODE elements from Eqs. 1 and 2, AGOS generates the following SolveODE and SolveAE methods that implement the numerical solution presented by Eq. 3.

```
void Solveode::solve(int iterations){
    for(i=1; i<iterations; i++)
        Vm[i] = dt* (-calc_I_ion()/Cm) + Vm[i-1];
}
double calc_I_ion(){
        return a*(Vm[i-1]-b)*(c-Vm[i-1])*Vm[i-1];
}
```

3 Tool Architecture

The translator tool comprises of three basic components: a Preprocessor for XML format, an Extractor of ODE conceptual elements, and a Code Generator. The components are organized as a pipeline. The Preprocessor reads an XML-based file (MathML or CellML) and extracts the content into an array of tree data structures. Every tree of this array is processed by the ODE extractor that identifies the ODE elements and stores them in appropriate data formats. At the end of the pipeline, the Code Generator combines the extracted information to a code template and generates the AGOS API. Fig. 2 presents the tool architecture where the relations between the basic components are illustrated.

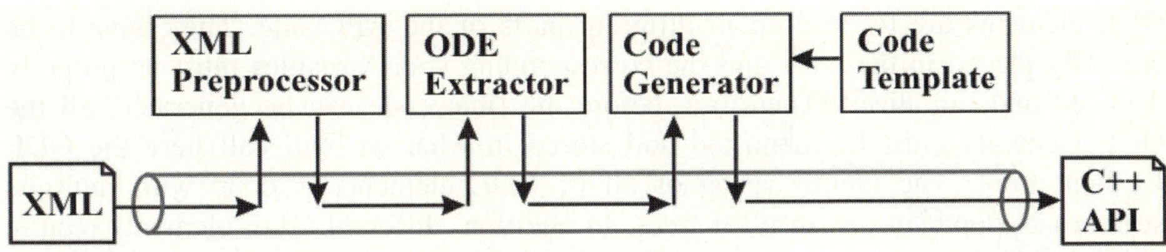

Fig. 2. AGOS Architecture

3.1 XML Preprocessing

The MathML description language uses prefix format on input, i.e. the operators precede the operands. Therefore, a tree is an appropriate structure to store the XML content as it facilitates the identification of the operands and operators. In addition, with the information stored in a tree it is easy to recover the equation formulation with a search in depth procedure. We use the DOM class library [7] to manipulate the XML input files. The Document Object Model (DOM) is an API for HTML and

XML files that provides a structural representation of the document, enabling programs and scripts to access and modify its content [7]. The information is extracted into a tree data structure with equation elements and XML tags. The DOM tree nodes contain information about each operand and operator, besides the equation type (if it is a differential equation or an algebraic one).

To illustrate the preprocessing step, Fig. 3 presents the corresponding Content MathML code and the generated tree of Eq. 1.

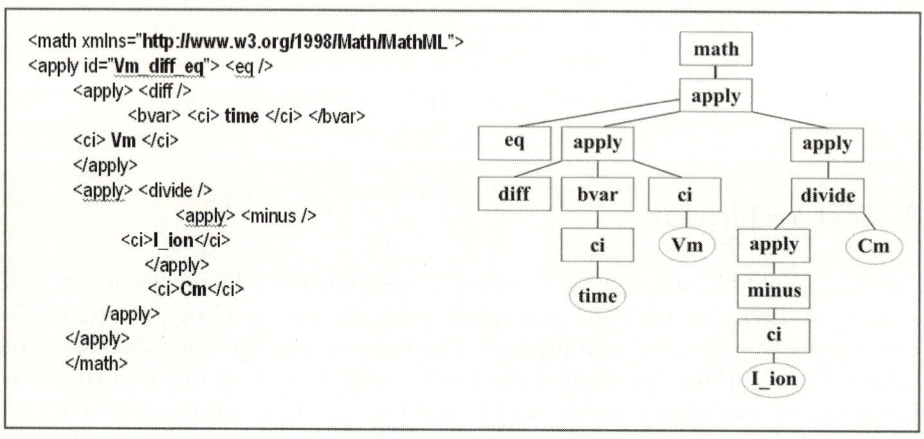

Fig. 3. Content MathML code and tree-like representation

3.2 Extracting ODE Elements

ODE elements are to be used in different parts of the API code. They have to be correctly placed in the code and the corresponding code variables must be properly declared and initialized. Therefore, before the final code can be generated, all the ODE elements must be identified and stored in what we will call here the ODE Element Pool. The identification of all of ODE elements is done with multiple searches in depth in the array of trees. In addition, different ODE elements require different data formats for storage and manipulation. Parameters, dependent and auxiliary variables are each stored in different linked lists. Examples of information stored here are the names, units and default values. The equations are stored in a linked list of trees. This way, the order between elements is preserved as well as information concerning the element type (operand or operator), element characteristic (infixed, prefixed, variable or constant), among others. Figure 5 illustrates the tree that corresponds to Eq. 1. During the creation of this data structures the XML tags are eliminated and the position of operands is standardized. Once the ODE elements are identified and stored in the appropriate data structures, the collection of these structures, i.e. the ODE Element Pool, contains all the necessary information for the Code Generator.

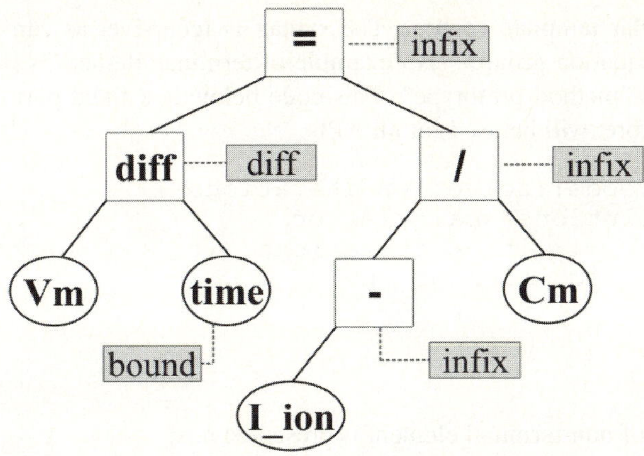

Fig. 4. The tree structure obtained from the MathML

3.3 Generating the AGOS Code

The adopted strategy for code generation is largely based on code templates. The syntactical structure of code templates can be described using formal grammar notation. The algorithm for code generation is inspired in a recursive algorithm for syntax analysis [8]. This algorithm fills in the C++ code template with data contained in the generated Pool of ODE elements. Next we illustrate the AGOS grammar.

```
<api> -> "Class header" "class body" <variables
        declaration>        <solution> <algebraic
        equation set> <GetVar>   <SetVar>      <Setup>
        <ParSet>

<variable declaration> -> "type" <variable> | "type"
        <variable>   <variables declaration>

<solution> ->  "method prototype"  <equation group>

<equation group>  ->  <equation>   |   <equation>
        <equation group>

<equation> ->  <dependent variable (t)> "="
        <discretization> "*"       <expression (t-dt)>
        "+" <dependent variable (t-dt)>

<algebraic equation set> -> <algebraic equation>   |
                <algebraic equation> <algebraic
        equation set>

<discretization> -> "d"<independent variable>
```

In the grammar, terminal symbols are enclosed by ("). The title of the terminal symbol indicates a piece of the code template. Non-terminal elements are enclosed by (<>). Such elements are defined elsewhere in the grammar or represent functions that fill in a particular template section. The syntax is recursive, as can be seen in the definition of <equation group>. An example of terminal element is presented below for the terminal "method prototype". This code below is a fixed part of the template code and, therefore, will be used for all APIs.

```
void Solveode::solve(int iterations){
    // solutions' calculation
    for(it_=1; it_ < iterations; it_++){
        // <equation group>
    }
}
```

An example of non-terminal element is presented next.

```
MMLVarListNode *cur = vlVariables;
fprintf(file,"//private variables\n");
fprintf(file,"//private: \n");
while(cur != NULL){
    fprintf(file,"\tdouble *%s;\t //%s \n",cur->name,
            cur->units);
    cur = cur->next;
}
```

The above code shows the implementation of the recursive definition of <variables declaration>. This part of the code generation uses the linked list structure that stores the dependent variables (linked list vlVariables) to dynamically generate the variable declaration of the AGOS API. The resulting code is:

```
//private variables
private:
            double *Vm;
```

4 Conclusions

In this work we described AGOS, a transformation tool that automatically generates executable code that solves and manipulates mathematical models described by initial value problems based on non-linear systems of ODEs and documented in the MathML or CellML standards. The support provided by this systematic transformation process aims on reducing the time during the various phases of scientific model development, implementation, debugging and use.

The AGOS Tool is available at [1], from where it is possible to download the API source-code. The AGOS API can also be used online via a web application, which uses the generated API to solve the ODE system and to visualize the results. Via a dynamic web form, that uses the reflexive AGOS methods, one is able to set up the ODE parameters and initial conditions of the specific submitted ODE system.

Acknowledgements. We thank the support provided the Brazilian Ministry of Science and Technology, CNPq (processes 506795/2004-7).

References

1. Fisiocomp. Laboratory of Computational Physiology. UFJF, Brazil (2005). http://www.fisiocomp.ufjf.br/
2. CellML biology, math, data, knowledge. Internet site address: http://www.cellml.org/
3. LI, J. and LETT, G.S.: Using MathML to Describe Numerical Computations, http://www.mathmlconference.org/2000/Talks/li/
4. CellML: mozCellML. http://www.cellml.org/tools/mozCellML/mozCellMLHelp/technical
5. W3C: Mathematical Markup Language Version 2.0, http://www.w3.org/TR/MathML2/
6. Keener,J., Sneyd, J.: Mathematical Physiology. Springer, 1 edition, 792p., (1998).
7. W3C, Document Object Model (DOM): http://www.w3.org/DOM/
8. Aho, A.V., Seit, R. and Ullman, J.D.: Compilers Addison Wesley, 500p., (1986).

Performance Comparison of Parallel Geometric and Algebraic Multigrid Preconditioners for the Bidomain Equations

Fernando Otaviano Campos, Rafael Sachetto Oliveira,
and Rodrigo Weber dos Santos

FISIOCOMP: Laboratory of Computational Physiology,
Department of Computer Science - Universidade Federal de Juiz de Fora (UFJF),
PO Box 15.064-91.501-970 - Juiz de Fora - MG - Brazil
`fernando.ocampos@terra.com.br`, `rsachetto@gmail.com`,
`rodrigo@dcc.ufjf.br`

Abstract. The purpose of this paper is to discuss parallel preconditioning techniques to solve the elliptic portion (since it dominates computation) of the bidomain model, a non-linear system of partial differential equations that is widely used for describing electrical activity in the heart. Specifically, we assessed the performance of parallel multigrid preconditioners for a conjugate gradient solver. We compared two different approaches: the Geometric and Algebraic Multigrid Methods. The implementation is based on the PETSc library and we reported results for a 6-node Athlon 64 cluster. The results suggest that the algebraic multigrid preconditioner performs better than the geometric multigrid method for the cardiac bidomain equations.

1   Introduction

The bidomain formulation [1] is currently the model that best reflects the electrical activity in the heart. The non-linear system of partial differential equations (PDEs) models both the intracellular and extracellular domains of cardiac tissue from an electrostatic point of view. The coupling of the two domains is done via non-linear models describing the current flow through the cell membrane.

Unfortunately, the bidomain equations are computationally very expensive. Efficient ways of solving the large linear algebraic system that arises from the discretization of the bidomain model have been a topic of research since 1994 [2]. Many different approaches among direct and iterative methods have been employed considering the problem's size and the computing resources available. However, iterative methods such as conjugate gradient (CG) are generally more scalable.

In previous works we have compared different parallel preconditioner methods for the Conjugate Gradient iterative algorithm. In [3] we have shown that preconditioners based on the Geometric Multigrid Method (GMG) performed better than the classical incomplete LU (ILU) preconditioners for 2D and 3D cardiac

V.N. Alexandrov et al. (Eds.): ICCS 2006, Part I, LNCS 3991, pp. 76–83, 2006.

electric propagation problems. In [4] the GMG preconditioners were compared to different Additive Schwarz (ASM) preconditioners. The results taken from a 16-node HP-Unix cluster indicated that the multigrid preconditioner was at least 13 times faster than the single-level Schwarz based techniques and requires at least 11% less memory.

In this paper we compare our previous parallel implementation of the GMG preconditioner [3], [4] to an Algebraic Multigrid (AMG) based parallel preconditioner [5]. We focus on the solution of the linear algebraic system associated with the elliptic part of the bidomain model, since this part dominates computation. We employ the CG method preconditioned with both, Geometric (GMG) and Algebraic (AMG) parallel multigrid (MG) techniques. The implementation is based on the PETSc C library [6] (which uses MPI) and is tested on problems involving thousands of unknowns. The results taken from a 6-node Athlon 64 Linux cluster indicate that the AMG preconditioner is at least 3 times faster than GMG and 117 times faster than the traditional ILU preconditioner.

2 Mathematical Formulation

The set of Bidomain equations[1] is currently one of the most complete mathematical models to simulate the electrical activity in cardiac tissue:

$$\chi \left(C_m \frac{\partial \phi}{\partial t} + f(\phi, t) \right) = \nabla.(\sigma_i \nabla \phi) + (\sigma_e \nabla \phi_e), \tag{1}$$

$$\nabla.((\sigma_e + \sigma_i)\nabla \phi_e) = -\nabla.(\sigma_i \nabla \phi), \tag{2}$$

$$\frac{\partial v}{\partial t} = g(\phi, \boldsymbol{\eta}), \quad \phi = \phi_i - \phi_e. \tag{3}$$

Where ϕ_e is the extracellular potential, ϕ_i the intracellular potential and ϕ is the transmembrane potential. Eq. (3) is a system of non-linear equations that accounts for the dynamics of several ionic species and channels (proteins that cross cell membrane) and their relation to the transmembrane potential. The system of (3) typically accounts for over 20 variables, such as ionic concentrations, protein channel resistivities and other cellular features. σ_i and σ_e are the intracellular and extracellular conductivity tensors, i.e. 3×3 symmetric matrices that vary in space and describe the anisotropy of the cardiac tissue. C_m and χ are the cell membrane capacitance and the surface-to-volume ratio, respectively.

Unfortunately, a solution of this large nonlinear system of partial differential equations (PDEs) is computationally expensive. One way to solve (1)-(3) at every time step is via the operator splitting technique [4]. The numerical solution reduces to a three step scheme which involves the solutions of a parabolic PDE, an elliptic PDE and a nonlinear system of ordinary differential equations (ODEs) at each time step. Rewriting equations (1)-(3) using the operator splitting technique (see [4] for more details) we get the following numerical scheme:

$$\varphi^{k+1/2} = (1 + \Delta t A_i)\varphi^k + \Delta t A_i(\varphi_e)^k; \tag{4}$$

$$\varphi^{k+1} = \varphi^{k+1/2} - \Delta t f(\varphi^{k+1/2}, \boldsymbol{\zeta}^k)/(C_m), \tag{5}$$

$$\zeta^{k+1} = \zeta^k + \Delta t g(\varphi^{k+1/2}, \zeta^k);$$
$$(A_i + A_e)(\varphi_e)^{k+1} = -A_i\varphi^{k+1}. \tag{6}$$

Where φ^k, φ_e^k and ζ^k discretizes ϕ, ϕ_e and η at time k Δ_t; A_i and A_e are the discretizations for $\nabla.((\sigma_i\nabla)/(\chi C_m)$ and $\nabla.((\sigma_e\nabla)/(\chi C_m)$, respectively. Spatial discretization was done via the Finite Element Method using a uniform mesh of squares and bilinear polynomials as previously described in [4].

Steps (4), (5) and (6) are solved as independent systems. Nevertheless, (4), (5) and (6) are still computationally expensive. One way of reducing the time spent on solving these equations is via parallel computing.

3   Parallel Multigrid Preconditioners

In the previous Section we presented a mathematical model for the electrical potential in the cardiac tissue. Many direct and iterative approaches have been employed in the solution of the linear systems that appear after the spatial discretization of the Bidomain equations. Direct factorization methods such as Cholesky or LU performed better than iterative methods for small simulation setups [7], i.e., when memory limitations were not a concern. However, for larger problems, for instance, the simulation of a whole three-dimensional (3D) heart in which the discretization leads to millions of nodes, an iterative method is mandatory. The preconditioned CG method has become the standard choice for an iterative solver of the bidomain formulation.

We used CG to solve the linear system and intended to compare both GMG and AMG as preconditioners. The solution of (4)-(6) was implemented in parallel using the PETSc C library, which uses MPI. CG is parallelized via linear domain decomposition. The spatial domain is decomposed into *proc* domains with equal sizes, where *proc* is the number of processors involved in the simulation. In addition, we compared the traditional block incomplete LU parallel preconditioner (ILU) against the multigrid ones.

The basic idea behind multigrid methods relies on the fact that simple iterative methods are very efficient at reducing high-frequency components of the residual, but are very inefficient with respect to the lower frequency components. In the classical MG theory [8], such iterative methods are called smoothers, since they smooth the error better than reduce its average. The multilevel solution to this problem is to project the residual onto a smaller space, a coarser grid of the problem, where the lower spatial frequency components can be handled more efficiently. The problem is now solved on the coarser grid and the residual is then projected back to the original space, the finer grid. This way, the residual on the finer grid has an approximation of the lower frequency components removed, and the convergence of the iterative method is faster.

In our GMG preconditioner implementation we successively generated coarser regular grids based on the finest regular grid G_0, i.e. the uniform element mesh. This procedure was repeated until the coarsest level, $G_{levels-1}$. For each grid pair, G_l and G_{l+1}, a prolongation rectangular matrix, P_l, was generated using

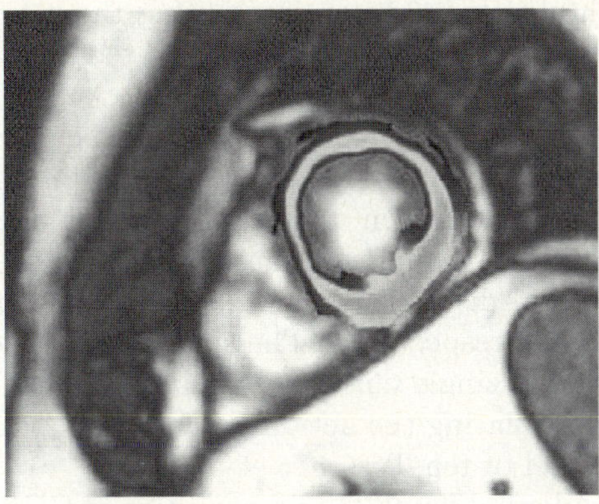

Fig. 1. Simulated electrical wave propagation overlapped to the Resonance Image

a bilinear interpolation scheme. For every grid level ($l = 0$ to $l = levels - 1$), a matrix A_l was generated by applying the finite element method to the particular grid. Further details of our GMG implementation can be found in [3].

The mainly difference between GMG and AMG preconditioners is the coarse grid selection. In the GMG scheme, just simple slices are made to the fine grid creating a coarse grid with half of the nodes in each direction. To select the coarse grid points in the AMG, we seek those unknowns u_i which can be used to represent the values of nearby unknowns u_j. It is done via the concepts of dependence and influence. We say that the point i depends on the point j or j influences i, if the value of the unknown u_j is important in determining the value of u_i from the ith equation. Based on measures of dependence and influence taken from the matrix coefficients, special heuristics generate the maximal subset of points of the coarse grid. Thus, the process of coarse matrices generation depends solely on the original finest-grid matrix. Different from the GMG, the finite element method is only used once in the AMG method, i.e. during the creation of the finest-grid matrix. All the other matrices are obtained algebraically.

In this work we adopted the parallel AMG code BoomerAMG [5] with its Falgout-coarsening strategy.

In both GMG and AMG preconditioners the smoother used for all but the coarsest level was an iterative method. For the coarsest level, we used a direct solver. This was not done in parallel, i.e., it was repeated on every processor, avoiding any communication.

4   Results

We performed several tests in order to compare the different preconditioners on a 6 node Linux Cluster, each node equipped with a AMD Athlon 64 $3\,GHz$ processor, $2\,GB$ of RAM and connected by $1\,Gbit/s$ Ethernet switch. In all tests, we simulated the cardiac electric propagation on a two-dimensional cut of the left ventricle obtained during the cardiac diastole phase by the resonance magnetic technique of a healthful person. After segmenting the resonance image, a two-dimensional mesh of 769×769 points was generated, that models the cardiac tissue, blood and torso.

All bidomain parameters were taken from [9]. The capacitance per unit area and the surface area - to- volume ratio are set to $1\,mF/cm^2$ and $2000/cm$, respectively. The interface between cardiac tissue and bath is modeled as described in [10]. All the other boundaries are assumed to be electrically isolated. The spatial and temporal discretization steps of the numerical model were set to $0.0148\,cm$ and $0.05\,ms$, respectively. The simulation was carried out for $5\,ms$, or 100 time steps, after a single current stimulus was introduced at a selected endocardial site. For simulating the action potential of cardiac cells we used the human ventricular model of ten Tusscher et al. [11].

The stop criterion adopted for all the preconditioned CG algorithms was based on the unpreconditioned and absolute L2 residual norm, $\|Ax_i - b\|^2 \leq tol$, where x_i was the solution at iteration i and tol was a tolerance which was set to 10^{-6}. Although this is not the most efficient stop criteria for the CG, it is the fairest one when comparing different preconditioning methods.

The performance measurements reported in this section, such as the execution time, CG number of iterations, average time per iteration, setup time and number of nonzeros in an particular grid are related to the solution of the elliptic part (2), since this part is responsible for around 80% of the whole simulation time. The memory usage is related to the whole model.

4.1 Parameter Tuning

We tuned the following parameters: *fill* for ILU; *levels* for GMG; *levels* and *strongthreshold* for AMG. Table 1 shows, for different numbers of processors, the optimal parameter values that yielded the fastest execution time. The parameter *fill* was varied from 0 to 4; GMG *levels* varied from 2 to 6; AMG *levels* from 6 to 16 and *strongthreshold* was set to 0.25, 0.50 and 0.75. In addition, all parameters were tuned for best execution time on 1, 2, 4 and 6 processors. A total of 160 simulations were performed during about one week of computation time.

Due to the long execution time demanded for the ILU preconditioner, just simulations on 6 processors were performed for this case. With this number of processors the optimal value of *fill* was 4.

For GMG, the optimal value of *levels* depended on *proc*. On a single processor, *levels* = 2 corresponded to the fastest execution. In parallel, however, since the coarsest grid is solved sequentially, the cost of fewer grid *levels* rivaled the gains of parallelism. Therefore, as *proc* increased, the optimal *levels* also increased to 3.

The AMG preconditioner performed better with the *strongthreshold* set to 0.25, i.e. the smallest experimented value. The choice of the *strongthreshold* value directly influences the number of grid points in each level, i.e. the number of non-zero elements of each matrix A_l. High *strongthresholds* generated rich coarse matrices in terms of the information that is kept from the finest level. This contributed towards faster convergence in terms of iteration count. However,

Table 1. Number of nonzeros on 1 and 6 processors

proc	levels	GMG	AMG
1	2	1329409	1722609
	8	-	8138
6	3	332929	683411
	8	-	8707

Table 2. Comparison between GMG and AMG on 1 and 6 processors

Type	proc	Time (s)	CG Iters.	Time / Iter
ILU		-	-	-
GMG	1	1867.05	1050	1.78
AMG		626.27	916	0.68
ILU		17502.88	175481	0.10
GMG	6	578.41	1290	0.45
AMG		141.12	900	0.16

this improvement did not result in faster execution times, since every level was considerably more computationally expensive.

AMG performed better with more *levels* (*levels* = 8 was the fastest) than GMG. Many factors may have contributed to this. It was shown before that AMG coarsening algorithms tend to coarsen in the direction of the dependence and perform better than the traditional geometric algorithms when the problem has anisotropic or discontinuous coefficients. Cardiac tissue is highly anisotropic and the conduction coefficients of the bidomain model reflect the cardiac tissue properties. Therefore, fewer levels in the GMG may be necessary in order to avoid loss of anisotropic information. In addition, the implementation of the AMG and GMG preconditioners differs in some aspects. The *smoothers* (the iterative methods used) are different. GMG uses a more robust and expensive method, a preconditioned CG, to *smooth* the residuals in all but the coarsest level. AMG uses a simple relaxation method. Thus, every GMG level is more computationally expensive than an AMG one. The direct methods used to solve the coarsest level are also different. GMG uses a more efficient and fast direct LU solver with nested dissection reordering. The AMG direct solver was very inefficient in handling large problems, i.e. coarsest matrices with less than 5 *levels*.

In summary, compare to GMG, AMG performed better with more levels and fewer non-zero elements in the coarsest matrix. Table 1 shows the number of non-zeros of the coarsest matrix of the simulations with 1 and 6 processors, for the AMG and GMG cases.

4.2   Performance Comparison and Parallel Speedup

Table 2 presents the execution time, total number of CG iterations and average time per iteration for all time steps (100 time steps) per processor for all preconditioners with the optimal parameters. Both GMG and AMG preconditioners

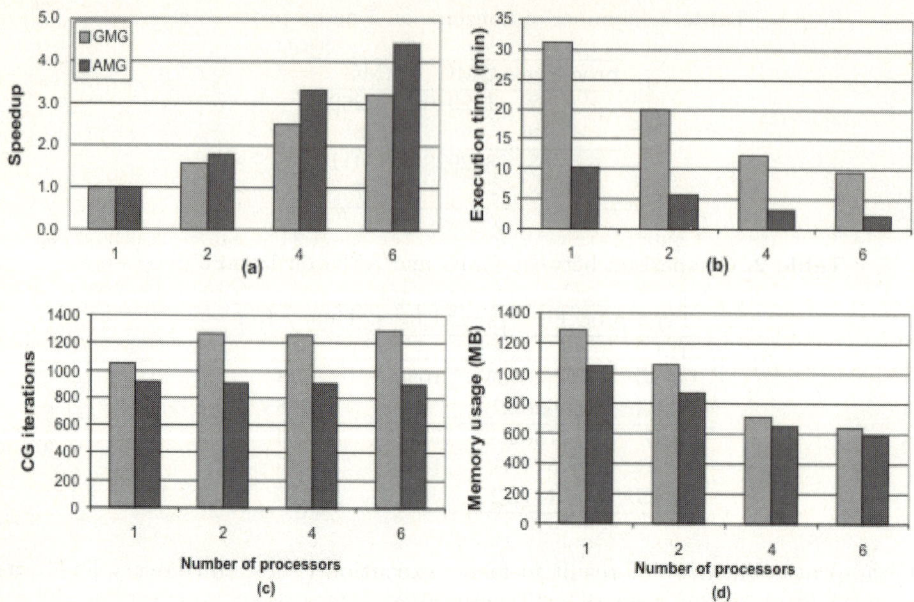

Fig. 2. Computational requirements for 2D anisotropic systole phase. (a) parallel speedup relative to single processor execution time for GMG and AMG preconditioners for different number of processors. (b) execution time, (c) total number of CG Iterations for 100 time steps and (d) memory usage.

achieved better performance results than ILU on 6 processors. GMG was 30.26 times faster than ILU. In the same conditions, AMG was 124.03 times faster than ILU. In addition, ILU took more than 17 thousands CG iterations to solve the problem, i.e. for the total 100 time steps. Although ILU presented a better average time/iteration than the multigrid methods, it needed much more CG iterations to solve the system. GMG and AMG converged with around one thousand iterations, i.e. 10 iterations per time step.

AMG was 2.98 (4.10) times faster than GMG on 1 (6) processors, and required 50% (54%) less on setup phase. When using the GMG preconditioner, the number of CG iterations increased as the number of processors scaled up from 1 to 6. The AMG method was more stable and the number of iterations did not increase when increasing the number of processors. Figure 2 shows the comparison between GMG and AMG preconditioners with 1, 2, 4 and 6 processors. Both multigrid preconditioners achieved reasonable parallel speedup (execution time on 1 processor/execution time) results, 3.23 for GMG and 4.44 for AMG on 6 processors. AMG performed better for all tests, it converged faster (took less CG iterations), required around 22% less memory with $proc \leq 2$ and 9% when $proc \geq 4$. Finally, according to the speedup results, AMG achieved a better scalability on 6 processors, which resulted from the smaller coarsest grid matrices. The direct method used in the coarsest grid is not parallelized. Thus, the larger matrices used by GMG in this level (see table 1) severely degrade the parallel speedup.

4.3   Conclusions

In this work, we employed the conjugate gradient algorithm for the solution of the linear system associated with the elliptic part of the bidomain equations and compared two different parallel multigrid preconditioners: the GMG and AMG. The results taken from a 6-node Athlon 64 Linux cluster indicate that the algebraic multigrid preconditioner is at least 3 times faster than GMG, requires at least 9% less memory and achieved a better scalability on 6 processors.

Acknowledgements. We thank the support provided by the Universidade Federal de Juiz de Fora (UFJF-Enxoval) and the Brazilian Ministry of Science and Technology, CNPq (processes 506795/2004-7 and 473872/2004-7).

References

1. Sepulveda N. G., Roth B. J., and Wikswo Jr. J. P.: Current injection into a two-dimensional anistropic bidomain. Biophysical J. **55** (1989) 987–999
2. Hooke N., Henriquez C., Lanzkron P., and Rose D.: Linear algebraic transformations of the bidomain equations: Implications for numerical methods. Math. Biosci. vol. 120, no. 2, pp. (1994) 127–45
3. Weber dos Santos R., PLANK G., BAUER S., and VIGMOND E. J.: Parallel Multigrid Preconditioner for the Cardiac Bidomain Model IEEE Trans. Biomed. Eng. **51(11)** (2004) 19601968
4. Weber dos Santos R., Plank G., Bauer S., and Vigmond E.: Preconditioning techniques for the bidomain equations. Lecture Notes In Computational Science And Engineering. **40** (2004) 571–580
5. Henson V. E., and Yang U. M.: BoomerAMG: a Parallel Algebraic Multigrid Solver and Preconditioner. Technical Report UCRL-JC-139098, Lawrence Livermore National Laboratory, (2000)
6. Balay S., Buschelman K., Gropp W., Kaushik D., Knepley M., McInnes L., Smith B., and Zhang H.: PETSc users manual. Technical report ANL-95/11 - Revision 2.1.15, Argony National Laboratory, (2002)
7. Vigmond E., Aguel F., and Trayanova N.: Computational techniques for solving the bidomain equations in three dimensions Trans. Biomed. Eng. IEEE. **49** (2002) 1260–9
8. Briggs W., Henson V., and McCormick S.: A Multigrid Tutorial. SIAM, Philadelphia, PA, Tech. Rep., (2000).
9. Jones, J. Vassilevski, P.: A parallel graph coloring heuristic. SIAM J. Sci. Comput. **14** (1993) 654–669
10. Krassowska W., and Neu. J. C.: Effective boundary conditions for syncytial tissues. IEEE Trans. Biomed. Eng **41** (1994) 143–150.
11. ten Tusscher K. H. W. J., Noble D., Noble P. J., and Panfilov A. V.: A model for human ventricular tissue J. Physiol. **286** (2004) 1573–1589

Simulating and Modeling Secure Web Applications

Ramon Nou, Jordi Guitart, and Jordi Torres

Barcelona Supercomputing Center(BSC)
Computer Architecture Department
Technical University of Catalonia
C/Jordi Girona 1-3, Campus Nord UPC
E-08034 Barcelona, Spain
{rnou, jguitart, torres}@ac.upc.edu

Abstract. Characterizing application servers performance becomes hard work when the system is unavailable or when a great amount of time and resources are required to generate the results. In this paper we propose the modeling and simulation of complex systems, such as application servers, in order to alleviate this limitation. Using simulations, and specifically coarse-grain simulations as we propose here, allows us to overcome the necessity of using the real system while taking only 1/10 of the time than that of the real system to generate the results. Our simulation proposal can be used to obtain server performance measurements, to evaluate server behavior with different configuration parameters or to evaluate the impact of incorporating additional mechanisms to the servers to improve their performance without the necessity of using the real system.

1 Introduction

We can view an Application Server based on the J2EE platform as a complex system, several things happen at the same time, and there are many levels involved; from threads to TCP/IP, including cryptographic and security issues. At the same time, all information that is confidential or has market value must be carefully protected when transmitted over the open Internet. Security between network nodes over the Internet is traditionally provided using HTTPS[1].

The increasing load that e-commerce sites must support and the widespread diffusion of dynamic web content and SSL increase the performance demand on application servers that host these sites. At the same time, the workload of Internet applications is known to vary over time, often in an unpredictable fashion, including flash crowds that cannot be processed. These factors sometimes lead to these servers getting overloaded. In e-commerce applications, which are heavily based on the use of security, such server behavior could translate to sizable revenue losses. These systems are known to be very complex to characterize. In our research, we need to spend four hours per test (including the start and stop of the server, the server warm up time, and several executions corresponding to several number of clients and different server parameters), but this time is

V.N. Alexandrov et al. (Eds.): ICCS 2006, Part I, LNCS 3991, pp. 84–91, 2006.

real time and exclusive, so resources are difficult to get. How could we test some hypotheses or ideas without using these resources? We propose to resolve this by estimating the performance measurements using some kind of performance model. With this approach we can get an approximation of the results using less resources and time.

Different approaches have been proposed in publications on performance analysis and prediction models for this type of e-business systems. Most of them exploit analytical models where analysis is based on Markov Chain Theory [2]. Queuing Networks and Petri Nets are among the most popular modeling formalisms that have been used. But due the complexity of todays e-business systems, the analytical methods for solving can not be used. Only by simplifying the system we can obtain manageable models. On the other hand, these systems cannot model, in an easy way, timeouts behavior. The simulation model is an abstract representation of the system elements and their interactions, and an alternative to analytical mathematical models. The main advantage of simulations is that it overcomes the limitation of complex theoretical models, while the methodological approach to simulation models and the analysis of simulation results is supported by statistical principles developed in the last 30 years. There are several works that simulate systems in order to extract information about them [3, 4], but in general, the number of proposals including modeling application servers and problems like the ones we are facing are scarce.

We are able to get results that would take a whole day on a real system (testing some parameter values), in less than an hour using only a desktop computer. This gives us the possibility to test several QoS policies in real time without having it implemented and running on a real machine.

The rest of the paper is organized as follows: Section 2 introduces the analyzed system, an application server with SSL security. Section 3 explains simulation environment and tools used. On Section 4 we describe the experimental environment. Inside Section 5 we explain all the blocks that build the simulated model. Section 6 compares simulation results with experimental results, and shows some interesting findings that can be obtained from simulated models. Some other experiments with simulation can be found on Report [5]. An extended version of this paper with more results can be found on Report [6].

2   Secure Dynamic Web Applications

The two components we are using on our systems are Dynamic web applications and SSL security. Dynamic web applications are multi-tiered applications, normally using a database. The client receives a HTML page with information gathered and computed from the database by the application server. Communication between client and server is protected using the SSL protocol ([7]). SSL increases the computation time necessary to serve a connection, due to the use of cryptography. The impact of SSL on server performance has been evaluated in [8]. It concludes that saturation of a server with SSL security provokes the degradation of the throughput. Further information about an admission control using SSL connections and its impact can be found in [9]. More information about the impact of using SSL on server performance can be found in [8].

3 Simulation Proposal

We are using a simulation tool and a performance analysis framework to generate the simulation.

Simulations are usually much more computationally intensive than analytic models. On the other hand, simulation models can be made as accurate as desired and focus over some specific target. To the best of our knowledge, there are not any simulation packages that are specifically oriented for these types of systems. We decided to use Objective Modular Network Testbed in C++ (OMNet++) [10]. OMNet++ offers us a way to build our modules with a programming language, so we can test our changes and fine-tune server model in a fast way.

In order to obtain the computation times of the different services, which will be used to construct the simulations model we propose using a performance analysis framework developed in our research center. This framework, which consists of an instrumentation tool called Java Instrumentation Suite (JIS [11]) and a visualization and analysis tool called Paraver [12], allows a fine-grain analysis of dynamic web applications. Further information about the implementation and its use for the analysis of dynamic web applications can be found in [11].

4 Benchmark and Platform

We simulate an e-business multi-tiered system composed of an application server, a database server and some distributed clients performing requests to the application server. This kind of environment has been frequently modeled [13], but we have added several new components to the simulation, which have not been considered before. First, a SSL protocol between the clients and the application server, and lastly, timeouts, which are harder to get on an analytical model.

We use Tomcat v5.0.19 [14] as the application server. In addition, in order to prevent server overload in secure environments, [9] we added to the Tomcat server a session-oriented adaptive mechanism that performs admission control based on SSL connections differentiation. Further details can be read on [9]. The client is using Httperf [15] to generate the requests of the workload from RUBiS benchmark [16] (RUBiS attempts to produce a workload similar to eBay auction site). The workload is divided into sessions, where every session issues several requests (burst and not-burst) with a thinktime between them.

5 Model Description

The system was modeled using five black boxes 1: A client that simulates Httperf and RUBiS (several others can be used, to represent other kind of scenarios), an

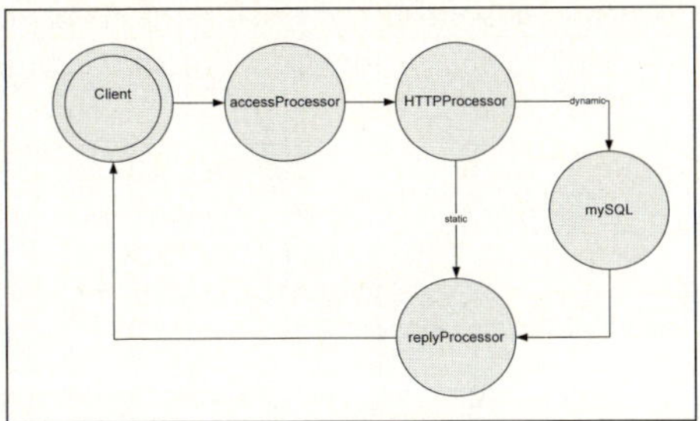

Fig. 1. Simulation modules

accessProcessor that simulates operating system backlog queue of connections and allow the setting up of admission policies. HTTPProcessor manages connections using SSL handshake and reusing SSL. HTTPProcessor also processes requests and sends to the MySQL database as needed. To conclude a replyProcessor gets the reply from MySQL or the HTTPProcessor and sends it to the client.

We do not want to build a system with a great level of detail; we do not include threads in our model even though we modeled a simplified HTTP 1.1 scheme. Our objective was to achieve an approximation of the real system values, without using a large processing time. This has some handicaps, because we cannot simulate or retrieve data for the response time (for example); we had not modeled all the components that affect it on a per-request level. We seek throughput data only so our model is enough for us, response time modeling will give more detail to cache-like mechanisms. We will show more specific details for some of these blocks. The code from the simulation framework can be found on [17] for more detail.

accessProcessor. Has a queue that sends requests to HTTPProcessor (and performs admission control). From here it is easy to control limitation policies on HTTPProcessor so we can quickly test several policies by just changing the code. Inside this block we had modeled typical Linux buffers behavior(backlog).

HTTPProcessor. Uses Processor Sharing scheduling. We are using a two pass processing mechanism: when a request arrives we process the connection (SSL process and request service) then we process static data as much as is needed. Next we see if the request needs further processing and we send it to the database block. These two phases were created to simulate the behavior of Httperf timeouts. Requests from this block can go to mySQL or to the client.

6   Evaluation

Here we include the validation of our approach when comparing the results obtained with real life systems. We also present some experiments showing the usefulness of our approach to evaluate system behavior when a real system is not available.

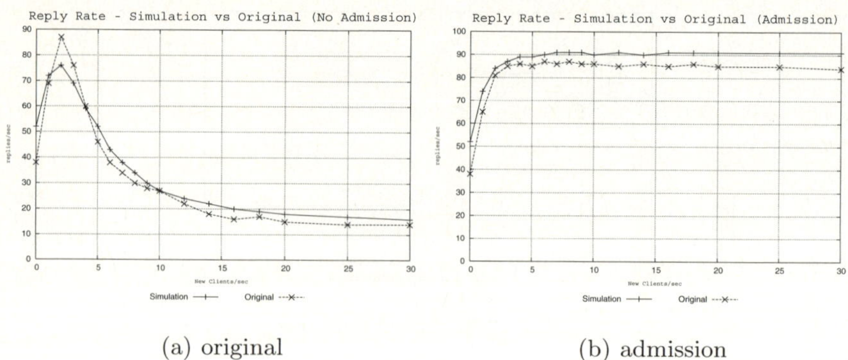

(a) original   (b) admission

Fig. 2. Throughput comparison between original Tomcat and with admission control in the real system versus when simulated

6.1   Model Validation

Comparison with the original Tomcat. Fig. 2.a shows the Tomcat throughput as a function of the number of new clients per second initiating a session with the server. When the number of clients needed to overload the server has been achieved, the server throughput degrades until approximately 20% of the maximum achievable throughput while the number of clients increases. The cause of this great performance degradation on server overload has been analyzed in [8]. When the server is overloaded, it cannot handle the incoming requests before the client timeouts expire; In this case, clients with expired timeouts are discarded and new ones are initiated, provoking the arrival of a great amount of new client connections that need the negotiation of a full SSL handshake, causing degradation of the server performance. In our simulation, as shown in Figure 2.a, the graph is a similar shape (with approximate values).

Comparison with Tomcat with Admission Control. Considering the described behavior, it makes sense to apply an admission control mechanism in order to improve server performance. Figure 2.b shows the Tomcat throughput as a function of the number of new clients per second initiating a session with the server. Notice that the server throughput increases linearly with respect to the input load (the server scales) until a determined number of clients hit the server. At this point , the throughput achieves its maximum value. Until this point, the server with admission control behaves in the same way as the original server. However, when the number of clients that would overload the server has been achieved, the admission control mechanism can avoid the throughput degradation, maintaining it in the maximum achievable throughput; in this case, the simulation is also able to achieve almost the same results than the real system, as shown in Figure 2.b.

6.2   Simulation Results

In the previous subsection, we have demonstrated that the simulation with a simple model, using a low level of detail, adapts well to the real system.

In addition, this kind of coarse-grain simulation allows us to run tests over hardware or systems that are not available and still obtain performance predictions. Its low complexity allows us to simulate long experiments without a high computational cost. Nevertheless, if it is necessary, this simulation model can be extended in any detail level in order to obtain other types of information.

In order to illustrate the possibilities that our simulation approach offers, in this section we numerate some experiments that cannot be performed in the real system (because it is not available or the execution time needed to run them is extremely high). In this way, we demonstrate that simulations can be very helpful to answer a great number of questions; confirming or discarding hypotheses and providing hints about server behavior with different system configurations.

Scalability of the application. Although our model is very simple, it can describe with great success the real model pattern. If we need to know the behavior of our application with a system with more CPUs, we can do it. However these kind of results with our simulation should be restricted to a certain number of CPUs. To enhance this number, we should increase complexity of the model to enable a more detailed behavior of contention between processors (for example to model thread behavior inside HTTPProcessors). With the same environment we can use simulation to provide us with a way to test several QoS admission policies, and how they will scale on the real system without using real resources. As an example we do not have machines available with more than 4 CPUs, so simulation fills this gap. Simulation is used to determine if a proposal is worth testing on a real system or not.

Extracting extra data from simulations. We can adapt the model to help us to extract other kind of data also. We can accomplish tests that would cost a lot of resources, while simulation also gives us more flexibility to add variables and count things like the maximum number of opened sessions on an instant. With this variable we could see how admission control affects the number of opened sessions. These kinds of statistical results can be achieved with ease by adding little code to the model.

Testing hypothesis involving system changes. Simulations were used to test how different algorithms, which change the order of requests, can improve server performance. The order should be changed on the backlog queue, so it requires several Linux Kernel modifications. Testing on a real system would involve rebooting machines (server is multi-user), modifying kernel internals, debugging and several other time consuming chores. With a simulation we were able to test these changes with a low cost (in time and resources). After changing the order of the backlog queue we obtained the results that we expected. Now we can start moving resources to implement our hypothesis on a real system, with a lower probability of failure.

7   Conclusions

In this paper we have presented a method to simulate an existing system and shown how we can obtain some experimental data without the need to use much resources or time. In this work, we used only data from a single CPU system run, and from its sample workload. Our model uses OMNET++ to the system behavior with all the detail we need, and predicts the shape of the real system graph. Although we do not model the system with deep enough detail to predict real numbers, we found that the results are very similar and good enough to make some predictions. In our model, every time we added a new feature, the results became more accurate, and we had achieved a great approximation on Figure 2.b.

We added to the normal simulations of application web servers some handicaps like timeouts and SSL, to show how it affects the server, and how it shapes the real life graphics. For more specific results we will need a more detailed model (i.e. OS contention with more than one CPU).

Using simulations gives us more flexibility and capability to extract data and run some tests that could be unaffordable with real resources. We can measure what throughput we could achieve (or at least an approximation) with the addition of more processors, threads or with a change of the number of clients. We did not need the real system for this, only a simulation. We can slightly modify the model to help us to decide what proposal of admission control is the best in a first approach. For example we can quickly analyze the number of rejected sessions(rejected sessions leads to lost revenue).

Simulations can be useful to test changes such as changing the backlog from FIFO to LIFO [18] and see how it improves performance. To conclude, coarsegrain simulations could give us basic guidelines to help us to test some changes in our systems and there are a great number of questions that can be answered by them. Thanks to the simulations we have obtained results such as those on 6.2. In these we can analyze the behavior of the system when we have more CPUs. Thanks to the validation we have seen that the obtained results come closer to the real ones. This can help us to evaluate these policies in a fast and quite reliable way and give us some hints to know if a proposal is good or bad.

Future work will include introducing simulations as a predictive tool inside an application server.

Acknowledgment

This work is supported by the Ministry of Science and Technology of Spain and the European Union (FEDER funds) under contract TIN2004- 07739-C02-01. Thanks to Lídia Montero from Dept. of Statistics and Operations Research, UPC for her help. For additional information please visit the Barcelona eDragon Research Group web site [17].

References

1. Rescorla, E.: HTTP over TLS. RFC 2818 (2000)
2. Bolch, G., Greiner, S., Meer, H.D., Trivedi, K.S.: Queueing networks and markov chains. Modelling and Performance Evaluation with Computer Science Applications. Wiley, New York (1998)
3. Stewart, C., Shen, K.: Performance modeling and system management for multi-component online services. NSDI (2005)
4. Uragonkar, B., G.Pacifi, Shenoy, P., Spreitzer, M., Tantawi, A.: An analytical model for multi-tier internet services and its applications. SIGMETRICS'05, Alberta, Canada (2005)
5. Nou, R.: Sorting backlog queue, impact over tomcat. Research Report UPC-DAC-RR-CAP-2005-30 (2005)
6. Nou, R., Guitart, J., Torres, J.: Simulating and modeling secure e-business applications. Research Report UPC-DAC-RR-2005-31 (2005)
7. Dierks, T., Allen, C.: The tls protocol, version 1.0. RFC 2246 (1999)
8. Guitart, J., Beltran, V., Carrera, D., Torres, J., Ayguadé, E.: Characterizing secure dynamic web applications scalability. IPDPS'05, Denver, Colorado, USA (2005)
9. Guitart, J., Beltran, V., Carrera, D., Torres, J., Ayguadé, E.: Session-based adaptative overload control for secure dynamic web application. ICPP-05, Oslo, Norway (2005)
10. WebPage: Omnet++. http://www.omnetpp.org (2005)
11. Carrera, D., Guitart, J., Torres, J., Ayguadé, E., Labarta, J.: Complete instrumentation requirements for performance analysis of web based technologies. ISPASS'03, pp. 166-176, Austin, Texas, USA (2003)
12. WebPage: Paraver. http://www.cepba.upc.es/paraver (2005)
13. Menacé, D.A., Almeida, V.A.F., Dowdy, L.W.: Performance by Design. Pearson Education (2004)
14. WebPage: Jakarta tomcat servlet container. http://jakarta.apache.org/tomcat (2005)
15. Mosberger, D., Jin, T.: httperf: A tool for measuring web server performance. Workshop on Internet Server Performance (WISP'98) (in conjunction with SIGMETRICS'98), pp. 59-67. Madison, Wisconsin, USA (1998)
16. Amza, C., Cecchet, E., Chanda, A., Cox, A., Elnikety, S., Gil, R., Marguerite, J., Rajamani, K., Zwaenepoel, W.: Specification and implementation of dynamic web site benchmarks. WWC-5, Austin, Texas, USA. (2002)
17. WebPage: Barcelona edragon research group. http://www.bsc.es/eDragon (2005)
18. N.Singhmar, Mathur, V., Apte, V., D.Manjunath: A combined lifo-priority scheme for overload control of e-commerce web servers. International Infrastructure Survivability Workshop, Lisbon, Portugal (2004)

A Treecode for Accurate Force Calculations*

Kasthuri Srinivasan** and Vivek Sarin

Department of Computer Science, Texas A&M University,
College Station, TX, USA
{kasthuri, sarin}@cs.tamu.edu

Abstract. A novel algorithm for computing forces in N-body simulations is presented. This algorithm constructs a hierarchical tree data structure and uses multipole expansions for the inverse square law. It can be considered as a multipole-based variant of Barnes-Hut algorithm [2]. The key idea here is the representation of forces (gravitational or electrostatic) through a multipole series by using the relationship between ultraspherical and Legendre polynomials. The primary advantage of this algorithm is the accuracy it offers along with the ease of coding. This method can be used in the simulation of star clusters in astrophysical applications in which higher order moments are expensive and difficult to construct. This method can also be used in molecular dynamics simulations as an alternative to particle-mesh and Ewald Summation methods which suffer from large errors due to various differentiation schemes. Our algorithm has $O(p^3 N \log N)$ complexity where p is typically a small constant.

1   Introduction

N-body simulations have become very important in theoretical and experimental analysis of complex physical systems. A typical N-body problem models physical domains where there is a system of N particles (or bodies), each of which is influenced by gravitational forces or electrostatic forces of all other particles. The naive brute force algorithm to compute the forces acting on each particle based on particle-particle interaction is an $O(N^2)$ algorithm. There are various approximation algorithms with lower complexity that have been proposed for these simulations. In the simulation of gravitational forces in astrophysics, Barnes-Hut algorithm [2] and its variants are used. For electrostatic force simulations occurring in molecular dynamics (MD), Fast Multipole algorithm (FMM) [5], particle-mesh methods like P3M [8] and Ewald Summation techniques [7] are widely applied.

For two particles of strengths q_1 and q_2, the force acting on q_1 due to q_2, is given by the inverse square law,

$$F = \frac{q_1 q_2}{|r_{12}|^2} \cdot \frac{r_{12}}{|r_{12}|}, \tag{1}$$

---

* This work has been supported in part by NSF under the grants NSF-CCR0113668 and NSF-CCR0431068.

** Corresponding author.

V.N. Alexandrov et al. (Eds.): ICCS 2006, Part I, LNCS 3991, pp. 92–99, 2006.

where $|r_{12}|$ is the distance between the particles. In this paper, the constant of proportionality (like the gravitational constant) is assumed to be 1. Even though the above law is applicable to both gravitational and electrostatic interactions, the required accuracy determines the algorithms to be used. The Barnes-Hut algorithm is an easy-to-code algorithm used in most astrophysical simulations where lower accuracy is often preferred (around 1%) over reduced execution time. In astrophysical simulations, the force is approximated by a multipole series expansion with respect to the center of mass. The first moments are large enough (since the masses are all positive) to approximate the forces. Therefore, the monopole moment and the first two moments in the multipole series provide enough accuracy. On the other hand, in MD simulations, higher accuracy is needed because the distance and time scales involved are very different from astrophysical problems. The Barnes-Hut scheme can be adapted to MD simulations by adding more terms to the multipole series. For electrostatic force simulations, however, the distribution of positive and negative charges impedes the numerical robustness of the Barnes-Hut algorithm [6]. Methods like Ewald Summation, particle-mesh based approaches like P3M, and FMM are preferred over Barnes-Hut algorithm for electrostatic problems.

In contrast, in the investigation of collisional astrophysical systems, namely star clusters and galactic nuclei, there is a need for high accuracy. Such applications require $|\delta E/E| \ll 0.04 N^{-1} t/t_{cr}$ where t and t_{cr} are time parameters, $|\delta E/E|$ is the relative energy accuracy and N is the number of stars in the system [4]. For a system of 10^3 stars, this means an upper bound of 10^{-5} on the error. In order to achieve such high accuracy, McMillan and Aarseth [3] use octupole terms in the multipole expansion which are quite cumbersome to construct. Moreover, increased accuracy is needed for larger systems in order to limit the cumulative errors on core-collapse time scales to acceptable levels. To our knowledge there is no prior work that uses higher order moments than octupoles.

In the case of electrostatic force interactions, particle-mesh algorithms like P3M and Ewald Summation techniques compete with FMM [9, 10]. Despite the superior asymptotic scaling of FMM when compared to earlier methods [$O(N)$ versus $O(N \log N)$], FMM has some disadvantages. Apart from the obvious difficulty in coding, FMM suffers from lack of energy conservation unless very high accuracy is employed [11]. In addition, special considerations regarding the degree of force interpolation must be taken into account to conserve momentum. Pollock and Glosi [10] discuss other advantages of P3M over FMM. Even in force interpolation methods used in particle-mesh algorithms and Ewald Summation methods, a general statement of accuracy seems to be difficult. The error analysis of force calculations in these methods are quite involved and they provide a fair comparison only if done at the same level of accuracy. The discretization methods introduce new sources of errors in addition to the ones originating from real and reciprocal space cutoffs. Furthermore, investigation of errors during force evaluation depends on several parameters such as mesh-size, interpolation orders and differentiation schemes. Deserno and Holm [12] illustrated remarkable differences in accuracy for these methods.

In this paper, we present a novel algorithm for accurate force calculations in N-body problems. This algorithm combines the ease of the Barnes-Hut method with a very high accuracy. Instead of using higher order moments, the algorithm uses additional terms of a multipole series to increase the accuracy. Our method can be used effectively in the calculation of star cluster interactions in which accuracy has been limited by octupole moments. Unlike particle-mesh and Ewald Summation based methods, our algorithm does not involve differentiation (either differentiation in Fourier space, analytic differentiation or discrete differentiation) to calculate the forces. Forces are directly computed through the series. This is especially useful where force interpolation is affected by conservation laws. Further, it may be of interest to note that the complexity of this algorithm is $O(p^3 N \log N)$, which is comparable to that of particle-mesh algorithms that use Fast Fourier Transforms.

The paper is organized as follows: Section 2 discusses the multipole expansions for forces calculations. Section 3 describes the algorithm, analyzes its complexity and discusses implementation issues. Section 4 presents numerical experiments. Conclusions are presented in Section 5.

2   Multipole Expansion for Forces

A multipole based tree code for computing potentials of the form $r^{-\lambda}$, $\lambda \geq 1$, was introduced in [1]. In this paper, we extend the result to evaluate the forces. The reader is advised to refer [1] for some of the discussions in this section. The key idea in computing the potentials of the form $r^{-\lambda}$ is the use of ultraspherical polynomials in a manner analogous to the use of Legendre polynomials for the expansion of the Coulomb potential r^{-1}. Ultraspherical polynomials (or Gegenbauer polynomials) are generalizations of the Legendre polynomials in terms of the underlying generating function. That is, if $x, y \in \Re$, then

$$\frac{1}{(1 - 2xy + y^2)^{\lambda/2}} = \sum_{n=0}^{\infty} C_n^{\lambda}(x) y^n, \tag{2}$$

where $C_n^{\lambda}(x)$ is an ultraspherical polynomial of degree n. They are also *higher dimensional* generalizations of Legendre polynomials [14, 15] in the sense that ultraspherical polynomials are eigenfunctions of the generalized angular momentum operator just as Legendre polynomials are eigenfunctions of the angular momentum operator in three dimensions. One can refer to [14] for a list of similarities between them. The above generating function can be used to expand $r^{-\lambda}$. In [1] we derived the following multipole expansion theorem for $r^{-\lambda}$ potentials.

Theorem 1 (Multipole Expansion for Potentials) [1]. *Suppose that k charges of strengths $\{q_i, i = 1, \ldots k\}$ are located at the points $\{Q_i = (\rho_i, \alpha_i, \beta_i), i = 1, \ldots, k\}$, with $|\rho_i| < a$. Then, for any point $P = (r, \theta, \phi)$ with $r > a$, the potential $\Phi(P)$ is given by*

$$\Phi(P) = \sum_{n=0}^{\infty} \sum_{m=0}^{\lfloor n/2 \rfloor} \frac{1}{r^{n+\lambda}} \mathbf{M_n^m} \cdot \mathbf{Y_{n,m}}(\theta, \phi), \tag{3}$$

where

$$\mathbf{M_n^m} = \sum_{i=1}^{k} q_i \rho_i^n B_{n,m}^{\lambda} \overline{\mathbf{Y_{n,m}}}(\alpha_i, \beta_i),$$

$$\mathbf{Y_{n,m}^T}(x, y) = \left[Y_{n-2m}^{-(n-2m)}(x, y), Y_{n-2m}^{-(n-2m)+1}(x, y), \ldots, Y_{n-2m}^{(n-2m)}(x, y) \right]$$

is a vector of spherical harmonics of degree $n - 2m$ *and*

$$B_{n,s}^{\lambda} = \frac{(\lambda)_{n-s}(\lambda - 1/2)_s}{(3/2)_{n-s} s!} (2n - 4s + 1). \tag{4}$$

Furthermore, for any $p \geq 1$,

$$\left| \Phi(P) - \sum_{n=0}^{p} \sum_{m=0}^{\lfloor n/2 \rfloor} \frac{\mathbf{M_n^m}}{r^{n+\lambda}} \cdot \mathbf{Y_{n,m}}(\theta, \phi) \right| \leq \frac{AB}{r^{\lambda-1}(r - a)} \left(\frac{a}{r} \right)^{p+1}, \tag{5}$$

where $A = \sum_{i=1}^{k} |q_i|$ *and* $B = \sum_{m=0}^{\lfloor n/2 \rfloor} \left| B_{n,m}^{\lambda} \right|.$

The multipole expansion for force computations can now be deduced as a corollary to the above theorem.

Corollary 1 (Multipole Expansion for Forces). *Suppose that* k *particles of strengths* $\{q_i,\ i = 1, \ldots k\}$ *are located at points whose position vectors are* $\{\rho_i, i = 1, \ldots, k\}$ *and let* $\{Q_i = (|\rho_i|, \alpha_i, \beta_i), i = 1, \ldots, k\}$ *denote their spherical coordinates with* $|\rho_i| < a$. *Then, for any vector* $\mathbf{r}$ *with coordinates* $P = (|\mathbf{r}|, \theta, \phi)$ *and* $|\mathbf{r}| > a$, *the* c^{th} *component of the force,* $F_c(P)$, *is given by*

$$F_c(P) = \sum_{n=0}^{\infty} \sum_{m=0}^{\lfloor n/2 \rfloor} \frac{1}{|\mathbf{r}|^{n+3}} \left[(\mathbf{r}_c)\mathbf{M_n^m} - {}_c\mathbf{V_n^m} \right] \cdot \mathbf{Y_{n,m}}(\theta, \phi) \tag{6}$$

where

$$\mathbf{M_n^m} = \sum_{i=1}^{k} q_i |\rho_i|^n B_{n,m}^3 \overline{\mathbf{Y_{n,m}}}(\alpha_i, \beta_i), \qquad {}_c\mathbf{V_n^m} = \sum_{i=1}^{k} \rho_{ci} \left(q_i |\rho_i|^n B_{n,m}^3 \overline{\mathbf{Y_{n,m}}}(\alpha_i, \beta_i) \right).$$

Here $\mathbf{r}_c$ *and* ρ_{ci} *are the* c^{th} *component of* $\mathbf{r}$ *and* ρ_i, *respectively. Furthermore, for any* $p \geq 1$, *the approximation* $\hat{F}_c^p(P)$ *obtained by truncating the expression in (6) after* p *terms, satisfies*

$$\left\| F_c(P) - \hat{F}_c^p(P) \right\|_2 \leq \frac{AB}{|\mathbf{r}|^2 (|\mathbf{r}| - a)} \left(\frac{a}{|\mathbf{r}|} \right)^{p+1} (|\mathbf{r}| + a),$$

where $A = \sum_{i=1}^{k} |q_i|$ *and* $B = \sum_{m=0}^{\lfloor n/2 \rfloor} \left| B_{n,m}^{\lambda} \right|.$

Proof. We note that $r_{12} = r - \rho_1$, and therefore, (1) can be written as

$$F(P) = \frac{q_1}{|r_{12}|^3} r - \frac{q_1}{|r_{12}|^3} \rho_1.$$

Using $\lambda = 3$ in (3) for $1/|r_{12}|^3$, the proof follows through the superposition of particles at $\{\rho_i, i = 1, \ldots, k\}$. The error bound can be obtained from (5) by using the triangle inequality $|r_{12}| \leq |r| + |\rho_1| \leq |r| + a$.

3 Treecode for Force Calculations

The treecode to compute the forces can be viewed as a variant of Barnes-Hut scheme that uses only particle-cluster multipole evaluations. The method works in two phases: the tree construction phase and the force computation phase. In the tree construction phase, a spatial tree representation of the domain is computed. At each step of this phase, if the domain contains more than one particle, it is recursively divided into eight equal sub-domains. This process continues until each sub-domain has at most one particle.

Each internal node in the tree computes and stores multipole series representation of the particles within its sub-domain. Note that we don't have the means to compute translations of multipole coefficients from child to parent nodes. We must compute the multipole coefficients at every internal node directly from the particles contained in the node. These coefficients are obtained using Corollary 1. Once the tree has been constructed, force at a point is computed using the multipole coefficients of a subset of the nodes in the tree, which are sufficiently far away from the point. A specific constant α is used to check if a node is far enough from the evaluation point. Given the distance of a point from the center of the sub-domain, d, and the side of the box, r, a point is considered to be far away from a node if the *multipole acceptance criterion* (MAC) defined as, d/r, is greater than α. For each particle, the algorithm proceeds by applying MAC to the root of the tree to determine whether an interaction can be computed; if not, the node is expanded and the process is repeated for each of its eight children. If the node happens to be a leaf, then the force is calculated directly by using the particles in the node. In order to improve the computational efficiency of the algorithm, the minimum number of particles in any leaf box can be set to a constant s. The pseudo-code for this method can be found in [1].

It can be seen from Corollary 1 that the complexity of computing the multipole coefficients at each level of the tree is $O(p^3 N)$, where N is the number of particles in the system and p is the multipole degree. Since there are $\log N$ levels in the tree for a uniform particle distribution, the total cost of computing the multipole coefficients is $O(p^3 N \log N)$. Similarly, it can be seen that the complexity of the force evaluation phase is $O(p^3 N \log N)$. Thus, the overall complexity for the algorithm is $O(p^3 N \log N)$.

The algorithm has been implemented in C++ programming language. The code first constructs the oct-tree and the multipole coefficients are calculated along with the tree construction. Standard Template Library (STL) data structures were used for the dynamically adaptive tree construction phase. The spherical symmetry of r^{-3} potentials, when expressed in spherical coordinates, leads to efficient computations. For example, spherical harmonics for r^{-1} potential is computed instead of computing the vector of spherical harmonics given in Theorem 1. Thus, computing $O(p^2)$ spherical harmonics for each particle is enough to compute $O(p^3)$ multipole coefficients, which results in significant time reduction for large systems. The constants defined in (4) are also precomputed and used in the force evaluation phase to reduce the overall computation time.

4   Experiments

In this section, we present results of numerical experiments performed on a 2.4 GHz, 512 MB Intel P4 PC running Red Hat 9.0. Due to space constraints we discuss two experiments only. In these tests, a direct summation code is the benchmark for comparing errors and execution time. For star-cluster applications, Mcmillan and Aarseth [3] compare accuracies for a problem with $N = 10^3$. They target median force error below 10^{-4} and achieve it using octupole moments for $MAC < 0.5$. Fig 1 shows the error in the force F incurred by truncating expression (6) after p terms. The relative error $\delta F/F$ is defined by

$$\left(\frac{\delta F}{F}\right)^2 = \frac{1}{N} \sum_{i=1}^{N} \left(\frac{|F_{tree} - F_{direct}|}{F_{scale}}\right)^2,$$

where F_{direct} is the force determined by direct summation, F_{tree} is the value returned by the tree algorithm and $F_{scale} = \sum_{ij} m_i/(r'_{ij})^2$ is a characteristic scale against which accuracy is measured. Here m_i's are the masses and the sums are taken over all particles in the system. Fig 1(a) illustrates the effect of MAC on error for a uniform distribution of 10^3 particles with random masses in the range $[0, 1]$. The maximum number of particles in leaf boxes were fixed at 10. For $MAC = 0.9$ we see that an accuracy of 10^{-5} is reached using multipole degree 2. Also, as MAC decreases, the error reduces proportionally. For a typical MAC between 0.6 and 0.7 used in the Barnes-Hut algorithm, we have an order of magnitude reduction in the error by increasing the multipole degree. In [3], for $MAC < 0.3$ the accuracy of the octupole component of the force deteriorates markedly. In our case, we see that dramatic reduction in the error is obtained as we increase the multipole degree. Also, Mcmillan et al. require $MAC < 0.5$ for accurate computations. Since the amount of direct computation is inversely proportional to MAC, it may be expensive to select such small MAC for large systems.

Fig 1(b) and Fig 2 demonstrate the effect of MAC on accuracy and execution time for 10^4 particles in random uniform distribution. Each particle carries a random positive or a negative charge. This test is carried out to verify the numerical robustness and consistency of our approach for electrostatic force interactions. As before, the multipole degree was allowed to vary between 2 and 4, and the number

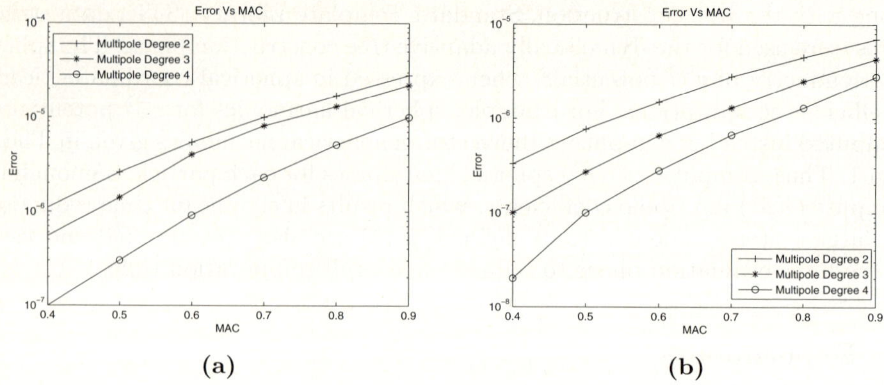

Fig. 1. Error Vs MAC with multipole degrees $p = 2, 3$ and 4 for (a) 1000 particles with positive masses (b) 10000 particles with positive and negative charges

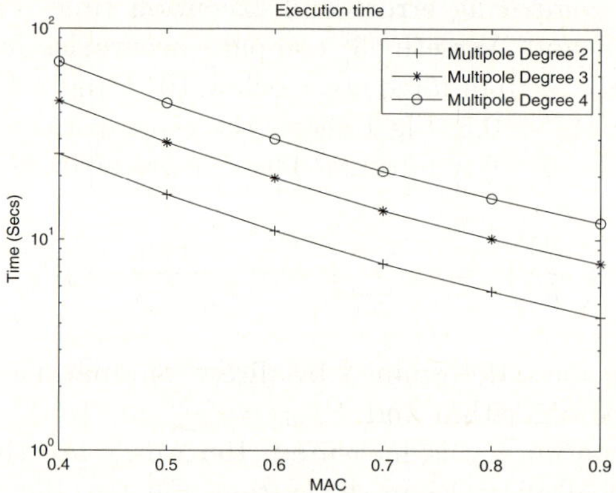

Fig. 2. Execution time for 10000 particles of positive and negative strengths

of particles per leaf box was fixed at 10. In Fig 2, we see a rapid increase in the execution time as we increase the multipole degree whereas the growth is smaller with respect to the MAC. In order to improve the accuracy for a small system, it is better to reduce the MAC than increase the multipole degree. However, tree-based force evaluation schemes are quite inefficient for small number of particles because the overhead involved in manipulating and traversing the tree data structure is fairly large. For $N > 10^4$, the treecode presented in this paper is expected to outperform the direct summation algorithm in execution time.

5   Conclusion

This paper presents an efficient algorithm for accurate calculation of forces using spherical coordinates. The proposed algorithm is superior to existing methods in which accuracy is limited by the choice of octupole moments and differentiation schemes. Our algorithm combines the ease of the Barnes-Hut algorithm with accurate force calculation, and can be used to compute gravitational forces in astrophysical simulations and electrostatic forces in molecular dynamics.

References

1. K. Kasthuri Srinivasan, H. Mahawar and V. Sarin. A Multipole Based Treecode Using Spherical Harmonics for the Potentials of the Form $r^{-\lambda}$. *Proceedings of the International Conference on Computational Science (ICCS)*, Lecture Notes in Computer Science, Springer-Verlag, Vol. 3514, pp. 107-114, Atlanta, GA, May 2005.
2. J. Barnes and P. Hut. A Hierarchical O(n log n) Force Calculation Algorithm. *Nature*, Vol. 324:446–449, (1986).
3. S.L.W. Mcmillan and S.J. Aarseth. An $O(N \log N)$ Integration Scheme for Collisional Stellar Systems. *The Astrophysical Journal*, Vol. 414, 200-212 (1993).
4. S. Pfalzner and P. Gibbon. Many-Body Tree Methods in Physics. The Cambridge University Press, Cambridge, Massachusetts (1996).
5. L. Greengard. *The Rapid Evaluation of Potential Fields in Particle Systems*. The MIT Press, Cambridge, Massachusetts (1988).
6. J. Board and K. Schulten. The Fast Multipole Algorithm. *IEEE, Computing in Science and Engineering*, Vol. 2, No. 1, January/February (2000).
7. P. Ewald. Die Berechnung Optischer und Elektrostatischer Gitterpotentiale. *Ann. Phys.*, Vol.64, 253-287 (1921).
8. R.W. Hockney J.W. Eastwood. Computer Simulation Using Particles, New York: McGraw-Hill (1981)
9. C. Sagui and T. Darden. Molecular Dynamics Simulation of Biomolecules: Long-Range Electrostatic Effects . *Annu. Rev. Biophys. Biomol. Struct* Vol. 28, 155-179 (1999).
10. E. Pollock and J. Glosli. Comments on PPPM, FMM and Ewald Method for Large Periodic Coulombic Syatems. In *Comp. Phys. Comm.*, Vol. 95, 93-110 (1996).
11. T. Bishop, R. Skeel and K. Schulten. Difficulties with Multiple Time Stepping and Fast Multipole Algorithm in Molecular Dynamics. *J. Comp. Chem.*, Vol. 18 1785-1791 (1997).
12. M. Deserno and C. Holm. How to Mesh up Ewald Sums: A Theoretical and Numerical Comparison of Various Particle Mesh Routines . *J.Comp.Phys*, Vol. 109 7678-7692 (1998).
13. E.J. Weniger. Addition Theorems as Three-Dimensional Taylor Expansions. *International Journal of Quantum Chemistry*, Vol.76, 280-295 (2000).
14. John Avery. Hyperspherical Harmonics, Kluwer Academic Publishers (1989)
15. Claus Müller. Analysis of Spherical Symmetries in Euclidean Spaces, Springer-Verlag (1991)

An Approximate Algorithm for the Minimal Cost Gateways Location, Capacity and Flow Assignment in Two-Level Hierarchical Wide Area Networks

Przemyslaw Ryba and Andrzej Kasprzak

Wroclaw University of Technology, Chair of Systems and Computer Networks
Wybrzeze Wyspianskiego 27, 50-370 Wroclaw, Poland
{przemyslaw.ryba, andrzej.kasprzak}@pwr.wroc.pl

Abstract. In the paper the problem of designing two-level hierarchical structure of wide area network is considered. The goal is to select gateways location, channel capacities and flow routes in order to minimize total cost of leasing capacities of channels of 2nd level network, subject to delay constraint. An approximate algorithm is proposed. Some results following from computational experiment are reported.

1 Introduction

Process of designing large wide area networks (WAN) containing hundreds of hosts and communication links is very difficult and demanding task. Conventional design procedures are suitable for small and moderate-sized networks. Unfortunately, when applied directly to large networks, they become very costly (from computational point of view) and sometimes infeasible. Problems with computational cost of huge wide area network design can be alleviated by building WAN as a hierarchical network [1]. In hierarchical networks, nodes are clustered using some nearness measures like geographical distance, traffic and reliability requirements. Communication networks of each cluster (1st level network) can be designed separately. In each cluster special communication nodes (gateways) are chosen. Function of gateways is to handle the traffic between nodes from the cluster they are located and nodes from other clusters. Gateways, with channels connecting them, form 2nd level network. Two-level networks can be grouped in 3rd level clusters and so on. Traffic between nodes in the same cluster uses paths contained in local communication network. Traffic between nodes in different 1st level networks is first sent to local gateway, then via 2nd level network of gateways is sent to gateway located in destination 1st level network to finally reach the destination node. Example of structure of hierarchical wide area network is presented in the Fig. 1.

In this paper problem of designing the two-level hierarchical wide area network connecting existing networks in order to minimize total cost of leasing capacities of channels of 2nd level network subject to delay constraint is considered. The Gateways Location, Capacity and Flow Assignment problem with cost criterion is formulated as follows:

V.N. Alexandrov et al. (Eds.): ICCS 2006, Part I, LNCS 3991, pp. 100–107, 2006.

given: topology of 1st level networks and 2nd level network, sets of potential gateways locations, set of potential 2nd level network channels capacities and costs (i.e. discrete cost-capacity function), traffic requirements,

minimize: leasing cost of channel capacities of 2nd level network,

over: gateways locations, 2nd level network channel capacities, multicommodity flow (i.e. routing),

subject to: multicommodity flow constraints, channel capacity constraints, delay constraint in hierarchical network.

Discrete cost-capacity function considered here is most important from the practical point of view for the reason that channels capacities can be chosen from the sequence defined by ITU-T (International Telecommunication Union – Telecommunication Standardization Section) recommendations. Such formulated problem is NP-complete as more general than the CFA problem with discrete cost-capacity function which is NP-complete [2, 3].

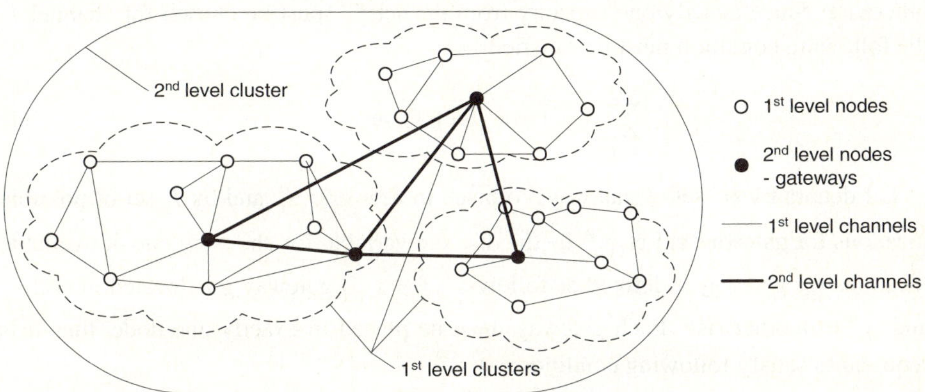

Fig. 1. Structure of the two-level hierarchical wide area network

Some algorithms for hierarchical network design can be found in [4, 5]. However, problem presented in [5] is limited to tree topology of hierarchical network. Algorithm for router location minimizing total cost of the network without constraints on quality indices and without assigning capacities to channels is presented in the paper [4]. Also, in algorithm for interconnecting two WANs presented in [6] assigning capacities to channels connecting the networks is not considered.

The problems considered in the literature presented above, do not take into account that the gateways location problem and capacity assignment problem should be considered simultaneously. Solving this problem is important from practical point of view due to the fact that it results in significant reduction of network exploitation cost. This paper joins problem of locating gateways in hierarchical wide area network with capacity assignment problem to minimize network cost and to satisfy quality demands. Thus, problem considered in the paper is more general and more important from practical point of view than problems, which can be found in the literature.

2 Problem Formulation

Consider a hierarchical WAN consisting of K networks on 1st level of hierarchy, each denoted by S_1^l, $l = 1,..., K$, and one 2nd level network denoted by S_2. Let N_1^l be set of nodes and L_1^l set of channels of 1st level network S_1^l. Let n be the total number of nodes in hierarchical network. Set of nodes N_2 consists of selected gateways and L_2 is set of channels connecting them. Let m be number of channels in 2nd level of hierarchical WAN and let p be total number of channels. For each channel in 2nd level network capacity must be chosen from the set of available capacities $C^i = \{c_1^i,...,c_{s(i)}^i\}$. Capacities in each set C^i are ordered in the following way: $c_1^i > c_2^i > ... > c_{s(i)}^i$. Let d_k^i be the cost of leasing capacity c_k^i for channel i.

Let x_k^i be the discrete variable for choosing one of available capacities for channel i defined as follows: $x_k^i = 1$, if the capacity c_k^i is assigned to channel i and $x_k^i = 0$ otherwise. Since exactly one capacity from the set C^i must be chosen for channel i, the following condition must be satisfied:

$$\sum_{k=1}^{s(i)} x_k^i = 1 \quad \text{for} \quad i = 1,..., m \tag{1}$$

Let denote by H^l set of gateways to place in network S_1^l and by J_g set of possible locations for gateway g. Let y_a^g be the discrete variable for choosing one of available locations for gateway g defined as follows: $y_a^g = 1$, if gateway g is located in node a and $y_a^g = 0$, otherwise. Each gateway must be placed in exactly one node, thus it is required to satisfy following condition:

$$\sum_{a \in J_g} y_a^g = 1, \ g \in H^l, \ l = 1,..., K \tag{2}$$

The cost d_k^i of leasing capacity c_k^i for channel i is defined as follows:

$$d_k^i = \sum_{a \in J_{g_1}} \sum_{b \in J_{g_2}} y_a^{g_1} y_b^{g_2} d_{ab}^{ik} \tag{3}$$

where g_1 and g_2 are gateways adjacent to channel i, d_{ab}^{ik} is cost of leasing capacity c_k^i of channel i between nodes a and b, which are possible locations of gateways g_1 and g_2 respectively.

Let $X_r^{'}$ be the permutation of values of variables x_k^i, $i = 1,..., m$ for which the condition (1) is satisfied, and let X_r be the set of variables which are equal to 1 in $X_r^{'}$. Similarly, let $Y_r^{'}$ be the permutation of values of all variables y_a^g for which condition (2) is satisfied and let Y_r be the set of variables which are equal to 1 in $Y_r^{'}$. The pair of sets (X_r, Y_r) is called a selection. Each selection (X_r, Y_r)

determines locations of gateways and channels capacities in the 2nd level of hierarchical WAN. Let $\Re$ be the family of all selections.

Let $d(X_r, Y_r)$ be the total cost of leasing capacities of channels of 2nd level network of hierarchical network, in which values of channels capacities are given by X_r and locations of gateways are given by Y_r.

$$d(X_r, Y_r) = \sum_{x_k^i \in X_r} x_k^i d_k^i \tag{4}$$

Then, the considered gateway location, flow and capacity assignment problem in hierarchical wide area network can be formulated as follows:

$$\min_{(X_r, Y_r)} d(X_r, Y_r) \tag{5}$$

subject to:

$$(X_r, Y_r) \in \Re \tag{6}$$

$$T(X_r, Y_r) \le T^{\max} \tag{7}$$

where $T(X_r, Y_r)$ denotes total average delay per packet in hierarchical network, given by Kleinrock's formula [7] and $T^{\max}$ its maximal acceptable value of that delay.

3 Algorithm

In this chapter we present an approximate algorithm for gateways location and capacity assignment minimizing cost of leasing capacities of channels of 2nd level network, subject to constraint on maximal admissible value of average delay per packet in hierarchical network.

Initially set Y_1 is calculated in such a way that total cost of leasing channels of 2nd level network is minimal. All channel capacities are set to maximal available values. If for such constructed network constraint (7) is violated, then one of gateways location is changed. Choice of gateway to change location is performed using exchange operation. After each change of gateway location new value of average delay per packet is compared with one already found. Location with higher value of average delay per packet is abandoned and removed from set of potential gateway locations. Operation is repeated until constraint (7) is satisfied. After that, capacities of selected channels is decreased to obtain network with lower leasing cost satisfying constraint (7) and value $d(X_r, Y_r)$ is compared with already found best solution d^*. If current cost of leasing channels capacities is lower, then d^* is updated. When constraint (7) is violated then operation of changing gateway locations is repeated. Algorithm terminates, when, for each gateway, the set of possible locations is empty or there is no channel which capacity can be decreased.

Algorithm finds heuristic solution after no more than $\left(\sum_{l=1}^{K} \sum_{g \in H^l} |J^g| \right) \left(\sum_{i=1}^{m} s(i) \right)$ iterations.

3.1 Exchange Operations

The purpose of exchange operation is to find a pair of variables x_k^i or variables y_a^g for substitution to generate a network with the least possible value of criterion (4). To take into account the constraint (7) we propose the auxiliary local criterion: $d(X_r, Y_r) + \alpha \cdot T(X_r, Y_r)$, where coefficient α converts value of average packets delay to cost expressed in [€/month]. Choice of variables x_k^i and x_j^i to exchange is made using modified criterion Δ_{kj}^i usually used in classical CFA problem, formulated as follow [2, 8]:

$$\Delta_{kj}^i = \begin{cases} \dfrac{\alpha}{\gamma}\left(\dfrac{f_i}{c_k^i - f_i} - \dfrac{f_i}{c_j^i - f_i}\right) + \left(d_k^i - d_j^i\right) & \text{for } f_i < c^i \\ \infty & \text{otherwise} \end{cases} \tag{8}$$

Choice of variables y_a^g and y_b^g to exchange may be evaluated by criterion δ_{ab}^g .

$$\delta_{ab}^g = \begin{cases} \dfrac{\alpha}{\gamma}\sum_{i \in L_1^l}\left(\dfrac{\tilde{f}_i}{c^i - \tilde{f}_i} - \dfrac{f_i}{c^i - f_i}\right) + \sum_{i:\langle g, g_2\rangle \in V^g}\left(d_{eb}^{ij} - d_{ea}^{ij}\right) & \text{for } \tilde{f}_i \leq c^i \\ \infty & \text{otherwise} \end{cases} \tag{9}$$

The second term of the sum estimates change of the leasing cost of channels adjacent to the gateway g of which location changes. The flow $\tilde{\mathbf{f}}$ is constructed as follows: the flow from all nodes in network S_1^l sent through gateway located in node a is moved from the routes leading to node a to the routes leading to node b. If the conditions $\tilde{f}_i \leq c^i$ are satisfied for every channel in the network S_1^l the flow $\tilde{\mathbf{f}}$ is feasible.

To exchange should be chosen the pair of variables for which the value of criterion δ_{ab}^g or Δ_{kj}^i is minimal.

3.2 Calculation Scheme of the Approximate Algorithm

Let J_g^r be the set of possible locations for gateway g in r-th iteration of the approximate algorithm. Let $J_g^1 = J_g$ for each gateway g. Let $Y_1 = \varnothing$, $d^* = \infty$, $r = 1$.

Step 1. Perform $x_1^i = 1$ for $i = 1, ..., m$. Compute Y_1, such that $d(X_1, Y_1) = \min\limits_{Y_r} d(X_1, Y_r)$

Step 2. Compute flow $\mathbf{f}$ using FD method [9], next compute $T(X_r, Y_r)$.

If $T(X_r, Y_r) > T^{\max}$ then go to step 3. Otherwise go to step 4.

Step 3. If $J_g^r = \varnothing$ for every g then go to step 6. Otherwise select the pair of variables

$y_a^g \in Y_r$ and $y_b^g : b \in J_g^r$ for which with the value of expression (9) is the greatest.

Perform $Y_{r+1} = \left(Y_r - \{y_a^g\}\right) \cup \{y_b^g\}$ Next compute $T(X_r, Y_{r+1})$.

If $T(X_r, Y_{r+1}) > T(X_r, Y_r)$ then perform $J_g^r = J_g^r - \{b\}$ and go to step 2.

Otherwise perform $J_g^{r+1} = J_g^r - \{a\}$ and $r = r+1$. Next go to step 2.

Step 4. If $d(X_r, Y_r) < d^*$ then perform $d^* = d(X_r, Y_r)$ and $(X_*, Y_*) = (X_r, Y_r)$.

Step 5. If $k = s(i)$ for every $x_k^i \in X_r$ then go to step 6. Otherwise select variable

$x_k^i \in X_r$, such that $k \le s(i)-1$, with greatest value of expression (8) and perform

$X_{r+1} = \left(X_r - \{x_k^i\}\right) \cup \{x_{k+1}^i\}$ and $r = r+1$. Go to step 2.

Step 6. Algorithm terminates. If $d^* = \infty$ then problem has no solution. Otherwise the selection (X_*, Y_*) associated with current value d^* is the near-optimal solution satisfying the constraints (6) and (7).

4 Computational Results

The presented approximate algorithm was implemented in C++ code. Extensive numerical experiments have been performed with this algorithm for many different hierarchical network topologies and for many possible gateways number and locations. The experiments were conducted with two main purposes in mind: first, to examine the impact of various parameters on solutions to find properties of the problem (5-7) important from practical point of view and second: to test the computational efficiency of proposed algorithm.

The dependence of the cost of leasing channel capacities d on maximal acceptable value of average delay per packet in hierarchical network $T^{\max}$ has been examined. In the Fig. 2 the typical dependence of d on the value $T^{\max}$ is presented for different

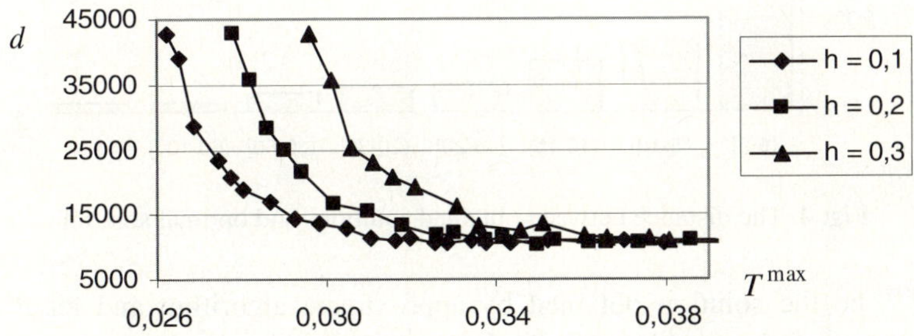

Fig. 2. The dependence of criterion function d on value $T^{\max}$

values of the average packet rates from external sources transmitted between each pair of nodes of the hierarchical network, which are denoted by h. It can be observed that the dependence of leasing cost of channel capacities d on T^{max} is decreasing function and that there exists such value $\hat{T}^{max}$, that the problem (5-7) has the same solution for each T^{max} greater or equal to $\hat{T}^{max}$.

Let G be the number of gateways, which must be allocated in some 1st level network of the hierarchical WAN. It has been examined how number of gateways G influences on the criterion function d. In the Fig. 3 typical dependence of leasing cost of channel capacities d on number of gateways G is presented.

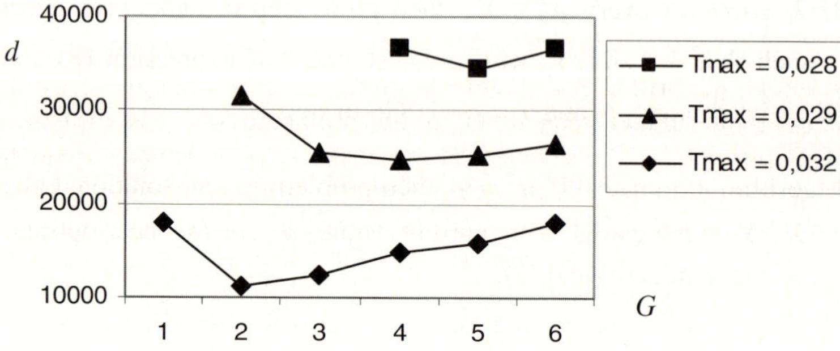

Fig. 3. The dependence of criterion function d on number of gateways G

Fig. 3 indicates that higher quality demands (lower values of constraint T^{max}) imply allocating higher number of gateways. For small number of gateways feasible solution cannot be found.. It can be also observed, that there exists such value of G for which the function $d(G)$ is minimal. Number of gateways G, for which value d is minimal, is higher for tighter constraint on T^{max}.

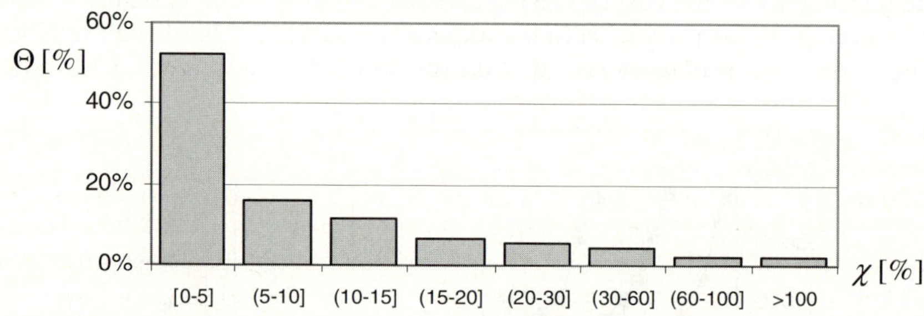

Fig. 4. The distance between obtained solutions and optimal solutions

Let d^{app} be the solution obtained by approximate algorithm and let d^{opt} be the optimal value of the problem (5-7). Let χ be the distance between approximate and optimal solutions: $\chi = \left| d^{app} - d^{opt} \right| / d^{opt} \cdot 100\%$. The value χ shows how the results obtained using the approximate algorithm are worse than the optimal solution.

Let

$$\Theta[a,b] = \frac{\text{number of solutions for which } \chi \in [a,b]}{\text{number of all solutions}} \cdot 100\%$$

denotes the fraction of solutions obtained from the approximate algorithm which are greater than optimal solutions between $a\%$ and $b\%$. The dependence Θ on divisions [0% - 5%], (5% - 10%), [10% - 15%), etc. is shown in the Fig. 4.

5   Conclusions

The approximate algorithm for solving the gateway location and network topology assignment problem in hierarchical network to minimize total cost of leasing channel capacities is presented. The considered problem is more general than the similar problems presented in the literature. It follows from computational experiments (Fig. 4) that more than 50% approximate solutions differ from optimal solutions at most 5%. Moreover, we noticed that the cost of the hierarchical WAN depends on the number of gateways. The presented approximate algorithm may be used whenever an optimal solution is not necessary and for large hierarchical networks.

This work was supported by a research project of The Polish State Committee for Scientific Research in 2005-2007.

References

1. Kleinrock L., Kamoun F.: Optimal Clustering Structures for Hierarchical Topological Network Design of Large Computer Networks. Networks 10 (1980) 221-248
2. Kasprzak, A.: Topological Design of the Wide Area Networks. Wroclaw University of Technology Press, Wroclaw (2001)
3. Markowski M., Kasprzak A.: An exact algorithm for host allocation, capacity and flow assignment problem in WAN. WITASI 2002, Kluwer Academic Publ. (2002) 73-82
4. Liu Z., Gu Y., Medhi D.: On Optimal Location of Switches/Routers and Interconnection. Technical Report, University of Missouri-Kansas City (1998)
5. Saha D. and Mukherjee A.: On the multidensity gateway location problem for multilevel high speed network. Computer Communications 20 (1997) 576-585
6. Liang S.C., Yee J.R.: Locating Internet Gateways to Minimize Nonlinear Congestion Costs. IEEE Transactions On Communications 42, (1994), 2740-50
7. Fratta, L., Gerla, M., Kleinrock, L.: The Flow Deviation Method: an Approach to Store-and-Forward Communication Network Design. Networks 3 (1973) 97-133
8. Markowski M., Kasprzak A.: The Web Replica Allocation and Topology Assignment Problem in Wide Area Networks: Algorithms and Computational Results. Lecture Notes in Computer Science 3483 (2005) 772-781
9. Walkowiak K.: A New Method of Primary Routes Selection for Local Restoration, Lectures Notes in Computer Science 3042 (2004) 1024-1035

Image-Based Robust Control of Robot Manipulators with Integral Actions

Min Seok Jie and Kang Woong Lee

School of Electronics, Telecommunication and Computer Engineering,
Hankuk Aviation University, 200-1, Hwajeon-dong, Deokyang-gu,
Koyang-city, Kyonggi-do, 412-791, Korea
Tel.: +82 2 3158 0166; Fax: +82 2 3159 9257
tomsey@korea.com, kwlee@mail.hangkong.ac.kr

Abstract. In this paper, we propose a robust visual feedback controller with integral action for tracking control of n-link robot manipulators in the presence of constant bounded parametric uncertainties. The proposed control input has robustness to the parametric uncertainty and reduces tracking error in the steady-state. The stability of the closed-loop system is shown by Lyapunov method. The effectiveness of the proposed method is shown by simulation and experiment results on the 5-link robot manipulators with two degree of freedom.

Keywords: robust control, visual feedback, integral action, robot manipulator.

1 Introduction

Applications of visual based robot control have been increased when the robot is working in unstructured environments. The use of visual feedback in these applications is an attractive solution for the position and motion control of robot manipulators. A visual servo control scheme can be classified in two configurations: fixed-camera, where the visual servoing camera is fixed with respect to the world coordinate frame and camera-in-hand, where the camera is mounted on the end-effector of the robot manipulator [1], [2]. In the camera-in-hand configuration, the camera supplies visual information of the object in the environment to the controller. The objective of this visual feedback control scheme is to move the robot manipulator in such a way that the projection of an object be at a desired position in the image plane obtained by the camera.

The manipulator vision system whose dynamics do not interact with the visual feedback loop can not achieve high performance for high speed tasks. In order to overcome this drawback, the controller must take into account the dynamics of robot manipulators [3]. The robot dynamics, however, includes parametric uncertainties due to load variations and disturbances. A visual servo controller taking into account robot dynamics must be robust to the parametric uncertainties and disturbances. Kelly [4] proposed an image-based direct visual servo controller for camera-in-hand robot manipulators, which is of a simple structure based on a transpose Jacobian term plus gravity compensation. In [5], a robust tracking controller has been designed to

V.N. Alexandrov et al. (Eds.): ICCS 2006, Part I, LNCS 3991, pp. 108 – 116, 2006.

compensate the uncertainty in the camera orientation and to ensure globally uniformly ultimate boundedness.

Robust control schemes require high feedback gains in order to reduce the tracking error. In practice, high feedback gains are limited because of hardware issues such as digital implementation, actuator saturation and noise contained in velocity measurements. Limitation of feedback gains induces large tracking errors. This problem can be overcome by integral control [6].

In this paper, we design a robust visual servo controller with integral action to compensate parametric uncertainties due to load variations or disturbances and to reduce tracking error. The closed-loop stability including the whole robot dynamics is shown by the Lyapunov method. The proposed robust visual servo controller is applied on a two degree of freedom 5-link robot manipulator to show the performance of the closed-loop system. The simulation and experimental results show convergence behavior of the image feature points.

This paper hereafter is organized as follows. In Sections 2 present the robot and camera models. The proposed robust control system with integral action is analyzed in Section 3. In Section 4 we introduce simulation and experiment results on a two degrees of freedom robot manipulators. The paper is finally summarized in Section 5.

2   Robot Model and Camera Model

In this work, we consider a robot manipulator with end effector which a camera is mounted on. The mathematical model of this system consists of the rigid robot dynamic model and the camera model.

In the absence of friction and disturbance, the dynamic equation of an n-link rigid robot manipulator can be expressed as [7]

$$M(q)\ddot{q} + C(q,\dot{q})\dot{q} + G(q) = \tau \tag{1}$$

where $q \in R^n$ is the vector of joint displacements, $\tau \in R^n$ is the vector of torques applied to the joints, $M(q) \in R^{n \times n}$ is the symmetric positive definite inertia matrix, $C(q,\dot{q})\dot{q} \in R^n$ is the vector of centripetal and Coriolis torques, and $G(q) \in R^n$ is the vector of the gravitational torques. The robot dynamic model (1) has the following properties.

Property 1: For the unknown constant parameter vector $\theta \in R^p$, the dynamic equation (1) can be expressed linearly

$$M(q)\ddot{q} + C(q,\dot{q})\dot{q} + G(q) = Y(q,\dot{q},\ddot{q})\theta = \tau \tag{2}$$

where $Y(q,\dot{q},\ddot{q}) \in R^{n \times p}$ is the known regression matrix.

The dynamic equation (1) can be changed to the error dynamic equation. Defining the tracking error as $\tilde{q} = q - q_d$, the error dynamic equation for the robot manipulator of (1) is given by

$$\ddot{\tilde{q}} = M^{-1}(q)[-M(q)\ddot{q}_d - C(q,\dot{q})\dot{q} - G(q) + \tau] \tag{3}$$

where q_d is the twice continuously differentiable desired trajectory.

In order to include integral action in the controller, let us define the new state vector

$$\sigma = \int_0^t \tilde{q}(\tau)d\tau \tag{4}$$

where $\sigma = [\sigma_1 \; \sigma_2 \; \cdots \; \sigma_n]^T$.

Defining the augmented state vector as $\zeta = [\sigma^T \; \tilde{q}^T \; \dot{\tilde{q}}^T]^T$, the augmented state equation is given by

$$\dot{\zeta} = A\zeta + BM^{-1}(q)\left[-Y(q,\dot{q},\ddot{q}_d)\theta + \tau\right] \tag{5}$$

where

$$A = \begin{bmatrix} 0 & I_{n\times n} & 0 \\ 0 & 0 & I_{n\times n} \\ 0 & 0 & 0_{n\times n} \end{bmatrix}, \; B = \begin{bmatrix} 0 \\ 0_{n\times n} \\ I_{n\times n} \end{bmatrix}, \; \text{and } Y(q,\dot{q},\ddot{q}_d)\theta = M(q)\ddot{q}_d + C(q,\dot{q})\dot{q} + G(q) \tag{6}$$

The motion dynamics of the image feature point described by the robot joint velocity as

$$\dot{\xi} = J(q,\xi,Z)\dot{q} \tag{7}$$

where $J(q,\xi,Z) = J_{img}(q,\xi,Z)\begin{bmatrix} {}^cR_w & 0 \\ 0 & {}^cR_w \end{bmatrix}\begin{bmatrix} I & 0 \\ 0 & T(q) \end{bmatrix} J_A(q) \tag{8}$

In robot control with visual feedback, the control problem is to design a controller to move the end-effector in such a way that the actual image features reach the desired ones specified in the image plane. Let us denote with ξ_d the desired image feature vector which is assumed to be constant. We define the image feature error as

$$\tilde{\xi} = \xi - \xi_d \tag{9}$$

Using (7), the time derivative of (9) can be expressed by

$$\dot{\tilde{\xi}} = J(q,\xi,Z)\dot{q} = J(q,\xi,Z)(\dot{\tilde{q}} + \dot{q}_d) \tag{10}$$

We take the desired joint velocity $\dot{q}_d$ as

$$\dot{q}_d = -J^+(q,\xi_d,Z)K_c\tilde{\xi} \tag{11}$$

where K_c is the positive definite matrix and $J^+(q,\xi_d,Z)$ is the pseudo inverse matrix defined by $J^+(q,\xi_d,Z) = \left[J^T(q,\xi_d,Z)J(q,\xi_d,Z)\right]^{-1}J^T(q,\xi_d,Z)$.

Substituting (11) into (10) leads to

$$\dot{\tilde{\xi}} = J(q,\xi,Z)\dot{\tilde{q}} - J(q,\xi,Z)J^+(q,\xi_d,Z)K_c\tilde{\xi} \tag{12}$$

3 Robust Control with Visual Feedback

In this section, we consider a robust visual feedback controller including integral action in order to compensate the bounded constant parametric uncertainties of robot manipulators. The proposed controller is given by

$$\tau = Y(q,\dot{q},\ddot{q}_d)\theta_o - M_o(q)K\zeta + J^+(q,\xi_d,Z)K_i\tilde{\xi} + \tau_N \tag{13}$$

where K and K_i is the symmetric positive definite gain matrices, τ_N is an additional nonlinear control input compensating for the bounded parametric uncertainties, $M_o(q)$ is the nominal matrix of $M(q)$, θ_o is the nominal value of the unknown parameter vector θ and $Y(q,\dot{q},\ddot{q}_d)\theta_o$ is defined as

$$Y(q,\dot{q},\ddot{q}_d)\theta_o = M_o(q)\ddot{q}_d + C_o(q,\dot{q})\dot{q} + G_o(q) \tag{14}$$

where $C_o(\cdot)$ and $G_o(\cdot)$ are the nominal matrices of $C(\cdot)$ and $G(\cdot)$, respectively.

Substituting the proposed input torque (13) into the equation (5) leads to

$$\dot{\zeta} = (A - BK)\zeta + BM^{-1}(q)[-Y(q,\dot{q},\ddot{q}_d)\tilde{\theta}$$
$$+ \tilde{M}(q)K\zeta + J^+(q,\xi_d,Z)K_i\tilde{\xi} + \tau_N] \tag{15}$$

where $\tilde{\theta} = \theta_o - \theta$, $\tilde{M}(q) = M(q) - M_o(q)$ and gain matrix K is chosen such that $(A - BK)$ is Hurwitz. We make the following assumptions to select an additional nonlinear input term and to prove the stability of the close-loop system.

Assumption 1: The norms of the following matrices can be bounded such that

$$\lambda_m \le \| M^{-1}(q) \| \le \lambda_M, \ \lambda_j \le \| J(q,\xi,Z) \| \le \lambda_J, \ \lambda_{j^+} \le \| J^+(q,\xi_d,Z) \| \le \lambda_{J^+},$$

$$K_m \le \| K \| \le K_M, \ K_{0m} \le \| K_0 \| \le K_{0M}, \ K_{cm} \le \| K_c \| \le K_{cM} \tag{16}$$

Assumption 2: There exist positive constants α_M, α_C and α_G such that

$$\| M(q) - M_o(q) \| \le \alpha_M \tag{17}$$
$$\| C(q,\dot{q}) - C_o(q,\dot{q}) \| \le \alpha_C \| \dot{q} \| \tag{18}$$
$$\| G(q) - G_o(q) \| \le \alpha_G \tag{19}$$

Assumption 3: There exist nonnegative constants β_1, β_2 and β_3 such that

$$\left\| Y(q,\dot{q},\ddot{q}_d)\tilde{\theta} \right\| \le \beta_1 + \beta_2 \|\zeta\| + \beta_3 \|\zeta\|^2 \tag{20}$$

where $\beta_1 = \alpha_M \| \ddot{q}_d \| + \alpha_C \| \dot{q}_d \|^2 + \alpha_G$, $\beta_2 = 2\alpha_C \| \dot{q}_d \|$, $\beta_3 = \alpha_C$.

These assumptions are hold since it is assumed that the desired q_d, $\dot{q}_d$ and $\ddot{q}_d$ belong to the compact set and parametric uncertainties of the robot manipulators are bounded.

We choose the nonlinear control input term τ_N as

$$\tau_N = \begin{cases} -\dfrac{\lambda_M \bar{\beta}_1(\zeta,\tilde{\xi})s}{\lambda_m \parallel s \parallel} - \dfrac{\bar{\beta}_2(\zeta,\tilde{\xi})s}{2\lambda_m \parallel s \parallel^2} & if \quad \lambda_M \bar{\beta}_1(\zeta,\tilde{\xi}) \parallel s \parallel \geq \mu \\[4mm] -\dfrac{\lambda^2{}_M \bar{\beta}_1{}^2(\zeta,\tilde{\xi})s}{\lambda_m \mu} - \dfrac{\bar{\beta}_2(\zeta,\tilde{\xi})s}{2\lambda_m \parallel s \parallel^2} & if \quad \lambda_M \bar{\beta}_1(\zeta,\tilde{\xi}) \parallel s \parallel < \mu \end{cases} \tag{21}$$

where $\mu > 0$ is a design parameter to be chosen, $s = B^T P \zeta$,

$\bar{\beta}_1(\zeta,\tilde{\xi}) = \beta_1 + \beta_2 \|\zeta\| + \beta_3 \|\zeta\|^2 + \alpha_M K_M + \lambda_{J+} K_M \parallel \tilde{\xi} \parallel$, and

$\bar{\beta}_2(\zeta,\tilde{\xi}) = \lambda_J K_{oM} \parallel \tilde{\xi} \parallel \parallel \zeta \parallel$.

Theorem 1: Suppose that assumption 1, 2 and 3 hold. Given the error dynamic equations (5) and (15), the proposed controller (13) with (21) ensures position tracking errors to be globally uniformly bounded.

Proof: Consider the Lyapunov function candidate

$$V = \zeta^T P \zeta + \frac{1}{2} \tilde{\xi}^T K_0 \xi \tag{22}$$

where K_0 is the positive definite matrix and $P = P^T > 0$ is the solution of the Lyapunov equation $(A - BK)^T P + P(A - BK) = -I$

Define the set Ω_c include the initial value

$$\Omega_c = \{(\zeta,\tilde{\xi}) \in R^{3n+2} \mid V(\zeta,\tilde{\xi}) \leq c\}, \ c > 0 \tag{23}$$

The time derivative of V along the trajectories of the equations (12) and (15) yields

$$\dot{V} = \dot{\zeta}^T P \zeta + \zeta^T P \dot{\zeta} + \tilde{\xi}^T K_0 \dot{\tilde{\xi}}$$

$$\leq -\parallel \zeta \parallel^2 - K_{om} \lambda_j \lambda_{j+} K_{cm} \parallel \tilde{\xi} \parallel^2 + 2\lambda_M \parallel s \parallel \bar{\beta}_1(\cdot)$$

$$+ \bar{\beta}_2(\cdot) + 2s^T M^{-1}(q)\tau_N \tag{24}$$

If $\lambda_M \bar{\beta}_1(\zeta,\tilde{\xi}) \parallel s \parallel \geq \mu$, we have

$$\dot{V} \leq -\parallel \zeta \parallel^2 - \eta \parallel \tilde{\xi} \parallel^2 \tag{25}$$

where $\eta = K_{om} \lambda_j \lambda_{j+} K_{cm}$.

If $\lambda_M \overline{\beta}_1(\zeta, \tilde{\xi}) \parallel s \parallel < \mu$, we have

$$\dot{V} \le - \parallel \zeta \parallel^2 - \eta \parallel \tilde{\xi} \parallel^2 + \frac{\mu}{2} \qquad (26)$$

For $\mu \le \dfrac{4\eta c}{a\lambda_{\min}(P)K_{om}}$ with $a > 1$

$$\dot{V} \le 0 \qquad (27)$$

If we define the set $\Omega_\mu = \{\lambda_M \overline{\beta}_1(\zeta, \tilde{\xi}) \parallel s \parallel < \mu\}$, from (27) the trajectory $\left(\zeta, \tilde{\xi}\right)$

will enter the set Ω_μ in infinite time and remain thereafter.

Therefore, the control law (13) with (21) guarantees the uniformly ultimate boundedness of the closed-loop system.

4 Simulation and Experimental Results

The proposed control method was implemented on 2-link robot manipulator manufactured by Samsung Faraman-AS1. The dynamic equation of the manipulators is given by

$$\begin{bmatrix} M_{11}(q) & M_{12}(q) \\ M_{12}(q) & M_{22}(q) \end{bmatrix} \begin{bmatrix} \ddot{q}_1 \\ \ddot{q}_2 \end{bmatrix} + \begin{bmatrix} 0 & C_{12}(q,\dot{q}) \\ C_{21}(q,\dot{q}) & 0 \end{bmatrix} \begin{bmatrix} \dot{q}_1 \\ \dot{q}_2 \end{bmatrix} + \begin{bmatrix} G_1(q) \\ G_2(q) \end{bmatrix} = \begin{bmatrix} \tau_1 \\ \tau_2 \end{bmatrix}$$

Setup for experiments is shown in Fig. 1. A motion control board(MMC) mounted in a main computer is used to execute the control algorithm. The feature signal is acquired by a image processing board(Matrox Meteor II) mounted on a main computer which processes the image obtained from a CCD camera and extracts the image feature. The CCD camera is mounted on the end-effector. The image obtained by the image processor has a 640×480 pixels resolution

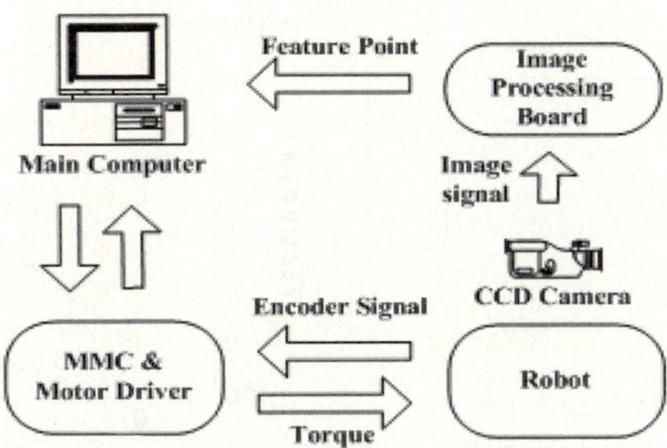

Fig. 1. Block diagram of vision systems

The variation interval of the parameter due to an unknown load is considered

$$0 \leq \Delta m_4 \leq 4.5 kg \ , \ 0 \leq \Delta I_2 \leq 0.005 Kgm^2$$

In the experimental test, we have considered one object feature point on a whiteboard. The white board was located at a distance $Z = 1m$ in front of the camera and parallel to the plane where the manipulator moves. The controller gain matrices K_c and K are chosen as $K_c = 10I$ and $K = [2I \ 50I \ 50I]$, respectively. The control gains K_i and K_o are chosen as $K_i = 7 \times 10^{-8}$ and $K_o = 7 \times 10^{-8}$, respectively. The initial positions of each joint are $q_1 = \pi/2(rad)$, $q_2 = \pi(rad)$. It was assumed that the initial feature point of the object was $\xi = [100 \ 100]^T$ pixels. The desired feature point coordinate was $\xi_d = [0 \ 0]^T$ pixels.

The simulation results are shown in Figs. 2, 3 and 4. Fig. 2 illustrates the trajectory of feature point on the image plane which shows the convergence to the desired feature

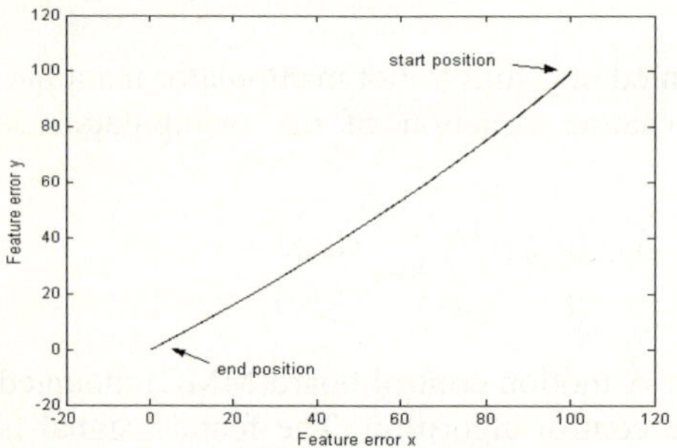

Fig. 2. Simulation results: Trajectories of feature errors using integral action(solid) and without integral action(dashed)

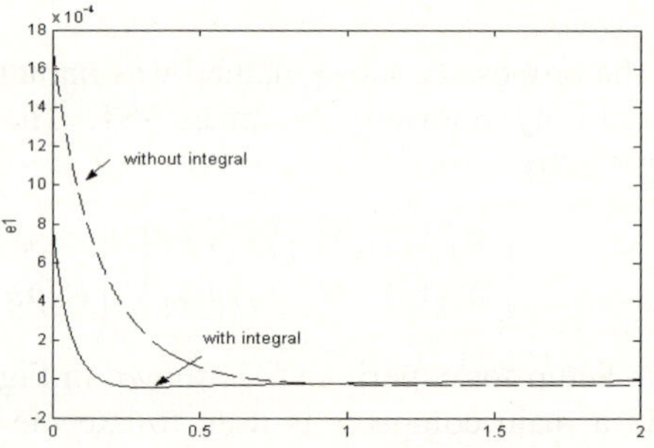

Fig. 3. Simulation results: Tracking errors of link 1 using integral action(solid) and without integral action(dashed)

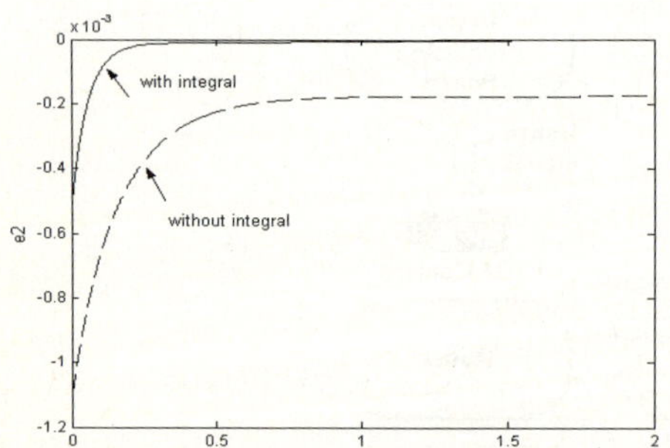

Fig. 4. Simulation results: Tracking errors of link 2 using integral action(solid) and without integral action(dashed)

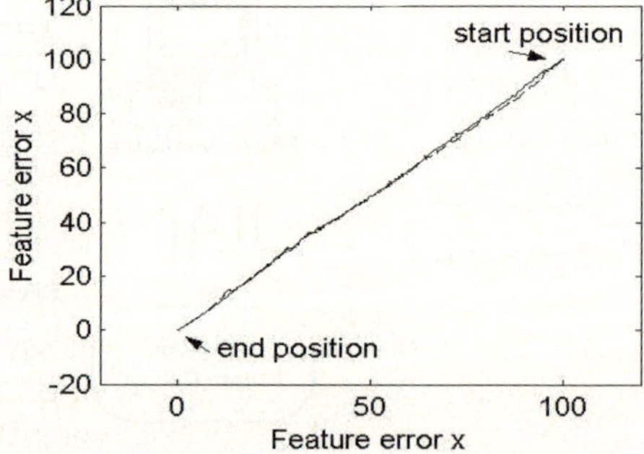

Fig. 5. Experimental results: Trajectories of feature errors using integral action(solid) and without integral action(dashed)

point. Fig. 3, 4 represents the tracking errors of joint 1 and 2. The results by the proposed method are compared to those without integral action. It is shown that better tracking performance results are archived by the robust controller with integral actions.

Figs. 5, 6 and 7 are experimental results that the proposed control algorithm applies to the 5-link Samsung Faraman robot. Experimental results show that the tracking performance is effective.

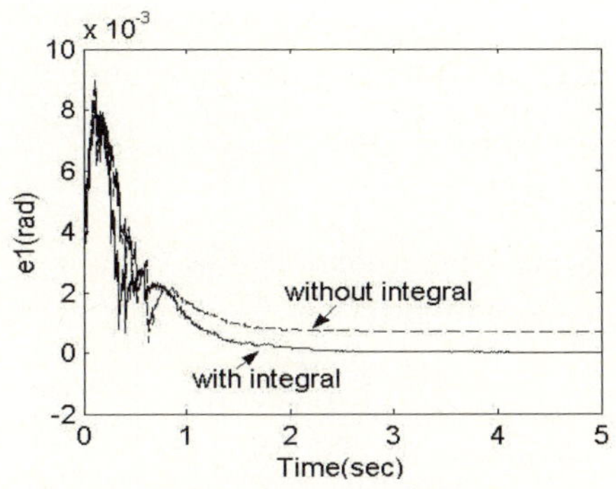

Fig. 6. Experimental results: Tracking errors of link 1 using integral action(solid) and without integral action(dashed)

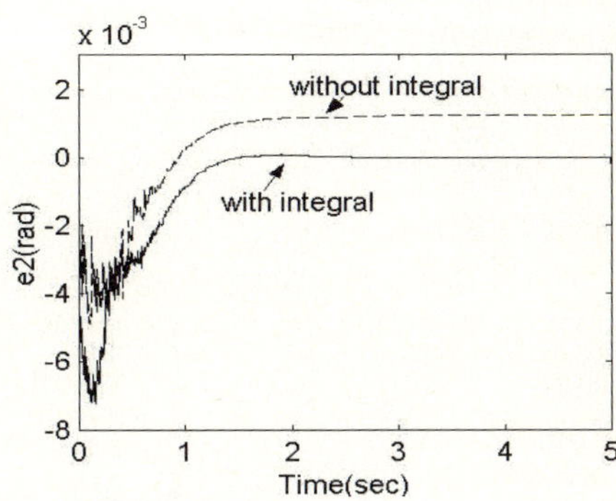

Fig. 7. Experimental results: Tracking errors of link 2 using integral action(solid) and without integral action(dashed)

5   Conclusions

In this paper, a robust controller with visual feedback for robot manipulators was proposed. The controller is of structure based on the image feature errors and the joint velocities fed back by the CCD camera and the encoder, respectively. The proposed controller with integral action reduces tracking error due to parametric uncertainties.

The ultimate uniform stability of the overall closed-loop system is proved by using the Lyapunov method. Simulation and Experiment results on a two degree of freedom manipulator have shown that the proposed control method has effectiveness to control robot manipulators with uncertainty.

References

1. Hashimoto, K.: VISUAL SERVOING. World Scientific (1993)
2. Espiau, E., Chaumette, F., Rives, P.: A new approach to visual servoing in robotics. IEEE Trans. Robotics and Automation, Vol. 8, No.3 (1992) 313-326
3. Hashimoto, K., Kimoto, T., Ebine, T., Kimura, H.: Manipulator control with image-based visual servo. IEEE International Conference on Robotics and Automation (1991) 2267-2272

4. Kelly, R., Carelli, R., Nasisi, O., Kuchen, B., Reyes, F.: Stable visual servoing of camera-in-hand robotic systems. IEEE/ASME Trans. Mechatronics, Vol. 5, No.1, (2000) 39-43
5. Zergeroglu, E., Dawson, D. M., Queiroz, M. S. de., Setlur, P.: Robust visual-servo control of robot manipulators in the presence of uncertainty. Journal of Robotic Systems, Vol. 20, issue, 2(2003) 93-106
6. Liu, G. J.,Goldenberg, A. A., Robust control of robot manipulators based on dynamic decomposition. IEEE Trans. Robotics and Automation, vol. 13, no. 5, (1997) 783-789
7. Spong, M. W., Vidyasagar, M.: Robot Dynamics and Control. Wiley, NewYork (1989)

Symmetric Runge-Kutta Methods with Higher Derivatives and Quadratic Extrapolation[*]

Gennady Yu. Kulikov[1], Ekaterina Yu. Khrustaleva[2],
and Arkadi I. Merkulov[2]

[1] School of Computational and Applied Mathematics,
University of the Witwatersrand, Private Bag 3,
Wits 2050, Johannesburg, South Africa
[2] Ulyanovsk State University, L. Tolstoy Str. 42,
432970 Ulyanovsk, Russia
gkulikov@cam.wits.ac.za, shabalkina@mail.ru,
merkul@vda.ru

Abstract. In this paper we study the symmetry of Runge-Kutta methods with higher derivatives. We find conditions which provide this property for the above numerical methods. We prove that the family of E-methods constructed earlier consists of symmetric methods only, which lead to the quadratic extrapolation technique in practice.

1 Introduction

Here, we study Runge-Kutta methods with higher derivatives applied to ordinary differential equations (ODEs) of the form

$$x'(t) = g\big(t, x(t)\big), \quad t \in [t_0, t_0 + T], \quad x(t_0) = x^0 \tag{1}$$

where $x(t) \in \mathbb{R}^n$ and $g : D \subset \mathbb{R}^{n+1} \to \mathbb{R}^n$ is a sufficiently smooth function. We remark that these numerical methods proved their efficiency. So, there are many papers published in this field (see [2], [3], [5], [7], [8], [11], [12], [13] and so on).

All methods of such sort can be represented as follows:

$$x_{ki} = x_k + \tau_k \sum_{j=1}^{l} \sum_{r=0}^{p_j} \tau_k^r a_{ij}^{(r)} g^{(r)}(t_{kj}, x_{kj}), \quad i = 1, 2, \ldots, l, \tag{2a}$$

$$x_{k+1} = x_k + \tau_k \sum_{j=1}^{l} \sum_{r=0}^{p_j} \tau_k^r b_j^{(r)} g^{(r)}(t_{kj}, x_{kj}), \quad k = 0, 1, \ldots, K-1, \tag{2b}$$

where $x_0 = x^0$, $t_{kj} \stackrel{\text{def}}{=} t_k + c_j \tau_k$, $g^{(r)}(t_{kj}, x_{kj})$ denotes the r-th derivative[1] of the right-hand side of problem (1) with respect to t evaluated at the point (t_{kj}, x_{kj}),

---

[*] This work was supported in part by the National Research Foundation of South Africa under grant No. FA2004033000016.

[1] Here and below the zero-derivative implies the original function.

V.N. Alexandrov et al. (Eds.): ICCS 2006, Part I, LNCS 3991, pp. 117–123, 2006.
© Springer-Verlag Berlin Heidelberg 2006

the coefficients $a_{ij}^{(r)}$, $b_j^{(r)}$ and c_j, $i, j = 1, 2, \ldots, l$, $r = 0, 1, \ldots, p_j$ are real numbers ($c_j \in [0, 1]$), and τ_k is a step size which may be fixed or variable. In general, the numbers p_j of the used derivatives are not equal.

When method (2) is implicit the important task is its correct implementation. In this case, we have to involve an additional iterative scheme and to keep a sufficient number of iteration steps, as indicated in [11], [12], in order to ensure high order convergence rate.

Here, we investigate the symmetry property of method (2). We find the conditions on the coefficients $a_{ij}^{(r)}$, $b_j^{(r)}$ and c_j which guarantee that method (2) is symmetric. For example, these conditions are used to prove the symmetry of E-methods developed in [11], [12]. Actually, the present study extends the results given in [4], [9] and [10] to the class of Runge-Kutta methods with higher derivatives. We just want to emphasize that symmetric numerical methods are of great importance when applied to reversible problems (see [6]) or when used as underlying methods in extrapolation algorithms (see the cited papers). However, we restrict ourselves to the quadratic extrapolation issue only in this paper.

2 Symmetric Runge-Kutta Methods with Higher Derivatives

It is clear that any particular Runge-Kutta method (2) with higher derivatives is determined uniquely by the set of coefficients $a_{ij}^{(r)}$, $b_j^{(r)}$ and c_j. Therefore, following Butcher's idea (see, for example, [1] or [4]), we represent method (2) by a partitioned tableau of the form

$$
\begin{array}{c|cccccccccc}
c_1 & a_{11}^{(0)} & a_{11}^{(1)} & \cdots & a_{11}^{(p_1)} & \cdots & a_{1l}^{(0)} & a_{1l}^{(1)} & \cdots & a_{1l}^{(p_l)} \\
c_2 & a_{21}^{(0)} & a_{21}^{(1)} & \cdots & a_{21}^{(p_1)} & \cdots & a_{2l}^{(0)} & a_{2l}^{(1)} & \cdots & a_{2l}^{(p_l)} \\
\vdots & \vdots & \vdots & \ddots & \vdots & \ddots & \vdots & \vdots & \ddots & \vdots \\
c_l & a_{l1}^{(0)} & a_{l1}^{(1)} & \cdots & a_{l1}^{(p_1)} & \cdots & a_{ll}^{(0)} & a_{ll}^{(1)} & \cdots & a_{ll}^{(p_l)} \\
\hline
 & b_1^{(0)} & b_1^{(1)} & \cdots & b_1^{(p_1)} & \cdots & b_l^{(0)} & b_l^{(1)} & \cdots & b_l^{(p_l)}
\end{array}
\tag{3}
$$

containing all the coefficients of a Runge-Kutta method with higher derivatives. The matrix A, consisting of the coefficients $a_{ij}^{(r)}$, is a real matrix of dimension $l \times (p_1 + p_2 + \ldots + p_l)l$, and the other coefficients form the real vectors b and c of dimensions $(p_1 + p_2 + \ldots + p_l)l$ and l, respectively, in tableau (3).

It is known that symmetric one-step methods are defined via their adjoint counterparts, as explained, for example, in [4, p. 219–222] or in [9]. Therefore our task now is to extend the appropriate results derived for coefficients of adjoint Runge-Kutta methods to the case of method (2) involving higher derivatives.

So, we start with

Theorem 1. *Let method (2) be an l-stage Runge-Kutta formula with higher derivatives, whose coefficients are given by tableau (3). Then its adjoint method* *is equivalent to an l-stage Runge-Kutta formula with higher derivatives of the form* (2) *whose coefficients* $a_{ij}^{(r)-}$, $b_j^{(r)-}$ *and* c_j^- *satisfy*

$$c_j^- = 1 - c_{l+1-j}, \tag{4a}$$

$$a_{ij}^{(r)-} = (-1)^r \left(b_{l+1-j}^{(r)} - a_{l+1-i,l+1-j}^{(r)} \right), \tag{4b}$$

$$b_j^{(r)-} = (-1)^r b_{l+1-j}^{(r)}, \quad r = 0, 1, \ldots, p_j, \ i, j = 1, 2, \ldots, l. \tag{4c}$$

Proof. It is well-known (see the cited papers) that in order to find the adjoint Runge-Kutta method with higher derivatives we have to interchange the pair (t_k, x_k) with the pair (t_{k+1}, x_{k+1}) and to replace τ with $-\tau$ in the increment function of the method (2). It follows from the fact that method (2) belongs to the class of one-step methods.

Having fulfilled this transformation we arrive at an l-stage Runge-Kutta method of the form

$$x_{ki} = x_k + \tau_k \sum_{j=1}^{l} \sum_{r=0}^{p_j} (-1)^r \tau_k^r \left(b_j^{(r)} - a_{ij}^{(r)} \right) g^{(r)}(t_{kj}, x_{kj}), \quad i = 1, 2, \ldots, l, \tag{5a}$$

$$x_{k+1} = x_k + \tau_k \sum_{j=1}^{l} \sum_{r=0}^{p_j} (-1)^r \tau_k^r b_j^{(r)} g^{(r)}(t_{kj}, x_{kj}), \quad k = 0, 1, \ldots, K - 1, \tag{5b}$$

where $t_{kj} = t_k + (1 - c_j)\tau_k$. If we now rearrange the nodes $c_i^- = 1 - c_{l+1-i}$ of method (5) to make them increasing; i.e., rearrange indexes by the rule $i \longleftrightarrow l+1-i$ and $j \longleftrightarrow l+1-j$ we will yield the conventional Runge-Kutta method with higher derivatives of the form (2) whose coefficients $a_{ij}^{(r)-}$, $b_j^{(r)-}$ and c_j^- satisfy formulas (4). The theorem is proved.

Papers [4, p. 219–222], [9] define a symmetric one-step method to be a method whose increment function coincides with the increment function of its adjoint method. Thus, Theorem 1 says that it is sufficient to require the following equalities:

$$c_j = 1 - c_{l+1-j}, \tag{6a}$$

$$a_{ij}^{(r)} = (-1)^r \left(b_{l+1-j}^{(r)} - a_{l+1-i,l+1-j}^{(r)} \right), \tag{6b}$$

$$b_j^{(r)} = (-1)^r b_{l+1-j}^{(r)}, \quad r = 0, 1, \ldots, p_j, \ i, j = 1, 2, \ldots, l. \tag{6c}$$

Note that formulas (6) will be necessary for the symmetry if method (2) is irreducible (see [14]). It is also important to understand that conditions (6b), (6c) imply

$$p_j = p_{l+1-j}, \quad j = 1, 2, \ldots, l; \tag{7}$$

i.e., equality (7) is required for any symmetric irreducible Runge-Kutta method with higher derivatives.

Our next goal is to show that symmetric irreducible Runge-Kutta methods of the form (2) do exist.

3   Symmetry of E-Methods with Higher Derivatives

To approach this, we refer to the family of Runge-Kutta methods with higher derivatives constructed by the collocation technique with multiple nodes (see [11], [12]). They have been called E-methods with higher derivatives and are of the form

$$x_{k+1/2} = x_k + \tau_k \sum_{r=0}^{p} \tau_k^r \left(a_1^{(r)} g_k^{(r)} + a_3^{(r)} g_{k+1}^{(r)} \right) + \tau_k a_2^{(0)} g_{k+1/2}^{(0)}, \tag{8a}$$

$$x_{k+1} = x_k + \tau_k \sum_{r=0}^{p} \tau_k^r \left(b_1^{(r)} g_k^{(r)} + b_3^{(r)} g_{k+1}^{(r)} \right) + \tau_k b_2^{(0)} g_{k+1/2}^{(0)} \tag{8b}$$

where $g_{k+i}^{(r)} \stackrel{\text{def}}{=} g^{(r)}(t_{k+i}, x_{k+i})$, $i = 0, 1/2, 1$, and p is the number of derivatives of the right-hand side of problem (1) used to compute the numerical solution. Evidently, method (8) belongs to the class of Runge-Kutta methods with higher derivatives, and its coefficients form the tableau

$$
\begin{array}{c|cccccccccc}
0 & 0 & 0 & \dots & 0 & 0 & 0 & 0 & \dots & 0 \\
1/2 & a_1^{(0)} & a_1^{(1)} & \dots & a_1^{(p)} & a_2^{(0)} & a_3^{(0)} & a_3^{(1)} & \dots & a_3^{(p)} \\
1 & b_1^{(0)} & b_1^{(1)} & \dots & b_1^{(p)} & b_2^{(0)} & b_3^{(0)} & b_3^{(1)} & \dots & b_3^{(p)} \\
\hline
 & b_1^{(0)} & b_1^{(1)} & \dots & b_1^{(p)} & b_2^{(0)} & b_3^{(0)} & b_3^{(1)} & \dots & b_3^{(p)}
\end{array}
\tag{9}
$$

and satisfy

$$
a_1^{(r)} = \frac{p+1}{r! 2^{p+r+2}} \sum_{i=0}^{p-r} \sum_{l=0}^{i+r} \sum_{j=0}^{p+1} \frac{(-1)^l (i+r)!}{l!(i+r-l)! j!(p+1-j)!(l+j+2)}
$$
$$
\times \sum_{q=0}^{i} \frac{(p+q)!}{q! 2^q}, \quad r = 0, 1, \dots, p, \tag{10a}
$$

$$
a_2^{(0)} = \frac{(p+1)!}{2} \sum_{l=0}^{p+1} \frac{(-1)^l}{l!(p+1-l)!(2l+1)}, \tag{10b}
$$

$$
a_3^{(r)} = \frac{(-1)^{r+1}(p+1)}{r! 2^{p+r+2}} \sum_{i=0}^{p-r} \sum_{l=0}^{i+r} \sum_{j=0}^{p+1} \frac{(-1)^j (i+r)!}{l!(i+r-l)! j!(p+1-j)!(l+j+2)}
$$
$$
\times \sum_{q=0}^{i} \frac{(p+q)!}{q! 2^q}, \quad r = 0, 1, \dots, p, \tag{10c}
$$

$$
b_1^{(r)} = a_1^{(r)} + (-1)^r a_3^{(r)}, \quad b_2^{(0)} = 2a_2^{(0)}, \quad b_3^{(r)} = (-1)^r a_1^{(r)} + a_3^{(r)}. \tag{10d}
$$

It was proven in [12] that method (8) with the coefficients calculated by formulas (10) is A-stable for any p. Thus, the E-methods with higher derivatives are a good means for a practical implementation when differentiation of the right-hand side of ODE (1) is not too complicated. Now we prove the symmetry of method (8).

Theorem 2. *E-method* (8) *with coefficients* (10) *is symmetric for any integer* $p \geq 0$.

Proof. To prove this theorem, it is sufficient to check the symmetry conditions (6), (7) for the coefficients in tableau (9) satisfying formulas (10) (see Section 2). It is obvious that (6a) and (7) hold. Then, by calculating the right-hand side of formula (6b) applied to the method (8) and using (10d), we obtain the following:

$$(-1)^r \left(b_3^{(r)} - a_3^{(r)} \right) = (-1)^r \left((-1)^r a_1^{(r)} + a_3^{(r)} - a_3^{(r)} \right) = a_1^{(r)},$$

$$(-1)^r \left(b_2^{(0)} - a_2^{(0)} \right) = (-1)^r \left((-1)^r a_2^{(0)} + a_2^{(0)} - a_2^{(0)} \right) = a_2^{(0)},$$

$$(-1)^r \left(b_1^{(r)} - a_1^{(r)} \right) = (-1)^r \left((-1)^r a_3^{(r)} + a_1^{(r)} - a_1^{(r)} \right) = a_3^{(r)}.$$

The condition (6b) also holds. In the same way, we easily check the last condition (6c). Theorem 2 is proved.

Next, we confirm this theoretical result with the numerical examples below which show that method (8) with coefficients (10) provide quadratic extrapolation for different values of the parameter p. We refer the reader to [4], [10] for particulars of quadratic extrapolation.

4 Quadratic Extrapolation

First, we apply formulas (10) to calculate coefficients of the following E-methods of orders 6 and 8, respectively; i.e., when $p = 1$ and $p = 2$:

$$x_{k+1/2} = x_k + \frac{131}{480}\tau_k g_k^{(0)} + \frac{23}{960}\tau_k^2 g_k^{(1)} - \frac{19}{480}\tau_k g_{k+1}^{(0)} + \frac{7}{960}\tau_k^2 g_{k+1}^{(1)} + \frac{4}{15}\tau_k g_{k+1/2}^{(0)}, \quad (11a)$$

$$x_{k+1} = x_k + \frac{7}{30}\tau_k \left(g_k^{(0)} + g_{k+1}^{(0)} \right) + \frac{1}{60}\tau_k^2 \left(g_k^{(1)} - g_{k+1}^{(1)} \right) + \frac{8}{15}\tau_k g_{k+1/2}^{(0)}; \quad (11b)$$

$$x_{k+1/2} = x_k + \frac{689}{2240}\tau_k g_k^{(0)} + \frac{169}{4480}\tau_k^2 g_k^{(1)} + \frac{17}{8960}\tau_k^3 g_k^{(2)} - \frac{81}{2240}\tau_k g_{k+1}^{(0)}$$
$$+ \frac{41}{4480}\tau_k^2 g_{k+1}^{(1)} - \frac{19}{26880}\tau_k^3 g_{k+1}^{(2)} + \frac{8}{35}\tau_k g_{k+1/2}^{(0)}, \quad (12a)$$

$$x_{k+1} = x_k + \frac{57}{210}\tau_k \left(g_k^{(0)} + g_{k+1}^{(0)} \right) + \frac{1}{35}\tau_k^2 \left(g_k^{(1)} - g_{k+1}^{(1)} \right)$$
$$+ \frac{1}{840}\tau_k^3 \left(g_k^{(2)} + g_{k+1}^{(2)} \right) + \frac{16}{35}\tau_k g_{k+1/2}^{(0)}. \quad (12b)$$

Second, we test E-methods (11) and (12) on the problem

$$x_1'(t) = 2t\, x_2(t)^{1/5} x_4(t), \quad x_2'(t) = 10t\, \exp\left\{ 5\left(x_3(t) - 1 \right) \right\} x_4(t), \quad (13a)$$

$$x_3'(t) = 2t\, x_4(t), \quad x_4'(t) = -2t\, \ln\left\{ x_1(t) \right\} \quad (13b)$$

where $x_i(0) = 1$, $i = 1, 2, 3, 4$. Problem (13) has the following exact solution:

$$x_1(t) = \exp\left\{ \sin(t^2) \right\}, \quad x_2(t) = \exp\left\{ 5\sin(t^2) \right\}, \quad x_3(t) = \sin(t^2) + 1, \quad x_4(t) = \cos(t^2).$$

Table 1. Global errors of the E-method (11) with q extrapolations applied to (13)

| | fixed step size | | | | |
q	$\tau = 1.00 \cdot 10^{-1}$	$\tau = 6.67 \cdot 10^{-2}$	$\tau = 4.44 \cdot 10^{-2}$	$\tau = 2.96 \cdot 10^{-2}$	$\tau = 1.98 \cdot 10^{-2}$
0	$1.1822 \cdot 10^{-01}$	$1.2221 \cdot 10^{-02}$	$1.0970 \cdot 10^{-03}$	$9.6250 \cdot 10^{-05}$	$8.4474 \cdot 10^{-06}$
1	$1.3139 \cdot 10^{-04}$	$4.1088 \cdot 10^{-06}$	$1.2647 \cdot 10^{-07}$	$5.1884 \cdot 10^{-09}$	$2.1093 \cdot 10^{-10}$
2	$1.8188 \cdot 10^{-06}$	$1.4897 \cdot 10^{-08}$	$2.0497 \cdot 10^{-10}$	$3.9575 \cdot 10^{-12}$	$3.6386 \cdot 10^{-13}$

Table 2. Global errors of the E-method (12) with q extrapolations applied to (13)

| | fixed step size | | | | |
q	$\tau = 1.00 \cdot 10^{-1}$	$\tau = 6.67 \cdot 10^{-2}$	$\tau = 4.44 \cdot 10^{-2}$	$\tau = 2.96 \cdot 10^{-2}$	$\tau = 1.98 \cdot 10^{-2}$
0	$2.2594 \cdot 10^{-03}$	$7.9717 \cdot 10^{-05}$	$2.9465 \cdot 10^{-06}$	$1.1041 \cdot 10^{-07}$	$4.4174 \cdot 10^{-09}$
1	$3.1320 \cdot 10^{-06}$	$2.8233 \cdot 10^{-08}$	$3.1639 \cdot 10^{-10}$	$6.1941 \cdot 10^{-12}$	$2.4070 \cdot 10^{-13}$
2	$2.2455 \cdot 10^{-08}$	$5.6611 \cdot 10^{-11}$	$6.6482 \cdot 10^{-13}$	$3.5671 \cdot 10^{-13}$	$1.4103 \cdot 10^{-13}$

Our aim in this section is to check that E-methods (11) and (12) are symmetric and, in fact, provide quadratic extrapolation in practice. To do this, we perform numerical integrations of problem (13) with 5 different fixed step sizes $\tau_k = \tau$ on the interval $[0, 3]$, as indicated in Tables 1 and 2. We use these methods without extrapolation ($q = 0$), with one extrapolation step ($q = 1$) and with two extrapolations ($q = 2$) as well. However, the underlying methods are implicit. Therefore we apply the Newton iteration with the trivial predictor in order to obtain the required numerical solution and keep the number of iteration steps per grid point high enough to ensure that the extrapolation works properly (see [10] for more detail). We choose integers 1,2,3 and so on as the extrapolation sequence. The exact solution mentioned above is used to calculate the global errors of the integrations in the sup-norm and, hence, to find the actual order of the extrapolation algorithms.

Tables 1 and 2 exhibit clearly that the quadratic extrapolation definitely works with E-methods (11) and (12); i.e., each step of the extrapolation algorithm raises the accuracy of the underlying method by two orders. This experiment confirms also that both E-methods are symmetric and the theoretical results given above are correct.

5   Conclusion

In this paper we have discussed the symmetry property of Runge-Kutta methods with higher derivatives. We have found the coefficients of adjoint Runge-Kutta methods with higher derivatives (Theorem 1) and determined the general conditions on the coefficients of the methods under consideration to be symmetric (formulas (6), (7)). The new result is an extension of the well-known formulas for conventional Runge-Kutta methods (see the cited papers) and, of course, covers this case. Finally, we have given some examples of symmetric Runge-Kutta methods with higher derivatives and confirmed this property with practical computations. From this point of view, the family of E-methods derived earlier in [11] and [12] looks attractive if the cost of the derivative calculation is not too high. However, more experiments are needed to identify practical properties of the E-methods in detail.

References

1. Butcher, J.C.: Numerical methods for ordinary differential equations. John Wiley and Son, Chichester, 2003
2. Fehlberg, E.: Eine methode zur fehlerverkleinerung bein Runge-Kutta verfahren. ZAMM. **38** (1958) 421–426
3. Fehlberg, E.: New high-order Runge-Kutta formulas with step size control for systems of first and second order differential equations. ZAMM. **44** (1964) T17–T19
4. Hairer, E., Nørsett, S.P., Wanner, G.: Solving ordinary differential equations I: Nonstiff problems. Springer-Verlag, Berlin, 1993
5. Hairer, E., Wanner, G.: Solving ordinary differential equations II: Stiff and differential-algebraic problems. Springer-Verlag, Berlin, 1996
6. Hairer, E., Wanner, G., Lubich, C.: Geometric numerical integration: structure preserving algorithms for ordinary differential equations. Springer-Verlag, Berlin, 2002
7. Kastlunger, K.H., Wanner, G.: Runge-Kutta processes with multiple nodes. Computing. **9** (1972) 9–24
8. Kastlunger, K.H., Wanner, G.: On Turan type implicit Runge-Kutta methods. Computing. **9** (1972) 317–325
9. Kulikov, G.Yu.: Revision of the theory of symmetric one-step methods for ordinary differential equations, Korean J. Comput. Appl. Math., **5** (1998) No. 3, 289–318
10. Kulikov, G.Yu.: On implicit extrapolation methods for ordinary differential equations, Russian J. Numer. Anal. Math. Modelling., **17** (2002) No. 1, 41–69
11. Kulikov, G.Yu., Merkulov, A.I., Khrustaleva, E.Yu.: On a family of A-stable collocation methods with high derivatives. In: Marian Bubak, Geert Dick van Albada, Peter M. A. Sloot, and Jack J. Dongarra (eds.): Computational Science — ICCS 2004. 4th International Conference, Krakow, Poland, June 6–9 2004. Proceedings, Part II. Lecture Notes in Computer Science, **3037** (2004) 73–80
12. Kulikov, G.Yu., Merkulov, A.I.: On one-step collocation methods with higher derivatives for solving ordinary differential equations. (*in Russian*) Zh. Vychisl. Mat. Mat. Fiz. 44 (2004) No. 10, 1782–1807; *translation in* Comput. Math. Math. Phys. **44** (2004) No. 10, 1696–1720
13. Nørsett, S.P.: One-step methods of Hermite type for numerical integration of stiff systems. BIT. **14** (1974) 63–77
14. Stetter, H.J.: Analysis of discretization methods for ordinary differential equations, Springer-Verlag, Berlin, 1973

A Note on the Simplex Method for 2-Dimensional Second-Order Cone Programming

Yu Xia[*]

The Institute of Statistical Mathematics,
4-6-7 Minami-Azabu, Minato-Ku,
Tokyo 106-8569, Japan
yuxia@ism.ac.jp

Abstract. We transform a 2-dimensional second-order cone program to a standard linear program in order to pivot a vector based on its three states. Related properties of the transformation are given. Based on the transformation, we interpret the simplex method and its sensitivity analysis for the 2-dimensional second-order cone programming, especially the state changes. Finally, we give some applications of the 2-dimensional second-order cone programming.

Keywords: Second-order cone programming, linear programming, simplex method, basis, sensitivity analysis.

1 Introduction

A second-order cone (SOC) in $\mathbb{R}^{n+1}$ is the set

$$\mathcal{Q}_{n+1} \stackrel{\text{def}}{=} \left\{ \mathbf{x} \in \mathbb{R}^{n+1} : x_0 \geq \sqrt{\sum_{i=1}^{n} x_i^2} \right\} .$$

We omit the subscript $n+1$ of $\mathcal{Q}$ when it is clear from the context. We write $\mathbf{x} \geq_{\mathcal{Q}} \mathbf{0}$ interchangeably with $\mathbf{x} \in \mathcal{Q}$, since $\mathcal{Q}$ induces a partial order. Other names of $\mathcal{Q}$ include Loréntz cone, ice-cream cone, and quadratic cone.

The standard form primal and dual second-order cone programming (SOCP) problem is the following:

$$
\begin{array}{llll}
\textbf{Primal} & & \textbf{Dual} & \\
\min & \mathbf{c}_1^T \mathbf{x} + \cdots + \mathbf{c}_n^T \mathbf{x} & \max & \mathbf{b}^T \mathbf{y} \\
\text{s.t.} & A_1 \mathbf{x}_1 + \cdots + A_n \mathbf{x}_n = \mathbf{b} , & \text{s.t.} & A_i^T \mathbf{y}_i + \mathbf{s}_i = \mathbf{c}_i \ (i = 1, \ldots, n) , \\
& \mathbf{x}_i \geq_{\mathcal{Q}_{N_i}} \mathbf{0} \ (i = 1, \ldots, n) ; & & \mathbf{s}_i \geq_{\mathcal{Q}_{N_i}} \mathbf{0} \ \ (i = 1, \ldots, n) .
\end{array}
\tag{1}
$$

[*] Supported in part by a dissertation fellowship from Rutgers University, U.S.A. and a postdoctoral fellowship for foreign researchers from JSPS (Japan Society for the Promotion of Science). I thank my Ph.D. supervisor Professor Farid Alizadeh for motivation and discussion of the paper and for financial support from his grants. I thank comments and suggestions of an anonymous referee.

V.N. Alexandrov et al. (Eds.): ICCS 2006, Part I, LNCS 3991, pp. 124–131, 2006.

Here $N_i \in \mathbb{N}$ is the dimension of variables $\mathbf{x}_i$ and $\mathbf{s}_i$ $(i = 1, \ldots, n)$; $\mathbf{y} \in \mathbb{R}^m$ is a dual variable; $\mathbf{b} \in \mathbb{R}^m$, $\mathbf{c}_i \in \mathbb{R}^{N_i}$, $A_i \in \mathbb{R}^{m \times N_i}$ $(i = 1, \ldots, n)$ are data.

SOCP is a useful tool for many practical applications and theoretical developments; see [1], [2] for a survey. Interior point methods have been extended to SOCP. On the other hand, simplex methods remain widely used practical procedures for linear programming (LP) because of the low running time of each step – linear vs cubic for that of interior point methods. These features are especially useful for " warm starting " and large scale computing. Parallel simplex methods and cluster computing grids further facilitate the very large scale application. Unfortunately, no simplex-like method possessed of the above merits exists for SOCP, because of the nonlinear nature of SOC. In this note, we consider a subclass of SOCP which simplex methods can be extended to. Observe that an LP model is a 1-dimensional SOCP model. When $N_i = 2(i = 1, \ldots, n)$, (1) is the following LP model, where we write the *jth* entry of $\mathbf{x}_i$ as $(x_i)_j$.

Primal

$$
\begin{aligned}
\min \quad & \mathbf{c}_1^T\mathbf{x} + \cdots + \mathbf{c}_n^T\mathbf{x} \\
\text{s.t.} \quad & A_1\mathbf{x}_1 + \cdots + A_n\mathbf{x}_n = \mathbf{b}, \\
& (x_i)_0 \geq (x_i)_1 \ (i = 1, \ldots, n), \\
& (x_i)_0 \geq -(x_i)_1 \ (i = 1, \ldots, n);
\end{aligned}
$$

Dual

$$
\begin{aligned}
\max \quad & \mathbf{b}^T\mathbf{y} \\
\text{s.t.} \quad & A_i^T\mathbf{y}_i + \mathbf{s}_i = \mathbf{c}_i \ (i = 1, \ldots, n), \\
& (s_i)_0 \geq (s_i)_1 \ (i = 1, \ldots, n), \\
& (s_i)_0 \geq -(s_i)_1 \ (i = 1, \ldots, n).
\end{aligned}
\tag{2}
$$

Note that the constraints $(x_i)_0 \geq 0 \, (i = 1, \ldots, n)$ can be obtained by adding $(x_i)_0 - (x_i)_1 \geq 0$ and $(x_i)_0 + (x_i)_1 \geq 0$ together. Nevertheless, it is not straightforward to extend the pivot rules for the simplex method to (2). Pivots for LP change variable states between positivity and zero — a variable enters the basis when it becomes zero and leaves the basis when it becomes positive. However, a vector associated with an SOC has three states: in the interior of the SOC (int $\mathcal{Q}$), in the boundary of the SOC (bd $\mathcal{Q}$), and zero.

In this note, we show how to transform (2) into a standard form LP model. With this, we interpret the simplex method and its sensitivity analysis for the 2-dimensional SOCP based on the original SOC variables of (2). We also give some applications of the 2-dimensional SOCP for motivations of developing efficient algorithms for it.

The rest of the paper is organized as follows. In § 2, we give a transformation that maps (2) into a standard form LP model. Properties of the transformation are also discussed. In § 3, we interpret the simplex method for the 2-dimensional SOCP and give the sensitivity analysis for the simplex method. In § 4, we present some applications of the 2-dimensional SOCP.

2   The Transformation

In this part, we give a transformation of a 2-dimensional second-order cone program to a standard form linear program. We also prove the equivalence of the standard form linear program to (2) and give its properties.

First, we observe that for any $\mathbf{x}_i \in \mathcal{Q}_1 (i \in \{1,\dots,n\})$, the state of $\mathbf{x}_i$ is determined by the number of active constraints:

$$(x_i)_0 + (x_i)_1 \geq 0 , \qquad \qquad ((+)\text{-constraint})$$
$$(x_i)_0 - (x_i)_1 \geq 0 . \qquad \qquad ((-)\text{-constraint})$$

That is, $\mathbf{x}_i = \mathbf{0}$ iff both $(+)$ and $(-)$-constraints are active; $\mathbf{x}_i \in \text{bd}\,\mathcal{Q}$ iff only one of the $(+)$ and $(-)$ constraints is active; $\mathbf{x}_i \in \text{int}\,\mathcal{Q}$ iff neither of the $(+1)$ and $(-)$ constraints is active.

The transformation. Define

$$\tilde{P} \stackrel{\text{def}}{=} \begin{pmatrix} \frac{\sqrt{2}}{2} & \frac{\sqrt{2}}{2} \\ \frac{\sqrt{2}}{2} & -\frac{\sqrt{2}}{2} \end{pmatrix} .$$

Then $\tilde{P}$ is an orthogonal matrix. In addition, $\tilde{P}^T = \tilde{P}$, $\tilde{P}^2 = I$.

Let $\mathbf{v}_i = [(v_i)_0, (v_i)_1]^T \stackrel{\text{def}}{=} \tilde{P}\mathbf{x}_i \ (i = 1,\dots,n)$. Then the following holds.

1. $\mathbf{x}_i \geq_{\mathcal{Q}} \mathbf{0}$ iff $\mathbf{v}_i \geq \mathbf{0}$.
   In addition, $\mathbf{x}_i \in \text{int}\,\mathcal{Q}$ iff $\mathbf{v}_i > \mathbf{0}$; $\mathbf{x}_i = \mathbf{0}$ iff $\mathbf{v}_i = \mathbf{0}$; $\mathbf{x}_i \in \text{bd}\,\mathcal{Q}$ iff one of $(v_i)_0$ and $(v_i)_1$ is positive and the other one is zero.
2. $|(v_i)_0|$ is the distance of $\mathbf{x}_i$ from the line $(x_i)_0 = -(x_i)_1$; $|(v_i)_1|$ is the distance of $\mathbf{x}_i$ from the line $(x_i)_0 = (x_i)_1$; $\|\mathbf{v}_i\|_2$ is the distance of $\mathbf{x}_i$ from $\mathbf{0}$.
   (a) $\mathbf{v}_i \geq \mathbf{0} \iff \mathbf{x}_i \in \mathcal{Q}$ and $\mathbf{v}_i$ are the distances of $\mathbf{x}_i$ from the two boundaries of SOC, i.e. the $(+)$ and $(-)$-constraints.
   (b) $(v_i)_0 \geq 0$ and $(v_i)_1 < 0 \iff \mathbf{x}_i \notin \mathcal{Q}$ and $-(v_i)_1$ is the distance of $\mathbf{x}_i$ from SOC.
   (c) $(v_i)_0 < 0$ and $(v_i)_1 \geq 0 \iff \mathbf{x}_i \notin \mathcal{Q}$ and $-(v_i)_0$ is the distance of $\mathbf{x}_i$ from SOC.
   (d) $\mathbf{v}_i \leq \mathbf{0} \iff \mathbf{x}_i \notin \mathcal{Q}$ and $\|\mathbf{v}_i\|_2$ is the distance of $\mathbf{x}_i$ from SOC.
3. $\sqrt{2}\mathbf{v}_i$ are the two eigenvalues of $\mathbf{x}_i$. The columns of $\frac{\sqrt{2}}{2}\tilde{P}$ are two eigenvectors of $\mathbf{x}_i$. In other words, $\left(\frac{\sqrt{2}}{2}\tilde{P}\right)(\sqrt{2}\mathbf{v}_i)$ is the spectral decomposition of $\mathbf{x}_i$ (see [1]).

Standard form linear program. Let P be a block diagonal matrix with each block being $\tilde{P}$. Let $\tilde{\mathbf{c}} \stackrel{\text{def}}{=} P\mathbf{c}$, $\tilde{A} \stackrel{\text{def}}{=} AP$. We consider the following standard form linear program:

Primal   **Dual**

$$
\begin{array}{llll}
\min & \tilde{\mathbf{c}}^T\mathbf{v} & \max & \mathbf{b}^T\mathbf{y} \\
\text{s.t.} & \tilde{A}\mathbf{v} = \mathbf{b}, & \text{s.t.} & \tilde{A}^T\mathbf{y} + \mathbf{w} = \tilde{\mathbf{c}}, \\
& \mathbf{v} \geq \mathbf{0}; & & \mathbf{w} \geq \mathbf{0}.
\end{array}
\tag{3}
$$

Complementary slackness. Let $(\mathbf{v}_i; \mathbf{w}_i) = (P\mathbf{x}; P\mathbf{s})$. Then

$$\mathbf{v}_i \cdot \mathbf{w}_i = \mathbf{0} \iff \mathbf{x}_i \circ \mathbf{s}_i = \mathbf{0} , \tag{4}$$

where $\mathbf{v}_i \cdot \mathbf{w}_i \stackrel{\text{def}}{=} \begin{bmatrix} (v_i)_0(w_i)_0 \\ (v_i)_1(w_i)_1 \end{bmatrix}$, $\mathbf{x}_i \circ \mathbf{s}_i \stackrel{\text{def}}{=} \begin{bmatrix} (x_i)_0(s_i)_0 + (x_i)_1(s_i)_1 \\ (x_i)_0(s_i)_1 + (x_i)_1(s_i)_0 \end{bmatrix} .$

Equivalence between (2) *and* (3). By duality theorem, a triple $(\mathbf{v}; \mathbf{w}; \mathbf{y})$ solves (3) iff

$$\tilde{A}\mathbf{v} = \mathbf{b}$$
$$\tilde{A}^T\mathbf{y} + \mathbf{w} = \tilde{\mathbf{c}}$$
$$\mathbf{v} \cdot \mathbf{w} = \mathbf{0}$$
$$\mathbf{v} \geq \mathbf{0}$$
$$\mathbf{w} \geq \mathbf{0} .$$

And $(\mathbf{x}; \mathbf{s}; \mathbf{y})$ is a solution to (2) iff it satisfies the following conditions [1].

$$A\mathbf{x} = \mathbf{b}$$
$$A^T\mathbf{y} + \mathbf{s} = \mathbf{c}$$
$$\mathbf{x} \circ \mathbf{s} = \mathbf{0}$$
$$\mathbf{x} \geq_{\mathcal{Q}} \mathbf{0}$$
$$\mathbf{s} \geq_{\mathcal{Q}} \mathbf{0} .$$

Together with (4) and the properties of P, we have that $(\mathbf{x}; \mathbf{s}; \mathbf{y})$ is a solution to (2) iff $(P\mathbf{x}; P\mathbf{s}; \mathbf{y})$ solves (3). In addition, they have the same objective value.

Strong duality. The properties of P and (4) also imply that the strong duality for the 2-dimensional SOCP holds if (i) both the primal and the dual have feasible solutions, or (ii) the primal has feasible solutions and the objective value is below bounded in the feasible region. For higher dimensional SOCP, neither (i) nor (ii) is sufficient for strong duality, see [1].

3 The Simplex Method

In this part, we interpret the simplex method for the 2-dimensional SOCP.

Basic solution. Without loss of generality, we assume A has full row rank; otherwise, either the linear constraints are inconsistent or some linear constraints are redundant. Let A_B be an $m \times m$ nonsingular matrix of A. Since P is nonsingular, $\tilde{A}_B$ is nonsingular, too. The constraints for which the column of the corresponding variable $\mathbf{v}_i$ belongs to $\tilde{A}_B$ are the basic constraints. Other constraints are the nonbasic constraints. The set of basic constraints is the basis. The corresponding vector $\mathbf{x}$ is the basic solution. If a basic solution $\mathbf{x}$ also satisfies $A\mathbf{x} = \mathbf{b}$ and $\mathbf{x} \geq_{\mathcal{Q}} \mathbf{0}$, $\mathbf{x}$ is called a basic feasible solution.

Let B_x represent the number of boundary blocks of a basic solution $\mathbf{x}$, I_x the number of its interior blocks, and O_x the number of its zero blocks. Then

$$B_{x^*} + 2I_{x^*} \leq m .$$

The simplex method. Next we interpret the primal simplex method with Bland's pivoting rule. Other simplex methods can be explained in a similar way.

1. Solve a phase I problem to get a basic feasible solution to the primal of (3). Assume the corresponding partition of the index set $\{i(k) \mid i \in \{1,\ldots,n\}, k \in \{-,+\}\}$ (i is the block index, k indicates the boundary constraints $(+)$ or $(-)$) is B and N, where $\tilde{A}_B \in \mathbb{R}^{m\times m}$ is nonsingular. Let $\mathbf{v}_B$ and $\mathbf{v}_N$ be the basic and nonbasic constraints, i.e. $\mathbf{v}_B = \tilde{A}_B^{-1}\mathbf{b} - \tilde{A}_B^{-1}\tilde{A}_N\mathbf{v}_N$.

2. If $\tilde{\mathbf{c}}_N - (\tilde{A}_B^{-1}\tilde{A}_N)^T\tilde{\mathbf{c}}_B \geq 0$, $\mathbf{x} = P\mathbf{v}$ is optimal for (2). Stop. Otherwise, there exists index $i(k)$ such that $\left(\tilde{\mathbf{c}}_N - (\tilde{A}_B^{-1}\tilde{A}_N)^T\tilde{\mathbf{c}}_B\right)_{i(k)} < 0$. That indicates that if $\mathbf{x}_i$ is moved away from the boundary k, the objective may be decreased.

3. Check the columns of $(\tilde{A}_B^{-1}\tilde{A}_N)_{i(k)}$ for such $i(k)$'s. If there exists an $i(k)$ such that $(\tilde{A}_B^{-1}\tilde{A}_N)_{i(k)} \leq \mathbf{0}$; then the problem is unbounded, i.e. $\mathbf{x}_i$ can be moved arbitrarily away from the boundary k to decrease the objective value infinitely. Otherwise, from the $i(k)$'s choose the smallest index $\bar{i}(\bar{k})$; from the indices $j(l)$'s with $(\tilde{A}_B^{-1}\tilde{A}_N)_{j(l),i(k)} > 0$ choose the smallest index $\bar{j}(\bar{l})$. Move $\mathbf{x}_{\bar{j}}$ to boundary $\bar{l}$, and move $\mathbf{x}_{\bar{i}}$ away from boundary $\bar{k}$ at a distance $(\tilde{A}_B^{-1}\mathbf{b})_{\bar{j}(\bar{l})}/(\tilde{A}_B^{-1}\tilde{A}_N)_{\bar{j}(\bar{l}),\bar{i}(\bar{k})}$.

4. Go to step 2 with the new basic, nonbasic constraints and coefficient matrix.

The state of the variable. In the above algorithm, each pivot affects at most two constraints: one active constraint becomes inactive, and one inactive constraint becomes active. Next, we consider how the state of $\mathbf{x}$ is affected by the pivots.

1. $\bar{i} = \bar{j}$

   In this case, $\mathbf{x}_{\bar{i}}$ must be moved from one boundary to the other boundary of $\mathcal{Q}$.

2. $\bar{i} \neq \bar{j}$

   In this case, the pivot affects two variables. The total number of active constraints for $\mathbf{x}_{\bar{i}}$ and $\mathbf{x}_{\bar{j}}$ is unchanged after the pivot. That total number can be only 1, 2, 3.

   (a) *The number is 1.* This means that the pivot makes an interior variable boundary and a boundary variable interior.

   (b) *The number is 2.* This means that after the pivot, a zero and an interior variable become two boundary variables, or vice versa.

   (c) *The number is 3.* This means that a zero variable is changed to a boundary variable, and a boundary variable is changed to a zero variable by the pivot.

Other methods for linear program, such as dual simplex algorithm, primal-dual simplex algorithm (see [3]) can also be applied to the 2-dimensional second-order cone program.

Sensitivity analysis. We can also perform sensitivity and parameter analysis on the 2-dimensional second-order cone programming.

Given a basic optimal solution for the primal-dual pair (2) with objective value ζ:

$$\zeta = \zeta^* - {\mathbf{w}_N^*}^T \mathbf{v}_N^*$$
$$\mathbf{v}_B = \mathbf{v}_B^* - \tilde{A}_B^{-1} \tilde{A}_N \mathbf{v}_N^* \, ,$$

where

$$\mathbf{v}_B^* = \tilde{A}_B^{-1} \mathbf{b}$$
$$\mathbf{v}_N^* = \mathbf{0}$$
$$\mathbf{w}_N^* = \tilde{A}_N^T \tilde{A}_B^{-T} \tilde{\mathbf{c}}_B - \tilde{\mathbf{c}}_N$$
$$\zeta^* = \tilde{\mathbf{c}}_B^T \tilde{A}_B^{-1} \mathbf{b} \, .$$

Next, we give the ranges of changes in cost coefficients $\mathbf{c}$ and right-hand side $\mathbf{b}$, under which current classification of optimal basic and nonbasic constraints remains optimal.

We first consider the change of right-hand side $\mathbf{b}$.

Assume $\mathbf{b}$ is changed to $\mathbf{b} + t\Delta\mathbf{b}$. The original basic constraints are now

$$\bar{\mathbf{v}}_B = \tilde{A}_B^{-1}(\mathbf{b} + t\Delta\mathbf{b}) \geq \mathbf{0} \, .$$

If $\bar{\mathbf{v}}_B \geq \mathbf{0}$, the current basis is still optimal. Solve $\bar{\mathbf{v}}_B \geq \mathbf{0}$ for t, we obtain

$$\left(\min_{j(l)\in B} -\frac{(\tilde{A}_B^{-1}\Delta\mathbf{b})_{j(l)}}{(\tilde{A}_B^{-1}\mathbf{b})_{j(l)}} \right)^{-1} \leq t \leq \left(\max_{j(l)\in B} -\frac{(\tilde{A}_B^{-1}\Delta\mathbf{b})_{j(l)}}{(\tilde{A}_B^{-1}\mathbf{b})_{j(l)}} \right)^{-1} \, .$$

And the objective will be $\zeta^* + t\tilde{\mathbf{c}}_B^T \tilde{A}_B^{-1} \Delta\mathbf{b}$.

Assume $\mathbf{c}$ is changed to $\mathbf{c} + t\Delta\mathbf{c}$. The dual nonbasic constraints are now

$$\bar{\mathbf{w}}_N = \mathbf{w}^* + \Delta\mathbf{w}_N \, ,$$

where

$$\Delta\mathbf{w}_N = \left(\tilde{A}_B^{-1} \tilde{A}_N \right)^T \Delta\tilde{\mathbf{c}}_B - \Delta\tilde{\mathbf{c}}_N \, .$$

The current basis remains optimal if $\bar{\mathbf{w}}_N \geq \mathbf{0}$. Solve $\bar{\mathbf{w}}_N \geq \mathbf{0}$ for t, we obtain

$$\left(\min_{i(k)\in N} -\frac{\Delta\mathbf{w}_{i(k)}}{\mathbf{w}_{i(k)}} \right)^{-1} \leq t \leq \left(\max_{i(k)\in N} -\frac{\Delta\mathbf{w}_{i(k)}}{\mathbf{w}_{i(k)}} \right)^{-1} \, .$$

And the objective will be $\zeta^* + t\Delta\tilde{\mathbf{c}}_B^T \tilde{A}_B^{-1} \mathbf{b}$.

Combining the above results, we can obtain that for simultaneous changes of right-hand side and cost coefficients.

4 Application

In this part, we give some applications of the 2-dimensional second-order cone programming.

Given a scalar u, its absolute value $|u|$ is equivalent to the optimal value of the following 2-dimensional SOCP model:

$$\min u_0$$
$$\text{s.t. } (u_0, u)^T \geq_{\mathcal{Q}} \mathbf{0}.$$

Therefore, given $A_i \in \mathbb{R}^{n_i \times m}$, $\mathbf{b}_i \in \mathbb{R}^{n_i}$ $(i = 1, \ldots, k)$, the L^1 and L^∞ norms associated with the affine transformation of a vector $A_i\mathbf{x} + \mathbf{b}_i$ can be formed as 2-dimensional SOCP models. Below are some examples.

Minimize the sum of weighted norms. Given some weights w_i $(i = 1, \ldots, k)$ for the k affine transformations of a vector $\mathbf{x}$, the problem of minimizing the weighted sum of L^1 or L^∞ norm can be formulated as a 2-dimensional SOCP model.

1. Minimize the sum of weighted L^1- norms.

$$\min \sum_{i=1}^{k} w_i \sum_{j=1}^{n_i} (u_{ij})_0$$
$$\text{s.t. } A_i\mathbf{x} + \mathbf{b}_i = \mathbf{u}_i \quad (i = 1, \ldots, k)$$
$$[(u_{ij})_0, (u_i)_j]^T \geq_{\mathcal{Q}} \mathbf{0} \quad (i = 1, \ldots, k; \ j = 1, \ldots, n_i).$$

We use $(u_i)_j$ to represent the *jth* entry of vector $\mathbf{u}_i = [(u_i)_1, \ldots, (u_i)_{n_i}]^T$. And $(u_{ij})_0$ is a variable not belonging to $\mathbf{u}_i$.

2. Minimize the sum of weighted $L^\infty-$ norms.

$$\min \sum_{i=1}^{k} w_i (u_i)_0$$
$$\text{s.t. } A_i\mathbf{x} + \mathbf{b}_i = \mathbf{u}_i \quad (i = 1, \ldots, k)$$
$$[(u_i)_0, (u_i)_j]^T \geq_{\mathcal{Q}} \mathbf{0} \quad (i = 1, \ldots, k; \ j = 1, \ldots, n_i).$$

Minimize the largest norm. The problem of minimizing the largest L^1 or L^∞ norm of the k norms can be cast as a 2-dimensional SOCP model.

1. Minimize the largest L^1- norm.

$$\min t$$
$$\text{s.t. } A_i\mathbf{x} + \mathbf{b}_i = \mathbf{u}_i \quad (i = 1, \ldots, k),$$
$$[(u_{ij})_0, (u_i)_j]^T \geq_{\mathcal{Q}} \mathbf{0} \quad (i = 1, \ldots, k; \ j = 1, \ldots, n_i),$$
$$\left[t, \sum_{j=1}^{n_i} (u_{ij})_0\right]^T \geq_{\mathcal{Q}} \mathbf{0} \quad (i = 1, \ldots, k).$$

2. Minimize the largest $L^\infty-$ norms.

The problem $\min \max_{1 \leq i \leq k} \|\bar{\mathbf{v}}_i\|_1$ can be formulated as the follow.

$$\min t$$
$$\text{s.t. } A_i\mathbf{x} + \mathbf{b}_i = \mathbf{u}_i \quad (i = 1, \ldots, k),$$
$$[t, (u_i)_j]^T \geq_{\mathcal{Q}} \mathbf{0} \quad (i = 1, \ldots, k; \ j = 1, \ldots, n_i).$$

Minimize the sum of r largest norms. Similarly as minimizing the sum of r largest Euclidean norms [1], the problem of minimizing the sum of r largest L^1 or L^∞ norms can be formulated as a 2-dimensional SOCP model.

1. Minimize the sum of r largest L^1- norm.

$$\min \sum_{i=1}^{r} u_i + rt$$
$$\text{s.t.} \quad A_i\mathbf{x} + \mathbf{b}_i = \mathbf{v}_i \quad (i = 1, \ldots, k)$$
$$[(v_{ij})_0, (v_i)_j]^T \geq_\mathcal{Q} \mathbf{0} \quad (i = 1, \ldots, k; \; j = 1, \ldots, n_j)$$
$$\left[t + u_i, \sum_{j=1}^{n_i} (v_{ij})_0\right]^T \geq_\mathcal{Q} \mathbf{0} \quad (i = 1, \ldots, k)$$

2. Minimize the sum of r largest $L^\infty-$ norm.

$$\min \sum_{i=1}^{r} u_i + rt$$
$$\text{s.t.} \quad A_i\mathbf{x} + \mathbf{b}_i = \mathbf{v}_i \quad (i = 1, \ldots, k)$$
$$[(v_i)_0, (v_i)_j]^T \geq_\mathcal{Q} \mathbf{0} \quad (i = 1, \ldots, k; \; j = 1, \ldots, n_j)$$
$$[t + u_i, (v_i)_0]^T \geq_\mathcal{Q} \mathbf{0} \quad (i = 1, \ldots, k)$$

Here, $\mathbf{v}_i = [(v_i)_1, \ldots, (v_i)_{n_i}]^T$. And t, u_i, $(v_{ij})_0$, $(v_i)_0$ are scalar variables.

5   Conclusion

We've transformed a 2-dimensional SOCP model into a standard form LP model and interpreted the simplex method and its sensitivity analysis for the model. Some applications have also been given.

References

1. F. Alizadeh and D. Goldfarb. Second-order cone programming. *Math. Program.*, 95(1, Ser. B):3–51, 2003.
2. Miguel Sousa Lobo, Lieven Vandenberghe, Stephen Boyd, and Hervé Lebret. Applications of second-order cone programming. *Linear Algebra Appl.*, 284(1-3):193–228, 1998.
3. Robert J. Vanderbei. *Linear programming: foundations and extensions.* Kluwer Academic Publishers, Boston, MA, 1996.

Local Linearization-Runge Kutta (LLRK) Methods for Solving Ordinary Differential Equations

H. De la Cruz[1,2], R.J. Biscay[3], F. Carbonell[3],
J.C. Jimenez[3], and T. Ozaki[4]

[1] Universidad de Granma, Bayamo MN, Cuba
[2] Universidad de las Ciencias Informáticas,
La Habana, Cuba
[3] Instituto de Cibernética, Matemática y Física,
La Habana, Cuba
[4] Institute of Statistical Mathematics,
Tokyo, Japan
hugo@uci.cu, biscay@icmf.inf.cu

Abstract. A new class of stable methods for solving ordinary differential equations (ODEs) is introduced. This is based on combining the Local Linearization (LL) integrator with other extant discretization methods. For this, an auxiliary ODE is solved to determine a correction term that is added to the LL approximation. In particular, combining the LL method with (explicit) Runge Kutta integrators yields what we call LLRK methods. This permits to improve the order of convergence of the LL method without loss of its stability properties. The performance of the proposed integrators is illustrated through computer simulations.

1 Introduction

The Local Linearization (LL) method (see, e.g., [9], [11], [15], [16], and references therein), also called exponentially fitted method [9], matricial exponentially fitted method [6], exponential Euler method [4] and piece-wise linearized method [16], is an explicit one-step integrator for solving ODEs. It is derived from the local linearization (first-order Taylor expansion) of the vector field of the equation at each time step. Theoretical and simulation results have demonstrated a number of dynamical properties of this approach, including A-stability and correct reproduction of phase portraits near hyperbolic equilibrium points and cycles [11]. Furthermore, its computational cost is relatively low, even for systems that when deal with standard methods require the use of either cumbersome implicit schemes or extremely small time steps.

However, a major drawback of the LL method is its low order of convergence, namely two. This has motivated the recent development of a class of higher order LL methods (called LLT integrators) that add a Taylor-based correction term to the LL solution [3]. In this way an arbitrary order of convergence is reached

V.N. Alexandrov et al. (Eds.): ICCS 2006, Part I, LNCS 3991, pp. 132–139, 2006.

without losing the stability properties of the LL method. But this is achieved at the expense of computing higher order derivatives of the vector field of the equation.

In the present paper an alternative approach is introduced for constructing higher order LL integrators. It is also based on the addition of a correction term to the LL approximation, but now this is determined as the solution of an auxiliary ODE. For the latter, any discretization scheme can be used; e.g. an explicit Runge-Kutta (RK) method, leading to what we call the Local Linearization-Runge Kutta (LLRK) methods. Computation of higher order derivatives is not required, and A-stability is insured. This approach can be thought of as a flexible framework for increasing the order of the LL solution as well as for stabilizing standard explicit integrators.

Likewise splitting and Implicit-Explicit Runge-Kutta (IMEX RK) methods (see, e.g., [1], [13]), the LLRK method is based on the representation of the vector field as the addition of two components. However, there are notable differences between these approaches: i) Typically in splitting and IMEX methods such a decomposition is global instead of local, and it is not based on a first-order Taylor expansion. ii) In contrast with IMEX and LLRK approaches, splitting methods construct an approximate solution by composition of the flows corresponding to the component vector fields. iii) IMEX RK methods are partitioned (more specifically, additive) Runge-Kutta methods that compute a solution $\mathbf{y} = \mathbf{u} + \mathbf{z}$ by solving certain ODE for $(\mathbf{u}, \mathbf{z})$, setting different RK coefficients for each block. LLRK methods also solve a partitioned system for $(\mathbf{u}, \mathbf{z})$ but a different one. In this, one of the blocks is linear and uncouple, and it is solved by the LL discretization. After inserting the (continuous time) LL approximation into the second block, this is treated as a non-autonomous ODE, for which any extant RK discretization can be used.

The paper is organized as follows. Section 2 reviews the standard, low order LL method, and its numerical implementation. Section 3 introduces a new class of higher order LL methods, including the Local Linearization-Runge Kutta (LLRK) methods, and discussed its computational aspects. Finally, Section 4 illustrates its performance through computer simulations.

2   Local Linearization Method

Let $\mathcal{D} \subset \mathbb{R}^d$ be an open set and $\mathbf{f} \in C^1\left(\mathbb{R} \times \mathcal{D}, \mathbb{R}^d\right)$. Consider the initial-value problem

$$\mathbf{x}'(t) = \mathbf{f}(t, \mathbf{x}(t)), \quad t_0 \leq t \leq T, \quad \mathbf{x}(t_0) = \mathbf{x}_0,$$

where $t_0, T \in \mathbb{R}$, $\mathbf{x}_0 \in \mathcal{D}$. Let $t_0 < t_1 < ... < t_N = T$ be a given partition of $[t_0, T]$, and denote $h_n = t_{n+1} - t_n$ and $\mathbf{\Lambda}_n = [t_n, t_{n+1}]$ for $n = 0, ..., N - 1$.

The LL discretization can be derived as follows (see, e.g., [11]). Define the local problems

$$\mathbf{x}'(t) = \mathbf{f}(t, \mathbf{x}(t)), \quad t \in \mathbf{\Lambda}_n, \tag{1}$$
$$\mathbf{x}(t_n) = \mathbf{x}_{t_n},$$

for given constants $\mathbf{x}_{t_n} \in \mathcal{D}$, $n = 0, ..., N-1$. These equations are approximated by linear ones on the basis of the first-order Taylor expansion of $\mathbf{f}(t, \mathbf{x})$ around $(t_n, \mathbf{x}_{t_n})$:

$$\mathbf{x}'(t) = \mathbf{L}(t_n, \mathbf{x}_{t_n})\,\mathbf{x}(t) + \mathbf{a}(t; t_n, \mathbf{x}_{t_n}), \quad t \in \Lambda_n, \quad \mathbf{x}(t_n) = \mathbf{x}_{t_n}, \tag{2}$$

where

$$\mathbf{L}(s, \mathbf{x}) = \mathbf{f}'_x(s, \mathbf{x}), \quad \mathbf{a}(t; s, \mathbf{x}) = \mathbf{f}(s, \mathbf{x}) - \mathbf{f}'_x(s, \mathbf{x})\,\mathbf{x} + \mathbf{f}'_t(s, \mathbf{x})(t - s). \tag{3}$$

Here, $\mathbf{f}'_x$ and $\mathbf{f}'_t$ denote the partial derivatives of $\mathbf{f}$ with respect to the variables $\mathbf{x}$ and t, respectively.

The problem (2) has a solution $\mathbf{y}^{LL}(\,.\,; t_n, \mathbf{x}_{t_n})$ that is explicit in terms of the fundamental matrix $\boldsymbol{\Phi}(t; t_n, \mathbf{x}_{t_n}) = \exp((t - t_n)\,\mathbf{f}'_x(t_n, \mathbf{x}_{t_n}))$ of the corresponding homogeneous linear system. Namely, for $t \in \Lambda_n$,

$$\mathbf{y}^{LL}(t; t_n, \mathbf{x}_{t_n}) = \boldsymbol{\Phi}(t; t_n, \mathbf{x}_{t_n})\mathbf{x}_{t_n} + \int_{t_n}^{t} \boldsymbol{\Phi}(u; t_n, \mathbf{x}_{t_n})\boldsymbol{\Phi}^{-1}(u; t_n, \mathbf{x}_{t_n})\mathbf{a}(u; t_n, \mathbf{x}_{t_n})\,du$$

$$= \mathbf{x}_{t_n} + (t - t_n)\,\varphi(t - t_n; t_n, \mathbf{x}_{t_n}), \tag{4}$$

where

$$\varphi(r; s, \mathbf{x}) = \frac{1}{r} \int_{0}^{r} e^{(r-u)\mathbf{L}(s,\mathbf{x})}(\mathbf{f}(s, \mathbf{x}) + \mathbf{f}'_t(s, \mathbf{x})(u - s))\,du.$$

The continuous-time *LL approximation* $\mathbf{y}^{LL}(t)$ on $t \in [t_0, T]$ is defined by concatenating the solutions (4) of said local linear problems starting at $\mathbf{y}^{LL}(t_0) = \mathbf{x}_0$:

$$\mathbf{y}^{LL}(t) = \mathbf{y}^{LL}(t; t_n, \mathbf{y}^{LL}(t_n)), \quad t \in \Lambda_n, \quad n = 0, ..., N-1.$$

Finally, the *LL discretization* is defined by evaluating the *LL approximation* at the discrete times $t = t_n$,

$$\mathbf{y}^{LL}_{t_n} = \mathbf{y}^{LL}(t_n), \quad n = 0, ..., N-1. \tag{5}$$

A number of schemes have been proposed for computing the LL discretization (see reviews in [3], [12]). An implementation that consists of computing just one matrix exponential is the following (see [10] for more details). For $t \in \Lambda_n$ and any $\mathbf{x}_{t_n} \in \mathcal{D}$, $\mathbf{y}^{LL}(t; t_n, \mathbf{x}_{t_n})$ is written as

$$\mathbf{y}^{LL}(t; t_n, \mathbf{x}_{t_n}) = \mathbf{x}_{t_n} + \mathbf{v}(t; t_n, \mathbf{x}_{t_n}). \tag{6}$$

In turn, $\mathbf{v}(t; t_n, \mathbf{x}_{t_n})$ can be obtained as a block of the matrix exponential $\exp((t - t_n)\mathbf{C}_n)$ according to the identity

$$\begin{bmatrix} \mathbf{F}(t; t_n, \mathbf{x}_{t_n}) & \mathbf{b}(t; t_n, \mathbf{x}_{t_n}) & \mathbf{v}(t; t_n, \mathbf{x}_{t_n}) \\ 0 & 1 & c(t; t_n, \mathbf{x}_{t_n}) \\ 0 & 0 & 1 \end{bmatrix} = e^{(t-t_n)\mathbf{C}_n}, \tag{7}$$

where

$$\mathbf{C}_n = \begin{bmatrix} \mathbf{f}'_x(t_n, \mathbf{x}_{t_n}) & \mathbf{f}'_t(t_n, \mathbf{x}_{t_n}) & \mathbf{f}(t_n, \mathbf{x}_{t_n}) \\ 0 & 0 & 1 \\ 0 & 0 & 0 \end{bmatrix} \in \mathbb{R}^{(d+2)\times(d+2)}, \tag{8}$$

and $\mathbf{F}\left(t; t_n, \mathbf{x}_{t_n}\right) \in \mathbb{R}^{d\times d}$, $\mathbf{b}\left(t; t_n, \mathbf{x}_{t_n}\right) \in \mathbb{R}^d$, $c\left(t; t_n, \mathbf{x}_{t_n}\right) \in \mathbb{R}$ are certain matrix blocks. Thus, the LL discretization can be obtained as

$$\mathbf{y}^{LL}_{t_{n+1}} = \mathbf{y}^{LL}_{t_n} + \mathbf{v}(t_{n+1}; t_n, \mathbf{y}^{LL}_{t_n}),$$

where $\mathbf{v}(t_{n+1}; t_n, \mathbf{y}^{LL}_{t_n})$ is computed through (7) with $t = t_{n+1}$.

A number of algorithms are available to compute the matrix exponential involved in this scheme, e.g. those based on stable Padé approximations with the scaling and squaring method, Schur decomposition, or Krylov subspace methods. The choice of one of them should be based on the size and structure of the Jacobian matrix $\mathbf{f}'_x$ ([8], [14]).

3 Local Linearization-Runge Kutta (LLRK) Methods

In this Section a modification of the LL method is introduced in order to improve its accuracy while retaining desirable stability properties. Specifically, in order to obtain a better approximation $\mathbf{y}_n\left(t\right)$ to the solution of the local problem (1) with initial condition $\mathbf{y}_{t_n}$ at $t = t_n$, consider the addition of a correction term $\mathbf{z}_n\left(t\right)$ to the LL solution,

$$\mathbf{y}_n\left(t\right) = \mathbf{y}^{LL}\left(t; t_n, \mathbf{y}_{t_n}\right) + \mathbf{z}_n\left(t\right), \quad t \in \Lambda_n, \tag{9}$$

where $\mathbf{y}^{LL}\left(t; t_n, \mathbf{y}_{t_n}\right)$ is defined by (4).

From the variation of constants formula (see, e.g., [5]) it follows that the solution $\mathbf{x}_n\left(t\right)$ of (1) starting from $\mathbf{y}_{t_n}$ at $t = t_n$ can be written as

$$\mathbf{x}_n\left(t\right) = \mathbf{y}^{LL}\left(t; t_n, \mathbf{y}_{t_n}\right) + \mathbf{r}_n\left(t\right),$$

where

$$\mathbf{r}_n\left(t\right) = \boldsymbol{\Phi}\left(t; t_n, \mathbf{y}_{t_n}\right) \int_{t_n}^{t} \boldsymbol{\Phi}^{-1}\left(u; t_n, \mathbf{y}_{t_n}\right) \mathbf{M}\left(u, \mathbf{x}_n(u); t_n, \mathbf{y}_{t_n}\right) du, \tag{10}$$

$\mathbf{M}(u, \mathbf{x}; t_n, \mathbf{y}_{t_n}) = \mathbf{f}(u, \mathbf{x}) - (\mathbf{L}(t_n, \mathbf{y}_{t_n})\mathbf{x} + \mathbf{a}(u; t_n, \mathbf{y}_{t_n}))$, and $\mathbf{L}, \mathbf{a}$ defined by (3).

By taking derivatives in (10) it is obtained that $\mathbf{r}_n\left(t\right)$ satisfies the initial-value problem

$$\mathbf{r}'_n\left(t\right) = \mathbf{g}_n(t, \mathbf{r}_n\left(t\right)), \quad t \in \Lambda_n, \quad \mathbf{r}_n\left(t_n\right) = 0, \tag{11}$$

where

$$\mathbf{g}_n(t, \mathbf{r}) = L(t_n, \mathbf{y}_{t_n})\mathbf{r} + \mathbf{M}(t, \mathbf{y}^{LL}(t; t_n, \mathbf{y}_{t_n}) + \mathbf{r}; t_n, \mathbf{y}_{t_n}).$$

Thus, an approximation $\mathbf{z}_n\left(t\right)$ to $\mathbf{r}_n\left(t\right)$ can be obtained by solving the initial-value problem (11) through any extant numerical integrator. In particular, we will focus on the approximation $\mathbf{y}_n(t)$ obtained from solving (11) by means of an explicit RK method. This will be called a *Local Linearization-Runge Kutta (LLRK) approximation*. Specifically, the choice of an s-stage explicit RK method with coefficients $\mathbf{c} = [c_i]$, $\mathbf{A} = [a_{ij}]$, $\mathbf{b} = [b_j]$ (see, [5]) leads to

$$\mathbf{z}_n(t) = (t - t_n) \sum_{j=1}^{s} b_j \mathbf{k}_j, \qquad t \in \mathbf{\Lambda}_n,$$

where

$$\mathbf{k}_i = \mathbf{g}_n\left(t_n + c_i(t - t_n), (t - t_n)\sum_{j=1}^{i-1} a_{ij}\mathbf{k}_j\right), \qquad i = 1, 2, ...s,$$

Finally, the *LLRK discretization* is defined by the recursion

$$\mathbf{y}_{t_{n+1}} = \mathbf{y}^{LL}(t_{n+1}; t_n, \mathbf{y}_{t_n}) + \mathbf{z}_n(t_{n+1}), \quad n = 0, 1, ..., N - 1, \tag{12}$$

starting at $\mathbf{y}_{t_0} = \mathbf{x}_0$.

When implementing the LLRK method (12), the required evaluations of $\mathbf{y}^{LL}(t; t_n, \mathbf{y}_{t_n})$ at $t = t_{n+1}$ and $t = t_n + c_i(t_{n+1} - t_n)$ can be computed by means of (6)-(8). We use the diagonal Padé approximation with the scaling and squaring method [14] to evaluate the matrix exponential involved in these expressions.

It should be noted some points that contribute to decrease the burden of the computation of (12). First, simplifications arise from the fact that $\mathbf{g}_n(t_n, \mathbf{0}) = \mathbf{0}$. Second, if (as usually) the Runge Kutta coefficients c_i are of the form $c_i = m_i a$ for some integer numbers m_i then (12) can be implemented in terms of a few powers of the same matrix exponential $\exp(ah_n C_n)$, where C_n is certain matrix.

Given $p \geq 2$, the notation LLRKp will indicate an LLRK method (12) obtained by using an order p RK method to compute $\mathbf{z}_n(t)$. Notice that any LLRKp discretization is A-stable because the LL discretization is so, and the former reduces to the latter for linear systems. Furthermore, it can be shown that any LLRKp method has order of convergence p if the needed evaluations of $\mathbf{y}^{LL}(\,.\,; t_n, \mathbf{y}_{t_n})$ are carried out with an error of order p.

4   Numerical Examples

The following example is taken from [2] in order to illustrate not only convergence issues of the LLRK discretization but also its dynamics around hyperbolic stationary points.

Example 1.

$$x_1' = -2x_1 + x_2 + 1 - \mu f(x_1, \lambda),$$
$$x_2' = x_1 - 2x_2 + 1 - \mu f(x_2, \lambda),$$

where $f(u, \lambda) = u(1 + u + \lambda u^2)^{-1}$.

For $\mu = 15$, $\lambda = 57$, this system has two stable stationary points and one unstable stationary point in the region $0 \leq x_1, x_2 \leq 1$. There is a nontrivial stable manifold for the unstable point which separates the basins of attraction for the two stable points (see [2]).

Figure 1 illustrates the performance of the LLRK3 scheme in this example. For comparison, Figure 1(a) presents the phase portrait obtained by the LL scheme with a very small step-size $\left(h = 2^{-13}\right)$, which can be regarded the exact solution for visualization purposes. Figures 1 (b)-(c)-(d) show the phase portraits obtained, respectively, by a third order explicit Runge-Kutta (RK3), the LL and the LLRK3 methods with step-size $h = 2^{-2}$.

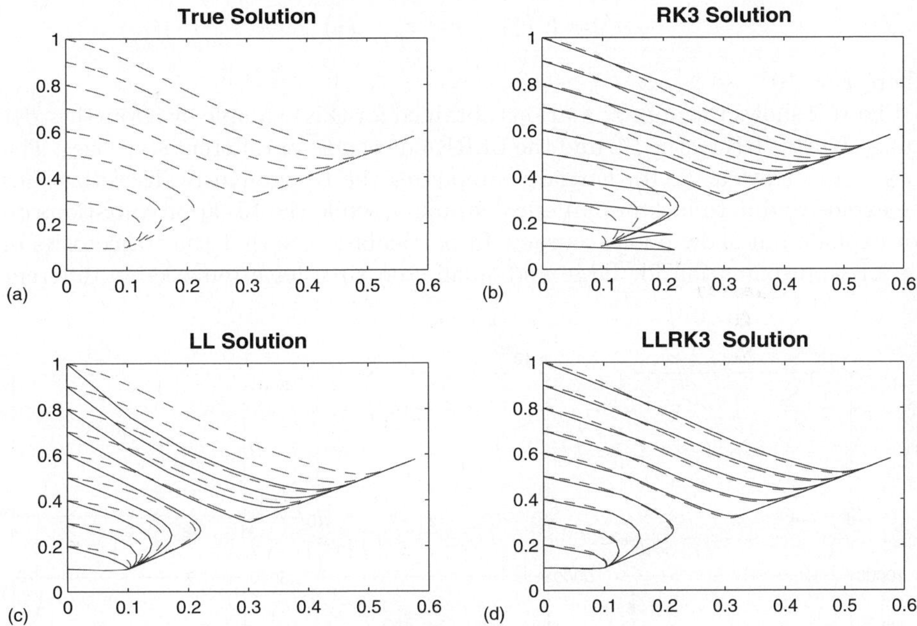

Fig. 1. (Example 1). a) Phase portrait obtained by the LL scheme with a very small step-size, $h = 2^{-13}$ (which can be thought of as the exact solution for visualization purposes). b), c), d) Continuous lines show the phase portraits obtained, respectively, by a third order Runge-Kutta (RK3), the LL and the LLRK3 methods with step-size $h = 2^{-2}$. For reference, the exact trajectories are also shown in each case as dashed lines.

It can be observed that the RK3 discretization fails to reproduce correctly the phase portrait of the underlying system near one of the point attractors. On the contrary, the exact phase portrait is adequately approximated near both point attractors by the LL and LLRK3 methods, the latter showing better accuracy. Also notice that the RK3 and LL discretizations do not approximate adequately the basins of attraction in the region shown in Figure 1. For instance, RK3 trajectories starting near $(0, 0.5)$ and LL trajectories starting near $(0, 0.6)$ go towards wrong point attractors in comparison with exact trajectories. In contrast, the attracting sets are much better represented by the phase portrait of the LLRK3 method. This demonstrates that, even for moderate values of p, the larger accuracy of LLRKp methods in comparison with the LL method can in practice leads to appreciable improvement in dynamical performance.

The next example illustrates the behavior of the LLRK4 method in a well-known stiff system that is frequently considered to test numerical integrators; namely, the Van der Pol equation (see, e.g., [7]).

Example 2.

$$x_1' = x_2,$$
$$x_2' = E\left(\left(1 - x_1^2\right) x_2 - x_1\right),$$

where $E = 10^3$.

Figure 2 shows the approximations obtained for this example by a fourth order Runge-Kutta (RK4), the LL and the LLRK4 methods for different step sizes. The trajectories start at $(2, 0)$. For large step sizes the Runge-Kutta discretizations are explosive due to lack of numerical stability, while the LL approximations do not explode but show poor accuracy. In particular, note that the trajectories of the LL approximation for large and small step sizes have remarkably different

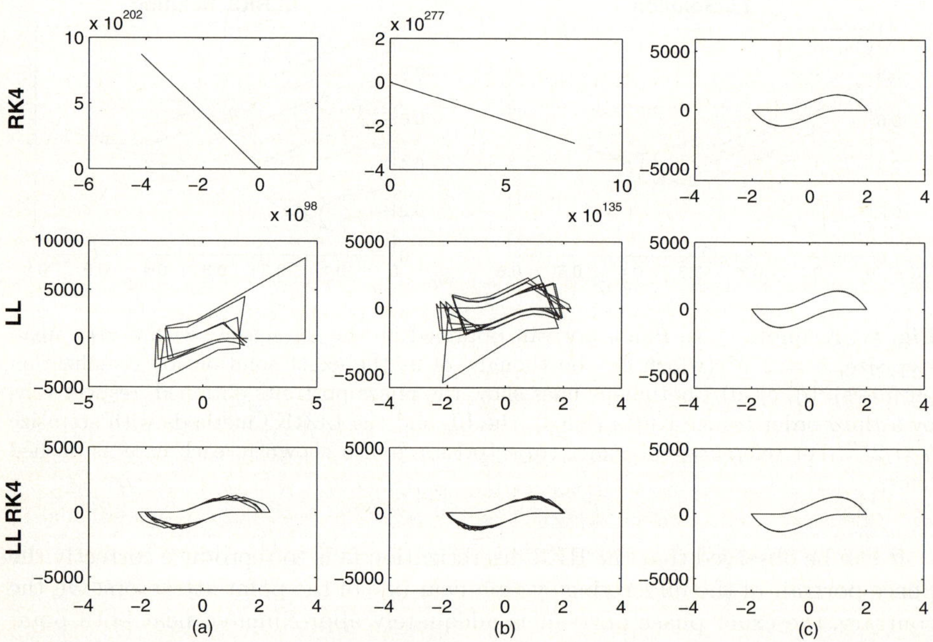

Fig. 2. (Example 2). Trajectories obtained for the example 2 by means of a fourth order Runge-Kutta (RK4), the LL and the LLRK4 methods for several step sizes: a) $h_1 = 0.00115$, b) $h_2 = 0.00095$, and c) $h_3 = 0.00025$.

ranges. In contrast, even for large step sizes the LLRK4 method achieves precise reconstructions of the underlying dynamics associated with a limit cycle.

References

1. Ascher, U. M., Ruuth, S. J. and Spiteri, R. J.: Implicit-Explicit Runge-Kutta methods for time-depending partial differential equations. Appl. Numer. Math. **25** (1995) 151-167
2. Beyn,W. J.: On the numerical approximation of phase portraits near stationary points. SIAM J.Numer. Anal. **24** (1987) 1095-1113
3. Biscay, R. J, De la Cruz, H., Carbonell, F., Ozaki, T. and Jimenez, J. C.: A Higher Order Local Linearization Method for Solving Ordinary Differential Equations. Technical Report, Instituto de Cibernetica, Matematica y Fisica, La Habana. (2005)
4. Bower, J. M. and Beeman, D.: The book of GENESIS: exploring realistic neural models with the general neural simulation system. Springer-Verlag (1995)
5. Butcher, J. C.: The Numerical Analysis of Ordinary Differential Equations. Runge-Kutta and General Linear Methods. John Wiley & Sons: Chichester (1987)
6. Carroll, J.: A matricial exponentially fitted scheme for the numerical solution of stiff initial-value problems. Computers Math. Applic. **26** (1993) 57-64.
7. Hairer, E. and Wanner, G.: Solving Ordinary Differential Equations II. Stiff and Differential-Algebraic Problems. Springer-Verlag: Berlin. Third edition (1996)
8. Higham, N. J.: The scaling and squaring method for the matrix exponential revisited. Numerical Analysis Report 452, Manchester Centre for Computational Mathematics. (2004)
9. Hochbruck, M., Lubich, C. and Selhofer, H.: Exponential integrators for large systems of differential equations. SIAM J. Sci. Comput. **19** (1998) 1552-1574
10. Jimenez, J. C.: A Simple Algebraic Expression to Evaluate the Local Linearization Schemes for Stochastic Differential Equations. Appl. Math. Lett. **15** (2002) 775-780
11. Jimenez, J. C., Biscay, R. J., Mora, C. M., and Rodriguez, L. M.: Dynamic properties of the Local Linearization method for initial-valued problems. Appl. Math. Comput. **126** (2002) 63-81
12. Jimenez, J. C. and Carbonell, F.: Rate of convergence of local linearization schemes for initial-value problems. Appl. Math. Comput. **171** (2005) 1282-1295
13. McLachlan, R.I and Quispel, G. R.W.: Splitting methods. Acta Numer. **11** (2002) 341-434
14. Moler, C. and Van Loan, C. F.: Nineteen dubious ways to compute the exponential of a matrix, twenty-five years later. SIAM Review. **45** (2003) 3-49
15. Ozaki, T.: A bridge between nonlinear time series models and nonlinear stochastic dynamical systems: a local linearization approach. Statist. Sinica. **2** (1992) 113-135
16. Ramos, J. I. and Garcia-Lopez, C. M. (1997).: Piecewise-linearized methods for initial-value problems. Appl. Math. Comput. **82** (1992) 273-302

The Study on the sEMG Signal Characteristics of Muscular Fatigue Based on the Hilbert-Huang Transform*

Bo Peng[1,2], Xiaogang Jin[2,**], Yong Min[2], and Xianchuang Su[3]

[1] Ningbo Institute of Technology, Zhejiang university,
Ningbo 315100, China
[2] AI Institute, College of Computer Science,
Zhejiang university, Hangzhou 310027, China
`xiaogangj@cise.zju.edu.cn`
[3] College of Software Engineering, Zhejiang university,
Hangzhou 310027, China

Abstract. Muscular fatigue refers to temporary decline of maximal power ability or contractive ability for muscle movement system. The signal of surface electromyographic signal (sEMG) can reflect the changes of muscular fatigue at certain extent. In many years, the application of signal of sEMG on evaluation muscular fatigue mainly focus on two aspects of time and frequency respectively. The new method Hilbert-Huang Transform(HHT) has the powerful ability of analyzing nonlinear and non-stationary data in both time and frequency aspect together. The method has self-adaptive basis and is better for feature extraction as we can obtain the local and instantaneous frequency of the signals. In this paper, we chose an experiment of the static biceps data of twelve adult subjects under the maximal voluntary contraction (MVC) of 80%. The experimental results proved that this method as a new thinking has an obvious potential for the biomedical signal analysis.

1   Introduction

Muscular fatigue [1] is an exercise-induced reduction in maximal voluntary muscle force. It may arise not only because of peripheral changes at the level of the muscle, but also because the central nervous system fails to drive the motoneurons adequately. The measurement of the muscular fatigue is according to the decline of the system maximal function. The adaptation of the neuromuscular system to heavy resistance exercise is a very complex result of many factors. This implies central and peripheral neural adaptations as well as chemical and morphological modifications of muscle tissue.

Surface electromyographic signal (sEMG) is an one-dimensional time series signal of neuromuscular system that recorded for skin surface, the time-frequency

* Supported by Zhejiang Provincial Natural Science Foundation of China under Grant No. Y105697 and Ningbo Natural Science Foundation (2005A610004).
** Corresponding author.

V.N. Alexandrov et al. (Eds.): ICCS 2006, Part I, LNCS 3991, pp. 140–147, 2006.

domain and nonlinear dynamical characters of it are sensitive to the intensity and state for muscular activities and therefore it is a valuable method for muscle functional evaluation. For many years, the analysis of sEMG concentrated on two main fields, time domain and the frequency domain [2]. The common use way of traditional time domain analysis is to regard the sEMG signal as the function of time, using some indicators like Integrated EMG (IEMG) or some statistical features such as Root Mean Square(RMS). In aspect of spectrum analysis, the typical method is the Fast Fourier Transform short as FFT. The spectrum we acquire from the FFT can show the distribution of the signals in different frequency component quantitatively. In order to portray the characteristic of the spectrum, researchers often use the following two indicators: Median Frequency (MF) and Mean Power Frequency (MPF) [3].

Most research works have found that the indicators of frequency analysis of the muscles in limbs and waist have good regularity in the condition of the static burthen. The main manifestation is the decline of the MPF or MF and the increase of power spectrum or the ratio of the low/high frequency during the process of fatiguing [4] [5]. However, under the dynamic condition, the alteration of the MPF and MF during the fatiguing process exist considerable difference from which can hardly get a universal conclusion. On account of above, we chose the static biceps data which have remarkable features of fatiguing to test our new method's performance in this paper.

2   The Method of Hilbert - Huang Transform

A new method [6] for analyzing nonlinear and non-stationary data has been developed in 1998 by N. E. Huang and others who lately made some amelioration on this method [7] in 1999. The key part of the method is the Empirical Mode Decomposition (EMD) method with which any complicated data set can be decomposed into a finite and often small number of Intrinsic Mode Functions (IMF) that admit well-behaved Hilbert transforms. This decomposition method is adaptive, and, therefore, highly efficient. Since the decomposition is based on the local characteristic time scale of the data, it is applicable to nonlinear and non-stationary processes. With the Hilbert transform, the IMF functions yield instantaneous frequencies as functions of time that give sharp identifications of imbedded structures. The final presentation of the results is the energy - frequency - time distribution, designated as the Hilbert spectrum. In this method, the main conceptual innovations are the introduction of IMF based on local properties of the signal, which makes the instantaneous frequency meaningful.

The method of Hilbert - Huang Transform (HHT) is composed of two parts: first one is the process of EMD which generate the products called IMF, and the second one it the traditional Hilbert transform and analysis on these IMF. IMF is a function that satisfies two conditions: (1) in the whole data set, the number of extremal and the number of zero crossings must either equal or differ at most by one; and (2) at any point, the mean value of the envelope defined by the local maxima and the envelope defined by the local minima is zero.

Find out all the local maxima and minima of the original signal. Once the extremal are identified, all the local maxima are connected by a cubic spline line as the upper envelope. Repeat the procedure for the local minima to produce the lower envelope. The upper and lower envelopes should cover all the data between them. Their mean is designated as m1, and the difference between the data and m_1 is the first component, h_1.

$$X(t) - m_1 = h_1 \tag{1}$$

Then, we have to judge whether h_1 be the IMF according to the requirements mentioned above. If not satisfy, the sifting process has to be repeated more times. In the second sifting process, h_1 is treated as the data, then,

$$h_1 - m_{11} = h_{11} \tag{2}$$

We can repeat this sifting procedure k times, until h_{1k} is an IMF, that is:

$$h_{1(k-1)} - m_{1k} = h_{1k} \tag{3}$$

Then, it is designated as:

$$c_1 = h_{1k} \tag{4}$$

The c_1 is the IMF component for the original signal. We can separate c_1 from the rest of the data by

$$X(t) - c_1 = r_1 \tag{5}$$

Since the residue, r_1, still contains information of longer period components, it is treated as the new data and subjected to the same sifting process as described above. This procedure can be repeated on all the subsequent r_j s, and the result is:

$$r_1 - c_2 = r_2, \cdots, r_{n-1} - c_n = r_n \tag{6}$$

The sifting process can be stopped, when the residue, r_n, becomes a monotonic function from which no more IMF can be extracted. By summing up equations (5) and (6), we finally obtain

$$X(t) = \sum_{i=1}^{n} c_i + r_n \tag{7}$$

Having obtained the intrinsic mode function components, we will have no difficulties in applying the Hilbert transform to each component, and computing the instantaneous frequency according to the following steps. For an arbitrary time series, $X(t)$, we can always have its Hilbert Transform, $Y(t)$, as

$$Y(t) = \frac{1}{\pi} P \int_{-\infty}^{+\infty} \frac{X(t')}{t - t'} \mathrm{d}t' \tag{8}$$

where P indicates the Cauchy principal value. With this definition, $X(t)$ and $Y(t)$ form the complex conjugate pair, so we can have an analytic signal, $Z(t)$, as

$$Z(t) = X(t) + \mathrm{i}Y(t) = a(t)\mathrm{e}^{\mathrm{i}\theta(t)} \tag{9}$$

in which we can also write as

$$a(t) = \left[X^2(t) + Y^2(t) \right]^{1/2} \tag{10}$$

$$\theta(t) = \arctan\left(\frac{Y(t)}{X(t)} \right) \tag{11}$$

and the instantaneous frequency defined as

$$\omega = \frac{\mathrm{d}\,\theta(t)}{\mathrm{d}\,t} \tag{12}$$

After performing the Hilbert transform on each IMF component, we can express the data in the following form:

$$X(t) = \sum_{j=1}^{n} a_j(t)\exp\left(\mathrm{i} \int \omega_j(t)\mathrm{d}\,t \right) \tag{13}$$

The equation also enables us to represent the amplitude and the instantaneous frequency as functions of time in a three-dimensional plot, in which the amplitude can be contoured on the frequency - time plane. This frequency- time distribution of the amplitude is designated as the Hilbert amplitude spectrum, $H(\omega; t)$, or simply Hilbert spectrum. If amplitude squared is more desirable commonly to represent energy density, then the squared values of amplitude can be substituted to produce the Hilbert energy spectrum just as well.

With the Hilbert spectrum defined, we can also define the marginal spectrum, $h(\omega)$, as

$$h(\omega) = \int_0^T H(\omega, t)\mathrm{d}\,t \tag{14}$$

In addition to the marginal spectrum, we can also define the instantaneous energy density level, IE, as

$$IE(t) = \int_\omega H^2(\omega, t)\mathrm{d}\,\omega \tag{15}$$

and the marginal energy spectrum ES, as

$$ES(\omega) = \int_0^T H^2(\omega, t)\mathrm{d}\,t \tag{16}$$

Although the HHT is a powerful method for analyzing nonlinear and non-stationary data, it has deficiencies also. Some research work has been done to ameliorate this [8]. However, the application of HHT in many fields have proved its performance exceed the traditional one-dimensional time series signals analysis methods [9][10][11][12].

3   Method and Result

3.1   Subjects

Twelve healthy adults of male university students, age ranging between 19 and 24, participates in this study. Subjects were all right hand dominant. They had no fierce movement before the test.

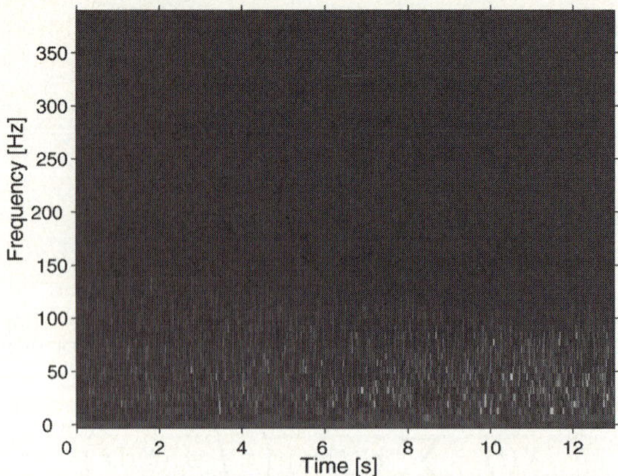

Fig. 1. The Hilbert spectrum of a sample sEMG (80%MVC) after the EMD process

3.2 Procedures

Subjects were first allowed to warm up the biceps. Then the wrists of the subjects were burdened with a weight of 80% MVC until they exhausted. During the procedures, signals were recorded with the sampling frequency of 1000Hz.

3.3 Results and Analysis

We deal with the data by the HHT, and Fig.1 shows the energy - frequency - time distribution of a sample sEMG (80% MVC) which we continue to use in Fig.2 and Fig.3, designated as the Hilbert spectrum. The brighter denote the higher energy and the darker denote the lower. Roughly, we can see from the Fig.1 the frequency distribution is wide at the beginning and concentrated to the lower frequency at the end, and in Fig.1, We can also see the energy became higher by the time variable for the right part of the spectrum is brighter than the

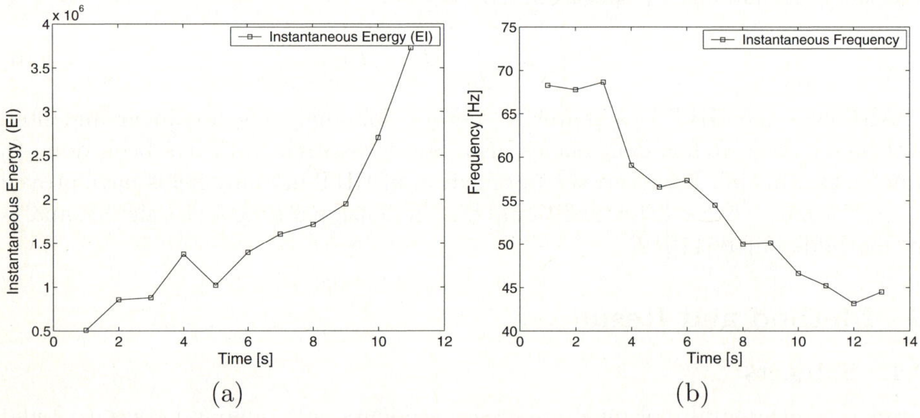

Fig. 2. (a)The instantaneous energy, IE the sample of the sample (b)The instantaneous frequency of the sample

left. We draw a conclusion qualitatively that during the process of the muscle
fatigue test, the frequency is going down and the energy is increasing. We get
more detailed and obvious features from further analysis below.

Fig.2(a) shows the changing trend of instantaneous energy (IE) by the factor
of time. We can see the energy approximately go up during the whole fatiguing
process. In Fig.2(b), we can see clearly that the instantaneous frequency is going
down during the fatiguing process. To calculate the instantaneous frequency,
please consult formulas. The ordinary way is to use the indicator MF or MPF
[3]. During the fatiguing course, the MF is move leftward, means the frequency
deceases generally. So, our results is consistent with the traditional approachs
and more convenient and evident [13].

In order to contrast, we extract the first second and last second of the signals
to test together, which we can see even more obviously the changes by fatiguing.
In the following analyzing, we use the average 80%MVC data of the twelve whole
subjects.

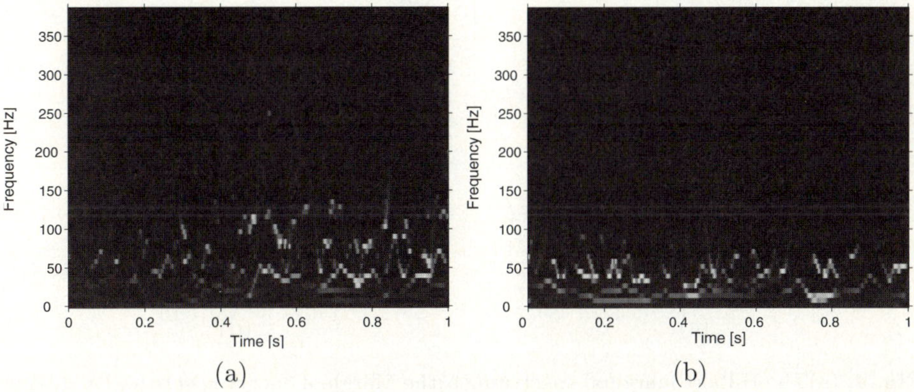

(a)   (b)

Fig. 3. The first second (a) and the last second (b)of the Hilbert spectrum of the
average twelve subjects' sEMG (80%MVC) after the EMD process

Fig.3(a) and Fig.3(b) show the Hilbert spectrum of the first second and the
last second of the average samples. Because the experiment was ended when the
subjects were exhausted, so we can consider the muscles were in the state of
fatigue in the last second. In contrast to Fig.3(a), frequency is concentrate to
the lower and the energy is higher in Fig.3(b). Although Fig.3(a) and Fig.3(b)
is the small sections of the Fig.1 in which they amplified the features of fatigue.

Fig.4(a) and Fig.4(b) show the energy's distribution on the frequency. We
can see the frequency of the last second is concentrate to the lower frequency
relatively and the energy is much higher than the first second's. This alteration
is even much remarkable in Fig.4(c) and Fig.4(d) which show the varying by the
time variable.

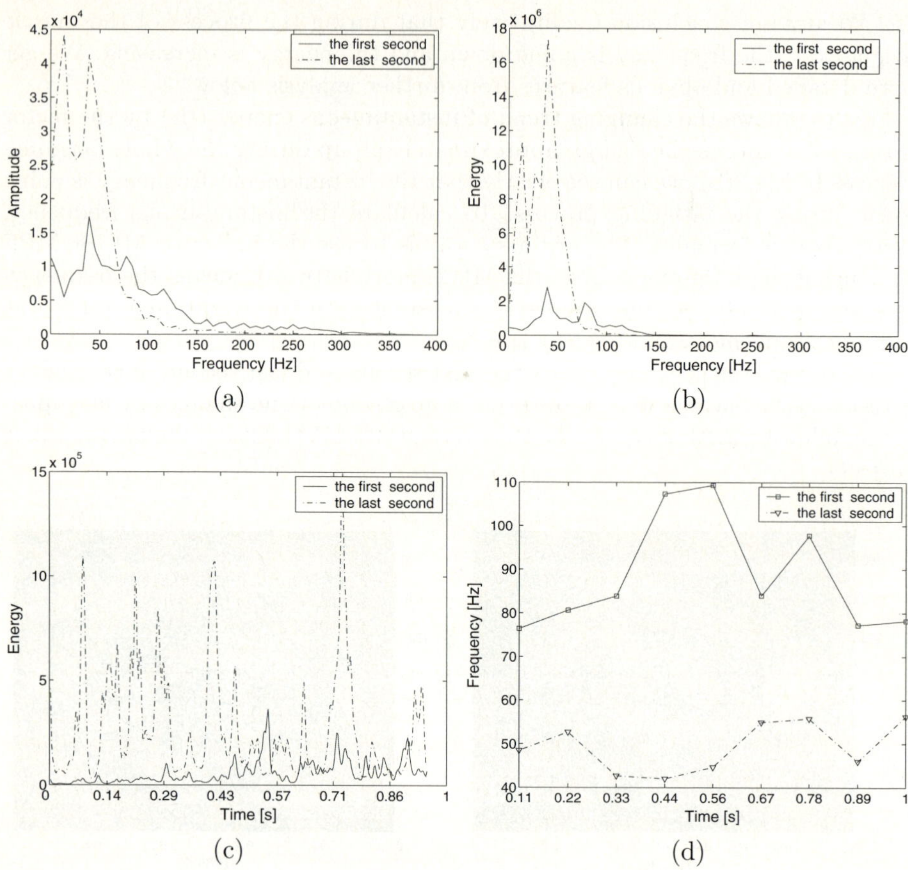

Fig. 4. (a)The Hilbert marginal spectrum,(b)the Marginal energy spectrum ES, (c)the instantaneous energy density level, IE, and(d) the instantaneous frequency of the first and the last second of average twelve subjects' sEMG (80%MVC)

4   Conclusion

In this paper, we adopted a new method-HHT, which is good at processing nonlinear and non-stationary data in two dimensions of both time domain and frequency domain to research the sEMG signal to find the features of muscular fatigue. We choose twelve adult university students as the experimental subjects. All the results we gained above are very consistent with the experiential conclusion that the frequency is going down and the energy is going up in the process of muscular fatiguing in [4][5]. The results also have proved that the HHT method has good performance and massive potential in analyzing the biomedical signals. However, the results in this paper are preliminary exploration, more works have to be done in the furture, even the HHT itself has some defects to ameliorate. Anyway, the HHT method has showed the extensive applicable foreground.

References

1. Gandevi S. C.: Spinal and supraspinal factors in human muscle fatigue. J. Physiological reviews. **81** (2001) 1725 -1789
2. Hagg G. M.: Interpretation of EMG spectral alternations and alternation indexes at sustained contraction. J. Appl. Physiol. **73** (1992) 1211-1217
3. Bilodeau M., Schindler-Ivens S., Williams D. M., Chandran R., Sharma S S.: EMG frequency content changes with increasing force and during fatigue in the quadriceps femoris muscle of men and women. J. Electromyography and Kinesiology. **13** (2003) 83-92
4. Mannion A. F., Connelly B., Wood K.: Electromyographic median frequency changes during isometric contraction of the back extensors to fatigue. Spine. **19** (1994) 1223-1229
5. Kazumi M., Tadashi M., Tsugutake S.: Changes in EMG parameters during static and dynamic fatiguing contractions. J. Electromyography and Kinesiology. **9** (1999) 9-46
6. Huang N. E., Shen Z., Long S. R.: The empirical mode decomposition and Hilbert spectrum for nonlinear and non-stationary time series analysis. J. Proc. R. Soc. London. **454** (1998) 899-995
7. Huang N. E., Shen Z., Long S. R.: A new view of non linear water waves: the Hilbert spectrum. J. Ann. Rev. Fluid Mech. **31** (1999) 417-457
8. Peng Z. K., Peter W .T., Chu F. L.: A comparison study of improved Hilbert-Huang transform and wavelet transform: Application to fault diagnosis for rolling bear. Mechanical Systems and Signal Processing. **19** (2005) 974-988
9. Dean G. D.: The Application of Hilbert-Huang Transforms to Meteorological Datasets. J. of Atmospheric and Oceanic Technology. Boston. **21** (2004) 599-612
10. Zhang R. R., VanDemark L., Liang L., Hu Y.: On estimating site damping with soil non-linearity from earthquake recordings. International Journal of Non-Linear, Mechanics. **39** (2004) 1501-1517
11. Phillips S. C., Gledhill R. J., Essex J. W., Edge C. M.: Application of the Hilbert-Huang Transform to the analysis of molecular dynamic simulations. J. Phys. Chem. A. **107** (2003) 4869-4876
12. Loh C. H., Wu T. C., Huang N. E.: Application of EMD+HHT method to identify near-fault ground motion characteristics and structural responses. BSSA. Special Issue of Chi-Chi Earthquake. **91** (2001) 1339-1357
13. Yang B. Z., Suh C. S.: Interpretation of crack-induced rotor non-linear response using instantaneous frequency. Mechanical Systems and Signal Processing. **18** (2004) 491-513

A New Approach for Solving Evolution Problems in Time-Parallel Way

Nabil R. Nassif[1], Noha Makhoul Karam[2], and Yeran Soukiassian[3]

[1] Mathematics Department,
American University of Beirut, Beirut Lebanon
[2] IRISA, Campus Beaulieu,
Université de Rennes I, Rennes, France
[3] Computer Science Department,
American University of Beirut, Beirut Lebanon

Abstract. With the advent of massively parallel computers with thousands of processors, a large amount of work has been done during the last decades in order to enable a more effective use of a higher number of processors, by superposing parallelism in time-domain, even though it is known that time-integration is inherently sequential, to parallelism in the space-domain[8]. Consequently, many families of predictor-corrector methods have been proposed, allowing computing on several time-steps concurrently[5], [6]. The aim of our present work is to develop a new parallel-in-time algorithm for solving evolution problems, based on particularities of a rescaling method that has been developed for solving different types of partial and ordinary differential equations whose solutions have a finite existence time[9]. Such method leads to a sliced-time computing technique used to solve independently rescaled models of the differential equation. The determining factor for convergence of the iterative process are the predicted values at the start of each time slice. These are obtained using "ratio-based" formulae. In this paper we extend successfully this method to reaction diffusion problems of the form $u_t = \Delta u^m + a u^p$, with their solutions having a global existence time when $p \le m \le 1$. The resulting algorithm **RaPTI** provides perfect parallelism, with convergence being reached after few iterations.

1 Introduction

Modelling complex phenomena in nature and sciences can often lead to solve time-dependent partial differential equations. In that context, and as the time-integration is inherently sequential, a lot of parallel algorithms have been first developed for the spatial discretization. However, with the advent of massively parallel computers with thousands of processors, the critical mesh resolution often remains far from the high capabilities of these supercomputers as detailed by Farhat and al in [8]. And one possible approach, in order to enable a more effective use of a higher number of processors, is to superpose parallelism in time-domain to the parallelism in the space-domain. One of the first (or maybe the first) has been suggested by Nievergelt [1] and led to multiple shooting methods.

V.N. Alexandrov et al. (Eds.): ICCS 2006, Part I, LNCS 3991, pp. 148–155, 2006.

Variants of this method were then developed by several authors, in particular by Chartier and Philippe in [2]. The basic underlining ideas in such algorithms are as follows:

1. Decompose the time integration interval into subintervals (time slices) and predict the initial values of the solution at the beginning of each slice.
2. Solve an initial value problem in each slice, and in a parallel way,
3. Force continuity of the solution branches, at the end points of successive slices, by means of an iterative (for example Newton-like) procedure.

The success of such parallel-time integration would necessarly depend on the number of iterations that should obviously be <u>much less</u> then the number of slices.

More recently, in 2001, Lions, Maday and Turinici proposed in [5] the "parareal algorithm" to solve evolution problems in a time-parallel way. This allows obtaining, in real time and using several processors, solutions that cannot be computed using one processor only. Algorithms has been implemented on linear and nonlinear parabolic problems and approximates successfully the solution later in time before having fully accurate approximations from earlier times. The main features of the method are the following:

1. Choice of a **coarse grid** that defines "large time-slices".
2. Prediction of the initial values of the solution on each time-slice, using for example a Euler-implicit scheme.
3. Iteration until convergence of the following process:
   - Solving independent evolution problems on each time-slice, using any method on a **fine grid**, leading to discontinuity "jump " between the end value of one slice and the predicted initial value on the next one.
   - Correction of the slices initial values by propagation of the jumps,

This scheme is inherently a multi-grid procedure. It combines coarse anf fine resolution in time, in the same spirit as what is done in space for domain decomposition methods, the coarse grid allowing us to propagate very quickly, on the whole interval, information reached on the fine grid up to any desired acuracy. Such algorithm has also received wide attention over the last few years: Maday and Bal proposed in [6] an improved version of the Parareal Algorithm (PA) which gives better answers for nonlinear problems and more importantly allows to tackle non-differentiable problems (whereas the former implementation was based on the linearization of the PDE). Farhat and Chandesris presented in [8] an original mathematical justification to the framework by using the theory of distribution. Many contributions, based on the former works, have been made during the 15th and 16th Domain Decomposition Conferences (Berlin 2003 and New York 2005). We cite mainly the following:

- In [13], Tromeur-Dervout and Guibert introduced adaptativity in the definition of the refinement of the time grid and the time domain splitting (in order to tackle stiff problems), and then proposed an adaptative parallel extrapolation algorithm (for very stiff problems).

– In [14] Farhat and al proposed a radical change of strategy in order to improve the performance of the parareal algorithm for second-order hyperbolic systems.

In this paper we present a new approach for parallel time integration. Our method is based on a rescaling procedure introduced by Nassif et al in [9] for problems that have finite existence time.

In section 2, we show that such procedure leads to a perfectly parallel-in-time alorithm in the case of a scalar ordinary differential equation of the form $y' = f(y)$, whereby coarse-grid time-slices can be explicitly found with exact initial predictions at the start of each slice.

In section 3, we generalize this rescaling approach to systems of differential equations that semi-discretize evolution problems of the form:

$$\frac{\partial u}{\partial t} = \Delta u^m + au^p.$$

This allows predictions of time-slices and simultaneously accurate initial values. We implement this method for linear PDE ($m = p = 1$) getting convergence (on the basis of a fast correcting procedure) in one or two iterations.

2 Perfect Time-Parallelism for a Scalar ODE

We consider the case of an elementary differential equation $y' = f(y)$ having **explosive solution in finite time**. One seeks $\{y; T_b\}$, $T_b < \infty$, $y:[0; T_b[\rightarrow \mathbb{R}$ such that:

$$\frac{dy}{dt} = f(y) \text{ for } 0 < t < T_b, \text{ with :} \tag{1}$$

$$y(0) = y_0 > 0, \text{ and } lim_{t \rightarrow T_b} |y(t)| = \infty.$$

The function f is such that:

$$f(y), f'(y) \geq 0, \ f(y_0) \neq 0, \ \int_{y_0}^{\infty} \frac{dy}{f(y)} < \infty. \tag{2}$$

The **Re-scaling method** consists in simultaneously introducing change of variables for the time t and the solution $y(t)$ in order to generate a sequence of *slices* of time intervals. On each of these subintervals, the computation of the solution is controlled by a preset threshold (or cut-off) value S. The change of variables is given by:

$$t = T_{n-1} + \beta_n s, \ y(t) = y_{n-1} + \alpha_n z(s) \tag{3}$$

where $y_n = y(T_n)$, and $\{\alpha_n\}$ and $\{\beta_n\}$ are sequences of normalizing parameters characteristic to the re-scaling method. The resolution of the initial problem (1) is then equivalent to that of rescaled models on each of time-slice, $[T_{n-1}, T_n]$, whereby we seek the pair $\{z, \{s_n\}\}$ where $z: [0; s_n] \rightarrow \mathbb{R}$, such that:

$$\frac{dz}{ds} = g_n(z) = \frac{\beta_n}{\alpha_n} f(y_{n-1} + \alpha_n z), \ 0 < s < s_n, \tag{4}$$

$$z(0) = 0, \; z(s_n) = S.$$

The criterion for limiting z not to out-grow pre-set threshold value S is only due to the nature of the solution which in this case is explosive. This condition must be adapted to other cases where the solution has a global existence time taking into account its behavior: bounded, unbounded, oscillatory, etc. Thus, in this particular case where the behavior is explosive, the choice of the sequence $\{\alpha_n\}$ is determined by $lim_{n\to\infty} y_n = \infty$ and $lim_{n\to\infty} T_n = T_b$, implying the consequent pertinent choices for $\{\alpha_n\}$, specifically: $\alpha_n = y_{n-1}$ when $f(y) = O(y^p)$ and $\alpha_n = 1$ when $f(y) = O(e^y)$. The underlying theory behind these choices is well detailed in [9] and [10]. The sequence $\{\beta_n\}$ is determined in order that the rescaled models (4) verify a **self-similarity concept**, formal or numerical. This leads into imposing $g_n(0) = g_1(0)$, $\forall n > 1$, implying for $\{\beta_n\}$, the choices: $\beta_n = \frac{\alpha_n}{f(y_{n-1})}$.

Perfect time-parallelism. Use of the re-scaling method for the purpose of parallel-time integration is based on the following principles:

1. The starting values of the solution on each time-slice are exactly determined by:

$$\forall n, \; y_n = y_{n-1} + \alpha_n S, \; \alpha_n = y_{n-1}, \text{ or } 1. \tag{5}$$

2. The size of the nth time slice $[T_{n-1}, T_n]$, is given by: $T_n - T_{n-1} = \beta_n s_n$, with the values of β_n also pre-determined by $\beta_n = \frac{\alpha_n}{f(y_{n-1})}$ and each s_n, computed independently of any other, given that the stopping criteria on each slice is the same one and in this scalar case is a function of S only.

Implementation of this method, with analysis of acceleration factors can be found in [15].

3   Extension to Partial Differential Equations

We first start by extending the re-scaling method to partial differential equations. This is illustrated on the case of finding $u : [0, T] \times \Omega \to \mathbb{R}$, $\Omega \subset \mathbb{R}^d$, $d = 1$ or 2, such that:

$$u_t - \Delta u^m = a u^p \tag{6}$$

for $m \leq p \leq 1$, homogeneous Dirichlet boundary conditions ($u(x,t) = 0$, $x \in \partial\Omega$, $t > 0$) and initial conditions ($u(x,0) = u_0(x) > 0$) allowing the solution to have various behaviors with respect to time ([7]). A first step consists in changing the variable u. Letting $v = u^m$ and $q = 1/m$, one has:

$$\frac{\partial v}{\partial t} = \frac{1}{q v^{q-1}} \Delta v + \frac{a}{q} v^{pq-q+1}, \; x \in \Omega \subset \mathbb{R}^d, \; t > 0 \tag{7}$$

with v verifying the same homogeneous Dirichlet boundary condition of u on $\partial\Omega$ and $v(x,0) = v_0 = u_0^m$. Re-scaling necessitates first semi-discretizing (7). In case we use Finite-Differences on the operator Δ, this leads to the following system of ordinary differential equations: Find $V : [0, T] \to \mathbb{R}^k$ such that:

$$\frac{dV}{dt} = -\frac{1}{q} D_W A V + F(V), \tag{8}$$

where $V(0) = V_0$ and $V(t) = (V_i(t))_{i=1,2,\ldots,k} \approx (v(x_i, t)_{i=1,2,\ldots,k}$,
$F(V) = \frac{a}{q} \left(f(V_i(t)^{pq-q+1}) \right)_{i=1,2,\ldots,k}$, $W(t) = \left(\frac{1}{[V_i(t)]^{q-1}} \right)_{i=1,2,\ldots,k}$,
$D_W = diag(W) \in \mathbb{R}^{k \times k}$ and A is a sparse symmetric positive matrix that discretizes the $-\Delta$ operator. Using the change of variables:

$$V(t) = V_{n-1} + D_{\alpha_n} Z(s), \text{ and } t = T_{n-1} + \beta_n s, \tag{9}$$

where $V_{n-1} = V(T_{n-1})$. Solving then the initial value problem (8) is equivalent to solving on each of the intervals $[T_{n-1}, T_n]$ to the rescaled models:

$$\frac{dZ}{ds} = \beta_n D_{\frac{1}{\alpha_n}}[-\frac{1}{q} D_W A V + F(V)], 0 < s < s_n, Z(0) = 0, \tag{10}$$

with the end of the slice s_n being determined by a condition depending on the behavior of the solution. For example, in case of explosive (unbounded) behavior, we use $||Z(s_n)||_\infty = S$. Otherwise, if $Z(s)$ cannot reach a threshold value S, we limit s_n to a maximum $\bar{s}$, $\forall n$. Thus, extending the choices used for α_n and β_n in the scalar case, it has been shown in [7] and [10] that suitable choices are given by $\alpha_n = V_{n-1} \in \mathbb{R}^k$ and (for the explosive case)

$$\beta_n = \frac{1}{||V_{n-1}^{pq-q}||_\infty} \in \mathbb{R} \tag{11}$$

Parallel-in-time implementation of the re-scaling method. Parallel-time algorithms require being able to determine: (i) the time slices (coarse-grid) and (ii) good predictions for the starting values of the solution at each of the time-slices. In the above scalar case, as the predicted values were the exact ones, parallelism was perfect. At present, the predicted values are only approximate, therefore requiring a correcting procedure that must be iteratively convergent. The basic idea of our predictions is inspired from the scalar case when $f(y) = O(y^p)$ and the sequence $\{y_n\}$ verifies $y_n = y_0(1 + S)^n = y_{n-1}(1 + S)$, $r = 1 + S$ being a transition ratio between two successive initial slice-values. This has been generalized in [7] and [10] to the vector case using the corresponding relation derived from the re-scaling method:

$$V_n = V_{n-1} + diag(\alpha_n)Z(s_n) \text{ where } \alpha_n = V_{n-1}, \text{ i.e.} \tag{12}$$

$$V_n = diag(V_{n-1})(e + Z(s_n)) = diag(e + Z(s_n))V_{n-1}, \tag{13}$$

where $e \in \mathbb{R}^k$ is the vector with all of its components equal to 1. By Defining the "transition" ratio vector:

$$r_n = (e + Z(s_n)). \tag{14}$$

the relation (13) becomes:

$$V_n = diag(r_n)V_{n-1}. \tag{15}$$

It follows from $\| Z(s_n) \|_\infty = S$ that $\| r_n \|_\infty = 1 + S$, which is invariant with respect to n and on the other hand is consistent with the scalar case ($k = 1$, $d = 1$). Thus, it appears that predicting starting values of the solution at each time-slice would result from a-priori information on r_n. For that purpose, we propose the following new approach to obtain time-parallelism through appropriate predictions of the transition ratio vectors r_n.

Parallel Algorithm: Ratio-Based Parallel Time Integration (RaPTI)

1. We run our method sequentially on n_s slices by computing the successive exact values of r_n, referred to as r_n^e, with a local slice computational tolerance ϵ_{tol}^l.
2. On following slices, $n > n_s$, statistical estimates techniques based on exact values r_n^e ($n \leq n_s$), allow us to predict values of r_n, denoted by r_n^p.
3. The previous step leads to compute **predicted initial values**, $V_n^p = diag(r_n^p) V_{n-1}^p$ for $n > n_s$. Note that for $n = n_s + 1$, $V_{n-1}^p = V_{n-1}^e$.
4. At this point, parallel computations of the solution can be executed on each nth slice, $n > n_s$, with a local computational tolerance ϵ_{tol}^l, with starting value V_{n-1}^p leading to an end value V_n^c. Knowing that the initial value of the $(n_s + 1)^{th}$ slice is exact, V_n^c is exact for $n = n_s + 1$.
5. For $n > n_s$, we define the sequence G_n of gaps at the end of the nth slice, as $G_n = V_n^c - V_n^p$.
6. For a given global computational tolerance $\epsilon_{tol}^g > \epsilon_{tol}^l$, we determine $n_{conv} > n_s$ such that for each nth slice, $n_s < n \leq n_{conv}$, $||G_n||_\infty \leq \epsilon_{tol}^g$, i.e. convergence is reached up to the n_{conv}^{th} slice.
7. We update n_s by n_{conv} and repeat steps 2 to 6 until the maximum time T, defining the interval $[0, T]$ of the evolution problem is reached.

Unlike the **Parareal** algorithm [5], our **RaPTI** procedure, **exludes any sequential computation** for $n > n_s$. Furthermore, since this new method is based on rescaling the variables, it generates dynamically the slices which may vary in size depending on the behavior of the solution. Obviously, the success of this method depends on the number of iterations (that repeat steps 2 to 6). This number should be **much less** than the number of slices needed to reach T. In fact, our tests have revealed cases of **perfect parallelism** where convergence has been reached in **one or two iterations**, particularly when the solution of the diffusion-reaction problem has a global time existence, i.e. $p \leq 1$. When $p > 1$, the solution **blows up in a finite time** and the **RaPTI** algorithm fails to give fast convergence unless we impose a gross global computational tolerance. A rule of thumb we have used in such situation consists in changing step 7 (correction step) of the algorithm to become:

7'. Repeat steps 2 to 6 using "Lions parareal algorithm"

This accelerates convergence at the expense of imposing sequential computations for predicting the new corrected values at begining of the slices.

Tests and Numerical results. Our experiments were run on the following problem:

$$\begin{cases} \frac{\partial u}{\partial t} - \Delta u = 3u^p & \text{for } x \in \Omega =\,]-1;1[\text{ and } t \geq 0 \\ u(x,t) = 0 & \text{for } x \in \partial\Omega \text{ and } t \geq 0 \\ u(x,0) = (1-x^2) & \text{for } x \in \Omega \end{cases}$$

We have used a threshold $S = 2$ to determine the sizes of the slices, a discretization space step of $h = 1/8$, a global tolerance $\epsilon_{tol}^g \geq \epsilon_{tol}^l = 5 \times 10^{-8}$ where ϵ_{tol}^l is the local tolerance used in computing the solution on every slice using a 4th order explicit Runge-Kutta method. Our results are summarized in the following tables where first column indicates the number of slices used for starting sequentially the computations:

1. $p = 1$; $\epsilon_{tol}^g = \frac{10^{-7}}{2}$.

n_s	Number of Slices	Iterations with **RaPTI**
2	16	1
2	32	1
2	64	1
2	128	1

2. $p = 0.8$; $\epsilon_{tol}^g = \frac{10^{-7}}{2}$.

			Number of Iterations	
n_s	Number of Slices	RaPTI	RaPTI with Parareal for correction (7')	
8	16	1	1	
8	32	1	1	
8	64	2	3	
8	128	2	3	

3. $p = 1.05$; **Results using RaPTI**.

		Iterations		
n_s	Number of Slices	$\epsilon_{tol}^g = \frac{10^{-4}}{2}$	$\epsilon_{tol}^g = \frac{10^{-3}}{2}$	$\epsilon_{tol}^g = \frac{10^{-2}}{2}$
2	8	5	1	1
2	16	13	6	1
2	32	29	22	2

4. $p = 1.05$; **Results with ratio-based predictions and corrections via Lions Parareal scheme**.

		Number of Iterations		
n_s	Number of Slices	$\epsilon_{tol}^g = \frac{10^{-4}}{2}$	$\epsilon_{tol}^g = \frac{10^{-3}}{2}$	$\epsilon_{tol}^g = \frac{10^{-2}}{2}$
2	8	3	1	1
2	16	3	3	1
2	32	5	4	3

4   Perspectives

We are presently testing the method for non-linear problems $(\forall m, p)$, with the **RaPTI** algorithm yielding the same super-convergence process when $p \leq m \leq 1$. We are also seeking a new prediction process to handle hyperbolic problems in which the solution exhibits quasi-periodic behavior.

References

1. **J.Nievergelt.** *Parallel methods for integration ordinary differential equations.* **Comm. ACM, 7:731-733, 1964**
2. **P.Chartier, B.Philippe.** *A parallel shooting technique for solving dissipative ODE's.* **Computing, vol.51, n3-4, 1993, p.209-236.**
3. **J. Erhel, S. Rault.** *Algorithme parallèle pour le calcul d'orbites.* **Springer-Verlag, (1989).**
4. **Fayad D., Nassif N., Cortas M.** *Rescaling technique for the Numerical computation of blowing-up solutions for semi-linear parabolic equations.* **2000,Manuscript.**
5. **J.L. Lions, Y.Maday, G.Turinici.** *Résolution d'EDP par un schéma en temps "pararéel".* **C.R.Acad.Sci.Paris, t.332, Srie 1, p.661-668. 2001.**
6. **Y. Maday, G. Bal.** *A parareal time discretization for non-linear pde's with application to the pricing of an American put.* **Recent developments in Domain Decomposition Methods (Zurich 2002), Lecture Notes in computational Science and Engineering, 23:189-202, Springer 2002.**
7. **Makhoul-Karam N.** *Résolution numérique d'équations paraboliques semi-linéaires à caractère d'explosion ou d'extinction par des méthodes de redimensionnement.* **Mémoire de DEA en MSI, 2002-2003.**
8. **Ch. Farhat, M.Chandesris.** *Time-decomposed parallel time-integrators.* **Int. J. Numer. Meth. Engng 2003. 58: 1397-1434.**
9. **N.R.Nassif, D.Fayad, M.Cortas.** *Slice-Time Computations with Re-scaling for Blowing-Up Solutions to Initial Value Differential Equations.* **V.S. Sunderam. et al. (Eds): ICCS 2005, LNCS 3514, pp. 58-65. Springer-Verlag 2005.**
10. **Cortas M.** *Méthode de re-dimensionnement (Rescaling technique) pour des équations aux dérivées ordinaires du 1er ordre à caractère explosif.* **Thèse, Université Bordeaux 1. Janvier 2005.**
11. **S.Vanderwalle.** *PARAREAL in a historical perspective (a review of space-time parallel algorithms).* **In the 16th International Conference on Domain Decomposition Methods, NYU 2005. Proceedings.**
12. **M.Gander, S.Vanderwalle.** *On the superlinear and linear convergence of the parareal algorithm.* **In the 16th International Conference on Domain Decomposition Methods, NYU 2005. Proceedings.**
13. **D.Guibert, D.Tromeur-Dervout.** *Adaptative Parareal for systems of ODEs.* **In the 16th International Conference on Domain Decomposition Methods, NYU 2005. Proceedings.**
14. **C.Farhat, J.Cortial, H.Bavestello, C.Dastillung.** *A time-decomposed time-parallel implicit algorithm for accelerating the solution of second-order hyperbolic problems.* **Berkley-Stanford computational fest, 7 mai 2005.**
15. **H.Arnaout, H.Mneimneh, Y.Soukiassian** *Parallel Algorithm for Sliced-Time Computation with Re-scaling for Blowing-up Solutions to Initial Value Problems.* **Master's Project. American University of Beirut. June 2005.**

CGO: A Sound Genetic Optimizer for Cyclic Query Graphs[*]

Victor Muntés-Mulero[1], Josep Aguilar-Saborit[1], Calisto Zuzarte[2],
and Josep-L. Larriba-Pey[1]

[1] DAMA-UPC, Computer Architecture Dept., Universitat Politècnica de Catalunya,
Campus Nord UPC, C/Jordi Girona Módul D6 Despatx 117 08034 Barcelona, Spain
{vmuntes, jaguilar, larri}@ac.upc.edu,
http://research.ac.upc.edu/DAC/DAMA-UPC
[2] IBM Canada Ltd, IBM Toronto Lab., 8200 Warden Ave.,
Markham, Ontario L6G1C7, Canada
calisto@ca.ibm.com

Abstract. The increasing number of applications requiring the use of
large join queries reinforces the search for good methods to determine
the best execution plan. This is especially true, when the large number
of joins occurring in a query prevent traditional optimizers from using
dynamic programming.

In this paper we present the Carquinyoli Genetic Optimizer (CGO).
CGO is a sound optimizer based on genetic programming that uses a subset of the cost-model of IBM®DB2®Universal Database^{TM}(DB2 UDB)
for selection in order to produce new generations of query plans. Our
study shows that CGO is very competitive either as a standalone optimizer or as a fast post-optimizer. In addition, CGO takes into account
the inherent characteristics of query plans like their cyclic nature.

1   Introduction

Query optimizers, which typically employ dynamic programming techniques [6],
have difficulties handling large join queries because of the exponential explosion
of the search space. Randomized search techniques remedy this problem by iteratively exploring the search space and converging to a nearly optimal solution.
Genetic algorithms [2] are a randomized search technique that models natural
evolution over generations using crossover, mutation and selection operations.

In this paper we present a genetic optimizer called the Carquinyoli[1] Genetic
Optimizer (CGO) that is coupled with the DB2 UDB optimizer's cost model.
CGO is an improvement over the work in [1] and [7] and proposes genetic oper-
ations that allow for cyclic plan graphs using genetic programming algorithms.

---

[*] Research supported by the IBM Toronto Lab Center for Advanced Studies and UPC
Barcelona. The authors from UPC want to thank Generalitat de Catalunya for its
support through grant GRE-00352.

[1] Name of a traditional Catalan cookie known for being very hard, which reminds us
of the complexity of the large join optimization problem.

V.N. Alexandrov et al. (Eds.): ICCS 2006, Part I, LNCS 3991, pp. 156–163, 2006.
© Springer-Verlag Berlin Heidelberg 2006

Our results show that the plans obtained by CGO equal those obtained by DB2 UDB for small queries, although the execution time of CGO is larger than that of DB2 UDB. When we turn to large queries, where DB2 UDB has to resort to heuristic approaches, CGO generates cheaper query plans, but the execution time of the optimizer is still larger. Note that DB2 UDB uses a greedy algorithm in those cases, since there is not enough memory to perform an exhaustive search using dynamic programming. In order to improve the optimization time, we propose a combined strategy that helps to improve the final plan obtained by DB2 UDB injecting the greedy plan in the initial population of CGO.

This paper is organized as follows. Section 2 introduces genetic optimization and describes CGO. Section 3 validates the genetic optimizer comparing it to DB2 UDB and proposes CGO as a post-optimizer for small databases. In sections 4 and 5, we explain the related work and draw some conclusions.

2   The Carquinyoli Genetic Optimizer

Inspired by the principles of genetic variation and natural selection, genetic programming performs operations on the members of a given population, imitating the natural evolution through several generations. Each member in the population represents a path to achieve a specific objective and has an associated cost. Starting with an initial population containing a known number of members, usually created from scratch, three operations are used to simulate evolution: *crossover operations*, which combine properties of the existing members in the population, *mutation operations*, which introduce new properties and *selection*, which discards the worst fitted members using a fitness function.

After applying these genetic operations, the population has been refreshed with new members. The new population is also called new *generation*. This process is repeated iteratively until a *stop condition* stops the execution. Once the stop condition is met, we take the best solution from the final population.

Query optimization can be reduced to a search problem where the DBMS needs to find the optimum execution plan in a vast search space. Each execution plan can be considered as a possible solution for the problem of finding a good access path to retrieve the required data. Therefore, in a *genetic optimizer*, every member in the population is a valid execution plan. Intuitively, as the population evolves, the average plan cost of the members decreases.

CGO. CGO is a sound genetic query optimizer based on the ideas of genetic programming outlined above. CGO assumes that there is a parser that transforms the SQL statements into a graph. This graph contains a vertex for each referenced relation and edges joining a pair of vertices when a join condition between attributes of these relations appear in the query statement.

CGO also assumes that a query execution plan (QEP) is a directed data flow graph, where leaf nodes represent the base access plans of the relations. Data flows from these nodes to the higher nodes in the graph. The non-leaf nodes process and combine the data from their input nodes using physical implementations of the relational operations of PROJECT, JOIN, etc., and the root node returns the final results of the query. The physical implementations of the operations used in the QEP are called plan operations and are described in more detail later in this section.

Cyclic Query Graphs. If we assume that a relation is never read twice, given N relations in a query we exactly need $N - 1$ join operations to join all the data during the query process. In a scenario without Cartesian products, this implies that, at least, we have $N - 1$ explicit join conditions in the SQL statement, joining all the relations accessed in the query process. Of course, the number of join conditions can be larger, forcing some join operations in the tree to have more than one join predicate. A number of join conditions larger than $N - 1$ implies cycles in the query graph.

In the presence of cyclic query graphs, two or more join predicates have to be used in some of the joins in the QEP. However, the same two conditions are not always merged together. Whether a join condition is located in a specific join operation depends on the order used to access the base relations.

One of the major enhancements of CGO is its capability to deal with cyclic query graphs. Our optimizer solves this problem merging the join predicates during the optimization process. Every join operation is associated with a single join predicate and is considered as an independent operation, giving versatility to our optimizer to handle cyclic query graphs.

Merging Join Conditions. Let us call $J_{i,j}$ a join operation with a single join predicate $Ri.a = Rj.b$. Every join operation has two input subtrees in the QEP, S_i and S_j, from where tuples are retrieved to perform the join process. We say that a relation R belongs to a subtree S, $R \in S$, if there is a scan operation in S having R as input relation.

An operation $J_{i,j}$ is merged with another existing single predicate join operation $J_{x,y}$ with subtrees S_x and S_y if

$$(R_i \in S_x \land R_j \in S_y) \lor (R_i \in S_y \land R_j \in S_x) \tag{1}$$

In fact, by *merge* we mean that both conditions $J_{i,j}$ and $J_{x,y}$ become a single operation with two join predicates.

Figure 1 shows an example where join operation $J5 = J_{1,3}$ has to be inserted in an existing partial plan. In step 1 we first check whether the join relations of $J5$ appear in the existing subtree. Since both join relations are scanned in the same subtree, we have to look for the node $J_{x,y}$ with subtrees S_x and S_y satisfying condition 1 (step 2). When node $J1 = J_{3,2}$ is checked, condition 1 is not satisfied. However, the optimizer detects that both input relations in $J5$ are in S_3 of $J1$. Therefore, $J2 = J_{1,5}$, the root node in S_3 of $J1$, is checked next. As $R_1 \in S_1 \land R_3 \in S_5$ is satisfied, $J2$ and $J5$ are merged in step 3. After the insertion $J2$, can be considered to have two join conditions or predicates, $R1 = R3$ and $R1 = R5$. However, as shown before, these two join predicates can be separated in other execution plan configurations. With this simple procedure, we allow CGO to handle cyclic query graphs, which previous genetic optimizers were not able to do.

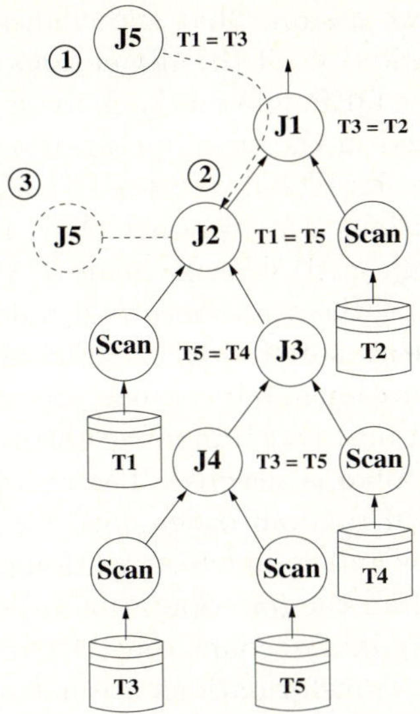

Fig. 1. Merging to join operations in an execution plan for a cyclic query graph

Plan Operations. The first aim of CGO is to outperform the most popular commercial optimizers when dealing with large join queries. Although commercial optimizers use a great variety of advanced techniques to improve the performance of DBMSs, for the sake of simplicity, CGO only considers a reduced set of these advanced techniques, such as prefetch or bit filters. The implementations considered by CGO are the *sequential scan* and the *index scan* for scan operations, the *Nested Loop Join*, *Merge-Scan Join* and *Hash Join* for join operations and two more unary operations, *Sort* and *Temp*, the first one for sorting data and the second one to materialize intermediate results in the QEP. Also, we reduced the language syntax to simplify our optimizer. Therefore, CGO can read simple SQL statements including multi-attribute selections, attribute constraints, equijoin conditions and sorting operations.

Cost Model. The cost model used to evaluate QEPs in CGO is an adaptation of the cost model in the DB2 UDB optimizer, in order to be comparable with existing optimizers. Overall, we use a fairly complete and comparable cost model based on CPU and I/O consumption.

Genetic Operations. In CGO, crossover operations randomly choose two QEPs in the population and produce two new trees preserving two subtrees from the parent plans and inserting the remaining operations one by one. Since each single-predicate join operation is handled separately, using this crossover method, the multiple join predicates in the QEP are automatically redistributed to the correct nodes and thus, we preserve the semantics of cyclic query graphs. A more detailed example can be found in [4].

Mutation operations are necessary to provide an opportunity to add new characteristics that are not represented in any of the execution plans in the population. In this paper we assume that the whole search space is accessible if CGO grants the exploration of all the dimensions of this search space: tree morphology, join order in the QEP, join methods to be used in the join operations and scan methods to be used in the scan operations. We propose four different kinds of mutation operations for CGO; (i) Swap (S): a join operation is randomly selected and its input relations are swapped. This mutation grants potential access to all tree morphologies; (ii) Change Scan (CS): CGO randomly chooses a scan operation and changes the scan method if indexes are available, making possible the access to all the scan methods; (iii) Change Join (CJ): the optimizer randomly chooses a join operation and it changes its implementation to one of the other available implementations; (iv) Random Subtree (RS): a subtree S from a randomly chosen execution plan is selected. The remaining join operations, not included in S are selected in random order until we have inserted all of them and, therefore, created a new and complete execution plan. This mutation grants the potential exploration of all the join operation orders.

S, CS and CJ do not take into account the occurrence of multiple join predicate operations since their transformations are not affected by the number of predicates. RS follows a construction method similar to crossover operations and thus, it allows for cyclic query graphs.

3   CGO Validation

In this section, we show that CGO is able to outperform a classical optimizer when this falls into heuristic algorithms because of a lack of memory, with large join queries. With this aim, we compare CGO with the optimizer of a well known commercial DBMS: the DB2 UDB optimizer. In order to verify the quality of the plans yielded by CGO we first compare the results obtained by CGO for queries based on the TPC-H Benchmark with query execution plans yielded by the DB2 UDB optimizer, that is, small join queries. Our results show, as expected, insignificant differences between the two optimizers as plotted in Figure 2 (left), although the optimization time of CGO is significantly larger than that of DB2 UDB. For more details, we refer the reader to [4]. Then, we compare CGO results for random-generated queries involving 20 and 30 relations, with those obtained by the commercial optimizer when it runs out of memory and uses its metaoptimizer, based on a sophisticated greedy join enumeration method. We always force the QEPs generated by CGO into the DB2 UDB optimizer, to avoid false results that are due to a possible bad integration of the DB2 UDB cost model. Finally, advanced features in DB2 UDB such as multidimensional clustering indexes, index ANDing and index ORing have been disabled.

3.1   Execution Details

We run two different test sets. In the first test set we generate 2 random databases and run 25 queries on each, where each query accesses 20 relations. In the second set, we create 4 random databases, running 25 random queries on each, where each query accesses 30 relations. Therefore, in total we execute 150 random queries in 6 randomly generated databases. For both, we run CGO on populations containing 250 members, performing 150 crossover operations and 160 mutation operations per generation, through 300 generations. We present all the results using the scaled cost, meaning the cost of the QEP generated by the DB2 UDB optimizer divided by the cost of the QEP generated by CGO. The tool used to generate the random databases and queries is detailed in [4].

In the first test, CGO obtained significantly cheaper plans. CGO obtained plans that were on average 3.83 times cheaper than those obtained by the DB2 UDB optimizer. Again, note that some optimizations of DB2 UDB were turned off and the greedy join algorithm was used by this optimizer. There were only 2 cases where the DB2 UDB plans were cheaper than those obtained by CGO. In the second test, which involved 30 relational joins, ignoring the outliers, the average scaled cost was 4.09 times cheaper with CGO. In the very best cases, CGO reaches improvements of three orders of magnitude.

3.2 Combining CGO and DB2 UDB: A Fast Post-optimizer

The experiments show the potential of CGO to find low-costed QEPs in front of an heuristic search-based optimizer. However, the optimization time required by CGO is larger than the time required by an heuristic search. The amount of computation time, saved during run time, can clearly compensate for the increasing optimization time since, usually, for large join queries the total execution time raises to several hours and even days. We are currently working to reduce the optimization time. Preliminary results based on very simple memory optimizations show improvements over the 95 % in optimization time reduction.

For this reason, queries that require long processing times will obtain a large benefit by using CGO alone; however, small and fast queries cannot afford long optimization times. Therefore, we propose to use CGO in a post-optimization phase. We obtain the QEP yielded by the DB2 UDB optimizer and force it as one of the QEP in the initial population of CGO.

Figure 2 (right) shows the total cost evolution for the best plan in the population across time. For 20 relations, we obtain, on average, QEP which are 1.71 times better than the initial plan generated by the DB2 UDB optimizer after 30 seconds of execution. The slight cost regressions for higher times and the values below 1 are a consequence of the complex integration of the DB2 UDB optimizer cost model into CGO, which makes it very hard to calculate the exact same cost. Therefore, although CGO always finds better plans, they are costed differently in DB2 UDB. However, it is important to notice that we can never get a worse execution plan than the one generated by the DB2 UDB optimizer since we can always preserve it in case CGO does not find anything better in a limited amount of time. Trends for 30 relations are similar.

We would like to remark that if CGO were using the exact same cost model as DB2 UDB, the genetic optimizer would benefit from it, thus amplifying the differences between both optimizers, since CGO currently wastes time trying to optimize some QEPs which cost is overestimated later by the cost model of DB2 UDB.

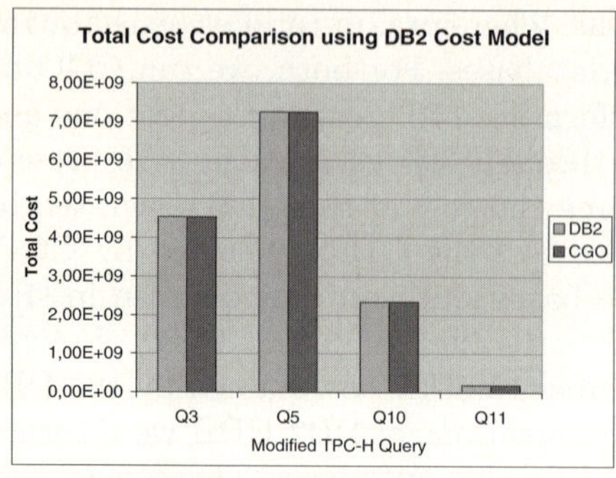
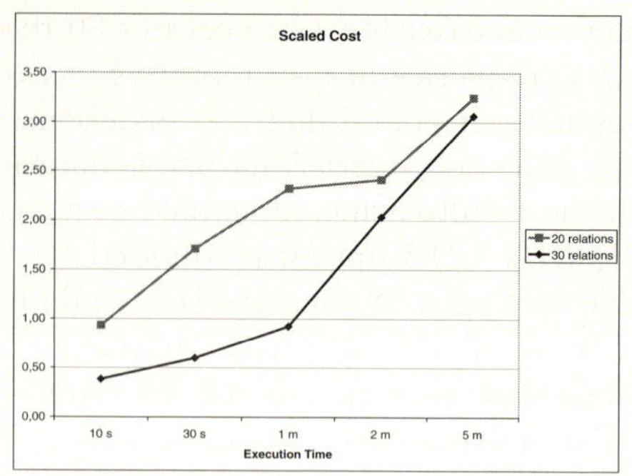

Fig. 2. (Left) Total Cost comparison for modified TPC-H queries using the cost model of the DB2 UDB optimizer. (Right) Evolution of the scaled cost of the best solution obtained by CGO across time.

4   Related Work

State-of-the-art query optimizers, which typically employ dynamic programming techniques [6], have difficulties handling large join queries because of the exponential explosion of the search space. In these situations, optimizers usually fall back to greedy algorithms. However, greedy algorithms, as well as other types of heuristic algorithms [9], do not consider the entire search space and thus may overlook the optimal plan, resulting in bad query performance, which may cause queries to run for hours instead of seconds. Randomized search techniques like Iterative Improvement or Simulated Annealing [3, 10, 7] remedy the exponential explosion of dynamic programming techniques by iteratively exploring the search space and converging to a nearly optimal solution.

Previous genetic approaches [1, 7] consider a limited amount of information per plan since these are transformed to chromosomes, represented as strings of integers. This lack of information usually leads to the generation of invalid plans that have to be repaired. A new crossover operation is proposed in [8] with the objective of making genetic transformations more aware of the structure of a database management system. Stillger proposed a genetic programming based optimizer that directly uses execution plans as the members in the population, instead of using chromosomes. However, mutation operations may lead to invalid execution plans that need to be repaired. Any of these previous approaches consider cyclic query graphs. A first genetic optimizer prototype was created for PostgreSQL [5], but its search domain is reduced to left-deep trees and mutation operations are deprecated, thus bounding the search to only those properties appearing in the execution plans of the initial population.

5   Conclusions

We presented CGO, a powerful genetic optimizer that is able to handle cyclic query graphs considering a rich variety of DBMS operations, reaching further than all previously proposed genetic optimizers. For the first time in the literature, we implemented a genetic optimizer that allows us to compare its results with the execution plans of a commercial DBMS, DB2 UDB.

The main conclusions of this paper are: (1) CGO is a sound optimizer able to compete with commercial optimizers for very large join queries. (2) CGO, used as a post-optimizer, significantly improves the execution plans obtained by DB2 UDB in a short amount of time. And (3) CGO is capable of finding the optimal or a near-optimal execution plan in a reasonable number of iterations for small queries.

Acknowledgment. IBM, DB2, and DB2 Universal Database are trademarks or registered trademarks of International Business Machines Corporation in the United States, other countries, or both. Other company, product, or service names may be trademarks or service marks of others.

References

1. Kristin Bennett, Michael C. Ferris, and Yannis E. Ioannidis. A genetic algorithm for database query optimization. In Rick Belew and Lashon Booker, editors, *Proceedings of the Fourth International Conference on Genetic Algorithms*, pages 400–407, San Mateo, CA, 1991. Morgan Kaufman.
2. J. Holland. Adaption in natural and artificial systems. The University of Michigan Press, Ann Arbor, 1975.
3. Yannis E. Ioannidis and Eugene Wong. Query optimization by simulated annealing. In *SIGMOD '87: Proceedings of the 1987 ACM SIGMOD international conference on Management of data*, pages 9–22, New York, NY, USA, 1987. ACM Press.
4. Victor Muntes, Josep Aguilar, Calisto Zuzarte, Volker Markl, and Josep Lluis Larriba. Genetic evolution in query optimization: a complete analysis of a genetic optimizer. Technical Report UPC-DAC-RR-2005-21, Dept. d'Arqu. de Computadors. Universitat Politecnica de Catalunya (http://www.dama.upc.edu), 2005.
5. PostgreSQL. http://www.postgresql.org/.
6. P. Griffiths Selinger, M. M. Astrahan, D. D. Chamberlin, R. A. Lorie, and T. G. Price. Access path selection in a relational database management system. In *Proceedings of the 1979 ACM SIGMOD international conference on Management of data*, pages 23–34. ACM Press, 1979.
7. Michael Steinbrunn, Guido Moerkotte, and Alfons Kemper. Heuristic and randomized optimization for the join ordering problem. *VLDB Journal: Very Large Data Bases*, 6(3):191–208, 1997.
8. Michael Stillger and Myra Spiliopoulou. Genetic programming in database query optimization. In John R. Koza, David E. Goldberg, David B. Fogel, and Rick L. Riolo, editors, *Genetic Programming 1996: Proceedings of the First Annual Conference*, pages 388–393, Stanford University, CA, USA, 28–31 July 1996. MIT Press.
9. A. Swami. Optimization of large join queries: combining heuristics and combinatorial techniques. In *SIGMOD '89: Proceedings of the 1989 ACM SIGMOD international conference on Management of data*, pages 367–376. ACM Press, 1989.
10. Arun Swami and Anoop Gupta. Optimization of large join queries. In *SIGMOD '88: Proceedings of the 1988 ACM SIGMOD international conference on Management of data*, pages 8–17, New York, NY, USA, 1988. ACM Press.

Multiscale Characteristics of Human Sleep EEG Time Series

In-Ho Song[1,2], In-Young Kim[2], Doo-Soo Lee[1], and Sun I. Kim[2]

[1] Department of Electrical and Computer Engineering, Hanyang University
[2] Department of Biomedical Engineering, College of Medicine, Hanyang University
Haengdang-dong, Seongdong-ku, Seoul, 133-791, South Korea
sunkim@hanyang.ac.kr

Abstract. We investigated the complexity of fluctuations in human sleep electroencephalograms (EEGs) by multifractals. We used human sleep EEG time series taken from normal, healthy subjects during the four stages of sleep and rapid eye movement (REM) sleep. Our findings showed that the fluctuation dynamics in human sleep EEGs could be adequately described by a set of scales and characterized by multifractals. Multifractal formalism, based on the wavelet transform modulus maxima, appears to be a good method for characterizing EEG dynamics.

1 Introduction

Since Babloyantz et al. reported that human sleep EEG signals had chaotic attractors during stage 2 and stage 3 sleep, nonlinear dynamic methods based on the concept of chaos have been used to analyze sleep EEG signals [1]. Fell et al. analyzed sleep EEG signals using the largest Lyapunov exponent (LLE) during sleep stages 1-4 and rapid eye movement (REM) sleep, and found significant differences in the values of LLE between the four stages and REM sleep [2] [3]. Kobayashi et al. found significant decrements in the mean correlation dimensions of EEGs from being awake to stage 3 sleep, and an increment during REM sleep [4]. However, these methods require a relatively large number of stationary data points to obtain reliable results [5] [6]. Indeed, it is difficult to meet this requirement in sleep EEGs, as well as in practical physiological systems [5] [6]. Recent research suggests that EEGs can be characterized in some situations by a single dominant timescale, indicating timescale invariance and a $1/f$ fractal structure [7] [8]. However, many physiological time series fluctuate in a complex manner and are inhomogeneous, suggesting that different parts of the signal have different scaling properties [9] [10]. A new method is needed to analyze sleep EEG signals, and one potential candidate is multifractal analysis based on wavelet transform modulus maxima (WTMM), which is a suitable method for obtaining information about dynamic systems without requiring a relatively large number of stationary data points [10] [11].

The aim of this study was to investigate the possibility that human sleep EEG signals can be characterized by a multifractal spectrum using WTMM. To

V.N. Alexandrov et al. (Eds.): ICCS 2006, Part I, LNCS 3991, pp. 164–171, 2006.

investigate whether physiological brain states according to sleep stages affect the phenomenon of multifractality in human sleep EEG dynamics, we compared the multifractal properties of sleep EEG signals obtained during the four sleep stages and REM sleep. Additionally, we used a surrogate data method to assess whether human sleep EEGs have long-range correlations.

2   Materials and Methods

2.1   Materials

The sleep EEG data files used in this study were selected from the sleep-EDF database consisting of physiological data obtained from four subjects with horizontal EOG, EEG, submental-EMG, oronasal airflow, etc. Fpz-Cz EEGs were sampled at 100 Hz, 12 bits per sample [12]. Sleep stages were scored according to the standard criteria of Rechtschaffen and Kales on 30 s epochs [12]. The file names used were sc4002e0, sc4012e0, sc4102e0 and sc4112e0. From these, 150 segments (30 awake, 30 stage 1, 30 stage 2, 30 slow wave sleep (SWS), and 30 REM) with a duration of 30 s (3000 points) were used. Band-pass filter settings were 0.53-40 Hz (12 dB/octave).

2.2   WTMM-Based Multifractal Formalism

As stated above, many physiological time series, including human sleep EEGs, have inhomogeneity and fluctuate in an irregular manner. To characterize sleep EEG signals and to extract local Hurst exponents from them, we used a WTMM-based multifractal formalism. The wavelet transform of sleep EEG signal, $f(x)$, is defined as

$$T_\psi[f](x_0, a) = \frac{1}{a} \int_{-\infty}^{\infty} f(x)\psi(\frac{x - x_0}{a})dx \tag{1}$$

where x_0 is the position parameter, a is the scale parameter, and $\psi(t)$ is the mother wavelet function [11]. In this study, the third derivative of the Gaussian function was used as the analyzing wavelet.

A partition function $Z_q(a)$ was defined as the sum of the power of order q of the local maxima of $|T_\psi[f](x_0, a)|$ at scale a [11].

$$Z_q(a) = \sum_{l \in L(a)} |T_\psi[f](x_{0l}(a), a)|^q \tag{2}$$

where $L(a)$ is the set of the maxima lines l existing at the scale a, and $x_{0l}(a)$ is the position, at the scale a, of the maximum belonging to the line l.

For small scales, it was expressed as

$$Z_q(a) \sim a^{\tau(q)} \tag{3}$$

where $\tau(q)$ is is the multifractal spectrum. For monofractal signals, $\tau(q)$ is a linear function: $\tau(q) = qh(q) - 1$, where $h(q) = d\tau(q)/dq = constant$, and is the global Hurst exponent [9] [10]. For multifractal signals, $\tau(q)$ is a nonlinear function and $h(q) = d\tau(q)/dq = not\ constant$ [9] [10]. For positive values of q, $Z_q(a)$ characterizes the scaling of the large fluctuations and strong singularities, whereas for negative values of q, $Z_q(a)$ characterizes the scaling of the small fluctuations and weak singularities.

Using a Legendre transform, the singularity spectrum $D(h)$ can be expressed as [9] [10] [11].

$$D(h) = q\frac{d\tau(q)}{dq} - \tau(q) \tag{4}$$

$D(h)$ can quantify the statistical properties of the different subsets characterized by different exponents, h. Nonzero $D(h)$ and $h = 0.5$ imply that the fluctuations in signal exhibit uncorrelated behavior. $h > 0.5$ corresponds to correlated behavior, while values of h in the range $0 < h < 0.5$ imply anticorrelated behavior [9] [10].

2.3   Surrogate Time Series and Statistical Analysis

To assess the presence of a long-range correlation in the human sleep EEG time series, we generated a surrogate time series by shuffling and integrating the human sleep EEG signal. The surrogate time series that was generated lost the long-range correlation; however, the distribution of the original human sleep EEG signal was preserved. To examine the differences between the mean values of the local Hurst exponents with maximum $D(h)$ for all sleep stages, one-way analysis of variance (ANOVA) and *post-hoc* analyses were performed. A paired-sample t test was used to compare the mean values of the original data and the surrogate data for each sleep stage. The significance level was 0.05.

3   Results

Multifractal spectra and singularity spectra were computed using the WTMM for the four sleep stages, REM sleep, and surrogate data. Ensemble averaged multifractal spectra, $\tau(q)$, for the four sleep stages, REM sleep and surrogate data were computed, and each singularity spectrum $D(h)$ was computed through a Legendre transform from the ensemble-averaged $\tau(q)$. Figure 1 shows the ensemble-averaged multifractal spectra for the awake stage and REM sleep with surrogate data. The spectra for the awake stage and its surrogate data are shown in the first panel of Fig. 1(a), and the spectra for REM sleep and its surrogate data are shown in the second panel of Fig. 1(b). The two shapes of $\tau(q)$ for the awake stage and its surrogate data, and for REM sleep and its surrogate data were different ($p < 0.0001$) in the range $q > 3$.

Likewise, Figure. 2 shows the ensemble-averaged multifractal spectra for the other sleep stages and their surrogate data. The spectra for stage 1 sleep and its surrogate data are shown in the first panel of Fig. 2(a). The second panel of Fig. 2(b) shows the spectra for stage 2 sleep and its surrogate data, and the third panel of Fig. 2(c) shows the spectra for SWS and its surrogate data. The two shapes of $\tau(q)$ between stage 1 sleep and its surrogate data, between stage 2 sleep and its surrogate data, and between SWS and its surrogate data, were all different ($p < 0.0001$) for q $(-10 < q < +10)$.

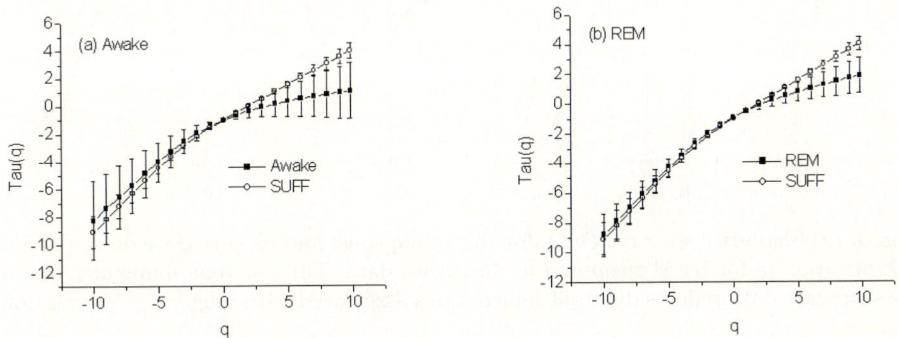

Fig. 1. Ensemble-averaged multifractal spectra for the awake stage and REM sleep with surrogate data. (a) $\tau(q)$ for the awake stage and its surrogate data. (b) $\tau(q)$ for REM sleep and its surrogate data.

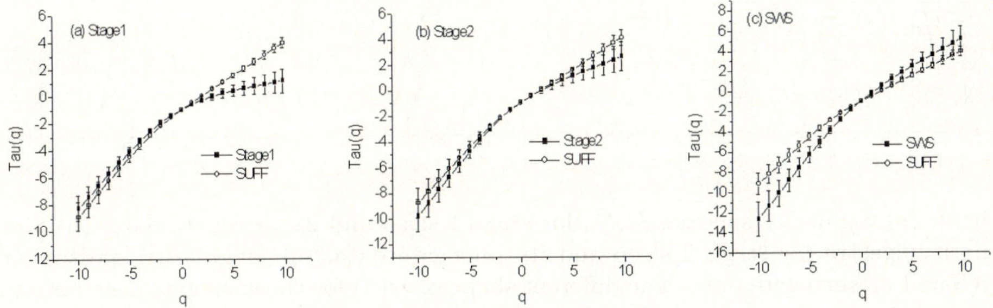

Fig. 2. Ensemble-averaged multifractal spectra for the other sleep stages and their surrogate data. (a) $\tau(q)$ for stage 1 sleep and its surrogate data. (b) $\tau(q)$ for stage 2 sleep and its surrogate data. (c) $\tau(q)$ for SWS and its surrogate data.

The singularity spectra for the four sleep stages, REM sleep and their surrogate data are shown in Fig. 3 and 4. The two shapes of $D(h)$ between each sleep stage and its surrogate data, and between REM sleep and its surrogate data, were significantly different. The range of local Hurst exponents with $D(h)$ greater than 0.75 was $0.24 < h < 1$ for each sleep stage and for REM sleep. However, the local Hurst exponents in the range $0.48 < h < 0.7$ corresponded to a $D(h)$ greater than 0.75 for the surrogate data.

Singularity spectra for the awake stage and its surrogate data are shown in the first panel of Fig. 3(a), and singularity spectra for REM sleep and its surrogate

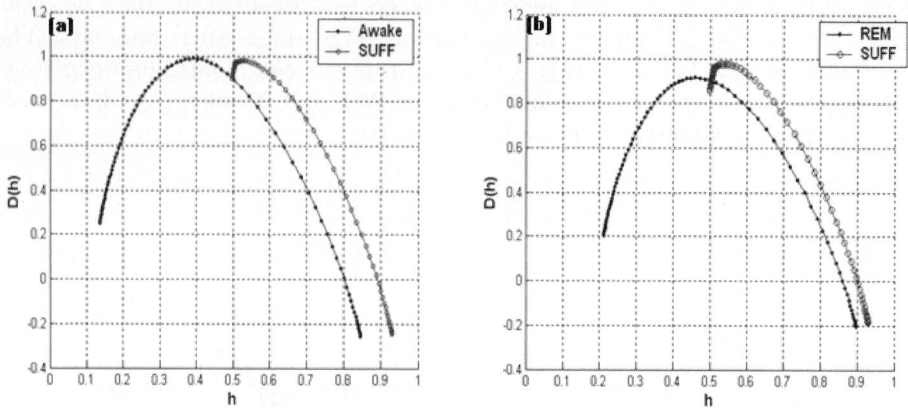

Fig. 3. (a) Sigularity spectra, $D(h)$, for the awake stage and its surrogate data. (b) Singularity spectra for REM sleep and its surrogate data. The different shape of $D(h)$ for the surrogate data reflects different fluatuations associated with long-range correlation.

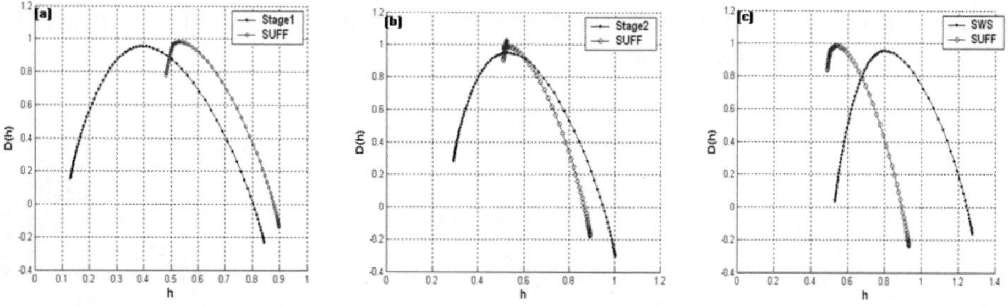

Fig. 4. (a) Sigularity spectra, $D(h)$, for stage 1 sleep and its surrogate data. (b) Singularity spectra for stage 2 sleep and its surrogate data. (c) Singularity spectra for SWS and its surrogate data. The different shape of $D(h)$ for the surrogate data reflects different fluatuations associated with long-range correlation.

data are shown in the second panel of Fig. 3(b). The mean values of local Hurst exponents with maximum $D(h)$ for the awake stage and for REM sleep were 0.39 and 0.46, respectively. The singularity spectra for stage 1 sleep and its surrogate data are shown in the first panel of Fig. 4(a). The second panel of Fig. 4(b) shows the singularity spectra for stage 2 sleep and its surrogate data, and the third panel of Fig. 4(c) shows the singularity spectra for SWS and its surrogate data. The mean values of local Hurst exponents with maximum $D(h)$ for stage 1, stage 2, and SWS were 0.39, 0.52 and 0.80, respectively. The results of the ANOVA are shown in Table 1.

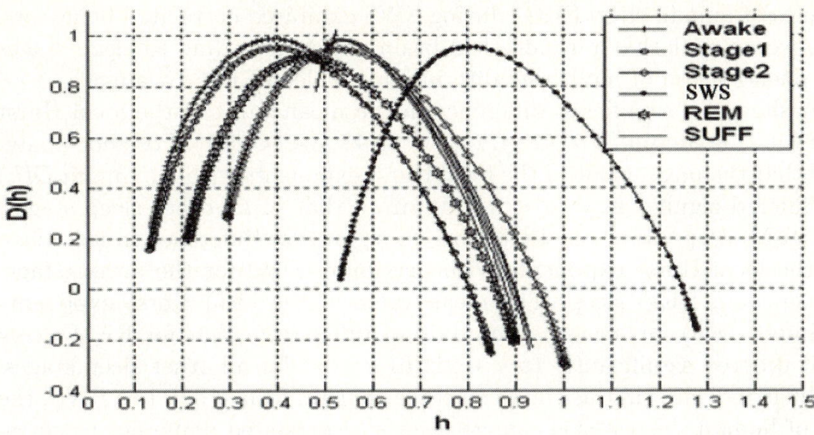

Fig. 5. Comparison of singularity spectra for each sleep stage, REM sleep and the surrogate data

Table 1. Comparison of local Hurst exponents with maximum $D(h)$. The results are presented as mean values $\pm$ SD. ANOVA was performed, followed by Sidak post hoc analysis.

Sleep stages and REM sleep	Mean $\pm$ Std	Sidak post hoc analysis
a) Awake	0.39 ± 0.29	a $<$ c $(p < 0.01)$; a $<$ d $(p < 0.001)$
b) Stage 1	0.39 ± 0.05	b $<$ c $(p < 0.01)$; b $<$ d $(p < 0.0001)$
c) Stage 2	0.52 ± 0.08	c $>$ a, b $(p < 0.01)$; c $<$ d $(p < 0.001)$
d) SWS	0.80 ± 0.11	d $>$ a, b, c, e $(p < 0.0001)$
e) REM	0.46 ± 0.12	e $<$ d $(p < 0.0001)$

4   Discussion and Conclusion

Figure 1 and 2 show the shapes of $\tau(q)$ for the four sleep stages and REM sleep were nonlinear, and Fig. 5 shows the shapes of $D(h)$ for the four sleep stages and REM sleep were broad, indicating that human sleep EEGs could not be characterized as monofractal and are therefore multifractal.

The subsets characterized by local Hurst exponents in the range $0.24 < h < 0.63$ were statistically dominant for the awake stage, stage 1 sleep, and REM sleep. We found that the dynamics of human sleep EEGs taken during the awake stage, stage 1 sleep, and REM sleep exhibited anticorrelated behaviors, indicating that large values are more likely to be followed by small values and vice versa. Statistically dominant subsets were characterized by local Hurst exponents of $0.38 < h < 0.71$ for stage 2 sleep. We found that the dynamics of human sleep EEGs during stage 2 sleep exhibited anticorrelated, correlated, and uncorrelated behaviors. For SWS, the subsets characterized by local Hurst exponents in the range $0.66 < h < 1$ were statistically dominant. We found that the dynamics of human sleep EEGs during SWS exhibited correlated behaviors. Therefore, we suggest that physiological brain states according to sleep stages affect the phenomenon of multifractality in human sleep EEG dynamics.

ANOVA showed a significant difference in the mean value of the local Hurst exponents with maximum $D(h)$ $(F(4, 145) = 35.17, p < 0.0001)$. *Post-hoc* analysis showed that the mean value of the local Hurst exponents with maximum $D(h)$ for SWS differed significantly $(p < 0.0001)$ from that of all other sleep stages, including REM sleep. However, there was no significant difference in the mean values of the local Hurst exponents with maximum $D(h)$ for the awake stage, stage 1 sleep, and REM sleep. The mean value of the local Hurst exponents with maximum $D(h)$ for stage 2 sleep did not differ from that for REM sleep; however, it differed significantly $(p < 0.01)$ from that for all other sleep stages. These findings indicate that as the sleep stage changes from stage 1 to SWS, the dynamics of human sleep EEGs change from anticorrelated dynamics to correlated dynamics. These findings are consistent with previous studies that show the activity of the human brain may become low as SWS sleep occurs.

The results observed after the surrogate test showed significant changes in the shape of $D(h)$ and in the values of the local Hurst exponents with maximum $D(h)$ for the four sleep stages and the REM sleep. The paired-sample t test showed significant differences between the awake stage and the surrogate data, and between the REM sleep and the surrogate data, for the range $q > 3$. These findings indicate that the scaling of the small fluctuations and weak singularities in human sleep EEGs during the awake stage and during REM sleep is similar to that in a simple random walk. However, the scaling of the large fluctuations and strong singularities in human sleep EEGs during the awake stage and REM sleep differ significantly from that in a simple random walk. The results of the paired-sample t test showed significant differences between stage 1 sleep and the surrogate data, between stage 2 sleep and the surrogate data, and between SWS and the surrogate data, for all q $(-10 < q < +10)$. These findings indicate that there are different fluctuations and different characteristics for the singularities between stage 1 sleep and the surrogate data, between stage 2 sleep and the surrogate data, and between SWS and the surrogate data. Therefore, we suggest that human sleep EEGs taken during the four sleep stages and REM sleep have long-range correlations because the surrogate data preserve the distributions of the original time series and remove long-range correlations.

In summary, we investigated the possibility that human sleep EEGs exhibit higher complexity than $1/f$ scaling. Our findings show the existence of inhomogeneities and multifractalities in human sleep EEGs during the four sleep stages and REM sleep. We found that human sleep EEGs taken during the awake stage and during REM sleep exhibited anticorrelated behaviors and had long-range correlations, whereas SWS EEGs exhibited correlated behaviors and had long-range correlations. We suggest that the complexity of human sleep EEGs can better be described by a set of scales rather than a single dominant scale.

Acknowlegements. This study was supported by a Nano-bio technology development project, Ministry of Science & Technology, Republic of Korea (2005-01249).

References

1. Acharya U, R., Faust, O., Kannathal, N., Chua, T., Laxminaryan, S.: Non-linear analysis of EEG signals at various sleep stages. Comput. Meth. Prog. Bio. **80** (2005) 37–45
2. Fell, J., Röschke, J., Beckmann, P.: Deterministic chaos and the first positive Lyapunov expoent: a nonlinear analysis of the human electroencephalogram during sleep. Boil. Cybernet. **69** (1993) 139–146
3. Fell, J., Röschke, J., Mann, K., Schäffner, C.: Discrimination of sleep stages: a comparison between spectral and nonlinear EEG measures. Electroencephalogr. Clin. Neurophysio. **98** (1996) 401–410
4. Kobayashi, T., Misaki,K., Nakagawa, H., Madokoro, S., Ihara, H., Tsuda, K., Umezawa, Y., Murayama, j., Isaki, K.: Non-linear analysis of the sleep EEG. Psychiatry Clin. Neurosci. **53** (1999) 159–161
5. Thomasson, N., Hoeppner,T.J., Webber Jr, C.L., Zbilut, J.P.: Recurrence quantification in epileptic EEGs. Phys. Lett. A. **279**. (2001) 94–101
6. Song, I.H., Lee, D.S., Kim, Sun.I.: Recurrence quantification analysis of sleep electroencephalogram in sleep apnea syndrome in humans. Neurosci. Lett. **366** (2004) 148–153
7. Watters, P.A.: Time-invariant long-range correlations in electroencephalogram dynamics. Int. J. Sys. Sci **31** (2000) 819–825
8. Linkenkaer-Hansen, K., Nikulin, V.V., Palva, S., Ilmoniemi, R.J., Palva, J.M.: Long-range temporal correlations and scaling behavior in human brain oscillations. J. Neurosci. **21** (2001) 1370–1377
9. Ivanov, P.Ch., Amaral, Luís A.N., Goldberger, A.L., Havlin, S., Rosenblum, M.G., Struzik, Z.R., Stanley, H.E.: Multifractality in human heartbeat dynamics. Nature. **399** (1999) 461–465
10. Ivanov, P.Ch., Amaral, Luís A.N., Goldberger, A.L., Havlin, S., Rosenblum, M.G., Stanley, H.E., Struzik, Z.R.: From $1/f$ noise to multifractal cascades in heartbeat dynamics. Chaos, **11** (2001) 641–652
11. Muzy, J.F., Bacry, E., Arneodo, A.: The multifractal formalism revisited with wavelet. Int. J. Bifurc. chaos. **4** (1994) 245–302
12. Mourtazaev, M.S., Kemp, B., Zwinderman, A.H., Kamphuisen, A.C.H.: Age and gender affect different characteristics of slow waves in the sleep EEG. Sleep **18** (1995) 557–564

A Hybrid Feature Selection Algorithm for the QSAR Problem

Marian Viorel Crăciun, Adina Cocu, Luminița Dumitriu, and Cristina Segal

Department of Computer Science and Engineering,
University "Dunărea de Jos" of Galați, 2 Științei, 800146, Romania
{mcraciun, cadin, lumi, csegal}@ugal.ro

Abstract. In this paper we discuss a hybrid feature selection algorithm for the Quantitative Structure Activity Relationship (QSAR) modelling. This is one of the goals in Predictive Toxicology domain, aiming to describe the relations between the chemical structure of a molecule and its biological or toxicological effects, in order to predict the behaviour of new, unknown chemical compounds. We propose a hybridization of the ReliefF algorithm based on a simple fuzzy extension of the value difference metric. The experimental results both on benchmark and real world applications suggest more stability in dealing with noisy data and our preliminary tests give a promising starting point for future research.

1 Introduction

Predictive Toxicology (PT) is one of the newest targets of the Knowledge Discovery in Databases (KDD) domain. Its goal is to describe the relationships between the chemical structure of chemical compounds and biological and toxicological processes. This kind of relationships is known as Structure-Activity Relationships (SARs) [1]. Because there is not a priori information about the existing mathematical relations between the chemical structure of a chemical compound and its effect, a SAR development requires close collaboration between researchers from Toxicology, Chemistry, Biology, Statistics and Artificial Intelligence – Data Mining and Machine Learning domains [2], in order to obtain reliable predictive models.

In real PT problems there is a very important topic which should be considered: the huge number of the chemical descriptors. Data Mining, as a particular step of the Knowledge Discovery in Databases (KDD) process [3] performs on the data subset resulted after the pre-processing procedure. Irrelevant, redundant, noisy and unreliable data have a negative impact, therefore one of the main goals in KDD is to detect these undesirable properties and to eliminate or correct them. This assumes operations of data cleaning, noise reduction and feature selection because the performance of the applied Machine Learning algorithms is strongly related with the quality of the data used.

Besides the removal of the problematic data, feature selection could also have importance in the reduction of the horizontal dimension and, consequently, of the hypothesis space of the data set: less attribute are more comprehensible, the smaller dimension of the input space allows faster training sessions, or even an improvement in predictive performance.

V.N. Alexandrov et al. (Eds.): ICCS 2006, Part I, LNCS 3991, pp. 172–178, 2006.

In the literature there are at least three broad trends in dealing with the problem of features selection: filter, wrapper and embedded methods. The filter based approach evaluate the quality of the attributes separately of a machine learning algorithm and taken into account the contribution of the attribute individually (e.g. the Information Theory based measures: Information Gain, Gain Ratio) or in the context of the other attributes from the training set (e.g. Relief [4], ReliefF [5], RReliefF [6]). Contrary, the wrapper methods use the machine learning algorithm with the purpose of quality evaluation of the attributes. The embedded techniques are nothing else than machine learning algorithms having the ability to extract most suited attributes learning the training data in the same time (e.g. Decision Trees, Bayesian Networks). In real world application wrapper and embedded methods seams to select the proper attributes offering superior performance. But, they are strongly dependent on the learner. On the other side, the filter techniques present an insight of the training data set and its statistical properties.

This paper presents an upgrading of the classical Relief algorithm using fuzzy theory [7]. This simple patch allows the algorithm to evaluate fuzzy attributes as well as the categorical and numerical ones. Also, if the fuzzyfication of the training data set is possible the algorithm will be less sensitive on noise. This Fuzzy Relief is tested and compared with few other methods using a well known data set and a toxicological set.

2 Distance and Fuzzy Sets

Usually, any similarity measure stands on a distance function or a metric. But when this is extended to measure the distance between two subsets of a metric space, the triangular inequality property is sometimes lost.

Let be $X \neq \Phi$ a set and $d: X \times X \rightarrow R_+$ a distance. Then, in this situation D could measure the distance between two sets:

$$D(A, B) = \begin{cases} \inf_{x \in A,\, y \in B} d(x, y), & \text{if } A \neq \Phi, B \neq \Phi \\ 0, & \text{if } A = \Phi \text{ or } B = \Phi \end{cases}, \quad A, B \subset X . \tag{1}$$

Nevertheless, even this is not properly a distance function; it might be used to measure the distance between two fuzzy sets:

$$d(A, B) = \int_0^1 D(A^\alpha, B^\alpha)\, d\alpha . \tag{2}$$

where A and B are two fuzzy sets and A^α, B^α are their α-cuts:

$$A^\alpha = \{t \in X \mid A(t) \geq \alpha\}, \ \alpha > 0 . \tag{3}$$

There are many ways to measure the distance between two fuzzy sets [8], [9]:

– using the difference between the centres of gravity,

$$d(A,B) = |CG(A) - CG(B)|, \ CG(A) = \frac{\int_U x\mu_A(x)dx}{\int_U \mu_A(x)dx} . \tag{4}$$

- Hausdorff distance,

$$D(A,B) = \sup_{\alpha \in [0,1]} \max\{|\, a_2(\alpha) - b_2(\alpha)\,|, |\, a_2(\alpha) - b_2(\alpha)\,|\} \ . \tag{5}$$

where α-cuts $A^{\alpha} = [a_1(\alpha),\ a_2(\alpha)]$ and $B^{\alpha} = [b_1(\alpha),\ b_2(\alpha)]$, $\alpha \in [0,1]$,
- C_{∞} distance,

$$C_{\infty} = ||A - B||_{\infty} = sup\{|A(u) - B(u)| : u \in U\} \ . \tag{6}$$

- Hamming distance,

$$H(A,B) = \int_U |\, A(x) - B(x)\,| \, dx \ . \tag{7}$$

With the support of one of these quasi-distance function, the similarity between fuzzy sets (linguistic variables) can be easily measured.

3 Fuzzy Relief

The Relief (Relief, ReliefF, RReliefF) family methods evaluates the contribution of the values of each attribute in the training data set to distinguish between most similar instances in the same class and in opposite (different) class. Using the difference function each attribute is scored being penalized if the values of the attribute are different for the instances in the same class and recompensed if the values of the attribute are different for the instances in the opposite class.

In order to extend the difference function to handle complex application with fuzzy attributes where the values might be overlapping, one could use one of the distance functions presented in the previous section. In this way, the evaluation of the similarity will be less noise sensitive and more robust. The similarity is determined using the distance (difference) function between the values $v^{(i,j)}$ and $v^{(i,k)}$ of the A_i attribute:

$$\mathrm{diff}\,(i, t_j, t_k) = \begin{cases} 0, & \text{if } A_i \text{ categorial and } v^{(i,j)} = v^{(i,k)} \\[2mm] 1, & \text{if } A_i \text{ categorial and } v^{(i,j)} \neq v^{(i,k)} \\[2mm] \dfrac{d(v^{(i,j)}, v^{(i,k)})}{\mathrm{Max}_i - \mathrm{Min}_i}, & \text{if } A_i \text{ numerical or fuzzy} \end{cases} \tag{8}$$

4 Experimental Results

The method proposed in the previous section is compared with few other classical methods: context dependent (such us ReliefF [5]) and myopic (such as Information Gain, Gain Ratio, and Chi Squared Statistics [11]) which evaluate the individual quality of each attribute.

The data sets used in the experiments were: the classical Iris set [12] and a real world toxicological data set provided by our partners in the framework of a national research project:

Iris data set
This well known data set is very often used as a benchmark and numerous publications reported very good results on it. It consists from 150 iris plant instances described by 4 attributes: sepal length, sepal width, petal length, and petal width. The plants are equal distributed in 3 classes: Iris Setosa, Iris Versicolour and Iris Virginica.

Toxicological data set
The set contains 268 organic chemical compounds characterized by 16 chemical descriptors and their toxic effect (the lethal dose) against 3 small mammals (rat, mouse, and rabbit). The compounds are grouped in 3 toxicity classes (low, medium, high) using the lethal doses. The class distribution is not equal this time. There are more chemical compounds in the most toxic class (156) than in the other two (72 and 40, respectively).

The data sets are artificially altered with different levels of noise between 0 and 25%. Not all the attributes were affected by noise. Only the values of 50% from the characteristics, the most predictive ones, discovered by the majority of the feature selection methods on the noiseless sets, were randomly modified. (e.g. for the Iris dataset the last two attributes are strongly correlated with the class and will be affected by noise, for the toxicology data set, the first eight descriptors are not altered).

The feature selection methods used in this set of experiments are:

Information Gain (IG) – evaluates the worth of an attribute A by measuring the information gain with respect to the class C:

$$Gain(A) = I(A;C) = H_C - H_{C|A} . \tag{9}$$

$$Gain(A) = -\sum_k p_{k.} \log p_{k.} - \sum_j p_{.j} \sum_k p_{k|j} \log p_{k|j} . \tag{10}$$

Gain Ratio (GR) – evaluates the worth of an attribute by measuring the gain ratio (the information gain normalized by the entropy of the attribute) with respect to the class:

$$GainR(A) = \frac{Gain(A)}{H_A} . \tag{11}$$

Chi squared statistics (Chi) – evaluates the worth of an attribute by computing the value of the chi-squared statistic with respect to the class.

ReliefF – evaluates the worth of an attribute by repeatedly sampling an instance and considering the value of the given attribute for the nearest instance of the same and different class.

Bayesian Networks (BN) – evaluates the importance of one attribute by observing how much the predictive performance of the model drops if we remove the corresponding variable. An important feature makes the prediction accuracy of the model to drop down when it is left out. On the other hand, if it removing a feature not affect to much the performance of the selected model, then it is less important. The predictive performance of the classifier is estimated with leave-one-out (LOO) cross validation method [13, 14].

FuzzyRelief – is the *Relief* algorithm working with fuzzy variables (after the fuzzyfication of the training data).

In order to compare the different filters for attribute selection, the well known Machine Learning method k-Nearest Neighbours (kNN), with k=10 and inverse distance weighting is used. The results of the experiments are obtained with *Weka* [11] and *B-Course* [15] software. Generally, the predictive performances of machine learning methods decrease and the importance of feature selection appears when the uncertainty (noise) in data increases.

The next two tables show the cross-validated (10-fold) performances of kNN in different circumstances. In both cases, first row contains the predictive performance obtained using all the available attributes for comparison purposes. The next rows include the kNN performance obtained in combination with different feature selection techniques: only half of the attributes, those with the higher ranks, are used for predictions.

Table 1 illustrates the predictive performance on the *Iris* dataset. The results of kNN working with or without feature selection methods are similar for both clean and very low noise level (5%) data. For noise level higher than 10%, all the attributes are a better choice in all situations. This is explainable taken into account the fact that the unaltered characteristics can compensate the noisy ones. The proposed hybrid selection technique, *FuzzyRelief*, demonstrates an analogous behaviour with almost all the other feature selection techniques (except *BN*).

Performances on the toxicological data are shown in Table 2. In this case the strength and flexibility of *FuzzyRelief* are more obvious. The ability of fuzzy techniques to deal with imperfect data (noise over 10%) in combination with a strong and well known feature selection method such as *ReliefF* yield to the appropriate attributes subset to describe the QSARs. The kNN accuracy in classification is enhanced.

Table 1. Prediction accuracy – Iris data set

Feature selection + kNN	Noise level					
	0%	5%	10%	15%	20%	25%
none+kNN	95%	92%	89%	90%	83%	81%
IG+kNN	95%	93%	82%	86%	79%	79%
GR+kNN	95%	93%	82%	86%	79%	73%
Chi+kNN	95%	93%	82%	86%	80%	79%
BN+kNN	95%	93%	86%	89%	80%	73%
ReliefF+kNN	95%	91%	83%	83%	79%	79%
FuzzyRelief+kNN	95%	93%	83%	83%	79%	79%

Table 2. Prediction accuracy – toxicological data set

Feature selection + kNN	Noise level					
	0%	5%	10%	15%	20%	25%
none +kNN	63%	56%	59%	57%	62%	56%
IG+kNN	57%	57%	57%	57%	62%	57%
GR+kNN	57%	57%	57%	57%	62%	57%
Chi+kNN	57%	57%	57%	57%	62%	57%
BN+kNN	59%	58%	57%	62%	63%	61%
ReliefF+kNN	65%	60%	59%	55%	61%	60%
FuzzyRelief+kNN	63%	59%	59%	64%	63%	62%

5 Conclusions

The results so far show the strength and the flexibility of *Fuzzy Relief* when possible uncertain data is used for training predictive data mining techniques. Its ranking performance proved in the case studies presented is comparable and some times better than the performances of other filter methods especially when the data is affected by noise. One of its drawbacks is the time needed to evaluate the similarity between the linguistic variables (between the fuzzy sets). Of course, depending on the target problem of the KDD process, the tradeoffs between the quality of the solutions and data, the dimensionality of the data sets and the available time will dictate the right strategy.

The future research will be focused in evaluating different similarity measures between the different fuzzy attributes and in testing the method on more real world data mining applications. Also, another future interest will be focused on evaluating different classifying methods in combination with *FuzzyRelief*.

Acknowledgement

This work was partially funded by the PRETOX Romanian project under the contract 407/2004 MENER.

References

1. Hansch, C. Hoekman, D., Leo, A., Zhang, L., Li, P., The expanding role of quantitative structure-activity relationship (QSAR) in toxicology, Toxicology Letters 79 (1995) 45-53
2. Y-t. Woo, A Toxicologist's View and Evaluation, Predictive Toxicology Challenge (PTC) 2000-2001
3. Fayad, U.M., Piatesky-Shapiro, G., Smyth, P., Uthurusamy, eds. Advances in Knowledge Discovery and Data Mining, (AAAI/MIT Press Menlo Park, CA), (1996)
4. Kira, K., Rendell, L. A., A practical approach to feature selection, In International Conference on machine learning, Morgan Kaufmann , (1992) 249-256
5. 5 Kononenko, I., Estimating attributes: Analysis and Extension of Relief. In Proc. of ECML'94, the Seventh European Conference in Machine Learning, Springer-Verlag, (1994) 171-182

6. Robnik Sikonja, M. and Kononenko, I., An adaptation of Relief for attribute estimation in regression, In Fisher, D., editor, Machine Learning: Proceedings of the Fourteenth International Conference (ICML'97), Morgan Kaufmann Publishers (1997) 296–304

7. Zadeh, L.A., Fuzzy Logic and Approximate Reasoning, Synthese, 30 (1975) 407-428

8. Pal, Sankar K., Shiu Simon C. K., Foundations of Soft Case-Based Reasoning, John Wiley and Sons, (2004)

9. Fuller, R., Introduction to Neuro-Fuzzy Systems, Advances in Soft Computing Series, Springer-Verlag Berlin, (1999)

10. Wilson, D.R., Martinez, T.R., Improved Heterogeneous Distance Functions, Journal of Artificial Intelligence Research (JAIR), 6-1 (1997) 1-34

11. Witten, I.H., Frank E., Data Mining: Practical Machine Learning Tools and Techniques with Java Implementations, Morgan Kaufmann Publishers, San Francisco, CA (1999)

12. Fisher, R.A., Iris Plant Database

13. Domingos P., Pazzani M., Beyond Independence: Conditions for the optimality of the simple bayeisan classifier, Proceeding of the Thirteenth International Conference on Machine Learning, (1996)

14. Elkan Charles, Naïve Bayesian Learning, Technical Report, University of California, (1997)

15. Myllymäki P., Silander T., Tirri H., Uronen P., B-Course: A Web-Based Tool for Bayesian and Causal Data Analysis. International Journal on Artificial Intelligence Tools, Vol 11, No. 3 (2002) 369-387

Sequential Probability Ratio Test (SPRT) for Dynamic Radiation Level Determination – Preliminary Assessment

Ding Yuan and Warnick Kernan

Remote Sensing Laboratory-Nellis
P. O. Box 98521, M/S RSL-11, Las Vegas, NV 89193-8521
{yuand, kernanwj}@nv.doe.gov

Abstract. A Sequential Probability Ratio Test (SPRT) algorithm for reliable and fast determination of a relative radiation level in a field environment has been developed. The background and the radioactive anomaly are assumed to follow the normal and Poisson distributions, respectively. The SPRT formulation has been derived and simplified based on these assumptions. The preliminary evaluation suggests that the algorithm, while offering confident estimations for the log-scaled radiation level, promises the additional advantage of reduction in sampling sizes, particularly in areas with a high radiation level.

1 Introduction

Reliable and fast estimation of the local relative radiation level with respect to that of the regional natural background, under prescribed confidence levels, is the central focus of many environmental radiological surveys. A local relative radiation level can be conventionally interpreted as a number of standard deviations away from the background radiation mean. The key issues are reliability and speed. In order to obtain a reliable estimate for a local relative radiation level, the conventional statistical wisdom teaches us that we need more samples. However, a high radiation anomaly may pose a significant health threat to the field surveyors. Conventional health wisdom tells us to leave the high radiation area as soon as possible. Therefore, we need a radiation level estimation algorithm that is both reliable for given confidence levels and fast for high radiation areas.

In our recent research, we tailored Wald's Sequential Probability Ratio Test (SPRT) [1] and developed an algorithm for the dynamic determination of the local relative radiation level. A preliminary experiment in the laboratory environment, using a common industrial radiation source, suggests that the algorithm is promising. It provided a confident estimate for the local radiation level, and reduced the sampling size requirement for high radiation spots.

2 Population Assumptions

Reconnaissance environmental radiation surveys are commonly performed by hand-held gamma-detection devices, such as Geiger counters. These gamma-detection devices produce readings of gross gamma ray counts, or the gross count (GC) detected during a given time interval [2].

V.N. Alexandrov et al. (Eds.): ICCS 2006, Part I, LNCS 3991, pp. 179–187, 2006.
© Springer-Verlag Berlin Heidelberg 2006

For a given high radiation anomaly, GC may be assumed to follow a Poisson Distribution [2] with PDF:

$$f_{\mu_1}(x) = P(X = x) = \frac{\mu_1^x\, e^{-\mu_1}}{x!}\ ,\ x = 0,\ 1, 2,\ ...,\ n \tag{1}$$

Where $\mu_1 = np$. n is a large number (number of atoms in the observed physical sample) and p is the decay rate for an individual particle for a given observation time interval. When n is large, it can be shown that:

$$f_{\mu_1}(x)\ \rightarrow\ N(\mu_1, \mu_1) \tag{2}$$

Where μ_1 is the mean GC of the physical sample.

For regional radiation background, we assume that GC follows a simple Normal distribution $N(\mu_0, \sigma_0^2)$ with probability distribution function (PDF):

$$f_{\mu_0, \sigma_0}(x) = \frac{1}{\sqrt{2\pi}\,\sigma_0}\, e^{-\frac{(x-\mu_0)^2}{2\sigma_0^2}}\ ,\quad -\infty < x < \infty \tag{3}$$

This different treatment is due to uncertainty about the number of counts in the background sample, and if it is large enough for Poison-Normal approximation that offers only one parameter for distribution fitting. Normal distribution, on the other hand, has two parameters to choose and should always make a better fit to the background samples.

3 Absolute and Discrete Radiation Levels

For measuring the absolute radiation level, the sample mean μ_1 is a natural choice. In a field survey, we are usually more interested in a discrete measure relative to the background radiation. For this reason, we shall introduce Log Departing Coefficient (LGC) as follows. Suppose that b (> 0) is a pre-selected log base, and ρ (>0) is a pre-selected scaling factor. Define (for $l \geq 1$):

$$\text{LGC} = l,\ \text{if } \mu_0 + b^{l-1}\rho\sigma_0 < \mu_1 \leq \mu_0 + b^l \rho\sigma_0 \tag{4}$$

For LGC$\geq$1, we have

$$\text{LGC} = \text{Ceiling}\left[\text{Log}_b\left[\frac{\mu_1 - \mu_0}{\rho\,\sigma_0}\right]\right] \tag{5}$$

Particularly, for a region with low radiation background, we may take b=2 and ρ=1, then:

$$\text{LGC} = \text{Ceiling}\left[\text{Log}_2\left[\frac{\mu_1 - \mu_0}{\sigma_0}\right]\right] \tag{6}$$

When LGC=0, we can say the sample mean is no more than two standard deviations from the regional mean. When LGC=1, we say that the sample mean is at least two standard deviations from the regional mean, but no more than four standard deviations from the regional mean, etc. For a region with higher radiation background, we may adjust factor ρ, making $\rho=0.5$ for compensating the growing speed of LGC. LGC defined this way can be used as a log-scaled discrete measure for the radiation departing level (L) from the regional background.

4 Alternative Hypotheses

Let's assume that we collect observations one at a time, using a handheld device such as a Geiger Counter, in an environmental reconnaissance radiological survey situation. We denote x_i the i'th observation, where the observation is the Gamma Gross Counts (GC) for fixed-length time intervals. Essentially for a sample, we have two possibilities:

(1) That the sample was from the background radiation population – the Null Hypothesis (H_0); and

(2) That the sample was from an anomaly population with higher radiation strength – the Alternative Hypothesis (H_1).

If we accept the null hypothesis H_0, then we continue our field survey. If we accept the alternative hypothesis H_1, then we mark the area as an anomaly for future detailed work and move on to the next spot.

Let α and β be the type I (false positive) and type II (false negative) errors associated with the decisions respectively,

$$\alpha = P\{\text{Accept } H_1 \text{ when } H_0 \text{ is true}\} = P(H_1 \mid H_0) \tag{7}$$

$$\beta = P\{\text{Accept } H_0 \text{ when } H_1 \text{ is true}\} = P(H_0 \mid H_1) \tag{8}$$

then $1-\alpha$ is the confidence level of accepting H_0 and $1-\beta$ is the confidence level of accepting H_1.

5 SPRT Basis

Wald's SPRT method [1] has the advantage of handling sequential sampling data. Let $\{x_1, x_2, x_3, ..., x_n\}$ be an (independently identically distributed) fresh data set collected since the last decision and a new decision is yet to be made. Assuming $f_0(x)$ and $f_1(x)$ are the PDFs for H_0 and H_1 respectively, we can construct a conventional Logarithmic Likelihood Ratio (LLR):

$$\Lambda_n = \text{LLR}(x_1, x_2, ..., x_n) = \text{Log} \frac{\prod_{i=1}^{n} f_1(x_i)}{\prod_{i=1}^{n} f_0(x_i)} = \sum_{i=1}^{n} \text{Log} \frac{f_1(x_i)}{f_0(x_i)} \tag{9}$$

The Wald's SPRT has the following form (although his original A and B are somehow reversed from what are commonly used today):

1) If $\Lambda_n < A$ then accept H_0, (and directly move on to next spot).

2) If $A \le \Lambda_n \le B$, request additional measurement made for the same spot.

3) If $\Lambda_n > B$ then accept H_1, (and mark anomaly etc., move on to the next spot).

Where A and B are two constants satisfying inequalities:

$$\mathrm{Log}\,\frac{\beta}{1-\alpha} \le A < B \le \mathrm{Log}\,\frac{1-\beta}{\alpha} \tag{10}$$

A common practice is simply to take:

$$A = \mathrm{Log}\,\frac{\beta}{1-\alpha} \tag{11}$$

$$B = \mathrm{Log}\,\frac{1-\beta}{\alpha} \tag{12}$$

It is clear that the parameters A and B are related only to the strengths (α, β), or confidences $(1\text{-}\alpha, 1\text{-}\beta)$, of the test, but are independent of the actual forms of the distribution functions.

6 SPRT Formulation

If using the PDFs discussed early for H_0 and H_1 respectively (using normal approximation for Poisson distribution), then

$$\mathrm{Log}\,\frac{f_1(x)}{f_0(x)} = \mathrm{Log}\,\frac{\sigma_0}{\sqrt{\mu_1}} + \frac{(x-\mu_0)^2}{2\,\sigma_0^2} - \frac{(x-\mu_1)^2}{2\,\mu_1} \tag{13}$$

Substitute these PDFs in the Log-Likelihood Ratio, we have:

$$\Lambda_n = \mathrm{LLR}(x_1,\, x_2,\, ...,\, x_n)$$

$$= \sum_{i=1}^{n}\left(\mathrm{Log}\,\frac{\sigma_0}{\sqrt{\mu_1}} + \frac{(x_i-\mu_0)^2}{2\,\sigma_0^2} - \frac{(x_i-\mu_1)^2}{2\,\mu_1}\right) \tag{14}$$

Denote sample mean and sum of squares respectively:

$$\bar{x} = \frac{1}{n}\sum_{i=1}^{n} x_i \tag{15}$$

$$S = \sum_{i=1}^{n}(x_i - \bar{x})^2 \tag{16}$$

Through some simplification, and using sample mean $\bar{x}$ for estimating μ_1, we should finally have:

$$\Lambda_n \overset{\Delta}{=} V_n^{(0)} - V_n^{(1)} \tag{17}$$

where

$$V_n^{(0)} = n \operatorname{Log} \sigma_0 + \frac{S + n(\bar{x} - \mu_0)^2}{2\sigma_0^2} \tag{18}$$

$$V_n^{(1)} = \frac{n}{2} \operatorname{Log} \bar{x} + \frac{S}{2\bar{x}} \tag{19}$$

7 Computational Details

We assume that the regional background population mean μ_0 and the standard deviations σ_0 have been estimated, and also also assume that the test strengths (α, β), or confidences $(1\text{-}\alpha, 1\text{-}\beta)$ are given. Subsequently constants A and B can be calculated using formula (11) and (12). For a new spot or physical sample, we take first $i_{\min}$ independent measures, then compute:

$$x^{(n)} = \frac{1}{n} \sum_{i=1}^{n} x_i \tag{20}$$

$$\mu_1 = x^{(n)} \tag{21}$$

$$S^{(n)} = \sum_{i=1}^{n} \left(x_i - x^{(n)}\right)^2 \tag{22}$$

$$V_n^{(1)} = \frac{n}{2} \operatorname{Log} \mu_1 + \frac{S^{(n)}}{2\mu_1} \tag{23}$$

$$V_n^{(0)} = n \operatorname{Log} \sigma_0 + \frac{S^{(n)} + n(x^{(n)} - \mu_0)^2}{2\sigma_0^2} \tag{24}$$

$$\Lambda_n = V_n^{(0)} - V_n^{(1)} \tag{25}$$

The decision and estimation for the local radiation level L are made as follows:

1) If $\Lambda_n < A$ then accept H_0 and assign $L = 0$, output both μ_1 and L;

2) If $A \leq \Lambda_n \leq B$, request additional measurement made for the same spot.

3) If $\Lambda_n > B$ then accept H_1 and compute:

$$L = \mathrm{LGC} = \operatorname{Ceiling}\left[\operatorname{Log}_b\left[\frac{\mu_1 - \mu_0}{\rho\,\sigma_0}\right]\right] \tag{26}$$

Now output both μ_1 and L, and move to the next physical sample or spot.

Although SPRT sampling sequence terminates with probability 1, it does require a stop limit. For case 2, we adapted Wald's stop limit[1]: if $n = i_{max}$ and we still have $A \le \Lambda_n \le B$, then:

$$\text{if } \Lambda_{i_{max}} \le \frac{A+B}{2}, \text{ we accept } H_0 \text{ and assign } L = 0;$$ (27)

$$\text{if } \Lambda_{i_{max}} > \frac{A+B}{2}, \text{ we accept } H_1 \text{ and compute } L = LGC$$ (28)

Apparently, all local radiation level Ls computed this way had confidence levels $(1-\alpha, 1-\beta)$.

This algorithm was implemented in Mathematica[3] for quick evaluation. In the implementation, we set $i_{min} = 5$, $i_{max} = 11$, α=0.05 and β=0.05.

8   Experimental Data

Experimental data were collected in a laboratory environment using a typical industrial Cs137 source. A device similar to a Geiger counter was used for measuring gamma GC. The counting interval of the device was set at half of a second. Figure 1 shows the half-second background gamma GC data measure of the lab ground. Background radiation has a mean $\mu_0 = 773$ and standard deviation $\sigma_0 = 33$. (This also suggests $33 = \sigma_0 \ne \sqrt{\mu_0} = 28$ for background radiation, and therefore anomaly source and background should be handled by different types of distributions.)

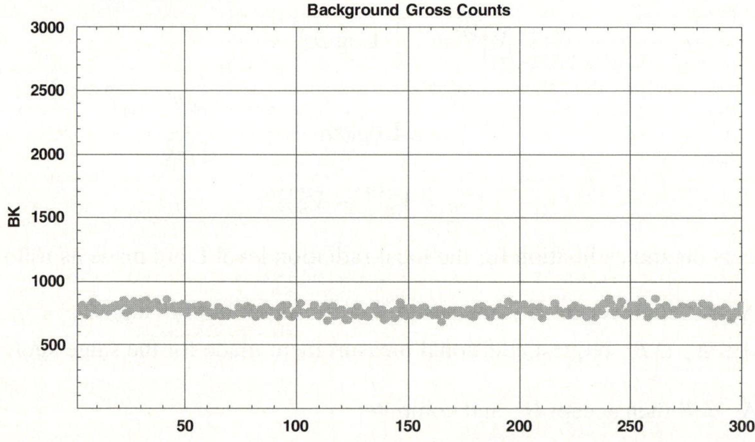

Fig. 1. Background gamma GC data, where the x-axis is the data-point position in the sequence

During the experiment, a common industrial source, Cs137, was placed on top of a table in the lab. The source gamma strength was measured at different distances from 1 to 50 feet for simulating sources of different strengths (Figure 2).

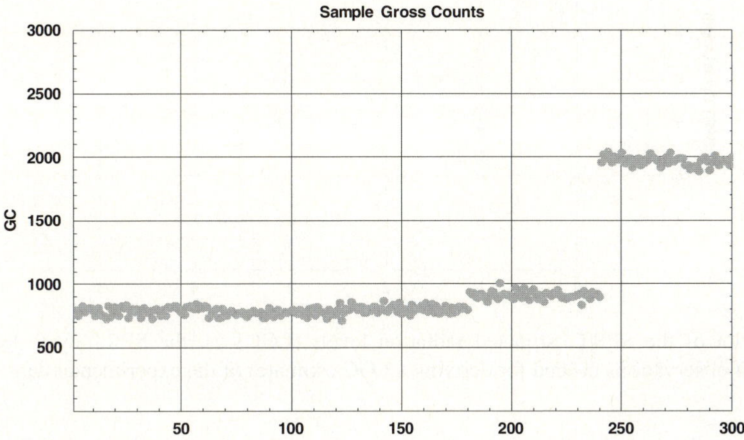

Fig. 2. Physical sample gamma GC data measured at different distances, where the x-axis is the data-point position in the sequence

9 Results and Analysis

The SPRT derived radiation levels (LGC) of the experimental data at different strengths (or distances) are shown in Figure 3. Comparing Figure 2 and 3, it is clear that the SPRT radiation level or LGC is exponentially correlated to the relative strength of the anomaly with respect to the background.

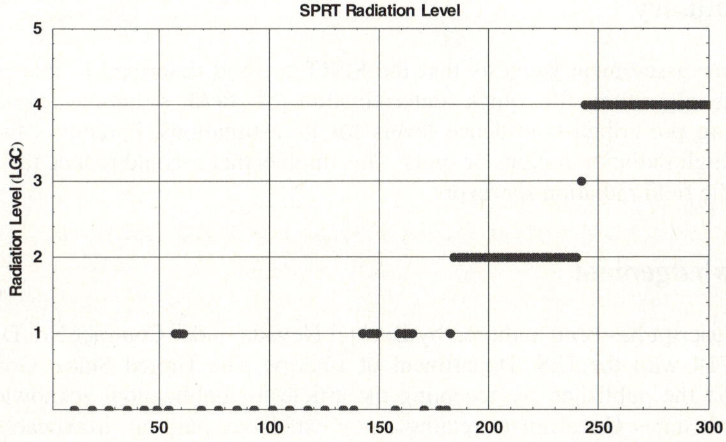

Fig. 3. SPRT derived anomaly radiation level (LGC) from the experimental data

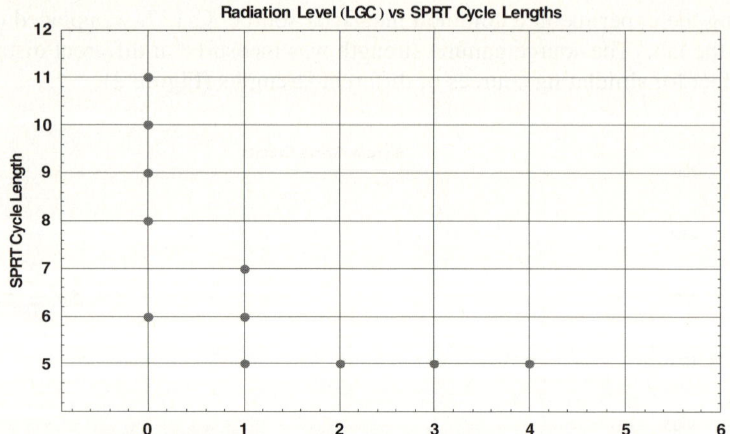

Fig. 4. Plot of the SPRT estimated radiation levels (LGC) vs. the SPRT cycle length (the number of observations needed for deriving a LGC estimate) of the experimental data

Further, the relationship between the estimated radiation level (LGC) and the SPRT cycle length, (i.e., the number of observations needed for deriving an estimate for radiation level) is analyze in Figure 4.

It is clear that for high-radiation anomaly ($LGC \geq 2$), the SPRT needs the minimum sample size $i_{min} = 5$ to derive the needed LGC with the given confidence levels. For low-radiation anomaly ($LGC = 1$), the SPRT needs no more than seven observations. For background confirmation ($LGC = 0$), the SPRT may need a longer observation cycle, anywhere from 6 to 11 observations. This is precisely a property we would like to have for a field radiation instrument, minimizing the time needed for high radiation anomaly sampling.

10 Summary

Preliminary assessment suggests that the SPRT method described in this paper is a promising algorithm for quick determination of field radiation levels. While maintaining prescribed confidence levels for its estimations, it reduces the sample sizes for high radiation regions or spots. This implies that it could reduce the exposure time for the field radiation surveyors.

Acknowledgement

This manuscript has been authored by Bechtel Nevada under Contract No. DE-AC08-96NV11718 with the U.S. Department of Energy. The United States Government retains and the publisher, by accepting the article for publication, acknowledge that the United States Government retains a non-exclusive, paid-up, irrevocable, world-wide license to publish or reproduce the published form of this manuscript, or allow others to do so, for United States Government purposes.

References

1. Wald, A., 2004, Sequential Analysis, Dover Ed., Dover Publications, Mineola, NY, USA.
2. Knoll, G. F., 1999, Radiation Detection and Measurement, 3rd Ed., John Wiley & Sons, Hoboken, NJ, USA.
3. Stephen Wolfram, 2003, The Mathmatica Book 5ed., Wolfram Media, Champaign, IL, USA.

Knowledge-Based Multiclass Support Vector Machines Applied to Vertical Two-Phase Flow

Olutayo O. Oladunni[1], Theodore B. Trafalis[1], and Dimitrios V. Papavassiliou[2]

[1] School of Industrial Engineering, The University of Oklahoma
202 West Boyd, CEC 124 Norman, OK 73019 USA
`{tayo, ttrafalis}@ou.edu`
[2] School of Chemical, Biological, and Materials Engineering, The University of Oklahoma
100 East Boyd, SEC T-335 Norman, OK 73019 USA
`dvpapava@ou.edu`

Abstract. We present a knowledge-based linear multi-classification model for vertical two-phase flow regimes in pipes with the transition equations of McQuillan & Whalley [1] used as prior knowledge. Using published experimental data for gas-liquid vertical two-phase flows, and expert domain knowledge of the two-phase flow regime transitions, the goal of the model is to identify the transition region between different flow regimes. The prior knowledge is in the form of polyhedral sets belonging to one or more classes. The resulting formulation leads to a Tikhonov regularization problem that can be solved using matrix or iterative methods.

1 Introduction

Multiphase flow in a pipe causes certain flow patterns to appear depending on pipe size, flow rates, fluid properties, and pipe inclination angle (when appropriate). Considerable progress has been made in defining such flow patterns [1-5], however there is no exact theory for their characterization. There is often disagreement about the transition boundaries between flow patterns making the selection of appropriate flow correlations for the description of the flow difficult. Therefore, it is important to develop a flow pattern model that minimizes the rate of misclassification errors (i.e., predicting the wrong flow regime for a given set of flow data) as well as extend the applicability of any new multiphase flow correlation to different pipe sizes, flow rates and fluid properties.

More recently, attempts have been made to identify multiphase flow regimes using non-linear classifiers derived from machine learning algorithms [6-8]. Trafalis et al. [9] employed a multi-classification support vector machine (MSVM) model, in which the superficial velocities and pipe sizes were used to detect flow regime transitions for two-phase flow in pipes. The model was data driven, and it outperformed the theoretical vertical flow correlation in terms of correct flow regime classification.

The primary objective of the present paper is to extend the model of [9] by exploring the effects of expert knowledge on the MSVM model. If one has prior information regarding a class or label, representing such knowledge and incorporating it into a classification problem can lead to a more robust solution. The vertical two-phase flow

V.N. Alexandrov et al. (Eds.): ICCS 2006, Part I, LNCS 3991, pp. 188–195, 2006.

regimes considered herein are bubble flow (class 1), intermittent flow (slug & churn flow, class 2), and annular flow (class 3). These flow regimes occur for a given set of parameters, (i.e., flow rates, fluid properties and pipe diameter). The theoretical correlations of McQuillan and Whalley [1], which were found to perform well for vertical two phase flow in [9], are used here as prior knowledge.

2   Linear Multiclassification Tikhonov Regularization Knowledge-Based Support Vector Machine: Pairwise Knowledge Set Formulation

Recently, prior knowledge formulation, incorporation, and solution have been studied in both the context of function approximations [10] and in the context of classifiers [11]. Here, we consider a problem of classifying data sets in R^n that are represented by a data matrix $A^i \in R^{m_i \times n}$, where $i = 1,..,k$ ($k \geq 2$ classes). Let A^i be an $m_i \times n$ matrix whose rows are points in the ith class, and m_i is the number of data in class i. Let A^j be an $m_j \times n$ matrix whose rows are points in the jth class, and m_j is the number of data in class j, and $y^{ij} = \pm 1$ for classes i and j, respectively. This problem can be modeled through the following optimization problem:

$$\min_{w,\gamma,\xi} \quad \frac{\lambda}{2}\sum_{i<j}^{k}\left\|w^{ij}\right\|^2 + \frac{1}{2}\sum_{i<j}^{k}\sum_{t=1}^{m_{ij}}(\xi_t^{ij})^2$$

$$s.t. \quad \xi_t^{ij} = y_t^{ij}(A_t^{ij}w^{ij} - \gamma^{ij}) - 1, \quad t = 1,....,m_{ij} \tag{2.1}$$

$$\text{where } m_{ij} = m_i + m_j, \quad A^{ij} = \begin{pmatrix} A^i \\ A^j \end{pmatrix}, \quad y^{ij} = \begin{pmatrix} y^i \\ y^j \end{pmatrix} = \begin{pmatrix} +1 \\ -1 \end{pmatrix}$$

Above, λ is the tradeoff constant, and ξ^{ij} are error slacks measuring the deviation of points from their respective bounding planes. Classification weights (w^{ij}, γ^{ij}) characterize the optimal separating planes

$$x^T w^{ij} = \gamma^{ij}, \quad i < j . \tag{2.2}$$

The locations of the optimal separating planes relative to the origin are determined by the value of γ^{ij}. Problem (2.1) is minimized parametrically with the tradeoff constant λ, which accounts for the tradeoff between minimum norm and minimum misclassification error.

Now, suppose that in addition to the classification problem, there is prior information belonging to one or more categories. The knowledge sets in n dimensional space are given in the form of a polyhedral set determined by the set of linear equalities and linear inequalities. The given knowledge sets are (see Fig. 1)

$$\{x \mid B^i x \leq b^i\} \text{ or } \{x \mid \overline{B}^i x = \overline{b}^i\}, \text{ belonging to class } i, \text{ and} \tag{2.3}$$

g_i or $\overline{g}_i$ are the number of prior knowledge (equality or inequality) constraints in class i.

$$\{x \mid B^j x \le b^j\} \text{ or } \{x \mid \overline{B}^j x = \overline{b}^j\}, \text{ belonging to class } j, \text{ and} \tag{2.4}$$

d_j or $\overline{d}_j$ are the number of prior knowledge (equality or inequality) constraints in class j.

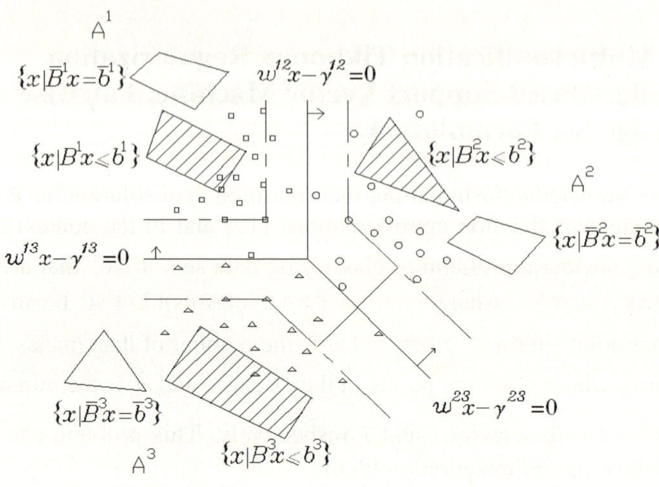

Fig. 1. A knowledge-based classification diagram

One can rewrite the constraints, and create an unconstrained optimization problem, called Linear Multi-classification Tikhonov Regularization Knowledge-Based Support Vector Machine ($L_M T_R KSVM$) [12, 13, 14].

$$\min_{w,\gamma,\xi,u,v} f_{L_M T_R KSVM}(w,\gamma,\xi,u,v) = \begin{bmatrix} \dfrac{\lambda}{2}\|w\|^2 + \dfrac{1}{2}\|Aw + E\gamma - e\|^2 + \\[2mm] \dfrac{1}{2}\left[\|B_u^T u + I_u w\|^2 + \|B_v^T v - I_v w\|^2\right] + \\[2mm] \dfrac{1}{2}\left[\|B_{bu}^T u + E_u \gamma + e_u\|^2 + \|B_{bv}^T v - E_v \gamma + e_v\|^2\right] \end{bmatrix}. \tag{2.5}$$

Where $u = [u^{ijT},....,u^{(k-1)kT},\overline{u}^{ijT},....,\overline{u}^{(k-1)kT}]^T$ is the vector of all multipliers referring to the ith class, and $v = [v^{ijT},....,v^{(k-1)kT},\overline{v}^{ijT},....,\overline{v}^{(k-1)kT}]^T$ is the vector of all multipliers referring to the jth class. ξ is a residual error vector accounting for the training data error, and vectors r,s,p,σ are residual error vectors accounting for the knowledge set error (i.e., violation of the knowledge constraints). The following matrices are defined for all $i < j$, $i,j = 1,....k$ classes (note that matrices describe the formation of the prior knowledge): Matrices $A^i \in R^{m_i \times n}$, $A^j \in R^{m_j \times n}$, $i < j$ are matrices whose rows belong to the ith and j^{th} class respectively; $e^i, e^j \in R^{m_i, m_j \times 1}$ are vectors of ones. Matrices

B_u and B_v are diagonal block matrices whose diagonals contain knowledge sets belonging to the ith (or j^{th}) class. The ith (or j^{th}) knowledge set is derived from the inequality and equality prior knowledge constraints. The diagonals of B_u, B_v are $B_{ij}^i, \overline{B}_{ij}^i \in R^{g_i \times n}$ ($B_{ij}^j, \overline{B}_{ij}^j \in R^{g_j \times n}$). Matrices B_{bu}, B_{bv} are matrices consisting of vectors $b_{ij}^i, \overline{b}_{ij}^i \in R^{g_i \times 1}$ ($b_{ij}^j, \overline{b}_{ij}^j \in R^{g_j \times 1}$) and each component of these vectors describes the bounds of the knowledge set. Matrices I_u, I_v are block matrices consisting of identity matrices $I_{uij}, \overline{I}_{uij} \in R^{n \times n}$ ($I_{vij}, \overline{I}_{vij} \in R^{n \times n}$). Matrices E_u, E_v are matrices consisting of entries $e_{uij}, \overline{e}_{uij} \in R$ ($e_{vij}, \overline{e}_{vij} \in R$), where $e_{uij}, \overline{e}_{uij}$ is equal to one. Vector e_u (e_v) is a vector of ones, where each entry corresponds to a vector $b_{ij}^i, \overline{b}_{ij}^i \in R^{g_i \times 1}$ ($b_{ij}^j, \overline{b}_{ij}^j \in R^{g_j \times 1}$). Each ith and j^{th} entry into B_u, B_{bu}, I_u, E_u, e_u, and B_v, B_{bv}, I_v, E_v, e_v completes a pairwise comparison. The formulation is applicable to $k \geq 2$ [14].

Problem (2.5) is the unconstrained $L_M T_R KSVM$ model which incorporates the knowledge sets (2.3) and (2.4) into the $L_M T_R SVM$ model (2.1). The tradeoff constant λ is also run through a range of values to achieve the best result. If its value increases then the minimum norm is achieved, but at the expense of having a higher training residual error and higher knowledge set residual error. If data is not available, the second term of the $L_M T_R KSVM$ model can be dropped. This result in classifiers based strictly on knowledge sets, and it is useful for situations in which only expert knowledge exists. The L_2 norm of all terms is considered because of its strong convexity of the objective function. Problem (2.5) is a convex unconstrained optimization problem that has a unique minimum point [14]. The decision function for classifying a point is given by:

$$D(x) = sign[x^T w^{ij} - \gamma^{ij}] = \begin{cases} +1, & \text{if point } (x) \text{ is in class } A^i \\ -1, & \text{if point } (x) \text{ is in class } A^j \end{cases}. \tag{2.6}$$

A new point x is classified according to the voting approach [15, 16]. For example, if sign [$x^T w^{ij} - \gamma^{ij}$] indicates that x belongs to class i, then the vote for class i is increased by one (otherwise j is increased by one). Finally, x belongs to the class with the largest vote. If two classes have identical votes, then select the one with the smallest index.

3 Vertical Two-Phase Flow Data and Prior Knowledge

Fifty percent of each data set described below were trained on the $L_M T_R KSVM$ model, and the other 50 % were used as test samples (for data source details see [9]).

<u>2D classification (data):</u> The 2D vertical flow data set uses two flow rates (superficial gas and liquid velocity) for one inch diameter pipes, with fluid properties at atmospheric conditions, to delineate three different flow regimes. There are 209 instances (points) and 2 attributes (features), 107 points were used as training samples and 102

points were used as test samples. The distribution of instances with respect to their class is as follows: 44 instances in class 1 (bubble flow), 102 instances in class 2 (intermittent flow), and 63 instances in class 3 (annular flow).

3D classification (data): The 3D vertical flow data set uses the pipe size in addition to the superficial gas and liquid velocity to delineate three different flow regimes. There are 424 instances, and 3 attributes, 206 data points used as training samples, 218 points used as test samples. The distribution of instances is 98 instances in class 1 (bubble flow), 217 instances in class 2 (intermittent flow), and 109 instances in class 3 (annular flow).

Prior Knowledge: In addition to the vertical flow data, prior knowledge is included to develop knowledge based classification models. Since the flow regime data are scaled by taking the natural logarithm of each instance, the prior knowledge also needs to be scaled. Below are the transition equations [1, 9] and their equivalent logarithmic transformations for 2D and 3D knowledge based classification:

- Bubble – intermittent flow transition

$$v_{LS} = 3.0 v_{GS} - 1.15 \left[\frac{g\sigma(\rho_L - \rho_G)}{\rho_L^2} \right]^{\frac{1}{4}} \tag{3.1}$$

$$\Rightarrow -0.9883\ln(v_{GS}) + \ln(v_{LS}) = 1.0608, \text{ For both 2D \& 3D classification}$$

- Bubble – dispersed bubble flow transition

$$v_{LS} \geq \frac{6.8}{\rho_L^{0.444}} \{g\sigma(\rho_L - \rho_G)\}^{0.278} \left(\frac{D}{\mu_L} \right)^{0.112} \tag{3.2}$$

$$\Rightarrow \begin{cases} \ln(v_{LS}) \geq 1.03345, \ 2D \text{ classification} \\ \ln(v_{LS}) \geq 1.4466 + 0.112\ln(D), \ 3D \text{ classification} \end{cases}$$

Above, D is the pipe diameter, g is the acceleration due to gravity, v_{GS}, v_{LS} are the gas and liquid superficial velocities, respectively, ρ_G, ρ_L are the gas and liquid densities, σ is the surface tension, and μ_G, μ_L are the gas and liquid viscosities, respectively. The correlations were selected based on the uncertainty of the transition lines as evidenced by the misclassification errors of the MSVM [9]. Also, as a result of the uncertainty, the threshold values of the vertical and horizontal correlations were deviated by a small deviation factor, $\pm\varepsilon$. This is in fact necessary because the correlation equations identify points on transition boundaries that have no distinct flow regime. So a small deviation of the thresholds facilitates the formation of bounds (deviated thresholds) for each flow regime. For instance equation (3.1), 2D case will be represented by

$$-0.9883\ln(v_{GS}) + \ln(v_{LS}) \geq 1.0608(1+\varepsilon) \rightarrow \text{Bubble}$$
$$-0.9883\ln(v_{GS}) + \ln(v_{LS}) \leq 1.0608(1-\varepsilon) \rightarrow \text{Intermitent} \tag{3.3}$$

4   Computational Results

In this section, the results of the analyzed data sets and prior sets described in section 3 are presented and discussed. The L_MT_RKSVM model is used to train the data sets with prior knowledge. To compare between the different models, a performance parameter (misclassification error) was defined as the fraction of misclassified points for a given data set

$$\beta = 1 - \left(\frac{Total\ number\ of\ correctly\ classified\ points}{Total\ number\ of\ observed\ points} \right). \tag{4.1}$$

The tradeoff constants considered are between $0 - 100$, and the deviation factors considered are within the interval $0.01 - 0.1$. The outputs (flow regimes) were coded as: 1 – bubble, 2 – intermittent (slug and churn) and 3 – annular.

Results of the 2D & 3D vertical flow data with prior knowledge information trained on the L_MT_RKSVM model and compared with the L_MT_RSVM model (2.1), are shown in Fig. 2 & 3. It should be noted that CPU time is measured in seconds. The theoretical correlations of McQuillan & Whalley [1] were also simulated to compare with the learning models. The error rate is 0.0163 for the 2D vertical flow data and 0.1227 for the 3D vertical flow data. All computations were performed using MATLAB [18]. In bold face are the lowest error rates for each tradeoff constant λ.

Fig. 2 presents the average test error rate results, based on three random samples for the 2D vertical flow data with prior knowledge as defined in section 4. The L_MT_RSVM and L_MT_RKSVM models generally report promising error rates. The error rate (0.0098) for the model with prior knowledge (L_MT_RKSVM) is the same as the one for the data driven model (L_MT_RSVM), but smaller than the error rate of the theoretical correlations for 2D vertical flow. Fig. 3 presents the average test error rate results, based on four runs with random samples for the 3D vertical flow data with prior knowledge. The model with prior knowledge (error rate of 0.0413) performs better than both the data driven model and the 3D vertical flow theoretical model.

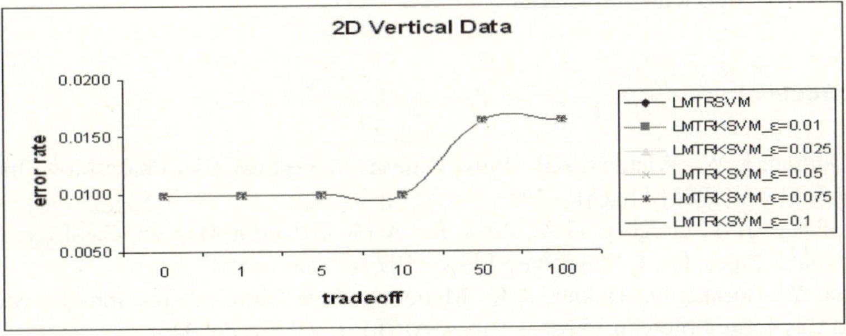

Fig. 2. Average test error rate for L_MT_RKSVM on 2D vertical flow data (varying tradeoff)

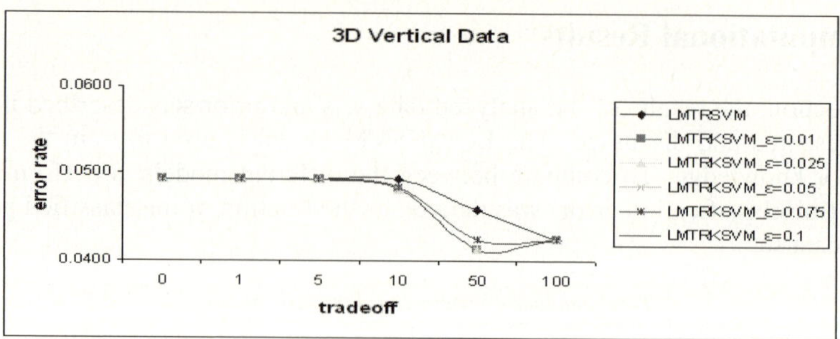

Fig. 3. Average test error rate for $L_M T_R KSVM$ on 3D vertical flow data (varying tradeoff)

Irrespective of the deviation factor, ε, the prior knowledge models appear to perform better or at least display equal level of performance (error rate) with the data driven model ($L_M T_R SVM$). For all data sets, the $L_M T_R KSVM$ performs better than the theoretical model. However, since most of the theoretical correlations are nonlinear models, a nonlinear classification of the data is likely to achieve better generalization ability and lower error rate (note that the present model assumes that the data are linearly separable).

5   Conclusion

In this paper, a knowledge-based multi-classification model called the Linear Multi-classification Tikhonov Regularization Knowledge-based Support Vector Machine ($L_M T_R KSVM$) was applied to vertical, two-phase flow data and comparisons were made with a data driven model ($L_M T_R SVM$) and a theoretical model. The $L_M T_R KSVM$ model using pipe size and superficial velocities as input training vector attributes of the flow regime produces a better overall misclassification error in comparison to theoretical models. Increasing the dimensionality of the classification problem to 3D for the vertical flow data still produces better error rates for $L_M T_R KSVM$ models. The theoretical model error rate is quite large and clearly out of the error rate interval of the $L_M T_R KSVM$ 3D vertical flow model. This highlights the strengths of the prior knowledge learning model.

References

1. McQuillan, K.W., Whalley, P.B.: Flow Patterns in Vertical Two-Phase Flow. Int. J. Multiphase Flow, (1985) 11, 161 – 175
2. Mandhane, J.M., Gregory, G.A., Aziz, K.: A Flow Pattern Map for Gas-Liquid Flow in Horizontal Pipes. Int. J. Multiphase Flow, (1974) 1, 537 – 553
3. Taitel, Y., Bornea, D., Dukler, A.E.: Modeling Flow Pattern Transitions for Steady Upward Gas-Liquid Flow in Vertical Tubes. AIChE J., (1980) 26, 345
4. Taitel, Y., Dukler, A.E.: A Model for Predicting Flow Regime Transitions in Horizontal and Near Horizontal Gas-Liquid Flow. AIChE J, (1976) 22, 47

5. Petalas, N., Aziz, K.: A Mechanistic Model for Multiphase Flow in Pipes. CIM98-39, Proceedings, 49th Annual Technical Meeting of the Petroleum Society of the CIM, Calgary, Alberta, Canada, (1998) June 8-10

6. Osman, E.A.: Artificial Neural Networks Models for Identifying Flow Regimes and Predicting Liquid Holdup in Horizontal Multiphase Flow. SPE 68219, (2000) March

7. Mi, Y., Ishii, M., Tsoukalas, L.H.: Flow Regime Identification Methodology with Neural Networks and Two-Phase Flow Models. Nuclear Engineering and Design, (2001), 204, 87 – 100

8. Ternyik, J., Bilgesu, H.I., Mohaghegh, S.: Virtual Measurement in Pipes, Part 2: Liquid Holdup and Flow Pattern Correlations. SPE 30976, (1995) September

9. Trafalis, T.B., Oladunni, O., Papavassiliou, D.V.: Two-Phase Flow Regime Identification with a Multi-Classification SVM Model. Industrial & Engineering Chemistry Research, (2005) 44, 4414 – 4426

10. Mangasarian, O.L., Shavlik, J.W., Wild, E.W.: Knowledge-Based kernel approximation. Journal of Machine Learning Research, 5, 1127-1141, (2004)

11. Fung, G., Mangasarian, O.L., Shavlik, J.W.: Knowledge-Based support vector machine classifiers. Neural Information Processing Systems 2002 (NIPS 2002), Vancouver, BC, December 10-12, (2002). ``Neural Information Processing Systems 15", S. Becker, S. Thrun and K. Obermayer, editors, MIT Press, Cambridge, MA, 2003, 521-528

12. Tikhonov, A.N., Arsenin, V.Y.: Solution of Ill-Posed Problems. Winston, Washington D.C., (1977)

13. Pelckmans, K., Suykens, J.A.K., De Moor, B.: Morozov, Ivanov and Tikhonov regularization based LS-SVMs. In Proceedings of the International Conference On Neural Information Processing (ICONIP 2004), Calcutta, India, Nov. 22-25, (2004)

14. Oladunni, O., Trafalis, T.B.: Linear Multi-classification Tikhonov Regularization Knowledge-based Support Vector Machine ($L_M T_R$KSVM), Technical Report, School of Industrial Engineering, University of Oklahoma, Norman, Oklahoma (2005)

15. Santosa, B., Conway, T., Trafalis, T.B.: Knowledge Based-Clustering and Application of Multi-Class SVM for Genes Expression Analysis. *Intelligent Engineering Systems through Artificial Neural Networks,* (2002), 12, 391 – 395

16. Hsu, C-W., Lin, C-J.: A Comparison of Methods for Multi-class Support Vector Machines. *IEEE Transactions on Neural Networks,* (2002) 13, 415 – 425

17. Lewis, J. M., Lakshmivarahan, S., Dhall, S.: Dynamic Data Assimilation. Cambridge University Press, (2006)

18. MATLAB User's Guide. The Math-Works, Inc., Natwick, MA 01760, (1994-2003). http://www.mathworks.com

Performance Improvement of Sparse Matrix Vector Product on Vector Machines

Sunil R. Tiyyagura[1], Uwe Küster[1], and Stefan Borowski[2]

[1] High Performance Computing Center Stuttgart,
Allmandring 30, 70550 Stuttgart, Germany
{sunil, kuester}@hlrs.de
[2] NEC HPCE GmbH, Heßbrühlstr. 21B, 70565 Stuttgart, Germany
sborowski@hpce.nec.com

Abstract. Many applications based on finite element and finite difference methods include the solution of large sparse linear systems using preconditioned iterative methods. Matrix vector multiplication is one of the key operations that has a significant impact on the performance of any iterative solver. In this paper, recent developments in sparse storage formats on vector machines are reviewed. Then, several improvements to memory access in the sparse matrix vector product are suggested. Particularly, algorithms based on dense blocks are discussed and reasons for their superior performance are explained. Finally, the performance gain by the presented modifications is demonstrated.

Keywords: Matrix vector product, Jagged diagonal storage, Vector processors.

1   Introduction

The main challenge facing computational scientists and engineers today is the rapidly increasing gap between sustained and peak performance of the high performance computing architectures. Even after spending considerable time on tuning applications to a particular architecture, this gap is an ever existing problem. Vector architecture provides a reasonable solution (at least up to certain extent) to bridge this gap [1]. The success of the Earth Simulator project [2] also emphasizes the need to look towards vector computing as a future alternative.

Sparse iterative solvers play an important role in computational science to solve linear systems arising from discretizing partial differential equations in finite element and finite difference methods. The major time consuming portions of a sparse iterative solver are the matrix vector product (MVP) and preconditioners based on domain decomposition like ICC, ILU, BILU, etc. The MVP algorithm depends on the storage format used to store the matrix non-zeros. Storage formats can be broadly classified as row, column or pseudo diagonal oriented formats. Commonly used examples for each of the formats are compressed row storage (CRS), compressed column storage (CCS) and jagged diagonal storage (JAD). A detailed discussion of different storage formats along with

V.N. Alexandrov et al. (Eds.): ICCS 2006, Part I, LNCS 3991, pp. 196–203, 2006.

corresponding conversions can be found in [3]. As row/column formats store the matrix non-zero entries of each row/column, the MVP algorithm in these cases is restricted to work on a single row/column in the innermost loop. This hinders the performance on vector machines if the number of non-zeros per row/column is less than the length of hardware vector pipeline. In case of NEC SX-8, a classical vector machine, the pipelines are 256 words long. Therefore, at least 256 non-zero entries per row/column are needed to have optimal performance on this architecture. If start-up time (pipeline depth and starting latencies) is additionally considered, the required number of entries for optimal performance is even larger.

Pseudo diagonal formats like the JAD storage are commonly used on vector machines as they fill up the vector pipelines and result in long average vector length. The length for the first few pseudo diagonals is equal to the size of the matrix and then decreases for the latter depending on the kind of problem. This helps in filling up the vector pipelines which results in superior performance over other formats for MVP. However, pseudo diagonal formats are not as natural as the row/column formats which makes their initial setup difficult. In the paper on hand, we focus on performance issues of the sparse MVP algorithm on vector machines. Different approaches to optimize the memory access pattern in this algorithm are addressed. The reduction of load/store operations for the result vector is regarded by modifying the algorithm and using vector registers. The more challenging problem of reducing the indirect memory access for the multiplied vector is considered by introducing a block based algorithm.

In Section 2 of the paper, we take a closer look at some recently proposed improvements for sparse storage formats. In Section 3, the changes decreasing the memory access in the result vector are explained. Then, the block based MVP algorithm is discussed. In Section 4, performance results are presented for the proposed changes. Section 5 summarizes the outcome of this paper.

2   Recent studies

There have been recent studies on optimizing sparse storage formats. It is worth understanding the implications of this work in the context of typical vector processing.

2.1   Optimizing the Storage

Transposed jagged diagonal format (TJAD) optimizes the amount of storage needed for a sparse matrix by eliminating the need for storing the permutation vector [4]. However, this forces a shift of indirect addressing from the multiplied vector to the result vector. As the result vector has to be both loaded and stored, this effectively doubles the amount of indirect addressing needed for MVP. This is a matter of concern on conventional vector architecture, but not much on cache based machines. Since this format is principally used on vector machines, this cannot be the desired alternative.

2.2 Improving Average Diagonal Length

Bi-jagged diagonal storage (Bi-JDS) combines both common JAD and TJAD to further increase the average length of diagonals [5]. The main idea is to store all the full length diagonals in JAD format and the remaining matrix data in TJAD format. TJAD has the disadvantage of increasing the indirect addressing as explained in the previous section. Setting up such a format within the scope of the whole iterative solver would cause changes to all algorithms that use the matrix data such as preconditioning. Furthermore, a lot of problems yield matrices with satisfactorily long average vector lengths in JAD format like matrices from structured finite element, finite difference and finite volume analysis. For problems involving surface integrals, it is common that the number of non-zeros in only some of the rows are extremely high. In this case, it may be advantageous to use such a scheme. This has to be extensively tested to measure its pros (long diagonals) against its cons (set-up cost and doubled indirect addressing).

2.3 Row Format on Vector Machines

Compressed row storage with permutation (CRSP) was introduced in [6]. Results for this format on Cray X1 show a performance of the MVP algorithm that is an order of magnitude higher than for common CRS format. The permutation introduced in the CRSP format adds a level of indirection to both the matrix and the multiplied vector, i.e. effectively two additional indirectly addressed memory loads per loop iteration. This is the overhead incurred because of permuting the rows in ascending order of their length. Although this results in tremendous performance improvement on Cray X1 due to caching, such an algorithm would perform poorly on conventional vector machines because of the heavy cost of indirect addressing involved (in absence of cache memory). The overhead of indirection in memory access is elaborated in the next section.

3 Improvements to the Algorithm

Here, several changes to the JAD MVP algorithm are proposed, which improve the performance on vector machines. In order to reduce memory access for the result vector, an algorithmic approach is to operate on more than one diagonal in the innermost loop. A technical alternative is the use of vector registers offered by vector machines. Finally, the more critical issue of reducing indirect memory access for the multiplied vector is addressed by considering block based algorithms. Before introducing the changes, a brief introduction to features of a vector architecture is provided.

3.1 Vector Processor

Vector processors like NEC SX-8 use a very different architectural approach than scalar processors. Vectorization exploits regularities in the computational structure to accelerate uniform operations on independent data sets. Vector arithmetic instructions are composed of identical operations on elements of vector operands located in vector registers.

For non-vectorizable instructions the NEC SX-8 processor also contains a cache-based superscalar unit. Since the vector unit is by far more powerful than the scalar unit, it is crucial to achieve high vector operation ratios, either via compiler discovery or explicitly through code and data (re-)organization. The vector unit has a clock frequency of 2 GHz and provides a peak vector performance of 16 GFlop/s (4 add and 4 multiply pipes working at 2 GHz). The total peak performance of the processor is 22 GFlop/s (including divide/sqrt unit and scalar unit). Table 1 gives an overview of the different processor units.

Table 1. NEC SX-8 processor units

Unit	No. of results per cycle	Peak (GFlop/s)
Add	4	8
Multiply	4	8
Divide/sqrt	2	4
Scalar		2
		Total = 22

3.2   Original JAD MVP Algorithm

The original JAD MVP algorithm is listed in Fig. 1. For simplicity, the permutation of the result vector is not shown. The performance limitations of this algorithm can be better understood in terms of an operation flow diagram shown in Fig. 2, which explains the execution of the vector instructions. It should be noted that there is only one load/store unit in the NEC SX-8 processor. In each clock cycle, two floating point (FP) operations per pipeline are possible : 1 add and 1 multiply. The main bottleneck of this algorithm is the indirect load of the multiplied vector (`vec`) which takes roughly 5 times longer than a directly addressed one on NEC SX-8. On superscalar architectures, this factor is in general even greater and more complicated to predict.

In 9 cycles (5 for indirectly addressed vector), the possible number of FP operations is 18 (add and multiply). But the effective FP operations in the innermost loop of the algorithm are only 2 (1 add and 1 multiply). Hence, the expected performance of this operation would be 2/18th of the vector peak.

```
for i=0, number_of_diagonals
  offset = jad_ptr[i]
  diag_length = jad_ptr[i+1] - offset
  for j=0, diag_length
    res[j] += mat[offset+j] * vec[index[offset+j]]
  end for
end for
```

Fig. 1. Original JAD MVP algorithm

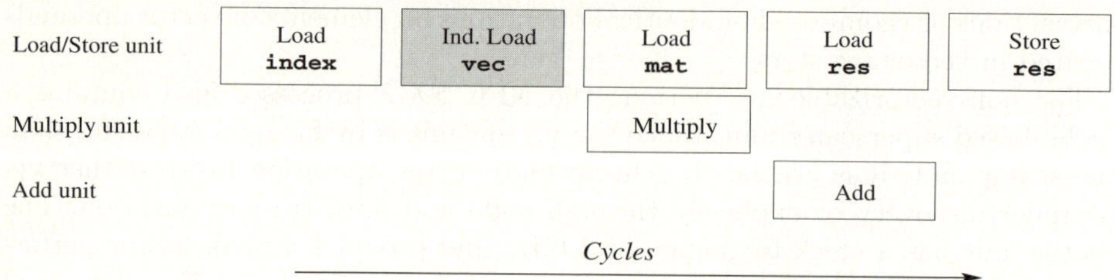

Fig. 2. Operation flow diagram for original JAD MVP algorithm

3.3   Working on Groups of Diagonals

Matrices originating from structured grids have groups of pseudo diagonals with equal length (stored in JAD format). One way to improve the performance is to operate on groups of equal length diagonals in the innermost loop instead of a single diagonal. This considerably saves load/store operations for the result vector (res). The accordingly modified algorithm is listed in Fig. 3. For simplicity, it works on utmost 2 diagonals of equal length. Extending this procedure to more diagonals improves the performance notably.

```
for i=0, number_of_diagonals
  offset = jad_ptr[i]
  dia_length = jad_ptr[i+1] - offset
  if( ((i+1)<number_of_diagonals) &&
      (dia_length==(jad_ptr[i+2]-jad_ptr[i+1])) )
    offset1 = jad_ptr[i+1]
    for j=0, diag_length
      res[j] +=   mat[offset+j] * vec[index[offset+j]]
              + mat[offset1+j] * vec[index[offset1+j]]
    end for
    i = i+1
  else
    for j=0, diag_length
      res[j] += mat[offset+j] * vec[index[offset+j]]
    end for
  end if
end for
```

Fig. 3. JAD MVP algorithm grouping at most 2 diagonals

3.4   Use of Vector Registers

Most vector machines provide a programmer interface to vector registers in order to temporarily store data, like the result vector (res). Using vector registers would need the user to strip mine the innermost loop. The resulting algorithm is listed in Fig. 4. This procedure does not depend on the grid

```
//Size of the hardware vector register
  strip = 256
  for j0=0, number_of_rows, strip
//Initialising the vector register
    for j=0, strip-1
      vregister[j]=0.0
    end for
//Performing the multiplication
    for i=0, number_of_diagonals
      offset = jad_ptr[i]
      diag_length = jad_ptr[i+1] - offset
      for j=j0, min(diag_length, j0+strip-1)
        vregister(j-j0) += mat[offset+j] * vec[index[offset+j]]
      end for
    end for
//Write results to memory
    for j=j0, min(number_of_rows,j0+strip-1)
      res[j] = vregister[j-j0]
    end for
  end for
```

Fig. 4. JAD MVP algorithm using vector registers

(structured/unstructured) and hence equally reduces the memory access for the result vector.

3.5   Operating on Dense Blocks

In the point based algorithms discussed so far, the major hurdle to performance of MVP is indirect memory addressing. To overcome this, block based computations exploit the fact that many problems have multiple physical variables per node. Thus, small blocks can be formed by grouping the equations at each node. Operating on such dense blocks considerably reduces the amount of indirect addressing required for MVP. This improves the performance dramatically on vector machines [7] and also remarkably on superscalar architectures [8]. The block based algorithm for MVP is listed in Fig. 5 (block size 2 for simplicity). The reduction in indirect addressing can be clearly seen in the corresponding operation flow diagram shown in Fig. 6. The multiplied vector is indirectly addressed only twice (`vec1` and `vec2`) instead of four times as it would be the case for the original algorithm. The needed index vector `index` is only loaded once and then incremented. To generalize, indirect memory addressing is reduced by a factor of block size (2 in this case).

4   Performance Results

The performance of the algorithm working on groups of diagonals is listed in Table 2. The case tested was a structured finite element problem with 8500

```
blksize = 2
for i = 0, number_of_diagonals
  offset = jad_ptr[i]
  diag_length = jad_ptr[i+1] - offset
  for j = 0, diag_length
    temp_mat = (offset+j)*blksize*blksize
    temp_vec = (index[offset+j])*blksize
    vec1 = vec(temp_vec)
    vec2 = vec(temp_vec+1)
    res(j*blksize)   +=   mat(temp_mat)   * vec1
                      + mat(temp_mat+1) * vec2
    res(j*blksize+1) +=   mat(temp_mat+2) * vec1
                      + mat(temp_mat+3) * vec2
  end for
end for
```

Fig. 5. Block based JAD MVP algorithm (for 2×2 blocks)

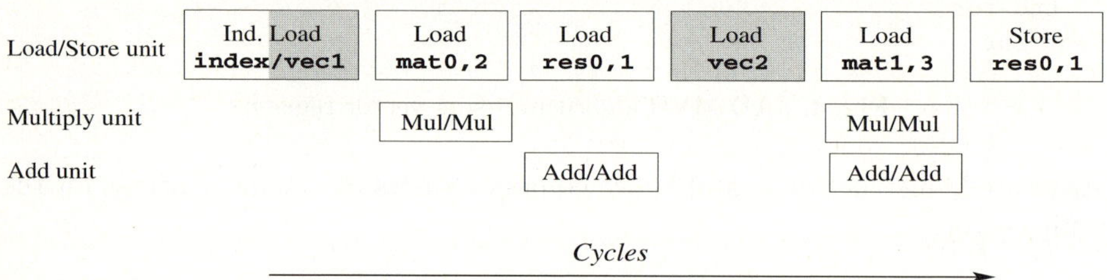

Fig. 6. Operation flow diagram for block based JAD MVP algorithm

hexahedral elements, 10065 nodes and 40260 unknowns. Increasing the group size improves the performance significantly. Grouping 9 diagonals results in a performance improvement of 78% over the original algorithm. Using vector registers, the performance of MVP was measured to be 2197 MFlop/s, about 23% improvement compared to the original algorithm.

Table 2. Performance of JAD MVP algorithm grouping diagonals

Max. group size	Performance (MFlop/s)
orig	1780
3	2511
5	2830
9	3181

Block based algorithms perform much better than the point based ones. For small test cases the performance of MVP for block sizes 4 and 5 is listed in Table 3. Operating on small dense blocks is a superior way to achieve a good percentage of the peak performance on vector machines (37% for block size 5). The performance improvement over the original algorithm is 235%. The improvement compared to the best performing point based algorithm introduced is 87%.

Table 3. Performance of block based JAD MVP algorithm

Block size	Performance (MFlop/s)
orig	1780
4	5804
5	5969

5 Summary

Several improvements to the matrix vector product algorithm based on jagged diagonal storage for vector machines have been suggested in the paper. Most of the memory access problems in this key operation of sparse iterative solvers have been addressed. Block based algorithms are necessary to achieve a good portion of the peak performance on vector machines and moreover should also benifit superscalar architectures. Further work is still required to look for efficient preconditioning methods based on dense blocks.

References

1. Oliker, L., Canning, A., Carter, J., Shalf, J., Ethier, S.: Scientific computations on modern parallel vector systems. In: Proceedings of the ACM/IEEE Supercomputing Conference 2004, Pittsburgh, USA (2004)
2. http://www.es.jamstec.go.jp/esc/eng/.
3. Saad, Y.: SPARSKIT: A basic toolkit for sparse matrix computations. Technical Report RIACS-90-20, NASA Ames Research Center, Moffet Field, CA (1994)
4. Montagne, E., Ekambaram, A.: An optimal storage format for sparse matrices. Information Processing Letters **90** (2004) 87–92
5. Hossain, S.: On efficient storage of sparse matrices. In: 2005 Istanbul Computational Science and Engineering Conference (ICCSE '05), ITU, Istanbul (2005)
6. D'Azevedo, E.F., Fahey, M.R., Mills, R.T.: Vectorized sparse matrix multiply for compressed row storage format. In: Proceedings of the 5th International Conference on Computational Science, Atlanta, USA, Springer-Verlag (2005)
7. Nakajima, K.: Parallel iterative solvers of geofem with selective blocking preconditioning for nonlinear contact problems on the earth simulator. GeoFEM 2003-005, RIST/Tokyo (2003)
8. Tuminaro, R.S., Shadid, J.N., Hutchinson, S.A.: Parallel sparse matrix vector multiply software for matrices with data locality. Concurrency: Practice and Experience **(3)10** (1998) 229–247

A New Reconstruction Algorithm in Spline Signal Spaces

Chen Zhao, Yueting Zhuang, and Honghua Gan

College of Computer Science, Zhejiang University,Hangzhou 310027, P.R. China
`csczhao@sohu.com`, `yzhuang@cs.zju.edu.cn`

Abstract. In this research letter, we introduce a reconstruction formula in spline signal spaces which is a generalization of former results in [11]. A general improved A-P iterative algorithm is presented. We use the algorithm to show reconstruction of signals from weighted samples and also show that the new algorithm shows better convergence than the old one. The explicit convergence rate of the algorithm is obtained.

1 Introduction

In the classical sampling problem, the reconstruction of f on $\mathbb{R}^d$ from its samples $\{f(x_j) : j \in J\}$, where J is a countable indexing set, is one of main tasks in many applications in signal or image processing. However, this problem is ill-posed, and becomes meaningful only when the function f is assumed to be bandlimited, or to belong to a shift-invariant space [1, 2, 3, 4, 8, 11, 12]. For a bandlimited signal of finite energy, it is completely characterized by its samples, and described by the famous classical Shannon sampling theorem. Obviously, the shift-invariant space is not a space of bandlimited function unless the generator is bandlimited.

In many real applications, sampling points are not always regular. For example, the sampling steps need to be fluctuated according to the signals so as to reduce the number of samples and the computational complexity. If a weighted sampling is considered, the system will be made to be more efficient [1, 2, 3, 4, 5, 11, 12]. It is well known that spline subspaces yield many advantages in their generation and numerical treatment so that there are many practical applications for signal or image processing. Therefore, the recent research of spline subspaces has received much attentions (see[3, 10, 11]).

For practical application and computation of reconstruction, Goh et al., showed practical reconstruction algorithm of bandlimited signals from irregular samples in [8], Aldroubi et al., presented a A-P iterative algorithm in [1, 2, 4]. We will improve and generalize the A-P iterative algorithm and also show that the new algorithm shows better than the old one for convergence rate. That is, we can easy control the convergence rate of the algorithm with our requirement. At the same time, we don't increase the number of the sampling point. But this algorithm is not perfect. Because we immolate(increase) computation complexity as soon as improve convergence rate of the algorithm.

V.N. Alexandrov et al. (Eds.): ICCS 2006, Part I, LNCS 3991, pp. 204–209, 2006.

2 Reconstruction Algorithm in Spline Spaces

By the special features of spline subspaces, we will present the new improved A-P algorithm and its convergence rate in spline spaces, which are more explicit. We introduce some notations and lemmas that will be used in this section.

The signal space $V_N = \{\sum_{k \in Z} c_k \varphi_N(\cdot - k) : \{c_k\} \in \ell^2\}$ is spline space generated by $\varphi_N = \chi_{[0,1]} * \cdots * \chi_{[0,1]}$ (N convolutions), $N \geq 1$.

Definition 2.1. A general bounded partition of unity(GBPU) is a set of function $\{\beta_{j_1}, \beta_{j_2}, \cdots, \beta_{j_r}\}$ that satisfy:

(1) $0 \leq \beta_{j_1}, \cdots, \beta_{j_r} \leq 1(\forall j_1 \equiv j_1(j), \cdots, j_r \equiv j_r(j) \in J)$, where J be countable separated index set.

(2) $supp\beta_{j_1} \subset B_{\frac{\delta}{r}}(x_{j_1}), \cdots, supp\beta_{j_r} \subset B_{\frac{\delta}{r}}(x_{j_r})$,

(3) $\sum_{j \in J}(\beta_{j_1} + \cdots + \beta_{j_r}) = 1$.

In fact, in the case of $r = 1$, the above GBPU definition is ordinary BPU definition be used in [1, 4].

We will assume that the weight function $\{\varphi_{x_j} : x_j \in X\}$ satisfy the following properties:

(i) $supp\varphi_{x_j} \subset B_{\frac{a}{r}}(x_j)$

(ii) there exist $M > 0$ such that $\int_{\mathbb{R}^d} |\varphi_{x_j}|dx \leq M$,

(iii) $\int_{\mathbb{R}^d} \varphi_{x_j}dx = 1$

The operator A and Q defined by $Af = \sum_{j \in J}\langle f, \varphi_{x_{j_1}}\rangle\beta_{j_1} + \cdots + \langle f, \varphi_{x_{j_r}}\rangle\beta_{j_r}$ and $Qf(x) = \sum_j f(x_{j_1})\beta_{j_1}(x) + \cdots + \sum_j f(x_{j_r})\beta_{j_r}(x)$, respectively.

The other definitions and notations can be found in [1, 4, 11, 12].

Lemma 2.1. [6] $\{\varphi_N(\cdot - k) : k \in \mathbb{Z}\}$ is Riesz basis for V_N, $A_N = \sum_k |\hat{\varphi}_N(\pi + 2k\pi)|^2$ and $B_N = 1$ are its lower and upper bounds, respectively.

Lemma 2.2. [4] If φ is continuous and has compact support, then for any $f \in V^p(\varphi) = \{\sum_{k \in \mathbb{Z}} c_k \varphi(\cdot - k) : (c_k) \in \ell^p\}$, the following conclusions (i)-(ii) hold:

(i) $\|f\|_{L^p} \approx \|c\|_{\ell^p} \approx \|f\|_{W(L^p)}$,

(ii) $V^p(\varphi) \subset W_0(L^p) \subset W_0(L^q) \subset W(L^q) \subset L^q(\mathbb{R})(1 \leq p \leq q \leq \infty)$.

Lemma 2.3. If $f \in V_N$, then for any $0 < \delta < 1$ we have $\|osc_\delta(f)\|_{L^2}^2 \leq (3N\delta)^2 \sum_{k \in \mathbb{Z}} |c_k|^2$, where $osc_\delta(f)(x) = \sup_{|y| \leq \delta} |f(x + y) - f(x)|$.

Lemma 2.4. [4] For any $f \in V^p(\varphi)$, the following conclusions (i)-(ii) hold:

(i) $\|osc_\delta(f)\|_{W(L^p)} \leq \|c\|_{\ell^p}\|osc_\delta(\varphi)\|_{W(L^1)}$,

(ii) $\|\sum_{k \in \mathbb{Z}} c_k \varphi(\cdot - k)\|_{W(L^p)} \leq \|c\|_{\ell^p}\|\varphi\|_{W(L^1)}$.

Lemma 2.5. *If* $X = \{x_n\}$ *is increasing real sequence with* $\sup_i (x_{i+1} - x_i) = \delta < 1$, *then for any* $f = \sum_{k \in \mathbb{Z}} c_k \varphi_N(\cdot - k) \in V_N$ *we have* $\|Qf\|_{L^2} \leq \|Qf\|_{W(L^2)} \leq (3 + \frac{2\delta}{r})\|c\|_{\ell^2}\|\varphi\|_{W(L^1)}$.

Proof. For $f = \sum_{k \in \mathbb{Z}} c_k \varphi_N(\cdot - k)$ we have

$$|f(x) - (Qf)(x)| \leq osc_{\frac{\delta}{r}}(f)(x).$$

From this pointwise estimate and Lemma 2.2, 2.4, we get

$$\|f - Qf\|_{W(L^2)} \leq \|osc_{\frac{\delta}{r}}(f)\|_{W(L^2)}$$
$$\leq \|c\|_{\ell^2}\|osc_{\frac{\delta}{r}}(\varphi_N)\|_{W(L^1)}.$$

By the results of [1] or [4] we know

$$\|osc_{\frac{\delta}{r}}(\varphi_N)\|_{W(L^1)} \leq 2(1 + \frac{\delta}{r})\|\varphi_N\|_{W(L^1)}.$$

Putting the above discussion together, we have

$$\|Qf\|_{L^2} \leq \|Qf\|_{W(L^2)} \leq \|f - Qf\|_{W(L^2)} + \|f\|_{W(L^2)}$$
$$\leq 2(1 + \frac{\delta}{r})\|c\|_{\ell^2}\|\varphi_N\|_{W(L^1)} + \|\sum_{k \in \mathbb{Z}} c_k \varphi_N(\cdot - k)\|_{W(L^2)}$$
$$\leq 2(1 + \frac{\delta}{r})\|c\|_{\ell^2}\|\varphi_N\|_{W(L^1)} + \|c\|_{\ell^2}\|\varphi_N\|_{W(L^1)}$$
$$\leq (3 + \frac{2\delta}{r})\|c\|_{\ell^2}\|\varphi_N\|_{W(L^1)}.$$

Theorem 2.1. *Let* P *be an orthogonal projection from* $L^2(\mathbb{R})$ *to* V_N. *If sampling set* $X = \{x_n\}$ *is a increasing real sequence with* $\sup_i(x_{i+1} - x_i) = \delta < 1$ *and* $\gamma = \dfrac{3N\delta}{r\sqrt{\sum_k |\hat{\varphi}_N(\pi + 2k\pi)|^2}} < 1$, *then any* $f \in V_N$ *can be recovered from its samples* $\{f(x_j) : x_j \in X\}$ *on sampling set* X *by the iterative algorithm*

$$\begin{cases} f_1 = PQf, \\ f_{n+1} = PQ(f - f_n) + f_n. \end{cases}$$

The convergence is geometric, that is,

$$\|f_{n+1} - f\|_{L^2} \leq \gamma^n \|f_1 - f\|_{L^2}.$$

Proof. By Lemma 2.1, Lemma 2.3 and properties of $\{\beta_{j1}, \cdots, \beta_{jr}\}$, we have

$$\|(I - PQ)f\|_{L^2}^2 = \|Pf - PQf\|_{L^2}^2 \leq \|P\|_{op}^2\|f - Qf\|_{L^2}^2 = \|f - Qf\|_{L^2}^2$$
$$\leq \|osc_{\frac{\delta}{r}}(f)\|_{L^2}^2 \leq (3N\frac{\delta}{r})^2 \sum_{k \in \mathbb{Z}} |c_k|^2 = (3N\frac{\delta}{r})^2\|c\|_{\ell^2}^2$$
$$\leq \left(\frac{3N\delta}{r\sqrt{\sum_k |\hat{\varphi}_N(\pi + 2k\pi)|^2}}\right)^2\|f\|_{L^2}^2.$$

Therefore

$$\|f_{n+1} - f\|_{L^2} = \|f_n + PQ(f - f_n) - f\|_{L^2} = \|PQ(f - f_n) - (f - f_n)\|_{L^2}$$
$$\leq \|I - PQ\|\|f - f_n\|_{L^2} \leq \cdots \leq \|I - PQ\|^n\|f - f_1\|_{L^2}.$$

Combining with the estimate of $\|I - PQ\|$, we can imply

$$\|f_{n+1} - f\|_{L^2} \leq \gamma^n\|f_1 - f\|_{L^2}.$$

Taking assumption $\gamma = \dfrac{3N\delta}{r\sqrt{\sum\limits_k |\hat{\varphi}_N(\pi+2k\pi)|^2}} < 1$, we know the algorithm is convergent.

In the following, we will show the new improved A-P iterative algorithm from weighted samples in spline subspace.

Theorem 2.2. *Let P be an orthogonal projection from $L^2(\mathbb{R})$ to V_N and weight function satisfy the following three conditions (i)-(iii):*

(i) *$supp\varphi_{x_j} \subset [x_j - \frac{a}{r}, x_j + \frac{a}{r}]$*
(ii) *there exist $M > 0$ such that $\int |\varphi_{x_j}(x)|dx \leq M$,*
(iii) *$\int \varphi_{x_j}(x)dx = 1$.*

If sampling set $X = \{x_n\}$ is a increasing real sequence with $\sup_i(x_{i+1} - x_i) = \delta < 1$ and we choose proper δ and a such that $\alpha = \dfrac{3N}{r\sqrt{\sum\limits_k |\hat{\varphi}_N(\pi+2k\pi)|^2}}(\delta +$

$a(3 + \frac{2a}{r})M) < 1$, then any $f \in V_N$ can be recovered from its weighted samples $\{\langle f, \varphi_{x_j}\rangle : x_j \in X\}$ on sampling set X by the iterative algorithm

$$\begin{cases} f_1 = PAf, \\ f_{n+1} = PA(f - f_n) + f_n. \end{cases}$$

The convergence is geometric, that is,

$$\|f_{n+1} - f\|_{L^2} \leq \alpha^n\|f_1 - f\|_{L^2}.$$

Proof. By $Pf = f$ and $\|P\|_{op} = 1$, for any $f = \sum\limits_{k\in\mathbb{Z}} c_k\varphi_N(\cdot - k) \in V_N$ we have

$$\|f - PAf\|_{L^2} = \|f - PQf + PQf - PAf\|_{L^2} \tag{1}$$
$$\leq \|f - Qf\|_{L^2} + \|Qf - Af\|_{L^2} \tag{2}$$

From the proof of Theorem 2.1, we have the following estimate for $\|f - Qf\|_{L^2}$:

$$\|f - Qf\|_{L^2} \leq (\dfrac{3N\delta}{r\sqrt{\sum\limits_k |\hat{\varphi}_N(\pi + 2k\pi)|^2}})\|f\|_{L^2}. \tag{3}$$

For the second term $\|Qf - Af\|_{L^2}$ of (2) we have the pointwise estimate

$$|(Qf - Af)(x)| \leq MQ(\sum\limits_{k\in\mathbb{Z}} |c_k|osc_{\frac{a}{r}}(\varphi_N)(x - k)).$$

From this pointwise estimate, Lemma 2.1, Lemma 2.3 and Lemma 2.5, it follows that:

$$\|Qf - Af\|_{L^2} \le M(3 + \frac{2a}{r})\|c\|_{\ell^2}\|osc_{\frac{a}{r}}(\varphi_N)\|_{W(L^1)} \tag{4}$$

$$\le M(3 + \frac{2a}{r})\frac{\|osc_{\frac{a}{r}}(\varphi_N)\|_{W(L^1)}}{\sqrt{\sum_k |\hat{\varphi}_N(\pi + 2k\pi)|^2}}\|f\|_{L^2} \tag{5}$$

$$\le M(3 + \frac{2a}{r})\frac{3Na}{r\sqrt{\sum_k |\hat{\varphi}_N(\pi + 2k\pi)|^2}}\|f\|_{L^2} \tag{6}$$

By combining (3) and (6), we can obtain

$$\|f - PAf\|_{L^2} \le \frac{3N}{r\sqrt{\sum_k |\hat{\varphi}_N(\pi + 2k\pi)|^2}}(\delta + a(3 + \frac{2a}{r})M)\|f\|_{L^2},$$

that is,

$$\|I - PA\|_{L^2} \le \frac{3N}{r\sqrt{\sum_k |\hat{\varphi}_N(\pi + 2k\pi)|^2}}(\delta + a(3 + \frac{2a}{r})M).$$

Similar to the procedure in the proof of Theorem 2.1, we have

$$\|f_{n+1} - f\|_{L^2} \le \alpha^n \|f_1 - f\|_{L^2}.$$

Remark 2.1. From the constructions of operator Q and A, we know why item r can appear in the convergence rate expression of the new improved algorithm. But r is not appear in the old algorithm. Hence this algorithm improves the convergence rate of the old algorithm. In addition, it is obvious that we can easily control the convergence rate through choosing proper r without changing sampling point gap δ. That is, when δ and a are proper given, we can obtain the convergence rate that we want through choosing proper r. We hope r be enough large. But we increase the computation complexity as soon as choose larger r. So we should choose proper r with our requirement.

3   Conclusion

In this research letter, we discuss in some detail the problem of the weighted sampling and reconstruction in spline signal spaces and provide a reconstruction formula in spline signal spaces, which is generalized and improved form of the results in [11]. Then we give general A-P iterative algorithm in general shift-invariant spaces, and use the new algorithm to show reconstruction of signals from weighted samples. The algorithm shows better convergence than the old one. We study the new algorithm with emphasis on its implementation and obtain explicit convergence rate of the algorithm in spline subspaces. Due to the limitation of the page number, we omit some numerical examples, proofs of lemma and theorem and will show their detail in regular paper.

References

1. A. Aldroubi. Non-uniform weighted average sampling and reconstruction in shift-invariant and wavelet spaces. *Appl. Comput. Harmon. Anal* 13(2002)156-161.
2. A. Aldroubi, H. Feichtinger. Exact iterative reconstruction algorithm for multivate irregular sampled functions in spline-like spaces: The L_p theory. *Proc. Amer. Math. Soc* 126(9)(1998)2677-2686.
3. A. Aldroubi, K. Gröchenig. Beurling-Landau-type theorems for non-uniform sampling in shift invariant spline spaces. *J. Fourier. Anal. Appl,* 6(1)(2000) 93-103.
4. A. Aldroubi and K. Gröchenig. Non-uniform sampling and reconstruction in shift-invariant spaces. *SIAM Rev* 43(4)(2001)585-620.
5. W. Chen, S. Itoh and J. Shiki. On sampling in shift invariant spaces. *IEEE Trans. Information. Theory* 48(10)(2002)2802-2810.
6. C. K. Chui. An introduction to Wavelet, Academic Press, New York,1992.
7. H. G. Feichtinger. Generalized amalgams, with applications to Fourier transform. *Can. J. of Math.,*42(3)(1990)395-409
8. S. S. Goh, I. G. H. Ong. Reconstruction of bandlimited signals from irregular samples. *Signal. Processing* 46(3)(1995)315-329.
9. K. Gröchenig. Localization of frames, Banach frames, and the invertibility of the frame operator, *J.Fourier.Anal.Appl* 10(2)(2004)105-132.
10. W. C. Sun and X. W. Zhou. Average sampling in spline subspaces. *Appl. Math. Letter,* 15(2002)233-237.
11. J. Xian, S. P. Luo and W. Lin, *Improved A-P iterative algorithm in spline subspaces. Lecture Notes in Comput. Sci,* 3037(2004) 58-64.
12. J. Xian, X. F. Qiang. *Non-uniform sampling and reconstruction in weighted multiply generated shift-invariant spaces. Far. East. J. Math. Sci,* 8(3)(2003), 281-293.

An Implicit Riemannian Trust-Region Method for the Symmetric Generalized Eigenproblem[*]

C.G. Baker[1,2], P.-A. Absil[3,4], and K.A. Gallivan[1]

[1] School of Computational Science, Florida State University, Tallahassee, FL, USA
[2] Computational Mathematics & Algorithms, Sandia National Laboratories,
Albuquerque, NM, USA
[3] Département d'ingénierie mathématique, Université catholique de Louvain,
1348 Louvain-la-Neuve, Belgium
[4] Peterhouse, University of Cambridge, UK

Abstract. The recently proposed Riemannian Trust-Region method can be applied to the problem of computing extreme eigenpairs of a matrix pencil, with strong global convergence and local convergence properties. This paper addresses inherent inefficiencies of an explicit trust-region mechanism. We propose a new algorithm, the Implicit Riemannian Trust-Region method for extreme eigenpair computation, which seeks to overcome these inefficiencies while still retaining the favorable convergence properties.

1 Introduction

Consider $n \times n$ symmetric matrices A and B, with B positive definite. The generalized eigenvalue problem

$$Ax = \lambda Bx$$

is known to admit n real eigenvalues $\lambda_1 \leq \ldots \leq \lambda_n$, along with associated B-orthonormal eigenvectors $v_1, \ldots, v_n$ (see [1]). We seek here to compute the p leftmost eigenvectors of the pencil (A, B). It is known that the leftmost eigenspace $\mathcal{U} = \text{colsp}(v_1, \ldots, v_p)$ of (A, B) is the column space of any minimizer of the generalized Rayleigh quotient

$$f : \mathbb{R}_*^{n \times p} \to \mathbb{R} : Y \mapsto \text{trace}\left((Y^T BY)^{-1}(Y^T AY)\right), \tag{1}$$

where $\mathbb{R}_*^{n \times p}$ denotes the set of full-rank $n \times p$ matrices.

This result underpins a number of methods based on finding the extreme points of the generalized Rayleigh quotient (see [2, 3, 4, 5, 6, 7] and references

[*] This work was supported by NSF Grant ACI0324944. The first author was in part supported by the CSRI, Sandia National Laboratories. Sandia is a multiprogram laboratory operated by Sandia Corporation, a Lockheed Martin Company, for the United States Department of Energy; contract/grant number: DE-AC04-94AL85000. The second author was partially supported by Microsoft Research through a Research Fellowship at Peterhouse, Cambridge.

V.N. Alexandrov et al. (Eds.): ICCS 2006, Part I, LNCS 3991, pp. 210–217, 2006.
© Springer-Verlag Berlin Heidelberg 2006

therein). Here, we consider the recently proposed [8, 9] Riemannian Trust-Region (RTR) method. This method formulates the eigenvalue problem as an optimization problem on a Riemannian manifold, utilizing a trust-region mechanism to find a solution. Similar to Euclidean trust-region methods [10, 11], the RTR method ensures strong global convergence properties while allowing superlinear convergence near the solution. However, the classical trust-region mechanism has some inherent inefficiencies. When the trust-region radius is too large, valuable time may be spent computing an update that may be rejected. When the trust-region radius is too small, we may reject good updates lying outside the trust-region. A second problem with the RTR method is typical of methods where the outer stopping criterion is evaluated only after exiting the inner iteration: in almost all cases, the last call to the inner iteration will perform more work than necessary to satisfy the outer stopping criterion.

In the current paper, we explore solutions to both of the problems described above. We present an analysis providing us knowledge of the model fidelity at every step of the inner iteration, allowing our trust-region to be based directly on the trustworthiness of the model. We propose a new algorithm, the Implicit Riemannian Trust-Region (IRTR) method, exploiting this analysis.

2   Riemannian Trust-Region Method with Newton Model

The RTR method can be used to minimize the generalized Rayleigh quotient (1). The right-hand side of this function depends only on $\mathrm{colsp}(Y)$, so that f induces a real-valued function on the set of p-dimensional subspaces of $\mathbb{R}^n$. (This set is known as the Grassmann manifold, which can be endowed with a Riemannian structure [4, 12].) The RTR method iteratively computes the minimizer of f by (approximately) minimizing successive models of f. The minimization of the models is done via an iterative process, which is referred to as the *inner iteration*, to distinguish it with the principal *outer iteration*. We present here the process in a way that does not require a background in differential geometry; we refer to [13] for the mathematical foundations of the technique.

Let Y be a full-rank, $n \times p$ matrix. We desire a correction S of Y such that $f(Y + S) < f(Y)$. A difficulty is that corrections of Y that do not modify its column space do not affect the value of the cost function. This situation leads to unpleasant degeneracy if it is not addressed. Therefore, we require S to satisfy some complementarity condition with respect to the space $\mathcal{V}_Y := \{YM : M \in \mathbb{R}^{p \times p}\}$. Here, in order to simplify later developments, we impose complementarity via B-orthogonality, namely $S \in \mathcal{H}_Y$ where

$$\mathcal{H}_Y = \{Z \in \mathbb{R}^{n \times p} : Y^T B Z = 0\}.$$

Consequently, the task is to minimize the function

$$\hat{f}_Y(S) := \mathrm{trace}\left(((Y + S)^T B(Y + S))^{-1}((Y + S)^T A(Y + S))\right), \quad S \in \mathcal{H}_Y.$$

The RTR method constructs a model m_Y of $\hat{f}_Y$ and computes an update S which approximately minimizes m_Y, so that the inner iteration attempts to solve the following problem:

$$\min m_Y(S), \quad S \in \mathcal{H}_Y, \quad \|S\|_2 \le \Delta,$$

where Δ (the *trust-region radius*) denotes the region in which we trust m_Y to approximate $\hat{f}_Y$. The next iterate and trust-region radius are determined by the performance of m_Y with respect to $\hat{f}_Y$. This performance ratio is measured by the quotient:

$$\rho_Y(S) = \frac{\hat{f}_Y(0) - \hat{f}_Y(S)}{m_Y(0) - m_Y(S)}.$$

Low values of $\rho_Y(S)$ (close to zero) indicate that the model m_Y at S is not a good approximation to $\hat{f}_Y$. In this scenario, the trust-region radius is reduced and the update $Y + S$ is rejected. Higher values of $\rho_Y(S)$ allow the acceptance of $Y + S$ as the next iterate, and a value of $\rho_Y(S)$ close to one suggests good approximation of $\hat{f}_Y$ by m_Y, allowing the trust-region radius to be enlarged.

Usually, the model m_Y is chosen as a quadratic function approximating $\hat{f}_Y$. In the sequel, in contrast to [9] where the quadratic term of the model was unspecified, we assume that m_Y is the *Newton model*, i.e., the quadratic expansion of $\hat{f}_Y$ at $S = 0$. Then, assuming from here on that $Y^T B Y = I_p$, we have

$$m_Y(S) = \mathrm{trace}\left(Y^T A Y\right) + 2\mathrm{trace}\left(S^T A Y\right) + \mathrm{trace}\left(S^T\left(AS - BS(Y^T A Y)\right)\right)$$
$$= \hat{f}_Y(0) + \mathrm{trace}\left(S^T \nabla \hat{f}_Y\right) + \frac{1}{2}\mathrm{trace}\left(S^T H_Y[S]\right),$$

where the gradient and the effect of the Hessian of $\hat{f}_Y$ are identified as

$$\nabla \hat{f}_Y = 2P_{BY} A Y \qquad H_Y[S] = 2P_{BY}\left(AS - BS(Y^T A Y)\right),$$

and where $P_{BY} = I - BY(Y^T BBY)^{-1}Y^T B$ is the orthogonal projector on the space perpendicular to the column space of BY.

Simple manipulation shows the following:

$$\hat{f}_Y(0) - \hat{f}_Y(S) = \mathrm{trace}\left(Y^T A Y - (I + S^T BS)^{-1}(Y + S)^T A(Y + S)\right)$$
$$= \mathrm{trace}\left((I + S^T BS)^{-1}(S^T BS(Y^T A Y) - 2S^T A Y - S^T AS)\right).$$

Consider the case where $p = 1$. The above equation simplifies to

$$\hat{f}_y(0) - \hat{f}_y(s) = (1 + s^T Bs)^{-1}\left(s^T Bs y^T A y - 2s^T A y - s^T As\right)$$
$$= (1 + s^T Bs)^{-1}\left(m_y(0) - m_y(s)\right),$$

so that

$$\rho_y(s) = \frac{\hat{f}_y(0) - \hat{f}_y(s)}{m_y(0) - m_y(s)} = \frac{1}{1 + s^T Bs}. \tag{2}$$

This allows the model performance ratio ρ_y to be constantly evaluated as the model minimization progresses, simply by tracking the B-norm of the current update vector.

3   Implicit Riemannian Trust-Region Method

In this section, we explore the possibility of selecting the trust-region as a sublevel set of the performance ratio ρ_Y. We dub this approach the Implicit Riemannian Trust-Region method.

3.1   Case $p = 1$

The analysis of ρ in the previous section shows that for the generalized Rayleigh quotient with $p = 1$, the performance of the model decreases as the iterate moves away from zero. However, in the case of the $p = 1$ generalized Rayleigh quotient, $\rho_y(s)$ has a simple relationship with $\|s\|_B$. Therefore, by monitoring the B-norm of the inner iterate, we can easily determine the value of ρ for a given inner iterate. Furthermore, the relationship between ρ and the B-norm of a vector, allows us to move along a search direction to a specific value of ρ. These two things, combined, enable us to redefine the trust-region based instead on the value of ρ.

The truncated conjugate gradient proposed in [9] for use in the simple RTR algorithm seeks to minimize the model m_Y within a trust-region defined explicitly as $\{s : \|s\|_2 \leq \Delta\}$. Here, we change the definition of the trust-region to $\{s : \rho_y(s) \geq \rho'\}$, for some $\rho' \in (0, 1)$. The necessary modifications to this algorithm are very simple. The definition of the trust-region occurs in three places: when detecting whether the trust-region has been breached; when constraining the update vector in the case that the trust-region was breached; and when constraining the update vector in the case that we have detected a direction of negative curvature. The new inner iteration is listed in Algorithm 1, with the differences highlighted.

Having stated the definition of the implicit trust-region, based on ρ, we need a mechanism for following a search direction to the edge of the trust-region. That is, at some outer step k and given s_j and a search direction d_j, we wish to compute $s = s_j + \tau d_j$ such that $\rho_{y_k}(s) = \rho'$. Given ρ' and denoting

$$\Delta_{\rho'} = \sqrt{\frac{1}{\rho'} - 1},$$ (3)

the desired value of τ is given by

$$\tau = \frac{-d_j^T B s_j + \sqrt{(d_j^T B s_j)^2 + d_j^T B d_j (\Delta_{\rho'}^2 - s_j^T B s_j)}}{d_j^T B d_j}.$$ (4)

A careful implementation precludes the need for any more matrix multiplications against B than are necessary to perform the iterations.

Another enhancement in Algorithm 1 is that the outer stopping criterion is tested during the inner iteration. This technique is not novel in the context of eigensolvers with inner iterations, having been proposed by Notay [14]. Our motivation for introducing this test is that, when it is absent, the final outer

---

Algorithm 1 (Preconditioned Truncated CG (IRTR))
Data: A,B symmetric, B positive definite, $\rho' \in (0,1)$, preconditioner M
Input: Iterate y, $y^T By = 1$
Set $s_0 = 0$, $r_0 = \nabla \hat{f}_y$, $z_0 = M^{-1} r_0$, $d_0 = -z_0$
for $j = 0, 1, 2, \ldots$

   Check κ/θ stopping criterion
   if $\|r_j\|_2 \leq \|r_0\|_2 \min\left\{\kappa, \|r_0\|_2^\theta\right\}$
      return s_j

   Check curvature of current search direction
   if $d_j^T H_y[d_j] \leq 0$
      Compute τ such that $s = s_j + \tau d_j$ satisfies $\boxed{\rho_y(s) = \rho'}$
      return s

   Set $\alpha_j = (z_j^T r_j)/(d_j^T H_y[d_j])$

   Generate next inner iterate
   Set $s_{j+1} = s_j + \alpha_j d_j$

   Check implicit trust-region
   if $\boxed{\rho_y(s_{j+1}) < \rho'}$
      Compute $\tau \geq 0$ such that $s = s_j + \tau d_j$ satisfies $\boxed{\rho_y(s) = \rho'}$
      return s

   Use CG recurrences to update residual and search direction
   Set $r_{j+1} = r_j + \alpha_j H_y[d_j]$
   Set $z_{j+1} = M^{-1} r_{j+1}$
   Set $\beta_{j+1} = (z_{j+1}^T r_{j+1})/(z_j^T r_j)$
   Set $d_{j+1} = -z_{j+1} + \beta_{j+1} d_j$

   Check outer stopping criterion
   $\boxed{\text{Compute } \|\nabla \hat{f}_{y+s_{j+1}}\|_2 \text{ and test}}$

end for.

---

step may reach a much higher accuracy than specified by the outer stopping criterion, resulting in a waste of computational effort. Also, while Notay proposed a formula for the inexpensive evaluation of the outer norm based on the inner iteration, we must rely on a slightly more expensive, but less frequent, explicit evaluation of the outer stopping criterion.

The product of this iteration is an update vector s_j which is guaranteed to lie inside of the ρ-based trust-region. The result is that the ρ value of the new iterate need not be explicitly computed, the new iterate can be automatically accepted, with an update vector constrained by model fidelity instead of a discretely chosen trust-region radius based on the performance of the last iterate. An updated outer iteration is presented in Algorithm 2, which also features an optional subspace acceleration enhancement à la Davidson [15].

Algorithm 2 (Implicit Riemannian Trust-Region Algorithm)
Data: A,B symmetric, B positive definite, $\rho' \in (0,1)$
Input: Initial subspace $\mathcal{W}_0$
for $k = 0, 1, 2, \ldots$

    Model-based Minimization
    Generate y_k using a Rayleigh-Ritz procedure on $\mathcal{W}_k$
    Compute $\nabla \hat{f}_{y_k}$ and check $\|\nabla \hat{f}_{y_k}\|_2$
    Compute s_k to approximately minimize m_{y_k} such that $\rho(s_k) \geq \rho'$ (Algorithm 1)

    Generate next subspace
    if *performing subspace acceleration*
        Compute new acceleration subspace $\mathcal{W}_{k+1}$ from $\mathcal{W}_k$ and s_k
    else
        Set $\mathcal{W}_{k+1} = \mathrm{colsp}(y_k + s_k)$
    end

end for.

3.2   A Block Algorithm

The analysis of Section 2 seems to preclude a simple formula for ρ in the case that $p > 1$. We wish, however, to have a block algorithm. The solution is to decouple the block Rayleigh quotient into the sum of p separate rank-1 Rayleigh quotients, which can then be addressed individually using the IRTR strategy. This is done as follows.

Assume that our iterates satisfy $Y^T AY = \Sigma = \mathrm{diag}(\sigma_1, \ldots, \sigma_p)$, in addition to $Y^T BY = I_p$. In fact, this is a natural consequence of the Rayleigh-Ritz process. Then given $Y = \begin{bmatrix} y_1 \ldots y_p \end{bmatrix}$, the model m_Y can be rewritten:

$$m_Y(S) = \mathrm{trace}\left(\Sigma + 2S^T AY + S^T(AS - BS\Sigma)\right)$$

$$= \sum_{i=1}^{p}\left(\sigma_i + 2s_i^T Ay_i + s_i^T(A - \sigma_i B)s_i\right) = \sum_{i=1}^{p} m_{y_i}(s_i).$$

It should be noted that the update vectors for the decoupled minimizations must have the original orthogonality constraints in place. That is, instead of requiring only that $y_i^T Bs_i = 0$, we require that $Y^T Bs_i = 0$ for each s_i. This is necessary to guarantee that the next iterate, $Y + S$, has full rank, so that the Rayleigh quotient is defined.

As for the truncated conjugate gradient, the p individual IRTR subproblems should be solved simultaneously, with the inner iteration stopped as soon as any of the iterations satisfy one of the inner stopping criteria (exceeded trust-region or detected negative curvature). If only a subset of iterations are allowed to continue, then the κ/θ inner stopping criterion may not be feasible.

The described method attempts to improve on the RTR, while retaining the strong global and local convergence properties of the RTR. The model fidelity guaranteed by the implicit trust-region mechanism allows for a straightforward

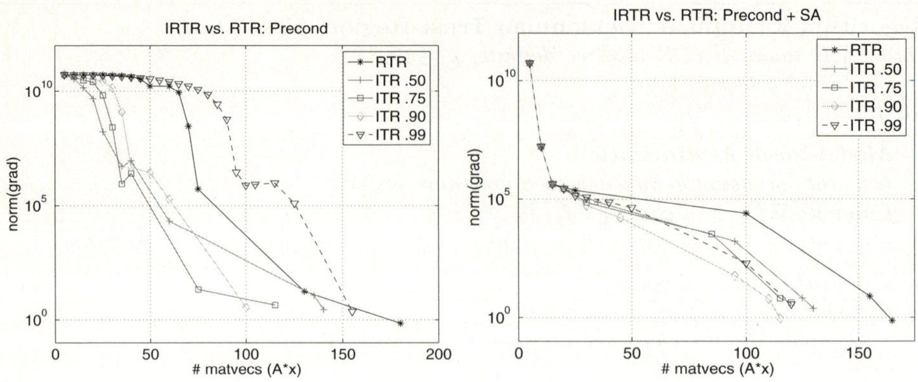

Fig. 1. Figures illustrating the efficiency of RTR vs. IRTR for different values of ρ'

proof of global convergence. Related work [16] presents the proofs of global convergence, along with a discussion regarding the consequences of early termination of the inner iteration due to testing the outer stopping criterion and an exploration of the RTR method in light of the ρ analysis presented here.

4   Numerical Results

The IRTR method seeks to overcome the inefficiencies of the RTR method, such as the rejection of computed updates and the limitations due to the discrete nature of the trust-region radius. We compare the performance of the IRTR with that of the classical RTR. The following experiments were performed in MATLAB (R14) under Mac OSX. Figure 1 considers a generalized eigenvalue problem with a preconditioned inner iteration. The matrices A and B are from the Harwell-Boeing collection BCSST24. The problem is of size $n = 3562$ and we are seeking the leftmost $p = 5$ eigenvalues. The inner iteration is preconditioned using an exact factorization of A. Two experiments are run: with and without subspace acceleration. When in effect, the subspace acceleration strategy occurs over the 10-dimensional subspace $\mathrm{colsp}([Y_k, S_k])$. The RTR is tested with a value of $\rho' = 0.1$, while the IRTR is run for multiple values of ρ'. These experiments demonstrate that the IRTR method is able to achieve a greater efficiency than the RTR method.

5   Conclusion

This paper presents an optimization-based analysis of the symmetric, generalized eigenvalue problem which explores the relationship between the inner and outer iterations. The paper proposes the Implicit Riemannian Trust-Region method, which seeks to alleviate inefficiencies resulting from the inner/outer divide, while still preserving the strong convergence properties of the RTR method. This algorithm was shown in numerical experiments to be capable of greater efficiency than the RTR method.

Acknowledgments. Useful discussions with Andreas Stathopoulos, Rich Lehoucq and Ulrich Hetmaniuk are gratefully acknowledged.

References

1. Stewart, G.W.: Matrix algorithms, Vol II: Eigensystems. Society for Industrial and Applied Mathematics, Philadelphia (2001)
2. Sameh, A.H., Wisniewski, J.A.: A trace minimization algorithm for the generalized eigenvalue problem. SIAM J. Numer. Anal. **19**(6) (1982) 1243–1259
3. Smith, S.T.: Optimization techniques on Riemannian manifolds. In: Hamiltonian and gradient flows, algorithms and control. Volume 3 of Fields Inst. Commun. Amer. Math. Soc., Providence, RI (1994) 113–136
4. Edelman, A., Arias, T.A., Smith, S.T.: The geometry of algorithms with orthogonality constraints. SIAM J. Matrix Anal. Appl. **20**(2) (1998) 303–353
5. Mongeau, M., Torki, M.: Computing eigenelements of real symmetric matrices via optimization. Comput. Optim. Appl. **29**(3) (2004) 263–287
6. Sameh, A., Tong, Z.: The trace minimization method for the symmetric generalized eigenvalue problem. J. Comput. Appl. Math. **123** (2000) 155–175
7. Knyazev, A.V.: Toward the optimal preconditioned eigensolver: locally optimal block preconditioned conjugate gradient method. SIAM J. Sci. Comput. **23**(2) (2001) 517–541
8. Absil, P.-A., Baker, C.G., Gallivan, K.A.: Trust-region methods on Riemannian manifolds with applications in numerical linear algebra. In: Proceedings of the 16th International Symposium on Mathematical Theory of Networks and Systems (MTNS2004), Leuven, Belgium, 5–9 July 2004. (2004)
9. Absil, P.A., Baker, C.G., Gallivan, K.A.: A truncated-CG style method for symmetric generalized eigenvalue problems. J. Comput. Appl. Math. **189**(1–2) (2006) 274–285
10. Moré, J.J., Sorensen, D.C.: Newton's method. In: Studies in numerical analysis. Volume 24 of MAA Stud. Math. Math. Assoc. America, Washington, DC (1984) 29–82
11. Nocedal, J., Wright, S.J.: Numerical Optimization. Springer Series in Operations Research. Springer-Verlag, New York (1999)
12. Absil, P.-A., Mahony, R., Sepulchre, R.: Riemannian geometry of Grassmann manifolds with a view on algorithmic computation. Acta Appl. Math. **80**(2) (2004) 199–220
13. Absil, P.-A., Baker, C.G., Gallivan, K.A.: Trust-region methods on Riemannian manifolds. submitted (2005)
14. Notay, Y.: Combination of Jacobi-Davidson and conjugate gradients for the partial symmetric eigenproblem. Numer. Linear Algebra Appl. **9**(1) (2002) 21–44
15. Sleijpen, G.L.G., Van der Vorst, H.A.: A Jacobi-Davidson iteration method for linear eigenvalue problems. SIAM J. Matrix Anal. Appl. **17**(2) (1996) 401–425
16. Baker, C.G., Absil, P.-A., Gallivan, K.A.: An implicit Riemannian trust-region method for the symmetric generalized eigenproblem. Technical Report FSU-SCS-2006-152, School of Computational Science, Florida State University (2006) http://scseprints.scs.fsu.edu.

Interval Arithmetic and Computational Science:
Performance Considerations

Alistair P. Rendell, Bill Clarke, and Josh Milthorpe

Department of Computer Science, Australian National University
Canberra ACT0200, Australia
`alistair.rendell@anu.edu.au`

Abstract. Interval analysis is an alternative to conventional floating-point computations that offers guaranteed error bounds. Despite this advantage, interval methods have not gained widespread use in large scale computational science applications. This paper addresses this issue from a performance perspective, comparing the performance of floating point and interval operations for some small computational kernels. Particularly attention is given to the Sun Fortran interval implementation, although the strategies introduced here to enhance performance are applicable to other interval implementations. Fundamental differences in the operation counts and memory references requirements of interval and floating point codes are discussed.

1 Introduction

The majority of science is concerned with physical phenomena, such as velocity, temperature, or pressure that are by their very nature continuous. Meanwhile computations of these quantities are performed using the discrete environment of the digital computer. To bridge this divide it is normal to approximate values to a specified precision using a finite set of machine-representable numbers. Inherent in this process is the concept of a rounding error, the effect of which can be hard to predict *a priori*.

Currently most scientific codes use IEEE 754 double precision arithmetic [1] and give relatively little or no attention to the effects of rounding errors. While this may have been okay in the past, on today's machines that are capable of multi teraflop ($>10^{12}$ operations) per second and with double precision arithmetic providing just 15-16 significant figures, it is easy to see the potential for rounding errors to compound and become as large as the computed quantities themselves.

Interval arithmetic is a fundamentally different approach to floating point computations that that was first proposed by R.E. Moore in 1965 [2], but is yet to achieve widespread use in computational science applications. The idea is to represent a floating point number by two floating point numbers corresponding to a lower and upper bound (referred to as the infima and suprema respectively). In concept it is identical to expressing the uncertainty in a quantity as 1.234 ± 0.001, except that the interval representation would be written as [1.233, 1.235] since this is easier to manipulate. The basic rules for interval addition and multiplication are as follows:

V.N. Alexandrov et al. (Eds.): ICCS 2006, Part I, LNCS 3991, pp. 218–225, 2006.

$$[\underline{x}, \overline{x}] + [\underline{y}, \overline{y}] = [\underline{x} + \underline{y}, \overline{x} + \overline{y}] \tag{1}$$

$$[\underline{x}, \overline{x}] \times [\underline{y}, \overline{y}] = [\min(\underline{x} \times \underline{y}, \underline{x} \times \overline{y}, \overline{x} \times \underline{y}, \overline{x} \times \overline{y}),$$

$$\max(\underline{x} \times \underline{y}, \underline{x} \times \overline{y}, \overline{x} \times \underline{y}, \overline{x} \times \overline{y})] \tag{2}$$

Division is slightly more complicated, particularly if the interval appearing in the denominator spans 0, and for details the reader is referred to [3]. Suffice it to say that anything that can be computed using floating point arithmetic can also be computed using intervals, although there are some fundamental differences between interval and standard floating point arithmetic. For instance interval arithmetic is not distributive [3], so the order in which interval operations are performed can lead to different interval results; the trick is to find the order of operations that give rise to the narrowest or sharpest possible interval result.

Interval computations can be used to bound errors from all sources, including input uncertainty, truncation and rounding errors. Given this apparent advantage it is perhaps somewhat surprising that intervals are not already widely use in computational science. The reason for this is arguably twofold; first difficulties associated with designing interval algorithms that produce narrow or sharp interval results, and second the performance penalty associated with use of intervals. In this paper we do not address the former, but instead focus on the performance penalty associated with performing interval operations on a computer.

There is a range of software products that are designed to support interval computations on popular architectures (see reference 4). Most of these (e.g. FILIB [5]) take the form of C++ template libraries and as such have obvious limitations for widespread use in computational science. A much more attractive alternative is provided by the Sun Studio Fortran compiler [6], as this has been extended to provide support for an interval data type. Technically this means that using this compiler it is possible to make an interval version of an existing computational science application code by simply changing all floating point data types to intervals. While we are investigating such things [7], in this paper the goal is to determine the likely performance penalty a typical user might see if he/she where to make such a switch.

2   Simple Interval Operations

Our first objective was to compare asymptotic performance for pipelined floating point and interval operations. This was done using the following simple Fortran loop:

```
do i = 1, size(input)
      result = result op input(i)
enddo
```

where `result` and `input` were either double precision floats or double precision intervals, and `op` was either addition or multiplication. This loop was placed in a separate procedure that was called many times, with care taken when initializing vector `input` to ensure that `result` did not over or underflow.

The benchmark code was compiled using version 8.1 of the Sun Fortran compiler (part of Sun Studio 10) with options:

```
-fast -xarch=v9b [-xia]
```

where `fast` is a macro corresponding to a number of different optimization flags, `v9b` requests a 64-bit binary, and `xia` indicates that the compiler should use interval support. Timing results for this simple benchmark run on a 900MHz UltraSPARC III Cu system are given in Table 1. As intervals require twice the storage of the equivalent floating point values results obtained using a given floating point vector length are compared with interval results obtained using half that vector length. Results for three different vector sizes are considered. For the small case the data resides in the 64KB level 1 cache, for the medium case the data resides in the 8MB level 2 cache, while for the large case data will be drawn from main memory.

Table 1. Performance (nsec) for summation and product benchmarks compiled using Sun Fortran 95 8.1 and run on a 900MHz UltraSPARC III Cu under Solaris 5.10

Name	—— Floating Point ——			——— Interval ———		
	Input Size	Sum	Product	Input Size	Sum	Product
Small	2000	1.7	1.7	1000	35.7	102
Medium	512000	1.7	1.7	256000	57.4	122
Large	4096000	7.2	7.0	2048000	58.6	124

The timing results for the floating point benchmarks are identical for both the small and medium vector size, reflecting the fact that the compiler has used prefetch instructions (evident in the assembly) to mask the cost of retrieving data from level 2 cache. For the large data set the performance drops markedly due to level 2 cache misses (confirmed using hardware performance counters). For the interval benchmark while there appears to be some cache effects, the most notably observation is the huge performance penalty associated with using intervals compared to floating point variables.

This prompts the obvious question, what should be the relative cost of performing an interval operation over the equivalent floating point operation? From equation 1 it would appear that an interval addition should cost at least twice that of a floating point addition – since two data items must be loaded into registers and two floating point additions performed (one on the infimas and one on the supremas). Some extra cost might also be expected since when adding the infimas the result must be rounded down, while when adding the supremas the result must be rounded up. For the interval product the expected performance is a little more difficult to predict; Equation 2 suggests the need to form 8 possible products and then a number of comparisons ino order to obtain the minimum and maximum of these. While we will return to this issue in more detail it does appear that a slowdown of 21 for the small interval addition benchmark and 60 for the small interval product benchmark over their equivalent floating point versions is excessive.

Indeed it is already clear from these simple results that for interval arithmetic to be widely used for large scale computational science applications the performance penalty associated with use of intervals needs to be substantially reduced. From above and via close inspection of the generated assembly there appears to be four main reasons why the Sun interval code is slower than its equivalent floating point code:

1. The inherent cost of interval arithmetic over equivalent floating point (at least 2 for addition and 8 for multiplication as discussed above);
2. The fact that the Sun compiler translates each interval operation into a function call;
3. A lack of loop unrolling due in part to the use of procedures to implement the basic interval operations; and
4. Need for additional instructions to change the rounding mode every time an interval operation is performed.

To quantify these effects Table 2 gives timing results for the floating point benchmark modified to run in an analogous fashion to the interval code. That is the compiler settings are first adjusted to prevent loop unrolling, then the benchmark is re-written so that each floating point addition or multiplication is performed in a separate function, and then additional calls are inserted into this separate function to switch rounding mode every time it is called. These results coupled with the larger inherent cost of interval operations explain the bulk of the performance differences between the floating point and interval benchmarks (i.e. the differences seen in Table 1).

Table 2. Performance (nsec) of modified sum and product floating-point benchmarks compiled using Sun Fortran 95 8.1 and run on a 900MHz UltraSPARC III Cu under Solaris 5.10. See text for details of modifications.

Size	Initial		No Unrolling		+Function Call		+Rounding	
	Sum	**Prod**	**Sum**	**Prod**	**Sum**	**Prod**	**Sum**	**Prod**
Small	1.7	1.7	4.5	5.6	24.5	24.5	28.9	26.9
Medium	1.7	1.7	4.5	5.6	29.4	30.7	33.8	32.4
Large	7.2	7.0	7.0	7.8	54.5	57.1	56.4	54.4

While the above analysis is based on observations from Sun's Fortran implementation of intervals it should be stressed that issues 2-4 are likely to occur for interval implementations based on C++ template libraries. Specifically these will invariably replace each interval operation by a function call which performs the basic operations under directed rounding, but in so doing the presence of the special function seriously limits the possibility of loop unrolling and other advanced compiler optimizations.

With the above performance data in mind, an obvious strategy when using intervals for large scale applications would be to rewrite selective kernels removing the use of separate function calls, minimizing the need for rounding mode to be switched, and generally enhancing the ability of the compiler to optimize the code. For example a modified version of the addition kernel would look like:

```
call round_up
sum%inf= -0.0d0
sum%sup= 0.0d0
do i = 1, size(input)
   sum%inf = sum%inf - input(i)%inf
   sum%sup = sum%sup + input(i)%sup
enddo
sum%inf = -sum%inf
call round_nearest
```

In this code the rounding mode is changed once before the loop and once at the end, and to enable only one rounding mode to be used one of the end points is negated before and after the loop (this trick is also used by the Sun compiler).

There is, however, one caveat that should be noted when attempting to hand optimizing any interval code. Specifically Sun implements an extended interval arithmetic [8] that guarantees containment for operations on infinite intervals such as $[-\infty,0]$. Dealing with infinite intervals adds a level of complexity that for the current purpose we will ignore. (Arguing that for the vast majority of computational science applications computations on infinity are not meaningful and in anycase if required this special condition can be handled by an "if test" that in most cases will be ignored given the underlying branch prediction hardware.)

Generating hand optimized code for the product benchmark is a little bit more complex as there are at least two alternative interval product formulations [7]. The first is a direct translation of the min/max operations given in Equation 2, while the second recognizes that depending on the signs of the two infima and two suprema it is usually only necessary to perform two multiplications in order to obtain the result. An exception arises when both operands span zero, in which case four multiplications are required. This scheme requires multiple if tests so will be referred to as the branching approach. For the purpose of generating hand optimized code for the product benchmark both options have been programmed.

At this point as well as considering the performance of the hand optimized sum and product kernels it is pertinent also to consider the performance of the Sun Fortran 95 intrinsic sum and product functions, since these perform identical functions to the original kernel. Thus in Table 3 we compare timings obtained from the basic interval code, with those obtained using the intrinsic functions and hand optimized versions of these routines. These show that the performance obtained using the intrinsic sum is significantly better than that obtained by the initial benchmark code. Indeed for the smallest vector size the intrinsic function outperforms the hand optimized code by 40%, but for the larger cases the hand optimized code is superior. The reason for this is that the hand optimized code has been compiled to include prefetch instructions, while the intrinsic function does not (evident from the assembly). Thus for the small benchmark where data is drawn from level 1 cache the intrinsic function shows superior performance, while for the larger benchmarks the processor stalls waiting for data to arrive in registers. For the hand optimized summation code the performance is now within a factor of four from the equivalent floating point code, and while this is larger than we might have expected, it is significantly better than our initial results.

For the product use of the intrinsic function is roughly twice as fast as the original code if the data is in cache. The hand optimized min/max approach appears to give the best performance, although it is still approximately 16 times slower than the equivalent floating point code. In comparison to min/max the branching approach is slightly slower, but this is probably not surprising given that exactly what branches are taken depends on the data values and if these are random (which they are by design in the benchmark used here) branch prediction is ineffective. It is therefore somewhat interesting to see (via disassembly of the relevant routine) that Sun appear to have implemented their interval product using a branching approach.

Table 3. Performance (nsec) comparison of initial summation and product benchmark with versions that uses Fortran intrinsic functions and versions that are hand optimised. Codes compiled using Sun Fortran 95 8.1 and run on a 900MHz (1.1nsec) UltraSPARC III Cu under Solaris 5.10.

	Summation Benchmark			Product Benchmark			
Name	Initial	Intrinsic	Hand	Initial	Intrinsic	Min/Max	Branch
Small	69.1	4.71	6.7	119	66.6	26.8	37.0
Medium	90.7	16.3	6.7	139	78.0	27.2	38.5
Large	92.3	73.0	36.6	141	136.0	56.6	48.5

Noting the relative difference between the two alternative product formulations it is of interest to consider the relative performance of these two schemes on other platforms, and in particular on an out-of-order processor with speculative execution. Thus the original Sun Fortran benchmark was re-written in C and run on an Intel Pentium 4 based system. As this processor has only a 512KB level 2 cache results are given in Table 4 for just the small and large benchmarks sizes. These show that the branching product formulation is now significantly faster than the min/max formulation. Comparing the Pentium 4 and UltraSPARC III Cu we find that the small summation benchmark runs faster on the Pentium 4 by a factor that is roughly equal to the clock speed ratio (2.6 faster compared to a clock speed ratio of 2.9). For the large summation benchmark this ratio is higher reflecting the greater memory bandwidth of the Pentium compared to the UltraSPARC. For the product the branching code is approximately 3.5 times faster than the min/max code on the UltraSPARC.

Table 4. Performance (nsec) of hand optimized C benchmark compiled using gcc 3.4.4 and run on a 2.6GHz (0.385nsec) Intel Pentium 4 system under the Linux 2.6.8-2-686-smp core

	Sum	Product	
Name	Hand	Min/Max	Branch
Small	2.6	35.2	7.6
Large	6.6	36.3	8.7

3 Compound Interval Operations

The previous section compared the performance of basic floating point and interval addition and multiplication operations. In this section we consider the dot-product and AXPY (Alpha times X plus Y) operations that form part of the level 1 Basic Linear

Algebra Subprogram (BLAS) library. As well as being slightly more complex, these operations are of interest since they form the building blocks for higher level BLAS operations, such as matrix multiplication. While the dot product and AXPY operations were explicitly coded in Fortran, dot product is also a Fortran90 intrinsic function and an interval version of AXPY is available as part of the Sun Performance Library. Thus in Table 5 we present timing results for both these operations obtained using basic Fortran code, Sun intrinsic functions or the Sun performance library, and hand optimized versions of these routines produced using a similar strategy to that outlined above for the simple benchmarks. These results again show that significant benefit can be gained from hand optimizing these routines.

Table 5. Performance (nsec) of various floating pint and interval level 1 BLAS benchmarks compiled using Sun Fortran 95 8.1 and run on a 900MHz (1.1nsec) UltraSPARC III Cu under Solaris 5.10. See text for details.

| | Floating Point | | Sun Interval | | Hand Interval | |
Name	Dot	AXPY	Dot	AXPY	Dot	AXPY
Small	3.4	6.3	81.4	250	18.1	34.1
Medium	3.3	6.6	108	306	18.4	85.0
Large	14.6	18.6	216	314	94.7	152

4   Conclusions and Discussions

Most software support for interval computation is in the form of libraries with all interval operations being compiled to library calls; this appears to be the approach that Sun have taken in providing Fortran support for intervals. The results obtained here show that such implementations are unlikely to result in performance levels that are acceptable for them to be used in large scale scientific computations. Better performance can be obtained by effectively inlining the interval operations, minimizing changes in rounding mode, and allowing the compiler to see and optimize as much of the time consuming interval code as possible. Such an approach does come with potential drawbacks, notably the containment problems alluded to above. For the bulk of computational science applications, however, we believe that this is not a major issue compared with the need to obtain much better performance from current interval implementations. (At least from the perspective of generating a few large scale proof of concept computation science applications that show a positive benefit from the use of intervals.)

While the performance of current interval implementations is likely to be too slow for them to find widespread use in computational science, it is of interest to consider architectural changes that might alter this conclusion. To this end in Table 6 we compare the floating point and memory operation mix required for floating point and interval based dot and AXPY operations (assuming that an interval multiplication requires 8 floating point multiplications). This shows that while on both the UltraSPARC III Cu and Pentium 4 processors the floating point versions are load/store limited, the interval versions tend to be floating point limited. Other non-trivial interval operations seems to exhibit similar behavior leading to us to the general conclusion that *interval arithmetic code requires a higher ratio of floating-point or flow-control operations to memory operations than does the corresponding floating point code.* Given that the performance of most modern computational science applications tend to be load store limited yet on chip real estate is becoming increasingly cheap (as evidenced by the development of larger on chip caches plus multithreaded and multi-core chips) raises the question as to whether with relatively little effort new chips could be designed to perform interval operations much faster.

Table 6. Comparison of operation mix for floating point and interval based Dot and AXPY level 1 BLAS operations (of length n) with hardware limits

| | | Floating-Point Ops | | Memory Ops | | Ratio |
		Add	Multiply	Load	Store	FP:Mem
Dot	Floating	n	n	2n	1	1:1
	Interval	2n	8n	4n	2	5:2
AXPY	Floating	n	n	2n	n	2:3
	Interval	2n	8n	4n	2n	5:3
Limit	UltraSPARC III Cu	1	1		1	2:1
	Pentium 4	1	1		1	2:1

Acknowledgements. This work was supported in part by the Australian Research Council through Discovery Grant DP0558228.

References

1. IEEE Standard 754-1985 for Binary Floating-point Arithmetic, IEEE, (1985). Reprinted in SIGPLAN 22, 9-25.
2. R.E. Moore, *"The Automatic Analysis and Control of Error in Digital Computation Based on the Use of Interval Numbers"*, in "Proc. of an Adv. Seminar Conducted by the Mathematics Research Center, US Army, University of Wisconsin, Madison", Ed L.B. Rall, John Wiley, (1965), pp 61-130. (Available at http://interval.louisiana.edu/Moores_early_papers/Moore_in_Rall_V1.pdf)
3. G.W. Walster, *"Interval Arithmetic: The New Floating-Point Arithmetic Paradigm"*, Oct 1997, Sun Microsystems white paper, available at http://www.mscs.mu.edu/~globsol/Papers/f90-support-clean.ps
4. Interval web page http://www.cs.utep.edu/interval-comp/main.html
5. M. Lerch, G. Tischler, and J.W. von Gudenberg, *filib++ - interval library specification and reference manual.* Techn. Rep. 279, Lehrstuhl für Informatik II, Universität Würzburg, 2001.
6. Fortran 95 interval arithmetic programming reference, Sun Studio 10. Tech. Rep. 819-0503-10, Sun Microsystems, Inc., Jan 2005.
7. J. Milthorpe, *"Using Interval Analysis to Bound Numerical Errors in Scientific Computing"*, honours thesis, Department of Computer Science, Australian National University, October 2005.
8. G.W. Walster, *Closed Interval Systems*, Sun Microsystems whitepaper, 2002, available at http://www.sun.com/software/sundev/whitepapers/closed.pdf.

Floating-Point Computation
with Just Enough Accuracy

Hank Dietz, Bill Dieter, Randy Fisher, and Kungyen Chang

University of Kentucky, Department of Electrical & Computer Engineering
`hankd@engr.uky.edu, dieter@engr.uky.edu,`
`randall.fisher@ieee.org, kchan0@engr.uky.edu`

Abstract. Most mathematical formulae are defined in terms of operations on real numbers, but computers can only operate on numeric values with finite precision and range. Using floating-point values as real numbers does not clearly identify the precision with which each value must be represented. Too little precision yields inaccurate results; too much wastes computational resources.

The popularity of multimedia applications has made fast hardware support for low-precision floating-point arithmetic common in Digital Signal Processors (DSPs), SIMD Within A Register (SWAR) instruction set extensions for general purpose processors, and in Graphics Processing Units (GPUs). In this paper, we describe a simple approach by which the speed of these low-precision operations can be speculatively employed to meet user-specified accuracy constraints. Where the native precision(s) yield insufficient accuracy, a simple technique is used to efficiently synthesize enhanced precision using pairs of native values.

1   Introduction

In the early 1990s, the MasPar MP1 was one of the most cost-effective supercomputers available. It implemented floating-point arithmetic using four-bit slices, offering much higher performance for lower precisions. Thus, Dietz collected production Fortran programs from various researchers and analyzed them to see if lower precisions could be used without loss of accuracy. The discouraging unpublished result: using the maximum precision available, static analysis could not *guarantee* that even one digit of the results was correct! The insight behind the current paper is that most results were acceptably accurate despite using insufficient precision. Why not deliberately use fast low precision, repeating the computation at higher precision only when a dynamic test of result accuracy demands it?

Bit-slice floating-point arithmetic is no longer in common use, but the proliferation of multimedia applications requiring low-precision floating-point arithmetic has produced DSP (Digital Signal Processor), SWAR (SIMD Within A Register)[1], and GPU (Graphics Processing Unit) hardware supporting *only* 16-bit or 32-bit floating-point arithmetic. Scientific and engineering applications often require accuracy that *native* multimedia hardware precisions cannot guarantee, but by using relatively slow (synthesized) higher-precision operations *only* to recompute values that did not meet accuracy requirements, the low cost and high performance of multimedia hardware can be leveraged.

V.N. Alexandrov et al. (Eds.): ICCS 2006, Part I, LNCS 3991, pp. 226–233, 2006.

Section 2 overviews our method for synthesizing higher precision operations using pairs of native values and gives microbenchmark results for native-pair arithmetic optimized to run on DSPs, SWAR targets, and GPUs. Section 3 presents a very simple compiler/preprocessor framework that supports speculative use of lower precision, automatically invoking higher precision recomputations only when dynamic analysis of the result accuracy demands it. Conclusions are summarized in Section 4.

2 Multi-precision Arithmetic Using Error Residuals

There are many ways to synthesize higher-precision operations [2]. The most efficient method for the target multimedia hardware is what we call *native-pair* arithmetic, in which a pair of native-precision floating point values is used with the `lo` component encoding the residual error from the representation of the `hi` component. We did not invent this approach; it is well known as *double-double* when referring to using two 64-bit doubles to approximate quad precision [3, 4]. Our contributions center on tuning the analysis, algorithms, data layouts, and instruction-level coding for the multimedia hardware platforms and performing detailed microbenchmarks to bound performance.

More than two values may be used to increase precision, however, successive values reduce the exponent range by at least the number of bits in the mantissa extensions. Ignoring this effect was rarely a problem given the number of exponent bits in an IEEE 754[5] compliant 64-bit binary floating-point `double`, but a 32-bit `float` has a 24-bit mantissa and only an 8-bit exponent. A `float` pair will have twice the native mantissa precision only if the exponent of the low value is in range, which implies the high value exponent must be at least 24 greater than the native bottom of the exponent range; thus, we have lost approximately 10% of the dynamic range. Similarly, treating four `float` as an extended-precision value reduces the effective dynamic range by at least 3×24, or 72 exponent steps – which is a potentially severe problem. Put another way, precision is limited by the exponent range to less than 11 `float` values.

One would expect, and earlier work generally assumes, that the exponents of the `lo` and `hi` components of a native-pair will differ by precisely the number of bits in the mantissa. However, values near the bottom of the dynamic range have a loss of precision when the `lo` exponent falls below the minimum representable value. A component value of 0 does not have an exponent per se, and is thus a special case. For non-zero component values, normalization actually ensures only that the exponent of `lo` is *at least* the number of component mantissa bits less than that of `hi`. Using `float` components, if the 25th bit of the higher-precision mantissa happens to be a 0, the exponent of `lo` will be at least 25 less – not 24 less. In general, a run of k 0 bits logically at the top of the lower-half of the higher-precision mantissa are absorbed by reducing the `lo` exponent by k. For this reason, some values requiring up to k bits more than twice the `native` mantissa precision can be precisely represented! However, this also means that, if the `native` floating-point does not implement denormalized arithmetic (many implementations do not[6, 7]), a run of k 0 bits will cause `lo` to be out of range (i.e., represented as 0) if an exponent of k less than that of `hi` is not representable; in the worst case, if the `hi` exponent is 24 above the minimum value and $k=1$, the result has only 25 rather than 48 bit precision. Earlier work[8] is oblivious to these strange numerical properties; our runtime accuracy checks are a more appropriate response.

Space does not permit listing our optimized algorithms in this paper. Although we used C for some development and testing, most compilers cannot generate good code for the routines in C because they tend to "optimize" the floating-point operations in a way that does not respect precision constraints. Further, significantly higher performance may be obtained by careful use of instruction set features and architecture-specific data layouts. The following subsections summarize the microbenchmark performance of our machine-specific, hand-optimized, assembly-level code for each target architecture.

2.1 Performance Using Host Processor Instructions

If native-pair operations using attached multimedia processors are too slow to be competitive with higher-precision operations on the host processor, then these operations should be performed on the host or can be divided for parallel execution across the host and multimedia hardware. Table 1 lists the official clock-cycle latencies for host processor native (X87) floating point operations using an AMD ATHLON[9] and INTEL PENTIUM 4[10].

Table 1. Performance, in clock cycles, of host processor instructions

type	processor	add	sub	mul	sqr	div	sqrt
32-bit `float`	ATHLON	4	4	4	4	16	19
32-bit `float`	PENTIUM 4	5	5	7	7	23	23
64-bit `double`	ATHLON	4	4	4	4	20	27
64-bit `double`	PENTIUM 4	5	5	7	7	38	38
80-bit extended	ATHLON	4	4	4	4	24	35
80-bit extended	PENTIUM 4	5	5	7	7	43	43

Native-pair operations constructed using these types are approximately an order of magnitude slower, so it is fairly obvious that pairing 32-bit `float` values is not productive. However, pairing 64-bit `double` values or 80-bit extended values is useful (although loading and storing 80-bit values is relatively inefficient). Thus, a host processor can effectively support at least five precisions, roughly corresponding to mantissas of 24, 53, 64, 106, and 128 bits with a separate sign bit.

2.2 DSP Targets

There are many different types of DSP chips in common use, most of which do not have floating-point hardware. Of those that do, nearly all support only precisions less than 64 bits. Our example case is the Texas Instruments TMS320C31[6], which provides non-IEEE 754 floating-point arithmetic using an 8-bit exponent and 24-bit mantissa, both represented in 2's complement. This DSP has specialized multiply-add support, which accelerates the multiply, square, and divide algorithms, but neither add nor subtract. Table 2 gives the experimentally-determined clock cycle counts for each of the `native` and `nativepair` operations.

Table 2. Cycle counts and instructions required for DSP operations

<table>
<tr><td colspan="7">(a) Opertaion cycle counts</td></tr>
</table>

type	add	sub	mul	sqr	div	sqrt
native	1	1	1	1	42	51
nativepair	11	11	25	19	112	119

(b) Instructions required

type	add	sub	mul	sqr	div	sqrt
native	1	1	1	1	33	39
nativepair	11	11	25	19	64	99

In general, an order of magnitude slowdown is incurred for `nativepair` operations, but divide and square root do better because they require executing many instructions for `native` operands. It is worth noting that the additional code size for `nativepair` operation sequences is modest enough to allow their use in embedded systems even if ROM space is tight; the number of instruction words for each operation is summarized in Table 2.

2.3 SWAR Targets and SWAR Data Layout

The most commonly used floating-point SWAR instruction sets are 3DNow![11, 7], SSE[12] (versions 1, 2, and 3[13] and the AMD64 extensions[14]), and ALTIVEC[15]. These instruction sets differ in many ways; for example, 3DNow! uses 64-bit registers while the others use 128-bit registers. However, there are a few common properties. The most significant commonality is that all of these SWAR instruction sets use the host processor memory access structures. Thus, the ideal data layout is markedly different from the obvious layout assumed in earlier multi-precision work.

Logically, the `hi` and `lo` parts of a `nativepair` may together be one object, but that layout yields substantial alignment-related overhead for SWAR implementations even if the `nativepair` values are aligned: different fields within the aligned objects have to be treated differently. The ideal layout separates the `hi` and `lo` fields to create contiguous, aligned, interleaved, vectors of the appropriate length. For example, 32-bit 3DNow! works best when pairs of `nativepair` values have their components interleaved as a vector of the two `hi` fields and a vector of two `lo` fields; for SSE and ALTIVEC, the vectors should be of length four. The creation of separate, Fortran-style, arrays of `hi` and `lo` components is not as efficient; that layout makes write combining ineffective, requires rapid access to twice as many cache lines, and implies address accesses separated by offsets large enough to increase addressing overhead and double the TLB/page table activity.

Given the appropriate data layout, for 3DNow! the primary complication is that the instruction set uses a two-register format that requires move instructions to avoid overwriting values. Table 3 the experimentally-determined cycle counts using the cycle count performance register in an AMD ATHLON XP.

All measurements were taken repeating the operation within a tight loop, which did allow some parallel overlap in execution (probably more than average for `swarnative` and less for `swarnativepair`). All counts given are for operations on two-element SWAR vectors of the specified types; for example, two `nativepair_add` operations are completed in 24 clock cycles. Although 3DNow! offers twice the 32-bit floating-point performance of the X87 floating-point support within the same processor, the

Table 3. Cycle counts for 3DNow! and SSE operations

	type	add	sub	mul	sqr	div	sqrt
3DNow!	swarnative	1	1	1	1	9	9
3DNow!	swarnativepair	24	28	27	14	57	40
SSE	float swarnativepair	51	50	148	129	173	199
SSE	double swarnativepair	45	48	48	42	50	-

X87 `double` arithmetic is faster than 3DNOW! `nativepair` for all operations except reciprocal and square root.

The SSE code is very similar to that used for 3DNOW!, differing primarily in data layout: there are four 32-bit values or two 64-bit values in each `swarnative` value. Thus, the float version produces twice as many results per `swarnativepair` operation. The code sequences were executed on an INTEL PENTIUM 4 and the cycle counter performance register was used to obtain the cycle counts in Table 3, which show that `float swarnativepair` does not compete well with host `double`, but `double swarnativepair` is very effective.

2.4 GPU Targets

DSP parts tend to be slow, but can function in parallel with a host processor; SWAR is fast, but does not work in parallel with the host. The excitement about GPU targets comes from the fact that they offer both the ability to operate in parallel with the host and speed that is competitive with that of the host.

Although there are many different GPU hardware implementations, all GPUs share a common assembly-language interface for vertex programs[16, 17] and for fragment (pixel-shading) programs[16, 17]. All GPUs use SWAR pixel operations on vectors of 4 components per register (corresponding to the red, green, blue, and alpha channels), with relatively inefficient methods for addressing fields within registers. Thus, the optimal data layout and coding for native-pair operations is very similar to that used for SSE. Oddly, the precision of GPU arithmetic is not standardized, ranging from 16-bit to 32-bit. For our latest experiments, we purchased a $600 NVIDIA GEFORCE 6800 ULTRA, which was then the fastest commodity GPU with roughly IEEE-compliant 32-bit floating point support; Table 4 shows the performance results .

Table 4. Relative cost of GPU operations

type	add	sub	mul	sqr	div	sqrt
swarnative	1	1	*	*	4	20
swarnativepair	11	11	18	10	35	28

To obtain the above numbers, it was necessary to resort to fractional factorial experimental procedures that timed combinations of operations and used arithmetic methods to extract times for individual operations. Each experiment was repeated 220 times to determine a 95% confidence interval for each time, which was then used to compute upper and lower bound times for the individual operations. Quirks of the NVIDIA GEFORCE 6800 ULTRA GPU and its assembler yielded inconsistent timing for some combinations of operations; there were insufficient consistent timings for the multiplication and squaring operations to determine the execution cost (probably about the same as add). All the other costs are listed in the above table relative to the cost of a `swarnative` add; in no case was the 95% confidence error greater than 1 unit. These results also are generally consistent with preliminary experiments we performed using an ATI RADEON 9800 XT with a less sophisticated timing methodology.

3   Compiler and Language Support

As mentioned earlier in this paper, traditional compiler technology is somewhat incompatible with the already awkward `nativepair` codings. Assembly-level coding is not viable for implementing complicated numerical algorithms. Implementation and benchmarking are beyond the scope of this paper, but we suggest that the compilation system should explicitly manage precision, including support for *speculative precision*.

Analysis and code generation for explicit precisions allows the compiler not only to maintain correctness while optimizing finite-precision computations, but also to select the fastest implementation for each operation individually – for example, using 3DNow! `swarnativepair` for square root and X87 double for other operations. Precision directives have been used to preprocess Fortran code to make use of an arbitrary-precision arithmetic package[8]. Better, the notation used in our SWARC[18] dialect of C can be extended: `int:5` specifies an integer of at least five bits precision, so `float:5` could specify a floating-point value with at least five mantissa bits. A less elegant notation can be supported using C++ templates. Requirements on dynamic range, support of IEEE 754 features like NAN and INFINITY, etc., are more complex and less commonly an issue; they can be specified using a more general, if somewhat awkward, syntax.

Speculative precision is based on specifying accuracy constraints. Accuracy requirements can be specified directly or, more practically, as both an accuracy required and a functional test to determine the accuracy of a result. The compiler would generate multiple versions of the speculative-precision code, one for each potentially viable precision. A crude but effective implementation can be created for C++ using a simple preprocessor with straightforward directives like:

#faildef *failure_code* *Defines **failure_code** as the code to execute when all precision alternatives fail; this definition can be used for many speculative blocks, not just one*

#specdef *name(item1, item2, ...)* *Defines name as the ordered sequence of types **item1**, **item2**, etc.; this definition can be used for many speculative blocks, not just one*

#speculate *name1 name2 ... Defines the start of a region of code which is to be speculatively executed for **name1**, **name2**, etc. taking each of the values specified in sequence*

#fail *Defines the position at which the failure action should be applied*

#commit *Defines the position at which the speculate region ends*

In practice, the accuracy check would be a much cheaper computation than the speculative computation; for example, Linpack and many other math libraries compute error terms that could be examined, but we prefer a short example for this paper. Suppose that there are both `float` and `double` versions of `sqrt()`, overloaded using the usual C++ mechanisms, and our goal is to use the cheapest version that can pass a simple accuracy test `mytest()`. Our short example could be coded as:

```
#faildef exit(1);
#specdef fd(float, double)
#speculate fd
  fd a = x; double b = sqrt(a); if (!mytest(b, x)) {
#fail
  } y = b;
#commit
```

Which would be preprocessed to create C++ code like:

```
#define faildef { exit(1); }
#define fd float
{ fd a = x; double b = sqrt(a);
if (!mytest(b, x)) {goto fail0_0;} y = b; } goto commit0;
#define fd double
fail0_0: ; { fd a = x; double b = sqrt(a);
if (!mytest(b, x)) { faildef } y = b; } commit0: ;
```

The syntax could be prettier, but this speculation mechanism has very little overhead and is general enough to handle many alternative-based speculations, not just speculation on types. Further, although the sample generated code simply tries the alternatives in the order specified (which would typically be lowest precision first), one could use a history mechanism resembling a branch predictor to intelligently alter the type sequence based on past behavior.

Another possible improvement is to optimize the higher-precision computations to incrementally improve the precision of the already-computed results, but this is much more difficult to automate. For example, using the `native` results as the initial top-halves of the `nativepair` computations may be cheaper than computing `nativepair` results from scratch, but the computation is changed in ways far too complex to be managed by a simple preprocessor.

4   Conclusions

Although floating-point arithmetic has been widely used for decades, the fact that it is a poor substitute for real numbers has continued to haunt programmers. Unexpected accuracy problems manifest themselves far too frequently to ignore, so specifying excessive precision has become the norm. Even the highest precision supported by the hardware sometimes proves insufficient. This paper suggests a better way.

Rather than statically fixing guessed precision requirements in code, we suggest a more dynamic approach: code to try the lowest potentially viable precision and try successively higher precisions only when the accuracy is (dynamically at runtime) determined to be inadequate. Thanks to demand in the multimedia community, lower-precision floating-point arithmetic is now often implemented with very high performance and very low cost. The techniques we have developed for extending precision using optimized native-pair arithmetic are commonly an order of magnitude slower than native. However, the large speed difference actually makes speculation more effective. Even if speculating lower precision usually fails to deliver the desired accuracy, an occasional success will reap a significant speedup overall. Only experience with a range of applications will determine just how often speculation succeeds.

References

1. Dietz, H.G., Fisher, R.J.: Compiling for SIMD within a register. In Chatterjee, S., Prins, J.F., Carter, L., Ferrante, J., Li, Z., Sehr, D., Yew, P.C., eds.: Languages and Compilers for Parallel Computing. Springer-Verlag (1999) 290–304
2. Bailey, D.H., Hida, Y., Jeyabalan, K., Li, X.S., Thompson, B.: Multiprecision software directory. http://crd.lbl.gov/~dhbailey/mpdist/ (2006)
3. Dekker, T.J.: A floating-point technique for extending the available precision. Numer. Math. **18** (1971) 224–242
4. Linnainmaa, S.: Software for doubled-precision floating-point computations. ACM Trans. Math. Softw. **7**(3) (1981) 272–283
5. IEEE: IEEE Standard for Binary Floating Point Arithmetic Std. 754-1985. (1985)
6. Texas Instruments: TMS320C3x User's Guide. (2004)
7. Advanced Micro Devices: 3DNow! Technology Manual. (2000)
8. Bailey, D.H.: Algorithm 719; Multiprecision translation and execution of FORTRAN programs. ACM Trans. Math. Softw. **19**(3) (1993) 288–319
9. Advanced Micro Devices: AMD Athlon Processor x86 Code Optimization Guide. (2002)
10. Intel: Intel Pentium 4 and Intel Xeon Processor Optimization Reference Manual. (2002)
11. Advanced Micro Devices: AMD64 Architecture Programmer's Manual Volume 5: 64-Bit Media and x87 Floating-Point Instructions. (2003)
12. Klimovitski, A.: Using SSE and SSE2: Misconceptions and reality. Intel Developer UPDATE Magazine (2001)
13. Smith, K.B., Bik, A.J.C., Tian, X.: Support for the Intel Pentium 4 processor with hyper-threading technology in Intel 8.0 compilers. Intel Technology Journal **08**(ISSN 1535-864X) (2004)
14. Advanced Micro Devices: AMD64 Architecture Programmer's Manual Volume 4: 128-Bit Media Instructions. (2003)
15. Freescale Semiconductor: AltiVec Technology Programming Interface Manual. (1999)
16. Microsoft: DirectX Graphics Reference. (2006)
17. Silicon Graphics, Inc: OpenGL Extension Registry. (2003)
18. Fisher, R.J., Dietz, H.G.: The Scc Compiler: SWARing at MMX and 3DNow. In Carter, L., Ferrante, J., eds.: Languages and Compilers for Parallel Computing. Springer-Verlag (2000) 399–414

Independent Component Analysis Applied to Voice Activity Detection

J.M. Górriz[1], J. Ramírez[1], C.G. Puntonet[3],
E.W. Lang[3], and K. Stadlthanner[3]

[1] Dpt. Signal Theory, Networking and communications, University of Granada, Spain
gorriz@ugr.es
http://www.ugr.es/~gorriz
[2] Dpt. Computer Architecture and Technology, University of Granada, Spain
[3] AG Neuro- und Bioinformatik, Universität Regensburg, Deutschland

Abstract. In this paper we present the first application of Independent Component Analysis (ICA) to Voice Activity Detection (VAD). The accuracy of a multiple observation-likelihood ratio test (MO-LRT) VAD is improved by transforming the set of observations to a new set of independent components. Clear improvements in speech/non-speech discrimination accuracy for low false alarm rate demonstrate the effectiveness of the proposed VAD. It is shown that the use of this new set leads to a better separation of the speech and noise distributions, thus allowing a more effective discrimination and a tradeoff between complexity and performance. The algorithm is optimum in those scenarios where the loss of speech frames could be unacceptable, causing a system failure. The experimental analysis carried out on the AURORA 3 databases and tasks provides an extensive performance evaluation together with an exhaustive comparison to the standard VADs such as ITU G.729, GSM AMR and ETSI AFE for distributed speech recognition (DSR), and other recently reported VADs.

1   Introduction

The demands of modern applications of speech communication are related to the need for increasing levels of performance in noise adverse environments. The new voice services including discontinuous speech transmission [1] or distributed speech recognition (DSR) over wireless and IP networks [2] are examples of such applications. These systems often require a noise reduction scheme working in combination with a precise VAD in order to palliate the harmful effect of the acoustic environment on the speech signal. Thus, numerous researchers have studied different strategies for detecting speech in noise and the influence of the VAD on the performance of speech processing systems [3, 4, 5, 6, 7]. Recently, an improved VAD using a long-term LRT test defined on the Bispectra coefficients [8] has shown significant improvements of the decision rule but with high computational cost. The latter VAD is based on a MO-LRT over a set of averaged bispectrum coefficients $\mathbf{y}_k$ which are assumed to be independent. This approach

V.N. Alexandrov et al. (Eds.): ICCS 2006, Part I, LNCS 3991, pp. 234–241, 2006.

is also assumed in the Fourier domain [7]. We use the latter algorithm in conjunction with the recursive PCA algorithm presented in [9] to build an efficient "on line" VAD, assessing its performance in an HMM-based speech recognition system.

The rest of the paper is organized as follows. Section 2 presents the Blind Source Separation (BSS) problem when dealing with Gaussian distributions. In section 3 we review the theoretical background related to MO-LRT applied to VAD showing the proposed signal model and analyzing the motivations for the proposed algorithm by showing the speech/non-speech correlation plots. Section 4 introduce a variation on the recursive PCA for the evaluation of the decision rule. Section 5 describes the experimental framework considered for the evaluation of the proposed statistical decision algorithm. Finally, in section we state some conclusions.

2   BSS in Gaussian Scenario

Let us denote by $\mathbf{x} = (x_1, \ldots, x_m)$ a zero m-dimensional random variable that can be observed and $\mathbf{s} = (s_1, \ldots, s_m)$ its m-dimensional statistically independent transform satisfying that:

$$\mathbf{s} = \mathbf{W}\mathbf{x} \tag{1}$$

where $\mathbf{W}$ is a constant (weight) square matrix. The BSS problem [10, 11] is to determine the previous matrix extracting the independent features s_i, $i = 1, \ldots, m$ and assuming $\mathbf{W}$ is constant. There are some ambiguities in the determination of the linear model (i.e. variances and order of independent components) but fortunately they are insignificant in most applications.

The multivariate Gaussian probability density function (pdf) of an $m \times 1$ random variable vector $\mathbf{x}$ is defined as:

$$p(\mathbf{x}) = \frac{1}{(2\pi)^{m/2} \det^{1/2}(\mathbf{R}(\mathbf{x}))} \exp\left[-\frac{1}{2}(\mathbf{x} - \mu)^T \mathbf{R}(\mathbf{x})^{-1}(\mathbf{x} - \mu)\right] \tag{2}$$

where μ is the mean vector and $\mathbf{R}(\mathbf{x})$ is the covariance matrix. It is assumed that $\mathbf{R}(\mathbf{x})$ is positive definite and hence $\mathbf{R}(\mathbf{x})^{-1}$ exists. The mean vector is defined as $[\mu]_i = E(x_i)$, $i = 1, \ldots m$ and the covariance matrix as $\mathbf{R}(\mathbf{x}) = E((\mathbf{x} - E(\mathbf{x}))(x - E(\mathbf{x}))^T))$. Uncorrelated jointly Gaussian variables (i.e. components of vector $\mathbf{x}$ in equation 2) satisfy that:

$$p(\mathbf{x}) = \prod_{i=1}^{m} p(x_i) \tag{3}$$

since its covariance matrix is diagonal. That is, they are statistically independent, then decorrelation and statistical independence are equivalent for jointly Gaussian variables. In VAD applications, the DFT coefficients of the incoming signal are usually assumed to be Gaussian independent somehow, using the model of overlapped MO-window (see section 3), they are not.

Principal Component Analysis (PCA) is a very efficient and popular tool for extracting uncorrelated components from a set of signals. PCA tries to find a linear transformation (Karhunen-Loéve) $\widetilde{\mathbf{s}} = \mathbf{W}^T\mathbf{x}$ into a new orthogonal basis (columns of $\mathbf{W}$) such that the covariance matrix in the new system is diagonal. Since $\mathbf{R}(\mathbf{x})$ is symmetric we can find $\mathbf{W}$ such that:

$$\mathbf{W}^T\mathbf{R}(\mathbf{x})\mathbf{W} = \mathbf{\Lambda} \tag{4}$$

Hence the PCA decorrelates the vector $\mathbf{x}$ also achieving statistical independence when it is multivariate Gaussian distributed[1].

3   PCA-MO-Likelihood Ratio Test

In a two hypothesis test (ω_0 =noise & ω_1 =speech in noise), the optimal decision rule that minimizes the error probability is the Bayes classifier. Given an J-dimensional observation vector $\hat{\mathbf{x}}$ to be classified, the problem is reduced to selecting the class (ω_0 or ω_1) with the largest posterior probability $P(\omega_i|\hat{\mathbf{x}})$. From the Bayes rule:

$$L(\hat{\mathbf{x}}) = \frac{p_{\mathbf{x}|\omega_1}(\hat{\mathbf{x}}|\omega_1)}{p_{\mathbf{x}|\omega_0}(\hat{\mathbf{x}}|\omega_0)} \begin{array}{c} > \\ < \end{array} \frac{P[\omega_0]}{P[\omega_1]} \Rightarrow \begin{array}{c} \hat{\mathbf{x}} \leftrightarrow \omega_1 \\ \hat{\mathbf{x}} \leftrightarrow \omega_0 \end{array} \tag{5}$$

In the LRT, it is assumed that the number of observations is fixed and represented by a vector $\hat{\mathbf{x}}$. The performance of the decision procedure can be improved by incorporating more observations to the statistical test. When $M = 2m + 1$ measurements $\hat{\mathbf{x}}_{-m}, \hat{\mathbf{x}}_{-m+1}, \ldots, \hat{\mathbf{x}}_m$ are available in a two-class classification problem, a MO-LRT can be defined by:

$$L_M(\hat{\mathbf{x}}_{-m}, \hat{\mathbf{x}}_{-m+1}, ..., \hat{\mathbf{x}}_m) = \frac{p_{\mathbf{x}_{-m}, \mathbf{x}_{-m+1}, ..., \mathbf{x}_m|\omega_1}(\hat{\mathbf{x}}_{-m}, \hat{\mathbf{x}}_{-m+1}, ..., \hat{\mathbf{x}}_m|\omega_1)}{p_{\mathbf{x}_{-m}, \mathbf{x}_{-m+1}, ..., \mathbf{x}_m|\omega_0}(\hat{\mathbf{x}}_{-m}, \hat{\mathbf{x}}_{-m+1}, ..., \hat{\mathbf{x}}_m|\omega_0)} \tag{6}$$

In order to evaluate the proposed MO-LRT VAD on an incoming signal, an adequate statistical model for the feature vectors in presence and absence of speech needs to be selected. The model selected is similar to that used by Sohn *et al.* [3] that assumes the DFT coefficients of the clean speech (S_j) and the noise (N_j) to be asymptotically independent Gaussian random variables. In our case, the decision rule is formulated over a sliding window consisting of $2m + 1$ observation vectors around the frame for which the decision is being made (see figure 1), then we have to assume they are at least jointly Gaussian distributed as in equation 2.

The PCA algorithm presented in the next section decorrelates the observed signals $\hat{\mathbf{x}}_k$ into a set of independent signal $\hat{\mathbf{s}}_\mathbf{k}$ hence the MO-LRT in equation 6 can be expressed as:

$$\ell_{l,m} = \sum_{k=l-m}^{l+m} \ln \frac{p_{\mathbf{s}_k|\omega_1}(\hat{\mathbf{s}}_k|\omega_1)}{p_{\mathbf{s}_k|\omega_0}(\hat{\mathbf{s}}_k|\omega_0)} \tag{7}$$

---

[1] Observe how any transformation of the kind $\widetilde{\mathbf{x}} = (\mathbf{VW})^T\mathbf{x}$ where $\mathbf{V}$ is a orthogonal matrix ($\mathbf{VV}^T = \mathbf{V}^T\mathbf{V} = I$) yields the same result.

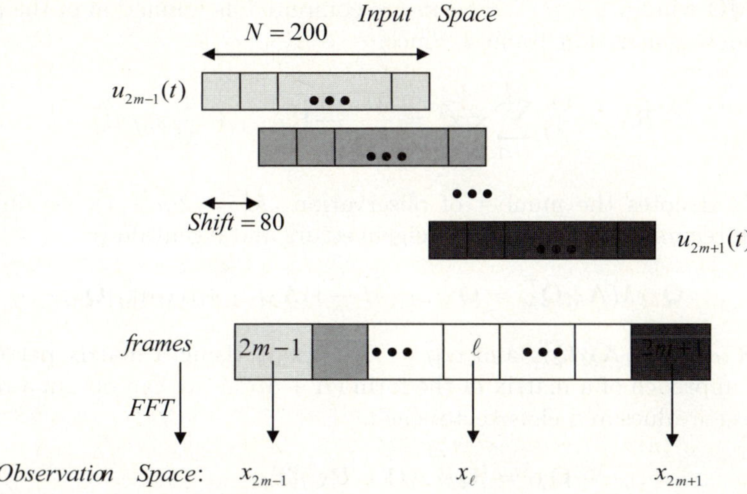

Fig. 1. Signal Model used in the PCA-MO-LRT. Observe how the use of overlapped windows introduces some correlation in the Observation Space then the assumption of statistical independence is not appropriate.

where l denotes the frame being classified as speech (ω_1) or non-speech (ω_0) and the pdf of the observations can be computed using:

$$p(\hat{\mathbf{s}}|\omega_0) = \prod_{j=0}^{J-1} \frac{1}{\pi \lambda_N(j)} \exp\left\{ -\frac{|s_j|^2}{\lambda_N(j)} \right\}$$
$$p(\hat{\mathbf{s}}|\omega_1) = \prod_{j=0}^{J-1} \frac{1}{\pi[\lambda_N(j)+\lambda_S(j)]} \exp\left\{ -\frac{|s_j|^2}{\lambda_N(j)+\lambda_S(j)} \right\}$$

(8)

where s_j represents the uncorrelated noisy speech DFT coefficients, J is the DFT resolution and $\lambda_N(j)$ and $\lambda_S(j)$ denote the variances of N_j and S_j, respectively.

By defining: $\Phi(k) = \ln \frac{p_{\mathbf{s}_k|\omega_1}(\hat{\mathbf{s}}_k|\omega_1)}{p_{\mathbf{s}_k|\omega_0}(\hat{\mathbf{s}}_k|\omega_0)}$, the LRT can be recursively computed:

$$\ell_{l+1,m} = \ell_{l,m} - \Phi(l-m) + \Phi(l+m+1)$$

(9)

and the decision rule is defined by:

$$\ell_{l,m} \quad \begin{array}{l} \geq \eta \quad \text{frame } l \text{ is classified as speech} \\ < \eta \quad \text{frame } l \text{ is classified as non - speech} \end{array}$$

(10)

where η is the decision threshold which is experimentally tuned for the best trade-off between speech and non-speech classification errors.

4 Recursive PCA Applied to VAD

In order to recursively evaluate the LRT over the set of uncorrelated signals in the current frame l we use a result in [9]. In the frame $l+1$ the PCA components for the MO-window $l - m, \ldots, l + m$ are computed as a function of the previous MO-window centered at frame l. Since:

$$\mathbf{R}_M = \frac{1}{M} \sum_{i=1}^{M} \mathbf{x}_i \mathbf{x}_i^T = \frac{M-1}{M} \mathbf{R}_{M-1} + \frac{1}{M} \mathbf{x}_M \mathbf{x}_M^T \qquad (11)$$

where M denotes the number of observation ($M = 2m + 1$), we obtain the following recursive formula for the eigenvectors and eigenvalues:

$$\mathbf{Q}_M M \mathbf{\Lambda}_M \mathbf{Q}_M^T = \mathbf{Q}_{M-1}[(M-1)\mathbf{\Lambda}_{M-1} + \alpha_M \alpha_M^T]\mathbf{Q}_M^T \qquad (12)$$

where $\mathbf{R}_M = \mathbf{Q}_M \mathbf{\Lambda}_M \mathbf{Q}_M^T$ and $\alpha_M = \mathbf{Q}_{M-1}^T \mathbf{x}_M$. Using a matrix perturbation analysis approach of a matrix in the form $(\mathbf{\Lambda} + \alpha \alpha^T)$, we can obtain a recursion for the eigenvalues and eigenvectors as [9]:

$$\begin{aligned} \mathbf{Q}_M &= [\mathbf{Q}_{M-1}(\mathbf{I} + \mathbf{P_V})]\mathbf{T}_M \\ \mathbf{\Lambda}_M &= [(1 - \lambda_M)\mathbf{\Lambda}_{M-1} + \mathbf{P_\Lambda}]\mathbf{T}_M^{-2} \end{aligned} \qquad (13)$$

where $\mathbf{T}_M$ is a diagonal matrix containing the inverses of the norms of each column of the matrix in brackets (top); $\mathbf{P_V}$ is an antisymmetric matrix whose $(i,j)^{th}$ entry is $\alpha_i \alpha_j / (\lambda_j + \alpha_j^2 - \lambda_i - \alpha_j^2)$ if $j \neq i$, and 0 if $j = i$; $\mathbf{P_\Lambda}$ is a diagonal matrix whose i^{th} diagonal entry is α_i^2; and λ_M is a memory depth parameter. Using the set of equations 13 we can obtain the uncorrelated components of the decision frame l from the decision frame $l - 1$ in a two step procedure: Let $\mathbf{R}_{M,l}$ denote the covariance matrix on the decision frame l of order $M = 2m + 1$.

1. From equation 13 obtain $\mathbf{R}_{M-1,l}$ using $\mathbf{R}_{M,l}$ and $\alpha = \alpha_{l-m}$.
2. From equation 13 obtain $\mathbf{R}_{M,l+1}$ using $\mathbf{R}_{M-1,l}$ and $\alpha = \alpha_{l+m}$.

With the aim of this recursion, the proposed VAD is computationally efficient enough to be used on a real time application and eludes the iterative eigenvector decomposition in each MO-window.

5   Experimental Framework

The ROC curves are used in this section for the evaluation of the proposed VAD. These plots completely describe the VAD error rate and show the trade-off between the speech and non-speech error probabilities as the threshold varies. The Aurora 3 Spanish SpeechDat-Car database was used in the analysis. This database contains recordings in a car environment from close-talking and hands-free microphones. Utterances from the close-talking device with an average SNR of about 25 dB were labeled as speech or non-speech for reference while the VAD was evaluated on the hands-free microphone. Thus, the speech and non-speech hit rates (HR1, HR0) were determined as a function of the decision threshold for each of the VAD tested. In figure 2 we observe an example of the VAD operation showing the components extracted by the recursive PCA and the correlation plot

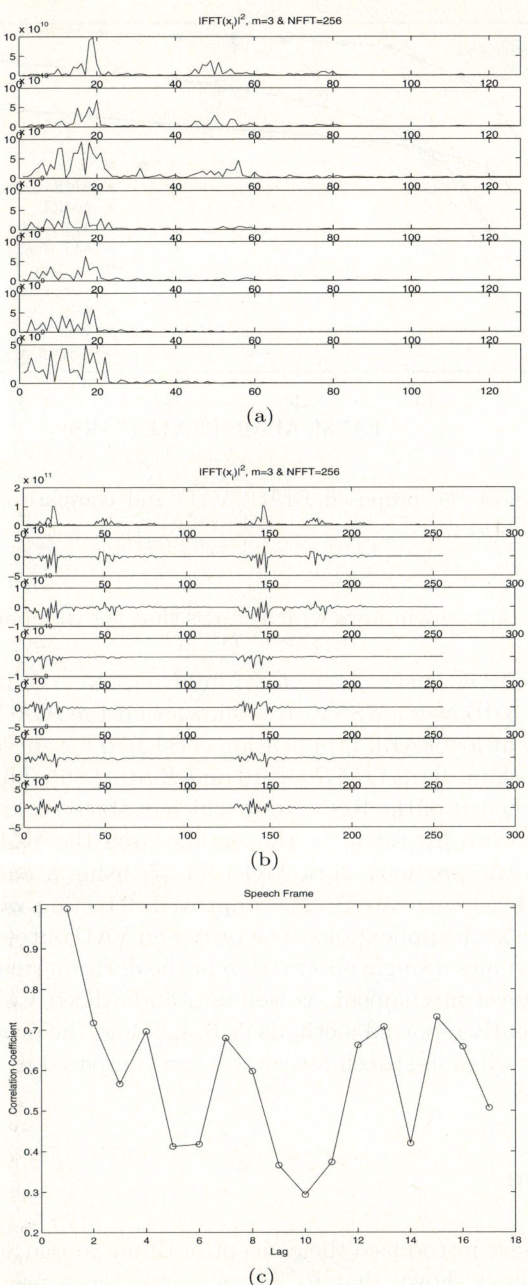

(a)

(b)

(c)

Fig. 2. Example of VAD operation for $m = 3$ and resolution $J = NFFT = 256$ (a) Set of observed DFT signals ($2m + 1 = 7$ windows) on speech frames. (b) Independent DFT signals on speech frames. (c) The correlation plot of the DFT observation vectors over the set of overlapped windows (m=8) reveals the non suitability of the previously proposed models (they are not independent as they are correlated).

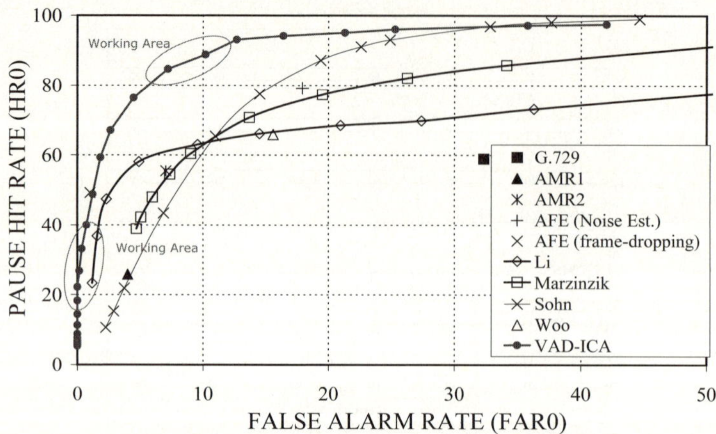

Fig. 3. ROC curves of the proposed BLRT VAD and comparison to standard and recently reported VADs

over the set of overlapped windows which shows that the independent assumption cannot be made.

Fig. 3 shows the ROC curves in the most unfavourable conditions (high-speed, good road) with a 5 dB average SNR. It is shown that the ICA-VAD obtains very good performance at low FAR0 (applications designed for speech detection) and at the common working areas ($FAR0 < 10$ and $HR0 > 80$). This is motivated by a shift-up and to the left of the ROC curve which enables working with improved speech and non-speech hit-rates. In the middle area the VAD reproduces the same accuracy as the previous work MO-LRT [7] using a completely different set of input signals (decorrelated). The improved detection rate over the latter method is major in such applications. The proposed VAD outperforms the Sohn's VAD [3], which assumes a single observation in the decision rule together with an HMM-based hangover mechanism, as well as standardized VADs such as G.729 and AMR and recently reported methods [5, 6, 4]. Thus, the proposed VAD works with improved speech/non-speech hit-rates when compared to the most relevant algorithms to date.

6   Conclusion

In this paper we have introduced the concept of Blind Source Separation in VAD applications. We have shown that PCA can be used as a preprocessing step in this scenario for improving the accuracy of LRT based VADs. The proposed algorithm is optimum in those scenarios ($FAR0 \rightarrow 0$) where the loss of speech frames could be unacceptable, causing a system failure. The VADs have been employed as a part of Speech Recognition Systems (i.e. ASR Systems) in the last years providing significant benefits. Hence the proposed VAD can be used to obtain relevant improvements over all standardized VADs in speech/pause detection accuracy and recognition performance.

Acknowledgements

This work has received research funding from the EU 6th Framework Programme, under contract number IST-2002-507943 (HIWIRE, Human Input that Works in Real Environments) and SESIBONN project (TEC2004-06096-C03-00) from the Spanish government. The views expressed here are those of the authors only. The Community is not liable for any use that may be made of the information contained therein.

References

1. ITU, "A silence compression scheme for G.729 optimized for terminals conforming to recommendation V.70," *ITU-T Recommendation G.729-Annex B*, 1996.
2. ETSI, "Speech processing, transmission and quality aspects (stq); distributed speech recognition; front-end feature extraction algorithm; compression algorithms," *ETSI ES 201 108 Recommendation*, 2000.
3. J. Sohn and et al., "A statistical model-based vad," *IEEE Signal Proccessing Letters*, vol. 16, no. 1, pp. 1–3, 1999.
4. M. Marzinzik and et al., "Speech pause detection for noise spectrum estimation by tracking power envelope dynamics," *IEEE Trans. on Speech and Audio Processing*, vol. 10, no. 6, pp. 341–351, 2002.
5. K. Woo and et al., "Robust vad algorithm for estimating noise spectrum," *Electronics Letters*, vol. 36, no. 2, pp. 180–181, 2000.
6. Q. Li and et al., "Robust endpoint detection and energy normalization for real-time speech and speaker recognition," *IEEE Trans. on Speech and Audio Proccessing*, vol. 10, no. 3, pp. 146–157, 2002.
7. J. Ramírez and et. al., "Statistical voice activity detection using a multiple observation likelihood ratio test," *IEEE Signal Processing Letters*, vol. 12, no. 10, pp. 689–692, 2005.
8. J. M. Górriz, J. Ramirez, J. C. Segura, and C. G. Puntonet, "An improved mo-lrt vad based on a bispectra gaussian model," *IEE Electronic Letters*, vol. 41, no. 15, pp. 877–879, 2005.
9. D. Erdogmus, Y. Rao, H. Peddaneni, A. Hegde, and J. Principe, "Recursive principal components analysis using eigenvector matrix perturbation," *EURASIP Journal on Applied Signal Processing*, vol. 13, pp. 2034–2041, 2004.
10. P. Comon, "Independent component analysis—a new concept?" *Signal Processing*, vol. 36, pp. 287–314, 1994.
11. A. J. Bell and T. J. Sejnowski, "An information maximisation approach to blind separation and blind deconvolution," *Neural Computation*, vol. 7, no. 6, pp. 1129–1159, 1995.

Characterizing the Performance and Energy Attributes of Scientific Simulations*

Sayaka Akioka[1], Konrad Malkowski[1], Padma Raghavan[1], Mary Jane Irwin[1],
Lois Curfman McInnes[2], and Boyana Norris[2]

[1] The Pennsylvania State University
{tobita, malkowsk, raghavan, mji}@cse.psu.edu
[2] Argonne National Laboratory
{mcinnes, norris}@mcs.anl.gov

Abstract. We characterize the performance and energy attributes of scientific applications based on nonlinear partial differential equations (PDEs). where the dominant cost is that of sparse linear system solution. We obtain performance and energy metrics using cycle-accurate emulations on a processor and memory system derived from the PowerPC RISC architecture with extensions to resemble the processor in the BlueGene/L. These results indicate that low-power modes of CPUs such as Dynamic Voltage Scaling (DVS) can indeed result in energy savings at the expense of performance degradation. We then consider the impact of certain memory subsystem optimizations to demonstrate that these optimizations in conjunction with DVS can provide faster execution time and lower energy consumption. For example, on the optimized architecture, if DVS is used to scale down the processor to 600MHz, execution times are faster by 45% with energy reductions of 75% compared to the original architecture at 1GHz. The insights gained from this study can help scientific applications better utilize the low-power modes of processors as well as guide the selection of hardware optimizations in future power-aware, high-performance computers.

1   Introduction

As microprocessors' clock frequencies have increased in recent years, their corresponding power consumption has also increased dramatically. The peak power dissipation and consequent thermal output often limit processor performance and hinder system reliability. Thus, due to thermal and cost concerns, architectural approaches that reduce overall energy consumption have become an important issue in high-performance parallel computing. The IBM BlueGene/L [17] is one example of power-aware supercomputer architectures, where lower-power, lower-performance processors are integrated into a massively parallel architecture, resulting in performance sufficient to claim the top spot in the Top500 list of supercomputer sites [27]. To fully exploit the performance potential of existing and future low-power architectures in scientific computations, the community must investigate the effects of hardware features on both performance and power.

* This work is supported by the National Science Foundation through grants ACI-0102537, CCF-0444345 and DMR-0205232.

V.N. Alexandrov et al. (Eds.): ICCS 2006, Part I, LNCS 3991, pp. 242–249, 2006.

Such work will have a two-fold impact: first, we will be able to target performance optimizations effectively; second, understanding the effects of new hardware features on realistic applications can help in the design of the next generation of power-aware architectures.

In this paper we profile a simulation of driven cavity flow [10] that uses fully implicit Newton-Krylov methods to solve the resulting system of nonlinear PDEs. We have selected this model problem because it has properties that are representative of many large-scale, nonlinear PDE-based applications in domains such as computational aerodynamics [1], astrophysics [16], and fusion [26]. The most time-consuming portion of the simulation is the solution of large, sparse linear systems of equations. Thus, we profile this kernel and interpret the impact of different memory hardware features and algorithm choices on the performance and power characteristics of the model problem.

The remainder of this paper is organized as follows. Section 2 summarizes related power-performance studies. Section 3 provides an overview of the driven cavity application, while Section 4 describes our approach to profiling its performance and power consumption. In Section 5 we present the results of these simulations. We conclude in Section 6 with observations of profiling trends and comments on future work.

2   Related Work

There are a number of studies on power-aware software optimization. The principal approach targets the use of dynamic voltage scaling (DVS) [25] to save central processing unit (CPU) energy. The basic idea of DVS is to reduce runtime energy as much as possible without a significant reduction in application performance; this technique allows the reduction of CPU idle time and consequently enables machines to spend more work time in a lower-power mode. Several studies investigate scheduling for DVS [9, 29, 15, 28, 18] or DVS-aware algorithms [19, 20]. However, we are not aware of any profiling and optimization approach that focuses explicitly on high-performance scientific computing.

There are few existing efforts to build power consumption models of applications [6, 11]. These approaches utilize performance counters on CPUs and actual power samples to build a model, thereby making this approach viable only when such counters and power measurements are available – usually not the case for newly released architectures. Rose et al. [13] are incorporating some temperature-based power estimates in the SvPablo tool, in addition to traditional hardware counter performance metrics. Feng et al. [14] are building a profiling tool that characterizes the power consumption and performance of scientific benchmarks, including SPEC CPU2000 [12] and the NAS parallel benchmarks [2]. Further investigation is needed to characterize and understand the behavior of real scientific applications.

3   A Model PDE-Based Application: Driven Cavity Flow

To explore power issues arising in the solution of large-scale nonlinear PDEs, we focus on the model problem of two-dimensional flow in a driven cavity, which has properties that are representative of our broader class of target applications. This problem is realistic yet sufficiently simple to provide a good a starting point for understanding the performance and energy attributes of PDE-based simulations.

The model is a combination of lid-driven flow and buoyancy-driven flow in a two-dimensional rectangular cavity. The lid moves with a steady and spatially uniform velocity and sets a principal vortex by viscous forces. The differentially heated lateral walls of the cavity invoke a buoyant vortex flow, opposing the principal lid-driven vortex. The nonlinear system can be expressed in the form $f(u) = 0$, where $f : R^n \to R^n$. We discretize this system using finite differences with the usual five-point stencil on a uniform Cartesian mesh, resulting in four unknowns per mesh point (two-dimensional velocity, vorticity, and temperature). Further details are discussed in [10]. Experiments presented in Section 5 employ a 64×64 mesh and the following nonlinearity parameters: lid velocity 10, Grashof number 600, and Prandtl number 1.

We solve the nonlinear system using an inexact Newton method (see, e.g., [23]) provided by the PETSc library [3], which (approximately) solves the Newton correction equation

$$f'(u^{k-1})\, \delta u^k = -f(u^{k-1}) \tag{1}$$

and then updates the iterate via $u^k = u^{k-1} + \alpha \cdot \delta u^k$, where α is a scalar determined by a line search technique such that $0 < \alpha \leq 1$. The overall time to solution is typically dominated by the time for repeatedly solving the sparse linearized systems (1) using preconditioned Krylov methods; such sparse solution comprises over 90% of the application execution time [5]. Consequently, the experiments in Section 5 focus on this important computational kernel. The dominance of sparse linear solves in overall time to solution is typical of many applications solving nonlinear PDEs using implicit and semi-implicit schemes.

4 Methodology

Our goal is to characterize the performance and power attributes of nonlinear PDE-based applications when low power modes, such as DVS, are used. An additional goal is to characterize the impact of memory subsystem optimization on performance and power. We use architectural emulation with cycle-accurate simulation of our application code to derive performance and power characteristics. For this purpose, we use SimpleScalar [8] and Wattch [7], which are two well-accepted tools in the computer architecture community. SimpleScalar is a cycle-accurate emulator of C code on a RISC architecture, which can be specified in full detail. Wattch is an extension of SimpleScalar and provides power consumption and energy metrics.

We start with a base architecture, denoted as **B**, to represent a processor and memory subsystem similar to that of the IBM BlueGene/L [17], designed specifically for power-aware scientific computing. **B** has two floating-point units (FPUs) and two integer arithmetic logic units (ALUs). The cache hierarchy consists of 32KB data / 32KB instruction Level 1 cache, 2KB Level 2 cache (L2), and 4MB unified Level 3 cache (L3). The L3 cache uses Enhanced Dynamic Random Access Memory (eDRAM) technology, similar to the BlueGene/L [17]. The main memory has 256MB of Double Data Rate SDRAM 2 (DDR2). The system is configured for a CPU frequency of 1000MHz with corresponding nominal V_{dd} voltages to model the effect of DVS. The frequency - V_{dd} pairs we use are: 400MHz-0.66V; 600MHz-0.84V; 800MHz-0.93V; 1000MHz-1.20V. We can then simulate the effects of DVS by scaling the frequency to 800MHz, 600MHz, and 400MHz with corresponding scaling of V_{dd}.

We employ a set of memory subsystem optimizations considered first in the evaluation of power and performance trade-offs of simple sparse kernels [21]:

- **W:** A wider memory bus of 128 bytes, as opposed to the 64-byte memory bus in the base architecture **B**.
- **MO:** Open memory page policy. This policy determines whether memory banks stay open or closed after a read/write operation. An open policy is more appropriate when there is some locality of data accesses.
- **LP:** Level 2 cache prefetching. After a read operation, the next line is prefetched. This optimization can improve performance if the data access pattern has locality.
- **MP:** Memory prefetching. This option is similar to **LP** above, but now the prefetching is done by the memory controller to avoid the latency of accessing the memory.
- **LM:** Load miss prediction logic. This feature can avoid L2 and L3 latencies when data is not present in the cache hierarchy.

We use labels of the form **B+W+MO** to specify a configuration that augments the basic architecture (**B**) with the wider data path (**W**) and memory-open (**MO**) features. Such memory subsystem optimizations are feasible and have been studied earlier in the architecture community. We evaluate the impact of these architectural configurations on the performance of sparse solvers in conjunction with DVS. More specifically, we focus on the seven configurations specified by: **B**, **B+W**, **B+W+MO**, **B+W+LP**, **B+W+MP**, **B+W+LM**, and **B+W+MO+LP+MP+LM** (or **All**).

5   Empirical Results

We simulate the driven cavity flow code introduced in Section 3 on SimpleScalar and Wattch to understand the performance and power attributes of its dominant computation, namely solving the sparse linear system given by Equation (1). We focus on observed performance and power metrics during the first iteration of Newton's method for different choices of the linear solver. We consider four popular schemes which are provided within PETSc [3]: the Krylov methods GMRES(30) and BiCG in combination with the incomplete factorization preconditioners ILU(0) and ILU(1) (see, e.g., [4,24]).

We begin by characterizing performance and power for **B**, at 1000MHz and when DVS is used to scale down to 800MHz, 600MHz, and 400MHz. Figure 1 shows the execution time in seconds (left), and the average system power per cycle in Watts (center) for the four sparse linear solution schemes, namely GMRES(30) + ILU(0), GMRES(30) + ILU(1), BiCG + ILU(0), and BiCG + ILU(1). The y axis represents values of the execution time or power at frequencies of 400MHz, 600MHz, 800MHz, 1000MHz for **B**. As expected, with frequency scaling down to lower values, the execution time for all methods increases. However, the relative increase for the same change in the frequency is different for different solvers. For example, BiCG+ILU(0) shows a greater increase in execution time than BiCG+ILU(1) when the frequency is scaled down from 600MHz

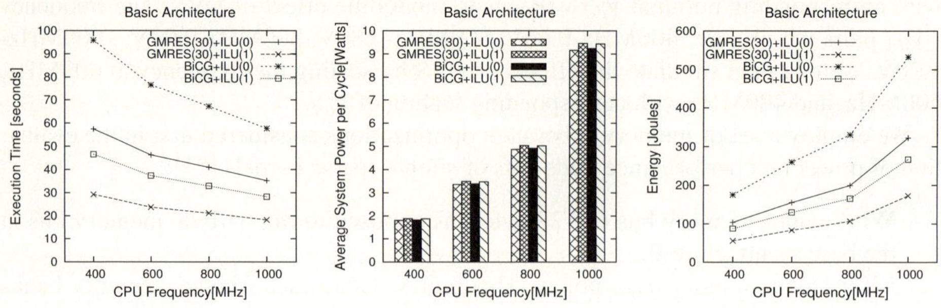

Fig. 1. Execution time in seconds (left), average system power per cycle in Watts (center), and energy in Joules (right) for one linear system solution with the base architecture at frequencies of 400MHz, 600MHz, 800MHz, and 1000MHz

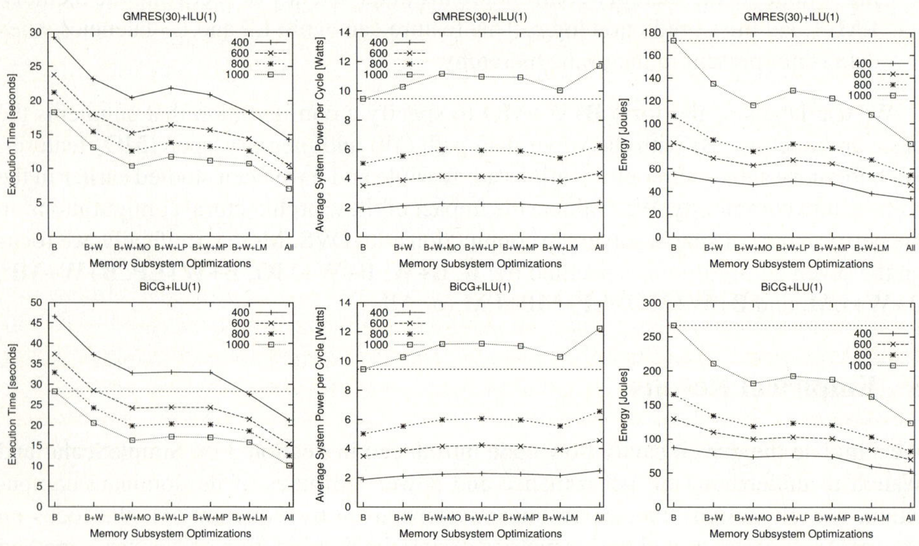

Fig. 2. Execution time in seconds (left), average system power per cycle in Watts (center), and total energy in Joules (right) for one linear system solution using GMRES(30)+ILU(1) (top) and BiCG+ILU(1) (bottom) at frequencies of 400MHz, 600MHz, 800MHz, and 1000MHz when memory subsystem optimizations are added sequentially starting with the base architecture

to 400MHz. The average power values are not substantially different across the solvers. This situation is not surprising as the underlying operations in the solvers are similar and therefore utilize the components of the architecture similarly. The corresponding energy values are shown in Figure 1 (right). As expected, the solvers consume substantially less energy at lower frequencies. However, such energy reductions through DVS come at the expense of degradation in the performance of the solvers, as indicated by substantial increases in execution time at the lowest frequency.

We next explore how the memory subsystem optimizations described in Section 4 help to improve performance. In Figure 2 (left), we focus on GMRES(30) + ILU(1) and BiCG + ILU(1). Each plot shows the execution time of a specific solver as the frequency is kept fixed and the architecture is changed to reflect the seven configurations, **B**, **B+W**, **B+W+MO**, **B+W+LP**, **B+W+MP**, **B+W+LM**, and **All**. These configurations are shown along the x axis, while the y axis values indicate execution time in seconds.

Again, the execution times for all architectural configurations and solvers increase as frequency scales down to lower values. The trends are similar for both solvers. More significantly, the execution times for **B** at 1000MHz are higher than the execution times when even the simplest optimization is added at lower frequency values down to 600MHz. A dotted flat line indicates the time for the solver on **B** at 1000MHz. When all optimizations are included, i.e., the configuration **All**, even at 400MHz the performance is improved over that observed for **B** at 1000MHz.

Figure 2 (center) shows the average system power per cycle in Watts corresponding to the execution times shown earlier in Figure 2 (left). Adding architectural optimizations increases the power incrementally. However, these increases are less significant than the substantial decreases obtained from DVS. Thus, system power values for the base architecture at 1000MHz are higher than at 400MHz, 600MHz and 800MHz consistently.

Figure 2 (right) shows the energy consumed by GMRES(30)+ILU(1) and BiCG+ILU(1). Once again we use a dotted line to denote the value for the base architecture at 1000MHz. Observe that with memory subsystem optimizations, the energy is reduced even without DVS. This effect, which is observed at all frequencies, is primarily due to improvements in execution time. For example, when using all optimizations at 600MHz, substantial reductions in energy are realized for both solvers. From these results, we conclude that the proper selection of memory subsystem optimizations in conjunction with DVS can reduce both execution time and energy consumed.

To quantify the relative improvements in execution time and energy, we define the following relative reduction (RR) metric. Consider a specific solution scheme, i.e., a specific combination of a Krylov method and preconditioner. Let $T_{f,q}$ denote the observed execution time at frequency f MHz for an architectural configuration labeled q. We define RR with respect to the execution time for **B** at 1000MHz, i.e., $T_{1000,B}$ as:

$$RR(T)_{f,q} = (T_{1000,B} - T_{f,q})/T_{1000,B}. \qquad (2)$$

Positive RR values indicate reductions in execution time, while negative values indicate increases. We can also define a corresponding RR metric for energy; we denote this for frequency f MHz and architectural configuration q as $RR(E)_{f,q}$.

Figure 3 shows $RR(T)$ and $RR(E)$ for each solver at each frequency for **All** relative to values for **B** at 1000MHz. RR values for the four solvers are indicated by a group of four bars at each frequency; the average value at each frequency is used in the line plot. Observe that on average, execution time is improved by approximately 20% at 400MHz with energy improvements of approximately 80%. Furthermore, if DVS is used to scale down to 600MHz, performance improvements are in excess of 45% on average with energy reductions of approximately 75%.

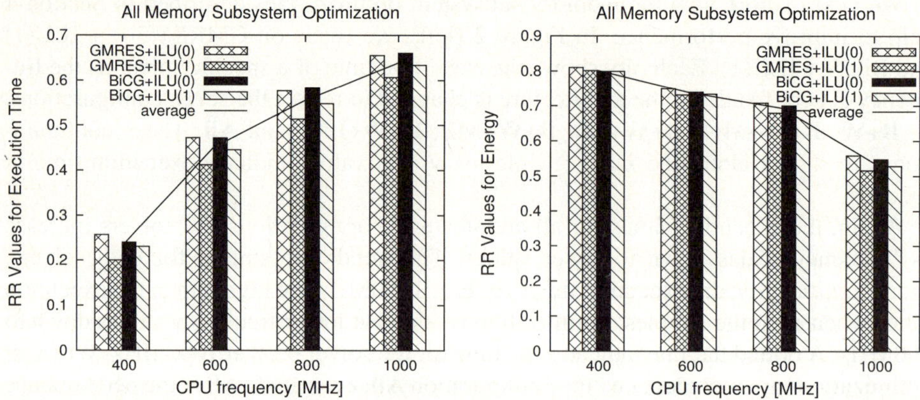

Fig. 3. Relative reductions (RR) in execution time (left) and energy (right) for one linear system solution with the architectural configuration **All** with respect to time/energy values for the basic architecture at 1000MHz. The line indicates the average reductions at each CPU frequency.

6   Conclusions

With the need for low-power, high-performance architectures, the combined performance and power characteristics of scientific applications on such architectures are an important area of study. In this paper, we observe the impact of memory subsystem optimizations and different linear solvers on the performance and power consumption of a nonlinear PDE-based simulation of flow in a driven cavity. Results using the SimpleScalar and Wattch simulators indicate that memory subsystem optimization plays a more important role than CPU frequency in reducing both execution time and power consumption. These results, taken together with our earlier studies [5,22] demonstrating the impact of different solution schemes on quality and performance, indicate that there is significant potential for developing adaptive techniques to co-manage performance, power and quality trade-offs for such scientific applications.

References

1. W. K. Anderson, W. D. Gropp, D. K. Kaushik, and et al. Achieving high sustained performance in an unstructured mesh CFD application. In *SC99*, 1999.
2. D. Bailey, T. Harris, W. Saphir, and et al. The NAS parallel benchmarks 2.0. Technical Report NAS-95-020, NASA Ames Research Center, 1995.
3. S. Balay, K. Buschelman, V. Eijkhout, and et al. PETSc users manual. Technical Report ANL-95/11 - Revision 2.3.0, Argonne National Laboratory, 2005. See `http://www.mcs.anl.gov/petsc`.
4. R. Barrett, M. Berry, T. F. Chan, and et al. *Templates for the Solution of Linear Systems: Building Blocks for Iterative Methods, Software, Environments, Tools*. SIAM, 1994.
5. S. Bhowmick, L. C. McInnes, B. Norris, and et al. The role of multi-method linear solvers in PDE-based simulations. In V. Kumar, M. L. Gavrilova, C. J. K. Tan, and P. L'Ecuyer, editors, *Lecture Notes in Computer Science, Computational Science and its Applications-ICCSA 2003*, volume 2677. Springer Verlag, 2003.

6. W. L. Bircher, M. Valluri, L. John, and et al. Runtime identification of microprocessor energy saving opportunities. In *ISLPED'05*, 2005.

7. D. Brooks, V. Tiwari, and M. Martonosi. Wattch: A framework for architectural-level power analysis and optimizations. In *ISCA'00*, 2000.

8. D. C. Burger and T. M. Austin. The SimpleScalar tool set, version 2.0. Technical Report 1342, UW Madison Computer Sciences, 1997.

9. J. Casmira and D. Grunwald. Dynamic instruction scheduling slack. In *2000 KoolChips Workshop*, 2000.

10. T. S. Coffey, C. T. Kelley, and D. E. Keyes. Pseudo-transient continuation and differential-algebraic equations. *SIAM J. Sci. Comput.*, 25(2), 2003.

11. G. Contreras and M. Martonosi. Power prediction for Intel XScale processors using performance monitoring unit events. In *ISLPED'05*, 2005.

12. Standard Performance Evaluation Corporation. The SPEC benchmark suite. `http://www.spec.org`.

13. L. A. de Rose and D. A. Reed. SvPablo: A multi-language architecture-independent performance analysis system. In *ICPP'99*, 1999.

14. X. Feng, R. Ge, and K. W. Cameron. Power and energy profiling of scientific applications on distributed systems. In *IPDPS'05*, 2005.

15. B. Fields, R. Bodik, and M. M. Hill. Slack: Maximizing performance under technological constraints. In *ISCA'02*, 2002.

16. B. Fryxell, K. Olson, P. Ricker, and et al. FLASH: An adaptive-mesh hydrodynamics code for modeling astrophysical thermonuclear flashes. *Astrophys. J. Suppl.*, 2000.

17. A. Gara, M. A. Blumrich, D. Chen, and et al. Overview of the Blue Gene/L system architecture. *IBM J. Res. & Dev.*, 49(2/3), 2005.

18. R. Ge, X. Feng, and K. W. Cameron. Performance-constrained, distributed DVS scheduling for scientific applications on power-aware clusters. In *SC05*, 2005.

19. C. Hsu and W. Feng. A power-aware run-time system for high-performance computing. In *SC05*, 2005.

20. N. Kappiah, V. W. Freeh, and D. K. Lowenthal. Just-in-time dynamic voltage scaling: Exploiting inter-node slack to save energy in MPI programs. In *SC05*, 2005.

21. K. Malkowski, I. Lee, P. Raghavan, and et al. Memory optimizations for tuned scientific applications: An evaluation of performance-power characteristics. Submitted to ISPASS'06.

22. L. McInnes, B. Norris, S. Bhowmick, and et al. Adaptive sparse linear solvers for implicit CFD using Newton-Krylov algorithms. volume 2, Boston, USA, June 17-20, 2003.

23. J. Nocedal and S. J. Wright. *Numerical Optimization*. Springer-Verlag, 1999.

24. Y. Saad. *Iterative Methods for Sparse Liner Systems*. SIAM, second edition, 2003.

25. G. Semeraro, D. H. Albonesi, S. G. Dropsho, and et al. Dynamic frequency and voltage control for a multiple clock domain microarchitecture. In *MICRO 2002*, 2002.

26. X. Z. Tang, G. Y. Fu, S. C. Jardin, and et al. Resistive magnetohydrodynamics simulation of fusion plasmas. Technical Report PPPL-3532, Princeton Plasma Physics Laboratory, 2001.

27. Top500.org. Top 500 supercomputer sites. `http://top500.org`, 2005.

28. W. Yuan and K. Nahrstedt. Energy-efficient soft real-time CPU scheduling for mobile multimedia systems. In *SOSP'03*, 2003.

29. D. Zhu, R. Melhem, and B. R. Childers. Scheduling with dynamic voltage/speed adjustment using slack reclamation in multi-processor real-time systems. In *RTSS'01*, 2001.

Computation of Si Nanowire Bandstructures on Parallel Machines Through Domain Decomposition

Tao Li[1], Ximeng Guan[1], Zhiping Yu[1], and Wei Xue[2]

[1] Institute of Microelectronics, Tsinghua University, Beijing 100084, China
{litaofrank99, gxm}@mails.tsinghua.edu.cn,
yuzhip@tsinghua.edu.cn
[2] Department of Computer Science and Technology, Tsinghua University,
Beijing 100084, China
xuewei@tsinghua.edu.cn

Abstract. This paper presents a methodology for calculating silicon nanowire (SiNW) bandstructures on parallel machines. A partition scheme is developed through domain decomposition and loading balance is considered for scheduling jobs on machines with different efficiencies. Using $sp^3d^5s^*$ tight-binding model, the Hamiltonian matrix of the SiNW is constructed and its eigenvalues are extracted with the Implicitly Restarted Arnoldi Method. Parallel calculation performance is tested on identical machines finally and a linear speedup is gained as the number of nodes increases.

1   Introduction

The future's scaling down of device size based on conventional silicon technology is predicted to face fundamental physical limits in the near future. In order to achieve the device miniaturization into nanometer scale, it is expected that conventional device structures should be modified or totally altered. With the rapid progress in nanofabrication technology, SiNWs with small diameters (<20 nm) have been synthesized and extensively studied for potential applications in nanoelectronics [1][2]. Due to the strong quantum confinement (QC) in ultrathin SiNWs, atomic bandstructure effects [3] are expected to play an important role on their device characteristics.

For modelling the nanoscale device characteristics, any realistic electronic model, whatever the underlying basis, must accurately reproduce the experimentally verified band gaps and effective masses for the relevant bands. The strength of tight-binding (TB) techniques is their ability to properly model such crucial material properties with a localized, atomistic orbital basis [4]. TB models have therefore developed into the primary choice for many researchers interested in the quantitative modelling of electronic structure on a nanometer scale. For the accurate description of the conduction subbands of the nanowire, it is necessary to include higher orbitals beyond the minimal sp^3 basis [4] and to employ the

V.N. Alexandrov et al. (Eds.): ICCS 2006, Part I, LNCS 3991, pp. 250–257, 2006.

$sp^3d^5s^*$ TB model, which was developed by Jancu *et al.* [5] and was found to describe quite well up to the two lowest conduction bands and as well as the valance bands of bulk Si.

Application of $sp^3d^5s^*$ TB model on SiNW bandstructures computation covers a broad spectrum of Brillouin zone and energy bands of interests [6]. In many practical cases, physical domains, which require a huge number of finite difference computational grids, are inevitable, and a complex nanowire cross-section is needed. In such circumstances, not only the PC memory size is too small to carry out the computations, but also a lot of CPU time is required. To facilitate such computational demand, tasks can be distributed to several machines so that each node is responsible for a smaller sub-task only. This idea underlies the present work of parallelizing the calculation of SiNW bandstructures. The implementation of the proposed parallel algorithm through domain decomposition on the Itanium-2 cluster is done by employing the commonly used message passing interface (MPI) library [7].

This paper is organized as follows. Section 2 presents the partition scheme through domain decomposition, followed by Section 3 which gives the algorithm for scheduling jobs on parallel machines with different efficiencies. In section 4, the computation for energy levels with $sp^3d^5s^*$ TB model on a single node machine is explained in detail. In Section 5, the bandstructures for SiNWs oriented along [112] direction with different cross-section shapes are computed and the performance is evaluated. Finally, a conclusion is given to close this paper.

2   Partition Scheme

A detailed calculation of the SiNW bandstructures in Brillouin zone needs a large number of finite difference computational grids, i.e. energy levels on a long list of k values are needed. Once the energy levels on each k point are carried out, the whole bandstructure could be charted by connecting the corresponding energy level respectively. Considering this, the partition scheme is straightforward developed through domain decomposition, and a master/slave program architecture is adopted for the parallel implementation.

On the first stage, a list of k values in the Brillouin zone are selected by the master node, then they are decomposed and distributed to the slave nodes in the cluster. When the energy levels computation is finished on one slave node, the results are sent back to the master and a new k value will be assigned. As shown in Fig. 1, this procedure continues until all points have been sent out and then the whole bandstructure is charted. In this partition scheme, communication occurs only between the master and slave nodes during the parallel computation, consequently, the partition scheme has a good scalability when grid size increases.

Another important aspect in parallel calculation is loading balance: for an urgent need for speed, jobs should be assigned to the slave nodes according to their computation abilities. In Sect. 3, We will show that this optimizing problem is polynomially solvable in our circumstance.

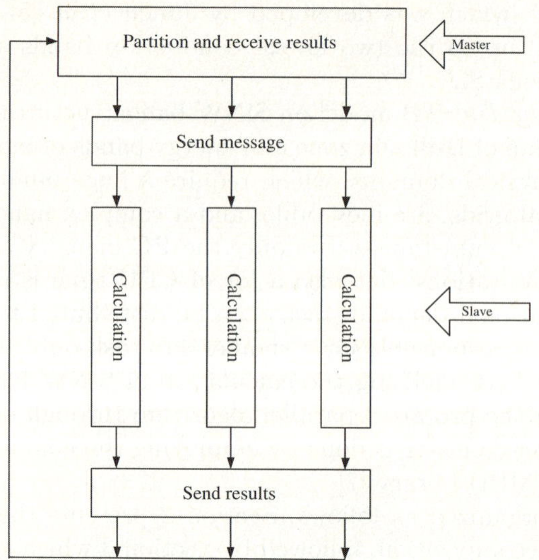

Fig. 1. Overview of the parallel calculation process

3   Scheduling Jobs on Parallel Machines

In this section, we consider the scheduling problem where n jobs $J_1, \ldots, J_n$ have to be processed by m machines with different efficiencies. Each job is described by a release date r_i, a processing time p_i and a cost function $f_i(t)$. This function represents the cost induced by J_i when it is finished at time t. The problem of minimizing the sum of the f_i functions consists of finding a set of completion times C_i for each job J_i such that:

- jobs start after their release date, i.e., $\forall t, C_i - p_i \geq r_i$,
- no more than m machines are used at any time t, i.e., $\forall t, |\{J_i||C_i - p_i \leq t < C_i\}| \leq m$,
- the objective function $\sum_i f_i(C_i)$ is minimal.

This problem, denoted as the $(P|r_i \sum f_i(C_i))$ in the standard scheduling terminology (e.g., [8]) is a generalization of several NP-hard scheduling problems. However, we refer to Brucker and Knust [9] for up to date complexity results on machine scheduling. As shown below, we study the specially practical case where:

- jobs are identical, i.e. differences between calculating energy levels of different $\boldsymbol{k}$ values are neglected,
- job release dates are the same and equal to zero,
- jobs are scheduled on uniform parallel machines (i.e., on machines that do not run at the same speed).

This optimizing problem is solved in this paper with a greedy algorithm. The initial list, which contains the time required to complete one job for each machine, is first established and the times are sorted from earliest to latest, then a job is assigned to the machine that corresponds to the first completion time in the list. This completion time is then increased by the time to complete an additional job on the machine, and the list is stored. This process is repeated until all the jobs have been sent out. When the procedure is finished, we successfully schedule these n jobs on m machines for the earliest completion time. This scheme satisfies the minimality property described in [10] and the algorithm is listed below. A time complexity analysis of this algorithm for scheduling a set of n jobs on m machines leads to $\Theta(mn)$.

```
algorithm GreedySchedule
   const
     n = the quantity of jobs;
     m = the quantity of nodes;
     Time: real array; {An array of size m, in which Time[i] stands
     for the time required to complete one job on machine i}
   Var
     Scheme: integer array; {An array of size m}
     Buffer: real array; {An array of size m}
     i, j: integer;
   begin
   i := 0;
   j := 0;
   repeat
     i := i + 1;
     Scheme[i] := 0;
     Buffer[i] := Time[i];
   until i = m;
   i := 0;
   repeat
     i := i + 1;
     j := FindMin(Buffer); {FindMin(Buffer) finds the smallest element
     in Buffer, and returns its index}
     Buffer[j] := Buffer[j] + Time[j];
     Scheme[j] := Scheme[j] + 1;
   until i = n;
   {Scheme is the array we want, in which Scheme[i] stands for the
     number of jobs assigned to machine i}
end.
```

4   Calculating Energy Levels with TB Model

In this section, the calculation of energy levels with $sp^3d^5s^*$ TB model on a single node machine is explained in detail. The Hamiltonian matrix is constructed first with TB model, and then the data structure is optimized to hold the Hamiltonian matrix which has billions of entries. At last, the eigenvalues of the Hamiltonian matrix, i.e. the energy levels, are extracted with the Implicitly Restarted Arnoldi Method.

4.1   Building Hamiltonian Matrix

In $sp^3d^5s^*$ TB model, the basis wave function of a crystal with a 3-D traditional symmetry is formed as a Bloch sum [6]:

$$\Phi_{\alpha l k} = \frac{1}{\sqrt[2]{N}} \sum_j e^{i(\boldsymbol{R}_j + \boldsymbol{r}_l)\cdot \boldsymbol{k}} \phi_\alpha(\boldsymbol{r} - \boldsymbol{R}_j - \boldsymbol{r}_l), \tag{1}$$

where ϕ_α is the atomic orbital of the state indexed by α, $\boldsymbol{k}$ is the 3-D wave vector, $\boldsymbol{R}_j$ denotes the position of the jth primitive cell and $\boldsymbol{r}_l$ is the relative position of the lth atom within the primitive cell [11]. The eigenfunction is then expressed as a linear combination of $\Phi_{\alpha l \boldsymbol{k}}$:

$$\Psi_{\boldsymbol{k}} = \sum_{\alpha,l} C_{\alpha l}\Phi_{\alpha l \boldsymbol{k}}. \tag{2}$$

The Hamiltonian of the system is expanded upon $\Phi_{\alpha l \boldsymbol{k}}$ to form a matrix [6].

Due to the nearest neighbor approximation introduced by the TB method, the Hamiltonian matrices of the SiNWs are always sparse, which usually range from 50000×50000 to 500000×500000. In this work, such a Hamiltonian matrix is constructed with basic sub-blocks in a pattern according to atom arrangement of a specific nanowire. As a result, we record the matrix in sub-blocks rather than other traditional ways of sparse matrix storage. As an example, we consider a [112] oriented SiNW with a cross-section size of 7.5nm×7.5nm. The size of the Hamiltonian matrix is 38400×38400. The number of nonzero entries of this gigantic matrix is approximately 1.92×10^6. If the sparse storage is applied, the memory needed to hold all the nonzero elements is 30M bytes. However, if the representation of this matrix is stored in sub-blocks, only a constant memory of 70k bytes is required at each node.

4.2   Eigenvalue Computation

Eigenvalues of the Hamiltonian matrix are solved using the Implicitly Restarted Arnoldi Method working on the Krylov subspace, and its related software package ARPACK, which is a collection of Fortran77 subroutines designed for large scale eigenvalue problems [12], is employed in this work. During the Arnoldi procedure, a matrix-vector multiplication is needed for solving the eigenvalues. However, there is no limit on the data structure for the matrix. In this work, a subblock-vector multiplication scheme is developed. As shown in Fig. 2, sub-blocks and the corresponding parts of the vector are multiplied, and then the results are collected and assembled into a new vector. Level 2 Basic Linear Algebra Subprograms (BLAS) are included to achieve an optimum performance for the multiplication.

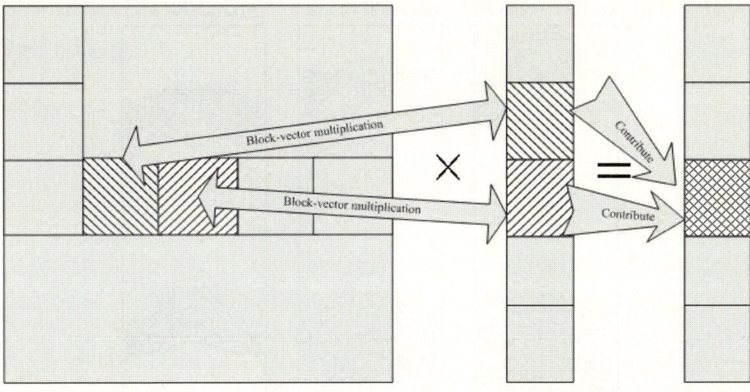

Fig. 2. The scheme of subblock-vector multiplication

As typically, only the energy levels near the bandgap (i.e., eigenvalues near the zero point) are cared in device simulation, the Hamiltonian matrix is not necessary to be diagonalized completely. However, the number of positive eigenvalues may not balance with the number of negative ones. This is not a satisfying solution as the information of both the conduction band and valance band is needed. To solve this problem, a matrix shift is carried out before matrix diagonalization. The calculated eigenvalues are then shifted back and stored.

5   Results

The cluster used for the bandstructure calculation consists of eight 1.0GHz Itanium-2 processors in four 2-CPU nodes connected through dual gigabit Ethernet. The MPI software used on this platform is MPICH-1.2.6 and the compilers are Intel Fortran and C/C++ compiler 9.0. The compilers switches used are: $-O2 - mcpu = itanium2 - mtune = itanium2$. Additionally, the Intel Math Kernel Library is included.

Bandstructures for three different SiNWs structures, including 56, 80, and 150 atoms in the supercell respectively, are computed. 24 points of k are chosen in the Brillouin zone, for each one of which 200 energy levels are selected. Finally, 100 energy levels on each k point are chosen and the bandstructures are charted in Fig. 3.

Fig. 4 shows the speedup and efficiency of the parallel calculation for different numbers of processors used. Here, the speedup is defined as the ratio of the run time using a single grid to the parallel run-time for calculating the bandstructures. From Fig. 4 we can see that a linear speedup is obtained and the overall performance of the parallel calculation is very good. Finally the machine time consumed for these three structures with different grid sizes is listed in Table 1.

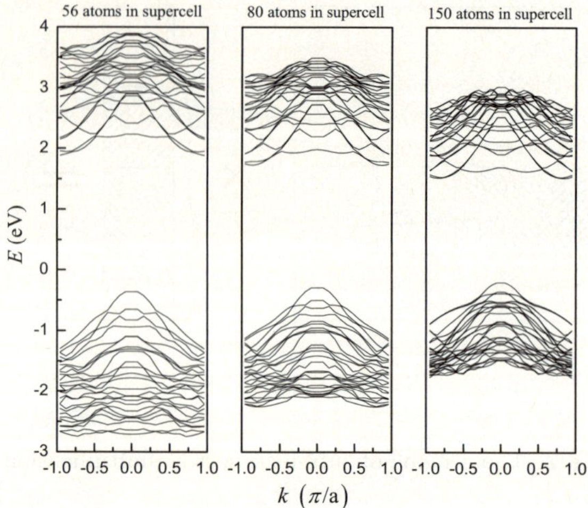

Fig. 3. Bandstructures of three [112] oriented SiNWs with different dimensions

Table 1. Machine time for bandstructure calculation

56 atoms		80 atoms		150 atoms	
#CPUs	Time (s)	#CPUs	Time (s)	#CPUs	Time (s)
1	1866	1	3384	1	7296
2	888	2	1573	2	3666
4	460	4	862	4	1819
8	299	8	421	8	906

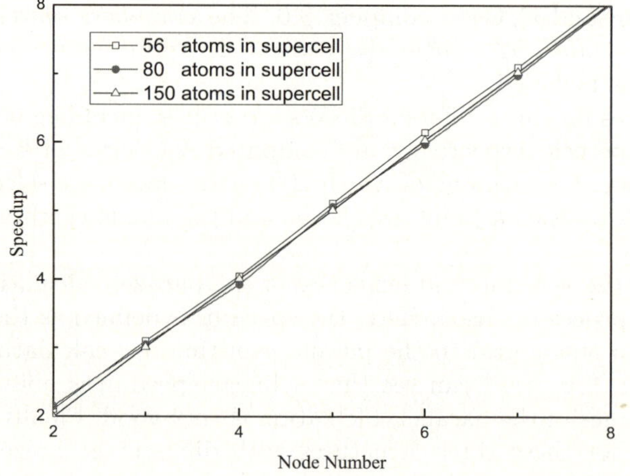

Fig. 4. Parallel computational speedup/efficiency

6   Conclusion

We have presented a parallel methodology of bandstructure calculation of SiNWs through domain decomposition in this paper. An algorithm has been developed for loading balance on heterogenous machines. Data structure has been optimized for storing the Hamiltonian matrix generated with $sp^3d^5s^*$ TB model, and the eigenvalues have been extracted with the Implicitly Restarted Arnoldi Method. Performance of the parallel methodology has been demonstrated on a cluster of identical parallel machines, and a linear speedup has been obtained.

Acknowledgement

This work is partially supported by a grant for Ph.D. education program from the Ministry of Education in China.

References

1. Cui, Y., Zhong, Z., Wang, D., Wang, W.U., Lieber, C.M.: High Performance Silicon Nanowire Field Effect Transistors. Nano Lett., Vol. 3 (2003) 149-152
2. Ma, D.D.D., Lee, C.S., Au, F.C.K., Tong, S.Y., Lee, S.T.: Small-Diameter Silicon Nanowire Surfaces. Science, (2003) 1874-1876
3. Zhao, X., Wei, C., Yang, L., Chou, M.Y.: Quantum Confinement and Electronic Properties of Silicon Nanowires. Phys. Rev. Lett., Vol. 92 (2004) 236805
4. Ko, Y.J., Shin, M., Lee, S., Park, K.W.: Effects of Atomistic Defects on Coherent Electron Transmission in Si Nanowires: Full Band Calculations. J. Appl. Phys., Vol. 89 (2001) 374-380
5. Jancu, J.-M., Scholz, R., Beltram, F., Bassani, F.: Empirical $spds^*$ Tight-binding Calculation for Cubic Semiconductors: General Method and Material Maramaters. Phys. Rev. B., Vol. 57 (1998) 6493-6507
6. Guan, X., Yu, Z.: Supercell Approach in Tight-Binding Calculation of Si and Ge Nanowire Bandstructures. Chin. Phys. Lett., Vol. 22 (2005) 2651-2654
7. Snir, M., Otto, S.W., Lederman, S.H-, Walker, D.W., Dongarra, J.: MPI The complete Reference. The MIT Press Cambridge, Massachussets (1996)
8. Cormen, T.H., Leiserson, C.E., Rivest, R.L., Stein, C.: Introduction to Algorithms. 2nd edn. The MIT Press, Massachussets (2001)
9. Brucker, P., Knust, S.: Complexity Results of Scheduling Problems. URL: http://www.mathematik.uni-osnabrueck.de/research/OR/class
10. Dessouky, M.I., Lageweg, B.J., Lenstra, J.K., van de Velde, S.L.: Scheduling Identical Jobs on Uniform Parallel Machines. Statist. Neerlandica, Vol. 44 (1990) 115-123
11. Yu, P.Y., Cadona, M.: Fundamentals of Semicondutors: Physics and Materials Properties. 3rd edn. New York: Springer, New York (2001)
12. Lehoucq, R., Maschhof, K.,Sorensen, D., Yang, C.: ARPACK. URL: http://www.cs.ucdavis.edu/ bai/ET/arnoldi_methods/overview_ARPACK.html

Semi-Lagrangian Scale Selective Two-Time-Level Scheme for Hydrostatic Atmospheric Model

Andrei Bourchtein[1], Ludmila Bourchtein[1], and Maxim Naumov[2]

[1] Institute of Physics and Mathematics, Pelotas State University, Brazil
`burstein@terra.com.br`
[2] Department of Computer Sciences, Purdue University, USA
`naumov@purdue.edu`

Abstract. A semi-Lagrangian scale selective finite difference scheme for hydrostatic atmospheric model is developed. The principal characteristics of the scheme are solution of the trajectory equations for advection, explicit first order approximation of physically insignificant adjustment terms and implicit time splitting discretization of the principal physical modes. This approach allows the use of large time steps, keeps practically the second order of accuracy and requires at each time step the amount of calculations proportional to the number of spatial grid points. The performed numerical experiments show computational efficiency of the proposed scheme and accuracy of the predicted atmospheric fields.

1 Introduction

Atmosphere parameters have a wide spectrum of spatial and temporal variations. Accordingly, the mathematical models of atmosphere (Euler or Navier-Stokes equations) contain solutions of different space and time scales. Analysis of the linearized equations reveals three principal types of atmospheric waves: acoustic, gravitational and inertial waves. Studying certain physical phenomenon one can try to filter the secondary effects still keeping all essential characteristics of the principal part. Analysis of the atmospheric data shows that weather systems are essentially defined by inertial motions while acoustic waves are practically neglectable. For large scale atmospheric dynamics the hydrostatic hypothesis is usually applied to filter out acoustic waves and simplify the governing equations. It has been proved to be effective simplification producing high quality atmospheric fields at reduced computational cost. Further attempts of simplification (Boussinesq, barotropic, quasi-geostrophic, etc. approximations) have not been succeeded in keeping the same level of the forecast accuracy.

Although gravity waves seems to be analytically inseparable from inertial ones and their contribution to weather systems is not neglectable, one can try to divide the entire spectrum of these waves in more relevant and insignificant parts by using appropriate numerical methods. This approach can be practically implemented if vertical decoupling takes place. In fact, expansion of the atmospheric fields by vertical normal modes reveals high heterogeneity in the distribution of available energy: few greatest vertical modes are responsible for about ninety percent of total

V.N. Alexandrov et al. (Eds.): ICCS 2006, Part I, LNCS 3991, pp. 258–266, 2006.

energy. This way, the general variability can be well predicted if the first vertical modes are approximated accurately while others are resolved more coarsely. Moreover, insignificant gravity waves contained in the smallest vertical modes can be approximated with even lower accuracy.

In this study, a scale separation is applied in the context of semi-Lagrangian semi-implicit (SLSI) method, which is currently the most efficient approach in numerical weather prediction and atmospheric modeling [8,11,15,16]. In this method the advective part is represented by equations of the trajectories of fluid particles, nonlinear terms are approximated explicitly and linear gravity waves implicitly along the above trajectories. At each time step the implicit terms require solution of 3D elliptic problem. Vertical decoupling transforms this problem into a set of 2D Helmholtz equations and eliminates necessity for implicit approximation of the insignificant vertical modes, but solution of the remaining elliptic problems is still expensive part of computations [4,7]. To overcome this difficulty, a time splitting method is applied to factorize each remaining 2D Helmholtz equation in a set of 1D problems, which are solved very effectively by direct Gelfand-Thomas algorithm. In order to reduce the splitting errors, which become great when time step exceeds the Courant-Friedrichs-Lewy (CFL) condition with respect to advection, the modified splitting proposed by Douglas et al. is applied [6,9,10,17]. This way, joining different numerical techniques we are able to construct computationally efficient and accurate SLSI model for the hydrostatic equations of the atmosphere.

2 Primitive Equations

Using time coordinate t, horizontal cartesian coordinates x, y and vertical coordinate $\sigma = p/p_s$, the governing equations of the hydrostatic atmosphere can be written as follows [11]:

$$\frac{du}{dt} = fv - G_x + N_u \ , \quad \frac{dv}{dt} = -fu - G_y + N_v \ , \tag{1}$$

$$G_{\ln \sigma} = -RT \ , \tag{2}$$

$$\frac{dP}{dt} = -D - \dot{\sigma}_\sigma \ , \quad \frac{dT}{dt} = \frac{RT_0}{c_p} \cdot \left(\frac{dP}{dt} + \frac{\dot{\sigma}}{\sigma} \right) + N_T \ . \tag{3}$$

Here $u, v, \dot{\sigma}, G, P, T$ are unknown functions, namely, u and v are the horizontal velocity components, $\dot{\sigma}$ is the vertical velocity component, $D = u_x + v_y$ is the horizontal divergence, $P = \ln p_s$, p and p_s are the pressure and surface pressure respectively, T is the temperature, $G = \Phi + RT_0 P$, $\Phi = gz$ is the geopotential, z is the height, $T_0 = const$ is the reference temperature profile.

The nonlinear terms N_u, N_v, N_T are expressed in the form

$$N_u = -R(T - T_0)P_x, \ N_v = -R(T - T_0)P_y \ , \ N_T = -\frac{R(T - T_0)}{c_p}\left(\frac{\dot{\sigma}}{\sigma} - D - \dot{\sigma}_\sigma \right).$$

The individual 3D derivative is

$$\frac{d\varphi}{dt} = \varphi_t + u\varphi_x + v\varphi_y + \dot{\sigma}\varphi_\sigma \quad , \quad \varphi = u, v, P, T$$

and the following parameters are used: f is the Coriolis parameter, g is the gravitational acceleration, R is the gas constant of dry air, c_p is the specific heat at constant pressure. Hereinafter the subscripts t, x, y, σ denote the partial derivatives.

3 Semi-Lagrangian Scale Selective Algorithm

The developed algorithm follows the general outline of the two-time-level SLSI method [11,15,16] with modifications imposed by application of vertical decoupling and horizontal splitting. First, the backward in time trajectory equations are solved:

$$\frac{d\mathbf{r}}{dt} = \mathbf{V} \ , \ \mathbf{r} = (x, y, \sigma) \ , \ \mathbf{V} = (u, v, \dot{\sigma}) \ , \ t \in [t_n, t_{n+1}] \ ; \ \mathbf{r}(t_{n+1}) = \mathbf{r}^a . \tag{4}$$

Here $\mathbf{r}^a$ are given coordinates of the arrival points chosen to be the grid points. One standard procedure for finding the departure points $\mathbf{r}(t_n) = \mathbf{r}^d$ is solving (4) by the fixed point iterations [11,15]

$$\Delta\mathbf{r}(s+1) = \tau\mathbf{V}^{n+1/2}(\mathbf{r}(s) - \Delta\mathbf{r}(s)/2) \ , \ \Delta\mathbf{r}(s) = \mathbf{r}^a - \mathbf{r}(s), \tag{5}$$

with velocity defined by the extrapolation formula at the intermediate time level $t_{n+1/2} = (n+1/2)\tau$: $\mathbf{V}^{n+1/2} = (3\mathbf{V}^n - \mathbf{V}^{n-1})/2$. Here s is the iteration number and τ is the time step. If the iteration convergence condition [14]

$$\tau \le \frac{2}{3V_d} \ , \ V_d = \max\left(|u_x|, |u_y|, |u_\sigma|, |v_x|, |v_y|, |v_\sigma|, |\dot{\sigma}_x|, |\dot{\sigma}_y|, |\dot{\sigma}_\sigma|\right) \tag{6}$$

is satisfied, then trilinear spatial interpolation of the velocity components to the trajectory points ensures finding the departure points with the second order of accuracy [13,15,16]. Using the maximum values of the wind component variations $V_d \approx 1.5 \cdot 10^{-4}\,\mathrm{s}^{-1}$, the maximum allowable time step obtained from (6) is $\tau \approx 70\,\mathrm{min}$.

The second stage consists of semi-Lagrangian forward-backward approximation of the prognostic equations (1),(3):

$$\frac{\hat{u}^{n+1} - u^n}{\tau} = f\hat{v}^{n+1} - \hat{G}_x^{n+1} + N_u^{n+1/2} \ , \ \frac{\hat{v}^{n+1} - v^n}{\tau} = -fu^n - \hat{G}_y^{n+1} + N_v^{n+1/2} \ , \tag{7}$$

$$\frac{\hat{P}^{n+1} - P^n}{\tau} = -D^n - \dot{\sigma}_\sigma^n, \ \frac{\hat{T}^{n+1} - T^n}{\tau} = \frac{RT_0}{c_p}\left(\frac{\hat{P}^{n+1} - P^n}{\tau} + \frac{\dot{\sigma}^n}{\sigma}\right) + N_T^{n+1/2} . \tag{8}$$

The nonlinear terms are found by extrapolation at the intermediate time level $t_{n+1/2}$:

$$N^{n+1/2} = \frac{N^{n+1/2} + N^{n+1/2}}{2} = \frac{1}{2}\left(\frac{3N^n - N^{n-1}}{2} + \frac{3N^n - N^{n-1}}{2}\right), \; N = N_u, N_v, N_T.$$

The superscripts $n-1$, n and $n+1$ denote the values at the past t_{n-1}, current t_n and new t_{n+1} time levels along the trajectory of air particles, that is,

$$\varphi^{n+1} = \varphi(t_{n+1}, \mathbf{r}(t_{n+1})), \; \varphi^n = \varphi(t_n, \mathbf{r}(t_n)), \; \varphi^{n-1} = \varphi(t_{n-1}, \mathbf{r}(t_{n-1})), \; \varphi = u, v, G, P, T, D, \dot{\sigma}.$$

Formulas (7),(8) can be solved in simple explicit way, but they have only the first order of accuracy and very restrictive CFL condition of stability:

$$\tau \leq \sqrt{2} h_g / c_g , \tag{9}$$

where h_g is a mesh size of spatial grid used for the gravity terms (that is, for the pressure gradient and divergence) and $c_g \approx 350 \, m/s$ is the maximum velocity of the gravity waves in the primitive system. On the spatial grid C with the main mesh size $h = 50 \, km$, the minimum gravity mesh size is $h_g = h/2 = 25 km$ [11]. Then the maximum allowable time step is about $100 \, sec$, which is very small as compared with accuracy requirements. If the coarser mesh size of 75km is used for approximation of the pressure gradient and divergence, then $\tau \approx 5 \, min$, which is still small.

The third stage consists of vertical decoupling and formulation of equations for implicit and more accurate approximation of the principal vertical modes. To this end, the equations for corrections to preliminary values are considered:

$$\frac{\delta u}{\tau} = f \frac{\delta v - \delta \hat{v}}{2} - \frac{\delta G_x - \delta \hat{G}_x}{2} \; , \; \frac{\delta v}{\tau} = -f \frac{\delta u + \delta \hat{u}}{2} - \frac{\delta G_y - \delta \hat{G}_y}{2} \; , \tag{10}$$

$$\frac{\delta P}{\tau} = -\frac{\delta D + \delta \hat{D}}{2} - \frac{\delta \dot{\sigma}_\sigma + \delta \hat{\dot{\sigma}}_\sigma}{2} \; , \; \frac{\delta T}{\tau} = \frac{RT_0}{c_p}\left(\frac{\delta P}{\tau} + \frac{\delta \dot{\sigma} + \delta \hat{\dot{\sigma}}}{2\sigma}\right) . \tag{11}$$

Here $\delta \hat{\varphi} = \hat{\varphi}^{n+1} - \varphi^n$, and unknown functions found by the formulas

$$\varphi^{n+1} = \hat{\varphi}^{n+1} + \delta \varphi \; , \; \varphi = u, v, G, P, T, D, \dot{\sigma}$$

coincide with solution of usual two-time-level SLSI scheme described in [12,16].

To avoid unnecessary corrections for insignificant vertical modes, the last system is vertically decoupled using the eigenvectors of the vertical structure matrix. To obtain these eigenvectors, the last two equations are simplified to the form

$$\left(\sigma \frac{\delta T}{\tau}\right)_\sigma = -\frac{RT_0}{c_p} \frac{\delta D + \delta \hat{D}}{2} \; .$$

and, by using the hydrostatic equation (2), temperature T is substituted by function G:

$$\left(\sigma \frac{\delta G_{\ln \sigma}}{\tau}\right)_\sigma = RT_0 \frac{R}{c_p} \frac{\delta D + \delta \hat{D}}{2} . \tag{12}$$

Equations (10), (12) form the closed system for three unknown functions. After vertical discretization on the Lorenz vertical K level grid [2,3,5], the discrete analogues of equations (10), (12) can be written as follows:

$$\frac{\delta \mathbf{u}}{\tau} = f \frac{\delta \mathbf{v} - \delta \hat{\mathbf{v}}}{2} - \frac{\delta \mathbf{G}_x - \delta \hat{\mathbf{G}}_x}{2}, \quad \frac{\delta \mathbf{v}}{\tau} = -f \frac{\delta \mathbf{u} + \delta \hat{\mathbf{u}}}{2} - \frac{\delta \mathbf{G}_y - \delta \hat{\mathbf{G}}_y}{2},$$

$$\frac{\delta \mathbf{G}}{\tau} = -RT_0 \mathbf{A} \frac{\delta \mathbf{D} + \delta \hat{\mathbf{D}}}{2}, \tag{13}$$

where $\mathbf{u}, \mathbf{v}, \mathbf{D}, \mathbf{G}$ are the vectors of order K and $\mathbf{A}$ is $K \times K$ matrix of the vertical structure. The distribution of variables on the Lorenz vertical grid and some natural approximations to vertical operators can be found in [2,3,5]. Since discretization on this grid is more straightforward for keeping conservation properties of the primitive equations it seems to be the most popular vertical grid for hydrostatic models [2,5].

It was proved in [5] that the matrix $\mathbf{A}$ has the spectral decomposition $\mathbf{A} = \mathbf{S} \Lambda \mathbf{S}^{-1}$ with the positive eigenvalue matrix $\Lambda = \mathbf{diag}[\lambda_1, \ldots, \lambda_K]$ and the matrix of eigenvectors (that is, vertical normal modes) $\mathbf{S}$. Of course, all these transformations and calculations related to finding of the vertical structure matrix and its spectral decomposition are made only once before numerical forecasting.

Multiplying the equations of system (13) on the left by $\mathbf{S}^{-1}$, one obtains K decoupled 2D systems

$$du_k = \tau f \frac{dv_k - d\hat{v}_k}{2} - \tau \frac{dG_{kx} - d\hat{G}_{kx}}{2}, \quad dv_k = -\tau f \frac{du_k + d\hat{u}_k}{2} - \tau \frac{dG_{ky} - d\hat{G}_{ky}}{2},$$

$$dG_k = -\tau c^2 \frac{dD_k + d\hat{D}_k}{2} . \tag{14}$$

Here $k = 1, \ldots, K$ is the index of vertical mode, $c_k = \sqrt{RT_0 \lambda_k}$ is the gravity wave speed of the k-th vertical mode and

$$d\varphi = \mathbf{S}^{-1} \cdot \delta \varphi , \quad d\hat{\varphi} = \mathbf{S}^{-1} \cdot \delta \hat{\varphi} , \quad c_k^2 = RT_0 \lambda_k , \tag{15}$$

that is, $d\varphi_k$ are the coefficients of expansion of physical corrections $\delta \varphi$ by the vertical normal modes $\mathbf{s}_k$, which compile the matrix $\mathbf{S}$, and analogously for $d\hat{\varphi}_k$. The eigenvalues λ_k (and c_k) are supposed to be numbered in decreasing order. From now on, we omit the subscript k, because it does not cause any ambiguity.

The last step of the algorithm is solution of (14) for the first principal I modes (other modes remain without change) by applying a time splitting technique. Each system (14) for $i = 1, \ldots, I$ is splitted into two subsystems solved successively for auxiliary corrections $d\tilde{\varphi}$ and final corrections $d\overline{\varphi}$:

$$d\tilde{u} = \tau f \frac{d\tilde{v} - d\hat{v}}{2} - \tau \frac{d\tilde{G}_x - d\hat{G}_x}{2} + h\,, \quad d\tilde{v} = 0\,, \quad d\tilde{G} = -\tau c^2 \frac{d\tilde{u}_x + d\hat{u}_x}{2}\,, \tag{16}$$

and

$$d\overline{u} - d\tilde{u} = 0\,, \quad d\overline{v} - d\tilde{v} = -\tau f \frac{d\overline{u} + d\hat{u}}{2} - \tau \frac{d\overline{G}_y - d\hat{G}_y}{2}\,, \quad d\overline{G} - d\tilde{G} = -\tau c^2 \frac{d\overline{v}_y + d\hat{v}_y}{2}\,. \tag{17}$$

Here h is an additional modification term by Douglas et al. [9,10]. It can be shown that (16),(17) is the second order approximation to (14) and the splitting error can be reduced with no penalty on simplicity of algorithm by choosing function h as follows:

$$h = -\frac{\tau^2 f^2}{4} du^* - \frac{\tau^2 f}{4}\left(dG^* - 2d\hat{G}\right)_y + \frac{\tau^2 c^2}{4} dv^*_{xy}\,, \quad d\varphi^* = \varphi^n - \varphi^{n-1}\,.$$

Each of the systems (16), (17) is transformed to 1D elliptic problem for G-corrections and the last is solved by simple Gelfand-Thomas algorithm. Found corrections are added to the values of the greatest modes and inverse vertical transformation returns the physical fields composed of the principal modes evaluated by semi-Lagrangian semi-implicit scheme and secondary modes calculated by equations (7),(8).

Applied scale separation allows to reduce the amount of computations at each time step and keep reasonably large time step. The linear stability analysis gives the approximate stability condition

$$\tau \leq \sqrt{2} h_g \big/ c_{I+1} \tag{18}$$

for the explicit modes (c_{I+1} is the maximum gravity wave speed of the explicit modes) and the trajectory convergence restriction (6) for the implicit modes. Since gravity speeds decrease rapidly as I increases, the explicit treatment of the smallest vertical modes does not cause strict limitation on the time step.

4 Numerical Tests

The described scheme was applied to 20-level hydrostatic model on horizontal grid with mesh size $h = 50\text{km}$. The first seven vertical modes with gravity wave speeds $c_1 = 343\,m/s$, $c_2 = 203\,m/s$, $c_3 = 122\,m/s$, $c_4 = 78.1\,m/s$, $c_5 = 54.3\,m/s$, $c_6 = 40.9\,m/s$, $c_7 = 32.6\,m/s$ were corrected and the remaining modes were treated explicitly with mesh size $h_g = 75\text{km}$ for the gravity wave terms. The greatest gravity speed of the explicit modes is $c_8 = c_{I+1} = 25.5\,m/s$ so that condition (18) allows to use time steps up to 60min, which is approximately equal to requirement (6) for the trajectory iterations. This way, the maximum time steps for traditional SLSI scheme (SLSIT) and described scale selective SLSI scheme (SLSIS) are practically coincide.

For evaluation of accuracy and computational efficiency of SLSIS scheme, its performance was compared with SLSIT scheme and with Eulerian leapfrog scheme

(LF). The last is a simple rather popular explicit scheme tested in different models of the atmosphere. The integrations were carried out on horizontal domain of $5000\text{x}5000\,\text{km}^2$ centered at Porto Alegre city ($30^0 S$, $52^0 W$) and the initial and boundary conditions were obtained from objective analysis and global forecasts of the National Centers for Environmental Prediction (NCEP).

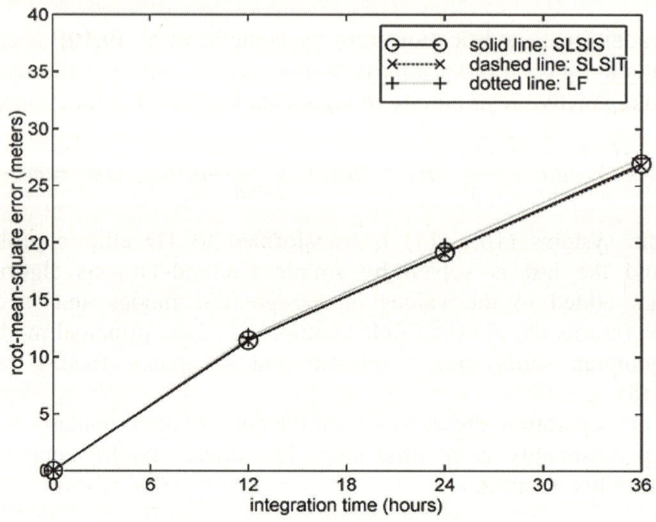

Fig. 1. Root-mean-square error of geopotential forecast at 500 hPa pressure level

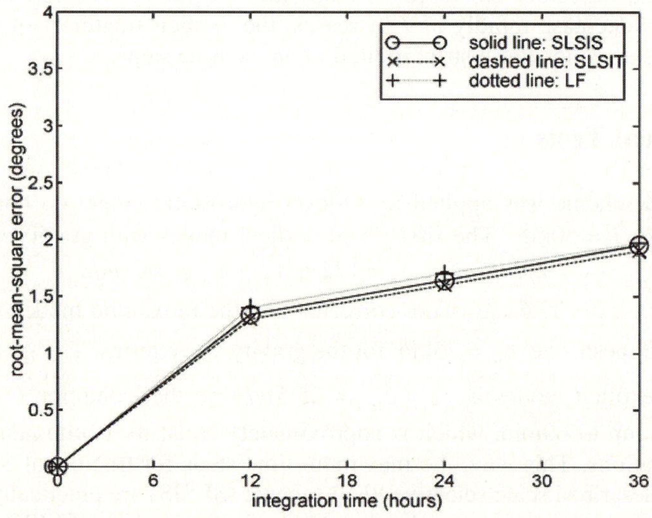

Fig. 2. Root-mean-square error of temperature forecast at 850 hPa pressure level

The 24-h forecasts were computed by using each of the above three schemes. The computational cost of one SLSIT forecast is about 18% of the LF computational time and the SLSIS scheme computation is about 42% faster than the SLSIT scheme. The root-mean-square differences between 24-h forecast and NCEP analysis are shown in Figs.1,2 for two elements: geopotential height at the 500hPa pressure level and temperature at the 850hPa pressure level. These characteristic elements are traditionally verified in the numerical weather prediction systems: the first reflects the dynamics of the middle atmosphere and the second is important for determination of humidity and cloud processes [1].

Another important skill measure for short-range forecasting is the correlation coefficient between predicted and observed tendency [1]. Evaluated for 12-h, 24-h and 36-h forecasts, this measure shows that the SLSIS and SLSIT forecasts are virtually coincide and the LF scheme has slightly lower accuracy, that is, it reveals relation between forecasts similar to that shown in Figs.1,2. Thus, performed evaluations confirm the validity of the applied method of scale separation and efficiency of the developed scheme.

Acknowledgements. This research was supported by Brazilian science foundation CNPq.

References

1. Anthes R.A., Kuo Y.H., Hsie E.Y., Low-Nam S., Bettge T.W.: Estimation of skill and uncertainty in regional numerical models. Q. J. R. Meteorol. Soc. **115** (1989) 763-806.
2. Arakawa A., Suarez M.J.: Vertical differencing of the primitive equations in sigma coordinates, Mon. Wea. Rev. **111** (1983) 34-45.
3. Bates J.R., Moorthi S., Higgins R.W.: A global multilevel atmospheric model using a vector semi-Lagrangian finite-difference scheme. Part I: Adiabatic formulation, Mon. Wea. Rev. **121** (1993) 244-263.
4. Bourchtein A.: Semi-Lagrangian semi-implicit space splitting regional baroclinic atmospheric model. Appl. Numer. Math. **41** (2002) 307-326.
5. Bourchtein A., Kadychnikov V.: Well-posedness of the initial value problem for vertically discretized hydrostatic equations. SIAM J. Num. An. **41** (2003) 195-207.
6. Browning G.L., Kreiss H.-O.: Splitting methods for problems with different timescales, Mon.Wea.Rev. **122** (1994) 2614-2622.
7. Burridge D.M., A split semi-implicit reformulation of the Bushby-Timpson 10 level model, Quart. J. Roy. Meteor. Soc. **101** (1975) 777-792.
8. Côté J., Gravel S., Methot A., Patoine A., Roch M., Staniforth A.: The operational CMC-MRB global environmental multiscale (GEM) model. Part I: Design considerations and formulation, Mon. Wea. Rev. **126** (1998) 1373-1395.
9. Douglas J., Kim S., Improved accuracy for locally one-dimensional methods for parabolic equations, Mathematical Models and Methods in Applied Science **11** (2001) 1563-1579.
10. Douglas J., Kim S., Lim H., An improved alternating-direction method for a viscous wave equation, in: Z. Chen, R. Glowinski, Kaitai L. (Eds.), Current Trends in Scientific Computing, Contemporary Mathematics, **329** (2003) 99-104.
11. Durran D.: Numerical Methods for Wave Equations in Geophysical Fluid Dynamics. Springer, New York (1999).

12. Hortal M.: The development and testing of a new two-time-level semi-Lagrangian scheme (SETTLS) in the ECMWF forecast model, Q.J.R.Met.Soc. **128** (2002) 1671-1687.
13. McDonald A.: Accuracy of multiply upstream, semi-Lagrangian advective schemes, Mon. Wea. Rev. **112** (1984) 1267-1275.
14. Pudykiewicz J., Benoit R., Staniforth A.: Preliminary results from a partial LRTAP model based on an existing meteorological forecast model. Atmos.-Ocean **23** (1985) 267-303.
15. Staniforth A., Côté J.: Semi-Lagrangian integration schemes for atmospheric models - A review. Mon. Wea. Rev. **119** (1991) 2206-2223.
16. Temperton C., Hortal M., Simmons A.J.: A two-time-level semi-Lagrangian global spectral model. Q. J. R. Meteorol. Soc. **127** (2001) 111-126.
17. Yakimiw E., Robert A.: Accuracy and stability analysis of a fully implicit scheme for the shallow water equations, Mon. Wea. Rev. **114** (1986) 240-244.

Identifying Cost-Effective Common Subexpressions to Reduce Operation Count in Tensor Contraction Evaluations

Albert Hartono[1], Qingda Lu[1], Xiaoyang Gao[1], Sriram Krishnamoorthy[1],
Marcel Nooijen[3], Gerald Baumgartner[4], David E. Bernholdt[6],
Venkatesh Choppella[1,7], Russell M. Pitzer[2], J. Ramanujam[5],
Atanas Rountev[1], and P. Sadayappan[1]

[1] Dept. of Computer Science and Engineering
[2] Dept. of Chemistry, The Ohio State University,
Columbus, OH 43210, USA
[3] Dept. of Chemistry, University of Waterloo,
Waterloo, Ontario N2L BG1, Canada
[4] Dept. of Computer Science
[5] Dept. of Electrical and Computer Engineering,
Louisiana State University,
Baton Rouge, LA 70803, USA
[6] Computer Sci. & Math. Div., Oak Ridge National Laboratory,
Oak Ridge, TN 37831, USA
[7] Indian Institute of Information Technology and Management,
Kerala, India

Abstract. Complex tensor contraction expressions arise in accurate electronic structure models in quantum chemistry, such as the coupled cluster method. Transformations using algebraic properties of commutativity and associativity can be used to significantly decrease the number of arithmetic operations required for evaluation of these expressions. Operation minimization is an important optimization step for the Tensor Contraction Engine, a tool being developed for the automatic transformation of high-level tensor contraction expressions into efficient programs. The identification of common subexpressions among a set of tensor contraction expressions can result in a reduction of the total number of operations required to evaluate the tensor contractions. In this paper, we develop an effective algorithm for common subexpression identification and demonstrate its effectiveness on tensor contraction expressions for coupled cluster equations.

1 Introduction

Users of current and emerging high-performance parallel computers face major challenges to both performance and productivity in the development of their scientific applications. For example, the manual development of accurate quantum chemistry models typically takes an expert several months of tedious effort; high-performance implementations can take substantially longer. One approach to address this situation is the use of automatic code generation to synthesize efficient parallel programs from the equations to be implemented, expressed in a very high-level domain-specific language. The

V.N. Alexandrov et al. (Eds.): ICCS 2006, Part I, LNCS 3991, pp. 267–275, 2006.

Tensor Contraction Engine (TCE) [3, 2] is such a tool, being developed through a collaboration between computer scientists and quantum chemists.

The first step in the TCE's code synthesis process is the transformation of input equations into an equivalent form with minimal operation count. Equations typically range from around ten to over a hundred terms, each involving the contraction of two or more tensors, and most quantum chemical methods involve two or more coupled equations of this type. This optimization problem can be viewed as a generalization of the matrix chain multiplication problem, which, unlike the matrix-chain case, has been shown to be NP-hard [6]. Our prior work focused on the use of single-term optimization (strength reduction or parenthesization), which decomposes multi-tensor contraction operations into a sequence of binary contractions, coupled with a global search of the composite single-term solution space for factorization opportunities. Exhaustive search (for small cases) and a number of heuristics were shown to be effective in minimizing the operation count [4].

Common subexpression elimination (CSE) is a classical optimization technique used in traditional optimizing compilers [1] to reduce the number of operations, where intermediates are identified that can be computed once and stored for use multiple times later. CSE is routinely used in the manual formulation of quantum chemical methods, but because of the complexity of the equations, it is extremely difficult to explore all possible formulations manually. CSE is a powerful technique that allows the exploration of the much larger algorithmic space than our previous approaches to operation minimization. However, the cost of the search itself grows explosively. In this paper, we develop an approach to CSE identification in the context of operation minimization for tensor contraction expressions. The developed approach is shown to be very effective, in that it automatically finds efficient computational forms for challenging tensor equations.

Quantum chemists have proposed domain-specific heuristics for strength reduction and factorization for specific forms of tensor contraction expressions (e.g., [7, 9]). However, their work does not consider the general form of arbitrary tensor contraction expressions. Single-term optimizations in the context of a general class of tensor contraction expressions were addressed in [6]. Approaches to single-term optimizations and factorization of tensor contraction expressions were presented in [4, 8]. Common subexpression identification to enhance single-term optimization was not considered in any of these approaches.

The rest of this paper is organized as follows. Section 2 provides a more detailed description of the operation minimization and the common subexpression elimination problem in the context of tensor contraction expressions. Section 3 describes our approach. Experimental results are presented in Section 4 and Section 5 concludes the paper.

2   Common Subexpressions and Operation Count Reduction

A tensor contraction expression comprises a sum of a number of terms, where each term might involve the contraction of two or more tensors. We first illustrate the issue of operation minimization for a single term, before addressing the issue of finding common subexpressions to optimize across multiple terms. Consider the following tensor contraction expression involving three tensors t, f and s, with indices x and z that have range V, and indices i and k that have range O. Distinct ranges for different indices is a characteristic of the quantum chemical methods of interest, where O and V correspond to the number of occupied and virtual orbitals in the representation of the molecule (typically $V \gg O$). Computed as a single nested loop computation, the number of arithmetic operations needed would be $2O^2V^2$.

$$r_i^x = \Sigma_{z,k} t_i^z f_z^k s_k^x \qquad \text{(cost=}2O^2V^2)$$

However, by performing a two-step computation with an intermediate I, it is possible to compute the result using $4OV^2$ operations:

$$I_z^x = \Sigma_k f_z^k s_k^x \qquad \text{(cost=}2OV^2); \qquad r_i^x = \Sigma_z t_i^z I_z^x \qquad \text{(cost=}2OV^2)$$

Another possibility using $4O^2V$ computations, which is more efficient when $V > O$ (as is usually the case in quantum chemistry calculations), is shown below:

$$I_i^k = \Sigma_z t_i^z f_z^k \qquad \text{(cost=}2O^2V); \qquad r_i^x = \Sigma_k I_i^k s_k^x \qquad \text{(cost=}2O^2V)$$

The above example illustrates the problem of single-term optimization, also called strength reduction: find the best sequence of two-tensor contractions to achieve a multi-tensor contraction. Different orders of contraction can result in very different operation costs; for the above example, if the ratio of V/O were 10, there is an order of magnitude difference in the number of arithmetic operations for the two choices.

With complex tensor contraction expressions involving a large number of terms, if multiple occurrences of the same subexpression can be identified, it will only be necessary to compute it once and use it multiple times. Thus, common subexpressions can be stored as intermediate results that are used more than once in the overall computation. Manual formulations of computational chemistry models often involve the use of such intermediates. The class of quantum chemical methods of interest, which include the coupled cluster singles and doubles (CCSD) method [7, 9], are most commonly formulated using the molecular orbital basis (MO) integral tensors. However the MO integrals are intermediates, derived from the more fundamental atomic orbital basis (AO) integral tensors. Alternate "AO-based" formulations of CCSD have been developed in which the more fundamental AO integrals are used directly, without fully forming the MO integrals [5]. However it is very difficult to manually explore all possible formulations of this type to find the one with minimal operation count, especially since it can depend strongly on the characteristics of the particular molecule being studied.

The challenge in identifying cost-effective common subexpressions is the combinatorial explosion of the search space, since single-term optimization of different product terms must be treated in a coupled manner. The following simple example illustrates the problem.

Suppose we have two MO-basis tensors, v and w, which can be expressed as a transformation of the AO-basis tensor, a, in two steps. Using single-term optimization to form tensor v, we consider two possible sequences of binary contractions as shown below, which both have the same (minimal) operation cost. Extending the notation above, indices p and q represent AO indices, which have range $M = O + V$.

Seq. 1: $I1_q^i = \sum_p a_q^p c_p^i$ (cost=$2OM^2$); $v_j^i = \sum_p I1_p^i d_j^p$ (cost=$2O^2M$)

Seq. 2: $I2_i^p = \sum_q a_q^p d_i^q$ (cost=$2OM^2$); $v_j^i = \sum_p I2_j^p c_p^i$ (cost=$2O^2M$)

To generate tensor w, suppose that there is only one cost-optimal sequence:

$$I1_q^i = \sum_p a_q^p c_p^i \qquad (cost=2OM^2); \qquad w_x^i = \sum_p I1_p^i e_x^p \qquad (cost=2OVM)$$

Note that the first step in the formation of w uses the same intermediate tensor $I1$ that appears in sequence 1 for v. Considering just the formation of v, either of the two sequences is equivalent in cost. But one form uses a common subexpression that is useful in computing the second MO-basis tensor, while the other form does not. If sequence 1 is chosen for v, the total cost of computing both v and w is $2OM^2 + 2O^2M + 2OVM$. On the other hand, the total cost is higher if sequence 2 is chosen ($4OM^2 + 2O^2M + 2OVM$). The $2OM^2$ cost difference is significant when M is large.

When a large number of terms exist in a tensor contraction expression, there is a combinatorial explosion in the search space if all possible equivalent-cost forms for each product term must be compared with each other.

In this paper, we address the following question: By developing an automatic operation minimization procedure that is effective in identifying suitable common subexpressions in tensor contraction expressions, can we automatically find more efficient computational forms? For example, with the coupled cluster equations, can we automatically find AO-based forms by simply executing the operation minimization procedure on the standard MO-based CCSD equations, where occurrences of the MO integral terms are explicitly expanded out in terms of AO integrals and integral transformations?

3   Algorithms for Operation Minimization with CSE

In this section, we describe the algorithm used to perform operation minimization, by employing single-term optimization together with CSE. The exponentially large space of possible single-term optimizations, together with CSE, makes an exhaustive search approach prohibitively expensive. So we use a two-step approach to apply single-term optimization and CSE in tandem.

The algorithm is shown in Fig. 2. It uses the single-term optimization algorithm, which is broadly illustrated in Fig. 1 and described in greater detail in our earlier work [4]. It takes as input a sequence of tensor contraction statements. Each statement defines a tensor in terms of a sum of tensor contraction expressions. The output is an optimized sequence of tensor contraction statements involving only binary tensor contractions. All intermediate tensors are explicitly defined.

The key idea is to determine the parenthesization of more expensive terms before the less expensive terms. The most expensive terms contribute heavily to the overall operation cost, and potentially contain expensive subexpressions. Early identification of these expensive subexpressions can facilitate their reuse in the computation of other expressions, reducing the overall operation count.

The algorithm begins with the *term set* to be optimized as the set of all the terms of the tensor contraction expressions on the right hand side of each statement. The set of intermediates is initially empty. In each step of the iterative procedure, the parenthesization for one term is determined. Single-term optimization is applied to each term in the term set using the current set of intermediates and the most expensive term is chosen to be parenthesized. Among the set of optimal parenthesizations for the chosen term, the one that maximally reduces the cost of the remaining terms is chosen. Once the term and its parenthesization are decided upon, the set of intermediates is updated and the corresponding statements for the new intermediates are generated. The procedure continues until the term set is empty.

SINGLE-TERM-OPT-CSE(E, is)

```
1   if E is a single-tensor expression
2     then return {⟨E,  ⟩}
3     else  \* E is a multiple-tensor contraction expression (i.e., E₁ * … * Eₙ) * \
4            {⟨p₁, is₁⟩, ⟨p₂, is₂⟩, …} ←
5                set of pairs of optimal parenthesization of E and its corresponding intermediate set
6                (the given intermediate set is is used to find effective common subexpressions)
7            return {⟨p₁, is₁⟩, ⟨p₂, is₂⟩, …}
```

Fig. 1. Single-term optimization algorithm with common subexpression elimination

OPTIMIZE$(stmts)$

```
 1   MSET ← set of all terms obtained from RHS expressions of stmts
 2   is ←    \* the set of intermediates * \
 3   while MSET ≠
 4   do Mheaviest ← the heaviest term in MSET
 5            (searched by applying SINGLE-TERM-OPT-CSE(Mᵢ, is) on each term Mᵢ ∈ MSET)
 6      PSET ← SINGLE-TERM-OPT-CSE(Mheaviest, is)
 7      ⟨pbest, isbest⟩ ← NIL
 8      profit ← 0
 9      for each ⟨pᵢ, isᵢ⟩ ∈ PSET
10      do cur_profit ← 0
11          for each Mᵢ ∈ (MSET − {Mheaviest})
12          do base_cost ← op-cost of optimal parenth. in SINGLE-TERM-OPT-CSE(Mᵢ, is)
13              opt_cost ← op-cost of optimal parenth. in SINGLE-TERM-OPT-CSE(Mᵢ, is ∪ isᵢ)
14              cur_profit ← cur_profit + (base_cost − opt_cost)
15          if ((⟨pbest, isbest⟩ = NIL) ∨ (cur_profit > profit)
16              then ⟨pbest, isbest⟩ ← ⟨pᵢ, isᵢ⟩
17                  profit ← cur_profit
18      stmts ← replace the term Mheaviest in stmts with pbest
19      MSET ← MSET − {Mheaviest}
20      is ← is ∪ isbest
21   return stmts
```

Fig. 2. Global operation minimization algorithm

4   Experimental Results

We evaluated our approach by comparing the optimized operation count of the MO-based CCSD T1 and T2 computations with the corresponding equations in which the occurrences of MO integrals are replaced by the expressions that produce them, referred to as the expanded form. Table 1 illustrates the characteristics of CCSD T1 and T2 equations. Fig. 3 shows the CCSD T1 equation, consisting of the computation of the MO

Table 1. Characteristics of input equations used in experiments

Equation	Number of terms	MO Integrals
CCSD T1	14	$v\_ooov, v\_oovv, v\_ovov, v\_ovvv$
CCSD T2	31	$v\_oooo, v\_ooov, v\_oovv, v\_ovoo, v\_ovov, v\_ovvv, v\_vvoo, v\_vvov, v\_vvvv$

$$\text{(1a)} \quad v\_ooov_{h3p1}^{h1h2} = (c\_mo_{h3}^{q1} * c\_mv_{p1}^{q2} * c\_om_{q3}^{h1} * c\_om_{q4}^{h2} * a\_mmmm_{q1q2}^{q3q4})$$

$$\text{(1b)} \quad v\_oovv_{p1p2}^{h1h2} = (c\_mv_{p1}^{q1} * c\_mv_{p2}^{q2} * c\_om_{q3}^{h1} * c\_om_{q4}^{h2} * a\_mmmm_{q1q2}^{q3q4})$$

$$\text{(1c)} \quad v\_ovov_{h2p2}^{h1p1} = (c\_mo_{h2}^{q1} * c\_mv_{p2}^{q2} * c\_om_{q3}^{h1} * c\_vm_{q4}^{p1} * a\_mmmm_{q1q2}^{q3q4})$$

$$\text{(1d)} \quad v\_ovvv_{p2p3}^{h1p1} = (c\_mv_{p2}^{q1} * c\_mv_{p3}^{q2} * c\_om_{q3}^{h1} * c\_vm_{q4}^{p1} * a\_mmmm_{q1q2}^{q3q4})$$

$$\text{(2)} \quad \text{residual}_{h1}^{p2} = 0.25 * (t\_vvoo_{h2h1}^{p2p1} * f\_ov_{p2}^{h2}) - 0.25 * (v\_ovov_{h1p2}^{h2p1} * t\_vo_{h2}^{p2})$$

$$+ 0.25 * (f\_vv_{p2}^{p1} * t\_vo_{h1}^{p2}) - 0.25 * (f\_oo_{h1}^{h2} * t\_vo_{h2}^{p1}) + 0.25 * f\_vo_{h1}^{p1}$$

$$- 0.25 * (t\_vo_{h2}^{p1} * t\_vo_{h1}^{p2} * t\_vo_{h3}^{p3} * v\_oovv_{p2p3}^{h2h3})$$

$$+ 0.25 * (t\_vvoo_{h2h1}^{p2p1} * t\_vo_{h3}^{p3} * v\_oovv_{p2p3}^{h2h3}) - 0.125 * (t\_vo_{h2}^{p1} * t\_vvoo_{h3h1}^{p2p3} * v\_oovv_{p2p3}^{h3h2})$$

$$- 0.125 * (t\_vo_{h1}^{p2} * t\_vvoo_{h2h3}^{p3p1} * v\_oovv_{p3p2}^{h2h3}) - 0.25 * (t\_vo_{h1}^{p2} * v\_ovvv_{p2p3}^{h2p1} * t\_vo_{h2}^{p3})$$

$$- 0.25 * (t\_vo_{h2}^{p1} * v\_ooov_{h1p2}^{h2h3} * t\_vo_{h3}^{p2}) - 0.25 * (t\_vo_{h2}^{p1} * t\_vo_{h1}^{p2} * f\_ov_{p2}^{h2})$$

$$+ 0.125 * (t\_vvoo_{h2h1}^{p2p3} * v\_ovvv_{p2p3}^{h2p1}) + 0.125 * (t\_vvoo_{h2h3}^{p2p1} * v\_ooov_{h1p2}^{h2h3})$$

Fig. 3. The input formulation of CCSD T1. For compactness, summations are implicit wherever the same index appears twice in a term.

integrals (Steps 1a–1d) and the expression for the single-excitation residual (Step 2). Whereas our examples above used rank-2 tensors for simplicity, the CCSD equations primarily involve rank-4 integral tensors.

The number of arithmetic operations depends upon O and V, which are specific to the molecule and quality of the simulation, but a typical range is $1 \leq V/O \leq 100$. To provide concrete comparisons, we set O to 10 and V to 100 or 500.

The CCSD computation proceeds through a number of iterations in which the AO integrals remain unchanged. At convergence, the amplitudes t_vo and t_vvoo attain values such that the residual vector in Step 2 of Fig. 3 is equal to zero and this typically takes 10–50 iterations. In different variants of CCSD, the MO integrals may also remain unchanged, or may change at each iteration, requiring the AO-to-MO transformation to be repeated. To represent these two cases, we use iteration counts of 10 and 1, respectively, to evaluate the different formulations obtained.

Tables 2 and 3 illustrate the results obtained by optimizing CCSD T1 and T2 equations with the algorithm described above. The total operation counts are shown for different (O,V) pairs, changing iteration counts, and choice of MO integrals to expand. We applied single-term optimization and CSE to the AO-to-MO calculation and the MO-basis expression separately, without expanding any MO integrals - this is representative of current implementations of coupled cluster methods. We report the operation count reduction using our approach relative to the optimized conventional two-step formulation as discussed above.

Table 2. Results of optimizing CCSD T1 with our algorithm

(O,V)	Iteration Count	Expanded Tensors	Total Operation Count	Reduction Factor
$(10,100)$	1	None	1.12×10^{10}	1
		v_ovvv	5.25×10^{9}	2.14
		v_ovvv, v_ooov, v_ovov	4.52×10^{9}	2.48
	10	None	1.40×10^{10}	1
		v_ovvv	1.20×10^{10}	1.17
		v_ovvv, v_ooov, v_ovov	1.18×10^{10}	1.19
$(10,500)$	1	None	5.36×10^{12}	1
		v_ovvv	1.59×10^{12}	3.37
		v_ovvv, v_ooov, v_ovov	1.51×10^{12}	3.55
	10	None	5.63×10^{12}	1
		v_ovvv	2.34×10^{12}	2.41
		v_ovvv, v_ooov, v_ovov	2.26×10^{12}	2.49

Table 3. Results of optimizing CCSD T2 with our algorithm

(O,V)	Iteration Count	Expanded Tensors	Total Operation Count	Reduction Factor
$(10,100)$	1	None	1.51×10^{11}	1
		v_vvvv	6.87×10^{10}	2.20
		v_vvvv, v_ovvv, v_vvov	5.40×10^{10}	2.80
	10	None	4.68×10^{11}	1
		v_vvvv	4.68×10^{11}	1
		v_vvvv, v_ovvv, v_vvov	4.67×10^{11}	1
$(10,500)$	1	None	2.85×10^{14}	1
		v_vvvv	2.72×10^{13}	10.48
		v_vvvv, v_ovvv, v_vvov	1.93×10^{13}	14.75
	10	None	4.22×10^{14}	1
		v_vvvv	1.76×10^{14}	2.40
		v_vvvv, v_ovvv, v_vvov	1.67×10^{14}	2.53

Among all the sixteen cases we have studied, twelve of them yield a reduction factor ranging from 2.14 to 14.75 and two of them have a reduction factor close to 1.2. We can conclude that our algorithm performs well in practice in most cases. The following observations can be made from the results in Tables 2 and 3.

- The benefits decrease with an increase of the iteration count;
- The benefits increase with increasing number of explicitly expanded terms; and
- The benefits are greater when the V/O ratio is large.

Fig. 4 shows an optimized formulation of the CCSD T1 equation in Fig. 3, when $(O,V) = (10,500)$ and the MO integrals v_ovvv, v_ooov, v_ovov are expanded. It may be seen that this form, with an operation-count reduction factor of 2.49, is significantly different from the original MO-basis formulation in Fig. 3. In this new formulation, the it arrays are the common subexpressions identified to reduce the operation count.

(1a) $\quad it\_1_{q2q3}^{q1h1} = (a\_mmmm_{q2q3}^{q4q1} * c\_om_{q4}^{h1})$

(1b) $\quad it\_2_{p1q1}^{h1h2} = (c\_mv_{p1}^{q2} * (c\_om_{q3}^{h1} * it\_1_{q1q2}^{q3h2}))$

(1c) $\quad v\_oovv_{p1p2}^{h1h2} = (c\_mv_{p1}^{q1} * it\_2_{p2q1}^{h2h1})$

(1d) $\quad it\_3_{h1}^{q1} = (c\_mv_{p1}^{q1} * t\_vo_{h1}^{p1})$

(1e) $\quad it\_4_{h3p1}^{h1h2} = (c\_mo_{h3}^{q1} * it\_2_{p1q1}^{h1h2})$

(1f) $\quad it\_5_{q2}^{q1} = (it\_1_{q2q3}^{q1h1} * it\_3_{h1}^{q3})$

(1g) $\quad it\_6_{p1}^{h1} = (v\_oovv_{p1p2}^{h1h2} * t\_vo_{h2}^{p2})$

(2) $\quad residual_{h1}^{p2} = 0.25 * f\_vo_{h1}^{p2} - 0.25 * (f\_oo_{h1}^{h2} * t\_vo_{h2}^{p1}) + 0.25 * (f\_vv_{p2}^{p1} * t\_vo_{h1}^{p2})$
$$+ 0.125 * (c\_vm_{q1}^{p1} * (it\_1_{q2q3}^{q1h2} * (c\_mv_{p1}^{q1} * (c\_mv_{p2}^{q1} * t\_vvoo_{h1h2}^{p2p1}))))$$
$$- 0.25 * ((f\_ov_{p1}^{h1} * t\_vo_{h2}^{p1}) * t\_vo_{h2}^{p1}) - 0.125 * ((t\_vo_{h3}^{p2} * v\_oovv_{p1p2}^{h1h2}) * t\_vvoo_{h2h3}^{p2p1})$$
$$- 0.125 * ((t\_vvoo_{h3h2}^{p1p2} * v\_oovv_{p1p2}^{h3h1}) * t\_vo_{h2}^{p1}) + 0.25 * (t\_vvoo_{h2h1}^{p2p1} * it\_6_{p2}^{h2})$$
$$- 0.25 * ((it\_6_{p1}^{h1} * t\_vo_{h2}^{p1}) * t\_vo_{h2}^{p1}) - 0.25 * (c\_vm_{q1}^{p1} * (c\_mo_{h1}^{q2} * it\_5_{q2}^{q1}))$$
$$- 0.25 * (c\_vm_{q1}^{p1} * (it\_3_{h1}^{q2} * it\_5_{q2}^{q1})) + 0.125 * (it\_4_{h1p2}^{h2h3} * t\_vvoo_{h3h2}^{p2p1})$$
$$- 0.25 * ((it\_4_{h2p1}^{h3h1} * t\_vo_{h3}^{p1}) * t\_vo_{h2}^{p1}) + 0.25 * (t\_vvoo_{h2h1}^{p2p1} * f\_ov_{p2}^{h2})$$

Fig. 4. The optimized formulation of CCSD T1. For compactness, summations are implicit wherever the same index appears twice in a term.

5 Conclusions

In this paper, we presented a coupled approach of utilizing single-term optimization and identification of common subexpressions to reduce the operation count in the evaluation of tensor contraction expressions. The benefits of the approach were shown by expanding the tensor contraction expressions in two representative computations, and demonstrating a reduction in the operation count for the composite computation.

Acknowledgments. This work has been supported in part by: the U.S. National Science Foundation through grants 0121676, 0121706, 0403342, 0508245, 0509442, 0509467, and 0541409; the Laboratory Directed Research and Development Program of Oak Ridge National Laboratory (ORNL); and a Discovery grant 262942-03 from the Natural Sciences and Engineering Research Council of Canada. ORNL is managed by UT-Battelle, LLC, for the US Dept. of Energy under contract DE-AC-05-00OR22725.

References

1. A. Aho, R. Sethi, and J. Ullman. *Compilers: Principles, Techniques, and Tools.* Addison-Wesley, 1986.
2. A. Auer, G. Baumgartner, D. Bernholdt, A. Bibireata, V. Choppella, D. Cociorva, X. Gao, R. Harrison, S. Krishanmoorthy, S. Krishnan, C. Lam, M. Nooijen, R. Pitzer, J. Ramanujam, P. Sadayappan, and A. Sibiryakov. Automatic code generation for many-body electronic structure methods: The Tensor Contraction Engine. *Molecular Physics*, 104(2):211–218, 20 January 2006.

3. G. Baumgartner, A. Auer, D. Bernholdt, A. Bibireata, V. Choppella, D. Cociorva, X. Gao, R. Harrison, S. Hirata, S. Krishnamoorthy, S. Krishnan, C. Lam, Q. Lu, M. Nooijen, R. Pitzer, J. Ramanujam, P. Sadayappan, and A. Sibiryakov. Synthesis of high-performance parallel programs for a class of ab initio quantum chemistry models. *Proceedings of the IEEE*, 93(2):276–292, February 2005.
4. A. Hartono, A. Sibiryakov, M. Nooijen, G. Baumgartner, D. Bernholdt, S. Hirata, C. Lam, R. Pitzer, J. Ramanujam, and P. Sadayappan. Automated operation minimization of tensor contraction expressions in electronic structure calculations. In *Proc. ICCS 2005 5th International Conference*, volume 3514 of *Lecture Notes in Computer Science*, pages 155–164. Springer, May 2005.
5. H. Koch, O. Christiansen, R. Kobayashi, P. Jørgensen, and T. Helgaker. A direct atomic orbital driven implementation of the coupled cluster singles and doubles (CCSD) model. *Chem. Phys. Lett.*, 228:233, 1994.
6. C. Lam, P. Sadayappan, and R. Wenger. On optimizing a class of multi-dimensional loops with reductions for parallel execution. *Parallel Processing Letters*, 7(2):157–168, 1997.
7. G. Scuseria, C. Janssen, and H. Schaefer. An efficient reformulation of the closed-shell coupled cluster single and double excitation (CCSD) equations. *The Journal of Chemical Physics*, 89(12):7382–7387, 1988.
8. A. Sibiryakov. Operation Optimization of Tensor Contraction Expressions. Master's thesis, The Ohio State University, Columbus, OH, August 2004.
9. J. Stanton, J. Gauss, J. Watts, and R. Bartlett. A direct product decomposition approach for symmetry exploitation in many-body methods. I. Energy calculations. *The Journal of Chemical Physics*, 94(6):4334–4345, 1991.

Prediction of Readthroughs Based on the Statistical Analysis of Nucleotides Around Stop Codons*

Sanghoon Moon, Yanga Byun, and Kyungsook Han[**]

School of Computer Science and Engineering, Inha University, Inchon 402-751, Korea
`jiap@inhaian.net, quska@inhaian.net, khan@inha.ac.kr`

Abstract. Readthrough is an unusual translational event in which a stop codon is skipped or misread as a sense codon. Translation then continues past the stop codon and results in an extended protein product. Reliable prediction of readthroughs is not easy since readthrough is in competition with standard decoding and readthroughs occur only at a tiny fraction of stop codons in the genome. We developed a program that predicts readthrough sites directly from statistical analysis of nucleotides surrounding all stop codons in genomic sequences. Experimental results of the program on 86 genome sequences showed that 80% and 100% of the actual readthrough sites were found in the top 3% and 10% prediction scores, respectively.

1   Introduction

Standard decoding of the genetic information is initiated from the start codon AUG and terminated by any of the three stop codons UAG, UAA and UGA. But the standard decoding rule can be changed occasionally by an event called 'recoding' [1]. A recoding event can occur during the elongation step (frameshift and hopping) or in the termination step (readthrough) of translation [2, 3].

In the case of readthrough, reading the stop codon is suppressed since stop codons are skipped or misread as sense codons. As a result, extended protein product is made from the readthrough process (Fig. 1). The codon usage of the local sequence surrounding stop codons is not random [4, 5], and in fact the sequence context around the stop codon is the major determinant that affects the efficiency of the translation termination [4, 5, 6, 7]. Namy *et al.* [8] showed 'poor termination contexts' that are rarely used at general termination sites. In *E. coli* and *S. cerevisiae*, the upstream sequence affects the termination efficiency [9]. The downstream sequence following the stop codon also affects the termination efficiency [4]. Bonetti *et al.* [5] demonstrated that translation termination is determined by synergistic interplay between upstream and downstream sequences.

There were previous computational and/or statistical approaches to finding readthrough sites in eukaryotes, prokaryotes and viruses [8, 10, 11, 12], but they are

---

[*] This work was supported by the Korea Science and Engineering Foundation (KOSEF) under grant R01-2003-000-10461-0.
[**] Corresponding author.

V.N. Alexandrov et al. (Eds.): ICCS 2006, Part I, LNCS 3991, pp. 276–283, 2006.

not fully automated. Some approaches find readthrough contexts that match the well-known readthrough context 'CAA UAG CAA' or 'CAR YYA' (Y is C or T, R is A or G), but our study shows that there are more readthrough sites not conforming to the context than those conforming to it.

We developed a fully automated program that finds readthrough sites directly from genome sequences with no prior knowledge. It considers SORF (semi open reading frame) and dORF (downstream open reading frame) in all three frames (-1, 0, +1) [10, 11]. We tested the program on 86 genome sequences from a number of organisms and successfully found readthrough sites. The rest of the paper presents the method and its experimental results in more detail.

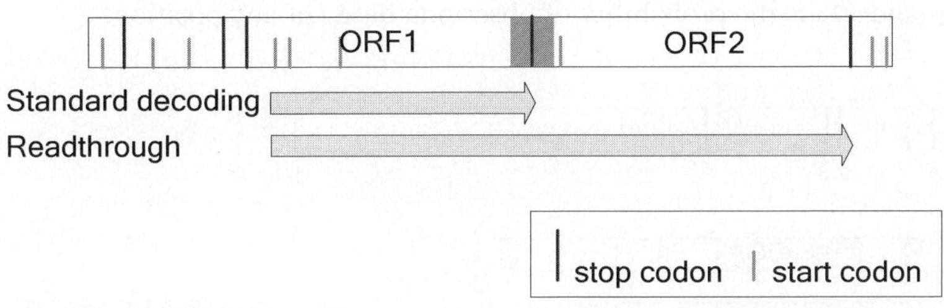

Fig. 1. In the standard decoding, translation is terminated at the stop codon. In the readthrough process, translation continues past the stop codon an extended protein product is produced.

2   Methods

2.1   Finding Readthrough Sites in Genome Sequences

Total 86 genome sequences were used as test data. 28 complete genome sequences of virus were obtained from the GenBank database [13]. 34 complete genome sequence and 24 complete coding sequences (CDSs) were obtained from the RECODE database [14].

We defined ORF1 and ORF2 in the same way as Namy *et al.* [11]. ORF1 is the area between the start codon and the first stop codon. ORF2 is the area between the first stop codon and the second stop codon. In the work by Namy *et al.* [11] the minimum length of ORF2 and the length between the first stop codon and the first start codon of ORF2 are fixed. However, different organisms have different lengths between the first stop codon and the first start codon of the second ORF. Our program allows the user to choose the length (Fig. 3D), so it is flexible to find readthrough sites of various organisms. Prediction of readthrough sites consists of five steps.

1. All genomic regions where two adjacent open reading frames (ORFs) are separated by a stop codon are identified (see Fig. 2).
2. For the second stop codon, 10 nucleotide positions (1 upstream nucleotide, stop codon and the following 6 nucleotides) are examined (colored region in Fig. 2).
3. Probabilities of A, C, G, and T are computed at each of the 10 positions, and the position specific score matrix (PSSM) is constructed from the probabilities (Fig.

3E). A score is computed for the second stop codon by equation 1 using PSSM and the probabilities of nucleotides.

4. If both ORFs do not have a start codon, the region is not considered as an open reading frame and excluded from candidates.

5. Scores are sorted in increasing order. Sites with lower scores are better ones than those with higher scores.

$$score = \sum_{k=0}^{9} \frac{p_{ki}}{p_i}, i \in \{A, C, G, T\} \tag{1}$$

where p_{ki} is the probability of observing base i at position k including the stop codon positions and p_i is the probability of observing base i at any position.

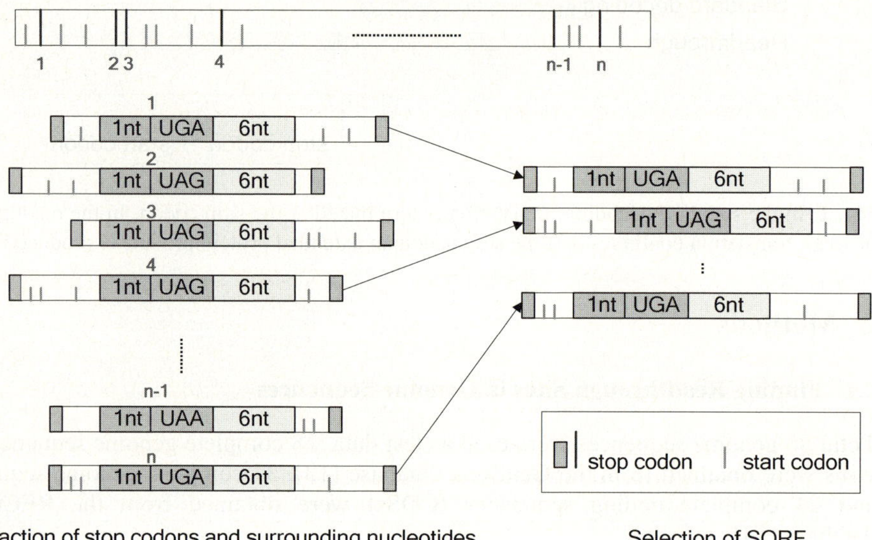

Fig. 2. Example of selecting semi open reading frames (SORF) in the genome sequence. SORFs 2, 3, and n-1 are excluded since start codons do not exist in both ORF1 and ORF2.

2.2 Implementation

The program for finding readthrough sites was implemented in Microsoft C#. As shown in Fig. 3A, stop codons of all three frames are examined by default, but the user can choose the frame that he/her wants to analyze. The user can also adjust the number of nucleotides surrounding a stop codon (Fig. 3B). Thus the user can compute the probability of the upstream and downstream nucleotides irrespective of the inclusion of stop codons (Fig. 3C). 10 in the first box of Fig. 3D indicates the minimum length of ORF1, 600 in the second box indicates the maximum length between the first stop codon and the first start codon of ORF2 and 10 in the third box means the minimum length of ORF2. From the parameters in Fig. 3A-D, the position specific score matrix is constructed (Fig. 3E). Based on our scoring scheme the scores are sorted in increasing order (Fig. 3F). Because our program uses the data grid control, the user can move data in Fig. 3E and F easily to Microsoft Excel or notepad with the copy and paste operations. The user can also sort data by column titles. Fig. 3G shows the graphical view of the genome sequence. The user can see ORFs in all frames and analyze them easily.

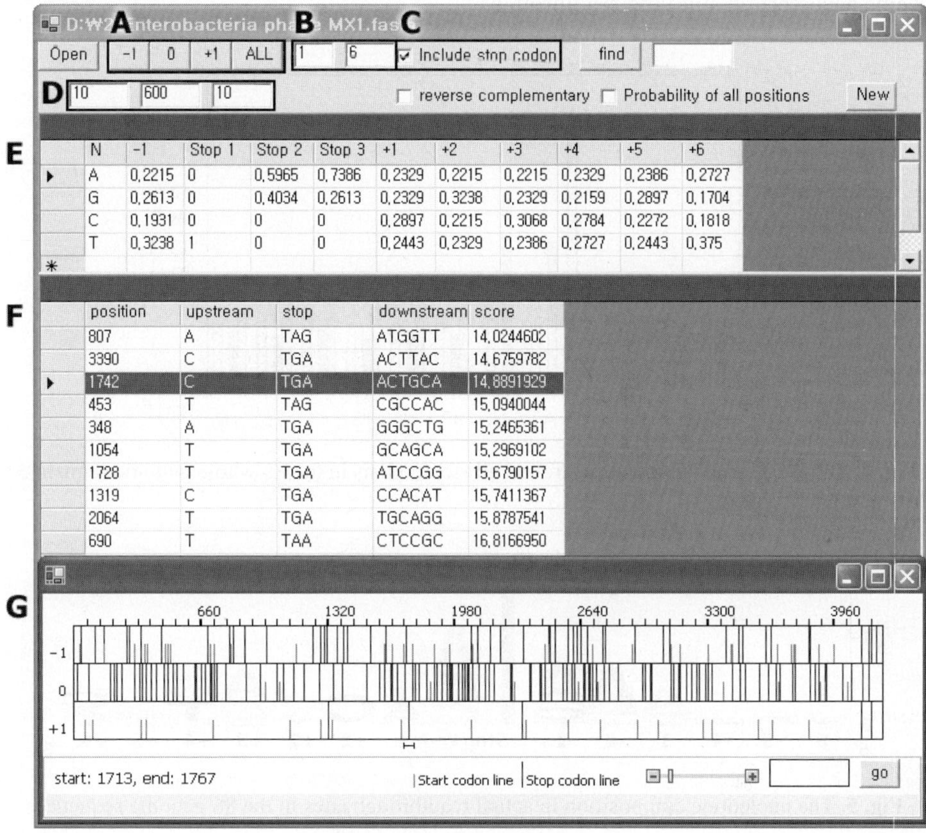

Fig. 3. Example of the user interface of our program

3   Results and Discussion

For the 86 genomic sequences we analyzed the average occurrence of each nucleotide at positions -3 to +6 from every stop codon. When we consider all stop codons in the genomic sequences, four nucleotides are observed in the positions with almost equal frequency (Fig. 4). However, the frequencies dramatically change for the stop codons in which readthroughs actually occur. Fig. 5 and Fig. 6 show sequence logos visualized by WebLogo [15] for the 86 genomic sequences and for other 26 sequences, respectively. The sequences in both data sets are known to have readthrough sites and obtained from RECODE.

Generally the role of the upstream codon for the stop codon efficiency is unclear. In Fig. 5, the nucleotides in -2 upstream codon (positions -4, -5, and -6) are distributed uniformly. In -1 upstream codon (positions -1, -2, and -3), nucleotide A is most abundant. T and A are abundant in Fig. 5. Interestingly G is very rare at position -1 (2% and 4% in Fig. 5 and 6, respectively). Most previous work on readthrough analyzed genomic sequences in terms of codons, but analysis in terms of nucleotides seems necessary in studying readthrough.

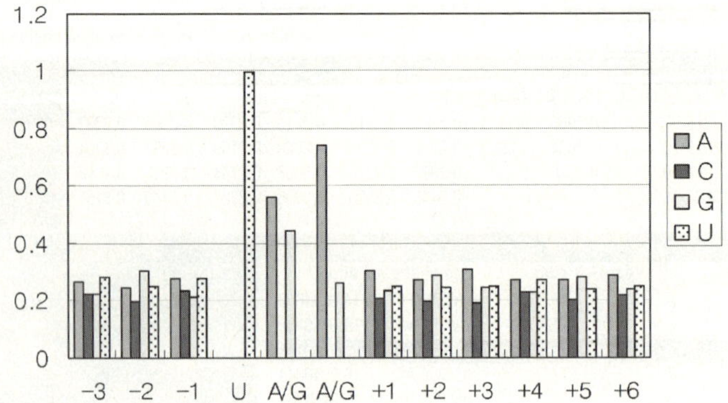

Fig. 4. The nucleotide composition around all stop codons in the 86 whole genome sequences

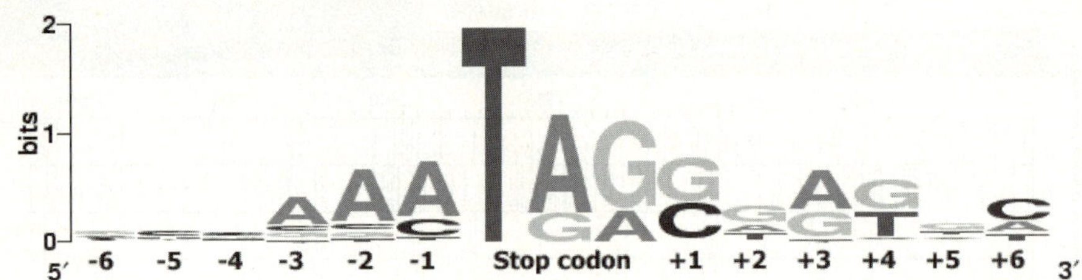

Fig. 5. The nucleotide composition in actual readthrough sites in the 86 genome sequences

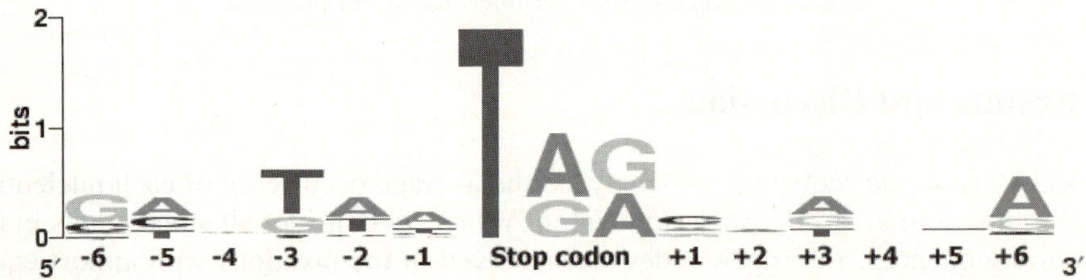

Fig. 6. The nucleotide composition in actual readthrough sites in 26 sequences from RECODE. The 26 sequences are not included in the 86 sequences used as test data.

Since G is very rare at -1 position, we examined the nucleotide at -1 position, a stop codon and the next 6 nucleotides downstream the stop codon. Tables 1-3 show the results of the analysis of stop codon readthrough. In the 86 sequences, total 23,735 (=6,416+7,343+10,154) stop codons were found. In Table 1, readthrough was not predicted for Middleburg virus (GI number: 28193965, accession number: AF339486). We obtained the complete CDS sequence of the virus from GenBank hyperlinked by RECODE, but the sequence was different from that in the RECODE database. We suppose the sequence (AF339486) does not have readthrough sites.

Table 1. Results of 24 complete CDSs from the RECODE database. Total number of stop codons: 6,416. *: no signal.

GI number	# of stop codons	Rank of the real RT in candidates (%)	GI number	# of stop codons	Rank of the real RT in candidates (%)
221091	190	4 (2.11%)	1841517	219	2 (0.91%)
332610	294	6 (2.04%)	2344756	573	1 (0.17%)
335192	153	5 (3.27%)	2582370	138	3 (2.17%)
393006	452	18 (3.98%)	3928747	191	3 (1.57%)
398066	376	10 (2.66%)	5714670	334	13 (3.89%)
409255	384	8 (2.08%)	6580858	373	11 (2.95%)
436017	286	2 (0.7%)	6580874	203	8 (3.94%)
533388	71	2 (2.82%)	7634686	394	5 (1.27%)
533391	63	3 (4.76%)	7634690	191	11 (5.76%)
786142	612	1 (0.16%)	8886896	305	4 (1.31%)
1016784	156	2 (1.28%)	10644290	147	4 (2.72%)
1236294	311	4 (1.29%)	28193965	191	*

Table 2. Results of 28 complete genomes from GenBank. Total number of stop codons: 7,343.

GI number	# of stop codons	Rank of the real RT in candidates (%)	GI number	# of stop codons	Rank of the real RT in candidates (%)
9625551	226	3 (1.33%)	20806010	366	2 (0.55%)
9625564	370	11 (2.97%)	20889313	136	2 (1.47%)
9629160	192	3 (1.56%)	20889365	337	3 (0.89%)
9629183	140	1 (0.71%)	22212887	382	5 (1.31%)
9629189	153	4 (2.61%)	25140187	311	4 (1.29%)
9635246	205	13 (6.34%)	50080143	128	6 (4.69%)
11072110	160	5 (3.13%)	30018246	214	3 (1.4%)
12018227	366	3 (0.82%)	30018252	201	7 (3.48%)
13357204	323	7 (2.17%)	33620701	839	15 (1.79%)
18254496	321	6 (1.87%)	38707974	221	5 (2.26%)
19881389	206	1 (0.49%)	39163648	174	2 (1.15%)
19919921	211	6 (2.84%)	50261346	154	1 (0.65%)
19919909	290	6 (2.07%)	51980895	228	1 (0.44%)
20153395	326	3 (0.92%)	66478128	163	3 (1.84%)

Table 3. Results of 34 complete genomes from RECODE. Total number of stop codons:10,154.

GI number	# stop codons	Rank of the real RT in candidates (%)	GI number	# stop codons	Rank of the real RT in candidates (%)
62128	323	6 (1.86%)	2801519	614	1 (0.16%)
218567	318	5 (1.57%)	3136264	308	9 (2.92%)
323338	195	6 (3.08%)	3396053	498	4 (0.8%)
331993	294	4 (1.36%)	3420022	334	9 (2.69%)
332140	222	4 (1.8%)	5442471	435	23 (5.29%)
333921	447	15 (3.36%)	5931707	368	10 (2.72%)
334100	423	11 (2.6%)	6018638	399	15 (3.76%)
335172	217	7 (3.23%)	6018642	163	4 (2.45%)
335243	374	4 (1.07%)	6143718	194	3 (1.55%)
408929	168	3 (1.79%)	6531653	178	2 (1.12%)
1335765	575	3 (0.52%)	7262472	144	2 (1.39%)
1685118	153	4 (2.61%)	7288147	438	9 (2.05%)
1752711	220	5 (2.27%)	7417288	151	1 (0.66%)
1902985	381	3 (0.79%)	10801177	246	20 (8.13%)
2213430	155	8 (5.16%)	30027702	151	3 (1.99%)
2231198	168	3 (1.79%)	30027703	218	1 (0.46%)
2801468	309	3 (0.97%)	30267510	195	4 (2.05%)

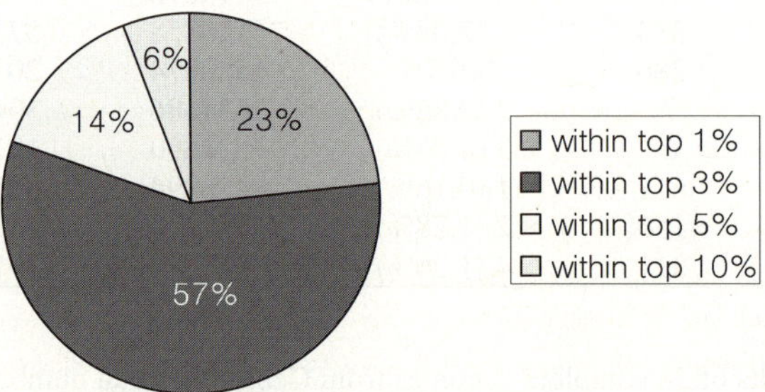

Fig. 7. The proportion of actual readthroughs in candidate readthroughs around all stop codons in the genome

It should be noted that only 0.36% (85 sites) of the total 23,735 stop codons are actual readthrough sites, and the actual sites were found with high prediction scores. 23% and 80% (=23%+57%) of the actual readthroughs were included in the top 1% prediction scores and 3% scores, respectively. All the actual readthroughs were found within the top 10% scores (see Fig. 7). Interestingly, only 23 of the 85 actual readthroughs conform to the well-known context 'CAA UAG CAA' or 'CAR YYA', suggesting that the well-known context is not sufficient for finding readthrough sites.

4   Conclusion

Finding readthrough genes is important because the readthrough process is associated with protein production and genetic control such as autoregulation. But prediction of readthrough is very difficult because readthrough process is in competition with the recognition as a sense codon and termination as a non-sense codon.

Based on codon preference and readthrough context, we have developed a program that predicts candidate readthrough sites. Our program focused on -1 upstream nucleotide, stop codon and 6 downstream nucleotides to find readthrough sites. The program does not require any prior knowledge or manual work. It finds candidate readthrough sites from the statistical analysis of genomic sequences only. The program was tested on 86 genome sequences and successfully predicted all known actual readthrough sites. If the user provides the information of the approximate ORF length, the prediction can be done more efficiently. We believe this is the first fully automated program capable of predicting readthrough sites.

References

1. Gesteland, R.F., Weiss, R.B., Atkins, J.F.: Recoding: re-programmed genetic decoding. Science 257 (1992) 1640-1641.
2. Gestealand, R.F., Atkins, J.F.: Recoding: dynamic reprogramming of translation. Annu. Rev. Biochem. 65 (1996) 741-768.
3. Namy, O., Rousset, J., Napthine, S., Brierley, I.: Reprogrammed genetic decoding in cellular gene expression. Mol. Cell 13 (2004) 157-169
4. Poole, E.S., Brown, C.H., Tate, W.P.: The identity of the base following the stop codon determines the efficiency of in vivo translational termination in Escherichia coli. EMBO Journal 14 (1995) 151-158
5. Bonetti, B., Fu, L., Moon, J., Bedwell, D.M.: The efficiency of translation termination is determined by a synergistic interplay between upstream and downstream sequences in Saccharomyces cerevisiae. Journal of Molecular Biology 251 (1995) 334-345
6. Namy, O., Hatin, I., Rousset, J.: Impact of the six nucleotides downstream of the stop codon on translation termination. EMBO reports 2 (2001) 787-793.
7. Harrell, L., Melcher, U., Atkins, J.F.: Predominance of six different hexanucleotide recoding signals 3' of read-through stop codons. Nucleic Acids Res. 30 (2002) 2011-2017.
8. Namy, O., Duchateau-Ngyen, G., Rousset, J.: Translational readthrough of the PDE2 stop codon modulates cAMP levels in Saccharomyces cerevisiae. Molecular Microbiology 43 (2002) 641-652
9. Mottagui-tabar, S., Tuite, M.F., Isaksson, L.A.: The influence of 5' codon context on translation termination in Saccharomyces cerevisiae. Eur. J. Biochem. 257 (1998) 249-254
10. Williams, I., Richardson, J., Starkey, A., Stansfield, I.: Genome-wide prediction of stop codon readthrough during translation in the yeast Saccharomyces cerevisiae. Nucleic Acids Res. 32 (2004) 6605-6616
11. Namy, O., Duchateau-Nguyen, G., Hatin, I., Denmat, S.H., Termier, M., Rousset, J.: Identification of stop codon readthrough genes in Saccharomyces cerevisiae. Nucleic Acids Research 31 (2003) 2289-2296
12. Sato, M., Umeki, H., Saito, R., Kanai, A., Tomita, M.: Computational analysis of stop codon readthrough in D.melanogaster. Bioinformatics 19 (2003) 1371-1380
13. Benson, D.A., Karsch-Mizrachi, I., Lipman, D.J., Ostell, J., Rapp, B.A., Wheeler, D.L.: GenBank. Nucleic Acids Res. 30 (2002) 17-20
14. Baranov, P., Gurvich, O.L., Hammer, A.W., Gesteland, R.F., Atkins, J.F.: RECODE. Nucleic Acids Res. 31 (2003) 87-89
15. Crooks, G.E., Hon, G., Chandonia, J., Brenner, S.E.: WebLogo: A sequencer logo generator. Genome Research 14 (2004) 1188-1190

Web Service for Finding Ribosomal Frameshifts*

Yanga Byun, Sanghoon Moon, and Kyungsook Han[**]

School of Computer Science and Engineering, Inha University, Inchon 402-751, Korea
`quska@inhaian.net, jiap@inhaian.net, khan@inha.ac.kr`

Abstract. Recent advances in biomedical research have generated a huge amount of data and software to deal with the data. Many biomedical systems use heterogeneous data structures and incompatible software components, so integration of software components into a system can be hindered by incompatibilities between the components of the system. This paper presents an XML web service and web application program for predicting ribosomal frameshifts from genomic sequences. Experimental results show that the web service of the program is useful for resolving many problems with incompatible software components as well as for predicting frameshifts of diverse types. The web service is available at http://wilab.inha.ac.kr/fsfinder2.

1 Introduction

A large-scale bioinformatics system often consists of several application programs dealing with a large volume of raw or derived data. For example, a data analysis program generates new data that may be modeled and integrated by other programs. However, the programs may not be interoperable due to the differences in data formats or running platforms. As a result, developing a bioinformatics system with a few application programs requires extra work to make the components compatible. In fact the difficulty in bioinformatics study comes more from the heterogeneity of the programs and data than from the quantity of the data.

Web service resolves some of these problems by exchanging messages between different applications developed by various programming languages. The basic process of web service is exchanging Simple Object Access Protocol (SOAP) messages described by XML (eXtensible Markup Language). When the server of web services receives the SOAP request with the parameters for calling the method of web service, the server returns the SOAP message in response to the method. An application program can use web services developed in different languages, whose results in turn can be a request message for another web service.

Recently various bioinformatics systems supporting web services have been developed [1]. European Bioinformatics Institute (EBI) provides several web services such as Dbfetch for biological database, ClustalW for multiple alignments of DNA and protein sequences, Fasta for nucleotide comparison [2]. Databases of DDBJ [3] and KEGG [4] give web service access. A tool named Taverna was developed for the bioinformatics workflows with several relevant biological web services [5].

---

[*] This work was supported by the Korea Science and Engineering Foundation (KOSEF) under grant R01-2003-000-10461-0.
[**] Corresponding author.

V.N. Alexandrov et al. (Eds.): ICCS 2006, Part I, LNCS 3991, pp. 284–291, 2006.

In previous work we developed a program called FSFinder (Frameshift Signal Finder) for predicting frameshifting [6]. Frameshifting is a change in reading frames by one or more nucleotides at a specific mRNA signal [7]. Frameshifts are classified into different types depending on the number of nucleotides shifted and the shifting direction. The most common type is a -1 frameshift, in which the ribosome slips a single nucleotide in the upstream direction. +1 frameshifts are much less common than -1 frameshifts, but have been observed in diverse organisms [8]. FSFinder is written in Microsoft C# and is executable on Windows systems only. To remove these limitations and to handle frameshifts of general type, we developed a web service and web application called FSFinder2. Users can predict frameshift sites of any type online from any web browser and operating system. By providing web service, FSFinder2 is more usable and compatible than the previous version of the program.

2   Systems and Method

Our system gives three ways to access: standalone application, web application and web service. The standalone application runs on Windows system only. The web application and service can be used with any web browser on any system. The web service is different from other web services in which the client simply makes a request based on the input format of the server and watches the web page returned in response to the request. On the other hand, the web service of our system is not only available through the web page but also enables to exchange soap messages in the XML format. This means that the user can use a web service without using the web page when he knows the input XML schema, output XML schema and the address of the web service. The user can also modify or reuse the reply message in the XML format to make it suitable to his system. The rest of this section describes the messages supported by the web service.

2.1   The Request SOAP Message

The 'FSFinder_run' message is the request message that includes the information of the input sequence and one or more user-defined models. Fig. 1A shows the elements of the request message. The 'components' element, which is a subelement of 'FSFinder_Model' can include any combination of four components: pattern type, secondary structure type, signal type and counter type [6]. The attribute 'spacer' is the space from the previous components. This attribute is needed from the second component. One number means fixed spacer and the range of numbers by 'nuber1~number2' means flexible spacer in the range.

Nucleotide characters (A, G, C, T or U, R, Y, M, K, S, W, H, B, V, D and N) and comma (,) are used to represent nucleotides. Comma marks the position of codons. It is optional and no comma represents any codon. The pattern component represents a pattern of nucleotides like the slippery site of the -1 frameshift model. Pattern type requires pattern characters, the match and exception nucleotide characters that represent the pattern characters. Repeated pattern characters are represented once. It finds the match sequence pattern with match characters without exception characters. The secondary structure component defines the size of pseudoknot parts in Fig. 2. The signal type can include any number of nucleotide signals. The counter type defines a count and the counter equation.

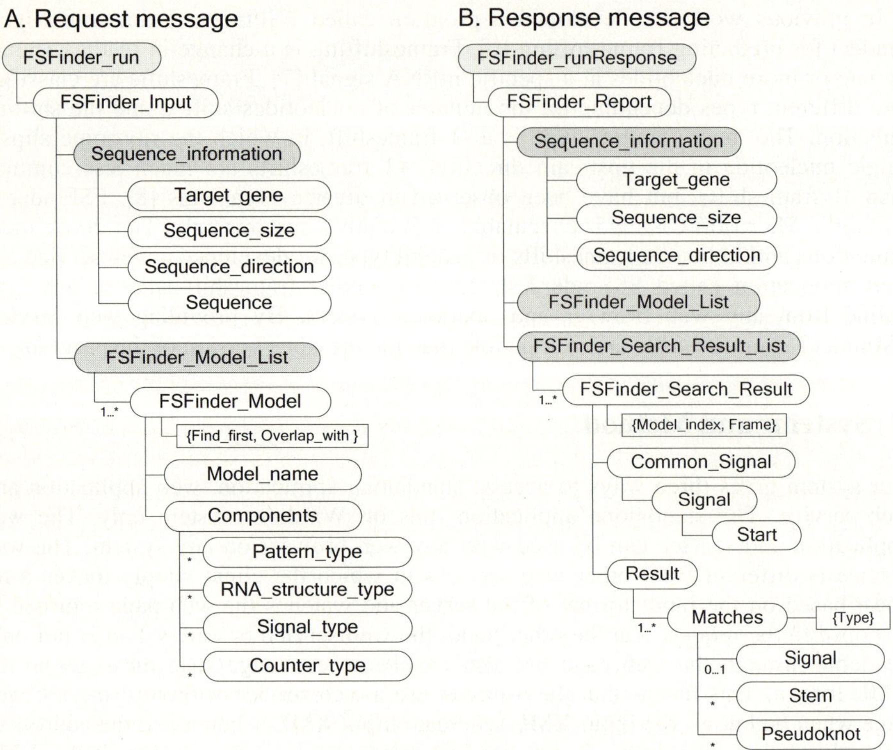

Fig. 1. The request message (left) and the response message (right) of the FSFinder2 web service. (A) The root element of the request message is 'FSFinder_run'. It requires one 'FSFinder_Input' element. 'Sequenece_information' and 'FSFinder_Model_List' are required as subelements of 'FSFinder_input'. 'Sequence_information' represents the input sequence. 'FSFinder_Model_List' includes one or more user-defined models. The element for the user-defined model is named 'FSFinder_Model'. This element requires overlapping open reading frames and the position of the component to be found first. (B) The root element of the response message is 'FSFinder_runResponse' and it has one child element named 'FSFinder_Report'. The child elements of 'FSFinder_Report' are 'Sequence_information', 'FSFinder_Model_list' and 'FSFinder_Search_Result_List'. 'Sequence_information' in the response message does not include the input sequence.

2.2   The Response SOAP Message

The response message includes the search results and the user-defined models specified in the request message. The structure of the response message is shown in Fig. 1B. The 'FSFinder_Search_Result_List' element has one or more 'FSFinder_Search_Result' elements. 'Model_index' and 'Frame' are the attributes of the element. 'Model_index' shows the index of the model from which the result came. 'Frame' is the frame where the search started. The search result includes one 'Common_Signal' and one 'Result' elements. The 'Result' element has 'Matches' elements in the same number and order as

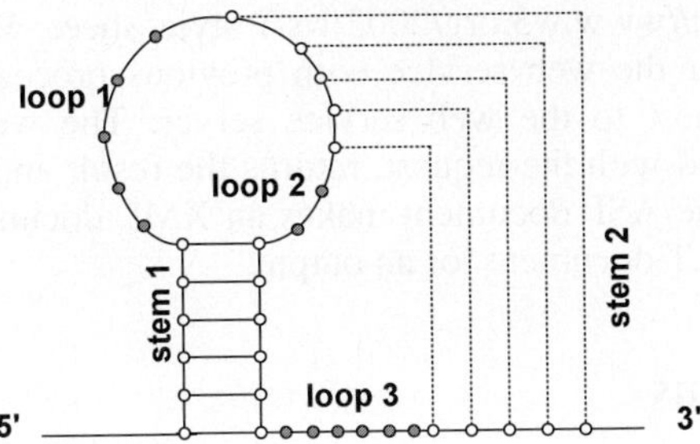

Fig. 2. Five parts of a pseudoknot structure. The size of each part can be defined by the subelements of 'RNA_Structure_type'.

specified in the components of the model. The 'Matches' element includes one signal or any number of stems and pseudoknots. 'Common_Signal' is the sequence fragment that has more than one match.

2.3 Web Application

Web application helps the user use the web service easily with a web browser (Fig. 3). Since the web application server sends an HTML document only in response to user request, all the user needs is a web browser that can read an HTML document. The web application server handles these requests with active server page (ASP). An ASP document generates an XML document and this XML document is shown as an HTML page with the current XSLT (the Extensible Stylesheet Language

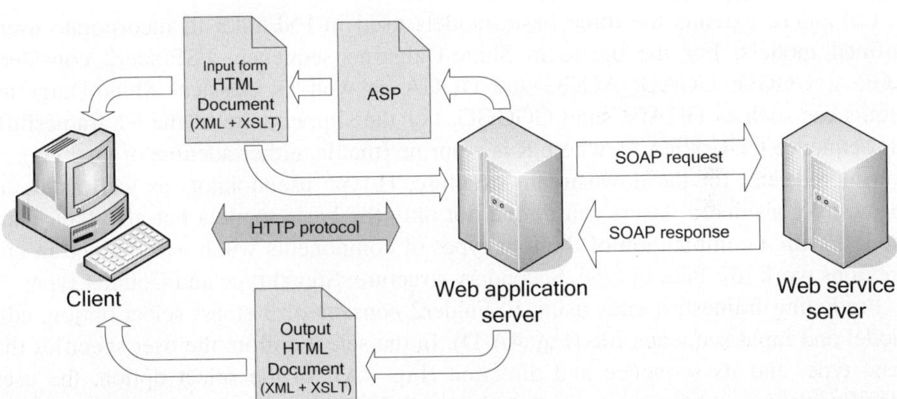

Fig. 3. The processes involved in the web application. The client connects to the web application server with HTML document using HTTP protocol. The web application server makes the request SOAP message and sends it. When the web service server sends back the result of the request, the web application server makes an XML document for the response SOAP message and returns the XML document in the current style sheet.

Transformations; http://www.w3.org/2002/ws/) style sheet. When the user finishes editing the request for the web service with previous process, the web application server sends the request to the web service server. The web service server calls 'FSFinder_run' method with the request, returns the result and sends back the SOAP response. Similarly, the ASP document makes an XML document and sends the output as the style of XSLT document for an output.

3   Implementations

FSFinder2 was implemented using XML, XSLT, ASP and JavaScript. If the user sends a query to the server after defining a new model, the computation is performed on the server side and the results are sent back to the user. Consequently FSFinder2 is independent of the operating systems and web browsers of his/her computer.

The XML schema of FSFinder2 described in the previous section can be downloaded at the web site. In the 'FSFinder in web' page, we provide the developer's guide containing information such as Web Services Description Language (WSDL), SOAP message and XML schema. Web service can be used only with the WSDL explanations. The results from any tools (applications, databases and so on) can be the input of this web service with making XML message as the form needed by this system. The output can also be input of any tools. Therefore, FSFinder2 guarantees high reusability and solves the problems from the differences of the implementation languages.

Three types of frameshift are considered as basic frameshifts, and their models are predefined: the most common -1 frameshifts, +1 frameshifts of the RF2 type, and +1 frameshift of the type found in the ornithine decarboxylase antizyme (ODC antizyme). The models for these frameshifts consist of a Shine-Dalgarno sequence, frameshift site, spacer and downstream secondary structure.

FSFinder2 extends the three basic models used in FSFinder to incorporate user-defined models. For the upstream Shine-Dalgarno sequence, FSFinder2 considers AGGA, GGGG, GGAG, AGGG and GGGA as well as classical Shine-Dalgarno sequences such as GGAGG and GGGGG. For the slippery site of the +1 frameshift, the sequence CUU URA C, where R is a purine (that is, either adenine or guanine), is considered, and for the downstream structure, H-type pseudoknots as well as stem-loops are considered. Users can define not only the basic models but also their own models with a combination of the four types of components witch was defined in our previous work [6]: Pattern type, Secondary structure, Signal type and Counter type.

Predicting frameshift sites using FSFinder2 consists of 3 steps: select option, edit model and input sequence file (Fig. 4A-D). In the select option, the user specifies the gene type, and its sequence and direction (Fig. 4A). In the select option, the user specifies the target gene type, whether the input DNA sequence is a complete genome or partial sequence, and its sense (+ or -). For the target gene type, the user can choose one of dnaX gene, oaz gene, prfB gene, other genes in bacteria and genes in viruses. Because it requires different methods for fining overlapping regions of ORFs, bacterial needs to be distinguished form viral genes.

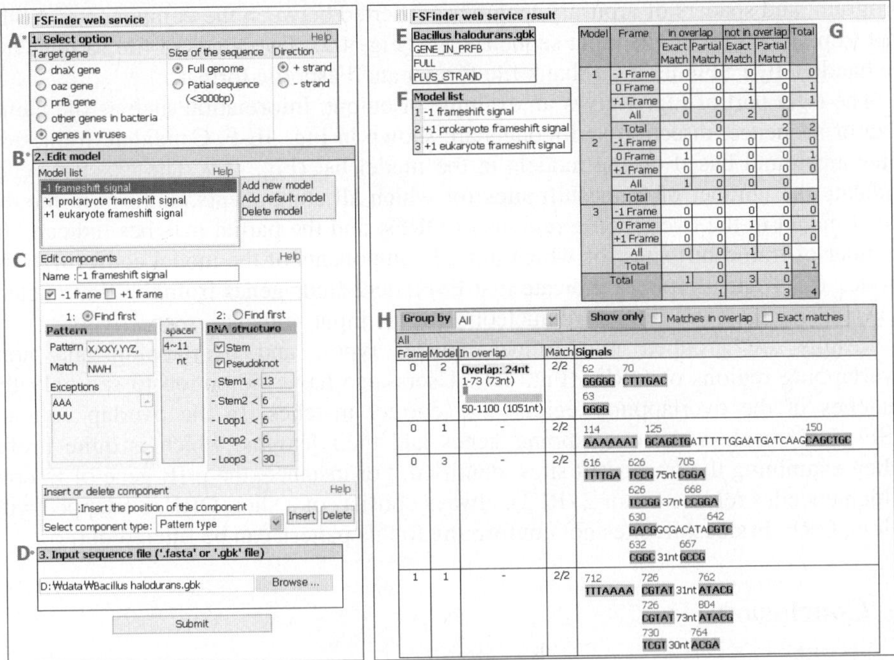

Fig. 4. The input page (left) and result page (right) of FSFinder2. Left: (A) The select option lets the user choose the type of genes expressed via frameshifts, the size of the sequence and its direction. (B, C) The user can define a new model by specifying its components and their locations. (D) The user selects the input sequence file. Right: (E) This box gives the file name of the input sequence, target gene, sequence size and direction. (F) This shows potential frameshift sites found by FSFinder2 for each model in the model list. (G) The results are separated into exact matches and partial matches in each of the overlapping and non-overlapping regions. (H) The results are grouped into model types, frames containing the frameshift sites and the overlapping regions of ORFs. FSFinder2 displays the locations and lengths of the overlapping ORFs. The match column shows the number of matched components out of the total components in the model. The signal column presents matched components and sequences. The matched parts in the signal are highlighted in the color defined by the user. Yellow, green, sky blue and red designate pattern type, RNA structure, signal type and counter type, respectively. The red numbers above the sequence designate the position of the component with respect to the first nucleotide of the sequence.

In the edit model, the user can define a new model of frameshift (Fig. 4B). After choosing 'Add new model' in the model list, the user defines the components of the new model (Fig. 4C). The check button of the -1 and +1 frames determines the type of frameshift. -1 frame is for finding -1 frameshift and +1 frame is for +1 frameshift. The component selected as 'find first' is searched earlier than any other component of the model. Below the 'find first' button, there are parameters of the component that the user can specify. After choosing the first component to search for, user can add a new component to the model. Each component can be changed before running the program, and spacers of arbitrary length are inserted between the components. As the last step, users choose the input sequence file (Fig. 4D). Two kinds of file formats can be handled by FSFinder2: Genbank file format and Fasta file format.

The right part of Fig. 4 gives an example of output. Information such as file name, size of sequence, direction and models are shown in Fig. 4E-F. Candidate frameshift sites are found based on the models in the model list (Fig. 4G). The exact matches indicate the number of frameshift sites for which all components conform to the defined model in the overlapping regions of ORFs, and the partial matches indicate the number of frameshift sites for which not all components of the model conform to the model. 0, +1 and -1 frames indicate that FSFinder2 finds genes from the first nucleotide, second nucleotide and third nucleotide of the input sequence, respectively.

Results are grouped into frames, model types, and overlapping and non-overlapping regions of ORFs (Fig. 4H). Users also have the option to see only the matches in the overlapping regions and correct matches. In the overlap column, FSFinder2 displays the overlapping genes and their lengths, which is quite useful when examining the frameshift sites identified. For instance, the prfB gene of E. coli, which encodes release factor 2 (RF2), always consists of a short ORF overlapped with a long ORF. Frameshift sites not conforming to this pattern can be filtered out.

4   Conclusion

We developed FSFinder2 to predict frameshift sites with a user-defined model. The web service and web application of FSFinder2 were implemented using XML, XSLT and JavaScript. If the user sends a query to the FSFinder2 server after setting parameters or defining a new model, all the computations are done on the server side. After computation, the server sends the results to the user. Consequently FSFinder2 is independent of the operating system or web browser of his/her computer. The web service provided by FSFinder2 is slightly different from most existing web applications in which a client makes a request based on the input format and watches the web page for a reply via the web browser. The FSFinder2 server and client exchange SOAP messages in the XML format according to the properties of the web service. If the user knows the input XML schema, output XML schema and address of the web service, the user can use the web service without using the web page. Since the reply message is sent in the XML format, the user can modify it to suit his/her system.

Experimental results of testing FSFinder2 on ~190 genomic and partial DNA sequences showed that it predicted frameshift sites efficiently and with greater sensitivity and specificity than other programs, because it focused on the overlapping regions of ORFs and prioritized candidate signals (For -1 frameshifts, sensitivity was 0.88 and specificity 0.97; for +1 frameshifts, sensitivity was 0.91 and specificity 0.94) [6, 9-11]. FSFinder2 has been successfully used to find unknown frameshift sites in the Shewanella genome. We believe FSFinder2 is the first program, guaranteed high reusability, capable of predicting frameshift signals of general type and that it is very useful for analyzing programmed frameshift signals in complete genome sequences. The web service of the FSFinder would be useful as not only the function of FSFinder2 itself but also intermediate of the cooperation with the other web services.

References

1. Stein, L.: Creating a bioinformatics nation. Nature, 417 (2002), 119–120
2. Pillai, S., Silventoinen,V., Kallio, K., Senger, M., Sobhany, S., Tate, J., Velankar, S., Golovin, A., Henrick, K., Rice, P., Stoehr, P., Lopez, R.: SOAP-based services provided by the European Bioinformatics Institute. Nucleic Acids Research, 33 (2005), 25-28
3. Miyazaki, S., Sugawara, H., Gojobori, T., Tateno, Y.: DNA Data Bank of Japan (DDBJ) in XML. Nucleic Acids Research, 31 (2003), 13-16
4. Kawashima, S., Katayama, T., Sato, Y. Kanehisa, M.: KEGG API: A Web Service Using SOAP /WSDL to Access the KEGG System. Genome Informatics, 14 (2003), 673-674
5. Oinn, T., Addis, M,. Ferris, J., Marvin, D., Senger, M., Greenwood, M., Carver, T., Glover, K., Pocock, M.R., Wipat, A., Li, P.: Taverna: a tool for the composition and enactment of bioinformatics workflows. Bioinformatics, 20 (2004), 3045-3054
6. Sanghoon, M., Yanga, B., Hong-Jin, K., Sunjoo, J., Kyungsook, H.: Predicting genes expressed via -1 and +1 frameshifts. Nucleic Acids Research, 32 (2004) 4884-4892
7. Baranov, P.V., Gesteland, R.F., Atkins, J.F.: Recoding: translational bifurca-tions in gene expression. Gene, 286 (2002), 187-201
8. Farabaugh, P.J.: Programmed translational frameshifting. Ann. Rev. Genetics, 30 (1996), 507-528
9. Hammell, A.B., Taylor, R.C., Peltz, S.W., Dinman, J.D.: Identification of putative programmed -1 ribosomal frameshift signals in large DNA databases. Genome Research, 9 (1999), 417-427
10. Bekaert, M., Bidou, L., Denise, A., Duchateau-Nguyen, G., Forest, J., Froidevaux, C., Hatin, Rousset, J., Termier, M.: Towards a computational model for -1 eukaryotic frameshifting sites. Bioinformatics, 19 (2003), 327-335
11. Shah, A.A. Giddings, M.C., Parvaz, J.B., Gesteland, R.F., Atkins, J.F., Ivanov, I.P.: Computational identification of putative programmed translational frameshift sites. Bioinformatics, 18 (2002), 1046-1053

A Remote Sensing Application Workflow and Its Implementation in Remote Sensing Service Grid Node

Ying Luo[1,4], Yong Xue[1,2,*], Chaolin Wu[1,4], Yincui Hu[1], Jianping Guo[1,4],
Wei Wan[1,4], Lei Zheng[1,4], Guoyin Cai[1], Shaobo Zhong[1], and Zhengfang Wang[3]

[1] State Key Laboratory of Remote Sensing Science, Jointly Sponsored by the Institute of
Remote Sensing Applications of Chinese Academy of Sciences and Beijing Normal University,
Institute of Remote Sensing Applications, Chinese Academy of Sciences, P.O. Box 9718,
Beijing 100101, China
[2] Department of Computing, London Metropolitan University, 166-220 Holloway Road,
London N7 8DB, UK
[3] China PUTIAN Institute of Technology, Shangdi Road, Beijing 100085, China
[4] Graduate School of the Chinese Academy of Sciences, Beijing, China
{jennyjordan@hotmail.com, y.xue@londonmet.ac.uk}

Abstract. In this article we describe a remote sensing application workflow in
building a Remote Sensing Information Analysis and Service Grid Node at
Institute of Remote Sensing Applications based on the Condor platform. The
goal of the Node is to make good use of physically distributed resources in the
field of remote sensing science such as data, models and algorithms, and
computing resource left unused on Internet. Implementing it we use workflow
technology to manage the node, control resources, and make traditional
algorithms as a Grid service. We use web service technology to communicate
with Spatial Information Grid (SIG) and other Grid systems. We use JSP
technology to provide an independent portal. Finally, the current status of this
ongoing work is described.

1 Introduction

Grid has been proposed as the next generation computing platform for solving large-
scale problems in science, engineering, and commerce [3][4]. There are many famous
Grid projects today: DataGrid, Access Grid, SpaceGRID, European Data Grid, Grid
Physics Network (GriPhyN), Earth System Grid (ESG), Information Power Grid,
TeraGrid, U.K. National Grid, etc. Of particular interest are SpaceGRID and ESG,
which focus on the integration of spatial information science and Grid.

The SpaceGRID project aims to assess how GRID technology can serve
requirements across a large variety of space disciplines, such as space science, Earth
observation, space weather and spacecraft engineering, sketch the design of an ESA-
wide Grid infrastructure, foster collaboration and enable shared efforts across space
applications. It will analyse the highly complicated technical aspects of managing,
accessing, exploiting and distributing large amounts of data, and set up test projects to
see how well the Grid performs at carrying out specific tasks in Earth observation,

---

* Corresponding author.

V.N. Alexandrov et al. (Eds.): ICCS 2006, Part I, LNCS 3991, pp. 292–299, 2006.

space weather, space science and spacecraft engineering. The Earth System Grid (ESG) is funded by the US Department of Energy (DOE). ESG integrates supercomputers with large-scale data and analysis servers located at numerous national labs and research centers to create a powerful environment for next generation climate research. This portal is the primary point of entry into the ESG.

The Condor Project has performed research in distributed high-throughput computing for the past 18 years, and maintains the Condor High Throughput Computing resource and job management software originally designed to harness idle CPU cycles on heterogeneous pool of computers [2]. In essence a workload management system for compute intensive jobs, it provides means for users to submit jobs to a local scheduler and manage the remote execution of these jobs on suitably selected resources in a pool. Condor differs from traditional batch scheduling systems in that it does not require the underlying resources to be dedicated: Condor will match jobs (*matchmaking*) to suited machines in a pool according to job requirements and community, resource owner and workload distribution policies and may vacate or migrate jobs when a machine is required. Boasting features such as check-pointing (state of a remote job is regularly saved on the client machine), file transfer and I/O redirection (i.e. remote system calls performed by the job can be captured and performed on the client machine, hence ensuring that there is no need for a shared file system), and fair share priority management (users are guaranteed a fair share of the resources according to pre-assigned priorities), Condor proves to be a very complete and sophisticated package. While providing functionality similar to that of any traditional batch queuing system, Condor's architecture allows it to succeed in areas where traditional scheduling systems fail. As a result, Condor can be used to combine seamlessly all the computational power in a community.

Grid workflows consist of a number of components, including computational models, distributed files, scientific instruments and special hardware platforms. Abramson *et al.* described an atmospheric science workflow implemented by web service [1]. Their workflow integrated several atmosphere models physically distributed. Hai studied the development of component-based workflow system, and he also studied the management of cognitive flow for distributed team cooperation [5]. We describe the Grid workflow in remote sensing science and show how it can be implemented using Web Services on a Grid platform in this paper. The workflow supports the coupling of a number of pre-existing legacy computational models across distributed computers. An important aspect of the work is that we do not require source modification of the codes. In fact, we do not even require access to the source codes. In order to implement the workflow we overload the normal file Input/Output (IO) operations to allow them to work in the Grid pool. We also leverage existing Grid middleware layers like Condor to provide access to control of the underlying computing resources.

In this paper we present remote sensing application workflow in building a Remote Sensing Information Analysis and Service Grid Node at Institute of Remote Sensing applications (IRSA) based on the Condor platform. The node is a special node of Spatial Information Grid (SIG) in China. The node will be introduced in Section 2. Several middleware developed in the node and the remote sensing workflow, with some detail of the functions of the various components will be demonstrated in Section 3. Finally, the conclusion and further development will be addressed in Section 4.

2 Remote Sensing Information Analysis and Service Grid Node

2.1 Spatial Information Grid (SIG)

SIG is the infrastructure that manages and processes spatial data according to users' demand. The goal of SIG is to build an application Grid platform for spatial information community, which can simplify and shield many complex technology and settings, facilitate SIG users, and make good use of resources physically distributed. There are many reasons why one might wish to have the SIG. First, the amount of spatial data is increasing amazingly. So that real time or almost real time processing needed by applications confronts much more difficulties in one single computer. Second, data, algorithm, and/or computing resources are physically distributed. Third, the resources may be "owned" by different organizations. Fourth, the use frequency of some resources is rather low.

A SIG at least contains: (1) A remote sensing Remote Sensing Information Analysis and Service Grid Node; (2) A data service node: traditional data base to a web service; (3) A management centre: resource register, find, trade, and management; (4) A portal: an entry to SIG user.

2.2 Remote Sensing Information Analysis and Service Grid Node

Remotely sensed data is one of the most important spatial information sources, so the research on architectures and technical supports of remote sensing information analysis and service grid node is the significant part of the research on SIG.

The aim of the node is to integrate data, traditional algorithm and software, and computing resource distributed, provide one-stop service to everyone on Internet, and make good use of everything pre-existing. The node can be very large, which contains many personal computers, supercomputers or other nodes. It also can be as small as just on personal computer. Figure 1 describes the architecture of Remote sensing information analysis and service Grid node. The node is part of the SIG, but it also can provide services independently. There're two entries to the node:

1. A portal implemented by JSP. User can visit it through Internet browses, such as Internet Explorer and others.
2. A portal implemented by web service and workflow technology. It is special for SIG. Other Grid resources and Grid systems can integrate with our node through this portal.

The node provides the application services such as aerosol optical depth retrieval, land surface temperature retrieval, soil moisture retrieval, surface reflectance retrieval, and vegetation index applications by MODIS data.

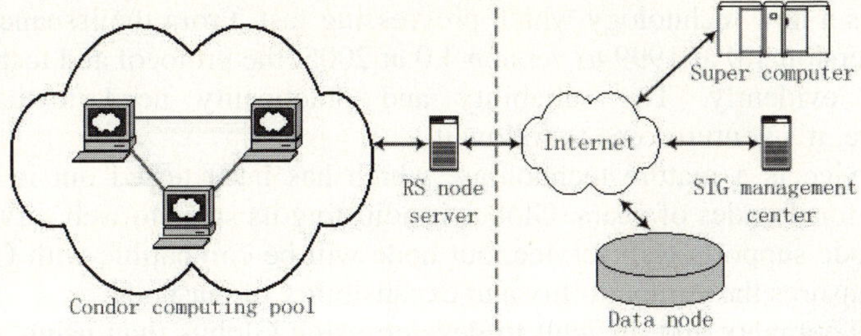

Fig. 1. Architecture of remote sensing information analysis and service Grid node

The remote sensing node server (Figure 2) responds for all the corresponding events about the services on it. It contains:

- Node container: provide the necessary grid work environment;
- Service/workflow: some functional entities or workflows will be encapsulated to services for Grid. Others functional entities called within the node will be encapsulated to workflows. By adding new services into the node container, the capability of the node can be augmented and in turn the function of the whole system can be improved incrementally.
- Middleware: traditional applications cannot be used on condor directly. So middleware is needed to provide an access to condor. The middleware stands for the division and mergence of sub-tasks, and monitor the condor computing pool and the process of sub-tasks running in the pool.
- Node management tool: it responds for issuance, register, and update to SIG management centre. It's up to the node management tool to trigger service and monitor the status of the service.

The node issues and registers its services to SIG manage centre periodically, responses calls of SIG, triggers services, and reports status. Receiving require from SIG manager, the node will find data form either local or remote data server according to the user's requirements, organize computing resource dynamically, trigger services, and monitor their running status.

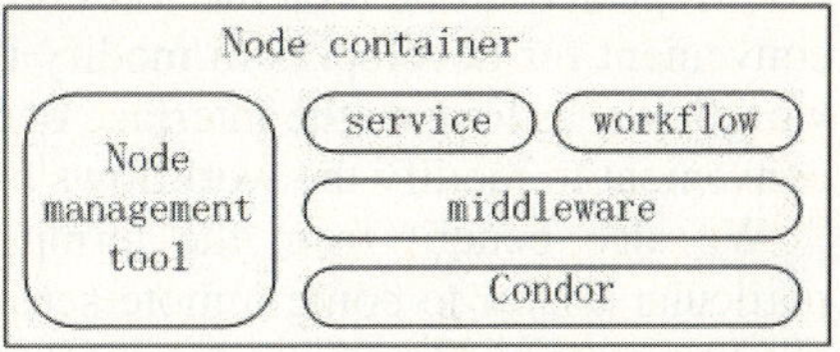

Fig. 2. Remote sensing node server

To decrease the total processing time of a task, the node will divide a task into several sub-tasks. The exact number and size of the sub-tasks is according to the current PC number and configure in the Condor computing pool and super computer. Only when the task is large enough, or on the user's request, the node triggers off the super computer to do a large sub-task. The method we trigger super computer is different with that of Condor computing pool. The Grid Service Spread (GSS) described by Wang *et al.* was used to implement it [6].

Implementing the node, we refer to use web service technology rather than Globus technology. The reasons are:

1. Globus is a new technology which progressing fast. From the issuance of Globus toolkit version 1.0 at 1999 to version 4.0 at 2005, the protocol and technology in it changed evidently. The reliability and practicality need textual research. Therefore, it's venturesome to follow it.
2. Web service is a mature technology, which has been tested out in the field of industry for decades of years. Globus is adjusting its steps to web service. So long as our node supports web service, our node will be compatible with Globus. Web service ensures the compatibility and expansibility of our node.
3. It's more complex and difficult to develop using Globus than using web service. There are some convenient tools available for web service. But few is for Globus.

3   Workflow Implementation with Web Service and Grid

Traditional method for link functional components is to program them into a static course, or involve human control. People have to realize all the permutation and combination of the components in order to deal with all possible cases. It need large amount of repeated work, and is discommodious to modify and extend. The workflow technology overcomes these shortages. Using the concept STEP, ACTION, STATE, and TRANSITION in finite state machine (FSM) for reference, workflow can skip among the components flexibly. So it's facility to organize components dynamically according the user request and the environment of the system.

As the temperature retrial example for demonstrating the implement of our node, which will be described in details in sections 3.1 and 3.2, we can cut-out any pre-processing components by configuring the workflow conditions and attributes. For example, if the primary data has been rectified outside our node, the workflow will skip the rectification component and turn into the next one.

The components called by workflows can be reused. So that people were set free from repeated work. For the loose coupling of the components and workflows, it's convenient for developers to modify the components themselves without harm to the workflows, as long as the interface of the components is unchanged. Furthermore, it is convenient to modify the workflows by adding or decreasing components in them.

We also benefit from the termination function of workflow technology. It is particular useful to some remote sensing applications which involve many steps and need long processing time. A workflow instance can be terminate by sending terminal signal artificially, if we don't want the instance execute anymore. An instance can stop itself when it finds out the components in it meet trouble or the required environment is not satisfied. But the workflow cannot stop a running component. It only can stop itself before or after the running component is finished.

Using workflow's status function, we can monitor the running status of a workflow and status of components in it. STATE has four meaning in the remote sensing node:

- State of the node: it describes the attributes of the node, such as busy, idle, current workload, current processing ability, service instance number in queue, etc.
- State of application/workflow instance: it presents whether the instance is finished, or which step is finished.
- State of components in an application: it has only two status: finished and underway.

- State of condor pool: it contains the detail configuration (operation system, IP address, architecture, CPU, and RAM) and running status (IDLE or BUSY) of the PCs in the condor computing pool, the number of the tasks submitted to condor.

3.1   An Implementation with Web Services on the Condor Platform

This implementation is for the connection to the SIG or other Grid application systems. Each application is implemented using web service technology such as SOAP, XML, and WSDL.

In our node, the workflow engine plays the role of a global coordinator to control different components to fulfil the functions requested by users according to a pre-defined process model. For example, Figure 3 presents the pre-defined process model of land surface temperature retrial. It involved the functional component of data format transfer, rectification, region selection, division, land surface temperature retrial, mergence, and return result to caller. The workflow will skip the functional components in front of the component whose input requirement has been satisfied by the primary data. Unfortunately, the system cannot recognize whether the data has been transferred or rectified without further information. It's necessary for the data node or someone else to describe these characters in a data metafile. The workflow will decide which components should be skipped by the metafile.

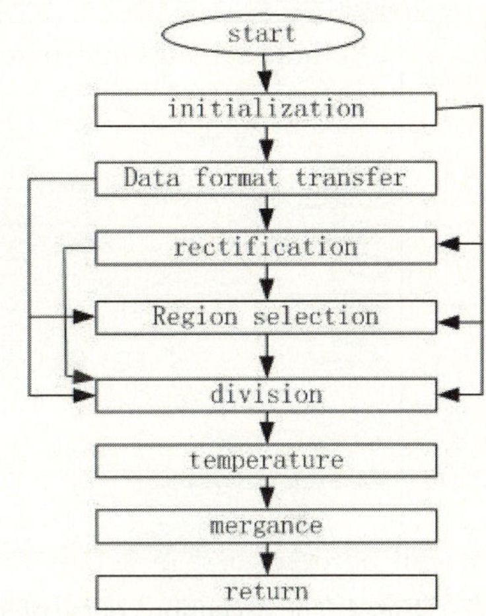

Fig. 3. Process model of land surface temperature retrieval

The division component calls the middleware shown in Figure 2. The middleware stands for checking the current status of condor computing pool, dividing data into pieces according to the number of the PCs in condor, generating all the files needed by Condor, packaging them, and transferring them to the location where condor can find them. The workflow submits the temperature component and data pieces to Condor. Then Condor stands for manage these sub-tasks. When Condor returns the results of sub-tasks, the mergence component will integrate them into the final result. And return component transfer this final result to caller. In this test, there were 5 PCs running operation system of Windows 2000 professional.

3.2   An Implementation with JSP on a Condor Platform

This implementation is for users to visit our node directly. The difference between the JSP implementation and web service one is:

1. It is not a web service application. There are no WSDL files and WSDD files for the JSP implementation. It only can be triggered and initialized artificially. It cannot be interact by other machines.

2. The implement with JSP can be man-machine interactive. Of course it can auto run, too. An application can be intervened when it is running.
3. It provides a data up load function before data format transfer component. So that users can process their data using the functions on our node.

Figure 4 shows the execution course of the land surface temperature retrieval application.

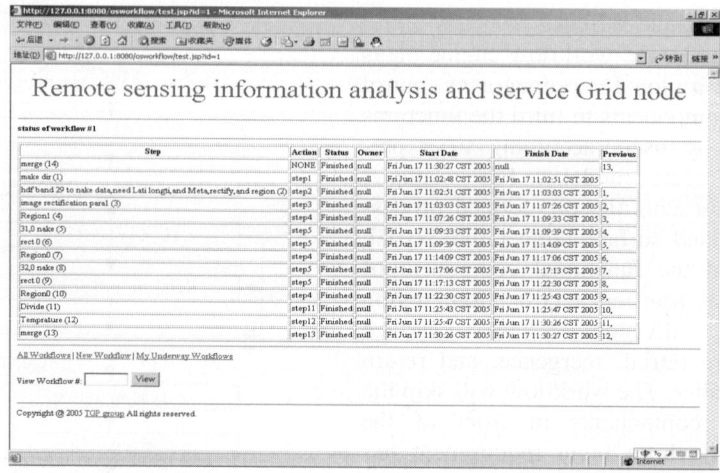

Step	Action	Status	Owner	Start Date	Finish Date	Previous
merge (14)	NONE	Finished	null	Fri Jun 17 11:30:27 CST 2005	null	13,
make dir (1)	step1	Finished	null	Fri Jun 17 11:02:48 CST 2005	Fri Jun 17 11:02:51 CST 2005	
hdf band 29 to nake data,need Lati longti,and Meta,rectify,and region (2)	step2	Finished	null	Fri Jun 17 11:02:51 CST 2005	Fri Jun 17 11:03:03 CST 2005	1,
image rectification paral (3)	step3	Finished	null	Fri Jun 17 11:03:03 CST 2005	Fri Jun 17 11:07:26 CST 2005	2,
Region1 (4)	step4	Finished	null	Fri Jun 17 11:07:26 CST 2005	Fri Jun 17 11:09:33 CST 2005	3,
31,0 nake (5)	step5	Finished	null	Fri Jun 17 11:09:33 CST 2005	Fri Jun 17 11:09:39 CST 2005	4,
rect 0 (6)	step5	Finished	null	Fri Jun 17 11:09:39 CST 2005	Fri Jun 17 11:14:09 CST 2005	5,
Region0 (7)	step4	Finished	null	Fri Jun 17 11:14:09 CST 2005	Fri Jun 17 11:17:06 CST 2005	6,
32,0 nake (8)	step5	Finished	null	Fri Jun 17 11:17:06 CST 2005	Fri Jun 17 11:17:13 CST 2005	7,
rect 0 (9)	step5	Finished	null	Fri Jun 17 11:17:13 CST 2005	Fri Jun 17 11:22:30 CST 2005	8,
Region0 (10)	step4	Finished	null	Fri Jun 17 11:22:30 CST 2005	Fri Jun 17 11:25:43 CST 2005	9,
Divide (11)	step11	Finished	null	Fri Jun 17 11:25:43 CST 2005	Fri Jun 17 11:25:47 CST 2005	10,
Temprature (12)	step12	Finished	null	Fri Jun 17 11:25:47 CST 2005	Fri Jun 17 11:30:26 CST 2005	11,
merge (13)	step13	Finished	null	Fri Jun 17 11:30:26 CST 2005	Fri Jun 17 11:30:27 CST 2005	12,

Fig. 4. The execution course of a temperature application workflow instance

4   Conclusions and Future Work

In this paper, we introduced our ongoing research on remote sensing information analysis and service Grid node. The node is service-oriented. The whole node is composed of many workflows. Each workflow stands for certain service. Some of the workflows were implemented with web service technology, so that they can be called by the SIG or other systems, which follow web service protocol. Web service technology endows our node with compatibility and machine-machine interaction. Other workflows are for JSP web site, so that user can visit our node directly. They only can be triggered artificially. The JSP technology endows our node with man-machine interaction. The execution course of the JSP workflows can be intervened by users. Workflow technology endows our node with easy modification and extensibility. SIG and Condor endows our node with the power of remote cooperation, resource share, and management physically distributed.

Our node is a demonstration to how to integrate data, traditional algorithms and models, and computing resource in order to provide one-stop service to users. We have implemented it mainly by workflow technology. We try to make a large virtual super computer by integrating cheap personal computers on Internet, and utilize them when they are left unused. Theoretically, the node can provide enough processing ability to anyone at anytime and anywhere if there are enough PCs left unused in it.

The services on the node seem to work well, but actually it is far away from intactness. There are still many problems to deal with. One problem is the long file transfer time between the PCs physically distributed. It weakens the decrease of the total processing time improved by using more PCs doing one task. To some multi-band image data, a feasible approach to deal with it is to transmit only the bands needed by the applications. The amount of the data may decrease a scalar grade.

Acknowledgement

This publication is an output from the research projects "Grid platform based aerosol fast monitoring modelling using MODIS data and middlewares development" (40471091) funded by the NSFC, China, "Dynamic Monitoring of Beijing Olympic Environment Using Remote Sensing" (2002BA904B07-2) and "Remote Sensing Information Processing and Service Node" (2003AA11135110) funded by the MOST, China and "Digital Earth" (KZCX2-312) funded by CAS, China.

References

[1] Abramson, D., Kommineni, J., McGregor J. L., and Katzfey, J., 2005, An Atmospheric Sciences Workflow and its implementation with Web Services. *Future Generation Computer Systems,* **Vol.21**, Issue 1, pp.69-78

[2] Basney, J., Livny, M., and Tannenbaum, T., 1997, High Throughput Computing with Condor, *HPCU news,* **Volume 1(2)**

[3] Foster, I., and Kesselman, C., 1997, Globus: a metacomputing infrastructure toolkit, Int. J. Supercomputer Application, **11 (2),** 115–128.

[4] Foster, I., and Kesselman, C. (Eds.), 1999, The Grid: Blueprint for a New Computing Infrastructure, (Morgan Kaufmann Publishers, USA).

[5] Hai, Z. G., 2003, Component-based workflow systems development, *Decision Support Systems,* **Volume 35**, Issue 4, Pages 517-536

[6] Wang, Y. G., Xue, Y., Jianqian Wang, Chaolin Wu, Yincui Hu, Ying Luo, Shaobo Zhong, Jiakui Tang, and Guoyin Cai, 2005, Java-based Grid Service Spread and Implementation in Remote Sensing Applications. *Lecture Notes in Computer Science,* **Vol.3516,** pp.496-503.

Predictive Analysis of Blood Gasometry Parameters Related to the Infants Respiration Insufficiency

Wieslaw Wajs[1], Mariusz Swiecicki[2], Piotr Wais[1], Hubert Wojtowicz[1], Pawel Janik[3], and Leszek Nowak[4]

[1] University of Mining and Metallurgy, Institute of Automatics,
Cracow, Poland
[2] University of Technology, Intitute of Computer Modelling,
Cracow, Poland
[3] Jagiellonian University, Faculty of Mathematics,
Astronomy and Applied Informatics,
Cracow, Poland
[4] Jagiellonian University, Faculty of Mathematics and Informatics,
Cracow, Poland

Abstract. The paper presents application of artificial immune system in time series prediction of the medical data. Prediction mechanism used in the work is basing on the paradigm stating that in the immune system during the response there exist not only antigene - antibody connections but also antigene - antigene connections, which role is control of antibodies activity. Moreover in the work learning mechanism of the artificial immune network, and results of carried out tests are presented.

1 Introduction

Biological phenomenons, and particularly organic processes are of a dynamic character. Diagnosis and therapy are of the same character. In approaching diagnosis and making therapeutic decision phenomenons happening in time are investigated [1]. Every doctor knows that it is rare to make a decision about diagnosis and treatment on the basis of only one clinical observation. Usually it is based on several patient's examinations, regular analysis of many biophysical and biochemical parameters, or imaging examination. On a particular level of these examinations preliminary diagnosis is determined, and later a final one. Basing on determined diagnosis, and on the evaluation of the sickness process dynamics, decision about the treatment is taken. The field of neonatology has made tremendous progress during the last twenty years. It's been made possible thanks to the progress in modern technologies of intensive medical care and progress in the applied sciences. Diagnostic examinations are based on micro - methods. Modern monitoring technologies, both non - invasive and invasive, aren't very strenuous for the patient, allow in an exact manner for continuous monitoring of gas exchange, lungs mechanics and functioning of circulatory system. Patophysiology of the infant respiration insufficiency and the role of the

V.N. Alexandrov et al. (Eds.): ICCS 2006, Part I, LNCS 3991, pp. 300–307, 2006.

surfactant deficiency in various forms of the respiration insufficiency was discovered. Physiology and patophysiology of the water - electrolyte economy of regular and premature neonates was also discovered. Initial stabilization of the infants state is a difficult task. It can take even few days. To achieve it the doctor analyses repeatedly many parameters related to the patient's health condition. To these parameters belong birth anamnesis, physiological examinations (body's weight, dieresis), results of additional examinations (biochemical, micro biological, imaging) and readings from the monitoring instruments (pulseximeter, cardio monitor, invasive measurement of the arterial pressure, respiratory mechanics monitor). It isn't rare when doctor appraises simultaneously over fifty variables. Analysis of this huge amount of data is hard and requires experience, all the more if the decision about the treatment should be made quickly. In the result of carried out data analysis doctor makes decision about the treatment expecting positive results, which are expressed by the desired changes in the results of additional examinations, readings of the monitoring instruments and physical examination. The whole process can be verified by comparing it to the model of respiratory insufficiency progress carried out by the doctor. Creation of this model is based on theoretical and empirical knowledge found in scientific books and papers and also on the doctor's experience. Out of necessity created model is of subjective character. Moreover results of analysis carried out by the doctor are always dependant of many external factors - recently read reports, weariness of the doctor etc. Intensification of gas exchange disorders are best reflected by arterial blood gasometry parameters examined in the context of currently used methods of ventilation support. For that reason as an output values four directly measured parameters of arterial blood gasometry: pH, pCO_2, pO_2, HCO_3 and oxygen parameter (pO_2/FiO_2) were chosen. Alkalis deficiency and hemoglobin oxygen saturation were omitted as derivatives of the four parameters mentioned above.

1.1   Parameters Prognosis

Below short characteristic of the parameters is presented:

pH - blood reaction is a parameter, which depends on the state of the respiratory system, but also on the circulatory system functioning, kidneys and metabolism of system. It is a general parameter, which can be treated as an indicator of the general state of the patient.

pCO_2 - partial pressure of the carbon dioxide is strictly correlated with the rate of the respiratory system efficiency. Above all it is dependent on minute ventilation. Its increase causes simultaneous decrease of pCO_2. Increase of the partial pressure of the carbon dioxide is usually the symptom of respiration disorders with lowered ventilation. However in serious respiration disorders with handicapped gas diffusion through the alveolar - capillary barrier also appear symptoms of handicapped exchange of carbon dioxide. Increase of CO_2 decreases blood's pH. pCO_2 of patients with supported ventilation depends mainly on the frequency of breaths (RR) and respiration capacity (TV).

pO_2 - partial pressure of oxygen is also strictly correlated with the rate of the respiratory system efficiency. Normal exchange of oxygen in lungs depends mainly on the proper ratio of ventilation to perfusion and on permeability of the alveolar - capillary barrier. Perturbation of the ventilation to perfusion ratio is observed usually in the case of atelectasis. Thickening of the alveolar - capillary barrier in the first place handicaps oxygen exchange, only further decrease of its permeability handicaps CO_2 exchange. pO_2 of patients with ventilation support depends on the oxygen concentration in the respiratory mixture (FiO_2) and on the rate of lung's dilation. The latter in the patients with ventilation support depends on the proper pressure in respiratory tracts (peak pressure of inspiration, average pressure in respiratory tracts). Oxygen deficiency in the organism causes increase of anaerobic metabolism and also increase of lactic acid and decrease of blood's pH production.

HCO_3 - hydrogen carbonate concentration in blood depends on system's metabolisms and kidney's functioning. In case of disproportions between transport of oxygen to tissues and metabolic requirements of system decrease of bicarbonate concentration and metabolic acidosis supervene.

Assessing particular clinical case one doesn't rely on only one, even complex parameter, usually one evaluates many various parameters. Because of that four parameters were chosen as inputs of artificial immune network. Exact classification of their values reflects direction and rate of changes - improvement or deterioration of respiratory system functioning, occurring in the natural progress of the illness and used treatment methods.

1.2   Source of Data

Research data used in the presented work, was gathered during few years long observation and hospitalization of patients on the Infant Intensive Care Ward of the Polish - American Institute of Pediatrics Collegium Medicum Jagiellonian University of Cracow. On this ward a computer database is used called Neonatal Information System (NIS), which stores information about all hospitalized infants.

2   An Artificial Immune Network for Signal Analysis

Main goal of our system is a prediction of signals and to achieve this goal several problems had to be solved. The first of them is connected with an algorithm of learning the immune network. The next problem is related to the structures of data, which are responsible for representation of signals. Solutions to these problems are presented below. The last paragraph presents results of signals prediction by our immune network.

2.1   Signal Representation

The input signal for the system, is interpreted as an antibody (Ab) so the task of immune network is to find an antigen Ag that will be suitable for Ab. The Ag-Ab representation will partially determine which distance measure shall be used to calculate their degree of interaction. Mathematically, the generalized shape of a molecule (m), either an antibody (Ab) or an antigen (Ag), can be represented by a set of real-valued coordinates $m =< m1, m2, ..., mL >$, which can be regarded as a point in an L-dimensional real-valued space.

$$D = \sqrt{\sum_{i=1}^{L}(ab_i - ag_i)^2} \tag{1}$$

The affinity between an antigen and an antibody is related to their distance that can be estimated via any distance measure between two vectors, for example the Euclidean or the Manhattan distance. If the coordinates of an antibody are given by $< ab1, ab2, ..., abL >$ and the coordinates of an antigen are given by $< ag1, ag2, ..., agL >$, then the distance (D) between them is presented in equation (1), which uses real-valued coordinates. The measure distances are called Euclidean shape-spaces.

2.2 Learning Algorithm

In this article, we are basing on an algorithm that was proposed in the papers [2, 3] by de Castro and Von Zuben. Our modified version of immune net algorithm [4] was adapted to cope with continuous signals prediction. The aforementioned algorithm is presented below. The learning algorithm lets building of a set that recognizes and represents the data structural organization. The more specific the antibodies, the less parsimonious the network (low compression rate), whilst the more generalist the antibodies, the more parsimonious the network with relation to the number of antibodies. The suppression threshold controls the specificity level of the antibodies, the clustering accuracy and network plasticity.

For each $Ag_j \in Ag$ do

- Determine its affinity $f_{i,j}$, i = 1,...,N, to all Ab_i. $f_{i,j} = 1/D_{i,j}$, i = 1,...,N
- A subset Ab(n) which contains the n highest affinity antibodies is selected;
- The n selected antibodies are going to clone proportionally to their antigenic affinity $f_{i,j}$, generating a set C of clones
- The set C is submitted to a directed affinity maturation process (guided mutation) generating a mutated set C^*, where each antibody k from C^* will suffer a mutation with a rate inversely proportional to the antigenic affinity $f_{i,j}$ of its parent antibody: the higher the affinity, the smaller the mutation rate:
- Determine the affinity $d_{k,j} = 1/D_{k,j}$ among Ag_j and all the elements of C^*:
- From C^*, re-select ξ % of the antibodies with highest $d_{k,j}$ and put them into a matrix M_j of clonal memory;
- Apoptosis: eliminate all the memory clones from M_j whose affinity $D_{k,j} > d$:
- Determine the affinity $s_{i,k}$ among the memory clones:
- Clonal suppression: eliminate those memory clones whose $s_{i,k} < s$:

- Concatenate the total antibody memory matrix with the resultant clonal memory M_j
- Determine the affinity among all the memory antibodies from Abm:
- Network suppression: eliminate all the antibodies such that $s_{i,k} < s$:
- Build the total antibody matrix $Ab \leftarrow [Ab(m); Ab(d)]$

Where:

Ab - available antibody repertoire ($Ab \in SN \times L, Ab = Ab(d) \cup Ab(m)$);

Ab(m) - total memory antibody repertoire ($Ab(m) \in Sm \times L, m \leq N$);

Ab(d) - d new antibodies to be inserted in Ab ($Ab(d) \in Sd \times L$);

Ag - population of antigens ($Ag \in Sm \times L$);

f_j - vector containing the affinity of all the antibodies Ab_i (i = 1,...N) with relation to antigen Ag_j. The affinity is inversely proportional to the Ag-Ab distance;

S - similarity matrix between each pair Ab_i - Ab_j, with elements $s_{i,j}$ (i,j = 1,...,N);

C - population of Nc clones generated from $Ab(C \in S_{NL})$;

C^* - population C after the affinity maturation process;

d_j - vector containing the affinity between every element from the set C^* with Ag_j;

ξ - percentage of the mature antibodies to be selected;

M_j - memory clone for antigen Agj (remaining from the process of clonal suppression);

M_j - resultant clonal memory for antigen Ag_j;

d - natural death threshold;

s - suppression threshold.

Artificial Immune System response algorithm:

```
NO_Ag = length(Ag);
Ab = Ag;
for i = NO_Ag : N_SegmentationWindowWidth - 1
   LenAg = length(Ab);
   [M, vbD, Dn ] = Answer_Net( M, Ab, i, ds );
     Ab=[Ab, M(1,i+1)];
end
```

The presented algorithm requires some explanations. There are important operations that are responsible for suppressive steps. The steps are called clonal suppression and network suppression, respectively. As far as a different clone is generated to each antigenic pattern presented, a clonal suppression is necessary to eliminate intra-clonal self-recognizing antibodies, while a network suppression is required to search for similarities between different sets of clones. After the learning phase, the network antibodies represent internal images of the antigens.

3   Example - Prediction of Blood Gasometry Parameters

In the given example use of the artificial immune network is presented in the prediction of the pH, pCO_2, pO_2 and HCO_3 parameters. Using a database, functioning for the few years on the Infant Intensive Care Ward of the Polish - American Institute of Pediatrics Collegium Medicum Jagiellonian University of Cracow, an artificial immune network was created, which task is prediction of the arterial blood gasometry parameters (pH, pCO_2, pO_2, HCO_3). In the process of training previous values of gasometry, respirator settings and surfactant administration were used as an input data.

Training process of the artificial immune network consists of two phases. First phase is a learning of the artificial immune network. This phase proceeds accordingly to the algorithm presented previously.

Training set is comprised of blood gasometry parameters time series starting from the time t. Time series are segmented as shown in Figure 1.

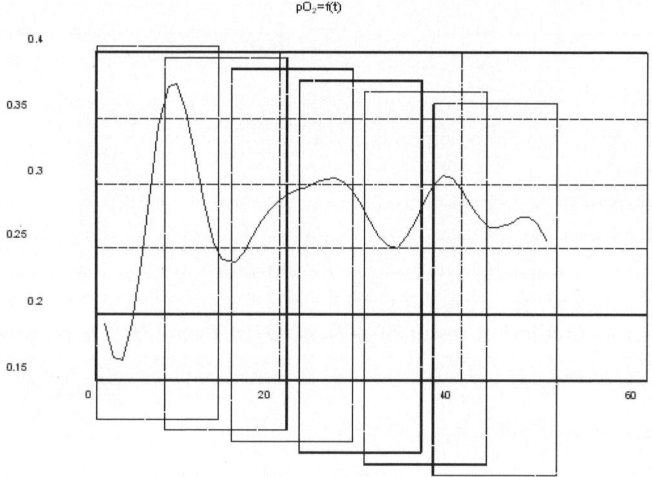

Fig. 1. pO2=f(t) time series segmentation process

Second phase tests the network's generalization abilities by presenting it input vectors from the test dataset.

3.1   Training Process

Dataset for the training process consisted of 480 samples, and parameters of the artificial immune network were set to the following values:

$$n = 4, \xi = 20\%, \sigma d = 0.8, \sigma s = 0.01$$

Stopping criterions of the training process were set to one hundred generations Ngen = 100. One of the more important parameters, which has a major influence on the immune network structure is a (σs) parameter. Changing value of the suppression threshold (σs) influences the generalization abilities of the artificial immune network.

3.2   Testing Process

Figure 2 presents example response of the network, for the time series, chosen for the infants blood gasometry parameters data set.

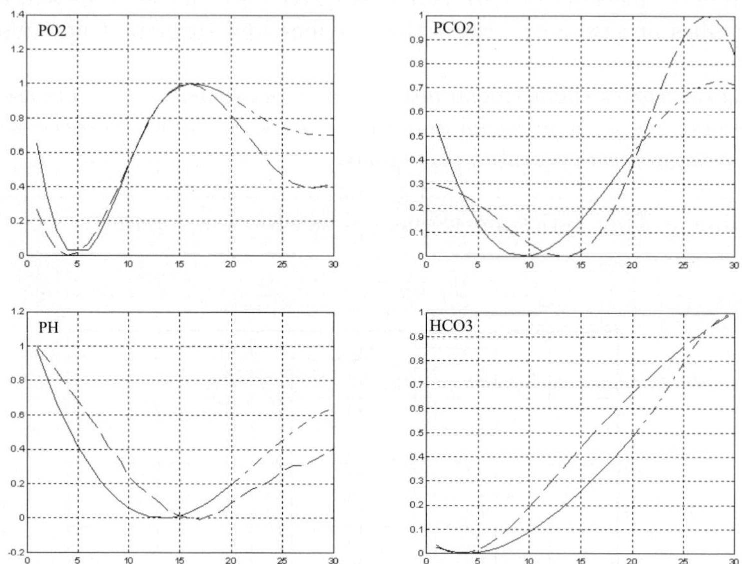

Fig. 2. Time series prediction result of pO_2, pCO_2, pH and HCO_3 parameters for one of the patients

Legend:

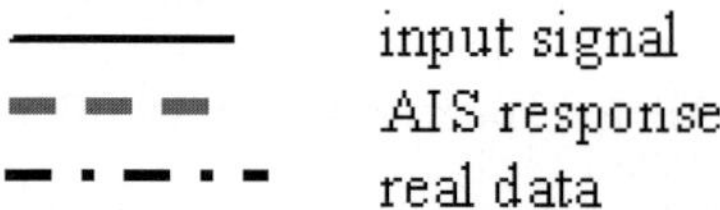

4   Summary

In the paper application of artificial immune network in time series prediction of medical data was presented. Prediction mechanism used in the work is basing on the paradigm stating that in the immune system during the response there exist not only antigene - antibody connections but also antigene - antigene connections, which role is control of antibodies activity. The mechanism turned out to be an interesting technique useful in prediction of the time series. Future work aims at increasing systems reliability by improving the algorithms used and acquring further data of the infants hospitalization.

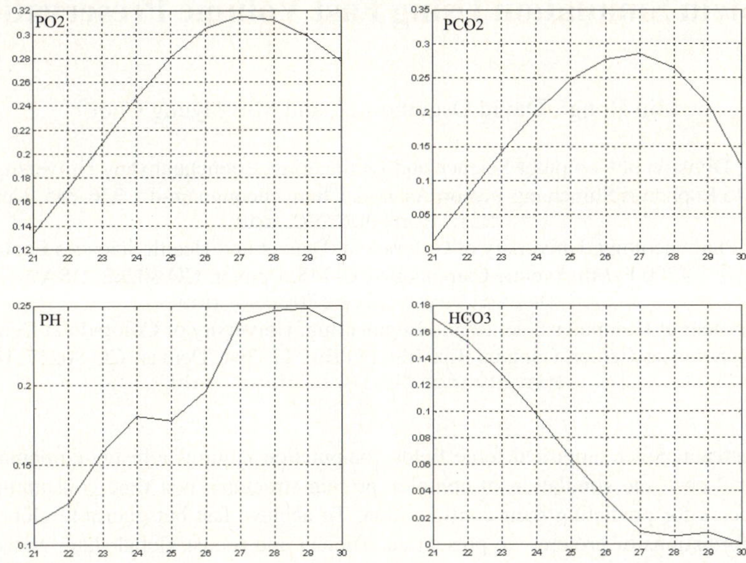

Fig. 3. Prediction error calculated for pO_2, pCO_2, pH and HCO_3 parameters

Acknowledgements

This work was partially supported by the Polish Government through a Grant No. 3T11E04227 from the State Committee of Scientific Research (KBN).

References

1. Kruczek, P. Assesment of neural networks methods usefulness in prediction of premature neonates respiration insufficiency, Doctoral Dissertation, Collegium Medicum, Jagiellonian University in Cracow, Cracow 2001.
2. De Castro, L. N., Von Zuben, F. J. (2000a) An Evolutionary Immune Network for Data Clustering, Proc. of the IEEE SBRN, pp. 84-89.
3. De Castro, L. N., Von Zuben, F. J. (2000b) The Clonal Selection Algorithm with Engineering Applications, GECCO'00 - Workshop Proceedings, pp. 36-37.
4. Wajs, W., Wais, P., Swiecicki, M., Wojtowicz, H. Artificial Immune System for Medical Data Classification, Proc. of the International Conference on Computational Science 2005, Springer LNCS 3516, pp. 810-812, 2005.

Protein Simulation Using Fast Volume Preservation

Min Hong[1], David Osguthorpe[2], and Min-Hyung Choi[3]

[1] Division of Computer Science and Engineering, Soonchunhyang University,
646 Eupnae-ri Shinchang-myeon Asan-si, Chungcheongnam-do, 336-745, Korea
`Min.Hong@UCHSC.edu`
[2] Pharmacology, University of Colorado at Denver and Health Sciences Center,
4200 E. 9th Avenue Campus Box C-245, Denver, CO 80262, USA
`David.Osguthorpe@UCHSC.edu`
[3] Department of Computer Science and Engineering, University of Colorado at Denver and
Health Sciences Center, Campus Box 109, PO Box 173364, Denver, CO 80217, USA
`Min-Hyung.Choi@cudenver.edu`

Abstract. Since empirical force fields computation requires a heavy computational cost, the simulation of complex protein structures is a time consuming process for predicting their configuration. To achieve fast but plausible global deformations of protein, we present an efficient and robust global shape based protein dynamics model using an implicit volume preservation method. A triangulated surface of the protein is generated using a marching cube algorithm in pre-processing time. The normal mode analysis based on motion data is used as a reference deformation of protein to estimate the necessary forces for protein movements. Our protein simulator provides a nice test-bed for initial screening of behavioral analysis to simulate various types of protein complexes.

1 Introduction

Since life and the development of all organisms are essentially determined by molecular interactions, the fundamental biological, physical, and chemical understanding of these unsolved detail behaviors of molecules are highly crucial. With the rapid accumulation of 3D (Dimensional) structures of proteins, predicting the motion of protein complexes is becoming of increasing interest. These structural data provide many insights on protein folding, protein-ligand interaction, protein-protein interaction and aid more rational approaches to assist drug development and the treatment of diseases. The analysis of deformation of proteins is essential in establishing structure-function relationships because a structure actually carries out a specific function by movement.

Advanced computer graphics hardware and software can offer real-time, interactive 3D and colorful protein complexes to the screen instead of lifeless chemical formulas. During the last decades, to discover and understand a wide range of biological phenomena, computer simulation experiments using quantum mechanics, molecular mechanics, and molecular dynamics simulation have opened avenues to estimate and predict the molecular level deformation [1, 2, 3]. Several applications such as Amber molecular dynamics [4] and CHARMM [5] have been developed. Quantum mechanics has been widely used for molecular modeling and it calculates the behavior of molecules at the electronic level. Although quantum mechanics provides accurate prediction for molecular simulation, it is limited to small sizes of molecules due to the expensive computational cost. Molecular mechanics calculates the energy of a

V.N. Alexandrov et al. (Eds.): ICCS 2006, Part I, LNCS 3991, pp. 308–315, 2006.

molecular system from atoms centered on the nuclear position and an empirical force field. Empirical force fields include bond related forces (bond stretching energy, bond angle bending energy, and torsional energy) and non-bond related forces (electrostatics interaction energy, hydrogen bonding energy, and van der Waals energy) to estimate the energy of the molecular system. Experimental data such as atomic geometry, bond lengths, and bond angles are obtained by X-ray crystal structures or NMR (Nuclear Magnetic Resonance) to set up the values of potential parameters. Molecular mechanics minimizes the given empirical energy functions using initial conditions and finds the minimum energy conformation. However, this method ignores kinetic energy of molecular system.

Although molecular mechanics reduces the computational cost substantially, the computational task is still time-consuming because proteins consist of large numbers of atoms. In addition, the motions of proteins are not static and dynamics are important for protein simulation. In a molecular dynamics simulation, the classical equations of motion for positions, velocities, and accelerations of all atoms are integrated forward in time using the well-known Newton equations of motion. It computes the time dependent deformation of protein structure. However, molecular dynamics simulation also depends on the atomic representation to estimate the evolution of conformations of protein complexes based on the interaction between atoms. Especially both electrostatics interaction forces and van der Waals forces require $O(N^2)$ computational complexity for performing the direct calculation. Thus its computational complexity is expensive for the high number of atoms which are required to predict the deformation of protein structure. The constraint enforcement for bond lengths or bond angles using nonlinear constraint equations at each time step such as SHAKE [6], and RATTLE [7] to eliminate high frequency vibration motions also require heavy computational cost for molecular dynamic simulation.

In this paper, instead of expensive bond lengths and angles constraints or non-bond related forces computations, we provide a new global shape based volume preservation for protein deformation. The complexity of our volume preservation algorithm is at a constant time and it requires virtually no additional computational burden over the conventional mass-spring dynamics model. Our method incorporates the implicit volume preservation constraint enforcement on a mass-spring system to represent protein complexes, so it provides an efficient platform for user manipulation at an interactive response rate. Therefore, the proposed method is well suitable for initial screening of behavioral analysis for complex protein structures.

2 Related Works

The modeling of complex proteins based on the molecular dynamics simulation has been one of the remarkable research areas. The canonical numerical integrations [8] have been used for molecular dynamics simulations using thermodynamic quantities and transport coefficients since they provide enhanced long-time dynamics [9, 10]. Nose-Hoover [11, 12] molecular dynamic is introduced using a constant-temperature and Martyna et al. [13] improved stiff system problems with Nose-Hoover chain dynamics. Jang and Voth [14] introduced simple reversible molecular dynamics algorithms for Nose-Hoover chain dynamics by extension of the Verlet algorithm. Constrained molecular dynamics simulation also has been used to remove numerical stiffness of bond length or bond angle potentials. In molecular dynamics, the length of the integration time step for numerical integration is dictated by the high frequency motions since hard springs can cause numerical instability. The high frequency motions of molecular dynamic simulation are less important than the low frequency motions which correspond to the global motions.

Ryckaert et al. [6] introduced the SHAKE algorithm based on the Verlet method to allow for bond constraints and is widely used for applying the constraints into molecular dynamic simulations. The SHAKE algorithm is a semi-explicit method. Although the SHAKE iteration algorithm is simple and has a low memory requirement, it requires a sufficiently small integration time step to converge to the solution. The adaptive relaxation algorithm for the SHAKE algorithm as presented by Barth et al. [15] iteratively determines the optimal relaxation factor for enhanced convergence. Andersen [7] proposed the velocity level Verlet algorithm, RATTLE, for a velocity level formulation of constraints. The SHAKE and RATTLE require solving a system of nonlinear constraint equations at each integration time step thus they require substantial computational cost. On the other hand, we applied the implicit volume-preserving constraint into protein simulations with our implicit constraint method to achieve the globally meaningful deformation of proteins.

3 Protein Structure

The protein consists of the spatial arrangement of amino acids which are connected to one another through bonds. Thus the protein is a set of atoms connected by bonds in 3D. Instead of using bonded and non-bonded empirical functions, we represent the bond connectivity by simple mass-spring system. Fig. 1 illuminates the structure of periplasmic lysine, arginine, ornithine binding protein (2LAO). The atoms (mass points) are connected by springs which are colored by green lines to propagate the energy. Although we generated springs for each amino acid, each amino acid is still isolated from other amino acids in the protein. Two or more amino acids are linked together by a dehydration synthesis to form a polypeptide. These characteristic chemical bonds are called peptide bonds. Therefore two amino acids are connected together by the carboxyl group and the amine group. Notice that a water molecule is removed during this process.

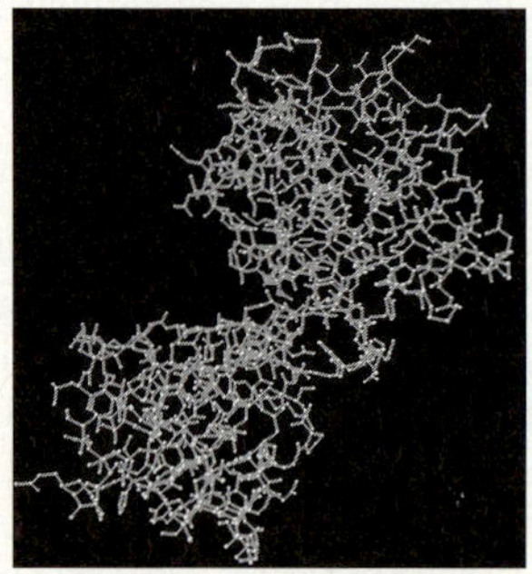

Fig. 1. Simple mass-spring system for periplasmic lysine, arginine, ornithine binding protein (2LAO)

4   Protein Surface

Since proteins interact with one another and other biological compounds through the surface, the protein surface is highly important to understand protein functions and deformations. The concept of a surface for protein involves the geometric properties of molecules, biophysics, and biochemistry. Usually the van der Waals surface is simply computed by overlapping the van der Waals spheres of each atom.

Lee and Richard [16] estimated the solvent accessible surface using a water molecule probe. The solvent accessible surface is determined by a water probe which exactly contacts the van der Waals spheres. Connolly [17] improved the solvent accessible surface using the molecular surface and reentrant surface and has been widely used to represent the surface of proteins. The atoms are idealized as spheres using the van der Waals radius and the Connolly surface is determined by the inward-facing surface of the water probe sphere as it rolls over the protein molecule.

Since fast and accurate computational geometry methods make it possible to compute the topological features of proteins, we applied the marching cube algorithm [18] to determine the surface of the protein. The marching cube algorithm has been widely used in applications of medical imaging to reconstruct 3D volumes of structures which can help medical doctors to understand the human anatomy present in the 2D slices and also applied to the description of the surface of biopolymers [19]. The 3D coordinates of every atom of the protein are obtained from PDB (Protein Data Bank) [20]. Initially, each atom is created using the van der Waals radius. However these standard values of the van der Waals radii create too many spatial cavities and tunnels. Thus we gradually increase the radius of atoms using a threshold until all atoms are overlapped with at least one of the other atoms.

The marching cube algorithm is one of a number of recent algorithms for surface generation with 3D volumetric cube data and it can effectively extract the complex 3D surface of protein. In addition, this algorithm provides the triangle mesh for the surface which is essential for using our volume preservation scheme. This algorithm detects not only the surface of the protein structure but also the cavities or tunnels of the protein. We can readily expand the basic principle of the marching cube algorithm to 3D.

We adopted the marching cube algorithm to generate the complex surface of proteins as a pre-processing stage. The high resolution of the triangulated mesh surface requires more computational cost and the resolution of the surface can be controlled using a threshold value which defines the size of cubes. The Fig. 2 shows the created refined surface of periplasmic lysine, arginine, ornithine binding protein (2LAO) and Adenylate kinase (1AKE). The surface nodes which create the surface of the protein are independently created from atoms. These surface nodes are connected by structure, bending, and sheer springs to preserve the physical properties. To represent the correct surface according to the deformation of the protein, the surface should be re-estimated by the marching cube algorithm at each time step, but it is a computationally expensive and painful task.

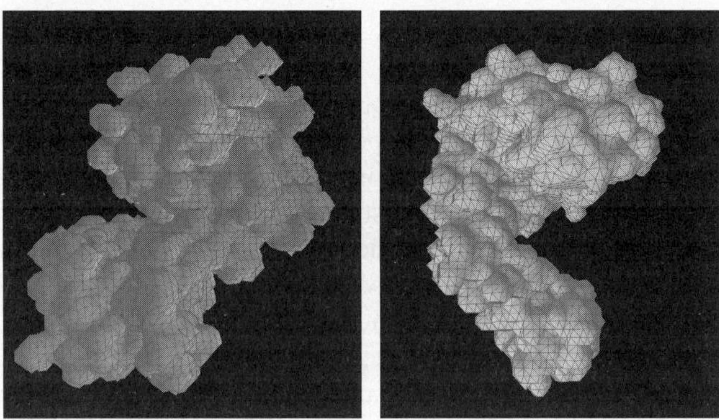

Fig. 2. The surface of proteins using marching cube algorithm: the left figure is periplasmic lysine, arginine, ornithine binding protein (2LAO) and the right figure is Adenylate kinase (1AKE)

5 Protein Volume

Volume changes of proteins on protein folding have been studied [21, 22] and their results show that volume changes are very small (less than 0.5 percentage from original volume) at normal pressure. To overcome the critical inherent drawback of volume loss of a mass-spring system, this paper proposes a real-time volume preservation method. Our volume preservation method maintains the global volume of a closed mesh structure and guarantees the constant volume constraint at every time step of a mass-spring simulation.

The total volume of a protein is estimated by summing the surface triangles of the protein using the Divergence Theorem. The surface of the protein is represented by flat triangular patches with coordinates (x, y, z) varying linearly on these patches. It is convenient to introduce the natural coordinates L_1, L_2, and L_3 and express the surface integral as

$$3V = \int \sum_{i=1}^{3} \left(x_i L_i N_x + y_i L_i N_y + z_i L_i N_z \right) dA \tag{1}$$

Note that the unit normal vector is constant on the triangular surface patch. The integral is easily evaluated using the following equation for integrating polynomials in L_i

$$\int L_1^{a_1} L_2^{a_2} L_3^{a_3} \, dxdy = 2A \frac{a_1! a_2! a_3!}{(a_1 + a_2 + a_3 + 2)!} \tag{2}$$

where a_1, a_2, and a_3 are non-negative integers, and A is the area of a triangle. We have the three cases: a_1, a_2, and a_3 are $a_1 = 1 \; a_2 = a_3 = 0$, $a_2 = 1 \; a_1 = a_3 = 0$, and $a_3 = 1 \; a_1 = a_2 = 0$. The three integrals we need are given by

$$\int L_1 dxdy = \int L_2 dxdy = \int L_3 dxdy = \frac{A}{3} \tag{3}$$

The total volume V can be obtained by

$$V = \sum_i \frac{A}{3}\left\{n_x(x_1 + x_2 + x_3) + n_y(y_1 + y_2 + y_3) + n_z(z_1 + z_2 + z_3)\right\} \tag{4}$$

where i is the volume contribution of surface triangle i. This volume must remain a constant over the entire simulation, so we cast this condition as a constraint in a dynamic system.

Let $\Phi(q,t)$ be the constraint to representing an algebraic constraint. To preserve the volume of object, the difference between V_0 (original volume) and V (current volume) should be 0 during the simulation.

$$\Phi(q,t) = V_0 - V = 0 \tag{5}$$

We applied the implicit constraint method [23] to maintain the volume constraint.

$$\Phi_q(q,t)M^{-1}\Phi_q^T(q,t)\lambda = \frac{1}{\Delta t^2}\Phi(q,t) + \Phi_q(q,t)(\frac{1}{\Delta t}\dot{q}(t) + M^{-1}F^A(q,t)) \tag{6}$$

where F^A are applied, gravitational and spring forces acting on the discrete masses, M is a diagonal matrix containing discrete nodal masses, λ is a vector containing the Lagrange multipliers and $\Phi_q = \dfrac{\partial V}{\partial q}$ is the Jacobian matrix. In equation (6), λ can be calculated by a simple division thus our volume preservation uses the implicit constraint method to preserve the protein volume with a mass-spring system at virtually no extra.

6 Force Generation for Protein Movement

Recently NMA (Normal Mode Analysis) [24, 25, 26] has been widely used to infer the dynamic motions of various types of proteins from an equilibrium conformation. NMA is a powerful theoretical method for estimating the possible motions of a given protein. To analyze the necessary external forces for protein movement, we used the Yale Morph Server [26] which applied NMA to achieve the movements of proteins as a reference deformation of the protein. We calculated the necessary forces based on the series of motion data files which are achieved from [26].

$$v_{n+1} = v_n + dt(F/m), \quad q_{n+1} = q_n + dtv_{n+1} \tag{7}$$

Equation (7) is the simple Euler integration to estimate the next status of positions and velocities. Here v is the velocity of atom, q is position of atom, F is the net force of system, m is the mass of atom, and dt is the integration time step. We already know the all position information of atoms for each time step, thus we can calculate the necessary force to move all atoms using equation (8).

$$v_{n+1} = (q_{n+1} - q_n)/dt, \quad F = m(v_{n+1} - v_n)/dt \tag{8}$$

Since the atoms of the protein and surface nodes are not connected to each other, when atoms are moved by estimated external forces in equation (8), the surface of the protein will not move. Instead of creating extra springs to follow the movement of atoms or regenerating the surface using the expensive marching cube algorithm at each time step, the surface nodes of the protein are moved according to the displacement of the closest atom to reduce computational requirements. Unlike the Yale Morph Server which only provides 2D motion of proteins with a fixed view point, our protein simulation provides 3D motion of protein deformation with a controllable view point.

7 Conclusion

Although we are using the current state-of-the-art computing power, it is still not computationally feasible to perform atomic molecular dynamics simulation for huge protein structures at an interactive response rate, partly due to the large conformation space. Instead of expensive and complicated force calculations at each time step to achieve the detail deformation of proteins, we applied the simple and fast mass-spring system. The stable and effective global shape based volume preservation constraint is applied to the protein simulation with the marching cube algorithm. Although our new global shape based volume preservation method sacrifices the micro level of dynamics in the protein movement, the proposed efficient platform of protein simulator can provide the globally correct (low-frequency motion) and biologically meaningful deformation of the protein complexes at an interactive level. Therefore, our simulator can be utilized as a nice test-bed for initial screening of behavioral analysis of protein complexes.

References

1. T. Schlick, Molecular modeling and simulation, Springer, New York, 2002.
2. A. R. Leach, Molecular modeling principles and applications, Addison Wesley Longman Limited, 1996.
3. L. Pauling, and E. B. Wilson, Introduction to quantum mechanics with applications to chemistry, Dover, New York, 1985.
4. Amber Molecular Dynamics, http://amber.scripps.edu/
5. CHARMM Development Project, http://www.charmm.org/
6. J. P. Ryckaert, G. Ciccotti, and H. J. C. Berendsen, Numerical integration of the Cartesian equations of motion of a system with constraints: Molecular dynamics of n-alkanes, Journal of Computational Physics, 23, pp. 327-341, 1977.
7. H. C. Andersen, Rattle: A velocity version of the SHAKE algorithm for molecular dynamics calculations, Journal of Computational Physics, 52, pp. 24-34, 1983.
8. D. Okunbor and R. D. Skeel, Canonical numerical methods for molecular dynamics simulations, Journal of Computational Chemistry, 15(1), pp. 72-79, 1994.
9. P. J. Channell, and J. C. Scovel, Symplectic integration of hamiltonian systems, Nonlinearity, 3, pp. 231-259, 1990.
10. D. Okunbor, Canonical methods for hamiltonian systems: Numerical experiments, Physics D, 60, pp. 314-322, 1992.
11. S. Nose, A molecular dynamics method for simulations in the canonical ensemble, Molecular Physics, 52(2), pp. 255-268, 1984.

12. W. G. Hoover, Canonical dynamics: Equilibrium phase-space distributions, Physical Review A, 31(3), pp. 1695-1697, 1985.

13. G. J. Martyna, M. L. Klein, and M. Tuckerman, Nosé–Hoover chains: The canonical ensemble via continuous dynamics, Journal of Chemical Physics, 97(4), pp. 2635-2643, 1992.

14. S. Jang, and G. A. Voth, Simple reversible molecular dynamics algorithms for Nosé–Hoover chain dynamics, Journal of Chemical Physics, 107(22), pp. 9514-9526, 1997.

15. E. Barth, K. Kuczera, B. Leimkuhler, and R. D. Skeel, Algorithms for constrained molecular dynamics, Journal of Computational Chemistry, 16, pp. 1192-1209, 1995.

16. B. Lee and F. M. Richards, The interpolation of protein structures: Estimation of static accessibility, Journal of Molecular Biology, 55, pp. 379-400, 1971.

17. M. L. Connolly, Solvent-accessible surfaces of proteins and nucleic acids, Science, pp. 709-713, 1983.

18. W. E. Lorensen, and H. E. Cline, Marching Cubes: A high resolution 3D surface construction algorithm, Proceedings of SIGGRAPH 1987, ACM Press / ACM SIGGRAPH, Computer Graphics Proceeding, 21(4), pp. 163-169, 1987.

19. A. H. Juffer, and H. J. Vogel, A flexible triangulation method to describe the solvent-accessible surface of biopolymers, Journal of Computer-Aided Molecular Design, 12(3), pp. 289-299, 1998.

20. The RCSB Protein Data Bank, http://www.rcsb.org/pdb/

21. P. E. Smith, Protein volume changes on cosolvent denaturation, Biophysical Chemistry, 113, pp. 299-302, 2005.

22. M. Gerstein, J. Tsai, and M. Levitt, The volume of atoms on the protein surface: Calculated from simulation, using voronoi polyhedra, Journal of Molecular Biology, 249, pp. 955-966, 1995.

23. M. Hong, M. Choi, S. Jung, S. Welch, and Trapp, J. Effective constrained dynamic simulation using implicit constraint enforcement. In IEEE International Conference on Robotics and Automation, 2005.

24. P. Dauber-Osguthorpe, D.J. Osguthorpe, P.S. Stern, J. Moult, Low-frequency motion in proteins - Comparison of normal mode and molecular dynamics of streptomyces griseus protease A, J. Comp. Phys., 151, pp. 169-189, 1999.

25. A. D. Schuyler, and G. S. Chirikjian, Normal mode analysis of proteins: A comparison of rigid cluster modes with Ca coarse graining, Journal of Molecular Graphics and Modelling, 22(3), pp. 183-193, 2004.

26. Yale Morph Server, http://molmovdb.mbb.yale.edu/

Third-Order Spectral Characterization of Termite's Emission Track

Juan-José González de-la-Rosa[1], I. Lloret[1], Carlos G. Puntonet[2],
A. Moreno[1], and J.M. Górriz[2]

[1] University of Cádiz, Electronics Area.
Research Group TIC168 - Computational Instrumentation and Industrial Electronics.
EPSA. Av. Ramón Puyol S/N. 11202, Algeciras-Cádiz, Spain
`juanjose.delarosa@uca.es`
[2] University of Granada, Department of Architecture and Computers Technology,
ESII, C/Periodista Daniel Saucedo. 18071, Granada, Spain
`carlos@atc.ugr.es`

Abstract. A higher-order frequency-domain characterization of termite activity (feeding and excavating) has been performed by means of analyzing diagonal slices of the bi-spectrum. Five sets of signals of different qualities were acquired using a high sensitivity probe-accelerometer. We conclude that it is possible to establish a third-order pattern (*spectral track*) associated to the termite emissions, and resulting from the impulsive response of the sensor and the body or substratum through which the emitted waves propagate.

1 Introduction

Termite detection has gained importance in the last decade mainly due to the urgent necessity of avoiding the use of harming termiticides, and to the joint use of new emerging techniques of detection and hormonal treatments, with the aim of performing an early treatment of the infestation. A localized partial infestation can be exterminated after two or three generations of these insects with the aid of hormones [1],[2].

User-friendly equipment is being currently used in targeting subterranean infestations by means of temporal analysis. An acoustic-emission (AE) sensor or an accelerometer is attached to the suspicious structure. Counting the hits produced by the insects and been registered by the accelerometer, the instrument outputs light signals. At the same time, the user can listen to the sounds and perform some pre-processing, like filtering or amplifying. A set of hits is defined as an acoustic event, which in fact constitutes the electronic *tracks* of the insects.

These instruments are based on the calculation of the root mean square (RMS) value of the vibratory waveform. The use of the RMS value can be justified both by the difficulty of working with raw AE signals in the high-frequency range, and the scarce information about sources and propagation properties of the AE waves. Noisy media and anisotropy makes even harder the implementation

V.N. Alexandrov et al. (Eds.): ICCS 2006, Part I, LNCS 3991, pp. 316–323, 2006.

of new methods of calculation and measurement procedures. A more sophisticated family of instruments make use of spectral analysis and digital filtering [3]. Both have the drawback of the relative high cost and their practical limitations.

In fact, the usefulness of the above prior-art acoustic techniques and equipments depends very much on several biophysical factors. The main one is the amount of distortion and attenuation as the sound travels through the soil ($\sim$600 dB m^{-1}, compared with 0.008 dB m^{-1} in the air) [1]. Furthermore, soil and wood are far from being ideal propagation media because of their high anisotropy, non-homogeneity and frequency dependent attenuation characteristics [3].

On the other hand, second order statistics and power spectra estimation (the second order spectrum) fail in low SNR conditions even with *ad hoc* piezoelectric sensors. Spectrum estimation and spectrogram extract time-frequency features, but ignoring phase properties of the signals. Besides, second-order algorithms are very sensitive to noise.

Other prior-art second-order tools, like wavelets and wavelet packets (time-dependent technique) concentrate on transients and non-stationary movements, making possible the detection of singularities and sharp transitions, by means of sub-band decomposition. The method has been proved under controlled laboratory conditions, up to a SNR=-30 dB [4].

As an alternative, higher order statistics (HOS), like the bi-spectrum, have proven useful for characterization of termites' emissions, using a synthetics of alarm signals and prior-known symmetrically-distributed noise processes [5],[6]. The conclusions of these works were funded in the advantages of cumulants; in particular, in the capability of enhancing the SNR of a signal buried in symmetrically distributed noise processes. The computational cost paid is the main drawback of the technique. Besides, in the practice the goal is to localize the infestation from weak evidences, in order to prevent greater destruction. For this reason it should be emphasized that non-audible signals have to be detected.

In this paper third-order spectra slices are used to characterize termite emissions. The results help the HOS researcher to better understand the higher-order frequency diagrams; in particular in the field of insect characterization by AE signal processing. The conclusions are based in records which were acquired within the surrounding perimeter of the infestation. The quality of the signals has been established using the criteria of audibility and the levels of quantization used in the digitalizing process by the data acquisition equipment. The accelerometer used is the SP1-L probe from AED2000 instrument, with a high sensitivity and a short band-width.

The paper is structured as follows: Section 2 summarizes the problem of acoustic detection of termites; Section 3 recalls the theoretical background of HOS, focussing on the computational tools. Experiments are drawn in Section 4, which is intended as a tool to interpret results from HOS-based experiments. Finally, conclusions are explained in Section 5.

2   Acoustic Detection of Termites

AE is defined as the class of phenomena whereby transient elastic waves are generated by the rapid (and spontaneous) release of energy from localized sources within a material, or the transient elastic wave(s) so generated. This energy travels through the material as a stress wave and is typically detected using a piezoelectric transducer, which converts the surface displacement (vibrations) to an electrical signal [4].

Termites use a sophisticated system of vibratory long distance alarm described, among others in, [1]. In [4] is is shown one of the impulses within a typical four-impulse burst (alarm signals) and its associated power spectrum of the specie (*reticulitermes lucifugus*). The carrier frequency of the drumming signal is defined as the main spectral component.

We are concerned about the spectral patterns of the signals; so we do not care about the energy levels. Besides, as a result of the HOS processing, the original energy levels of the signals are lost, but not as the extent of the levels of parasitic noise which are coupled to the signals. Furthermore, the amplitudes of the noise processes, which could be coupled to the waveform under study, are significatively reduced if the probability density function of the stochastic process is symmetrically distributed [5].

Signals from feeding an excavating do not exhibit a concrete time pattern (bursts and impulses equally spaced). They comprise random impulse events that provoke the same response of the sensor and the traversed media. The main handicap is the low intensity of the involved levels. We are concerned about detecting activity by rejecting noise.

In the following we settle the mathematical foundations of HOS for the characterization process.

3   Higher-Order Statistics (HOS)

The motivation of the poly-spectral analysis is three fold: (a) To suppress Gaussian noise processes of unknown spectral characteristics; the bi-spectrum also suppress noise with symmetrical probability distribution, (b) to reconstruct the magnitude and phase response of systems, and (c) to detect and characterize nonlinearities in time series.

Before cumulants, non-Gaussian processes were treated as if they were Gaussian. Cumulants and their associated Fourier transforms, known as poly-spectra [7], reveal information about amplitude and phase, whereas second order statistics (variance, covariance and power spectra) are phase-blind [8].

The relationship among the cumulants of stochastic signals, x_i, and their moments can be calculated by using the *Leonov-Shiryayev* formula. The second-, third-, and fourth-order cumulants are given by [5], [8] equation 1:

$$Cum(x_1, x_2) = E\{x_1 \cdot x_2\}. \tag{1a}$$

$$Cum(x_1, x_2, x_3) = E\{x_1 \cdot x_2 \cdot x_3\}. \tag{1b}$$

$$
\begin{aligned}
Cum(x_1, x_2, x_3, x_4) =\; & E\{x_1 \cdot x_2 \cdot x_3 \cdot x_4\} \\
& - E\{x_1 \cdot x_2\}E\{x_3 \cdot x_4\} \\
& - E\{x_1 \cdot x_3\}E\{x_2 \cdot x_4\} \\
& - E\{x_1 \cdot x_4\}E\{x_2 \cdot x_3\}.
\end{aligned}
\tag{1c}
$$

In the case of non-zero mean variables x_i have to be replaced by $x_i\text{-}E\{x_i\}$.

Let $\{x(t)\}$ be a rth-order stationary random real-valued process. The rth-order cumulant is defined as the joint rth-order cumulant of the random variables $x(t)$, $x(t+\tau_1),\ldots,\,x(t+\tau_{r-1})$,

$$
\begin{aligned}
& C_{r,x}(\tau_1, \tau_2, \ldots, \tau_{r-1}) \\
& = Cum[x(t), x(t + \tau_1), \ldots, x(t + \tau_{r-1})].
\end{aligned}
\tag{2}
$$

The second-, third- and fourth-order cumulants of zero-mean $x(t)$ can be expressed using equations 1 and 2, via:

$$
C_{2,x}(\tau) = E\{x(t) \cdot x(t + \tau)\}.
\tag{3a}
$$

$$
C_{3,x}(\tau_1, \tau_2) = E\{x(t) \cdot x(t + \tau_1) \cdot x(t + \tau_2)\}.
\tag{3b}
$$

$$
\begin{aligned}
& C_{4,x}(\tau_1, \tau_2, \tau_3) \\
& = E\{x(t) \cdot x(t + \tau_1) \cdot x(t + \tau_2) \cdot x(t + \tau_3)\} \\
& \quad - C_{2,x}(\tau_1)C_{2,x}(\tau_2 - \tau_3) \\
& \quad - C_{2,x}(\tau_2)C_{2,x}(\tau_3 - \tau_1) \\
& \quad - C_{2,x}(\tau_3)C_{2,x}(\tau_1 - \tau_2).
\end{aligned}
\tag{3c}
$$

We assume that the cumulants satisfy the bounding condition given in equation 4:

$$
\sum_{\tau_1=-\infty}^{\tau_1=+\infty} \cdots \sum_{\tau_{r-1}=-\infty}^{\tau_{r-1}=+\infty} |C_{r,x}(\tau_1, \tau_2, \ldots, \tau_{r-1})| < \infty.
\tag{4}
$$

The higher-order spectra are usually defined in terms of the rth-order cumulants as their $(r\text{-}1)$-dimensional Fourier transforms

$$
\begin{aligned}
& S_{r,x}(f_1, f_2, \ldots, f_{r-1}) \\
& = \sum_{\tau_1=-\infty}^{\tau_1=+\infty} \cdots \sum_{\tau_{r-1}=-\infty}^{\tau_{r-1}=+\infty} C_{r,x}(\tau_1, \tau_2, \ldots, \tau_{r-1}) \\
& \qquad \cdot \exp[-j2\pi(f_1\tau_1 + f_2\tau_2 + \cdots + f_{r-1}\tau_{r-1})].
\end{aligned}
\tag{5}
$$

The special poly-spectra derived from equation 5 are power spectrum ($r=2$), bi-spectrum ($r=3$) and try-spectrum ($r=4$). Only power spectrum is real, the others are complex magnitudes. Poly-spectra are multidimensional functions which comprise a lot of information. As a consequence, their computation may be impractical in some cases. To extract useful information one-dimensional slices of cumulant sequences and spectra, and bi-frequency planes are employed in non-Gaussian stationary processes [6].

Once summarized the foundations of the experiment, hereinafter we present que results obtained by means of the tools described here.

4 Experimental Results

The data acquisition stage took place in a residential area of the "Costa del Sol" (Málaga-Spain), at the beginning of the reproductive season of the termites. The sensors were attached (plunged) in the soil surrounding affected trees (above the roots). We worked under the hypothesis of having a 500 m-radius of an affected circular perimeter.

The probe SP1-L from the equipment AED2000 has been taken as the reference with a twofold purpose. First, we analyze the power spectra of signals with different qualities, to decide wether it is possible to use second-order spectra for identification purposes. Secondly, we settle down a higher-order detection criterion in the frequency domain which it is supposed to enhance detection.

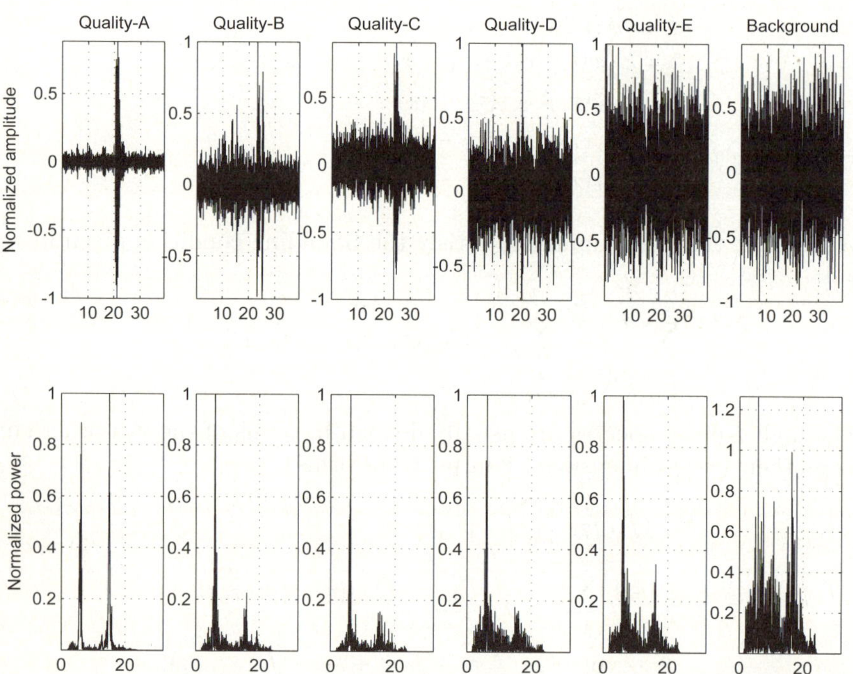

Fig. 1. Average power spectra of six categories of emissions in comparison to the the background. The first row of graphs contains one signal of each quality type. Abscissae of time sequences (ms); abscissae of spectra (kHz).

Signals have been previously high-pass filtered in order to suppress low frequency components which would mask the higher-frequency components. A 5th-order *Butterworth* digital filter with cut-off frequency of 2 kHz is used.

Each graph is the result of averaging a number of 15, 2500-sample registers from *reticulitermes lucifugus* feeding activity. Figure 1 shows the average power spectra of six signal categories according to their amplitude quantization levels. These AE signals were acquired by the sensor SP1-L, using a sampling frequency of 64 kHz and a resolution of 16 *bits*. These spectra are compared with the background in order to establish an identification criterion. Vibrations of qualities D and E are inaudible.

Quality-A (Q-A) signals' amplitudes belong to the quantization levels interval $[25000, 30000]$. Q-B levels are in $[15000, 20000]$. Q-C levels are in $[10000, 15000]$. Quality-D levels are in the interval $[5000, 10000]$. Finally, Q-E impulses are completely buried in the background. We are sure about the infestation, so the AE events in this series are due to termite activity, probably in a 3 meter-radius subset within the area under study.

The two main frequency components in the spectra of Figure 1 appear at 6 and 15 kHz, respectively, which are associated with the frequency response of the sensor, to the features of the sounds produced by the termite species and to the characteristics of the substratum, through which the emissions propagate. We conclude that using the probe SP1-L we can detect an infestation by interpreting the power spectra diagrams. This is due to the differences that the emissions exhibit in comparison with the flatter shape of the background spectra (sixth column of graphs in Figure 1).

The calculation of 3th-order spectra is performed with a twofold purpose. The first objective is to enhance the detection criteria in the frequency domain. The second purpose is to use more economic sensors, with a lower sensitivity and a higher band-width. Figure 2 shows the average diagonal bi-spectra associated with the signals acquired with the probe SP1-L. A maximum lag $\chi = 512$ was selected to compute the third-order auto-cumulants of the signals. The bottom bi-spectrum characterizes the background sound (the most unfavorable, with an amplitude lower in four orders of magnitude than Q-A).

The main frequency component in Figure 2 (6 kHz) permits to establish a detection criterion based on the identification of this maximum value. It is remarkable that this frequency is also associated with the sensor. Another sensor would display another bi-spectrum shape. For this reason the proposed method of insect detection is based on the prior characterization of the transducer. On the other hand, the magnitudes of the bi-spectra in Figure 2 don't suffer a dramatic attenuation from high to low levels (two orders from Q-A to Q-E). This fact reinforces the criterion of identification, in the sense that it is the noise (symmetrically distributed) which is mainly reduced in the higher-order computation.

On the basis of these results we establish the conclusions related to the identification criterion proposed.

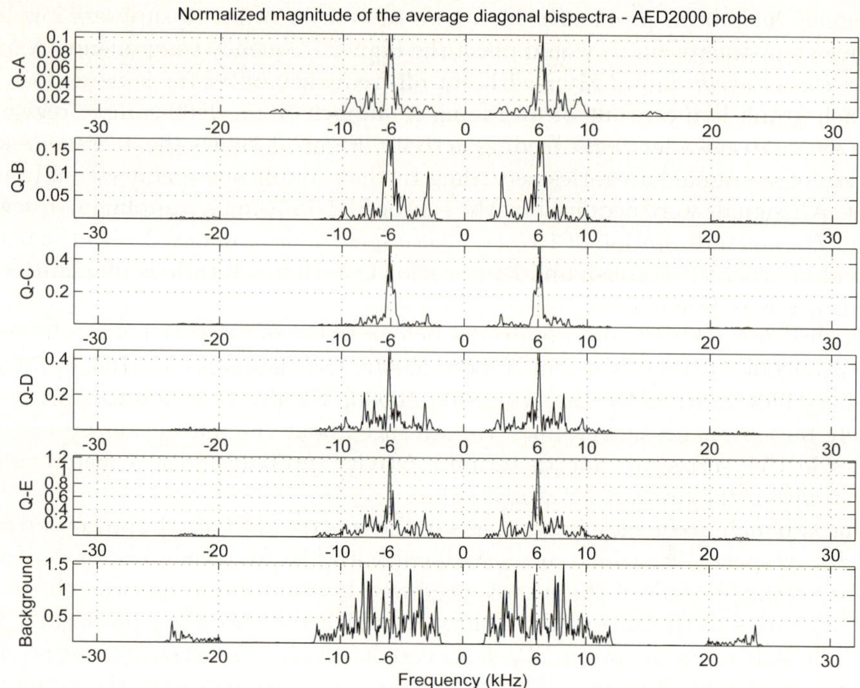

Fig. 2. Average diagonal bi-spectra of signals acquired with the probe SP1-L from AED2000 in comparison to the background sounds, for a maximum lag, $\chi = 512$. Each bi-spectrum results from averaging a number of 15, 2500-data registers.

5   Conclusions

In this work it has been shown that the diagonal slices of the bi-spectrum are valid and convenient tools for obtaining decision criteria to distinguish a possible infestation, based on the feeding activities of the termites. We have funded this conclusion on three arguments:

First, higher-order cumulants and spectra, as defined herein, enable the signal analysis procedure to have access to waveform information that is typically unavailable when using prior art (second-order) methods. In particular, we remark the enhancement of the frequency diagrams. This is due to the rejection exerted on symmetrically distributed noise processes. In fact, non-Gaussian processes are completely characterized by means of HOS.

Secondly, the potentially valuable information contained in an AE signal (most part of its spectrum) is related to the impulses. The average spectrum reveals amplitude information (the resonance peaks) but phase information is not shown. Higher-order spectra are arrangements of complex numbers and contain this additional information which can be valuable in a pattern recognition or identification criterion context.

Finally, using different sensors the criterion changes the frequency *set-point*. Besides, the probability of a false alarm is very low, considering the fact that we had to provide, intentionally, the worst case of background noise. Repeatability has been estimated in a 75 per cent.

Future work is focussed on reducing the computational complexity of HOS in two directions. On one side, we are using compact functions, like *FFT* and *FFTshift*. Secondly, we have to adopt a compromise between the maximum lag (χ) and the resolution, in order to save storage memory and time. These actions are oriented to implement the algorithms in a digital signal processor, in an autonomous hand-instrument.

Acknowledgement

The authors would like to thank the *Spanish Ministry of Education and Science* for funding the projects DPI2003-00878, TEC2004-06096 and PTR1995-0824-OP.

References

1. Röhrig, A., Kirchner, W.H., Leuthold, R.H.: Vibrational alarm communication in the african fungus-growing termite genus macrotermes (isoptera, termitidae). Insectes Sociaux **46** (1999) 71–77
2. De la Rosa, J.J.G., Lloret, I., Puntonet, C.G., Górriz, J.M.: Wavelets transforms applied to termite detection. In: IEEE Third International Workshop on Intelligent Data Acquisition and Advanced Computing Systems: Technology and Applications (IDAACS'2005). Proceedings, Budapest-Hungary, Institute of Computer Information Technologies (2005) oral presentation.
3. Mankin, R.W., Fisher, J.R.: Current and potential uses of acoustic systems for detection of soil insects infestations. In: Proceedings of the Fourth Symposium on Agroacoustic. (2002) 152–158
4. De la Rosa, J.J.G., Puntonet, C.G., Lloret, I., Górriz, J.M.: Wavelets and wavelet packets applied to termite detection. Lecture Notes in Computer Science (LNCS) **3514** (2005) 900–907 Computational Science - ICCS 2005: 5th International Conference, GA Atlanta, USA, May 22-25, 2005, Proceedings, Part I.
5. De la Rosa, J.J.G., Puntonet, C.G., Lloret, I.: An application of the independent component analysis to monitor acoustic emission signals generated by termite activity in wood. Measurement (Ed. Elsevier) **37** (2005) 63–76 Available online 12 October 2004.
6. De la Rosa, J.J.G., Lloret, I., Puntonet, C.G., Górriz, J.M.: Higher-order statistics to detect and characterise termite emissions. Electronics Letters **40** (2004) 1316–1317 Ultrasonics.
7. Nykias, C.L., Mendel, J.M.: Signal processing with higher-order spectra. IEEE Signal Processing Magazine (1993) 10–37
8. Mendel, J.M.: Tutorial on higher-order statistics (spectra) in signal processing and system theory: Theoretical results and some applications. Proceedings of the IEEE **79** (1991) 278–305

Parallel Optimization Methods Based on Direct Search

Rafael A. Trujillo Rasúa[1,2], Antonio M. Vidal[1],
and Víctor M. García[1]

[1] Departamento de Sistemas Informáticos y Computación
Universidad Politécnica de Valencia
Camino de Vera s/n, 46022 Valencia, España
{rtrujillo, avidal, vmgarcia}@dsic.upv.es
[2] Departamento de Técnicas de Programación,
Universidad de las Ciencias Informáticas,
Carretera a San Antonio de los Baños Km 2 s/n,
La Habana, Cuba
trujillo@uci.cu

Abstract. This paper is focused in the parallelization of Direct Search Optimization methods, which are part of the family of derivative-free methods. These methods are known to be quite slow, but are easily parallelizable, and have the advantage of achieving global convergence in some problems where standard Newton-like methods (based on derivatives) fail. These methods have been tested with the Inverse Additive Singular Value Problem, which is a difficult highly nonlinear problem. The results obtained have been compared with those obtained with derivative methods; the efficiency of the parallel versions has been studied.

1 Introduction

The general unrestricted optimization problem can be stated as follows: Given a continuous function $f : \Re^n \longrightarrow \Re$, find the minimum of this function. The most popular optimization methods use derivative or gradient information to build descent directions. However, there are many situations where the derivatives may not exist or cannot be computed, or are too expensive to compute. In such situations the derivative-free methods can be the only resource. These methods were studied first by Hooke and Jeeves [5] in 1961. In 1965 another important method of this kind was discovered, the Nelder-Mead simplex method [11].

All these methods are known to be fairly robust and locally convergent, but they are also quite slow, compared with derivative methods. For some time, this caused a lack of interest in these methods. However, with the advent of parallel computing these methods were again popular, since they are easily parallelizable [13, 2, 6].

This paper address the parallelization of a subset of these methods, termed in [9] as "Direct Search methods", where these methods are deeply described and studied. For the work described in this paper, different versions of these Direct

V.N. Alexandrov et al. (Eds.): ICCS 2006, Part I, LNCS 3991, pp. 324–331, 2006.

Search methods have been implemented and parallelized; in this paper the best versions are presented, and, as expected, the parallel versions obtain substantial time reductions over the sequential versions. Still more remarkable is the overall robustness of the direct search methods. These methods have been tested with a very difficult problem, the Inverse Additive Singular Value Problem (IASVP), which is a problem where Newton-like methods usually fail to converge, even using globalization techniques as Armijo's rule [7] . As will be shown, the Direct Search methods converge where Newton's method fails, although they are still substantially slower than Newton's method.

The next section is devoted to the description of the Direct Search methods and the sequential versions implemented. Next, the parallel versions are presented, followed by the numerical results of the sequential and parallel methods. Finally the conclusions shall be given.

2   Direct Search Methods

We will present two different direct search methods; both belong to the class of *Generating Set Search* (GSS), as defined by Kolda et. al. in [9]. These methods start from an initial point x_0, an initial step length Δ_0 and a set of directions spanning $\Re^n$: $D_0 = \{d_i\}_{i=1,\ldots p}, d_i \in \Re^n$, so that every vector in $\Re^n$ can be written as a nonnegative linear combination of the directions in D_k.

The driving idea of the GSS methods is to find a decrease direction among those of D_k. To do that, in each iteration k the objective function f is evaluated along the directions in D_k. At that stage, the actual point shall be x_k and the step length is Δ_k; therefore, $f(x_k + \Delta_k d_i)$ is computed, for $i = 1, ..., p$, until a direction d_i is found such that $f(x_k + \Delta_k d_i) < f(x_k)$. If no such direction is found, the step length is decreased and the function is evaluated again along the directions.

When an acceptable pair (Δ_k, d_i) is found, the new point is updated:

$(x_{k+1} = x_k + \Delta_k d_i)$, a new step length Δ_{k+1} is computed and the set of directions is possibly modified or updated. This procedure shall be repeated until convergence (that is, when the step length is small enough).

This general algorithmic framework can be implemented in many ways.Among the versions that we have implemented, we have chosen the two versions described below, since they give the best results.

2.1   Method GSS1

In our first version, the direction set chosen as

$$D_k = \{\pm e_i\} \cup \{(1,1,\cdots,1),(-1,-1,\cdots,-1)\}, \tag{1}$$

that is, the set of the coordinate axes and the vectors $(1,1,\cdots,1)$ and $(-1,-1,\cdots,-1)$ which very often accelerate the search when the initial point is far from the optimum.

Another characteristic is the strategy proposed in [6] of doubling the step length when the same descent direction is chosen in two consecutive iterations.

A distinct characteristic of our method (not proposed before, as far as we know) is the appropriate rotations of the set of directions. The evaluations are carried out in the order in which the directions are located in D_k. Therefore, if the descent directions are located in the first positions of the vector, the algorithm should find the optimum faster. To accomplish that, we propose that if in the iteration k the first descent direction is d_i, it means that $d_1, ..., d_{i-1}$ are not descent directions and, most likely, will not be descent directions in the next iterations. Therefore, in our algorithm, these directions are displaced to the end of the D_k set; that is, the new set would be: $(d_i, d_{i+1}, ..., d_n, d_1, ..., d_{i-1})$.

2.2 Method GSS2

This method follows a similar strategy to the described by Hooke and Jeeves in [5]. Here the initial direction set is $D_0 = \{\pm e_i\}$, and, unlike in GSS1, all the directions in D_k are explored. Then, choosing all the descent directions: $d_{i,1}^{(k)}, d_{i,2}^{(k)}, ...d_{i,j}^{(k)}$, a new descent direction is built:

$$d = d_{i,1}^{(k)} + d_{i,2}^{(k)} + ... + d_{i,j}^{(k)} \qquad (2)$$

which probably will cause a greater descent. If this is not true, the algorithm will choose among $d_{i,1}^{(k)}, d_{i,2}^{(k)}, ...d_{i,j}^{(k)}$ the direction with larger descent.

Clearly, this algorithm should need less iterations for convergence than GSS1, but the cost of each iteration is larger, since the function will be evaluated along all the directions.

3   Parallelization of the Direct Search Methods

The simplicity of the direct search methods and the independence of the function evaluations along each direction makes the parallelization of these methods a relatively easy task. The only serious problem to solve is the load imbalance that can suffer the algorithm GSS1. We will describe first the simpler parallel version of GSS2, and then two different options for GSS1.

3.1   Parallel Method PGSS2

The $2n$ function evaluations needed for each iteration of GSS2 can be distributed among p processors, so that each one carries out approximately $\frac{2n}{p}$ evaluations.The parallel algorithm PGSS2 would start by distributing the initial point x_0, the initial step length Δ_0 and a set of approximately $\frac{2n}{p}$ directions. In each iteration each processor would evaluate the function along each direction, and would return the descent directions found and the corresponding function values to the "Root" processor. This processor would form the new direction (as in (2)), would evaluate the function along this direction and obtain the new point x_{k+1} and the new step length Δ_{k+1}, which would be broadcasted to the other processors.

In this algorithm PGSS2 there is hardly any load imbalance, and few communications, so that the speed-up is expected to be close to the optimum.

3.2 Parallel Method PGSS1

The parallel version of the algorithm GSS1 follows the same data distribution of GSS2. However, the underlying strategy must be different. Let us recall that in the GSS1 algorithm the function is evaluated sequentially along each direction until a descent direction is found.

As in the parallel version of GSS2, the directions are distributed among the processors. If a processor finds among its directions a descent one, it should warn the rest of processors to stop searching. The strategy chosen for this parallel algorithm is to select a number of evaluations m that any processor should perform before reporting to the other processors. If after m evaluations there has been no success, they must perform m evaluations more, and repeat the process until a descent direction is found or they run out of directions.

It might happen that when the processors broadcast their results, several descent directions are found. If this happens, the direction of larger descent is chosen. This can cause that the parallel algorithm PGSS1 can make less iterations than the sequential version GSS1.

3.3 Parallel Method Master-Slave MSPGSS1

In the section 3.2 the algorithm described will perform a number of innecesary evaluations, depending on the value of m, and maybe some innecesary communications (if m is too small). As an attempt to improve it, we implemented an asynchronous parallel version, using a Master-Slave scheme.

The Master processor controls all the search, while the slaves perform the evaluations of the function along the directions. The directions are distributed among the slaves; after each evaluation, the slave sends the value with a non-blocking message to the master processor. The non-blocking message allows that the slave goes on with its work. When the master detects a descent, sends to all the slaves a Success message; then, computes the new point and step length, and broadcasts it to all the slaves.

A drawback of this algorithm is that there is one processor less to perform evaluations, although this allows a better control and overlaps computing time with communication time.

4 Experimental Results

4.1 Sequential Experiments

All these algorithms have been implemented and applied to a difficult problem, the Inverse Additive Singular Value Problem (IASVP), which can be defined as:

Given a set of matrices $A_0, A_1, ..., A_n \in \Re^{m \times n}$ $(m \geq n)$ and a set of real numbers $S^* = \{S_1^*, S_2^*, ..., S_n^*\}$, where $S_1^* > S_2^* > ... > S_n^*$, find a vector $c = [c_1, c_2, ..., c_n]^t \in \Re^n$, such that S^* are the singular values of

$$A(c) = A_0 + c_1 A_1 + ... + c_n A_n. \tag{3}$$

This problem is usually formulated as a nonlinear system of equations, and is solved with different formulations of Newton's method; several formulations are analyzed in [3]. However, it can be formulated as well as an optimization problem, minimizing the distance between the desired singular values S^* and the singular values of $A(c)$.

The sequential algorithms were implemented in C, using Blas [4] and LAPACK [1], and were executed in a 2GHz Pentium Xeon with 1 GByte of RAM and with operating system Red Hat Linux 8.0. For the experiments carried out with sequential algorithms, random square matrices were generated with sizes $n = 5, 10, 15, 20, 25, 30, 40, 50$; the solution vector c^* was chosen randomly as well, and the initial guesses $c_i^{(0)}$ were taken perturbing the solution with different values δ: $c_i^{(0)} = c^* + \delta, i = 1, \cdots, n$.

To obtain a fair appreciation of the performance of the Direct Search methods applied to the IASVP, the results of these methods were compared to Newton's method. Since there was convergence problems, Armijo's rule used to improve convergence in Newton's method. However, it became clear that Newton's method is very sensitive to the distance from the initial point to the solution. When the perturbation δ used was small, $\delta = 0.1$ or smaller, Newton's method converged always. However, when $\delta = 1.1$ it only converged in the case with $n = 5$, and when $\delta = 10.1$ it did not converge in any test. Meanwhile, the Direct Search methods converged in all the cases, although in a quite large number of iterations.

Table 1. Number of iterations to convergence, $\delta = 0.1$

n	5	10	15	20	25	30	40	50
GSSI	82	630	2180	967	827	6721	7914	7178
GSSII	51	669	611	372	322	2237	1368	7292
Newton	3	3	7	5	7	7	5	6

In the table 2 are shown the execution times of GSS1, GSS2 and Newton's method for the case $\delta = 0.1$, when Newton's method converges. GSS1 is faster than GSS2, but Newton's method is much faster than Direct Search methods, when it converges.

Table 2. Execution Times (seconds), $\delta = 0.1$

n	5	10	15	20	25	30	40	50
$GSSI$	0,03	1,26	3,37	2,80	4,10	45,00	139,76	237,48
$GSSII$	0,02	0,50	2,96	4,12	6,77	86,21	170,34	2054,88
$Newton$	0,00	0,00	0,01	0,01	0,02	0,04	0,08	0,25

These results show that the application of the sequential Direct Search methods to high complexity problems is not practical, since the execution times are too high. However, they show a very interesting property; at least for this problem, they are far more robust than Newton's method when the initial guess is far away from the solution.

4.2 Parallel Experiments

The tests were carried out in a cluster with 20 2GHz Intel Xeon biprocessors, each one with 1 Gbyte of RAM, disposed in a 4x5 mesh with 2D torus topology and interconnected through a SCI network. The parallel methods were implemented using MPI [12]. The sizes of the problems were conditioned by the high cost of the sequential solution. Due to this, the sizes of the problems were not too large: $n = 72, 96, 120, 144$. However, it must be noted that Newton's algorithm did not converge in these cases, even taking a initial guess very close to the real solution. The tables 3,4,5 summarize the execution times of the parallel algorithms.

Table 3. Execution Times (seconds) PGSS1

$size\backslash processors$	1	2	4	6	8	10	16
72	763	307	136	80	59	64	32
96	3929	1982	621	362	336	266	187
120	10430	4961	1891	943	669	605	364
144	36755	17935	5626	3187	2209	1599	1207

Table 4. Execution Times (seconds) PGSS2

$size\backslash processors$	1	2	4	6	8	10	16
72	278	141	72	50	38	33	23
96	2710	1400	710	482	375	313	210
120	5020	2581	1322	897	703	560	377
144	14497	7349	3749	2537	1948	1565	1028

Table 5. Execution Times (seconds) MSPGSS1

$size\backslash processors$	1	2	4	6	8	10	16
72	763	762	214	133	95	108	58
96	3929	3940	1228	767	467	482	263
120	10430	10614	3494	2108	1459	1134	958
144	36755	36786	12258	6244	4375	3630	2286

It must be observed that the sequential algorithm is the same for PGSS1 and MSPGSS1, and therefore the times are identical. These times are quite similar as well to the times of the MSPGSS1 with two processors, since there is only a slave processor performing the evaluations.

The Master-slave algorithm performs worse than the other two. The fact of having one less processor performing evaluations does not seem to be the reason, since with 10 or 16 processors the influence should be less important.

The comparison of the algorithms PGSS1 and PGSS2 shows interesting trends; for these new sizes the sequential algorithm GSS1 deteriorates quite fast when the problem size grows, unlike in the smaller test cases (See Table 2). With only one processor, the algorithm GSS2 is twice faster for large problems.

However, when the number of processors increases the situation changes; the PGSS2 algorithm gives good speedups, as expected, but the PGSS1 algorithm obtains great advantages when the number of processors grows, so that for large number of processors it obtains execution times close to those of PGSS2. It can be seen that the algorithm PGSS1 obtains speed-ups over the theoretical maximum, which is the number of processors. This happens because the parallel algorithm does not follow the same search strategy than the sequential one; this causes that the number of iterations and the number of function evaluations is different for both algorithms.

The algorithm PGSS2 gives the best results overall.

5   Conclusions

Three parallel Direct search methods have been implemented and described. Some new strategies have been applied:

1. In the GSS1 method, the new strategy of rotating the serch directions complements the strategy described in [6] of doubling the time step. This increases the probabilities of a descent direction being selected in two consecutive iterations, decreasing the average execution time.
2. In the GSS2 method, the addition of all the descent directions generates easily a new descent direction.
3. A Master-slave asynchronous algorithm was designed and implemented. It obtains quite good speed-ups, but the overall time execution is, with the present version, worse than the other two versions.

Finally, we would like to remark the great robustness shown by these methods in the IASVP problem, which is a really testing problem. For this problem, the convergence radius shown by these methods is far larger than the shown by Newton method, which, when converges, is much faster.

This shows that the Direct Search methods could be very useful for solution of difficult problems; if not for the full procedure, they could be used to obtain a good initial approximation for Newton's method, composing therefore a hybrid method. As shown, the long execution times of the Direct Search methods can be alleviated using parallel versions of these methods.

Acknowledgement

This work has been supported by Spanish MCYT and FEDER under Grant TIC2003-08238-C02-02 .

References

1. Anderson E., Bai Z., Bishof C., Demmel J., Dongarra J.: LAPACK User Guide; Second edition. SIAM (1995)
2. Dennis J.E. and Torczon V.: Direct search methods on parallel machines; SIAM J. Optim., 1, (1991) 448-474
3. Flores F., García V.M., and Vidal A.M.: Numerical Experiments on the solution of the Inverse Additive Singular Value Problem; Computational Science - 5th International Conference, Atlanta, GA, USA (2005)
4. Hammarling S., Dongarra J., Du Croz J., Hanson R.J.: An extended set of fortran basic linear algebra subroutines; ACM Trans. Mat. Software (1988)
5. Hooke R. and Jeeves T.A.: Direct Search solution of numerical and statistical problems; Journal of the Association for Computing Machinery, (1961) 212-229
6. Hough P.D., Kolda T.G. and Torczon V.: Asyncronous parallel pattern for nonlinear optimization. SIAM J. Sci. Comput., 23, (2001) 134-156
7. C. T. Kelley: Iterative methods for linear and nonlinear equations; SIAM (1995).
8. Kolda T.G. and Torczon V.: On the Convergence of Asyncronous Parallel Pattern Search; Tech. Rep. SAND2001-8696, Sandia National Laboratories, Livermore, CA, (2002); SIAM J. Optim., submitted.
9. Kolda T. G., Lewis R. M. and Torczon V.: Optimization by Direct Search: New perspective on Some Clasical and Modern Methods; SIAM Review, Vol. 45, No. 3 (2003) 385-442
10. Kumar V., Gramar A., Gupta A. and Kerypis G.: Introduction to Parallel Computing: Desing and analysis of Algorithms; Redwood City, CA (1994)
11. Nelder J.A. and Mead R.: A simplex method for function minimization; The Computer Journal, 7 (1965) 308-313
12. Snir M., Otto S., Huss-Lederman S., Walker D. and Dongarra J.: MPI: The Complete Reference; MIT Press, (1996)
13. Torczon V.: Multi-Directional Search: A Direct Search Algorithm for Parallel Machines; Technical Report 90-7, Department of Computational and Applied Mathematics, Rice University, Houston, TX (1990), Author's 1989 Ph.D. dissertation.

On the Selection of a Transversal to Solve Nonlinear Systems with Interval Arithmetic

Frédéric Goualard and Christophe Jermann

LINA, FRE CNRS 2729 – University of Nantes – France
2, rue de la Houssinière – BP 92208 – F-44322 Nantes cedex 3
{Frederic.Goualard, Christophe.Jermann}@univ-nantes.fr

Abstract. This paper investigates the impact of the selection of a transversal on the speed of convergence of interval methods based on the nonlinear Gauss-Seidel scheme to solve nonlinear systems of equations. It is shown that, in a marked contrast with the linear case, such a selection does not speed up the computation in the general case; directions for researches on more flexible methods to select projections are then discussed.

1   Introduction

The extensions to interval arithmetic [10] of Newton and nonlinear Gauss-Seidel methods [13] do not suffer from lack of convergence or loss of solutions that cripple their floating-point counterparts, which makes them well suited to solve systems of highly nonlinear equations.

For the linear case, it is well known that reordering equations and variables to select a transversal is paramount to the speed of convergence of first-order iterative methods such as Gauss-Seidel [3, 5]. Transversals may also be computed [14, 7, 4] in the nonlinear case when using nonlinear Gauss-Seidel methods [12] and when solving the linear systems arising in Newton methods (e.g., preconditioned Newton-Gauss-Seidel, aka Hansen-Sengupta's method [6]).

Interval-based nonlinear Gauss-Seidel (INLGS) methods are of special importance because they constitute the basis for interval constraint algorithms [4] that often outperform extensions to intervals of numerical methods.

We show in this paper that, in the general case, it is not possible to choose statically at the beginning of the computation a good transversal when using an INLGS method. We also present evidences that reconsidering the choice of the transversals after each Gauss-Seidel outer iteration is potentially harmful since it may delay the splitting of domains when the INLGS method is floundering.

Section 2 gives some background on interval arithmetic and its use in the algorithm based on nonlinear Gauss-Seidel that is used in this paper; Section 3 describes previous works on the selection of a good transversal, and presents experimental evidences that such a choice may actually be baseless; Section 4 explores alternative ways to select a good set of projections by either choosing more than n projections for a system of n equations on n variables, or by reconsidering the choice of a transversal dynamically during the solving process;

V.N. Alexandrov et al. (Eds.): ICCS 2006, Part I, LNCS 3991, pp. 332–339, 2006.

Lastly, Section 5 delves into all the experimental facts presented so far to propose new directions of research for speeding up the solving of systems of nonlinear equations with INLGS-based algorithms.

2 An Interval Nonlinear Gauss-Seidel Method

Interval arithmetic [10] replaces floating-point numbers by closed connected sets of the form $I = [\underline{I}, \overline{I}] = \{a \in \mathbb{R} \mid \underline{I} \leqslant a \leqslant \overline{I}\}$ from *the set* $\mathbb{I}$ *of intervals*, where $\underline{I}$ and $\overline{I}$ are floating-point numbers. In addition, each n-ary real function ϕ with domain $\mathcal{D}_\phi$ is extended to an interval function Φ with domain $\mathcal{D}_\Phi$ in such a way that the *containment principle* is verified:

$$\forall A \in \mathcal{D}_\phi \, \forall I \in \mathcal{D}_\Phi : A \in I \implies \phi(A) \in \Phi(I)$$

Example 1. The *natural interval extensions* of addition and multiplication are defined by:

$$I_1 + I_2 = [\underline{I_1} + \underline{I_2}, \overline{I_1} + \overline{I_2}]$$
$$I_1 \times I_2 = [\min(\underline{I_1 I_2}, \underline{I_1}\overline{I_2}, \overline{I_1}\underline{I_2}, \overline{I_1 I_2}), \max(\underline{I_1 I_2}, \underline{I_1}\overline{I_2}, \overline{I_1}\underline{I_2}, \overline{I_1 I_2})]$$

Then, given the real function $f(x, y) = x \times x + y$, we may define its natural interval extension by $f(x, y) = x \times x + y$, and we have that $f([2, 3], [-1, 5]) = [3, 14]$.

Implementations of interval arithmetic use outward rounding to enlarge the domains computed so as not to violate the containment principle, should some bounds be unrepresentable with floating-point numbers [8].

Many numerical methods have been extended to interval arithmetic [11, 13]. Given a system of nonlinear equations of the form:

$$f_1(x_1, \ldots, x_n) = 0$$
$$\ddots \tag{1}$$
$$f_n(x_1, \ldots, x_n) = 0$$

and initial domains $I_1, \ldots, I_n$ for the variables, these methods are usually embedded into a *branch-and-prune* algorithm BaP that manages a set of boxes of domains to tighten. Starting from the initial box $D = I_1 \times \cdots \times I_n$, BaP applies a numerical method "prune" to tighten the domains in D around the solutions of System (1), and bisects the resulting box along one of its dimensions whose width is larger than some specified threshold ε. The BaP algorithm eventually returns a set of boxes whose largest dimension has a width smaller than ε and whose union contains all the solutions to Eq. (1)—note that the boxes returned may contain zero, one, or more than one solution.

The *interval nonlinear Gauss-Seidel method* is a possible implementation for prune. It considers the n *unary projections*:

$$f_1^{(1)}(x_1, I_2, \ldots, I_n) \;= 0$$
$$\ddots \tag{2}$$
$$f_n^{(n)}(I_1, \ldots, I_{n-1}, x_n) = 0$$

and uses any unidimensional root-finding method to tighten the domain of each variable x_i in turn. Unidimensional Newton leads to the *Gauss-Seidel-Newton method* [12], whose extension to intervals is the *Herbort-Ratz method* [7].

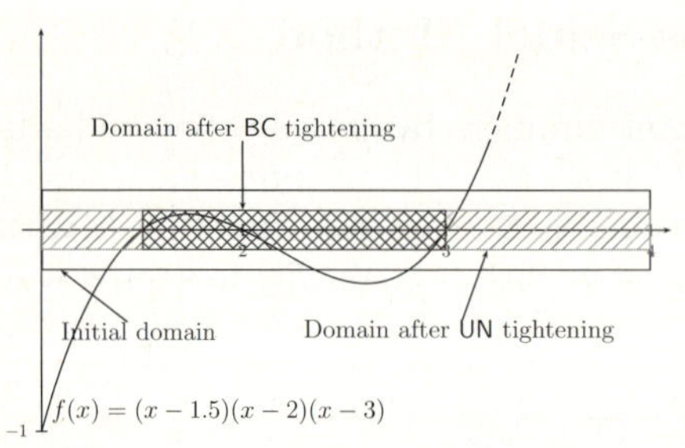

Fig. 1. Comparison of UN and BC

Let UN be the elementary step performed by one unidimensional Newton application to the projection $f_i^{(j)}$, where i and j may be different [12]. As soon as D is moderately large, it is very likely that each projection constraint will have many "solutions" that are not solutions of the original real system, and whose discarding slows down the computation. The Newton method will also fail to narrow down the domain of some x_i if there is more than one solution to the corresponding projection constraint for the current box D, thereby demanding more splittings in the BaP algorithm. Achieving the right balance between the amount of work required by the prune method and the number of splittings performed overall is the key to the maximum efficiency of BaP. In this very situation, experimental evidences show that trying harder to narrow down the domain of x_i pays off [2]. A way to do it is to ensure that the canonical intervals $[I_j, I_j^+]$ and $[\overline{I_j^-}, \overline{I_j}]$, whose bounds are two consecutive floating-point-numbers, are solutions of $f_i^{(j)}(I_1, \ldots, I_{j-1}, x_j, I_{j+1}, \ldots, I_n) = 0$. Let BC be an algorithm that ensures this property (called *box consistency [2] of x_j w.r.t. the constraint $f_i = 0$ and D*) for a projection $f_i^{(j)}$. A simple method to implement it uses a dichotomic process to isolate the leftmost and rightmost solutions included in D of each projection constraint.

Example 2. Consider the constraint $f(x) = (x - 1.5)(x - 2)(x - 3) = 0$ and the domain $I = [1, 4]$ for x (See Fig. 1). The UN method leaves I unchanged because the derivative of f over the initial domain contains 0 while BC narrows down I to $I' = [1.5, 3]$, which is the smallest interval included in I that contains all the solutions to the interval constraint $f(x) = 0$.

3   Static Selection of a Transversal

When System (1) is linear, it is well known that one should reorder f_is and x_js such that the coefficient matrix becomes strictly diagonal dominant [12]. Many authors have noticed that nonlinear Gauss-Seidel (be it on intervals or not) is equally sensitive to such a reordering. When System (1) is nonlinear, one may exchange rows and columns in its *incidence matrix*[1] so as to obtain a transversal

---

[1] The incidence matrix M associated to System (1) is the zero-one matrix where M_{ij} is 1 if and only if x_j occurs in f_i.

of n pairs (f_i, x_i) corresponding to the unary projections in System (2) that will hopefully maximize the convergence rate of INLGS.

Several authors have attempted to compute good transversals for nonlinear problems:

- Sotiropoulos *et al.* [14] select a transversal for a polynomial system at the beginning of the computation by looking at the syntactic structure of the equations (variables with the largest degree in the system, . . .), and by using numerical considerations only to break ties. In their paper, the static transversal is used in an interval Newton-Gauss-Seidel algorithm;
- Herbort and Ratz [7] compute the Jacobian J of System (1) w.r.t. the initial box D and select projections according to whether the corresponding entry in the Jacobian straddles zero or not. Their method is not completely static since they recompute the Jacobian after each iteration of INLGS (the choice of projections is not completely reconsidered, though). In addition, it theoretically allows for the choice of more than n projections;
- Goualard [4] determines an $n \times n$ matrix of weights W from the Jacobian J of Eq. (1) w.r.t. the initial box D corresponding to the distance of each interval J_{ij} to zero. He then computes a maximum weight perfect matching in the bipartite weighted graph associated to W, which gives a set of n projections on which to apply INLGS.

Table 1. Selecting a transversal vs. using all projections

Problems Name (n, #*sols*)	HH	HG	HS	HB	GSA
1 Barton (5,1)	NA	881	—	881	**378**
2 Bronstein (3,4)	8593	6712	**4430**	6712	10204
3 Broyden-banded (100,1)	**4500**	**4500**	NA	**4500**	30780
4 Broyden-tridiag. (10,2)	**12714**	1192697	13334	13385	31917
5 Combustion (10,1)	39324	8711299	NA	69546	**2581**
6 Extended Crag-Levy (8,36)	61707	7532	—	7532	**6612**
7 Extended Powell (10,32)	272977	268485	—	267591	**25130**
8 Grapsa-Vrahatis (3,2)	NA	246755	299085	17880	**15978**
9 MAT (3,1)	9639	**8590**	—	8657	18744
10 Moré-Cosnard (100,1)	**1200**	**1200**	NA	**1200**	100000
11 Robot (8,16)	NA	36179	NA	36179	**8510**
12 Troesch (10,1)	**260**	**260**	—	**260**	728
13 Yamamura (5,5)	**690**	4984061	**690**	**690**	3450

Number of calls to BC *to find all solutions up to a precision of* 10^{-8}

We tested the above heuristics on thirteen classical problems [1, 9, 14]. Eight problems are polynomial, and five are not. The heuristics of Herbort and Ratz (**HH**), Goualard (**HG**), and Sotiropoulos *et al.* (**HS**) served to compute a transversal of n projections used in an INLGS algorithm where the univariate root-finding method is BC. The initial box is the one published in the papers cited. Table 1 presents the overall number of calls to BC needed to compute a set of solution boxes with width less than 10^{-8}. The dash entries correspond to the cases where an heuristics is not applicable (non polynomial problems for **HS**). An "NA" entry signals that the problem could not be solved in less than 5 minutes on an AMD Athlon™ XP 2600+. Each problem is identified by its name followed by its dimension and its number of solutions between parentheses. Heuristics **HB** corresponds to the choice of a maximum weight perfect matching computed from a matrix W where W_{ij} is the *relative reduction*[2] performed by applying BC on $f_i^{(j)}$ for the initial box. This heuristics serves as a benchmark of what could be the best choice of a transversal, assuming the efficiency of a projection does not vary during the solving process. Column **GSA** gives the results when using all possible projections (that is, at most n^2). The boldfaced entries correspond to the smallest number of calls to BC per problem.

Analysis of Table 1. Whatever the heuristics chosen, selecting statically (or semi-statically for **HH**) a transversal of n projections is better than using the at most n^2 projections (**GSA**) for only 7 problems out of 13. We have monitored the amount of reduction obtained by applying BC on each of the possible projections during the whole solving process for every problem[3]. Upon close study, it is possible to separate the problems into two categories: those for which exactly n projections reduce the domains much better than the others, and those for which no such dichotomy can be made. The results for **GSA** follow closely that division: if there indeed exists a good transversal, the heuristics usually fare better than **GSA** because they are likely to choose it; if there is no good transversal, **GSA** is better because it avoids selecting a bad set of n projections, and makes the most out of the capability of all the projections to tighten the domains. This is in line with the conclusions by Goualard [4]. For those problems having a good transversal (2, 3, 4, 9, 10, 12, 13), the results of the heuristics vary widely, and some problems cannot be solved in the allotted time. At this point, the only reason that comes to mind is that the heuristics did choose a bad transversal and were stuck with it since the choice is not reconsidered dynamically. We will see in the next section that another reason may explain this situation. The **HB** heuristics also appears a good choice for problems with a good transversal since it never flounders on the seven presented here. However, the fact that it is not always the best method shows that the first reduction of a projection does not measure its overall efficiency, i.e., the efficiency of a projection varies during the solving process. In addition, it is slow on problems without a transversal. On the other hand, we cannot rely on **GSA** to solve problems with a good transversal since it is too computationally expensive to handle n^2 projections instead of just n of them, especially for problems with a dense incidence matrix (e.g., Moré-Cosnard).

---

[2] The relative reduction is defined by $(\mathrm{w}(I_j^b) - \mathrm{w}(I_j^a))/\mathrm{w}(I_j^b)$ where $\mathrm{w}(I_j^b)$ (resp. $\mathrm{w}(I_j^a)$) is the width of the domain of x_j before (resp. after) applying BC on $f_i^{(j)}$.

[3] All the log files containing detailed statistics for the problems presented are available at `http://interval.constraint.free.fr/problem-statistics.tar.gz`

The problems without a good transversal may have no projection that is consistently better than the others; alternatively, it is possible that there exists a set of n or more good projections whose composition varies with the domains of the variables during the solving process. The next section investigates this matter to find out whether it is possible to optimize the computation for all problems.

4   Beyond the Choice of a Static Transversal

In order to assess conclusively whether a dynamic recomputation of a transversal may lead to good performances for all problems, we have decided to consider the **HB** heuristics only. Obviously, the results obtained are then only a benchmark of the level of performances that is theoretically attainable, since the **HB** heuristics works as an oracle able to tell us in advance what are the best reductions possible at some point in the computation.

Table 2 presents the results of our tests:

- $\mathbf{HB}_1^n$ is the **HB** heuristics presented in the previous section;
- $\mathbf{HB}_\infty^n$ corresponds to the dynamic recomputation of a transversal in the same way as **HB** after each outer iteration of INLGS;
- $\mathbf{HB}_1^{n+k}$ statically selects the best projection per variable (n) and then the best projection for each equation not already covered in the first selection ($0 \leqslant k \leqslant n - 1$). It removes the *transversality constraint* that may yield suboptimal projections, avoiding also the use of costly matching algorithms;
- $\mathbf{HB}_\infty^{n+k}$ recomputes dynamically a set of $n + k$ projections according to $\mathbf{HB}_1^{n+k}$ after each outer iteration of INLGS.

Table 2. Dynamic and static selection of projections

Problems Name $(n, \#sols)$	$\mathbf{HB}_1^n$	$\mathbf{HB}_\infty^n$	$\mathbf{HB}_1^{n+k}$	$\mathbf{HB}_\infty^{n+k}$	**GSA**
1 Barton (5,1)	881	**140**	1470	299	378
2 Bronstein (3,4)	6712	**6011**	7961	6308	10204
3 Broyden-banded (100,1)	**4500**	**4500**	**4500**	6041	30780
4 Broyden-tridiag. (10,2)	**13385**	41019	**13385**	48941	31917
5 Combustion (10,1)	69546	**950**	47848	1309	2581
6 Extended Crag-Levy (8,36)	7532	7544	63947	**5252**	6612
7 Extended Powell (10,32)	267591	**12137**	415969	18412	25130
8 Grapsa-Vrahatis (3,2)	17880	**6575**	421945	9341	15978
9 MAT (3,1,5)	8657	8324	12554	**7589**	18744
10 Moré-Cosnard (100,1)	**1200**	**1200**	**1200**	1201	100000
11 Robot (8,16)	36179	5711	2219189	**3422**	8510
12 Troesch (10,1)	**260**	**260**	**260**	265	728
13 Yamamura (5,5)	**690**	710	**690**	931	3450

Number of calls to BC *to find all solutions up to a precision of* 10^{-8}

Analysis of Table 2. **GSA** is no longer the best method on any of the problems, leading to the conclusion that it is always worthwhile to consider subsets of good projections. The dynamic heuristics ($\mathbf{HB}_\infty^n$ and $\mathbf{HB}_\infty^{n+k}$) lead to the least number of calls to BC for the majority of the problems, which is not so surprising since they guarantee to know in advance for the current iteration the projections for which BC will tighten variables domains the most. The fact that $\mathbf{HB}_\infty^n$ is slightly better than $\mathbf{HB}_\infty^{n+k}$ is an evidence that most problems that do not have a static transversal may still have a dynamic one. Note, however, that the transversality constraint might lead to choose some of the less effective projections. This may explain why $\mathbf{HB}_\infty^{n+k}$ is better than $\mathbf{HB}_\infty^n$ on Problems 6, 9, and 11. While dynamic heuristics are often better, they sometimes lead to poorer performances (e.g., for Problems 4 and 13). A close look at the computation logs for these problems reveals that the intervals of time between two bisections in $\mathbf{HB}_\infty^n$ and $\mathbf{HB}_\infty^{n+k}$ are larger than in $\mathbf{HB}_1^n$ and $\mathbf{HB}_1^{n+k}$, which is quite natural since the better choice of the projections leads to more reductions, and then to more iterations before the quiescence of INLGS and the necessity to split. This is good when the reductions are significant; on the other hand, such a behavior is undesirable when the reductions performed are small: it would then be much better to stop the iterations and bisect the domains altogether.

5   Conclusions

We have seen in Section 3 that heuristics that select a transversal only once at the beginning are doomed to fail on many problems for which such a static transversal does not exist. However, we have found in Section 4 that it is usually possible to isolate dynamically a set of good projections whose composition evolves during the solving process; as said before, the method we have used to find the elements of this set is only a benchmark of the optimizations attainable since it requires to find in advance what are the projections with which BC will tighten the domains of the variables the most. What is more, we may expect that the cost of recomputing a good transversal dynamically by any heuristics, however cheap, may more than offset the benefit of not having to consider n^2 projections for all problems but the ones with the densest incidence matrix.

There is still hope however: we have seen in Table 2 that a static heuristics like $\mathbf{HB}_1^n$ works as well as the best dynamic heuristics for the problems that have a good static transversal. A direction for future researches is then to identify beforehand whether a problem has such a good static transversal, and revert to **GSA** if it does not. More generally, it would be interesting to determine whether a projection will be always good, always bad, or of varying interest in order to use it in an appropriate way. Along these lines, our preliminary experiments suggest that Artificial Intelligence-based methods such as *reinforcement learning* approaches [15] used to solve the *Nonstationary Multi-Armed Bandit Problem* are well-suited to tackle the dynamics behind the behavior of the projections during the solving process.

References

1. P. I. Barton. The equation oriented strategy for process flowsheeting. Dept. of Chemical Eng. MIT, Cambridge, MA, 2000.
2. F. Benhamou, D. McAllester, and P. Van Hentenryck. CLP(Intervals) revisited. In *Procs. Intl. Symp. on Logic Prog.*, pages 124–138, Ithaca, NY, November 1994. The MIT Press.
3. I. S. Duff. On algorithms for obtaining a maximum transversal. *ACM Trans. Math. Software*, 7(3):315–330, September 1981.
4. F. Goualard. On considering an interval constraint solving algorithm as a free-steering nonlinear gauss-seidel procedure. In *Procs. 20th Annual ACM Symp. on Applied Computing (Reliable Comp. and Applications track)*, volume 2, pages 1434–1438. Ass. for Comp. Machinery, Inc., March 2005.
5. A. J. Hughes Hallett and L. Piscitelli. Simple reordering techniques for expanding the convergence radius of first-order iterative techniques. *J. Econom. Dynam. Control*, 22:1319–1333, 1998.
6. E. R. Hansen and S. Sengupta. Bounding solutions of systems of equations using interval analysis. *BIT*, 21:203–211, 1981.
7. S. Herbort and D. Ratz. Improving the efficiency of a nonlinear-system-solver using a componentwise newton method. Research report 2/1997, Institut für Angewandte Mathematik, Universität Karslruhe (TH), 1997.
8. T. J. Hickey, Q. Ju, and M. H. Van Emden. Interval arithmetic: from principles to implementation. *J. ACM*, 48(5):1038–1068, September 2001.
9. INRIA project COPRIN: Contraintes, OPtimisation, Résolution par INtervalles. The COPRIN examples page. Web page at `http://www-sop.inria.fr/coprin/logiciels/ALIAS/Benches/benches.html`.
10. R. E. Moore. *Interval Analysis*. Prentice-Hall, Englewood Cliffs, N. J., 1966.
11. A. Neumaier. *Interval methods for systems of equations*, volume 37 of *Encyclopedia of Mathematics and its Applications*. Cambridge University Press, 1990.
12. J. M. Ortega and W. C. Rheinboldt. *Iterative solutions of nonlinear equations in several variables*. Academic Press Inc., 1970.
13. H. Ratschek and J. Rokne. Interval methods. In *Handbook of Global Optimization*, pages 751–828. Kluwer Academic, 1995.
14. D. G. Sotiropoulos, J. A. Nikas, and T. N. Grapsa. Improving the efficiency of a polynomial system solver via a reordering technique. In *Procs. 4th GRACM Congress on Computational Mechanics*, volume III, pages 970–976, 2002.
15. R. S. Sutton and A. G. Barto. *Reinforcement Learning: An Introduction*. The MIT Press, 1998.

Landscape Properties and Hybrid Evolutionary Algorithm for Optimum Multiuser Detection Problem

Shaowei Wang[1], Qiuping Zhu[1], and Lishan Kang[2]

[1] School of Electronic Information, Wuhan University, Wuhan, Hubei, 430079, P.R. China
[2] State Key Laboratory of Software Engineering, Wuhan University, Wuhan, Hubei, 430072, P.R. China.
{shwwang, qpzhu, kang}@whu.edu.cn

Abstract. Optimum multiuser detection (OMD) for CDMA systems is an NP-complete problem. Fitness landscape has been proven to be very useful for understanding the behavior of combinatorial optimization algorithms and can help in predicting their performance. This paper analyzes the statistic properties of the fitness landscape of the OMD problem by performing autocorrelation analysis, fitness distance correlation test and epistasis measure. The analysis results, including epistasis variance, correlation length and fitness distance correlation coefficient in different instances, explain why some random search algorithms are effective methods for the OMD problem and give hints how to design more efficient randomized search heuristic algorithms for it. Based on these results, a multi-start greedy algorithm is proposed for multiuser detection and simulation results show it can provide rather good performance for cases where other suboptimum algorithms perform poorly.

1 Introduction

Multiple access interference (MAI) is the main factor limiting performance in CDMA systems. While optimum multiuser detection (OMD) [1], which is based on the maximum likelihood sequence estimation rule, is the most promising technique for mitigating MAI, its computational complexity increases exponentially with the number of active users, which leads to its implementation impractical.

From a combinatorial optimization viewpoint, the OMD is an NP-complete problem [2]. Randomized search heuristics (RSH) are effective methods for such kinds of problems, so many RSH based multiuser detectors have been studied and exhibit better performance than that of the other linear or nonlinear detectors. Earlier works on applying RSH to OMD problem can be found in [3][4][5][6][7].The essence of optimum multiuser detection is to search for possible combinations of the users' entire transmitted bit sequence that maximizes the logarithm likelihood function (LLF) derived from the maximum likelihood sequence estimation rule [1], which is called fitness function or objective function in the RSH multiuser detectors. Comparing with so much emphasis on the implementation details and the performance analysis of these algorithms, little attention has been paid on the analysis of statistic characteristics of the OMD problem in terms of combinatorial optimization.

V.N. Alexandrov et al. (Eds.): ICCS 2006, Part I, LNCS 3991, pp. 340–347, 2006.

Fitness landscape has been proven to be a power concept in combinatorial optimization theory [8][9][10]. Moreover, the concept can be used to understand the behavior of heuristic algorithms for combinatorial optimization problems and to predict their performance. In this paper, we formulate the fitness landscapes of OMD problem by taking the LLF as the objective function and analyze their local and global characteristics as well as gene interaction properties [11][12]. With the analysis results, we propose a multi-start greedy (MSG) multiuser detector and compare its performance with other detectors.

The remainder of this paper is organized as follows. In Sect.2, we state the OMD problem and construct its fitness landscape. Statistic properties of the fitness landscape, including autocorrelation function, fitness distance correlation and epistasis variance, are described in Sect.3 with analysis results. In Sect.4, we propose the MSG algorithm and compare its performance with others. A short conclusion is given in Sect.5.

2   Optimum Multiuser Detection Problem and Fitness Landscape

2.1   Optimum Multiuser Detection Problem

Assume a binary phase shift keying (BPSK) transmission through an additive-white-Gaussian-noise (AWGN) channel shared by K active users with packet size M in an asynchronous DS-CDMA system (synchronous system is the special case of the asynchronous one). The sufficient statistics for demodulation of the transmitted bits $\mathbf{b}$ are given by MK matched filter outputs $\mathbf{y}$ [13]

$$\mathbf{y} = \mathbf{RAb} + \mathbf{n} \tag{1}$$

here $\mathbf{A}$ is the $MK \times MK$ diagonal matrix whose $k + iK$ diagonal element is the kth user's signal amplitude A_k and $i = 1, 2, ..., M$. $\mathbf{R} \in \mathbb{R}^{MK \times MK}$ is the signature correlation matrix and can be written as

$$\mathbf{R} = \begin{bmatrix} \mathbf{R}[0] & \mathbf{R}^{T}[1] & \mathbf{0} & ... & \mathbf{0} & \mathbf{0} \\ \mathbf{R}[1] & \mathbf{R}[0] & \mathbf{R}^{T}[1] & ... & \mathbf{0} & \mathbf{0} \\ \mathbf{0} & \mathbf{R}[1] & \mathbf{R}[0] & ... & \mathbf{0} & \mathbf{0} \\ \vdots & \vdots & \vdots & \vdots & \vdots & \vdots \\ \mathbf{0} & \mathbf{0} & \mathbf{0} & ... & \mathbf{R}[1] & \mathbf{R}[0] \end{bmatrix} \tag{2}$$

where $\mathbf{R}[0]$ and $\mathbf{R}[1]$ are $K \times K$ matrices defined by

$$R_{jk}[0] = \begin{cases} 1, & \text{if } j = k; \\ \rho_{jk,} & \text{if } j < k; \\ \rho_{kj,} & \text{if } j > k; \end{cases} \tag{3}$$

$$R_{jk}[1] = \begin{cases} 0, & \text{if } j \geq k; \\ \rho_{jk,} & \text{if } j < k. \end{cases} \qquad (4)$$

ρ_{jk} denotes the partial crosscorrelation coefficient between the jth user and the kth user.

The optimum multiuser detection problem is to generate an estimation sequence $\hat{\mathbf{b}} = [\hat{b}_1(1), \hat{b}_2(1), ..., \hat{b}_K(1), ..., \hat{b}_1(M), \hat{b}_2(M), ..., \hat{b}_K(M)]^T$ to maximum the objective function

$$f(\mathbf{b}) = 2\mathbf{y}^T \mathbf{A}\mathbf{b} - \mathbf{b}^T \mathbf{A}\mathbf{R}\mathbf{A}\mathbf{b} \qquad (5)$$

which means to search 2^{MK} possible bit sequence exhaustively. It is a combinatorial optimization problem and is proven to be NP-complete [2].

2.2   Fitness Landscape of OMD Problem

The notion of fitness landscape comes from biology, where it is used as a framework for thinking about evolution. Landscape theory has emerged as an attempt to devise suitable mathematical structures for describing the static properties of landscape as well as their influence on the dynamics of adaptation [10].

In formal terms, a fitness landscape consists of three ingredients: A set $\mathbf{S}$ of possible solutions, the notion of element distance d in $\mathbf{S}$ and a fitness function $f : \mathbf{S} \to \mathbb{R}$. So the fitness landscape of OMD problem can be defined as follows.

1. The solution space $\mathbf{S} : \{-1,1\}^{MK}$, where K is the number of active users and M is the packet size.

2. If $\mathbf{b}_i$ and $\mathbf{b}_j$ are two elements in $\mathbf{S}$, $i, j \in \mathbb{N}$, the distance between them is the Hamming distance of the two binary vetors: $d(\mathbf{b}_i, \mathbf{b}_j) = \sum_{n=1}^{MK} b_i^n \oplus b_j^n$.

3. The fitness function is the LLF defined in equation (5).

3   Statistic Properties Analysis of Fitness Landscape

3.1   Random Walk Correlation, Fitness Distance Correlation and Epistasis Correlation

3.1.1   The Random Walk Correlation Function

To measure the ruggedness of a fitness landscape, Weinberger [8] suggests the use of random walk correlation function

$$r(s) = \frac{\langle f(x_t) f(x_{t+s}) \rangle - \langle f \rangle^2}{\langle f^2 \rangle - \langle f \rangle^2} \tag{6}$$

where $\langle \cdot \rangle$ is expectation and f is fitness. x_t is a time series and $f(x_t)$ defines the correlation of two points s steps away along random walk through the fitness landscape, $-1 \le r(s) \le 1$. Then the normalized correlation length l is defined as

$$l = -N \times \ln(|r(1)|) \tag{7}$$

for $r(1) \ne 0$, N is the dimension of $\mathbf{S}$.

3.1.2 The Fitness Distance Correlation

Fitness distance correlation (FDC) coefficient is another measure for problem difficulty for heuristic algorithms such as evolutionary algorithms [9]. Denote the shortest distance between a possible solution and the global optimum solution as d, the FDC coefficient ρ is defined as

$$\rho = \frac{\langle fd \rangle - \langle f \rangle \langle d \rangle}{\sqrt{(\langle f^2 \rangle - \langle f \rangle^2)(\langle d^2 \rangle - \langle d \rangle^2)}} \tag{8}$$

where $-1 \le \rho \le 1$.

3.1.3 The Epistasis Correlation

Denote $\hat{f}$ as the estimation fitness value based on gene decomposition [11], epistasis variance $EpiN$ and epistasis correlation $EpiC$ are defined as [12]

$$EpiN = \frac{\sqrt{\sum_{x \in S}(f(x) - \hat{f}(x))}}{\sqrt{\sum_{x \in S} f^2(x)}} \tag{9}$$

$$EpiC = \frac{\sum_{x \in S}(f(x) - \bar{f})(\hat{f}(x) - \bar{\hat{f}})}{\sqrt{\sum_{x \in S}(f(x) - \bar{f})^2}\sqrt{\sum_{x \in S}(\hat{f}(x) - \bar{\hat{f}})^2}} \tag{10}$$

where $\bar{f} = \langle f \rangle$, $\bar{\hat{f}} = \langle \hat{f} \rangle$. $0 \le EpiN \le 1$, $0 \le EpiC \le 1$. Generally, the lower the value of

$EpiN$, the higher the value of $EpiC$, and the weaker the epistasis of the fitness function.

3.2 Numerical Results and Discussions

Experiments have been conducted to estimate the coefficients discussed above. E_b is the bit energy, N_0 is the two-sided power spectrum density of Gaussian noise, and K is the number of active users. All results are obtained by performing 10^6 Monte Carlo runs. Table 1 shows the numerical result of 9 different cases.

Table 1. Numerical results of different cases

Case	K	E_b / N_0	l	ρ	EpiN	EpiC
1	10	3	2.313	-0.911	0.109	0.982
2	10	6	2.317	-0.910	0.110	0.980
3	10	9	2.314	-0.930	0.111	0.979
4	20	3	2.351	-0.870	0.123	0.962
5	20	6	2.225	-0.869	0.124	0.962
6	20	9	2.327	-0.869	0.124	0.958
7	30	3	2.249	-0.846	0.128	0.951
8	30	6	2.310	-0.849	0.128	0.945
9	30	9	2.355	-0.834	0.128	0.944

Though the fitness landscape of OMD problem is dynamic because of the changes of E_b / N_0 and K, from Table 1, we can see the statistics of the fitness landscape (l, ρ, EpiN and EpiC) vary within a narrow range. The FDC coefficient ρ keeps negative value close to -1, which means the LLF is appropriate as the cost function for heuristics based multiuser detector. The normalized correlation length l is almost the same in all cases, which gives hints how to select the neighborhood size in heuristics. For example, $k = 2$ is appropriate for the $k - opt$ multiuser detector [6]and may give the best tradeoff between performance and computational complexity, because l is close to 2.

Table 1 also shows that EpiN and EpiC change very little when E_b / N_0 and K vary within a wide range and the value of EpiN is relatively small, which means the epistasis of the LLF is very weaker and the fitness value of a possible solution is almost determined independently by each bit of the solution vector. Therefore greedy heuristic method may be very efficient local search algorithm for OMD problem [3][5].

4 Multi-Start Greedy Algorithm for Multiuser Detection

4.1 Multi-Start Greedy Algorithm

The MSG algorithm consists of three phases. Phase 1, generate n initial solutions distributed in the whole solution space randomly. This phase is consider as multi-start process which imitates the population of evolutionary algorithms (EAs). Phase 2, perform greedy local search algorithm on each solution and produce n local optima.

This phase is a local search process. Phase 3, choose the best one (based on their fitness) from these local optima as the global optimum solution.

The greedy local search algorithm in phase 2 can be described as follows. Without loss the generality, for a current solution vector $\mathbf{b}^{t-1} = [b_1^{t-1}, b_2^{t-1}, ..., b_N^{t-1}]^T$, where N is the length of the solution vector, denote the vector with only the jth bit different from $\mathbf{b}^{t-1}$ by $\mathbf{b}_{difj}^{t-1} = [b_1^{t-1}, ..., -b_j^{t-1}, ..., b_N^{t-1}]^T$. Calculate the gain g_j^{t-1}

$$g_j^{t-1} = f(\mathbf{b}_{difj}^{t-1}) - f(\mathbf{b}^{t-1}) \tag{11}$$

for $j = 1, 2, ..., N$, here f is defined in equation (5). Then each bit of the next solution vector $\mathbf{b}^t$ can be updated as

$$b_k^t = \begin{cases} -b_k^{t-1} & \text{if } k = \arg\{\max_j g_j^{t-1}\} \text{ and } g_j^{t-1} > 0 \\ b_k^{t-1} & \text{otherwise} \end{cases} \tag{12}$$

This process is repeated until a local optimal is obtained (there are no better solutions that are in its neighborhood).

4.2 Simulation Results

The BER performance of the conventional detector (CD), evolutionary programming detector (EP) [4], gradient guided detector (GGD) [5], parallel interference cancellation (PIC) detector [14] and the proposed MSG is illustrated in Fig.1 by the curves of BER versus E_b/N_0. The number of users is 10 and 20 ($K = 10, 20$) in Fig.1(a) and (b) respectively and the packet size is 3 ($M = 3$). It is obvious that MSG detector outperforms other detectors.

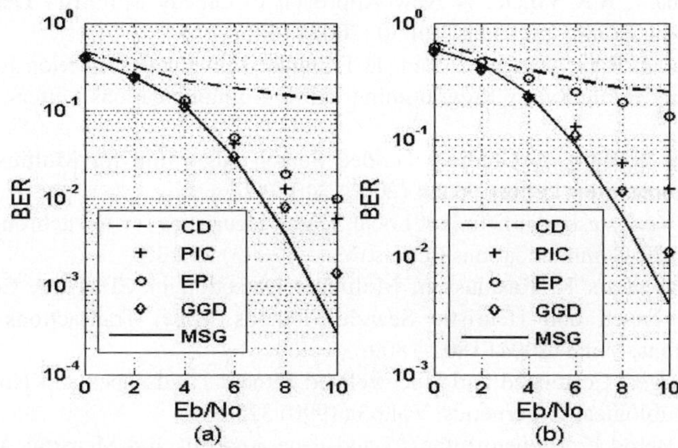

Fig. 1. BER against E_b/N_0 performance of MSG, GGD, EP, PIC and CD for $M = 3$ (a) $K = 10$ and (b) $K = 20$

The computational complexity of MSG can be determined easily. The initialization is to generate n random vector and requires computations of $O(n)$. Each local optimal requires $n \times (n-1)/2$ fitness value computations. So the computational complexity of a local optimal is $O(n^2)$ and the computational complexity of MSG is $O(n^3)$. A fast greedy heuristic is proposed in [15]. With this method, a local optimal can be obtained in $O(n)$. Then the computational complexity of MSG can be reduced to $O(n^2)$.

5 Conclusions

This paper studies the important statistic properties of the fitness landscape of the OMD problem and proposes a multi-start greedy algorithm based multiuser detector. The analysis results about the fitness landscape explain why some heuristics multiuser detector are effective and give hints how to design more efficient randomized search heuristics for the OMD problem. The multi-start greedy algorithm which is founded on these analysis results can provide good performance for cases where other suboptimum algorithms perform poorly.

Acknowledgement. This work is supported by the National Natural Science Foundation of China (No. 60473081).

References

1. S. Verdu: Minimum Probability of Error for Asynchronous Gaussian Multiple-access Channels. IEEE Transactions on Information Theory,Vol.32 (1986) 85-96.
2. S. Verdu: Computational Complexity of Optimal Multiuser Detection. Algorithmica, vol.4 (1989) 303–312.
3. A. AlRustamani, B.R Vojcic: A New Approach to Greedy Multiuser Detection. IEEE Transactions on Communications,Vol.50 (2002) 1326-1336.
4. H.S.Lim, M.V.C.Rao, Alan W.C. Tan, H.T.Chuah: Mulatiuser Detection for DS-CDMA Systems Using Evolutionary Programming. IEEE Communications Letters, Vol.7 (2003) 101-103.
5. Jun Hu, Rick S.Blum: A Gradient Guided Search Algorithm for Multiuser Detection. IEEE Communications Letters, Vol.4 (2000) 340-342.
6. H.S.Lim, B. Venkatesh: An Efficient Local Search Heuristics for Asynchronous Multiuser Detection. IEEE Communications Letters, Vol.7 (2003) 299-30.
7. Peng Hui Tan, Lars K. Rasmussen: Multiuser Detection in CDMA-A Comparison of Relaxations, Exact, and Heuristic Search Methods. IEEE Transactions on Wireless Communicatons, Vol.3 (2004) 1802-1809.
8. E.D. Weinberger: Correlated and Uncorrelated Fitness Landscapes and How to Tell the Difference. Biological Cybernetics, Vol.63 (1990) 325-336.
9. Peter Merz, Bernd Freisleben: Fitness Landscape Analysis and Memetic Algorithms for the Quadratic Assignment Problem. IEEE Transactions on Evolutionary Computation, Vol.4 (2000) 337-352.

10. P.F. Stadler: Fitness Landscape. Lecture Notes in Physics, Springer-Verlag Berlin Heidelberg, (2002) 183-204.
11. C. Reeves, C. Wright: An experimental design perspective on genetic algorithms. Foundations of Genetic Algorithms 3, L.D. Whitley and M. D. Vose, Eds. San Mateo, CA: Morgan Kaufmann, (1995) 7–22.
12. Bart Naudts, Leila Kallel: A Comparison of Predictive Measure of Problem Difficulty in Evolutionary Algorithms. IEEE Transactions on Evolutionary Computation, Vol.4 (2000) 1-15.
13. S. Verdu: Multiuser Detection. Cambridge University Press, Cambridge, U.K. (1998).
14. Dariush Divsalar, Marvin K. Simon and Dan Raphaeli: Improved Parallel Interference Cancellation for CDMA. IEEE Transactions on Communications, Vol.46 (1998) 258-268.
15. P. Merz and B. Freisleben: Greedy and local search heuristics for unconstrained binary quadratic programming. Journal of Heuristics, vol. 8 (2002)197–213.

A Parallel Solution of Hermitian Toeplitz Linear Systems[*][**]

Pedro Alonso[1], Miguel O. Bernabeu[1], and Antonio M. Vidal[1]

Universidad Politécnica de Valencia, cno. Vera s/n, 46022 Valencia, Spain
{palonso, mbernabeu, avidal}@dsic.upv.es

Abstract. A parallel algorithm for solving complex hermitian Toeplitz linear systems is presented. The parallel algorithm exploits the special structure of Toeplitz matrices to obtain the solution in a quadratic asymptotical cost. Our parallel algorithm transfors the Toeplitz matrix into a Cauchy–like matrix. Working on a Cauchy–like system lets to work with real arithmetic. The parallel algorithm for the solution of a Cauchy–like matrix has a low amount of communication cost regarding other parallel algorithms that work directly on the Toeplitz system. We use a message–passing programming model. The experimental tests are obtained in a cluster of personal computers.

1   Introduction

In this work we propose a parallel algorithm for the solution of

$$Tx = b \ , \tag{1}$$

with $T = (t_{ij}) = (t_{|i-j|})_{0 \le i,j < n} \in \mathbb{C}^{n \times n}$ hermitian and being $b, x \in \mathbb{C}^n$ the independent and the solution vectors, respectively.

There exist the so called *fast* algorithms to obtain the solution to this problem with an order of magnitude lower than the classical algorithms. These algorithms are based on the *displacement rank* property of the *structured* matrices. Nevertheless, there are two main drawbacks. Firstly, if the Toeplitz matrix is not strongly regular, *fast* algorithms can break down or can produce poor results regarding the accuracy of the solution. The other one deals with its parallelization because due to the dependency between operations causes a large number of point–to–point and broadcast–type communications when a message passing model programming is used [1].

The parallel algorithm presented in this paper transforms the Toeplitz matrix into a another type of structured matrix called Cauchy–like. This reduces significantly the execution time with the number of processors thanks to the use of just only one broadcast–type communication per iteration. In addition,

---

[*] Supported by Spanish MCYT and FEDER under Grant TIC 2003-08238-C02-02.
[**] Funded by Secretaría de Estado de Universidades e Investigación del Ministerio de Educación y Cultura of Spain.

V.N. Alexandrov et al. (Eds.): ICCS 2006, Part I, LNCS 3991, pp. 348–355, 2006.

working on Cauchy–like matrices avoids the algorithm to break down although it can still produce inaccuracy results. However, a refinement technique can be incorporated to improve the precision of the solution like it is done in [2] for the real non–symmetric case. Furthermore, we have used a blocking factor that minimizes the execution time overlapping computational and communication operations.

The parallel algorithm constitutes a particular improvement for hermitian matrices. The more general non–hermitian case can be found in [3]. Other related works based on the same idea applied to the real symmetric case was presented in [4].

Standard libraries like LAPACK [5] and ScaLAPACK [6] are used to achieve a more easy and portable implementation. The `fftpack` library [7, 8] is also used together with our own implementation by using the *Chirp-z* factorization [9].

The next two sections includes a brief revision of the mathematical background and the sequential algorithm. Afterward, Sect. 4 and Sect. 5 shows the parallel algorithm and the experimental results, respectively.

2   Rank Displacement and Cauchy–Like Matrices

The concept of *rank displacement* [10] describes the special structure of *structured* matrices. The definition uses the *displacement equation* of a given $n \times n$ matrix. If the rank r of the *displacement equation* is considerably lower than n ($r \ll n$), it is said that the matrix is *structured*.

Given the Toeplitz matrix of (1) its displacement equation can be defined in several ways. A useful form for our purposes is

$$F\,T - T\,F = G\,H\,G^* \,, \tag{2}$$

where $F = Z + Z^T$, being Z the one position down shift matrix, and being $G \in \mathbb{C}^{n \times 4}$ and $H \in \mathbb{C}^{4 \times 4}$ the so called *generator pair*.

It is said that matrix $C = (c_{ij})_{1 \leq i,j \leq n}$ is a hermitian Cauchy matrix, if for a complex vector $\lambda = (\lambda_i)_1^n$, the matrix

$$\nabla_\lambda C = ((\lambda_i - \lambda_j)c_{ij})_1^n \ , \ (\lambda_i - \lambda_j)c_{ij} = 1 \ , \tag{3}$$

has a very low rank with respect to n. If $(\lambda_i - \lambda_j)c_{ij} \neq 1$, matrix $\nabla_\lambda C$ is said to be a Cauchy–like matrix.

Both Toeplitz and Cauchy–like matrices as defined in (1) and (3), respectively, are structured matrices. Furthermore, there exists a direct relation between both classes of matrices by means of linear transformations. However, it is possible to avoid working in the complex plane in the hermitian case as follows.

Let S be the normalized Discrete Sine Transform (DST-I) [11, 12] matrix

$$S = \sqrt{\frac{2}{n+1}} \left(\sin \frac{ij\pi}{n+1} \right) \, , i,j = 1, \ldots, n \, , \tag{4}$$

where $S = S^T$ and $SS^T = S^T S = I$, being I the identity matrix, and let P be a *odd–even* permutation matrix so $P \left(x_1\ x_2\ x_3\ x_4\ \ldots \right) = \left(x_1\ x_3\ \ldots\ x_2\ x_4\ \ldots \right)$. Let $\Re(T)$ and $\Im(T)$ be the real and imaginary part of T, then

$$PS\Re(T)SP^T = \begin{pmatrix} M_1 & 0 \\ 0 & M_2 \end{pmatrix}, \quad \text{and} \quad PS\Im(T)SP^T = \begin{pmatrix} 0 & -M_3^T \\ M_3 & 0 \end{pmatrix},$$

where M_1, M_2 and M_3 are all real symmetric Cauchy–like matrices of size $\lceil n/2 \rceil \times \lceil n/2 \rceil$, $\lfloor n/2 \rfloor \times \lfloor n/2 \rfloor$ and $\lceil n/2 \rceil \times \lfloor n/2 \rfloor$, respectively.

Being matrix $\mathcal{D}$ defined as

$$\mathcal{D} = \begin{pmatrix} I_{\lceil n/2 \rceil} & \\ & iI_{\lfloor n/2 \rfloor} \end{pmatrix},$$

where $i = \sqrt{-1}$, the following transformation allows to convert a hermitian Toeplitz matrix into a real symmetric Cauchy–like matrix

$$C = \mathcal{D}PSTSP^T\mathcal{D}^* = \begin{pmatrix} M_1 & -M_3^T \\ M_3 & M_2 \end{pmatrix}. \tag{5}$$

Applying the previous transformation to the displacement equation (2) the displacement equation of the real Cauchy–like matrix in (5) is obtained

$$\Lambda C - C \Lambda = \hat{G} H \hat{G}^T, \tag{6}$$

where

$$\hat{G} = \left(\Re(\mathcal{G})\ \Im(\mathcal{G}) \right) \text{ and } \mathcal{G} = \frac{1}{\sqrt{n+1}}\mathcal{D}PSG_{:,1:2}. \tag{7}$$

Thus, the solution of a hermitian Toeplitz linear system (1) is approached by solving the real symmetric Cauchy–like system

$$C\hat{x} = \hat{b}, \tag{8}$$

where $\hat{b} = \mathcal{D}PSb$ and $\hat{x} = \mathcal{D}PSx$ so the solution of (1) is $x = \mathcal{D}^*P^T S\hat{x}$.

The Cauchy–like linear system (8) is solved performing the factorization $C = LDL^T$ with L lower triangular and D diagonal. This factorization can be done in a "fast" way due to the displacement rank property of structured matrices.

3   Triangular Factorization of Cauchy–Like Matrices

Gohberg, Kailath and Olshevsky [13] proposed an algorithm (GKO) to factorize non–hermitian Cauchy–like matrices. Following the same idea is not hard to derive a fast algorithm for the triangular factorization of real symmetric Cauchy–like matrices that exploits its displacement rank property.

Given the following partitions of matrices C and Λ (6),

$$C = \begin{pmatrix} d & l^T \\ l & C_1 \end{pmatrix}, \Lambda = \begin{pmatrix} \lambda_1 & 0 \\ 0 & \Lambda_1 \end{pmatrix} ,$$

where $\left(d\ l^T \right)^T$ is a one dimensional array, and being $X = \begin{pmatrix} 1 & 0 \\ l/d & I_{n-1} \end{pmatrix}$, if the transformation $X^{-1}(.)X^{-T}$ is applied to (6), we have

$$(X^{-1}\Lambda X)(X^{-1}CX^{-T}) - (X^{-1}CX^{-T})(X^T\Lambda X^{-T}) =$$

$$\begin{pmatrix} \lambda_1 & 0 \\ \frac{\Lambda_1 l - \lambda_1 l}{d} & \Lambda_1 \end{pmatrix} \begin{pmatrix} d & 0 \\ 0 & C_{sc} \end{pmatrix} - \begin{pmatrix} d & 0 \\ 0 & C_{sc} \end{pmatrix} \begin{pmatrix} \lambda_1 & \frac{l^T \Lambda_1 - \lambda_1 l^T}{d} \\ 0 & \Lambda_1 \end{pmatrix} = (X^{-1}\hat{G})H(X^{-1}\hat{G})^T .$$

Equating element $(2,2)$ of the previous expression it is obtained

$$\Lambda_1\ C_{sc} - C_{sc}\ \Lambda_1 = \hat{G}_1\ H\ \hat{G}_1^T ,$$

that is, the displacement equation of the Schur complement C_{sc} of C with respect to its first element d. Matrix $\hat{G}_1 = X^{-1}\hat{G}$ is the *generator* of C_{sc}. This property about that the Schur complements of a structured matrix are also structured is used to derive a triangular factorization fast algorithm.

The computation of the first column of C and the computation of $\hat{G}_1$ defines the first step of the algorithm. The LDL^T factorization of C is obtained repeating this process n iterations on the successive arising Schur complements (Alg. 1).

Algorithm 1 (LDL^T factorization of a real symmetric Cauchy–like matrix): *Given $\hat{G}$, H, Λ (6) and diagonal entries of C, this algorithm returns the unit lower triangular factor L and the diagonal matrix D such that $C = LDL^T$.*

> **for** $j = 1, \ldots, n$
> $\quad D_{j,j} = C_{j,j}$
> $\quad$ **for** $i = j+1, \ldots, n$
> $\qquad$ 1. $L_{i,j} = (\hat{G}_{i,:}\ H\ \hat{G}_{j,:}^T)/(D_{j,j}\ (\lambda_i - \lambda_j))$
> $\qquad$ 2. $C_{i,i} \leftarrow C_{i,i} - D_{j,j}L_{i,j}^2$
> $\qquad$ 3. $\hat{G}_{i,:} \leftarrow \hat{G}_{i,:} - L_{i,j}\hat{G}_{j,:}$
> $\quad$ **end for**
> **end for**

All elements of each Schur complement of C can be computed by solving the Lyapunov equation shown in step 1 of Alg. 1 except diagonal entries. That is why these elements must be computed a priori before calling Alg. 1. There exist several fast algorithms to do that like the one in [14]. However, we have developed a new one that lets to obtain diagonal entries of the Cauchy–like matrix arising from $C = ST_LS$ where T_L is lower triangular Toeplitz. This is useful to obtain diagonal entries of different symmetric Cauchy–like matrices of the form $C = SMS$ where M can be either symmetric or non–symmetric or even the product of two Toeplitz matrices. Furthermore, the algorithm uses a minimal amount of memory avoiding to use of some vectors arising in the description of the algorithm by other authors.

P_0	$\hat{G}_0$	$L_{0,0}$
P_1	$\hat{G}_1$	$L_{1,0}\ L_{1,1}$
P_2	$\hat{G}_2$	$L_{2,0}\ L_{2,1}\ L_{2,2}$
P_0	$\hat{G}_3$	$L_{3,0}\ L_{3,1}\ L_{3,2}\ L_{3,3}$
P_1	$\hat{G}_4$	$L_{4,0}\ L_{4,1}\ L_{4,2}\ L_{4,3}\ L_{4,4}$
$\vdots$	$\vdots$	$\vdots\quad\vdots\quad\vdots\quad\vdots\quad\vdots\quad\ddots$

Fig. 1. Example of data distribution with 3 processors

4   The Parallel Algorithm

The central part of the parallel algorithm falls in the factorization of C. The generator $\hat{G}$ is partitioned in n/η blocks of size $\eta \times 4$ and distributed using the BLACS model. Under this model a *logical* unidimensional mesh with $p \times 1$ processors is built. Blocks of $\hat{G}$ are cyclically distributed among the p processors such that block $\hat{G}_k$, $k = 0,\ldots,n/\eta - 1$, belongs to processor $P_{j \bmod p}$. The ith row of $\hat{G}$ belongs to block $\hat{G}_{i/\eta}$. Matrix Λ and the diagonal of C are both partitioned and distributed in the same way. Alg. 2 is the parallel version of Alg. 1. An example of distribution of $\hat{G}$ and L with 3 processors can be seen in Fig. 1.

Algorithm 2 (Parallel LDL^T factorization of a symmetric Cauchy–like matrix): *Given $\hat{G}$, H, Λ (6) and the diagonal of C cyclically distributed for a given block size $\eta \geq 1$, this algorithm returns the unit lower triangular factor L and the diagonal matrix D such that $C = LDL^T$ both partitioned in blocks of η rows cyclically distributed as the argument matrices.*

Each processor P_j, $j = 0,\ldots,p-1$, performs:

    for $k = 0,\ldots,n/\eta - 1$

      Let $C_{sc}^k = \begin{pmatrix} C_1 & C_3^T \\ C_3 & C_2 \end{pmatrix}$, $C_1 \in \mathrm{IR}^{\eta \times \eta}$, be the Schur complement of C

      with respect to the leading submatrix of order $k\eta$.

      if *$(\hat{G}_k \in P_j)$*

         Compute $C_1 = L_{kk}D_kL_{kk}^T$ by using Alg. 1 and broadcast $\hat{G}_k$ and Λ_k.

      else *Receive $\hat{G}_k$ and Λ_k.*

      end if

      for $i = k+1,\ldots,n/\eta - 1$

        if *$(\hat{G}_i \in P_j)$*

           Compute L_{ik} and update $\hat{G}_i$ and the diagonal entries of C_{ii}.

        end if

      end for

    end for

Updating blocks $\hat{G}_i$ and diagonal entries of C_{ii}, $k < i < n/\eta$, can be easily derive from the operations described in Alg. 1 if only η iterations are applied. In each step of Alg. 2 the following factorization is obtained

$$C_{sc}^k = \begin{pmatrix} L_{k,k} \\ L_{k+1:n/\eta-1,k} \end{pmatrix} \begin{pmatrix} D_{k,k} & 0 \\ 0 & 0 \end{pmatrix} \begin{pmatrix} L_{k,k}^T & L_{k+1:n/\eta-1,k}^T \end{pmatrix} + \begin{pmatrix} 0 & 0 \\ 0 & C_{sc}^{k+1} \end{pmatrix},$$

being C_{sc}^{k+1} implicitly known throughout the generator.

The main advantage of Alg. 2 is that it performs only one broadcast per iteration on the contrary that other parallel algorithms for the solution of the same problem. The size and the number of messages are both a function of η. Different values of η changes the overlapping between communications and computations and also the weight of both factors in the total cost of the algorithm. The optimum value of η depends on the machine and is experimentally tuned.

The complete parallel algorithm is summarized in Alg. 3.

Algorithm 3 (Parallel solution of a hermitian Toeplitz linear system):
Given a hermitian Toeplitz matrix $T \in \mathbb{C}^{n \times n}$ and a right hand side vector $b \in \mathbb{C}^n$, this algorithm returns the solution vector x of the linear system $Tx = b$.
On each processor P_j, for $j = 0, \ldots, p-1$:

1. *Every processor computes matrices $\hat{G}$ (7), Λ (6) and vector $\hat{b}$ (8), and distribute them cyclically by blocks of size $\eta \times 4$.*
2. *Apply Alg. 2 to obtain $C = LDL^T$.*
3. *Solve $LDL^T \hat{x} = \hat{b}$ in parallel by using PBLAS routines.*
   P_0 gathers $\hat{x}$ from the rest of processors and computes $x = \mathcal{D}^* P^T S \hat{x}$.

Another improvement used is the efficient computation of the DST (4). The time used by the `fftpack` routine to apply a DST to a vector highly depends on the value of the prime numbers in which $n + 1$ is factorised. This routine is faster as these prime numbers are small. We have implement a DST routine based on the *Chirp–z* factorization used in the computation of DFT's whose computational cost is independent of the size of these prime numbers (Table 1).

5   Experimental Results

The experimental results have been obtained in a cluster of 10 two–processor boards with two Intel Xeon at 2 GHz. and with 1 Gb. of RAM. The

Table 1. Excution time of $\hat{G}$ with and without using the *Chirp–z* factorization

n	6998	7498	7998	8498	8998	9998	10498	10998
p. d.	$3 \cdot 2333$	7499	$19 \cdot 421$	$3 \cdot 2833$	8999	$3^2 \cdot 11 \cdot 101$	10499	$17 \cdot 647$
`fftpack`	0.64	2.93	0.33	0.96	4.25	0.45	5.76	0.64
Chirp–z	0.36	0.38	0.41	0.60	0.64	0.72	0.77	0.81

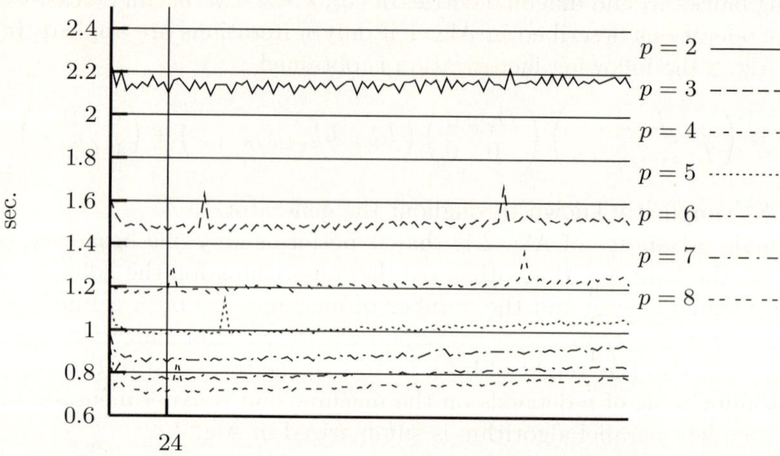

Fig. 2. Time versus different values of η and number of processors for $n = 12000$

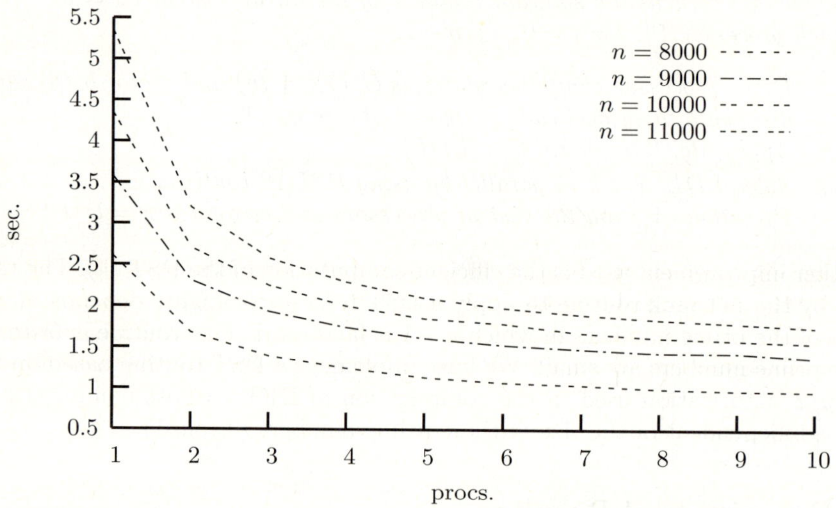

Fig. 3. Time for different size problems and different number of processors.

interconnection network is a SCI with a 2D torus topology. In the experiments, each MPI process is mapped onto only one processor of each node.

The first test consists of tuning the block size η used in the partition of $\hat{G}$. Fig. 2 shows the execution time of Alg. 2, that is the second step of Alg. 3, so it can be concluded that there exists a range of values for η that can be used to get the best performance. We have chosen a size of $\eta = 24$ independently of the number of processors in our target machine. A more detailed study with other problem sizes shows that this is always the best choice despite of some a very slight variation of time in the different values of η.

Once the optimal value of η is fixed, the next experiment deal with the total cost of the parallel algorithm with different problem sizes. Figure 3 shows a large reduction in time achieved with the increment in the number of processors. This is specially significant because this result means the parallel algorithm can be useful in applications with real time constraints.

6   Conclusions

We have presented a parallel algorithm that exploits the displacement structure of Toeplitz and Cauchy–like matrices as the sequential algorithms do. The algorithm does not break down if the Toeplitz matrix is not strongly regular, uses real arithmetic and reduces the execution time with the use of up to 10 processors. This is a challenge taking into account how difficult it is to improve the performance of this type of fast algorithms due to the operation dependency and the large cost of a message communication regarding the time of a flop.

References

1. Alonso, P., Badía, J.M., Vidal, A.M.: Parallel algorithms for the solution of toeplitz systems of linear equations. LNCS **3019** (2004) 969–976
2. Alonso, P., Badía, J.M., Vidal, A.M.: An efficient and stable parallel solution for non–symmetric Toeplitz linear systems. LNCS **3402** (2005) 685–692
3. Alonso, P., Vidal, A.M.: An efficient parallel solution of complex toeplitz linear systems. Lecture Notes in Computer Science (to appear in 2006)
4. Alonso, P., Vidal, A.M.: The symmetric–toeplitz linear system problem in parallel. LNCS **3514** (2005) 220–228
5. Anderson, E., Bai, Z., Bischof, C., Demmel, J., Dongarra, J., Croz, J.D., Greenbaum, A., Hammarling, S., McKenney, A., Ostrouchov, S., Sorensen, D.: LAPACK Users' Guide. Second edn. SIAM, Philadelphia (1995)
6. Blackford, L., et al.: ScaLAPACK Users' Guide. SIAM, Philadelphia (1997)
7. Swarztrauber, P.: Vectorizing the FFT's. Academic Press, New York (1982)
8. Swarztrauber, P.: FFT algorithms for vector computers. Parallel Computing **1** (1984) 45–63
9. Loan, C.V.: Computational Frameworks for the Fast Fourier Transform. SIAM Press, Philadelphia (1992)
10. Kailath, T., Sayed, A.H.: Displacement structure: Theory and applications. SIAM Review **37** (1995) 297–386
11. Bojanczyk, A.W., Heinig, G.: Transformation techniques for toeplitz and toeplitz-plus-hankel matrices part I. transformations. Technical Report 96-250, Cornell Theory Center (1996)
12. Bojanczyk, A.W., Heinig, G.: Transformation techniques for toeplitz and toeplitz-plus-hankel matrices part II. algorithms. Technical Report 96-251, Cornell Theory Center (1996)
13. Gohberg, I., Kailath, T., Olshevsky, V.: Fast Gaussian elimination with partial pivoting for matrices with displacement structure. Mathematics of Computation **64** (1995) 1557–1576
14. Thirumalai, S.: High performance algorithms to solve Toeplitz and block Toeplitz systems. Ph.d. thesis, Graduate College of the University of Illinois at Urbana-Champaign (1996)

Speech Event Detection Using Support Vector Machines

P. Yélamos[1], J. Ramírez[1], J.M. Górriz[1],
C.G. Puntonet[2], and J.C. Segura[1]

[1] Dept. of Signal Theory, Networking and Communications,
University of Granada, Spain
`javierrp@ugr.es`
[2] Dept. of Architecture and Computer Technology,
University of Granada, Spain

Abstract. An effective speech event detector is presented in this work for improving the performance of speech processing systems working in noisy environment. The proposed method is based on a trained support vector machine (SVM) that defines an optimized non-linear decision rule involving the subband SNRs of the input speech. It is analyzed the classification rule in the input space and the ability of the SVM model to learn how the signal is masked by the background noise. The algorithm also incorporates a noise reduction block working in tandem with the voice activity detector (VAD) that has shown to be very effective in high noise environments. The experimental analysis carried out on the Spanish SpeechDat-Car database shows clear improvements over standard VADs including ITU G.729, ETSI AMR and ETSI AFE for distributed speech recognition (DSR), and other recently reported VADs.

1 Introduction

With the advent of wireless communications, new speech services are being deployed with the development of modern robust speech processing technology. An important obstacle affecting these systems is the environmental noise and its harmful effect on the system performance. Most of the noise reduction algorithms often require a precise voice activity detector (VAD). The detection task is not as trivial as it appears since the increasing level of background noise degrades the classifier effectiveness.

Since their introduction in the late seventies [1], Support Vector Machines (SVMs) marked the beginning of a new era in the learning from examples paradigm. SVMs have attracted recent attention from the pattern recognition community due to a number of theoretical and computational merits derived from the Statistical Learning Theory [2,3] developed by Vladimir Vapnik at AT&T. Enqing [4] applied SVMs to the VAD problem showing promising results when the standardized ITU-T G.729 VAD [5] speech features were used as the inputs to the classification module. Later, this VAD was incorporated to a variable low bit-rate speech codec [6] using the local cosine transform. Recently, Qi *et al.*

V.N. Alexandrov et al. (Eds.): ICCS 2006, Part I, LNCS 3991, pp. 356–363, 2006.

[7] has extended these ideas to the problem of classifying speech into voiced, unvoiced and silence frames. Again the SVM-based classifier operated on the G.729 speech features including the full-band energy difference, the low-band energy difference, the spectral distortion and the zero-crossing rate. This paper shows an effective SVM-based speech event detector for low-delay speech processing. The proposed method combines a noise robust speech processing feature extraction process together with a trained SVM model for classification. The results show improvements in speech/pause discrimination when compared to standardized VADs [5, 8, 9] and other recently published VAD methods [10, 11, 12, 13].

2   Support Vector Machines

SVMs have recently been proposed for pattern recognition in a wide range of applications by its ability for learning from experimental data. The reason is that SVMs are much more effective than other conventional parametric classifiers. In SVM-based pattern recognition, the objective is to build a function $f : R^N \longrightarrow \{\pm 1\}$ using training data that is, N-dimensional patterns $\mathbf{x}_i$ and class labels y_i:

$$(\mathbf{x}_1, y_1), (\mathbf{x}_2, y_2), ..., (\mathbf{x}_\ell, y_\ell) \in R^N \times \{\pm 1\} \tag{1}$$

so that f will correctly classify new examples $(\mathbf{x}, y)$.

Hyperplane classifiers are based on the class of decision functions:

$$f(\mathbf{x}) = \text{sign}\{(\mathbf{w} \cdot \mathbf{x}) + b\} \tag{2}$$

It can be shown that the optimal hyperplane is defined as the one with the maximal margin of separation between the two classes. The solution $\mathbf{w}$ of a constrained quadratic optimization process can be expanded in terms of a subset of the training patterns called support vectors that lie on the margin:

$$\mathbf{w} = \sum_{i=1}^{\ell} \nu_i \mathbf{x}_i \tag{3}$$

Thus, the decision rule depends only on dot products between patterns:

$$f(\mathbf{x}) = \text{sign}\{\sum_{i=1}^{\ell} \nu_i (\mathbf{x}_i \cdot \mathbf{x}) + b\} \tag{4}$$

The use of kernels in SVM enables to map the data into some other dot product space (called feature space) F via a nonlinear transformation $\Phi : R^N \longrightarrow F$ and perform the above linear algorithm in F. Figure 1 illustrates this process where the 2-D input space is mapped to a 3-D feature space where the data is linearly separable. The kernel is related to the Φ function by $k(\mathbf{x}, \mathbf{y}) = (\Phi(\mathbf{x}) \cdot \Phi(\mathbf{y}))$ In the input space, the hyperplane corresponds to a nonlinear decision function whose form is determined by the kernel. There are three common kernels that are used by SVM practitioners for the nonlinear feature mapping:

- Polynomial

$$k(\mathbf{x}, \mathbf{y}) = [\gamma(\mathbf{x} \cdot \mathbf{y}) + c]^d \tag{5}$$

- Radial basis function (RBF)

$$k(\mathbf{x}, \mathbf{y}) = \exp(-\gamma \|\mathbf{x} - \mathbf{y}\|^2) \tag{6}$$

- Sigmoid

$$k(\mathbf{x}, \mathbf{y}) = \tanh(\gamma(\mathbf{x} \cdot \mathbf{y}) + c) \tag{7}$$

Thus, the decision function is nonlinear in the input space

$$f(\mathbf{x}) = \text{sign}\{\sum_{i=1}^{\ell} \nu_i k(\mathbf{x}_i, \mathbf{x}) + b\} \tag{8}$$

and the parameters ν_i are the solution of a quadratic programming problem that are usually determined by the well known Sequential Minimal Optimization (SMO) algorithm [14]. Many classification problems are always separable in the feature space and are able to obtain better results by using RBF kernels instead of linear and polynomial kernel functions [15, 16].

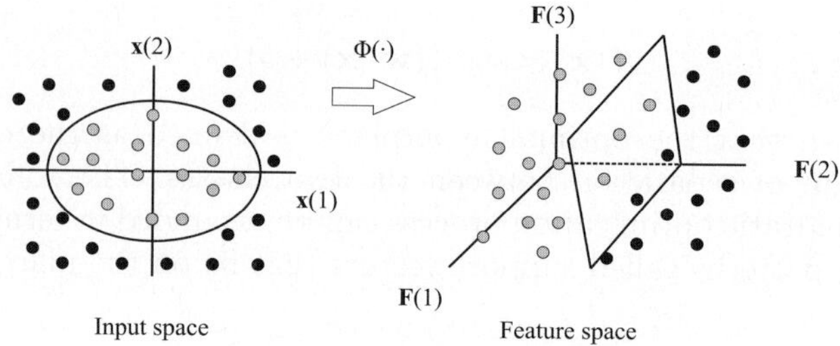

Fig. 1. Effect of the map from input to feature space where the separation boundary becomes linear

3   Speech Event Detection

The first step in defining the classification rule is the training process on the training data set and the associated class labels. The signal is preprocessed and a feature vector is extracted for training. Once the SMV model has been trained, the proposed speech event detector is described as follows: i) the input signal is decomposed into speech frames and feature extraction is conducted for classification, and ii) the speech features $\mathbf{x}$ are processed by the SVM decision function f defined in equation 8.

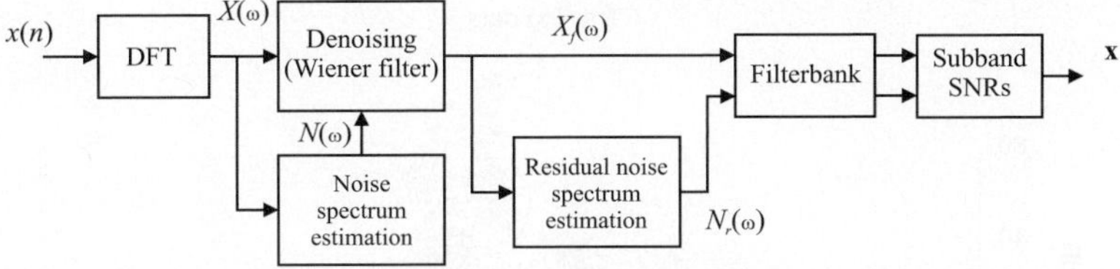

Fig. 2. Block diagram of the proposed SVM-based VAD

3.1　Feature Extraction

The algorithm for feature extraction is stated as follows. The input signal $x(n)$ sampled at 8 kHz is decomposed into 25-ms overlapped frames with a 10-ms window shift. The current frame consisting of 200 samples is zero padded to 256 samples and power spectral magnitude $X(\omega)$ is computed through the discrete Fourier transform (DFT). A denoising process based on a two-stage Wiener filter is applied to improve the performance of the VAD in high noise environments. It is described as follows:

1. Spectral subtraction.

$$S_1(\omega) = L_s X_f(\omega) + (1 - L_s)\max(X(\omega) - \alpha N(\omega), \beta X(\omega)) \tag{9}$$

2. First WF design and filtering.

$$\begin{aligned}
\mu_1(\omega) &= S_1(\omega)/N(\omega) \\
W_1(\omega) &= \mu_1(\omega)/(1 + \mu_1(\omega)) \\
S_2(\omega) &= W_1(\omega)X(\omega)
\end{aligned} \tag{10}$$

3. Second WF design and filtering.

$$\begin{aligned}
\mu_2(\omega) &= S_2(\omega)/N(\omega) \\
W_2(\omega) &= \max(\mu_2(\omega)/(1 + \mu_2(\omega)), \beta) \\
X_f(\omega) &= W_2(\omega)X(\omega)
\end{aligned} \tag{11}$$

where $L_s = 0.99$, $\alpha = 1$ and $\beta = 10^{(-22/10)}$ is selected to ensure a -22dB maximum attenuation for the filter in order to reduce the high variance musical noise that normally appears due to rapid changes across adjacent frequency bins. Once the input signal has been denoised, a filterbank reduces the dimensionality of the feature vector to a representation including broadband spectral information suitable for detection. Thus, the signal and the residual noise is passed through a K-band filterbank which is defined by

$$E_B(k) = \sum_{\omega=\omega_k}^{\omega_{k+1}} X_f(\omega); \quad N_B(k) = \sum_{\omega=\omega_k}^{\omega_{k+1}} N(\omega) \tag{12}$$
$$\omega_k = \tfrac{\pi}{K}k \qquad k = 0, 1, ..., K - 1$$

and the subband SNRs are computed as

$$\mathrm{SNR}(k) = 20\log_{10}\left(\frac{E_B(k)}{N_B(k)}\right) \qquad k = 1, 2, ..., K - 1 \tag{13}$$

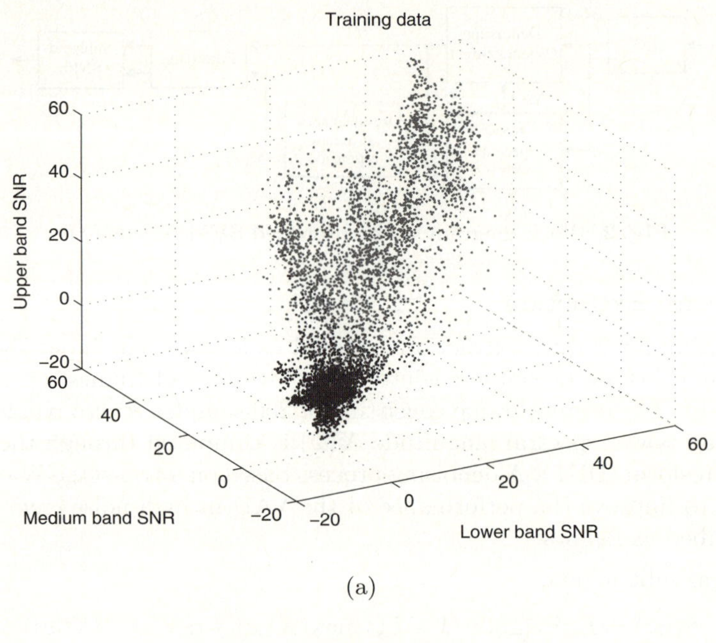

(a)

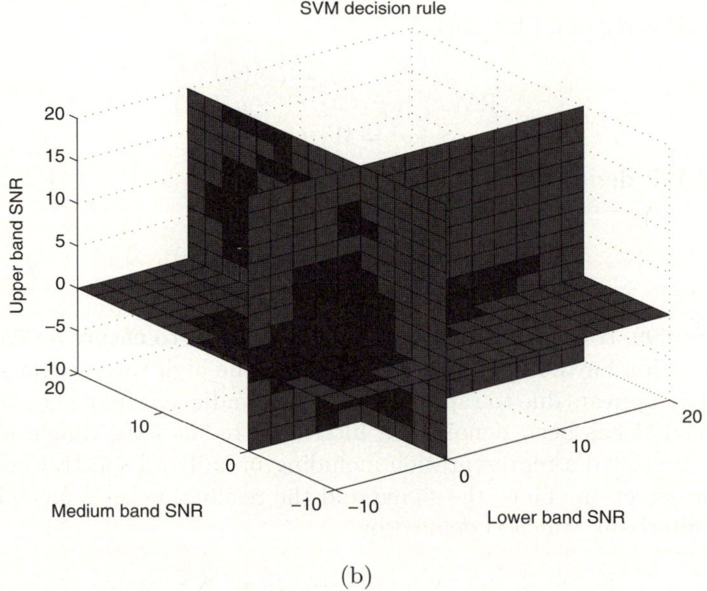

(b)

Fig. 3. Classification rule in the input space after training a 3-band SVM model. a) Training data set, b) SVM classification rule.

3.2   Training

The SVM model has been trained using LIBSVM software tool [17]. A training set consisting of 12 utterances of the AURORA 3 Spanish SpeechDat-Car (SDC) was used. This database contains 4914 recordings using close-talking and distant microphones from more than 160 speakers. The files are categorized into three noisy conditions: quiet, low noisy and highly noisy conditions, which represent different driving conditions with average SNR values between 25dB, and 5dB. The recordings used for training the SVM are selected to deal with different noisy conditions. Fig. 3.1.a shows the training data set in the 3-band input space. After the training process, the SVM decision rule defined by equation 8 is graphically shown in Fig 3.1.b where the non-speech and speech classes are clearly distinguished in the 3-D space. Note that, the SVM model learns how the signal is masked by the noise and automatically defines the decision rules in the input space.

The SVM formulation is based on C-Support Vector Classification [18, 3] while the decision rule is defined by equation 8. An RBF kernel is used and the training process consists in finding the solution of a primal problem

$$\min_\alpha \quad \tfrac{1}{2}\alpha^T \mathbf{Q}\alpha - \mathbf{e}^T\alpha$$
$$0 \le \alpha_i \le C, \quad i = 1, 2, ..., \ell \tag{14}$$
$$\text{subject} \quad \text{to} \quad \mathbf{y}\alpha = 0$$

by using LIBSVM [17], where $\mathbf{e}= [1\ 1\ ...\ 1]$, $C > 0$ is the upper bound and $Q_{ij} = y_i y_j k(\mathbf{x}_i, \mathbf{x}_j)$. After this process, the support vectors $\mathbf{x}_i$ and coefficients α_i required to evaluate the decision rule defined in equation 8 are selected where $\nu_i = y_i\alpha_i$. Note that, b can be used as a decision threshold for the VAD in the sense that the working point of the VAD can be shifted in order to meet the application requirements.

4   Experimental Framework

This section analyzes the proposed VAD and compares its performance to other algorithms used as a reference. The analysis is based on the ROC curves, a frequently used methodology to describe the VAD error rate. The AURORA subset of the original Spanish SDC database [19] was used again in this analysis. The non-speech hit rate (HR0) and the false alarm rate (FAR0= 100-HR1) were determined for each noisy condition being the actual speech frames and actual speech pauses determined by hand-labelling the database on the close-talking microphone. Thus, the objective is to work as close as possible to the ideal [0%,100%] point where both speech and non-speech are determined with no error. Preliminary experiments determined that increasing the number of subbands up to four subbands improved the performance of the proposed VAD by shifting the ROC curves in the ROC space.

Fig. 4 compares the ROC curve of the proposed VAD to frequently referred algorithms [11, 12, 13, 10] for recordings from the distant microphone high noisy conditions. The working points of the ITU-T G.729, ETSI AMR and AFE VADs are also included. The results show improvements in detection accuracy. Thus, among all the VAD examined, our VAD yields the lowest false alarm rate for a fixed non-speech hit rate and also, the highest non-speech hit rate for a given false alarm rate. The benefits are especially important over ITU-T G.729 [5], which is used along with a speech codec for discontinuous transmission, and over the Li's algorithm [12], that is based on an optimum linear filter for edge detection. The proposed VAD also improves Marzinzik's VAD [13] that tracks the power spectral envelopes, and the Sohn's VAD [10], that formulates the decision rule by means of a model-based statistical likelihood ratio test.

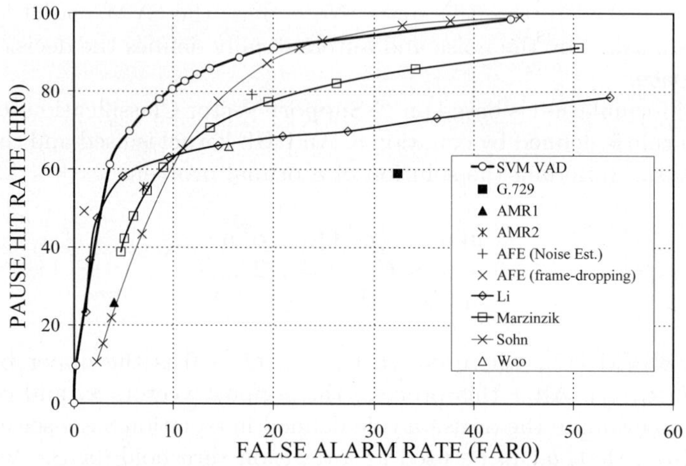

Fig. 4. Comparative results to other VAD methods

5   Conclusions

This paper has shown an effective speech event detector combining spectral noise reduction and support vector machine learning tools. The use of kernels enables defining a non-linear decision rule in the input space which is defined in terms of subbands SNRs. It is also shown the ability of SVM tools to learn how the speech is masked by the acoustic noise. With these and other innovations the proposed method has shown to be more effective than VADs that define the decision rule in terms of an average SNR values. The proposed algorithm also outperformed ITU G.729, ETSI AMR1 and AMR2 and ETSI AFE standards and recently reported VAD methods in speech/non-speech detection performance.

Acknowledgements

This work has been funded by the European Commission (HIWIRE, IST No. 507943) and the Spanish MEC project TEC2004-03829/FEDER.

References

1. Vapnik, V.: Estimation of Dependences Based on Empirical Data. Springer-Verlag, New York (1982)
2. Vapnik, V.: The Nature of Statistical Learning Theory. Springer-Verlag, Berlin (1995)
3. Vapnik, V.: Statistical Learning Theory. John Wiley and Sons, Inc., New York (1998)
4. Enqing, D., Guizhong, L., Yatong, Z., Xiaodi, Z.: Applying support vector machines to voice activity detection. In: 6th International Conference on Signal Processing. Volume 2. (2002) 1124–1127
5. ITU: A silence compression scheme for G.729 optimized for terminals conforming to recommendation V.70. ITU-T Recommendation G.729-Annex B (1996)
6. Enqing, D., Heming, Z., Yongli, L.: Low bit and variable rate speech coding using local cosine transform. In: Proc. of the 2002 IEEE Region 10 Conference on Computers, Communications, Control and Power Engineering (TENCON '02). Volume 1. (2002) 423–426
7. Qi, F., Bao, C., Liu, Y.: A novel two-step SVM classifier for voiced/unvoiced/silence classification of speech. In: International Symposium on Chinese Spoken Language Processing. (2004) 77–80
8. ETSI: Voice activity detector (VAD) for Adaptive Multi-Rate (AMR) speech traffic channels. ETSI EN 301 708 Recommendation (1999)
9. ETSI: Speech processing, transmission and quality aspects (STQ); distributed speech recognition; advanced front-end feature extraction algorithm; compression algorithms. ETSI ES 201 108 Recommendation (2002)
10. Sohn, J., Kim, N.S., Sung, W.: A statistical model-based voice activity detection. IEEE Signal Processing Letters **16** (1999) 1–3
11. Woo, K., Yang, T., Park, K., Lee, C.: Robust voice activity detection algorithm for estimating noise spectrum. Electronics Letters **36** (2000) 180–181
12. Li, Q., Zheng, J., Tsai, A., Zhou, Q.: Robust endpoint detection and energy normalization for real-time speech and speaker recognition. IEEE Transactions on Speech and Audio Processing **10** (2002) 146–157
13. Marzinzik, M., Kollmeier, B.: Speech pause detection for noise spectrum estimation by tracking power envelope dynamics. IEEE Transactions on Speech and Audio Processing **10** (2002) 341–351
14. Platt, J.: Fast Training of Support Vector Machines using Sequential Minimal Optimization. In: Advances in Kernel Methods - Support Vector Learning. MIT Press (1999) 185–208
15. Clarkson, P., Moreno, P.: On the use of support vector machines for phonetic classification. In: Proc. of the IEEE Int. Conference on Acoustics, Speech and Signal Processing. Volume 2. (1999) 585–588
16. Ganapathiraju, A., Hamaker, J., Picone, J.: Applications of support vector machines to speech recognition. IEEE Transactions on Signal Processing **52** (2004) 2348–2355
17. Chang, C., Lin, C.J.: LIBSVM: a library for support vector machines. Technical report, Dept. of Computer Science and Information Engineering, National Taiwan University (2001)
18. Cortes, C., Vapnik, V.: Support-vector network. Machine Learning (1995)
19. Moreno, A., Borge, L., Christoph, D., Gael, R., Khalid, C., Stephan, E., Jeffrey, A.: SpeechDat-Car: A Large Speech Database for Automotive Environments. In: Proceedings of the II LREC Conference. (2000)

BRUST: An Efficient Buffer Replacement for Spatial Databases

Jun-Ki Min

School of Internet-Media Engineering
Korea University of Technology and Education
Byeongcheon-myeon, Cheonan, Chungnam,
Republic of Korea, 330-708
jkmin@kut.ac.kr

Abstract. This paper presents a novel buffer management technique for spatial database management systems. Much research has been performed on various buffer management techniques in order to reduce disk I/O. However, many of the proposed techniques utilize the temporal locality of access patterns. In spatial database environments, there exists not only the temporal locality, where a recently access object will be accessed again in near future, but also spatial locality, where the objects in the recently accessed regions will be accessed again in the near future. Thus, in this paper, we present a buffer management technique, called BRUST, which utilizes both the temporal locality and spatial locality in spatial database environments.

1 Introduction

Spatial database management is an active area of research over the past ten years [1] since the applications using the spatial information such as geographic information systems (GIS), computer aided design (CAD), multimedia systems, satellite image databases, and location based service (LBS), have proliferated. In order to improve the performance of spatial database management systems (SDBMSs), much of the work on SDBMSs has focused on spatial indices [2, 3] and query processing methods [4, 5].

Since data volume is larger than current memory sizes, it is inevitable that disk I/O will be incurred. In order to reduce disk I/O, a buffer is used. Since efficient management of the buffer is closely related to the performance of databases, many researchers have proposed diverse and efficient buffer management techniques.

Well known buffer management techniques utilize temporal locality, where recently accessed data will be accessed in the near future. With SDBMSs, there also exists spatial locality which is the property where objects in recently accessed regions will be accessed in the near future. Therefore, spatial locality should be also considered in buffer management techniques. However, traditional buffer management techniques consider the temporal locality only.

V.N. Alexandrov et al. (Eds.): ICCS 2006, Part I, LNCS 3991, pp. 364–371, 2006.

In this paper, we present a novel buffer management algorithm, called *BRUST* (Buffer Replacement Using Spatial and Temporal locality). In BRUST, using a spatially interesting point (SIP), the buffer management selects a victim which is replaced with a newly access object.

We implemented BRUST and performed an experimental study over various buffer sizes and workloads. The experimental results confirm that BRUST is more efficient than the existing buffer replacement algorithms on SDBMSs environments.

2 Related Work

When the buffer is full, buffer management methods find a victim to be replaced with a newly loaded object by analyzing the access pattern using the buffer replacement algorithm. There is a long and rich history of the research performed on buffer management. The core of buffer management techniques is the buffer replacement algorithm. In this section, we present the diverse buffer replacement algorithms.

2.1 Traditional Buffer Replacement Algorithm

The most well known buffer replacement algorithm among the various buffer replacement algorithms is LRU [6] (Least Recently Used). The LRU buffer replaces the object which has not been accessed for the longest time (i.e., least recently accessed object). Since the LRU algorithm is based on the simple heuristic rule such that the recently accessed object will be accessed in the near future, the LRU algorithm cannot support diverse data access patterns efficiently.

To overcome this problem, LRU-k [7] was proposed. LRU-k keeps track of the times for the last k references to a object, and the object with the least recent last k-th access will then be replaced. Of particular interest, LRU-2 replaces the object whose penultimate (second to last) access is least recent. LRU-2 improves upon LRU because the second to last access is a better indicator of the interarrival time between accesses than the most recent access. LRU-k keeps k-access history for each object and so the process of finding a victim is expensive.

Thus, John and Shasha [8] proposed 2Q which behaves like LRU-2 but is more efficient. 2Q handles the buffer using two separate queues: A1IN and AM. A1IN acts as a first-in-first-out queue and AMQ acts as an LRU queue. When an object not used in the near past is loaded, the object is inserted into A1IN. Otherwise, the object is inserted into AMQ. In 2Q, the history of object replacement is maintained by A1OUT. A1OUT does not contain the object itself. Thus, using the contents of A1OUT, the decision of whether the object was used in the near past or not is made. A disadvantage of 2Q is that the performance of 2Q is determined by the sizes of the queues: A1IN, AM, A1OUT.

The frequency counter and recency history are the major indications of temporal locality. LRFU [9] integrates the two indications. In LRFU, each object x in the buffer has the following value $C(x)$.

$$C(x) = \begin{cases} 1 + 2^{-\lambda}C(x) & \text{if x is referenced at t time} \\ 2^{-\lambda}C(x) & \text{otherwise} \end{cases} \tag{1}$$

In the above formula, λ is a tunable parameter. For newly loaded objects, $C(x)$ is 0. When a buffer replacement is required, LRFU selects the object whose $C(x)$ value is smallest as the victim. In LRFU, when λ approaches 1, LRFU gives more weight to more recent reference. Thus, the behavior of LRFU is similar to that of LRU. When λ is equal to 0, C(x) simply counts the number of accesses. Thus, LRFU acts as LFU. Therefore, the performance of LRFU is determined by λ.

Practically, it is hard to determine the optimal values for tunable parameters such as λ of LRFU and the queue sizes of 2Q. Megiddo and Modha suggested ARC [10] which dynamically changes the behavior of the algorithm with respect to the access pattern. Like 2Q, ARC separates a buffer whose size is c into queues: B1 whose size is p and B2 whose size is $c - p$. B1 and B2 are LRU queues. A newly accessed object is loaded into B1, and an accessed object which is in B1 or B2 is moved into B2. The behavior of ARC is determined by parameter p. If a hit occurs on B1, p increases. Otherwise, p decreases. Note that p is not the actual size of B1 but target size of B1. So, the size of B1 may not be equal to p. When a buffer replacement occurs, if the size of B1 is greater than p, a victim is chosen from B1. Otherwise, a victim is chosen from B2. In this approach, the incremental ratio of p may vary according to the learning rate. In other words, the main goal of eliminating tunable parameters is not accomplished.

Also, LFU-k [11] which is a generalization of LFU has been proposed. And, TNPR [12] which estimates the interval time of accesses for each object was presented. In addition, for buffering an index instead of data, ILRU [13] and GHOST [14] has been suggested.

2.2   Buffer Management Techniques for SDBMSs

The buffer replacement algorithms presented in Section 2.1 consider only temporal locality. However, some buffer management techniques for SDBMSs has been proposed.

Papadopoulos and Manolopoulous proposed LRD-Manhattan [15]. In general, a spatial object is represented as an MBR (Minimum Bounded Rectangle) to reduce the computational overhead. Probabilistically, when a point is selected in a unit space, the probability of a large sized object that contains the point is greater than that of a small sized object in the uniform assumption [16]. In other words, large sized spatial objects may be accessed more than small sized spatial objects. LRD-Manhattan computes the average of the access density (i.e., access ratio of each object) and the normalized MBR size of each spatial object. Then, LRD-Manhattan selects a spatial object whose average value is the minimum among all objects in the buffer as a victim.

Recently, ASB [17] which considers the LRU heuristic and the sizes of spatial objects together was proposed. In this technique, a buffer consists of two logically separated buffers, like ARC. The buffer B1 is maintained using the LRU heuristic and the buffer B2 is maintained using the sizes of spatial objects. A newly accessed object is loaded into B1. When the size of the buffer B1 is insufficient, the least recently used object is moved into the buffer B2. When buffer replacement is required, the object whose MBR size is smallest is selected from B2 as a victim. Also, the sizes of B1 and B2 are incrementally changed with respect to the property of a newly accessed object (see details in [17]).

The techniques presented above utilize a static property (i.e., the size of MBR). Thus, these techniques do not suggest efficient buffer management with respect to dynamic access patterns.

3   Behavior of BRUST

In spatial database environments, not only the temporal locality but also spatial locality [18] where spatial queries converge on a specific area of the work space within a certain period exists since user's interest focus on a certain area. In other words, if a certain area is frequently accessed, then the spatial objects near that area have high probability of being accessed.

When considering only temporal locality, a spatial object in the frequently accessed area may be a victim. In this case, the efficiency of the buffer is degraded. Also, if we only consider spatial locality, some access patterns such as random access and liner scan are not efficiently supported.

Therefore, in contrast to the previous techniques, the proposed buffer management technique, BRUST, considers temporal locality and spatial locality together. The basic heuristic rule of BRUST is that a spatial object near the frequently used area will be used in the near future.

First, in order to utilize the spatial locality, we estimate a spatially interesting point (SIP) with respect to the spatial locations of accessed objects. The initial location of a SIP is the center point of workspace. Let the currently estimated SIP be $< x_{sip}, y_{sip} >$ and a spatial object whose center point p $< x_p, y_p >$ is be accessed. Then a SIP should be modified in order the reflect the change of user's interest. The following formula is for the estimation of a new SIP.

$$
\begin{aligned}
x_{sip} &= x_{sip} + w_x \cdot (x_{sip} - x_p) \\
y_{sip} &= y_{sip} + w_y \cdot (y_{sip} - y_p)
\end{aligned}
\tag{2}
$$

During the estimating phase of a SIP, we consider outliers since, in general, a user interesting location is gradually changed. Thus, as presented in equation (2), the weight factors w_x and w_y are multiplied. As distance between the current SIP and a newly accessed object increases, the probability that the location of a newly accessed object is a user interesting location dramatically decreases. Thus, we use a simple decrease function $(1/e)^\lambda$. As described in equation (3), w_x decreases as the distance between the current SIP and a newly accessed object increases.

$$
w_x = \left(\tfrac{1}{e}\right)^{|x_{sip} - x_p| / Domain_x}
\tag{3}
$$
$$
\text{where } e \text{ is the base of natural logarithm,}
$$
$$
\text{and } Domain_x \text{ is the length of workspace on x-coordinate}
$$

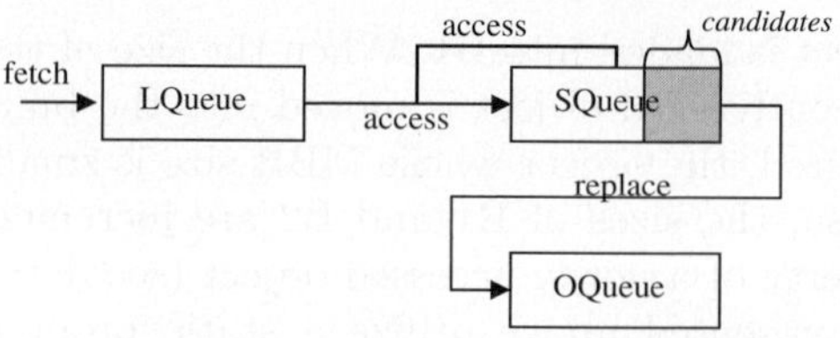

Fig. 1. The behavior of BRUST

The behavior of BRUST is described in Figure 1. As depicted in Figure 1, there are three queues in BRUST. Among the queues, LQueue contains the objects which have been accessed once recently. SQueue contains the objects which have been accessed at least twice recently. Conceptually, LQueue maintains the recently accessed objects and SQueue maintains the frequently accessed objects. The sizes of LQueue and SQueue are adaptively changed like ARC. OQueue does not contain objects like AOUT queue in 2Q but keeps the replacement history. The size of OQueue is 30% of the size of SQueue.

If a buffer replacement is required, BRUST finds the farthest object from a SIP in SQueue and makes the farthest object a victim. To find out the farthest object in SQueue, all objects in SQueue should be evaluated. It degrades the efficiency of the buffer management. Thus, in BRUST, a victim is selected from the portion of SQueue, called *candidates*. The gray box in Figure 1 represents *candidates*.

Now we have an important question which concerns the size of *candidates*. The basic idea of the problem is that the size of *candidates* is changed with respect to the access pattern.

Note that BRUST selects a object in *candidates* as a victim. The case that a replaced object in the near past is re-referenced means the influence of temporal locality is larger than that of spatial locality. Recall that OQueue exists in BRUST in order to keep the replace history. Thus, if a newly access object is found in OQueue, BRUST reduces the size of *candidates* in order to reduce the effect of spatial locality. Otherwise, BRUST increases the size of *candidates*.

4   Experiments

In this section, we show the effectiveness of BRUST compared with the diverse buffer management techniques: LRU, 2Q, ARC and ASB. As mentioned earlier, LRU, 2Q and ARC consider only temporal locality. And ASB consider temporal locality and the property of s spatial object (i.e., the size of MBR). We evaluated the hit ratio of BRUST using the real-life and synthetic data over various sized buffers. However, due to the space limitation, we show only the experimental result of the real-life data when buffer sizes are 5%, 10% and 20% of total size of objects. The real-life data in our experiments were extracted from TIGER/Line data of US Bureau of the Census [19]. We used the road segment data of Kings county of the California State. The number of spatial objects is 21,853 and the size of work space is 840,681×700,366.

To evaluate the effectiveness of BRUST, we made 3 kinds of access pattern. First, we made an uniform access pattern (termed *Uniform*) where all spatial objects have same access probability. In order to measure the effect of temporal locality, we made a temporally skewed access pattern (termed *Time Skew*) using Zipf distribution. In temporally skewed access pattern, 80% of the references accesses 20% of spatial objects. Finally, we made a spatially skewed access pattern (termed *Spatial Skew*). In this pattern, 90% of the references access the spatial objects which are in a 10% sized region of the work space. Thus, in the spatially skewed access pattern, temporal locality and spatial locality appear intermixedly. In each access pattern, 1,000,000 accesses of spatial objects occur.

The following figures presents the experimental results over diverse sized buffers.

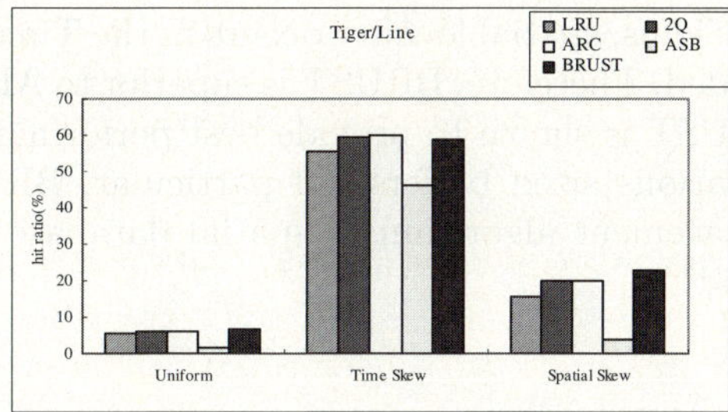

Fig. 2. The results on the 5% sized buffer

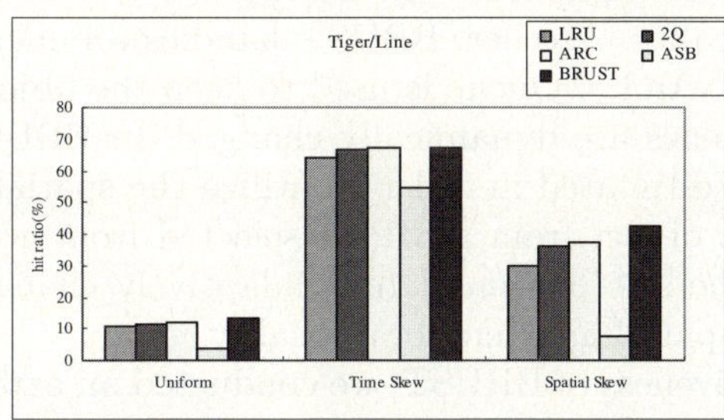

Fig. 3. The results on the 10% sized buffer

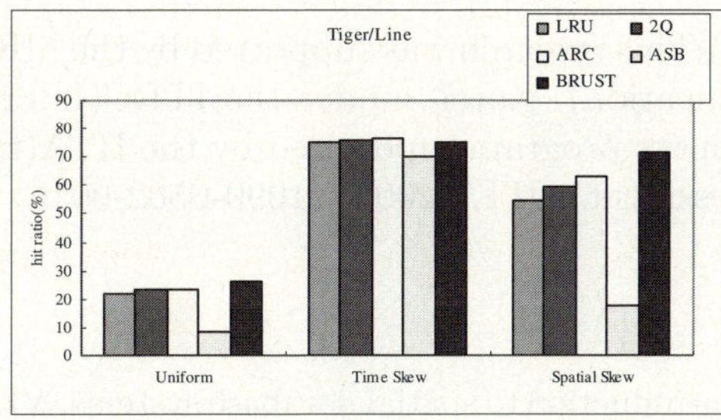

Fig. 4. The results on the 20% sized buffer

As shown in Figure 2, Figure 3, and Figure 4, BRUST shows the best hit ratio over most of all cases. Of particular, BRUST outperforms LRU, 2Q, ARC, and ASB on the *Time Skew* workload over diverse sized buffers even though BRUST considers temporal locality and spatial locality together.

ASB shows the worst performance over most of cases since a victim is selected with respect to a static property (i.e., MBR size). LRU does not show the most efficient performance but not show the worst performance over all cases. 2Q and ARC show good performance on the *Time Skew* workload since 2Q and ARC is devised for the temporally skewed accesses.

BRUST shows the better performance than ARC on the *Time Skew* workload with the real-life data. Recall that BRUST dynamically changes the sizes of SQueue and OQueue like ARC. In addition, BRUST adaptively changes the sizes of *candidates* with respect to the effect of a SIP. In the real-life data set, locations of spatial objects are clustered. Thus, spatial locality occurs in the Time Skew workload although it is not intended. Therefore, BRUST is superior to ARC.

Consequently, BRUST is shown to provide best performance over diverse access patterns with various sized buffers. Of particular, BRUST is superior to the other buffer replacement algorithm in spatial database environments (i.e., *Spatial Skew* workload).

5   Conclusion

In this paper, we present a novel buffer management technique, called BRUST (Buffer Replacement Using Spatial and Temporal locality) which consider spatial and temporal locality together. BRUST handles a buffer using two queues: LQueue and SQueue. And, OQueue is used to keep the object replacement history. The sizes of queues are dynamically changed. In BRUST, a spatial interesting point (SIP) is estimated in order to utilize the spatial locality. A victim, which is the farthest object from a SIP, is selected from a portion of SQueue, called *candidates*. The size of *candidate* is adaptively changed with respect to the influence of the spatial and temporal locality.

To show the effectiveness of BRUST, we conducted an extensive experimental study with the synthetic and real-life data. The experimental results demonstrate that BRUST is superior to existing buffer management techniques in spatial databases environments.

Acknowledgement. This research was supported by the MIC(Ministry of Information and Communication), Korea, under the ITRC(Information Technology Research Center) support program supervised by the IITA(Institute of Information Technology Assessment) (IITA-2005-C1090-0502-0016)

References

1. Guting, R.H.: An introduction to spatial database systems. VLDB **3** (1994) 357–399
2. Guttman, A.: The R-tree: A Dynamic index structure for spatial searching. In: Proceedings of ACM SIGMOD Conference. (1984) 47–57

3. Brinkhoff, T., Kriegel, H., Scheneider, R., Seeger, B.: The R*-tree: An Efficient and Robust Access Method for Points and Rectangles. In: Proceedings of ACM SIGMOD Conference. (1990) 322–331
4. Min, J.K., Park, H.H., Chung, C.W.: Multi-way spatial join selectivity for the ring join graph. Information and Software Technology **47** (2005) 785–795
5. Papadias, D., Mamoulis, N., Theodoridis, Y.: Processing and Optimization of Multiway Spatial Join Using R-Tree. In: Proceedings of ACM PODS. (1999) 44–55
6. Effelsberg, W.: Principles of Database buffer Management. ACM TODS **9** (1984) 560–595
7. O'Neil, E.J., Neil, P.E.O., Weikum, G.: The LRU-K Page Replacement algorithm for database disk buffering. In: Proceedings of ACM SIGMOD Conference. (1993) 297–306
8. Johnson, T., Shasha, D.: 2Q: a Low Overhead High Performance Buffer Management Replacement Algorithm. In: Proceedings of VLDB Conference. (1994) 439–450
9. D. Lee, J.C., Kim, J.H., Noh, S.H., Min, S.L., Cho, Y., Kim, C.S.: LRFU: A Spectrum of Policies that subsumes the Least Recently Used and Least Frequently Used Policies. IEEE Tans. Computers **50** (2001) 1352–1360
10. Megiddo, N., Modha, D.S.: ARC: A Self-tuning, Low Overhead Replacement Cache. In: Proceedings of USENIX FAST Conference. (2003)
11. Sokolinsky, L.B.: LFU-K: An Effective Buffer Management Replacement Algorithm. In: Proceedings of DASFAA. (2004) 670–681
12. Juurlink, B.: Approximating the Optimal Replacement Algorithm. In: ACM CF Conference. (2004)
13. Sacco, G.M.: Index Access with a Finite Buffer. In: Proceedings of VLDB Conference. (1987)
14. Goh, C.H., Ooi, B.C., Sim, D., Tan, K.: GHOST: Fine Granularity Buffering of Index. In: Proceedings of VLDB Conference. (1999)
15. Papadopoulos, A., Manolopoulos, Y.: Global Page Replacement in Spatial Databases. In: Proceedings of DEXA. (1996)
16. Kamel, I., Faloutsos, C.: On Packing R-Trees. In: Proceedings of CIKM. (1993) 490–499
17. Brinkhoff, T.: A Robust and Self-tuning Page Replacement Strategy for Spatial Database Systems. In: Proceedings of DEXA. (2002) 533–552
18. Ki-Joune, L., Robert, L.: The Spatial Locality and a Spatial Indexing Method by Dynamic Clustering in Hypermap System. In: Proceedings of SSD. (1990) 207–223
19. Bureau, U.C.: UA Census 2000 TIGER/Line Files. (http://www.census.gov/geo/www/tiger/tigerua/ua_tgr2k.html)

Effects of O_3 Adsorption on the Emission Properties of Single-Wall Carbon Nanotubes: A Density Functional Theory Study

B. Akdim[1], T. Kar[2], D.A. Shiffler[3], X. Duan[1], and R. Pachter[1,*]

[1] Air Force Research Laboratory, Materials and Manufacturing Directorate
Wright-Patterson Air Force Base, OH, USA
Brahim.Akdim@wpafb.af.mil, Xiaofeng.Duan@wpafb.af.mil,
Ruth.Pachter@wpafb.af.mil
[2] Department of Chemistry and Biochemistry, Utah State University, Logan, UT, USA
tkar@wpafb.af.mil
[3] Air Force Research Laboratory, Directed Energy Directorate
Kirtland Air Force Base, NM, USA
Donald.Shiffler@wpafb.af.mil

Abstract. In this study, we report density functional theory calculations to examine the effects of O_3 adsorption on the field emission properties of capped C(5,5) single-wall carbon nanotubes. Structural changes, adsorption energies, and the first ionization potential for possible adsorption sites are discussed, including an applied field in the calculations. The results suggest a suppression of the emission upon O_3 adsorption, explained by the charge transfer, while the favored adsorption for the etched structures rationalizes enhancement due to sharper tips upon opening of the carbon nanotube when ozonized, consistent with experimental observations.

1 Introduction

Applications of single-wall carbon nanotubes (SWCNTs) could be enhanced, or adversely affected, by surface adsorption, through non-covalent or covalent interactions (covalent surface chemistry was recently reviewed [1]). In this work we focus on the effects of surface adsorption on the field emission properties of carbon nanotubes, which attracted considerable attention for use in applications such as microwave amplifiers, X-ray sources, and flat panel displays, summarized [2], and improved upon [3]. It is notable, however, that despite significant technological developments, there are still unanswered questions, because the I-V characteristics often deviate from Fowler-Nordheim's relationship:

$$I = \alpha E_{eff}^2 \exp(-\beta / E_{eff}) \tag{1}$$

where α is a constant related to the geometry, E_{eff} is the effective field at the emitter tip, and β is proportional to the work function.

---

* Corresponding author.

V.N. Alexandrov et al. (Eds.): ICCS 2006, Part I, LNCS 3991, pp. 372–378, 2006.

Indeed, changes in the field emission of SWCNTs due to geometrical and local field effects were previously examined theoretically [4], while in our ongoing interest to enhance field emission characteristics of SWCNTs, we investigated the effects of Cs surface adsorption [5,6], which decreases the ionization potential (IP) of the SWCNT. Most recently, enhanced field emission was observed for HfC-coated carbon nanotubes, also attributed to the lower work function of HfC [7].

On the other hand, upon O_2 adsorption, a decrease in the output current was noted experimentally [8], consistent with our density functional theory (DFT) IP results [9], demonstrating a suppression of the field emission, particularly when including the effects of an electric field. At the same time, the effects of O_3 adsorption at a SWCNT tip have not yet been addressed theoretically, even though ozonized carbon nanotubes are often present in samples due to purification procedures [1]. The field emission characteristics upon ozonation were explored experimentally [10], suggesting that the observed enhancement with a longer time treatment by ozone of the sample, may be due to an opening of the carbon nanotube upon etching, resulting in sharper tips, thus leading to a higher emission current. In this study we carried out DFT calculations to study the effects of O_3 on the adsorption and field emission properties of SWCNTs.

2   Computational Details

2.1   Models

Calculations were performed by an all-electron linear combination of atomic orbitals DFT approach, applying DMOL3 [11], with the Perdew-Burke-Ernzerhof (PBE) [12] exchange-correlation functional within the generalized gradient approximation (GGA). A double numerical polarized basis set was used. The studied C(5,5) models are shown in Figure 1 (70 C; 10 H, saturating the SWCNT at one end), with the following adsorption sites: physisorption (config.-a); adsorption@h—h (config.-b); adsorption@h—p (config.-c); etched structure of adsorption@h—p (config.-d). An electric field was directed from above the tip to mimic the emission environment.

2.2   Computational Efficiency

DFT-based electronic structure calculations for molecular/periodic systems scale as ca. $O(N^3)$, where N is the number of basis functions (bf). We carried out timings for carbon nanotube model compounds, for one SCF iteration applying Gaussian03 [13] (defined as: molecular geometry; guess density matrix; evaluate integrals; solve KS equation) on a Compaq ES45, with up to 8100 bf. Calculations were carried out on 8 processors, and assuming linear scalability in this case, an $O(N^{3.45})$ scaling was noted. The serial efficiency of about 13% for an 180 atoms model (2700 bf's), when running on a 2GFlop processor, was consistent with NWChem results [14].

The bottlenecks in DFT include the electronic Coulomb problem, (x-c) quadrature, and the diagonalization of the Kohn-Sham (KS) matrix. Recent advances in achieving "linear scaling" in DFT consist, for example, of methods to tackle the Coulomb problem, based on the FMM, as reviewed [15], also by Beck [16], summarizing real-space techniques in DFT. An $O(N)$ real space technique has been applied by Bernholc's group [17].

3 Results and Discussion

3.1 Structural Parameters

Structural parameters of configurations (a)-(d) (Figure 1) are listed in Table 1. The increase in the C—C bond length at the adsorption site by 15% and 16%, for the h—h and h—p sites, respectively, demonstrated a weakening of the bonds upon O_3 adsorption, being similar to our results for sidewall ozonation [18], but larger than for O_2 [9], which is consistent with the well known reactivity of ozone [19]. In addition, a decrease of the bond angle in O_3 by 16% upon adsorption was notable. In the etched configuration, the dissociation of O_3 forms two adsorption sites (config.-d), an etched and an epoxide-type structure, for which an increase in the C—C bond distance of about 10% was noted. Small structural changes were shown upon physisorption, and the distance of O_3 to the SWCNTs was calculated to be ca. 3.2Å. These results are comparable to the results for the adsorption at the sidewall of a SWCNT [18].

Table 1. Structural parameters

	Field (eV/Å)	Pristine	config.-a	config.-b	config.-c	config.-d
O—O—O°	0.0		117	100	101	
	0.5			100	101	
	1.0			100	100	
O—O (Å)	0.0		1.30	1.46	1.47	
	0.5			1.47	1.48	
	1.0			1.46	1.50	
C—O (Å)	0.0		3.19	1.45	1.44	epoxide.43 ketone: 1.22
	0.5			1.46	1.47	epoxide: 1.45 ketone: 1.23
	1.0			1.48	1.51	epoxide: 1.48 ketone: 1.25
C—C (Å)	0.0	1.45 (h—p) 1.40 (h—h)	1.45 (h—p) 1.40 (h—h)	1.62 (h—h)	1.66 (h—p)	epoxide: 1.60 ketone: 2.75
	0.5	1.46 (h—p) 1.41 (h—h)		1.62 (h—h)		epoxide: 1.57 ketone: 2.76
	1.0	1.46 (h—p) 1.41 (h—h)		1.62 (h—h)	1.66 (h—p)	epoxide: 1.55 ketone: 2.76

3.2 Adsorption Energies

Our results showed a strong chemisorption of O_3, with binding energies of -1.4 eV and -0.74 eV for the h—h and h—p adsorption sites, respectively (Table 2). Most notable is the stronger adsorption upon etching, namely for config.-d, characterized by a higher adsorption energy (E_{ad}=-2.77 eV), as compared to O_2 adsorption (E_{ad}=-1.17 eV) [9]. These results are consistent with the O_3 reactivity, and may explain the enhancement of field emission for the etched sharp tips of the ozone treated carbon

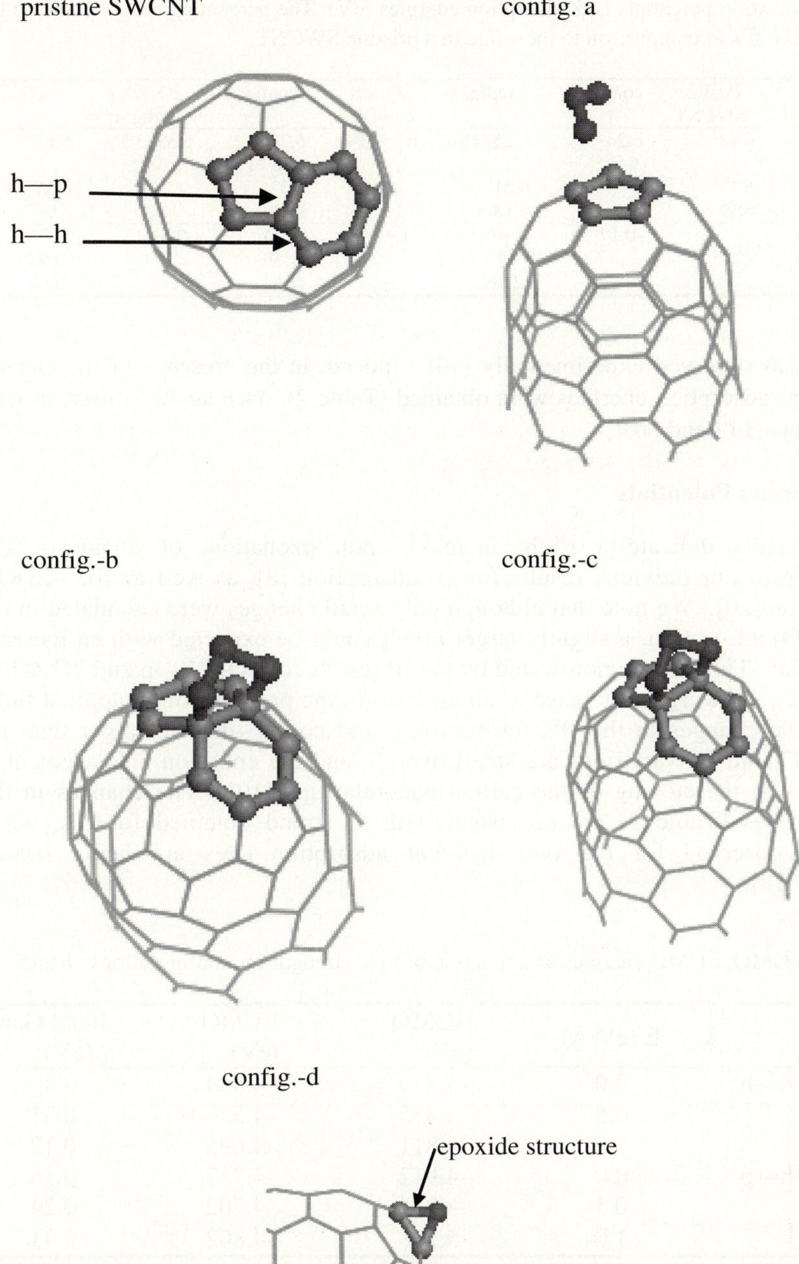

Fig. 1. C(5,5) SWCNT models

Table 2. Ionization potentials and adsorption energies (eV). The percentage values indicate the increase of the IPs in comparison to the value in a pristine SWCNT.

eV	Field (eV/Å)	Pristine SWCNT	config.-a	config.-b	$2O_3$@h-h config.-b'	config.-c	$2O_3$@h-p config.-c'	config.-d
IP	0.0	6.17	6.24	6.25 (1%)	6.34 (3%)	6.28 (1.6%)	6.35 (3%)	6.31
	0.5	6.17		6.21		6.32		6.55
	1.0	6.17		6.18		6.60		6.90
Ead	0.0		-0.17	-1.40	-1.36	-0.74	-0.63	-2.77
	0.5			-1.72		-1.50		-3.93
	1.0			-2.20		-2.83		-5.62

nanotubes, as observed experimentally [10]. Indeed, in the presence of the electric field, higher adsorption energies were obtained (Table 2), once again, consistent with the experimental trend [10].

3.3 Ionization Potentials

The IP results indicate a slight increase upon ozonation, of about 1—2%, comparable to our previous results for O_2 adsorption [8], as well as for –COOH modification [20]. We note that although only small changes were calculated in the IPs upon O_3 adsorption, a slightly larger effect could be expected with an increase in O_3 uptake. This was demonstrated by the IP results for $2O_3$@h—p and $2O_3$@h—h (Figure 2), showing an increase of about 3%. In the presence of an applied field, although the changes in the IPs for config.-c and config.–d were larger than for config.-b (Table 2), variations are small overall, and the emission enhancement is mostly due to the etching of the carbon nanotube tips [10]. The changes in the HOMO values (Table 3) are consistent with the trend obtained for IPs, while variations observed for the two different adsorption sites are being further examined [18].

Table 3. HOMO, LUMO energies as a function of field strength for configurations –b and –c

	E (eV/Å)	HOMO (eV)	LUMO (eV)	Band-Gap (eV)
O_3@h—h	0.0	-4.879	-4.484	0.40
	0.5	-4.835	-4.526	0.31
	1.0	-4.811	-4.695	0.12
O_3@h—p	0.0	-4.912	-4.737	0.16
	0.5	-4.944	-4.702	0.24
	1.0	-5.208	-4.802	0.41

Mulliken population analyses show charge transfer to the adsorbed O_3 moiety, explaining the increase of the IP, as compared to pristine SWCNTs (Figure 2). A larger charge transfer was obtained for two O_3 adducts [18].

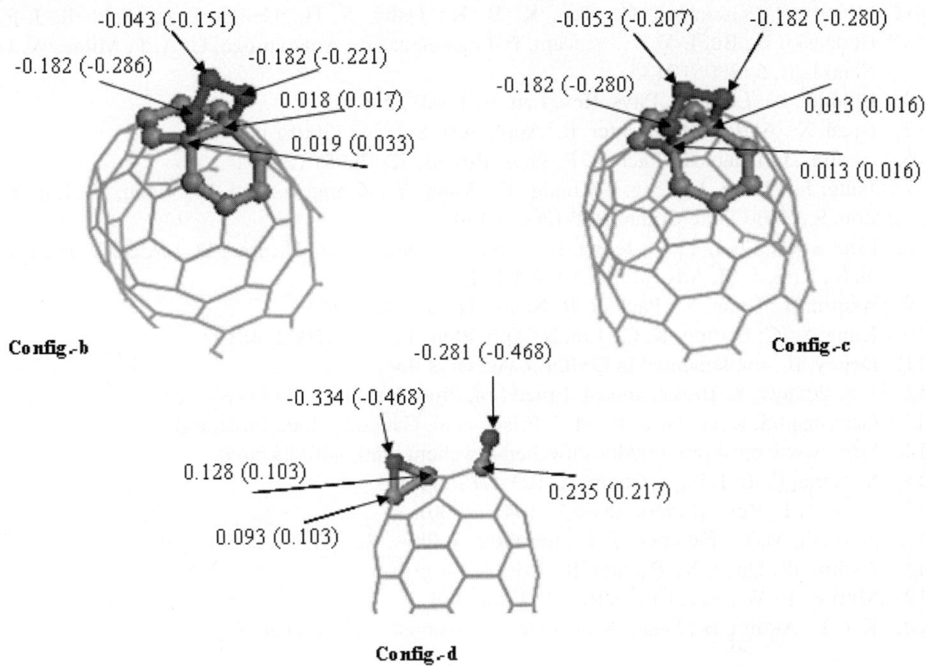

Fig. 2. Mulliken partial atomic charges (e) for O_3 adsorption @h—h and h—p sites; 0 (1) eV/Å applied electric field (corresponding values are given in parentheses)

4 Conclusions

In this study, we report DFT calculations to examine the effects of O_3 adsorption on the field emission properties of capped C(5,5) single-wall carbon nanotubes. The first ionization potential results, including an applied field in the calculations, for possible adsorption sites, have shown a suppression of the emission upon O_3 adsorption, explained in terms of charge transfer, while the favored adsorption for the etched structures rationalizes the enhancement due to sharper tips upon opening of the carbon nanotube when treated with ozone for a long period of time, consistent with experimental observations.

Acknowledgements

The support of the Wright Brothers Institute, Dayton, Ohio, and help provided by the Major Shared Resource Center for High Performance Computing, ASC/HP, are gratefully acknowledged.

References

1. Banerjee, S.; Hemraj-Benny, T.; Wong, S. S. Adv. Mater. 17 (2005) 17
2. Cheng, Y., Zhou, O., C. R. Physique 4 (2003) 1021

3. Minoux, E. Groening, O., Teo, K. B. K., Dalal, S, H., Gangloff, L., Schnell, J.-P., Hudanski, L., Bu, I. Y. Y., Vincent, P. Legagneux, P., Amaratunga, G. A. J., Milne, W. I., Nano Lett. 5 (2005) 2135
4. Buldum, A., Lu, J. P., Phys. Rev. Lett. 91 (2003) 236801
5. Duan, X., Akdim, B., Pachter, R., Appl. Surf. Sci. 243 (2005) 11
6. Akdim, B., Duan, X., Pachter, R. Phys. Rev. B, 72 (2005) 121402
7. Jiang, J., Zhang, J., Feng, T., Jiang, T., Wang, Y., Zhang, F., Dai, L., Wang, X., Liu, X., Zou, S., Solid State Comm. 135 (2005) 390
8. Lim, S. C., Choi, Y. C., Jeong, H. J., Shin, Y. M., An, K. H., Bae, D. J., Lee, Y. H., Lee, N. S., Kim, J. M. Adv. Mater. 13 (2001) 1563
9. Akdim, B.; Duan, X.; Pachter, R. Nano Lett. 3 (2003) 1209
10. Kung, S.–C; Hwang, K. C.; Lin, N. App. Phys. Lett. 80 (2002) 4819
11. Delley, B., implemented in DMOL3, Accelrys, Inc.
12. J. P. Perdew, K. Burke, and M. Ernzerhof, Phys. Rev. Lett. 77 (1996) 3865
13. Gaussian 03, Revision B.05, M. J. Frisch et al. Gaussian, Inc.: Pittsburgh, PA, 2003
14. http://www.emsl.pnl.gov/docs/nwchem/nwchem.html, authors cited.
15. Scuseria, G. E. J. Phys. Chem. A. 103 4782 (1999).
16. Beck, T. L. Rev. Modern Phys. 72 1041 (2000).
17. Nordelli, M. B., Fattebert, J.-L., Bernholc, J. Phys. Rev. B. 64, 245463 (2001).
18. Akdim, B., Duan, X., Pachter, R., work in progress.
19. Murray, R. W., Acc. Chem. Res. 1 (1968) 313
20. Kar, T., Akdim, B.; Duan, X.; Pachter, R., manuscript in preparation.

Developing Metadata Services for Grid Enabling Scientific Applications

Choonhan Youn[*], Tim Kaiser, Cindy Santini, and Dogan Seber

San Diego Supercomputer Center, University of California at San Diego
9500 Gilman Drive
La Jolla, CA 92093-0505
{cyoun, tkaiser, csantini, seber}@sdsc.edu

Abstract. In a web-based scientific computing, the creation of parameter studies of scientific applications is required to conduct a large number of experiments through the dynamic graphic user interface, without paying the expense of great difficulty of use. The generated parameter spaces which include various problems are incorporated with the computation of the application in the computational environments on the grid. Simultaneously, for the grid-based computing, scientific applications are geographically distributed as the computing resources. In order to run a particular application on the certain site, we need a meaningful metadata model for applications as the adaptive application metadata service used by the job submission service. In this paper, we present how general XML approach and our design for the generation process of input parameters are deployed on the certain scientific application as the example and how application metadata is incorporated with the job submission service in SYNSEIS (SYNthetic SEISmogram generation) tool.

1 Introduction

The Grid project [1] is essentially a giant research effort among loosely federated groups to build a seamless computing infrastructure for distributed computing, for example, the cyberinfrastructure for geosciences, GEON [2], national computational grids, TeraGrid [3]. The ultimate vision of Grid computing is that it will be able to provide seamless access to politically and geographically distributed computing resources: a researcher somewhere potentially has access to all the computing power he/she needs to solve a particular problem (assuming he/she can afford this), and all resources are accessed in a homogeneous manner using Grid technologies [4].

In a grid computing environments, scientific applications are deployed into the Grid as resources. For interacting with applications on the Grid, the information of actual applications is necessary. For example, an actual application may be wrapped by XML objects as the application proxy, which can be used to invoke the application, either directly or through submission to a queuing system without modifying the application source code. Thus, we need a general purpose set of schemas that describes how to use a particular application. Having a general application description mechanism allows

---

[*] This research was funded by NSF grants EAR-0353590 and EAR 0225673.

V.N. Alexandrov et al. (Eds.): ICCS 2006, Part I, LNCS 3991, pp. 379–386, 2006.
© Springer-Verlag Berlin Heidelberg 2006

user interfaces to be developed independently of the service deployment. And the application metadata may be discovered and bound dynamically through XML database for storing and querying. For this, we reuse and extend the Application XML Descriptors developed by the Community Grids Lab, Indiana University, Bloomington (More detailed description is available from [5]) as the information repository.

Going beyond simple submission and management of running jobs, for the repeated execution of the same application with different input parameters resulting in different outputs, the creating process of the parameter studies manually is a tedious, time-consuming, and error-prone. Scientists want to simplify the complex parameter study process for obtaining the solutions from their applications using wide varieties of input parameter values. Those parameter studies of high-performance distributed, scientific applications have been a more challenging problem for conducting a large number of experimental simulations. The parameter spaces are related to the individual problem sizes which is obtained through several successive stages on the portal.

In this paper, we describe our designs and initial efforts for building interacting metadata services for the parameter studies and the discovery of applications in SYNSEIS application tool. Using GEON grid environments [6] and national-scale TeraGrid supercomputer centers [3] for high performance computing, the SYNSEIS application tool is developed as a portlet object that can provide the hosting environments interacting with the codes and the data retrieval systems [7]. It is built using a service-based architecture for reusability and interoperability of each constituent service components which are exposed as Web services as well. The SYNSEIS targets a well-tested, parallel finite difference code, e3d developed by Lawrence Livermore National Laboratory [8], with a wide variety of input parameters. Input file data formats in e3d application are described in the various forms depending on the user's selection. We obviously want to support a more scaleable system with common data formats that may be shared between the legacy input file data formats which the code uses. The common data format may be translated in a variety of the legacy input formats. Also, from the portal architecture point of view, it becomes possible to develop general purpose tools for manipulating the common data elements and a well-defined framework for adding new applications. Using the meta-language approach, XML, we present XML schema and our design for related data services for describing the code input parameters. This XML object simply interacts with the hosting environments of the code, converting it to the legacy input data formats, and then allowing the code to be executed, submitted to batch queuing systems, and monitored by users.

2 Related Work

We briefly review here some motivating examples for the parameter studies. Nimrod [9] a tool for managing the execution of parameterized simulations on distributed workstations and combining the advantages of remote queue management systems with those of distributed applications. This tool builds a simple description of the parameters and the necessary control scripts about a particular application for running the code and generates a job for each set from the parameter creation. And then this job is submitted to the remote host and any required files are also transferred to the host.

Unlike Nimrod, users have the ability to access multiple job submission environments including any combination of queue systems such as PBS, LSF, Condor, and so on. In order to continue the job submission autonomously without the continued presence of the parameter study tool, ILab [10] has constructed the GUI for controlling parameter studies using the perl script and Tk tool kit. After creating input files from the parameter studies the job launching process is initiated.

Nimrod and ILab's parameter studies tool is restricted to the parameterization of the input files. On the contrary, ZENTURIO [11] uses a direct-based language to annotate arbitrary files and specify arbitrary application parameters as well as performance metrics. Using this language-based approach, a large number of experiments are potentially generated and submitted to the host.

QuakeSim project developed by the Community Grids Lab, Indiana University, Bloomington [12] is science portal for the earthquake simulation modeling codes. It provides the unifying hosting environment for managing the various applications. The applications are wrapped as XML objects that can provide simple interactions with the hosting environments of the codes, allowing the codes to be executed, submitted to batch queuing systems, and monitored by users through the browser. For support interactions between simulation codes, there are common formats, well-defined XML tags for making legacy input and output parameters using Web service approach.

3 XML Schema for Code Input

XML is an important information technology as it can build and organize metadata to describe resources and the raw data generated either by codes or by scientific instruments. This metadata will enable more precise search methods as envisaged by the Semantic Web. XML has the advantage of being human-readable and hierarchically organized, but is verbose and thus not ideal for very large datasets. Instead, it is more often useful to have the XML metadata description point to the location of the data and describe how that data is formatted, compressed, and to be handled. It may also be transmitted easily between distributed components by using Web service. XML may be used to encode and provide the data structure to code input data files in a structured way. We may thus be quite specific about which definitions of location, resolution, geology, the external source, surfaces, volumes, stations, layers, or other parameters we are actually using within a particular XML document. We do not expect that our definitions will be a final standard adopted by all, but it is useful to qualify all our definitions.

When we examined the inputs for the e3d application which is using in SYNSEIS tool currently, it became apparent that the data may be split into two portions in Figure 1: the code-basic data definitions, and code-optional data definitions for incorporating various code parameters, such as number of stations, and layers. We highlight the major elements here. We structure our XML dialect definitions as being composed of the following:

- Grid Dimension: describes the location, dimension, and the grid spacing for the grid.
- Time Stepping: includes the number of time steps and the time step increment.
- Velocity Model: includes various parameters needed to characterize the griddled velocity model such as p, s, r, or attenuation coefficients.
- Source: includes type, location, amplitude, frequency, fault parameters.
- Seismogram Output: includes location, output name, and mode for writing the seismogram in SAC format [16].
- Image Output: includes the number of time steps for producing a series of mapview images through the surface grid nodes.
- Volume Output: includes the number of time steps for outputting individual data volumes at selected time steps.
- Layer: includes parameters for the crustal model format.

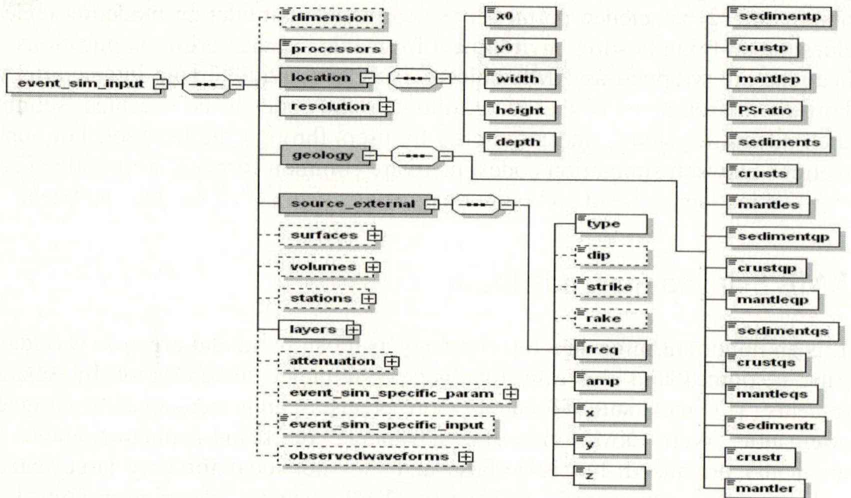

Fig. 1. The main block diagram of the e3d XML Schema

Note that we are not modifying the codes to take directly the XML input. Rather, we use XML descriptions as an independent format that may be used to generate input files for the codes in the proper legacy format. This XML schema simply defines the information necessary to implement the input files in a particular application.

4   Implementing User Interfaces

In the design of user interface, it is very difficult to be faithful to one's functions as well as be beautiful. If the user interface design is too functional, it is stiff and formal. And if that design places great emphasis on the beauty, it is easy to be decorative as well. So, our goal of user interface design is to get the benefit at both functions and the beauty at the same time.

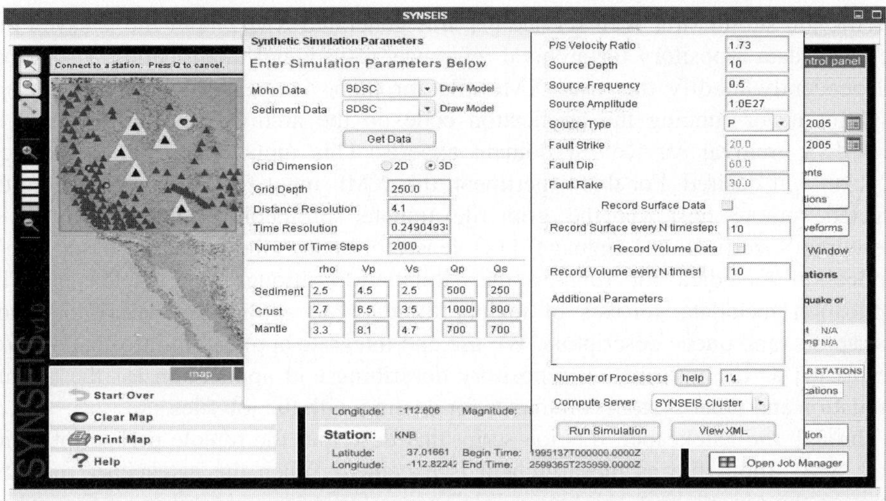

Fig. 2. SYNSEIS's interactive graphic user interface

More or less the e3d application has more complex parameter creation process. We have developed to construct our SYNSEIS Graphic User Interface using the flash application [13] as shown in Figure 2 for minimizing the difficulty of building a set of parameterized input files. Our aim in our SYNSEIS tool is also to provide users with a nice and a variety of user-centric and dynamic environments on the web for representing the input parameters of the earthquake experiments. In Figure 2, as one of scenarios, users are able to select some certain area in US map using the mapping tool and compose the location, several event points and station locations by using the data retrieval Web service [7], and the selected points for the experiment within the boundary area. And then another window specifies other input parameters to allow users to do the graphical selection of the appropriate parameter data fields and designate the set of values by using data model Web service [7]. If users set up the "Distance Resolution" field in Figure 2, "Time Resolution" and "Source Frequency" have been parameterized automatically because of unsuccessful run of the scientific program under the consideration. After the text selection of the appropriate fields, finally the XML input file is generated for running the experiments. Because the parameter creation process is integrated within SYNSEIS GUI, its use is quite easy, intuitive, and trivial.

5   Interactions of the Metadata Service for the Application

We describe how SYNSEIS system exports the XML input file to the specific input data formats of the code and integrate the metadata for the application. In Figure 3, we may express the input file for e3d application using XML format. XML generator collects and composes some data needed for running the code from the user interface, for example, map services, data model Web service, and data descriptors. Using our specific job submission Web service [7], this generated XML input file is saved into the XML data repository for archival reference, re-use, and modification. Users may independently modify this user XML file for their own experiment purposes and resubmit it for running the application code on the archival session provided by SYNSEIS archival service for domain experts. This input XML file is therefore recyclable, if desired. For this experiment, this XML input file is transferred into the targeted remote host via the grid file transfer protocol. Simultaneously, RSL (Resource Specification Language) [14] generator creates the job script based on the application metadata for running the Globus job through the gatekeeper. This application metadata consists of mainly three parts: application descriptors, host descriptors, and queue descriptors. We use and integrate application information Web service [5] as the information repository describing e3d application for the seismic simulation and other system commands for dealing with the job files.

Through the gatekeeper, this job script that describes the remote perl script which takes XML input file as the argument is executed. When the gatekeeper runs this script on the remote host, the "Input File" generator creates several input files which are for actually running the code and the queue script in a remote certain directory for this experiment. At this stage, we must export the XML to the legacy input format for a particular application.

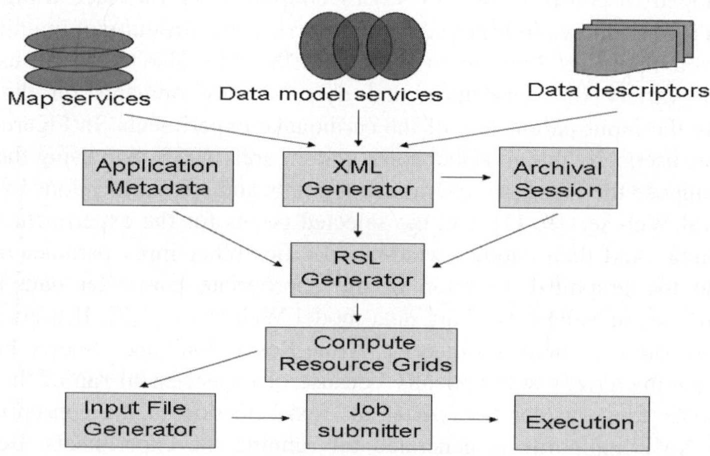

Fig. 3. Processing steps for interactions with the metadata for the application

We assume that a cluster of machines has the job scheduler such as PBS, or Condor. That is, in order for the job submitter to launch jobs run, we need the local or remote compute environments that may require any of Globus, PBS, and other job scheduler. This job scheduler which is accessed from Globus interface [15] is used for queuing and starting jobs. Finally, this batch queue script containing PBS directives followed by shell commands is submitted to the job scheduler by executing the remote perl script.

6 Conclusion and Future Directions

We have built an application-specific tool called SYNSEIS using Web services approach. Since this architecture is based on the service-oriented, those useful services can be plugged into any general frameworks and be put together on the workbench. At the implementation phase of those services, some computing services are required for integrating with the metadata services on the framework. For example, we provide a job submission Web Service to run the relevant applications in a computational and hosting environment, TeraGrid or GEONgrid using Grid technologies, especially Globus interface, including a file transfer. So, we have presented the application XML metadata for constructing the job script for running the code within this service. This application metadata for describing the actual application which we suggested can be used by application web service, having providing the application interfaces as the application proxy. The application descriptor schema contains the "HostBinding" element that indicates the host descriptor, which describes the hosting environment, especially the location, type, parameters of the queuing system. In our application tool, more system commands are required for doing the job and data handling. So, for a more general way, we will keep extending this schema to put the system environments.

In the design of the user interface, we have also implemented the easy-to-use parameterization process for complex earthquake simulations using SYNSEIS SWF (the file format used by Macromedia Flash) code. Through the dynamic graphic user interface, users can select and compose the data items. And then XML input data object is generated for the user experiments on the computational environments. Currently, since we provide simply the event-station pair and point source case for this experiment, we will take into consideration providing event-multiple stations situations, a line source implementation, and multiple seismogram plots additionally.

On the back end, at the time of submitting the job that contains users' XML file, the remote perl script consists of mainly two parts: the converting the XML file into the legacy input file and the job launching process. Because its process is tightly coupled, if both parts are not run successfully, for example, the converting process is done, but the job submission is failed, the remote perl script running is not useful even if the converting process is successful. For more modularity and reusability, those steps are redesigned. Creating the input files and the job directory for this experiment from the XML file will be generated by using Web service. For example, the QuakeSim portal [12] has the capability for exporting and importing the XML input file into the legacy format between applications running. Using the common XML format, QuakeSim applications are able to communicate each other via the Data hub. The current job submission is used for the job scheduler directly. That is, the job is submitted to the job scheduler directly. As the effective way, in order to get full functionalities provided by Globus interface, the launching and monitoring job will be performed through the Globus metacomputing middleware.

References

1. Berman, F., Fox, G., and Hey, T.: Grid Computing: Making the Global Infrastructure a Reality. Wiley, 2003
2. Cyberinfrastructure for the Geosciences: http://www.geongrid.org
3. TeraGrid project: http://www.teragrid.org
4. Fox, G. C., Gannon, D., and Thomas, M. A Summary of Grid Computing Environments. *Concurrency and Computation: Practice and Experience*, Vol. 14, No. 13-15, pp 1035-1044, 2002
5. Youn, C., Pierce, M., and Fox, G., "Building Problem Solving Environments with Application Web Service Toolkits" ICCS 2003 Workshop on Complex Problem Solving Environments for Grid Computing, LNCS 2660, pp. 403-412, 2003
6. C.Youn, C. Baru, K. Bhatia, S. Chandra, K. Lin, A. Memon, G. Memon and D. Seber. GEONGrid Portal: Design and Implementations. *GCE 2005: Workshop on Grid Computing Portals* held in conjunction with SC 2005, Seattle, WA, USA, November 12-18, 2005
7. C. Youn, T. Kaiser, C. Santini and D. Seber. Design and Implementation of Services for a Synthetic Seismogram Calculation Tool on the Grid. *ICCS 2005: 5th International Conference*, Atlanta, GA, USA, May 22-25, 2005, Proceedings, Part 1, LNCS 3514, pp. 469-476, 2005
8. Larsen, S.: e3d: 2D/3D Elastic Finite-Difference Wave Propagation Code. Available from http://www.seismo.unr.edu/ftp/pub/louie/class/455/e3d/e3d.txt
9. Abramson, D., Sosic, R.,Giddy, J., Hall, B.: Nimrod: A Tool for Performing Parametised Simulations using DistributedWorkstations. The 4th IEEE Symposium on High Performance Distributed Computing, Virginia, August 1995 IEEE Computer Society Press, Silver Spring, MD, 1995, pp. 520–528
10. M. Yarrow, K.M. McCann, R. Biswas, R.F. Van der Wijngaart, An Advanced User Interface Approach for Complex Parameter Study Process Specification on the Information Power Grid, in Proceedings of the First IEEE/ACM International Workshop on Grid Computing, Bangalore, India, December 2000, Lecture Notes in Computer Science, Vol. 1971, Springer, London, UK 2000, 146 – 157
11. R. Prodan, T. Fahringer, ZENTURIO: a grid middleware-based tool for experiment management of parallel and distributed applications. Journal of Parallel and Distributed Computing, Vol. 64 (6), Academic Press, Orlando, FL, USA, 2004, pp. 693 – 707
12. M. Pierce, C. Youn and G. Fox, Interacting Data Services for Distributed Earthquake Modeling, ICCS 2003 Workshop on Computational Earthquake Physics and Solid Earth System Simulation, LNCS 2659, pp. 863-872, 2003
13. Allaire, J.: Macromedia Flash MX—A next-generation rich client. March 2002. Available from http://www.macromedia.com/devnet/mx/flash/whitepapers/richclient.pdf
14. RSL v1.0. See http://www-fp.globus.org/gram/rsl_spec1.html
15. Gregor von Laszewski, Ian Foster, Jarek Gawor, and Peter Lane. A Java Commodity Grid Kit," Concurrency and Computation: Practice and Experience, vol. 13, no. 8-9, pp. 643-662, 2001
16. SAC – Seismic Analysis Code: http://www.llnl.gov/sac/

In Silico Three Dimensional Pharmacophore Models to Aid the Discovery and Design of New Antimalarial Agents

Apurba K. Bhattacharjee*, Mark G. Hartell, Daniel A. Nichols, Rickey P. Hicks, John E. van Hamont, and Wilbur K. Milhous

Department of Medicinal Chemistry, Division of Experimental Therapeutics,
Walter Reed Army Institute of Research, 503 Robert Grant Avenue,
Silver Spring, MD 20910, U.S.A.
apurba.bhattacharjee@na.amedd.army.mil

Abstract. Malaria is one of the most dangerous diseases affecting primarily poor people of tropical and subtropical regions. The search for novel drugs against specific parasites is an important goal for antimalarial drug discovery. This study describes how 3D pharmacophores for antimalarial activity could be developed from known antimalarials and be used as screening tools for virtual compound libraries to identify new antimalarial candidates with examples of indolo[2,1-b]quinazoline-6,12-diones (tryptanthrins) that exhibited *in vitro* activity below 100 ng/mL. These models mapped on the potent analogues and also onto other well-known antimalarial drugs of different chemical classes including quinolines, chalcones, rhodamine dyes, Pfmrk CDK inhibitors, malarial KASIII inhibitors, and plasmepsin inhibitors. The pharmacophores allowed search and identification of new antimalarials from in-house multi-conformer 3D CIS database and enabled custom designed synthesis of new potent analogues that are found to be potent against *in vitro* W2, D6, and TM91C235 strains of *P. falciparum*.

1 Introduction

Malaria, one of the most severe of the human parasitic diseases, causes about 500 million infections worldwide and approximately 3-5 million deaths every year. The search for novel antimalarial drugs against specific parasitic targets is an important goal for antimalarial drug discovery. [1] The situation is rapidly worsening mainly due to non-availability of effective drugs and development of drug resistance of a large number of non-immune people in areas where malaria is frequently transmitted. [1] Chloroquine, mefloquine, and other frontline drugs for the treatment and prevention of malaria, are becoming increasingly ineffective. [1] Artemisinin analogues such as artesunate and arteether were later introduced and found to be quite effective, particularly against drug-resistant *P. falciparum*, but observations of drug-induced and dose-related neurotoxicity in animals have raised concern about the safety of these compounds for human use.[2,3] Therefore, much effort and attention are needed for discovery and development of new and less toxic antimalarial drugs.

Indolo[2,1-b]quinazoline-6,12-dione, known as tryptanthrin, is an alkaloid isolated from the Taiwanese medicinal plant *Strobilanthes cusia*, and its substituted

V.N. Alexandrov et al. (Eds.): ICCS 2006, Part I, LNCS 3991, pp. 387–394, 2006.

derivatives (Table 1) were recently studied at the Walter Reed Army Institute of Research, Silver Spring, Maryland, U.S.A.[4] and found to have remarkable *in vitro* antimalarial activity against *P. falciparum*, both sensitive and multidrug-resistant strains. The more potent analogs exhibit IC50 values in the range 0.43 to 10 ng/mL, about one one-thousandth of the concentrations necessary to inhibit bacteria. Furthermore, the compounds are also found to be highly potent against strains of *P. falciparum* that are up to 5000-fold resistant to atovoquone, 50-fold resistant to chloroquine, and 20-fold resistant to mefloquine. Thus, this novel class of compounds has opened a new chapter for study in the chemotherapy of malaria.

Computer assisted molecular modeling (CAMM) has made remarkable progress in recent years in the design and discovery of new potential bioactive chemical entities.[5] The current advances in these methodologies allow direct applications ranging from accurate *ab initio* quantum chemical calculations of stereoelectronic properties, generation of three-dimensional pharmacophores, virtual database searches to identify potent bioactive agents, to simulate and dock drug molecules in the binding pockets of proteins that are crucial and directly related to biomedical research.

In continuation of our efforts to design and discover new antimalarial therapeutic agents using computational methodologies [6,7], we present here our reported [4] 3D chemical feature based pharmacophore model and its utility as a tool for 3D database searches for identification of new structurally different classes of compounds for further study.

2   Results and Discussion

Three-dimensional pharmacophore for antimalarial activity was found to contain two hydrogen bond acceptor (lipid) functions and two aromatic hydrophobic functions at a specific geometric positions (Fig. 1) of the tryptanthrins was reported by us earlier. [5] The pharmacophore was developed using the CATALYST methodology [8] from a set of 17 structurally diverse tryptanthrin derivatives including the parent compound as the training set shown in Table 1. The correlation between the experimental and predicted activity of these compounds were found to be as R = 0.89. The more potent analogs of the training set (Table 1) such as **1** and **4** map well with the pharmacophore (Fig. 2a & 2b) whereas, the less potent analogues such as **11** and **16** do not map adequately with the hypothesis (Fig. 2c & 2d). Cross-validation of the generated pharmacophore was carried out by preparing a "test set" of 15 additional tryptanthrin compounds (Table 1) that were tested for *in vitro* antimalarial activity against D6 and W2 clones of *P. falciparum* identical to the original training set. The predicted and the experimental IC50 values for the test set tryptanthrins are found reasonably well reproduced [4], correlation as R = 0.92, within the limits of uncertainty [3], thus demonstrating the predictive power of the original pharmacophore. As observed in the training set, the more potent analogues of the test set such as **4'** and **11'** map well (Fig. 3a & 3b) with the pharmacophore whereas, the less potent analogues of the test set do not map adequately.

To further examine the validity of the pharmacophores, its features were mapped onto a series of eight antimalarial drugs that are currently used in the United States [3], viz., quinine, chloroquine, mefloquine, primaquine, hydroxychloroquine, pyrimethamine, sulfadoxine, and doxycycline (Fig. 4). The mapping indicates that quinine completely satisfies the requirement with the pharmacophore (Fig. 4a) whereas, the other drugs map in varying degrees. It may be worthwhile to mention here that quinine and other quinoline-containing antimalarials including chloroquine have shown varying capacity to inhibit malaria heme polymerase extracted from *P. falciparum* tropozoites. [9] In particular, the interaction of quinine with heme has been well documented. [10] Since the pharmacophore maps well on quinine and in varying degrees on the other quinoline-containing antimalarials, it may be reasonable to speculate that the tryptanthrin compounds may target heme polymerase from the *P. falciparum* tropozoites. Preliminary results from NMR experiments on select tryptanthrin analogues have shown positive indications toward inhibition of the hemin polymerization process (unpublished results). In addition, we have also mapped the pharmacophore on a few recently reported potent antimalarials such as the chalcones11 and rhodacyanine dyes. [12] Surprisingly again, the more potent chalcone analogues such as, 2',3',4'-trimethoxy-4-trifluoromethyl and 2',4'-dimethoxy-4-ethyl chalcones (Fig 5a & 5b) and the rhodacyanine dye MKT-077 and its *para* analogue map well on to the tryptanthrin pharmacophore (Fig 6a & Fig 6b). In addition, the pharmacophore led to a search of the in-house multiconformer 3D databases that led to the identification of a fairly potent plasmepsin inhibitor: 6-diethylamino-2,3-diphenylbenzofuran, two excellent Pfmrk inhibitors from the tryptanthrin classes, such the 8-nitro- and 7-chloro-4-azatryptanthrins, and a moderately good inhibitor, 4-(benzenesulfonamido)-salicylic acid for the malarial KASIII enzyme. [4]

Table 1. Structure of Tryptanthrins

#	1-	2-	3-	4-	X	1'	2'	3'	4'
Training set:									
1	CH	CH	CH	-N=	C=O	CH	CH	C-Cl	CH
2	CH	CH	C-F	-CH	C=O	CH	CH	C-Cl	CH
3	CH	-N=	CH	CH	C=O	CH	CH	C-CH$_2$ CH$_3$	CH
4	CH	CH	C-N <(CH$_2$)$_4$> N-CH$_3$	CH	C=O	CH	CH	C-Cl	CH
5	CH	-N=	C-CH$_3$	-N=	C=O	CH	CH	CH	CH
6	CH	CH	CH	-CH	C=O	CH	CH	CH	CH
7	CH	-N=	CH	-CH	C=O	CH	CH	CCH- C$_7$H$_{16}$	CH

Table 1. (*continued*)

8	-N=	CH	CH	-N=	C=O	CH	CH	CH	CH
9	CH	CH	CH	C-OCH$_3$	C=C-phenyl	CH	CH	C-F	CH
10	CH	CH	C-F	CH	C=O	CH	C-N<(CH$_2$)$_4$>N-CH$_3$	C-F	CH
11	CH	CH	CH	CH	-O-	CH	CH	CH	CH
12	CH	-N=	CH	CH	-S-	CH	CH	CH	CH
13	CH	CH	CH	-N=	C=O	CH	C-Cl	CH	CH
14	-N=	C-OH	-N=	C-OH	C=O	CH	CH	C-I	CH
15	CH	CH	CH	CH	C=indole	CH	CH	CH	CH
16	CH	CH	CH	CH	C=C-C=C-phenyl	CH	CH	CH	CH
17	CH	CH	CH	CH	C-dioxane	CH	CH	C-Br	CH

Test set :

1[*]	CH	-N=	CH	CH	C=O	CH	CH	C-C$_8$H$_{17}$	CH
2'	CH	CH	CH	-N=	C=O	CH	CH	C-Cl	CH
3'	CH	CH	CH	CH	C=O	CH	CH	CH	C-Cl
4'	CH	CH	C-S-C$_2$H$_4$OH	CH	C=O	CH	CH	C-Cl	CH
5'	CH	-N=	CH	CH	C=O	CH	CH	C-C$_4$H$_9$	CH
6'	CH	-N=	CH	CH	C=O	CH	CH	C$_2$H$_5$	CH
7'	CH	CH	CH	C-OCH$_3$	C=O	CH	CH	CH	CH
8'	CH	-N=	CH	CH	C=O	CH	CH	C-CHOCH$_3$C-(CH$_3$)$_2$	CH
9'	CH	CH	CH	CH	C=O	CH	CH	C-OCF$_3$	CH
10'	CH	CH	CH	-N=	C=O	CH	CH	C-I	CH
11'	CH	CH	CH	C-OCH$_3$	C=O	CH	CH	C-I	CH
12'	CH	CH	C-piperidine	CH	C=O	CH	CH	C-Cl	CH

Table 1. (*continued*)

13'	CH	-N=	CH	CH	C=O	CH	CH	C-Br	CH
14'	CH	CH	CNCH$_3$(CH$_2$)$_2$OH	CH	C=O	CH	CH	C-Cl	CH
15'	CH	C-CH$_3$	CH	CH	C=O	CH	CH	CH	CH

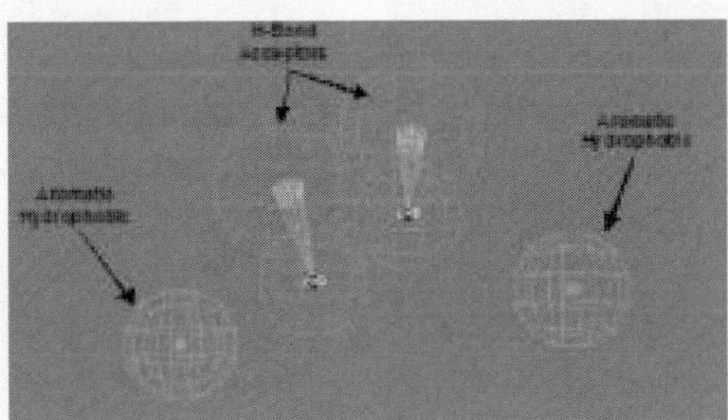

Fig. 1. Pharmacophore for antimalarial activity of the tryptanthrins

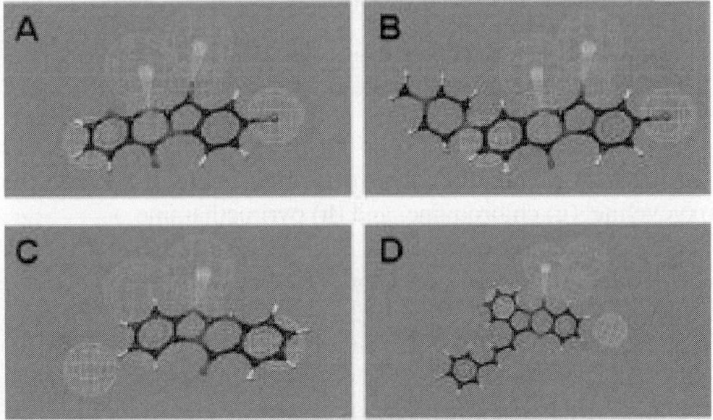

Fig. 2. Mapping of the more potent analogues: (a) **1** and (b) **4**, and less potent analogues: (c) **11** and (d) **16**, onto the pharmacophore model

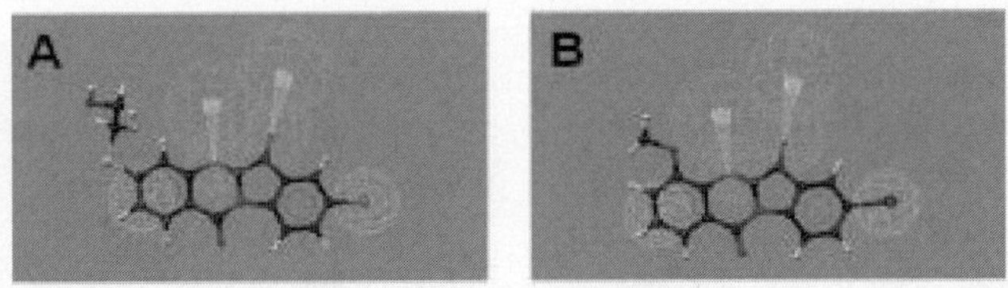

Fig. 3. Mapping of the more potent analogues: (a) **4'** and (b) **11'** of the test set onto the pharmacophore

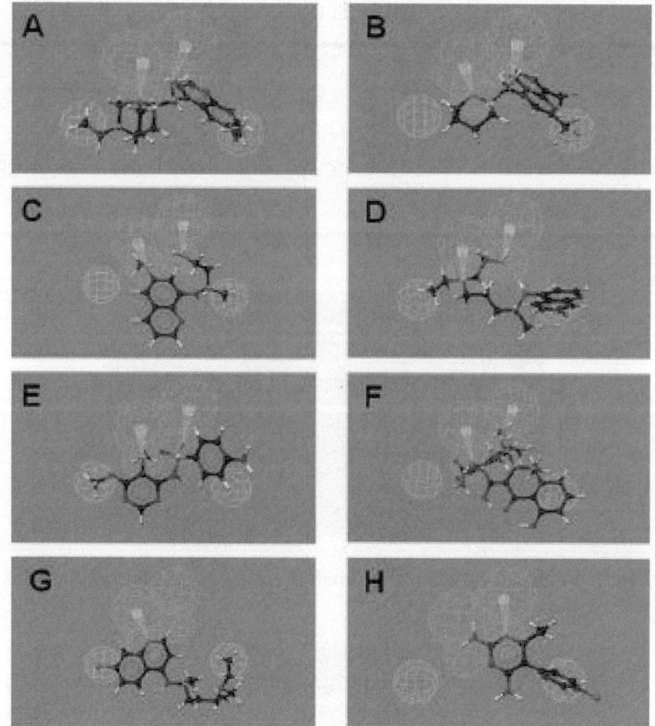

Fig. 4. Mapping of the pharmacophore onto eight commonly used antimalarial drugs in the United States: (a) quinine, (b) mefloquine, (c) primaquine, (d) hydroxyxhloroquine, (e) sulfadoxine, (f) doxycycline, (g) chloroquine, and (h) pyrimethamine

Fig. 5. Mapping of the pharmacophore on (a) 4,-trifluoromethyl-2',3',4'-trimethoxychalcone, and (b) 4-ethyl-2',4'-dimethoxychalcone

Fig. 6. Mapping of the pharmacophore on (a) rhodacyanine dye MKT-077 and (b) *para*-rhodacyanine dye MKT-077

3   Conclusions

The study demonstrates how the chemical features such as the hydrogen bond donors, acceptors, hydrophobicity etc. of potent compounds can be organized to develop pharmacophores for activity and utilize them for virtual *in silico* search of new class of compounds to open new chapters for chemotherapeutic study. Specifically, in this investigation we have shown how molecular characteristics from a set of diverse tryptanthrin derivatives may be organized to be both statistically and mechanistically significant for potent antimalarial activity that may have universal applicability. In addition, the resulting model can be used to unravel a possible rationale for the target-specific antimalarial activity of these compounds. The validity of the pharmacophore extends to structurally different class of compounds, and thereby provides a powerful template from which novel drug candidates may be identified for extended study. Since the identity of the target for antimalarial activity of these compounds is unknown, the three-dimensional QSAR pharmacophore should aid in the design of well-tolerated target-specific antimalarial agents.

4   Experimental Section

Biological testing and procedure adopted for the pharmacophore generation utilizing the CATALYST [8] methodology is described earlier. [5]

Acknowledgement

The work was funded under a Military Infectious Disease Research Program grant. The opinions expressed herein are the private views of the authors and are not to be considered as official or reflecting the views of the Department of Army or the Department of Defense.

References

1. Malaria Foundation International, http://www.malaria.org/, and sites given therein.
2. Vroman, J.A., Gaston, M.A., Avery, M.A.: Current Progress in the Chemistry, Medicinal Chemistry and Drug Design of Artemisinin Based Antimalarials. Curr. Pharm. Design. 5 (1999) 101-138.
3. Bhattacharjee, A.K., Karle, J.M.: Stereoelectronic Properties of Antimalarial Artemisinin Analogues in Relation to Neurotoxicity. Chem. Res. Toxicol. 12 (1999) 422- 428.
4. Bhattacharjee, A.K., Hartell, M.G., Nichols, D.A., Hicks, R.P., Stanton, B., van Hamont, J.E., Milhous, W.K.: Structure-activity relationship study of antimalarial indolo [2,1-b]quinazoline-6-12-diones (tryptanthrins). Three dimensional pharmacophores modeling and identification of new antimalarial candidates. European J. Med. Chem. 39 (2004) 59-67.
5. Buchwald, P., Bodor, N.: Computer-aided drug design: the role of quantitative structure-property, structure-activity and structure-metabolism relationships (QSPR, QSAR, QSMR). Drug Future. 27 (2002) 577-588.

6. Bhattacharjee, A.K., Kyle, D.E., Vennerstrom, J.L., Milhous, W.K.: A 3D QSAR Pharmacophore Model and Quantum Chemical Structure Activity Analysis of Chloroquine(CQ)-Resistance Reversal. J. Chem. Info. Comput. Sci. 42 (2002) 1212-1220.
7. Bhattacharjee, A.K., Geyer, J. A., Woodard, C.L., Kathcart, A.K., Nichols, D.A., Prigge, S.T., Li, Z., Mott, B.T., Waters, N.C.: A Three Dimensional In Silico Pharmacophore Model for Inhibition of *Plasmodium Falciparum* Cyclin Dependent Kinases and Discovery of Different Classes of Novel Pfmrk Specific Inhibitors. J. Med. Chem. 47 (2004) 5418-5426.
8. CATALYST Version 4.5 software, 2000, Accelrys Inc., San Diego, CA.
9. Slater, A.F.G., Cerami, A.: Inhibition of chloroquine of a novel haem polymerase enzyme activity in malaria trophozoites. Nature 355 (1992) 167-169.
10. Meshnick, S.R.: In Malaria Parasite Biology, Pathogenesis and Protection: From quinine to qinghaosu: historical perspectives. Sherman, I.W. (eds.): ASM Press, Washington, D.C. (1998) 341-353.
11. Liu, M., Wilairat, P., Go, M.L.: Antimalarial alkoxylated and hydroxylated chalcones: structure-activity relationship analysis. J. Med. Chem. 44 (2001) 4443-4452.
12. Takasu, K.; Inoue, H.; Kim, H.K.; Suzuki, M.; Shishido, T.; Wataya, Y.; Ihara, M. Rhodacyanine dyes as antimalarials. 1. Preliminary evaluation of their activity and toxicity. J. Med. Chem. 45 (2002) 995-998.

Fuzzy Logic Speech/Non-speech Discrimination for Noise Robust Speech Processing

R. Culebras, J. Ramírez, J.M. Górriz, and J.C. Segura

Dept. of Signal Theory, Networking and Communications
University of Granada, Spain

Abstract. This paper shows a fuzzy logic speech/non-speech discrimination method for improving the performance of speech processing systems working in noise environments. The fuzzy system is based on a Sugeno inference engine with membership functions defined as combination of two Gaussian functions. The rule base consists of ten fuzzy if then statements defined in terms of the denoised subband signal-to-noise ratios (SNRs) and the zero crossing rates (ZCRs). Its operation is optimized by means of a hybrid training algorithm combining the least-squares method and the backpropagation gradient descent method for training membership function parameters. The experiments conducted on the Spanish SpeechDat-Car database shows that the proposed method yields clear improvements over a set of standardized VADs for discontinuous transmission (DTX) and distributed speech recognition (DSR) and also over recently published VAD methods.

1 Introduction

The deployment of new wireless speech communication services finds a serious implementation barrier in the harmful effect of the acoustic noise present in the operating environment. These systems often benefits from voice activity detectors (VADs) which are frequently used in such application scenarios for different purposes.

Detecting the presence of speech in a noisy signal is a problem affecting numerous applications including robust speech recognition [1, 2], discontinuous transmission (DTX) in voice communications [3, 4] or real-time speech transmission on the Internet [5]. The classification task is not as trivial as it appears, and most of the VAD algorithms often fail in high noise conditions. During the last decade, numerous researchers have developed different strategies for detecting speech in a noisy signal [6, 7, 8, 9] and have evaluated the influence of the VAD effectiveness on the performance of speech processing systems [10, 11, 12, 13].

Since its introduction in the late sixties [14], fuzzy logic has enabled defining the behavior of many systems by means of qualitative expressions in a more natural way than mathematical equations. Thus, an effective approach for speech/non-speech discrimination in noisy environments is to use these techniques that enables describing the decision rule by means of if-then clauses which

V.N. Alexandrov et al. (Eds.): ICCS 2006, Part I, LNCS 3991, pp. 395–402, 2006.

are selected based on the knowledge of the problem. Beritelli [15] showed a robust VAD with a pattern matching process consisting of a set of six fuzzy rules. However, no specific optimization was performed at the signal level since the system operated on feature vectors defined by the popular ITU-T G.729 speech coding standard [4]. This paper shows an effective VAD based on fuzzy logic rules for low-delay speech communications. The proposed method combines a noise robust speech processing feature extraction process together with a trained fuzzy logic pattern matching module for classification. The experiments conducted on speech databases shows that the proposed method yields improvements over different standardized VADs for DTX and distributed speech recognition (DSR) and other VAD methods.

2   Fuzzy Logic

Fuzzy logic [16] is much closer in spirit to human thinking and natural language than traditional logic systems. Basically, it provides an effective means of capturing the approximate, inexact nature of the real world. Viewed in perspective, the essential part of a fuzzy logic system is a set of linguistic rules related by the dual concepts of fuzzy implication and the compositional rule of inference.

Fuzzy logic consists of a mapping between an input space and an output space by means of a list of if-then statements called rules. These rules are useful because they refer to variables and the adjectives that describe these variables. The mapping is performed in the fuzzy inference stage, a method that interprets the values in the input vector and, based on some set of rules, assigns values to the output.

Fuzzy logic starts with the concept of a fuzzy set. A fuzzy set F defined on a discourse universe U is characterized by a membership function $\mu_F(x)$ which takes values in the interval $[0, 1]$. A fuzzy set is a generalization of a crisp set. A membership function provides the degree of similarity of an element in U to the fuzzy set. A fuzzy set F in U may be represented as a set of ordered pairs of a generic element x and its grade of membership function: $F = \{(x, \mu_F(x))|x \in U\}$.

The concept of linguistic variable was first proposed by Zadeh who considered them as variables whose values are not numbers but words or sentences in a natural or artificial language. A membership function $\mu_F(x)$ is a curve that defines how each point in the input space is mapped to a membership value (or degree of membership) between 0 and 1. The most commonly used shapes for membership functions are triangular, trapezoidal, piecewise linear and Gaussian. Membership functions were chosen by the user arbitrarily in the past, based on the user's experience. Now, membership functions are commonly designed using optimization procedures. The number of membership functions improves the resolution at the cost of greater computational complexity. They normally overlap expressing the degree of membership of a value to different attributes.

Fuzzy sets and fuzzy operators are the subjects and verbs of fuzzy logic. Fuzzy logic rules based on if-then statements are used to formulate the conditional statements that comprise fuzzy logic. A single fuzzy if-then rule assumes the form if x is F then y is G where F and G are linguistic values defined by fuzzy

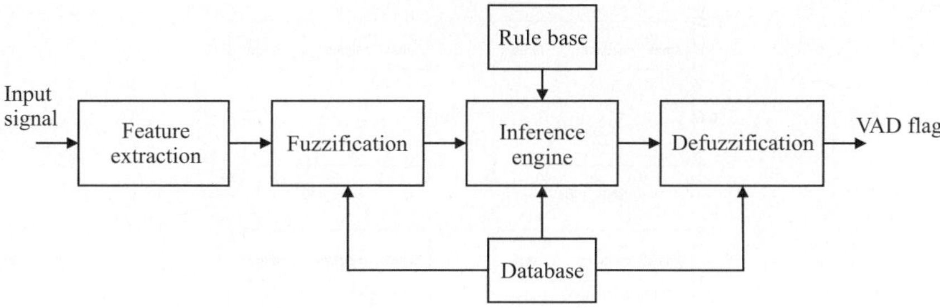

Fig. 1. Feature extraction

sets. The if-part of the rule is called the antecedent or premise, while the then-part of the rule is called the consequent or conclusion. Interpreting an if-then rule involves distinct parts: *i*) evaluating the antecedent (which involves fuzzifying the input and applying any necessary fuzzy operators), and *ii*) applying that result to the consequent.

3   Voice Activity Detection

Figure 1 shows the basic configuration of a fuzzy logic VAD which comprises five principal components: *i*) the feature extraction process that prepares discriminative speech feature for the fuzzy logic classifier, *ii*) the fuzzification interface performs a scale mapping, that transfers the range of values into the corresponding universe of discourse and performs the function of fuzzification, that converts input data into suitable linguistic variables viewed as labels of fuzzy sets, *iii*) the knowledge base comprises a knowledge of the application domain and the objective of the VAD. It consists of a database, which provides necessary definitions which are used to define linguistic VAD rules and a linguistic (fuzzy) rule base, which characterizes the VAD goal by means of a set of linguistic rules and the user experience, *iv*) the decision making logic is the kernel of the fuzzy logic VAD. It has the capability of simulating human decision making based on fuzzy concepts and of inferring actions employing fuzzy implication and the inference rules, and *v*) the defuzzification interface performs a scale mapping, which converts the range of output values into the corresponding universe of discourse, and defuzzification, which yields a nonfuzzy VAD flag.

3.1   Feature Extraction

The feature vector consists of the Zero Crossing Rates (ZCR) defined by:

$$\text{ZCR} = \frac{\sum_{n=1}^{N-1} |\operatorname{sign}(x(n)) - \operatorname{sign}(x(n-1))|}{2} \tag{1}$$

and the K-band signal-to-noise ratios (SNR) that are calculated after a previous denoising process by means of a uniformly distributed filter bank on the discrete Fourier spectrum of the denoised signal.

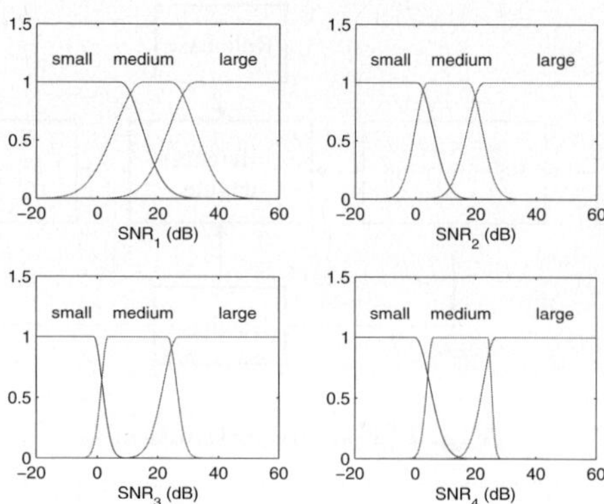

Fig. 2. Membership functions for subband SNRs

3.2   Inference Engine

A Sugeno inference engine was preferred over Mamdani's method since: i) it is computationally efficient, ii) it works well with linear techniques, iii) it works well with optimization and adaptive techniques, iv) it has guaranteed continuity of the output surface and v) it is well-suited to mathematical analysis. Once the inputs have been fuzzified, we know the degree to which each part of the antecedent has been satisfied for each rule. The input to the fuzzy operator is two or more membership values from fuzzified input variables. Any number of well-defined methods can fill in for the AND operation or the OR operation. We have used the product for AND, the maximum for OR and the weighted average as the defuzzification method. Finally, the output of the system is compared to a fixed threshold η. If the output is greater than η, the current frame is classified as speech (VAD flag= 1) otherwise it is classified as non-speech or silence (VAD flag= 0). We will show later that modifying η enables the selection of the VAD working point depending on the application requirements.

3.3   Membership Function Definition

The initial definition of the membership functions is based on the expert knowledge and the observation of experimental data. After the initialization, a training algorithm updates the system in order to obtain a better definition of the membership functions. Two-sided Gaussian membership functions were selected. They are defined as a combination of Gaussian functions

$$f(x; \mu_1, \sigma_1, \mu_2, \sigma_2) = f_1(x; \mu_1, \sigma_1) f_2(x; \mu_2, \sigma_2)$$

$$f_i(x; \mu_i, \sigma_i) = \begin{cases} \exp\left(-\frac{(x-\mu_i)^2}{2\sigma_i^2}\right) & x \le \mu_i \\ 1 & x > \mu_i \end{cases} \tag{2}$$

where the first function specified by σ_1 and μ_1, determines the shape of the leftmost curve while the second function determines the shape of the rightmost

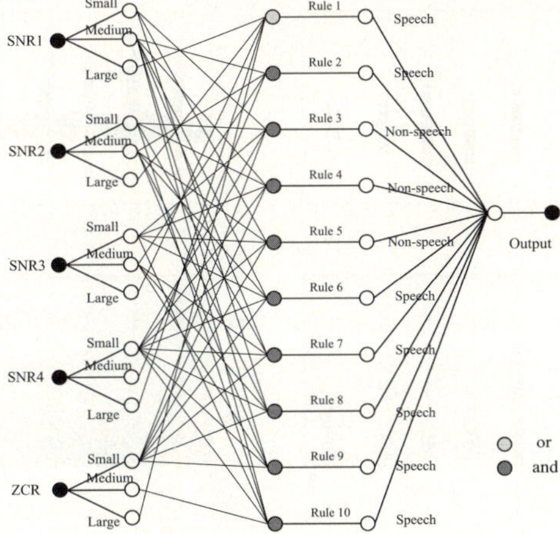

Fig. 3. Rule base

curve. Figure 2 shows the membership functions for the $K = 4$ subband SNRs that are used together with the ZCR as inputs of the fuzzy logic VAD.

3.4   Rule Base

The rule base consists of ten fuzzy rules which were trained using ANFIS [17]. It applies a combination of the least-squares method and the backpropagation gradient descent method for training membership function parameters to emulate a given training data set. An study of the better conditions for the training processed was carried out using utterances of the Spanish SpeechDat-Car (SDC) database [18]. This database contains 4914 recordings using close-talking (channel 0) and distant microphones (channel 1) from more than 160 speakers. The files are categorized into three noisy conditions: quiet, low noisy and highly noisy conditions, which represent different driving conditions. Four different training sets were used: *i*) quiet ch1, *ii*) low ch1 *iii*) high ch1, and *iv*) a combination of utterances from the three previous subsets. Training with data from the three categories yielded the best results in speech/pause discrimination.

Figure 4 shows the operation of the VAD on an utterance of the SpeechDat-Car database recorded with the close talking and distant microphones. It shows the effectiveness of the fuzzy logic pattern matching for discriminating between speech and non-speech even in a high noise environments.

4   Experiments

This section analyzes the proposed VAD and compares its performance to other algorithms used as a reference. The analysis is based on the ROC curves [19], a frequently used methodology to describe the VAD error rate. The Spanish

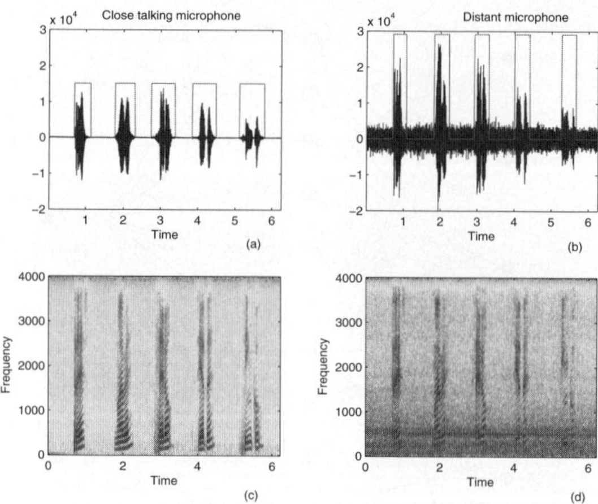

Fig. 4. Operation of the fuzzy VAD on an utterance of the Spanish SpeechDat-Car database. VAD decision for utterances recorded with close talking (a) and distant microphones (b) and associated spectrograms.

SDC database [18] was used. The non-speech hit rate (HR0) and the false alarm rate (FAR0= 100-HR1) were determined for each noisy condition being the actual speech frames and actual speech pauses determined by hand-labelling the database on the close-talking microphone. Figure 5 shows the ROC curves of the proposed VAD and other frequently referred algorithms [20, 21, 19, 6] for recordings from the distant microphone in quiet and high noisy conditions. The working points of the ITU-T G.729, ETSI AMR and ETSI AFE VADs are also included. The results show improvements in speech detection accuracy over standard VADs and a representative set of recently reported VAD algorithms [20, 21, 19, 6].

5   Conclusion

This paper showed the effectiveness of fuzzy logic concepts for robust speech/ non-speech discrimination. The fuzzy system is based on a Sugeno inference engine with membership functions defined as combination of two Gaussian functions. The rule base consists of ten fuzzy if then statements defined in terms of the denoised subband signal-to-noise ratios (SNRs) and the zero crossing rates (ZCRs). Its operation is optimized by means of a hybrid training algorithm combining the least-squares method and the backpropagation gradient descent method for training membership function parameters. With these and other innovations the proposed algorithms unveils significant improvements over ITU-T G.729, ETSI AMR and ETSI AFE standards as well as over VADs that define the decision rule in terms of averaged subband speech features.

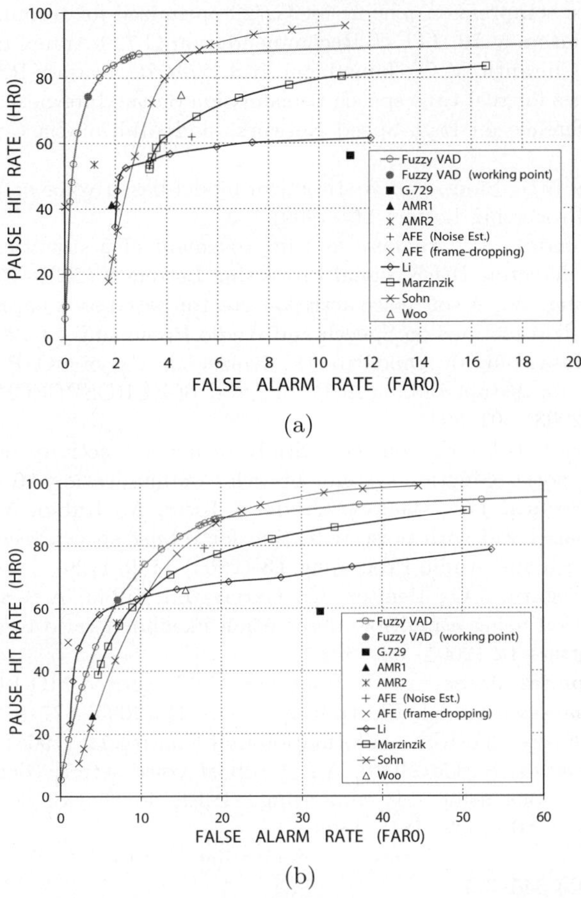

Fig. 5. Comparative results. *a*) Quiet ch1, *b*) High ch1.

Acknowledgements

This work has been funded by the European Commission (HIWIRE, IST No. 507943) and the Spanish MEC project TEC2004-03829/FEDER.

References

1. Karray, L., Martin, A.: Towards improving speech detection robustness for speech recognition in adverse environments. Speech Communition (2003) 261–276
2. Ramírez, J., Segura, J.C., Benítez, M.C., de la Torre, A., Rubio, A.: A new adaptive long-term spectral estimation voice activity detector. In: Proc. of EUROSPEECH 2003, Geneva, Switzerland (2003) 3041–3044
3. ETSI: Voice activity detector (VAD) for Adaptive Multi-Rate (AMR) speech traffic channels. ETSI EN 301 708 Recommendation (1999)

4. ITU: A silence compression scheme for G.729 optimized for terminals conforming to recommendation V.70. ITU-T Recommendation G.729-Annex B (1996)
5. Sangwan, A., Chiranth, M.C., Jamadagni, H.S., Sah, R., Prasad, R.V., Gaurav, V.: VAD techniques for real-time speech transmission on the Internet. In: IEEE International Conference on High-Speed Networks and Multimedia Communications. (2002) 46–50
6. Sohn, J., Kim, N.S., Sung, W.: A statistical model-based voice activity detection. IEEE Signal Processing Letters 16 (1999) 1–3
7. Cho, Y.D., Kondoz, A.: Analysis and improvement of a statistical model-based voice activity detector. IEEE Signal Processing Letters 8 (2001) 276–278
8. Gazor, S., Zhang, W.: A soft voice activity detector based on a Laplacian-Gaussian model. IEEE Transactions on Speech and Audio Processing 11 (2003) 498–505
9. Armani, L., Matassoni, M., Omologo, M., Svaizer, P.: Use of a CSP-based voice activity detector for distant-talking ASR. In: Proc. of EUROSPEECH 2003, Geneva, Switzerland (2003) 501–504
10. Bouquin-Jeannes, R.L., Faucon, G.: Study of a voice activity detector and its influence on a noise reduction system. Speech Communication 16 (1995) 245–254
11. Ramírez, J., Segura, J.C., Benítez, C., de la Torre, A., Rubio, A.: An effective subband osf-based vad with noise reduction for robust speech recognition. IEEE Trans. on Speech and Audio Processing 13 (2005) 1119–1129
12. Ramírez, J., Segura, J.C., Benítez, C., García, L., Rubio, A.: Statistical voice activity detection using a multiple observation likelihood ratio test. IEEE Signal Processing Letters 12 (2005) 689–692
13. Górriz, J., Ramírez, J., Segura, J., Puntonet, C.: Improved MO-LRT VAD based on bispectra gaussian model. Electronics Letters 41 (2005) 877–879
14. Zadeh, L.A.: Fuzzy algorithm. Information and Control 12 (1968) 94–102
15. Beritelli, F., Casale, S., Cavallaro, A.: A robust voice activity detector for wireless communications using soft computing. IEEE Journal of Selected Areas in Communications 16 (1998) 1818–1829
16. Mendel, J.: Fuzzy logic systems for engineering: A tutorial. Proceedings of the IEEE 83 (1995) 345–377
17. Jang, J.S.R.: ANFIS: Adaptive-network-based fuzzy inference systems. IEEE Transactions on Systems, Man, and Cybernetics 23 (1993) 665–685
18. Moreno, A., Borge, L., Christoph, D., Gael, R., Khalid, C., Stephan, E., Jeffrey, A.: SpeechDat-Car: A Large Speech Database for Automotive Environments. In: Proceedings of the II LREC Conference. (2000)
19. Marzinzik, M., Kollmeier, B.: Speech pause detection for noise spectrum estimation by tracking power envelope dynamics. IEEE Transactions on Speech and Audio Processing 10 (2002) 341–351
20. Woo, K., Yang, T., Park, K., Lee, C.: Robust voice activity detection algorithm for estimating noise spectrum. Electronics Letters 36 (2000) 180–181
21. Li, Q., Zheng, J., Tsai, A., Zhou, Q.: Robust endpoint detection and energy normalization for real-time speech and speaker recognition. IEEE Transactions on Speech and Audio Processing 10 (2002) 146–157

Classification of Surimi Gel Strength Patterns Using Backpropagation Neural Network and Principal Component Analysis

Krisana Chinnasarn[1], David Leo Pyle[2], and Sirima Chinnasarn[3]

[1] Department of Computer Science, Burapha University, Thailand
[2] c/o School of Chemical Engineering and Analytical Science,
The University of Manchester, UK
[3] Department of Food Science, Burapha University, Thailand

Abstract. This paper proposes two practically and efficiently supervised and unsupervised classifications for surimi gel strength patterns. An supervised learning method, backpropagation neural network with three layers of 17-34-4 neurons for each later, is used. An unsupervised classification method consists of the data dimensionality reduction step via the PCA algorithm and classification step using correlation coefficient similarity measure. In the similarity measure step, each surimi gel strength pattern is compared with the surimi eigen-gel patterns, produced by the PCA step. In this paper, we consider a datum pattern as a datum dimension. The training data sets (12 patterns or 12 data dimensions) of surimi gel strength are collected from 4 experiments having different fixed setting temperature at $35^{o}C$, $40^{o}C$, $45^{o}C$, and $50^{o}C$, respectively. Testing data sets (48 patterns) are including original training set and their added Gaussian noise with 1, 3 and 5 points, respectively. From the experiments, two proposed methods can classify all testing data sets into its proper class.

Keywords: Backpropagation neural network, Principal component analysis, surimi eigen-gel patterns, backprogpagation neural network.

1   Introduction

Surimi is a concentrate of the myofibrillar proteins, primarily myosin and actin, of fish muscle, that is mechanically deboned, water-washed fish meat. In order to manufacture surimi-based products, the raw or frozen surimi is ground with salt and other ingredients, and then passed through thermal processing to set the shape and develop the texture [8]. During the thermal process, myofibrillar protein is induced to form the actomyosin gel, mainly derived from the myosin portions, which gives the unique elastic textural properties of the products [4]. Surimi is very well-known in food processing topics. Applications or surimi based products are Crap-analog or Kani-kamabogo, Kamaboko, Chikuwa, Satsumage, fish sausage and Shell fish-analog, etc. [8]

V.N. Alexandrov et al. (Eds.): ICCS 2006, Part I, LNCS 3991, pp. 403–410, 2006.
© Springer-Verlag Berlin Heidelberg 2006

Generally, surimi gel is formed following two heating-steps, a preheating process below 50^oC (setting step) prior to cooking at 80^oC or 90^oC to produce an opaque, highly elastic and strengthened gel [8]. The setting period is an important step in surimi gel development. This is because the setting at low temperature prior to heating at a higher temperature allows slow ordering of the protein molecule resulting in good gelation, fine structure and great elasticity [6] [8]. Different temperatures in the setting step result in different conformation of the proteins leading to different structure, and strength of the final gel [7].

The qualities of surimi gel, notably gel strength and structure, can be affected during gelation by various physical conditions such as high pressure, heating temperature and heating period during setting. In this paper, we emphasize on only surimi gel strength patterns which obtained from various setting temperatures and times. From the preliminary experiments, it can be observed that all patterns of surimi gel strength have similar shape. The final gel strengths increased to the maximum value and then decreased to the plateau in all setting tempertures. It is difficult to determine classes of each surimi gel strenth. In order to determine their proper classes, both supervised and unsupervised classification learning methods are applied to identify the qualities of surimi gel pattern.

The paper is organized as follows: Section 2 explains Cod surimi material, backpropagation neural network, principal surimi gel strength construction, and similarity method; Some experimental results are presented in Section 3 and conclusions in Section 4, respectively.

2   Materials and Methods

2.1   Materials

Cod (Gadus morhua) fillets were purchased from Frosts Fish LTD, Reading, UK. The fillets were skinned, rinsed with clean water, blended in a Lab Micronizer (Waring Commercial), washed by the ratio of water: minced fish at 3:1 (v/w). Then, the minced fish was dewatered by a basket centrifuge, then mixed with 6% sugar and 0.2% tetrasodium pyrophosphate and stored in an air blast freezer at -18^oC as a frozen surimi sample.

2.2   Gel Preparation

The frozen surimi was thawed at 4^oC until the temperature of surimi reached 0^oC and then, mixed with 2.5% salt by using a Lab Micronizer (Waring Commercial). The sol obtained was stuffed into stainless-steel cylinders of 2.5 cm. inner diameter and 2.5 cm. length. Surimi gels were prepared by heat setting at 35^oC, 40^oC, 45^oC and 50^oC for 0, 5, 10, 15, 20, 25, 30, 35, 40, 45, 50, 55, 60, 120, 180, 240, and 300 minutes in a water bath (Grant Y28, type VFP). Then, the gels were cooked at 90^oC for 20 minutes, and cooled in ice water. The obtained surimi gels were stored at 4^oC for 24 hours before analysed.

2.3   Gel Strength Analysis

The gel strengths of surimi gels were tested by using a TA-XT2 Texture Analyser (Stable Micro Systems Ltd., Surrey, UK.). All the cooked gels were compressed to 15 mm at the speed of 1.1 mm/second using a cylindrical shaped probe of 2.5 cm in diameter. Then, the changes in applied force were recorded. The gel strength was obtained from the peak force multiplied by the compression distance at the peak.

2.4   Proposed Classification Methods

The objective of the paper is to classify each surimi gel strength pattern into a proper class. We propose both supervised and unsupervised learning methods to identify the class or quality of surimi gel patterns.

Backpropagation Neural Network. Artificial neural network or neural network is a parallel processing mechanism which similar to human brain learning system. There are both supervised, unsupervised and reinforcement learning systems [5]. The first and simple neural network structure is called the *Perceptron neural network*. Perceptron consists of input neurons, connection weight, adding function, activation function, and output as illustrated in Fig. 1(**a**).

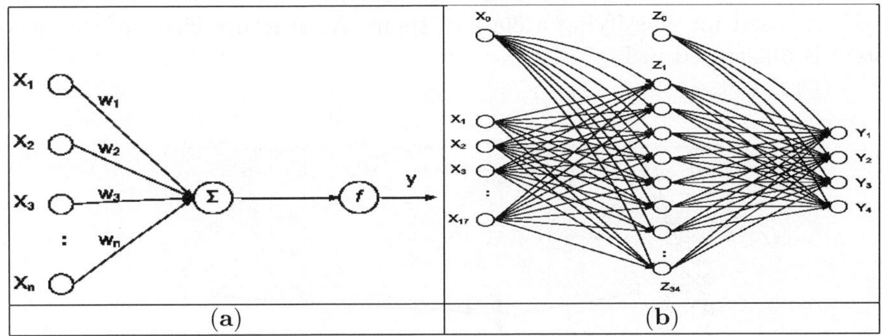

Fig. 1. (**a**) Perceptron neural network and (**b**) Backpropagation neural network used in the paper

where $[x_1, x_2, \cdots, x_n]^T$ are the input vectors, $[w_1, w_2, \cdots, w_n]^T$ are the connection weights, β is a bias, $f()$ is an activation function and y is an output. y can be written as follows:

$$net = \sum(x_1w_1 + x_2w_2 + \cdots + x_nw_n) + \beta \tag{1}$$

$$y = f(net) \tag{2}$$

The backpropagation neural network [5] with three layers, input-hidden- output, is used. The neurons in each layer are 17-34-4 for input-hidden-output layer, respectively. Neural network learning algorithm starts with the random weight w_{hi} and w_{jh}. The input vectors $[x_1, x_2, \cdots, x_{17}]^T$ are fed to the hidden layer Z using the weight w_{hi}. Then hidden vectors $[z_1, z_2, \cdots, z_{34}]^T$ are fed again by the weight w_{jh} in order to produce the output vectors $[y_1, \cdots, y_4]^T$. Backpropagation network structure is shown in Fig. 1(**b**). Let t be a desired target. The objective function is a summed squared error (SSE) as follows:

$$E = \frac{1}{2} \sum (y - t)^2 \tag{3}$$

And the activation function is a Sigmoidal function as follows:

$$f(net) = \frac{1}{1 + e^{-net}} \tag{4}$$

Principal Component Analysis to Surimi Gel Pattern. The second method for classifying each surimi gel strength is the Principal component analysis and the similarity measure method. An identifiable system consists of two main parts, the construction of surimi gel eigen-patterns and similarity analysis. First, the PCA method is used for searching an eigen-pattern by returning the principal direction for each class of surimi gel quality. Second, correlation coefficient similarity method between the eigen gel pattern and surimi gel strength pattern is used for classifying a class of them. A structure chart of the overall system is illustrated in Figure 2.

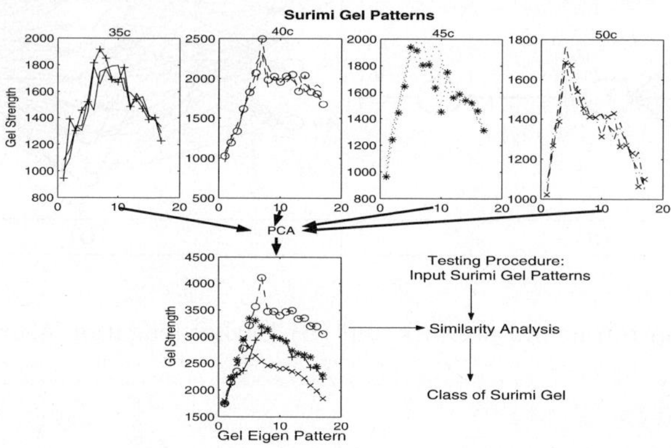

Fig. 2. System Structures for Surimi Gel Pattern Identification

Constructing principal direction or eigen-pattern is presented in this section [2]. Constructing procedures are composed of two main steps, surimi gel strength matrix formation and computing an eigen-pattern. First, the surimi gel strength matrix formation step, input data for modeling surimi gel eigen-patterns are surimi gel strength from 3 replications, detailed in Section 2.2. Then from the previous conditions, three input vectors of surimi gel strength were obtained for each setting temperature. And they are formed as a row major matrix with a dimension of 3 by 17.

Second, computing an eigen-pattern step, the principal component analysis can be carried out. The calculation of surimi gel eigen-pattern is given by the solution to the following step [3]:

1. Compute the mean for each row $mean_i$
2. Generate the zero-mean matrix $\mathbf{Mz}$ for each row i,

$$\mathbf{Mz}_{i,j} = \mathbf{M}_{i,j} - mean_i \qquad (5)$$

   where $1 \leq i \leq 17$.
3. Compute the covariance matrix $\mathbf{Cov}$ for $\mathbf{Mz}$,

$$\mathbf{Cov}(\mathbf{Mz}) = \mathbf{E}[(\mathbf{Mz})(\mathbf{Mz})^T] \qquad (6)$$

   where $\mathbf{E}[.]$ is an expectation value and $(\mathbf{Mz})^T$ is a transpose matrix of $\mathbf{Mz}$. Then the dimensions of $\mathbf{Cov}(\mathbf{Mz})$ are 3 by 3.
4. Compute an eigenvalue, d, and eigenvector, v, for the covariance matrix $\mathbf{Cov}$.
5. Project $\mathbf{M}$ onto new axis by eigen-vector v,

$$\mathbf{P} = v^T \mathbf{M} \qquad (7)$$

   where v^T is a transpose matrix of v

Similarity Measure. The main objective of the classification problem is to find a natural grouping in a set of data [3]. In other words, we want to say that the samples in class c_i are more like one another than like samples in other class c_j, where $i \neq j$.

Correlation Coefficient Analysis, ρ. The correlation coefficient ρ between two random variable x and y can be described as:

$$\rho(x,y) = \frac{cov(x,y)}{\sqrt{var(x)var(y)}} \qquad (8)$$

where $cov(x,y)$ is the covariance between two random variables and $var(.)$ is the variance of a random variable. Correlation coefficient values are within $[-1, \ldots, 1]$. If x and y are completely correlated, $\rho(x,y)$ is 1 or -1. If they are absolutely uncorrelated, $\rho(x,y)$ is 0.

3   Experimental Results

Some simulations have been made on surimi Gel Strength patterns, which contain 17 data points for each pattern, details were given in Section 2.2. Testing sets (12 data) include the original training set and their added Gaussian noise with 1, 3 and 5 points, respectively. Then, in total, the testing sets have 48 patterns. Graphical representation of all testing sets are illustrated in

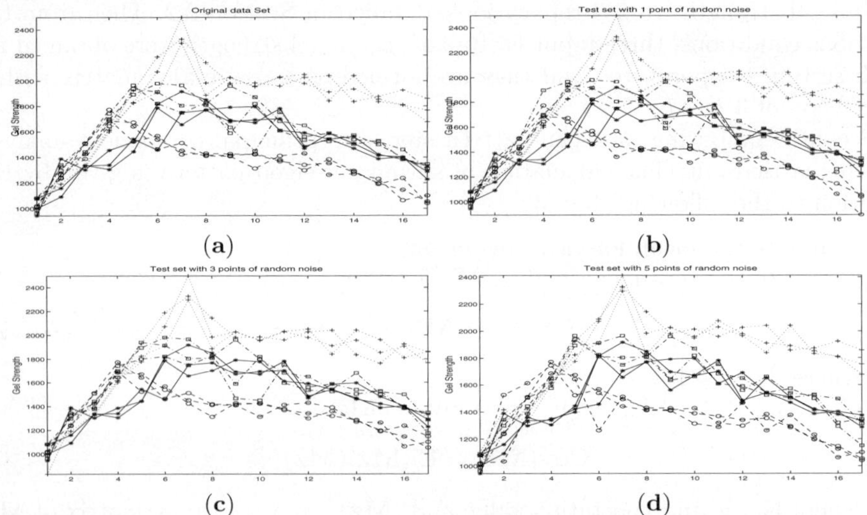

Fig. 3. (a) Original surimi gel patterns created by 35°C, 40°C, 45°C and 50°C, (b) Original surimi gel added one point of Gaussign noise, (c) Original surimi gel added three points of Gaussign noise and (d) Original surimi gel added five points of Gaussign noise

Figure 3 (a), (b), (c), and (d), respectively. It can be seen that it is very difficult to identify a class for input pattern by hand. How can we do that?

3.1   Backpropagation Neural Network

In the learning step, we set $\epsilon \leq 0.00001$. The learning is terminated after 35 epochs. Tables 1(a), (b), (c), and (d) display the probability of each surimi gel strength pattern to be each expected class. Probability values from Tables 1(a), (b), (c), and (d) confirm that all testing set patterns can be classified into a proper class even they have some noises. For example from Table 1(d), degree of similarity between testing pattern 1 to expected class 1, class 2, class 3, and class 4 are 0.9957, 0.0000, 0.0000, and 0.0000, respectively. Then, it is classified to class 1 which is correct.

3.2   Principal Component Analysis

Tables 2(a), (b), (c) and (d) display the degree of similarity between testing data sets and the principal direction of surimi gel strength using correlation coefficient analysis. Degree of similarity among them from Tables 2(a), (b), (c) and (d) confirm that all testing set patterns can be classifed into a proper class even they have some noises. For example from Table 2(c), degree of similarity between testing pattern 3 to expected class 1, class 2, class 3, and class 4 are 0.9462, 0.8789, 0.7765, and 0.4154, respectively. Then it is classified to class 1 which is correct.

Table 1. Probability of the surimi gel strength pattern using backpropagation neural network (**a**) original surimi gel patterns, (**b**) added 1 point of random gaussian noise, (**c**) added 3 points of random gaussian noise and (**d**) added 5 points of random gaussian noise

Testing pattern	Expected classes				Expected classes				Its Actual class
	Class 1	Class 2	Class 3	Class 4	Class 1	Class 2	Class 3	Class 4	
1	**0.9998**	0.0000	0.0000	0.0000	**0.9999**	0.0000	0.0000	0.0000	1
2	**0.9999**	0.0013	0.0000	0.0000	**0.9998**	0.0015	0.0000	0.0000	1
3	**1.0000**	0.0000	0.0000	0.0000	**1.0000**	0.0000	0.0000	0.0000	1
4	0.0000	**0.9994**	0.0008	0.0000	0.0000	**0.9994**	0.0008	0.0000	2
5	0.0001	**0.9996**	0.0000	0.0000	0.0001	**0.9996**	0.0000	0.0000	2
6	0.0000	**0.9984**	0.0001	0.0000	0.0000	**0.9814**	0.0029	0.0000	2
7	0.0000	0.0000	**0.9995**	0.0000	0.0000	0.0000	**0.9998**	0.0000	3
8	0.0000	0.0000	**0.9999**	0.0000	0.0000	0.0000	**0.9999**	0.0000	3
9	0.0000	0.0008	**0.9990**	0.0000	0.0000	0.0008	**0.9990**	0.0000	3
10	0.0000	0.0000	0.0002	**0.9999**	0.0000	0.0000	0.2592	**0.9930**	4
11	0.0000	0.0000	0.0005	**1.0000**	0.0000	0.0000	0.0011	**1.0000**	4
12	0.0002	0.0000	0.0000	**1.0000**	0.0000	0.0000	0.0248	**1.0000**	4
	(a)				(b)				
1	**0.9989**	0.0000	0.0000	0.0000	**0.9957**	0.0000	0.0000	0.0000	1
2	**0.9643**	0.0000	0.0000	0.0000	**1.0000**	0.0001	0.0000	0.0000	1
3	**1.0000**	0.0000	0.0000	0.0000	**1.0000**	0.0000	0.0000	0.0000	1
4	0.0000	**0.9992**	0.0056	0.0000	0.0000	**0.9960**	0.0000	0.0000	2
5	0.0000	**0.9997**	0.0000	0.0000	0.0000	**0.9996**	0.0000	0.0000	2
6	0.0000	**0.9984**	0.0002	0.0000	0.0000	**0.9988**	0.0001	0.0000	2
7	0.0002	0.0074	**0.9984**	0.0000	0.0000	0.0000	**0.9992**	0.0000	3
8	0.0000	0.0000	**1.0000**	0.0000	0.0000	0.0000	**0.9995**	0.0000	3
9	0.0000	0.0005	**0.9991**	0.0000	0.0000	0.0005	**0.9972**	0.0000	3
10	0.0004	0.0000	0.0000	**0.9995**	0.0000	0.0000	0.0351	**1.0000**	4
11	0.0000	0.0000	0.0000	**1.0000**	0.0000	0.0000	0.0000	**1.0000**	4
12	0.0002	0.0000	0.0001	**0.9999**	0.0000	0.0000	0.0027	**1.0000**	4
	(c)				(d)				

Table 2. Correlation coefficient (ρ) between (**a**) original surimi gel patterns, (**b**) added 1 point of random gaussian noise, (**c**) added 3 points of random gaussian noise and (**d**) added 5 points of random gaussian noise and the principal surimi gel eigen-patterns

Testing pattern	Expected classes				Expected classes				Its Actual class
	Class 1	Class 2	Class 3	Class 4	Class 1	Class 2	Class 3	Class 4	
1	**0.9575**	0.8661	0.7968	0.4126	**0.9572**	0.8421	0.8006	0.4142	1
2	**0.9585**	0.8258	0.8662	0.5848	**0.9581**	0.8264	0.8761	0.5949	1
3	**0.9534**	0.8567	0.7852	0.4321	**0.9449**	0.8334	0.7677	0.4182	1
4	0.8717	**0.9805**	0.7685	0.3218	0.8722	**0.9806**	0.7691	0.3227	2
5	0.8807	**0.9896**	0.7594	0.3489	0.8806	**0.9898**	0.7635	0.3531	2
6	0.8678	**0.9894**	0.7307	0.3105	0.8706	**0.9857**	0.7365	0.3229	2
7	0.7763	0.6666	**0.9659**	0.8288	0.7266	0.5869	**0.9246**	0.8297	3
8	0.8874	0.7663	**0.9653**	0.7679	0.8893	0.7673	**0.9657**	0.7679	3
9	0.7992	0.7661	**0.9549**	0.7524	0.7993	0.7670	**0.9549**	0.7524	3
10	0.5423	0.3323	0.8013	**0.9819**	0.5730	0.3794	0.8212	**0.9741**	4
11	0.4523	0.2820	0.7529	**0.9717**	0.4556	0.2649	0.7368	**0.9587**	4
12	0.4919	0.3613	0.8201	**0.9713**	0.4353	0.2792	0.7715	**0.9652**	4
	(a)				(b)				
1	**0.9565**	0.8656	0.8003	0.4076	**0.9425**	0.8575	0.7916	0.4205	1
2	**0.9659**	0.8174	0.8384	0.5187	**0.9501**	0.8440	0.7805	0.4328	1
3	**0.9462**	0.8789	0.7765	0.4154	**0.9338**	0.8443	0.7172	0.3690	1
4	0.8617	**0.9751**	0.7702	0.3253	0.8698	**0.9777**	0.7382	0.2794	2
5	0.8993	**0.9875**	0.7875	0.3741	0.9005	**0.9853**	0.7622	0.3499	2
6	0.8635	**0.9848**	0.6980	0.2759	0.8597	**0.9826**	0.7261	0.3158	2
7	0.6937	0.6066	**0.9313**	0.8130	0.8094	0.6932	**0.9713**	0.8010	3
8	0.8871	0.7564	**0.9542**	0.7791	0.8875	0.7346	**0.9383**	0.7427	3
9	0.8061	0.7788	**0.9432**	0.7446	0.8568	0.8045	**0.9708**	0.7488	3
10	0.5288	0.3145	0.7921	**0.9635**	0.3667	0.1332	0.6376	**0.9196**	4
11	0.4302	0.2484	0.7185	**0.9584**	0.3782	0.2118	0.6627	**0.9181**	4
12	0.4894	0.4065	0.8076	**0.9281**	0.5128	0.4503	0.8260	**0.9138**	4
	(c)				(d)				

4   Concluding Remarks

This paper reports some applications of Backpropagation neural network and PCA (Principal Component Analysis). Backpropagation neural network is the supervised learning method. But PCA is an unsupervised classification method based on data information. Throughout the paper, the identification of surimi gel patterns using Backpropagation neural network and eigen-pattern together with the correlation coefficient has been outlined. Good identification is obtained from the Backpropagation neural network but we need to know a class of each pattern in advance. On the other hand, the advantages of the PCA method developed in this paper are as follows. (1) Each class or principal direction of surimi gel strengths is obtained by an unsupervised method, the PCA. Thus, for further identification of the surimi gel strength class, we can easily do this by passing surimi gel strength into the PCA. Then a new class of surimi gel strengths is created. In other words, the principal direction of surimi gel strengths is dynamic based on data characteristics. (2) It can be concluded that the correlation coefficient method gives an efficient result in the case where the testing patterns have very similar shape.

References

1. Chen.T, Amari.S.-I. and Lin.Q. A Unified Algorithm for Pricipal and Minor Component Extraction *Neural Networks*, Vol. 11, pp. 385-390, 1998.
2. Chinnasarn.K, Chinnasarn.S, and Pyle.D.L., Surimi Gel Pattern Identification Using Eigen-Pattern and Similarity Analysis, accepted to the 9th *National Computer Science and Engineering Conference: NCSEC2005*, Bangkok, Thailand.
3. Duda.R.O, Hart.P.E., and Stork.D.G. *Pattern Classification*, 2nd edition, John Wiley & Sons, Inc., 2001.
4. Hall, G.M. and Ahmad,N.H. Surimi and fish-mince products. In G.M. Hall (ed). *Fish Processing Technology*. London: Blackie Acadimic & Professional. 1997.
5. Haykin.S. *Neural Network a Comprehensive foundation*. 2nd, Prentice Hall,1999.
6. Hermansson, A.M. Physico-chemical aspects of soy proteins structure formation. *Journal of Texture Studies*. 9: pp 33-58, 1978.
7. Hossain, M.I., et al. Contribution of the polymerization of protein by disulfide bonding to increased Gel Strength of walleye pollack surimi gel with preheating time. *Fisheries Science*. 67 : pp 710-717, 2001.
8. Lanier, T.C. and Lee, C.M. *Surimi Technology*. New York : Marcel Dekker, 1992.

Optimal Matching of Images Using Combined Color Feature and Spatial Feature

Xin Huang, Shijia Zhang, Guoping Wang, and Heng Wang

Human-Computer Interaction & Multimedia Lab,
Department of Computer Science and Technology,
Peking University, 100871 Beijing, P.R. China
{hx, zsj, wgp, hengwang}@graphics.pku.edu.cn

Abstract. In this paper[1] we develop a new image retrieval method based on combined color feature and spatial feature. We introduce an ant colony clustering algorithm, which helps us develop a perceptually dominant color descriptor. The similarity between any two images is measured by combining the dominant color feature with its spatial feature. The optimal matching theory is employed to search the optimal matching pair of dominant color sets of any two images, and the similarity between the query image and the target image is computed by summing up all the distances of every matched pair of dominant colors. The algorithm introduced in this paper is well suited for creating small spatial color descriptors and is efficient. It is also suitable for image representation, matching and retrieval.

1 Introduction

Color is a widely used low-level feature in content-based image retrieval systems [1, 4, 5, 6], because of its characteristic of invariance with respect to image scaling and orientation. Smith [1] proposed a method to quantize colors into 166 bins in the *HSV* color space. Zhang [2] gave a new dividing method to quantize the color space into 36 non-uniform bins. It has been observed that the color quantization schemes have a major and common drawback. That is similar colors might be quantized to different bins in the histogram, thus increasing the possibility of retrieving dissimilar images.

Besides color histogram, another commonly used method is to apply clustering based techniques in quantizing the color space. Ma et al. utilized a vector quantization called Generalized Lloyd algorithm (GLA) [3] to quantize the *RGB* color space. Mojsilovic [4] proposed a new quantization scheme in the *Lab* space based on spiral lattice. However, the problem of how to extract semantic information from the image still remains the biggest obstacle in the content-based image retrieval system [13, 14]. Rogowitz performed psychophysical experiments [6] analyzing human perception of image content, showing that visual features have a significant correlation with

---

[1] The research was supported by No. 2004CB719403 from The National Basic Research Program of China (973 Program), No. 60473100□No. 60573151 from National Natural Science Foundation of China.

V.N. Alexandrov et al. (Eds.): ICCS 2006, Part I, LNCS 3991, pp. 411–418, 2006.

semantically relevant information. Mojsilovic indicated that even with the absence of semantic cues, "semantically correct retrievals" [5] can also be achieved by perceptually based features. By exploiting the fact that the human eye cannot perceive a large number of colors at the same time, nor is it to distinguish close colors well, we aim to create a small color descriptor, which is suitable for image representation, matching and retrieval.

In this paper we introduce a color feature extraction method based on ant colony clustering algorithm, which models the behavior of ants' collecting corpses and is self-organizing. The algorithm extracts perceptually dominant colors as the basis for image matching. The spatial information of dominant colors is then taken into account in order to enlarge feature space and increase the retrieval precision. Similarity metric between any two images is established by using an optimal matching algorithm in graph theory.

2 Dominant Color Feature Extraction

Ant colony clustering algorithm [7, 8, 9] has been proposed and applied in various areas since 1990s, while it models the ants' behavior of piling corpses. Researchers found that the ants can assemble the ant corpses into several piles in their studies. Deneubourg proposed a model that explains the ants' behavior of piling corpses, which is commonly called BM (Basic Model) [7] to describe the ants' clustering activity. The general idea is that when an unloaded ant encounters a corpse, it will pick it up with a probability that increases with the degree of isolation of the corpse; when an ant is carrying a corpse, it will drop the corpse with a probability that increases with the number of corpses in the vicinity. The picking and dropping operations are biased by the similarity and density of data items within the ants' local neighborhood.

The step of dominant colors extraction based on ant colony clustering is as follows. First an input image is transformed into $CIELAB$ color space. We get the training sequence consisting of M source vectors: $T = \{x_1, x_2, \ldots, x_M\}$. The source vector that is three-dimensional consists of L, a, b value in $CIELAB$ color space. Then we utilize the ant colony clustering algorithm [7, 8] to extract the dominant colors from the training sequence T. The first step is to randomly project training sequence T onto a plane, and a few virtual ants are generated, randomly placed on the plane. Then the density measure of each ant is computed [8]. Each ant acts according to its current state and corresponding probability. Finally several clustering centers are visually formed through the ants' collective actions. The algorithm is ended with a few clustering dominant colors generated. After using the ant colony clustering algorithm, we extract the dominant color set denoted as $C = \{c_1, c_2, \cdots c_K\}$, $P = \{p_1, p_2, \cdots p_K\}$, where each dominant color $c_i = \{L_i, a_i, b_i\}$ is a three-dimensional Lab color value, and p_i is the corresponding size percentage. In our experiments the number of dominant colors K is assigned the value 16.

3   Combined Color Feature and Spatial Feature

In the procedure of color clustering, we only consider the color feature of each image. Thus it may lose color distribution information and lead to false retrieval. In order to prevent this problem, we introduce the color spatial information to enlarge the feature space of dominant colors. Moment [10, 11] is a simple and effective way for representing the spatial feature in images. It has the prominent property of being invariant to image rotation, shift and scale. We use the centroid of the dominant colors and the second-order central moment [11] to represent spatial features. The centroid represents the location of each dominant color, and the second-order central moment indicates the mass distributing information of dominant colors.

In Statistics, moment represents fundament distributing properties of random variables. The $p+q$ th-order moments of a bounded function $f(x, y)$ with two variables is defined as:

$$M_{pq} = \int \int x^p y^q f(x, y) dx dy \tag{1}$$

where p and q are nonnegative integers [11].

Suppose the size of an image is $m \times n$. After extracting dominant color features by ant colony clustering, the dominant color set is denoted as $C = \{c_1, c_2, \cdots c_K\}$, the $p+q$ th-order moments of dominant color c_i can be defined as follows:

$$M_{pq}^i = \sum_{x=0}^{m-1} \sum_{y=0}^{n-1} x^p y^q f(x, y, i) \tag{2}$$

If the pixel at coordinates (x, y) belongs to the dominant color c_i, then $f(x, y, i) = 1$; otherwise $f(x, y, i) = 0$.

Then the centroid coordinates $(\overline{x}_i, \overline{y}_i)$ of each dominant color can be computed using the first-order moment, $\overline{x}_i = \dfrac{M_{10}^i}{M_{00}^i}, \overline{y}_i = \dfrac{M_{01}^i}{M_{00}^i}$. The $j+k$ th-order central moments of dominant color c_i can be defined as:

$$\mu_{jk}^i = \sum_{x=0}^{m-1} \sum_{y=0}^{n-1} (x - \overline{x}_i)^j (y - \overline{y}_i)^k f(x, y, i) \tag{3}$$

We use the second-order central moment μ_{11} to describe the mass distributing feature of dominant colors.

4   Similarity Measure

In order to define the similarity metric between two images, we first give the formula of computing the distance between two dominant colors c_i and c_j. Both the color feature and spatial feature are considered in defining the distance. According to the dominant color feature and spatial feature defined in section 2 and section 3, we compute four corresponding distances in section 4.1.

4.1 Distance Computation

Distance $dCc(c_i, c_j)$ is the color difference of c_i and c_j in *CIELAB* color space. Distance $dPt(c_i, c_j)$ is the area percentage difference of c_i and c_j. Distance $dCt(c_i, c_j)$ is the centroid coordinates difference of c_i and c_j. Distance $d\mu(c_i, c_j)$ is the second-order central moment difference of c_i and c_j. The formulas of four distances are defined as follows:

$$dCc(c_i, c_j) = \sqrt{(L_i - L_j)^2 + (a_i - a_j)^2 + (b_i - b_j)^2} \tag{4}$$

$$dPt(c_i, c_j) = |P_i - P_j| \tag{5}$$

$$dCt(c_i, c_j) = \sqrt{(\overline{x}_i - \overline{x}_j)^2 + (\overline{y}_i - \overline{y}_j)^2} \tag{6}$$

$$d\mu(c_i, c_j) = |\mu_i - \mu_j| \tag{7}$$

We define the overall distance between c_i and c_j as follows:

$$D(c_i, c_j) = w_1 dCc(c_i, c_j) + w_2 dPt(c_i, c_j) + w_3 dCt(c_i, c_j) + w_4 d\mu(c_i, c_j) \tag{8}$$

w_i is the weight assigned to the corresponding distance. We have assigned different weight to each distance, which is shown in Table 1 in Appendix. According to the weight in each group of Table 1, the performance of image retrieving is evaluated and the result is presented in Fig. 4. From Fig. 4, we can see that the weights assigned the values $w_1 = 0.4$, $w_2 = 0.4$, $w_3 = 0.15$, $w_4 = 0.05$, achieve the best retrieving performance in our experiments. From analysis of Fig. 4, we can see the color feature (the dominant color and its area percentage) is still the most significant part in defining the distance, and the centroid also has more obvious influence on retrieval precision when compared with the second-order central moment.

$D(c_i, c_j)$ is a normalized value so that the value of similarity between c_i and c_j can be defined as:

$$Sim(c_i, c_j) = 1 - D(c_i, c_j) \tag{9}$$

4.2 Optimal Matching

Given two images, a query image A and a target image B, each of them has the dominant color set $C^a = \{c_1^a, c_2^a, \cdots c_K^a\}$ and $C^b = \{c_1^b, c_2^b, \cdots c_K^b\}$ respectively, where K is the number of dominant colors of each image. In order to compute the similarity of the two images, we first have to search the optimal matching dominant colors between the two dominant color sets C^a and C^b.

We use the optimal matching method in graph theory [12] to solve the problem. We construct the bipartite graph as $G = \{C^a, C^b, E\}$, where C^a and C^b are dominant color sets of two images. $E = \{e_{i,j}\}$ is the edge sets, where a weight $w_{i,j}$ is assigned to the edge $e_{i,j}$ in G. $w_{i,j}$ is the value of similarity between two dominant colors c_i^a and c_j^b, computed by formula (9). Given the weighted bipartite graph G (An example is shown in Fig.1), the *Kuhn-Munkres* algorithm [12] can be used to solve the optimal matching problem. This algorithm has been applied in some research such as content-based video retrieval [16] and document similarity search [17]. The computational complexity of *Kuhn-Munkres* algorithm is $O(K^3)$. Based on the optimal matching theory, the similarity measure of the query image and the target image can be computed by the sum of all distances between every matched pair of dominant colors. Then the retrieval result is ranked according to the value of similarity.

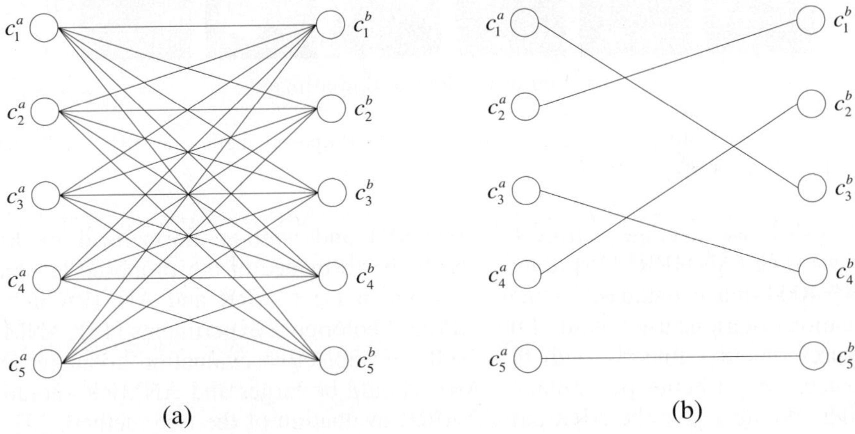

(a)   (b)

Fig. 1. (a) A bipartite graph of two dominant color sets. (b) The optimal matching result.

5   Experimental Results

We have developed a content-based image retrieval system called PKUQBIC to validate the efficiency of proposed algorithms and techniques in our paper. The image database consists of 4000 images, distributed into 28 different categories. We present the retrieval result of the proposed algorithm in this paper and compare it with other two clustering based algorithms [3] and [4]. The proposed algorithm in this paper is called *CSOP* (Color-Spatial Optimal Matching). From Fig. 2 we can see the method proposed in this paper is well defined, and achieves much better retrieving results than the other two methods.

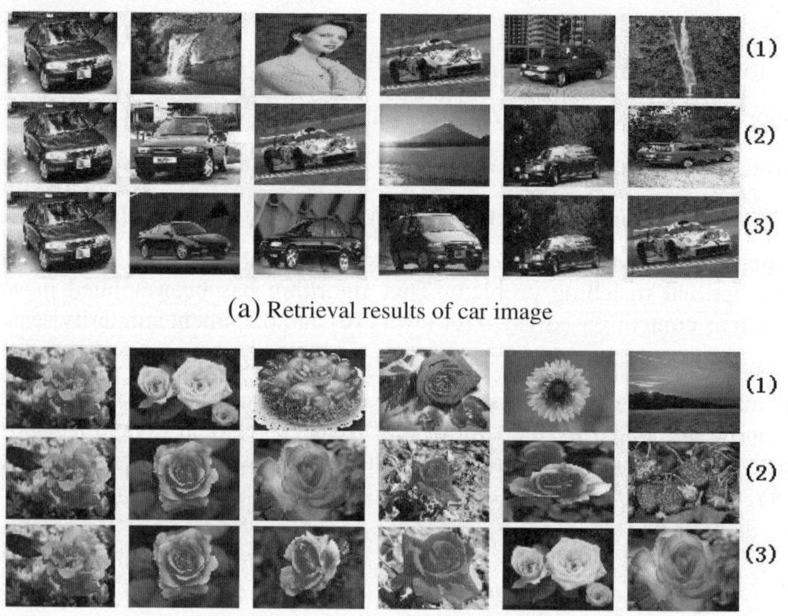

(a) Retrieval results of car image

(b) Retrieval results of flower image

Fig. 2. Retrieval results of three methods, with (1) Proposed method in [3], (2) Proposed method in [4], (3) *CSOP* method

We also use average retrieval rate (ARR) and average normalized modified retrieval rank (ANMRR) [15] to evaluate the performance of our proposed technique in the 4000-image database, which is shown in Fig.3. ARR and ANMRR are the evaluation criterions used in all of the MPEG-7 color core experiments [15]. ANMRR measure coincides linearly with the results of subjective evaluation about retrieval accuracy. To get better performance, ARR should be larger and ANMRR should be smaller. We also give the ARR and ANMRR evaluation of the two methods [3] and [4] in order to compare them with *CSOP* algorithm. From Fig.3 we can see that *CSOP* gets a significant improvement in retrieval performance compared with the other two

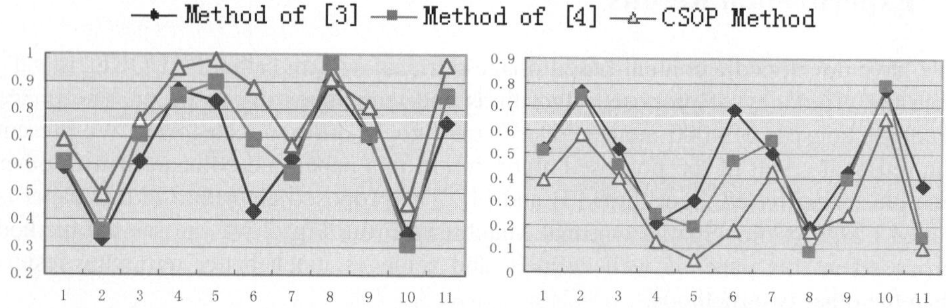

Fig. 3. ARR performance (left) and ANMRR performance (right) of the three methods

methods. The horizontal axes in Fig.3 (including Fig.4) denote corresponding image category, listed as: 1-fruit, 2-cup, 3-building, 4-sky, 5-face, 6-car, 7-hill, 8-fire, 9-bird, 10-dog, 11-sea, and the vertical axes denote the ARR and ANMRR performance.

6 Conclusion

Along with the fact that visual features have a significant correlation with semantic information of image, this paper proposes an ant colony clustering scheme to extract the dominant color features that well match human perception of images. Spatial feature combined with the dominant color feature is taken into account to measure the similarity. Besides we develop a perceptually based image similarity metric based on optimal dominant color matching algorithm, which is used to search the optimal matching pair of dominant color sets of any two images. The future work is to extend the proposed algorithm *CSOP* to include other spatial information such as the texture feature or shape feature to measure similarity, and a larger image database should be employed to evaluate the performance of the proposed scheme.

Acknowledgements

The authors would like to thank Mr. Yuxin Peng from Institute of Computer Science & technology of Peking University and Ms. Jianying Hu from IBM T.J. Watson Research Center, for their advice and help during the research.

References

1. J. R. Smith, S. F. Chang, A fully automated content-based image query system, Proceedings of ACM Multimedia, pp. 87 − 98, 1996
2. L. Zhang, F. Lin, and B. Zhang, A CBIR method based on color-spatial feature, IEEE Regin10 Annual International Conference, pp. 166-169, 1999
3. W.Y. Ma, Y. Deng, and B.S. Manjunath, Tools for texture/color base search of images, Proc. SPIE, vol. 3016, pp. 496–505, 1997
4. A. Mojsilovic, J. Hu, and E. Soljanin, Extraction of perceptually important colors and similarity measurement for image matching, retrieval, and analysis, IEEE Trans. Image Processing, vol. 11, pp. 1238-1248, Nov. 2002
5. A.Mojsilovic, J.Kovacevic, J.Hu, etc., Matching and retrieval based on the vocabulary and grammar of color patterns, IEEE Trans. Image Processing, vol. 9, pp. 38–54, Jan. 2000
6. B. Rogowitz, T. Frese, J. Smith, C. A. Bouman, and E. Kalin, Perceptual image similarity experiments, Proc. SPIE, 1997
7. J.L. Deneubourg, S. Goss, N. Frank, The dynamics of collective sorting: robot-like ants and ant-like robots, Proc. Of the 1st Int. Conference on Simulation of Adaptive Behavior: From Animals to Animats, MIT Press/ Bradford Books, Cambridge, MA, pp.356-363, 1991
8. E. Lumer and B. Faieta, Diversity and adaption in populations of clustering ants, In Proc. of the 3rd International conference on Simulation of Adaptive Behaviour: From Animals to Animats 3, MIT Press, Cambridge, MA, pp. 501−508, 1994
9. R.S. Parpinelli, H.S. Lopes, and A.A. Freitas, Data mining with an ant colony opti-mization algorithm, IEEE Trans. on Evolutionary Computing, vol. 6, pp. 321–332, Aug. 2002
10. Yap, P.- T., Paramesran, R., Seng-Huat Ong, Image analysis by Krawtchouk moments, IEEE Transactions on Image Processing, Vo. 12, pp. 1367 − 1377, 2003

11. Kenneth R. Castleman, Digital Image Processing, Prentice-Hall International, Inc. 1996
12. L. Lovász and M. D. Plummer, Matching Theory. Amsterdam, The Netherlands: North Holland, 1986
13. Wei Jiang, Guihua Er, and Qionghai Dai, Multilayer semantic representation learning for image retrieval, International Conf. on Image Processing, vol. 4, pp. 2215-2218, 2004
14. Feng Jing, Mingjing Li, Hong-Jiang Zhang, and Bo Zhang, A unified framework for image retrieval using keyword and visual features, IEEE Trans. on Image Processing, vol. 14, pp. 979-989, 2005
15. B.S Manjunath, J-R Ohm, V.V. Vasudevan, and A. Yamada, Color and texture descritors, IEEE Trans. Circuits Syst. for Video Technol., Vol. 11, pp. 703-715, June 2001
16. Peng Y.X, Ngo C.W, Dong Q.J, Guo ZM, Xiao JG, An approach for video retrieval by videoclip, Journal of Software, 14(8):1409~1417, 2003
17. X. J. Wan, Y. X. Peng, A new retrieval model based on texttiling for document similarity search[J], Comput. Sci. & Technol, VOL.20, NO.4, 2005

Appendix

Different weights are used to combine the color feature and spatial feature to compute the distance between any two images. We construct Table 1, in which five typical weight groups (T1, T2, T3, T4, T5) are assigned to coordinate the color feature and spatial feature. Retrieving performance of *CSOP* assigned with the five weight groups is evaluated using the 4000-image database of PKUQBIC, shown in Fig. 4. We can see weight group T3 achieves the best retrieving performance. The appendix shows that using combined features performs better than using either mainly spatial feature or only color feature.

Table 1. Five typical weight groups

	w_1	w_2	w_3	w_4
T1	0.1	0.1	0.4	0.4
T2	0.3	0.3	0.2	0.2
T3	0.4	0.4	0.15	0.05
T4	0.4	0.4	0.1	0.1
T5	0.5	0.5	0	0

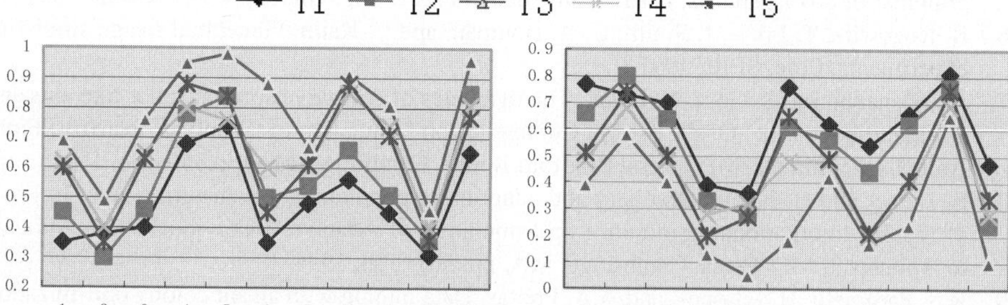

Fig .4. ARR performance (left) and ANMRR performance (right) according to different weights assigned in Table 1

A Novel Network Intrusion Attempts Prediction Model Based on Fuzzy Neural Network*

Guiling Zhang and Jizhou Sun

Department of Electronic Information Engineering, Tianjin University, 300072, China
glzhang808@sohu.com, jzsun@tju.edu.cn

Abstract. Identifying the intrusion attempts of the monitored systems is extremely vital for the next generation intrusion detection system. In this paper, a novel network intrusion attempts prediction model (FNNIP) is developed, which is based on the observation of network packet sequences. A new fuzzy neural network based on a novel BP learning algorithm is designed and then applied to the network intrusion attempts predicting scheme. After given the analysis of the features of the experimental data sets, the experiment process is detailed. The experimental results show that the proposed Scheme has good accuracy of predicting the network intrusion attempts by observing the network packet sequences.

Keywords: Intrusion Attempt Prediction, Intrusion Detection, Fuzzy Neural Network.

1 Introduction

Intrusion detection techniques have become an important research area in modern information security system. Denning first proposes a model for building a real time intrusion detection expert system by analyzing the profiles representing the system behavior from audit records [1]. S. Forrest and others propose an intrusion detection model based on short sequences of system calls [2]. W. Lee and others build a data-mining framework for intrusion detection system [4, 5]. H. Jin and coworkers extends the W. Lee's work by applying fuzzy data mining algorithm to intrusion detection [3]. Paper [6] and paper [7] apply neural networks to building intrusion detection models. The above methods are only able to detect intrusions after the attacks have occurred, either partially or fully, which makes it difficult to contain or stop the attack in real time [9]. So it is necessary to tap intrusion attempts prediction techniques into IDS.

N. Ye and others proposed two methods of forecasting normal activities in computer systems for intrusion detection. Their method provides performance improvement on intrusion detection [8]. L. Feng and coworkers apply a plan recognition method for predicting the anomaly events and the intensions of possible intruders to a computer system based on the observation of system call sequences [9].

This paper extends their works by observing network packet sequences to predict intrusion attempts. We develop a new intrusion attempts prediction model based on

---

* This research was supported by the National High Technology Development 863 program of China under Grant No. 2002AA142010.

V.N. Alexandrov et al. (Eds.): ICCS 2006, Part I, LNCS 3991, pp. 419–426, 2006.

fuzzy neural network (FNNIP) for network intrusion detection system. Only the critical features in the network connection records are employed to describe the actions of the normal or attacks. The experimental results show that the proposed Scheme has good accuracy of predicting the network intrusion attempts by observing the network packet sequences.

The rest of this paper is organized as follows: How to construct the structure of the proposed FNNIP is discussed in section 2. In this section, a new BP learning algorithm is also designed for the proposed scheme. Section 3 details the description of the network traffic behaviors. Some experiments are described in section 4. In section 5, some conclusions and future works are given.

2 The Structure of Fuzzy Neural Network for Network Intrusion Attempts Prediction

The structure of the proposed fuzzy neural network based network intrusion attempts prediction (FNNIP) system is designed in Fig.1 [12].

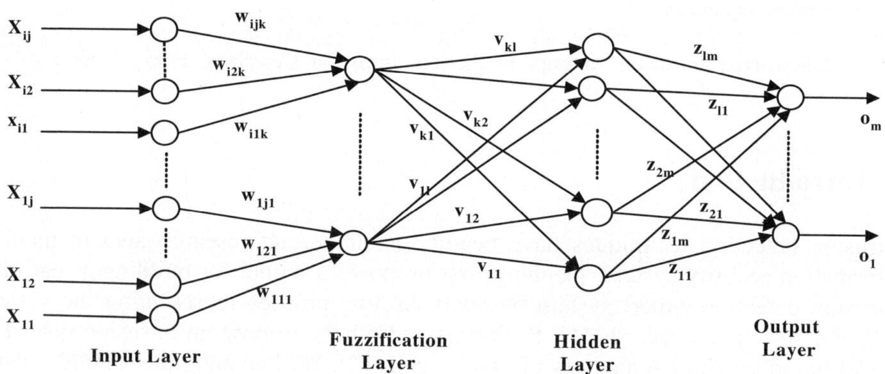

Fig. 1. The Structure of The Proposed FNNIP

The input layer consists of N input vectors $\mathbf{x}_i, i = 1, 2, \cdots, N-1, N$; each vector $\mathbf{x}_i$ has t_j elements. The total of the input nodes are $N \times t_j$.

The fuzzification layer has $t_f = N$ nodes. Each node in the fuzzification layer represents a fuzzy membership function for the t_j input variable in vector $\mathbf{x}_i$. The fuzzy membership function used is described in equation (1).

$$\mu_{ij_k}(x_{ij}) = e^{-(\sum_{j=1}^{t_j} w_{ijk} x_{ij})^2} \tag{1}$$

Where μ_{ijk} is the kth output of the fuzzification layer, w_{ijk} is the weight between input nodes x_{ij} and the kth fuzzification nodes.

For each node in the hidden layer, the inputs are μ_{ijk} and the outputs h_l are as follows [13]:

$$h_l = f(net_l) \qquad net_l = \sum_{k=1}^{t_f} v_{kl} \mu_{ijk}(x_{ij}) \tag{2}$$

Where v_{kl} are the weights between the fuzzification layer and the hidden layer.

In the output layer, each output o_m is as follows:

$$o_m = f(net_m) \quad net_m = \sum_{l=0}^{t_h} z_{lm} h_l \tag{3}$$

Where t_h is the number of nodes in the hidden layer, z_{lm} are the weights between the hidden layer and the output layer and $f(net) = \dfrac{1}{1+e^{-x}}$.

The error function of the FNNIP is:

$$E = \tfrac{1}{2} \sum_m (t_m - o_m)^2 \tag{4}$$

Where t_m is the expected output for the mth output node o_m. The weights between hidden layer and output layer are adjusted by BP algorithm as follows [12]:

$$z_{lm}(n) = z_{lm}(n-1) + \Delta z_{lm}(n) \tag{5}$$

$$\text{Where} \qquad \Delta z_{ilm}(n) = \eta \delta_m h_l + \alpha \Delta z_{lm}(n-1) \tag{6}$$

Where n is the iteration count, $\delta_m = t_m - o_m$. The weights between fuzzification layer and hidden layer are adjusted by the following equations [12]:

$$v_{kl}(n) = v_{kl}(n-1) + \Delta v_{kl}(n) \tag{7}$$

$$\text{Where} \qquad \Delta v_{kl}(n) = \eta \delta_{kl} \mu_{ijk} + \alpha \Delta v_{kl}(n-1) \tag{8}$$

$$\delta_{kl} = (\prod_{k=1}^{t_f} \mu_{ijk})(\sum_{m=1}^{t_o} \delta_m z_{lm} / \mu_{ijk}) \tag{9}$$

Where t_o is the total number of the output nodes. Similarly, the weights between input layer and fuzzification layer are adjusted as follows [12, 13]:

$$w_{ijk}(n) = w_{ijk}(n-1) + \Delta w_{ijk}(n) \tag{10}$$

$$\text{Where} \quad \Delta w_{ijk}(n) = \eta \delta_{kl} \mu_0 + \alpha \Delta w_{ijk}(n-1) \tag{11}$$

$$\mu_0 = \frac{1}{t_j} \sum_{j=1}^{t_j} \exp(-(w_{ijk} x_{ijk})^2) \tag{12}$$

In the formula above, "η" is a learning rate, which controls the rate of convergence, in the beginning of training it has the larger value, then decrease quickly [13]. We set $\eta = b_2(1 - t/t_m)$, b_2 is a constant value and in the range of [0,1] (it is set to 0.3 in this paper), t_m is the maximal training number in advance, t is the tth training. The momentum constant "α" (kept at 0.6 throughout) is able to add to speed up the training process and avoid local minima. The initial weights are randomly selected in the interval [-1, +1]. The training process is continued till $E(n) < \varepsilon$ at all points or the number of iteration reaches its maximum (e.g., 5000). The value of ε is taken to be $1 \times e^{-4}$ during training.

3   The Description of the Network Traffic Behaviors

In this paper, we assume that the current state of network connection record dependents on the latest t_j continuous network connection records sequences. So, the network intrusion prediction method is defined as:

Definition 1. Given the latest t_j continuous network connection records sequences $\mathbf{X}$, each record $\mathbf{x_i}$ in $\mathbf{X}$ has N-1 connection features and 1 label feature (normal or abnormal). The next network connection record $\mathbf{x_{next}}$ is predicted by the given latest t_j serial network connection records, and the label feature (normal or abnormal) may also be predicted:

$$\mathbf{X} = (\mathbf{x_1}, \mathbf{x_2}, \cdots, \mathbf{x}_{t_j}) \Rightarrow \mathbf{x}_{next} \tag{13}$$

According to definition 1, it is assumed that the next network action can be correctly predicted by the latest continuous network actions.

To improve the performance of the FNNIP, we should reduce the dimensions of the input vector. Some important features for intrusion prediction system are employed. Before we analyze the data set, data standardization must be performed. The data standardization process is described as follows [3]:

$$x'_{ik} = \frac{x_{ik} - \overline{x}_k}{s_k}, \quad (i=1,2,\ldots,n, k=1,2,\ldots,m) \tag{14}$$

Where $\overline{x}_k$ and s_k is the mean value and standard deviation of one feature or the kth dimension of data set [3].

$$\overline{x}_k = \frac{1}{n}\sum\nolimits_{i=1}^{n} x_{ik} \tag{15}$$

$$s_k = \sqrt{\frac{1}{n}\sum\nolimits_{i=1}^{n} (x_{ik} - \overline{x}_k)^2} \tag{16}$$

This procedure transforms the mean of the set of feature values to zero and the standard deviation to one. But the $x_{ik}^{'}$ may not be in the interval [0,1] and it is processed as follows [3]:

$$x_{ik}^{''} = \frac{x_{ik}^{'} - \min_{1 \le i \le n}\{x_{ik}^{'}\}}{\max_{1 \le i \le n}\{x_{ik}^{'}\} - \min_{1 \le i \le n}\{x_{ik}^{'}\}}, (k = 1,2,\ldots,m) \tag{17}$$

To select the important features, we calculate the correlation coefficient between x_i and x_j as follows [3]:

$$r_{ij} = \frac{\sum\nolimits_{k=1}^{m} \left|x_{ik} - \overline{x}_i\right|\left\|x_{jk} - \overline{x}_j\right|}{\sqrt{\sum\nolimits_{k=1}^{m} (x_{ik} - \overline{x}_i)^2} \cdot \sqrt{\sum\nolimits_{k=1}^{m} (x_{jk} - \overline{x}_j)^2}} \tag{18}$$

It is strong negative correlation between x_i and x_j when $r_{ij} = -1$ and it is strong positive correlation between x_i and x_j when $r_{ij} = 1$.

In this paper, we compute the correlation coefficients between two features by the soft ware SPSS [3]. If the $r_{ij} \ge 0.7$, we select only one of them to represent these two fields. The selected features are describe as follows:

$$\mathbf{x} = (x_1, x_2, \cdots, x_N) = (duration, service, flag, src\_bytes, dst\_bytes, wrong\_$$
$$fragment, urgent, hot, num\_failed\_log ins, su\_attempted, num\_root, num\_$$
$$file\_creations, num\_access\_files, count, srv\_count, serror\_rate, srv\_serror$$
$$\_rate, rerror\_rate, srv\_rerror\_rate, same\_srv\_rate, srv\_diff\_host\_rate, \tag{19}$$
$$dst\_host\_count, dst\_host\_srv\_count, dst\_host\_same\_src\_port\_rate,$$
$$dst\_host\_srv\_diff\_host\_rate, label)$$

4 Experiments

4.1 Training the FNNIP

The proposed FNNIP are evaluated by the famous KDD'99 data sets [10]. There are four main categories attacks in the KDD'99 data sets: DoS, R2L, U2R and Probing.

The training data is randomly collected from the original raw KDD'99 data sets, which contains all of four categories attacks (20%) and normal records (80%).

We select only 17 features from the total 41 features of each connection record and an additional label feature as the inputs of the FNNIP (N=18). The latest 5 (t_j =5 for the FNNIP) continuous connection records are employed to predict the next connection records. So, the total input nodes of the FNNIP are 90. A slide window (window size is 5 records and the step is 1) employed covers the training data and its outputs are passed to the FNNIP. The total nodes of fuzzification and the total nodes of output layer are 18, respectively. The total nodes of hidden layer are selected as H=30,40,50, respectively.

During training process, the training data set is exposed to the proposed FNNIP and the weights of the FNNIP are initialized randomly. To adjust the weights, the actual outputs are compared with desired targets, respectively. If the output and target match (in the same range), no change is made to the weights. However, if the output differs from the target (in the different range) a change must be made to the specific weight. The training process is continued till $E(n) < \varepsilon$ at all points or the number of iteration reaches its maximum (e.g., 10000).

4.2   Testing the FNNIP

Four groups test data sets are generated from the KDD'99 test data set. It contains a total of 24 attacks in training data and additional 14 types attacks in the test data only. The extracted test data sets contain 80% normal connection records and 20% attacks records but not including the training data set. The testing data is also normalized as described in section 3.2. We use the Euclidean distance to measure the similarity between the active output $\mathbf{o}_o = \{o_{o1}, o_{o2}, \cdots, o_{on-1}\}$ and the corresponding records $\mathbf{o}_t = \{o_{t1}, o_{t2}, \cdots, o_{tn-1}\}$ in the test data (except the label feature):

$$d = \sqrt{\sum_{i=1}^{n-1} (o_{oi} - o_{ti})^2} \tag{20}$$

If d is little than a specific value (e.g. 0.3) then the result of prediction of connection record is right, otherwise, the result of prediction is wrong.

For the label feature, if its value of the prediction is larger than 0.5, the prediction record is abnormal; otherwise, the prediction record is normal.

The hit occurs only when the connection record is correctly predicted and its corresponding label is rightly predicted.

The FNNIP presented in this paper is evaluated using ROC charts. The experimental results are showed in Fig.2 to Fig.5, which are the average results for the four groups testing data set and when the Euclidean distance satisfies d<0.3.

Given a specific number of hidden nodes for the FNNIP, we can obtain a pair of hit rate and false alarm for specific type of attacks. Fig.2 to Fig.5 show that the performance of the FNNIP increases with different numbers of hidden nodes, but the hit rate can only improve slightly after the number of hidden node is greater than 40. Fig.2 is the ROC of prediction of the DoS attacks. The hit-rate is greater than 80% when false alarm rate is 10% and the number of hidden nodes of FNNIP is greater than 40. If we allow the false alarm greater than 17%, the hit-rate is greater than 90%.

Fig.3 is the ROC of prediction of the Probing attacks. The hit-rate is greater than 80% when false alarm rate is 5% and the number of hidden nodes of FNNIP is greater than 40. If we agree that the false alarm greater than 15%, the hit-rate is greater than 90%.

Fig.4 is the ROC of prediction of the U2R attacks. The hit-rate is greater than 80% when false alarm rate is 5% and the number of hidden nodes of the FNNIP is greater than 50%. If we permit the false alarm greater than 20%, the hit-rate is greater than 90%.

Fig.5 is the ROC of predicting the R2L attacks. From Fig.5, The hit-rate is between 70% and 75% when false alarm rate is 10% and the number of hidden nodes of the FNNIP is greater than 40. If we allow the false alarm greater than 30%, the hit-rate may be greater than 90%.

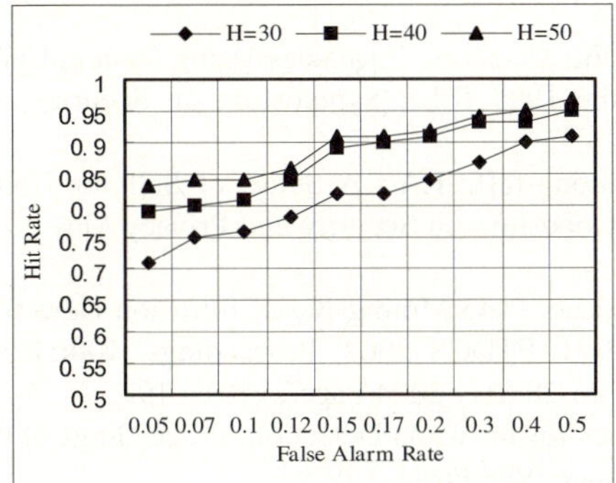

Fig. 2. The ROC for the DoS attacks

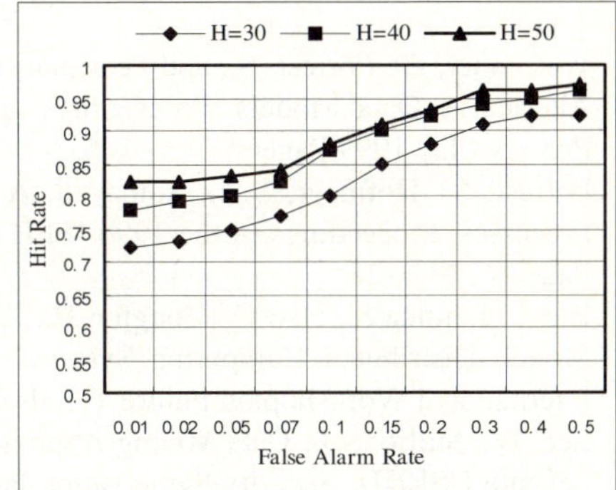

Fig. 3. The ROC for the Probing attacks

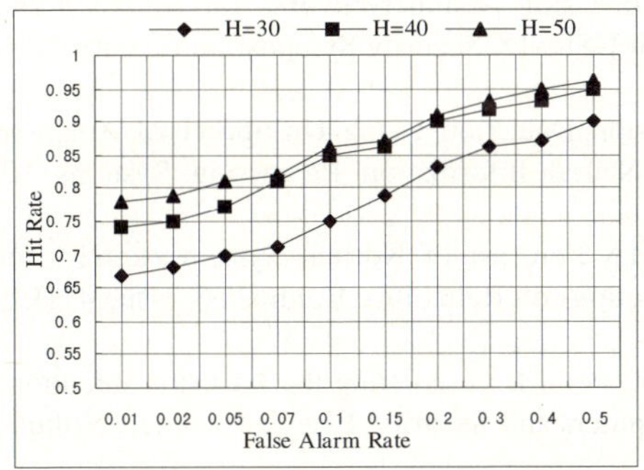

Fig. 4. The ROC for the U2R attacks

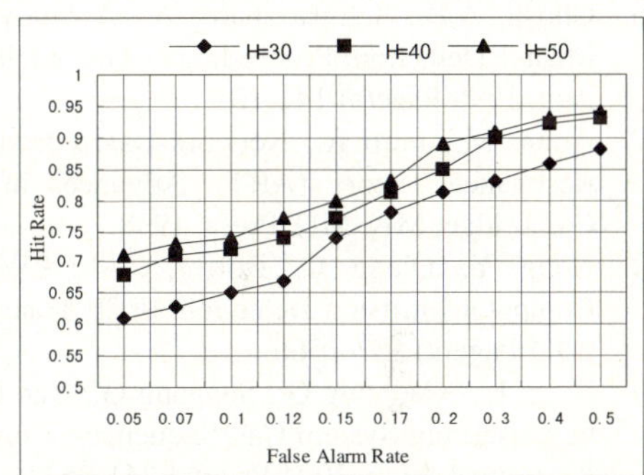

Fig. 5. The ROC for the R2L attacks

From Fig.2 to Fig.5, we can conclude that the DoS and R2L attacks are more flexible than the other two type attacks during their intrusion procedure. It is more difficult to predict the Probing and U2R attacks correctly.

5   Conclusions and Future Work

A novel method of predicting intrusion intentions based on fuzzy neural network has been investigated in this paper. The proposed FNNIP use the latest continuous connection records and their labels ("normal" or "intrusion") to predict the next record and the corresponding label ("normal" or "intrusion"). It is evaluated by the famous KDD'99 dataset and the experiment results demonstrate that the proposed FNNIP has acceptable ability of predicting different type of attacks in network connection records.

However there are some issues to be studied in the future. First, how much is the optimal number of the latest continuous records to predict the next records? Second, what features are more important to the FNNIP prediction scheme? Final, how much is optimal number of hidden nodes for FNNIP corresponding to the different parameters?

References

1. Warrender, C., Forrest, S., and Pearlmutter, B.: Detecting Intrusions Using System Calls: Alternative Data Models. Proceedings of the 1999 IEEE Symposium on Security and Privacy May 1999 Page(s) 133 - 145.
2. Forrest, S., Hofmeyr, S.A.; Somayaji, A., Longstaff, T.A.: A Sense of Self for UNIX Processes. Proceedings of the 1996 IEEE Symposium on Security and Privacy May 1996 Page(s) 120 - 128.
3. Hai J., Jianhua S., Hao C., Zongfen H.: A Fuzzy Data Mining Based Intrusion Detection Model. Distributed Computing Systems, 2004. FTDCS 2004. Proceedings. 10th IEEE International Workshop on Future Trends of 26-28 May 2004 Page(s) 191 - 197.
4. Lee, W., Stolfo, S.J.: Data Mining Approaches for Intrusion Detection. Proceedings of the Seventh USENIX Security Symposium, January 1998 Page(s) 79-93.
5. Wenke L., Stolfo, S.J.; Mok, K.W.: A Data Mining Framework for Building Intrusion Detection Models. Proceedings of the 1999 IEEE Symposium on Security and Privacy May 1999 Page(s) 120 - 132.
6. Ghosh, A.K., Schwartzbard, A.: A Study in Using Neural Networks for Anomaly and Misuse Detection. Proceedings of the Eighth USENIX Security Symposium (Security'99), Aug. 1999 Page(s) 141-151.
7. Amini M., Jalili R.: Network-Based Intrusion Detection Using Unsupervised Adaptive Resonance Theory (ART). Advances in Neural Information Processing Systems 10, Cambridge, MA: MIT Press, 1998.
8. Nong, Y., Qiang, C., Borror, C.M.: EWMA Forecast of Normal System Activity for Computer Intrusion Detection. IEEE Transactions on Reliability Volume 53, Issue 4, Dec. 2004 Page(s) 557 - 566.
9. Feng, L., Xiaohong G., Sangang G., Yan G., Peini L.: Predicting the Intrusion Intentions by Observing System Call Sequences. Computers and Security, Elsevier Science, Volume 23, Issue 3, May, 2004 Page(s) 241-252.
10. KDD Cup 1999 data. http://kdd.ics.uci.edu/databases/kddcup99/kddcup99.htm.
11. Mukkamala S., Sung A. H.: Feature Selection for Intrusion Detection Using Neural Networks and Support Vector Machines. Journal of the Transportation Research Board, Transportation Research Record No 1822, 2003 Page(s) 33-39.
12. Dash, P.K., Pradhan, A.K., Panda, G.: A Novel Fuzzy Neural Network Based Distance Relaying Scheme. IEEE Transactions On Power Delivery, VOL. 15, NO. 3, July 2000 Page(s) 902 - 907.
13. Yuan, F., Wu, H., and Yu, G.: Web Users' Classification Using Fuzzy Neural Network. Knowledge-Based Intelligent Information and Engineering Systems: 8th International Conference, Kes 2004, Wellington, New Zealand, September, 2004.

On the Random Sampling Amplitude Error

Shouyuan Yang[1], Zhanjie Song[2,*], and Xingwei Zhou[3,**]

[1] Department of Mathematics, Naikai University and LPMC,
Tianjin 300071, China
yshouy@sina.com
[2] School of Science, Tianjin University,
Tianjin 300072, China
zhanjiesong@tju.edu.cn
[3] Department of Mathematics, Naikai University and LPMC,
Tianjin 300071, China
xwzhou@naikai.edu.cn

Abstract. The main purpose of this paper is to examine the distribution of the random amplitude error for the sampling problem in diverse situations, and specific formulas are given, which reveal the connection between the random errors of the sampled values and the amplitude error caused by them. The information loss error is also included as a special case.

1 Introduction and Preliminaries

Sampling theories are now widely used in many areas, especially in digital signal processing and transmitting. The most important feature of all sampling theorems is that a continuous signal can be recovered from a sequence of sampled values. The most famous sampling theorem which is usually attributed to Shannon stated that

$$f(t) = \sum_{n \in \mathbb{Z}} f\left(\frac{n}{2\sigma}\right) \frac{\sin 2\pi\sigma(t - n/2\sigma)}{2\pi\sigma(t - n/2\sigma)}$$

for any σ-bandlimited signals $f(t)$, i.e., $f(t) \in L^2(\mathbb{R})$ and its Fourier transform $\hat{f}(\xi) := \int_{\mathbb{R}} f(t)e^{-i2\pi\xi t}dt$ supported on $[-\sigma, +\sigma]$, where $L^2(\mathbb{R})$ denote the space of all square integrable signals. The classical sampling theorem has been extended in many ways during the last five decades. The most important extension may be nonuniform sampling and sampling in other signal spaces, such as spline-like (shift-invariant) spaces and wavelet subspaces, e.g., see [1, 4, 5, 10]. Higher dimensional sampling is also considered by many researchers because of its wide application in image processing and many other areas.

* Supported by the National Natural Science Foundation of China (60476042 and 60572113), the Liuhui Center for Applied Mathematics.

** Supported by the Natural Science Foundation of China under Grant 60472042 and the Research Fund for the Doctoral Program of Higher Education.

V.N. Alexandrov et al. (Eds.): ICCS 2006, Part I, LNCS 3991, pp. 427–434, 2006.
© Springer-Verlag Berlin Heidelberg 2006

Now let us introduce some notations. We use $\mathrm{I\!R}^d$ and $\mathbb{Z}^d$ to denote the d-dimensional Euclidean space and unit lattice, respectively. $L^{(}\mathrm{I\!R}^d)$ and $l^2(\mathbb{Z}^d)$ denote the space of all square integrable signals defined on $\mathrm{I\!R}^d$ and the space of all square summable sequences defined on $\mathbb{Z}^d$, respectively. With the inner product $\langle f, g \rangle = \int_{\mathrm{I\!R}^d} f(t)\overline{g(t)}dt$, $L^2(\mathrm{I\!R}^d)$ constitutes a Hilbert space. Obviously, the sampling problem would be meaningless if no restriction is imposed on the signal space and the set of sampling points. Throughout this paper we assume that the signal space $V \subseteq L^2(\mathrm{I\!R}^d)$ and the set of sampling points $X := \{t_j\}_{j \in J} \subseteq \mathrm{I\!R}^d$ satisfy the following conditions:

i). There exists a sequence $\{s_n : n \in \mathbb{Z}^d\}$ of functions in V which is called a sampling sequence of V such that

$$f(t) = \sum_{n \in \mathbb{Z}^d} f(t_n)s_n(t) \tag{1}$$

for any $f \in V$, where the convergence is in the $L^2(\mathrm{I\!R}^d)$-sense. In particular, if there exist $s \in V$ such that $\{s(\cdot - t_n) : n \in \mathbb{Z}^d\}$ constitutes a sampling sequence of V, then s is said to be a sampling function.

ii). The sampling operator $\mathrm{S}_X : V \to l^2(X)$ defined by $\mathrm{S}_X f = (f(t_j))_{j \in J}$ is a bounded linear operator, i.e.,

$$\sum_{j \in J} |f(t_j)|^2 \leq B\|f\|_2^2, \quad \text{for all } f \in V,$$

where B is a constant independent of f.

It is worthwhile pointing out that so far all the sampling theorems either include the above conditions as a assumption or include other assumptions from which the above conditions can be obtained as a conclusion. Here we list the sampling sequence or sampling functions for several well-known sampling problems:

I) Uniform sampling for band-limited functions. The signal space is B_σ, which consists of all σ-bandlimited signals defined on $\mathrm{I\!R}$, the system $\{\mathrm{sinc}\, 2\pi\sigma(\cdot - n/2\sigma)\}_{n \in \mathbb{Z}}$ constitutes a sampling sequence of B_σ, where $\mathrm{sinc}\, t := \sin t/t$. Hence $\mathrm{sinc}\, 2\pi\sigma(\cdot)$ is a sampling function of B_σ, the reconstruction formula is exactly the Shannon sampling theorem.

II) If $\sigma = 1$, $\{t_n\}_{n \in \mathbb{Z}}$ is a sequence of real numbers such that $|t_n - n| \leq L < 1/4$ for all n, then by Kadec's $\frac{1}{4}$-theorem (e.g., see [11]), the sequence $\{G_n(t)\}_{n \in \mathbb{Z}}$ constitutes a sampling sequence of B_σ, where

$$G_n(t) := \frac{G(t)}{G'(t_n)(t - t_n)}, \qquad G(t) := t \prod_{n \in \mathbb{Z}} \left(1 - \frac{t^2}{t_n^2}\right).$$

III) If the signal space is a spline-like space $V^2(\varphi)$ defined as follows

$$V^2(\varphi) := \left\{ \sum_{n \in \mathbb{Z}^d} c_n \varphi(\cdot - n) : c = (c_n) \in l^2(\mathbb{Z}^d) \right\},$$

where φ satisfies

$$0 < c \le G_\varphi(\xi) = \sum_{j \in \mathbb{Z}^d} |\hat\varphi(\xi + j)|^2 \le C, \quad \text{a.e. } \xi \in \mathbb{R}^d \tag{2}$$

and some decay and smoothness condition, e.g., φ is continuous and satisfies

$$\|\varphi\|_{W(L^p(\mathbb{R}^d))} := \left(\sum_{k \in \mathbb{Z}^d} \sup_{t \in [0,1]^d} |\varphi(t+k)|^p \right)^{1/p} < \infty,$$

then the function s determined by

$$\hat s(\xi) = \frac{\hat\varphi(\xi)}{\sum_{j \in \mathbb{Z}^d} \varphi(j) e^{2\pi i j \cdot \xi}} \tag{3}$$

is a sampling function of $V^2(\varphi)$, and $\{s(\cdot - n)\}_{n \in \mathbb{Z}^d}$ is a sampling sequence, e.g., see [1, 10]. If the sampling points are not uniformly distributed, we can also construct a sampling sequence of $V^2(\varphi)$. Indeed, if we let $\tilde\varphi$ be determined by $\hat{\tilde\varphi}(\xi) = \hat\varphi(\xi)/G_\varphi(\xi)$, where $G_\varphi(\xi)$ defined in (2), then $K(x,y) := \sum_{j \in \mathbb{Z}^d} \varphi(x-j)\tilde\varphi(y-j)$ is a reproducing kernel (e.g., see [16]), namely,

$$f(t) = \langle f, K(t, \cdot) \rangle, \quad \text{for all } t \in \mathbb{R}^d, \ f \in V^2(\varphi). \tag{4}$$

If the sampling points $\{t_j\}$ are dense enough, then $\{K(t_j, \cdot)\}$ constitutes a frame for $V^2(\varphi)$, and its dual frame $\{\widetilde{K(t_j, \cdot)}\}$ is what we try to find, e.g., see [1, 13, 14, 15].

IV) Let φ be a scaling function (e.g., see [5, 8, 9]) satisfying (2) and certain decay and smoothness condition, $\{V_m : m \in \mathbb{Z}\}$ be the multi-resolution analysis generated by φ (e.g., see [5]). If s be the function determined by (3), then for each m the system $\{s_{m,n} : n \in \mathbb{Z}\}$ constitutes a sampling basis of V_m, where $s_{m,n} = \varphi(2^m \cdot -n)$. The reconstruction formula is

$$f(t) = \sum_{n \in \mathbb{Z}^d} f\left(\frac{n}{2^m}\right) s_{m,n}(t), \quad \text{for all } f \in V_m.$$

There are several type of errors which occur in in the real application of sampling theorems, e.g., see [6]. In [3], Atreas et al examined the truncation error of the reconstruction formula in wavelet subspaces. It was not long before Yang et al extended their results to higher dimensional cases and spline-like spaces, e.g., see [12, 14]. In this paper we shall investigate the random amplitude error for the above sampling expansions. Specifically, let $f(t_j)$ be the true value of the signal f at the sample t_j, and $\widetilde{f(t_j)}$ be the sampled value obtained by apparatus, of course it cannot be absolutely precise, since it is often noised by a random error. Let $\lambda(t_j)$ be the relative error defined by

$$\lambda(t_j) := \frac{\widetilde{f(t_j)} - f(t_j)}{f(t_j)}$$

if $f(t_j) \neq 0$, otherwise $\lambda(t_j) = \mathrm{sgn}(\widetilde{f(t_j)} - f(t_j)) \cdot \infty$, where $\mathrm{sgn}(\cdot)$ denotes the sign function, i.e., $\mathrm{sgn}(x) = 1$ if $x > 0$, $\mathrm{sgn}(x) = -1$ if $x < 0$, and $\mathrm{sgn}(0) = 0$, and $0 \cdot \infty = 0$ in the definition of $\lambda(t_j)$ by convention. Since the relative error is determined by the inertia of the sampling apparatus and many other unknown factors, it is impossible to find out its precise value, so we assume that all $\lambda(t_j)$'s are independent and identically distributed (i.i.d.) random variables with finite first moments. The amplitude error is defined by

$$\mathrm{Am}\, f(t) := \mathrm{Rec}\, f(t; \cdots, \widetilde{f(t_j)}, \cdots) - f(t) \,,$$

where $\mathrm{Rec}\, f(t; \cdots, \widetilde{f(t_j)}, \cdots)$ denotes the signal reconstructed from the sequence $\{\widetilde{f(t_j)}\}$ of measured samples.

2   Random Amplitude Error Estimation

In this section we assume that the $L^2(\mathrm{I\!R}^d)$-norm of the original signal $f(t)$ is finite, and then examine the distribution of the amplitude error in terms of this norm. We assume henceforth that the relative errors $\lambda(t_j)$ are i.i.d. random variables with $\mathrm{E}[\lambda(t_j)] = 0$ and $\mathrm{E}[|\lambda(t_j)|] = \delta < \infty$ if no other assumptions are claimed, where $\mathrm{E}[X]$ denotes the expectation (mean) of the random variable X.

2.1   Uniform Sampling

Without loss of generality, we assume that the unit lattice $\mathrm{Z\!\!Z}^d$ and the signal space $V \subseteq L^2(\mathrm{I\!R}^d)$ satisfy the conditions i) and ii) given in Section 1. Let $\{s(\cdot - j) : j \in \mathrm{Z\!\!Z}^d\}$ be a sampling sequence of the signal space V. Then we have the following reconstruction formula

$$f(t) = \sum_{j \in \mathrm{Z\!\!Z}^d} f(j) s(t - j) \,, \tag{5}$$

and the amplitude error can be rewrite as

$$\mathrm{Am}\, f(t) = \mathrm{Rec}\, f(t; \cdots, \widetilde{f(t_j)}, \cdots) - f(t)$$
$$= \sum_{j \in \mathrm{Z\!\!Z}^d} \widetilde{f(j)} s(t - j) - \sum_{j \in \mathrm{Z\!\!Z}^d} f(j) s(t - j)$$
$$= \sum_{j \in \mathrm{Z\!\!Z}^d} \lambda(j) \cdot f(j) s(t - j) \,. \tag{6}$$

Hence we have

$$\mathrm{E}[\mathrm{Am}\, f(t)] = \sum_{j \in \mathrm{Z\!\!Z}^d} \mathrm{E}[\lambda(j)] \cdot f(j) s(t - j) = 0 \tag{7}$$

and

$$\mathrm{E}[|\operatorname{Am} f(t)|] \leq \mathrm{E}\left[\sum_{j \in \mathbb{Z}^d} |\lambda(j)| \cdot |f(j)s(t-j)|\right]$$

$$= \delta \sum_{j \in \mathbb{Z}^d} |f(j)s(t-j)|$$

$$\leq \delta \left(\sum_{j \in \mathbb{Z}^d} |f(j)|^2\right)^{1/2} \cdot \left(\sum_{j \in \mathbb{Z}^d} |s(t-j)|^2\right)^{1/2}$$

$$\leq \delta \cdot B^{1/2} \cdot \|f\|_2 \cdot \left(\sum_{j \in \mathbb{Z}^d} |s(t-j)|^2\right)^{1/2}. \tag{8}$$

Now by the Chebyshev's inequality, from (7) and (8) we get that

$$\operatorname{Prob}\{|\operatorname{Am} f(t)| < \epsilon\} \geq 1 - \frac{\delta \cdot B^{1/2}}{\epsilon}\|f\|_2 \cdot \left(\sum_{j \in \mathbb{Z}^d} |s(t-j)|^2\right)^{1/2}. \tag{9}$$

Hence we have proved the first part of the following theorem.

Theorem 1. *Let the set of sampling points $\mathbb{Z}^d$ and the signal space V satisfy the conditions given in Section 1, and assume that there exists a sampling function $s \in V$ such that the reconstruction formula (5) holds for all t. If s decays fast enough such that $\sum_{j \in \mathbb{Z}^d} |s(t-j)|^2 < \infty$, then the amplitude error satisfies (9). In particular, if $|s(t)| \leq C(1 + |t|^\alpha)^{-1/2}$, where $\alpha > d$, then we have*

$$\operatorname{Prob}\{|\operatorname{Am} f(t)| < \epsilon\} \geq 1 - \frac{C \cdot \delta \cdot B^{1/2} \cdot \alpha^{1/2} \cdot 2^{(\alpha+d)/2}}{\epsilon \cdot (\alpha - d)^{1/2}}\|f\|_2 \tag{10}$$

for all t.

Proof. Only inequality (10) needs to prove. For each $j \in \mathbb{Z}^d$, let Q_j be a closed ball centred at j with radius $1/2$. For $y \in Q_j$ and $t \in Q_j^c$, where Q_j^c denotes the complement of Q_j, direct calculations show that

$$\frac{1 + |t-y|^\alpha}{1 + |t-j|^\alpha} \leq \frac{1 + (|t-j| + |y-j|)^\alpha}{1 + |t-j|^\alpha}$$

$$\leq \frac{1 + (|t-j| + 1/2)^\alpha}{1 + |t-j|^\alpha}$$

$$\leq \frac{1 + |t-j|^\alpha \left(1 + \frac{1}{2|t-j|}\right)^\alpha}{1 + |t-j|^\alpha}$$

$$\leq 2^\alpha.$$

For $t \in Q_j$, we can also prove that $(1 + |t - y|^\alpha)/(1 + |t - j|^\alpha) \leq 2^\alpha$. Hence $(1 + |t - j|^\alpha)^{-1} \leq 2^\alpha \cdot (1 + |t - y|^\alpha)^{-1}$ for all $y \in Q_j$ and all $t \in \mathbb{R}^d$, and therefore

$$\sum_{j \in \mathbb{Z}^d} |s(t - j)|^2 \leq C^2 \sum_{j \in \mathbb{Z}^d} (1 + |t - j|^\alpha)^{-1}$$

$$\leq C^2 \cdot 2^\alpha \sum_{j \in \mathbb{Z}^d} |Q_j|^{-1} \int_{Q_j} (1 + |t - y|^\alpha)^{-1} \, dy$$

$$\leq C^2 \cdot 2^{\alpha+d} \frac{d}{S_d} \int_{\mathbb{R}^d} (1 + |t - y|^\alpha)^{-1} \, dy$$

$$\leq C^2 \cdot 2^{\alpha+d} \cdot d \cdot \left(\frac{1}{d} + \frac{1}{\alpha - d} \right), \tag{11}$$

where $|Q_j|$ and S_d denote the volume of the closed ball Q_j and the area of the d-dimensional unit sphere, respectively. The inequalities (9) and (11) lead to the conclusion immediately.

Note that for band-limited sampling theorems the sampling function can be obtained by dilating the function $\text{sinc}(\cdot)$, therefore, obviously satisfies the decay condition required in Theorem 1; for sampling theorems in the spline-like spaces, the decay of the sampling function is guaranteed by the decay of the generator φ. Indeed, Yang has proved that in the spline-like spaces the asymptotic rate of decay of the sampling function is the same as that of the generator (see [12]). As for the wavelet subspaces, it can be viewed as a spline-like space generated by the dilated scaling function, so the amplitude error estimate obtained in spline-like spaces can be easily extended to wavelet subspaces.

2.2   Nonuniform Sampling

Now let us consider the general case. Let the signal space $V \subseteq L^2(\mathbb{R}^d)$ and the set of sampling points $\{t_j : j \in J\}$ satisfy the conditions i) and ii) given in Section 1. Then the amplitude error can be rewrite as

$$\text{Am} \, f(t) = \sum_{j \in J} \lambda(t_j) \cdot f(t_j) s(t - t_j).$$

By the same techniques we can prove the following results.

Theorem 2. *Let the set of sampling points $\{t_j : j \in J\}$ and the signal space V satisfy the conditions given in Section 1. If the sampling sequence $\{s_j \in V : j \in J\}$ satisfies that $\sum_{j \in J} |s_j(t)|^2 < \infty$ uniformly, then we have*

$$\text{Prob}\{| \text{Am} \, f(t)| < \epsilon\} \geq 1 - \frac{\delta \cdot B^{1/2}}{\epsilon} \|f\|_2 \cdot \left(\sum_{j \in J} |s_j(t)|^2 \right)^{1/2} \tag{12}$$

for all t. In particular, if $|s_j(t)| \leq C(1+|t-t_j|^\alpha)^{-1/2}$ for all $j \in J$, where $\alpha > d$, and the set of sampling points $\{t_j : j \in J\}$ are separated, i.e., $\inf_{j,l \in J, j \neq l} |x_j - x_l| = \mu > 0$, then we have

$$\text{Prob}\{|\operatorname{Am} f(t)| < \epsilon\} \geq 1 - \frac{C \cdot \delta \cdot B^{1/2} \cdot \alpha^{1/2} \cdot 2^{(\alpha+d)/2}}{\epsilon \cdot \mu^{d/2} \cdot (\alpha - d)^{1/2}} \|f\|_2 \tag{13}$$

for all t.

We point out that for nonuniform sampling in spline-like spaces, the decay of the sampling sequence $\{s_j : j \in J\}$ is also guaranteed by the decay of the generator φ, e.g., see [7]. Secondly, the constant B appearing in condition ii) in Section 1 depends on the density of the samples, and its existence is guaranteed by the separateness of the samples.

3   Random Information Loss Error Estimation

If the relative errors are binary valued, namely, $\lambda(t_j)$ either takes the value 1 or takes the value 0, no other value is allowed, then the corresponding amplitude error is called the information loss error in [2]. In that paper, the error caused by the missing of some sampled data are considered, where $\lambda(t_j) = 1$ for the sampling points t_j at which the sampled values $f(t_j)$ are missing and $\lambda(t_j) = 0$ otherwise. In the present paper we assume that the missing occurs randomly, and $\lambda(t_j)$ are i.i.d. random variables with $\text{Prob}\{\lambda(t_j) = 1\} = p$ and $\text{Prob}\{\lambda(t_j) = 0\} = 1 - p$. Since the following results are just special cases of Theorem 2, so we omit its proof.

Theorem 3. *Let the set of sampling points $\{t_j : j \in J\}$ and the signal space V satisfy the conditions given in Section 1, and $\lambda(t_j)$ be the corresponding relative errors with all the properties stated above. If the sampling sequence $\{s_j \in V : j \in J\}$ satisfies that $\sum_{j \in J} |s_j(t)|^2 < \infty$ uniformly, then we have*

$$\text{Prob}\{|\operatorname{Am} f(t) - p| < \epsilon\} \geq 1 - \frac{p \cdot B^{1/2}}{\epsilon} \|f\|_2 \cdot \left(\sum_{j \in J} |s_j(t)|^2\right)^{1/2} \tag{14}$$

for all t. In particular, if $|s_j(t)| \leq C(1+|t-t_j|^\alpha)^{-1/2}$ for all $j \in J$, where $\alpha > d$, and the set of sampling points $\{t_j : j \in J\}$ are separated, i.e., $\inf_{j,l \in J, j \neq l} |x_j - x_l| = \mu > 0$, then we have

$$\text{Prob}\{|\operatorname{Am} f(t) - p| < \epsilon\} \geq 1 - \frac{C \cdot p \cdot B^{1/2} \cdot \alpha^{1/2} \cdot 2^{(\alpha+d)/2}}{\epsilon \cdot \mu^{d/2} \cdot (\alpha - d)^{1/2}} \|f\|_2 \tag{15}$$

for all t.

References

1. A. Aldroubi, and K. Gröchenig, Nonuniform sampling and reconstruction in shift-invariant spaces. SIAM Rev. **43**(4)(2001) 585-620
2. N.treas, N.Bagis and C.Karanikas, The information loss error and the jitter error for regular sampling expansions, Sampling Theory in Signal and Image Processing, **1**(3)(2003) 261-276
3. N. Atreas, J. J. Benedetto, and C. Karanikas, Local sampling for regular wavelet and Gabor expansions(to appear)
4. John J. Benedetto, Irregular sampling and frames, in: C. K. Chui (Ed.), Wavelets: A Tutorial in Theory and Applications. (1992) 445-507
5. I. Daubechies, Ten Lectures on Wavelets. CBMS-NSF Series in Applied Math. SIAM Philadelphia. 1992.
6. H.Feichtinger and K.Gröchenig, Error analysis in regular and irregular sampling theory, Appl. Anal. **50** (1993) 167-189.
7. K. Gröchenig, Localization of frames, Banach frames, and the invertibility of the frame operator. J. Fourier Anal. Appl.(to appear)
8. S. Mallat, A Wavelet Tour of Signal Processing. Academic Press, Boston, 1998
9. Y. Meyer, Ondelettes Et Opérateurs. Hermann, Paris, 1990
10. Michael Unser, Sampling–50 years after Shannon. Procdings of the IEE. **88** (4) (2000) 569-587
11. R.M.Yang, An Introduction to Nonharmonic Analysis, Academic Press, New York.
12. S. Y. Yang, Local error estimation for samling problems, Appl. Math. Comp. **158** (2004) 561-572.
13. S.Y.Yang, Wavelet Frames, Local Sampling Problems and Applications, PhD thesis, 2004.
14. S.Y.Yang and W.Lin, Local sampling problems, Lecture Notes in Computer Science, **3037** (2004) 81-88.
15. S.Y.Yang, The local property of several operators on sampling, Applicable anal. **83** (9) (2004) 905-913.
16. K. Yao, Application of reproducing kernel Hilbert spaces–bandlimited signal models. Inform. and control. **11** (1967) 429-444

Enhancing 3D Face Recognition by Combination of Voiceprint*

Yueming Wang, Gang Pan, Yingchun Yang, Dongdong Li, and Zhaohui Wu

Department of Computer Science and Engineering
Zhejiang University, Hangzhou, 310027, P.R. China
{ymingwang, gpan}@zju.edu.cn

Abstract. This paper investigates the enhancement of identification performance when using voice classifier to help 3D face recognition. 3D face recognition is well known for its being superior to 2D due to the invariance in illumination, make-ups and pose. However, it is still challenged by expression variance. The partial ICP method we used for 3D face recognition could implicitly and dynamically extract the rigid parts of facial surface and be able to get much better performance than other methods in 3D face recognition under expression changes. This work serves to further improve the performance of recognition by combining a voiceprint classifier into partial ICP method. We implement 9 combination schemes, and experiments on database of 360 models with 40 subjects, 9 3D face scans with four different kinds of expression and 9 sessions of utterance for each subject, shows improvement of performance is very promising.

1   Introduction

Identity recognition has been received much attention during past two decades and biometric technologies, such as fingerprint, iris, face and voice, has been studied extensively. Though biometric technologies based on fingerprint and iris can offer greater accuracy, they require much greater explicit cooperation from the user. For example, fingerprint requires that the subject cooperate in making physical contact with the sensor surface. Recognition through face and voice are nature methods for their unlimited feature. Thus it appears that there is significant potential application-driven demand for improved performance in face and voice recognition system.

Most efforts have been made for face recognition from 2D images[1], but only a few approaches exploited 3D information[5]. Although the 2D face recognition system has good performance under constrained conditions, it is still challenged by changes in illumination, pose and expression [1, 6]. Because the 2D image is only a projection of the 3D human face essentially, the system performance can be improved by utilizing 3D information which is the explicit representation of facial surface[6]. However, facial expression is still a big challenge even

---

* The authors are grateful for the grants from NSF of China (60503019, 60525202) and Program for New Century Excellent Talents in University (NCET-04-0545).

using 3D data in face recognition because in fact facial surface is a non-rigid object. "Different expressions between the gallery and probe sets degrade rank-one recognition rates in 3D face by as much as 33%", reported by [3].

Our previous work has explored facial expression effects in 3D face recognition using partial ICP [7]. The partial ICP method could partly overcome the problem of facial expression variance by dynamically extracting the rigid parts of facial surfaces to match in face identification. This paper will evaluate the enhancement of performance by partial ICP when combining a voice classifier into it. Since Gaussian Mixture Models(GMM) provide a robust speaker representation for the difficult task of speaker identification using corrupted , unconstrained speech [8], we use GMM-based method as a voiceprint component in combination.

The paper is organized as follows: Sec. 2 and Sec. 3 introduce the partial ICP and GMM classifiers respectively. Sec. 4 describe the combining schemes used in the experiments. The experimental results and conclusions are in Sec. 5 and Sec. 6 respectively.

2   3D Face Recognition Component

Our previous work has provided detailed description about partial ICP in 3D face recognition [7]. We briefly review it as follows.

The famous ICP algorithm, developed by Besl and Mckay [4], is used to register the point sets by an iterative procedure which is widely used in field of 3D rigid object registration. Let point set $P_1 = \{p^1_1, \cdots, p^1_M\}$ and point set $P_2 = \{p^2_1, \cdots, p^2_N\}$. The ICP algorithm can be summarized as:

1. $P_2(0) = P_2$, l=0
2. **Do**
3.   **For** each point p^2_i in $P_2(l)$
4.   Find the closest point y_i in P_1
5.   **End For**
6.   The closest points y_i form a new point set $Y(l)$ where the pairs of points
7. $\{(p^2_1, y_1), \cdots, (p^2_N, y_N)\}$ describe the correspondences between P_1 and
8. $P_2(l)$.
9.   **If** registration error E between P_1 and $P_2(l)$ is too large
10.   Compute transformation $T(l)$ between $(P_2(l),\ Y(l))$,
11.   Apply transformation $P_2(l+1) = T(l) \bullet P_2(1)$,$l$=$l$+1
12.   **Else**
13.   Stop
14.   **End If**
15. **While** $\|P_2(l+1) - P_2(l)\|$ >threshold

where point y_k in set $Y(l)$ denotes the closest point in P_1 to the point $p^2_k(l)$ in $P_2(l)$ and the registration error between P_1 and $P_2(l)$ is

$$E = \frac{1}{N} \sum_{k}^{N} \|y_k - p^2_k(l)\|^2 \tag{1}$$

For convergence of ICP, the coarse registration usually is carried out before the iterative process.

In general, when utilization of ICP in 3D face recognition, two facial surfaces are registered by the above method. Then the value of E computed in the last time of iterative steps is treated as dissimilarity measure of two faces.

When matching two facial surfaces with different expressions, the difference between the pairs of nearest points may become large due to shape deformation which may have a large effect when performing least-squares minimization and E is no longer accurate as a dissimilarity metric. If only those pairs of points with relatively less deformation are selected as input of calculation of E, the registration error E may be still able to distinguish different subjects while remain small when matching models of same subjects with different expression.

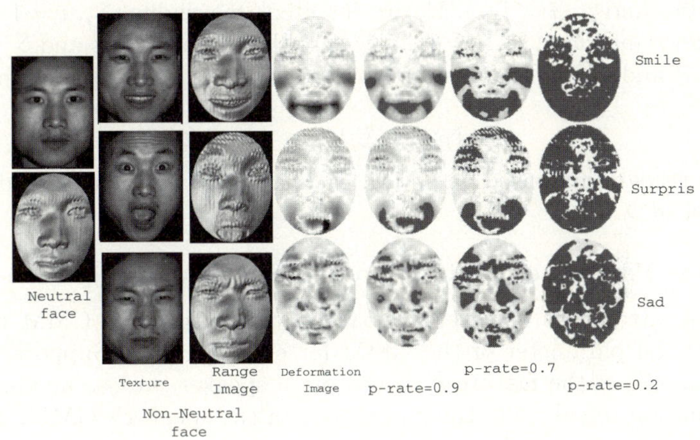

Fig. 1. Discarded area in facial surface with different *p-rate*(in red)

While the traditional ICP-based method in 3D face recognition uses all point pairs in computing transformation $T(l)$ and E [3], we do it by selecting parts of the point pairs. After sorting the distances of pairs of points in increasing order, we reject the worst $n\%$ of pairs based on distance in each pair. That is, only first $(1-n\%)$ part of distances and corresponding point pairs from sorted distances are chosen to compute transformation E and $T(l)$. Considering the last E that is used as dissimilarity measure of matching, discarding $n\%$ of pairs means removing those points in non-rigid region of facial surface. Thus, it is an implicit method to extract points in rigid parts of facial surface to register and match and the rigid parts extracted are varied according to deformation of facial surface among different matching models. We denote it *partial ICP* for 3D face recognition approach and call $(1-n\%)$ *p-rate*.

Fig. 1 shows some deformation images in which the darker indicates more deformation and the lighter means less deformation and areas in red indicate regions discarded by setting certain *p-rate*.

3   Speaker Recognition Component

3.1   GMM Description

A Gaussian mixture density is a weighted sum of M component densities which provide a smooth approximation to the underlying long-term sample distribution of observations obtained from utterances by a given speaker [8]. It can be formalized as:

$$p(\overrightarrow{x}|\lambda) = \sum_{i=1}^{M}(p_i * b_i(\overrightarrow{x}))$$

(2)

$$b_i(\overrightarrow{x}) = \frac{1}{(2\pi)^{D/2}|\Sigma_i|^{1/2}}exp\{-\frac{1}{2}(\overrightarrow{x}-\overrightarrow{u_i})'\Sigma_i^{-1}(\overrightarrow{x}-\overrightarrow{u_i})\}$$

(3)

where $\overrightarrow{x}$ is a D-dimensional random vector, $b_i(\overrightarrow{x})$, $i = 1, ..., M$, are the component densities and p_i, $i = 1, ..., M$, are the mixture weights, Σ_i, $i = 1, ..., M$, are the covariance matrices, $\overrightarrow{u_i}$, $i = 1, ..., M$, are the mean vectors and $\sum_{i=1}^{M} p_i = 1$. Given p_i, $\overrightarrow{u_i}$ and Σ_i, a Gaussian mixture density can be completely parameterized as:

$$\lambda = \{p_i, \overrightarrow{u_i}, \Sigma_i\}, i = 1, ..., M.$$

(4)

In speaker recognition based on GMM, each speaker has a corresponding GMM represented by λ.

3.2   GMM Parameter Estimation and Speaker Recognition

Given some utterances of several speakers, the purpose of GMM training is to estimate the parameter of the GMM for each speakers. Suppose vector set $X = \{\overrightarrow{x_1}, ..., \overrightarrow{x_T}\}$ is the feature vectors from a speaker, we use maximum likelihood estimation to calculate the parameters of the speaker's GMM. The GMM likelihood function is as

$$p(X|\lambda) = \prod_{t=1}^{T} p(\overrightarrow{x_t}|\lambda)$$

(5)

For $p(X|\lambda)$ is a nonlinear function of λ, we use an expectation-maximization algorithm(EM) to compute the parameters [10].

Suppose S speakers labelled as $1, ..., S$ are trained, each has corresponding GMM with parameters $\lambda_1, ..., \lambda_S$ and X are feature vector from a test speaker, identification can be performed by finding the speaker model with label $\hat{S}$ which has the maximum posteriori probability. It can be formalized as:

$$\hat{S} = arg(max_{1\leq k\leq S}P_r(\lambda_k|X)) = arg(max_{1\leq k\leq S}\frac{p(X|\lambda_k)P_r(\lambda_k)}{p(X)})$$

(6)

4   Combining Schemes

Following fusion schemes are used in our experiments:

1. Minimum, Maximum, Average and Product scheme
2. Sum, Weighted Sum and Scores Difference-based Weighted Sum
3. "Naive"-Bayes Combination(NB) and Behavior-Knowledge Space

Let $x \in R^n$ be a feature vector, $\{1, 2, ..., c\}$ be the label set of c classes, $D_i(x) = [d_{i,1}(x), ..., d_{i,c}(x)]^T, i = 1, 2$, be the output of the classifier D_i and $d_{i,j}(x)$ be the degree of "support" given by classifier D_i to the hypothesis that x comes from class j. The fusion result can be presented as

$$\hat{D}(x) = \zeta(D_1(x), D_2(x)) \tag{7}$$

where ζ denotes fusion scheme. Thus, the class label $\hat{L_D}$ of feature vector x can be decided by:

$$\hat{L_D} = arg(max_{1 \leq i \leq c}(d_i)), d_i \in \hat{D} \tag{8}$$

The decision profile(DP) of two classifier's outputs can be organized as:

$$DP(x) = \begin{bmatrix} d_{1,1}(x), ..., d_{1,j}(x), ...d_{1,c}(x) \\ d_{2,1}(x), ..., d_{2,j}(x), ...d_{2,c}(x) \end{bmatrix} \tag{9}$$

• Minimum, Maximum, Average and Product Scheme

$$Minimum : \hat{D}(x) = \{d_i(x)|d_i(x) = min(d_{1,i}(x), d_{2,i}(x))\}, \tag{10}$$

$$Maximum : \hat{D}(x) = \{d_i(x)|d_i(x) = max(d_{1,i}(x), d_{2,i}(x))\}, \tag{11}$$

$$Average : \hat{D}(x) = \{d_i(x)|d_i(x) = (d_{1,i}(x) + d_{2,i}(x))/2\}, \tag{12}$$

$$Product : \hat{D}(x) = \{d_i(x)|d_i(x) = d_{1,i}(x) * d_{2,i}(x)\}, \tag{13}$$

where, $d_{1,i}(x) \in D_1(x), d_{2,i}(x) \in D_2(x), i = 1, ..., c, min(\bullet, \bullet), max(\bullet, \bullet)$ denotes the smaller and larger element of two given elements respectively.

• Sum, Weighted Sum and Scores Difference-Based Weighted Sum

$$Sum : \hat{D}(x) = \{d_i(x)|d_i(x) = d_{1,i}(x) + d_{2,i}(x)\}, \tag{14}$$

$$WeightedSum : \hat{D}(x) = \{d_i(x)|d_i(x) = w_1 * d_{1,i}(x) + w_2 * d_{2,i}(x)\}, \tag{15}$$

where,

$$w_i = \frac{1 - 2E_i}{2 - 2\sum_{j=1}^{2}(E_j)}, i = 1, 2, j = 1, 2, i \neq j \tag{16}$$

and E_j is the error rate of classifier j.

Scores Difference-based Weighted Sum (SDWS) is a trained fusion method based on Weighted Sum rule which using training data and the error rates of classifiers to compute the weights. The detail of SDWS can be seen in [11].

• "Naive"-Bayes Combination(NB) and Behavior-Knowledge Space.

These two schemes tested in our experiments are both trained fusion methods. Using the label results of different classifiers on training data, both "Naive"-Bayes Combination(NB) and Behavior-Knowledge Space schemes compute the probability of each kind of labels combination. When given a testing model, NB and BKS decide the class of the model by indexing into the probability using current labels combination. The detailed description of these techniques can be found in [9].

5   Experimental Results

5.1   Data Acquisition

The data used in our experiments include two parts for partial ICP classifier and GMM classifier respectively. One is facial range data set and the other is the utterance set. The database consists of total 40 different persons.

The facial range data set includes 360 scans in which there are 9 scans for each subject, 2 scans with smile expression, 2 scans with surprise expression, 2 scans with sad expression, 3 scans with neutral expression. All face models are acquired by InSpeck 3D MEGA Capturor DF. Fig. 2 shows some examples of face models. We put one neutral expression face model for each subject into gallery and the other 320 scans are put into probe.

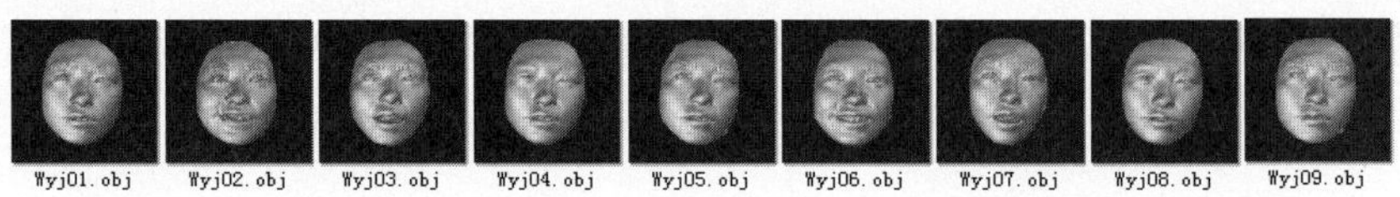

Fig. 2. Face models acquired by InSpeck 3D MEGA Capturor DF

The speech data of each person is divided into 9 sessions with different text. One session is used to train GMM to get the subject's model parameter λ. In the other 8 sessions, 10 prompts are asked to read per session for each subject which are used for testing.

5.2   Results

In the paper, two experiments are conducted to evaluate the enhancement of combining GMM classifier into partial ICP. For those fusion schemes that must be trained, we used leave-one-out method to train them before combination. Before combination, the output of each classifier are normalized so that each matching process issues a support score in [0,1].

Firstly, 5 different values of *p-rate* { 0.6, 0.7, 0.8, 0.9, 1} and 9 fusion methods are tested in our experiments. Rank-1 recognition rates are shown in table 1.

From the table, it can be seen that:

(1) Rank-1 recognition rates can be enhanced remarkly by combining GMM classifier into partial ICP classifier with following fusion method: Minimum, Average, Product, Sum, Weighted Sum and SDWS. The largest improvement of performance is done by SDWS. An average improvement of SDWS, about 5.283%, is reached.
(2) While the schemes Average, Product, Sum and Weighted Sum achieve similar improvement of performance, the methods, Maximum, NB and BKS, have some decline in rank-1 recognition rate. With each *p-rate*, we do not see any evidence, in this study, of increasing recognition rate with these three fusion schemes under leave-one-out training method when necessary.

Table 1. Recognition performance varied by different *p-rate* and fusion schemes

Fusion Scheme	Rank-1 Recognition Rate				
	p-rate=100%	*p-rate*=90%	*p-rate*=80%	*p-rate*=70%	*p-rate*=60%
partial ICP	90.63%	94.06%	96.56%	95%	94.69%
GMM	91.25%				
Minimum	91.87%	96.88%	95.31%	96.88%	96.88%
Maximum	91.25%	92.19%	94.69%	93.13%	93.13%
Average	96.25%	96.88%	97.5%	98.12%	98.12%
Product	96.88%	96.88%	97.5%	98.12%	98.12%
Sum	96.25%	96.88%	97.5%	98.12%	98.12%
Weighted Sum	96.25%	96.88%	97.5%	97.81%	98.12%
SDWS	97.19%	97.5%	98.44%	98.44%	98.44%
NB	90.94%	91.87%	94.37%	92.81%	92.5%
BKS	90.63%	91.87%	94.37%	92.81%	92.5%

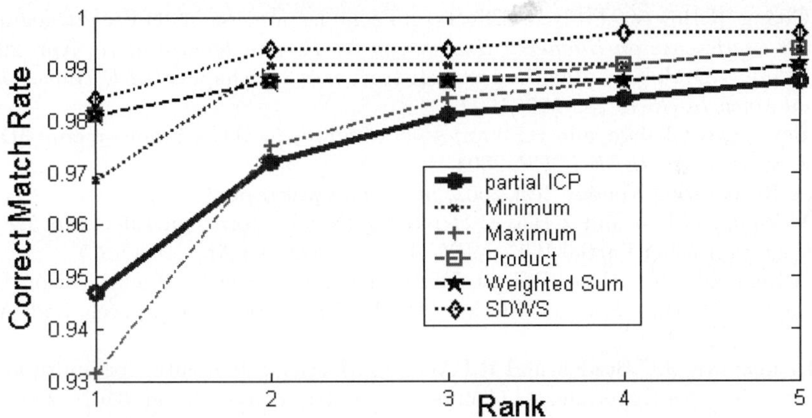

Fig. 3. Performance results by different fusion methods

(3) When *p-rate*=60% or 70%, the rank-1 recognition rates are maintained better than other *p-rate*. With SDWS fusion method, the experiment yields 98.44% rank-1 recognition rate.

Secondly, setting *p-rate* = 60%, we also plot several Cumulative Match Characteristics (CMC) Curves with different fusion methods, shown in Fig. 3.

From the figure, when considering first 5 best matching models instead of only rank-one model in recognition, identification rate are above 99% using almost all fusion methods except BKS and NB(Experimental results of fusion scheme Average and Sum are not shown in Fig. 3 because their curve will be overlayed by others).

6   Conclusion

The paper explores the improvement of identity recognition rate when combining a voice classifier, GMM-based method, into a 3D face classifier, partial ICP-based method. Several fusion methods are tested and the experimental results are compared with each other. 6 fusion methods can improve the performance of identification. The best average improvement of rank-1 recognition rate, about 5.283%, demonstrates that voice classifier can enhance the performance of 3D face recognition remarkly.

References

1. W.Zhao, R.Chellappa, P.J.Phillips, A.Rosenfeld. Face recognition: a literature survey. *ACM Computing Surveys*, 35(4):399-458, 2003
2. W. Zhao, R. Chellappa. Illumination-insensitive face recognition using symmetric shape-form-shading. *Proc. IEEE ICCV*, 1:286-293, 2000.
3. K.Chang, K.Bowyer, P.Flynn. Effects on Facial Expression in 3D Face Recognition, *SPIE Conference on Biometric Technology for Human Identification*, Apr. 2005.
4. P.J.Besl, N.D.McKay, A method for registration of 3-D shapes, *IEEE Trans.Pattern Anal.Mach.Intell.* 14:239-256, 1992.
5. K.Bowyer, K.Change, and P.Flynn, A short survey fo 3D and multi-modal 3D+2D face recognition, *IEEE ICPR*, 2004.
6. Face Recognition Vendor Test 2002, http://www.frvt.org/.
7. Y.M.Wang, G.Pan and Z.H.Wu, Exploring Facial Expression Effects in 3D Face Recognition using Partial ICP, *IEEE ACCV* , 2006. to Appear. Oral.
8. D.A.Reynolds, R.C.Rose, Robust Text-independent Speaker Identification Using Gaussion Mixture Speaker Models, IEEE Transactions on Speech and Audio Processing,3-1:1995.
9. L.I.Kuncheva, J.C.Bezdek and R.P.W.Duin, Decision templates for multiple classifier fusion: An Experimental Comparison. Pattern Recognition, 34(2), 2001, 299-314.
10. A.Dempster, N.Laird and D.Rubin, Maximum liklihood from incomplete data via the EM algorithm, J.Royal Stat.Soc., vol.39, 1977, 1-38.
11. D.D. Li, Y.C. Yang and Z.H. Wu, Combining Voiceprint and Face Biometrics for Speaker Identification Using SDWS. To appeared in Interspeech 2005.

Physical Modeling of Laser-Induced Breakdown of Glass

Jaemyoung Lee[1,*], Michael F. Becker[2], and Taikyeong T. Jeong[2]

[1] Korea Polytechnic University,
2121 Jungwang Shihung Kyuggi, Korea(ROK) 429-793
lee@kpu.ac.kr
[2] Department of Electrical and Computer Engineering
University of Texas at Austin

Abstract. We made a physical model for investigation of laser-induced breakdown of glass. To estimate the laser energy absorption through electron heating, we derive a power transfer rate equation and an electron number density equation for a steady state as a function of temperature and electric field. Numerical simulations using the derived equations show that the laser power absorption dependence of glass on temperature and electric field strength.

1 Introduction

Laser ablation has emerged as one of promising techniques for material processing such as thin film depositions, nanoparticle fabrications, etc. Laser ablation has proven its advantages in material processing because it does not contaminates materials during the process. The foundations of laser ablation, however, lie in the old field of laser-material interactions where many materials were irradiated with high power laser pulses. Therefore, studies about the laser-material interactions have been required to investigate the laser-material interactions.

In this paper, we form a physical model to analyze the laser-induced breakdown of glass, provide simple equations for numerical analysis using experimental results published from other groups, and show simulation results about energy absorption, electron number density in terms of temperature and electric field using the derived equations.

2 Theoretical Background

The interaction between the laser field and free electrons can explain the high power laser breakdown mechanism. Electron avalanche theory [1, 2, 3] has been accepted to explain the high power laser breakdown of transparent(wide band gap) solids at both visible and near-infrared wavelengths. This theory assumes that breakdown starts above the critical laser intensity where the energy gain

---

* Corresponding author.

V.N. Alexandrov et al. (Eds.): ICCS 2006, Part I, LNCS 3991, pp. 443–448, 2006.

rate from the laser field by a few electrons exceeds the energy loss rate due to lattice phonon scattering. These starting electrons with density n_0 increase their kinetic energies in the high electric field, and initiate impact ionization leading to an exponential increase of the free electron density.

S.C. Jones et al. [3] found discrepancies in this theory for explaining laser induced breakdown of wide band gap materials such as ultrapure alkali halides and SiO_2 at visible and near-infrared wavelengths. In one experiment, they measured a lattice temperature that increased gradually as the laser intensity increased and found that breakdown occurred around the melting temperature. They observed that the lattice temperature gradually increased with increasing laser intensity. They explained this phenomenon as a multiphoton absorption process rather than impact ionization process.

Lattice heating was caused by simultaneous free-electron mediated energy transfer from the laser field to the lattice, *free-electron heating* [4, 5]. B. Gorshkov, et al. [5], investigated carrier generation processes, both impact ionization and multiphoton absorption, for laser-induced breakdown and concluded that the breakdown mechanism begins to change from an impact ionization process to a multiphoton generation process at near infrared wavelengths such that, at visible wavelengths, the multiphoton absorption mechanism is dominant in wide band gap materials.

Pulse width is also an important parameter because electron density increases exponentially with time for impact ionization while linearly for multiphoton absorption. At wavelengths longer than 2 μm, the electron multiplication rate and energy transfer rate approach the dc limit, since as the laser frequency decreases, the momentum relaxation rate becomes larger than the laser frequency. In this region, the photon field of the laser is assumed to be a sinusoidal electric field in which the electrons move. This implies that photon energy is small compared to the electron kinetic energy such that energy change from the field can be continuous in time. At short wavelengths(< 500 nm), impact ionization and free-electron heating are less dominant because the impact ionization threshold moves to higher electric fields, making multiphoton absorption the dominant mechanism. At these wavelengths, even though the multiphoton absorption mechanism is dominant, free electron heating is another efficient mechanism to contribute to lattice heating and melting at laser intensities below the impact ionization threshold.

3   Physical Model for Numerical Analysis

In the case of glass, D. Arnold et al. [6], showed that neither the quantum-mechanical approach nor the standard classical approach followed their exact derivation of the power transfer rate from the laser field to electrons at a wavelength of 1 μm. The power transfer rate at this wavelength lay between these two approaches.

In our numerical simulation of irradiation with a 1 μm laser, the exact power transfer rate of D. Arnold and E. Cartier [6] was used. This approach eliminated

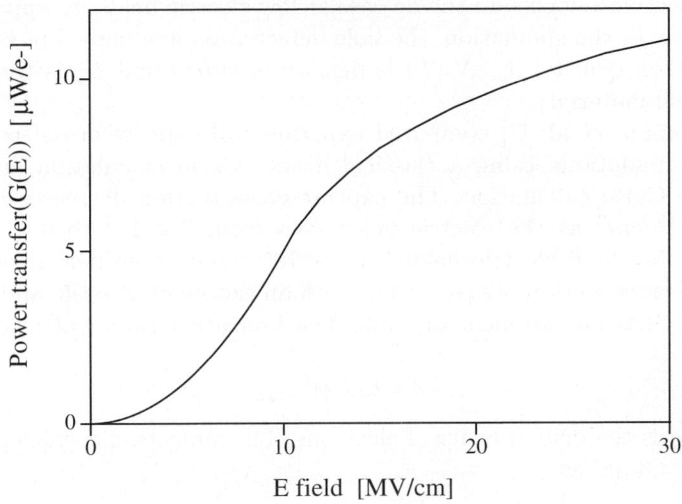

Fig. 1. Power transfer per electron versus E field

additional complex calculations, such as the laser field to free electron interaction equation, to get the power transfer rate. Equation 1 was devised to imitate their result for energy transfer from electrons to the lattice. In Figure 1, the equation 1 is plotted.

$$G(E) = 5 \times 10^{-14} \times E^2, \quad if \quad E < 10^7 [V/cm]$$
$$= 3 \times ln(E + 210^6) - 39.7, \quad if \quad E > 10^7 [V/cm] \tag{1}$$

The evolution of the free-electron density, $n_e(t)$, is essential to modeling a microsphere ablated by a high power laser pulse. Generation and recombination processes are included in the ionization rate equation as shown below,

$$\frac{\partial n}{\partial t} = P_{gen} - P_{loss} + P_{therm} \tag{2}$$

where P_{gen} is laser induced ionization generation rate, P_{loss} is loss rate due to electron hole recombination, and P_{therm} is thermal ionization rate. These equations can be described by

$$P_{gen} = \gamma(E) \times n_e \tag{3}$$
$$P_{loss} = \sigma(E) \times v_e \times n_e \times n_h \tag{4}$$
$$P_{therm} = N_c(T) \exp\left(-\frac{E_g}{2kT}\right) \tag{5}$$

where $\gamma(E)$ is the electric field dependent ionization coefficient, E is electric field, $\sigma(E)$ is the Coulombic capture cross section with field dependence. The quantity

v_e is the effective velocity. In the dc case at low electric fields, v_e approaches the drift velocity. In the simulation, the hole density, n_h, is assumed to be the same as the electron density, n_e. $N_c(T)$ is density of states and E_g is the band gap energy of the material.

D. Buchanan, et al. [7] compared experimental capture cross sections with numerical simulations using a classical Monte Carlo calculation and a quantum Monte Carlo calculation. The capture cross section decreases from 10^{-12} to 3×10^{-15} cm^2 as the electric fields goes from 2×10^5 to 3×10^6 V/cm. Above the threshold electric field($\approx 1.2 \times 10^6$ V/cm), the field dependence of the capture cross section is a power law with an exponent of -1.5, and below this threshold, it has an exponent of -3.0. The ionization rate, $\gamma(E)$, is expressed as,

$$\gamma = \alpha(E)V_{drift} \tag{6}$$

where V_{drift} is the drift velocity of electrons. The ionization coefficient, $\alpha(E)$, is given by D. Du [8] as

$$\alpha(E) = \frac{eE}{U_i} exp\left(-\frac{E_i}{E(1 + E/E_p) + E_{kT}}\right) \tag{7}$$

The constants E_{kT}, E_p, and E_i are threshold electric fields for electrons to overcome the decelerating effects of thermal, phonon, and ionization scattering, respectively. For the laser fluences used in this simulation, the ionization rate, equation 2, reaches steady state in much less than 1 ns. Therefore, it is solved for a steady state carrier density at each time step of the simulation. Solving

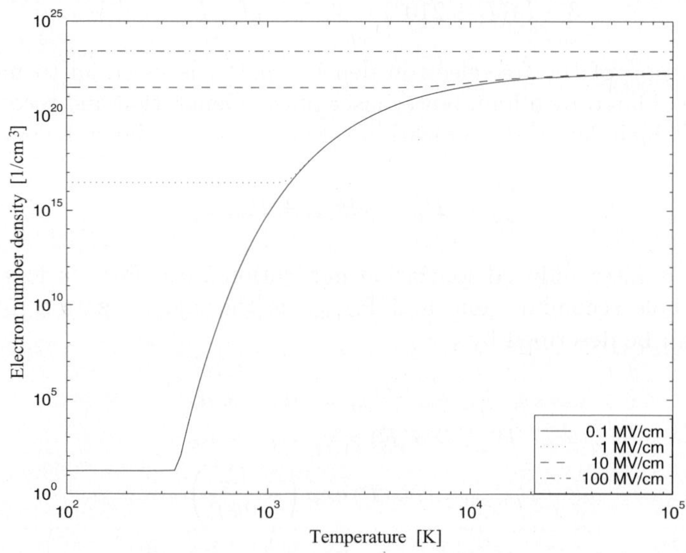

Fig. 2. Electron number density vs. temperature for four values of laser E field

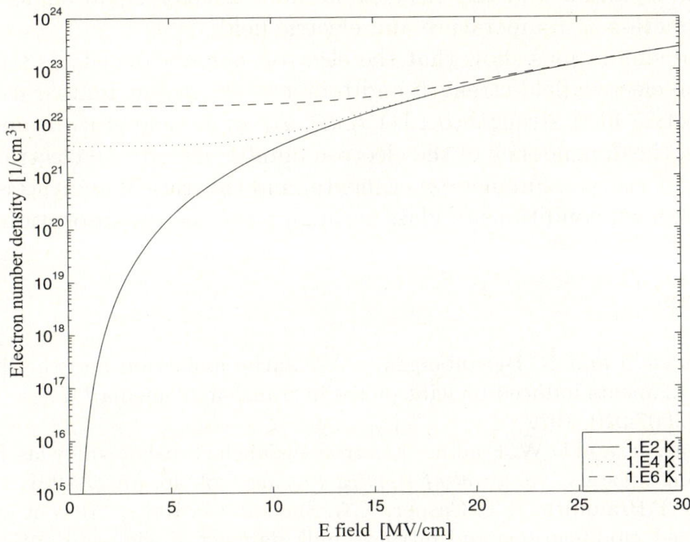

Fig. 3. Electron number density vs. laser E field for three temperatures

these equation for the electron number density, $n(E, T)$ results in the following equation,

$$n(E, T) = \frac{1}{2\sigma(E)v_{drift}} \left(\alpha(E)V_{drift} + \sqrt{(\alpha(E)V_{drift})^2 + [2\sigma(E)V_{drift}n(T)]^2} \right)$$

(8)

where $n(T)$ is the thermal generation given by eq. 5.

The electron number density, given by eq. 8 is shown in Figs. 2, and 3. In our simulation, the average charge number, Z, was obtained by dividing the electron number density by the atom number density which was calculated by dividing the mass density by the average atomic mass.

Figure 3 shows that thermal generation of electrons is dominant in the low electric field region, while in the low temperature region electrons are excited by the laser electric field. In the shockwave experiment reported in this dissertation, the highest electric field was 8 MV/cm which corresponds to the laser energy at the highest enhancement factor with an applied fluence of 12 J/cm^2. Temperature in this simulation was up to a few 10,000 K. For this case, the maximum electron number density was about 3 - 5 10^{22} cm^{-3}.

4   Conclusion

To investigate the laser power absorption via electron heating, we form a physical model of laser energy absorption of glass through electron and derive the power transfer rate equation and the electron number density equation for a steady state as a function of temperature and electric field.

Numerical simulations show that the electron number density is strongly depends on the electric field strength and temperature at low temperature(100K) and low electric filed strength($0.1MV/cm$). At high temperature and electric field regions, the dependency of the electron number density on electric field and temperature decreases. Further experimental and theoretical analysis is required for the breakdown conditions of glass through laser energy absorption.

References

1. E. Yablonovitch and N. Bloembergen, "Avalanche ionization and the limiting diameter of filaments induced by light pulses in transparent media," *Phys. Rev. Lett.*, vol. 29, pp.907-910, 1972
2. L.H. Holway, Jr. and D.W. Fradin, "Electron avalanche breakdown by laser radiation in insulating crystals," *Journal of Applied Physics*, vol. 46, pp.279-291, 1975
3. S.C. Jones, P.Braunlich, R.T. Casper, X.A. Shen and P. Kelly, "Recent progress on laser-induced modifications and intrinsic bulk damage of wide-gap optical materials," *Opt. Eng.*, vol. 28, pp.1039-1068, 1989
4. A. Epifanov, "Avalanche ionization induced in solid transparent dielectrics by strong laser pulses," *Sov. Phys.-JETP*, vol. 40, pp.897-9028, 1975
5. B. Gorshkov, A. Epifanov and A. Manenkov, "Avalanche ionization produced in solids by large radiation quanta and relative role of multiphoton ionization in laser-induced breakdown," *Sov. Phys.-JETP*, vol. 49, pp.309-315, 1979
6. D. Arnold and E. Cartier, "Theory of laser-induced free-electron heating and impact ionization in wide-band-gap solids," *Physical review B*, vol. 46, pp.15102-15115, 1992
7. D.A. Buchanan, M.V. Fischetti and D.J. DiMaria, "Coulombic and neutral trapping centers in silicon dioxide," *Physcal Review B*, vol. 43, pp.1471-1456, 1991
8. D. Du, X. Liu, G. Korn, J. Squier and G. Mourou, "Laser-induced breakdown by impact ionization in SiO_2 with pulse widths from 7 ns to 150 fs," *Appl. Phys. Lett.*, vol. 64, pp.3071-3073, 1994

An Enhanced Speech Emotion Recognition System Based on Discourse Information

Chun Chen, Mingyu You, Mingli Song, Jiajun Bu, and Jia Liu

College of Computer Science, YuQuan Campus, ZheJiang University
Hangzhou, P.R.CHINA, 310027
{chenc, roseyoumy, brooksong, bjj, liujia}@zju.edu.cn

Abstract. There are certain correlation between two persons' emotional states in communication, but none of previous work has focused on it. In this paper, a novel conversation database in Chinese was collected and an emotion interaction matrix was proposed to embody the discourse information in conversation. Based on discourse information, an enhanced speech emotion recognition system was presented to improve the recognition accuracy. Some modifications were performed on traditional KNN classification, which could reduce the interruption of noise. Experiment result shows that our system makes 3% - 5% relative improvement compared with the traditional method.

1   Introduction

Researches on understanding and modelling human emotions have attracted increasing attention in the artificial intelligence field. As a major indicator of human emotions, speech plays an important role in detecting affective states. Accurate emotion recognition from speech signals will benefit the human-machine interaction [1]. It has broadly potential applications in areas such as education, consumer service and entertainment.

There are lots of researches that attempt to characterize emotional states of human speech. Most previous efforts involving speech emotion recognition have tended to focus on either lexical [2, 3] or acoustic information [4, 5, 6]. These studies usually used non-natural speech, including short isolated utterances, expressed by professional actors. Systems embedding lexical features into emotion recognition assumed that certain words correlated with emotional states. But as we know, the relationship between words and emotions is fuzzy and sometimes confused. One word may indicate several possible emotions and one emotional state can be conveyed by different words. Besides, lexical information always needs manual transcription for each utterance, which is difficult to be realized automatically.

On the other hand, researches on emotion detection of spoken dialog system tried to classify more natural occurring emotion of actual users. In order to achieve better performance, lexical and acoustic feature sets were augmented with additional features such as context and discourse information[7, 8, 9, 10]. Contextual factors really have influences on emotion identification, which can

V.N. Alexandrov et al. (Eds.): ICCS 2006, Part I, LNCS 3991, pp. 449–456, 2006.

be found in the work of Crystal[11, 12, 13]. Liscombe[7] et al. included prosodic context, lexical context and discourse context as contextual features in emotion prediction and increased classification accuracy by 2.6%. Contributions made by context information indicate that emotion is expressed in a language environment. Traditional operation extracting prosodic features from isolated utterances is inconsistent with emotion perception of human beings. Contextual features focus on utterances of single person, but dialogs appear frequently in daily life. Many studies paid attention to dialog system and added discourse information into emotion recognition[8, 9, 10]. Ang et al.[8] included discourse features such as turn location within the conversation and dialog acts of the current turn (repeat, repair, neither). Lee et al.[9] labeled dialog acts as rejection, repeat, rephrase, ask-start over or none of the above. The addition of discourse information added approximately 3% relative improvement over using lexical and prosodic features alone. But the discourse features considered were only based on the categorization of users' responses when interacting with machine. Few work paid attention to the dialog between two persons. But it is known that the emotions of two talkers have certain correlation.

This paper presents research on simultaneously recognizing two talkers' emotional states using discourse information. In addition to extracting typical acoustic features, we combined emotional correlation between two persons in dialog into speech emotion recognition. Using this extended method, we observed an increase in prediction accuracy.

2   Conversation Corpus and Discourse Information

In the field of emotion recognition, there are ongoing debates concerning how to define basic emotion categories. Different researchers have different opinions and some psychologists even argue against the categorical labels for human emotions. In this paper, we just focus on the archetypal emotions - happy, sad, fear, angry, surprise and neutral. Besides, how to obtain amount of realistic data for research is another hard task. Most studies in speech emotion recognition asked subjects to simulate certain emotions with neutral semantic content[14]. Since those data sets are limited to utterances of single person in archetypal emotions, results based on them may still have distance to real-life scenarios. On the other hand, it's hard for real data to cover all the emotion categories needed and the noise in real environment is also a problem. With aforementioned analysis, we should turn to sources containing conversations and happening in a relatively quiet background. The living theater and broadcast drama are suitable sources for dialog corpus in controlled environment. They also embody most emotional correlation of conversation in daily life. As the beginning of dialog corpus collection, we tried to extract 4000 conversations from Chinese drama lasting hundreds of hours totally. Our data corpus covers the six archetypal emotions mentioned above and contains dialogs covering man to man, man to woman and woman to woman.

In table 1, we mark two persons having conversation as A and B. A represents the first person and B denotes the second one. The left part of table 1 presents

Table 1. Distribution of The Other's Emotion When Given One Person's Emotional State (The abbreviated emotion labels in line are same to those in column)

A	Ne.(%)	An.	Fe.	Ha.	Sa.	Su.	B	Ne.	An.	Fe.	Ha.	Sa.	Su.
Neutral	26.67	22.22	6.67	6.67	11.11	26.67	Neutral	23.53	27.45	13.73	11.76	5.88	17.65
Angry	32.56	46.51	2.33	2.33	4.65	11.63	Angry	20.83	41.67	12.5	0	8.33	16.67
Fear	41.18	35.29	5.88	0	11.76	5.88	Fear	50	16.67	16.67	16.67	0	0
Happy	30	0	5	40	5	20	Happy	16.67	5.56	0	44.44	0	33.33
Sad	20	26.67	0	0	40	13.33	Sad	29.41	11.76	11.76	5.88	35.29	5.88
Surprise	36	32	0	24	4	4	Surprise	48	20	4	16	8	4

A's emotion distribution when given the emotional state of B. Similarly, emotion distribution of B is shown at the right part of table 1. As an example, we look into the first line of part A. This line exhibits A's emotion distribution when the emotional state of B is neutral. From the listed numbers, we find that when B is neutral, the probability of A on neutral is 26.67%. Probabilities of A on angry, fear, happy, sad and surprise are 22.22%, 6.67%, 6.67%, 11.11% and 26.67%, respectively. Different probability on each emotion category implies that B will affect the emotional state of A in dialog and vice versa. For example, when B is happy, A is likely to be also happy considering the corresponding high probability. We were wondering if we could embed this information into speech emotion recognition in order to improve the recognition accuracy. A detailed experiment is designed in the following section.

3   An Enhanced Speech Emotion Recognition

Based on the analysis above, we propose an enhanced speech emotion recognition system in Figure 1. The system is composed of two parts - training and

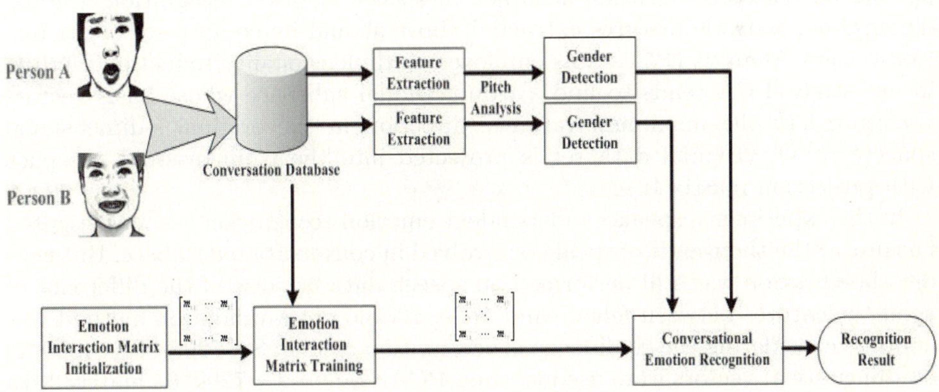

Fig. 1. System Overview

testing. In the off-line training phase, we collected 4000 dialogs into the conversation database. The corpus includes thousands of people instead of two in Figure 1 which is just for simplification. An initialized emotion interaction matrix was trained to embody the emotion correlation between people in dialog in the training course. When new data came into the system for testing, acoustic features were extracted and gender classification was performed based on pitch analysis. Then a method combining emotion interaction matrix was proposed to recognize emotional states of both persons using those acoustic features. The whole process can be divided into three steps mentioned below.

3.1   Acoustic Features Extraction

The conversation corpus we collected is sampled at 16kHZ frequency and 16 bits resolution with monophonic Windows PCM format. In this study, we extracted 48 prosodic and 16 formant frequency features. Prosody is mainly related to the rhythmic aspects of speech, and believed to be the primary indicator of speakers' emotion state. The extracted prosodic features include: max, min, mean, median of Pitch (Energy); mean, median of Pitch (Energy) rising/ falling slopes; max, mean, median duration of Pitch (Energy) rising/ falling slopes; mean, median of Pitch (Energy) plateaux at maxima/ minima; max, mean, median duration of Pitch (Energy) plateaux at maxima/ minima. Here, if the first derivative is approximately zero and the second derivative is positive, the point belongs to a plateau at a local minimum. If the second derivative is negative, it belongs to a plateau at a local maximum. We also investigated formant frequency features which are widely used in speech processing applications. Statistical properties including max, min, mean, median of the first, second, third, and fourth formant were extracted.

3.2   Dimensionality Reduction

Because high dimensional data can dramatically raise computational complexity and decrease classification accuracy in speech emotion recognition, the 64-dimensional acoustic features extracted above should be compressed. Principal Component Analysis (PCA) was employed as dimensionality reduction method in our study. PCA tends to find a t-dimensional subspace whose basis vectors correspond to the maximum variance direction in the original s-dimensional space($t \ll s$). Original data set is projected into the t-dimensional subspace with projection matrix W_{PCA}.

In the experiment, speaker independent emotion recognition was investigated because of the thousands of speakers involved in conversation database. But gender classification was still performed on speech data because of the difference of acoustic features between female and male. 10-fold cross-validation method was adopted considering the confidence of recognition results. So, 7200(90%*4000*2) 64-dimensional vectors were used to train PCA. We used a 7200*64 matrix X to represent these vectors. After X was normalized and mean-subtracted, we got matrix Y. $Y^T Y$ formed a covariance matrix M which was 64*64. Eigenvalues and eigenvectors were computed for M. Eigenvectors corresponding to the largest t eigenvalues were selected to create the PCA projection matrix W_{PCA}. t was the number of eigenvalues that guaranteed energy E was greater than 0.9. Here energy E was defined in equation(1):

$$E_t = \sum_{j=1}^{t} \lambda_j / \sum_{j=1}^{64} \lambda_j \tag{1}$$

where λ_j was the jth eigenvalue. In our experiment, t equaled to 27. 3600(7200/2) training conversations and 400(800/2) testing ones were both projected into subspace using W_{PCA}.

3.3   Emotion Recognition Based on Discourse Information

Having the low dimensional features, K-Nearest-Neighbor(KNN) was adopted to classify the data into six emotional states. K-Nearest-Neighbor is a simple classification which range the testing data into the class most of its k nearest neighbors belonging to. It is the classical implementation of KNN. We made some modifications to KNN for the sake of our enhanced recognition system. K nearest neighbors were calculated for testing utterance which was same to KNN classification. In the process of K nearest neighbors' calculation, Euclidean distance was adopted as the distance measurement between feature vectors of training utterance and testing utterance. K nearest neighbors belonged to M classes $\{C_1, C_2, \cdots, C_M\}$. The probability of belonging to C_i was defined by: $P_i = N_i/N_k$ where N_i denoted the number of nodes belonging to C_i and N_k equaled to k which was the total number of nodes. So the probabilities for M classes was $\{P_1, P_2, \cdots, P_M\}$ and we used P_{m_1} to be the highest probability and P_{m_2} to be the second. In classical KNN, C_{m_1} with P_{m_1} is selected as the recognition result for test utterance. However, we'd like to make our decision based on P_{m_1} and P_{m_2} instead of depending on P_{m_1} alone.

Different from conventional methods, the emotional states of two persons in dialog were recognized together in our system. Let us use M_A to stand for the emotion interaction matrix of A when given emotional state of B and M_B to stand for that of B. $P_{m_1}^A$ and $P_{m_2}^A$ were the largest two distribution probabilities of person A's neighbors and $P_{m_1}^B$ and $P_{m_2}^B$ were those of B's. There were four situations based on different $P_{m_1}^A$, $P_{m_2}^A$, $P_{m_1}^B$ and $P_{m_2}^B$ listed below. We used a constant Th to represent the threshold of comparison.

(1) $P_{m_1}^A - P_{m_2}^A \geq Th$ and $P_{m_1}^B - P_{m_2}^B \geq Th$
In this case, we believed $C_{m_1}^A$ corresponding to $P_{m_1}^A$ and $C_{m_1}^B$ corresponding to $P_{m_1}^B$ were outstanding ones among those candidates. So we just chose $C_{m_1}^A$ as the emotion recognition result for person A and $C_{m_1}^B$ for person B.

(2) $P_{m_1}^A - P_{m_2}^A < Th$ and $P_{m_1}^B - P_{m_2}^B \geq Th$
In this case, $C_{m_1}^B$ corresponding to $P_{m_1}^B$ was selected as the recognition result of person B just as situation(1). But for person A, there wasn't such class with prominent performance, in other words, it was not sure which class should be

selected as the result. In our system, we selected $C_{m_i}^A$ corresponding to $P_{m_i}^A$ defined in equation(2) as the recognition result for person A.

$$P_{m_i}^A = \arg \max_{i=1}^{2}(P_{m_i}^A \times (M_A)_{m_i^A m_1^B} \times P_{m_1}^B) \tag{2}$$

Here, we embedded emotion interaction information mentioned above into the speech emotion recognition system. Such method could save those candidates in $\{C_1, C_2, \cdots, C_M\}$ which might have lower probabilities because of noise. As human beings, we also use this rule in emotion perception. If we are not sure about the other's emotional state, we'd like to judge it by our experience on what emotion would be most likely.

(3) $P_{m_1}^A - P_{m_2}^A \geq Th$ and $P_{m_1}^B - P_{m_2}^B < Th$

This case is similar to situation(2) excepting for using matrix M_B instead of M_A. Here we omit the detailed operations.

(4) $P_{m_1}^A - P_{m_2}^A < Th$ and $P_{m_1}^B - P_{m_2}^B < Th$

In this case, both A and B could not find sure recognition results. When recognizing emotional state of person A, $C_{m_i}^A$ corresponding to $P_{m_i}^A$ defined in equation(3) was selected as the result.

$$P_{m_i}^A = \arg \max_{i,j=1}^{2}(P_{m_i}^A \times (M_A)_{m_i^A m_j^B} \times P_{m_j}^B) \tag{3}$$

Similarly, person B's recognition result depended on $P_{m_j}^B$ defined in equation(4).

$$P_{m_j}^B = \arg \max_{i,j=1}^{2}(P_{m_i}^A \times (M_B)_{m_i^A m_j^B} \times P_{m_j}^B) \tag{4}$$

4   Experiment Result

In our experiment, we set k to 10 in K-Nearest-Neighbor searching and Threshold Th to 20%. In order to evaluate the performance of our new method, we

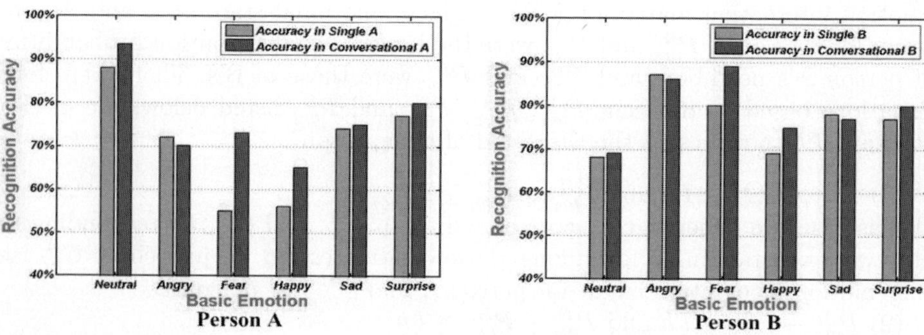

Fig. 2. Recognition Accuracy for person A and person B in our method (Accuracy in Conversational X) and traditional method (Accuracy in Single X)

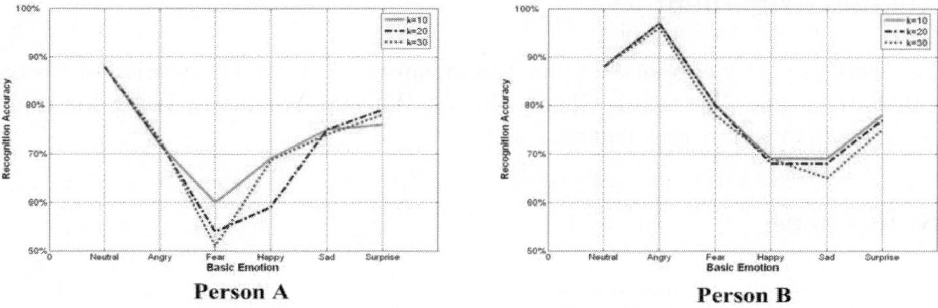

Fig. 3. Recognition Accuracy on Different Choice of k

also included a traditional emotion recognition process. In the traditional way, emotions of two persons in dialog were recognized separately and classical KNN classification was employed. The recognition accuracy of six emotions using our enhanced emotion recognition method and the traditional method is shown in Figure 2. From the figure, we can find out that the enhanced speech emotion recognition system outperforms traditional method on almost all of the basic emotions. On the average, our system is observed 5% relative improvement on person A and 3% on person B compared with traditional method. For person A, traditional method has different performance on different emotion. The accuracy on emotion fear is 30% lower than that on neutral, which will impact on the system performance. Our method balances the performance on each emotional state and improve the recognition accuracy totally.

Besides, the performance on different choice of k in K-Nearest-Neighbor searching is compared in Figure 3. None of the k achieves outstanding result compared with other choices. But $k = 10$ always has acceptable performance, especially on the emotion fear of person A. In addition to performance, simple computation is also an advantage of $k = 10$.

5 Conclusion and Future Work

This paper presents an enhanced speech emotion recognition system based on discourse information between human beings. Instead of hurriedly choosing one class as the recognition result, all possible classes were investigated. Experiment result shows that the enhanced method makes improvements at almost all of the emotional states and balanced the performance on every emotion. As we expected, interaction information used in the communication of humans did help the emotion recognition of computers.

The emotion interaction matrix we collected is only a beginning work. Because of the small conversation database, the correlation of emotional states between talking people can not be well indicated. More efforts should be put on the collection of conversation database and the discovery of emotional relationship between talking people.

Acknowledgement

The work is partly supported by National Natural Science Foundation of China (60203013). And We thank Cheng Jin, Qi Wu and Weiguang Wang for their generous help to our experiment and paper.

References

1. A. Mehrabian, "Communication without words", Psychology Today, 2(4), pp. 53-56, 1968
2. Z. J. Chuang and C. H. Wu, "Emotion recognition from textual input using an emotional semantic network", in Proceedings of ICSLP, Denver, Colorado, USA, 2002, pp. 2033C2036.
3. B. Schuller, G. Rigoll and M. Lang, "Speech emotion recognition combining acoustic features and linguistic information in a hybrid support vector machine-belief network architecture ", in Proc. IEEE International Conference on Acoustics, Speech, and Signal Processing, Volume 1, pp. 577-580, May 2004.
4. D. Ververidis, C. Kotropoulos and I. Pitas, "Automatic emotional speech classification", in Proc. IEEE International Conference on Acoustics, Speech, and Signal Processing, Volume 1, pp. 593-596, May 2004.
5. C. Lee, S. Narayanan, and R. Pieraccini, "Recognition of negative emotions from the speech signal", in Proc. Automatic Speech Recognition Understanding, Dec. 2001.
6. J. Yuan, L. Shen, and F. Chen, "The acoustic realization of anger, fear, joy, and sadness in chinese", in Proceedings of ICSLP, Denver, Colorado, USA, 2002, pp. 2025C2028.
7. J. Liscombe, G. Riccardi and D. Hakkani-Tür, "Using Context to Improve Emotion Detection in Spoken Dialog Systems", In the Proceedings of EUROSPEECH'05, September, 2005.
8. J. Ang, R. Dhillon, A. Krupski, E. Shriberg, and A. Stolcke, "Prosody-based automatic detection of annoyance and frustration in human-computer dialog", in Proceedings of ICSLP, Denver, Colorado, USA, 2002, pp. 2037C 2039.
9. C. M. Lee and S. Narayanan, "Towards detecting emotions in spoken dialogs", IEEE Transactions on Speech and Audio Processing, in press, 2004.
10. A. Batliner, K. Fischer, R. Huber, J. Spilker, and E. Nöth, "How to find trouble in communication", Speech Communication, vol. 40, pp. 117C143, 2003.
11. D. Crystal, "Prosodic Systems and Intonation in English", Cambridge University Press, 1969
12. D. Crystal, "The English Tone of Voice", Edward Arnold, 1975
13. D. Crystal, "The Cambridge Encyclopaedia of the English Language", Cambridge University Press, 1995
14. M. You, C. Chen and J. Bu, "CHAD: A CHINESE AFFECTIVE DATABASE", in Proc. Affective Computing and Intelligent Interaction, pp. 542 - 549, 2005

Simulation of Time-Multiplexing Cellular Neural Networks with Numerical Integration Algorithms

V. Murugesh and K. Murugesan

Department of Computer Science and Engineering,
National Institute of Technology, Tiruchirappalli – 620 015, Tamil Nadu, India
{murugesh, murugu}@nitt.edu

Abstract. A novel approach to simulate Cellular Neural Networks (CNN) is presented in this paper. The approach, time-multiplexing simulation, is prompted by the need to simulate hardware models and test hardware implementations of CNN. For practical applications, due to hardware limitations, it is impossible to have a one-to-one mapping between the CNN hardware processors and all the pixels of the image. This simulator provides a solution by processing the input image block by block, with the number of pixels in a block being the same as the number of CNN processors in the   hardware. The algorithm for implementing this simulator is presented along with popular numerical integration algorithms. Some simulation results and comparisons are also presented.

1   Introduction

Cellular Neural Networks (CNNs) are analog, time-continuous, nonlinear dynamical systems and formally belong to the class of recurrent neural networks. Since their introduction in 1988 (by Chua and Yang [2-3]), it has been the subject of intense research. Initial applications include image processing, signal processing, pattern recognition and solving partial differential equations etc.

Lee and Pineda de Gyvez [4] introduced Euler, Improved Euler and Fourth-Order Runge-Kutta algorithms in time-multiplexing CNN simulation. In this article, we consider the same problem (discussed by Chi-Chien Lee and Jose Pineda de Gyvez [4]) but presenting a different approach using the algorithms such as Euler, RK-Gill and RK-Butcher with more accuracy.

2   Cellular Neural Networks

The basic circuit unit of CNN is called a cell. It contains linear and nonlinear circuit elements. Any cell, $C(i, j)$, is connected only to its neighbouring cells i.e. adjacent cells interact directly with each other. This intuitive concept is called neighbourhood and is denoted as $N(i, j)$. Cells not in the immediate neighbourhood have indirect effect because of the propagation effects of the dynamics of the network. Each cell has a state x, input u, and output y. The state of each cell is bounded for all time $t > 0$ and, after the transient has settled down, a cellular

V.N. Alexandrov et al. (Eds.): ICCS 2006, Part I, LNCS 3991, pp. 457–464, 2006.

neural network always approaches one of its stable equilibrium points. This fact is relevant because it implies that the circuit will not oscillate. The dynamics of a CNN has both output feedback (A) and input control (B) mechanisms. The first order nonlinear differential equation defining the dynamics of a cellular neural network cell can be written as follows

$$C\frac{dx_{ij}}{dt} = -\frac{1}{R}x_{ij}(t) + \sum_{C(k,l)\in N(i,j)}A(i,j;k,l)y_{kj}(t) + \sum_{C(k,l)\in N(i,j)}B(i,j;k,l)u_{kl} + I \tag{1}$$

$$y_{ij}(t) = \frac{1}{2}\left(\left|x_{ij}(t)+1\right| - \left|x_{ij}(t)-1\right|\right)$$

where x_{ij} is the state of cell $C(i,j)$, $x_{ij}(0)$ is the initial condition of the cell, C is a linear capacitor, R is a linear resistor, I is an independent current source, $A(i,j;k,l)y_{kl}$ and $B(i,j;k,l)u_{kl}$ are voltage controlled current sources for all cells $C(k,l)$ in the neighbourhood $N(i,j)$ of cell $C(i,j)$, and y_{ij} represents the output equation.

It is to be noted from the summation operators that each cell is affected by its neighbour cells. $A(.)$ acts on the output of neighbouring cells and is referred to as the feedback operator. $B(.)$ in turn affects the input control and is referred to as the control operator. Specific entry values of matrices $A(.)$ and $B(.)$, are application dependent, are space invariant and are called cloning templates. A current bias I and the cloning templates determine the transient behavior of the cellular nonlinear network.

For software simulation purposes, equation (1) is solved within each cell, in a discretized form, to simulate its state dynamics. One common way of processing a large complex image is using a raster CNN approach discussed by Murugesh and Murugesan [7]. This approach implies that each pixel of the image is mapped onto a CNN processor. That is, we have an image processing function in the spatial domain that can be expressed as:

$$g(x,y) = T(f(x,y)) \tag{2}$$

where $f(.)$ is the input image, $g(.)$ the processed image, and T is an operator on $f(.)$ defined over the neighbourhood of (x,y). From the hardware implementation's point of view, this is a very exhaustive approach. For practical applications, in the order of 2,50,000 pixels, the hardware would require an enormous amount of processors which would make its implementation unfeasible. An alternative to this scenario is time-multiplexing simulation.

3   Time-Multiplexing Simulation

Image sizes may be in the order of thousands of pixels which mean a large CNN array is required. This is because the mapping between the image and CNN is one-to-one and each pixel in the image has a corresponding cell in the CNN array. Practically this is unfeasible. Multiplexing the image processing operator is a suitable method to overcome this problem. In this approach, the image is processed block by block and the block size is equal to the used CNN array dimension. This approach leads to two errors in the calculation of border pixels of any block since they are calculated without the effect of their neighbours. The following two equations state these errors using a neighbourhood radius equal to one:

$$\varepsilon_{ij}^{B} = \sum_{i=1}^{i=3} b_{ij+1} sign\left(u_{ij+1}\right) \tag{3}$$

$$\varepsilon_{ij}^{A} = \sum_{i=1}^{i=3} a_{ij+1} y_{ij+1}(t) \tag{4}$$

Equation (4) represents the error ε_{ij}^{A} caused in the calculation of a border cell C_{ij} in the current block due to loosing the feedback effect of neighbouring cells and neglecting the feed-forward effect. Equation (3) represents the error ε_{ij}^{B} caused in the calculation of a border cell C_{ij} due to loosing the feed-forward effect of its neighbouring cells and neglecting the feedback effect. Both the equations (3) and (4) are written for two horizontally adjacent blocks.

3.1 Overlap and Belt Approaches

To calculate the border cells more accurately, the following two approaches are used:

(i)   To eliminate the error ε_{ij}^{B} , a belt of width equal to the neighbourhood radius of CNN from the original image is used around the block, as shown in the Fig. 1(a).

(ii) An overlap between every two adjacent blocks $block_{i}$ and $block_{i+1}$ is used to minimize the error ε_{ij}^{A} , as shown in Fig. 1(b). This overlap is proportional to twice the neighbourhood radius of the CNN. The result of simulating $block_{i}$ is stored except the outer row or column in the overlap area between the two blocks that belongs to $block_{i+1}$ is set to the final state of $block_{i}$.

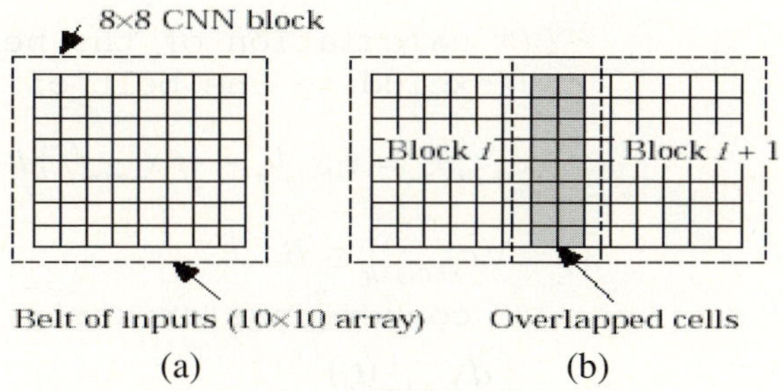

Fig. 1. (a) Belt of inputs   (b) Overlapped pixels

For the purpose of understanding the overall idea of this simulation approach better, the simplified algorithm is presented below:

```
Algorithm: (Time-Multiplexing CNN simulation)
```

$$B = \left(C_{ij} \setminus i = 1,...,block\_x \wedge j = 1,...,block\_y \right)$$

$$P \subset B = \text{ set of border cells (lower left corner)}$$

```
overlap = number of cell overlaps
belt = width of input cells
M = number of rows of the image
N = number of columns of the image
for (i=0; i<M; I+=block_x - overlap)
    for(j=0;j<N;j+=block_y - overlap)
        {
/* load initial conditions for the cells in the block
    except for those in the borders */

        for ( p=-belt,q<block_x+belt;p++)
            for(q=-belt;q<block_y+belt;q++) {
```

$$x_{i+p,j+j}(t_n) = \begin{cases} u_{ij} \\ 1 \forall C_{i+p,j+j} \in B \\ -1 \end{cases}$$

```
            } /*end for */
            /* if the block is all white or black
            don't Process it*/
```

$$\text{if}\left(x_{i+p,j+q} = -1 \vee x_{i+p,j+q} = 1 \forall c_{i+p,j+q} \in B \right)$$

```
                {
                    Obtain the final states from
                    memory
                    continue;
                }
            do { /*normal raster simulation*/
                for (p=0; p<block_x; p++) {
                    for (q=0; q<block_y; q++)

                {/* calculation of the next state
                    excluding the belt of inputs */
```

$$x_{i+p,j+q}(t_{n+1}) = x_{i+p,j+q}(t_n) + \int_{t_n}^{t_{n+1}} f\left(x_{i+p,j+q}(t_n) \right) dt$$

$$\forall c_{i+p,j+q} \in B$$

```
                /* convergence criteria */
```

$$\text{if} \left(\frac{dx_{i+p,j+q}(t_n)}{dt} = 0 \right) \text{ and}$$

$$y_{kl} = \pm 1$$

$$\forall C(k,l) \in N_r(i+p,j+q) \{$$

```
converged_cells++;
    } /* end for */
    /* update the state values */
```

$$x_{i+p,j+q}(t_n) = x_{i+p,j+q}(t_n+1) \forall c_{i+p,j+q} \in B ;$$

```
}while(converged_cells<(block_x*blovk_y));
/* store new state values excluding the ones
     corresponding to the border cells*/
```

$$A \leftarrow x_{ij} \forall C_{ij} \in B \setminus P$$

```
} /*end for */
```

4 Numerical Integration Algorithms

The CNN is described by a system of nonlinear differential equations. Therefore, it is necessary to discretize the differential equation for performing simulations. For computational purpose, a normalized time differential equations describing CNN is used by Nossek et al. [10].

$$f'(x(\pi\tau)) := \frac{dx_{ij}(\pi\tau)}{dt} = -x_{ij}(\pi\tau) + \sum_{C(k,l)\in N_r(i,j)} A(i,j;k,l)\, y_{kl}(\pi\tau)$$

$$+ \sum_{C(k,l)\in N_r(i,j)} B(i,j;k,l)\, u_{kl} + I$$

$$y_{ij}(\pi\tau) = \frac{1}{2}\left(\left| x_{ij}(\pi\tau)+1 \right| - \left| x_{ij}(\pi\tau)-1 \right| \right) \tag{5}$$

Where τ is the normalized time. For the purpose of solving the initial-value   problem, well established numerical integration techniques are used. These methods can be derived using the definition of the definite integral

$$x_{ij}((n+1)\tau) - x_{ij}(\pi\tau) = \int_{\tau_n}^{\tau_{n+1}} f'(x(\pi\tau))\, d(\pi\tau) \tag{6}$$

Three of the most widely used Numerical Integration Algorithms are used in Time-Multiplexing Simulation described here. They are the Euler's Algorithm; RK-Gill Algorithm discussed by Oliveira [9] and the RK-Butcher Algorithm discussed by Murugesan et al. [5], [6] and Murugesh and Murugesan [7], [8].

5 Simulation Results and Comparisons

All the simulation reported here are performed using a SUN BLADE 1500 workstation, and the simulation time used for comparisons is the actual CPU time used.

The input image format is the bitmap format (xbm), which is commonly available and easily convertible from popular image formats like GIF or JPEG.

Using actual numbers can easily show how much improvement is achieved. The size of Fig. 2(a) is 355×400 (1,42,000 pixels), and an Averaging template is used for simulation comparisons. First, using the raster CNN simulator discussed by Murugesh and Murugesan [7], the simulation took 200.42 seconds. Next, with the regular time-multiplexing simulator (with overlapping and input belt) the simulation took 342.28 seconds. Finally, the time-multiplexing with the time-saving scheme performed the same simulation in 240.20 seconds, almost a 33% improvement from the regular time-multiplexing. The size of two dimensional window of 10×10, with two column overlapping is used. It may be noted that this algorithm maintains all the edges of the original one.

(a)

(b)

Fig. 2. (a) Original image (b) After Averaging Template using Time-Multiplexing simulation

Since speed is one of the main concerns in the simulation, finding the maximum step size that still yields convergence for a template can be of helpful in speeding up the system. The speed-up can be achieved by selecting an appropriate step size Δt for that particular template. Even though the maximum step size may slightly vary from one image to another, the values in the Fig. 3 serve as a good reference for step size comparison. These results were obtained by trial and error over more than 100 simulations on a diamond figure. If the step size is chosen too small, the simulation might take much iteration; hence, it will take longer time to achieve convergence.

On the other hand, if the step size is taken too large, it might not converge at all or it would converge to erroneous steady state values (beyond step size 5); the latter remark can be observed for the Euler and RK-Gill algorithm which is plotted in Fig. 4. Hence, the speed of convergence of RK-Butcher algorithm for large step size is much faster than Euler and RK-Gill algorithms.

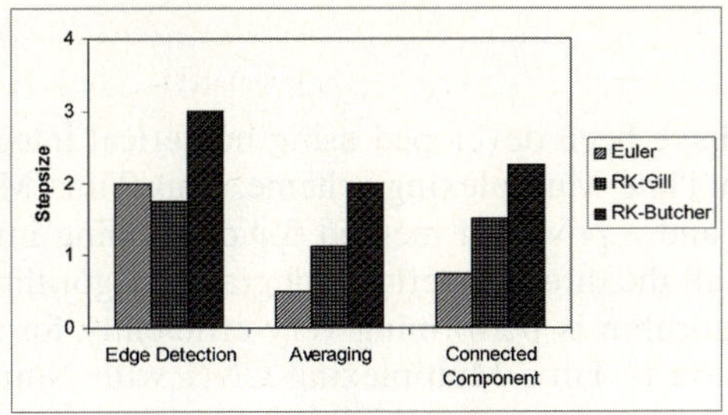

Fig. 3. Maximum step size for three different templates

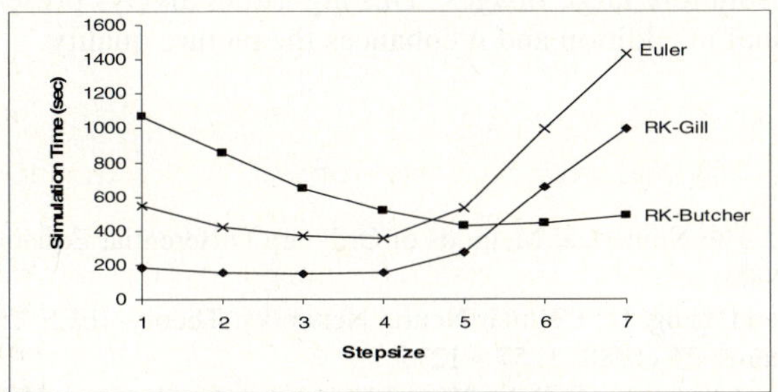

Fig. 4. Simulation time comparison using Edge Detection template

The results of Fig. 4 were obtained by simulating a small image of size 16×16 (256 pixels) using Edge Detection template on a diamond figure.

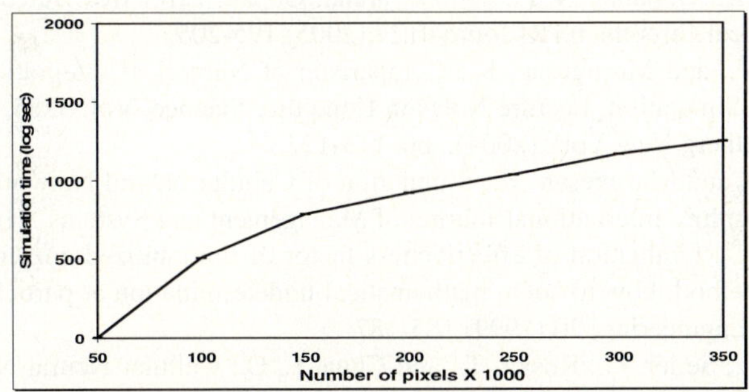

Fig. 5. CPU performance (SUN BLADE 1500 work station) for distinct sizes in number of pixels

Simulation time computations are shown in Fig. 5 using an Averaging template for images of sizes about 2,50,000 pixels and it is observed from Fig. 5, the simulation time will increase if the number of pixels chosen is too large in a log linear fashion.

6 Conclusion

Versatile algorithms have been developed using numerical integration algorithms for simulating CNN with Time-Multiplexing scheme. This Time Multiplexing algorithm is a very simple one and a powerful method for developing image processors using CNN. In fact, among all the three numerical integration algorithms, the one developed using RK-Butcher algorithm is performing very efficiently for solving this problem. The notable observation is Time Multiplexing CNN with Numerical Integration always converges for a larger step size, which in turn results in lowest simulation time for any given image size. For a given step size, the convergence time of this algorithm is log linear for all larger size images and this is an additional attractive feature as we deal with high resolution/ large images. This algorithm always preserves the edges as given in the original in addition and it enhances the picture quality.

References

1. Butcher, J. C.: The Numerical Methods of Ordinary Differential Equations, John Wiley & Sons, U.K. (2003).
2. Chua, L. O., and Yang, L.: Cellular Neural Networks: Theory, IEEE Transactions on Circuits and Systems, 35 (1988) 1257 – 1272.
3. Chua, L. O., and Yang, L.: Cellular Neural Networks: Applications, IEEE Transactions on Circuits and Systems, 35 (1988) 1273 – 1290.
4. Chi-Chien Lee and Jose Pineda de Gyvez: Time-Multiplexing CNN Simulator, IEEE International Symposium on Circuits and Systems, 6 (1999) 407-410.
5. Murugesan, K., Sekar, S., Murugesh, V. and Park, J. Y.: Numerical Solution of an Industrial Robot Arm Control Problem using the RK-Butcher algorithm, International Journal of Computer Applications in Technology, 19 (2004) 132-138.
6. Murugesan, K., Gopalan, N. P., and Devarajan Gopal.: Error free Butcher algorithms for linear Electrical Circuits, ETRI Journal, 27 (2005) 195-205.
7. Murugesh, V., and Murugesan, K.: Comparison of Numerical Integration Algorithms in Raster CNN Simulation, Lecture Notes in Computer Science, Vol. 3285, Springer-Verlag, Berlin Heidelberg New York (2004), pp. 115-122.
8. Murugesh,V. and Murugesan, K.: Simulation of Cellular Neural Networks using the RK-Butcher algorithm, International Journal of Management and Systems, 21(2005)65-78.
9. Oliveira, S. C.: Evaluation of effectiveness factor of immobilized enzymes using Runge-Kutta-Gill method: how to solve mathematical undetermination at particle center point?", Bio Process Engineering, 20 (1999) 185-187.
10. Nossek, J. A., Seiler, G., Roska. T., and Chua. L. O.: Cellular Neural Networks: Theory and Circuit Design, International Journal of Circuit Theory and Applications, 20 (1992) 533-553.

Dynamics of POD Modes in Wall Bounded Turbulent Flow

Giancarlo Alfonsi[1] and Leonardo Primavera[2]

[1] Dipartimento di Difesa del Suolo, Università della Calabria
Via P. Bucci 42b, 87036 Rende (Cosenza), Italy
alfonsi@dds.unical.it
[2] Dipartimento di Fisica, Università della Calabria
Via P. Bucci 33b, 87036 Rende (Cosenza), Italy
lprimavera@fis.unical.it

Abstract. The dynamic properties of POD modes of the fluctuating velocity field developing in the wall region of turbulent channel flow are investigated. The flow of viscous incompressible fluid in a channel is simulated numerically by means of a parallel computational code based on a mixed spectral-finite difference algorithm for the numerical integration of the Navier-Stokes equations. The DNS approach (Direct Numerical Simulation of turbulence) is followed in the calculations, performed at friction Reynolds number $Re_\tau = 180$. A database representing the turbulent statistically steady state of the flow through 10 viscous time units is assembled and the Proper Orthogonal Decomposition technique (POD) is applied to the fluctuating portion of the velocity field. The dynamic properties of the most energetic POD modes are investigated showing a clear interaction between streamwise-independent modes and quasi-streamwise modes in the temporal development of the turbulent flow field.

1   Introduction

The hypothesis incorporated in all turbulence theories that have been formulated in the last decades is that of the local isotropy of the small turbulent scales (Kolmogorov [1]), i.e. the postulate that the small-scale structures of turbulent flows possess universal statistical properties independent of the large scales. Local isotropy has also been enforced in most of the existing *SGS* (subgrid-scale) closures within the *LES* (Large Eddy Simulation) approach to turbulence modelling.

The verification of the hypotheses that provide the basis for turbulence theories is an issue of remarkable relevance for both theoreticians and modellers. Several researchers provided evidence of the "-5/3" velocity spectrum in the inertial range, while the issue of the universal statistical properties of small turbulent scales in both inertial and dissipation ranges is still controversial. One of the difficulties encountered is that of reaching values of the Reynolds number able to assure the development of a sufficiently broad spectrum of scales for the possible establishment of local isotropy of the small scales.

The properties of the velocity field has been studied numerically in homogeneous shear flow with constant mean shear by Pumir & Shraiman [2]. They showed that for

V.N. Alexandrov et al. (Eds.): ICCS 2006, Part I, LNCS 3991, pp. 465–472, 2006.

values of the Taylor microscale Reynolds number of order 100 the value of the derivative skewness of the velocity fluctuation in the direction of the mean flow along the direction of the mean velocity gradient, is of order 1. Garg & Warhaft [3] studied experimentally the properties of the small-scale velocity field in homogeneous shear flow with constant mean shear. They found that there is a significant skewness (of order 1) of the derivative of the longitudinal velocity fluctuation in the direction of the mean gradient. Third-order transverse structure functions of the longitudinal velocity were found to have a scaling range, showing the existence of anisotropy at both inertial and dissipation scales. Shen & Warhaft [4] continued the experiments. They showed that the fifth moment of the derivative of the longitudinal fluctuation in the direction of the mean gradient, is of order 10 with no diminution with the Reynolds number. Moreover, fifth- and seventh-order inertial subrange skewness structure functions are of order 10 and 100 respectively. These results show that there exists velocity anisotropy in both inertial and dissipation ranges, for the Reynolds number tested.

A more recent approach to turbulence modelling, that differs from the most used RANS (Reynolds Averaged Navier-Stokes equations) and LES (Large Eddy Simulation), involves methods for the reduction of the turbulent phenomenon to a system with a limited number of degrees of freedom. The Proper Orthogonal Decomposition (POD) is a technique that permits the extraction of appropriately-defined modes of the flow from the background flow, that can be subsequently projected onto the system of the Navier-Stokes equations to obtain a low-order model of a given turbulent flow (Podvin & Lumley [5], Omurtag & Sirovich [6]). A relevant issue in this context is the investigation of the dynamic properties of the flow modes as extracted with the POD technique.

The present work addresses the issue of the dynamic characteristics of the coherent structures of turbulence in moderately turbulent channel flow, educed by applying the Proper Orthogonal Decomposition to the fluctuating portion of the velocity field. The POD modes are calculated from a numerical database assembled with the use of a parallel computational code for the numerical integration of the Navier-Stokes equations in the case of the plane channel at friction Reynolds number $Re_\tau = 180$.

2   Methods

The simulations have been performed with a parallel computational code based on a mixed spectral-finite difference technique. The unsteady Navier-Stokes equations for incompressible fluids with constant properties in three dimensions and non-dimensional conservative form, is considered ($i \, \& \, j = 1,2,3$):

$$\frac{\partial u_i}{\partial x_i} = 0 \tag{1a}$$

$$\frac{\partial u_i}{\partial t} + \frac{\partial}{\partial x_j}\left(u_i u_j\right) = -\frac{\partial p}{\partial x_i} + \frac{1}{Re_\tau}\frac{\partial^2 u_i}{\partial x_j \partial x_j} \tag{1b}$$

where $u_i(u,v,w)$ are the velocity components in the cartesian coordinate system $x_i(x,y,z)$. Equations (1) are nondimensionalized by the channel half-width δ for lenghts, wall shear velocity $u_\tau = \sqrt{\tau_w/\rho}$ for velocities, ρu_τ^2 for pressure and δ/u_τ for time, being $Re_\tau = (u_\tau \delta/v)$ the friction Reynolds number.

The fields are admitted to be periodic in the streamwise (x) and spanwise (z) directions, and equations (1) are Fourier transformed accordingly. The nonlinear terms in the momentum equation are evaluated pseudospectrally by anti-transforming the velocities back in physical space to perform the products (FFTs are used). A dealiasing procedure is applied to avoid errors in transforming the results back to Fourier space. In order to have a better spatial resolution near the walls, a grid-stretching law of hyperbolic-tangent type has been introduced for the grid points along y, the direction orthogonal to the walls. For the time advancement, a third-order Runge-Kutta algorithm has been implemented and the time marching procedure is accomplished with the fractional-step method. No-slip boundary conditions at the walls and cyclic conditions in the streamwise and spanwise directions have been applied to the velocity. More detailed descriptions of the numerical scheme, of its reliability and of the performance obtained on the parallel computers that have been used, can be found in Alfonsi *et al.* [7] and Passoni *et al.* [8],[9],[10].

By recalling the wall formalism, one has: $x_i^+ = x_i u_\tau/v = x_i/\delta_\tau$, $t^+ = tu_\tau^2/v = tu_\tau/\delta_\tau$, $\delta^+ = \delta/\delta_\tau$, $u^+ = \overline{u}/u_\tau$, $Re_\tau = u_\tau \delta/v = \delta/\delta_\tau = \delta^+$, where $\overline{u}$ is streamwise velocity averaged on a x-z plane and time, $\delta_\tau = v/u_\tau$ is the viscous length and δ/u_τ the viscous time unit. The characteristic parameters of the numerical simulations are the following. Computing domain: $L_x = 2\pi\delta$, $L_y = 2\delta$, $L_z = \pi\delta$; $L_x^+ = 1131$, $L_y^+ = 360$, $L_z^+ = 565$. Computational grid: $N_x = 96$, $N_y = 129$, $N_z = 64$. Grid spacing: $\Delta x^+ = 11.8$, $\Delta y_{center}^+ = 4.4$, $\Delta y_{wall}^+ = 0.87$, $\Delta z^+ = 8.8$. It can be verified that there are 6 grid points in the y direction within the viscous sublayer ($y^+ \leq 5$). The Kolmogorov spatial microscale, estimated using the criterion of the average dissipation rate per unit mass across the width of the channel, results $\eta^+ \approx 1.8$. After the insertion of appropriate initial conditions, the initial transient of the flow in the channel has been simulated, the turbulent statistically steady state has been reached and then calculated for a time $t = 10\,\delta/u_\tau$ ($t^+ = 1800$). 20000 time steps have been calculated with a temporal resolution of $\Delta t = 5\times10^{-4}\,\delta/u_\tau$ ($\Delta t^+ = 0.09$).

In Table 1 predicted and computed values of a number of mean-flow variables are reported (U_b and U_c are the bulk mean velocity and the mean centerline velocity respectively, while Re_b and Re_c are the related Reynolds numbers). The predicted values of U_c/U_b and C_f are obtained from the correlations suggested by Dean [11] [$U_c/U_b = 1.28(2Re_b)^{-0.0116}$; $C_f = 0.073(2Re_b)^{-0.25}$] while the computed skin friction coefficient [$C_f = (2\tau_w/\rho U_b^2)$; $\tau_w = \mu(\partial U/\partial y)_{wall}$] is calculated using the value of the shear stress at the wall actually obtained in the computations (a finite difference routine is used).

Table 1. Predicted vs. computed mean-flow variables

Predicted variables

Re_τ	Re_b	Re_c	U_b/u_τ	U_c/u_τ	U_c/U_b	C_f
180	2800	3244	15.56	18.02	1.16	8.44×10^{-3}

Computed variables

Re_τ	Re_b	Re_c	U_b/u_τ	U_c/u_τ	U_c/U_b	C_f
178.74	2786	3238	15.48	17.99	1.16	8.23×10^{-3}

The Proper Orthogonal Decomposition is a technique that can be applied for the extraction of the coherent structures from a turbulent flow field (Berkooz *et al.* [12], Sirovich [13]). By considering an ensemble of temporal realizations of a velocity field $u_i(x_j,t)$ on a finite domain D, one wants to find which is the most similar function to the elements of the ensemble, on average. This problem corresponds to find a deterministic vector function $\varphi_i(x_j)$ such that (*i* & *j=1,2,3*):

$$\max_\psi \frac{\left\langle \left| \left(u_i(x_j,t), \psi_i(x_j) \right) \right|^2 \right\rangle}{\left(\psi_i(x_j), \psi_i(x_j) \right)} = \frac{\left\langle \left| \left(u_i(x_j,t), \varphi_i(x_j) \right) \right|^2 \right\rangle}{\left(\varphi_i(x_j), \varphi_i(x_j) \right)}. \tag{2}$$

A necessary condition for problem (2) is that $\varphi_i(x_j)$ is an eigenfunction, solution of the eigenvalue problem and Fredholm integral equation of the first kind:

$$\int_D R_{ij}(x_l, x_l')\varphi_j(x_l')dx_l' = \int_D \left\langle u_i(x_k,t)u_j(x_k',t) \right\rangle \varphi_j(x_k')dx_k' = \lambda\varphi_i(x_k) \tag{3}$$

where $R_{ij} = \left\langle u_i(x_k,t)u_j(x_k',t) \right\rangle$ is the two-point velocity correlation tensor. To each eigenfunction $\varphi_i^{(n)}(x_j)$ is associated a real positive eingenvalue $\lambda^{(n)}$ and every member of the ensemble can be reconstructed by means of the modal decomposition $u_i(x_j,t) = \sum_n a_n(t)\varphi_i^{(n)}(x_j)$. The contribution of each mode to the kinetic energy content of the flow is given by $E = \int_D \left\langle u_i(x_j,t)u_i(x_j,t) \right\rangle dx_j = \sum_n \lambda^{(n)}$, being E the turbulent kinetic energy in the domain D. In the present work the POD is used for the analysis of the fluctuating portion of the velocity field. The two homogeneous directions are handled in Fourier space so that the optimal representation of the velocity field in the statistical sense outlined above is sought in the direction normal to the solid walls.

3   Results

As a result of the decomposition, $3N_y(387)$ POD modes and correspondent eigenvalues have been determined for each wavenumber index pair (m,n). Table 2 reports the individual fraction of the turbulent kinetic energy and the cumulative energies of the velocity fluctuations of the first 10 most energetic modes of the decomposition (m and n are the wavenumbers along x and z, respectively and q is the generic POD mode.

Table 2. Energy content of the first 10 eigenfunctions

Index	Mode (m,n,q)	Energy fraction	Energy sum
1	(0,1,1)	0.03220	0.03220
2	(0,2,1)	0.02173	0.05393
3	(0,2,2)	0.01535	0.06929
4	(1,1,1)	0.01508	0.08437
5	(1,2,1)	0.01454	0.09891
6	(0,3,1)	0.01197	0.11089
7	(1,3,1)	0.01196	0.12286
8	(1,2,2)	0.01160	0.13446
9	(0,4,1)	0.01053	0.14499
10	(1,4,1)	0.00972	0.15472

About 6.9% of the energy resides in the first three streamwise-independent modes (m=0). The first mode that exhibits a streamwise dependence is the fourth.

Figure 1 shows two surfaces of constant streamwise velocity (the light surface is positive, the dark surface is negative streamwise velocity) reconstructed from the first most energetic eigenfunction. The visualization shows two structures elongated in the streamwise direction. Flow representations (not reported) of the second and third eigenfunctions have shown streamwise-independent structures similar to those of Figure 1, with appropriate repetitions according to the values of m and n. Figure 2 shows surfaces of constant streamwise velocity reconstructed from the fourth most energetic eigenfunction of the decomposition. This is the first streamwise-dependent eigenfunction. The visualization shows couples of bean-shaped quasi-streamwise flow structures, where one of the structure of each couple is lower with respect to the first (more displaced toward the center of the channel). Flow representations (not reported) of the fifth eigenfunction have shown bean-shaped structures aligned in the streamwise direction, similar to those of Figure 2.

Figure 3 shows the flow structure – identified in terms of surfaces of constant streamwise velocity – formed by the sum of the first five most energetic POD modes (three x-independent structures of the type shown in Figure 1 and two x-dependent turbulent structures of the type shown in Figure 2) at $t^+ = 72$. Two dominant structures elongated in the streamwise direction are visible. With particular reference

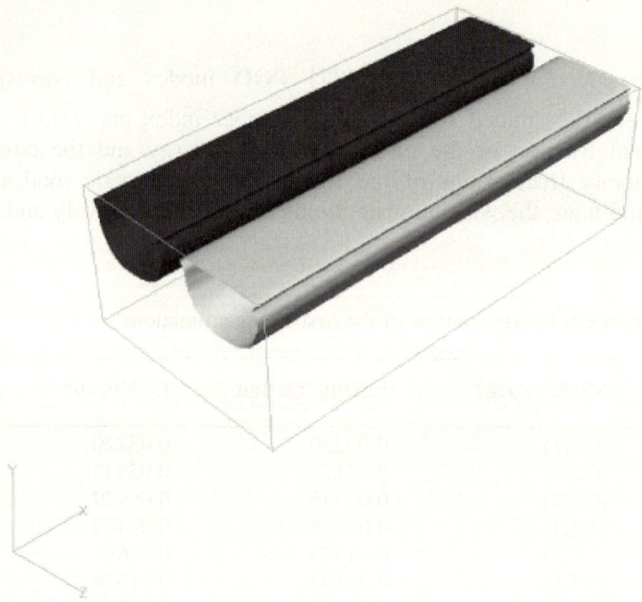

Fig. 1. Surfaces of constant x-velocity reconstructed from the first POD mode

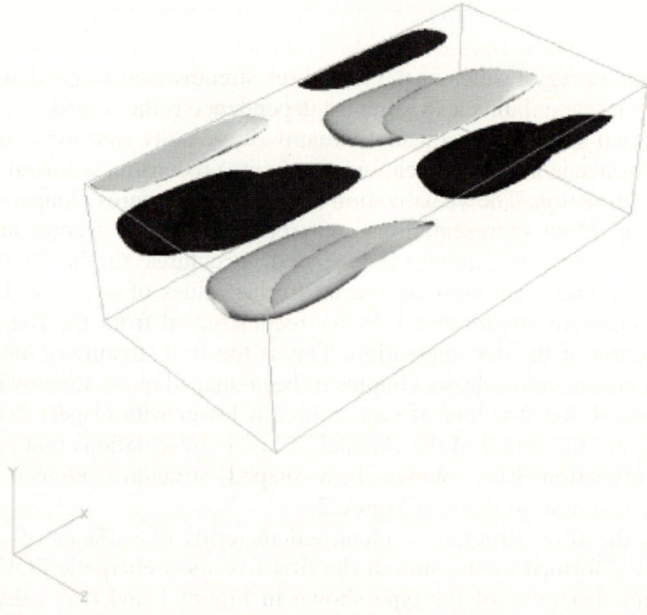

Fig. 2. Surfaces of constant x-velocity reconstructed from the fourth POD mode

Fig. 3. Surfaces of constant x-velocity reconstructed from the first five POD modes at $t^+ = 72$

to the positive (light) surface it clearly appears that the shape of the basic streamwise-elongated structure is altered because of the interaction with the travelling bean-shaped quasi-streamwise modes. This is the basic mechanism of evolution in time of the turbulent flow structures, that has been revealed due to the fact that the flow phenomena have been described in terms of dominant (most energetic) turbulent structures (the POD modes in this context).

4 Concluding Remarks

The analysis of the flow field of numerically simulated turbulent channel flow is performed in terms of flow modes determined with the POD technique. The dynamic properties of the POD modes are investigated, revealing the basic mechanism of evolution in time of the turbulent structures. This mechanism consists in the mutual interaction of two different types of modes, streamwise-independent structures and bean-shaped, quasi-streamwise modes aligned in the streamwise direction.

References

1. Kolmogorov A.N.: The local structure of turbulence in incompressible viscous fluid for very large Reynolds numbers. *Dokl. Akad. Nauk. SSSR* **30** (1941) 301
2. Pumir A. & Shraiman B.I.: Persistent small scale anisotropy in homogeneous shear flows. *Phys. Rev. Lett.* **75** (1996) 3114

3. Garg S. & Warhaft Z.: On the small scale structure of simple shear flow. *Phys. Fluids* **10** (1998) 662

4. Shen X. & Warhaft Z.: The anisotropy of the small scale structure in high Reynolds number ($R_\lambda \approx 1000$) turbulent shear flow. *Phys. Fluids* **12** (2000) 2976

5. Podvin B. & Lumley J.L.: A low-dimensional approach for the minimal flow unit. *J. Fluid Mech.*, **362** (1998) 121

6. Omurtag A. & Sirovich L.: On low-dimensional modeling of channel turbulence. *Theor. Comp. Fluid Dyn.* **13** (1999) 115

7. Alfonsi G., Passoni G., Pancaldo L. & Zampaglione D.: A spectral-finite difference solution of the Navier-Stokes equations in three dimensions. *Int. J. Num. Meth. Fluids* **28** (1998) 129

8. Passoni G., Alfonsi G., Tula G. & Cardu U.: A wavenumber parallel computational code for the numerical integration of the Navier-Stokes equations. *Parall. Comp.* **25** (1999) 593

9. Passoni G., Cremonesi P. & Alfonsi G.: Analysis and implementation of a parallelization strategy on a Navier-Stokes solver for shear flow simulations. *Parall. Comp.* **27** (2001) 1665

10. Passoni G., Alfonsi G. & Galbiati M.: Analysis of hybrid algorithms for the Navier-Stokes equations with respect to hydrodynamic stability theory. *Int. J. Num. Meth. Fluids* **38** (2002) 1069

11. Dean R.B.: Reynolds number dependence of skin friction and other bulk flow variables in two-dimensional rectangular duct flow. *J. Fluids Eng.* **100** (1978) 215

12. Berkooz G., Holmes P., Lumley J.L.: The Proper Orthogonal Decomposition in the analysis of turbulent flows. *Ann. Rev. Fluid Mech.* **25** (1993) 539

13. Sirovich L.: Turbulence and the dynamics of coherent structures. Part I: coherent structures. Part II: symmetries and transformations. Part III: dynamics and scaling. *Quart. Appl. Math.* **45** (1987) 561

Score Evaluation Within the Extended Square-Root Information Filter

Maria V. Kulikova[1] and Innokenti V. Semoushin[2]

[1] School of Computational and Applied Mathematics,
University of the Witwatersrand, Private Bag 3, Wits 2050,
Johannesburg, South Africa
mkulikova@cam.wits.ac.za

[2] Ulyanovsk State University, 42 Leo Tolstoy Str., 432970 Ulyanovsk, Russia
i.semushin@ulsu.ru
http://staff.ulsu.ru/semoushin/

Abstract. A newly developed algorithm for evaluating the Log Likelihood Gradient (score) of linear discrete-time dynamic systems is presented, based on the extended Square-Root Information Filter (eSRIF). The new result can be used for efficient calculations in gradient-search algorithms for maximum likelihood estimation of the unknown system parameters. The theoretical results are given with the examples showing the superior perfomance of this computational approach over the conventional one.

1 Introduction

Consider the discrete-time linear dynamic stochastic system

$$x_{t+1} = F_t x_t + G_t w_t, \qquad t = 0, 1, \ldots, N \tag{1}$$

$$z_t = H_t x_t + v_t, \qquad t = 1, 2, \ldots, N \tag{2}$$

with the system state $x_t \in \mathrm{I\!R}^n$, the state disturbance $w_t \in \mathrm{I\!R}^q$, the observed vector $z_t \in \mathrm{I\!R}^m$, and the measurement error $v_t \in \mathrm{I\!R}^m$, such that the initial state x_0 and each w_t, v_t of $\{w_t : t = 0, 1, \ldots\}$, $\{v_t : t = 1, 2, \ldots\}$ are taken from mutually independent Gaussian distributions with the following expectations:

$$\mathbf{E}\left\{\begin{bmatrix} x_0 \\ w_t \\ v_t \end{bmatrix}\right\} = \begin{bmatrix} \bar{x}_0 \\ 0 \\ 0 \end{bmatrix}, \quad \mathbf{E}\left\{\begin{bmatrix} (x_0 - \bar{x}_0) \\ w_t \\ v_t \end{bmatrix}\begin{bmatrix} (x_0 - \bar{x}_0) \\ w_t \\ v_t \end{bmatrix}^T\right\} = \begin{bmatrix} P_0 & 0 & 0 \\ 0 & Q_t & 0 \\ 0 & 0 & R_t \end{bmatrix}$$

and $\mathbf{E}\left\{w_t w_{t'}^T\right\} = 0$, $\mathbf{E}\left\{v_t v_{t'}^T\right\} = 0$ if $t \neq t'$. Assume the system is parameterized by a vector $\theta \in \mathrm{I\!R}^p$ of unknown system parameters. This means that all the above characteristics, namely F_t, G_t, H_t, $\bar{x}_0$, $P_0 \geq 0$, $Q_t \geq 0$ and $R_t > 0$, can depend upon θ (the corresponding notations $F_t(\theta)$, $G_t(\theta)$ and so on, are suppressed for the sake of simplicity).

V.N. Alexandrov et al. (Eds.): ICCS 2006, Part I, LNCS 3991, pp. 473–481, 2006.

Models like (1), (2), together with the associated Kalman filter, have been intensively used in many application fields such as control, communications, and signal/image processing. To apply the Kalman filter/smoother in the case of parametric uncertainty, it is necessary to identify the model, i. e., to estimate the unknown system parameters (or directly the Kalman filter/smoother parameters) from the available measurements $Z_1^t = (z_1, z_2, \ldots, z_t)$, $t = 1, 2, \ldots, N$.

A very general and well known approach to parameter estimation is the maximum likelihood (ML) principle. The ML method for estimating θ requires the maximization of the Log Likelihood Function (LLF) $L_\theta\left(Z_1^N\right)$, which is the logarithm of the joint probability density of $z_1, z_2, \ldots, z_N$, with respect to θ. It incorporates various gradient-search optimization algorithms. In this context, it is very important to find a means for efficient computation of Log Likelihood Gradient (LLG) known as the "score".

The first solutions to the problem [1, 2] obtained by the direct differentiating the Kalman filtering equations are time consuming as they require the computation of p vector *filter sensitivity equations* and p matrix *Riccati-type sensitivity equations*, both run recursively in the forward time direction. An alternative approach was developed by Yared [3] who assumed a steady-state Kalman filter and used an adjoint filter, resulting in one additional backward pass in time. It works faster than the forward "differentiated" filter although at the expence of increased storage requirements. Wilson and Kumar simplified Yared's results [4] by moving from derivatives of Kalman variables to those of the original system matrices. They reduced the problem to solving a time-invariant matrix equation, an algebraic Riccati equation, a time-invariant Kalman filter equation, and the backward adjopint filter equation.

Segal and Weinstein developed LLG formulas for both the discrete and continuous-time systems [5, 6] using Fisher's identity and employing the Kalman smoother. Their approach enables to compute not only the LLG, but (approximately) the log-likelihood Hessian and the Fisher information matrix of θ as well. The continuous-time formula in [5] uses the Itô integral, which cannot be instrumented. For continuous-time case, Leland developed the improved formulas for the LLG [7, 8] free of this limitation.

Despite this remarkable success it is useful to develop alternative algorithms. The most important reason is that the previously mentioned methods suffer from some limitations mainly stemming from the use of the conventional Kalman filter (KF) implementation. This implementation — in terms of covariance matrices — is particularly sensitive to roundoff errors [12]. As a consequence, any method for evaluating the log-likelihood gradient based on the conventional KF implementation, inherits this drawback and so cannot be considered as numerically stable. The alternative solutions to evaluate the LLG can be found in alternative KF implementation methods developed for dealing with the problem of numerical instability in many papers, for example [9, 10, 11], and described in [12].

In this paper we derive a new algorithm for evaluating the LLG based upon the extended Square-Root Information Filter (eSRIF) recently proposed in [11]. Unlike the Conventional Kalman filter (CKF), the eSRIF avoids numerical instabilities arising from roundoff errors and has also the added feature of being better suited to parallel and to very large scale integration (VLSI) implementations (see [11]). Thus, inheriting these advantages, we expect our method to outperform the conventional KF mechanization for accuracy. These expectations will be verified by two examples of ill-conditioned problems.

The paper is organized as follows. In Section 2 we present a new algorithm for evaluating the LLG based upon the eSRIF. The comparison of the developed algorithm and the conventional approach in terms of sensitivity to roundoff errors is given in Section 3. Section 4 presents some numerical results and finally, Section 5 concludes the paper.

2 Log Likelihood Gradient Evaluation

The Log Likelihood Function (LLF) of system (1), (2) is given by

$$L_\theta\left(Z_1^N\right) = -\frac{1}{2}\sum_{t=1}^{N}\left\{\frac{m}{2}\ln(2\pi) + \ln(\det(R_{e,t})) + e_t^T R_{e,t}^{-1} e_t\right\}$$

where $Z_1^N = (z_1, z_2, \ldots z_N)$ is N-step measurement history, $e_t \overset{\text{def}}{=} z_t - H_t \hat{x}_t$ is the zero-mean innovation sequence whose covariance is determined as $R_{e,t} \overset{\text{def}}{=} \mathbf{E}\left\{e_t e_t^T\right\} = H_t P_t H_t^T + R_t$. The matrix P_t is the error covariance matrix of the time updated estimate $\hat{x}_t$ of the state vector generated by the Kalman filter.

Let $l_\theta(z_t)$ denote the negative LLF for the t-th measurement z_t in system (1), (2), given measurement history $Z_1^{t-1} \overset{\text{def}}{=} \{z_1, z_2, \ldots, z_{t-1}\}$, then

$$l_\theta(z_t) = \frac{1}{2}\left\{\frac{m}{2}\ln(2\pi) + \ln(\det(R_{e,t})) + e_t^T R_{e,t}^{-1} e_t\right\}. \tag{3}$$

From (3) one can easily obtain the expression for the LLG. Let $R_{e,t} = R_{e,t}^{T/2} R_{e,t}^{1/2}$ where $R_{e,t}^{1/2}$ is a square-root factor of the matrix $R_{e,t}$ and $\bar{e}_t$ are the normalized innovations, i. e. $\bar{e}_t = R_{e,t}^{-T/2} e_t$. Taking into account that the matrix $R_{e,t}^{1/2}$ is triangular, we can write down the expression

$$\frac{\partial}{\partial\theta_i}\left[\ln(\det(R_{e,t}^{1/2}))\right] = \mathbf{tr}\left[R_{e,t}^{-1/2}\,\frac{\partial\left(R_{e,t}^{1/2}\right)}{\partial\theta_i}\right]. \tag{4}$$

Hence,

$$\frac{\partial l(z_t)}{\partial\theta_i} = \mathbf{tr}\left[R_{e,t}^{-1/2}\,\frac{\partial\left(R_{e,t}^{1/2}\right)}{\partial\theta_i}\right] + \bar{e}_t^T\frac{\partial\bar{e}_t}{\partial\theta_i}, \qquad i = 1, 2, \ldots, p \tag{5}$$

as it follows directly from (3) and (4).

According to the goal pursued by this research and stated in Section 1, we consider the eSRIF presented in [11]. For convenience, we reformulate it in the following form: given $P_0^{-T/2}$ and $P_0^{-T/2}\hat{x}_0 = P_0^{-T/2}\bar{x}_0$; calculate

$$O_t \begin{bmatrix} R_t^{-T/2} & -R_t^{-T/2}H_tF_t^{-1} & R_t^{-T/2}H_tF_t^{-1}G_tQ_t^{T/2} & -R_t^{-T/2}z_t \\ 0 & P_t^{-T/2}F_t^{-1} & -P_t^{-T/2}F_t^{-1}G_tQ_t^{T/2} & P_t^{-T/2}\hat{x}_t \\ 0 & 0 & I_q & 0 \end{bmatrix}$$

$$= \begin{bmatrix} R_{e,t}^{-T/2} & 0 & 0 & -\bar{e}_t \\ -P_{t+1}^{-T/2}K_{p,t} & P_{t+1}^{-T/2} & 0 & P_{t+1}^{-T/2}\hat{x}_{t+1} \\ * & * & * & * \end{bmatrix} \tag{6}$$

where O_t is any orthogonal transformation such that the matrix on the right-hand side of formula (6) is block lower triangular. The matrix $P_t^{1/2}$ is a square-root factor of P_t, i. e. $P_t = P_t^{T/2}P_t^{1/2}$, $P_t^{1/2}$ is upper triangular. Similarly, we define $P_{t+1} = P_{t+1}^{T/2}P_{t+1}^{1/2}$, $R_t = R_t^{T/2}R_t^{1/2}$, $Q_t = Q_t^{T/2}Q_t^{1/2}$ and $R_{e,t} = R_{e,t}^{T/2}R_{e,t}^{1/2}$. For convenience we shall also write $A^{T/2} = (A^{1/2})^T$, $A^{-1/2} = (A^{1/2})^{-1}$ and $A^{-T/2} = (A^{-1/2})^T$. Additionally, $K_{p,t} = F_tP_tH_t^TR_{e,t}^{-1}$.

It is easy to see that the LLG (5) in terms of eSRIF (6) is given by

$$\frac{\partial l(z_t)}{\partial \theta_i} = -\,\mathbf{tr}\left[R_{e,t}^{1/2}\frac{\partial\left(R_{e,t}^{-1/2}\right)}{\partial \theta_i}\right] + \bar{e}_t^T\frac{\partial \bar{e}_t}{\partial \theta_i}, \qquad i = 1, 2, \ldots, p. \tag{7}$$

To establish our algorithm for efficient evaluation of LLG (7) we prove the following result.

Lemma 1. *Let*

$$QA = L \tag{8}$$

where Q is any orthogonal transformation such that the matrix on the right-hand side of formula (8) is lower triangular and A is a nonsingular matrix. If the elements of A are differentiable functions of a parameter θ then the upper triangular matrix U in

$$Q'_\theta Q^T = U^T - U \tag{9}$$

is, in fact, the upper triangular part of the matrix $QA'_\theta L^{-1}$.

Having applied Lemma 1 to eSRIF (6), we obtain the following algorithm for computing LLG (7).

Algorithm LLG-eSRIF

I. For each θ_i, $i = 1, 2, \ldots, p$, calculate

$$
O_t
\begin{bmatrix}
\dfrac{\partial}{\partial \theta_i}\left(R_t^{-T/2}\right) & \dfrac{\partial}{\partial \theta_i}\left(S_t^{(1)}\right) & \dfrac{\partial}{\partial \theta_i}\left(S_t^{(2)}\right) & \left|\ \dfrac{\partial}{\partial \theta_i}\left(S_t^{(3)}\right)\right. \\
0 & \dfrac{\partial}{\partial \theta_i}\left(S_t^{(4)}\right) & \dfrac{\partial}{\partial \theta_i}\left(S_t^{(5)}\right) & \left|\ \dfrac{\partial}{\partial \theta_i}\left(S_t^{(6)}\right)\right. \\
0 & 0 & 0 & 0
\end{bmatrix}
=
\begin{bmatrix}
X_i & Y_i & M_i & L_i \\
N_i & V_i & W_i & K_i \\
* & * & * & *
\end{bmatrix}
$$

where O_t is the same orthogonal transformation as in (6) and

$$
S_t^{(1)} = -R_t^{-T/2} H_t F_t^{-1}, \; S_t^{(2)} = R_t^{-T/2} H_t F_t^{-1} G_t Q_t^{T/2}, \; S_t^{(3)} = -R_t^{-T/2} z_t,
$$

$$
S_t^{(4)} = P_t^{-T/2} F_t^{-1}, \quad S_t^{(5)} = -P_t^{-T/2} F_t^{-1} G_t Q_t^{T/2}, \quad S_t^{(6)} = P_t^{-T/2} \hat{x}_t.
$$

II. For each θ_i, $i = 1, 2, \ldots, p$, compute the matrix

$$
J_i =
\begin{bmatrix}
X_i & Y_i & M_i \\
N_i & V_i & W_i
\end{bmatrix}
\begin{bmatrix}
R_{e,t}^{-T/2} & 0 & 0 \\
-P_{t+1}^{-T/2} K_{p,t} & P_{t+1}^{-T/2} & 0 \\
* & * & *
\end{bmatrix}^{-1}.
$$

III. For each θ_i, $i = 1, 2, \ldots, p$, we split the matrices J_i obtained at Step II as follows:

$$
J_i = \underbrace{\left[\ \overbrace{\left[L_i + D_i + U_i\right.}^{m+n+q} \Big| \quad\ *** \quad\ \right]}_{m+n} \quad \left.\right\} m+n
$$

where L_i, D_i and U_i are the strictly lower triangular, diagonal and strictly upper triangular parts of matrix J_i, respectively.

IV. For each θ_i, $i = 1, 2, \ldots, p$, compute the following quantities:

$$
\begin{bmatrix}
\dfrac{\partial R_{e,t}^{-T/2}}{\partial \theta_i} & 0 \\[2ex]
-\dfrac{\partial\left(\tilde{P}_{t+1}^{-T/2} K_{p,t}\right)}{\partial \theta_i} & \dfrac{\partial \tilde{P}_{t+1}^{-T/2}}{\partial \theta_i}
\end{bmatrix}
= \left[L_i + D_i + U_i^T\right]
\begin{bmatrix}
R_{e,t}^{-T/2} & 0 \\
-\tilde{P}_{t+1}^{-T/2} K_{p,t} & \tilde{P}_{t+1}^{-T/2}
\end{bmatrix}
$$

$$
\frac{\partial \bar{e}_t}{\partial \theta_i} = \left[\frac{\partial R_{e,t}^{-T/2}}{\partial \theta_i} - X_i\right] R_{e,t}^{T/2} \bar{e}_t + Y_i F_t \hat{x}_t - L_i,
$$

$$
\frac{\partial S_{t+1}^{(6)}}{\partial \theta_i} = \left[\frac{\partial\left(\tilde{P}_{t+1}^{-T/2} K_{p,t}\right)}{\partial \theta_i} + N_i\right] R_{e,t}^{T/2} \bar{e}_t + \left[\frac{\partial \tilde{P}_{t+1}^{-T/2}}{\partial \theta_i} - V_i\right] F_t \hat{x}_t + K_i.
$$

Table 1. Comparison of Rounded Solutions to Problem 1 evaluated at the point $\theta = 1$

Filter Implementation	Solution	
	Exact Answer	Rounded Answer
"differentiated" Conventional Covariance Filter	$(P_1)'_\theta\big\|_{\theta=1} = \begin{bmatrix} \dfrac{e^2}{1+e^2} & 0 \\ 0 & 1 \end{bmatrix}$	$(P_1)'_\theta\big\|_{\theta=1} \overset{r}{=} \begin{bmatrix} 0 & 0 \\ 0 & 1 \end{bmatrix}$
"differentiated" Conventional Information Filter	$(P_1^{-1})'_\theta\big\|_{\theta=1} = -\begin{bmatrix} \dfrac{1+e^2}{e^2} & 0 \\ 0 & 1 \end{bmatrix}$	$(P_1^{-1})'_\theta\big\|_{\theta=1} \overset{r}{=} -\begin{bmatrix} \dfrac{1}{e^2} & 0 \\ 0 & 1 \end{bmatrix}$
Algorithm LLG-eSRIF	$\left(P_1^{-1/2}\right)'_\theta\big\|_{\theta=1} = -\dfrac{1}{2}\begin{bmatrix} \dfrac{\sqrt{1+e^2}}{e} & 0 \\ 0 & 1 \end{bmatrix}$	$\left(P_1^{-1/2}\right)'_\theta\big\|_{\theta=1} \overset{r}{=} -\dfrac{1}{2}\begin{bmatrix} \dfrac{1}{e} & 0 \\ 0 & 1 \end{bmatrix}$

V. Finally, compute the LLG according to (7).

Remark 1. Since, the matrices in (7) are triangular, only the diagonal elements of $R_{e,t}^{1/2}$ and $\dfrac{\partial\left(R_{e,t}^{-1/2}\right)}{\partial\theta_i}$ need to be computed. Hence, the Algorithm LLG-eSRIF allows the $m \times m$-matrix inversion of $R_{e,t}$ to be avoided in the evaluation of LLG.

3 Ill-Conditioned Example Problems and Comparison

To illustrate and compare the performance of the presented Algorithm LLG-eSRIF and the conventional approach, i.e. a straightforward differentiation of the KF ("differentiated" KF), two simple test problems have been constructed.

Problem 1. Given:

$$P_0 = \begin{bmatrix} \theta & 0 \\ 0 & \theta \end{bmatrix}, H = \begin{bmatrix} 1, 0 \end{bmatrix}, R = e^2\theta \quad \text{and} \quad F = I_2, \ Q = 0, \ G = \begin{bmatrix} 0, 0 \end{bmatrix}^T$$

where θ is an unknown parameter, I_2 is an identity 2×2 matrix, $0 < e << 1$; to simulate roundoff we assume $e + 1 \neq 1$ but $e^2 + 1 \overset{r}{=} 1$.

Calculate: $(P_1)'_\theta$ at the point $\theta = 1$.

For the $\theta = 1$, this example illustrates the initialization problems that result when $H_t P_t H_t^T + R_t$ is rounded to $H_t P_t H_t^T$ (see, for example, [9]). The exact answer and the rounded answer, using $e^2 + 1 \overset{r}{=} 1$ in all calculations, are given for the "differentiated" CKF, "differentiated" Conventional Information filter (CIF) and the new Algorithm LLG-eSRIF developed in this paper (see Table 1).

Table 2. Comparison of Rounded Solutions to Problem 2 evaluated at the point $\theta = 1$

Filter Implement.	Solution	
	Exact Answer	Rounded Answer
"diff." CKF	$(P_1)'_\theta\big|_{\theta=1} = \dfrac{1}{2+e^2}\begin{bmatrix} 1+e^2 & -1 \\ -1 & 1+e^2 \end{bmatrix}$	$(P_1)'_\theta\big|_{\theta=1} \stackrel{r}{=} \dfrac{1}{2}\begin{bmatrix} 1 & -1 \\ -1 & 1 \end{bmatrix}$
"diff." CIF	$(P_1^{-1})'_\theta\big|_{\theta=1} = -\dfrac{1}{e^2}\begin{bmatrix} 1+e^2 & 1 \\ 1 & 1+e^2 \end{bmatrix}$	$(P_1^{-1})'_\theta\big|_{\theta=1} \stackrel{r}{=} -\dfrac{1}{e^2}\begin{bmatrix} 1 & 1 \\ 1 & 1 \end{bmatrix}$
Algorithm LLG-eSRIF	$\left(P_1^{-1/2}\right)'_\theta\big|_{\theta=1} = -\dfrac{1}{2}\begin{bmatrix} \sqrt{\dfrac{2+e^2}{1+e^2}} & \dfrac{1}{e\sqrt{1+e^2}} \\ 0 & \dfrac{\sqrt{1+e^2}}{e} \end{bmatrix}$	$\stackrel{r}{=} -\dfrac{1}{2}\begin{bmatrix} \sqrt{2} & \dfrac{1}{e} \\ 0 & \dfrac{1}{e} \end{bmatrix}$

It can be seen that all algorithms except the "differentiated" CKF give the nonsingular result. A singular answer will lead to a zero gain if a second measurement of the same type is processed.

Problem 2. Given: same as Problem 1, but $H = \begin{bmatrix} 1, 1 \end{bmatrix}$.
      Find: $(P_1)'_\theta$ at the point $\theta = 1$.

The exact solution and the rounded one for Problem 2 are summarized in Table 2. With this more general type of ill-conditioning, only Algorithm LLG-eSRIF gives a nonsingular result.

4   Numerical Results

To substantiate the above theoretical result experimentally, we consider the example taken from [10]. Our simulation experiments are broken down into the following steps. Step 1 is planned to compute the negative LLF; in so doing, we compare our simulation results with those produced by the Conventional Information Filter (CIF). Step 2 is intended to compute the LLG; in this case, we compare the result generated by the Algorithm LLG-eSRIF with those produced by the "differentiated" CIF.

Example 1. Let the test system (1), (2) be defined as follows:

$$x_{t+1} = \begin{bmatrix} 1 & \Delta t \\ 0 & e^{-\Delta t/\tau} \end{bmatrix} x_t + \begin{bmatrix} 0 \\ 1 \end{bmatrix} w_t,$$

$$z_t = \begin{bmatrix} 1 & 0 \end{bmatrix} x_t + v_t$$

where τ is a parameter which needs to be estimated.

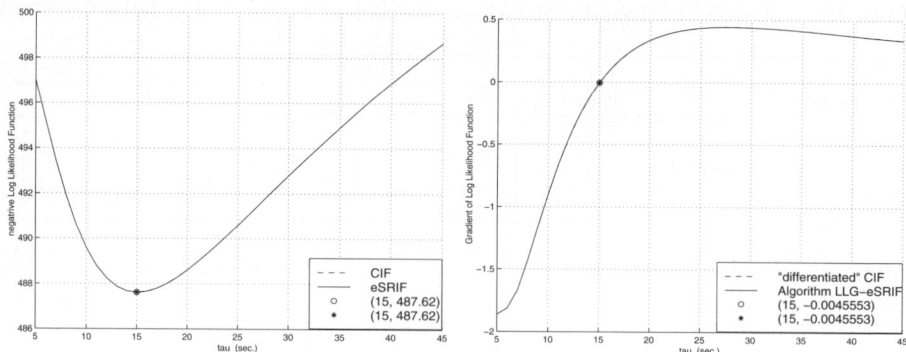

Fig. 1. The negative LLF and LLG calculated by the Conventional Information Filter, the extended Square-Root Information Filter, the "differentiated" Conventional Information Filter and Algorithm LLG-eSRIF, respectively.

The results of Step 1 and Step 2 are shown in Fig. 1. For the test problem, $\tau^* = 15$ was chosen as the true value of parameter τ. As can be seen, all two algorithms for evaluatinf the LLG produce exactly the same result and give the same zero point. Besides, it is readily seen that the estimate $\hat{\tau}_{\min}$ minimizing the negative LLF coincides with the estimate $\hat{\tau}_{\mathrm{grad}}$ at which the LLG is zero. Moreover, all estimates fit the true value of parameter τ, i.e. $\tau^* = 15$.

5   Conclusion

In this paper, the new algorithm for evaluating the Log Likelihood Gradient (score) of linear discrete-time dynamic systems has been developed. The necessary theory has been given and substantiated by the computational experiments. Two ill-conditioned example problems have been constructed to show the superior perfomance of the Algorithm LLG-eSRIF over the conventional approach.

References

1. G. C. Goodwin and R. L. Payne, *Dynamic System Identification: Experiment Design and Data Analysis.* New York: Academic, 1977.
2. R. L. Goodrich and P. E. Caines, Linear System Identification from Non-stationary Cross-sectional Data, *IEEE Trans. Automat. Contr.,* vol. AC-24, pp. 403–411, June 1979.
3. K. I. Yared, On Maximum Likelihood Identification of Linear State Space models, Mass. Inst. Technol., Cambridge, MA, *Rep. LIDS-TH-920,* 1979.
4. D. A. Wilson and A. Kumar, Derivative Computations for the Log Likelihood Function, *IEEE Trans. Automat. Contr.,* vol. AC-27, pp. 230–232, Feb. 1982.
5. M. Segal and E. Weinstein, A New Method for Evaluating the Log-Likelihood Gradient (Score) of Linear Dynamic Systems, *IEEE Trans. Automat. Contr.,* vol. AC-33, pp. 763–766, Aug. 1988.

6. M. Segal and E. Weinstein, A New Method for Evaluating the Log-Likelihood Gradient, the Hessian, and the Fischer Information Matrix for Linear Dynamic Systems, *IEEE Trans. Automat. Contr.*, vol. AC-35, pp. 682–687, May 1989.

7. R. P. Leland, A New Formula for the Log-Likelihood Gradient for Continuous-Time Stochastic Systems, *IEEE Trans. Automat. Contr.*, vol. AC-40, pp. 1295–1300, July 1995.

8. R. P. Leland, An Improved Log-Likelihood Gradient for Continuous-Time Stochastic Systems with Deterministic Input, *IEEE Trans. Automat. Contr.*, vol. AC-41, pp. 1207–1210, Aug. 1996.

9. P.G. Kaminski, A.E. Bryson, S.F. Schmidt, Discrete Square Root Filtering: A survey of Current Techniques. *IEEE Trans. on Aut. Cont.* **V. AC-16.**(6), P. 727–735, 1971.

10. G.J. Bierman, M.R. Belzer, J.S. Vandergraft, D.W. Porter, Maximum likelihood estimation using square root information filters. *IEEE Trans. on Autom. Contr.* **35**(12), P. 1293–1298, 1990.

11. P. Park, T. Kailath, New square-root algorithms for Kalman filtering. *IEEE Trans. on Autom. Contr.* **40**(5), P. 895–899, 1995.

12. M.S. Grewal, A.P. Andrews, *Kalman Filtering: Theory and Practice.* Prentice-Hall, Englewood Cliffs, New Jersey, 2001.

An Improved Algorithm for Sequence Pair Generation

Mingxu Huo and Koubao Ding[*]

Dept. of Information and Electronic Engineering, Zhejiang University,
Hangzhou 310027, P.R. China
`{huomingxu, dingkb}@zju.edu.cn`

Abstract. Sequence Pair is an elegant representation for block placement of IC design, and the procedure to generate the SP from an existing placement is necessary in most cases. An improved generation algorithm is proposed instead of the existing methods that are either difficult or inefficient to be implemented. The algorithm simplifies the definition of relation between blocks and avoids employing complicated graph operations. The time complexity of the algorithm is $O(n^2)$ and can be reduced to $O(n \log n)$, where n is the number of blocks. The experimental results of the algorithm show its superiority in running time.

1 Introduction

The floorplanning and block placement problems become increasingly important in physical design of Integrated Circuits (IC), where floorplanning can be regarded as placement with soft module blocks. Such problems are usually solved in two phases, i.e., the initial constructive phase, and iterative improvement one [1]. They are complex combinatorial optimization problems and most of their sub-problems are NP-Complete or NP-Hard [2]. Therefore heuristic approaches such as Simulated Annealing (SA) algorithm [3] are widely used to generate good layouts at the iterative stage, where representations of the placement is one of the crucial factors in evaluation of the costs.

In contrast with the so-called flat or absolute representations where the blocks are specified in terms of absolute coordinates on a plane without grids, a large number of representations of geometrical topological relations of blocks were proposed, e.g., slicing [4], mosaic [5], compacted [6] and P*-admissible representations [2, 7, 8, 9]. The P*-admissible representations can represent the most general floorplans and contain a complete structure for searching an optimal solution, among which the Sequence Pair (SP) is most favored and widely researched recently.

The SP related efforts with SA algorithms in the literature usually randomly generate an initial SP and then pack it to evaluate its cost. However, Force Directed Relaxation [10, 11] and other analytical algorithms that have been studied for a long time can construct better placements than the random ones. In addition, with the occurrence of incremental physical design [12] or Engineering Change Orders (ECO), it is imperative to find an effective and efficient approach to generate the SP from an arbitrary existing placement as an initial configuration.

---

[*] Corresponding author.

V.N. Alexandrov et al. (Eds.): ICCS 2006, Part I, LNCS 3991, pp. 482–489, 2006.

The original method to generate SP is "Gridding" [2], which is so complicated that it can hardly be implemented and the time complexity is assumed to be O (n^3), where n is the number of blocks. The generation procedure in Parquet [13] uses dynamic programming algorithm to find transitive closure graphs (TCG), which runs in O (n^2) time. In [14], Huo and Ding proposed a much faster generation method and an algorithm that runs much faster than the TCG algorithm from Parquet. As we know, the time complexity of traversing a graph is O (n^2) or O (n + e), where e is the number of edges, although it was reported (without proof or experimental results) by Kodama et. al. that the time complexity of the proposed "Fast-Gridding" algorithm by means of tracing a constraint graph in [15] is O (n log n).

In this paper, an improved algorithm to generate SP is proposed, which determines the position of each block on a generation plane based on the relations of every two blocks, and the time complexity is O (n^2). A faster O (n log n) algorithm is also proposed.

2   Sequence Pair

The topological relations of any two non-overlap module blocks are horizontal and vertical, i.e., left to, right to, above and below [16], as shown in Fig. 1. Diagonal relations can be simply degenerated by preferring horizontal relations to vertical ones, as in Fig. 1(c) where b_i is assumed to be left to b_j, unless there is a chain of vertical relations, which is considered as a vertical relation, as shown in Fig. 1(d). For example, in Fig. 2 block 8 is both left to and above block 3, but there is a vertical relation chain of them, i.e., block 8 is above block 6 and block 6 above block 3, so block 8 is said to be above block 3 instead of left to it. Such relations cannot be determined by just check two blocks, so that many procedures calculate indirect relations from the transitive closure of immediate relations, which is usually time-consuming.

An HV-Relation-Set (HVRS) for a set of blocks is a set of horizontal or vertical relations for all block pairs [17], and a Feasible-HVRS involves all the relations of blocks excluding non-realizable relations, e.g. {b_i is left to b_j, b_j is left to b_k, b_k is left to b_i} is not a Feasible-HVRS for blocks b_i, b_j and b_k.

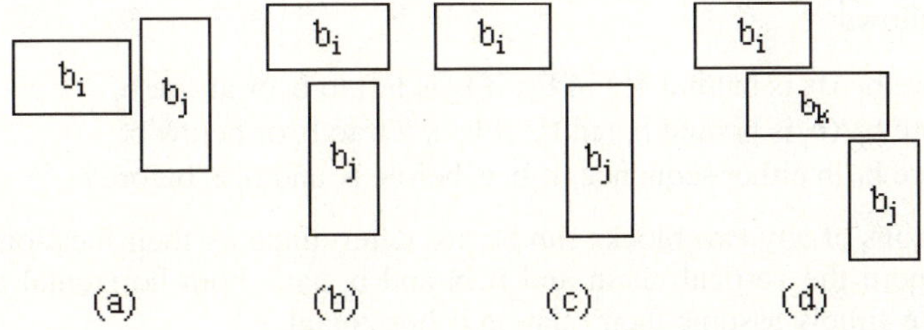

Fig. 1. Possible relations between two blocks b_i and b_j. (a) b_i is left to b_j; (b) b_i is above b_j; (c) b_i is diagonal to b_j and assumed to be left to b_j; (d) b_i is diagonal to b_j but a vertical relation chain exists between b_i and b_j, i.e., b_i is above b_k and b_k is above b_j, so b_i is above b_j.

A Sequence Pair (SP) is an ordered pair of sequences Γ_+ and Γ_-, each of which is a sequence of n block names [2], e.g., the SP from Fig. 2 is $(\Gamma_+, \Gamma_-) = (154806273, 012453687)$. A sequence pair corresponds to a Feasible-HVRS as follows:

- b_i is left to b_j (b_j is right to b_i): if $(\Gamma_+, \Gamma_-) = (\ldots b_i \ldots b_j \ldots, \ldots b_i \ldots b_j \ldots)$
- b_i is below b_j (b_j is above b_i): if $(\Gamma_+, \Gamma_-) = (\ldots b_j \ldots b_i \ldots, \ldots b_i \ldots b_j \ldots)$

The time complexity of the original evaluation procedure "Packing" is $O(n^2)$ [2], and it was later sped up to $O(n \log \log n)$ [18].

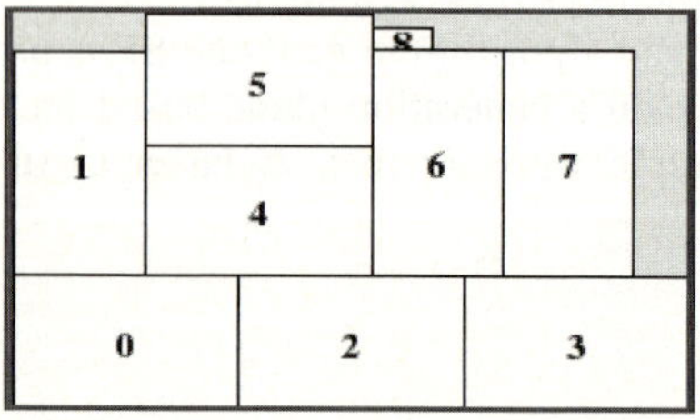

Fig. 2. A placement of the MCNC apte benchmark with SP (1 5 4 8 0 6 2 7 3, 0 1 2 4 5 3 6 8 7)

3   Generation Algorithms

3.1   Algorithm Embedding

According to the generation method Embedding from [14], algorithm Embedding can be proposed naturally. To determine the Feasible-HVRS of n blocks, each block needs to be compared with every other n-1 blocks, which implies the $O(n^2)$ time complexity. However, not all the comparisons are necessary and some of them can be removed to accelerate the procedure, e.g., if the information of b_i above b_j and b_j above b_k is acquired, b_i is absolutely above b_k.

Resolve relations and ordering criteria in sequences
The orders of blocks in both sequences can be determined from relations of blocks b_i and b_j, as follows:

- b_i is before b_j (b_j is behind b_i) in Γ_+: if b_i is left to b_j or above b_j
- b_i is before b_j (b_j is behind b_i) in Γ_-: if b_i is left to b_j or below b_j
- b_i is before b_k in either sequence: if b_i is before b_j and b_j is before b_k.

The relations of any two blocks can be just determined by their locations and sizes. Here we ignore the vertical chain and if b_i and b_j have both horizontal and vertical relations. We simply assume their relation is horizontal.

Sort the blocks according to their x coordinates
The order of blocks to be embedded must be determined, in order to eliminate ambiguities and achieve a correct result. In this algorithm, we sort the blocks

according to their x coordinates with introspective sorting and selection algorithms that performs O (n log n) time complexity in worst cases as discussed in [19].

Insert block names into linked lists according to the sorted order

Two linked lists listx and listy are used to represent the Γ_+ and Γ_- respectively, which are generated separately by inserting the block names. To insert block b_i into a list, we traverse the list from its header, check the relations of b_i with every visited block b_j, and stop to insert b_i before b_j, if it is found that b_j should be behind b_i in the list. However, if no such block exists, b_i is appended to list because it should be current list tail.

Program section of Embedding for in C++ (only for listx)

```
sort_blocks(xx) // sort by x coordinates
listx.push_back(xx[0]);// insert the first block
for(idx=1; idx<n; ++idx ) {
  i = xx[idx]; //get the block name
  for(pos=listx.begin();pos!=listx.end();++pos) {
    j = *pos;
    r = relation(i,j);
    if(r == LEFT || r == ABOVE) { // i precedes j
      listx.insert(pos, i);// insert i before j
      break;
    }
  }
  if(pos == listx.end()){ // i is currently list tail
    listx.push_back(i);
  }
}
```

The procedure "relation" in the algorithm compares block b_i and block b_j only, so it runs for constant time. The time complexity of the outer for loop is O (n), so is the inner one. And thus the total time complexity of algorithm Embedding is O (n^2).

Sometimes a placement may correspond to multiple sequence pairs, e.g., the placement in Fig. 2 has another sequence pair (1 5 4 8 0 6 7 2 3, 0 1 2 3 4 5 6 7 8) as well as the one discussed in this paper. We pack the generated SP again and check whether the locations of all the blocks in the new placement match those in the old one, in order to validate the result.

A sequence pair obtained by Embedding imposes horizontal relation on any pair of blocks only when the relation of the pair cannot be coded as vertical.

Lemma. **Embedding is correct when vertical relation chains exists.**

Proof: Suppose the relation of b_i and b_j is diagonal and there exists a chain of vertical relations between them. Without loss of generality, b_i is supposed to be left to and above b_j, and only one block b_k is in the chain, i.e., b_i is above b_k and b_k is above b_j, as illustrated in Fig. 1 (d). If the relation of b_i and b_j is checked first, the incorrect result will be returned, i.e., b_i is left to b_j, according to the ordering criteria. Now that the relation of x coordinates of the three blocks is $x_i < x_k < x_j$, and the order to embed blocks is according to their x coordinates, b_k is embedded in advance of b_j and the correct vertical relation of b_i and b_k is obtained first. Accordingly, the correct vertical relation of b_k and b_j is obtained later, so the relation of b_i and b_j is implicated correctly, i.e., a vertical one, instead of being mistakenly resolved.

***Theorem.* The Sequence Pair generated by algorithm Embedding is correct.**

Proof: If no diagonal relations exist, the relation of every two blocks is unique and the orders of blocks in both sequences can be correctly determined. If diagonal relations exist but there are no chains of vertical relations, since we regard such diagonal relations as horizontal ones, the relation of every pair of blocks is also unique and the result is correct. The last case is there are both diagonal relations and vertical relation chains in a placement. According to the preceding Lemma, the algorithm is correct too. Since the relations of every pair of blocks are determined without ambiguity, the order in either sequence of SP is correct.

3.2   Reducing Running Time to O (n log n)

Since both Γ_+ and Γ_- are ordered sequences, we can use balanced binary search trees to store them. Make use of two binary trees treex and treey to represent Γ_+ and Γ_- respectively, and generate treex and treey separately by inserting the blocks into the trees. We name this algorithm LogAlgo.

Program section of LogAlgo algorithm in C++ (for treex only)

```
sort_blocks(xx)
//ltspx is a function object
Ltlogsp ltspx(LEFT, ABOVE);
//treex is a binary tree with
//ordering criteria ltspx
Btree treex(ltspx);
for(idx=0; idx<n; ++idx ) {
  i = xx[idx].idx;
  treex.insert(i);
}
```

Class Ltlogsp invokes procedure relation (i, j) that functions the same with the one of algorithm Embedding when a comparison of two blocks b_i and b_j is needed. The insertion operation of Btree includes searching on a binary tree and keeping it balanced, whose time complexity is O (log n). There are n blocks to be inserted to the tree in the loop, so the total time complexity is now reduced to O (n log n) from O (n^2).

3.3   Experimental Results

After sorting, the order of blocks of the placement in Fig. 1 to be embedded is: 0 1 4 5 2 6 8 3 7. Γ_+ and Γ_- can be generated separately and simultaneously. Table 1 shows the running process of the algorithm Embedding to generate the listx from the placement in Fig. 1 (the results for listy and for other benchmarks is omitted), and the number of comparisons is 34. Table 2 shows the process of algorithm LogAlgo, with 20 comparisons. Fig. 3 demonstrates the balanced binary tree treex when block 3 and block 7 is inserted respectively.

Table 1. Run of the algorithm after inserted block 0 in list listx. A, B, L and R stand for relations between two blocks of above, below, left to and right to respectively. List listx is the list before (without block name in parenthesis) and after intertion of current block.

#	Comparisons	ListX	#	Comparisons	ListX
1	A0	(1) 0	**6**	R1, R5, R4, R0, A2	1 5 4 0 (6) 2
4	R1, A0	1 (4) 0	**8**	R1, R5, R4, R0, A6	1 5 4 0 (8) 6 2
5	R1, A4	1 (5) 4 0	**3**	R1, R5, R4, R0, R8, B6, R2	1 5 4 0 8 6 2 (3)
2	R1, B5, B4, R0	1 5 4 0 (2)	**7**	R1, R5, R4, R0, R8, R6, R2, A3	1 5 4 0 8 6 2 (7) 3

Table 2. Run of the algorithm after inserted block 0 in Btree treex. The inorder traversal results of treex are listed in column TreeX.

#	Comparisons	TreeX	#	Comparisons	TreeX
1	A0	(1) 0	**6**	R4, R0, A2	1 5 4 0 (6) 2
4	A0, R1	1 (4) 0	**8**	R4, A6, R0	1 5 4 0 (8) 6 2
5	A4, R1	1 (5) 4 0	**3**	R4, B6, R2	1 5 4 0 8 6 2 (3)
2	B4, R0	1 5 4 0 (2)	**7**	R4, R6, R2, A3	1 5 4 0 8 6 2 (7) 3

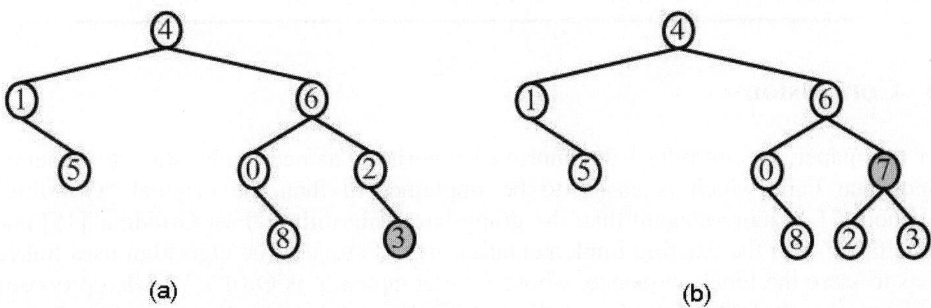

Fig. 3. the Btree treex (a) after block 3 and (b) block 7 was inserted, respectively. The newly inserted node is in shade. Please note that after block 7 was inserted, the tree rotated to keep balanced.

Experiments on all the MCNC benchmark[1] circuits and the some of the GSRC benchmark[2] circuits are performed in comparison with the generation routine from Parquet [13] on running time. We first generate the placements of the benchmarks with the Simulated Annealing algorithm using the LCS evaluation algorithm [18], and save the placements and their corresponding sequence pairs. Then we take the placements as input and generate their sequence pairs as output with the programs respectively. The accumulative running time of the generation 1000 times are listed in

[1] See: http://www.cse.ucsc.edu/research/surf/GSRC/MCNCbench.html
[2] See: http://www.cse.ucsc.edu/research/surf/GSRC/GSRCbench.html

Table 3. It can be seen that our algorithms runs much faster than the routine from Parquet, and the more complex the placement is, the more speed improvement our program achieves. In addition, as the number of blocks grows larger than 33, the LogAlgo algorithm with O (n log n) time complexity outperforms the Embedding evidently. When there are less than 33 blocks, additional time is needed for the LogAlgo algorithm to keep the tree balanced.

Table 3. Running time in seconds of our programs and Paquet on MCNC and GSRC benchmarks

Benchmark	*block #*	*Parquet*	*Embedding*	*LogAlgo*
apte	9	0.0580	0.0100	0.0120
xerox	10	0.0740	0.0110	0.0130
hp	11	0.0840	0.0120	0.0130
ami33	33	0.7369	0.0470	0.0440
ami49	49	1.8127	0.0780	0.0670
n50a	50	1.7257	0.0830	0.0670
n100a	100	7.4579	0.2570	0.1380
n200a	200	35.8016	0.8989	0.2940
n300a	300	92.4549	1.9057	0.4349

4   Conclusions

In this paper, we introduce an improved algorithm named Embedding to generate Sequence Pair, which is easier to be implemented than the original "Gridding" method [2], is more elegant than the graph-based algorithm "Fast-Gridding"[15] and runs faster than the existing implementation from Parquet. The algorithm uses linked lists to store the block sequence, whose time complexity is O (n^2). We also proposed an algorithm LogAlgo based on balanced binary search trees that costs O (n log n) running time. When a placement is composed with more than 33 blocks the LogAlgo reveals its superiority.

References

1. Goto, S., An efficient algorithm for the two-dimensional placement problem in electrical circuit layout,IEEE Trans. On Circuits and Systems, (1998), 28(1): 12 - 18
2. Murata, H., et al., Rectangle-packing-based module placement. 1995 IEEE/ACM ICCAD. Digest of Technical Papers, (1995):. 472-479.
3. Kirkpatrick, S., C.D. Gelatt, and M.P. Vecchi, optimization by simulated annealing. Science, 1983. 220(4598): (1983) 671-680.
4. Otten, R.H.J.M., Automatic floorplan design. ACM IEEE 9th DAC Proceedings, (1982): 261-267.

5. Hong, X.L., et al., A non-slicing floorplanning algorithm using corner block list topological representation, in 2000 IEEE Asia-Pacific Conference On Circuits And Systems - Electronic Communication Systems. (2000). 833-836.
6. Pei-Ning, G., C. Chung-Kuan, and T. Yoshimura, An O-tree representation of non-slicing floorplan and its applications. Proceedings 1999 DAC, (1999): 268-273.
7. Nakatake, S., et al., Module packing based on the BSG-structure and IC layout applications. IEEE TCAD Of Integrated Circuits And Systems, (1998). 17(6): 519-530.
8. Jai-Ming, L. and C. Yao-Wen, TCG: a transitive closure graph-based representation for non-slicing floorplans. Proceedings of the 38th DAC, (2001): 764-769.
9. Lin, J.M. and Y.W. Chang, TCG-S: Orthogonal coupling of P*-admissible representations for general floorplans, in 39th DAC, Proceedings. (2002). 842-847.
10. M. Sarrafzadeh and M. Wang, NRG: Global and detailed placement, Proc. IEEE ICCAD , 1997. (1997):532–537
11. N. Quinn, and M. Breuer, A force-directed component placement procedure for printed circuit boards, IEEE Trans. CAS, vol.CAS-26, no.1979(6):.377-388
12. Cong, J. and M. Sarrafzadeh, Incremental physical design. Proceedings International Symposium on Physical Design, 2000. ISPD-2000, (2000): 84-92.
13. Adya, S.N. and I.L. Markov, Fixed-outline floorplanning through better local search. Proceedings 2001 IEEE ICCD: VLSI in Computers and Processors. 2001, (2001): 328-334.
14. Huo, MX and Ding, KB, A quick generation method of sequence pair for block placement, COMPUTATIONAL SCIENCE - ICCS 2005, PT 3 3516: 954-957
15. Kodama, C., Fujiyoshi, K. and Koga, T., A novel encoding method into sequence-pair,. Proceedings ISCAS 2004, 2004 (5): 329 – 332
16. Onodera, H., Y. Taniguchi, and K. Tamaru, Branch-and-bound placement for building block layout. 28th ACM/IEEE DAC. Proceedings 1991, (1991): 433-439.
17. Murata, H., et al., A mapping from sequence-pair to rectangular dissection. Proceedings of the ASP-DAC '97. 1997, (1997): 625-633.
18. Xiaoping, T. and D.F. Wong, FAST-SP: a fast algorithm for block placement based on sequence pair. Proceedings of the ASP-DAC 2001. 2001, (2001): 521-526.
19. D. R. Musser, "Introspective Sorting and Selection Algorithms", Software Practice and Experience, 1997, 27(8):983

Implicit Constraint Enforcement for Rigid Body Dynamic Simulation

Min Hong[1], Samuel Welch[2], John Trapp[2], and Min-Hyung Choi[3]

[1] Division of Computer Science and Engineering, Soonchunhyang University,
646 Eupnae-ri Shinchang-myeon Asan-si, Chungcheongnam-do, 336-745, Korea
`Min.Hong@UCHSC.edu`
[2] Department of Mechanical Engineering, University of Colorado at Denver and Health
Sciences Center, Campus Box 109, PO Box 173364, Denver, CO 80217, USA
`{Sam, jtrapp}@carbon.cudenver.edu`
[3] Department of Computer Science and Engineering, University of Colorado at Denver and
Health Sciences Center, Campus Box 109, PO Box 173364, Denver, CO 80217, USA
`Min-Hyung.Choi@cudenver.edu`

Abstract. The paper presents a simple, robust, and effective constraint enforcement scheme for rigid body dynamic simulation. The constraint enforcement scheme treats the constraint equations implicitly providing stability as well as accuracy in constrained dynamic problems. The method does not require ad-hoc problem dependent parameters. We describe the formulation of implicit constraint enforcement for both holonomic and non-holonomic cases in rigid body simulation. A first order version of the method is compared to a first order version of the well-known Baumgarte stabilization.

1 Introduction

Rigid body simulation is highly important in the modeling of various physical systems and has been comprehensively studied in electrical and mechanical engineering. Rigid body dynamics involves a variety of holonomic constraints such as ball-and-socket, hinges, sliders, universal joints, and contact constraints. In addition, non-holonomic constraints have been used in simulations of robots and cars. In rigid body dynamics, constraint enforcement is critical to guarantee successful simulations since small but accumulating constraint drift could cause instabilities. Constraint satisfaction through the use of Lagrange multipliers represents a challenging problem numerically as the resulting system of equations is a mixed ODE algebraic system. Early workers solved this problem by differenting the constraint equations replacing the algebraic equation with an ODE but the solutions exhibited numerical drift in the constraint error. Baumgarte's stabilization method [1] was introduced to reduce this numerical drift using parameter dependent stabilization terms and is still widely used today due to it's simplicity.

In this paper, we propose an implicit holonomic and non-holonomic constraint enforcement method for rigid body dynamics that provides numerical stability and accuracy without requiring ad-hoc stabilization parameter terms while providing the same asymptotic computational cost as the Baumgarte method. The approach is to implicitly expand the algebraic constraint in a Taylor series to the same order as the order of

V.N. Alexandrov et al. (Eds.): ICCS 2006, Part I, LNCS 3991, pp. 490–497, 2006.

the numerical method used to integrate the ODE system. The Taylor expansion is implicit in the state variables thus the method has desirable stability characteristics.

2 Related Work

The seminal work of Baumgarte [1] included second order feedback terms to stabilize constraints and has been used successfully for many applications [2, 3, 4, 6]. Numerous improved methods have appeared over the decades since Baumgarte presented his method but Baumgarte's stabilization is widely used due to its simplicity and the ease with which it is understood. The main drawback to Baumgarte stabilization is that the required parameters must be chosen in an *ad-hoc* manner. Another approach is post-stabilization in which a correction is made to the state variables in order to minimize the constraint error. An example of this is Cline and Pai [5] who proposed a post-stabilization approach for rigid body simulation. Post-stabilization requires additional computational load to reinstate the accuracy in the constraint and it can cause error in the motion of objects under complicated situations since the constraint drift reduction is performed independently from the conforming dynamic motions [7].

3 Implicit Holonomic Constraint Enforcement

In this section a simple implicit constraint enforcement method is presented and compared to Baumgarte's stabilization. It will be shown that a certain selection of Baumgarte parameters will result in the two methods being nearly identical for the holonomic case. A stability analysis will be presented and used as a guide to select parameters for Baumgarte's stabilization technique. Let Φ be the vector of holonomic constraints. The equations of motion for the holonomically constrained system are written as

$$M\ddot{r} + \Phi_r^T \lambda = F \tag{1}$$

$$J\dot{\omega} + \Phi_\pi^T \lambda = T - \tilde{\omega}J\omega \tag{2}$$

$$\Phi(r,e) = 0 \tag{3}$$

$$\dot{e} = \frac{1}{2}G(e)\omega \tag{4}$$

Where r are the center of mass coordinates, e are the Euler parameters associated with the orientation of the rigid bodies, Φ_r and Φ_π are the constraint jacobians associated with r and the orientation, respectively and λ is a vector containing the Lagrange multipliers. M and J are inertia matrices and $\tilde{\omega}$ is the skew-symmetric matrix associated with the angular velocity ω. The matrix defining the time derivative of the Euler parameters and the angular velocity is given by:

$$G(e) = \left\{ \begin{array}{cccc} -e1 & e0 & e3 & -e2 \\ -e2 & -e3 & e0 & e1 \\ -e3 & e2 & -e1 & e0 \end{array} \right\} \tag{5}$$

492   M. Hong et al.

Equations (1), (2) and (4) are discretized to first order as:

$$\dot{r}(t+\Delta t) = \dot{r}(t) + \Delta t M^{-1}F - \Delta t M^{-1}\Phi_r^T \lambda \tag{6}$$

$$r(t+\Delta t) = r(t) + \Delta t \dot{r}(t+\Delta t) \tag{7}$$

$$\omega(t+\Delta t) = \omega(t) + \Delta t J^{-1}T - \Delta t J^{-1}\tilde{\omega}(t)J\omega(t) - \Delta t J^{-1}\Phi_\pi^T \lambda \tag{8}$$

$$e(t+\Delta t) = e(t) + \frac{\Delta t}{2}G(e)\omega(t+\Delta t) \tag{9}$$

The basic philosophy of the implicit constraint enforcement technique we use is to expand the algebraic constraint function in a Taylor series to the same order as the discrete equations. The resulting expansion is:

$$\Phi(t+\Delta t) = \Phi(t) + \left\{ \Phi_r \dot{r}(t+\Delta t) + \Phi_\pi \omega(t+\Delta t) \right\} \Delta t = 0 \tag{10}$$

Equations (6) through (8) may be used to eliminate the new time velocity terms to obtain the linear system that must be solved for the Lagrange multiplier:

$$\left\{ \Phi_r M^{-1}\Phi_r^T + \Phi_\pi J^{-1}\Phi_\pi^T \right\} \lambda = \frac{\Phi}{\Delta t^2} + \frac{1}{\Delta t}\left(\Phi_r \dot{r} + \Phi_\pi \tilde{\omega} \right) + \Phi_r M^{-1}F + \Phi_\pi J^{-1}(T - \tilde{\omega}J\tilde{\omega}) \tag{11}$$

In this equation and in all equations that follow terms without explicit time dependence indicated are evaluated at old time. Once this system is solved for the Lagrange multipliers equations (6) through (9) are used to update the other variables. Note that this linear system is symmetric positive definite. Baumgarte's stabilization technique may be used for this system in the following way. First the constraint is stabilized by adding two terms to the second derivative with parameters α and β as follows:

$$\ddot{\Phi} + 2\alpha\dot{\Phi} + \beta^2\Phi = 0 \tag{12}$$

A similar procedure is used to obtain the linear system to be solved for the Lagrange multipliers:

$$\left\{ \Phi_r M^{-1}\Phi_r^T + \Phi_\pi J^{-1}\Phi_\pi^T \right\} \lambda = \beta^2\Phi + 2\alpha\left(\Phi_r \dot{r} + \Phi_\pi \tilde{\omega} \right) + \Phi_r M^{-1}F$$
$$+ \Phi_\pi J^{-1}(T - \tilde{\omega}J\tilde{\omega}) + \left(\Phi_r \dot{r} \right)_r \dot{r} + \left(\Phi_\pi \tilde{\omega} \right)_\pi \tilde{\omega} \tag{13}$$

As with the simple implicit method, once this system is solved for the Lagrange multipliers equations (6) through (9) are used to update the other variables. Baumgarte stabilization is often criticized because selection of the parameters α and β is *ad-hoc* and problem dependent. Greenwood [9] suggests that both parameters should be proportional to the time step and a cursory examination of equations (12) and (10) indicates that the choices $\alpha = 0.5\Delta t^{-1}$ and $\beta = \Delta t^{-1}$ would result in our method and Baumgarte's stabilization differing by only the last two terms in equation (12). In what follows we analyze the stability of these schemes and use the result to guide us in a selection for the parameters α and β. In order to analyze the stability the schemes described above we lump the center of gravity and orientation variables into a single vector q and we consider a linearized system with linearized constraints with constraint gradient given by

$$\Phi_q(q) = B \tag{14}$$

The dynamic system will have only constraint forces as we wish to isolate the effect of constraint enforcement on stability. The discretized homogeneous equations of motion may be written as:

$$\dot{q}(t+\Delta t) = \dot{q}(t) - \Delta t M^{-1} B^T \lambda \tag{15}$$

$$q(t+\Delta t) = q(t) + \Delta t \dot{q}(t+\Delta t) \tag{16}$$

Note that the mass matrix and inertia matrix are now lumped together. The linear system that must be solved for the Lagrange multiplier becomes.

$$B\,M^{-1}B^T\lambda = \frac{1}{\Delta t^2} Bq(t) + \frac{1}{\Delta t} B\dot{q}(t) \tag{17}$$

The Lagrange multipliers may now be eliminated and the resulting linear state-space form of the equations of motion is

$$\begin{Bmatrix} q(t + \Delta t) \\ \dot{q}(t + \Delta t) \end{Bmatrix} = X \begin{Bmatrix} q(t) \\ \dot{q}(t) \end{Bmatrix} \tag{18}$$

where

$$X = \begin{bmatrix} I - M^{-1}\,B^T\,A^{-1}\,B & \Delta t\left(I - M^{-1}\,B^T\,A^{-1}B\right) \\ \dfrac{-1}{\Delta t}M^{-1}\,B^T\,A^{-1}\,B & I - M^{-1}\,B^T\,A^{-1}B \end{bmatrix} \tag{19}$$

where $A = B\,M^{-1}B^T$. The condition for stability of this system of equations is

$$\rho(X) \le 1 \tag{20}$$

where $\rho(X)$ is the spectral radius of the matrix X. A similar result may be derived for the Baumgarte method with the result

$$\begin{Bmatrix} q(t + \Delta t) \\ \dot{q}(t + \Delta t) \end{Bmatrix} = \Lambda \begin{Bmatrix} q(t) \\ \dot{q}(t) \end{Bmatrix} \tag{21}$$

Where

$$\Lambda = \begin{bmatrix} I - \Delta t^2 \beta^2 M^{-1}\,B^T\,A^{-1}\,B & \Delta t\left(I - 2\Delta t\alpha M^{-1}\,B^T\,A^{-1}B\right) \\ -\Delta t \beta^2 M^{-1}\,B^T\,A^{-1}\,B & I - 2\Delta t\alpha M^{-1}\,B^T\,A^{-1}B \end{bmatrix} \tag{22}$$

The condition for stability for the Baumgarte method is

$$\rho(\Lambda) \le 1 \tag{23}$$

Note that the choices $\alpha = 0.5\Delta t^{-1}$ and $\beta = \Delta t^{-1}$ would result in the Baumgarte stabilization method having the same spectral radius as our implicit method as the differing terms were discarded in the linearization. In order to investigate the stability of the methods further we consider the simple problem of a planar double pendulum with links modeled by point masses. The equations of motion are written using Cartesian coordinates thus we have two constraints that force the link lengths to be constant. The link lengths considered are unity as are the point masses. The constraints are linearized about a co-linear position and the resulting constraint jacobian for this linearized system is

$$\mathbf{B} = \begin{bmatrix} 2 & 0 & 0 & 0 \\ -2 & 0 & 2 & 0 \end{bmatrix}$$

(24)

For this case, $\rho(X) \leq 1$ for all Δt thus the implicit constraint enforcement method is unconditionally stable. The stability behavior of the Baumgarte's stabilization depends on the selection of the parameters α and β. For example, if we select $\alpha, \beta = 10$ the resulting stability limit is $\Delta t \leq 0.83$. If we select parameters $\alpha = \Delta t^{-1}$ and $\beta = \Delta t^{-1}$ which give us a critically damped response with time constant proportional to Δt we will find that the system is unstable. Therein lies the difficulty in parameter selection for Baumgarte's stabilization technique. Improper choices for the parameters α and β often results in time step limitations due to stability or in the time constants associated with constraint satisfaction being too slow. One of the advantages of the simple implicit constraint enforcement scheme described in this paper is that the search for a good choice of parameters need not be undertaken and, as we will show, the method produces accurate results. For the comparisons in the rest of this paper we will use the parameter selections $\alpha = 0.5\Delta t^{-1}$ and $\beta = \Delta t^{-1}$ for the reasons cited above.

We note that these parameter selections specify a damping coefficient of 0.5 and a time constant on the order of the time step for constraint satisfaction.

4 Implicit Non-holonomic Constraints Enforcement

Non-holonomic constraints typically restrict velocities for points on objects. In the development that follows the implicit constraint enforcement technique is developed using functional dependences on the center of gravity coordinates only. Extension to dependence on orientation coordinates follows identical development. The equations of motion for the non-holonomic constraint can be written as

$$^{N}\Phi_{r}(r)\dot{r} = v_{o}$$

(25)

Following our earlier philosophy the constraint written at new time is

$$^{N}\Phi_{r}(r(t+\Delta t))\,\dot{r}(t+\Delta t) = v_{o}(t+\Delta t)$$

(26)

Next a Taylor series expansion is written to the same order as the discrete equations:

$$^{N}\Phi_{r}(r(t+\Delta t))\,\dot{r}(t+\Delta t) = {}^{N}\Phi_{r}(r(t))\,\dot{r}(t) + \left({}^{N}\Phi_{r}(r(t))\,\dot{r}(t)\right)_{r}(r(t+\Delta t) - r(t))$$
$$+ \left({}^{N}\Phi_{r}(r(t))\,\dot{r}(t)\right)_{\dot{r}}(\dot{r}(t+\Delta t) - \dot{r}(t))$$

(27)

where the subscript r indicates partial differentiation with respect to r. After simplification we have:

$$^{N}\Phi_{r}(r(t+\Delta t))\,\dot{r}(t+\Delta t)=\left(^{N}\Phi_{r}(r(t))\,\dot{r}(t)\right)_{r}(r(t+\Delta t)-r(t))+\,^{N}\Phi_{r}(r(t))\,\dot{r}(t+\Delta t) \quad (28)$$

Equation (7) may be used to eliminate $r(t+\Delta t)$ to obtain

$$^{N}\Phi_{r}(r(t+\Delta t))\,\dot{r}(t+\Delta t)=\Delta t\left(^{N}\Phi_{r}(r(t))\,\dot{r}(t)\right)_{r}\dot{r}(t+\Delta t)+\,^{N}\Phi_{r}(r(t))\,\dot{r}(t+\Delta t) \quad (29)$$

The first term may be written to second order replacing $\dot{r}(t+\Delta t)$ with $\dot{r}(t)$ resulting in

$$^{N}\Phi_{r}(r(t+\Delta t))\,\dot{r}(t+\Delta t)=\Delta t\left(^{N}\Phi_{r}(r(t))\,\dot{r}(t)\right)_{r}\dot{r}(t)+\,^{N}\Phi_{r}(r(t))\,\dot{r}(t+\Delta t) \quad (30)$$

Equation (6) may be used to eliminate $\dot{r}(t+\Delta t)$ and the result is substituted into Equation (32) to obtain the linear system that must be solved for the Lagrange multiplier:

$$^{N}\Phi_{r}M^{-1}\,^{N}\Phi_{r}^{T}\lambda=\frac{1}{\Delta t}\,^{N}\Phi_{r}\dot{r}+\,^{N}\Phi_{r}M^{-1}F+\left(^{N}\Phi_{r}\dot{r}\right)_{r}\dot{r}-\frac{v_{o}(t+\Delta t)}{\Delta t} \quad (31)$$

Note that we obtain the desirable symmetric positive definite linear system as we did with the holonomic case. The development for the case with orientation variable dependence of the constraints is identical. Once this system is solved for the Lagrange multipliers equations (6) through (9) are used to update the other variables. Baumgarte's stabilization for the non-holonomic case begins with the constraint function written as $g(t) = 0$. Baumgarte's stabilization may be expressed as (Greenwood [8]).

$$\dot{g} + 2\alpha g + \beta^{2}\int_{o}^{t}gdt = 0 \quad (32)$$

For our case we write

$$g(t)=\,^{N}\Phi_{r}(r)\dot{r}-v_{o}=0 \quad (33)$$

and Baumgarte's stabilization becomes

$$^{N}\Phi_{r}M^{-1}\,^{N}\Phi_{r}^{T}\lambda=-\dot{v}_{o}+\left(^{N}\Phi_{r}\dot{r}\right)_{r}\dot{r}+\,^{N}\Phi_{r}M^{-1}F+2\alpha\left(^{N}\Phi_{r}\dot{r}-v_{o}\right)+\beta^{2}\int_{0}^{t}\left(^{N}\Phi_{r}\dot{r}-v_{o}\right)dt \quad (34)$$

Note again the appearance of extra terms in when using Baumgarte's stabilization and in particular the appearance of the integral that must be calculated. In the results that follow we evaluate the integral with a first-order method consistent with the method used to integrate the ODE's.

5 Experiments

To verify the accuracy of the proposed holonomic and non-holonomic implicit constraint enforcement, we performed a three-link rigid bar simulation in Fig. 1. Three separate rigid bars are connected by three spherical joints. The non-holonomic

496   M. Hong et al.

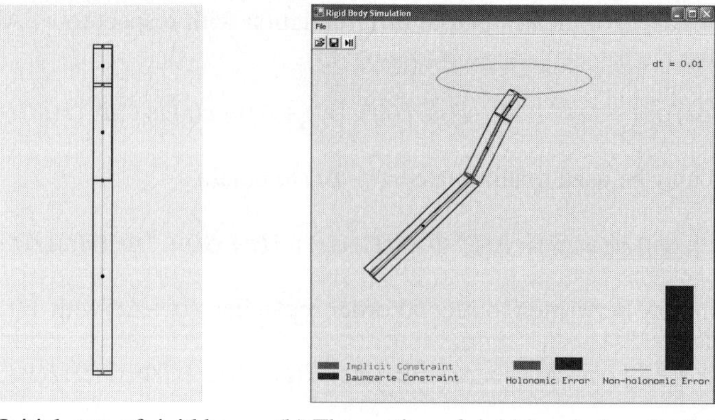

(a) Initial state of rigid bar   (b) The motion of rigid-bar during simulation

Fig. 1. The three-link rigid bar simulation with holonomic and non-holonomic constraints. The green circle indicates the trajectory of the point where the non-holonomic constraint is applied.

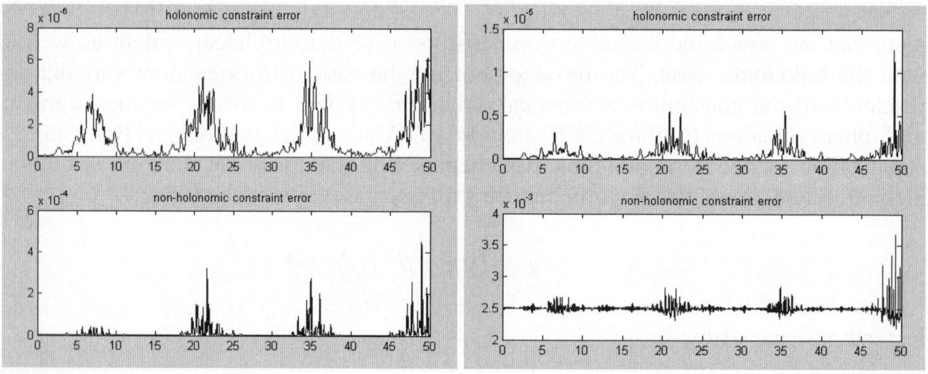

(a) Implicit constraint error using $\Delta t = 0.01$ (b) Baumgarte constraint error using $\Delta t = 0.01$

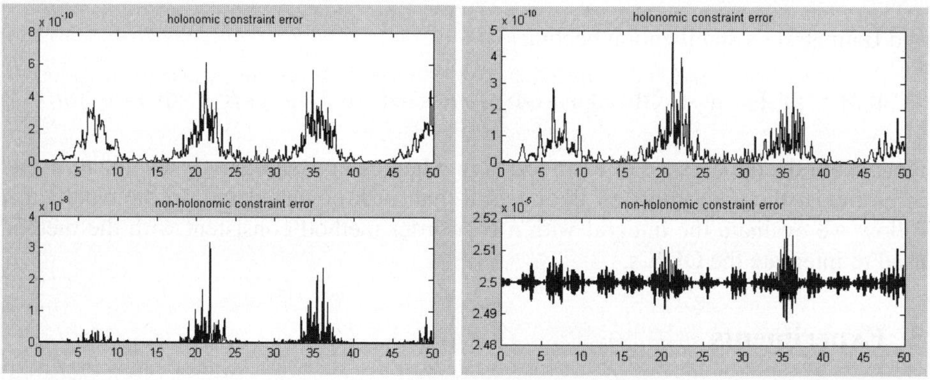

(c) Implicit constraint error using $\Delta t = 0.001$ (d) Baumgarte constraint error using $\Delta t = 0.001$

Fig. 2. Comparison of holonomic and non-holonomic constraint errors for three-link rigid bar simulation

constraint is applied on the top end of the first link. The point in the end of this link has a circular velocity prescribed. The constraint is integrable but we treat it as a non-holonomic constraint for illustration purposes. The prescribed velocity results in the circular movement of the end point of the link in the x-y plane.

We recorded the holonomic and non-holonomic constraint errors over a time duration of 50 seconds and the results are shown in Fig 2. The Baumgarte stabilization parameters used are $\alpha = 0.5\Delta t^{-1}$ and $\beta = \Delta t^{-1}$. The results shown indicate somewhat better results for the non-holonomic constraint errors with our method and similar results for the holonomic constraint errors. These results are typical of what we have observed and we note that if we do not include the integral in the Baumgarte method the results are less accurate.

6 Conclusion

This paper describes an implicit constraint enforcement method for rigid body simulations with both holonomic and non-holonomic constraints. Our implicit constraint enforcement method has several advantages. This method is implicit thus it has desirable stability characteristics. The method does not require problem dependent parameters and does not require an integral computation for the non-holonomic case. The method is simple to implement, requiring fewer terms than Baumgarte stabilization. The method results in a symmetric positive definite linear system identical to that in Baumgarte stabilization. The method is easily extended to second order and the second order method will appear in future work.

References

1. J. Baumgarte, Stabilization of constraints and integrals of motion in dynamical systems, Computer Methods in Applied Mechanics, pp. 1-16, 1972.
2. D. Baraff, Linear-time dynamics using Lagrange multipliers, In proceedings of SIGGRAPH 1996, ACM Press / ACM SIGGRAPH, Computer Graphics Proceedings, pp. 137-146, 1996.
3. R. Barzel, and A. Barr, A modeling system based on dynamic constraints, Computer Graphics, 22(4), pp. 179-188, 1988.
4. J. C. Platt, and A. Barr, Constraint methods for flexible models, Computer Graphics, 22(4), pp. 279-288, 1988.
5. M. B. Cline, and D. K. Pai, Post-stabilization for rigid body simulation with contact and constraints, In proceedings of the IEEE International Conference on Robotics and Automation, pp. 3744-3751, 2003.
6. H. Jeon, M. Choi, and M. Hong, Numerical stability and convergence analysis of geometric constraint enforcement in dynamic simulation systems, In proceedings of the International Conference on Modeling, Simulation and Visualization Methods, pp. 207-213, 2004.
7. M. Hong, M. Choi, S. Jung, S. Welch, and J. Trapp, "Effective constrained dynamic simulation using implicit constraint enforcement," in *Proceedings of the 2005 IEEE International Conference on Robotics and Automation*, pp. 4531-4536, 2005.
8. D. Greenwood, Advanced Dynamics, Cambridge, 2003

Heat Diffusion – Searching for the Accurate Modeling

Malgorzata Langer[1], Janusz Wozny[1], Malgorzata Jakubowska[2], and Zbigniew Lisik[1]

[1] Institute of Electronics, Technical University of Lodz,
90-924 Lodz, ul. Wolczanska 211/215, Poland
{malgorzata.langer, jwozny, zbigniew.lisik}@p.lodz.pl
[2] Institute of Electronic Materials Technology, Wolczynska Str. 133
01-919 Warsaw, Poland
malgorzata.jakubowska@itme.edu.pl

Abstract. The authors study three approaches which allow to model the steady state heat conduction in a 2D multiphase composite. The subject under investigating is the thermal conductivity of a c-BN composite thick film with possible inclusions of air bubbles. To define the thermal conductivity we have utilized (i) a commercial program ANSYS, for which a random structure has been externally generated, (ii) a cellular automata (CA) based model and (iii) a modified cellular automata based model where we have taken into account a thermal contact resistance between adjacent grains of c-BN.

1   Introduction

Up-to-now the whole electronics, that is based on silicon can work in the low temperature regime (max. 473K), and yet one should be aware that it will be soon when the device should meet the demand to take as huge power density, as 400 W/cm^2 [1]. That is why there are intensive works kept on designing new and improving known materials with extremely high heat conductivity to be capable taking such values over. The good and verified thermal model for mixtures, conglomerates and other composites becomes an issue. Such materials as cubic boron nitride (c-BN), gallium nitride (GaN), silicon carbide (SiC), and some others are to widen the temperature range for electronic devices. One hopes that the mentioned materials may allow increasing the temperature range even up to about 900K. The Table 1 introduces cubic boron nitride features among which the good heat conductivity, electrical insulation capabilities, the resistance against high temperature and chemical agents' impact make one being interested in this material. As with CVD (Chemical Vapor Deposition) or PVD (Physical Vapor Deposition) it is impossible obtaining layers thicker than 3000 Å (they delaminate [2]) and yet, it is difficult to obtain crystalline structure that is wanted in electronics, the conception of thick film technology – screen printing was realized and the c-BN films, 20 – 150 µm thick, with a high adhesion to the substrate were deposited.

The tested films were deposited on alumina (Al_2O_3 96%), with a screen printing technique through 325 mesh screen, then fired in a thick film belt furnace in 850^{0}C, 10 minutes. The printed thick film composition ('electronic paste'), based on c-BN is a dispersive mixture of inorganic powders in a solution of organic resins. It consists of three basic components:

V.N. Alexandrov et al. (Eds.): ICCS 2006, Part I, LNCS 3991, pp. 498 – 505, 2006.

- the active phase that determines the electrical and thermal properties of the layer, i.e. the grains of cubic boron nitride, 2-5µm big, obtained with method shown in [3];
- the auxiliary phase (bonding) – bismuth-boron-silicate glass;
- the organic vehicle

The obtained film consists of c-BN grains surrounded by glass, with closed and open pores filled with air (Fig. 1)

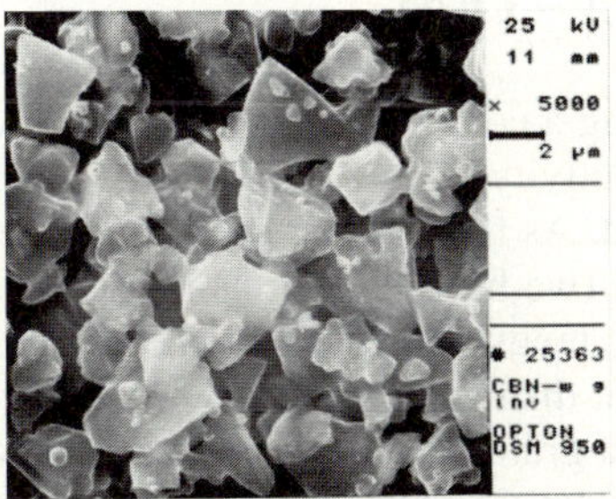

Fig. 1. SEM image with inverted contrast; a sample of the c-BN composite (magnified 5000x)

Table 1. Components and material properties of a c-BN composite sample

Material	thermal conductivity $[W/m^2K]$	density $[g/cm^3]$	volume fraction $[\%]$
c-BN	1300	3.48	85.67
glass	12	6.64	4.32
air	0.00271	0.0013	10

It is obvious the mutual ratio of these phases influences the thermal conductivity of the film. The measurements for the film 100 µm thick (made with the method described in [4]) result with values in the range 100-200 W/mK, depending on the c-BN grain fraction, the purity, and the volume of pores. Comparing this number to the c-BN thermal conductivity, one can see that although the thermal conductivity of c-BN is very high, the effective thermal conductivity of a composite film is much lower due to the presence of the glass, air inclusions and the fact that there are grains of c-BN only and not a continuous layer of a monocrystal. Since the gas pores are very small, one can neglect the convection and treat them as a material of low thermal conductivity. So the conductivity of the composite layer is affected by the size and the shape of grains, their orientation and the volume of c-BN phase. Since the thickness of a layer is a dozen of grains one should consider the shape, the size and the position of every grain.

There is a lack of references, where thermal modeling of composites deals with different sizes and irregular shapes of grains. For instance in [5] every particle with the high thermal conductivity is considered as a circle or sphere of randomly varying radius and position. In [6] and [7] every node represents an element with the high or low conductivity. No thermal resistance of contacts was considered between neighboring elements. The effective thermal conductivity of the layer was governed by the existence of percolation path, mainly. A (random) resistors network has been generated first [5], [6], [7] and then the solution was obtained with the use of various approaches.

We have gone one step lower where the heat conduction inside grains and between them should be properly modeled to get accurate effective conductivity of the film.

2 The First Approach – FEM

The first idea is always using a commercial software, and we have done it to calculate an effective thermal conductivity. The ANSYS software, based on the FEM (Finite Element Method), was used. As the number of grains may be higher than few dozens so we prepare a special file (we have done it using Scilab 3.0 [8], but it may be done in any other programming language). This is the script that covers input data for geometry: information about distribution function of the grain size, permissible angles between adjacent edges of a grain and the final size of a composite sample. Thus, the structure is completely stochastic. The basic assumption is that the grains cannot overlap one another. Thus, after generating, any grain is to be placed in the structure in such a location where it does not overlap with other grains and its position is as low as possible (Fig. 2 shows an example how the geometry is constructed).

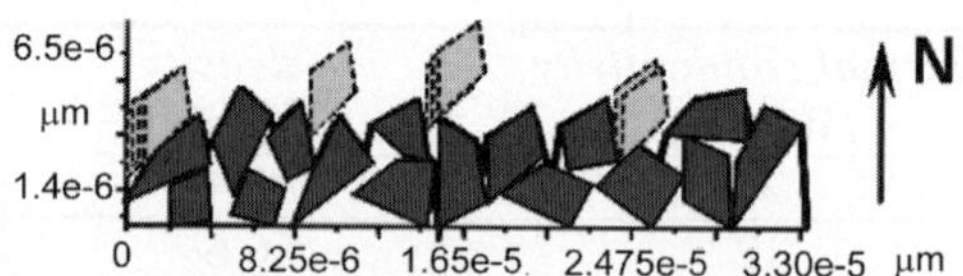

Fig. 2. A midstep of the geometry generation process. We have to place a grain in the lowest possible location.

An example of such an input file's result is presented in Fig. 3.

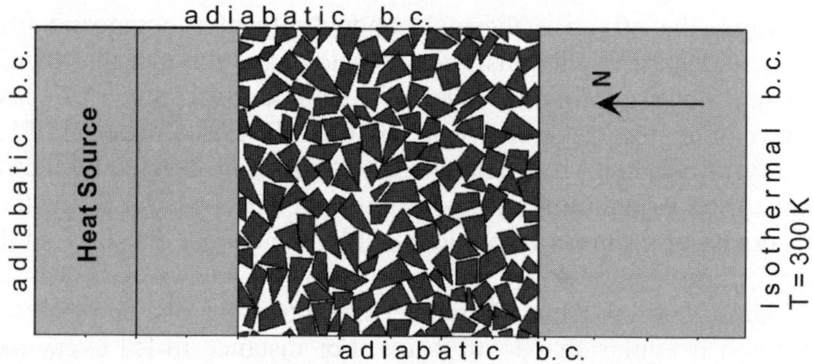

Fig. 3. The geometry used for defining the effective thermal conductivity of a composite layer

Additional (grey) layers work as buffers between the film and the boundary conditions and guarantee a space for a heat source. Setting boundary conditions as presented in Fig. 3 assures that the entire heat flux must flow trough the composite layer. The task for ANSYS is to mesh the domain, to solve it, and to export some results Fig. 4

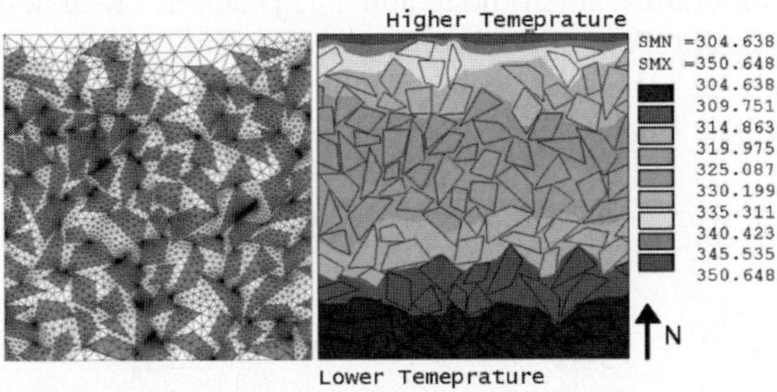

Fig. 4. The mesh and temperature distribution of the composite layer

Since one knows the dissipated power, dimensions of the film and average temperatures on both sides of it, one can calculate its effective thermal conductivity. For the film in Fig. 4 the volume fraction of c-BN was ca. 50% and no air was considered. The effective thermal conductivity amounts to ca. 64 $Wm^{-1}K^{-1}$.

The mesh consists of $12 \cdot 10^3$ nodes. It took about 1 min (2800+ Athlon PC) to obtain these results. So one can reasonably expect that also 3D simulation can be performed.

Of course there are some issues that must be considered. First, the presented method of generating geometry of the structure doesn't allow getting high (ca. > 70%) c-BN volume ratios. Then, if we get a large number of grains comparing to the glass phase, the space between them can be very narrow which will cause problems with mesh generation. Thus a case when two grains have a common edge should be also taken into account. Summarizing, a "simple" problem of creating the composite structure evolves into modeling of composite fabrication process, not discussed here. However, the presented results show that it is possible to obtain the effective thermal conductivity with this method.

3   Standard Cellular Automata (CA) Approach

To solve the steady-state heat diffusion differential equation there is no need to utilize any sophisticated commercial software. One can use a cellular automata, which allows to model behaviors of physical systems [9],[10], and also to mimic a diffusion process. On a homogeneous 2D square lattice the temperature at a given node is obtained as a mean value of temperatures at 4 neighboring nodes (1) [11]. This is enough to mimic the diffusion. When the lattice is not uniform or the environment is not homogeneous one has to use a weighted mean (1). Such a case is presented in

Fig. 5. The space between nodes can be chosen freely as well as thermal conductivity of each cell. Then, the only problem is to obtain the weights.

$$T(i, j) = \sum_{k=1}^{4} p_k T_k (i, j) \ .$$

(1)

where $T(i,j)$ is the temperature at (i,j) node and $T_k(i,j)$ denotes k-th neighbour of (i,j) node, p_k is a proper weight. A sum of all p_k's is always one.

The weights can be obtained using finite difference representation of the differential equation. The final result is (2) [11].:

$$T_{i,j} = \left[p_e T_{i+1,j} + p_w T_{i-1,j} + p_n T_{i,j+1} + p_s T_{i-1,j+1} + S_{i,j} HGEN \right] \frac{1}{t}$$

$$p_e = \frac{\Delta y}{2\Delta x}\left(k_{i,j} + k_{i,j-1}\right); p_w = \frac{\Delta y}{2\Delta x}\left(k_{i-1,j} + k_{i-1,j-1}\right); p_n = \frac{\Delta x}{2\Delta y}\left(k_{i-1,j} + k_{i,j}\right); \ .$$

(2)

$$p_s = \frac{\Delta x}{2\Delta y}\left(k_{i-1,j-1} + k_{i,j-1}\right); t = p_e + p_w + p_n + p_s$$

where HGEN is a heat generation rate, and $S_{i,j}$ is the heat generation area.

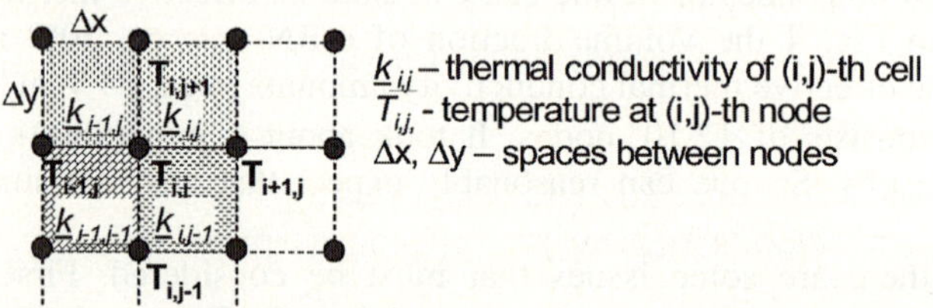

Fig. 5. Every node has its own temperature which depends on temperatures of adjacent cells. Such local relation is typical for cellular automata.

For the isothermal boundary condition (b.c.) we fix the temperature at the boundary node. If the edge is adiabatic, an image of the node which is nearest to the edge is placed, the other side of boundary is to obtain a reflective property of adiabatic b.c.

The structure and the temperature distributions are shown in Fig. 6. The structure resolution is 100x100 pixels, and each pixel represents one cell. The shade indicates material.

To calculate the steady-state distribution we needed ca. 10 minutes (Athlon XP, 2800+). The effective thermal conductivity was found about 90 $Wm^{-1}K^{-1}$.

The volume ratios were respectively: cBN - 83% , glass - 7%, air - 10%. This is much closer to the values placed in Table 1, but also close to the measurements. However the lower bound of glass phase is still limited by the size of one pixel. For 100 pixels we cannot decrease it below 1% if any. And each grain must be bounded by the glass. To decrease the volume fraction of the glass one should have to increase the resolution which would increase time of calculation. For 200x200 pixels 2 hours (Athlon XP, 2800+) were necessary to find the final solution.

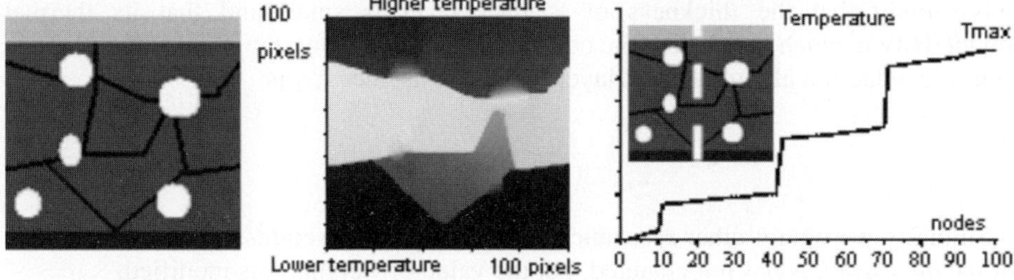

Fig. 6. The 100x100 pixels structure, the temperature distribution and the temperature profile along dashed line. The size of each pixel was assumed 1μm×1μm.

So we have to deal with two problems: we have to (i) decrease time of computation by lowering the resolution (keeping in mind the accuracy), (ii) decrease the lower limit of the glass volume .This can be done by a slight modification of the CA

4 Modified Cellular Automata Approach

Having in mind the temperature distribution (Fig. 6), and the temperature profile, one can see that the temperature steps correspond to the glass regions and the profile reminds a profile with contact thermal resistances. Thus one can eliminate pixels which surround each grain by introducing contact resistances.

The formulas (2) were derived from the finite difference, but they can also be obtained by connecting adjacent nodes by thermal resistors. Introducing thermal contact resistances the resistances can be calculated from the conductivity and dimensions of a cell Fig. 7.

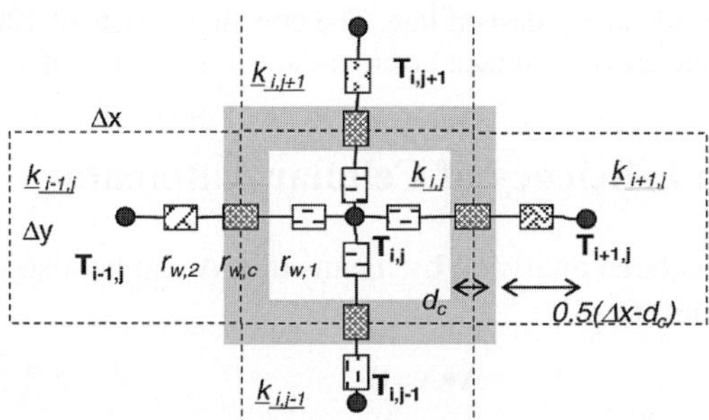

Fig. 7. The way of introducing contact thermal resistance. d_c – thickness of the contact. Please note that here the node lies at the center of the cell.

The resistance of the contact is calculated as the resistance of a layer with the thickness d_c.

$$r_{w,c} = \frac{1}{k_c} \frac{d_c}{\Delta y \cdot [1m]} \, . \tag{3}$$

Assuming that the thickness of a glass layer is small and that its thermal conductivity is much lower than the conductivity of c-BN one can assume that there is no heat conduction along the joint layer and the resistance $r_{w,1}$ is:

$$r_{w,1} = \frac{1}{k_{i,j}} \frac{\Delta x}{2(\Delta y - d_c) \cdot [1m]} \ . \tag{4}$$

Similarly we obtain other resistances and then proper weights. So the final recipe for the diffusive CA has not changed, but the value of weights was modified.

Two temperature profiles in two composite films are presented in Fig 8 . The composition of the films a) and b) is: c-BN – 91% and 88%, glass – 1% and 3%, air – 8% and 8% of volume ratio. The effective thermal impedance was about 773 $Wm^{-1}K^{-1}$ and 515 $Wm^{-1}K^{-1}$, respectively. We can see that for the b) structure the steps are much higher since the thickness of the glass film is higher. Even a small change of glass content may change the effective conductivity significantly. For this method of modeling there is no bottom limit for glass fraction.

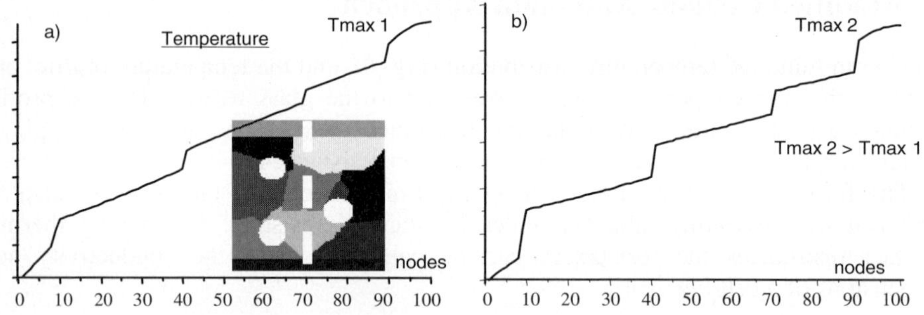

Fig. 8. Temperature profile along dashed line. The domain consists of 100x100 pixels of a size 1μm×1μm each. The thickness of contact layer was a) 0.1 and b) 0.3 of a cell size.

5   Discussion on Efficiency of Cellular Automata

The system which has been analyzed by means of CA can be also solved by forming a system of linear equations:

$$A \bullet X = B \ . \tag{5}$$

For the domain of 100x100 pixels X – is a vector of 10^4 unknown temperatures. A is a $10^4 \times 10^4$ sparse matrix. B is a 10^4 elements vector. When using CA we need at least one 100x100 matrix where we store temperatures and four 100x100 matrices for weights. So we do not need a $10^4 \times 10^4$ matrix. Even if it's a sparse matrix we have to remember indeces of non zero elements. The simplest iterative methods of solving (5) are the Jacobi and Gauss-Seidel algorithms [12]. In both cases we have not obtained shorter time of computation in comparison to CA algorithm. It doesn't mean that CA methods are fast enough. The speed is a problem, but a problem which can be solved. First we can use a non-uniform grid: finer near grain boundaries and courser inside a grain. This will affect only weights. Moreover, different levels of mesh could be introduced so an approximated solution could be found first, and next, using more accurate mesh a detail solution would be obtained.

6   Conclusions

The purpose of this study was to investigate an efficient method of accurate modeling of composites' thermal properties. We have considered: (i) an approach which utilizes ANSYS software, (ii) a cellular automata method.

Using the first approach, the result comes very fast but if we have the geometry already prepared. And the basic problem is to generate the structure. We should consider not only the position of each grain but also a possible sticking one grain to another if we need to lower the glass fraction. So should be ready to face up a problem of modeling fabrication process.

Of course this can be the case for CA approach. However we can use a scanned and reviewed image of the composite layer as input data. Since a CA can easily deal with complex boundaries we are not restricted to simple shapes which are necessary for ANSYS. The CA is an easy approach and very intuitive.

References

1. J.L., German R.M., Hems K.F., Guiton Th.A., Injection Molding AlN for Thermal Management Applications, Bull. Am. Ceram. Soc. 75[8], 1996, pp. 61-65.
2. Mirkarimi P.B., McCarty K.F., Medlin D.L., Review of Advances In Cubic Boron Nitride Symthesis, Materials Science and Engineering, R21 (1997), pp. 47-100
3. Giellissee P. Niculescu H., Temblay J., Achmatowicz S., Jakubowska M., & others, High Thermal Conductivity Cubic Boron Nitride Thick Films, Proc. of 2001 Int. Symp. on Microelectronics, Oct. 9-11, 2001, Baltimore, pp.379-383.
4. Achmatowicz S., Zwierkowska E., Wyżkiewicz I., Łobodziński W., New Approach to Thermal Conductivity of Thin Films Measurements by Means of Comparative Method, Proc. of 35th Int. Symp. On Microel., 2003, pp. 655-660
5. Gerenrot D., Berlyand L., Philips J.: Random Network Model for Heat Transfer in High Contrast Composite Materials, IEEE Trans. on Advanced Packaging, vol. 26, No. 4, Nov. 2003, 410416
6. Devpura A., Prasher R. S.: Percolation Theory Applied to the Analysis of Thermal Interface Materials in Flip-Chip Technology, Itherm 2000, May 24-26, 2000, Las Vegas, NV, 21-28
7. Staggs J.E.J.: "Estimating the thermal conductivity of chars and porous residues using thermal resistor networks" Fire Safety Journal 37 (2002) 107–119
8. http://scilabsoft.inria.fr/
9. Wolf-Gladrow D.: Lattice Gas Cellular Automata and Lattice Boltzmann Models: An Introduction, Lecture notes in mathematics 1795, Springer-Verlag 2000
10. Wolfram S.: A new kind of science, Wolfram Media (May, 2002)
11. Haji-Sheikh A. Monte Carlo Methods in In Handbook of Numerical Heat Transfer, John Wiley & Sons, New York, 1988, pp. 673-722
12. Jon H. Mathews, Kurtis D. Fink Numerical Methods Using Matlab 3rd ed., Prentice Hall '99

Parallel Exact and Approximate Arrow-Type Inverses on Symmetric Multiprocessor Systems

George A. Gravvanis and Konstantinos M. Giannoutakis

Department of Electrical and Computer Engineering, School of Engineering,
Democritus University of Thrace, 12, Vas. Sofias street, GR 671 00 Xanthi, Greece
{ggravvan, kgiannou}@ee.duth.gr

Abstract. In this paper we present new parallel inverse arrow-type matrix algorithms based on the concept of sparse factorization procedures, for computing explicitly exact and approximate inverses, on symmetric multiprocessor systems. The parallel implementation of the new inversion algorithms is discussed and numerical results are presented, using the simulation tool of Multi-Pascal.

1   Introduction

Sparse matrix computations, which have inherent parallelism, are of central importance, because of the applicability to real-life problems and are the most time-consuming part in computational science and engineering computations. Hence research efforts were focused on the production of efficient parallel computational methods and related software suitable for multiprocessor systems, [1, 2, 5, 11, 12, 14].

Until recently, direct methods have been effectively used, but the increase of size, even with the use of modern computer systems, has become a barrier to such methods, [2, 3]. Additionally, the solution of sparse linear systems, because of its applicability to real-life problems, is obtained by iterative methods, which are in competitive demand after the emergence of Krylov subspace methods, [5, 11, 14].

An important achievement over the last decades is the appearance and use of Explicit Preconditioned Methods, [4], for solving sparse linear systems, and the preconditioned form of a linear system $Au = s$ is $MAu = Ms$, where M is preconditioner, [1, 2, 5, 11, 14]. The preconditioner M has therefore to satisfy the following conditions: (i) MA should have a "clustered" spectrum, (ii) M can be efficiently computed in parallel and (iii) finally "$M \times$ vector" should be fast to compute in parallel, [2, 5, 11, 12, 14].

Hence, the derivation of parallel methods was the main objective for which several families of parallel inverses of an arrow-type matrix, are proposed. The main motive for the derivation of the parallel exact and approximate inverse matrix algorithms is that they result in parallel direct methods and in parallel iterative methods in conjunction with parallel preconditioned conjugate gradient-type schemes respectively, for solving arrow-type linear systems on multiprocessor systems.

V.N. Alexandrov et al. (Eds.): ICCS 2006, Part I, LNCS 3991, pp. 506–513, 2006.

2 Parallel Exact and Approximate Inverses

Let us consider the arrow - type linear system, i.e.,

$$Au = s, \tag{1}$$

where A is a sparse arrow - type $(n \times n)$ matrix of the following form:

$$
A \equiv
\begin{bmatrix}
b_1 & c_1 & & & & & p_1 \\
a_2 & & & & & & \\
& & & 0 & & & \\
& & & & & & \\
& 0 & & & & p_{n-2} & \\
& & & & & c_{n-1} & \\
v_1 & & & & v_{n-2} & a_n & b_n
\end{bmatrix} . \tag{2}
$$

Arrow-type matrices occur in practice for example, in the course of the Lanczos method for solving the eigenvalue problem for large sparse matrices, in the eigenstructure problems of arrowhead matrices which arise from applications in molecular physics and in the application of the finite element or finite difference method over a region by removing part of the region. This class of arrow-type linear systems captures the class of linear systems, which can be considered as a special tridiagonal linear systems, which occur when solving certain constant-coefficient elliptic partial differential equations by the Fourier method, or using finite difference methods to solve linear constant-coefficient boundary value problems, [6, 7, 8].

Let us now assume the factorization of the coefficient matrix A, i.e.

$$A = LU, \tag{3}$$

where L and U are sparse strictly lower and upper triangular matrices of the same profile as the matrix A, [6, 7, 8]. The elements of the L and U decomposition factors were computed by the so-called **A**rrow - **t**ype **A**pproximate **LU**-type **Fa**ctorization algorithm (henceforth **ATALUFA** algorithm), [6]. The memory requirements of the **ATALUFA** algorithm are $O(5n)$ words and the computational work is $O(6n)$ multiplicative operations, [6].

Let $M = (\mu_{i,j})$, $i, j \in [1, n]$, be the exact inverse of the coefficient matrix A. The elements of $M = (\mu_{i,j})$ can be computed by solving recursively the following systems, [6]:

$$ML = U^{-1} \quad \text{and} \quad UM = L^{-1}. \tag{4}$$

The **E**xact **I**nverse **A**rrow - **t**ype **M**atrix algorithm (henceforth called the **EIATM** algorithm) for computing the elements of the exact inverse, has been presented in [6]. Similarly, by retaining δl elements next to the main diagonal, the elements of the approximate inverse $M^{\delta l} = (\mu_{i,j})$, $i \in [1, n]$, $j \in [\max(1, i - \delta l + 1), \quad \min(n, i + \delta l - 1)]$, were computed by the **B**anded

Approximate **I**nverse **A**rrow - type **M**atrix algorithm (henceforth called the **BAIATM** algorithm), [6].

For the parallelization of the **EIATM** and the **BAIATM** algorithms on symmetric multiprocessor systems, the **ATALUFA** algorithm was used as a "front-end" computational procedure. The elements of the matrix M were computed in a sequence, as shown in Eq. 5 (for $n = 8$ and $\delta l = 3$ without loss of generality), because of the dependence of the elements of the inverse during the construction. The values of the parentheses at the superscript of elements (e.g. $\mu_{i,j}^{(k)}$), indicate that the element $\mu_{i,j}$ was computed at the k-th step of the algorithm, while the elements with the same superscript (i.e. (k)) were computed concurently. The pattern for the sequence of computations was considered as an anti-diagonal motion, starting from the element $\mu_{n,n}$ down to $\mu_{1,1}$, where the elements on an anti-diagonal were computed in parallel.

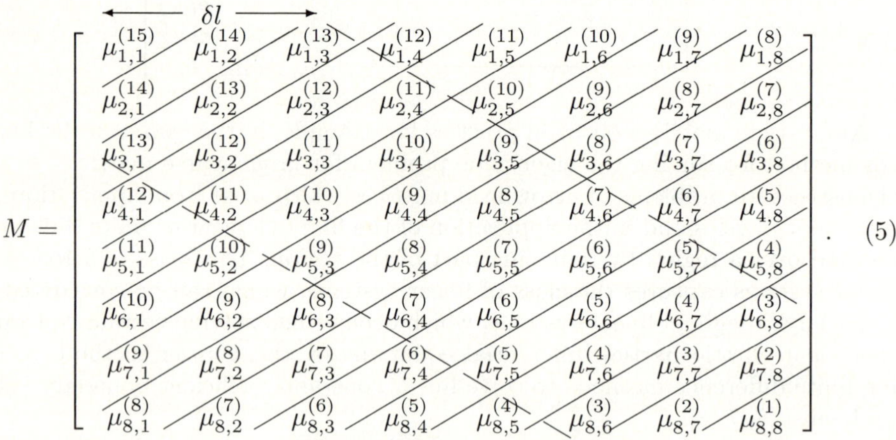

$$ M = \begin{bmatrix}
\mu_{1,1}^{(15)} & \mu_{1,2}^{(14)} & \mu_{1,3}^{(13)} & \mu_{1,4}^{(12)} & \mu_{1,5}^{(11)} & \mu_{1,6}^{(10)} & \mu_{1,7}^{(9)} & \mu_{1,8}^{(8)} \\
\mu_{2,1}^{(14)} & \mu_{2,2}^{(13)} & \mu_{2,3}^{(12)} & \mu_{2,4}^{(11)} & \mu_{2,5}^{(10)} & \mu_{2,6}^{(9)} & \mu_{2,7}^{(8)} & \mu_{2,8}^{(7)} \\
\mu_{3,1}^{(13)} & \mu_{3,2}^{(12)} & \mu_{3,3}^{(11)} & \mu_{3,4}^{(10)} & \mu_{3,5}^{(9)} & \mu_{3,6}^{(8)} & \mu_{3,7}^{(7)} & \mu_{3,8}^{(6)} \\
\mu_{4,1}^{(12)} & \mu_{4,2}^{(11)} & \mu_{4,3}^{(10)} & \mu_{4,4}^{(9)} & \mu_{4,5}^{(8)} & \mu_{4,6}^{(7)} & \mu_{4,7}^{(6)} & \mu_{4,8}^{(5)} \\
\mu_{5,1}^{(11)} & \mu_{5,2}^{(10)} & \mu_{5,3}^{(9)} & \mu_{5,4}^{(8)} & \mu_{5,5}^{(7)} & \mu_{5,6}^{(6)} & \mu_{5,7}^{(5)} & \mu_{5,8}^{(4)} \\
\mu_{6,1}^{(10)} & \mu_{6,2}^{(9)} & \mu_{6,3}^{(8)} & \mu_{6,4}^{(7)} & \mu_{6,5}^{(6)} & \mu_{6,6}^{(5)} & \mu_{6,7}^{(4)} & \mu_{6,8}^{(3)} \\
\mu_{7,1}^{(9)} & \mu_{7,2}^{(8)} & \mu_{7,3}^{(7)} & \mu_{7,4}^{(6)} & \mu_{7,5}^{(5)} & \mu_{7,6}^{(4)} & \mu_{7,7}^{(3)} & \mu_{7,8}^{(2)} \\
\mu_{8,1}^{(8)} & \mu_{8,2}^{(7)} & \mu_{8,3}^{(6)} & \mu_{8,4}^{(5)} & \mu_{8,5}^{(4)} & \mu_{8,6}^{(3)} & \mu_{8,7}^{(2)} & \mu_{8,8}^{(1)}
\end{bmatrix} . \qquad (5) $$

If we consider that the command **forall** is responsible for process creation and execution, then the algorithm for the **P**arallel implementation of the **EIATM** algorithm (henceforth called the **PEIATM** algorithm) is:

```
For k = 1 to n
     forall l = 1 to k
          call inverse(n − l + 1, n − k + l)
for k = n − 1 downto 1
     forall l = 1 to k
          call inverse(l, k − l + 1)
```

while the algorithm for the **P**arallel implementation of the **BAIATM** algorithm (henceforth called the **PBAIATM** algorithm) is:

```
For k = 1 to δl
     forall l = 1 to k
          call inverse(n − l + 1, n − k + l)
m = 2
```

for $k = (\delta l + 1)$ **to** n
    forall $l = m$ **to** $(k - m + 1)$
       call $inverse(n - l + 1, n - k + l)$
    if $(k - \delta l) \bmod 2 = 0$ **then**
       $m = m + 1$
$m = m - 1$
for $k = (n - 1)$ **downto** $(\delta l + 1)$
    forall $l = m$ **to** $k - m + 1$
       call $inverse(l, k - l + 1)$
    if $(k - \delta l) \bmod 2 = 1$ **then**
       $m = m - 1$
for $k = \delta l$ **downto** 1
    forall $l = 1$ **to** k
       call $inverse(l, k - l + 1)$

where the function $inverse(i,j)$ computes the element $\mu_{i,j}$ of the exact inverse, and can be described as follows, [6]:

function $inverse(i,j)$
if $(i \geq j)$ **then**
    if $(i = n)$ **then**
       if $(j = n)$ **then**
          $\mu_{n,n} = 1/w_n$
       else
          if $(j = n - 1)$ **then**
             $\mu_{n,n-1} = (-d_j \cdot \mu_{n,n})/w_j$
          else
             $\mu_{n,j} = (-d_j \cdot \mu_{n,j+1} - e_j \cdot \mu_{n,n})/w_j$
    else
       if $(i = j)$ **then**
          if $(j = n - 1)$ **then**
             $\mu_{n-1,n-1} = (1 - d_{n-1} \cdot \mu_{i,n})/w_j$
          else
             $\mu_{i,j} = (1 - d_j \cdot \mu_{i,j+1} - e_j \cdot \mu_{i,n})/w_j$
       else
          $\mu_{i,j} = (-d_j \cdot \mu_{i,j+1} - e_j \cdot \mu_{i,n})/w_j$
else
    if $(j = n)$ **then**
       if $(i = n - 1)$ **then**
          $\mu_{n-1,n} = -g_{n-1} \cdot \mu_{n,n}$
       else
          $\mu_{i,n} = -g_i \cdot \mu_{i+1,n} - h_i \cdot \mu_{n,n}$
    else
       $\mu_{i,j} = -g_i \cdot \mu_{i+1,j} - h_i \cdot \mu_{n,j}$

It should be mentioned that the parallel construction of exact inverses on distributed systems has been studied and implemented in [9], and is under further investigation.

3   Numerical Results

For the parallel implementation of the **PEIATM** and **PBAIATM** algorithm described above, the simulation tool of *Multi-Pascal* has been utilized, where a time unit of the simulated time is approximately equivalent to one microsecond of the real execution time on a general purpose multiprocessor, [13]. The architecture platform is that of a shared memory system consisted of 512 processors.

The relative speedups and efficiency, for several values of the order n and number of processors for the **PEIATM** algorithm are presented in Table 1. In Fig. 1 and 2 the relative speedups versus the order n for several number of processors and the relative speedups versus the number of processors for several values of the order n, are presented respectively for the **PEIATM** algorithm.

Table 1. Speedups and efficiency for the **PEIATM** algorithm, for various values of the order n and number of processors

n	no_proc=2		no_proc=4		no_proc=8		no_proc=13	
	Speedup	Effi/ncy	Speedup	Effi/ncy	Speedup	Effi/ncy	Speedup	Effi/ncy
15	1.825	0.913	2.867	0.717	3.505	0.438	3.645	0.280
20	1.856	0.928	2.975	0.744	3.898	0.487	4.155	0.320
40	1.894	0.947	3.083	0.771	4.931	0.616	5.361	0.412
60	1.902	0.951	3.102	0.776	5.494	0.687	5.957	0.458
80	1.905	0.953	3.106	0.777	5.870	0.734	6.314	0.486

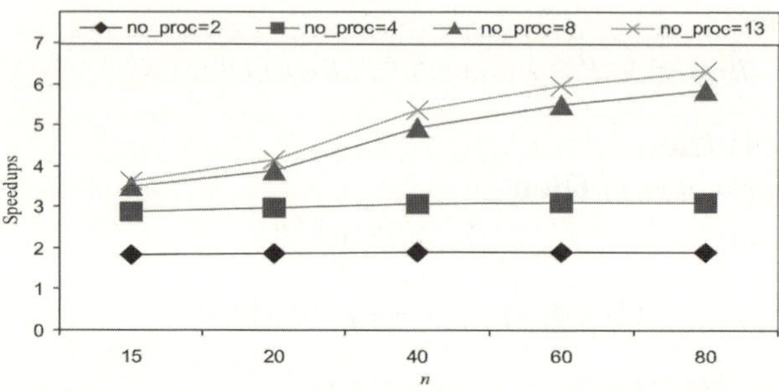

Fig. 1. Speedups versus the order n for the **PEIATM** algorithm for several number of processors

The relative speedups and efficiency for several values of the order n, the retention parameter δl and number of processors for the **PBAIATM** algorithm are presented in Table 2. In Fig. 3 and 4 the relative speedups versus the retention parameter δl for several number of processors with $n = 100$, and the relative speedups versus the order n for several number of processors with $\delta l = n/2$, are presented respectively for the **PBAIATM** algorithm.

It should be noted that the restrictions imposed by the developing environment used, did not allow us to fully explore cases for $n > 80$ for the exact inverse

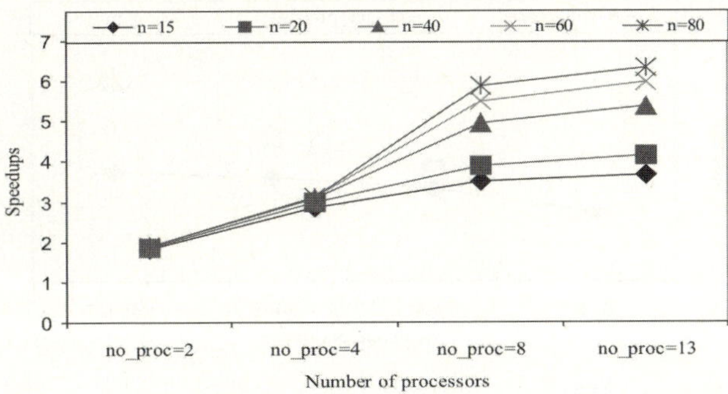

Fig. 2. Speedups versus the number of processors for the **PEIATM** algorithm for several values of the order n

Table 2. Speedups and efficiency for the **PBAIATM** algorithm, for various values of the order n, δl and number of processors

n	δl	no_proc=2		no_proc=4		no_proc=5	
		Speedup	Efficiency	Speedup	Efficiency	Speedup	Efficiency
20	2	1.104	0.552				
	4	1.477	0.739	1.760	0.440	1.760	0.352
	8	1.637	0.819	2.201	0.550	2.307	0.461
	$n/2$	1.669	0.835	2.376	0.594	2.421	0.484
40	2	1.107	0.554				
	4	1.484	0.742	1.784	0.446	1.784	0.357
	8	1.650	0.825	2.258	0.565	2.370	0.474
	$n/2$	1.793	0.897	2.809	0.702	2.943	0.589
60	2	1.108	0.554				
	4	1.486	0.743	1.792	0.448	1.792	0.358
	8	1.654	0.827	2.275	0.569	2.389	0.478
	$n/2$	1.783	0.892	2.953	0.738	3.151	0.630
80	2	1.108	0.554				
	4	1.487	0.744	1.796	0.449	1.795	0.359
	8	1.656	0.828	2.283	0.571	2.398	0.480
	$n/2$	1.768	0.884	3.118	0.780	3.262	0.652
100	2	1.108	0.554				
	4	1.487	0.744	1.798	0.450	1.798	0.360
	8	1.657	0.829	2.288	0.572	2.403	0.481
	$n/2$	1.762	0.881	3.218	0.805	3.351	0.670

algorithm and for $n > 100$ for the approximate inverse algorithm. Additionally, the number of processors $no\_proc = 13$ for the exact inverse and $no\_proc = 5$ for the approximate inverse, was the maximum number of processors allowed by the simulation environment in each case (for $\delta l = 2$ the maximum number was $no\_proc = 2$).

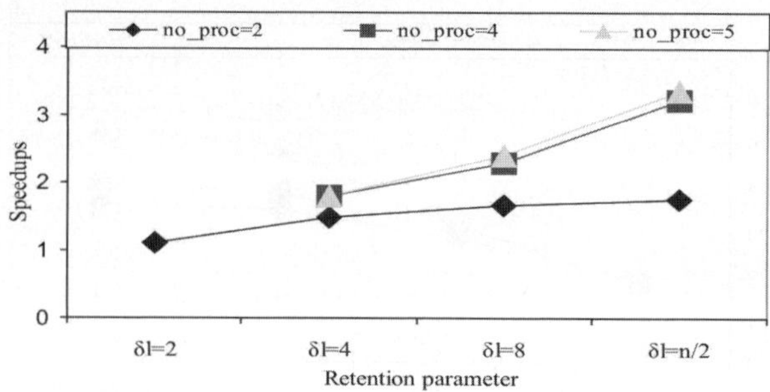

Fig. 3. Speedups versus the retention parameter δl for the **PBAIATM** algorithm for several number of processors with $n = 100$

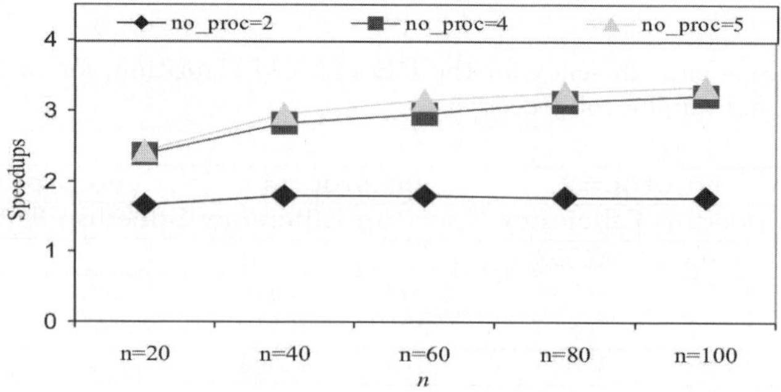

Fig. 4. Speedups versus the order n for the **PBAIATM** algorithm for several number of processors with $\delta l = n/2$

We observe that the relative speedups and efficiency are increasing as the order n increases for various values of the number of processors. Although the relative speedups increase when adding more processors on the system, the efficiency is decreasing, but remains high for large values of n.

Numerical results for such arrow-type linear systems have been presented in [6, 7, 8] for solving boundary value problems and in [10] for solving problems considered in reliability engineering and in particular performability evaluation of multitasking and multiprocessor systems.

4   Conclusion

The design of parallel explicit exact and approximate inverses results in efficient parallel direct and iterative methods for solving arrow-type linear systems on multiprocessor systems. Despite the restrictions of the simulated environment used as the experimental platform, the experimental results obtained tend to the upper theoretical bounds.

Finally, further parallel algorithmic techniques will be investigated in order to improve the speedups and efficiency of the parallel construction of explicit exact and approximate inverses, on symmetric multiprocessor systems.

References

[1] Akl, S.G.: Parallel Computation: Models and Methods. Prentice Hall (1997)

[2] Dongarra, J.J., Duff, I., Sorensen, D., van der Vorst, H.A.: Numerical Linear Algebra for High-Performance Computers. SIAM (1998)

[3] Duff, I.: The impact of high performance computing in the solution of linear systems: trends and problems. J. Comp. Applied Math. **123** (2000) 515-530

[4] Evans, D.J.: Preconditioning Methods: Theory and Applications. Gordon and Breach Science Publishers (1983)

[5] Gravvanis, G.A.: Explicit Approximate Inverse Preconditioning Techniques. Archives of Computational Methods in Engineering **9(4)** (2002) 371-402

[6] Gravvanis, G.A.: Explicit isomorphic iterative methods for solving arrow-type linear systems. Inter. J. Comp. Math. **74(2)** (2000) 195-206

[7] Gravvanis, G.A.: Solving parabolic and nonlinear 1D problems with periodic boundary conditions. CD-ROM Proceedings of the European Congress on Computational Methods in Applied Sciences and Engineering 2000

[8] Gravvanis, G.A.: Parallel preconditioned algorithms for solving special tridiagonal systems. Proceedings of Dynamical Systems and Applications **III**, (1999) 241-248 Dynamic Publishers

[9] Gravvanis, G.A.: Parallel matrix techniques. In: K. Papailiou, D. Tsahalis, J. Periaux, C. Hirsch, M. Pandolfi eds. Computational Fluid Dynamics **I** (1998) 472-477 Wiley

[10] Gravvanis, G.A., Platis, A.N., Giannoutakis, K.M., Violentis, J.B., Lipitakis, E.A.: Performability evaluation of multitasking and multiprocessor systems by explicit approximate inverses. Proceedings of the International Conference on Parallel and Distributed Processing Techniques and Applications, H.R. Arabnia and Youngsong Mun eds. **III** (2003) 1324-1331 CSREA Press

[11] Grote, M.J., Huckle, T.: Parallel preconditioning with sparse approximate inverses. SIAM J. Sci. Comput. **18** (1977) 838-853

[12] Grote, M.J., Simon, H.D.: Parallel preconditioning and approximate inverses on the connection machine. In: R.F. Sincovec, D.E. Keyes, L.R. Petzold and D.A. Reed, eds. Parallel Processing for Scientific Computing **2** (1993) 519-523 SIAM

[13] Lester, B.P.: The Art of Parallel Programming. Prentice-Hall Int. Inc (1993)

[14] Saad, Y., van der Vorst, H.A.: Iterative solution of linear systems in the 20th century. J. Comp. Applied Math. **123** (2000) 1-33

A Permutation-Based Differential Evolution Algorithm Incorporating Simulated Annealing for Multiprocessor Scheduling with Communication Delays

Xiaohong Kong[1,2], Wenbo Xu[1], and Jing Liu[1]

[1] School of Information Technology, Southern Yangtze University,
Wuxi 214122, China
nancykong@hist.edu.cn, xwb@sytu.edu.cn
[2] Henan Institute Of Science and Technology,
Xinxiang, Henan 453003,China

Abstract. Employing a differential evolution (DE) algorithm, we present a novel permutation-based search technique in list scheduling for parallel program. By encoding a vector as a scheduling list and differential variation as s swap operator, the DE algorithm can generate high quality solutions in a short time. In standard differential evolution algorithm, while constructing the next generation, a greedy strategy is used which maybe lead to convergence to a local optimum. In order to avoid the above problem, we combine differential evolution algorithm with simulated annealing algorithm which relaxes the criterion selecting the next generation. We also use stochastic topological sorting algorithm (STS) to generate an initial scheduling list. The results demonstrate that the hybrid differential evolution generates better solutions even optimal solutions in most cases and simultaneously meet scalability.

1   Introduction

Given parallel program modelled by a directed acyclic graph (DAG), the objective of scheduling the tasks to multiprocessors is minimizing the completion time or makespan while satisfying the precedence constraints. The problem is NP-hard even simplified model with some assumptions and becomes more complex under realistic application such as arbitrary task execution and communication times. Due to the intractability, many classical heuristics have been proposed to find out sub-optimal solution of the problem, the idea behind these heuristic algorithms is to tradeoffs the solution quality and the complexity [1-5]. Recently meta-heuristics search approaches have also made some accomplishment on solving the problem [1][2][3].

Since DE was first introduced to minimizing possibly nonlinear and non-differentiable continuous space functions [6], it has been successfully applied in a variety of applications [7]. In this paper, we exploit a hybrid differential evolution algorithm to construct the solution for parallel program scheduling with the permutation-based solution presentation.

V.N. Alexandrov et al. (Eds.): ICCS 2006, Part I, LNCS 3991, pp. 514–521, 2006.

2   The Multiprocessor Scheduling with Communication Delays

It is popular to model the multiprocessor scheduling using a directed acyclic graph (DAG), which can be defined by a tuple $G = (V, E, C, W)$, where $V = \{n_j, j = 1 : v\}$ is the set of task nodes and $v = |V|$ is the number of nodes, E is the set of communication edges and $e = |E|$ is the number of edges,C is the set of edge communication costs, and W is the set of node computation costs. The value $c(n_i, n_j) \in C$ corresponds to the communication cost incurred while the task n_i and n_j are scheduled to different processors, which is zero if both nodes are assigned on the same processor. The value $w(n_i) \in W$ is the execution time of the node $n_i \in V$. The edge $e_{i,j} = (n_i, n_j) \in E$ represents the partial order between tasks ni and n_j, which dictate that a task cannot be executed unless all its predecessors have been completed their execution.

The target system M is consisted of m identical or homogeneous processors with local memory connected by an interconnection network with a certain topology. When scheduling tasks to machines, we assume every task of a parallel program can be executed on any processor and only on one processor non-preemptively and the system executes computation and communication simultaneously.

Scheduling the graph G to M is to find out pairs of (task, processor) which optimize the scheduling length or completion time. Most scheduling algorithms are based on the so-called list scheduling strategy. The basic idea of list scheduling is to make a scheduling list (a sequence of nodes for scheduling) by assigning them some priorities, and then assign the tasks to processor according to some rule such as the earliest start time first.

3   Differential Evolution Algorithm

Differential evolution (DE) is one of the latest evolutionary optimization methods proposed by Storn and Price [6]. Like other evolution algorithms, mutant operator, Crossover operator, selection operator are introduced to generate a next generation, but DE has its advantages such as simple concept, immediately accessible for practical applications, simple structure, ease of use, speed to get the solutions, and robustness, parallel direct search method [6].

At the heart of the DE method is the strategy that the weighted difference between two vectors selected randomly is exerted on the perturbed vector to generate a trial vector, then the trial vector and the assigning vector exchange some elements according to probability, better individuals are selected as members of the generation $G+1$.

For example, one version DE/rand/2 updates according to the following formulates:

(1) Initial population,$X_{i,G}$, $i = 0, 1, 2, \cdots, NP - 1, NP$ is the number of population.
(2) Evolution operation, for every $X_{i,G}$, denote running vector.

Mutation

A mutation vector v is generated according to

$$V_{i,G+1} = X_{r1,G} + F1 * (X_{r2,G} - X_{r3,G}) + F_2 * (X_{r4,G} - X_{r5,G}) \qquad (1)$$

$r1, r2, r3, r4$ and $r5 \in [0, NP - 1]$ re mutually different integer and different from running index i; $F_1, F_2 \in [0, 2]$ are constant factor which controls the amplification of the differential variation $(X_{r2,G} - X_{r3,G})$ and $(X_{r4,G} - X_{r5,G})$.

Crossover

The trial vector is formed,

$$U_{i,G+1} = (u_{0i,G+1}, u_{1i,G+1}, \ldots, u_{(D-1)i,G+1}) \qquad (2)$$

$$\text{where }, u_{ji} = \begin{cases} v_{ji,G+1} & j =< n >_D, < n+1 >_D, \ldots, < n+L-1 >_D \\ x_{ji,G} & \text{otherwise} \end{cases} \qquad (3)$$

where $<>_D$ denote the modulo function with modulus D. The starting index n in (2) is a randomly chosen integer from the interval $[0, D-1]$. The integer L is drawn from the interval $[0, D-1]$ with the probability

$$Pr(L = v) = (CR)^v \qquad (4)$$

$CR \in [0, 1]$ is the crossover probability and constitutes a control variable for the DE scheme. The values of both n, D and L can refer to literature [6].

Selection

In order to determine whichever of $U_{i,G+1}$ and $X_{i,G}$ is transferred into the next generation, the fitness values of the two are comparedand the better is preserved.

(3) Stop Criterion

This process is repeated until a convergence occurs.

4   Applying DE Heuristic to Scheduling Problem

The DE algorithm with few control variables is robust, easy to use and lends itself very well to parallel computation [6]. However, the continuous nature of the algorithm limited DE to apply to combinatorial optimization problems. In order to use it in parallel program scheduling problem, we must re-define the operations in following way as to take into account the precedence relations between tasks.

4.1   Redefining the DE

Defining the vector: Every vector in differential evolution algorithm is represented by a feasible permutation of tasks, a tasks list satisfying topology order.

Defining of differential variation: In our proposed algorithm, the differential variations $(X_{r2,G} - Xr3, G)$ and $(X_{r4,G} - X_{r5,G})$ is defined as a set of Swap

Operator on task nodes in scheduling list[8]. Consider a normal solution sequence of multiprocessor scheduling with n nodes, here we define Swap Operator $SO(n_i, n_j)$ as exchanging node n_i and node n_j in scheduling list. Then we define $\widetilde{S} = S + SO(n_i, n_j)$ as a new solution on which operator $SO(n_i, n_j)$ acts. For example,

$$\{n_1, n_2, n_3, n_4, n_5, n_6\} + SO(n_1, n_3) = \{n_3, n_2, n_1, n_4, n_5, n_6\} \tag{5}$$

Plus operation between two SOs: Swap Set SS is a set of Swap Operators.

$$SS = (SO_1, SO_2, SO_3, \cdots, SO_n) \tag{6}$$

When Swap Set acts on a solution, all the Swap Operators of the swap Set act on the solution in order. i.e.

$$SS = \{(n_i^k, n_j^k), i, j \in \{1, 2, \cdots, N\}, k \in \{1, 2, \cdots, \}\} \tag{7}$$

which represents that node n_i^k and n_j^k are swapped firstly, and n_i^2 and n_j^2 are swapped secondly, and so forth. Define plus operation between SO_1 and SO_2 as the union of the two swap operators, denote

$$SO_1 + SO_2 \tag{8}$$

so Swap Set operation can be described by the following formula[8]:

$$\begin{aligned}
\widetilde{S} = S + SS &= S + (SO_1, SO_2, SO_3, \cdots, SO_n) \\
&= ((SO_1, SO \cdots) + SO_2) + \ldots + SO_n
\end{aligned} \tag{9}$$

The plus sign "+" above means continuous swap operations on one solution.

Plus operation between two SSs: If several Swap Sets have the same results as a single Swap Set acting on one solution, we define the operator "$\oplus$" as merging several Swap Sets into a new swap Set[8]. For instance, there is two Swap Sets SS_1 and SS_2, SS_1 and SS_2 act on one solution S in order, and there is another Swap Set $S\widetilde{S}$ acting on the same solution S, then get the same solution $\widetilde{S}$, call that $S\widetilde{S}$ is equivalent to $SS_1 \oplus SS_2$.

Minus operation between two vectors: Suppose there are two vectors, A and B, a Swap Set SS which can act on B to get vector A, i.e. we can swap the nodes in B according to A from left to right to get SS. So there must be an equation $A = B + SS$. We define the minus operation between vectors A and B as a SS, that is. $A = B + SS \Leftrightarrow A - B = SS$.

Updating: On the basis of above, Formula (1) has already no longer been suitable for the scheduling problem. We update it as follows:

$$V_{i,G+1} = X_{r1,G} + (X_{r2,G} - X_{r3,G}) \oplus (X_{r4,G} - X_{r5,G}) \tag{10}$$

4.2 Stochastic Topological Sorting Algorithm

On the basis of above, the initial vectors, initial scheduling list, must satisfy the precedence constrains. The topological sorting algorithm (TS) can serve the purpose, but the TS has two fatal disadvantages, one of which is that it is based on the depth-first search so that the topological orders generated by the TS algorithm cannot cover the whole feasible solutions, the other of which is that the topological order is fixed because it is subject to the storage structure of the DAG in the computer. In [9], we devise a novel sorting algorithm, a stochastic topological sorting algorithm (STS). The STS algorithm will be used in this paper to generate an initialized population.

4.3 Crossover Operation

In our algorithm, crossover operators adopted do not exchange the values of the elements but the order the elements appear in vectors to avoid permutation infeasibility when using permutation-based DE for the scheduling. We lend this idea from partially mapped crossover (PMX) [10], so that the order the activities appear in the multidimensional vectors other than the values of the elements are changed during updating process. The strategy of PMX that performs swapping elements is illustrated in Fig.1.

Here, the mutation vector and the running vector, respectively, resemble parent 1 and parent 2 in PMX, except that the two vectors should not be incorporated into a vector. Element(called element 1) of mutation vector will be determined according to formula(3) to see if the activity represented by the element will be moved to another placement or if the element will be swapped with another one (called element 2), which is the element that element 1 maps in the running vector[11]. When the placements of the two elements satisfy the predecessors-successors constraints, the crossover operation takes place.

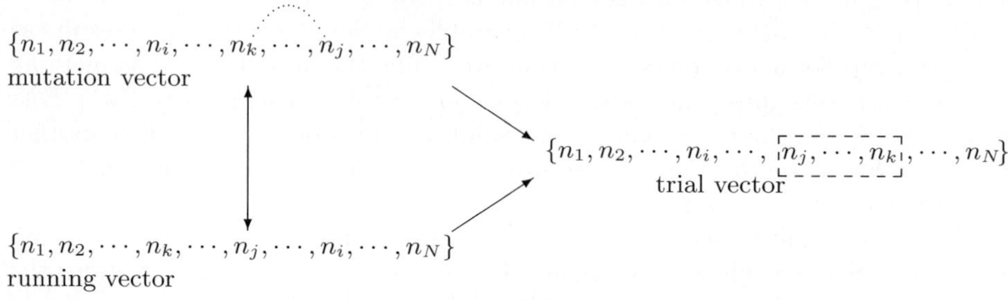

Fig. 1. An example of the partially mapped crossover operator in DE vectors

4.4 Simulated Annealing Selection Operation

Once a new solution is generated, a decision must be made whether or not to accept the newly derived vector as next generation. In standard DE, a greedy strategy is utilized to determine trial and running vector, which means the better of fitness of the two survive into next generation. The greedy criterion can converge fairly fast, but it runs the risk of becoming trapped by a local minimum[6].

The mechanism of Simulated Annealing is introduced into DE[12], which effectively optimize the objective function while moving both uphill and downhill. According to SA algorithm, selection strategy not only accepts better solutions (decreasing scheduling length) but also some worse solutions(increasing scheduling length). The Metropolis criterion decides on acceptance probability of worse solutions[12].

5   Experiment Results

Benchmark instances from website (http://www.mi.sanu.ac.yu/ tanjad/) are used to test the validity of our approach with three different selection strategies. In order to demonstrate the efficiency of the SA select operation, we test the benchmark instances using three strategies, SA selection operation (DE+SA), standard DE(STDE), the trial vector directing into next generation(DE+RAND) respectively.

In our experiments, we select the ogra100 problem with different edge densities , which is defined as the number of edges divided by the number of edges of a completely connected graph with the same number of nodes [5]. For each density, the mean deviations from the optimal schedule for 10 runs are recorded in Table1 for DE with and without SA selection strategy respectively. DE with SA always has comparable performance on the results of DE without SA whether using different processors or different densities.

Table 1. Results of the DE algorithm with different selection strategies compared against optimal solutions (% deviations) for the random task graphs with three densities using 2, 4, 6, and 8 processors.(amplification factor F1=F2=1, Iteration number equals 20).

Graph density	No. of processors											
	STDE				DE+RAND				DE+SA			
	2	4	6	8	2	4	6	8	2	4	6	8
60	3.75	8.00	9.25	10.88	2.88	7.13	12.88	11.25	2.88	6.75	8.63	6.79
70	2.75	6.38	7.25	7.25	2.63	8.50	12.13	9.38	1.63	6.75	6.79	6.75
80	1.63	5.00	7.63	4.63	1.38	8.13	10.25	7.38	1.00	4.00	4.75	4.75
90	1.00	1.12	5.48	5.50	1.13	3.50	5.50	5.50	0.50	1.63	4.00	4.00
Avg. Dev.	2.28	5.12	7.40	7.06	2.00	6.81	10.18	8.38	1.50	4.78	6.04	5.57

At the same time, we also investigate how the density of graphs affects the scheduling results. Because the lengths of optimal schedules are given depend onlyon task number, it appears that dense graphs spent less communication costs. Fig.2 is the result. Based on same tasks number and processors, less deviation is achieved for dense-task graphs.

Finally, we test the average deviation under different task numbers, and the result is illustrated in Fig.3. With the tasks number increasing, the communication cost between tasks assigned to different processors has a vital proportion to

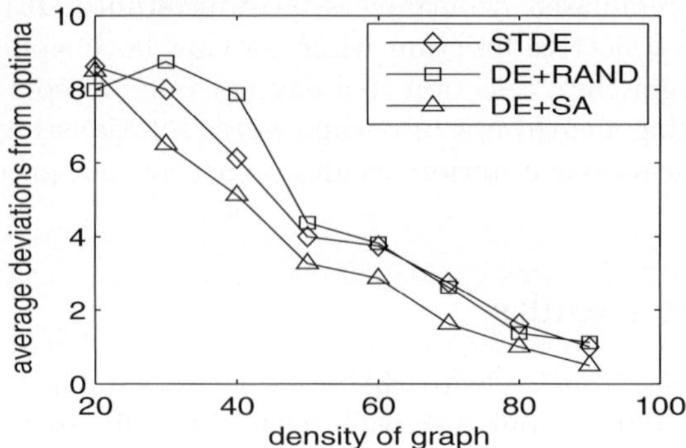

Fig. 2. The average % deviations from the optimal of the solutions generated by the DE algorithm for the random task graphs with different densities using 2 processors

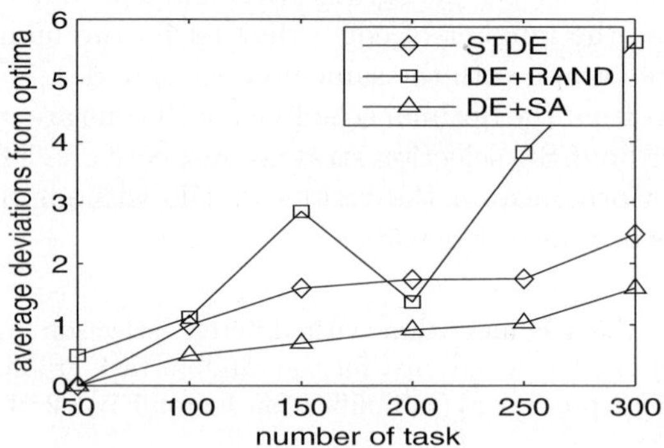

Fig. 3. The average % deviations from the optimal of the solutions generated by the DE algorithm for the random task graphs with different tasks using 4 processors

the overall cost. A conclusion can be observed that the DE algorithm with SA selection strategy for scheduling has a good scalability.

From the values above it is evident that a hybrid strategy of DE incorporating temperature-based acceptance criterion of SA is preferable to greedily selecting manipulation because the use of probability acceptance to inferior solutions in SA enhances the solution diversity in search process. The effect is always desirable due to the advantage of the DE with SA to avoid the local optima.

6   Summary

We have presented a differential evolution algorithm for multiprocessor DAG scheduling. As can be see from the previous results of the performance-testing experiment, in most cases the permutation-based DE method can find near-optimal schedule, especially the DE method with SA selection technique. In practice, we combine the DE technique of searching in global space with the SA capacity to jump out of local optimum in selecting an optimal scheduling list.

References

1. I. Ahmad, M. K. Dhodhi: Multiprocessor Scheduling in a Genetic Paradigm. Parallel Computing,Vol. 22 (1996) 395-406
2. Y.-K. Kwok, I. Ahmad: Efficient Scheduling of Arbitrary Task Graphs to Multiprocessors Using a Parallel Genetic Algorithm. J. Parallel and Distributed Computing,Vol 47,(1997) 58-77
3. T. Davidovic, P. Hansen, N. Mladenovic: Permutation Based Genetic, Tabu and Variable Neighborhood Search Heuristics for Multiprocessor Scheduling with Communication Delays. Asia-Pacific Journal of Operational Research,Vol. 22, (2005) 297-326
4. T. Davidovic, T. G. Crainic: New Benchmarks for Static Task Scheduling on Homogeneous Multiprocessor Systems with Communication Delays. Publication CRT-2003-04, Centre de Recherche sur les Transports, Universite de Montreal (2003)
5. T. Davidovic, P. Hansen, N. Mladenovic: Variable Neighborhood Search for Multiprocessor Scheduling Problem with Communication Delays. Proc. MIC2001, 4th Metaheuristic International Conference, J. P. de Sous, Porto, Portugal (2001)
6. R. Storn, K. Price: Differential Evolution-a Simple and Efficient Heuristic for Global Optimization over Continuous Space. Journal of Global Optimization, Vol. 11,(1997) 341-359
7. R. Storn: System Design by Constraint Adaptation and Differential Evolution. IEEE Transaction on Evolutionary Computation,vol. 3 (1999) 22-34
8. KANG-PING WANG, LAN HUANG,CHUN-GUANG ZHOU,WE1 PANG: Particle Swarm Optimization for Traveling Salesman Problem. Proceedings of the Second International Conference on Machine Learning and Cybernetics,vol. 5 (2003) 1583-1585
9. Wenbo Xu, Jun Sun: Efficient Scheduling of Task Graphs to Multiprocessors Using a Simulated Annealing Algorithm. DCABES 2004 Proceedings,Vol .1 (2004) 435-439
10. D.E. Goldberg: Genetic Algorithms in Search, Optimization, and Matching Learning. Addison-Wesley Publishing Company, Inc.,Reading (1989)
11. Hong Zhang, Xiaodong Li, Heng Li: Particle Swarm Optimization-Based Schemes for Resource-Constrained Project Scheduling. Automation in Construction,Vol. 14 (2005) 393-404
12. S.Kirkpatrick, C.Gelatt, M.Vecchi: Optimization by Simulated Annealing. Science,vol. 220 (1983) 671-680

Accelerating the Viterbi Algorithm for Profile Hidden Markov Models Using Reconfigurable Hardware

Timothy F. Oliver, Bertil Schmidt, Yanto Jakop, and Douglas L. Maskell

School of Computer Engineering, Nanyang Technological University, Singapore 639798
{yanto_jakop, tim.oliver}@pmail.ntu.edu.sg,
{asbschmidt, asdouglas}@ntu.edu.sg

Abstract. Profile Hidden Markov Models (PHMMs) are used as a popular tool in bioinformatics for probabilistic sequence database searching. The search operation consists of computing the Viterbi score for each sequence in the database with respect to a given query PHMM. Because of the rapid growth of biological sequence databases, finding fast solutions is of highest importance to research in this area. Unfortunately, the required scan times of currently available sequential software implementations are very high. In this paper we show how reconfigurable hardware can be used as a computational platform to accelerate this application by two orders of magnitude.

1   Introduction

Profile Hidden Markov Models (PHMMs) have been introduced to molecular biology to statistically describe protein families [6,8]. This statistical description can be used for sensitive and selective protein sequence database searching [5]. The scan operation consists of aligning each sequence in the database to a query PHMM using the well-known Viterbi algorithm [12]. This type of database search is widely used in biological research. Examples include searching for trans-membrane domains [9] and Spin/SSTY homologues [11], to name just a few. However, due to the quadratic time complexity of the Viterbi algorithm this search can take hours or even days depending on the database size, query PHMM length, and hardware used. Therefore, several parallel solutions for PHMM database searching have been developed on coarse-grained architectures such as clusters [13] and grids [2] as well as on fine-grained architectures such as SIMD boards [3,10] and graphics cards [7].

In this paper we show how re-configurable field-programmable gate array (FPGA)-based hardware platforms can be used to accelerate PHMM database scanning by two orders of magnitude. Since there is a large overall FPGA market, this approach has a relatively small price/unit and also facilitates upgrading to FPGAs based on state-of-the-art technology. We present a high-speed implementation on a Virtex II XC2V6000. The implementation is also portable to other FPGAs.

This paper is organised as follows. In Section 2, we introduce the Viterbi algorithm used to align a PHMM to a sequence. The parallel algorithm and its mapping onto a reconfigurable platform are explained in Section 3. The performance is evaluated and compared to previous implementations in Section 4. Section 5 concludes the paper with an outlook to further research topics.

V.N. Alexandrov et al. (Eds.): ICCS 2006, Part I, LNCS 3991, pp. 522–529, 2006.

2 Viterbi Algorithm for Profile Hidden Markov Models

Biologists have characterized a growing resource of protein families that share common function and evolutionary ancestry. PHMMs have been identified as a suitable mathematical tool to statistically describe such families and PHMM databases such as PFAM [1] have been created. The general transisiton structure of a PHMM is shown in Figure 1. It consists of a linear sequence of nodes. Each node has a match (M), insert (I) and delete state (D). Between the nodes are transitions with associated probabilities. Each match state and insert state also contains a position-specific table with probabilities for emitting a particular amino acid. Both transition and emission probabilities can be generated from a multiple sequence alignment of a protein family.

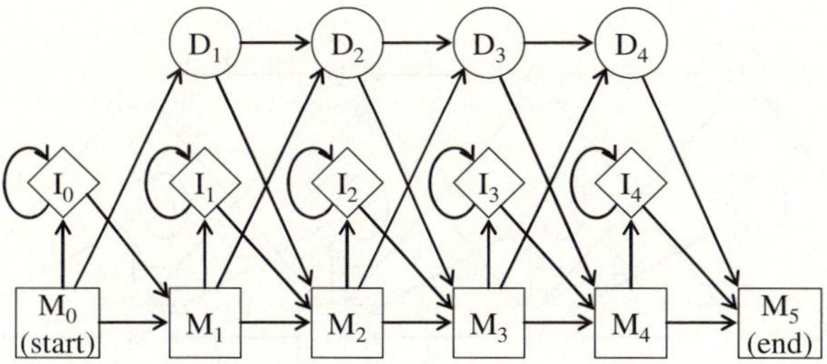

Fig. 1. The transition structure of a PHMM of length 4. Squares represent match states, circles represent delete states and diamonds represent insertions.

A PHMM can be aligned to a given protein sequence to determine the probability that the sequence belongs to the modeled protein family. The most probable path through the PHMM generating a sequence equal to the given sequence determines a similarity score. The well-known Viterbi algorithm can compute this score by dynamic programming (DP). The computation is given by the following recurrence relations.

$$M(i,j) \;=\; e(M_j,s_i) \;+\; \max \begin{cases} M(i-1,j-1)+tr(M_{j-1},M_j) \\ I(i-1,j-1)+tr(I_{j-1},M_j) \\ D(i-1,j-1)+tr(D_{j-1},M_j) \end{cases}$$

$$I(i,j) \;=\; e(I_j,s_i) \;+\; \max \begin{cases} M(i-1,j)+tr(M_j,I_j) \\ I(i-1,j)+tr(I_j,I_j) \end{cases}$$

$$D(i,j) \;=\; \max \begin{cases} M(i,j-1)+tr(M_{j-1},D_j) \\ D(i,j-1)+tr(D_{j-1},D_j) \end{cases}$$

where $tr(state1,state2)$ is the transition cost from $state1$ to $state2$ and $e(M_j,s_i)$ is the emission cost of amino acid s_i at state M_j. $M(i,j)$ denotes the score of the best path matching subsequence $s_1...s_i$ to the submodel up to state j, ending with s_i being emitted by state M_j. Similarly $I(i,j)$ is the score of the best path ending in s_i being emitted by I_j, and, $D(i,j)$ for the best path ending in state D_j. Initialization and termination are given by $M(0,0)=0$ and $M(n+1,m+1)$ for a sequence of length n and a PHMM of length m. By adding jump-in/out costs, null model transitions and null model emission costs the equation can easily be extended to implement Viterbi local scoring (see e.g. [4]).

An alignment example is illustrated in Figures 2, 3, and 4. A PHMM with transition scores is given in Figure 2. The emission scores of the M-states are given in Figure 3. The I-states emission scores in this example are set to zero, i.e. $e(I_j,s_i) = 0$ for all i ,j. The Viterbi DP matrix for computing the global alignment score of the sequence HEIKQ and the given PHMM is shown in Figure 4. The three values M, I, D at each position are displayed as $_D M^I$. A traceback procedure starting at $M(6,5)$ and ending at $M(0,0)$ (shaded cells in Figure 4) delivers the optimal path through the given PHMM emitting the sequence HEIKQ.

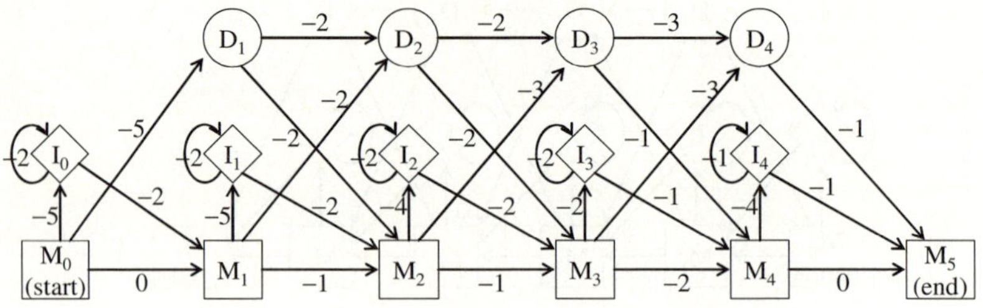

Fig. 2. The given PHMM of length 4 with transition scores

	A	C	D	E	F	G	H	I	K	L	M	N	P	Q	R	S	T	V	W	Y
M_1	-1	-1	-1	-1	1	-1	3	1	-1	-1	-1	-1	-1	-1	-1	-1	-1	-1	-1	-1
M_2	1	0	0	1	0	1	0	0	1	0	0	0	0	0	0	0	0	0	0	1
M_3	2	0	0	0	0	2	0	0	0	0	0	0	0	0	0	0	0	0	0	0
M_4	-1	-1	1	-1	-1	-1	1	-1	2	-1	-1	-1	-1	-1	-1	-1	1	-1	-1	-1

Fig. 3. Emission scores of the M-states for the PHMM in Figure 2

	0	1	2	3	4	5
∅	$*0*$	$_{-5}{-}\infty^{-\infty}$	$_{-7}{-}\infty^{-\infty}$	$_{-9}{-}\infty^{-\infty}$	$_{-12}{-}\infty^{-\infty}$	
H	$_{-\infty}{-}\infty^{-5}$	$_{-\infty}3^{-7}$	$_{1}{-}7^{-\infty}$	$_{-1}{-}9^{-\infty}$	$_{-4}{-}9^{-\infty}$	
E	$_{-\infty}{-}\infty^{-7}$	$_{-\infty}{-}8^{-2}$	$_{-4}3^{-2}$	$_{0}{-}1^{-4}$	$_{-3}{-}3^{-6}$	
I	$_{-\infty}{-}\infty^{-9}$	$_{-\infty}{-}6^{-4}$	$_{-6}{-}4^{-1}$	$_{-3}2^{-3}$	$_{-1}{-}2^{-5}$	
K	$_{-\infty}{-}\infty^{-11}$	$_{-\infty}{-}10^{-6}$	$_{-8}{-}5^{-3}$	$_{-5}{-}3^{0}$	$_{-4}2^{-3}$	
Q	$_{-\infty}{-}\infty^{-13}$	$_{-\infty}{-}12^{-8}$	$_{-10}{-}8^{-5}$	$_{-7}{-}5^{-2}$	$_{-6}{-}2^{-2}$	-2

Fig. 4. The Viterbi DP matrix for computing the global alignment score of the protein sequence HEIKQ and the given PHMM

3   Mapping onto a Reconfigurable Platform

The three values of I, D, and M of any cell in the Viterbi DP matrix can only be computed if the values of all cells to the left and above have been computed. But the calculations of the values of diagonally arranged cells parallel to the minor diagonal are independent and can be done simultaneously. Assuming we want to align a subject sequence to a query PHMM on a linear array of processing elements (*PEs*) this parallelization is achieved by mapping the Viterbi calculation as follows: one PE is assigned to each node of the query PHMM. The subject sequence is then shifted through the linear chain of PEs (see Fig. 5). If l_1 is the length of the subject sequence and l_2 is the length of the query PHMM, the comparison is performed in l_1+l_2-1 steps on l_1 PEs, instead of the $l_1 \times l_2$ steps required on a sequential processor.

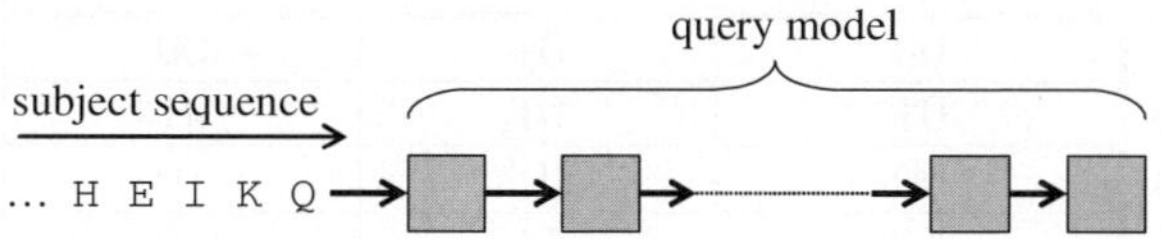

Fig. 5. Systolic sequence comparison on a linear processor array

Figure 6 shows our design for each individual PE. It contains registers to store the following temporary DP matrix values: $M(i-1,j-1)$, $I(i-1,j-1)$, $D(i-1,j-1)$ (upper-left cells) and $M(i-1,j)$, $I(i-1,j)$, $D(i-1,j)$ (upper cells). The values $M(i,j-1)$, $I(i,j-1)$, and $D(i,j-1)$ are stored in the left neighbour. Each PE holds the emission probabilities $e(M_j,s_i)$ and $e(I_j,s_i)$ for the corresponding PHMM node in two look-up-tables (LUTs). The look-ups of $e(M_j,s_i)$ and $e(I_j,s_i)$ and their addition are done in one clock cycle. The results are then passed to the next PE in the array together with the sequence character.

The data width (dw) is scaled to the required precision (usually $dw = 24$ bits is sufficient). The LUT depth is scaled to hold the number of emission scores per node (usually 20 for aminoacid sequences). The emission width (ew) is scaled to accommodate the dynamic range required by the emission score (usually $ew=16$ is sufficient). The look-up address width (lw) is scaled in relation to the LUT depth. All numbers are represented in 2-complement form. Furthermore, the adders in our PE design use saturation arithmetic.

In order to achieve high clock frequencies fast saturation arithmetic is crucial to our design. Therefore, we have added two tag bits to our number representation. These two tags encode the following cases: number (00), +max (01), −max (10), and not-a-number (NaN) (11). The tags of the result of an addition and maximum operation are calculated according to Table 1 and 2. Our representation has the advantage that result tags can be computed in a very simple and efficient way: if any of the operand's tags is set in an addition, a simple bit-wise OR operation suffices. Otherwise, the tags will be set according to the overflow bit of the performed addition.

Table 1. Computation of result tags in the case of an addition

add	number (00)	+max (01)	−max (10)	NaN (11)
number (00)	00[a]	01	10	11
+max (01)	01	01	11	11
−max (10)	10	11	10	11
NaN (11)	11	11	11	11

[a]except the case that the result produces an overflow, then the result tag is 01 (if MSB is set) or 10 (if MSB is not set)

Table 2. Computation of result tags in the case of a maximum operation

max	number (00)	+max (01)	−max (10)	NaN (11)
number (00)	00	01	00	11
+max (01)	01	01	01	11
−max (10)	00	01	10	11
NaN (11)	11	11	11	11

Assuming, we are aligning the subject sequence $S = s_1...s_M$ of length M to a query PHMM of length K on a linear processor array of size K using the Viterbi algorithm. As a preprocessing step, the transition and emission probabilities of states M_j, I_j, and D_j are loaded into PE j, $1 \leq j \leq K$. S is then completely shifted through the array in $M+K-1$ steps as displayed in Figure 5. In iteration step k, $1 \leq k \leq M+K-1$, the values $M(i,j)$, $I(i,j)$, and $D(i,j)$ for all i, j with $1 \leq i \leq M$, $1 \leq j \leq K$ and $k=i+j-1$ are computed in parallel in all PEs within a single clock cycle. For this calculation, PE j, $2 \leq j \leq K$, receives the values $M(i,j-1)$, $I(i,j-1)$, $D(i,j-1)$ and s_i from its left neighbor $j-1$, while all other required values are stored locally. Thus, it takes $M+K-1$ steps to compute the alignment score with the Viterbi algorithm. However, notice that after the last character of S enters the array, the first character of a new subject sequence can be input for the next iteration step. Thus, all subject sequences of the database can be pipelined with only one-step delay between two different sequences.

So far we have assumed a processor array equal in size of the query model length. In practice, this rarely happens. Since the length of the HMMs may vary, the computation must be partitioned on the fixed size processor array. The query model is usually larger than the processor array. For sake of clarity we firstly assume a query sequence of length K and a processor array of size N where K is a multiple of N, i.e. $K=p \cdot N$ where $p \geq 1$ is an integer. A possible solution is to split the computation into p passes:

The first N nodes of the query model are assigned to the processor array and the corresponding emission and transition scores are loaded. A number of database sequences to be aligned to the query model then cross the array; the M-, I-, and D-value computed in PE N in each iteration step are output. In the next pass the following N nodes of the query model are loaded into the array. The data stored previously is loaded together with the corresponding subject sequences and sent again through the processor array. The process is iterated until the end of the query model is reached.

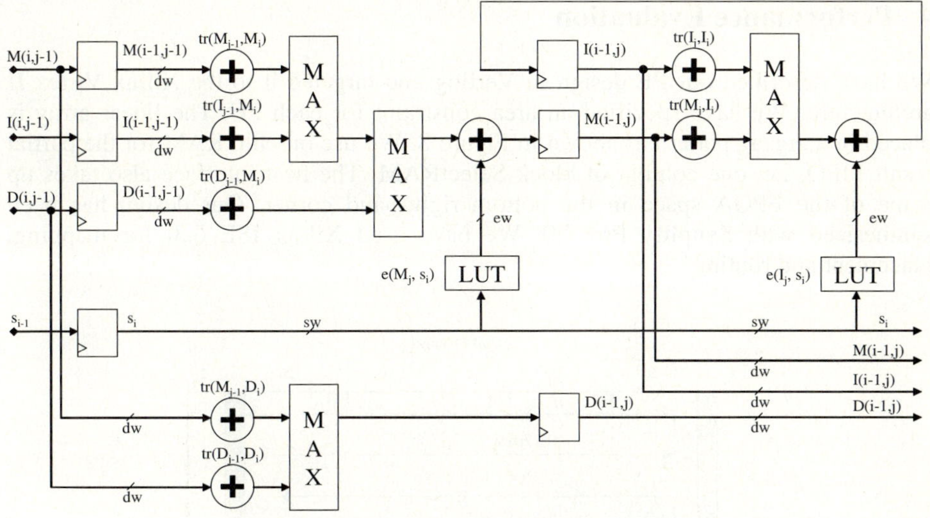

Fig. 6. Schematic diagram of our PE design

The database sequences are passed in from the host one by one through a first-in first-out (FIFO) buffer. The database sequences have been pre-converted to LUT addresses. For query lengths longer than the PE array, the intermediate results are stored in a FIFO. The FIFO depth is sized to hold the longest sequence in the database. The database sequence is also stored in the FIFO. On each consecutive pass an LUT offset is added to address the emission table corresponding to the model of the next iteration step within the PEs. Figure 7 illustrates this solution for 4 PEs.

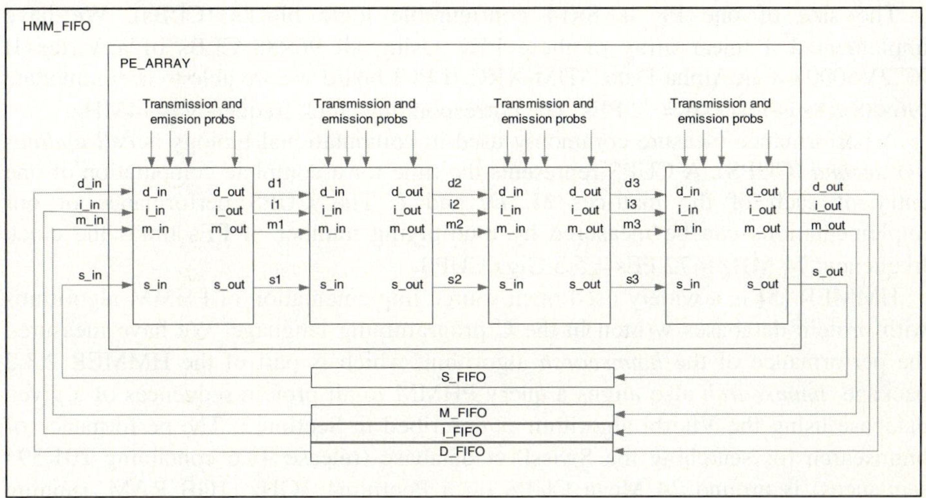

Fig. 7. System implementation

4 Performance Evaluation

We have described our PE design in Verilog and targeted it to the Xilinx Virtex II architecture. We have specified an area constraint for each PE. The linear array is placed in a zigzag pattern as shown in Figure 8. We use on-chip RAM for the partial result FIFO, i.e. one column of block SelectRAM. The host interface also takes up some of the FPGA space in the bottom right-hand corner. Our design has been synthesized with Synplify Pro 7.0. We have used Xilinx ISE 6.3i for mapping, placement and routing.

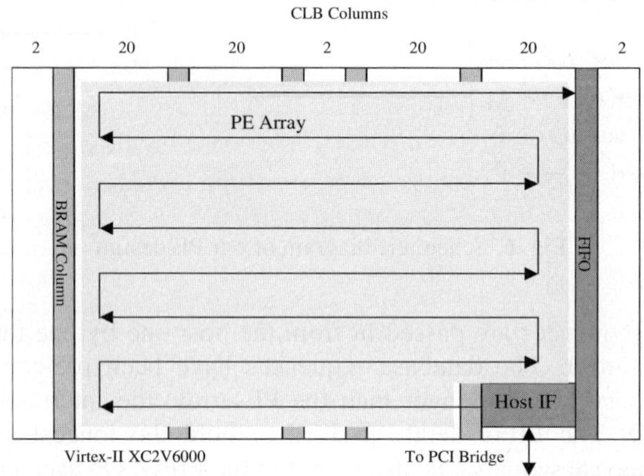

Fig. 8. System Floor plan in the XC2V6000 on the Alpha-Data ADM-XRC-II Board

The size of one PE is 8×14 configurable logic blocks (CLBs). We have implemented a linear array of these PEs. Using all 96×88 CLBs of a Virtex II XC2V6000 on an Alpha-Data ADM-XRC-II PCI board we are able to accommodate (96×88)/(8×14) =12×6 = 72 PEs. The corresponding clock frequency is 74MHz.

A performance measure commonly used in computational biology is *cell updates per second* (CUPS). A CUPS represents the time for a complete computation of one entry of each of the matrices M, D, and I. The CUPS performance of our implementations can be measured by multiplying number of PEs times the clock frequency: 74 MHz × 72 PEs = 5.3 Giga CUPS.

HMMER [5] is a widely used open source implementation of PHMM algorithms with protein databases written in the C programming language. We have measured the performance of the *hmmsearch* algorithm, which is part of the HMMER 2.3.2 package. *hmmsearch* also aligns a query PHMM to all protein sequences of a given database using the Viterbi algorithm as described in Section 2. The performance of hmmsearch for searching the SwissProt database (release 48.6 containing 201,594 sequences) is around 24 Mega CUPS on a Pentium4 3GHz, 1GB RAM, running Linux 2.6.11. Hence, our FPGA implementation achieves a speedup of 220.

5   Conclusions and Future Work

In this paper we have demonstrated that re-configurable hardware platforms provide a cost-effective solution to high performance biological sequence database searching with PHMMs. We have described a partitioning strategy to implement database scans using the Viterbi algorithm on a fixed-size processor array with varying query model lengths. Our PE design and partitioning strategy outperforms available sequential desktop implementations by two orders of magnitude. Our future work includes extending our design to compute local alignments between a sequence and a PHMM and making our implementation available to biologists as a webserver.

References

1. Bateman, A., et al: The PFAM Protein Families Database, *Nucleic Acid Research*, 32: 138-141 (2004)
2. Chukkapalli, G., Guda, C., Subramaniam, S.: SledgeHMMER: a web server for batch searching the pfam database, *Nucleic Acid Research* 32 (July), W542-544 (2004)
3. Di Blas, A. et al: The UCSC Kestrel Parallel Processor, *IEEE Transactions on Parallel and Distributed Systems* 16 (1) 80-92 (2005)
4. Durbin, R., Eddy, S., Krogh, A., Mitchison, G.: Biologcial Sequence Analysis, Probabilistic models of proteins and nucleic acids, *Cambridge University Press* (1998)
5. Eddy, S.R.: HMMER: Profile HMMs for protein sequence analysis, http://hmmer.wustl.edu (2003)
6. Eddy, S.R.: Profile Hidden Markov Models, *Bioinformatics* 14, 755-763 (1998)
7. Horn, D.R., Houston, M., Hanrahan, P.: ClawHMMER: A Streaming HMMer-Search Implementation, *ACM/IEEE Conference on Supercomputing* (2005)
8. Krogh A., Brown, M., Mian, S., Sjolander, K., Hausler, D.: Hidden Markov Models in computational biology: Applications to protein modeling, *Journal of Molecular Biology* 235, 1501-1531 (1994)
9. Narukawa, K., Kadowaki, T.: Transmembrane regions prediction for G-protein-coupled receptors for hidden markov models, *Proc. 15th Int. Conf. on Genome Informatics* (2004)
10. Schmidt, B., Schröder, H.: Massively Parallel Sequence Analysis with Hidden Markov Models, *International Conference on Scientific & Engineering Computation*, World Scientific, Singapore (2002)
11. Staub, E., Mennerich, D., Rosenthal, A.: The Spin/Ssty repeat: a new motif identified in proteins involved in vertebrate development from gamete to embryo, *Genome Biology* 3, 1 (2001)
12. Viterbi, A.J.: Error bounds for convolutional codes and an asymptotically optimum decoding algorithm, *IEEE Transactions on Information Theory* 13, 2, 260-269 (1967)
13. Zhu, W., Niu, Y., Lu, J., Gao, G.R.: Implementing Parallel Hmm-Pfam on the EARTH Multithreaded Architecture, *2nd IEEE Computer Society Bioinformatics Conference*, 549-550 (2003)

Benchmarking and Adaptive Load Balancing of the Virtual Reactor Application on the Russian-Dutch Grid

Vladimir V. Korkhov[1,2] and Valeria V. Krzhizhanovskaya[1,2]

[1] University of Amsterdam, Faculty of Science, Section Computational Science
[2] St. Petersburg State Polytechnic University, Russia
{vkorkhov, valeria}@science.uva.nl

Abstract. This paper addresses a problem of porting a distributed parallel application to the Grid. As a case study we use the Virtual Reactor application on the Russian-Dutch Grid testbed. We sketch the Grid testbed infrastructure and application modular architecture, and concentrate on performance issues of one of the core parallel solvers on the Grid. We compare the performance achieved on homogeneous resources with that observed on heterogeneous computing and networking infrastructure. To increase the parallel efficiency of the solver on heterogeneous resources we developed an adaptive load balancing algorithm. We demonstrate the speedup achieved with this technique and indicate the ways to further enhance the algorithm and develop an automated procedure for optimal utilization of Grid resources for parallel computing.

Keywords: Grid, benchmarking, adaptive load balancing, heterogeneous resources, parallel distributed application, Virtual Reactor, PECVD.

1 Introduction

The importance of fully integrated simulators is recognized by various research groups and scientific software companies [1]. Our Virtual Reactor application [2,3] was developed for simulation of plasma enhanced chemical vapour deposition (PECVD) reactors, multiphysics systems spanning a wide range of spatial and temporal scales. Simulation of three-dimensional flow with chemical reactions and plasma discharge in complex geometries is one of the most challenging and demanding problems in computational science and engineering, requiring both high-performance and high-throughput computing. This application serves as a test-case driving and validating the development of the Russian-Dutch computational Grid (RDG) for distributed high performance simulation [4]. The Virtual Reactor is particularly suitable for "gridification" since it can be decomposed into a number of functional components. Moreover, this application requires large parameter space exploration, which can be efficiently organized on the Grid. For that we rely upon the Nimrod-G parameter sweep middleware [5]. Our work on porting the Virtual Reactor to the Grid started within the CrossGrid EU project [6]. Some results of these efforts were reported in [2]. The RDG Grid is the successor of the CrossGrid as it is based on CrossGrid software and serves as a testbed for the Virtual Reactor.

In this paper we present the results of ongoing work on porting the Virtual Reactor application to the Russian-Dutch Grid. We demonstrate the results of benchmarking

V.N. Alexandrov et al. (Eds.): ICCS 2006, Part I, LNCS 3991, pp. 530–538, 2006.

of one of the parallel solvers on the RDG, indicate the bottleneck of the parallel algorithm used on heterogeneous Grid resources, and propose a generic approach for adaptive workload balancing that takes into account the processors power and interprocessor communications. Further we show the results of implementation of the load balancing algorithm, and conclude the paper with discussion and future plans.

2 Russian-Dutch Grid Testbed Infrastructure

Generally a site within a Grid testbed can be of one of the four types depending on homo- or heterogeneity of underlying resources: homogeneous worker nodes on uniform (I) or non-uniform (II) links; and heterogeneous nodes on uniform (III) or non-uniform (IV) links. Currently the Russian-Dutch Grid testbed consists of five sites with different infrastructures: Amsterdam (3 nodes, 4 processors) and St. Petersburg (4 nodes, 6 processors) – type IV; Novosibirsk (3 processors) – type II; Moscow1 (12 nodes, 24 processors) and Moscow2 (14 processors) – type I.

The Russian-Dutch Grid testbed is built with the CrossGrid middleware [6] based on the LCG-2 distributions, and sustains the interoperability with the CrossGrid testbed. More information on the RDG testbed can be found in [4].

3 Porting of the Virtual Reactor to the Grid

The Virtual Reactor application includes the basic components for reactor geometry design; computational mesh generation; plasma, flow and chemistry simulation; editors of chemical processes and gas properties connected to the corresponding databases; pre- and postprocessors, visualization and archiving modules [2]. This is schematically shown in Fig. 1, where we emphasize the *simulation* components.

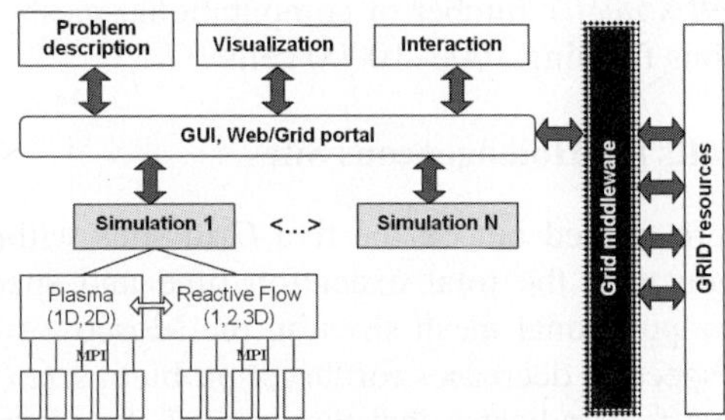

Fig. 1. Functional scheme of the Virtual Reactor application

The aim of our research is to virtualize separate components of the application to run them as services and combine on the Grid. The core components are modules simulating gas flow, chemical reactions and film deposition processes occurring in a PECVD reactor. The details on numerical methods and parallel algorithm employed in the solver are described in [7]. The most important features are the following: for stability reasons, implicit schemes were applied, thus forcing us to use a sweep-type algorithm for solving equations in every "beam" of computational cells in each spatial direction of the Cartesian mesh. A special parallel algorithm was developed with beams distribution among the processors with communications exploiting a synchronous Master-Slave model [7]. The algorithm was implemented using the MPI message passing interface. In the testbed we use generic MPICH-P4 built binaries that can be executed on all the testbed machines using Globus job submission service.

In order to make a Grid application more efficient it is necessary to perform initial benchmarking of the application modules on Grid resources to reveal existing bottlenecks in the application architecture and possible mismatch with the Grid environment. The results of this benchmarking activity are described in the next section.

4   Benchmarking of the Virtual Reactor on the Grid

The tests performed on the CrossGrid testbed showed that most of the interactive components of the Virtual Reactor do not set restrictions on the environment and can be effectively run on distributed Grid resources. Here we concentrate on benchmarking of the *simulation* modules. Each simulation consists of two basic components: one for plasma simulation and another for reactive flow simulation (see Fig. 1). These two components exchange only a small amount of data every hundred or thousand time steps, therefore the network bandwidth is not critical for their communication. Next, we focus on benchmarking the individual parallel solvers, starting from a 2D reactive flow solver. To measure the dependency of solver performance upon the input data, multiparameter variation (of the computational mesh size, number of simulation time steps, and number of processors) has been applied. We started from a light-weighted problem not simulating the chemical and plasma processes, with a simplified reactor geometry consisting of a single block that allows easy tracking of parameter influence on the execution time. In these tests, a single-block topology was used. The block was subdivided into a (*ncell* x *ncell*) number of computational mesh cells, with *ncell* running from 40 to 100, thus forming 1600–10000 cells.

4.1   Benchmark Results for Homogeneous Sites

The measurements were carried out on the five Grid sites within the RDG testbed. Figures 2 and 3 demonstrate the total execution time and speedup of the parallel solver for different computational mesh sizes on the Moscow site of Type I (homogeneous cluster). The speedup decreases for larger problem size (with more computational mesh cells). This fact indicates that the ratio of the interprocess communications bandwidth to the processor performance is not high enough for the light-weighted problems with relatively small number of operations per computational cell. To get insight into the computation/communication relations within the solver we measured the communication time for different types of the problem and for varying mesh sizes. We observed that for light-weighted problems interprocess communication time grows super-linearly with increasing the mesh size, although the amount of data transferred is linearly proportional to the number of mesh cells. This behaviour shall be studied further by extensive benchmarking of the network links. Some peculiarities in the communication time can be seen in Fig. 4: (1) The communication time grows non-monotonically with the number of processors, but drops down on every processor with an even number; and (2) The time of MPI Receive calls is an order of magnitude higher for the first few processors.

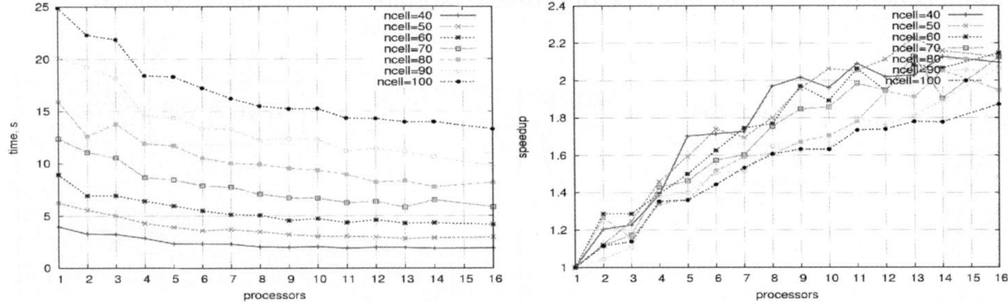

Fig. 2. Dependency of the total execution time on the number of processors

Fig. 3. Speedup achieved by the solver for different computational mesh sizes

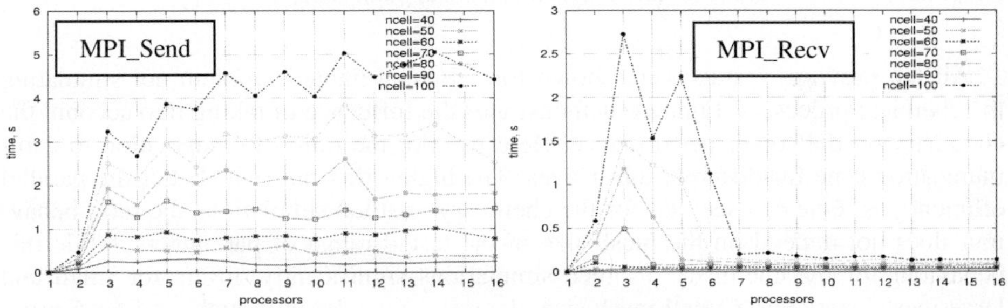

Fig. 4. Dependency of the communication time on the number of processors for different computational mesh sizes

These results reflect the topology, network and nodes features of the tested Grid site:

(1) Since the site consists of two-processor nodes, the network channels work more efficiently for data transfers between the Master and a Slave processor if a connection was already established with another Slave processor on the same node. This can be explained by implementation of the MPI library which saves network resources while opening and maintaining connections for concurrent processes on the same node.

(2) The "peaks" of the MPI Receive time for the first few processors (see Fig. 4 right) are caused by the constraints on the portions of data that could be accommodated at once. The constraining factors could be the network bandwidth distribution, the processor cache size, the memory available on the node, or a combination of these factors.

In Figure 5 the total execution time is presented along with the contributions of calculation and communication. For a smaller mesh (Fig. 5 left), the communication time makes a relatively small contribution into the total execution time even for a large number of processors involved. For a larger mesh (Fig. 5 right), communication makes up to 30% of the execution time. This result confirms that the network bandwidth is not sufficient for this type of problem (see also explanations to Fig. 3).

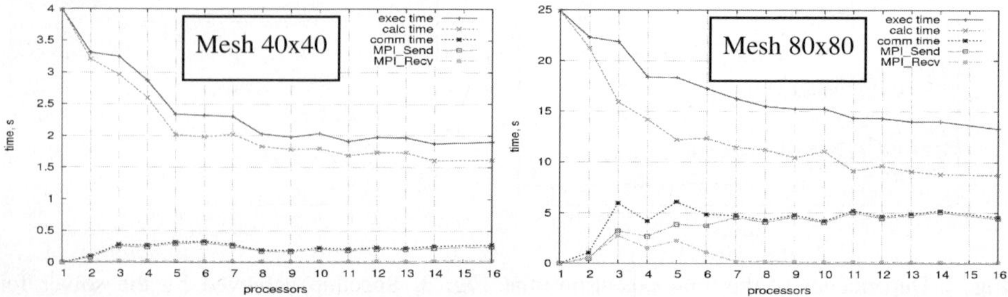

Fig. 5. Total execution time and contributions of the calculation and communication depending on the number of processors for different computational mesh sizes

All the previous results were shown for a light-weighted problem not simulating the chemical processes. Figure 6 demonstrates the influence of taking into account the chemistry on the solver performance. Here we plot the ratio of computation to communication time for different mesh sizes. The higher this ratio, the better the parallel efficiency is. One can see that for the chemistry-enabled simulations the ratio behaviour does not depend on the mesh size in the tested range of parameters, while this behaviour for the chemistry-disabled simulations significantly differs for small and large mesh sizes. For a small mesh size, the ratio stays decently high, and for 6 processors and more it reaches the level of the chemistry-enabled simulations. For a larger mesh, the computation/communication ratio for the no-chemistry simulations is very low, thus diminishing the overall parallel efficiency.

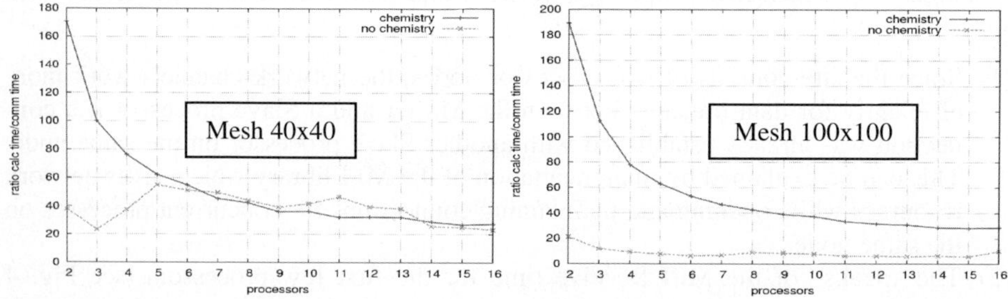

Fig. 6. The ratio of computation to communication time for chemistry-enabled and chemistry-disabled simulations

4.2 Heterogeneous Sites: Load Balancing and Benchmarking

The parallel algorithm was originally developed for homogeneous computer clusters with equal processor power, memory and interprocessor communication bandwidth. In case of submitting equal portions of a parallel job to the nodes with different performance, all the fast processors have to wait at the barrier synchronization point till the slowest ones get the job done. The same problem occurs if the network connection from the Master processor to some of the Slave processors is much slower than to the others. As we have shown in the previous section, for communication-bound simulations (chemistry-disabled simulation with large computational meshes), the communication time on low-bandwidth networks is of the order of calculation time, therefore the heterogeneity of the interprocessor communication links is a hindrance as considerable as the diversity of the processor power. One of the natural ways to adapt the solver to the heterogeneous Grid resources is to distribute the portions of job among the processors proportionally to the processor performance and network connections.

The issue of load balancing in Grid environment is addressed by a number of research groups. Generally studies on load balancing consider distribution of processes to computational resources on the system/library level with no modifications in the application code [11]. Less often load balancing code is included into the application source code to improve performance in specific cases [12]. Some research projects concern load balancing techniques that use source code transformations to improve the execution of the application [10]. We employ the application-centric approach where the balancing decisions are taken by the application itself, however the algorithm and the code estimating available resources and suggesting the optimal load balancing of a parallel job is generic and can be employed in any parallel application to be executed on heterogeneous resources.

We developed a mechanism for estimating the "weight" of a processor according to its processing power and network connection to the Master processor. The values of the weights determine how much work will be executed by each processor. Similar approach was used in [9] for heterogeneous computer clusters, however the same tools can not be used in Grid environments, where the weights shall be calculated every time the solver is started on a new set of dynamically assigned processors.

The link bandwidth between the Master and Slave processors is estimated using MPI_Send transfers of a predefined data block (MPI buffer size is 10^6 of MPI_DOUBLEs) in the beginning of the solver execution, after the resources were allocated. The CPU power was obtained by a function from the *perfsuite* library [8]. The node weights were calculated as follows:

$$weight_i = c\_CPU \cdot CPUweight_i + c\_NET \cdot NETweight_i;$$

$$CPUweight_i = CPU_i \Big/ \sum_j CPU_j \; ; \; NETweight_i = MPI\_SendTime_i^{-1} \Big/ \sum_j MPI\_SendTime_j^{-1}$$

The main factor in distributing the load was the processor power (c_CPU=1.0), and to take into account the influence of the network connections we introduced the cn parameter (cn=c_NET=[0.0 … 2.0]). The value of cn=0 means that the diversity of communication links is not taken into account, and cn=1 means that influence of the links bandwidth is considered as important as the processor power.

To illustrate the approach described above, we present the results obtained for a light-weighted problem of chemistry-disabled simulation of a real reactor geometry with 10678 cells on St. Petersburg Grid site, which is heterogeneous in both CPU power and network connections of the nodes (Type IV). There are two 3 GHz nodes and two dual 450 MHz nodes. One of the dual nodes is placed in a separate network segment with 10 times lower bandwidth (10 Mbit/s against 100 Mbit/s in the main segment). Figures 7 and 8 illustrate the speedup achieved by applying the workload balancing technique with different values of the network influence parameter cn. The speedup was calculated as the ratio of the execution time without load balancing to that applying the balancing algorithm. The most noticeable speedup is observed for 3 and 4 processors in the considered resource configuration. This is explained by the fact that in these cases the solver was run on equal-performance processors connected with the network links of different bandwidth (Type II infrastructure). Figure 8 shows that in this case the speedup grows linearly with the increase of the network influence coefficient cn. The slowdown observed on 2 processors with our balancing algorithm is discussed in the next section (item 4).

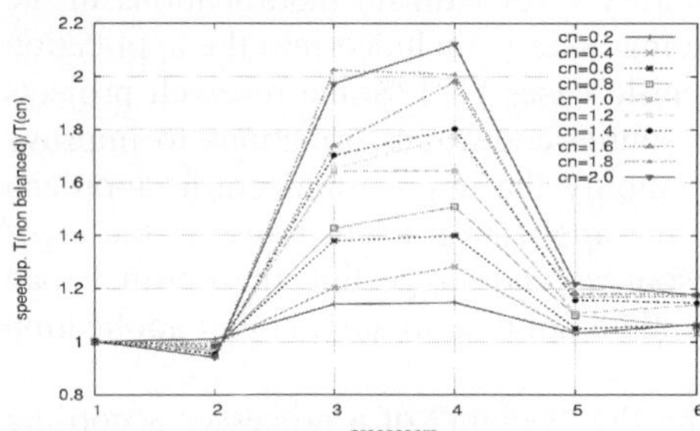 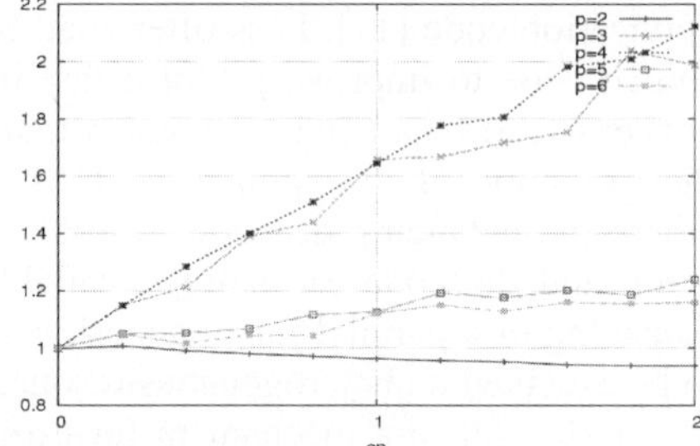

Fig. 7. Speedup of the load-balanced version compared to the non-balanced solver

Fig. 8. Dependency of speedup on the network influence parameter cn

5 Discussion

Analysis of the results achieved with the workload balancing algorithm suggests that the following issues shall be addressed in order to optimize the balancing technique:

1. The type of resources assigned to the parallel solver shall play a role in choosing the c_CPU and cn coefficients: for the Type II resources (homogeneous worker nodes with heterogeneous interconnections) c_CPU shall be set to 0.0, as only the network heterogeneity shall be compensated by load balancing. For the Type III (heterogeneous worker nodes with uniform interconnections) cn shall be 0.0; and only for the Type IV (heterogeneous nodes with heterogeneous interconnections) both c_CPU and cn parameters shall be adjusted optimally. This can be done automatically by enriching the weighting algorithm with a function analyzing the CPU and network responses of the nodes participating in the simulation.

2. To choose optimal values of the network weighting coefficient cn for the Type IV resources, for each particular problem to be simulated we shall analyze the ratio of

the computation to communication time. This can be also theoretically estimated as a function of the CPU power to the network bandwidth ratio.

3. To measure the interprocess communication rate, we sent a fixed amount of data from the Master to each Slave processor. However the response of the communication channels to increasing amount of data is not scaled linearly. For the slower networks this tendency is even more pronounced. This brings us to a conclusion that the amount of data sent to measure the links performance shall be close to the amount really transferred within the solver for every particular problem, mesh size, geometry and number of processors in a parallel job.

4. To calculate the weight of the Master processor, we used a fixed artificial value of the *MPI_SendTime* for this processor. Often it was much lower than the values of measured connections to the Slaves. It caused assigning excessive load for the Master processor, which slowed down the simulation because the Master shall perform co-ordination and execute some additional functions. A simple solution would be to dynamically set this parameter to the value of a Slave processor with the fastest link to the Master.

7 Conclusions and Future Work

In this paper we addressed the issue of porting a cluster-based problem solving environment to the Grid using as a test case a distributed parallel Virtual Reactor on the Russian-Dutch Grid testbed. We illustrated the performance issues that occur while porting computational components from homogeneous cluster environment to the Grid. To adapt the parallel programs to the heterogeneous Grid resources, we developed a generic workload balancing technique that takes into account specific parameters of the Grid resources dynamically assigned to a parallel job. We plan to enhance the algorithm and create a library for automatic load balancing on the Grid..

Benchmarking the components of a distributed application allowed us to evaluate their performance dependencies. Applying these results to improve Grid resource management for the Virtual Reactor is another direction of our future work.

Acknowledgments. The authors would like to thank Irina Shoshmina, Breanndán Ó Nualláin and the RDG Grid deployment team for their assistance. The research was conducted with financial support from the NWO/RFBR projects 047.016.007 and 047.016.018, and from the Virtual Laboratory for e-Science project (www.vl-e.nl).

References

1. www.cfdrc.com, www.fluent.com, www.semitech.us, www.softimpact.ru
2. V.V. Krzhizhanovskaya et al. *Grid -based Simulation of Industrial Thin-Film Production.* Simulation: Transactions of the Society for Modeling and Simulation International, V. 81, No. 1, pp. 77-85 (2005)
3. V.V. Krzhizhanovskaya et al. *A 3D Virtual Reactor for Simulation of Silicon-Based Film Production.* Proceedings of the ASME/JSME PVP Conference. ASME PVP-Vol. 491-2, pp. 59-68, PVP2004-3120 (2004)
4. Project "High performance simulation on the Grid" http://grid.csa.ru/
5. Nimrod-G: http://www.csse.monash.edu.au/~davida/nimrod/

6. CrossGrid EU Science project: http://www.eu-CrossGrid.org
7. V.V. Krzhizhanovskaya et al. *Distributed Simulation of Silicon-Based Film Growth.* Proceedings of the 4th PPAM conference, LNCS, V. 2328, pp. 879-888. Springer-Verlag 2002
8. R. Kufrin. *PerfSuite: An Accessible, Open Source Performance Analysis Environment for Linux.* 6th International Conference on Linux Clusters. Chapel Hill, NC. (2005)
9. J.D. Teresco et al. *Resource-Aware Scientific Computation on a Heterogeneous Cluster.* Computing in Science & Engineering, V. 7, N 2, pp. 40-50, 2005
10. R. David et al. *Source Code Transformations Strategies to Load-Balance Grid Applications.* LNCS vol. 2536, pp. 82-87, Springer-Verlag, 2002
11. A. Barak et al. *The MOSIX Distributed Operating System, Load Balancing for UNIX,* LNCS, vol. 672, Springer-Verlag, 1993
12. G. Shao et al. *Master/Slave Computing on the Grid.* Proceedings of Heterogeneous Computing Workshop, pp 3-16, IEEE Computer Society (2000)

Improved Prediction Methods for Wildfires Using High Performance Computing: A Comparison*

Germán Bianchini, Ana Cortés, Tomàs Margalef, and Emilio Luque

Departament d'Informàtica, E.T.S.E, Universitat Autònoma de Barcelona,
08193-Bellaterra (Barcelona) Spain

Abstract. Recently, dry and hot seasons have seriously increased the risk of forest fire in the Mediterranean area. Wildland simulators, used to predict fire behavior, can give erroneous forecasts due to lack of precision for certain dynamic input parameters. Developing methods to avoid such parameter problems can improve significantly the fire behavior prediction. In this paper, two methods are evaluated, involving statistical and uncertainty schemes. In each one, the number of simulations that must be carried out is enormous and it is necessary to apply high-performance computing techniques to make the methodology feasible. These techniques have been implemented in parallel schemes and tested in Linux cluster using MPI.

1   Introduction

Forest fires are a very serious hazard that, every year, cause significant damage around the world from the ecological, social, economic and human point of view [7]. These hazards are particularly dangerous when meteorological conditions are extreme with dry and hot seasons. An example of fire effects under severe conditions is summer 2003. That year, the temperatures in the Mediterranean area were extremely high and there was also a lack of precipitations. A relevant example of fire effects under these conditions was Portugal. In this country alone, 420,000 hectares were burned and, as a consequence, 20 people died.

To reduce the negative effects of fire it is crucial to improve current fire risk assessment methods. These methods base their recommendations on the results provided by a certain wildfire simulator. However, in most cases, the results provided by simulation tools do not match real propagation. Thus, such simulation tools are not wholly useful, since predictions are not reliable. One of the most common sources of deviation in fire simulation spread from that of real-fire propagation is imprecision in input simulator parameters. There are certain parameters that could be defined as static factor parameters such as the terrain slope and the vegetation type because they remain invariable through time. However,

* This work has been supported by the MEyC-Spain under contract TIN 2004-03388 and by the European Commission under contract EVG1-CT-2001-00043 SPREAD.

V.N. Alexandrov et al. (Eds.): ICCS 2006, Part I, LNCS 3991, pp. 539–546, 2006.

there are other parameters, which can be referred to as dynamic parameters, which typically are very difficult to determine as the fire progresses. Some examples of this kind of parameters are the moisture content in the vegetation and wind conditions [12].

One of the most common ways of approaching this problem, consists in searching for an optimal set of input parameters in order to obtain the best simulation results [4]. However, there are some authors that reject the idea that there is only one optimum parameter set in a simulator calibration. They consider that there are multiple parameter sets that may be acceptable in simulating the system under study [2, 11]. GLUE is one of the most relevant methods within this category. The main drawback of this method is caused by its random way of working. To overcome this difficulty, we propose the S^2F^2M method. S^2F^2M applies statistical analysis by simulating the fire propagation, considering a wide range of parameter combinations to determine different scenarios.

The GLUE and S^2F^2M methods are described in section 2 and 3 respectively. Since both strategies need to execute a large number of simulations, we have used the parallel scheme master-worker, implemented with an MPI [8] as a message-passing library and executed on a PC Linux cluster. The implementation details are described in section 4. Section 5 provides a comparison study between both strategies based on real field fires. And finally, the main conclusions are reported in section 6.

2   GLUE Methodology

The GLUE method of Beven and Binley [2] is a Monte Carlo simulation based approach to model conditioning and uncertainty estimation. It rejects the idea that there is only one optimum parameter set in a model calibration. It considers that there are multiple parameter sets and even multiple model structures that may be acceptable in simulating the system under study. Therefore, it is possible to evaluate the relative likelihood of a given model and parameter set in reproducing the available data to test the models. Then, uncertainty in the predictions may be estimated by calculating a likelihood weighted cumulative distribution of a predicted variable based on the simulated values from all the retained simulations (those with a likelihood value greater than zero). Thus, for any model predicted variable, Z:

$$P(\hat{Z}_t < z) = \sum_{i=1}^{N} L = \left[M(\Theta_i) | \hat{Z}_{t,i} < z \right]$$

where $P(\hat{Z}_t < z)$ are prediction quantiles, $\hat{Z}_{t,i}$ is the value of variable Z at time t simulated by model $M(\Theta_i)$ with parameter set Θ_i and likelihood $L[M(\Theta_i)]$. Then, the accuracy in estimating such prediction quantiles will depend on having a suitable sample of models to represent the behavioral part of the model. In this framework the parameter values are treated as a set with their associated likelihood value so that any interactions between parameter values in fitting the available observations are included implicitly in the conditioning process.

In our case, we use a fuzzy measure of goodness of fit. Initial prior likelihoods were set to zero for all the parameter sets. The updating of likelihoods from one time step to the following one consisted in averaging of the prior and the current likelihoods. In order to make the uncertainty limits converge when the actual rate of spread did not change, this average can be raised, optionally, to a power p ($p \leq 1$):

$$L_p(M(\Theta_i)) = \frac{[L_0(M(\Theta_i) + L(M(\Theta_i)|Y)]^p}{C}$$

Where $L_0(M(\Theta_i)$ is the prior likelihood of the model M with the parameter set Θ_i; $L(M(\Theta_i)|Y)$ is the goodness of fit of the of the model M with the parameter set Θ_i to the latest observations Y; $L_p(M(\Theta_i)$ is the posterior likelihood of the model M with the parameter set Θ_i; and C is a constant which ensures the sum of the posterior likelihoods of all the parameters to 1.

3 S^2F^2M Methodolgy

The methodology of S^2F^2M is based on statistics. When there are a lot of significant factors involved (i.e. weather, wind speed, slope, etc.), the best strategy is to use a **factorial experiment**. A factorial experiment is one in which the factors vary at the same time [10] (for example, wind conditions, moisture content and vegetation parameters). A **scenario** represents each particular situation that results from a set of values.

For a given time interval, we want to know whether a portion of the terrain (called a cell) will be burnt or not. If n is the total number of scenarios and n_A is the number of scenarios in which the cell was burned, we can calculate the **ignition probability** as:

$$P_{ign}(A) = n_A/n$$

The next step is to generalize this reasoning and apply it to some cell sets. In this manner we obtain a matrix with values representing the probability of each cell catching fire.

In this way, after calculation of the ignition probability (P_{ign}) for each cell, we can compare the real case against our matrix. Then, we find a key P_{ign} which defines an area similar to the real situation. So, we can use this value to predict in a next time the possible fire behavior.

S^2F^2M uses a forest fire simulator as a black box which needs to be fed with different parameters in order to work. A particular setting of the set of parameters defines an individual scenario. These parameters correspond to the parameters proposed in the Rothermel model [12].

For each parameter we define a range and an increment value, which are used to move throughout the interval. For a given parameter i (which we will refer to as *Parameter_i*) the associated interval and increment is expressed as:

[Inferior_threshold_i, Superior_threshold_i], Increment_i

Then, for each parameter i, it is possible to obtain a number C_i (parameter domain cardinality), which is calculated as follows:

$$C_i = \frac{((Superior\_threshold\_i - Inferior\_threshold\_i) + Increment\_i)}{Increment\_i}$$

Finally, from each parameter's cardinality it is possible to calculate the total number of scenarios obtained from variations of all possible combinations.

$$\#Scenarios = \prod_{i=1}^{p} C_i$$

where p is the number of parameters.

4   Implementation

Both methods described above have been implemented in two operational systems that incorporate a simulation kernel and apply a methodology to evaluate the fitness function. This system has been developed on a PC Linux cluster using MPI as the message passing library.

4.1   The Simulator

The S^2F^2M and GLUE system use as a simulation core the wildland simulator proposed by Collin D. Bevins, which is based on the fireLib library [5]. **fireLib** is a library that encapsulates the BEHAVE fire behavior algorithm [1]. In particular, this simulator uses a cell automata approach to evaluate fire spread. The terrain is divided into square cells and a neighborhood relationship is used to evaluate whether a cell will be burnt and at what time the fire will reach those cells.

As inputs, this simulator accepts maps of the terrain, vegetation characteristics, wind and the initial ignition map.

The output generated by the simulator consists of a map of the terrain in which each cell is labeled with its ignition time.

4.2   The Fitness Function

To evaluate and compare the systems' responses, we defined a fitness function. Since both systems use an approximation based on cells, the fitness function was specified as follows:

$$Fitness = \frac{(\#cells \bigcap -\#IgnitionCells)}{(\#cells \bigcup -\#IgnitionCells)}$$

where, $\#cells \bigcap$ represents the number of cells in the intersection between the simulation results and the real map, $\#cells \bigcup$ is the number of cells in the union

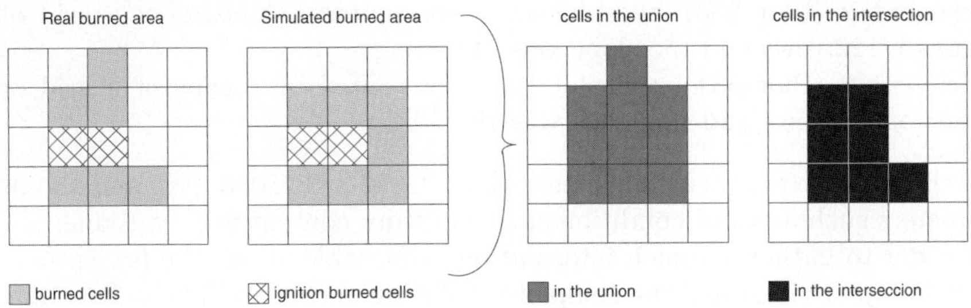

Fig. 1. Calculating the fitness for a 5 x 5 cell terrain

of the simulation results and the real situation, and $\#IgnitionCells$ represent the number of burned cells before starting the simulation.

Figure 1 shows an example of how to calculate this function for a terrain made up of 5x5 cells. In this case, the fitness function is $(7-2)/(10-2) = 0.7124$.

A fitness value equal to one corresponds to the perfect prediction because it means that the predicted area is equal to the real burned area. On the other hand, a fitness equal to zero indicates the maximum error, because in this case the simulation did not coincide with reality at all.

4.3   Parallelisation

GLUE and S^2F^2M have to make a large quantity of calculations because they use a sequential simulator as a kernel [5]. For this reason they need to make a simulation for each resulting combination of parameters ($\#Scenarios$), giving as a result a very time consuming simulation.

Using multiple computational resources working in parallel to obtain the desired efficiency is a solution. We believe a master-worker architecture is suitable to achieve this aim, because a main processor can calculate each combination of parameters and send them to a set of workers. These workers carry out the simulation and return the map to the master. This resulting map indicates which cells are burned and which are not. GLUE implementation follows the same scheme as S^2F^2M. For more information about the S^2F^2M implementation, see [3].

5   Experimental Results

To compare the systems we used three experiments in the field. These burns took place in Serra da Lousã (Gestosa, Portugal (40°15'N, 8°10'O)) , at an altitude of between 800 and 950 m above sea level. The burns were part of the SPREAD project [9]. In the Gestosa field experiments [6], terrain was divided into dedicated plots in order to carry out different sorts of tests and measurements. We worked with plots 520, 533 and 534, which had the following characteristics:

Experiment 1 (Plot 520): the plot was represented by means of a grid of 89 columns x 109 rows and the slope was 18°.

544   G. Bianchini et al.

Experiment 2 (Plot 533): the plot was represented by means of a grid of 95 columns x 123 rows and the slope was 21°.
Experiment 3 (Plot 534): the plot was represented by means of a grid of 75 columns x 126 rows and the slope was 19°.

In the three experiments each cell was 3.28083 x 3.28083 feet, and the other parameters such as wind conditions and moisture content were variable.

In order to gather as much information as possible about the fire-spread behavior, a camera recorded the complete evolution of the fire. The video obtained was analyzed and several images were extracted every 2 minutes in the first and second experiment, and every 1 minute in the third. From the images the corresponding fire contours were obtained and converted to cell format in order for S^2F^2M and GLUE to interpret them.

Experiments were conducted on a cluster (16 processors) of homogenous Pentium 4, 1.8 Ghz (SUSE Linux 8.0) connected by 100 Mb/sec network.

5.1 Experiment 1

To make comparisons we fixed the initial time to 6 minutes (when it is possible to do the first prediction) and a limit value of 14 minutes. In figure 2 we can observe that the prediction proposed by S^2F^2M is always better or, in the worse case, equal to the GLUE predictions. The highest fitness value (0.7624) is reached at time 12. Also it is possible to observe that in time 6, 10 and 14, the fitness value becomes significantly lower. However, that value is still above GLUE fitness and above the best individual case of S^2F^2M. On the left figure it is possible to see the real propagation.

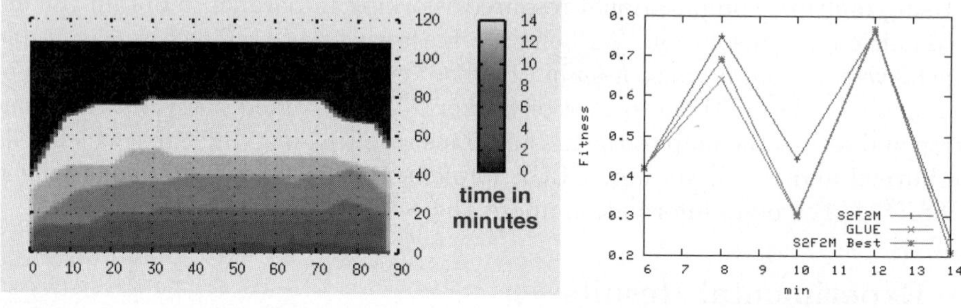

Fig. 2. Real spread for Plot 520 (axes in feet). Comparison between the methods.

5.2 Experiment 2

The second experiment had a similar duration to previous burning. Figure 3 shows the real propagation and the resulting fitness after applying the methods studied. This is an interesting case to analyze, because in it our method has a lower fitness than the others on three of four points. Such a situation has an explanation, which is related to the method. If we look at figure 4, in this simple

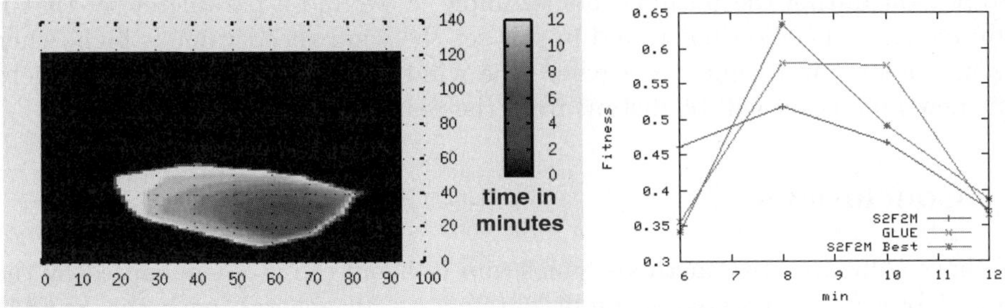

Fig. 3. Real spread for Plot 533 (axes in feet). Comparison between the methods.

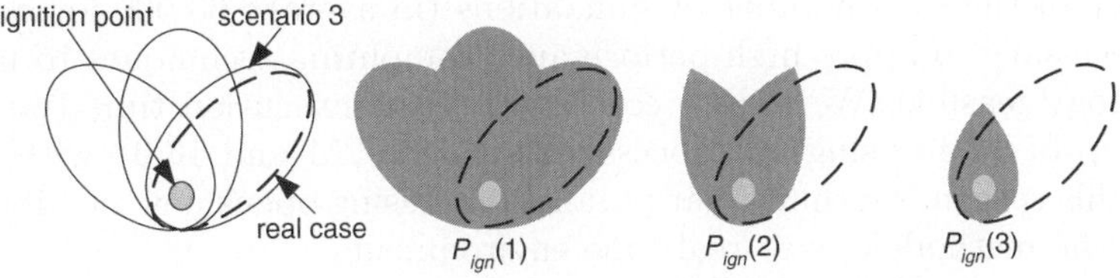

Fig. 4. Explanation

example we have only three scenarios. Scenario number 3 has a similar behavior to the real case, but when we choose $P_{ign}(1)$ or $P_{ign}(2)$ or $P_{ign}(3)$, we discover that fitness on each case is lower than those individual case.

5.3 Experiment 3

Finally, figure 5 shows the fitness function obtained on plot 534 (the shortest experiment). It is possible to identify clearly that the S^2F^2M, as in the first experiment, produces the best predictions.

In general, we saw that fitness function increases in certain intervals and decreases in others. The reason could be because of quick weather changes which

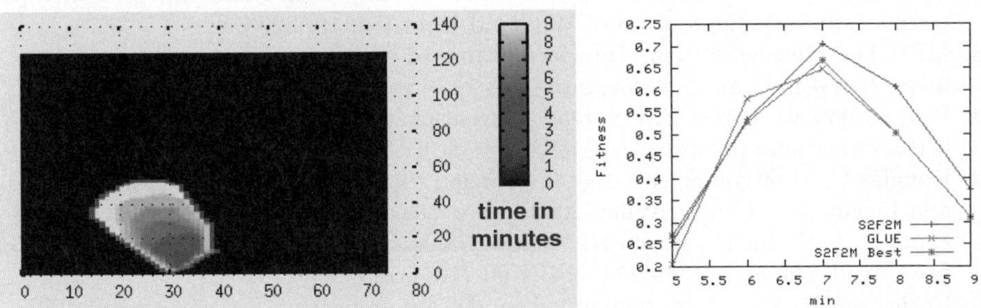

Fig. 5. Real spread for Plot 534 (axes in feet). Comparison between the methods

are present during the burning. For example, if we find a parameter set that is suitable for a specific time z, and in this set the wind speed value is high, when we use this set in a time $z + 1$ where the wind speed value changes brusquely, this new prediction will be distant from the real situation.

6   Conclusions

We have compared two methods which aim to avoid the problem caused by the lack of precision for dynamic input parameters. In general, with the S^2F^2M method we obtain a better prediction that using GLUE. In some case we saw that the S^2F^2M prediction was slightly worse, but importantly we know the reason why, and we can look for a solution to improve the method.

Because of the high number of simulations (in average 60,000 for each step) it was necessary to apply high-performance computing techniques to make the methodology feasible. We reduce considerably the execution time (we reached a speed-up of 14.43 using 16 processors with S2F2M and 13.94 with GLUE), and, for this reason, we think that parallel processing opens new possibilities for applying the methodology to real-time environments.

References

1. Andrews P. L. "BEHAVE: Fire Behavior prediction and modeling systems - Burn subsystem, part 1". General Technical Report INT-194. Odgen, UT, US Department of Agriculture, Forest Service, Intermountain Research Station; 1986.
2. Beven K., Binley A. 1992. "The future of distributed models: model calibration and uncertainty prediction". Hydrological Processes 6:279-298.
3. Bianchini G., Cortés A., Margalef T., Luque E. "S^2F^2M - Statistical System for Forest Fire Management". LNCS 3514, pp. 427-434. 2005.
4. Abdalhaq B., Bianchini G., Cortés A., Margalef T., Luque E.: "Improving Wildland Fire Prediction on MPI Clusters". LNCS 2840, pp. 520-528, 2003.
5. Collins D. Bevins, "FireLib User Manual & Technical Reference", 1996. http://www.fire.org.
6. ADAI - CEIF (Center of Forest Fire Studies) http://www.adai.pt/ceif/Gestosa/
7. Morgan P., Hardy C., Swetnam T. W., Rollins M. G., Long D. G. 2001. Mapping fire regimes across time and space: Understanding coarse and fine-scale fire patterns. International Journal of Wildland Fire, Vol. 10: 329-342.
8. MPI: The Message Passing Interface Standard http://www-unix.mcs.anl.gov/mpi/
9. Project Spread, Forest Fire Spread Prevention and Mitigation http://www.adai.pt/spread/
10. Douglas C. Montgomery, George C. Runger, "Probabilidad y Estadística aplicada a la Ingeniería", Limusa Wiley 2002 ISBN: 968-18-5914-6
11. Piñol P., Salvador R., Beven K. "Model Calibration and uncertainty prediction of fire spread". 2002, pp. 99- 111. ISBN 90-77017-72-0
12. Rothermel R. C., "A mathematical model for predecting fire spread in wildland fuels", USDA FS, Ogden TU, Res. Pap. INT-115, 1972.

Support Vector Machine Regression Algorithm Based on Chunking Incremental Learning

Jiang Jingqing[1, 2], Song Chuyi[2], Wu Chunguo[1, 3], Marchese Maurizio[4],
and Liang Yangchun[1,4,*]

[1] College of Computer Science and Technology, Jilin University, Key Laboratory of Symbol
Computation and Knowledge Engineering of Ministry of Education, Changchun 130012, China
[2] College of Mathematics and Computer Science,
Inner Mongolia University for Nationalities, Tongliao 028043, China
[3] The Key Laboratory of Information Science & Engineering of
Railway Ministry/The Key Laboratory of Advanced Information Science and Network
Technology of Beijing, Beijing Jiaotong University, Beijing 100044, China
[4] Department of Information and Communication Technology,
University of Trento, Via Sommarive 14, 38050, Povo (TN) Italy

Abstract. On the basis of least squares support vector machine regression
(LSSVR), an adaptive and iterative support vector machine regression algorithm
based on chunking incremental learning (CISVR) is presented in this paper.
CISVR is an iterative algorithm and the samples are added to the working set in
batches. The inverse of the matrix of coefficients from previous iteration is used
to calculate the regression parameters. Therefore, the proposed approach permits
to avoid the calculation of the inverse of a large-scale matrix and improves the
learning speed of the algorithm. Support vectors are selected adaptively in the iteration to maintain the sparseness. Experimental results show that the learning
speed of CISVR is improved greatly compared with LSSVR for the similar training accuracy. At the same time the number of the support vectors obtained by the
presented algorithm is less than that obtained by LSSVR greatly.

1 Introduction

The support vector machine (SVM) is a novel learning method that is constructed
based on statistical learning theory. The support vector machine has been studied
widely since it was presented in 1995. It has been applied to pattern recognition
broadly and its excellent performance has been shown in function regression problems. Training a standard support vector machine requires the solution of a large-scale
quadratic programming problem. This is a difficult problem when the number of the
samples exceeds a few thousands. Many algorithms for training the SVM have been
studied. Osuua [1] proposed a decomposition algorithm and the quadratic programming problem for standard SVM is divided into a serial small-scale quadratic programming sub-problem. Focusing on the problem of the working set selection,
Joachims [2] presented a SVM^{Light} algorithm to implement the decomposition
algorithm in [1] efficiently. A sequential minimal optimization algorithm (SMO) was
proposed by Patt [3]. It transformed the quadratic programming problem for standard

---

* Corresponding author.

V.N. Alexandrov et al. (Eds.): ICCS 2006, Part I, LNCS 3991, pp. 547–554, 2006.

SVM to the minimization quadratic programming problem that could be solved analytically. Suykens [4] suggested a least squares support vector machine (LSSVM) in which the inequality constrains were replaced by equality constrains. By this way, solving a quadratic programming was converted into solving linear equations. The efficiency of training SVM is improved greatly and the difficulty of training SVM is cut down. Suykens [5] studied the LSSVM for function regression further. Hao [6] proposed a chunking incremental learning algorithm for LSSVM to deal with classification problem. In this paper, an adaptive and iterative support vector machine regression algorithm based on chunking incremental learning (CISVR) is presented. The support vectors are selected adaptively in the iteration to maintain the sparseness and the samples are added to working set in batches.

2   Least Squares Support Vector Machine for Regression (LSSVR)

According to [5], let us consider a given training set of l samples $\{x_i, y_i\}_{i=1}^{l}$ with the ith input datum $x_i \in R^n$ and the ith output datum $y_i \in R$. The aim of support vector machine model is to construct the decision function that takes the form:

$$f(x, w) = w^T \varphi(x) + b \tag{1}$$

where the nonlinear mapping $\varphi(\cdot)$ maps the input data into a higher dimensional feature space. In least squares support machine for function regression the following optimization problem is formulated

$$\min_{w, e} \; J(w, e) = \frac{1}{2} w^T w + \gamma \sum_{i=1}^{l} e_i^2 \tag{2}$$

subject to the equality constraints

$$y_i = w^T \varphi(x_i) + b + e_i, \quad i = 1,\ldots, l \tag{3}$$

This corresponds to a form of ridge regression. The Lagrangian is given by

$$L(w, b, e, \alpha) = J(w, e) - \sum_{i=1}^{l} \alpha_i \{ w^T \varphi(x_i) + b + e_i - y_i \} \tag{4}$$

with Lagrange multipliers α_k. The conditions for the optimality are

$$\begin{cases} \dfrac{\partial L}{\partial W} = 0 \rightarrow w = \sum_{i=1}^{l} \alpha_i \varphi(x_i) \\[2mm] \dfrac{\partial L}{\partial b} = 0 \rightarrow \sum_{i=1}^{l} \alpha_i = 0 \\[2mm] \dfrac{\partial L}{\partial e_i} = 0 \rightarrow \alpha_i = \gamma e_i \\[2mm] \dfrac{\partial L}{\partial \alpha_i} = 0 \rightarrow w^T \varphi(x_i) + b + e_i = 0 \end{cases} \tag{5}$$

for $i = 1,...,l$. After eliminating e_i and w, we could have the solution by the following linear equations

$$\begin{bmatrix} 0 & \vec{1}^T \\ \vec{1} & \Omega + \gamma^{-1}I \end{bmatrix}\begin{bmatrix} b \\ a \end{bmatrix} = \begin{bmatrix} 0 \\ y \end{bmatrix} \tag{6}$$

where $y = [y_1,..., y_l]^T, \vec{1} = [1,...,1]^T, \alpha = [\alpha_1,...,\alpha_l]^T$ and the Mercer condition

$$\Omega_{kj} = \varphi(x_k)^T \varphi(x_j) = \psi(x_k, x_j) \quad k, j = 1,...,l \tag{7}$$

is applied. Set $A = \Omega + \gamma^{-1}I$. If A is a symmetric and positive-definite matrix, A^{-1} exists. Solving the linear equations (6) we obtain the solution

$$\alpha = A^{-1}(y - b\vec{1}) \qquad b = \frac{\vec{1}^T A^{-1} y}{\vec{1}^T A^{-1} \vec{1}} \tag{8}$$

Substituting w in Eq. (1) with the first equation of Eqs. (5) and using Eq. (7) we have

$$f(x, w) = y(x) = \sum_{i=1}^{l} \alpha_i \psi(x, x_i) + b \tag{9}$$

where α_i and b are the solution to Eqs. (6). The kernel function $\psi(\cdot)$ can be chosen as linear function $\psi(x, x_i) = x_i^T x$, polynomial function $\psi(x, x_i) = (x_i^T x + 1)^d$ or radial basis function $\psi(x, x_i) = \exp\{-\|x - x_i\|_2^2 / \sigma^2\}$.

3 Adaptive and Iterative Least Squares Support Vector Machine Regression Algorithm Based on Chunking Incremental Learning

3.1 Chunking Increment Procedure

According to Eq. (6), set

$$A_N = \Omega + \gamma^{-1}I \qquad \overline{\alpha}_N = \alpha \qquad \overline{y}_N = y \tag{10}$$

where N is the number of samples in current working set. Eq. (8) can be rewritten as

$$\overline{\alpha}_N = A_N^{-1}(\overline{y}_N - b\vec{1}) \qquad b = \frac{\vec{1}^T A_N^{-1} \overline{y}_N}{\vec{1}^T A_N^{-1} \vec{1}} \tag{11}$$

$\vec{1} = (1,...,1)^T$. When K new coming samples $(x_{N+1}, y_{N+1}), (x_{N+2}, y_{N+2}),...,(x_{N+K}, y_{N+K})$ are added to the current working set, we could calculate the parameters according to Eq. (12)

$$\overline{\alpha}_{N+K} = A_{N+K}^{-1}(\overline{y}_{N+K} - b\vec{1}) \qquad b = \frac{\vec{1}^T A_{N+K}^{-1} \overline{y}_{N+K}}{\vec{1}^T A_{N+K}^{-1} \vec{1}} \tag{12}$$

where $\vec{1} = (1,...,1)^T$, $\overline{\alpha}_{N+K} = (\overline{\alpha}_N, \alpha_{N+1},...,\alpha_{N+K})$, $\overline{y}_{N+K} = (\overline{y}_N, y_{N+1},..., y_{N+K})$,

$$A_{N+K} = \begin{bmatrix} A_N & Q \\ Q^T & S \end{bmatrix} \qquad Q = \begin{bmatrix} \Omega_{1,N+1} & \Omega_{1,N+2} & \cdots & \Omega_{1,N+K} \\ \cdots & \cdots & \cdots & \cdots \\ \Omega_{N,N+1} & \Omega_{N,N+2} & \cdots & \Omega_{N,N+K} \end{bmatrix}$$

$$S = \begin{bmatrix} \Omega_{N+1,N+1} & \Omega_{N+1,N+2} & \cdots & \Omega_{N+1,N+K} \\ \cdots & \cdots & \cdots & \cdots \\ \Omega_{N+K,N+1} & \Omega_{N+K,N+2} & \cdots & \Omega_{N+K,N+K} \end{bmatrix} + \gamma^{-1} I$$

According to the algorithm in [6], the matrix A_{N+K}^{-1} in Eq. (12) could be calculated from matrix A_N^{-1} and the inverse of a small $K{\times}K$ matrix, that is

$$A_{N+K}^{-1} = \begin{bmatrix} A_N^{-1} & 0 \\ 0 & 0 \end{bmatrix} + \begin{bmatrix} -A_N^{-1}Q \\ I \end{bmatrix} [S - Q^T A_N^{-1} Q]^{-1} \begin{bmatrix} -Q^T A_N^{-1} & I \end{bmatrix} \tag{13}$$

where 0 is a matrix whose elements are all zero. I is a unit matrix with K rows and K columns. In this way the calculation for the inverse of a large-scale matrix could be avoided.

3.2 Decrement Procedure

The number of support vectors will increase with the chunking increment procedure. To maintain the sparseness of support vectors, a decrement procedure is implemented after the chunking increment procedure. A support vector is omitted in this procedure. Meanwhile, a trained sample in the working set corresponding to the discarded support vector is also omitted. Form [7], $A_{l-1}^{-1} = (\hat{a}_{ij})_{i,j \neq k}$ can be calculated from $A_l^{-1} = (\tilde{a}_{ij})$, $A_l = (a_{ij})$ and $A_{l-1} = (a_{ij})_{i,j \neq k}$ in the decrement procedure, that is

$$\hat{a}_{ij} = \tilde{a}_{ij} - \frac{1}{a_{kk}} \tilde{a}_{ik} \tilde{a}_{kj}, \quad i, j \neq k \tag{14}$$

where A_{l-1} is a matrix obtained from A_l by omitting the kth row and the kth column.

3.3 Steps of CISVR Algorithm

Set the training sample set $T = \{s_i \mid s_i = (x_i, y_i), x_i \in R^n, y_i \in R, i = 1,2,\cdots,l\}$. The form of the regression function is

$$f(x,\alpha,b) = \sum_{i \in \mathbf{W}} \alpha_i \psi(x, x_i) + b \equiv f(x) |_{\tilde{\mathbf{w}}} \tag{15}$$

where α and b are the regression parameters, W is named working set whose elements are the training samples selected to calculate the regression parameters, and $\tilde{w}$

is the regression parameters set which is decided by working set W. Set θ is the precision in training and testing, the precision in stop criterion is ε.

Steps of CISVR algorithm are as follows:

Initialization: set $W = \{(x_1, y_1),...,(x_N, y_N)\}$ and calculate A^{-1} analytically. Calculate $\tilde{W}$ and $f(x)|_{\tilde{W}}$ from Eqs. (8) and (9). Set k=0.

for $i = N+1,...,l$ **do**

    adaptive learning

      1.   read a sample $s_i = (x_i, y_i)$

      2.   **if** $\left|f(x_i)|_{\tilde{W}} - y_i\right| > \theta$ **and** $s_i \notin W$ **then**

      3. $W = W \cup \{s_i\}$, k=k+1

      4.   **end if**

      5.   **if** k=K **then**

      6.   calculate $\tilde{W}$ by chunking increment procedure

      7.   find the minimization support vector $\left|\alpha_{i^*}\right| = \min_{s_i \in W}\{\left|\alpha_i\right|\}$

      8. $\hat{W} = W \setminus \{s_{i^*}\}$ $//\hat{W}$ is temporary working set

      9.   calculate $(\hat{W})\tilde{}$ and temporary regression function $f(x)|_{\hat{W}}$

                $// (\hat{W})\tilde{}$ is the temporary regression parameters set corresponding to the

                $//$ temporary working set $\hat{W}$

     10.   read a sample s_{i+1}

     11.   **if** $\left|f(x_{i+1})|_{(\hat{W})\tilde{}} - y_{i+1}\right| \le \theta$ **then**

     12. $W = \hat{W}$ $\tilde{W} = (\hat{W})\tilde{}$

     13.   **end if**

     14.   k=0

     15. **end if**

end for

while the stop criterion is false **do**

    for $i = 1,...,l$ **do**

      adaptive learning

    end for

end while

The stop criterion is related to the objective value. The formulation of the objective function is $J(w,e)|_W = \dfrac{1}{2}\left\|w\right\|^T_{w \in \tilde{W}} + \dfrac{\gamma}{2}\sum_{s_i \in W} e_i^2$, where $w = \sum_{s_i \in W} \alpha_i \varphi(x_i)$, $e_i = \dfrac{1}{\gamma}\alpha_i$. The meaning of the defined stop criterion is that the procedure ends when the relative error of objective values in the two adjacent iterations is smaller than a given precision ε. In the decrement procedure, the minimization support vector is omitted because it has least effect on the performance of the regression function. The matrix A^{-1} in the current iteration is obtained from that in the previous iteration in both chunking increment and decrement procedure. In this way, it is possible on one hand to avoid calculating the inverse for a large-scale matrix and on the other hand to improve the learning speed of the procedure.

4 Numerical Experiments

In order to examine the efficiency of CISVR algorithm and compare CISVR with LSSVR algorithm, numerical experiments are performed using two kinds of data sets. One kind of data set is composed of the simply elementary functions which include $f(x) = \sin(x)$ and $f(x) = x^2$. These functions are used to test the regression ability for the known function. The other kind of data set is composed of Mackey-Glass (MG) system and simple function $f(x) = \sin c(x)$. The MG system is a blood cell regulation model established in 1977 by Mackey and Glass. It is a chaos system $\dfrac{dx}{dt} = \dfrac{a \cdot x(t-\tau)}{1 + x^{10}(t-\tau)} - b \cdot x(t)$ described in [8], where $\tau = 17$ $a = 0.2$ $b = 0.1$ $\Delta t = 1$ $t \in (0,400)$. The embedded dimensions are $n = 4,6,8$ respectively. The sample function is $f(x) = \begin{cases} 1 & x = 0 \\ \dfrac{\sin(x)}{x} & x \neq 0 \end{cases}$. An RBF kernel function $\psi(x_i,x_j) = \exp(-\|x_i - x_j\|^2 / (2\sigma^2))$ is employed in these two algorithms. The parameters γ and σ are showed in Tab.1. The other parameters are as follows: $\theta = 0.01, \varepsilon = 0.01$. The comparison between LSSVR

Table 1. Parameters used in algorithm

		sin	square	sinc	MG system4	MG system6	MG system8
γ		50000	30000	5000	50000	50000	50000
σ	LSSVR	1.0	1.0	2.0	2.0	2.0	2.0
	CISVR	1.0	1.0	1.0	1.0	1.0	1.0

and CISVR are showed in Tab.2, where the third column is the number of support vectors, the forth column is the seconds for training, and the fifth and seventh columns are the regression accuracy for training and testing, respectively. The regression accuracy is a ratio that is the number of samples whose relative error is smaller than θ to the number of samples in the working set (testing set). The sixth and eighth columns are the mean square error for training and testing, respectively. It can be seen from Tab.2 that the learning speed of CISVR is much faster than LSSVR. Moreover,

Table 2. Comparison between CISVR algorithm and standard LSSVR

Dataset $l \times n$	Algorithm name	# of SVs	Train time (CPU s)	Accuracy (train%)	MSE (train)	Accuracy (test%)	MSE (test)
sin	LSSVR	3000	1465.05	99.93	3.58e-009	99.87	4.19e-009
3000×1	CISVR	56	4.44	99.96	9.31e-007	99.70	1.04e-006
square	LSSVR	3000	1463.44	99.97	2.62e-005	99.97	2.49e-005
3000×1	CISVR	132	10.65	97.76	3.29e-003	97.73	2.85e-003
sinc	LSSVR	3000	1448.28	99.93	8.23e-011	99.80	1.14e-010
3000×1	CISVR	53	4.984	99.83	2.87e-008	99.56	3.18e-008
MG system4	LSSVR	6000	11048.69	100	1.44e-008	100	1.49e-008
6000×4	CISVR	16	52.35	100	6.65e-007	100	6.91e-007
MG system6	LSSVR	6000	12954.34	100	8.06e-009	100	8.48e-009
6000×6	CISVR	14	55.64	100	9.67e-007	100	1.02e-006
MG system8	LSSVR	6000	13104.99	100	3.30e-009	100	3.28e-009
6000×8	CISVR	10	54.92	100	1.08e-006	100	1.13e-006

the number of support vectors is less than that obtained by LSSVR for the similar regression accuracy.

5   Discussion and Conclusion

In this paper we propose an adaptive and iterative support vector machine regression algorithm based on the chunking incremental learning and the least square support vector machine regression algorithm. The samples are added to the working set in batches. The support vectors are selected adaptively in the iteration and the sparseness of support vectors is maintained. Meanwhile, the inverse of matrix A in the previous iteration is used to calculate the regression parameters. Therefore, the proposed approach can avoid calculating the inverse of a large-scale matrix, and at the same time, substantially improve the learning speed compared to that of LSSVR for the similar regression accuracy.

Acknowledgment

The authors are grateful to the support of the National Natural Science Foundation of China (60433020), the science-technology development project for international collaboration of Jilin Province of China (20050705-2), the doctoral funds of the National Education Ministry of China (20030183060), Graduate Innovation Lab of Jilin University (503043), and "985" Project of Jilin University. The last two authors would like to thank the support of the EuMI School and the Erasmus Mundus programme of the European Commission.

References

1. Osuna E., Freund R., Girosi F.: An Improved Training Algorithm for Support Vector Machines. IEEE Workshop on Neural Networks and Signal Processing, Amelia Island, (1997) 276-285
2. Joachims T.: Making Large-scale Support Vector Machine Practical. In Advances in Kernel Methods-Support Vector Learning, Cambridge, Massachusetts: The MIT Press, (1999) 169-184.
3. Platt J.C.: Fast Training of Support Vector Machines Using Sequential Minimal Optimization. In Advances in Kernel Methods-Support Vector Learning, Cambridge, Massachusetts: The MIT Press, (1999) 185-208.
4. Suykens J.A.K., Vandewalle J.: Least Squares Support Vector Machine Classifiers. Neural Processing Letters, Vol.9 (1999) 293–300.
5. Suykens J.A.K., Lukas L., Wandewalle J.: Sparse Approximation Using Least Squares Support Vector Machines. In Proc. of the IEEE International Symposium on Circuits and Systems (ISCAS 2000), Geneva, Switzerland, (2000) 757-760.
6. Hao Z.F., Yu S., Yang X.W., Hu R., Zhao F., Liang Y.C.: Online LS-SVM Learning for Classification Problems Based on Incremental Chunk. Lecture Notes in Computer Science, Vol.3173 (2004) 558-564.
7. Cauwenberghs G., Poggio T.: Incremental and Decremental Support Vector Machine Learning. In Advances in Neural Information Processing Systems, Cambridge, MA: MIT Press, Vol.13 (2001) 426-433.
8. Flake G.W., Lawrence S.: Efficient SVM Regression Training with SMO. Machine Learning, Vol.46 (2002) 271–290.

Factorization with Missing and Noisy Data

Carme Julià, Angel Sappa, Felipe Lumbreras, Joan Serrat, and Antonio López

Computer Vision Center and Computer Science Department,
Universitat Autònoma de Barcelona,
08193 Bellaterra, Spain
{cjulia, asappa, felipe, joans, antonio}@cvc.uab.es

Abstract. Several factorization techniques have been proposed for tackling the *Structure from Motion* problem. Most of them provide a good solution, while the amount of missing data is within an acceptable ratio. Focussing on this problem, we propose an incremental multiresolution scheme able to deal with a high rate of missing data, as well as noisy data. It is based on an iterative approach that applies a classical factorization technique in an incrementally reduced space. Information recovered following a coarse-to-fine strategy is used for both, filling in the missing entries of the input matrix and denoising original data. A statistical study of the proposed scheme compared to a classical factorization technique is given. Experimental results obtained with synthetic data and real video sequences are presented to demonstrate the viability of the proposed approach.[1]

1   Introduction

Structure From Motion (SFM) consists in extracting the 3D shape of a scene as well as the camera motion from trajectories of tracked features. Factorization is a method addressing to this problem. The central idea is to express a matrix of trajectories W as the product of two unknown matrices, namely, the 3D object's shape S and the relative camera pose at each frame M: $W_{2f \times p} = M_{2f \times r} S_{r \times p}$, where f and p are the number of frames and feature points respectively and r the rank of $W_{2f \times p}$. These factors can be estimated thanks to the key result that their rank is small and due to constraints derived from the orthonormality of the camera axes. The *Singular Value Decomposition* (SVD) is generally used when there are not missing entries. Unfortunately, in most of the real cases not all the data points are available, hence other methods need to be used.

In the seminal approach Tomasi and Kanade [1] propose an initialization method in which they first decompose the largest full submatrix by the factorization method and then the initial solution grows by one row or by one column at a time, unveiling missing data. The problem is that finding the largest full submatrix of a matrix with missing entries is a NP-hard problem. Jacobs [2]

---

[1] This work has been supported by the Government of Spain under the CICYT project TRA2004-06702/AUT. The second author has been supported by The Ramón y Cajal Program.

V.N. Alexandrov et al. (Eds.): ICCS 2006, Part I, LNCS 3991, pp. 555–562, 2006.

treats each column with missing entries as an affine subspace and shows that, for every r-tuple of columns, the space spanned by all possible completions of them must contain the column space of the completely filled matrix. Missing entries are recovered by finding the least squares regression onto that subspace. However, this approach is strongly affected by noise on the data. An incremental SVD scheme of incomplete data is proposed by Brand [3]. The main drawback of that scheme is that the final result depends on the order in which the data are encountered. Brandt [4] proposes a different technique that addresses the affine reconstruction under missing data by means of an EM algorithm. Although the feature points do not have to be visible in all views, the affine projection matrices in each image must be known. A method for recovering the most reliable imputation, addressing the SFM problem, is provided by Suter and Chen [5]. They propose an iterative algorithm to employ this criterion to the problem of missing data. Their aim is not to obtain the factors M and S, but the projection onto a low rank matrix to reduce noise and to fill in missing data. Wiberg [6] introduces the *Alternation* technique to solve the factorization with missing data. Since then, several variants of this approach have been proposed in the literature. In [7], Buchanan and Fitzgibbon summarize different factorization approaches with missing data and propose the *Alternation/Damped Newton Hybrid*, which combines the *Alternation* strategy with the *Damped Newton* method.

One disadvantage of the above methods is that the result depends on the percentage of missing data. They give a good factorization while the amount of missing points is reduced, which is not common in real image sequences, unfortunately. Additionally to this problem, when real sequences are considered, the presence of noisy data needs to be taken into account in order to evaluate the performance of the factorization technique. Addressing to these problems, we propose to use an iterative multiresolution scheme, which incrementally fill in missing data. A statistical study of the performance of the proposed scheme is carried out considering different percentages of missing and noisy data. The key point of the implemented approach is to work with a reduced set of feature points along a few number of consecutive frames. Thus, the 3D reconstruction corresponding to the selected feature points and the camera motion of the used frames are obtained. Missing entries of the trajectory matrix are recovered, while, at the same time, noisy data are filtered.

This paper is organized as follows. Section 2 contains a brief review of Alternation factorization techniques for the case where there are missing data. Section 3 presents the incremental multiresolution scheme used to factorize a matrix of trajectories that has a large amount of missing data. The error function is defined in section 4. Section 5 contains results obtained with synthetic and real data. Conclusions and future work are given in section 6.

2   Alternation Technique

Let be $W_{2f \times p}$ the matrix of trajectories of p feature points, tracked over f frames—also denoted as W, or input matrix. The goal of Alternation is to find the best rank r approximation to W, where $r < 2f, p$. That is, to compute the matrix factors M and S such that minimize the cost function:

$$\|W - MS\|_F^2 \tag{1}$$

where $\| \cdot \|$ is the Frobenius matrix norm [8]. In the case of missing data:

$$\|W - MS\|_F^2 = \sum_{i,j} |W_{ij} - (MS)_{ij}|^2 \tag{2}$$

where i and j correspond to the index pairs where W_{ij} is defined.

The algorithm starts with an initial random $2f \times r$ matrix M_0 and repeats the next steps until the product $M_k S_k$ converges to W:

$$S_k = (M_{k-1}^t M_{k-1})^{-1} M_{k-1}^t W \qquad M_k = W S_k (S_k^t S_k)^{-1} \tag{3}$$

As pointed out in [9], the most important advantage of this 2-step algorithm is that these equations are the matrix versions of the normal equations. That is, each M_k and S_k is the least-squares solution of a set of equations of the form $W = MS$. Besides, since the updates of M given S (and analogously in the case of S given M) can be independently done for each row of S, missing entries in W correspond to omitted equations. Due to that fact, with a few data points the method would fail to converge, but this happens only with large amounts of missing data.

In the application of affine SFM, the last row of S should be filled with ones $S = \begin{bmatrix} X & 1 \end{bmatrix}^t$, where X are the 3D recovered coordinates of the feature points. Through the paper, an Alternation for SFM with motion constraints (AM) approach has been used. Hence, at each iteration k, it is used the fact that M is the motion matrix—the relative camera pose at each frame. Therefore, given M_{k-1}, and before computing S_k, we impose the orthonormality of the camera axes at each frame.

3   Proposed Approach

We propose an iterative multiresolution approach using the AM previously described. We will refer it as Incremental Alternation with Motion constraints (IAM). Essentially, our basic idea is to generate sub-matrices with a reduced density of missing points. Thus, the AM could be used for factoring these sub-matrices and recovering their corresponding 3D shape and motion. The proposed technique consists of two stages, which are fully explained below.

3.1   Observation Matrix Splitting

Let $W_{2f \times p}$ be the observation matrix of p feature points tracked over f frames containing missing entries. Let k be the index denoting the current iteration number. In a first step, W is split into $k \times k$ non-overlapped sub-matrices, each one of them defined as $W_{k(i,j)}$, $i \in (0, \lfloor \frac{2f}{k} \rfloor]$, $j \in (0, \lfloor \frac{p}{k} \rfloor]$. For the sake of presentation simplicity, hereinafter a sub-matrix in the current iteration level k will be referred as W_k (assuming $k > 1$, since $k = 1$ is simply the AM method).

Although the idea is to focus the process in a small area (sub-matrix W_k), with a reduced density of missing data, recovering information from a small patch can be easily affected from noisy data, as pointed out in [5]. Hence there is a trade off between the size of a sub-matrix and the confidence of its recovered data. In order to improve the confidence of recovered data a multiresolution approach is followed. In a second step, and only when $k > 2$, four W_{2k} overlapped sub-matrices, with twice the size of W_k are computed as illustrated in Fig.1. The idea of this enlargement process is to study the behavior of feature points contained in W_k when a bigger region is considered. Other strategies were tested in order to compute in a fast and robust way sub-matrices with a reduced density of missing entries (e.g. quadtrees, ternary graph structure), but they do not give the desired and necessary properties of overlapping.

Since generating four W_{2k} for every W_k is a computationally expensive task, a simple and more direct approach is followed. It consists in splitting the input matrix W in four different ways, by shifting W_{2k} half of its size (i.e., W_k) through rows, columns or both at the same time. When all these matrices are considered together, the overlap between the different areas is obtained.

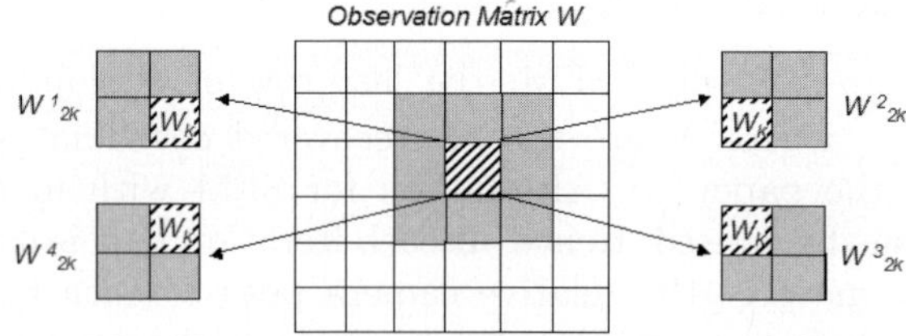

Fig. 1. W_{2k} overlapped matrices of the observation matrix W, computed during the first stage (section 3.1), at iteration $k = 6$

3.2   Sub-matrices Processing

At this stage, the objective is to recover missing data by applying AM at every single sub-matrix. Independently of their size hereinafter sub-matrices will be referred as W_i.

Given a sub-matrix W_i, the AM gives its corresponding M_i and S_i matrices. Their product could be used for computing an approximation error ε_i such as equation (2). In case the resulting error is smaller than a user defined threshold σ, every point in W_i is kept in order to be merged with overlapped values after finishing the current iteration. Additionally, every point of W_i is associated with a weighting factor, defined as $\frac{1}{\varepsilon_i}$, in order to measure the goodness of that value. These weighting factors are later on used for merging data on overlapped areas. Otherwise, the resulting error is higher than σ, computed data are discarded.

Finally, when every sub-matrix W_i has been processed, recovered missing data are used for filling in the input matrix W. In case a missing datum has been recovered from more than one sub-matrix (overlapped regions), those recovered data are merged by using their corresponding normalized weighting factors. On the contrary, when a missing datum has been recovered from only one sub-matrix, this value is directly used for filling in that position.

Once recovered missing data were used for filling in the input matrix W, the iterative process starts again (section 3.1) splitting the new matrix W (the input one merged with recovered data) either by incrementing k one unit or, in case the size of sub-matrices W_k at the new iteration stage is quite small (the smaller W_k size was set to 5×5), by setting $k = 2$. This iterative process is applied until one of the following conditions is true: a) the matrix of trajectories is totally filled; b) at the current iteration no missing data were recovered; c) a maximum number of iterations is reached.

4   The Error Function

As pointed out in [5], the cost function defined by equation (2) could be ambiguous and in some cases contradictory. That is because that formula only takes into account the recovered values corresponding to known features, but it ignores how the rest of entries are filled. Therefore, in order to compare the proposed IAM scheme with the classical AM, an error function that considers all the features of the sequence, including missing points, is used (see equation (1)). Unfortunately, this is only possible when we have access to the whole information. Hence, when real data are considered, in order to perform a comparison by using the proposed error function, a full matrix should be selected. This matrix is used as input and missing data are randomly removed.

As it will be presented in short, noise is added to the input matrix W. However, since the aim is also to study how the data are filtered, the entries of the matrix of trajectories given by the product MS are compared with the corresponding elements in the input matrix W.

5   Experimental Results

As it was previously mentioned we want to do an statistical study about the filtering capability of the IAM scheme compared to the classical AM. At the same time, the robustness to missing data will be considered. In order to perform a comparison that help us to infer some conclusions, different levels of Gaussian noise—standard deviation (also denoted as std) with values from $\frac{1}{8}$ to 1, both in a synthetic and a real case—are added into the 2D feature point trajectories and different amounts of missing data are considered—from 20% up to 80%. Statistical results are obtained by applying: a) AM over the input matrix W; b) AM after filling in missing data with the proposed IAM. Although this strategy consists of two parts (IAM+AM), for simplicity, we referred it as IAM.

For each setting (level of noise, amount of missing data) 100 attempts are repeated and the number of *convergent* cases—those in which the error value is smaller than a threshold—obtained with each approach is computed. Due to the fact that for each setting the error takes a different range of values, a unique threshold is difficult to define. Therefore, it is defined for each particular setting, by the mean of the inliers error values ϵ [2] obtained from AM and IAM. Notice that more *convergent* cases does not necessary mean that a better performance is achieved for that setting. The idea is compare AM and IAM, not the different settings. Experiments using both synthetic and real data are presented below.

5.1   Synthetic Object

Synthetic data are randomly generated by distributing 35 3D feature points over the whole surface of a cylinder, see Fig.2 (left). The cylinder is defined by a radius of 100 and a height of 400; it rotates over its axis. The corresponding input matrices are obtained using different number of frames. In order to obtain a low percentage of missing data, a few frames are taken and the resulting input matrices are quite small. The IAM performs worse in these cases. However, the goal is to show its performance for a great amount of missing data, when other factorization techniques tend to fail. Additionally, the small size of the input matrices does not help to see the denoising capability of both AM and IAM.

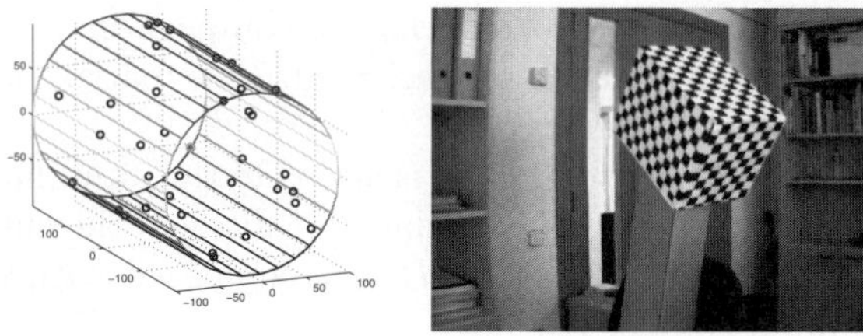

Fig. 2. (left) Synthetic object used to test the proposed approach. (right) Input object used for the real case.

As shown in Fig. 3 (left), the case of free-noisy data differs considerably from the others. In particular, for the IAM approach, a high average of convergent cases is obtained, no matter the ratio of missing data. Notice that with IAM and for 80% of missing data, a ratio of convergence of 100 is obtained. That does not mean a better result that working with a low percentage of missing data. As mentioned above, the idea is not to compare with the other settings, but with the AM, which has a ratio of convergence of about 2 for the same setting.

---

[2] The inliers ϵ are defined as $|\epsilon| < q_3 + 1.5\Delta q$—where q_3 is the value of the third quartile and Δq is the interquartile distance $q_3 - q_1$.

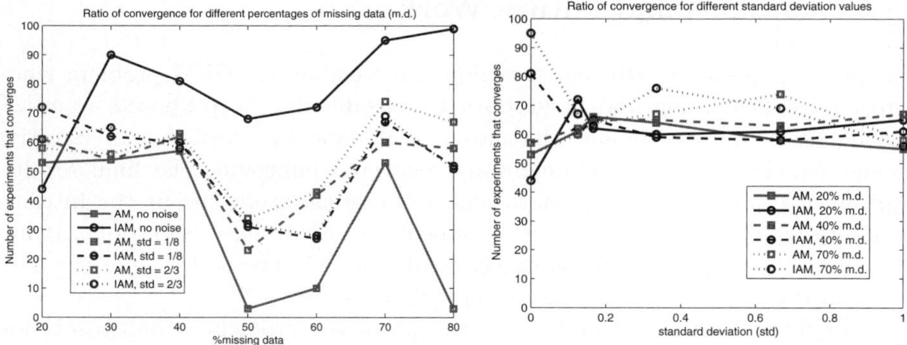

Fig. 3. Synthetic case. (left) Ratio of convergence for the different percentages of missing data, fixing different standard deviation (*std*) values. (right) The same for different *std* values, fixing various percentages of missing data. Zero *std* means no noisy data.

5.2   Real Object

Experimental results with a real video sequence of 101 frames with a resolution of 640×480 pixels are presented. The studied object is shown in Fig. 2 (right). A single rotation around a vertical axis is performed. Feature points are selected by means of a corner detector algorithm and 87 points over the object to be studied are considered. An iterative feature tracking algorithm has been used. More details about corner detection and tracking algorithm can be found in [10]. Different ratios of missing data are obtained by randomly removing data; the removed data are recorded in order to compute the error value (1).

Again, in Fig. 4 (left) it seems that for the case of no noise and a percentage of missing data from 40 up to 60, both AM and IAM performs worse than for other percentages. As shown in Fig. 4 (right), the number of convergent cases is in general higher applying IAM than AM.

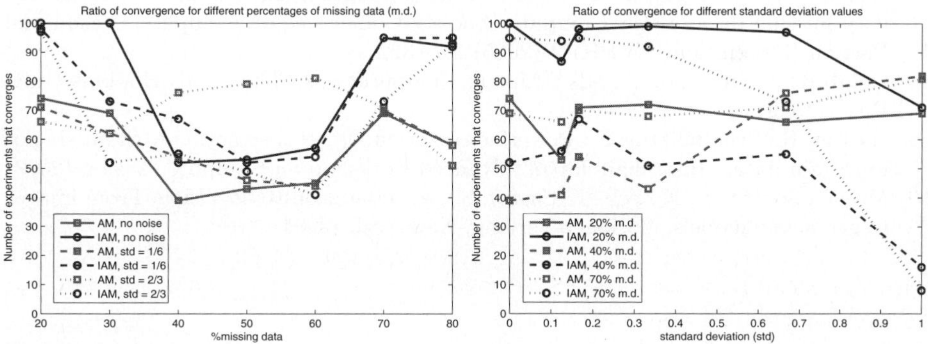

Fig. 4. Real case. (left) Ratio of convergence for the different percentages of missing data, fixing different standard deviation (*std*) values. (right) The same for different *std* values, fixing various percentages of missing data. Zero *std* means no noisy data.

6   Conclusions and Future Work

This paper presents an efficient technique for tackling the SFM problem when a high ratio of missing and noisy data is considered. The proposed approach exploits the simplicity of an Alternation technique by means of an iterative scheme. Missing data are incrementally recovered improving the final results. Noise have been added to the data and a statistical study about the filtering capability of AM compared to our incremental strategy have been done. It has been shown that, in most of the cases, results of IAM are better than the ones of AM in the sense of number of convergent cases.

In the future, we would like to use the proposed icremental multiresolution scheme with other classical factorization techniques. Additionally, other functions that consider the goodness of the obtained M and S and not only the recovered elements of W will be studied.

References

1. Tomasi, C., Kanade, T.: Shape and motion from image streams: a factorization method. Full report on the orthographic case (1992)
2. Jacobs, D.: Linear fitting with missing data for structure-from-motion. Computer vision and image understanding, CVIU (2001) 7–81
3. Brand, M.: Incremental singular value decomposition of uncertain data with missing values. In: Proceedings, ECCV. (2002) 707–720
4. Brandt, S.: Closed-form solutions for affine reconstruction under missing data. In: Proceedings Statistical Methods for Video Processing Workshop, in conjunction with ECCV. (2002) 109–114
5. Chen, P., Suter, D.: Recovering the missing components in a large noisy low-rank matrix: Application to sfm. IEEE Transactions on Pattern Analysis and Machine Intelligence **26** (2004)
6. Wiberg, T.: Computation of principal components when data is missing. In: Proceedings Second Symposium of Computational Statistics. (1976) 229–326
7. Buchanan, A., Fitzgibbon, A.: Damped newton algorithms for matrix factorization with missing data. IEEE Computer Society Conference on Computer Vision and Pattern Recognition (CVPR) **2** (2005) 316–322
8. Golub, G., Van Loan, C., eds.: Matrix Computations. The Johns Hopkins Univ. Press (1989)
9. Hartley, R., Schaffalitzky, F.: Powerfactorization: 3d reconstruction with missing or uncertain data. Australian-Japan advanced workshop on Computer Vision (2003)
10. Ma, Y., Soatto, J., Koseck, J., Sastry, S.: An invitation to 3d vision: From images to geometric models. Springer-Verlang, New York (2004)

An Edge-Based Approach to Motion Detection*

Angel D. Sappa and Fadi Dornaika

Computer Vison Center
Edifici O Campus UAB
08193 Barcelona, Spain
{sappa, dornaika}@cvc.uab.es

Abstract. This paper presents a simple technique for motion detection in steady-camera video sequences. It consists of three stages. Firstly, a coarse moving edge representation is computed by a set of arithmetic operations between a given frame and two equidistant ones (initially the nearest ones). Secondly, non-desired edges are removed by means of a filtering technique. The previous two stages are enough for detecting edges corresponding to objects moving in the image plane with a dynamics higher than the camera's capture rate. However, in order to extract moving edges with a lower dynamics, a scheme that repeats the previous two stages at different time scales is performed. This temporal scheme is applied over couples of equidistant frames and stops when no new information about moving edges is obtained or a maximum number of iterations is reached. Although the proposed approach has been tested on human body motion detection it can be used for detecting moving objects in general. Experimental results with scenes containing movements at different speeds are presented.

1 Introduction

A number of techniques for motion detection have been proposed during last years (e.g., [1], [2], [3]). An extensive survey of the current state of the art in image change detection is given in [4]. The most common approaches compute a background image and then threshold the difference between each frame and this estimated background. This difference will automatically unveil moving objects (foreground) present in the scene. Background modeling and subtraction approaches have been extensively used and mainly rely on the use of color or luminance information (e.g. [5], [6]). [7] utilizes color and edge information in order to improve the quality and reliability of the results. It requires several frames to compute an initial estimation of the background image. Since the background is exposed to permanent changes, it has to be updated periodically. Typical approaches update background model by means of Gaussian mixtures [8].

In contrast to iterative updating algorithms, [9] proposes a background estimation algorithm that utilizes a global optimization to identify the periods of time in which

---

* This work was supported by the Government of Spain under the CICYT project TIN2005-09026 and The Ramón y Cajal Program.

V.N. Alexandrov et al. (Eds.): ICCS 2006, Part I, LNCS 3991, pp. 563–570, 2006.

background content is visible in a small block of the image. Since foreground regions are excluded, no bias towards the foreground color will occur in the reconstructed background. The main drawback of background-modeling techniques appears when moving objects always overlap the same area.

On the contrary to previous approaches, the difference between consecutive images was also used to detect motion. For instance, [10] and [11] propose techniques based on the difference between consecutive frames. In [10], moving objects are detected by a combination of three edge maps: *a)* a background edge map, *b)* an edge map computed from the difference of two consecutive frames, and *c)* an edge map from the current frame. In both approaches an interframe scheme that only considers two consecutive frames is proposed; therefore, objects moving with a low dynamics are only detected by both an edge labeling process and a parameters' tuning process.

Our work is closely related to the work presented in [10]. However, it is more advantageous than [10] since: *(1)* efficiency is higher due to the fact that there is no need to compute a background model, *(2)* moving objects are directly extracted by means of their moving edges without tuning any user defined parameter and *(3)* complex scenes, containing objects with different dynamics, can be processed. The proposed technique is based on the use of arithmetic operations between the current frame and other two equidistant ones (backward and forward along the video sequence). It allows handling scenes containing bodies moving at different speeds.

The proposed technique consists of three stages. Firstly, a coarse representation of moving edges is computed. Secondly, that representation is filtered given rise to an image only containing those objects moving with a speed higher than the camera's capture rate. Finally, these two stages are applied iteratively in order to extract all the moving objects present in the current frame. The paper is organized as follows. Section 2 introduces the moving edge detection stage. Section 3 presents the filtering stage, proposed to remove non-moving edges generated in noisy regions. The iterative process is presented in section 4. Experimental results are presented in section 5 and conclusions are finally given in section 6.

2   Moving Edge Detection

Given a video sequence defined by f frames, the algorithm starts by computing their corresponding edges by means of the Canny edge detector [12]. These segmented frames, E_i, contain all the edges of the input frames, F_i. At this first stage the objective is to extract a coarse description of those edges defining moving objects.

In order to detect those edges, a set of arithmetic operations is applied over three consecutive frames $\{n\text{-}m, n, n+m\}$. The philosophy of this first stage is to detect moving edges based on the fact that they will be placed at different positions when consecutive frames are considered. Firstly, the signed differences between edges extracted from a central frame and edges corresponding to two nearest neighbors are computed $(DE_l = \lfloor E_n - E_{n-1} \rfloor$ and $DE_r = \lfloor E_n - E_{n+1} \rfloor)$. From these differences, only positive pixels are considered; pixels with a negative value are set to zero. Each one of these new images (DE_l , DE_r) essentially contains moving edges together with some background edges occluded by the non-overlapped difference (DE_l , DE_r). The latter will be called δ edges, see Fig. 3(*bottom*). The amount of δ edges depends on the speed of the moving objects in the image plane. In addition to the previous edges, the new images also contain edges generated by noisy data or by small differences in the edge representation computed by the Canny edge detector (edges are quite sensitive to light variations). All these non-moving edges will be removed during the next stage by a filtering algorithm, while δ edges are easily removed by merging the computed images (DE_l, DE_r) through an AND logical operation:

$$ME = DE_l \cap DE_r \tag{1}$$

δ edge removal stage is one of the differences with respect to [10], where occluded edges are removed by using an edge map generated by combining background edges and the edges of the current frame.

Fig. 1. (*left*) Original frame. (*right*) Edge representation computed by the Canny edge detector.

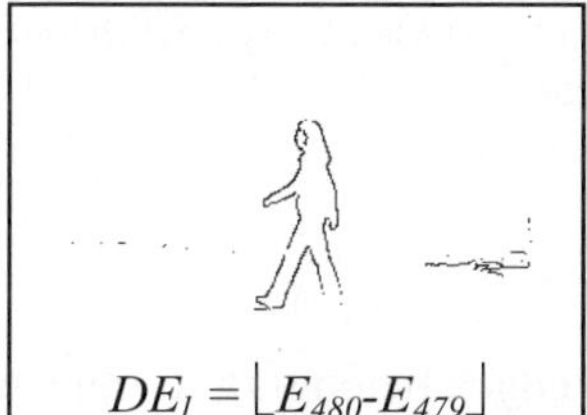  

Fig. 2. (*left*) Edges computed by subtracting to the central frame the previous one. (*center*) Result after subtracting the next one. (*right*) Final edge representation *ME*, computed from DE_l and DE_r.

As mentioned above, an image *ME* still contains edges belonging to non-moving objects generated by noisy data. They are removed next by a filtering stage. Fig. 2(*right*) shows an illustrations of the resulting *ME* image, corresponding to Fig. 1, computed after merging Fig. 2(*left*) with Fig. 2(*center*). Notice that at this particular sequence, motion is performed with a low dynamics—a walking displacement; hence, there are not unveiled δ edges in Fig. 2(*left*) neither in Fig. 2(*center*). Notice that DE_l, DE_r and therefore *ME,* contain some edges corresponding to noisy data from Fig. 1(*right*), which will be removed next.

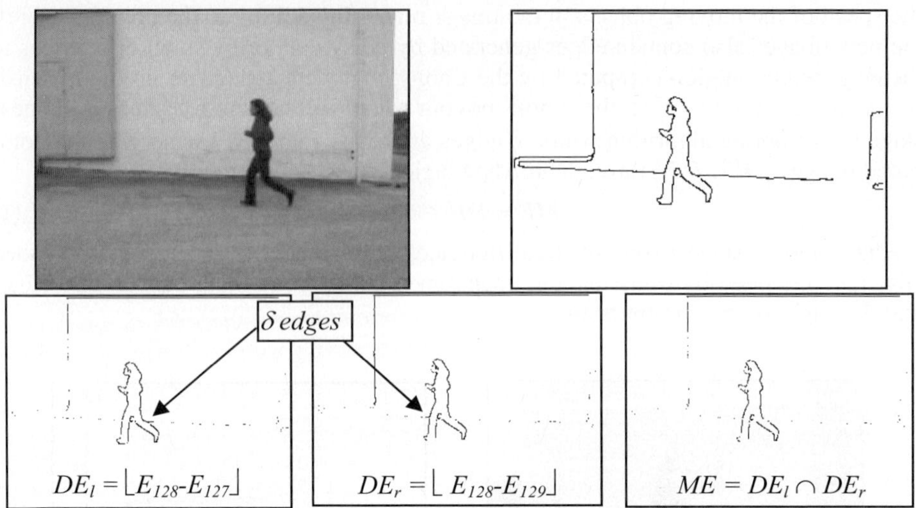

Fig. 3. (*top-left*) Original frame. (*top-right*) Edge representation computed by the Canny edge detector. (*bottom-left*) Edges computed by subtracting to the central frame the previous one. (*bottom-center*) Result after subtracting the next one. (*bottom-right*) Final edge representation *ME*, computed from DE_l and DE_r.

Fig. 3 shows the result obtained with a scene containing a movement having higher dynamics. Differently to the previous case, Fig. 3(*bottom-left*) and Fig. 3(*bottom-center*) show some δ edges. The final edge representation is shown in Fig. 3(*bottom-right*), again there are some edges corresponding to noisy data.

3 Non-moving Edge Removal

The outcome of the previous stage is an image containing edges belonging to objects moving with a speed, in the image plane, higher than the camera capture rate. In addition, that image contains edges belonging to non-moving objects, which are originated due to the fact that the random noise created in one frame is different from the one created in other frames. These differences generate slight changes in the edge position (or new edges), which make that even stationary background edges are not removed when the differences between the current frame and its neighbors is computed (DE_l, DE_r) (see Fig. 2(*right*) and Fig. 3(*bottom-right*)). The objective at this stage is to remove all these non-moving edges.

As shown in [10] and [11], an easy and robust way to extract a noiseless edge representation of moving edges is to apply the Canny operator over the difference of two original frames $\zeta(|F_n - F_{n-1}|)$, instead of performing the difference of the computed edges. It is because Gaussian convolution, included in the Canny operator, suppresses the noise in the luminance difference by smoothing it:

$$\zeta(|F_n - F_{n-1}|) = \theta\,(\,\nabla G * |F_n - F_{n-1}|\,) \tag{2}$$

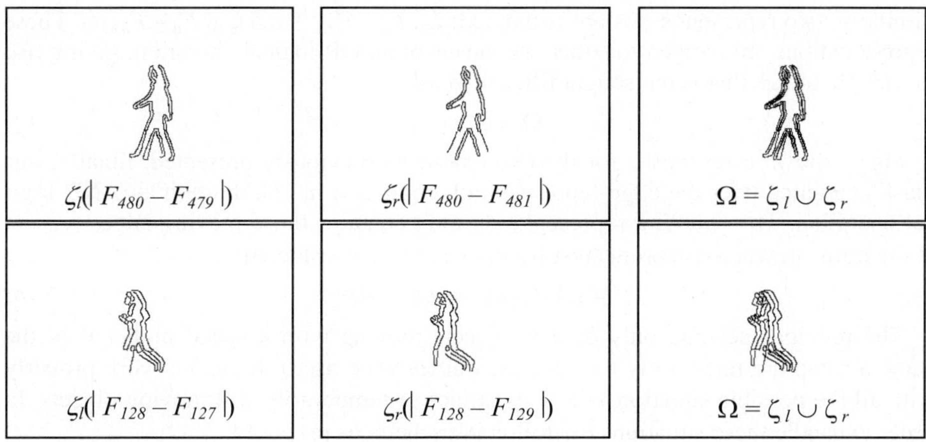

Fig. 4. (*top*) Filter mask for merged edges (*ME*) shown in Fig. 2(*right*). (*bottom*) Filter mask for merged edges (*ME*) shown in Fig. 3(*bottom-right*).

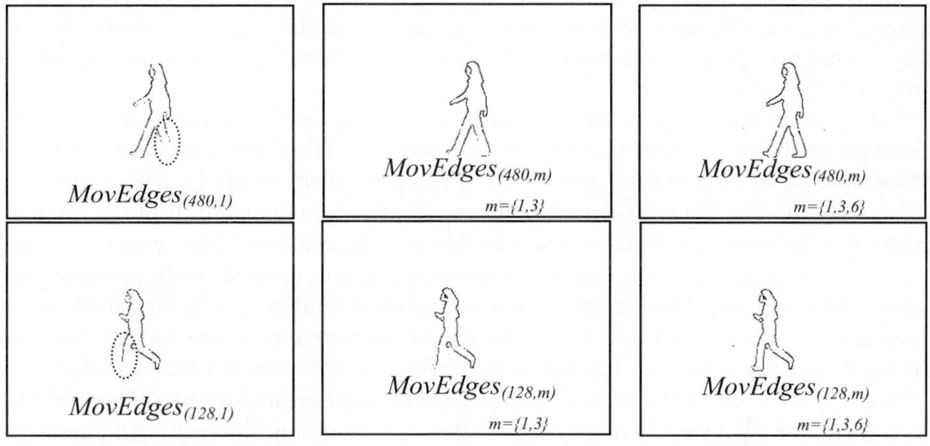

Fig. 5. (*top*) Moving edges extracted from frame 480, Fig. 1(*top-right*), after two iterations. (*bottom*) Moving edges extracted from frame 128, Fig. 3(*top-right*), after two iterations.

where the edges of the original input frames difference, $\zeta(|\,F_n - F_{n-1}\,|)$, are computed by the Canny edge detector by performing a gradient operation ∇ on the Gaussian convoluted image $G*F$, followed by applying the nonmaximum suppression to the gradient magnitude to thin the edges and the thresholding operation with hysteresis to detect and link them (θ). This strategy has already been used in [10] to extract the edges of moving objects by merging this representation with other two edge map representations (edges from the background, automatically or manually computed, and edges from the current frame). In the current implementation we propose to take advantage of this noiseless edge representation and to use it as a filtering mask.

Similarly, two representations are computed: $\zeta_l(|F_n - F_{n-1}|)$ and $\zeta_r(|F_n - F_{n+1}|)$. These representations are merged together, by means of an OR logical operation, giving rise to a single image that is the sought filtering mask:

$$\Omega = \zeta_l \cup \zeta_r \tag{3}$$

Fig. 4 shows filter masks for the two examples previously presented. Finally, this mask is applied over the edge representation computed in (*1*), through an AND logical operation. The resulting representation only contains those moving edges present at the frame n, when its two nearest frames ($n\pm1$) are considered:

$$MovEdges_{(n,1)} = \Omega \cap ME \tag{4}$$

The previous scheme only detects objects moving with a speed higher than the camera's capture rates (only two nearest frames were used). It cannot work properly with all the possible situations—low dynamics or temporarily still moving objects. In order to handle these situations the following scheme is proposed.

4 Detecting Moving Objects

The previous stages can easily be extended by considering not only the two nearest frames but a combination of two frames equidistant to the one under study. In this way, an iterative process has been proposed to detect all the spectra of moving edges present in the scene.

Let E_n be the edges extracted from frame n by using the Canny operator. The technique presented in previous sections, is now used by taking into account a couple of frames placed at m backward and forward positions from n ($m>1$). Again, moving edges computed by (*1*) are filtered by means of (*3*), also computed from the frame n together with both $n\pm m$ frames. The variable m is incremented after every iteration and the computed moving edges, $MovEdges_{(n,m)}$, are merged with previous results—OR operation. This iterative process is applied until no new information about moving edges is extracted or a maximum number of iterations is reached. In this case the algorithm stops and moving objects are defined by the extracted moving edges.

In order to speed up the process, in the current implementation the variable m has been increased by a step of three frames after each iteration (*m += 3*). An attractive point of the proposed scheme, when human motion is considered, is that this iterative approach allows detecting all body parts independently of their particular dynamics. Human body displacement (e.g. walking, running) is a good example of a movement involving different dynamics. Its particularity, over other rigid moving objects, is that in spite of the fact that the center of gravity could have associated a constant velocity, each body part has a different non-constant velocity; this velocity, for example during a walking period, is temporarily null for the foot that is in contact with the floor. Hence, detection of human body displacement is an attractive topic, where, up to our knowledge none of those algorithms based on the use of only two consecutive frame differences is able to efficiently detect without further considerations.

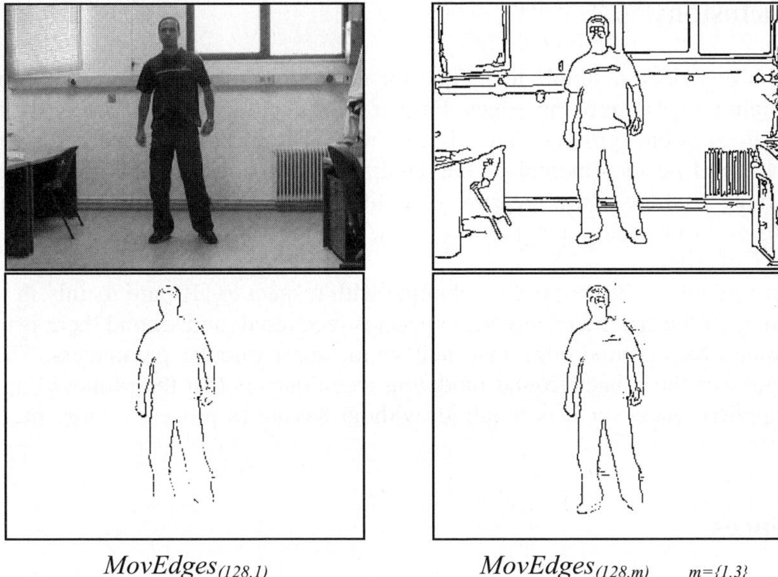

$MovEdges_{(128,1)}$ $MovEdges_{(128,m)}$ $m=\{1,3\}$

Fig. 6. (*top-left*) Original frame. (*top-right*) Edge representation computed by the Canny edge detector. (*bottom-left*) Moving edges extracted after one iteration. (*bottom-right*) Moving edges extracted after two iterations.

5   Experimental Results

The proposed technique has been tested with several video sequences depicting body motion having different dynamics. In the paper two different illustrations have been used (one with low dynamics and the other with high dynamics). Fig. 5 shows final results of both illustrations. Fig. 5(*top-left*) has been obtained after filtering Fig. 2(*right*) with the mask presented in Fig. 4(*top-right*). While Fig. 5(*top-center*) and Fig. 5(*top-right*) present moving edges obtained after two and three iterations respectively—edges corresponding to the highlighted region in Fig. 5(*top-left*) have been recovered when frames further than one position were considered. Fig. 5(*bottom-left*) has been obtained after filtering Fig. 3(*bottom-right*) with the mask presented in Fig. 4(*bottom-right*). Similarly, the highlighted region corresponds to the body part with lowest dynamics. Fig. 5(*bottom-center*) and Fig. 5(*bottom-right*) present moving edges obtained after two and three iterations respectively.

Fig. 6 presents results obtained with an indoor video sequence. Fig. 6(*top-left*) shows an original frame, while its corresponding edges, computed by Canny edge detector, are presented in Fig. 6(*top-right*). Moving edges obtained after one and two iterations are presented in Fig. 6(*bottom*).

Finally, although at the current implementation segmenting the bodies' region is not addressed, they can be easily handled by detecting regions bounded by the first and last edge points along rows and columns [10]. Extracted points will define the moving body regions.

6 Conclusions

This paper described a simple technique for recovering moving objects by extracting their defining edges—moving edges. Further works will consider labeling those non-moving edges as background edges. In this way, if it necessary, a background representation could be incrementally generated; moreover after computing a full background image, where some measure of confidence is reached, the algorithm could switch from moving edge detection to a background subtraction approach, probably reducing CPU time.

Improvements of the proposed technique with respect to [10] are mainly in two aspects. First, all the spectra of moving objects is recovered; and second there is no need to generate a background edge map neither to tune particular parameters. The main advantage over those background modeling techniques is that the proposed approach can be applied whenever it is required, without having to process a large part of the video.

References

1. Kim, J., Kim, H.: Efficient Region-Based Motion Segmentation for a Video Monitoring System. Pattern Recognition Letters **24** (2003) 113-128
2. Sheikh, Y., Shah, M.: Bayesian Object Detection in Dynamic Scenes. IEEE Int. Conf. on Computer Vision and Pattern Recognition, San Diego, CA, USA, June 2005
3. Kellner, M., Hanning, T.: Motion Detection Based on Contour Strings. IEEE Int. Conf. on Image Processing, Singapore, October 2004
4. Radke, R., Andra, S., Al-Kofahi, O., Roysam, B.: Image Change Detection Algorithms: A Systematic Survey. IEEE Trans. On Image Processing, **14** (2005) 294-307
5. Bichsel, M.: Segmenting Simply Connected Moving Objects in a Static Scene. IEEE Trans. on Pattern Analysis and Machine Intelligence **16** (1994) 1138-1142
6. Francois, A., Medioni, G.: Adaptive Color Background Modeling for Real-Time Segmentation of Video Streams. Int. Conf. on Imaging Science, Systems and Technology, Las Vegas, NA, June 1999
7. Jabri, S. , Duric, Z., Wechsler, H., Rosenfeld, A.: Detection and Location of People in Video Images Using Adaptive Fusion of Color and Edge Information. 15th. Int. Conf. on Pattern Recognition, Barcelona, Spain, Sep. 2000
8. Lee, D.: Effective Gaussian mixture learning for video background subtraction. IEEE Trans. on Pattern Analysis and Machine Intelligence **27** (2005) 827-832
9. Farin, D., de With, P., Effelsberg, W.: Robust Background Estimation for Complex Video Sequences. IEEE Int. Conf. on Image Processing, Barcelona, Spain, Sep. 2003.
10. Kim, C., Hwang, J.: Fast and Automatic Video Object Segmentation and Tracking for Content-Based Applications. IEEE Trans. on Circuits and Systems for Video Technology **12** (2002) 122-129
11. Marqués, F., Molina, C.: Object Tracking for Content-Based Functionalities. SPIE Vis. Commun. Image Processing, vol. 3024, San Jose, CA, Feb. 1997
12. Canny, J.F.: A Computational Approach to Edge Detection. IEEE Trans. on Pattern Analysis and Machine Intelligence **6** (1986) 679-698

A Dominating Set Based Clustering Algorithm for Mobile Ad Hoc Networks

Deniz Cokuslu, Kayhan Erciyes, and Orhan Dagdeviren

Izmir Institute of Technology
Computer Eng. Dept., Urla, Izmir 35340, Turkey
{denizcokuslu, kayhanerciyes, orhandagdeviren}@iyte.edu.tr

Abstract. We propose a new Connected Dominating Set (CDS) based algorithm for clustering in Mobile Ad hoc Networks (MANETs). Our algorithm is based on Wu and Li's [14] algorithm, however we provide significant modifications by considering the degrees of the nodes during marking process and also provide further heuristics to determine the color of a node in the initial phase. We describe, analyze and measure performance of this new algorithm by simulation and show that it performs better than Wu and Li's [14] algorithm especially in the case of dense networks.

1   Introduction

MANETs do not have any fixed infrastructure and consist of wireless mobile nodes that perform various data communication tasks. MANETs have potential applications in rescue operations, mobile conferences, battlefield communications etc. Conserving energy is an important issue for MANETs as the nodes are powered by batteries only[1].

Clustering has become an important approach to manage MANETs. In large, dynamic ad hoc networks, it is very hard to construct an efficient network topology. By clustering the entire network, one can decrease the size of the problem into small sized clusters. Clustering has many advantages in mobile networks. Clustering makes the routing process easier, also, by clustering the network, one can build a virtual backbone which makes multicasting faster. However, the overhead of cluster formation and maintenance is not trivial. In a typical clustering scheme, the MANET is firstly partitioned into a number of clusters by a suitable distributed algorithm. A Cluster Head (CH) is then allocated for each cluster which will perform various tasks on behalf of the members of the cluster. The performance metrics of a clustering algorithm are the number of clusters and the count of the *neighbor nodes* which are the adjacent nodes between clusters that are formed [1].

In this study, we search various graph theoretic algorithms for clustering in MANETs and propose a new distributed algorithm. *Dominating Set based Clustering* which is a fundamental approach and related work in this area are reviewed in Section 2. We illustrate our algorithm and sample results in Section 3 and the Conclusion Section provides the overview.

V.N. Alexandrov et al. (Eds.): ICCS 2006, Part I, LNCS 3991, pp. 571–578, 2006.
© Springer-Verlag Berlin Heidelberg 2006

2 Background

2.1 Clustering Using Dominating Sets

A dominating set is a subset S of a graph G such that every vertex in G is either in S or adjacent to a vertex in S[2]. Dominating sets are widely used in clustering networks[8]. Dominating sets can be classified into three main classes, Independent Dominating Sets (IDS), Weakly Connected Dominating Sets (WCDS) and Connected Dominating Sets (CDS)[4].

- *Independent Dominating Sets:* IDS is a dominating set S of a graph G in which there are no adjacent vertices. Fig. 1.a shows a sample independent dominating set where black nodes show cluster heads.
- *Weakly Connected Dominating Sets (WCDS):* A weakly induced subgraph $(S)_w$ is a subset S of a graph G that contains the vertices of S, their neighbors and all edges of the original graph G with at least one endpoint in S. A subset S is a weakly-connected dominating set, if S is dominating and $(S)_w$ is connected [5]. Black nodes in Fig. 1.b show a WCDS example.
- *Connected Dominating Sets:* A connected dominating set (CDS) is a subset S of a graph G such that S forms a dominating set and S is connected. Fig. 1.c shows a sample CDS.

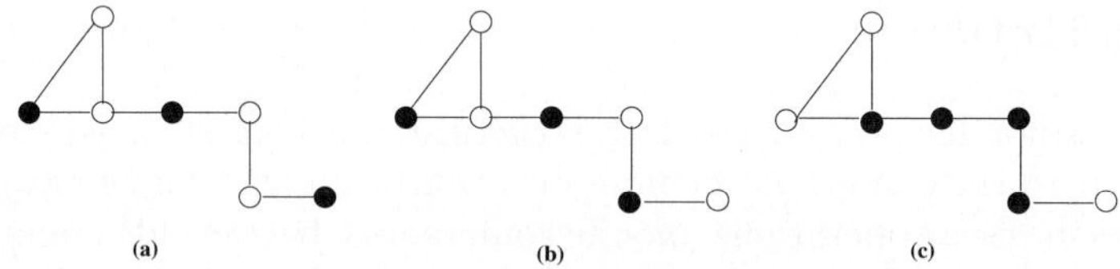

Fig. 1. (a)IDS (b)WCDS (b)CDS

2.2 Dominating Set Algorithms

Various algorithms exist for clustering in IDS, WCDS and CDS.

Clustering Using IDS: Baker and Ephremides [6] proposed an independent dominating set algorithm called *highest vertex ID*. A very similar algorithm to the highest id algorithm is the *lowest id algorithm* by Gerla and Tsai [7]. Gerla and Tsai developed another algorithm to find the independent dominating sets called the *highest degree algorithm*. Although these algorithms are considered as important algorithms, Chen et al. [8] proposed that these algorithms are not working correctly for some graphs. To solve this incorrect operation, Chen et al. developed the *k-distance independent dominating set* algorithm.[9].

Clustering Using WCDS: Although independent dominating sets are suitable for constructing optimum sized dominating sets, they have some deficiencies such as lack of direct communication between cluster heads. In order to obtain the connectivity between cluster heads, WCDSs can be used to construct clusters.

The WCDS was first proposed for clustering in ad hoc networks by Chen and Liestman [10] called *zonal clustering.*

Clustering Using CDS: CDSs have many advantages in network applications such as ease of broadcasting and constructing virtual backbones [11], however, when we try to obtain a connected dominating set, we may have undesirable number of cluster heads. So, in constructing connected dominating sets, our primary problem is to find a minimal connected dominating set. Guha and Khuller [12] proposed two centralized greedy algorithms for finding suboptimal connected dominating sets. Das and Bharghavan [13] provided distributed implementations of Ghua and Khuller's algorithms [12]. Wu and Li [14], improved Das and Bhraghavan's distributed algorithm as a localized distributed algorithm for finding connected distributed sets in which each node only needs to know its distance-two neighbor [13]. Then Wu and Dai, proposed an extended version of this algorithm which uses more general rules in order to cluster graphs[18]. Xinfang Yan, Yugeng Sun, and Yanlin Wang [15] proposed a heuristic algorithm for minimum connected dominating set which uses uptime and power levels of the nodes as heuristics. Peng-Jun Wan, Khaled M. Alzoubi and Ophir Frieder [16] proposed a distributed algorithm for finding a CDS which constructs the dominating set using the Maximal Independent Sets. Hui Liu, Yi Pan and Jiannong Cao [17], improved Wu and Li's algorithm [14] by adding a third phase elimination. In the additional third phase, the algorithm searches redundant cluster heads. A cluster head is eliminated if it is dominated by two of its cluster head neighbors. Our algorithm is based on Wu and Li's CDS Algorithm [14].

Wu and Li CDS Algorithm: Wu and Li CDS Algorithm [14] is a step wise operational distributed algorithm, in which every node has to wait for others in lock state in the algorithm. In this algorithm, initially each vertex marks itself $WHITE$ indicating that it is not dominated yet. In the first phase, a vertex marks itself $BLACK$ if any two of its neighbors are not connected to each other directly. In the second phase, a $BLACK$ marked vertex v changes its color to $WHITE$ if either of the following conditions is met:

1. $\exists u \in N(v)$ which is marked $BLACK$ such that $N[v] \subseteq N[u]$ and $id(v) < id(u)$;
2. $\exists u, w \in N(v)$ which is marked $BLACK$ such that $N(v) \subseteq N(u) \bigcup N(w)$ and $id(v) = min\{id(v), id(u), id(w)\}$;

3 The Distributed Dominating Set Based Clustering Algorithm

3.1 General Idea

We propose a distributed algorithm which finds a minimal connected dominating set in a MANET. We developed our algorithm based on Wu's CDS Algorithm but we add some extra heuristics. First, we determine some situations that a

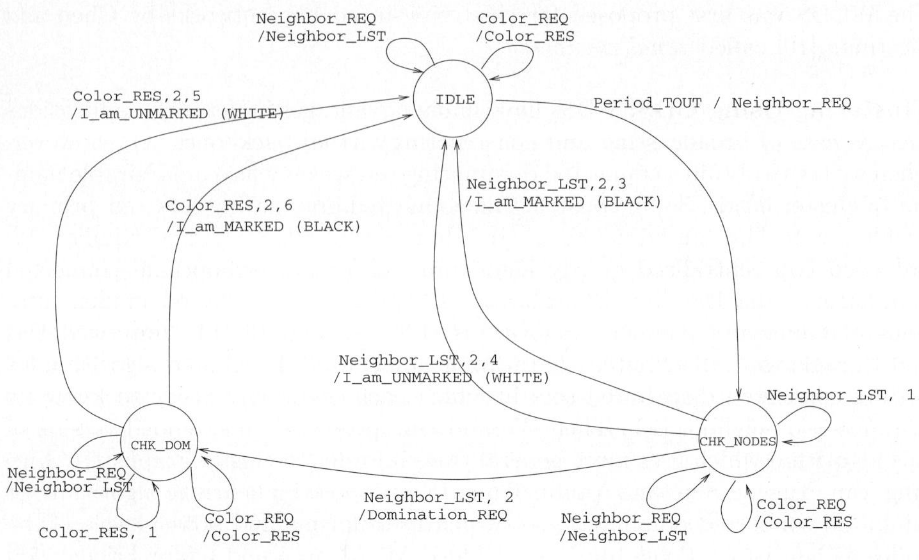


```
1) count < Neighbor_number
2) count = Neighbor_number
3) Having at least one isolated neighbor
4) All neighbors are connected OR i am isolated
5) One of the four pruning rules is true
6) None of the four pruning rules is true
```

Fig. 2. Finite State Machine of the Clustering Algorithm

node cannot change its color after the first phase. We also consider the degree of a node when marking it.

3.2   Algorithm

We assume that each node has a unique *node_id* and knows its adjacent neighbors. Each node has a *color* indicating whether the node is in the dominating set or not. The *color* is set to *BLACK* if the node is in the dominating set, or *WHITE* if the node is not in the dominating set. Color *GRAY* is used to indicate that the node is marked after the first phase, but it will change its color after the second phase as either *WHITE* or *BLACK*. The finite state diagram for the algorithm can be seen in Fig. 2. *Period_TOUT* message triggers the algorithm and is sent periodically. *Neighbor_REQ* message is sent to collect a list of distance-2 neighbors. *Neighbor_LST* message includes a list of adjacent neighbors of sending node. *Color_REQ* message is used to collect a node's neighbor's colors after the first phase. *Color_RES* message includes sender's color after the first phase. Each node is in the *IDLE* state and colored as *UNDEFINED_COLOR* initially. When all *Neighbor_LST* messages are collected in the *CHK_NODES* state, the node checks the following heuristics to determine if it will be among the ones whose color will not alter after the first phase:

- If the node has at least one isolated neighbor, it changes its color to *BLACK* and its state to *IDLE*.
- If all neighbors of the node are directly connected to each other or if the node is an isolated node, it changes its color to *WHITE* and its state to *IDLE*.

If the node is not suitable for one of these two coloring heuristics, then it changes its color to *GRAY* and its state to *CHK_DOM*. When the node switches to state *CHK_DOM*, it multicasts a *Color_REQ* message to its neighbors. Then it waits until all its neighbors send their colors. When the node v collects all color information, it starts to apply the following rules:

1. $\exists u \in N(v)$ which is marked *BLACK* such that $N[v] \subseteq N[u]$;
2. $\exists u, w \in N(v)$ which is marked *BLACK* such that $N(v) \subseteq N(u) \bigcup N(w)$;
3. $\exists u \in N(v)$ which is marked *GRAY* such that $N[v] \subseteq N[u]$ and $degree(v) < degree(u)$ OR $(degree(v) = degree(u)$ AND $id(v) < id(u))$;
4. $\exists u, w \in N(v)$ which is marked *GRAY* OR *BLACK* such that $N(v) \subseteq N(u) \bigcup N(w)$ and $degree(v) < min\{degree(u), degree(w)\}$ OR $degree(v) = min\{degree(u), degree(w)\}$ AND $id(v) < min\{id(u), id(w)\}$;

If one of these rules is true, then the node v changes its color to *WHITE*, else it changes its color to *BLACK*. After the node determines its permanent color, it changes its state to *IDLE*. At any state, a node can receive request messages to help other nodes run their algorithms. These messages are *Neighbor_REQ* and *Color_REQ*. In such a case, the node prepares the required information requested in the received message and continues to its current operation. No state changes are performed in these cases.

3.3 An Example Operation

We obtained the resulting connected dominating set in Fig. 3 by using our algorithm. This section explains the algorithm step by step by using the sample graph in Fig. 3. Runtime of the algorithm is explained for nodes 1 and 8, but we remark again that each node is running the algorithm concurrently.

- *Execution of algorithm at node 1:* When node 1 times out the period, it sets its *color* to *UNDEFINED_COLOR*, sends a *Neighbor_REQ* message to all of its neighbors and changes its state to *CHK_NODES*. It waits at that state until all response messages are collected from its neighbors. Once responses are collected, node 1 prepares a list of its directly unconnected neighbors

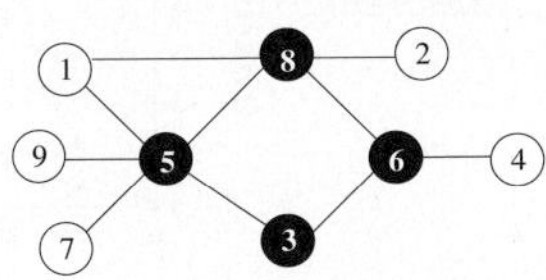

Fig. 3. Example

by looking at *Neighbor_LST* messages. In this stage, it finds out that all of its neighbors are directly connected, therefore the node 1 sets its color to *WHITE* and its state to *IDLE*.

– *Execution of algorithm at node 8:* When node 8 times out the period, it sets its *color* to *UNDEFINED_COLOR*, sends a *Neighbor_REQ* message to all of its neighbors and changes its state to *CHK_NODES*. Once responses are collected, node 8 checks if one of the heuristics is suitable for it, at that stage, it finds out that the node 8 has an isolated neighbor (*node2*), so it sets its color to *BLACK* and its state to *IDLE*.

3.4   Analysis

Theorem 1. *Time complexity of the clustering algorithm is $\Theta(4)$.*

Proof. Every node executes the distributed algorithm by the exchange of 4 messages. Since all these communication occurs concurrently, at the end of this phase, the members of the CDS are determined, so the time complexity of the algorithm is $\Theta(4)$.

Theorem 2. *Message complexity of the clustering algorithm is $O(n^2)$ where n is the number of nodes in the graph.*

Proof. For every mark operation of a node, 4 messages are required (*Neighbor_REQ, Neighbor_LST, Color_REQ, Color_RES*). Assuming every node has

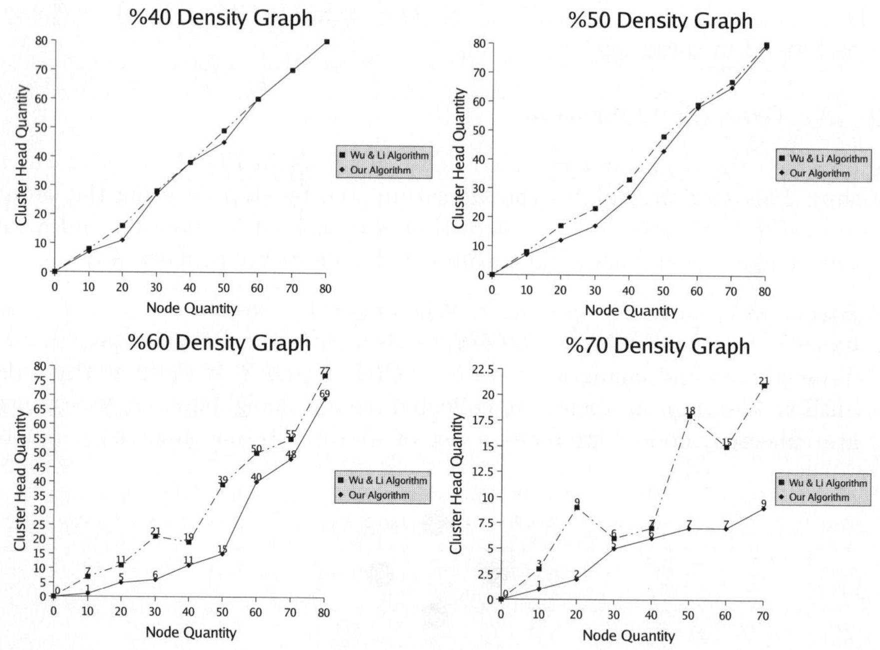

Fig. 4. Test Results

n-1 adjacent neighbors, total number of messages sent is $4(n - 1)$. Since there are n nodes, total number of messages in the system is $n(4(n - 1))$ Therefore messaging complexity of our algorithm has an upperbound of $O(n^2)$.

3.5   Results

We implemented the Dominating Set Based Clustering Algorithm with POSIX threads in RedHat. Each thread is treated as a mobile node. Graph is constructed randomly with different densities. Fig. 4 displays the cluster head numbers at graphs with 40, 50, 60 and 70 per cent densities. The results are taken at ranging from 10 to 90 nodes in a random graph. It can be seen that our modified algorithm performs significantly better than Wu and Li's[14] algorithm at higher densities.

4   Conclusions

In this paper, we proposed a distributed algorithm with significant modifications on Wu and Li's algorithm for constructing a CDS. We showed the implementation of the algorithm and analyzed its time and message complexities. We showed that we can decrease the size of the CDS by adding some heuristics to Wu and Li's algorithm. We also showed that this improvement can be significant especially when the number of nodes in the MANET is large. Therefore, we can conclude that the algorithm can be preferable in dense mobile networks. Clustering described in this study can be used for various distributed tasks in MANET's such as routing and multicast communications. We are planning to experiment various *total order multicast* algorithms in such an environment where message ordering is provided by the cluster heads on behalf of the ordinary nodes of the MANET.

References

1. Nocetti, F., B., Gonzalez, J., S. and Stojmenovic, I. : Connectivity Based K-Hop Clustering in Wireless Networks, Telecommunication Systems, (2003), 205-220.
2. West, D. : Introduction to Graph Theory, Second edition, Prentice Hall, Upper Saddle River, N.J., (2001).
3. Chen, Y., Z., P. and Liestman, A., L. : Approximating Minimum Size Weakly-Connected Dominating Sets for Clustering Mobile Ad Hoc Networks, Proc. of 3'rd ACM Int l. Symp. Mobile Ad Hoc Net. and Comp., (2002), 16572.
4. Haynes, T., W., Hedetniemi, S., T. and Slater, P., J. : Domination in Graphs, Advanced Topics, Marcel Dekker, Inc., (1998).
5. Chen, Y., Z., P., Liestman, A., L. and Jiangchuan, L. : Clustering Algorithms for Ad Hoc Wireless Networks, Nova Science Publishers, (2004).
6. Baker, D. and Ephremides, A. : The Architectural Organization of a Mobile Radio Network via a Distributed Algorithm, Communications, IEEE Transactions, (1981), 29(11), 1694-1701.
7. Gerla, M. and Jack T., C., T. : Multicluster, Mobile, Multimedia Radio Network, Wireless Networks, 1(3), (1995), 255-265.

8. Chen, G., Nocetti, F.,G., Gonzalez and J.S., Stojmenovic, I. : Connectivity Based K-Hop Clustering in Wireless Networks, System Sciences, Proc. of the 35th Annual Hawaii International Conference, (2002), 2450-2459.
9. Ohta, T., Inoue, S. and Kakuda, Y. : An Adaptive Multihop Clustering Scheme for Highly Mobile Ad Hoc Networks, Proc. of 6th ISADS'03, (2003).
10. Chen, Y., P. and Liestman, A., L. : A Zonal Algorithm for Clustering Ad Hoc Networks, International Journal of Foundations of Computer Science, (2003), 14(2), 305-322.
11. Stojmenovic, I., Seddigh M. and Zunic, J. : Dominating Sets and Neighbor Elimination-Based Broadcasting Algorithms in Wireless Networks, IEEE Transactions on Parallel and Distributed Systems, (2002), 13, 14-25.
12. Guha S. and Khuller, S. : Approximation Algorithms for Connected Dominating Sets, Springer Verlag New York, LLC, ISSN: 0178-4617, (1998).
13. Das, B. and Bharghavan, V. : Routing in Ad-Hoc Networks Using Minimum Connected Dominating Sets, Communications, ICC97 Montreal, 'Towards the Knowledge Millennium', IEEE International Conference, (1997), 1, 376-380.
14. Wu, J. and Li, H. : A Dominating-Set-Based Routing Scheme in Ad Hoc Wireless Networks, Springer Science+Business Media B.V., Formerly Kluwer Academic Publishers B.V. ISSN: 1018-4864, (2001).
15. Yan, X., Sun, Y. and Wang, Y. : A Heuristic Algorithm for Minimum Connected Dominating Set with Maximal Weight in Ad Hoc Networks, Proc. of the Grid and Cooperative Computing Second International Workshop, (2003), 719-722.
16. Wan, P., J., Alzoubi, K., M. and Frieder, O. : Distributed Construction of Connected Dominating Set in Wireless Ad Hoc Networks, Springer Science+Business Media B.V., Formerly Kluwer Academic Publishers B.V., , (2002), 9(2), 141-149.
17. Liu, H., Pan, Y. and Cao, J. : An Improved Distributed Algorithm for Connected Dominating Sets in Wireless Ad Hoc Networks, Parallel and Distributed Processing and Applications, Proc. of the ISPA 2004, (2004), 340.
18. Wu, J. and Dai, F. : An Extended Localized Algorithm for Connected Dominating Set Formation in Ad-Hoc Wireless Networks, IEEE Transactions on Parallel and Distributed Systems, (2004), 15(10).

MVRC Heuristic for Solving the Multi-Choice Multi-Constraint Knapsack Problem

Maria Chantzara and Miltiades Anagnostou

School of Electrical & Computer Engineering, National Technical University of Athens
9 Heroon Polytechniou Str, 157 73 Zografou, Athens, Greece
{marhantz, miltos}@telecom.ntua.gr

Abstract. This paper presents the heuristic algorithm Maximizing Value per Resources Consumption (MVRC) that solves the Multi-Choice Multi-Constraint Knapsack Problem, a variant of the known NP-hard optimization problem called Knapsack problem. Starting with an initial solution, the MVRC performs iterative improvements through exchanging the already picked items in order to conclude to the optimal solution. Following a three step procedure, it tries to pick the items with the maximum Value per Aggregate Resources Consumption. The proposed algorithm has been evaluated in terms of the quality of the final solution and its run-time performance.

1 Introduction

One of the most studied combinatorial optimization problems is the *Knapsack Problem* (KP). Numerous problems of different fields such as capital budgeting, cargo loading and resource allocation are modeled as a variant of it. Due to the high applicability of this NP-hard problem, it has been widely studied. The objective of the original KP is to optimize resource allocation, or more precisely, how to distribute a fixed amount of resources among several actions in order to obtain maximum payoff. The 0-1 KP considers a knapsack of specific capacity and a set of items; each of them has specific weight and value. The objective is to determine which items should be placed in the knapsack so as to maximize the total value of the items contained in it without exceeding its capacity. Another variant of the 0-1 KP refers to the case that there are multiple constraints regarding resources and is called *Multi-Dimension or Multi-Constraint KP* (MDKP). Another one is the *Multi-Choice KP* (MCKP). In this case the items are divided into groups and the objective is to pick exactly one item of every group in order to maximize the total value.

A combination of the MDKP and MCKP variants is the *Multi-Choice Multi-Constraint Knapsack Problem* (MMKP). Its formal definition is: There are n groups of items. Group i contains l_i items. Each item ij has a particular value v_{ij} and weights $r_{ij} = (r_{1ij},...,r_{mij})$ regarding the m resource constraints: $C = (C_1,...,C_m)$. The objective is to pick exactly one item from each group in order to have maximum total value of the collected items, subject to the m resource constraints. With decision variable: $x_{ij} \in \{0,1\}$, the MMKP is formulated as follows. The objective function to be maximized is the total value of the picked items:

V.N. Alexandrov et al. (Eds.): ICCS 2006, Part I, LNCS 3991, pp. 579–587, 2006.

$$TotalValue = \sum_{i=1}^{i=n} \sum_{j=1}^{j=l_i} v_{ij} x_{ij} \qquad (1)$$

The constraints are:

$$\sum_{i=1}^{i=n} \sum_{j=1}^{j=l_i} r_{kij} x_{ij} \leq C_k \, , k = 1,...,m \qquad (2)$$

$$\sum_{j=1}^{j=l_i} x_{ij} = 1, i = 1,...,n \qquad (3)$$

This paper presents the heuristic Maximizing Value per Resources Consumption (MVRC) for solving the MMKP. It is an improvement of the heuristic HEU [1] which is proved to be one of the best known algorithms in terms of performance (solution optimality vs. computation time). Starting with an initial solution, MVRC performs iterative improvements though exchanging the already picked items in order to conclude to the optimal solution. Unlike HEU that tries to minimize the resources consumption, MVRC picks the items with the maximum Value per Aggregate Resources Consumption. Finally, MVRC solves the MMKP instances in less time that is important for cases that require real-time decision making. We have applied the proposed heuristic to the MMKP instance referring to the discovery of the information sources for acquiring data on behalf of the context-aware services. Regarding this problem, each requested type of info corresponds to a group, and the available quality levels of the same context data correspond to the items of the group. The resource constraints refer to the cost and the latency bound for obtaining the data. The value that needs to be maximized is the expected utility of the picked items. The formulation of context source discovery as MMKP has been analyzed in [2]. In this case, the values do not follow monotone feasibility order, since items with higher cost correspond to higher expected utility, while items with higher response time to lower.

The rest of the paper is organized as follows: the literature review related to MMKP is reported in Section 2. In Section 3, the proposed heuristic MVRC is detailed and its worst-case complexity is computed. In Section 4 the evaluation of the MVRC performance is presented. Finally, Section 5 provides some conclusive remarks.

2 Related Work

In [3] and recently in [4], reviews of the KPs and literature on methods to solve them are presented. The proposed exact algorithms for solving the different variants are based on the "branch-and-bound" approach or dynamic programming techniques, that are capable of producing optimal solutions but they require much time. Therefore, heuristics providing near-optimal solutions within low computation time are developed. We are particularly interested in the literature about the MMKP variant. The exact algorithm for the MMKP is a branch-and-bound [5] that performs a complete enumeration keeping the best solution found so far in order to find the optimal one. If a partial solution cannot improve on the best, it is abandoned.

However, this approach cannot be applied when the selection should be done real-time, due to its high complexity and therefore, heuristics are developed.

One of the first heuristics for solving the MMKP is presented in [6]. It uses the concept of graceful degradation from the most valuable items based on Lagrange Multipliers. This algorithm fails to find a solution when the resources are extremely short and the feasible solutions are very few. In [1] the heuristic HEU is detailed. This approach is based on the concept of "aggregate resources" proposed in [7] that converts the multiple resource dimensions into only one through penalizing the use of resources. In fact, it applies a large penalty for a heavily used resource and a small penalty for a lightly used resource. The proposed method starts from an initial pick of the least valuable items, finds a feasible solution by exchanging items based on the Toyoda concept, while it ensures that there is improvement regarding the resources consumption. It finally upgrades the feasible solution in terms of solution value through iterative exchanges resulting to another feasible solution. The HEU has been applied to the QoS management problem, where the QoS levels follow monotone feasibility order since a QoS level with higher utility requires more resources. In order to have better solutions for the MMKP instances where some higher-valued items consume less resources than lower-valued, the authors of [8] have proposed another method. This one applies a transformation technique to map the multi-dimensional resource to single dimension and constructs convex hulls to reduce the search space of solutions. Comparing it with the HEU showed that even though it produces solutions with significant lower value especially for the correlated data sets, it concludes to the solution in significant reduced time. As a result, it can be stated that it mostly fits to cases where run-time performance is of greater interest than the solution quality. Finally, the authors of [9] propose a set of algorithms for finding the solution of an MMKP instance. The first one constructs an initial feasible solution through a greedy approach, the second one improves the quality of the initial solution using a complementary procedure and the third one searches for the best feasible solution over a set of neighborhoods according to the "guided local search" method. The evaluation of the performance of this approach showed that it outperforms the HEU in terms of the solution quality but for the run-time performance no metrics are given.

3   The MVRC Heuristic Algorithm

Before describing the MVRC heuristic, we introduce some notations. Considering item ij with value v_{ij} and resources usage $r_{ij} = (r_{1ij},...,r_{mij})$, we define the *Aggregate Resources Consumption* (ARC): $ARC_{ij} = \dfrac{r_{1ij} * C_1 + ... + r_{kij} * C_k}{\sqrt{C_1^2 + ... + C_k^2}}$, and the *Value-per unit of ARC* (V-ARC): $V - ARC_{ij} = v_{ij} \big/ ARC_{ij}$. Assuming that we have the problem's solution described by the vector S=$(1j_1,2j_2,..., ij_i,..,nj_n)$, denoting the items picked per each group, and resources consumption $R_S = (R_1,R_2,...R_k)$ where $R_k = \sum\limits_{i=1}^{i=n} r_{kij_i}$. If the already picked item ij_i of group i is exchanged by the item ij, the new solution is: S'=$(1j_1,2j_2,...,ij,...,nj_n)$. The resources consumption becomes $R_{S'}=(R'_1,R'_2,...,R'_k)$.

For the new solution S′, we define the *Aggregate Resources Requirements* (ARR):
$ARR_{ij} = \dfrac{(R'_1 - R_1)*R_1 + ... + (R'_k - R_k)*R_k}{\sqrt{R_1^2 + ... + R_k^2}}$, and the *Value Update-per unit of ARR*

(VU-ARR): $VU - ARR_{ij} = \dfrac{v_{ij} - v_{ij_i}}{ARR_{ij}}$. For the solution S, we also define the *feasibility*

factor of each resource k as follows: $F_k = {R_k}\big/{C_k}$. If $F_k \leq 1$, the resource k is called

feasible; otherwise it is infeasible. In the same respect, if $F_k \leq 1$ for each k=1,2,…,m, the solution S is called feasible; otherwise it is called infeasible. Moreover for the picked items in solution S, we define the *feasibility factor of each item* ij_i of group i

for each resource k, as follows: $FI_{kj_i} = {r_{kij_i}}\big/{R_k}$. Finally, the sum of the values of the

picked items is the *solution value*: $V_S = \displaystyle\sum_{i=1}^{i=n} v_{ij_i}$.

The MVRC heuristic follows three steps in order to reach the final solution. Given the MMKP input instance, Step1 produces an initial solution. Step 2 produces a feasible solution through iteratively exchanging the already picked items. In Step 3 the feasible solution is improved by iteratively picking items with higher values.

Fig. 1. MVRC heuristic

The MVRC steps are described in more detail below:

STEP 1: We find the initial solution by selecting the item with the highest V-ARC of each group. In case the solution is a feasible Step 3 follows; otherwise Step 2 follows.
STEP 2: In this step, we exchange one of the items of the current solution in order to find a feasible solution. The decision about the exchange is the outcome of the following sub-steps. Firstly, the resource with the highest feasibility factor is found (step 2.1) and in relation to this resource, the picked item with the highest feasibility factor is determined (step 2.2). Then, we exchange this item with an item of the same group that has lower V-ARC (step 2.3). In case the solution which has come out of this exchange remains infeasible, Step 2 is repeated; otherwise, Step 3 follows. If Step 2 fails to find any feasible solution, the algorithm terminates without a solution.
STEP 3: In this step, we exchange one of the items of the current solution with another of the same group in order to increase the solution value. The criterion is the maximization of the VU-ARR ratio so that the new solution remains feasible. For deciding on the item to insert in the knapsack, we identify the following cases. Having two candidate items, with aggregate resources requirements ARR_1 and ARR_2 respectively: (i) If both ARR_1, $ARR_2 \leq 0$, the item that maximizes the solution upgrade is preferred. (ii) If both ARR_1, $ARR_2 > 0$, the one with the maximum VU-

ARR is preferred. (iii) If $ARR_1 \leq 0$ and $ARR_2 > 0$, the one that maximizes solution value upgrade is preferred. In case, a new exchange can be performed, Step 3 is repeated; otherwise, the algorithm terminates and the current solution is the optimal.

In the following, we present the upper bounds of the computational complexity of the three steps of the MVRC heuristic for the MMKP input instance consisting of n groups, m resources and l items per group. It is assumed that in each group the items are arranged in non-decreasing order. In Step 1, for every item the ratio V-ARC is computed with complexity $O(nml)$. For the sorting of the items according to the V-ARC ratios we use a simple algorithm with complexity $O(nl!)$. For the feasibility check the complexity is $O(m)$. Therefore, the complexity of Step 1 is $O(nml+ nl!+m)$. The step 2.1 requires m comparisons and is $O(m)$. The step 2.2 is $O(n)$. The step 2.3 requires constant time since the items are already ordered. The Step 2 can be repeated at most $n(l-1)$. Thus, the complexity of Step 2 is $O(n^2(l-1)m)$. In Step 3, computing the VU-ARR has $O(m)$ complexity, while finding the item to be exchanged may require $n(l-1)$ computations. Thus, the complexity is $O(n(l-1)m)$. Since the feasible upgrades could be $n(l-1)$ at most, the complexity of Step 3 is $O(n^2(l-1)^2m)$. Finally, we conclude that the complexity of the MVRC heuristic is $O(nml+ nl!+m+n^2(l-1)m+n^2(l-1)^2m)$.

4 Evaluation

In order to test the performance of the proposed algorithm, we implemented the MVRC algorithm along with the HEU and the exhaustive approach. The HEU is one of the best known heuristics that has been evaluated comparatively to all proposed approaches of the literature. The exhaustive approach is an exact algorithm that computes all possible combinations and checks their feasibility in order to find the optimal one (its complexity is $O\left(ml^{\,n}\right)$). We applied the three algorithms to instances of different sizes under the scope to test both the quality of the solution and the run-time performance. The implementation of the algorithms is done in Java 1.4., while we ran the algorithms on a 1.4 GHz Pentium Fujitsu Siemens Lifebook with 512MB running Linux. The data sets were generated according to the state-of-the-art pattern described in [8] that generated both correlated and uncorrelated data sets. Concerning the correlated data instances, that are harder to solve [10], the value of an item depends on its weights, while concerning the uncorrelated, the value and the weights are independent. The metrics measuring the performance of the algorithms are:

1. *Percentage (%) optimality of the solution* that measures the quality of the produced solution. Assuming that V_{opt} is the optimal one and V_i is the one we wish to compare, the percentage optimality is defined as follows:
   $$\%Opt = V_i / V_{opt} * 100$$
2. *Computation time* describing the required time to execute the algorithm.
3. *Number of updates* that are performed till the final solution is found, showing how quick the algorithm converges to the final solution. Unlike the computation time, this metric does not depend on the implementation of the algorithm.

We recorded these metrics for all algorithms with the increase in the number of groups, items per group, resource constraints. For each size of data set, we generated

10 instances and applied the algorithms to all of them. However, we report the average of the performance metrics of the 10 test instances of the specific data set size. The graphs of figures 2-4 depict the % optimality of the solutions for small-sized data sets, since we are unable to run the exhaustive approach for large-sized data instances due to memory and computation time requirements. The graphs of figures 5-7 depict the computation time for the two algorithms to find a solution for large-sized data instances. Finally, the figures 8-10 depict the number of updates respectively.

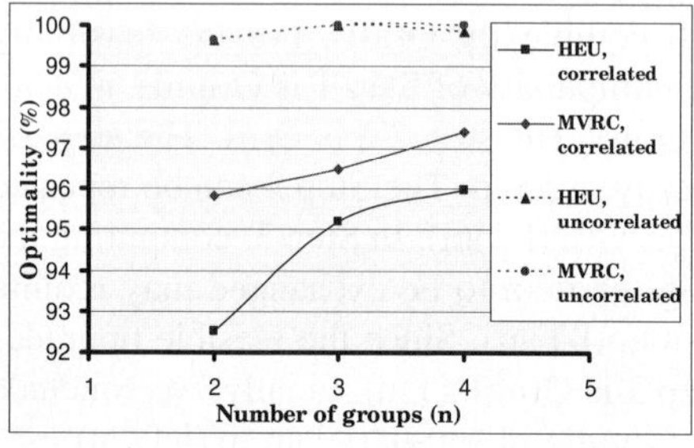

Fig. 2. Optimality in relation to the number of groups (l=20, m=2)

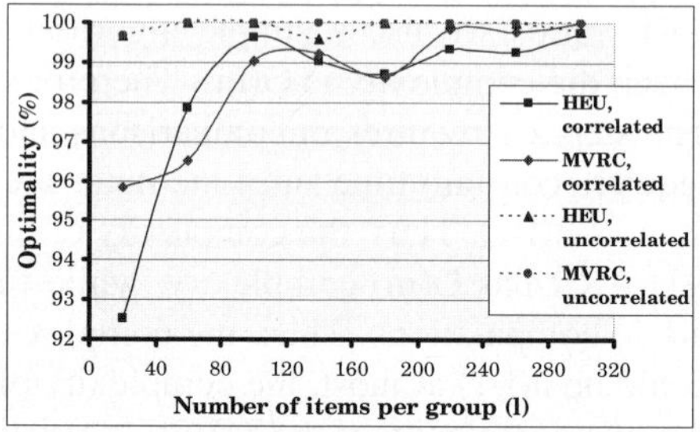

Fig. 3. Optimality in relation to the number of items per group (n=2, m=2)

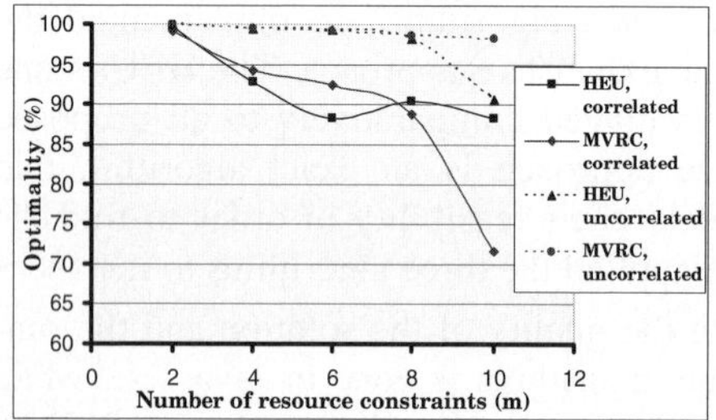

Fig. 4. Optimality in relation to the number of res. constraints (n=2, l=20)

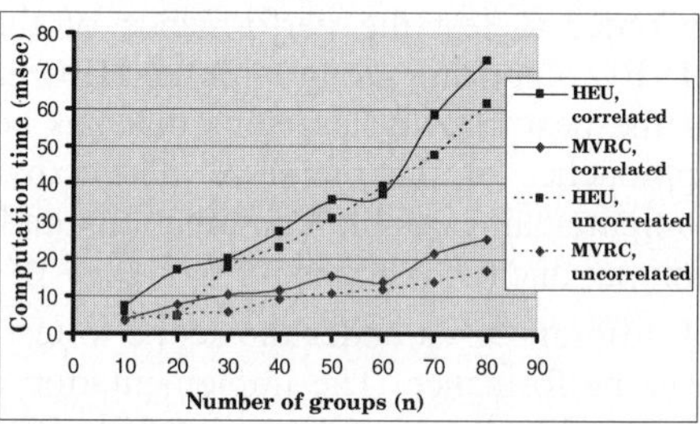

Fig. 5. Computation time in relation to the number of groups (l=10, m=2)

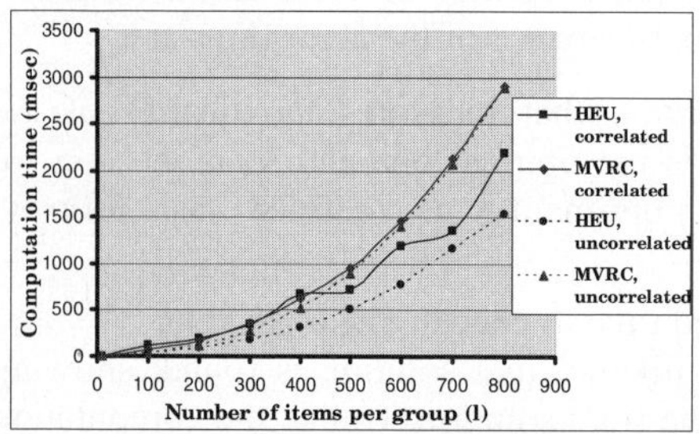

Fig. 6. Computation time in relation to the number of items per group (n=10, m=2)

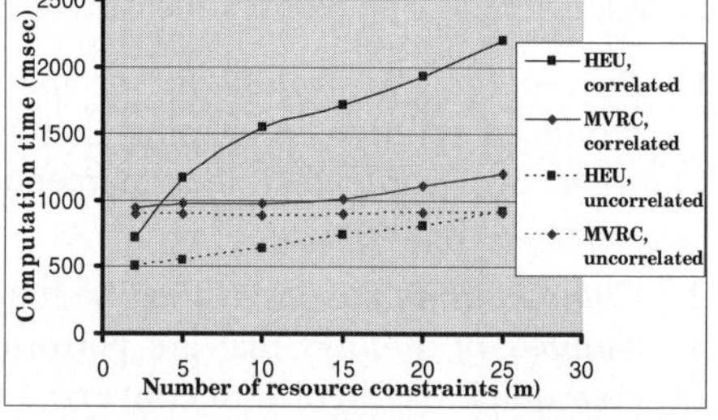

Fig. 7. Computation time in relation to the number of resource constraints (n=10, l=500)

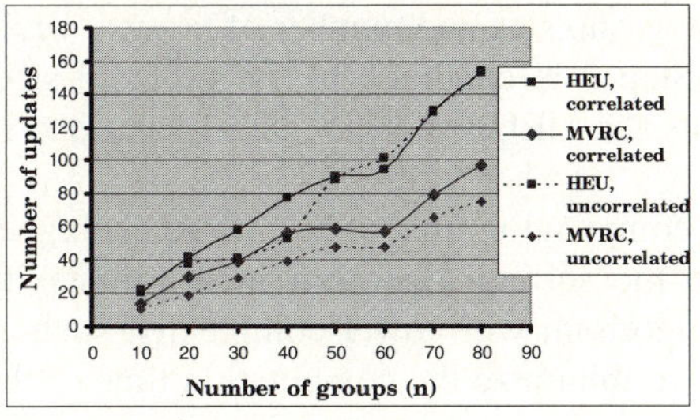

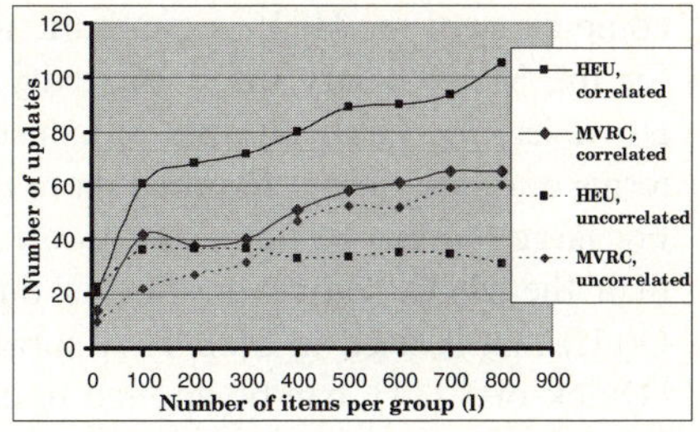

Fig. 8. Number of updates in relation to the number of groups (l=10, m=2)

Fig. 9. Number of updates in relation to the number of items per group (n=10, m=2)

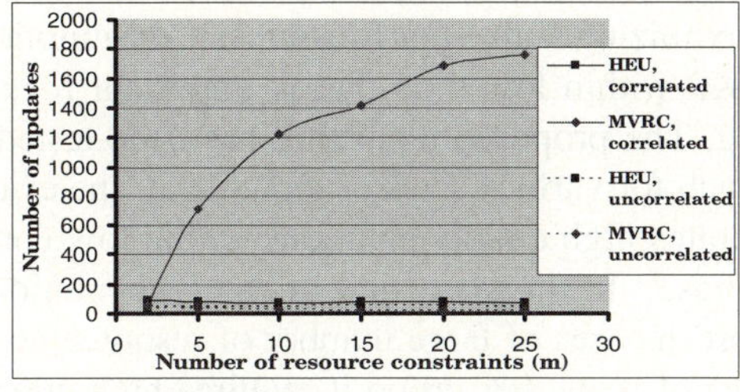

Fig. 10. Number of updates in relation to the number of resource constraints (n=10, l=500)

Observing the above figures we conclude to the following evaluation results:

- Testing the algorithms with small-sized data instances showed that both algorithms provide feasible solutions for every case. However, using large-sized input instances the algorithms showed that MVRC fails to find feasible solutions for large number of resource constraints. In fact, for n=10, l=500, m=20, the failure rate is 50% while for m=25, it becomes 70%. For the other large-sized input instances, no failure is noticed.

- Regarding the quality of the solutions, we conclude that both algorithms produce solutions with value very close to the one produced by the exhaustive approach. For the smaller data instances, the increase of the problem size result in the increase of optimality, while for larger data instances the optimality tends to stabilize. Regarding the quality of the solutions for large data sets, we observe that the solutions produced by the two algorithms are quite close.

- It may happen that smaller data sets take longer than larger data sets. This happens because of the fact that the data sets are produced randomly. If the data set consists of very few feasible solutions it might take less time to get the final solution.

- The performance for the uncorrelated data sets is better than the correlated ones.

- The MVRC outperforms the HEU in terms of solution optimality. It also performs less updates till finding the final solution, and requires lower computation time. As the data sizes increase, the increase rates of computation time and performed updates get higher for the HEU, while for the MVRC it tends to remain stable. The complexity of the HEU is quadratic to the group size and number of groups, while for the MVRC only the complexity of Step 3 is quadratic to the problem size. However, for large number of constraints the HEU produces better solutions in terms of quality but it requires more time.

- For large number of items per group, the computation time of the MVRC is higher than the HEU. However, this is due to the sorting algorithm with complexity $O(nl!)$ that is used in Step 1. A sorting algorithm with lower complexity, such as "Quicksort" [11], can be applied in order to minimize the computation time of the MVRC.

5 Conclusions

We presented the Maximizing Value per Resources Consumption   heuristic algorithm for solving the MMKP within low time that is important for cases that require real-time decision making. The proposed algorithm has been tested against the HEU and the exhaustive approach for various sizes of input data. The evaluation results showed that the MVRC generates high quality solutions within low computation time. As the input data sizes increase, the increase rate of the computation time remains stable. However, it falls short in cases of large number of resource constraints. Moreover, in cases of large number of items per group it requires high computation time, but this can be improved if we utilize sorting algorithms with lower complexity.

References

1. Khan, S., Kin, F. Li, Manning, E., Akbar, M.: Solving the Knapsack Problem for Adaptive Multimedia Systems. Studia Informatica (2002), Special Issue on Combinatorial Problems. Vol. 2. No. 1, pp. 154-174.
2. Chantzara, M., Anagnostou, M.: Evaluation and Selection of Context Information. In Proceedings of the 2nd International Workshop on Modelling and Retrieval of Context (MRC 2005), Edinburgh, Scotland, (July 31- August 1 2005). CEUR Workshop Proceedings, ISSN 1613-0073.
3. Martello, S., Toth, P.: Knapsack Problems: Algorithms and Computer Implementations. Wiley-Inrescience Series in Discrete Mathematics and Optimization, 1990, John Wiley & Sons, Chichester, England.
4. Kellerer, H., Pferschy, U., Pisinger, D.: Knapsack Problems. Springer, 2004 ,ISBN: 3-540-40286-1, p. 546.
5. Khan, S.: Quality Adaptation in a Multi-Session Adaptive Multimedia System: Model, Algorithms and Architecture. PhD Thesis, Department of Electronical and Computer Engineering, University of Victoria, Canada (1998).
6. Moser, M., Jokanovic, D., Shiratori, N.: An Algorithm for the Multidimensional Multiple-Choice Knapsack Problem. IECE Trans Fundamentals Electron (1997), Vol.80. pp. 582-589.
7. Toyoda, Y.: A Simplified Algorithm for Obtaining Approximate Solutions to Zero-one Programming Problems. Management Science (1975), Vol. 21, No. 12, pp. 1417-1427.

8. Akbar, M., Rahman, M., Kaykobad, M., Manning, E., Shoja, G.: Solving the Multidimensional Multiple-choice Knapsack Problem by Constructing Convex Hulls. Computers & Operations Research, Elsevier (May 2006). Vol. 33, No. 5, pp. 1259-1273.
9. Hifi, M., Micrafy, M., Sbihi, A.: Heuristic Algorithms for the Multiple-choice Multidimensional Knapsack Problem. Journal of the Operational Research Society (December 2004). Vol. 55, pp. 1323-1332.
10. Pisinger, D.: Where are the hard knapsack problems?. Computers and Operations Research (September 2005). Vol. 32, No. 9, pp. 2271-2284.
11. Cormen, T., Leiserson, C., Rivest, R., Stein, C.: Introduction to Algorithms. MIT Press and McGraw-Hill.

FACT: A New Fuzzy Adaptive Clustering Technique

Faezeh Ensan, Mohammad Hossien Yaghmaee, and Ebrahim Bagheri

Department of Computing, Faculty of engineering
Ferdowsi University of Mashhad, Mashhad, Iran
Fa_En93@stu-mail.um.ac.ir, hyaghmae@um.ac.ir,
Eb_ba63@stu-mail.um.ac.ir

Abstract. Clustering belongs to the set of mathematical problems which aim at classification of data or objects into related sets or classes. Many different pattern clustering approaches based on the pattern membership model could be used to classify objects within various classes. Different models of Crisp, Hierarchical, Overlapping and Fuzzy clustering algorithms have been developed which serve different purposes. The main deficiency that most of the algorithms face is that the number of clusters for reaching the optimal arrangement is not automatically calculated and needs user intervention. In this paper we propose a fuzzy clustering technique (FACT) which determines the number of appropriate clusters based on the pattern essence. Different experiments for algorithm evaluation were performed which show a much better performance compared with the typical widely used K-means clustering algorithm.

Keywords: Fuzzy Clustering, Unsupervised Classification, Adaptive Pattern Categorization, Fuzzy C-means.

1 Introduction

Clustering belongs to the set of mathematical problems which aim at classification and assignment of data or objects to related sets or classes. The act of classification could well be applied through supervised or unsupervised learning methods [1]. In the Supervised model, Patterns are learnt using some familiar, previously classified data. Multi-layered Perceptron - MLP, Support Vector Machine - SVM, and Decision Trees are illustrious examples of such learning algorithms. This type of learning may be called the learning by example methodology. On the other hand, and in the unsupervised method, mainly named clustering, entities are classified in homogeneous classes so that neighboring patterns are assembled in similar collections. In this approach object-class association is not previously known and clusters are formed based on some object similarity criteria.

As unsupervised learning models, such as clustering, can semi-consciously detect well separated classes amongst available data based on their intrinsic features, they have been extensively used in different scientific fields. Their use can significantly vary from Medicine and its application to disease detection, to Intrusion Detection Systems (IDS) for network activity division into two typical types of intrusive and

V.N. Alexandrov et al. (Eds.): ICCS 2006, Part I, LNCS 3991, pp. 588–594, 2006.

non-intrusive. New applications of clustering have been found in data (web) mining and adaptive systems where user characteristics modeling, session detection and etc can be achieved through modified clustering algorithms. Pattern recognition can also be an important field of clustering application.

Most clustering techniques assume a well defined distinction between the clusters so that each pattern can only belong to one cluster at a time. This supposition can neglect the natural ability of objects existing in multiple clusters. For this reason and with the aid of fuzzy logic, fuzzy clustering can be employed to overcome this weakness. The membership of a pattern in a given cluster can vary between 0 and 1. In this model one single pattern can have different degrees of membership in various clusters. A pattern belongs to the cluster where it has the highest membership value.

In this paper we aim to propose a fuzzy clustering technique which is capable of detecting the most appropriate number of clusters based on a density factor. This algorithm is completely insensitive to the initial number of employed clusters; however the initial value should always be lower than the optimal cluster number. Although a very low number of initial clusters will increase the computation time and CPU usage but it will prevent the algorithm from choosing the incorrect number of clusters. The method discovers the number of clusters by intelligently splitting capable clusters and creating new cluster centers through outlier detection.

The paper is organized in the following sections: Section 2 will introduce the proposed fuzzy clustering heuristic. Sections 3 presents experimental results obtained from the algorithm implementation and the following section will conclude and provide related works in the field.

2   FACT Heuristic

Although Fuzzy C-means algorithm shows strengths in many areas but it lacks the ability to determine the appropriate number of clusters for pattern classification and requires the user to define the correct number of clusters. Many applications of clustering like pattern recognition or intrusive data classification require the clustering algorithm to decide on the proper number of clusters, as the correct number of classes is not a priori known. In the proposed heuristic we devise an algorithm which exploits a modified version of Fuzzy C-means in which U (Membership Degree Matrix) is not randomly initialized. The other two strengths of this heuristic is that it is based on a fuzzy split-outlier detector and a Cluster Density Criterion (CDC). The fuzzy split algorithm was to some extent inspired by [8]. Some fuzzy clustering algorithms such as [9] are based on the minimization of the objective function value as their ultimate goal. This criterion serves as a great factor for the algorithms with a predefined number of clusters; however in heuristics which have an adaptive approach to cluster number assessment, this factor cannot be used. This is because the objective function will decrease with the increase of the number of clusters and hence causes further cluster splitting which results in an incorrect number of clusters (the number of clusters will most likely end up being identical to the number of available patterns). For this reason using the objective functions as the basis for successful split assessment is unreasonable. We define and apply CDC for split success comparison. FACT is comprised of 3 main steps which are further explained in the following paragraphs:

Step 1 – Initialization
The existing version of the Fuzzy C-means is applied to the set of available patterns by setting the initial cluster number and m to 2. The outputs of this step are the preliminary values for U and CDC.

Step 2 – Outlier Detection
a)   Cluster Member Assignment
Every pattern in the fuzzy clustering algorithm has a membership degree in all available clusters. The process of pattern to cluster assignment is done through allocating the pattern to the cluster in which it has the highest membership degree. Matrix M, $[m_{ij}]_{c*n}$ is defined as follows:

$$M_{ij} = \begin{cases} U_{ij}, & if \quad \overset{c}{\underset{i=1}{Max}} U_{ij} = U_{ij} \\ 0, & else \end{cases} \tag{6}$$

b)   Local Outlier Detection
In this sub step the candidates in each cluster to be the final outliers over all of the patterns are selected. This process selects the pattern with the lowest non-zero membership value in vector M_i where i shows the current cluster (7).

$$Candidate_i = \overset{n}{\underset{j=1}{Min}}(M_{ij}) \quad where \quad M_{ij} \neq 0 \tag{7}$$

c)   Final Outlier Selection and Splitting
The pattern with the lowest value in the Candidate vector (OP) is selected as the ultimate outlier. The coordinates of OP are used as the basis for the center of a new cluster. Let OP = {op_1, op_2... op_r} be the outlier point, the new cluster center will be calculated using (Eq.8):

$$Center (c+1) = OP + \lambda \tag{8}$$

Where $\lambda = (\lambda_1, \lambda_2... \lambda_r) \sim 0$.

Having calculated the value of the new cluster center, the previous composition of pattern classifications can be altered and rearranged based on c+ 1 cluster. Matrix U is updated using (Eq. 4) where the upper bound of k is c+1.The modified version of fuzzy C-means is now tuned using the calculated U and c+1 number of clusters and is used to create the new cluster composition. After having split the cluster formation into a new arrangement, the CDC will be updated (Eq. 9). The value obtained from the division of the new CDC to the former CDC is multiplied by a coefficient, α, which is between 0 and 1. To show that splitting has improved the clustering, θ_{t+1} should be larger than θ_t and thus the splitting procedure is confirmed and stabilized. The value for α is usually set to 0.2. The θ is named the Feedback Control Parameter (FCP) which controls the system behavior using a feedback from the prior iteration.

$$CDC_{t+1} = \sum_{i=1}^{c+1}\sum_{j=1}^{n} M_{ij} \tag{9}$$

$$\theta_{t+1} = \alpha(\frac{CDC_{t+1}}{CDC_t}) + (1-\alpha)\theta_t \qquad (10)$$

If the splitting has been unsuccessful 2.c is repeated with the next pattern in the Candidate vector.

Step 3 – Test
If none of the patterns available in the Candidate vector can serve as a successful splitting point for improving the current cluster arrangement, the algorithm will terminate with the current composition on hand else it will increase the number of clusters by one unit and resume algorithm execution from 2.a.

3   Experimental Results

Four main pattern sets were used in the first set of experiments. Wisconsin Breast Cancer Databases containing 699 patterns were cleaned to be used in the comparison procedure. The patterns were 9 dimensional data with 2 main classes (malignant and benign). Pima Indians Diabetes Database was the second pattern set used which included 768 patterns with 8 attributes for each pattern. The training was done to test positive or negative diabetes tests. The patterns were initially obtained from the National Institute of Diabetes and Digestive and Kidney Diseases. The third pattern set was the Liver-disorders Database from the BUPA Medical Research Ltd. This pattern set consisted of 345 patterns each with 7 numeric-valued attributes. The Statlog Project Heart Disease Database is made up of 270 patterns which are used to classify normal and abnormal patients using 13 different traits. The four pattern sets were taken from the UCI Machine Learning Repository at [10]. The pattern sets were divided into two parts for train and test purposes. The exact division is shown in table 1.

Table 1. The number of Train and Test patterns used in each pattern set

Pattern Set	Patterns	Train Patterns			Test patterns		
		Sum	*First class*	*Second class*	*Sum*	*First class*	*Second class*
Wisconsin Breast Cancer	683	400	303	197	183	141	42
Pima Indians Diabetes	768	500	318	182	268	182	86
Liver-disorders	345	250	110	140	95	35	60
Statlog Project Heart Disease Database	270	220	124	96	50	26	23

For the sake of clarity and to show K-means' dependency to the number of clusters, different variations of cluster numbers were created for k-means performance evaluation ranging from 2 to 34 clusters. Figure 1 depicts K-means performance under different circumstances. As it can be inferred from the diagram, different number of clusters employed, can significantly affect the final outcome and hence be evaluated as a negative impact on the overall algorithm performance.

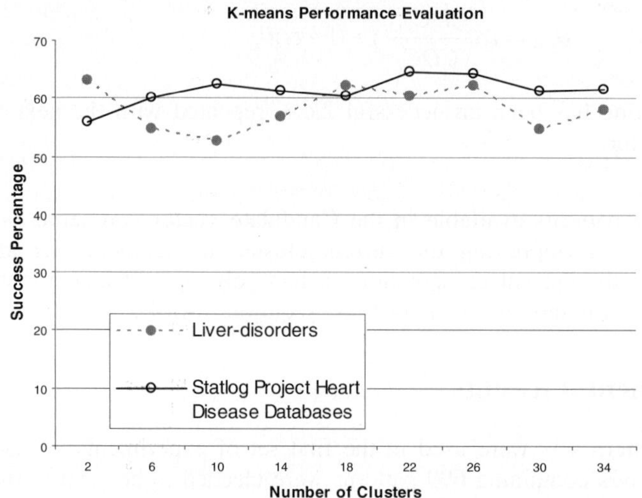

Fig. 1. K-means Performance Evaluation

The proposed heuristic was examined under several different criteria and compared with the K-means algorithm. As the K-means algorithm clusters data using a predefined number of classes and this is different from what the proposed heuristic does, the appropriate comparison model should have been devised. In our analysis, FACT was first applied to the pattern sets for clustering. This step detected the available classes (c) in the pattern sets. K-means was then executed with three different initial cluster numbers of c-1, c and c+1. Although the evaluations were initially done based upon the number of FACT detected classes but the optimal number of classes known from the omniscient were also applied to the K-means algorithm to compare the optimal success ratio of both heuristics. As K-means provides different clustering compositions each time it is run due to its sensitivity to the initial state; it was executed 100 runs on every pattern set with the specified cluster number and the average results were used. Table 2 compares the performance of both heuristics based on the percentage of correct pattern classification. As it can be clearly seen, the FACT algorithm outperforms the K-means algorithm in 3 of the pattern bases and reaches optimality in the Liver-disorders pattern set. The important point is that due to the differences in the essence of the algorithms the number of appropriate clusters for the FACT might differ from the optimal number of clusters for K-means, but even with the selection of the best number of clusters in K-means, FACT still shows better performance.

The next experiment was done on pattern sets which were statistically created to form well-defined class boundaries. Each pattern set had colonies of patterns consisting of 200 objects. 4, 5 and 7 colonies had been integrated into the pattern sets forming pattern sets with 800 (PS1), 1000 (PS2) and 1400 (PS3) patterns. To compare the performance of each clustering model, the Quadratic Error (QE) factor introduced in [11] was used. Let $X_i = (x_i1, x_i2, \ldots, x_in)$ be the members of cluster I and cc_i be the centroid of the ith cluster, QE is defined as the average of the mean squared distances of each pattern to the cluster centroids as shown in (Eq.11).

$$QE = \frac{1}{n}\sum_{i=1}^{n}\frac{\sum_{p_l \in C_i}\|p_l - cc_i\|^2}{|C_i|} \tag{11}$$

Table3 compares the values for the QE factor obtained from different heuristics. The smaller the value for the Quadratic Error is, the higher the inter object relationship in one cluster would be. FACT related QE values show much better performance for the proposed algorithm. Figure 2 shows one of the devised pattern sets depicting the cluster centers chosen using each algorithm which reveals a better centroid placement strategy in the FACT algorithm.

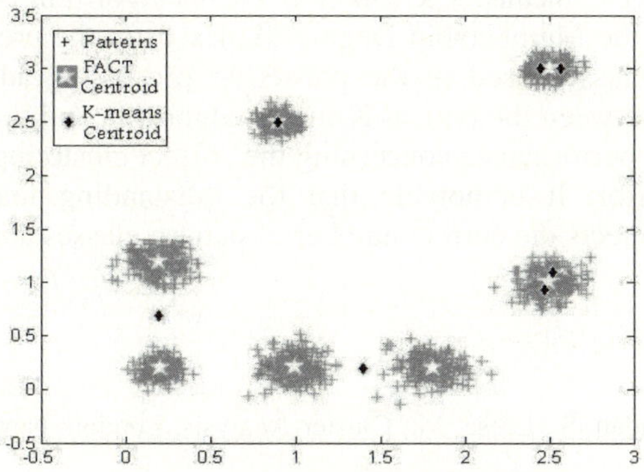

Fig. 2. PS3 and Centroid Placement Strategy

Table 2. The comparison of the two heuristics based on the correct classifications rate.1 and 2 show the number of detected clusters using the FACT algorithm and the success percentage achieved, respectively. The number of clusters used to reach the best success percentage and the success percentage achieved in K-means are also shown in 3 and 4.

Pattern Set	FACT		K-means			
	Number of Clusters[1]	Success Percentage[2]	Number of Clusters	Success Percentage	Optimal Number of Clusters[3]	Optimal Success Percentage[4]
Wisconsin Breast Cancer	8	100.00	7 8 9	98.57 98.36 98.91	2	100.00
Pima Indians Diabetes	68	71.31	67 68 69	67.67 65.10 66.10	14	69.96
Liver-disorders	2	63.16	2 3 4	63.16 63.16 53.68	2	63.16
Statlog Project Databases	9	70.00	8 9 10	59.50 60.66 60.66	18	64.66

Table 3. QE-based Comparison

Pattern Set	Number of Clusters	Quadratic Error of FACT	Quadratic Error of K-means
PS1	4	0.005934	0.006044
PS2	5	0.0064	0.0306
PS3	7	0.0075	0.0321

4 Conclusion

In this paper we have proposed a fuzzy adaptive clustering algorithm which modifies the well known fuzzy C-means. The Fuzzy C-means algorithm is altered so that it is initialized based on the Membership Degree Matrix from the previous iteration. The number of pattern classes used in the clustering process is adaptively calculated. Comparisons done between the typical K-means algorithm and the proposed heuristic demonstrate a better performance concerning the correct clustering percentage and the Quadratic Error factor. It is notable that the outstanding feature of the FACT algorithm is that it detects the correct number of pattern classes adaptively.

References

1. Everitt, B.S., Landau, S., Leese, M., Cluster Analysis, London: New York, Halsted Press, 1993.
2. Hansen, P., Mladenovic, N., J-Means: a new local search heuristic for minimum sum-of-squares clustering, Pattern Recognition 34 (2), 2001, pp.405–413.
3. MacQueen, J., Some methods for classification and analysis of multivariate observations, Proceedings of the Fifth Berkeley Symposium on Mathematical Statistics and Probability, Vol.2, 1967, pp.281-297.
4. Sanjiv, K.B., Adaptive K-Means Clustering, FLAIRS Conference 2004
5. A survey of recent advances in hierarchical clustering algorithms The Computer Journal, Volume 26, Issue 4, 1983, pp. 354-359.
6. Barthélemy, J.P., Brucker, F., NP-hard approximation problems in overlapping clustering, Journal of Classification 18, 2001, pp.159-183.
7. Bezdek, J.C., Pattern Recognition with Fuzzy Objective Function Algorithms, Plenum Press, New York, 1981.
8. Guan, Y., Ghorbani, A., and Belacel, N., K-means+: An autonomous clustering algorithm, in submission.
9. Belacel, N., Hansen, P., and Mladenovic, N., Fuzzy J-means: A new heuristic for fuzzy clustering, Pattern Recognition 35, 2002, pp.2193–2200.
10. http://www.ics.uci.edu/~mlearn/MLSummary.html
11. Bacao, F., Lobo, V., Painho, M., Self-Organizing Maps as efficient initialization procedures and substitutes for k-means clustering, International Conference on Computational Science, 2005

Algorithm for K Disjoint Maximum Subarrays

Sung Eun Bae and Tadao Takaoka

Department of Computer Science and Software Engineering
University of Canterbury, Christchurch, New Zealand
{seb43, tad}@cosc.canterbury.ac.nz

Abstract. The maximum subarray problem is to find the array portion that maximizes the sum of array elements in it. For K disjoint maximum subarrays, Ruzzo and Tompa gave an $O(n)$ time solution for one-dimension. This solution is, however, difficult to extend to two-dimensions. While a trivial solution of $O(Kn^3)$ time is easily obtainable for two-dimensions, little study has been undertaken to better this. We first propose an $O(n + K \log K)$ time solution for one-dimension. This is equivalent to Ruzzo and Tompa's when order is considered. Based on this, we achieve $O(n^3 + Kn^2 \log n)$ time for two-dimensions. This is cubic time when $K \leq n/\log n$.

1   Introduction

The maximum subarray problem determines an contiguous array elements that sum to the maximum value with respect to all possible array portions within the input array. When the input array is two-dimensional, we find a rectangular subarray with the largest possible sum.

In the sales database, the maximum subarray problem may be applied to identify certain group of consumers most interested in a particular product. This is also used in the analysis of long genomic DNA sequences to identify a biologically significant portion. In graphics, we can find the brightest area within the image after subtracting the average pixel value from each pixel.

For the one-dimensional case, we have an optimal $O(n)$ time sequential solution, known as Kadane's algorithm [5]. A simple extension of this solution can solve the two-dimensional problem in $O(m^2 n)$ time for an $m \times n$ array $(m \leq n)$, which is cubic when $m = n$ [6]. The subcubic time algorithm is given by Tamaki and Tokuyama [10], which is further simplified by Takaoka [9].

Finding K maximum sums is a natural extension. We can define two categories depending on whether physical overlapping of solutions is allowed or not.

For K overlapping maximum subarrays, significant improvements have been made since the problem was first discussed in [1] and [3]. Recent development by Cheng et al. [7] and Bengtsson and Chen [4] established an optimal solution of $O(n + K \log K)$ time. For two-dimensions, $O(n^3)$ time is possible [2, 7].

The goal of the K-disjoint maximum subarray problem is to find K maximum subarrays, which are disjoint from one another. Ruzzo and Tompa's algorithm [8] finds all disjoint maximum subarrays in $O(n)$ time for one-dimension.

V.N. Alexandrov et al. (Eds.): ICCS 2006, Part I, LNCS 3991, pp. 595–602, 2006.

To the best of Authors' knowledge, little study has been undertaken on this problem for higher dimensions. Particularly, an algorithm for the two-dimensional case may be used to select brightest spots in graphics, and such a technique may be also applied to motion detection and video compression.

In this paper, we discuss the difficulty involved in extending Ruzzo and Tompa's algorithm [8] to two-dimensions and design an alternative algorithm for one-dimension that is more flexible to extend to higher dimensions. Based on the new framework, we present an $O(m^2 n + K m^2 \log n)$ time solution for two-dimensions where (m, n) is the size of the input array. This is cubic time when $m = n$ and $K \leq n/\log n$.

2 Problem Definition and Difficulty in Two-Dimensions

2.1 Problem Definition

For a given array $a[1..n]$ containing mixture of positive and negative real numbers, the maximum subarray is the consecutive array elements of the greatest sum. The definition of K disjoint maximum subarrays is given as follows.

Definition 1 (Ruzzo and Tompa [8]). *The k-th maximum subarray is the consecutive subarray that maximizes the sum of array elements* disjoint *from the $(k-1)$ maximum subarrays*

In addition, we impose sorted order on K maximum subarrays.

Definition 2. *The k-th maximum subarray is not greater than the $(k-1)$-th maximum subarray.*

It is possible for a subarray of zero sum adjacent to a subarray of positive sum to create an overlapping subarray with tied sums. To resolve this, we select the one with smaller area if there are subarrays of tied sums. Another subtle problem arises with the value of K. For $k < K$, it is possible that the k-th maximum subarray becomes non-positive. We may stop the process at this point even if the K-th maximum is yet to be found. A non-positive maximum subarray is essentially a single negative array element, which is trivial to find. Let $\bar{K}$ be the maximum number of positive disjoint maximum subarrays. Theoretically $1 \leq \bar{K} \leq n/2$ and is data dependent. Throughout this paper, we assume that K, the number of maximum subarrays we wish to find, is not greater than $\bar{K}$.

Example 1. $a = \{3,51,-41,-57,52,59,-11,93,-55,-71,21,21\}$. From the array a, the maximum subarray is 193, $a[5] + a[6] + a[7] + a[8]$ if the index of first element is 1. We denote this by $192(5, 8)$. The second and third maximum subarrays are $54(1, 2)$ and $42(11, 12)$. The fourth is $-41(3, 3)$, so $\bar{K} = 3$.

A trivial solution may be repeated application of Kadane's algorithm [5, 6]. When the first maximum subarray is found in $O(n)$ time, we replace the element values within the solution with $-\infty$. The second and subsequent maximum subarrays are found by repeating this process. This is $O(Kn)$ time. Ruzzo and Tompa's algorithm [8] takes $O(n)$ time for $\bar{K}$ disjoint maximum subarrays, but requires sorting if Definition 2 needs to be met.

Algorithm 1. Maximum subarray for one-dimension

1: If the array becomes one element, return its value.
2: Let M_{left} be the solution for the left half.
3: Let M_{right} be the solution for the right half.
4: Let M_{center} be the solution for the center problem.
5: $M \leftarrow \max\{M_{left}, M_{right}, M_{center}\}$.

2.2   Problems in Two-Dimensions

For an (m, n) array, $a[1..m][1..n]$, we wish to find K disjoint maximum subarrays which are in rectangular shape. We denote a subarray of sum x with coordinates of top-left corner (r_1, c_1) and bottom-right corner (r_2, c_2) by $x(r_1, c_1)|(r_2, c_2)$. In the following example, we compute $K = 4$ disjoint maximum subarrays.

Example 2.

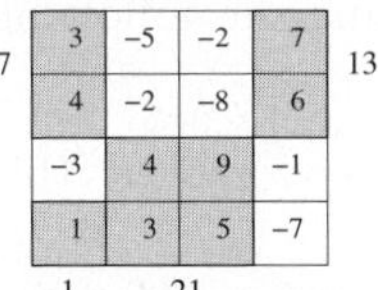

For $K = 4$, K disjoint maximum subarrays are $21(3,2)|(4,3)$, $13(1,4)|(2,4)$, $7(1,1)|(2,1)$ and $1(4,1)|(4,1)$.

In this example, $\bar{K} = 4$. When $K > 4$, the subsequent subarrays will be comprised of a single negative array element such as $-1(3,4)|(3,4)$, $-2(1,3)|(1,3)$ etc.

As is in one-dimension, a trivial solution for finding K disjoint maximum subarrays is repeated application of Kadane's algorithm. This is $O(Km^2n)$ time or $O(Kn^3)$ time for $m = n$. In the worst case, where $\bar{K} = n^2/2$, we have $O(n^5)$ time for $K = \bar{K}$.

For more efficient solution, it is natural to consider extending Ruzzo and Tompa's algorithm [8]. While we omit the details of this algorithm, it seems difficult to extend to two-dimensions as we have to organize the scanning in two directions such that the rectangular subarray may be found.

In the following section, we present another algorithm for one-dimension. This algorithm provides solid framework to extend to the two-dimensional case.

3   One-Dimensional Case

For a one-dimensional array $a[1..n]$, we compute the prefix sum s such that $s[x] = \sum_{i=1}^{x} a[i]$. We assume $s[0] = 0$.

Algorithm 1 shows the outer framework. In this algorithm, the center problem is to obtain an array portion that crosses over the central point with maximum sum, and can be solved in the following way. Note that the prefix sums once computed are used throughout recursion. We assume that n is power of 2 without loss of generality.

$$M_{center} = \max_{\substack{n/2 \leq i \leq n \\ 0 \leq j < n/2}} \{s[i] - s[j]\} = \max_{n/2 \leq i \leq n} \{s[i]\} - \min_{0 \leq j < n/2} \{s[j]\} \qquad (1)$$

The recursive computation of this algorithm can be conceived as a tournament-like selection process, which we describe in the following.

Algorithm 2. Build tournament for $a[f..t]$

---

procedure buildtree($node, f, t$) **begin**

1: $(from, to) \leftarrow (f, t)$
2: **if** $from = to$ **then**
3:    $(L, M, G) \leftarrow (s[f-1], a[f], s[f])$ // *node* is a leaf
4: **else** // *node* is an internal node
5:    create two children *left* and *right*
6:    buildtree($left, f, \frac{f+t}{2} - 1$) //build left subtree
7:    buildtree($right, \frac{f+t}{2}, t$)) //build right subtree
8:    $L \leftarrow \min\{L_{left}, L_{right}\}$, $R \leftarrow \max\{G_{left}, G_{right}\}$
9:    $M \leftarrow \max\{M_{left}, M_{center}, M_{right}\}$ where $M_{center} \leftarrow G_{right} - L_{left}$

end

3.1 Tournament

We construct a binary tree bottom-up where each node contains the following attributes.

- $(from,to)$: the coverage, i.e., the range of array elements covered by this node.
- L: the minimum prefix sum within $(from - 1, to - 1)$. Abbreviates "least".
- G: the maximum prefix sum within the coverage. Abbreviates "greatest".
- M: the maximum sum found within the coverage. Refer to Lemma 1.
- (noL, noG): boolean values initially both *false*. Discussed in Section 3.2.

We denote the left child of an internal node x by x_{left} and the right child by x_{right}. Variables of the child node are given with subscript such as L_{left}, meaning that L of x_{left}. Throughout this paper, we call this tree the *tournament*, or simply T. The root of T will be referred to as $root(T)$.

Based on Algorithm 1, we design Algorithm 2 that recursively builds T. Note that the computation of M_{center} at line 9 is due to Eq. 1. After *buildtree(root,1,n)* is processed, the value of M at $root(T)$ is the maximum sum in the array $a[1..n]$.

We let each L, G and M carry two indices such as $M.from$ and $M.to$ to indicate that M is the sum of array elements $a[M.from]..a[M.to]$. If M_{center} is chosen for M, $M.from = L_{left}.to + 1$ and $M.to = G_{right}.to$. The following two facts are easily observed. Proofs are omitted.

Lemma 1. *When a node x has coverage $(from, to)$, its M is the maximum subarray inside this coverage. i.e., $from \leq M.from \leq M.to \leq to$.*

Lemma 2. *The maximum subarray of $a[1..n]$ is M at $root(T)$.*

When there is a tie during computation, such as $L_{left} = L_{right}$, we select the one that will result in M with smaller physical size. For example, when $L_{left} = L_{right}$, we select L_{right} as M_{center} will have smaller physical size by subtracting L_{right}. Similarly, when $G_{left} = G_{right}$, we select G_{left}. For the same reason, we choose the one with smaller physical size in computing $M = \max\{M_{left}, M_{right}, M_{center}\}$.

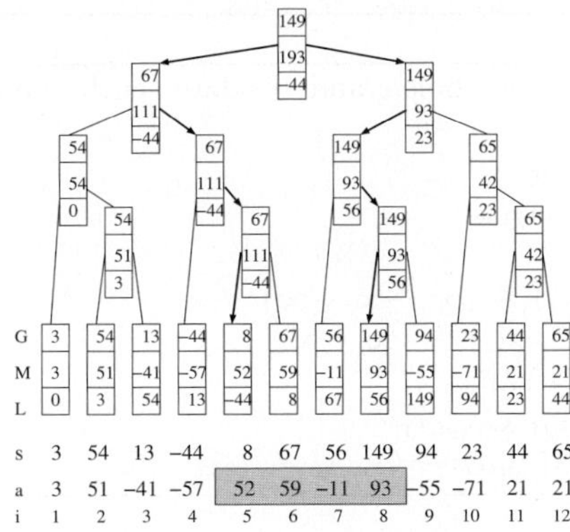

Fig. 1. Tournament T	Fig. 2. Subarray deletion

Fig. 1 shows the example in Section 2.1 computed by the tournament. Each node shows a 3-tuple of (L, M, G). The value of M at root, 193 represents the maximum sum. The figure omits the location $(M.from, M.to)$ which is $(5, 8)$.

3.2 Delete a Subarray

We discuss how we delete a subarray from the tournament, so that the tree will produce a maximum subarray that is disjoint from the deleted portion. In the following description, we use a term *hole* to refer to the portion to be deleted or has been deleted.

With the index *holeBegin* and *holeEnd*, the location of the hole, we trace the tree from the root to find subtrees whose coverage is inside the hole. In Fig. 2, dark subtrees are those to be deleted. These subtrees can be deleted by removing the link a,b and c. Since we only delete the link, deleting each subtree takes $O(1)$ time. Actual memory deallocation will be done during the post-processing.

After deletion is done, there are nodes that no longer have two children. Such nodes include 2,4,6. When a node has one child missing, we assume that this node receives a 3-tuple $(\infty, -\infty, -\infty)$ from the missing child.

The maximum subarray in the range of (u, v) is determined by node 1. We want it to be disjoint from the hole. If M_{center} becomes M at node 1, we have $(M.from, M.to) = (L_{left}.to + 1, G_{right}.to)$. This M is a "superarray" of the hole as it covers the hole. In general, a node with coverage that encompasses the whole range of the hole can "potentially" produce M_{center} overlapping the hole as M_{center} becomes a superarray of the hole. Node 0 is another node that has such potential, however, if L_{left} comes from region II,III or IV, M_{center} at node 0 can be disjoint from the hole. So we can not simply disable computing M_{center} at such nodes. We resolve this issue by Algorithm 3. The objective of the following algorithm is to ensure that M_{center} is a subarray disjoint from the hole. If no subarray disjoint from the hole can be obtained for M_{center}, we set $M_{center} = -\infty$ to represent no value.

Algorithm 3. Update tournament T

Recursively trace from $root(T)$ downwards the hole, and update each node x.

  1: **if** $holeEnd$ is in left subtree **then** $noG \leftarrow$ true
  2: **if** $holeBegin$ is in right subtree **then** $noL \leftarrow$ true
     //Recursively update (L, M, G)
  3: **if** x_{left} was deleted **then** $(L_{left}, M_{left}, G_{left}) \leftarrow (\infty, -\infty, -\infty)$
  4: **if** x_{right} was deleted **then** $(L_{right}, M_{right}, G_{right}) \leftarrow (\infty, -\infty, -\infty)$
  5: **if** noL **then** $L \leftarrow L_{right}$ **else** $L \leftarrow \min\{L_{left}, L_{right}\}$
  6: **if** noG **then** $G \leftarrow G_{left}$ **else** $G \leftarrow \max\{G_{left}, G_{right}\}$
  7: $M \leftarrow \max\{M_{left}, M_{right}, M_{center}\}$, where $M_{center} \leftarrow G_{right} - L_{left}$

When a node has the flag noL set, it means that this node will not use L_{left} for updating L. Similarly, noG means G_{right} is not used for updating G. However, L_{left} and G_{right} are still used to compute M_{center} regardless of the flags. Basically these flags block propagating L_{left} and G_{right} to the parent node.

If we delete the hole $(5, 8)$ from Fig. 1, and update T as described above, the second maximum subarray $54(1, 2)$ will be obtained from the root.

We propose the following lemma holds. If we use the *minimum hole inclusive tree* (MHIT), the smallest subtree that contains the hole (as shown in Fig. 2), as the basis, it can be inductively proved. We omit the proof due to limited space.

Lemma 3. *After deleting the hole and applying Algorithm 3, $root(T)$ produces maximum subarray M that is disjoint from the hole.*

3.3 Analysis

We find the first maximum subarray by building T in $O(n)$ time. To compute the next maximum subarray, we regard the previous solution as a hole and perform the deletion and flag updates as described in Section 3.2. Deleting nodes involves traversing two paths from the root to $holeBegin$ and $holeEnd$. Since the height of the tree is $O(\log n)$, the time for the next maximum subarray is bounded by $O(\log n)$. To obtain K disjoint maximum subarrays in sorted order, the total time is therefore $O(n + K \log n)$. Note that $O(n + K \log n) = O(n + K \log K)$ for any integer K according to [4, 7]. For ranking K disjoint maximum subarrays, this is equivalent to Ruzzo and Tompa's algorithm [8] which requires extra time for sorting.

4 Two-Dimensional Case

In this section, we extend algorithm for one-dimension to two-dimensions.

4.1 Strip Separation

An easy way to extend an algorithm for the one-dimensional case to two-dimensions is *strip separation* technique, such that the two-dimensional array $a[1..m]$

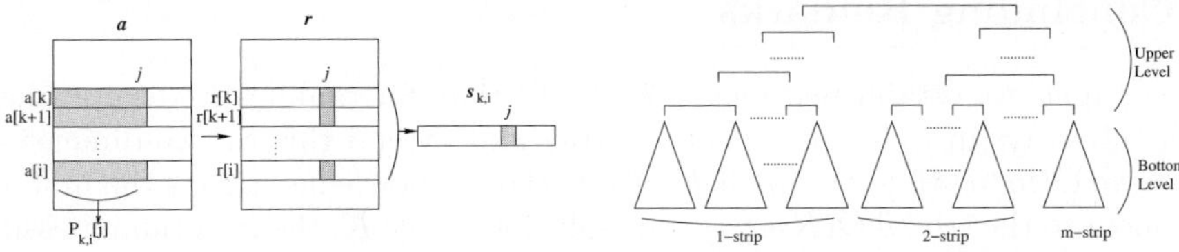

Fig. 3. Separating strips from a two-dimensions

Fig. 4. Two-level tournament

$[1..n]$ is separated into $m(m+1)/2$ strips. As shown in Fig. 3, a strip $s_{k,i}$ is the prefix sum array of a portion $P_{k,i}[1..n]$. We call a strip consisting of x rows a $x-strip$.

We pre-process the row-wise prefix sum $r[1..m][1..n]$, such that $r[i][j] = a[i][1] + a[i][2] + .. + a[i][j]$ in $O(mn)$ time. Then $s_{k,i}[j]$ is computed by $r[k][j] + ..r[i][j]$. The time for strip separation is $O(m^2 n)$.

4.2   Two-Level Tournament

We organize a two-level tournament as shown in Fig. 4. The bottom-level is composed of $O(m^2)$ tournaments of each strip. A tournament of $s_{k,i}$ has $M_{k,i}$, the maximum sum of the strip, at the root. Since each strip is regarded as a one dimensional problem, $M_{k,i}$ is given as $x(l,j)$, from which we obtain the original rectangular subarray $x(k,l)|(i,j)$. The upper-level is a tournament where all $M_{k,i}$ of bottom-level tournaments participate. There are $O(m^2)$ participants. The winner of the upper-level is the maximum subarray for two-dimensions. As building a tournament is linear time, we spend $O(m^2 n)$ time for the bottom-level tournaments, and $O(m^2)$ time for the upper-level.

4.3   Next Maximum

Suppose $x^*(k^*, l^*)|(i^*, j^*)$ is selected for the first maximum by the upper-level. Let us consider a strip $s_{k,i}$. If its row-wise coverage (k,i) is disjoint from (k^*, i^*), this strip definitely produces $M_{k,i}$ disjoint from the first maximum. Otherwise, it is possible that this strip produces an overlapping $M_{k,i}$. There are $O(m^2)$ such strips. By creating a hole (l^*, j^*) in tournaments of such a strip and updating the tree as per Section 3.2, we ensure that all $M_{k,i}$ are disjoint from $(k^*, l^*)|(i^*, j^*)$.

Now that all winners of each bottom-level tournament have become disjoint from the first maximum, we are safe to re-build the upper-level to select the second maximum. Subsequent disjoint maximum subarrays are found by repeating these steps. Each maximum subarray is computed by $O(m^2 \log n)$ time overlap removal and $O(m^2)$ time upper-level re-build. The latter time is absorbed into the former. For K maxima, it is $O(Km^2 \log n)$ time.

Including the time for initial setting, the total time is $O(m^2 n + Km^2 \log n)$. For $K \leq \frac{n}{\log n}$, we have a cubic time $O(m^2 n)$.

5   Concluding Remarks

In this paper, we established $O(n + K \log K)$ time for ranking K disjoint maximum subarrays in a one-dimensional array and extend this to two-dimensions to achieve $O(m^2 n)$ time for small K. To Authors' knowledge, this is the first improvement to the trivial $O(Km^2 n)$ time solution. Since $\bar{K}$, the maximum possible K, can be as large as $mn/2$ depending on the data, reduction of the factor K is significant. It will be an interesting question to determine $\bar{K}$ in advance.

In the current form of our algorithm, in fact, the upper-level does not require a tournament. Linear time maximum selection algorithm can be used instead without increasing the complexity. It is, however, expected that the two-level tournament may provide a solid structural platform for further improvement.

The second term of the complexity, $O(Km^2 \log n)$, seems possible to improve. Currently, we update $O(m^2)$ bottom-level tournament trees on computation of each maximum subarray. If $M_{k,i}$ of a bottom-level tournament is no longer positive, we may discard such a tree to reduce the size of the bottom-level. By doing so, if $\bar{K} = mn/2$, only $O(m)$ bottom-level tournaments of 1-strip will remain. Also if a hole (l^*, j^*) has been already created in a bottom-level tournament in the previous computation, we may skip creating it again. As no more than $O(n)$ holes can be made in each strip, this will result in the second term not exceeding $O(m^2 n \log n)$ even if $K > n$. If such an idea is incorporated, the overall complexity may be reduced to $O(m^2 n + \min(K, n) \cdot m^2 \log n)$.

References

1. Bae, S.E., Takaoka, T.: Algorithms for the problem of K maximum sums and a VLSI algorithm for the K maximum subarrays problem. ISPAN 2004, Hong Kong, 10-12 May, (2004) 247–253. IEEE Comp.Soc. Press.
2. Bae, S.E., Takaoka, T.: Improved algorithms for the K-maximum subarray problem for small K. COCOON 2005, Kunming,China, 16-19 Aug. (2005) 621–631
3. Bengtsson, F., Chen, J.: Efficient algorithms for the k maximum sums. ISAAC 2004, Hong Kong, 20-22 Dec.,(2004) 137–148. Springer-Verlag
4. Bengtsson, F., Chen, J.: A note on ranking k maximum sums. research report 2005:08 (2005) Luleå University of Technology
5. Bentley, J.: Programming pearls: algorithm design techniques. Commun. ACM, Vol.**27(9)** (1984) 865–873.
6. Bentley, J.: Programming pearls: perspective on performance. Commun. ACM, Vol.**27(11)** (1984) 1087–1092.
7. Cheng, C.H., Chen, K.Y., Tien, W.C., Chao, K.M.: Improved algorithms for the k maximum sums problems. ISAAC 2005, Sanya, Hainam, China, 19-21 Dec.(2005) 799–808.
8. Ruzzo, W.L., Tompa, M.: A linear time algorithm for finding all maximal scoring subsequences. ISMB'99, Heidelberg, Germany, 6-10 Aug.(1999) 234–241.
9. Takaoka, T.: Efficient algorithms for the maximum subarray problem by distance matrix multiplication. Elec. Notes in Theoretical Comp. Sci., Vol.**61** (2002) Elsevier
10. Tamaki, H., Tokuyama, T.: Algorithms for the maximum subarray problem based on matrix multiplication. SODA 1998, San Francisco, 25-27 Jan. (1998) 446–452.

An Evolutionary Approach in Information Retrieval

T. Amghar, B. Levrat, and F. Saubion

LERIA, Université d'Angers
2, Bd Lavoisier 49045 Angers, France
{amghar, levrat, saubion}@info.univ-angers.fr

Abstract. One critical step in information retrieval is the skimming of the returned documents, considered as globally relevant by an Information retrieval system as responses to a user's query. This skimming has generally to be done in order to find the parts of the returned documents which contain the information satisfying the user's information need. This task may be particularly heavy when only small parts of the returned documents are related to the asked topic. Therefore, our proposition here is to substitute an automatic extraction and recomposition process in order to provide the user with synthetic documents, called here composite documents, made of parts of documents extracted from the set of documents returned as responses to a query. The composite documents are built in such a way that they summarize as concisely as possible the various aspects of relevant information for the query and which are initially scattered among the returned documents. Due to the combinatorial cost of the recomposition process, we use a genetic algorithm whose individuals are texts and that aims at optimizing a satisfaction criterion based on similarity. We have implemented several variants of the algorithm and we proposed an analysis of the first experimental results which seems promising for a preliminary work.

1   Introduction

One critical step in information retrieval is the skimming of the returned documents, considered as globally relevant by an Information retrieval system [1, 10, 6] as responses to a user's query. This skimming has generally to be done in the aim to find the parts of the returned documents which contain the information satisfying the user's information need. This task may be particularly heavy in the case where only small parts of the returned documents are related to the asked topic. For example, it may be the case when a general document base or a set of press agency dispatches is asked for technical questions: some documents may incidentally contain relevant information but, as they have not been written to satisfy these kinds of topics, only small parts of them can eventually contain the suitable information. Therefore, our proposition here is to furnish a response to a query not directly in terms of documents of the base taken in their wholes, but rather by producing new documents, extracted from the initially furnished

V.N. Alexandrov et al. (Eds.): ICCS 2006, Part I, LNCS 3991, pp. 603–610, 2006.

documents as responses. The built documents, called composite documents, are made of parts of the previous ones, called here segments, and are intended to synthesize or summarize the information relevant for the query. For that purpose, we use an automatic extraction and recomposition process which provide the user with this synthetic documents made of parts of documents relevant for his information need initially scattered among the returned documents. So, the user may focus his attention only on interesting information avoiding the boring task of exploring non relevant parts in the returned documents.

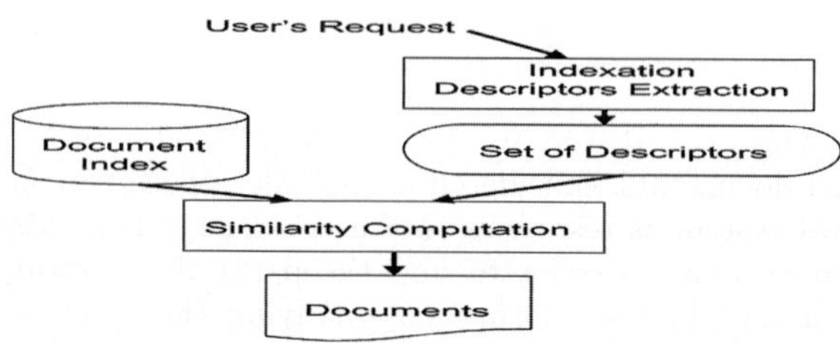

Fig. 1. Information retrieval process

The combinatorial aspect of the problem and its modeling characteristics led us to consider resolution paradigms that are commonly used in the combinatorial optimization community. Moreover, viewing the generation of composite documents as the result of a succession of segmentations and recombinations of documents until a satisfaction criterion threshold is reached make this task suitable for a treatment by evolutionary algorithms, and more precisely by genetic algorithms to solve combinatorial problems, a genetic algorithm manages a pool (population) of configurations (individuals) of the problem and its purpose is to generate the best possible configuration, a solution if this notion is appropriate. In our context, the characterization of what constitutes a solution is a rather difficult task, since it is not only strongly related to the satisfaction degree of the end user, but depends also on what the initial materials are (i.e.the documents base from which it is built). What is searched for is not appropriately said a solution but rather what is the best individual we can built from initial data. So stated, our problem is merely an optimization rather than a solution finding one. In this sense our objective will be to provide good answers with regards to a given multipurpose quality function, taking into account different evaluation points of view. In this genetic algorithm approach, individuals consists of parts of documents and are evaluated according to a fitness function that takes into account its semantics appropriateness with regards to the initial query of the user. The population of documents evolves thanks to recombinations and mutations that correspond to exchanges of text parts between two individuals and the random changes of some these parts. These basic genetic operators produce new individuals that are added to the current population. In order to converge toward a good individual, a selection stage in necessary and consists in eliminating the less performing individuals with regards to the fitness function from the current population . This process is repeated until a maximum number of iterations has been reached. Experiments have been performed on a specific corpus and provide very promising results.

2   Background

2.1   Information Retrieval

The vectorial model [9] constitutes an alternative to the binary model which responds to their major defaults (lack of partial results, lack of a similarity evaluation criterion and lack of ranking of the returned documents). It has been shown as sufficiently simple and efficient compared with other more elaborated models like the extended boolean one [8] or the generalized vector one [11] which are computationally more costly. This is enough to justify this choice as a model for our information retrieval system, since our works mainly bears on the set of the returned documents to a user's query by an information retrieval system, and it doesn't matter what precisely is its underlying model. Let us sketch however here some of the underlying principles of the this model sufficient to the comprehension of our work. The main characteristic we use of this model is that it rends it possible to represent the contribution of the terms to the general semantics of a text by means of non-binary weights. Thus queries and documents representations are expressed in terms of vectors of weights. Each vector component expresses the semantic weight of the term associated to the component and renders both the representativity of the term for the question or the document and its ability to distinguish documents the ones from the others. The application of the vectorial algebra makes it possible to model the semantic similarity between documents or between documents and queries as a vectorial distance expressed, for example, by the angles between the concerned vectors. In this case, this measurement of similarity can be chosen as the cosine between the two vectors in the space of terms considered as having N for its dimension if there are N terms in the documents base. This cosine similarity criterion has a real value between 0 and 1, collinearity corresponds to the value 1 and indicates the highest similarity degree interpreted as the highest semantic similarity between the compared entities (a query and a document or a document and another document). The weights of the terms t pertaining to a document o are defined in the following way :

Definition 1. *Weight of term t in an object o*

$$w_o(t) = f_o(t)(1 - log(\tfrac{N}{f(t)}))$$

where $f_o(t)$ is the normalized frequency of the term t in the object o, $f_o(t) = occur_o(t)/occur_o(t_{max})$ where t_{max} is the term with the highest occurrence number in the object o, and $occur_o(t)$ is the number of times that the term t appears in the object o, $f(t)$ is the number of documents containing the term t and N the total number of documents of the corpus.

The similarity between a request q and a document d or between two documents is also defined as we have previously said as a cosine measurement of the angle between the two associated vectors. More formally, let $\vec{o} \equiv (o_1, \ldots, o_N)$ the vector corresponding to the object o (document or request), with o_i indicating the weight associated with the ith component of o (this component is associated with the ith term), then the similarity between the two objects (in fact here, a document d and a request q) in this model is calculated as follows :

Definition 2. *Similarity*

$$sim(d, q) = \frac{\vec{d} . \vec{q}}{|\vec{d}| . |\vec{q}|}$$

Segmentation of Documents. The segmentation consists in cutting out documents provided by the document retrieval system in order to extract paragraphs containing elementary pieces of information since, most of the time, a complete document may include different topics. Our purpose is then to recombine these paragraphs into a more relevant composite document. Segmentation can be considered from several points of view : we may either exploit the layout of the text (morphosyntaxic criteria like punctuation) or detect semantic variations with methods like Text Tiling[3] or Decision Tree. As mentioned in the introduction, we want to reduce the computational effort as much as possible and we have chosen to use a morphosyntaxic segmentation of texts into paragraphs. In the experimental section, we will see that this type of segmentation provides promising results and can be achieved very quickly.

2.2   Genetic Algorithms

Genetic algorithms (GA)[4, 5] are mainly based on the notion of adaptation of a population (i.e., a set of individuals) to a criterion using evolution operators like crossover [2]. If individuals are considered as potential solutions to a given problem, applying a genetic algorithm consists in generating better and better individuals with respect to the problem by selecting, crossing, and mutating them. This approach reveals very useful for problems with huge search spaces. We give a general framework for GAs and we refer the reader to [7] for a survey. A GA consists of the following components:

- a representation of the individuals,
- an evaluation function *eval*: the evaluation function rates each potential solution with respect to the given problem,
- genetic operators that define the composition of the children: two different operators will be considered: crossover allows to generate new individuals (the offsprings) by crossing individuals of the current population (the parents), mutation arbitrarily alters one or more genes of a selected individual,
- parameters: population size p_{size} and probabilities of crossover p_c and mutation p_m.

We now precise how this general algorithmic scheme will be instantiated to fit our problem.

3 Designing a Genetic Algorithm for Composite Documents Generation

As mentioned in the introduction, our purpose is to produce composite documents in order to answer to a given query. The general process consists in using first a document retrieval system in order to select a sets of relevant initial documents, to cut them into basic paragraphs and then to combined them, thanks to a specific GA, to obtain a new composite document, that is expected to contain all pertinent informations with regards to the initial query. This full process can be thus schemed as in figure 2. We only depict here the GA part that we have designed for this problem.

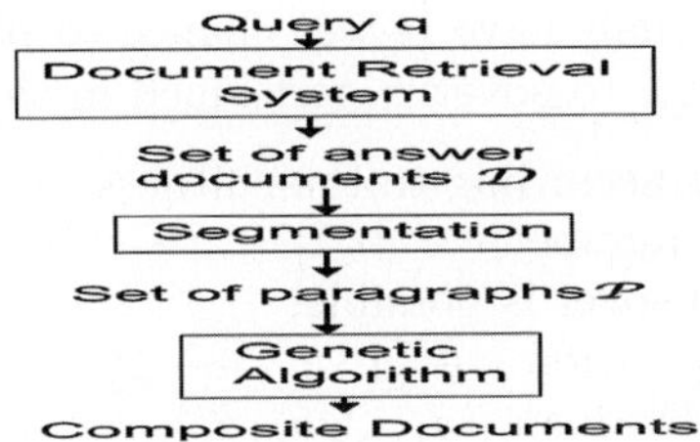

Fig. 2. General process

Representation. A crucial point for the design of an efficient evolutionary solving procedure relies on the representation of the search space. In our context, the initial ordered set of documents $\mathcal{D}$ generated by the retrieval system, will be cut into a set of basic paragraphs $\mathcal{P}$ from elements of $\mathcal{D}$ (see section 2.1). Therefore, our search space is $2^{\mathcal{P}}$. Individuals will be documents of a given size (here, composed of 50 paragraphs) and their set is denoted I. A population π is a subset of I.

Evaluation. In the composition of documents, we have to consider several criteria of evaluation. Of course, in order to assess the proximity between a text and the user's query we use a measure of similarity. Here we do not want to evaluate paragraphs independently and we only allow our algorithm to use global similarity of a document. We may also use a specific indicator to rate an individual according to the order of the documents provided by the document retrieval process (i.e., the order of initial documents in $\mathcal{D}$). We first define a function that computes the average rank of an individual i with regards to the ranks in $\mathcal{D}$ of the documents its paragraphs come from : $rank(i) = \frac{1}{n}\Sigma_{k=1}^{n}r_k$ where r_k is the rank in $\mathcal{D}$ of the document that originally contains the kth paragraph of i. Using the definition of similarity given in definition 2, we define an evaluation function with regards to a given query q. This function takes into account both similarity and rank $Eval1_q(i) = -log(\frac{1}{rank(i)})(sim(i,q) - rank(i))$. We will also introduce a simple similarity evaluation $Eval2_q(i) = sim(i,q)$ in order to compare different versions of the algorithm.

Population Management. The initial population is randomly generated from the set of all paragraphs in order to generate a set of 100 individuals. At each step, crossover and mutation are applied to generate new individuals that are added to the current population. The 100 best individuals of the current population are then selected in order to generate the next population and to keep a constant size. This process is repeated until a maximum number of iteration has been reached (this number is fixed here to 100).

Genetic Operators. Crossover operators [2] are used in the evolution process to combine individuals from the population and produce new ones. Here we focus on a basic crossover operator, the single point crossover, that takes as input two individuals, the parents, and produces two children by exchanging the values of the parents. The parents are classically selected according to their fitness (i.e., the best evaluated individuals have better chance to participate to a crossover operation). More precisely, crossover is performed in the following way:

- Select two individuals according to their fitness
- Generate randomly a number $r \in [0, 1]$
- If $r > p_c$ then the crossover is possible;
    - Select a random position $p \in \{1, \ldots, n - 1\}$
    - From selected individuals $(a_1, ..., a_p, a_{p+1}, ..., a_n), (b_1, ..., b_p, b_{p+1}, ..., b_n)$, two new individuals $(a_1, ..., a_p, b_{p+1}, ..., b_n)$, $(b_1, ..., b_p, a_{p+1}, ..., a_n)$ are added to the current population.

The mutation is defined as:

- For each individual i and for each paragraph k in i, generate a random number $r \in [0, 1]$,
- if $r > p_m$ then mutate k (randomly exchange the paragraph with any element from $\mathcal{P}$).

For the experiments, p_c is set to 0.2 and $p_m = 0.02$ (empirical settings determined by experimentation).

4   Experiments

The previous GA has been implemented and connected with an existing document processing system.

Extraction of Documents. In our approach, we first retrieve, from a corpus of texts, documents corresponding to the user's query with a document search system (see figure 1). These documents are ordered according to their global similarity with the initial query. We use here the corpus of documents of TREC (Text REtrieval Conference : http://trec.nist.gov/) which consists of columns of several newspapers written in ASCII code. Our document retrieval system is a PERL program named PIRES (Perl Information REtrieval System : http://www.info.univ-angers.fr/pub/robin/). This program is based on the classical vectorial model and the documentary research is implemented thanks to a measure of similarity.

Genetic Algorithm. According to the different evaluation functions provided in the evaluation section, we have implemented three different versions of the GA that correspond to the three possible evaluations:

- GA_1 : evaluation $Eval1_q$ including *rank*
- GA_2 : evaluation $Eval2_q$ using only the similarity
- GA_3 : blind GA with no evaluation function in order to compare our results with a full random document composition process.

Remind that the main parameters are : individual size $= 50$, population size $= 100$, probability of crossover $= 0.2$, probability of mutation $= 0.02$ and number of generations $= 100$.

Comparison Criteria. Our purpose is to test and compare the three versions of the genetic algorithm. Our experimental process is the following (according to fig. 2) : we fix a query and we extract a set $\mathcal{D}$ of documents from our corpus with PIRES. These documents serve as basis for the text tiling process in order to obtain our set of paragraphs $\mathcal{P}$. After the runs of the GA, the paragraphs are evaluated by hand in order to rate from a user point of view their pertinence with regards to the query. Based on this pertinence evaluation we define a first experimental evaluation criterion (Recall) which is the rate of relevant paragraphs appearing in an individual $NbPert_i$ with regards to the total number of relevant paragraphs which were available in the corpus $NbPert_{Total}$. A second criterion is the similarity between a individual and the initial query $sim(i, q)$. The last criterion (Recall Back) evaluates the covering ability of an individual i with respect to the initial set of documents $\mathcal{D}$, i.e. we calculate the average rank r_k of the paragraphs k in i with regards to the initial order in $\mathcal{D}$.

Experimental Results. We test two different queries : Query1 *Oil platforms attacks during the gulf war* and Query2 *Brain drain migrations*. We first give the average results for the three versions of the GA described above. We chose to run each algorithm 100 times and to compute the average values of the criteria previously described over the 15 best individuals obtained after a full run of 100 generations.

	Query1			Query2		
	GA_1	GA_2	GA_3	GA_1	GA_2	GA_3
Recall	0.0072	0.0058	0.0047	0.9256	0.7606	0.6086
Similarity	0.9978	0.9534	0.9306	0.9999	0.9999	0.8721
Recall Back	0.0220	0.0090	0.0068	0	0	0

This results show the improvements for all the three evaluation measures of GA_1 and GA_2 compared with GA_3. It is interesting to see GA_1 appears to be the more efficient for the three evaluation criteria (it has the particular ability to better cover the initial set of documents (recall back criterion)). Results on the query 2 are due to the extreme parsimony of relevant information for it in the returned documents. Best individuals of each GA versions contain the same relevant segment duplicated so as to constitute a document. This explains the high value of the similarity measure and the value zero for the recall back measure since this segment belong to the first relevant document initially returned by PIRES. We have performed several tests with other queries and observed similar results.

5   Conclusion

In this paper we have presented a new approach to generate composite documents that provide the user a better overview and a faster exploitation of a set of documents he gets from an information retrieval system. Our techniques based on a GA could be more efficiently tuned and extended in several ways. We could introduce more sophisticated evaluation criteria such as the intrinsic consistency of a composite document, with regards to the relationships of its different pieces of text. The genetic framework could be used to naturally handle such a multi-objective optimization problem.

References

1. R. Baeza-Yates and B. Ribeiro-Neto. *Modern Information Retrieval.* ACM Press/Addison-Wesley, 1999.
2. D.E. Goldberg. *Genetic Algorithms for Search, Optimization, and Machine Learning.* Reading, MA:Addison-Wesley, 1989.
3. Marti A. Hearst. Texttiling: Segmenting text into multi-paragraph subtopic passages. *Computational Linguistics*, 23(1):33–64, 1997.
4. John H. Holland. *Adaptation in Natural and Artificial Systems.* University of Michigan Press, 1975.
5. K. A. De Jong. *An analysis of the behavior of a class of genetic adaptive systems.* Phd thesis, University of Michigan, 1975.
6. Michael Lesk. *Practical Digital Libraries: Books, Bytes, and Bucks.* Morgan Kaufmann, 1997.
7. Z. Michalewicz. *Genetic algorithms + data structures = evolution programs (3rd, extended ed.).* Springer, 1996.
8. Gerard Salton, Edward A. Fox, and Harry Wu. Extended boolean information retrieval. *Commun. ACM*, 26(11):1022–1036, 1983.
9. Gerard Salton and Michael Lesk. Computer evaluation of indexing and text processing. *J. ACM*, 15(1):8–36, 1968.
10. Ian H. Witten, Alistair Moffat, and Timothy C. Bell. *Managing Gigabytes: Compressing and Indexing Documents and Images.* Van Nostrand Reinhold, 1994.
11. S. K. Michael Wong, Wojciech Ziarko, and P. C. N. Wong. Generalized vector space model in information retrieval. In *SIGIR*, pages 18–25, 1985.

An Index Data Structure for Searching in Metric Space Databases

Roberto Uribe[1], Gonzalo Navarro[2], Ricardo J. Barrientos[1],
and Mauricio Marín[1,3]

[1] Computer Engineering Department, University of Magallanes, Chile
[2] Computer Science Department, University of Chile
[3] Center for Quaternary Studies, CEQUA, Chile

Abstract. This paper presents the Evolutionary Geometric Near-neighbor Access Tree (EGNAT) which is a new data structure devised for searching in metric space databases. The EGNAT is fully dynamic, i.e., it allows combinations of insert and delete operations, and has been optimized for secondary memory. Empirical results on different databases show that this tree achieves good performance for high-dimensional metric spaces. We also show that this data structure allows efficient parallelization on distributed memory parallel architectures. All this indicates that the EGNAT is suitable for conducting similarity searches on very large metric space databases.

1 Introduction

Searching for similar objects into a large collection of objects stored in a metric-space database has become an important problem. For example, a typical query for these applications is the *range query* which consists on retrieving all objects within a certain distance from a given query object. From this operation one can construct other ones such as the nearest neighbors. Applications can be found in voice and image recognition, and data mining problems.

Similarity can be modeled as a metric space as stated by the following definitions.

Metric Space. A metric space is a set X in which a distance function is defined
$d : X^2 \rightarrow R$, such that $\forall\, x, y, z \in X$,
1. $d(x,y) \geq 0$ *and* $d(x,y) = 0$ iff $x = y$.
2. $d(x,y) = d(y,x)$.
3. $d(x,y) + d(y,z) \geq (d(x,z)$ (triangular inequality).

Range query. Given a metric space *(X,d)*, a finite set $Y \subseteq X$, a query $x \in X$, and a range $r \in R$. The results for query x with range r is the set $y \in Y$, such that $d(x,y) \leq r$.

The distance between two database objects in a high-dimensional space can be very expensive to compute and in many cases it is certainly the relevant performance metric to optimize; even over the cost secondary memory operations.

V.N. Alexandrov et al. (Eds.): ICCS 2006, Part I, LNCS 3991, pp. 611–617, 2006.

For large and complex databases it then becomes crucial to reduce the number of distance calculations in order to achieve reasonable running times. This makes a case for the use of parallelism.

The distance function encapsulates the particular features of the application objects which makes the different data structures for searching general purpose strategies [4]. Well-known data structures for metric spaces are BKTree [3], MetricTree [9], GNAT [2], VpTree [12], FQTree [1], MTree [5], SAT [6], Slim-Tree [8]. Some of them are based on clustering and others on pivots. The EGNAT proposed in this paper is based on clustering [10].

In the case of pivots based strategies a set of (usually random) objects are selected from the database and distances are calculated to organize the pivots in a, for example, tree fashion. A search query is executed by calculating distances between the query object and the pivots so that the search space is reduced by applying the triangular inequality to discard tree branches.

The strategies based on clustering divide the space in areas, where each area has a center point. Information is stored in each area so that it allows easy discarding of the whole area by just comparing the query with the center point. The strategies based on clustering are better suited than pivots ones for high-dimensional metric spaces.

Voronoi Diagrams. Consider a set of point $\{c_1, c_2, \ldots, c_n\}$(centers). A Voronoi Diagram is defined as the subdivision of the plane in n areas, one for each c_i, such that q is in the area c_i if and only if the euclidean distance holds $d(q, c_i) < d(q, c_j)$ for each c_j, with $j \neq i$.

The EGNAT is based on the concepts of Voronoi Diagrams and is an extension of the GNAT proposed in [2], which in turn is a generalization of the *Generalized Hyperplane Tree* (GHT) [9]. Basically the tree is constructed by taking two selected points (the two children of the root) and distributing the remaining points according with how close in distance they are to one of the two points. This is repeated recursively in each sub-tree.

In the GNAT k points, instead of two, are selected to divide the space $\{p_1, p_2, \ldots, p_k\}$, where every remaining point is assigned to the closet one among the k points.

Most data structures and algorithms for searching in metric-space databases were not defined to be dynamic ones [4]. However, some of them allow insertion operations in an efficient manner once the whole tree has been constructed from an initial set of points (objects). Deletion operations, however, are particularly complicated because in this strategies the invariant that supports the data structure can be easily broken with a sufficient number of deletions, which makes it necessary to re-construct from scratch the whole tree from the remaining points. Experimental results about these issues can be found in [7].

When we consider the use of secondary memory we find in the literature a few strategies which are able to cope efficiently with this requirement. Among the well-know strategies are the *M-Tree* [5] which has a similar performance to the GNAT in terms of number of accesses to disk and overall size of the data structure.

2   Evolutionary Geometric Near-Neighbor Access Tree

The construction of the initial EGNAT is performed using the GNAT method proposed by [2], that is

1. Select k points called *centers*, $p_1, \ldots, p_k$.
2. Associate every remaining point with the nearest center. The set of points associated with every center p_i is denoted by D_{p_i}.
3. For each pair of centers (p_i, p_j), the following range is calculated,

$$\text{range}\,(p_i, D_{p_j}) = \left[\min\{d\,(p_i, D_{p_j})\}, \max\{d\,(p_i, D_{p_j})\}\right].$$

4. The tree is constructed recursively for each element in D_{p_i}.

Every set D_{p_i} represents a sub-tree whose root is p_i.

Additionally, the EGNAT [10] is data structure in which the nodes are created as buckets in which the only information is the distance to their father. This allows a significant reduction in space used in disk and also allows good performance in terms of number of distance evaluations. When a node becomes full of objects it evolves from a bucket to a GNAT node by re-inserting all objects in the bucket to the new GNAT sub-tree node.

Searching in the EGNAT is performed recursively as follows,

1. Assume that we are interested in retrieving all objects with distance $d \leq r$ to the query object q (range query). Let P be the set of centers of the current node in the search tree.
2. Choose randomly a point p in P, calculate the distance $d(q, p)$. If $d(q, p) \leq r$, add p to the output set result.
3. $\forall\, x \in P$, if $[d(q, p) - r, d(q, p) + r] \cap \text{range}(p, D_x)$ is empty, the remove x from P.
4. Repeat steps 2 and 3 until processing all points (objects) in P.
5. For all points $p_i \in P$, repeat recursively the search in D_{p_i}.

3   EGNAT Performance (Sequentially and in Parallel)

In figure 1 we show results comparing the EGNAT with the M-Tree for both number of distance calculations and access to secondary memory. These results show that EGNAT is more efficient than the M-Tree. The results shown in the figure were obtained for the best tunning parameters for each data structure (19, 20, 0.4, 0.1, details in references).

We also tested the suitability of the EGNAT data structure for supporting query processing in parallel. We evenly distributed the database among P processors of a 10-processors cluster of PCs. Queries are processed in batches as we assume an environment in which a high traffic of queries is arriving to a broker machine. The broker routes the queries to the processors in a circular manner. We takes batches as we use the BSP model of computing for performing the parallel computations [11].

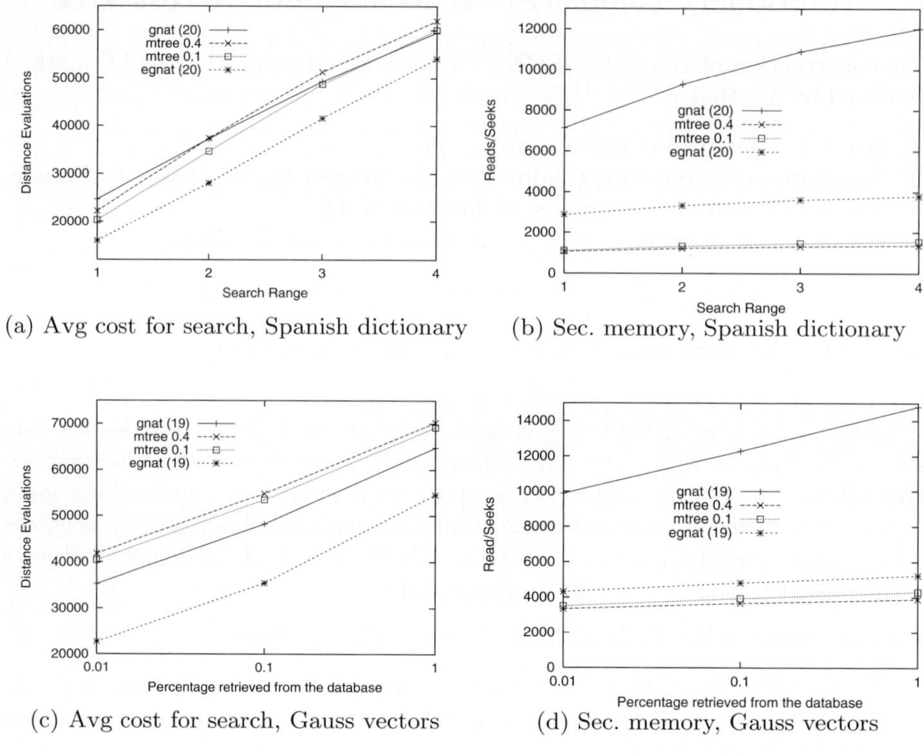

(a) Avg cost for search, Spanish dictionary   (b) Sec. memory, Spanish dictionary

(c) Avg cost for search, Gauss vectors   (d) Sec. memory, Gauss vectors

Fig. 1. Results for the local index approach

In the bulk-synchronous parallel (BSP) model of computing [11], any parallel computer (e.g., PC cluster, shared or distributed memory multiprocessors) is seen as composed of a set of P processor-local-memory components which communicate with each other through messages. The computation is organized as a sequence of *supersteps*. During a superstep, the processors may only perform sequential computations on local data and/or send messages to other processors. The messages are available for processing at their destinations by the next superstep, and each superstep is ended with the barrier synchronization of the processors.

We used two approaches to the parallel processing of batches of queries. In the first case, an independent EGNAT is constructed in the piece of database stored in each processors. Queries in this case start at any processor at the beginning of each superstep. The first step in processing a particular query is to send a copy of it to all processors including itself. At the next superstep the searching algorithms is performed in the respective EGNAT and all solutions found are reported to the processor that originated the query. We call this strategy the *local index* approach.

In the second case, we assume that a single EGNAT has been constructed considering the whole database. The first levels of the tree are kept duplicated in every processor. The size of this tree is large enough to fit in main memory.

Downwards the tree branches or sub-trees are evenly distributed onto secondary memory of the processors. A query stars at any processor and the sequential algorithms is used for the first levels of the tree. The copies of the query "travel" to other processors to continue the search in the sub-trees stored in remote secondary memory. Note that queries can divided in multiple copies according to the tree paths that contains valid results. Thus these copies are processed in parallel when are sent to different processors. We call this strategy the *global index* approach.

Results for databases formed by natural language text, a large set of points formed with Gaussian distribution and a collection of images are shown in the figure 2. A total of 10,000 queries are processed. The results show that the local index approach achieves good efficiency in parallel. In particular, because of the relatively larger cost of disk access with respect to communication cost, we observed super-linear speedups in the results. The speedups for the global index approach were very similar.

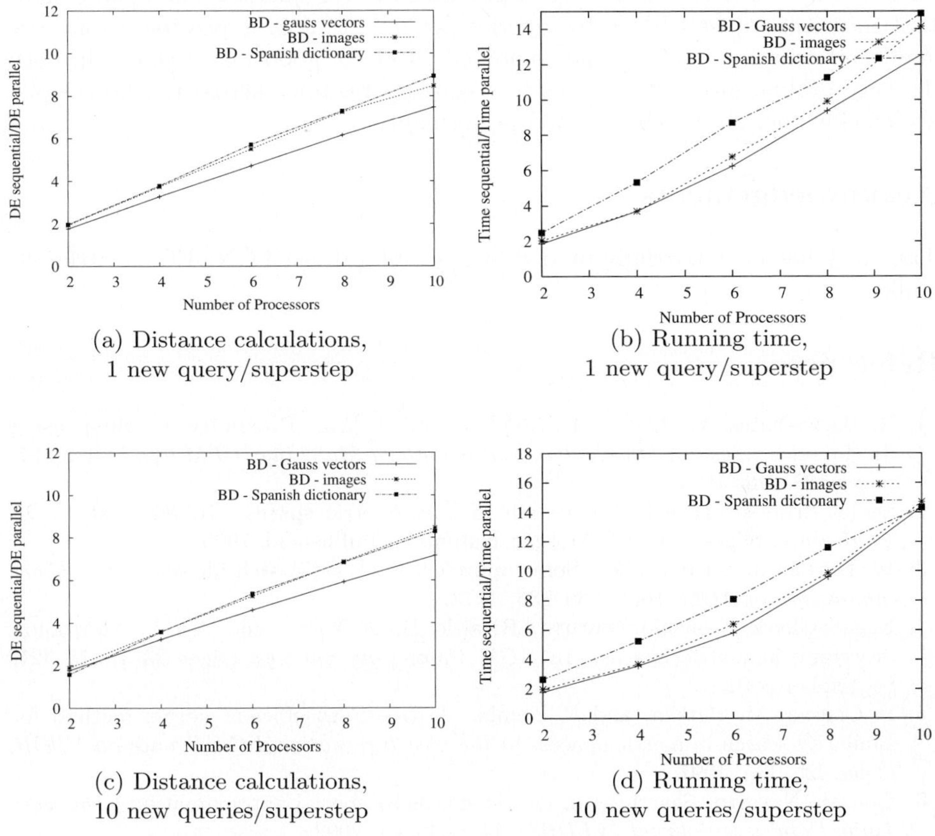

(a) Distance calculations,
1 new query/superstep

(b) Running time,
1 new query/superstep

(c) Distance calculations,
10 new queries/superstep

(d) Running time,
10 new queries/superstep

Fig. 2. Results for the local index approach

4   Conclusions

We have described the EGNAT data structure and shown its performance both sequentially and in parallel.

The results show that this data structure is a good choice for systems large enough that tree nodes has to be stored in secondary memory. It also allows insert and delete operations to take place once the tree has been constructed.

For the sequential case the results with different databases show that EG-NAT is more efficient than the well-known M-Tree both in number of distance evaluations required to solve range queries and amount of accesses to secondary memory.

For the parallel setting, the results show that the EGNAT admits an efficient parallelization. The results for running time show that it is feasible to significantly reduce the running time by the inclusion of more processors. This because of distance calculations takes in parallel during solutions of batches of queries. We emphasize that for use of parallel computing to be justified we must put ourselves in a situation of a very high traffic of user queries. The results show that in practice just with a few queries per unit time it is possible to achieve good performance. That is, the combined effects of good load balance in both distance evaluations and accesses to secondary memory across the processors, are quickly enough to achieve good performance.

Acknowledgements

This work has been partially funded by research project FONDECYT 1040611, Chile.

References

1. R. Baeza-Yates, W. Cunto, U. Manber, and S. Wu. Proximity matching using fixedqueries trees. In *5th Combinatorial Pattern Matching (CPM'94)*, LNCS 807, pages 198–212, 1994.
2. Sergei Brin. Near neighbor search in large metric spaces. In *the 21st VLDB Conference*, pages 574–584. Morgan Kaufmann Publishers, 1995.
3. W. Burkhard and R. Keller. Some approaches to best-match file searching. *Communication of ACM*, 16(4):230–236, 1973.
4. Edgar Chvez, Gonzalo Navarro, Ricardo Baeza-Yates, and Jos L. Marroqun. Searching in metric spaces. In *ACM Computing Surveys*, pages 33(3):273–321, September 2001.
5. P. Ciaccia, M. Patella, and P. Zezula. M-tree : An efficient access method for similarity search in metric spaces. In *the 23st International Conference on VLDB*, pages 426–435, 1997.
6. Gonzalo Navarro. Searching in metric spaces by spatial approximation. *The Very Large Databases Journal (VLDBJ)*, 11(1):28–46, 2002.
7. Gonzalo Navarro and Nora Reyes. Fully dynamic spatial approximation trees. In *the 9th International Symposium on String Processing and Information Retrieval (SPIRE 2002)*, pages 254–270, Springer 2002.

8. Caetano Traina, Agma Traina, Bernhard Seeger, and Christos Faloutsos. Slim-trees: High performance metric trees minimizing overlap between nodes. In *VII International Conference on Extending Database Technology*, pages 51–61, 2000.

9. J. Uhlmann. Satisfying general proximity/similarity queries with metric trees. In *Information Processing Letters*, pages 40:175–179, 1991.

10. R. Uribe. A space-metric data structure for secondary memory. Master's thesis, Computer Science Department, University of Chile, Santiago, Chile, Abril 2005.

11. L.G. Valiant. A bridging model for parallel computation. *Comm. ACM*, 33:103–111, Aug. 1990.

12. P. Yianilos. Data structures and algoritms for nearest neighbor search in general metric spaces. In *4th ACM-SIAM Symposium on Discrete Algorithms (SODA'93)*, pages 311–321, 1993.

Unsplittable Anycast Flow Problem:
Formulation and Algorithms

Krzysztof Walkowiak

Chair of Systems and Computer Networks, Faculty of Electronics,
Wroclaw University of Technology, Wybrzeze Wyspianskiego 27, 50-370 Wroclaw, Poland
Tel.: (+48)713203539; Fax.: (+48)713202902
`Krzysztof.Walkowiak@pwr.wroc.pl`

Abstract. Our discussion in this article centers on a new optimization problem called *unsplittable anycast flow problem* (UAFP). We are given a directed network with arc capacities and a set of anycast requests. *Anycast* is a one-to-one-of-many delivery technique that allows a client to choose a content server of a set of replicated servers. In the context of unsplittable flows, anycast request consists of two connections: upstream (from the client to the server) and the downstream (in the opposite direction). The objective of UAFP is to find a subset of the requests of maximum total demand for which upstream and downstream connection uses only one path and the capacity constraint is satisfied. To our best survey, this is the first study that addresses the UFP (unsplittable flow problem) in the context of anycast flows. After formulation of UAFP, we propose several heuristics to solve that problem. Next, we present results of simulation evaluation of these algorithms.

Keywords: UFP, connection-oriented network, anycast.

1 Introduction

In this paper we consider a new version of the Unsplittable Flow Problem (UFP) for anycast flows. Anycast is a *one-to-one-of-many* technique to deliver a packet to one of many hosts. Anycast paradigm becomes popular, since there is a need to facilitate the distribution of popular content (electronic music, movies, books, software) in the Internet. One of exemplary techniques that apply anycast traffic is Content Delivery Network (CDN). For more details on anycast flows and CDNs refer to [3], [8-10]. In this work we consider on anycast unsplittable flows. Several connection-oriented (c-o) techniques apply unsplittable flows, e.g. MultiProtocol Label Switching (MPLS), Asynchronous Transfer Mode (ATM), optical network [2]. In c-o networks an anycast demand consists of two connections: one from the client to the server (upstream) and the second one in the opposite direction (downstream). Upstream connection is used to send user's requests. Downstream connection carries requested data. Consequently, each anycast demand is defined by a following triple: client node, upstream bandwidth requirement and downstream bandwidth requirement. In contrast, a unicast demand is defined by the following triple: origin node, destination node and bandwidth requirement. According to observations of real networks and user behavior,

V.N. Alexandrov et al. (Eds.): ICCS 2006, Part I, LNCS 3991, pp. 618–625, 2006.

there is asymmetry in anycast flows, since usually more data is received by clients then is sent to content servers Therefore, bandwidth of upstream connection is typically much lower than bandwidth of downstream connection.

The considered *unsplittable anycast flow problem* (UAFP) can be formulated as follows

Given	network topology, anycast traffic demand pattern, location of replica servers, link capacity
Minimize	volume of un-established anycast requests
Over	selection of replica server, routing (path assignment)
Subject to	anycast connection-oriented flow constraints, capacity constraints

To the best of the author's knowledge, this is the first work that addresses the UFP in the context of anycast flows. The objectives of this paper are twofold: mathematical formulation of the UAFP and development of effective heuristics for the UAFP. Evaluation of proposed algorithms through simulation experiments is also provided.

2 Related Work

The most intuitive approach to solve unicast UFP is the *greedy algorithm* (GA), which proceed all connections in one pass and either allocate the processed request to the shortest path or reject the request if such a feasible path does not exist, i.e. origin and destination node of the connection do not belong to the same component of considered graph [5]. A modification of GA called *bounded greedy algorithm* (BGA) works as follows [5], [7]: Let L be a suitable chosen parameter. Reject the request if there is no feasible path of the length at most L hops. Otherwise accept the request. Another version of GA is *careful BGA* (cBGA) proposed in [7]. Online algorithms can also solve the UFP. Several such algorithms were developed in the context of dynamic routing in MPLS networks [6]. MPLS supports the explicit mode, which enables the source node of the LSP to calculate the path. The main goal of dynamic routing is to minimize the number of rejected calls or the volume of rejected calls. The most common approach to dynamic routing is the shortest path first (SPF) algorithm based on an administrative weight (metric).

3 Problem Formulation

In this section we will formulate the UAFP. We consider an offline UAFP, in which we know a priori all requests that are to be located in the network. To mathematically represent the problem we introduce the following notations

Sets:

V set of vertices representing the network nodes.

A set of arcs representing network directed links.

P set of connections in the network. A connection can be of two types: downstream or upstream.

Π_p the index set of candidate routes (paths) π_p^k for connection p. Route π_p^0 is a "null" route, i.e. it indicates that connection p is not established. For upstream connection, paths are between client node and server. Analogously, downstream candidate paths connect server and client node.

X_r set of variables x_i^k, which are equal to one. X_r determines the unique set of currently selected routes.

Constants:

δ_{pa}^k equal to1, if arc a belongs to route k realizing connection p; 0 otherwise

Q_p Volume (estimated bandwidth requirement) of connection p

c_a capacity of arc a

$\tau(p)$ index of the connection associated with connection p. If p is a downstream connection $\tau(p)$ must be an upstream connection and vice versa.

$o(\pi)$ origin node of route π. If π is a "null", then route, $o(\pi)=0$.

$d(\pi)$ destination node of route π. If π is a "null", then route, $d(\pi)=0$.

Variables:

x_p^k 1 if route $k \in \Pi_p$ is selected for connection p and 0 otherwise.

f_a flow of arc a.

The UAFP can be formulated as follows

$$LF = \min_{X_r} \quad \sum_{p \in P} x_p^0 Q_p \tag{1}$$

subject to

$$\sum_{k \in \Pi_p} x_p^k = 1 \qquad \forall p \in P \tag{2}$$

$$x_p^k \in \{0,1\} \quad \forall p \in P, \forall k \in \Pi_p \tag{3}$$

$$f_a = \sum_{p \in P} \sum_{k \in \Pi_p} \delta_{pa}^k x_p^k Q_p \quad \forall a \in A \tag{4}$$

$$f_a \leq c_a \quad \forall a \in A \tag{5}$$

$$\sum_{k \in \Pi_p} x_p^k d(\pi_p^k) = \sum_{k \in \Pi_{\tau(p)}} x_{\tau(p)}^k o(\pi_{\tau(p)}^k) \qquad \forall p \in P \tag{6}$$

$$X_r = \left(\bigcup_{i,k:x_i^k=1} \left\{ x_i^k \right\} \right) \tag{7}$$

The objective function (1) is a lost flow (*LF*). Function *LF* is as a sum of all demands (connections) that are not established (variable x_p^0 is 1). It should be noted that also an equivalent objective function could be applied, in which we maximize the total volume of established connections. Condition (2) states that the each connection can use only one route or is not established. Therefore, we index the subscript of variable x_p^k starting from 0, i.e. variable x_p^0 indicates whether or not connection p is established. If it is established, $x_p^0 = 0$ and one of variables x_p^k ($k>0$), which is equal to 1, indicates the selected path. Constraint (3) ensures that decision variables are binary ones. (4) is a definition of an arc flow. Inequality (5) denotes the capacity constraint. Equation (6) guarantees that two routes associated with the same anycast demand connect the same pair of nodes. Furthermore, (6) assures that if upstream connection is not established ($x_p^0 = 0$ and the "null" route is selected) the downstream connection of the same anycast demand is not established, and vice versa. Finally, (7) is a definition of a set X called a selection that includes all variables x, which are equal to 1. Each selection denotes for each connection either the selected route or indicates that the particular connection is not established. Note, that we call a connection p *established* in selection X if $x_p^0 \notin X$.

4   Algorithms

The first, quite intuitive, approach to solve UAFP can be a modification of heuristics developed for unicast UFP. Now, we shortly discuss how to adapt GA (greed algorithm) to anycast flows. We refer to this new algorithm as AGA (anycast GA). AGA proceeds all anycast requests in a one pass. Requests can be sorted accordingly to selected criterion (e.g. bandwidth requirement). For each request we perform the following procedure. First, we construct a residual network for downstream connection of considered anycast demand, i.e. we remove from the network all arcs that have less residual capacity than volume of downstream connection. Next, we find shortest paths between each server and client node. If none path exists, the demand is rejected. Otherwise, we select a shortest path among all found paths. We repeat the same procedure for upstream connection. The residual network is computed again. However, in this case we try to find a shortest path from the client node to the server node already selected for downstream connection. This follows from the asymmetry of anycast flows discussed in Section 1, because the downstream connection is more important than upstream connection.

Now we propose two new offline algorithms for the UAFP: Anycast Greedy Algorithm with Preemption (AGAP) and Anycast Greedy Algorithm with Preemption and Flow Deviation (AGAPFD). Both algorithms use the *preemption* mechanism, which consists in removing from the network already established connections in order to enable establishment of other connections to minimize the objective function.

We apply some temporary variables in both algorithms. Sets H and F are selections including decision variables x equal to 1. Sets B and D include indexes of connections. Operator *first(B)* returns the index of the first connection in set B. Operator *sort(H)* returns indexes of connections included in H ordered according to their paths' length given by the metric CSPF [1] starting with the longest. We select CSPF metric due to its effectiveness in many dynamic routing problems. Operator *AGA(H,i,j)* returns either the pair of indexes of calculated routes: downstream and upstream according to AGA or a pair of zeros, if a pair feasible routes does not exist for connections i and j associated with the same anycast demand. Operator $ue(H)$ returns indexes of connections, which are not established in H, while $es(H)$ returns indexes of connections established in H.

Algorithm AGAP

Step 1. Let H denote an initial solution, in which none connection is established. Let $B:=sort(H)$.

Step 2. Set $i:=first(B)$ and $\{d,u\}:=AGA(H,i,\tau(i))$. Calculate $B:=B-\{i\}$, $B:=B-\{\tau(i)\}$).

   a) If $d>0$ set $H:=\left(H-\left\{x_i^0\right\}\right)\cup\left\{x_i^u\right\}$, $H:=\left(H-\left\{x_{\tau(i)}^0\right\}\right)\cup\left\{x_{\tau(i)}^d\right\}$. Go to step 2c.

   b) If $d=0$ go to step 3.

   c) If $B=\varnothing$ then stop the algorithm. Otherwise go to step 2.

Step 3. Set $D:=es(sort(H))$.

   a) Set $j:=first(D)$. Set $D:=D-\{j\}$ and $F:=\left(H-\left\{x_j^m\right\}\right)\cup\left\{x_j^0\right\}$, where $x_j^m\in H$. Next set $D:=D-\{\tau(j)\}$ and $F:=\left(F-\left\{x_{\tau(j)}^m\right\}\right)\cup\left\{x_{\tau(j)}^0\right\}$, where $x_{\tau(j)}^m\in H$.

   b) Find $\{d,u\}:=AGA(F,i,\tau(i))$. If $d>0$ set $H:=\left(H-\left\{x_i^0\right\}\right)\cup\left\{x_i^u\right\}$, $H:=\left(H-\left\{x_{\tau(i)}^0\right\}\right)\cup\left\{x_{\tau(i)}^d\right\}$ and go to step 2.

   c) If $d=0$ then go to step 3d.

   d) If $D=\varnothing$ then go to step 2c. Otherwise go to step 3a.

The idea of AGAP is as follows. We start with an "empty" solution – none connection is established. We process connections one-by-one sorted according to selected criterion. If the AGA can find a pair of feasible paths for current anycast demand, we establish upstream and downstream connections (step 2a). Otherwise (step 2b), we go back to already established connections and preempt each of these demands (step 3a) trying to establish again considered anycast connections (step 3b-d).

Algorithm AGAPFD(α,β)

Step 1. Let X_1 denote a feasible initial solution given by AGA. Sort all connections in X_1 according to their bandwidth requirements starting with the heaviest. Set $j:=1$.

Step 2. Set $H:=X_j$. Let $B:=sort(H)$. Let l denote the number of connections established in H. Set $k:=0$.

   a) Set $i:=first(B)$. Calculate $B:=(B-\{i\})$, $B:=B-\{\tau(i)\}$), $F:=\left(H-\left\{x_i^m\right\}\right)\cup\left\{x_i^0\right\}$ where $x_i^m\in H$, $F:=\left(F-\left\{x_{\tau(i)}^m\right\}\right)\cup\left\{x_{\tau(i)}^0\right\}$, where $x_{\tau(i)}^m\in H$.

   b) Set $H:=F$ and $k:=k+1$.

   c) If $k>\alpha\cdot l$ then go to step 3. Otherwise go to step 2a.

Step 3. Let $F:=AFDNB(H)$ be a selection of route variables calculated according to the Anycast Flow Deviation for Non-bifurcated flows algorithm [10].

Step 4. Set $H:=F$ and $B:=ue(H)$.

   a) Set $i:=first(B)$ and find $\{d,u\}:=AGA(H,i,\tau(i))$. Calculate $B:=B-\{i\}$, $B:=B-\{\tau(i)\}$), $H:=\left(H-\left\{x_i^0\right\}\right)\cup\left\{x_i^u\right\}$ and $H:=\left(H-\left\{x_{\tau(i)}^0\right\}\right)\cup\left\{x_{\tau(i)}^d\right\}$.

   b) Set $H:=F$.

c) If $B = \varnothing$ then go to step 5. Otherwise go to step 4a.

Step 5. If $j \geq \beta$ stop the algorithm. Otherwise set $j := j+1$, $X_j := H$ and go to step 2.

The algorithm has two input parameters that can be calibrated. Parameter $\alpha \in [0,1]$ is used to find set for preemption of α established connections. The second parameter of the GAPFD algorithm - β - is a number of iterations for which the main loop of the algorithm is repeated. The main idea of the AGAPFD algorithm is as follows. We start with a feasible solution X_1, found by AGA. Next, in step 2 we remove from the network a number of connections. We find $\alpha \cdot l$ connections with the longest routes computed according to the CSPF metric [1] assigned to each arc, where l denotes the number of established connections in a given selection. Next, we remove these connections. The major goal of preemption is to remove connections having routes using the most congested arcs. In Step 3 we re-optimize routes of established connections in order to change the allocation of arcs' flows and enable creation of as many as possible of un-established connections, what should yield improvement of UAFP objective function. We use the Anycast Flow Deviation for Non-bifurcated flows (AFDNB) algorithm proposed in [10]. Note that AFDNB uses as objective the network delay function, which includes the capacity constraint (5) as a penalty function. This guarantees feasibility of obtained selection. Furthermore, former studies shows that the delay function provides proportional allocation of network flows – more open capacity is left for other demands [4]. Since AGA processes connections sequentially, there is no chance to change the route of already established connections. Running of AFDNB in step 4 eliminates this constraint and enables re-optimization of already established routes. In Step 4 we process all un-established connections using AGA. The main loop of the algorithm (Steps 2-5) is repeated β times.

5 Results

Algorithms AGA, AGAP, AGAPFD were coded in C++. The network on which we conduct our experiment consists of 36 nodes and 144 directed links [10]. We test 10 demand patterns for 5 various locations of 2, 3, or 4 content servers. Each of 10 demand patterns includes 360 anycast requests generated randomly. Due to asymmetry of anycast flows, we assume that upstream bandwidth is 0.1 of the downstream bandwidth. To compare results we apply *competitive ration* performance indicator. The competitive ration is defined as the difference between result obtained for a particular algorithm and the minimum value of objective function yielded by the best algorithm. For instance, if in a test consisting of simulations of various algorithms the minimum value of lost flow is 2000 and the considered algorithm yields 2500; the competitive ration is calculated as follows: (2500-2000)/2500=20%. For presentation of aggregate results we apply the *aggregate competitive ration*, which is a sum of competitive rations over all considered experiments.

The first objective of numerical experiments was evaluation of various metrics and orderings for AGA. We have tested two link metrics: HOP and CSPF [1]. Demands were sorted according to the bandwidth requirement of downstream connection in decreasing and increasing order. In Table 1 we report aggregate competitive ration of AGA for these cases. We can watch that the best result provides CSPF metric and demand sorted in decreasing order.

Table 1. Aggregate competitive ration of AGA using ordering CDLI and various metrics for experiment C

Servers	Decreasing Hop	Decreasing CSPF	Increasing HOP	Increasing CSPF
2	3.5%	3.0%	152.7%	165.3%
3	14.6%	3.0%	415.9%	426.4%
4	52.8%	7.9%	1495.4%	1536.6%

The next goal of experiments was tuning of AGAPFD. Due to initial trial runs we decided to set the number of iterations (parameter β) to 10. We run simulations for the following values of parameter α={0.0; 0.05; 0.1; 0.2; 0.3; 0.4; 0.5; 0.6; 0.7; 0.8; 0.9; 1.0}. In Fig. 1 we report comparison of results obtained for tested values of α. Due to results presented in Fig. 1 we decide to set α=0.3 for further simulations.

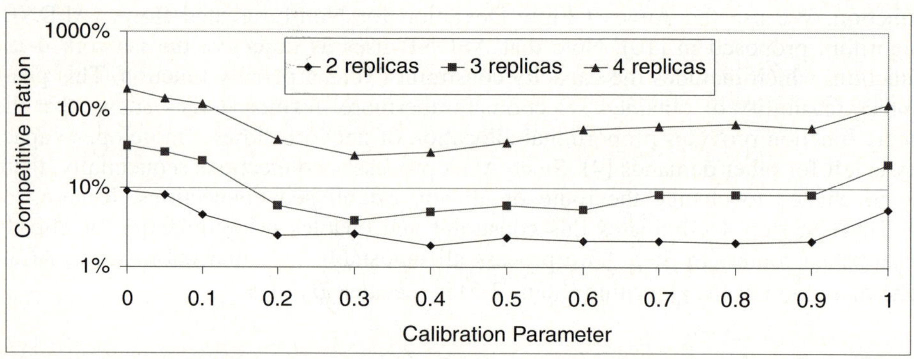

Fig. 1. Aggregate competitive ration of parameter α for various numbers of content servers

Finally, we run experiments to compare performance of AGA, AGAP and AGAPFD(0.3,10). In Table 2 we present aggregate competitive ration for these algorithms. We can see that AGAPFD outperforms other methods – in all cases it can find the best result. The difference between AGAP and AGA increases with the number of content servers.

The main drawback of AGAPFD is the calculation time – AGAPFD needs about 25 times more decision time than AGA. Most of the time is spent on running AFDNB algorithm. AGAP needs about 3-4 times more than AGA. However, it should be noted that we consider the offline UAFP. Consequently, the decision time is not the most important criterion in our considerations.

Table 2. Aggregate competitive ration of tested algorithms

Servers	AGA	AGAP	AGAPFD(0.3,10)
2	6.50%	6.02%	0.00%
3	31.00%	27.42%	0.00%
4	156.64%	112.90%	0.00%

6 Conclusion

In this paper, we have formulated the Unsplittable Anycast Flow Problem. To our best knowledge, this is the first survey that addresses the UFP in the context of anycast flows. We have developed three heuristics solving offline version of UAFP. Through simulations we have tuned and compared the algorithms. The UAFP problem is motivated by service provider needs for fast deployment of bandwidth guaranteed services enabling fast and effective distribution of popular content over Internet. Results of our work can be applied for optimization of flows of Content Delivery Networks located in connection-oriented environment, e.g. MPLS network.

Acknowledgements. This work was supported by a research project of the Polish State Committee for Scientific Research carried out in years 2005-2007.

References

1. Crawley, E., Nair, R., Jajagopalan, B., Sandick, H.: A Framework for QoS-based Routing in the Internet. RFC2386 (1998)
2. Grover, W.: Mesh-based Survivable Networks: Options and Strategies for Optical, MPLS, SONET and ATM Networking. Prentice Hall PTR, Upper Saddle River, New Jersey (2004)
3. Hao, F., Zegura, E., Ammar, M.: QoS routing for anycast communications: motivation and an architecture for DiffServ networks. IEEE Communication Magazine, 6 (2002), 48-56
4. Kasprzak, A.: Designing of Wide Area Networks. Wroclaw Univ. of Tech. Press, (2001)
5. Kleinberg, J.: Approximation algorithms for disjoint paths problems. PhD thesis, MIT, Cambridge, (1996)
6. Kodialam, M., Lakshman, T.: Minimum Interference Routing with Applications to MPLS Traffic Engineering. In Proceedings of INFOCOM (2000), 884-893
7. Kolman, P., Scheideler, C.: Improved bounds for the unsplittable flow problem. In Proc. of the Symposium on Discrete Algorithms (2002), 184-193
8. Markowski, M., Kasprzak, A.: The web replica allocation and topology assignment problem in wide area networks: algorithms and computational results. Lectures Notes in Computer Science, Vol. 3483 (2005), 772-781
9. Peng, G.: CDN: Content Distribution Network. Technical Report, sunysb.edu/tr/rpe13.ps.gz, (2003)
10. Walkowiak, K.: Heuristic Algorithm for Anycast Flow Assignment in Connection-Oriented Networks. Lectures Notes in Computer Science, Vol. 3516 (2005), 1092-1095

Lagrangean Heuristic for Anycast Flow Assignment in Connection-Oriented Networks

Krzysztof Walkowiak

Chair of Systems and Computer Networks, Faculty of Electronics,
Wroclaw University of Technology, Wybrzeze Wyspianskiego 27, 50-370 Wroclaw, Poland
Krzysztof.Walkowiak@pwr.wroc.pl

Abstract. In this work we address the problem of anycast flow assignment. *Anycast* is a one-to-one-of-many delivery technique that allows a client to choose a content server of a set of replicated servers. We formulate an optimization problem of anycast flows assignment in a connection-oriented network, which is 0/1 and NP-complete. Thus, we propose a new effective heuristic algorithm based on Lagrangean relaxation technique. To our best survey, this is the first study that applies the Lagrangean relaxation to anycast flow problem. We evaluate the performance of the proposed scheme by making a comparison with its counterpart using a sample network topology and different scenarios of traffic demand patterns and replica location. Obtained results show advantage of the Lagrangean heuristic over a previously proposed algorithm.

1 Introduction

The influence of the Internet can be noticed in many areas of people's life, and in the future its importance will grow. Currently, most of transmissions issued in the Internet are unicast – two individual systems connect *one-to-one* to exchange data. However, new approaches are developed to overcome problems now being encountered in computer networks, e.g. network latency, congestion, growing costs. One of the most interesting approaches that can facilitate the development of new services in the Internet and enable QoS (Quality of Service) guarantees is the *anycast* paradigm. Anycast is a *one-to-one-of-many* technique to deliver a packet to one of many hosts. Significance of anycast transmission will grow with popularity of the Internet and development of new services, e.g. distribution of electronic music, movies, books. One of the most famous technologies that applies anycast traffic is Content Delivery Network (CDN). CDN uses many content servers offering the same content replicated in various locations. User-perceived latency and the other QoS parameters can be easily and inexpensively improved by the CDN.

In this work we concentrate on problems of CDN design in MPLS (Multiprotocol Label Switching) environment. The MPLS proposed by the Internet Engineering Task Force (IETF) in [8] is a networking technology that enables traffic engineering and QoS performance for carrier networks. MPLS is a connection-oriented (c-o) technique and is widely applied in backbone networks. Thus, MPLS should be addressed in designing of CDNs. The considered problem can be formulated as follows:

V.N. Alexandrov et al. (Eds.): ICCS 2006, Part I, LNCS 3991, pp. 626–633, 2006.
© Springer-Verlag Berlin Heidelberg 2006

Given	network topology, traffic demand pattern, location of replica servers, link capacity
Minimize	network delay
Over	selection of replica server, routing (path assignment)
Subject to	connection-oriented flow constraints, capacity constraints

The considered problem was formulated and solved in [9]. In this paper we propose a Lagrangean relaxation method combined with subgradient optimization to iteratively solve the problem by a heuristic algorithm ANBFD [9] using as initial solutions results given solving dual problems. Lagrangean relaxation with subgradient optimization has been used successfully for solving various communication network routing problems [3], [7]. Our starting point is the algorithm proposed in [3]. However, we modified this approach according to anycast flow constraints.

Due to limited size of the paper we cannot present more information on anycast flow, CDN and networks optimization. For more information refer to [1], [4-6], [7-9].

2 Anycast Flow Assignment Problem

Since we consider a connection-oriented network, we model the network flow as non-bifurcated multicommodity flow. In c-o networks an anycast demand must consist of two connections: one from the client to the server (upstream) and the second one in the opposite direction (downstream). Upstream connection is used to send user's requests. Downstream connection carries requested data. Consequently, each anycast demand is defined by a following triple: client node, upstream bandwidth requirement and downstream bandwidth requirement. In contrast, a unicast demand is defined by a following triple: origin node, destination node and bandwidth requirement. To establish a unicast demand a path satisfying requested bandwidth and connecting origin and destination nodes must be calculated. Optimization of anycast demands is more complex. The first step is the server selection process. Next, when the server node is selected, both paths: upstream and downstream can be calculated analogously to unicast approach. However, the main constraint is that both connections associated with a particular anycast demand must connect the same pair of network nodes.

To mathematically represent the problem, we introduce the following notations:

V set of n vertices representing the network nodes.

A set of m arcs representing network directed links.

P set of q connections in the network. A connection can be of two types: downstream or upstream.

Π_p the index set of candidate routes (paths) π_p^k for connection p. For upstream connection paths are between client node and a node hosting a content server. Analogously, downstream candidate paths connect a server and a client node.

X_r set of route selection variables x_p^k, which are equal to one. X_r determines the unique set of currently selected routes.

c_a capacity of arc a.

Q_p bandwidth requirement of connection p.

$\tau(p)$ index of the connection associated with connection p. If p is a downstream connection $\tau(p)$ must be an upstream connection and vice versa.

628   K. Walkowiak

δ_{pa}^k binary variable, which is 1 if arc a belongs to route π_p^k and is 0 otherwise.

$o(\pi)$ origin node of route π.

$d(\pi)$ destination node of route π.

T total arrive rate of messages in the network

The anycast non-bifurcated flow assignment (AFA1) problem is as follows

$$\min_{X_r} \quad \frac{1}{T} \sum_{a \in A} \frac{f_{ar}}{(c_a - f_{ar})} \tag{1}$$

subject to

$$f_{ar} = \sum_{a \in P} \sum_{k \in \Pi_p} \delta_{pa}^k x_p^k Q_p \quad \forall a \in A \tag{2}$$

$$\sum_{k \in \Pi_p} x_p^k = 1 \qquad \forall p \in P \tag{3}$$

$$0 \le f_{ar} \le c_a \qquad \forall a \in A \tag{4}$$

$$\sum_{k \in \Pi_p} x_p^k d(\pi_p^k) = \sum_{k \in \Pi_{\tau(p)}} x_{\tau(p)}^k o(\pi_{\tau(p)}^k) \qquad \forall p \in P \tag{5}$$

$$x_p^k \in \{0,1\} \qquad \forall p \in P; k \in \Pi_p \tag{6}$$

$$X_r = \left(\bigcup_{p,k: x_p^k = 1} \left\{ x_p^k \right\} \right) \tag{7}$$

Variable r denotes the index of set X_r including information on currently selected routes given by variables x equal to 1. The goal is to minimize the network delay [2], [7]. Condition (3) states that each connection can use only one route. (4) is a capacity constraint. Equation (5) guarantees that two routes associated with the same anycast demand connect the same pair of nodes. Constraint (6) ensures that decision variables are binary ones. Note that the constraints set doesn't include a constraint guaranteeing that upstream (downstream) connection ends (starts) in a node with a replica server. This follows from the construct of set Π_p that contains only feasible candidate routes, i.e. each route connects the client node and a replica node. The AFA problem given by (1-7) is NP-complete. This follows from the fact that the unicast non-bifurcated flow assignment is NP-complete [8]. AFA is 0/1 optimization problem. Complexity of AFA problem is a result of huge solution space.

Accordingly to [3] we transform the AFA1 problem into an equivalent formulation called AFA2, which is better suited for a Lagrangean relaxation procedure. Since the objective function (1) is not decreasing with f_a, we can replace equality (2) with an inequality leading to the following problem

$$\varpi = \min_{X_r} \quad \frac{1}{T} \sum_{a \in A} \frac{f_{ar}}{(c_a - f_{ar})} \tag{8}$$

subject to (3-7) and

$$\sum_{p \in P} \sum_{k \in \Pi_p} \delta_{pa}^k x_p^k Q_p \leq f_a \quad \forall a \in A \tag{9}$$

3 Lagrangean Relaxation

The major idea of the Lagrangean Relaxation (LR) decomposition algorithm is to consider the dual problem of by relaxing constraints to obtain a simpler subproblem iteratively to drive towards the optimal solution of the original problem. Therefore, a suitable constraint set is chosen to be relaxed. In our case we use the approach proposed in [3] and relax the constraint (9) using a vector of positive Lagrangean multipliers λ_a, $a \in A$. Consequently, the following Lagrangean relaxation of problem AFA2, called LR_AFA is formulated as

$$\varphi(\lambda) = \min \quad (\frac{1}{T} \sum_{a \in A} \frac{f_{ar}}{(c_a - f_{ar})} + \sum_{a \in A} \lambda_a (\sum_{p \in P} \sum_{k \in \Pi_p} \delta_{pa}^k x_p^k Q_p - f_{ar}) \tag{10}$$

subject to (3-6).

The set of feasible solutions for the problem AFA is a subset of the set of feasible solutions for the Lagrangean relaxation of AFA. Since we assume that

$$0 \leq \lambda_a \quad \forall a \in A \tag{11}$$

in any feasible solution of AFA, the term $\sum_{a \in A} \lambda_a (\sum_{p \in P} \sum_{k \in \Pi_p} \delta_{pa}^k x_p^k Q_p - f_{ar})$ is non-positive and thus, the value of the objective function (10) never exceeds the value of the objective function in problem AFA. Thus, accordingly to Lagrangean relaxation theory, whenever problem AFA has a feasible solution, $\varphi(\lambda) \leq \varpi$. Consequently, for each vector of multipliers λ, $\varphi(\lambda)$ is a lower bound of delay function. The best lower bound can be obtained for the vector of multipliers λ^* for which $\varphi(\lambda^*) = \max_\lambda \varphi(\lambda)$.

The objective function of Lagrangean problem can be rewritten as

$$\varphi(\lambda) = \min \quad \left\{ \frac{1}{T} \sum_{a \in A} \frac{f_{ar}}{(c_a - f_{ar})} - \lambda_a f_{ar} \right\} + \left\{ \sum_{a \in A} \lambda_a (\sum_{p \in P} \sum_{k \in \Pi_p} \delta_{pa}^k x_p^k Q_p) \right\} \tag{12}$$

Since there are no coupling constraints between variables f_a and variables x_p^k, the considered problem can be divided into two independent subproblems:

Subproblem 1

$$\varphi_1(\lambda) = \min \left\{ \frac{1}{T} \sum_{a \in A} \frac{f_{ar}}{(c_a - f_{ar})} - \lambda_a f_{ar} \right\} \tag{13}$$

subject to (4).

Subproblem 2

$$\varphi_2(\lambda) = \min \left\{ \sum_{a \in A} \lambda_a (\sum_{p \in P} \sum_{k \in \Pi_p} \delta^k_{pa} x^k_p Q_p) \right\} \tag{14}$$

subject to (3) and (5-6).

According to formulation of objective function we can separate subproblem 1 into $m=|A|$ subproblems and each of such subproblems is solved by

$$f_{ar} = \begin{cases} (c_a - \sqrt{c_a/(T \cdot \lambda_a)}) & \text{when} \quad \lambda_a > 1/(T \cdot c_a) \\ 0 & \text{otherwise} \end{cases} \tag{15}$$

Subproblem 2 can be separated into $q=|P|$ subproblems, one for each connection

$$\varphi_2^p = \min \left\{ \sum_{k \in \Pi_p} x^k_p \alpha^k_p \right\} \tag{16}$$

subject to (3), (5-6) and

$$\alpha^k_p = \sum_{a \in A} \lambda_a \delta^k_{pa} Q_p \tag{17}$$

α^k_p denotes the length of route $k \in \Pi_p$. The minimum for each connection p is attained when for the shortest route calculated under a metric $\lambda_a Q_p$. It means that the subproblem 2 consists of a number of shortest path problems. Comparing to the unicast version of considered problem addressed in [3], in the anycast flow problem to find a shortest path we must take into account all content servers. Another modification is the coupling between upstream and downstream connections given by (5). Therefore, upstream and downstream connections must be processed jointly. Consequently, for each pair of connections we calculate a shortest path using Dijkstra under the metric $\lambda_a Q_p$ to each of content servers. Next, we select a pair of paths that satisfy constraint (5) and have the smallest sum of lengths. Such a procedure guarantees optimal solution of subproblem 2. If the set of candidate routes includes only a subset of all possible routes, we must find a pair of routes for which length is the smallest.

4 Subgradient Search

Given multiplier's vector λ we can solve the Lagrangean relaxation problem by solving two subproblems as presented above. Let $x^k_p(\lambda)$ and $f_{ar}(\lambda)$ be an optimal solution of Lagrangean relaxation for a fixed vector of multipliers λ. The corresponding subgradient of the dual function (10) at λ is given by

$$\gamma_a(\lambda) = (\sum_{p \in P} \sum_{k \in \Pi_p} \delta_{pa}^k x_p^k(\lambda) Q_p) - f_{ar}(\lambda) \qquad \forall a \in A \tag{18}$$

The multipliers are updated as follows

$$\lambda_a^{i+1} = \max(1, \lambda_a^i + t_i \gamma_a^i) \qquad \forall a \in A \tag{19}$$

The step-size, t_k, can be given as [3], [7]

$$t_i = \rho(\bar{\varphi} - \varphi(\lambda^i)) / \left\| \gamma^i \right\|^2 \tag{20}$$

$\bar{\varphi}$ denotes un upper bound on the dual function, which can be calculated using a heuristic algorithm that can find a feasible solution of the primal problem. Note that ρ is commonly used in the range $0 \leq \rho \leq 2$ [7].

Lagrangean Relaxation Heuristic (LRH) Algorithm

Step 0. Select an initial λ^0, Set $\rho := 2$, $\rho_{maxiter} = 5$, $\rho_{min} := 0.005$, $i_{max} = 100, .i := 0$, $\rho_{iter} := 0$, $\varpi^{best} := \propto$, $\lambda^* := \lambda^0$. Using a heuristic, compute an overestimate $\bar{\varphi}$ of $\phi(\lambda^*)$ or set $\bar{\varphi}$ to an arbitrary large value.

Step 1. Set $i := i+1$, $\rho_{iter} := \rho_{iter} + 1$. Given λ^i, solve the LR_AFA as decoupled subproblems to obtain $\phi(\lambda^i)$, x^i, f^i.

Step 2a. If $(\phi(\lambda^i) > \phi(\lambda^*))$ then $\lambda^* := \lambda^i$ and $\rho_{iter} := -1$.

Step 2b. Use x^i, f^i to compute feasible primal objective ϖ of the problem AFA2 using ANBFD[9] algorithm. If $\varpi < \varpi^{best}$ then $\varpi^{best} := \varpi$, $x^{best} := x^i$, $f^{best} := f^i$, $\bar{\varphi} := \varpi^{best}$.

Step 2c. If $(\rho_{iter} > \rho_{maxiter})$ then $\rho := \max\{\rho/2, \rho_{min}\}$ and $\rho_{iter} := 0$.

Step 3. Use decoupled solutions x^i, f^i to compute: subgradient $\gamma^i(\lambda^i)$ (refer to (18)); step size t_i (refer to (19)); multipliers λ^{i+1} (refer to (20)).

Step 4. If $i > i_{max}$, stop. Otherwise go to 1.

5   Results

We now describe our simulation setup and scenarios. Algorithms LRH and ANBFD were coded in C++. Next, we run a number of tests to evaluate effectiveness of this algorithm for various simulation scenarios. The network on which we conduct our experiment consists of 36 nodes and 144 directed links [9]. We run experiments for various scenarios of replica location and number of replicas. According to observations of real networks and user behavior, more data is received by clients then is sent to replicas Therefore, we assume that upstream bandwidth is 0.1 of the downstream bandwidth. The total demand is calculated as a sum of all downstream and all upstream bandwidth requirements. We assume that there are 10 anycast demands for each node in the network. Therefore, the total number of anycast demands is $10n = 360$. Bandwidth requirements of demands are generated randomly.

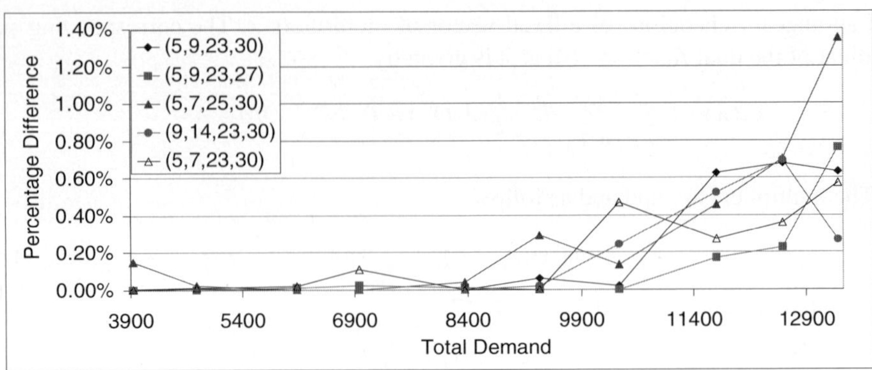

Fig. 1. Percentage difference between LRH and ANBFD for various replica location

In Table 1 we present results of algorithms LRH and ANBFD for the following four locations of replica servers: (5,23), (5,23,30), and (5,9,23,30). Fig. 1 shows the percentage difference in results of LRH and ANBFD obtained for various location of 4 replicas. If we compare LRH against ANBFD, we can notice that for relatively low loads both algorithms perform similarly. However, when the volume of all connections increases, LRH yields better results than ANBFD. This follows from the fact that LRH can search larger solution space, while ANBFD have a tendency of quick convergence to local minimum and does not have mechanisms to enlarge the number of tested combinations. Another, quite obvious, observation is that increasing the number of replicas decreases the network delay. In addition, if there are more replicas in the network, demands with higher bandwidth requirements can be satisfied.

Table 1. Performance of LRH and ANBFD for various replica location and demands

Total Demand	(5,23)		(5,23,30)		(5,9,23,30)	
	ANBFD	LRH	ANBFD	LRH	ANBFD	LRH
3933	2.4320	2.4320	1.8383	1.8380	1.3240	1.3240
4781	2.8480	2.8480	1.9786	1.9775	1.4140	1.4139
6125	4.1142	4.1033	2.3472	2.3458	1.5582	1.5582
6956	6.0474	6.0474	2.6890	2.6864	1.6384	1.6384
8359			3.3583	3.3503	1.8801	1.8801
9351			4.1990	4.1854	2.1807	2.1793
10400			6.0389	6.0248	2.5018	2.5013
11708			20.9596	20.7720	3.1085	3.0891
12588					3.6577	3.6331
13332					4.2587	4.2319

In Table 2 we report the decision time of both algorithms. Certainly, LRH is much slower than ANBFD, because in each of 100 iterations LRH must run ANBFD algorithm. However, execution time of ANBFD is not deterministic – it depends on situation in the network. Therefore, for (5,23), (5,23,30) and (5,9,23,30) cases, the execution time of LRH is about 94, 85 and 92 times longer than for ANBFD, respectively.

Table 2. Average decision time of LRH and ANBFD in seconds

Algorithm	(5,23)	(5,23,30)	(5,9,23,30)
ANBFD	0.883	1.188	0.844
LRH	83.035	101.369	77.680

5　Conclusions

In this paper, we have addressed the anycast flow assignment problem in connection-oriented networks. We have developed a new heuristic algorithm based on the Lagrangean-relaxation method combined with subgradient optimization to iteratively solve the problem by a heuristic ANBFD proposed in [9]. Based on experimental results, our conclusion is that LRH outperforms ANBFD, especially for congested networks. The main overhead of LRH is much higher computational cost. However the considered problem is offline and consequently the decision time isn't the most important criterion in our study.

Acknowledgements. This work was supported by a research project of the Polish State Committee for Scientific Research carried out in years 2005-2007.

References

1. Awerbuch, B., Brinkmann, A., Scheideler C.: Anycasting in adversarial systems: routing and admission control. Lecture Notes in Computer Science, LNCS 2719 (2003), 1153-1168
2. Fratta, L., Gerla, M., Kleinrock, L.: The Flow Deviation Method: An Approach to Store-and-Forward Communication Network Design. Networks Vol. 3 (1973), 97–133
3. Gavish, B., Huntler, S.: An Algorithm for Optimal Route Selection in SNA Networks. IEEE Trans. Commun., Vol. COM-31, 10 (1983), pp. 1154-1160
4. Hao, F., Zegura, E., Ammar, M.: QoS routing for anycast communications: motivation and an architecture for DiffServ networks. IEEE Communication Magazine, 6 (2002), 48-56
5. Markowski, M., Kasprzak, A.: The web replica allocation and topology assignment problem in wide area networks: algorithms and computational results. Lectures Notes in Computer Science, Vol. 3483 (2005), 772-781
6. Peng, G.: CDN: Content Distribution Network. sunysb.edu/tr/rpe13.ps.gz, (2003)
7. Pióro, M., Medhi, D.: Routing, Flow, and Capacity Design in Communication and Computer Networks. Morgan Kaufman Publishers (2004)
8. Rosen, E., Viswanathan, A., Callon, R.: Multiprotocol Label Switching Architecture. RFC 3031 (2001)
9. Walkowiak, K.: Heuristic Algorithm for Anycast Flow Assignment in Connection-Oriented Networks. Lectures Notes in Computer Science, Vol. 3516 (2005), 1092-1095

Low Complexity Systolic Architecture
for Modular Multiplication over GF(2^m)

Hyun-Sung Kim[1] and Sung-Woon Lee[2]

[1] Kyungil University, School of Computer Engineering,
712-701, Kyungsansi, Kyungpook Province, Korea
[2] Tongmyong University, Dept. of Information Security,
Busan, Korea

Abstract. The modular multiplication is known as an efficient basic operation for public key cryptosystems over GF(2^m). Various systolic architectures for performing the modular multiplication have already been proposed based on a standard basis representation. However, they have high hardware complexity and long latency. Thereby, this paper presents a new algorithm and architecture for the modular multiplication in GF(2^m). First, a new algorithm is proposed based on the LSB-first scheme using a standard basis representation. Then, bit serial systolic multiplier is derived with a low hardware complexity and small latency. Since the proposed architecture incorporates simplicity, regularity, and modularity, it is well suited to VLSI implementation and can be easily applied to modular exponentiation architecture. Furthermore, the architecture will be utilized for the basic architecture of crypto-processor.

1 Introduction

The arithmetic operations in the finite field have several applications in error-correcting codes[1], cryptography[2, 3], digital signal processing[4], and so on. Information processing in such areas usually requires performing multiplication, inverse/division, and exponentiation. Among these operations, the modular multiplication is known as the basic operation for public key cryptosystems over GF(2^m) [2- 4]. Exponentiation is computed efficiently by the sequences of modular multiplications. And division and inverse can be regarded as a special case of exponentiation because $B^{-1} = B^{\,2^m-2}$ [5, 6].

Recently, three types of multipliers over GF(2^m) have been proposed that are easily realized using VLSI techniques. These are normal, dual, and standard basis multipliers, which have their own distinct features. The normal and dual basis multipliers need basis conversion, while the standard does not. In the following, we restrict our attention to the standard basis multiplier.

Numerous architectures for modular multiplication in GF(2^m) have been proposed in [7-10] over the standard basis. In 1984, Yeh et al. proposed two systolic array architectures with the LSB-first modular multiplication [7]. Wang et al. in [8] proposed two systolic architectures with the MSB-first fashion with less control problems as compared to [6]. Jain et al. proposed another multiplier architecture [9]. Its latency is smaller than those of other standard-basis multipliers, but there are broadcast lines in

V.N. Alexandrov et al. (Eds.): ICCS 2006, Part I, LNCS 3991, pp. 634 – 640, 2006.

the circuit. Wu et al. in [10] proposed bit-level systolic arrays with a simple hardware complexity with the MSB-first modular multiplication.

This paper proposes a new algorithm and its architecture for the modular multiplication with a bit-serial systolic architecture. The proposed algorithm supports the LSB-first scheme and the proposed architecture has a good time and area complexity compared to the previous multipliers.

2 Modular Multiplication

A finite field GF(2^m) has 2^m elements and it is assumed that all the (2^m-1) non-zero elements of GF(2^m) are represented using the standard basis. Let $A(x)$ and $B(x)$ be two elements in GF(2^m) and $F(x)$ be the primitive polynomial, where $A(x) = a_{m-1}x^{m-1} + a_{m-2}x^{m-2} + \cdots + a_1x + a_0,$ and $B(x) = b_{m-1}x^{m-1} + b_{m-2}x^{m-2} + \cdots + b_1x + b_0$, where a_i and b_i, $\in$ GF(2) ($0 \leq i \leq m$-1). A finite field of GF(2^m) elements is generated by a primitive polynomial of degree m over GF(2). Let $F(x)$ be an irreducible polynomial that generates the field and is expressed as $F(x) = x^m + f_{m-1}x^{m-1} + \cdots + f_1x + f_0$. If α is the root of $F(x)$, then $F(\alpha) = 0$, and $F(\alpha) \equiv \alpha^m = f_{m-1}\alpha^{m-1} + \cdots + f_1\alpha + f_0$, where $f_i \in$ GF(2) ($0 \leq i \leq m$-1). To compute the AB operation, the following equation is the common LSB-first algorithm

$$P = AB \bmod F(x)$$
$$= b_0[A\alpha^0 \bmod F(x)] + b_1[A\alpha \bmod F(x)] + b_2[A\alpha^2 \bmod F(x)] + \cdots \qquad (1)$$
$$+ b_{m-1}[A\alpha^{m-1} \bmod F(x)]$$

A new algorithm for the modular multiplication is derived from the equation 1 that is suitable for systolic array implementation. The modular reduction is necessary because of the operation $[A\alpha^i \bmod F(x)]$ on the i-th step. We just try to concentrate on the modular reduction from the first term of the above equation, $b_0 A \bmod F(x)$. The subsequent terms in the above equation are accumulated until reaching the end. The procedure of the new algorithm is as follows:

First,
$$P_1 = b_0[A\alpha^0 \bmod F(x)]$$
$$= [\sum_{k=0}^{m-1} a_k b_{m-1}\alpha^k]\alpha^0 \bmod F(x) \qquad (2)$$

However, in this step we can overlap the modular multiplication of $[A\alpha \bmod F(x)]$ for the next step operations. For the better understanding, we will use the symbol $A^{(i)}$ for the i-th snapshot for A, which has the value of $A\alpha^i \bmod F(x)$. Thereby, our algorithm computes two operations as follows

$$\begin{cases} P^{(1)} = b_0 A^{(0)} = [\sum_{k=0}^{m-1} a^{(0)}{}_k b_{m-1}\alpha^k]; \\ \\ A^{(1)} = A^{(0)}\alpha \bmod F(x) = [\sum_{k=0}^{m-1} a^{(0)}{}_k\alpha^k]\alpha \bmod F(x) = [\sum_{k=0}^{m-1} a^{(1)}{}_k\alpha^k]; \end{cases} \qquad (3)$$

where

$$\begin{cases} p_k^{(1)} = a_k^{(0)} b_{m-1} \ (k = 0, \cdots , m\text{-}1); \\ a_k^{(1)} = a_{m-1}^{(0)} f_k + a_{k-1}^{(0)} \ (k = 1, \cdots , m\text{-}1); \\ a_0^{(1)} = a_{m-1}^{(0)} f_0 \ ; \end{cases}$$

(4)

In the general case,

$$\begin{cases} P^{(i)}= P^{(i-1)} +b_{i-1}A^{(i-1)} = [\sum_{k=0}^{m-1} (p^{(i-1)}{}_k + a^{(i-1)}{}_k b_{m-1})\alpha^k]; \\ A^{(i)} = A^{(i-1)}\alpha \bmod F(x) =[\sum_{k=0}^{m-1} a^{(i-1)}{}_k \alpha^k]\alpha \bmod F(x)=[\sum_{k=0}^{m-1} a^{(i)}{}_k \alpha^k] \ ; \end{cases}$$

(5)

where

$$\begin{cases} p_k^{(i)} = p_k^{(i-1)} + a_k^{(i-1)} b_{m-1} \ (k = 0, \cdots , m\text{-}1); \\ a_k^{(i)} = a_{m-1}^{(i-1)} f_k + a_{k-1}^{(i-1)} \ (k = 1, \cdots , m\text{-}1); \\ a_0^{(i)} = a_{m-1}^{(i-1)} f_0 \ ; \end{cases}$$

(6)

Finally,

$$P^{(m)}= P^{(m\text{-}1)} +b_{m\text{-}1}A^{(m\text{-}1)} = [\sum_{k=0}^{m-1} a^{(i-1)}{}_k b_{m-1} \alpha^k];$$

(7)

where $p_k^{(m)} = p_k^{(m-1)} + a_k^{(m-1)} b_{m-1} \ (k = 0, \cdots , m\text{-}1).$

Thus the multiplication of two elements A and B in GF(2^m) can be computed using the above new recursive algorithm. From the algorithm, we can derive an efficient systolic multiplier by following the procedures in [8-9].

3   Systolic Modular Multiplier

This section proposes a serial systolic multiplier. Fig.1 shows the dependence graph for our new multiplication over GF(2^4). The inputs A and F enter the array in parallel from the top row, while B is from the leftmost column. The output P is transmitted from the bottom row of the array in parallel.

Fig 2 (a) shows a basic cell in Fig. 1 for the general case where the circuit function is primarily governed by the following recurrence equation:

$$\begin{cases} p_k^{(i)} = p_k^{(i-1)} + a_k^{(i-1)} b_{m-1} \ (k = 0, \cdots , m\text{-}1); \\ a_k^{(i)} = a_{m-1}^{(i-1)} f_k + a_{k-1}^{(i-1)} \ (k = 1, \cdots , m\text{-}1); \\ a_0^{(i)} = a_{m-1}^{(i-1)} f_0 \ ; \end{cases}$$

(8)

where the k-th bit ($p_k^{(i)}$) of P_i is the intermediate result of the product. Fig. 2 (b) shows a processing element for the case that the cell is located in the last row.

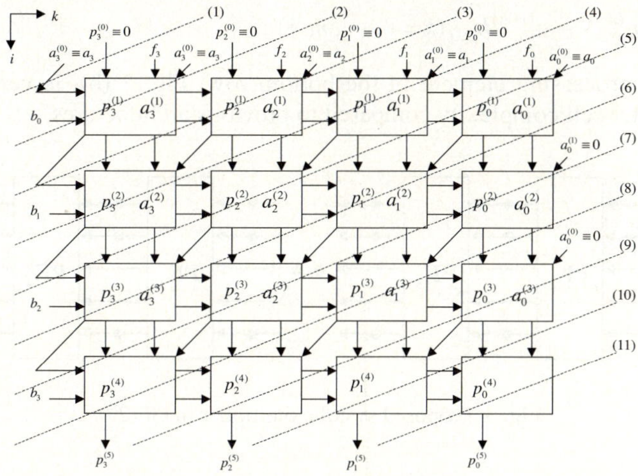

Fig. 1. Dependence graph in GF(2^4)

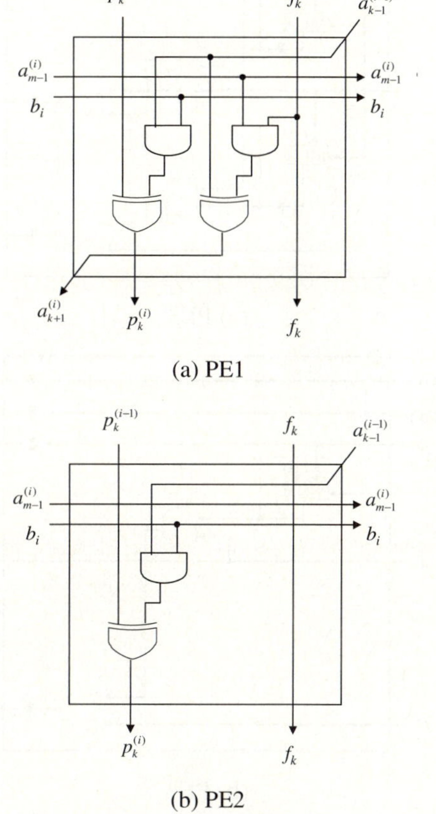

(a) PE1

(b) PE2

Fig. 2. Processing elements

$$p_k^{(m)} = p_k^{(m-1)} + a_k^{(m-1)} b_{m-1} \ (k = 0, \cdots, m\text{-}1) \ ; \tag{9}$$

Note that the processing element in the bottom row, Fig. 2 (b), is very simple and reduces the total cell complexity compared to previous architectures.

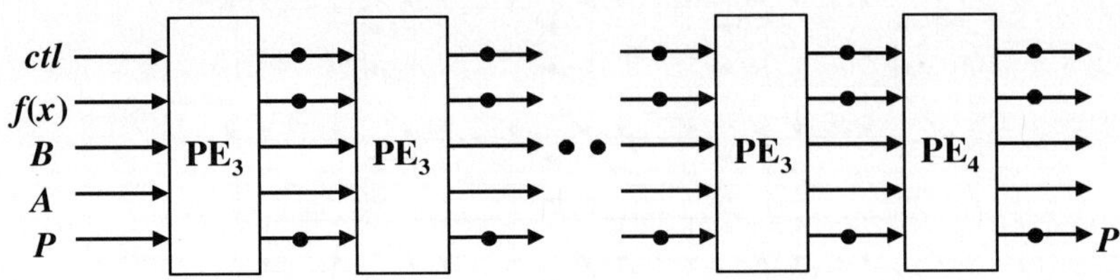

Fig. 3. Proposed systolic multiplier in GF(2^m)

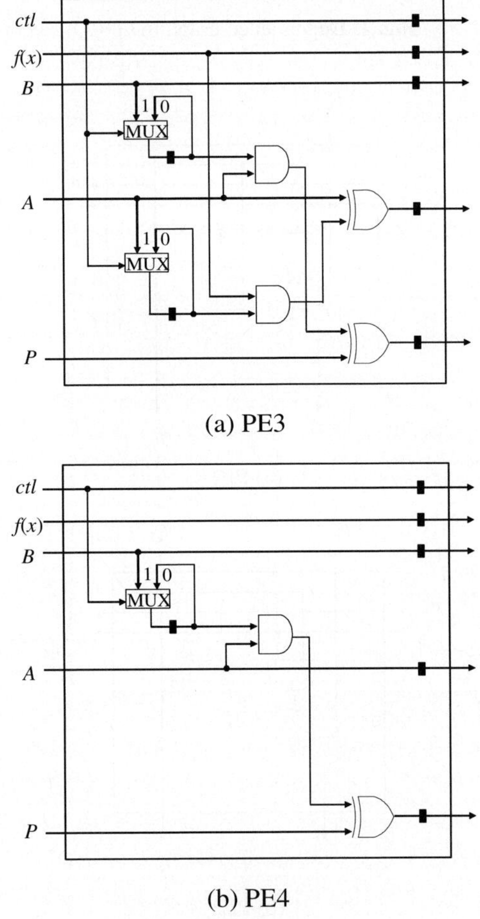

(a) PE3

(b) PE4

Fig. 4. Processing elements

An optimal systolic array among the several systolic arrays is derived by the projection vector $[1,0]^T$ and schedule vector $[2,1]^T$ from the dependency graph in Fig. 1. Fig. 3 shows the proposed bit-serial systolic multiplier. The dot ("•") on the data flow means the buffer for one time step delay. Let $I = (I_{ctl}, I_{f(x)}, I_B, I_A, I_P)$ denote the layout of input values for the systolic array. $I_{ctl}, I_{f(x)}, I_B, I_A,$ and I_P are data sequences of control signals, $f(x)$, B, A, and P, respectively. Each sequence is as follows:

$$I_{ctl} = \{1, 0, 0, \dots, 0\}$$
$$I_{f(x)} = \{\,\bullet\,, f_{m-1}, f_{m-2}, \dots, f_0\}$$
$$I_B = \{b_{m-1}, b_{m-2}, \dots, b_0\}$$
$$I_A = \{a_{m-1}, a_{m-1}, a_{m-2}, \dots, a_0\}$$
$$I_P = \{0, 0, 0, \dots, 0\}$$

The " • " of the digit sequences means a time delay of input data. Only the first bit among m bits of the control signal has the value 1 as the values of a and b should be held for m clock times.

Fig 4 (a) shows a basic cell in Fig. 3 for the general case where the circuit function is primarily governed by the equation 8. Fig. 4 (b) shows a processing element for the case that the cell is located in the last row using the equation 9.

The structure of a systolic array is similar to Yeh $et\ al.$ in [4] except the input scheduling and the structure of the last processing element. Since the horizontal path of each cell only requires two delay elements, except for the last cell, the latency is $3m$-2.

Table 1. Comparison of the bit-serial systolic architectures in GF(2^m)

Circuit Item	Yeh et al. [7]	Wang et al. [8]	Proposed	
No. of cells	m	m	m	
Function	$AB+C$	$AB+C$	$AB+C$	
Throughput	$1/\,m$	$1/\,m$	$1/\,m$	
Latency	$3m$	$3m$	$3m$-2	
Critical path	$T_{AND} + T_{XOR\text{-}2}$	$T_{AND} + T_{XOR\text{-}3}$	$T_{AND} + T_{XOR\text{-}2}$	
Cell complexity	2 AND 2 XOR 7 latches 2 MUX	2 AND 1 XOR$_3$ 7 latches 2 MUX	PE3	2 AND 2 XOR 7 latches 2 MUX
			PE4	1 AND 1 XOR 2 latches 1 MUX
Algorithm fashion	LSB	MSB	LSB	

4 Analysis

Our multiplier was simulated and verified using the ALTERA MAX+PLUS II simulator. Table 1 shows comparisions between the proposed architecture and the related circuits. We will give a comparison of systolic architectures with Yeh et al.'s in [7] and Wang et al.'s in [8].

Before the comparison, it was assumed that AND and XOR_i denote a 2-input AND gate and i-input XOR, respectively. T_{AND} and T_{XORi} are the propagation delay of a 2-input AND gate and i-input XOR gate, respectively. Latches are 1-bit latch and MUX is a 2-input 1-output switch. Table 1 shows that our multiplier has a good area and time complexity compared with the previous architectures.

5 Conclusion

This paper has explored a new algorithm for computing the modular multiplication into a low-complexity systolic architecture in $GF(2^m)$. A comparison between related systolic architectures reveals that the new systolic architecture has lower property than the conventional architectures for the hardware complexity and the latency. Furthermore, it can be used as the basic architecture for computing an inverse/division operation. Moreover, the architecture has a simplicity, regularity, and modularity. Thereby, it is well suited to VLSI implementation and it can be easily utilized for the crypto-processor chip design.

References

1. W.W.Peterson and E.J.Weldon, *Error-correcting codes*, MIT Press, MA, 1972.
2. D.E.R.Denning, *Cryptography and data security*, Addison-Wesley, MA, 1983.
3. A.Menezes, *Elliptic Curve Public Key Cryptosystems*, Kluwer Academic Publishers, Boston, 1993.
4. I.S.Reed and T.K.Truong, "The use of finite fields to compute convolutions," *IEEE Trans. Inform. Theory*, 21, 1975, pp.208-213.
5. H.S.Kim, *Bit-Serial AOP Arithmetic Architecture for Modular Exponentiation*, PhD. Thesis, Kyungpook National University, 2002.
6. S.W.Wei, "VLSI architectures for computing exponentiations, multiplicative inverses, and divisions in $GF(2^m)$," *IEEE Trans. Circuits and Systems,* 44, 1997, pp.847-855.
7. C.S.Yeh, S.Reed, and T.K.Truong, "Systolic multipliers for finite fields $GF(2^m)$," *IEEE Trans. Comput.,* vol.C-33, Apr. 1984, pp.357-360.
8. C.L.Wang and J.L.Lin, "Systolic Array Implementation of Multipliers for Finite Fields $GF(2^m)$," *IEEE Trans. Circuits and Systems*, vol.38, July 1991, pp796-800.
9. S. K. Jain and L. Song, "Efficient Semisystolic Architectures for finite field Arithmetic," *IEEE Trans. on VLSI Systems*, vol. 6, no. 1, Mar. 1998.
10. C.W.Wu and M.K.Chang, "Bit-Level Systolic Arrays for Finite-Field Multiplications," *Journal of VLSI Signal Processing*, vol. 10, pp.85-92, 1995.
11. S.Y.Kung : 'VLSI Array Processors,' Prentice-Hall, 1987.
12. K.Y.Yoo : 'A Systolic Array Design Methodology for Sequential Loop Algorithms,' PhD. thesis, Rensselaer Polytechnic Institute, New York, 1992.

A Generic Framework for Local Search:
Application to the Sudoku Problem

T. Lambert[1,2], E. Monfroy[1,3,*], and F. Saubion[2]

[1] LINA, Université de Nantes, France
Firstname.Name@lina.univ-nantes.fr
[2] LERIA, Université d'Angers, France
Firstname.Name@univ-angers.fr
[3] Universidad Santa María, Valparaíso, Chile
Firstname.Name@inf.utfsm.cl

Abstract. Constraint Satisfaction Problems (CSP) provide a general framework for modeling many practical applications. CSPs can be solved with complete methods or incomplete methods. Although some frameworks has been designed to formalized constraint propagation, there are only few studies of theoretical frameworks for local search. In this paper, we are concerned with the design of a generic framework to model local search as the computation of a fixed point of functions and to solve the Sudoku problem. This work allows one to simulate standard strategies used for local search, and to design easily new strategies in a uniform framework.

1   Introduction

Sudoku literally means *single number* in Japanese and has reached recently an international popularity. The success of this great puzzle game probably comes from the simplicity of its rules : place digits between 1-9 on a 9×9 grid such that each digit appears once in each row, column and each 3x3 sub-grid. This problem can obviously be considered as a Constraint Satisfaction Problem (CSP), usually defined by a set of variables associated to domains of possible values and by a set of constraints.

Local search techniques [1] have been successfully applied to various combinatorial optimization problems (scheduling, timetabling, transportation ...). In the CSP solving context, local search algorithms are used either as the main resolution technique or in cooperation with other resolution processes (e.g., constraint propagation) [3, 6]. Unfortunately, the definitions and the behaviors of these algorithms are often strongly related to specific implementations and problems.

Usual constraint propagation techniques have already been used to solve Sudoku and in this paper, we are interested in using various local search techniques to solve this problem. Since we want to test different resolution heuristics, our purpose is to use a generic framework based on basic functions in order to provide uniform framework which help to better understand existing local search algorithms and to design new ones.

---

* The author has been partially supported by the Chilean National Science Fund through the project FONDECYT N°1060373.

V.N. Alexandrov et al. (Eds.): ICCS 2006, Part I, LNCS 3991, pp. 641–648, 2006.

In [7] we have extended the mathematical framework of K.R. Apt [2] to take into account hybridizations of constraint propagation with local search algorithms. The purpose of this paper is to focus on the modeling of basic local search processes and then to improve this previous work by providing a more comprehensive definition to local search algorithms. To obtain a finer definition of local search, we propose a specific computation structure and define the basic functions that will be used iteratively on this structure to create a local search process. Processes are abstracted at the same level by some homogeneous functions called reduction functions. The result of local search is then computed as a fixed point of this set of functions.

The paper is organized as follows : in Section 2 we recall basic definitions related to CSP and local search algorithms. In Section 3 we describe our framework by defining its main components. In Section 4, we provide the operational semantics as the computation of a fixed point of functions over the ordered structure, which corresponds to the instantiation of the generic iteration algorithm proposed by K.R. Apt. Popular local search algorithms and strategies are then designed and applied to solve Sudoku in Section 5 before concluding in Section 6.

2   Solving CSP with Local Search

A CSP is a tuple (X, D, C) where $X = \{x_1, \cdots, x_n\}$ is a set of variables taking their values in their respective domains $D = \{D_1, \cdots, D_n\}$. A constraint $c \in C$ is a relation $c \subseteq D_1 \times \cdots \times D_n$. In order to simplify notations, D will also denote the Cartesian product of D_i and C the union of its constraints. A tuple $d \in D$ is a solution of a CSP (X, D, C) if and only if $\forall c \in C, d \in c$. In this paper, we always consider finite domains.

Given an optimization problem (which can be minimizing the number of violated constraints and thus trying to find a solution to the CSP), local search techniques [1] aim at exploring the search space, moving from a sample to one of its neighbors. These moves are guided by a fitness function that evaluates the benefit of such a move in order to reach a local optimum.

For the resolution of a CSP (X, D, C), the search space can be often defined as the set of possible tuples of $D = D_1 \times \cdots \times D_n$ and the neighborhood is a mapping $\mathcal{N} : D \to 2^D$. This neighborhood function defines indeed the possible moves from a sample of D to one of its neighbors and therefore fully defines the exploration landscape. The fitness (or evaluation) function $eval$ is related to the notion of solution and can be defined as the number of constraints c that are not satisfied by the current sample. The problem to solve is then a minimization problem. Given a sample $d \in D$, two basic cases can be identified in order to continue the exploration of D: intensification (choose $d' \in \mathcal{N}(d)$ such that $eval(d') < eval(d)$) and diversification(choose any other neighbor d'). Any local search algorithm is based on the management of these basic heuristics by introducing specific control features. Here, we abstract moves and neighborhood by functions computing over a given ordered structure.

3   A Computational Framework

In this section, we define the computation structure we will use to represent local search states, together with the functions that will be required to model local search as a fixed point computation.

3.1   The Computation Structure

As we have seen, local search acts usually on a structure which corresponds to points of the search space. Here, we propose a more general and abstract definition based on the notion of sample, already suggested.

Definition 1 (Sample). *Given a CSP (X, D, C), a sample function is a function $\varepsilon : D \to 2^D$. By extension, $\varepsilon(D)$ denotes the set $\{\varepsilon(d) \mid d \in D\}$.*

Generally, $\varepsilon(d)$ is restricted to d and $\varepsilon(D) = D$, but it can also be a scatter of tuples around d, or a box of tuples covering d. Indeed, the search space D is abstracted by $\varepsilon(D)$ to be used by the local search. Note that in any case, $\varepsilon(D)$ is finite since we consider D to be finite.

The general process of local search can be abstracted by two stages: generate the neighborhood of the current sample and move from the current sample to one of the previously computed neighbor.We define then a local search path as a tuple of samples which represents the path already constructed by the local search process.

Definition 2 (Local Search Path). *A local search path p is a finite sequence $(s_1, \cdots, s_n)$ such that $\forall 1 \leq i \leq n, s_i \in \varepsilon(D)$.*

We denote by $P_{\varepsilon(D)}$ the set of all possible local search paths on $\varepsilon(D)$. Given a tuple $p = (s_1, \cdots, s_n) \in \varepsilon(D)^n$, and an element $s \in \varepsilon(D)$, we denote $p' = p \oplus s$ the tuple $(s_1, \cdots, s_n, s)$. To simplify notation, we denote $s \in p$ the fact that a sample s is a component of a path p.

We now define orderings on this structure. From a practical point of view, a local search process aims at building a finite path whose length is either determined by the fact that a solution has been reached or that a maximum number of iterations have been performed. Therefore, our orderings take into account these two main aspects of local search.

Definition 3 (Ordering on paths). *Given $p=(s_1, \cdots, s_n)$ and $p'=(s'_1, \cdots, s'_m)$ two paths of $P_{\varepsilon(D)}$, $p \sqsubseteq p'$ iff $s'_m \in Sol_{\varepsilon(D)}$ or $s_n \notin Sol_{\varepsilon(D)}$ and $m \geq n$.*

In order to handle simultaneously paths and neighborhoods, we first define the notion of local search configuration.

Definition 4 (Local Search Configuration). *A local search configuration $\mathcal{C}_{\varepsilon(D)}$ is a pair (p, V) where $p = (s_1, \cdots, s_n)$ is a local search path and $V \subseteq 2^{\varepsilon(D)}$.*

$\mathcal{C}$ is the set of all configurations. The ordering on path is obviously extended to configurations taking into account set inclusion for the neighborhood.

3.2   Reduction Function Definitions

Our definition of reduction functions is based on K.R. Apt´s framework [2].

Definition 5 (Reduction function on a structure). *Given a partial ordering $(D, \sqsubseteq)$, a reduction function f is a function from D to D which satisfies the following properties:*

- $\forall x \in D, x \sqsubseteq f(x)$ *(inflationary)*
- $\forall x, y \in D, x \sqsubseteq y \Rightarrow f(x) \sqsubseteq f(y)$ *(monotonic)*

As previously described, the basic components to build a single local search path are move and neighborhood computation.

Definition 6 (Move Function). *A move function is a function $\mu : \mathcal{C} \to \mathcal{C}$ such that $\mu(p, V) = (p', \emptyset)$ where $p' = p \oplus s$ with $s \in V$ if $p = p'' \oplus s'$ and $s' \notin Sol_{\varepsilon(D)}$ and $V \neq \emptyset, p' = p$ otherwise.*

Definition 7 (Neighborhood Function). *A neighborhood function is a function $\nu : \mathcal{C} \to \mathcal{C}$ with $\nu(p, V) = (p, V \cup V')$ such that $V' \subseteq \varepsilon(D)$ and $V' \cap V = \emptyset$.*

Move and neighborhood functions are reduction functions.

3.3   Restricting Functions to Match a Practical Framework

As mentioned before, we must remark that stop conditions are always added in local search algorithms in order to insure termination. These conditions are basically based on a maximum number of allowed search steps or on a notion of solution (if this notion is available). In the context of CSP solving, this notion of solution has been clearly defined and is taken into account in the definition of our computation structure. Here the maximum number of operations will be defined by a maximum number σ of steps in each path (maximal length of a path). Given a move or neighborhood function $f: \mathcal{C} \to \mathcal{C}$, we define its restriction $f^\sigma: \mathcal{C} \to \mathcal{C}$ as: $f^\sigma(p, V) = f(p, V)$ if $|p| \leq \sigma$ and $f^\sigma(p, V) = (p, V)$ otherwise.

We must insist on the fact that, after these practical restrictions, we only consider P as the set of all possible local search paths of size σ. Therefore, this set is finite and $\mathcal{C}$ is also finite. Note that only the restriction concerning σ is required to insure finiteness of the structures which is needed to fit the generic iteration framework (section 4).

4   Local Search as a Fixed Point of Reduction Functions

In our framework, local search will be described as a fixed point computation on the previously ordered structure.

4.1 Chaotic Iterations

In [2], K.R. Apt proposed the chaotic iteration framework, a general theoretical framework for computing limits of iterations of a finite set of functions over a partially ordered set. In this paper, we do not recall all the theoretical results of K.R. Apt, but we just give the **GI** algorithm for computing fixed point of functions. Consider a finite set F of functions, and d an element of a partially ordered set $\mathcal{D}$.

GI: Generic Iteration Algorithm

$d := \perp$;
$G := F$;
While $G \neq \emptyset$ do
       choose $g \in G$;
       $G := G - \{g\}$;
       $G := G \cup update(G, g, d)$;
       $d := g(d)$;
Endwhile

where $\perp$ is the least element of the partial ordering $(\mathcal{D}, \sqsubseteq)$, G is the current set of functions still to be applied ($G \subseteq F$), and for all G, g, d the set of functions $update(G, g, d)$ from F is such that:

P1 $\{f \in F - G \mid f(d) = d \wedge f(g(d)) \neq g(d)\} \subseteq update(G, g, d)$.
P2 $g(d) = d$ implies that $update(G, g, d) = \emptyset$.
P3 $g(g(d)) \neq g(d)$ implies that $g \in update(G, g, d)$

Suppose that all functions in F are reduction functions as defined before and that $(\mathcal{D}, \sqsubseteq)$ is finite (note that finiteness is important as is has already been mentioned for our structure) then every execution of the **GI** algorithm terminates and computes in d the least common fixed point of the functions from F (see [2]).

We now use the **GI** algorithm to compute the fixed point of our functions. The algorithm is thus feed with:

- a set of move and neighborhood functions, that compose the set F,
- $\perp = \emptyset \in P_{\varepsilon(D)}$ to instantiate initial d,
- the ordering that we use is the ordering $\sqsubseteq$ on $P_{\varepsilon(D)}$.

In our context, the algorithm terminates and computes the least common fixed point of the functions from F, i.e., the result of the whole local search. Inspired by [2], the proof partially relies on an invariant $\forall f \in F - G, f(d) = d$ of the "while" loop in the algorithm. This invariant is preserved by our characterization of the update function. Moreover, since we keep a finite partial ordering and a set of monotonic and inflationary functions, the results of K.R. Apt can be extended here.

5   Solving the Sudoku

As mentioned in the introduction, the Sudoku problem consists in filling a 9×9 grid so that every row, every column, and every 3×3 box contains the digits from 1 to 9. Although Sudoku, when generalized to n^2 x n^2 grids to be filled in by numbers from 1 to n^2 is NP-complete, the popular 9×9 grid with 3×3 regions is not difficult to solve with a simple computer program. Therefore, in order to increase the difficulty, we consider 16×16 grids (published under the name "super Sudoku"), 25×25 and 36×36 grids. On this problem, we will show that our generic framework allows us to easily define local search algorithms and to combine and compare them.

5.1   CSP Model

Consider a n^2 x n^2 problem, an instinctive formalization considers a set of n^4 variables whose correspond, to all the cells to fill in. Using this, the set of related constraints is defined by AllDiff global contraints [8] representing : all digits appears only once in each row, once in each column and once in each $n \times n$ square the grid has been subdivided.

This model formalizes Sudoku problems in such a way that the CSP to solve has, for instance, 1296 variables and 108 constraints for the 36×36 grid. Usual approaches consider grids with a subset of cells already filled in and complete it (i.e., the real game for players), or aim at generating grids such that only one solution can be reach. In this paper we consider empty grids and our goal is to generate full ones. Even if Constraint Programming is useful for small problems, complete techniques are intractable for bigger instances. Indeed, grids with already filled positions (i.e., instantiated variables) are easier and we want to attack larger and more difficult problem since we consider local search strategies because of their ability in solving large scale problems.

We consider only grids such that all different digits appear once in each $n \times n$ sub-square. This corresponds to enforce some constraints of the problem directly in the encoding structure and it implies a restriction of the neighborhood to all possible swaps between cells of a same square. The *eval* function is related to the notion of solution and is defined for each constraint alldiff c as a *cost* of violation (number of assignments to correct to satisfy the constraint). It is equal to 0 if the constraint is satisfied.

Concerning LS methods, on one hand, Tabu search (TS) [4] has been successfully applied to solve CSP [5]. Basically, this algorithm forbids moving to a sample that was visited less than l steps before. To this end, the list of the last l visited samples is memorized. On the other hand, we consider a basic descent technique with random walks RW where random moves are performed according to a certain probability p. According to our model, we only have now to design functions of the generic algorithm of Section 4.1 to model strategies. Neighborhood functions are functions $C \to C$ such that $(p, V) \mapsto (p, V \cup V')$ with different conditions:

$$FullNeighbor : V' = \{s \in D | s \notin V\}$$
$$TabuNeighbor : V' = \{s \in D | \not\exists k, \ n - l \leq k \leq n, s_k = s\}$$
$$DescentNeighbor : p = (s_1, \ldots, s_n) \text{ and } V' = s \subset D \text{ s.t. } \not\exists s' \in V \text{ s.t } eval(s') <$$
$$eval(s_n)$$

Move functions are functions $C \to C$ such that $(p, V) \mapsto (p', \emptyset)$ with different conditions:

$$BestMove : p' = p \oplus s' \text{ and } eval(s') = min_{s'' \in V} eval(s'').$$
$$ImproveMove : p = p'' \oplus s_n \text{ and } p' = p \oplus s \text{ s.t. } eval(s') < eval(s_n).$$
$$RandomMove : p' = p \oplus s' \text{ and } s' \in V.$$

We can precise here the input set of function F for algorithm GI.

$$Tabusearch : \{TabuNeighbor; BestNeighBor\}$$
$$Randomwalk : \{FullNeighbor; BestNeighBor; RandomNeighbor\}$$
$$TabuSearch + Descent : \{TabuNeighbor; DescentNeighbor$$
$$; ImproveNeighBor; BestNeighBor\}$$
$$Randomwalk + Descent : \{FullNeighbor; BestNeighBor$$
$$; RandomNeighbor; DescentNeighbor$$
$$; ImproveNeighBor\}$$

Remark that the different algorithms correspond here to different sets of input functions and different behaviours of the *choose* function in the GI algorithm. This choose function select alternatively neighborhood function and move functions. For the Random Walk algorithm, given a probability parameter p, we have to introduce a quota of p *BestMove* functions and $1 - p$ *RandomMove* used in GI. Concerning Tabu Search, we use here a *TabuNeighbor* with $l = 10$ and *BestMove* functions to built our Tabu Search algorithm. At last, we combine a descent strategy by adding *DescentNeighbor* and *ImproveMove* to the previous sets in order to design algorithms in which a Descent is first applied in order to reach more quickly to a good configuration.

5.2 Experimentation Results

In Table 1. we compare results of the tabu search and random walks associated with descent on different instances of Sudoku problem. Note that we have evaluated the difficulty of the problem thanks to a classic complete method with propagation and split, we obtained a more than one day cpu time cost for a 36×36 grid. At the opposite, by a simple formalization of the problem and thanks to a function application model, we are able to reach to a solution with classical local search algorithms and this starting from an empty grid. For each method and for each instance, 2000 runs have been performed except for 36×36 problem, 500 runs. Results we obtain by adding descent in a Tabu Search or in a Random Walk method, allows us to reduce the computation time to reach a solution. We may remark that the hybrid strategies combining several move and neighborhood functions provide better results. Our framework allows us to tune easily the balance between the different basic functions that characterize basic solving strategies and thus to design various algorithms in a single generic algorithm, which is not so clear if the methods are considered from a pure algorithmic point of view.

Table 1. Results of Sudoku problem by different search approaches

n^2 x n^2	TabuSearch			RandomWalk		
	16x16	25x25	36x36	16x16	25x25	36x36
cpu time Avg	3,14	115,08	3289,8	3,92	105,22	2495
deviations	1,28	52,3	1347,4	1,47	49,3	1099
mvts Avg	405	3240	22333	443	2318	13975
n^2 x n^2	Descent + TabuSearch			Descent +RandomWalk		
	16x16	25x25	36x36	16x16	25x25	36x36
cpu time Avg	2,34	111,81	2948	2,41	82,94	2455
deviations	1,42	55,04	1476	1,11	36,99	1092
mvts Avg	534	3666	20878	544	2581	14908

6 Conclusion

In this paper, we have proposed a framework for modeling CSP resolution with local search techniques. This framework provides a computational model as the computation of a fixed point of functions over a partial ordering, inspired the initial works of K.R. Apt [2]. The generic algorithm has been used to solve the Sudoku problem and allow us to compare different resolution approach in a very uniform framework where strategies can be easily designed.

This framework could be extended in order to include complete resolution mechanisms (constraint propagation, domain splitting) and even other meta-heuristics such as evolutionary algorithms. It could also be used for experimental studies as it provides a uniform description framework for various methods in an hybridization context.

References

1. E. Aarts and J.K. Lenstra, editors. *Local Search in Combinatorial Optimization.* John Wiley and Sons, 1997.
2. K. Apt. From chaotic iteration to constraint propagation. In ICALP '97, number 1256 in LNCS, pages 36–55. Springer, 1997. invited lecture.
3. F. Focacci, F. Laburthe, and A. Lodi. Local search and constraint programming. In Handbook of Metaheuristics, Kluwer, 2002.
4. F. Glover and M. Laguna. *Tabu Search.* Kluwer Academic Publishers, 1997.
5. J.K. Hao and R. Dorne. Empirical studies of heuristics local search for constraint solving. In *CP'96 LNCS 1118*, pages 184–208, 1996.
6. N. Jussien and O. Lhomme. Local search with constraint propagation and conflict-based heuristics. *Artificial Intelligence*, 139(1):21–45, 2002.
7. E. Monfroy, F. Saubion, and T. Lambert. On hybridization of local search and constraint propagation. In ICLP 2004, number 3132 in LNCS, pages 180–194, 2004.
8. J.C. Régin. A filtering algorithm for constraint of difference in csps. In *National Conference of Artificial Intelligence*, pages 362–367, 1994.

C-Means Clustering Applied to Speech Discrimination

J.M. Górriz[1], J. Ramírez[1], I. Turias[2],
C.G. Puntonet[3], J. González[3], and E.W. Lang[4]

[1] Dpt. Signal Theory, Networking and communications, University of Granada, Spain
gorriz@ugr.es
http://www.ugr.es/~gorriz
[2] Dpt. Computer Science, University of Cádiz, Spain
[3] Dpt. Computer Architecture and Technology, University of Granada, Spain
[4] AG Neuro- und Bioinformatik, Universität Regensburg, Deutschland

Abstract. An effective voice activity detection (VAD) algorithm is proposed for improving speech recognition performance in noisy environments. The proposed speech/pause discrimination method is based on a hard-decision clustering approach built over a set of subband log-energies. Detecting the presence of speech frames (a new cluster) is achieved using a basic sequential algorithm scheme (BSAS) according to a given "distance" (in this case, geometrical distance) and a suitable threshold. The accuracy of the Cl-VAD algorithm lies in the use of a decision function defined over a multiple-observation (MO) window of averaged subband log-energies and the modeling of noise subspace into cluster prototypes. In addition, time efficiency is also reached due to the clustering approach which is fundamental in VAD real time applications, i.e. speech recognition. An exhaustive analysis on the Spanish SpeechDat-Car databases is conducted in order to assess the performance of the proposed method and to compare it to existing standard VAD methods. The results show improvements in detection accuracy over standard VADs and a representative set of recently reported VAD algorithms.

1 Introduction

The emerging wireless communication systems are demanding increasing levels of performance of speech processing systems working in noise adverse environments. These systems often benefits from using voice activity detectors (VADs) which are frequently used in such application scenarios for different purposes. Speech/non-speech detection is an unsolved problem in speech processing and affects numerous applications including robust speech recognition [1, 2], discontinuous transmission [3, 4], real-time speech transmission on the Internet [5] or combined noise reduction and echo cancellation schemes in the context of telephony [6]. The speech/non-speech classification task is not as trivial as it appears, and most of the VAD algorithms fail when the level of background noise increases. During the last decade, numerous researchers have developed different

V.N. Alexandrov et al. (Eds.): ICCS 2006, Part I, LNCS 3991, pp. 649–656, 2006.

650   J.M. Górriz et al.

strategies for detecting speech on a noisy signal [7] and have evaluated the influence of the VAD effectiveness on the performance of speech processing systems [8]. Most of them have focussed on the development of robust algorithms with special attention on the derivation and study of noise robust features and decision rules [9, 10, 11, 7]. The different approaches include those based on energy thresholds, pitch detection, spectrum analysis, zero-crossing rate, periodicity measure or combinations of different features.

The speech/pause discrimination can be described as an unsupervised learning problem. Clustering is one solution to this case where data is divided into groups which are related "in some sense". Despite the simplicity of clustering algorithms, there is an increasing interest in the use of clustering methods in pattern recognition [15], image processing [16] and information retrieval [17, 18]. Clustering has a rich history in other disciplines [12] such as machine learning, biology, psychiatry, psychology, archaeology, geology, geography, and marketing. Cluster analysis, also called data segmentation, has a variety of goals. All related to grouping or segmenting a collection of objects into subsets or "clusters" such that those within each cluster are more closely related to one another than objects assigned to different clusters. Cluster analysis is also used to form descriptive statistics to ascertain whether or not the data consist of a set of distinct subgroups, each group representing objects with substantially different properties.

The essay is organized as follows: in section 2 we describe a suitable signal model to detect the presence of speech frames in noisy environments. In the following section 3, we apply cluster analysis to form "descriptive statistics" transforming the noise sample set into a soft-noise model with low dimensional feature. A complete experimental framework is shown in section 4. Finally we state some conclusions and acknowledgements in the last part of the paper.

2   A Suitable Model for VAD

Let $x(n)$ be a discrete time signal. Denote by y_j a frame of signal containing the elements:

$$\{x_i^j\} = \{x(i + j \cdot D)\}; \quad i = 1 \ldots L \tag{1}$$

where D is the window shift and L is the number of samples in each frame. Consider the set of $2 \cdot m + 1$ frames $\{y_{l-m}, \ldots y_l \ldots, y_{l+m}\}$ centered on frame y_l, and denote by $Y(s, j)$, $j = l - m, \ldots l \ldots, l + m$ its Discrete Fourier Transform (DFT) resp.:

$$Y_j(\omega_s) \equiv Y(s, j) = \sum_{n=0}^{N_{FFT}-1} x(n + j \cdot D) \cdot exp\left(-j \cdot n \cdot \omega_s\right). \tag{2}$$

where $\omega_s = \frac{2\pi \cdot s}{N_{FFT}}$, $0 \leq s \leq N_{FFT} - 1$ and N_{FFT} is the number of points or resolution used in the DFT (if $N_{FFT} > L$ then the DFT is padded with zeros). The log-energies for the l-th frame, $E(k, l)$, in K subbands ($k = 0, 1, ..., K - 1$), are computed by means of:

$$E(k, l) = \log \left(\frac{K}{N_{FFT}} \sum_{s=s_k}^{s_{k+1}-1} |Y(s, l)|^2 \right) \tag{3}$$

$$s_k = \left\lfloor \frac{N_{FFT}}{2K} k \right\rfloor \quad k = 0, 1, ..., K - 1,$$

where an equally spaced subband assignment is used and $\lfloor \cdot \rfloor$ denotes the "floor" function. Hence, the signal log-energy is averaged over K subbands obtaining a suitable representation of the input signal for VAD [19], the observation vector at frame l, $\mathbf{E}(l) = (E(0, l), \ldots, E(K - 1, l))^T$. The VAD decision rule is formulated over a sliding window consisting of $2m+1$ observation (feature) vectors (log-energies) around the frame for which the decision is being made (l), as we will show in the following sections. This strategy consisting on "long term information" provides very good results using several approaches for VAD such as [13, 14] etc.

3   C-Means Clustering over the Feature Vectors

C-means clustering is a method for finding clusters and cluster centers in a set of unlabeled data [20]. The number of cluster centers (prototypes) C is a priori known and the C-means iteratively moves the centers to minimize the total within cluster variance. Given an initial set of centers the C-means algorithm alternates two steps [21]:

- for each cluster we identify the subset of training points (its cluster) that is closer to it than any other center;
- the means of each feature for the data points in each cluster are computed, and this mean vector becomes the new center for that cluster.

3.1   Noise Modeling

In our algorithm, this procedure is applied to a set of initial pause frames (log-energies) in order to characterize the noise space. Then we call this set of clusters noise prototypes [1]. Each observation vector ($\mathbf{E}$ from equation 3) is uniquely labeled, by the integer $i \in \{1, \ldots, N\}$, and uniquely assigned to a prespecified number of prototypes $C < N$, labeled by an integer $c \in \{1, \ldots, C\}$. The dissimilarity measure between observation vectors is the squared Euclidean distance:

$$d(\mathbf{E}_i, \mathbf{E}_j) = \sum_{k=0}^{K-1} (E(k, i) - E(k, j))^2 = ||\mathbf{E}_i - \mathbf{E}_j||^2 \tag{4}$$

and the loss function to be minimized is defined as:

$$W(C) = \frac{1}{2} \sum_{c=1}^{C} \sum_{C(i)=c} \sum_{C(j)=c} ||\mathbf{E}_i - \mathbf{E}_j||^2 = \sum_{k=1}^{C} \sum_{C(i)=c} ||\mathbf{E}_i - \bar{\mathbf{E}}_k||^2, \tag{5}$$

---

[1] The word cluster is assigned to different classes of labeled data, that is $\mathbf{K}$ is fixed to 2 (noise and speech frames).

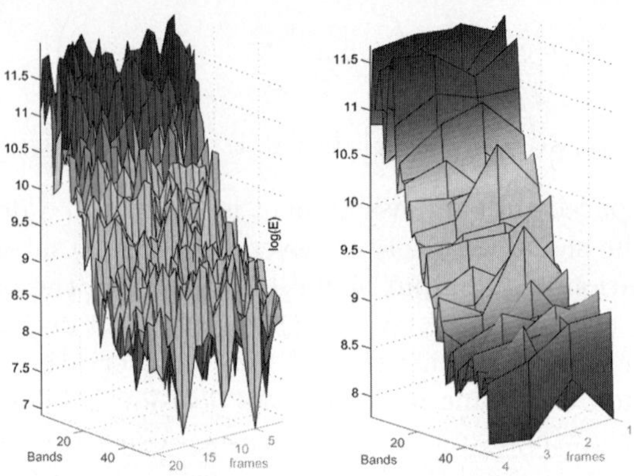

Fig. 1. a) 20 log Energies of noise frames, computed using $N_{FFT} = 256$, averaged over 50 subbands. b) Clustering approach to the latter set of log Energies using hard decision **C**-means (C=4 prototypes).

where $C(x)$ denotes the prototype associated to observation x and

$$\bar{\mathbf{E}}_c = (\bar{E}(1,c), \ldots, \bar{E}(K,c))^T \tag{6}$$

is the mean vector associated with the c-th prototype. Thus, the loss function is minimized by assigning the N observations to the C prototypes in such a way that within each prototype the average dissimilarity of the observations is minimized. Once convergence is reached, N K-dimensional pause frames are efficiently modeled by C K-dimensional noise prototype vectors denoted by $\bar{\mathbf{E}}_c^{opt}$, $c = 0, \ldots, C - 1$. In figure 1 we observed how the complex nature of noise can be simplified (smoothed) using a clustering approach. The clustering approach speeds the decision function in a significant way since the dimension of feature vectors is reduced substantially $(N \to C)$.

3.2   Soft Decision Function for VAD

In order to classify the second labeled data (log energies of speech frames) we use a BSAS using a MO window centered at frame l, as shown in section 2. For this purpose let consider the same dissimilarity measure, a threshold of dissimilarity γ and the maximum clusters allowed $\mathbf{K} = 2$.

Let $\hat{\mathbf{E}}(l)$ be the decision feature vector that is based on the MO window as follows:

$$\hat{\mathbf{E}}(l) = \max\{\mathbf{E}(i)\}, \quad i = l - m, \ldots, l + m \tag{7}$$

This selection of the feature vector describing the actual frame is useful as it detects the presence of voice beforehand (pause-speech transition) and holds the detection flag, smoothing the VAD decision (as a hangover based algorithm [11, 10] in speech-pause transition).

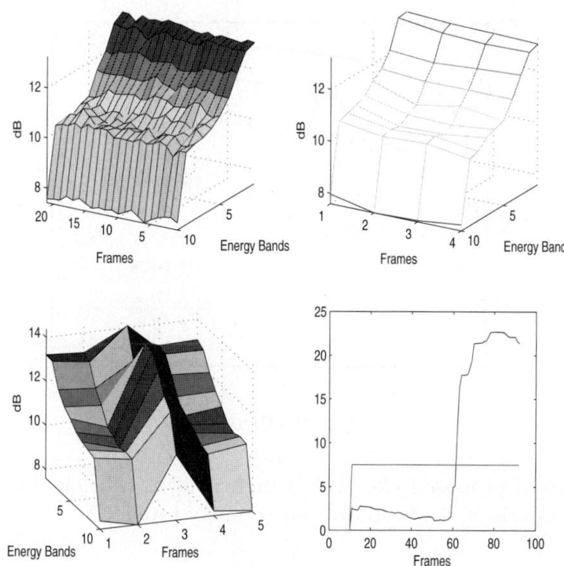

Fig. 2. Step of the algorithm. The frame selected is classified as speech frame (VAD=1) as is shown in the decision function a) Noise log-energy subbands. b) **C**-means centers prototypes. c) comparison between noise prototypes and the log-energy of the current frame. d) decision function and threshold versus frames.

Finally, the presence of a new cluster (speech frame detection) is satisfied if:

$$||\hat{\mathbf{E}}(l) - < \bar{\mathbf{E}}_c > ||^2 > \gamma \tag{8}$$

where $< \bar{\mathbf{E}}_c >$ is the averaged noise prototype and γ is the decision threshold. The set of noise prototypes are updated in pause frames (not satisfying equation 8)) in a competitive manner (only the closer noise prototype is moved towards the current feature vector):

$$\bar{\mathbf{E}}_{c'} = arg_{min}\left(||\bar{\mathbf{E}}_c - \hat{\mathbf{E}}(l)||^2\right) \quad \Rightarrow \quad \bar{\mathbf{E}}_{c'}^{new} = \alpha \cdot \bar{\mathbf{E}}_{c'}^{old} + (1 - \alpha) \cdot \hat{\mathbf{E}}(l) \tag{9}$$

where α is a normalized constant with value close to one for a soft decision function (i.e. we selected in simulation $\alpha = 0.99$).

In figure 2 we show an step detail in the algorithm. We display the noise log energy model (top-left), the clustering **C**-means approach, the log-energy of current frame (frame=3) included in the noise prototypes ($C = 4$) and the decision rule versus time.

4   Experimental Framework

Several experiments are commonly conducted to evaluate the performance of VAD algorithms. The analysis is normally focused on the determination of misclassification errors at different SNR levels [11], and the influence of the VAD

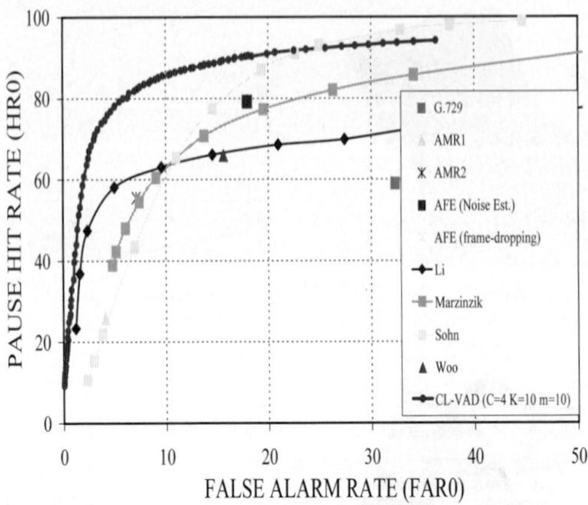

Fig. 3. ROC curves of proposed Cl-VAD in high noisy conditions for $m = 10$, $K = 10$ and $C = 8$ and comparison to standard and recently reported VADs

decision on speech processing systems [8, 1]. The experimental framework and the objective performance tests conducted to evaluate the proposed algorithm are described in this section.

The ROC curves are used in this section for the evaluation of the proposed VAD. These plots describe completely the VAD error rate and show the trade-off between the speech and non-speech error probabilities as the threshold γ varies. The Spanish SpeechDat- Car database [22] was used in the analysis. This database contains recordings in a car environment from close-talking and hands-free microphones. Utterances from the close-talking device with an average SNR of about $25dB$ were labeled as speech or non-speech for reference while the VAD was evaluated on the hands-free microphone. Thus, the speech and non-speech hit rates $(HR1, HR0)$ were determined as a function of the decision threshold γ for each of the VAD tested. Figure 3 shows the ROC curves in the most unfavorable conditions (high-speed, good road) with a 5 dB average SNR. It was shown that increasing the number of observation vectors m improves the performance of the proposed Cl-VAD. The best results are obtained for $m = 10$ while increasing the number of observations over this value reports no additional improvements. The proposed VAD outperforms the Sohn's VAD [7], which assumes a single observation likelihood ratio test (LRT) in the decision rule together with an HMM-based hangover mechanism, as well as standardized VADs such as G.729 and AMR [4, 3]. It also improve recently reported methods [7, 10, 9, 11]. Thus, the proposed VAD works with improved speech/non-speech hit rates when compared to the most relevant algorithms to date.

5   Conclusions

A new VAD for improving speech detection robustness in noisy environments is proposed. The proposed Cl-VAD is based on noise modeling using **C**-means clustering and benefits from long term information for the formulation of a soft decision rule. The proposed Cl-VAD outperformed Sohn's VAD, that defines the LRT on a single observation, and other methods including the standardized G.729, AMR and AFE VADs, in addition to recently reported VADs. The VAD performs an advanced detection of beginnings and delayed detection of word endings which, in part, avoids having to include additional hangover schemes or noise reduction blocks. Obviously it also will improve the recognition rate when it is considered as part of a complete speech recognition system. The experimental work on this part is on the way. In addition a soft decision based clustering approach for modeling noise prototypes and decision function is currently on progress.

Acknowledgements

This work has received research funding from the EU 6th Framework Programme, under contract number IST-2002-507943 (HIWIRE, Human Input that Works in Real Environments) and SESIBONN and SR3-VoIP projects (TEC2004-06096-C03-00, TEC2004-03829/TCM) from the Spanish government. The views expressed here are those of the authors only. The Community is not liable for any use that may be made of the information contained therein.

References

1. L. Karray and A. Martin, Towards improving speech detection robustness for speech recognition in adverse environments, 2003, Speech Communitation, number 3,pages 261-276.
2. J. Ramírez, J. C. Segura, M. C. Benítez, A. de la Torre and A. Rubio, A New Adaptive Long-Term Spectral Estimation Voice Activity Detector,Proc. of EUROSPEECH 2003, 2003, Geneva, Switzerland, September, pages 3041-3044.
3. ETSI, Voice Activity Detector (VAD) for Adaptive Multi-Rate (AMR) Speech Traffic Channels, 1999, ETSI EN 301 708 Recommendation.
4. ITU, A silence compression scheme for G.729 optimized for terminals conforming to recommendation V.70, 1996, ITU-T Recommendation G.729-Annex B.
5. A. Sangwan, M. C. Chiranth, H. S. Jamadagni, R. Sah, R. V. Prasad, V. Gaurav, VAD Techniques for Real-Time Speech Transmission on the Internet, IEEE International Conference on High-Speed Networks and Multimedia Communications, 2002, pages 46-50.
6. F. Basbug, K. Swaminathan and S. Nandkumar, Noise Reduction and Echo Cancellation Front-End for Speech Codecs, 2003, IEEE Transactions on Speech and Audio Processing, 11, num 1, pages 1-13.
7. J. Sohn, N. S. Kim and W. Sung, A statistical model-based voice activity detection, 1999, IEEE Signal Processing Letters,vol 16,num 1, pages 1-3,.
8. R. L. Bouquin-Jeannes and G. Faucon, Study of a voice activity detector and its influence on a noise reduction system, 1995, Speech Communication, vol 16, pages 245-254.
9. K. Woo, T. Yang, K. Park and C. Lee, Robust voice activity detection algorithm for estimating noise spectrum, 2000, Electronics Letters, vol 36, num 2, pages 180-181.

10. Q. Li, J. Zheng, A. Tsai and Q. Zhou, Robust endpoint detection and energy normalization for real-time speech and speaker recognition, 2002, IEEE Transactions on Speech and Audio Processing, vol 10, num 3, pages 146-157.
11. M. Marzinzik and B. Kollmeier, Speech pause detection for noise spectrum estimation by tracking power envelope dynamics, 2002, IEEE Transactions on Speech and Audio Processing, vol 10, num 6, pages 341-351.
12. Fisher, D. 1987. Knowledge acquisition via incremental conceptual clustering. Machine Learning 2:139–172.
13. J.M. Górriz, J. Ramírez, J.C. Segura and C.G. Puntonet, Improved MO-LRT VAD based on bispectra Gaussian model, 2005, Electronics Letters, vol 41, num 15, pages 877-879.
14. J. Ramírez, José C. Segura, C. Benítez, L. García and A. Rubio, Statistical Voice Activity Detection using a Multiple Observation Likelihood Ratio Test, 2005, IEEE Signal Processing Letters, vol 12, num 10, pages 689-692.
15. Anderberg, M. R. 1973. Cluster Analysis for Applications. Academic Press, Inc., New York, NY.
16. Jain, A. K. and Flynn, P. J. 1996. Image segmentation using clustering. In Advances in Image Understanding. A Festschrift for Azriel Rosenfeld, N. Ahuja and K. Bowyer, Eds, IEEE Press, Piscataway, NJ, 65-83.
17. Rasmussen, E. 1992. Clustering algorithms. In Information Retrieval: Data Structures and Algorithms, W. B. Frakes and R. Baeza-Yates, Eds. Prentice-Hall, Inc., Upper Saddle River, NJ, 419-442
18. Salton, G. 1991. Developments in automatic text retrieval. Science 253, 974-980.
19. J. Ramírez, José C. Segura, C. Benítez, A. de la Torre, A. Rubio, An Effective Subband OSF-based VAD with Noise Reduction for Robust Speech Recognition, 2005, In press IEEE Trans. on Speech and Audio Processing.
20. J. B. MacQueen, Some Methods for classification and Analysis of Multivariate Observations, Proceedings of 5-th Berkeley Symposium on Mathematical Statistics and Probability, Berkeley, University of California Press, 1:281-297 (1967)
21. T. Hastie, R. Tibshirani and J. Friedman The Elements of Statistical Learning Data Mining, Inference, and Prediction Series: Springer Series in Statistics 1st ed. 2001. ISBN: 0-387-95284-5
22. A. Moreno, L. Borge, D. Christoph, R. Gael, C. Khalid, E. Stephan and A. Jeffrey, SpeechDat-Car: A Large Speech Database for Automotive Environments, Proceedings of the II LREC Conference, 2000.

An Improved Particle Swarm Optimization Algorithm for Global Numerical Optimization

Bo Zhao

Jiangsu Electric Power Research Institute Corporation Limited
Nanjing 210063, Jiangsu, China
zhaobozju@163.com

Abstract. This paper presents an improved particle swarm optimization algorithm (IPSO) for global numerical optimization. The IPSO uses more particles' information to control the mutation operation. A new adaptive strategy for choosing parameters is also proposed to assure convergence of the IPSO. Meanwhile, we execute the IPSO to solve eight benchmark problems. The results show that the IPSO is superior to some existing methods for finding the best solution, in terms of both solution quality and algorithm robustness.

1 Introduction

Particle swarm optimization (PSO) is one of the evolutionary computation techniques based on the social behavior metaphor [1]. It is initialized with a population of random solutions, conceptualized as particles. Each particle in PSO flies through the search space with a velocity according to its own and the whole population's historical behaviors. The particles have a tendency to fly toward better search areas over the course of a search process. In recent years, PSO has been widely used for numerical optimization and many other engineering problems. Global optimization problems arise in almost every field of science, engineering, and business. Now, the PSO is becoming one of the popular methods to address them due to its simplicity of implementation and ability to quickly converge to a reasonably good solution. But, in global optimization problems, the major challenge is that the POS may be trapped in the local optima of the objective function. The issue is particularly challenging when the dimension is high and there are numerous local optima. Therefore, some improved methods have been proposed, such as the fully informed particle swarm [2], a hybrid of GA and PSO [3], a cooperative approach to PSO [4], a hierarchical particle swarm optimizer [5], a multiagent-based PSO [6], and so on.

In the standard PSO algorithm, the neighborhood of a particle consist of all particles, so that the global best position, i.e., the best solution found so far, directly influences its behavior. Hence, for social science context, a PSO system combines a social-only component model and a cognition-only model. The social-only model component suggests that individuals ignore their own experience and adjust their behavior according to the successful beliefs of individuals in the neighborhood. On the other hand, the cognition-only model component treats individuals as isolated beings. The real strength of the particle swarm derives from the social interactions

V.N. Alexandrov et al. (Eds.): ICCS 2006, Part I, LNCS 3991, pp. 657–664, 2006.

among
particles as they search the space collaboratively. So, in this paper, an improved adaptation strategy with enhanced social influence is proposed for PSO algorithm. This adaptation strategy is similar to the social society in that a group of leaders, i.e. several fittest individuals in a population, play a major role in reproduction process. This is different from the standard PSO with only one leader, the fittest individual in a generation, is selected to the reproduction process. The improved particle swarm optimization (IPSO) has been tested with some benchmark functions. The experimental results show that the IPSO achieves a good performance for test functions, which illustrate that the IPSO overcomes the problem of premature convergence of the standard PSO method in some degree.

2   Improved Particle Swarm Optimization (IPSO)

2.1   Standard Particle Swarm Optimization Approach (PSO)

PSO is developed through simulation of bird flocking in two-dimension space. According to the research results for a flock of birds, birds find food by flocking. The observation leads the assumption that information is shared inside flocking. Moreover, according to observation of behavior of human groups, behavior of each individual (agent) is also based on behavior patterns authorized by the groups such as customs and other behavior patterns according to the experiences by each individual. The position of each agent is represented by XY-axis position and the velocity is expressed by v_x (the velocity of X-axis) and v_y (the velocity of Y-axis). Modification of the agent position is realized by using the position and the velocity information [7].

Searching procedures by PSO based on the above concept can be described as follows: a flock of agents optimizes a certain objective function. Each agent knows its best value so far (*pbest*) and its position. The information is corresponding to personal experiences of each agent. Moreover, each agent knows the best value so far in the group (*gbest*) among *pbests*. The information is corresponding to knowledge of how the other agents around them have performed. Namely, each agent tries to modify its position using the following information:

- The distance between the current position and *pbest*, p_t.
- The distance between the current position and *gbest*, $\hat{p}_t$.

The modification can be represented by the concept of velocity. Velocity of each agent can be modified by the following equation:

$$v_{t+1} = c_0 v_t + c_1 r_1(t) \times (p_t - x_t) + c_2 r_2(t) \times (\hat{p}_t - x_t),$$

(1)

where c_0, c_1 and c_2 are positive constant coefficients, r_1 and r_2 are uniformly distributed random numbers in [0,1], v_t is the current velocity of the particle at iteration t, x_t is current position of the particle at iteration t, v_{t+1} is the modified velocity at iteration $t+1$.

Using the above equation (1), a certain velocity that gradually gets close to *pbests* and *gbest* can be calculated. The current position (searching point in the solution space) can be modified by the following equation:

$$x_{t+1} = x_t + v_{t+1}, \tag{2}$$

where x_{t+1} is the modified position at iteration $t+1$.

2.2 Improved Particle Swarm Optimization Approach (IPSO)

Due to the shortage of the social interactions among particles in the standard PSO algorithm, an improved adaptation strategy with enhanced social interactions is proposed for particle swarm optimization (PSO) algorithm in this section. This adaptation strategy uses more particles' information to control the mutation operation and extends the original formulas of the PSO method, which can search the global optimal solution more effectively.

The third term added to the right-hand-side of the velocity equation (1) is derived from the successes of the others; it is considered a "social influence" term. It is found that when this effect is removed from the algorithm, performance is abysmal. So the social interaction is an important factor to improve the PSO performance. To enhance the social interactions in the algorithm, this paper proposes a new method of improved PSO using some fittest particles' information to modify particle's position and velocity. Namely, at ith iteration, we rearrange the particles in descending order according to their fitness and select the last n particles to modify particle's position and velocity. Let $\hat{p}_{i,t}$ denote current position of the particle i in these particles at iteration t. The updating equations of IPSO method can be described in the following:

$$v_{t+1} = c_0 v_t + c_1 r_1(t) \times (p_t - x_t) + \sum_{i=1}^{n} c_{2,i} r_{2,i}(t) \times (\hat{p}_{i,t} - x_t), \tag{3}$$

$$x_{t+1} = x_t + v_{t+1}, \tag{4}$$

Each particle of IPSO method modifies its position and velocity using the best solution particle achieved and several *gbest* of neighborhood particles. It is similar to the social society in that the group of leaders could make better decisions. However, in standard PSO, only one *gbest* of neighborhood particles is employed. This process using some neighborhood particles can be called 'intensifying' and 'enhancing' the social influence. Based on this understanding, we should intensify these particles that could lead individuals to better fitness. This reinforces the exploitation and exploration of PSO. As a particle swarm population searches over time, individuals are drawn toward one another's successes, with the usual result being clustering of individuals in optimal regions of the space.

2.3 IPSO Algorithm's Convergence

The form of the recurrence relation of particle's position can be derived as follows: substituting equation (3) into (4) and from equation $v_t = x_t - x_{t-1}$, we have the following equation:

$$x_{t+1} = \left(1 + c_0 - c_1 r_1(t) - \sum_{i=1}^{n} c_{2,i} r_{2,i}(t)\right) x_t - c_0 x_{t-1} + c_1 r_1(t) p_t + \sum_{i=1}^{n} c_{2,i} r_{2,i}(t) \hat{p}_{i,t}, \tag{5}$$

which is a non-homogeneous recurrence relation that can be solved using standard recursive techniques. This recurrence relation can be written as a matrix-vector product,

$$\begin{bmatrix} x_{t+1} \\ x_t \\ 1 \end{bmatrix} = \begin{bmatrix} 1 + c_0 - \eta_0 - \eta_1 & -c_0 & \eta_0 p_t + \eta_1 p \\ 1 & 0 & 0 \\ 0 & 0 & 1 \end{bmatrix} \begin{bmatrix} x_t \\ x_{t-1} \\ 1 \end{bmatrix}, \tag{6}$$

where $\eta_0 = c_1 * r_1(t)$, $\eta_1 = \sum_{i=1}^{n} c_{2.i} * r_{2.i}(t)$, $\eta_1 p = \sum_{i=1}^{n} c_{2.i} * r_{2.i}(t) * \hat{p}_{i,t}$. The characteristic polynomial of the equation (6) is,

$$(\lambda - 1)\left(\lambda^2 - (1 + c_0 - \eta_0 - \eta_1)\lambda + c_0\right),$$

which has a trivial root of $\lambda_1 = 1.0$, and two other solutions (7) and (8),

$$\lambda_2 = \frac{(1 + c_0 - \eta_0 - \eta_1) + \gamma}{2}, \tag{7}$$

$$\lambda_3 = \frac{(1 + c_0 - \eta_0 - \eta_1) - \gamma}{2}, \tag{8}$$

where

$$\gamma = \sqrt{(1 + c_0 - \eta_0 - \eta_1)^2 - 4c_0}, \tag{9}$$

Note that λ_1, λ_2 and λ_3 are both eigenvalues of the equation (6). The explicit form of the recurrence relation (5) is then given by equation,

$$x_t = k_1 \lambda_1^t + k_2 \lambda_2^t + k_3 \lambda_3^t = k_1 + k_2 \lambda_2^t + k_3 \lambda_3^t, \tag{10}$$

where k_1, k_2 and k_3 are constants determined by the initial conditions of the system at each iteration.

An important aspect of the behavior of a particle concerns whether its trajectory (specified by x_t) converges or diverges. The conditions under which the sequence $\{x_t\}_{t=0}^{+\infty}$ will converge are determined by the magnitude of the values λ_2 and λ_3 as computed using equations (7) and (8). From equation (9), it is clear that there are two cases:

1) Case A: $(1 + c_0 - \eta_0 - \eta_1)^2 < 4c_0$

In this case, r will be a complex number with a non-zero imaginary component. A complex r results in λ_2 and λ_3, being complex numbers with non-zero imaginary components as well. Consider the value of x_t in the limit, thus equation (10) becomes,

$$x_t = k_1 + k_2 \cdot \|\lambda_2\|^t \cdot e^{j\theta t} + k_3 \cdot \|\lambda_3\|^t \cdot e^{j\sigma t}, \tag{11}$$

where $\|\lambda_2\|^t \cdot e^{j\theta t}$ and $\|\lambda_3\|^t \cdot e^{j\sigma t}$ are the exponent expression of the trivial roots. Clearly, equation (11) explains the sequence $\{x_t\}_{t=0}^{+\infty}$ will converge when $\max\left(\|\lambda_2\|,\|\lambda_3\|\right) < 1$, so, $\lim\limits_{t\to\infty} x_t = k_1 + k_2 \cdot \lambda_2^t + k_3 \cdot \lambda_3^t = k_1$.

2)   *Case B:* $\left(1 + c_0 - \eta_0 - \eta_1\right)^2 \geq 4c_0$

In this case, r, λ_2 and λ_3, will be real numbers, from equation (10), if $\max\left(\|\lambda_2\|,\|\lambda_3\|\right) < 1$, then the sequence $\{x_t\}_{t=0}^{+\infty}$ converges.

From the analysis of above two cases, we obtain the convergence condition of the sequence $\{x_t\}_{t=0}^{+\infty}$ is $\max\left(\|\lambda_2\|,\|\lambda_3\|\right) < 1$.

One popular choice of updating parameters is $c_0 = 0.7298$, $c_1 = 1.49618$ and $\sum_{i=1}^{n} c_{2.i} = 1.49618$ [8]. On account of $\eta_0 = c_1 * r_1(t)$ and $\eta_1 = \sum_{i=1}^{n} c_{2.i} * r_{2.i}(t)$, so $\eta_0 \in (0, 1.49618)$, $\eta_1 \in (0, 1.49618)$ and $\eta_0 + \eta_1 \in (0, 2.9924)$ .When $\eta_0 + \eta_1 \in (0, 0.0212]$, from equation (9), it will imply a real-valued γ, which corresponds to case B, then $\gamma \geq 0$ and $\max(\|\lambda_2\|,\|\lambda_3\|) = \dfrac{1.7298 - \eta_0 - \eta_1 + \gamma}{2} < 1$. Similarly, $\eta_0 + \eta_1 \in (0.0212, 2.9924)$ will result in a complex r value, which corresponds to case A and from equations (7) and (8) we can assure $\|\lambda_2\| = \|\lambda_3\| < 1$. The above analysis shows that the choices of parameters satisfy the convergence conditions and will assure convergence of the sequence $\{x_t\}_{t=0}^{+\infty}$.

IPSO method differs from PSO method is that η_1 is the sum of the coefficients of n *gbest* particles. The degree of importance of each particle is weighed through particle's fitness value. The better the particle's fitness value is, the more important the particle's influence is. So the paper proposed the rule of updating parameters chosen as follows:

$$c_{2,i} = \dfrac{\dfrac{1}{\hat{f}_i}}{\sum\limits_{i=1}^{n} \dfrac{1}{\hat{f}_i}} \cdot 1.49618 , \tag{12}$$

where $\hat{f}_i$ is the *gbest* particle's value. This adaptive strategy of updating parameters in IPSO can assure convergence of IPSO method and enhance the global convergence capability of IPSO method.

3   Numerical Experiments and Results

In order to verify the effectiveness and efficiency of the proposed IPSO method, eight benchmark functions have been used in Table 1.

Table 1. Comparisons of generalization capability with published results

Function Name	Function Expression	Dimensions (N)	Initial Range
Schwefel	$F_1(x) = \sum_{i=1}^{N} \left(-x_i \sin\sqrt{|x_i|}\right)$	30	(-500,500)
Rastrigin	$F_2(x) = \sum_{i=1}^{N} \left[x_i^2 - 10\cos(2\pi x_i) + 10\right]$	30	(-5.12,5.12)
Ackley	$F_3(x) = -20\exp\left(-0.2\sqrt{\frac{1}{n}\sum_{i=1}^{N} x_i^2}\right) - \exp\left(\frac{1}{n}\sum_{i=1}^{N}\cos(2\pi x_i)\right) + 20 + \exp(1)$	30	(-32,32)
Griewank	$F_4(x) = \frac{1}{4000}\sum_{i=1}^{N} x_i^2 - \prod_{i=1}^{N}\cos\left(\frac{x_i}{\sqrt{i}}\right) + 1$	30	(-600,600)
Rosenbrock	$F_5(x) = \sum_{i=1}^{N-1}\left[100(x_{i+1} - x_i)^2 + (1 - x_i)^2\right]$	30	(-30,30)
Sphere	$F_6(x) = \sum_{i=1}^{N} x_i^2$	30	(-100,100)
Schwefel	$F_7(x) = \sum_{i=1}^{N}|x_i| + \prod_{i=1}^{N}|x_i|$	30	(-10,10)
Schwefel	$F_8(x) = \sum_{i=1}^{N}\left(\sum_{j=1}^{i} x_j\right)^2$	30	(-100,100)

Some parameters must be assigned to before the IPSO method is used to solve problems in following experiments. Since the test problem dimensions are high, a moderate population size is set to 70 and the maximal generation is set to 1000. To evaluate uncertain value combinations of n of the IPSO method, we have been executed 50 times to solve the above test function problem under various value combinations. The results show that the best solution can be obtained by the IPSO method when $n=4$. Hence, in the following study, we always choose $n=4$.

Owing to the randomness in IPSO, the algorithm is executed 50 independent times on each test function. Table 2 shows the optimal results of IPSO. From Table 2, we see that the mean function values are equal or close to the optimal ones, and the standard deviations of the function values are relatively small, except for function F_2. These results indicate that the proposed IPSO method can find optimal or close-to-optimal solutions, and its solution quality is quite stable.

Table 3 summarizes the optimal results as obtained by some existing methods. These existing algorithms include standard particle swarm optimization algorithm (PSO), conventional genetic algorithm (CGA), orthogonal genetic algorithm with quantization (QGA/Q), fast evolutionary programming (FEP), evolutionary optimization (EO), enhanced simulated annealing (ESA) and conventional evolutionary programming with Gaussian mutation operator (CEP/GMO). Since each of these existing algorithms was executed to solve some of the above-mentioned functions,

Table 3 has included all of the available results for comparison. From Table 3, these results show that the optimal solutions determined by the IPSO lead to high quality solutions, which confirms that the IPSO is well capable of determining the global or close-to-optimal solutions. Firstly, Table 3 compares IPSO with PSO and CGA. For all numerical optimization functions, IPSO gives better solutions than PSO and CGA. Compared IPSO with PEP and OGA/Q, we see that FEP and OGA/Q obtain good performance on numerical optimization problem, even better solutions in some optimization functions. But the IPSO method is simple and has not complicated operator, not as PEP and OGA/Q. However, in eight benchmark functions, except for F_2, the optimal solutions obtained by the IPSO are nothing less than these solutions obtained by PEP and OGA/Q. It indicates that IPSO can, in general, give good mean solution quality. At the same time, compared IPSO with EO and ESA, as can been seen, IPSO obtains better solutions in the available results for comparison, and displays a good performance in solving these global numerical optimization problems.

Table 2. Optimal results of IPSO

Test function	Mean function value	Standard deviation of function value	Globally minimal function value
F_1	-12569.487	7.056×10^{-6}	-12569.5
F_2	11.5312	0.1242	0
F_3	8.253×10^{-16}	7.413×10^{-17}	0
F_4	0	0	0
F_5	9.584×10^{-1}	2.761×10^{-1}	0
F_6	4.096×10^{-96}	1.721×10^{-96}	0
F_7	1.654×10^{-8}	7.357×10^{-9}	0
F_8	0	0	0

Table 3. Comparison between IPSO and some existing algorithms

	Mean function value							
	IPSO	PSO	CGA	OGA/Q	FEP	EO	ESA	CEP/GMO
F_1	**-12569.487**	-9903.8	-9094.75	**-12569.45**	**-12554.5**	--	--	--
F_2	**11.5312**	26.8639	22.967	**0**	$\mathbf{4.6 \times 10^{-2}}$	46.47	--	120.00
F_3	$\mathbf{8.25 \times 10^{-16}}$	7.99×10^{-15}	2.697	$\mathbf{4.44 \times 10^{-16}}$	$\mathbf{1.8 \times 10^{-2}}$	--	--	9.10
F_4	**0**	0.022	1.258	**0**	$\mathbf{1.6 \times 10^{-2}}$	0.40	--	2.52×10^{-7}
F_5	$\mathbf{9.58 \times 10^{-1}}$	16.770	150.79	$\mathbf{7.52 \times 10^{-1}}$	--	1911.59	17.10	86.70
F_6	**0**	4.83×10^{-48}	4.96	**0**	$\mathbf{5.7 \times 10^{-4}}$	9.88	--	3.09×10^{-7}
F_7	$\mathbf{5.40 \times 10^{-22}}$	1.65×10^{-8}	7.93×10^{-1}	**0**	$\mathbf{8.1 \times 10^{-3}}$	--	--	1.99×10^{-3}
F_8	$\mathbf{2.70 \times 10^{-11}}$	0.324	18.82	**0**	$\mathbf{1.6 \times 10^{-2}}$	--	--	17.60

Figure 1 shows the evolution process of F_1 obtained using IPSO, PSO and CGA respectively. It is clear for the figure that the solution by IPSO is converged to high quality solutions at the early iterations. Similar results are obtained for F_2 to F_8. According to the above analysis, considering together more particles' information to control the mutation operation, the IPSO method performs better than the PSO, both in the quality of the solution discovered and in the velocity of convergence, and simulation results show that IPSO outperforms CGA and PSO.

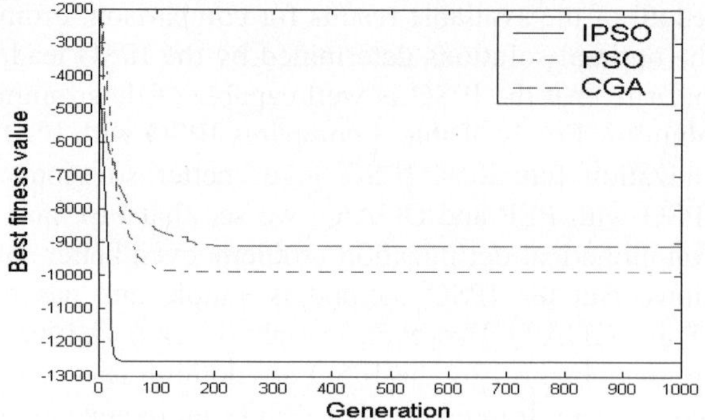

Fig. 1. Optimization procedure by IPSO, PSO and CGA

4 Conclusions

An improved particle swarm optimization algorithm (IPSO) has been developed for determination of the optimal or close-to-optimal solutions of global numerical optimization problem. The IPSO approach uses more particles' information to control the mutation operation. The convergence property of IPSO is analyzed using standard results from the dynamic system theory and guidelines for proper algorithm parameter selection are derived. A new adaptive strategy for choosing parameters is also proposed to assure convergence of IPSO. Meanwhile, it was found from experimental results that IPSO could find higher quality solutions reliably with the faster converging characteristics than some existing methods on the problem studies.

References

1. Kennedy, J., Eberhart, R. C.: Particle swarm optimization. In: Proceedings of IEEE International Conference on Neural Networks. (1995) 1942-1948
2. Mendes, R., Kennedy, J.: The fully informed particle swarm: simpler, maybe better. IEEE Trans. Evolutionary Computation. 3 (2004) 204-210
3. Juang, C.F.: A hybrid of genetic algorithm and particle swarm optimization for recurrent network design. IEEE Trans. System, Man and Cybernetics-Part B. 2 (2004) 997-1006
4. van den Bergh, F., Engelbrecht, P.: A cooperative approach to particle swarm optimization,. IEEE Trans. Evolutionary Computation. 3 (2004) 225-239
5. Janson, S., Middendorf, M.: A hierarchical particle swarm optimizer and its adaptive variant. IEEE Trans. System, Man and Cybernetics-Part B, 6 (2005) 1272-1282
6. Zhao, B., Guo, C.Y., Cao, Y.J.: A multiagent-based particle swarm optimization approach for optimal reactive power dispatch. IEEE Trans. Power Systems. 2 (2005) 1070-1078
7. Kennedy, J., Eberhart, R.C.: Swarm Intelligence. Morgan Kaufmann Publishers (2001)
8. Clerc, M., Kennedy, J.: The Particle Swarm-explosion, stability, and convergence in a multidimensional complex space. IEEE Trans. on Evolutionary Computation. 1 (2002) 58-73

Pores in a Two-Dimensional Network of DNA Strands – Computer Simulations

M.J. Krawczyk[*], and K. Kułakowski

Faculty of Physics and Applied Computer Science,
AGH University of Science and Technology,
al. Mickiewicza 30, 30-059 Kraków, Poland
Tel.: (48 12) 6173518; Fax.: (48 12) 6340010
gos@fatcat.ftj.agh.edu.pl

Abstract. Formation of random network of DNA strands is simulated on a two-dimensional triangular lattice. We investigate the size distribution of pores in the network. The results are interpreted within theory of percolation on Bethe lattice.

PACS Code: 05.10.Ln, 87.14.Gg, 87.15.Aa.

Keywords: DNA junctions, DNA networks, computer simulations.

1   Introduction

Besides their extremely important biological function, molecules of DNA are useful in many areas, as biotechnology, nanotechnology and electronics. Possibility of design of molecules of exactly known composition and length opened an area for using DNA for construction of different structures. Appropriate choice of DNA sequences allows to obtain not only linear chains but also branched DNA molecules. These synthetical branches are analogs of branched DNA molecules (Holliday junctions) discovered in living cells. Addition of sticky ends to such junctions enables a construction of complex molecular systems. Holliday junctions occurring naturally are present between two symmetrical segments of DNA strands, and because of that, the branch point is free to migrate. Such a behaviour is not advantageous when DNA strands are used for a construction of networks. Elimination of symmetry in synthetically made DNA molecules fixes the branch points location. Synthetical junctions give an opportunity for building complex structures from branches prepared specially for given purposes; this enables to construct junctions with different numbers of arms. Addition of sticky ends to three- or four-armed branches allows to construct two-dimensional lattices, and using six-armed branches – for a construction of 3D lattices [1, 2]. Different kinds of branched molecules can be connected with linear strands or to each other. Appropriate choice of DNA and other allows for a construction of lattices of specific parameters.

---

[*] Corresponding author.

V.N. Alexandrov et al. (Eds.): ICCS 2006, Part I, LNCS 3991, pp. 665–672, 2006.
© Springer-Verlag Berlin Heidelberg 2006

One of the most important problems occurring during the network formation is that the angles between the arms of branched junctions are variable. This problem can be solved by using branches with four different sticky ends. Besides a stabilization of the edges, this solution eliminates improper connections between arms. Second important problem with DNA network formation is that the polymer strands are flexible. They can cyclize on itself what prevents further growth of the system [2]. To construct periodic lattices, such as e.g. crystals, which can be used in nanoelectronics, polymers must be rigid. Construction of crystals is possible since DNA double-crossover motifs were developed [2]. As the DNA double helix width is about 2 nm and one helical repeat length is about 3.4 nm, lattices made from DNA strands can be useful in the emerging nanotechnology. It is possible to use DNA to form crystals with edge length of more than 10 μm and a uniform thickness of about 1-2 nm [4]. One of main advantages of DNA structures is a possibility of their modifications by binding different ions and groups of ions or fluorescent labels [3, 2]. This is useful in electronic devices where nanometer-scale precision is extremely important.

Here we are interested in a design of random networks of DNA molecules. This aim is motivated as follows: the above mentioned limitations and conditions of the production of DNA lattices and crystals can be found to be too severe and in consequence too costful for some applications, where periodic structures are not necessary. In this case, properties of networks of DNA can be relevant for these potential applications of various two-dimensional structures, e.g. of polymers and liquid crystals [6, 7]. On the other hand, process of formation of a random network is close to some model processes of current interest for theories of disordered matter [8]. Then, interpretation of our numerical results allows to comment to what extent assumptions of these theories are minimal. As it is discussed below, our results suggest that some conclusions of the percolation theory in uncorrelated Bethe lattice can apply to correlated structures.

We are going to investigate the size distribution of empty holes (pores) in a random network. This aim should allow to evaluate the network in a role of a kind of molecular sieve [9]. Whilst disorder in such a sieve is obviously a disadvantage when compared to periodic structures [2], it seems to us worthwhile to evaluate consequences of this disorder.

In subsequent section the model is described and some details are given on its numerical realization. In Section III we show the results on the size distribution of the pores, obtained for a triangular two-dimensional lattice. Short discussion closes the text.

2   Model

In our model we initially consider two kinds of molecules: linear DNA strands and DNA-junctions, which connect ends of the strands. The latter can be of three or four arms; they are called Y- and X-junctions. We assume that the length of the linear molecules is much larger, than the size of the junctions. Also, we assume that the number of junctions is so much larger than the number of strands, that *i)* practically any collision of two ends of the strands leads to their fixed connection, *ii)* the junctions do not influence the motion of the strands in other way, than as in *i)*. Summarizing, our assumptions on the DNA-junctions (enough small and enough numerous) allow to neglect them during the simulation.

To speed up numerical calculations, the algorithm is based on a periodic lattice. It is straightforward to apply the square lattice in the case of X-junctions, and the triangular lattice in the case of Y-junctions. The length of the linear DNA strand is chosen to be an integer multiple of the size of the lattice cell. An additional assumption is that, no more than arbitrary chosen number M of different DNA strands are placed at the same lattice node, at the same time. This limits the density of the system from above, to reproduce the steric effect. We have only three model parameters: the length L of the strands, their number N, and the maximal number M of strands at one lattice cell. A strand cannot rotate, but only moves along its axis. The direction of the latter is random, with equal probabilities 1/2 at each time step. Initial set of the strand positions is selected to be entirely random (positions and orientations), with only the above mentioned limitation of the density. We choose periodic boundary conditions, to limit an influence of the borders. Subsequent positions are controlled by the assumption that in the one cell can not be more molecules that arbitrary chosen number M. Possible orientations of the strands are horizontal or vertical in the case of square lattice. In the case of triangular lattice, three orientations of the strands are equally possible, along all three sides of the triangle. If the number of ends of the strands at a given cell of the lattice is equal to or more than two (but not more than M, which is forbidden by the code), these strands become connected to a DNA-junction and to each other and they cannot move anymore. If at any time step two different strands are placed at the same cells only one of them, chosen randomly, can be connected to the DNA-junction. In one simulation step all molecules which are not immobile can move with equal probability to the right or left if they are oriented horizontally, and up or down if they are vertical in the case of square lattice. In the case of triangular lattice the strands can move in two possible directions, in accordance with their orientation. At each time step each strand can move by the length equal to the size of the one lattice cell. A movement is accepted if, as its consequence, the number of strands at all lattice cells is not more then M.

All results are obtained as average value from ten independent simulations, and the lattice size was 512×512. Each simulation was made for 100 time steps. Duration of the simulation was chosen so as to a number of free strands was in range of few percent (2-3). In all simulations we studied distribution of the pores, i.e. the lattice nodes which are not occupied by the DNA strands as a function of the model parameters. The size distribution of the pores was investigated using the Hoshen-Kopelman algorithm [10].

3   Results

The results presented in Figs. 1-4 are obtained for the case of triangular lattice with $M = 3$. The number N of molecules introduced to the system varies from 10^4 (Fig. 1) to 4×10^4 (Fig. 4). The plot shown in Fig. 1 shows that there are large pores in the system, but their number is exactly one for each size s; in other words, the system is diluted. For larger N we see that the pore size distribution is close to a power law, i.e. $N_g(s) \propto s^{-\tau}$, with $\tau \approx 1.6$ for $N = 2 \times 10^4$ and $\tau \approx 2.0$ for $N = 3 \times 10^4$. This is seen in Figs. 2 and 3. A further increase of the number

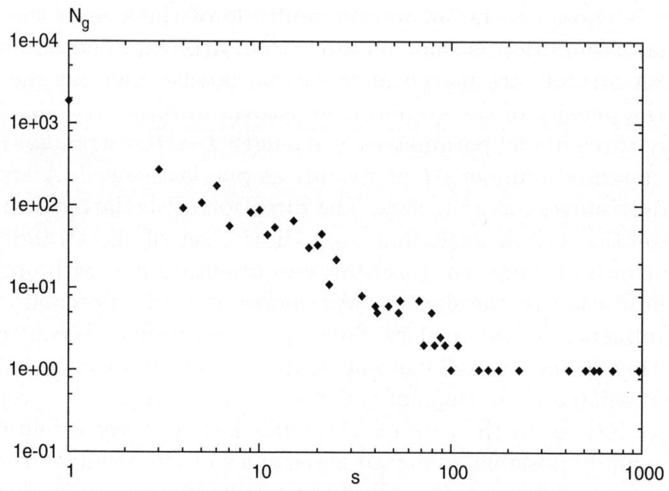

Fig. 1. Pore size distribution $N_g(s)$ for $L = 7$, $N = 10^4$, $M = 3$

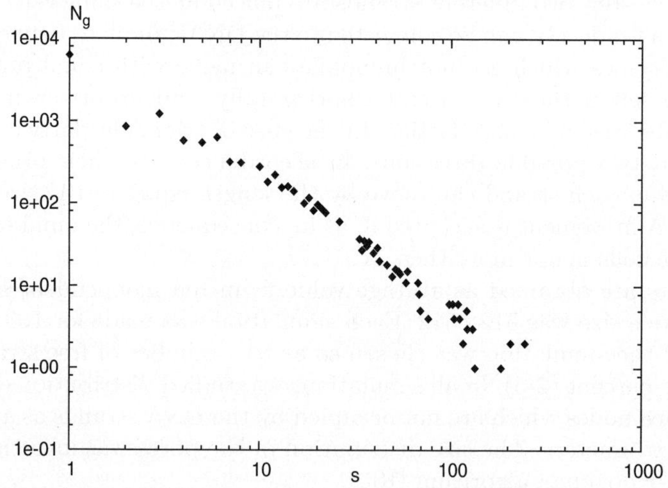

Fig. 2. Pore size distribution $N_g(s)$ for $L = 7$, $N = 2 \times 10^4$, $M = 3$

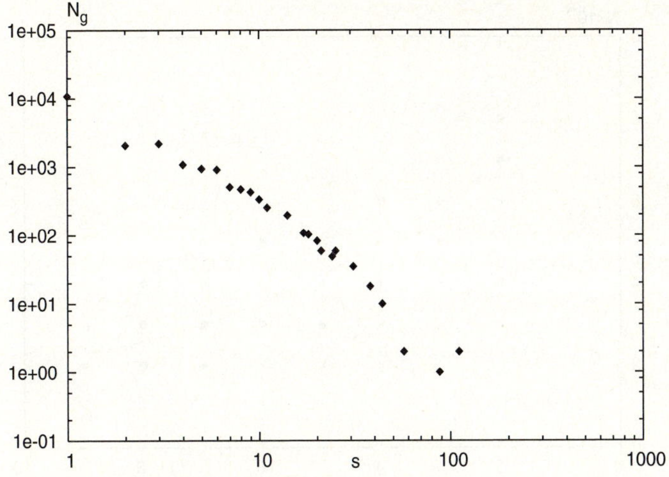

Fig. 3. Pore size distribution $N_g(s)$ for $L = 7$, $N = 3 \times 10^4$, $M = 3$

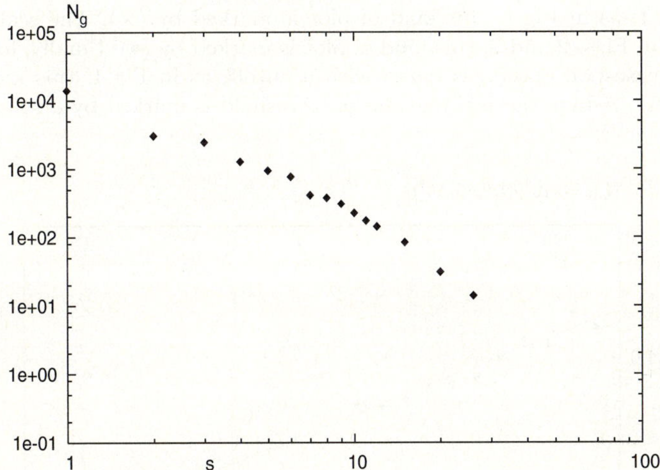

Fig. 4. Pore size distribution $N_g(s)$ for $L = 7$, $N = 4 \times 10^4$, $M = 3$

N of molecules leads to a cutoff of the scale-free character of the plot (Fig. 4). The size of maximal pores is then reduced from 100 or more to about 30.

Similar results are obtained for other strand lengths L. As a rule, for low values of N the shape of the function $N_g(s)$ is close to the one in Fig. 1. For larger N we see a power-law function, as in Figs. 2 and 3. When N increases further, large pores vanishes. This common behaviour is summarized in Fig. 5, where a kind of phase diagram is presented. Assuming that the scale-free power-law function is proper at only one value of $L_c(N)$, where the concentration of the strands is at the percolation threshold, we marked approximate position of this

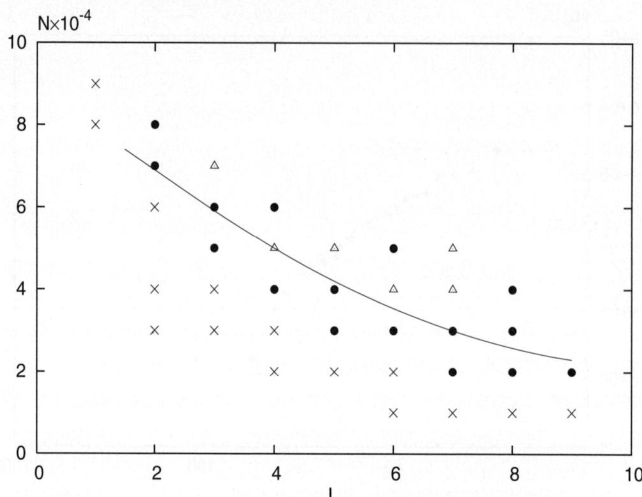

Fig. 5. Phase diagram (L,N) for $M = 3$. Different kinds of plots of $N_g(s)$ are marked with different signs. In the case of small N, a part of plot for large s is a horizontal line at $N_g = 1$, as in Fig.1; this kind of plot is marked by ($\times$). The scale-free power functions, as in Figs. 2 and 3; this kind of plot is marked by ($\bullet$). Finally, for the largest N, in many cases we observe a curve with a cutoff, as in Fig.4; this kind of plot is marked by ($\triangle$). A hypothetical percolation threshold is marked by a solid line.

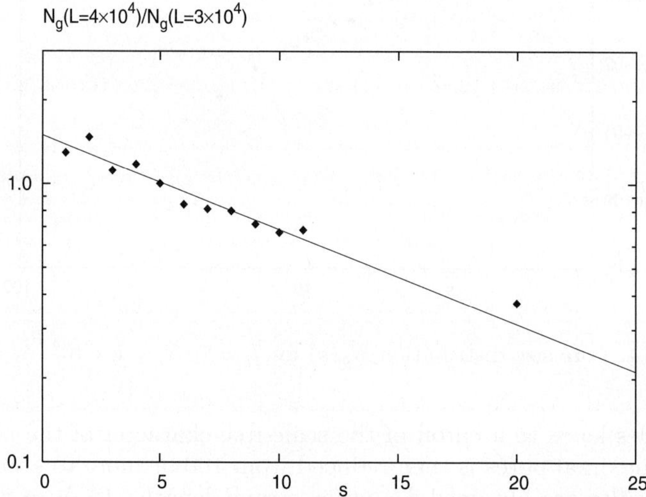

Fig. 6. $N_g(L = 4 \times 10^4)/N_g(L = 3 \times 10^4)$ against s. The obtained points follow a straight line, in accordance with the theory of Bethe lattice [8].

threshold with a schematic solid line. This plot is qualitative: the only conclusion from our data is that L_c decreases with N.

4   Discussion

As it was noted in the Introduction, our results appear to be close to what could be expected for an uncorrelated percolation in Bethe lattice [8]. For $M = 3$, the percolation threshold is $p_c = 1/(M - 1) = 0.5$. By symmetry, we can interchange occupied cells and empty cells. Then, the pore distribution for $p > p_c$ is equivalent to the cluster size distribution below the percolation threshold, i.e. for $p < p_c$, where the giant component is absent. At the percolation threshold, we expect that the average number of strands at the network nodes is $p_c M = 1.5$. Actually, we obtain the average number of strands equal to 1.39, 1.48, 1.52 and 1.53 for $L = 7$ and $N = 1, 2, 3$ and 4×10^4, respectively. This means that, keeping the analogy with the uncorrelated percolation in the Bethe lattice, we are close to the percolation threshold. Theoretical value of the exponent τ in the two-dimensional Bethe lattice is $187/91 \approx 2.05$ [8]. Taking into account our numerical errors, this is not in contradiction with our evaluations. Moreover, the cluster size distribution above the percolation threshold in the Bethe lattice theory fulfils the equation $N_g(s, L)/N_g(s, L_c) \propto exp(-cs)$. This formula is checked in Fig. 6 with data from Figs. 3 and 4. Again, the obtained data are not in contradiction with theory.

However, our network is neither uncorrelated, nor the Bethe lattice. Correlations appear because the strands are allowed to move until finally attached to some other strands. From this point of view, the system is analogous to the diffusion-limited aggregation (DLA), where the density of the growing cluster decreases in time. The difference is that in DLA, the cluster grows from some nucleus, which is absent in our simulation. On the other hand, loops are not prevented here, unlike in the Bethe lattice. The similarity of our results to the latter model allows to suspect that in random structure near the percolation threshold the contribution to the cluster size distribution from accidental loops remains small.

The same calculations are performed also for the two-dimensional square lattice. Here the power-law like character of the pore size distribution is less convincing. It seems likely that this additional difficulty is due to the fact that the triangular lattice is closer to the off-lattice model than the square lattice. In general, the lattice introduces numerical artifacts, which can be expected to disturb the system for small values of L. In the limit of large L, it is only the anisotropy of the lattice which is different from the ideal off-lattice limit. This numerically induced anisotropy allows for three orientations for the triangular lattice and two orientations for the square lattice. In an off-lattice calculation, the number of orientations would be infinite.

References

1. Seeman N.C., TIBTECH **17** (1999) 437-443
2. Seeman N.C. et al., Nanotechnology **9** (1998) 257-273
3. Ito Y., Fukusaki E., J.Mol.Catal.B:Enzymatic **28** (2004) 155-166
4. Niemeyer Ch.M., Current Opinion in Chemical Biology **4** (2000) 609-618
5. Mao Ch. et al., Nature **397** (1999) 144-146

6. Clarke S. M. et al., Macromol.Chem.Phys. **198** (1997) 3485-3498
7. Sheming Lu et al., Cryst. Res. Technol. **39** (2004) 89-93
8. Stauffer D., Aharony A., Introduction to Percolation Theory, 2nd edition, Taylor and Francis, Routledge 2003
9. Polarz S,. Smarsly B., J.Nanoscience and Nanotechnology **2** (2002) 581-612
10. Hoshen J., Kopelman R., Phys. Rev. B **14** (1976) 3438-3445

Efficient Storage and Processing of Adaptive Triangular Grids Using Sierpinski Curves

M. Bader and Ch. Zenger

Dept. of Informatics, TU München, 80290 München, Germany

Abstract. We present an algorithm to store and process fully adaptive computational grids requiring only a minimal amount of memory. The adaptive grid is specified by a recursive decomposition of triangular grid cells; the cells are stored and processed in an order that is given by Sierpinski's space filling curve. A sophisticated system of stacks is used to ensure the efficient access to the unknowns. The resulting scheme makes it possible to process grids containing more than one hundred million cells on a common workstation, and is also inherently cache efficient.

1 Introduction

One of the most common approaches to modelling and simulation in Computational Science is based on partial differential equations (PDEs) and their numerical discretisation with finite element or similar methods. One of the key requirements in the generation of the respective computational grids, is the possibility for adaptive refinement. Unfortunately, introducing adaptive refinement most often leads to a trade-off between memory requirements and computing time. This is basically a result of the need to obtain the neighbour relationships between grid cells both during the grid generation and during the computation on the grids. Storing these neighbour relations explicitly allows arbitrary, unstructured meshes, but requires a considerable memory overhead. It is not uncommon for codes using unstructured grids to use up more than 1 kilobyte of memory per unknown, a large part of which is due to the explicit storage of the grid structure. In this paper, however, we want to address a situation where memory should be saved as far as possible, which requires the use of a strongly structured grid. The structure helps to save most of the memory overhead, but the neighbour relations now have to be computed, instead. In a strongly adaptive grid, this can be a rather time-consuming task.

In that context, Guenther[3], Mehl[4], and others have recently introduced an approach that allows full adaptivity using recursively substructured rectangular grids (similar to octrees). To store such adaptive grids requires only a marginal amount of memory. To efficiently process the grids they presented a scheme that combines the use of space-filling curves and a sophisticated scheme of stacks. The stack-like access even leads to excellent cache efficiency. Moreover, parallelisation strategies based on space-filling curves are readily available (see [8], for example).

In this paper, we will present a similar approach for grids resulting from a recursive splitting of triangles. Such grid generation strategies are, for example,

V.N. Alexandrov et al. (Eds.): ICCS 2006, Part I, LNCS 3991, pp. 673–680, 2006.

used by Stevenson[6] or Behrens et.al.[1]. Behrens even uses a space-filling curve (Sierpinski curve) for parallelisation, and observes an overall benefit which is a result of the locality properties induced by the space-filling curve. In this article, we will present an algorithm to efficiently process these kinds of triangular grids. The algorithmic scheme will be similar to that used by Guenther[3] and Mehl[4]: it is also based on the use of stack structures, and shows equally promising properties with respect to cache efficiency.

2   Recursively Structured Triangular Grids and Sierpinski Curves

For description of the grid generation, we stick to the simplest case and use a right, isosceles triangle as the computational domain. The computational grid is then constructed in a recursive process. Starting with a grid cell that consists of the entire computational domain, we recursively split each triangle cell into two congruent subcells. This recursive splitting is repeated until the desired resolution of the grid has been reached. The grid may be adaptive and may even contain hanging nodes—only in section 5 will we place some restrictions on the structure of the grid, which result from the application of a multigrid scheme.

Figure 1 shows a simple grid constructed by this scheme. The respective substructuring tree is shown next to it. Note that a respective uniformly refined recursive construction is used to define the so-called *Sierpinski curve*[5]. Thus, we can use the Sierpinski curve to generate a linear order on the grid cells (see Figure 1). This corresponds to a depth-first traversal of the substructuring tree. To store the grid structure therefore requires only one bit per node of the construction tree: this bit indicates whether a node (i.e. a grid cell) is a leave or whether it is adaptively refined.

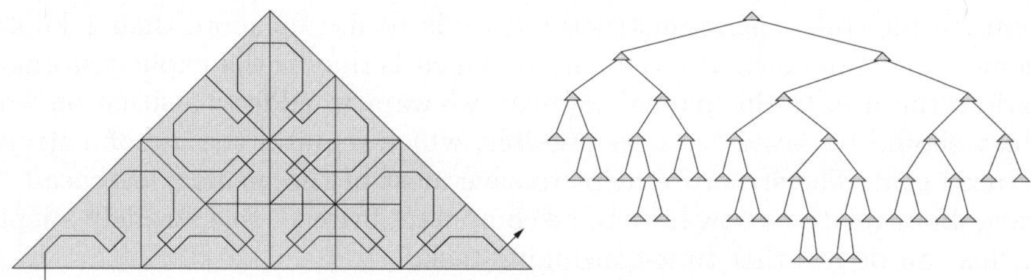

Fig. 1. Recursive construction of the grid on a triangular computational domain

There are several straightforward extensions to this basic scheme, which offer a quite flexible treatment of reasonably complicated computational domains[1]:

1. Instead of only one initial triangle, a simple grid of several initial triangular cells may be used.

2. The cells can be arbitrary triangles instead of right, isosceles ones as long as the structure of the recursive subdivision is not changed: one leg of each triangular cell will be defined as the *tagged edge* and take the role of the hypotenuse.
3. The subtriangles do not need to be real subsets of the parent triangle: whenever a triangle is subdivided, the tagged edge can be replaced by a linear interpolation of the boundary (compare Figure 2).

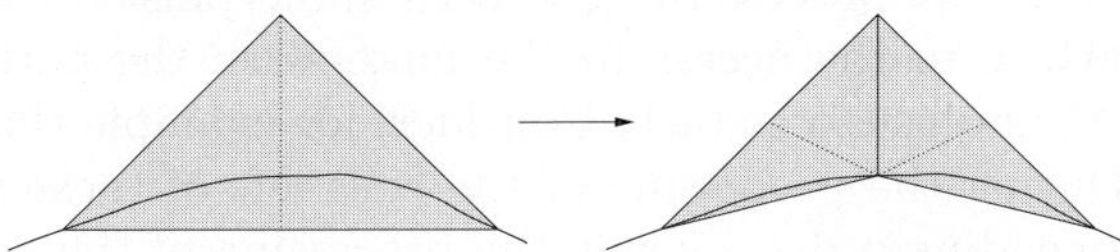

Fig. 2. Subdividing triangles at boundaries

3   Discretisation

After specifying the grid generation strategy, the discretisation of the given PDE is the next step in the simulation pipeline. Consider, for example, a discretisation using linear finite elements on the triangular grid cells. The discretisation will generate an element stiffness matrix and a corresponding right-hand side for each grid cell. Accumulation of these local systems will lead to a global system of equations for the unknowns, which are placed on the nodes of the grid.

Sticking to our overall principle to save memory as far as possible, we assume that storing either the local or the global systems of equations is considered to be too memory-consuming. Instead, we assume that it is possible to compute the stiffness matrix on the fly or even hardcode it into the software. Then our memory requirements are constricted to a minimal amount of overhead to store the recursive grid structure plus the inevitable memory we need to store the values of the unknowns.

Iterative solvers typically contain the computation of the matrix-vector product between stiffness matrix and the vector of unknowns as one of the fundamental subtasks. In a classical, node-oriented approach, this product is evaluated line-by-line using a loop over the unknowns. This requires, for each unknown, to access the unknowns of all neighbouring nodes. However, in a recursively structured grid, as it is described in section 2, not all neighbours will be easily accessible: a neighbour may well be part of an element that lies in an entirely different subtree. Therefore, the grid should be processed in a cell-oriented way, instead. Proceeding through the grid cell-by-cell, we can compute the local contributions of the respective element stiffness matrices, and accumulate the local contributions to compute a correction for the unknowns.

4   Cache Efficient Processing of the Computational Grid

In such a cell-oriented processing of grids, the problem is no longer to access all neighbours of the currently processed unknown. Instead, we need to access all unknowns that are situated within the current element. A straightforward solution would be store the indices of the unknowns for each element, but the special structure of the grids described in section 2 offers a much better option: it requires that the elements are processed along the Sierpinski curve.

Then, as we can see in Figure 3, the Sierpinski curve divides the unknowns into two halves – one half lying to the left of the curve, the other half lying to the right. We can mark the respective nodes with two different colours, for example red and green. Now, if we process the grid cells in Sierpinski order, we recognize an interesting pattern in the access to the unknowns: the pattern is perfectly compatible with the access to a stack. Consider, for example the unknowns 5 to 10 in Figure 3. During the processing of the cells left of these unknowns, they are accessed in ascending order. During the processing of the cells to the right, they are accessed in descending order. In addition, the unknowns 8, 9, and 10 are in turn placed on top of the respective stack.

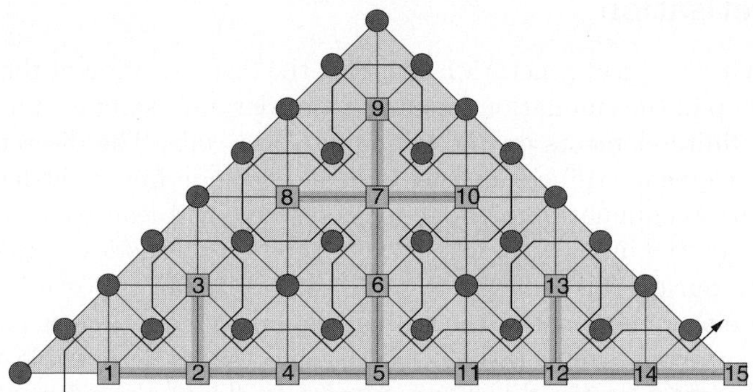

Fig. 3. Marching through a grid of triangular element in Sierpinski order. The nodes to the left and right of the curves are accessed in an order that motivates the use of a stack to store intermediate results.

In the final algorithm, a system of four stacks is required to organise the access to the unknowns:

- one read stack that holds the initial values of the unknowns;
- two helper stacks, a *green* stack and a *red* stack, that hold intermediate results for the unknowns of the respective colour;
- and finally a write stack to store the updated values of the unknowns.

In addition, whenever we move from a processed cell to the next one, two unknowns can always be directly reused, because they are adjacent to the common edge. Thus, we only have to deal with the two remaining unknowns – one in the cell that was just left, the other one in the entered cell –, both lying in a corner opposite to the common edge.

The remaining unknown in the exited cell will either be put onto the write stack (if its processing is already complete) or onto the helper stack of the correct colour (if it still has to be processed by other cells). To decide whether the processing is completed, we can simply use a counter for the number of accesses. To decide the colour of the helper stack, we need to know whether an unknown is lying to the left of the curve or to its right. This decision can be easily made, if we know where the Sierpinski curve enters and leaves the triangle element. As the Sierpinski curve always enters and exits a grid cell at the two nodes adjacent to the hypotenuse, there are only three possible scenarios:

1. the curve enters through the hypotenuse – then it can only leave across the opposite leg (it will not go back to the cell it just left);
2. the curve enters through the adjacent leg and leaves across the hypotenuse;
3. the curve enters and leaves across the opposite legs of the triangle element.

Figure 4 illustrates these three scenarios, and also shows that obtaining the colouring of a node is then trivial.

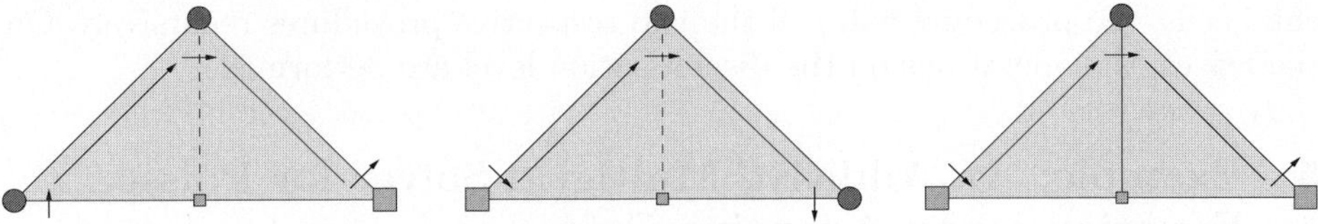

Fig. 4. Three scenarios to determine the colouring of the edges: nodes to the left of the curve are *red* (circles), nodes to the right are *green* (boxes)

Next, we have to decide how to obtain the remaining unknown in the entered grid cell. It will be either be taken from the read stack, if it has not been used before, or otherwise from the respective coloured helper stack. This decision solely depends on whether the unknown has already been accessed before. Therefore, we consider whether the adjacent triangle cells have already been processed or not. For two out of the three neighbour cells, this property is already known: the cell adjacent to the entering edge has already been processed; the cell adjacent to the exit edge has not. The third cell can be either *old* (processed already) or *new* (not yet processed). Consequently, we split each of our existent three scenarios according to this additional criterion, and obtain six new scenarios, such as illustrated in Figure 5. The strategy where to obtain the remaining unknown is then straightforward:

– if at least one of the adjacent edges is marked as *old*, we have to fetch the unknown from the respective coloured stack;
– if both adjacent edges are marked as *new*, we have to fetch the unknown from the read stack.

This strategy works fine, even though it does not consider all elements adjacent to the node. Combinations that would make the scheme fail cannot occur due to the special recursive construction of the Sierpinski curve. Figure 5 also shows that the scenarios for the two respective subcells are always determined.

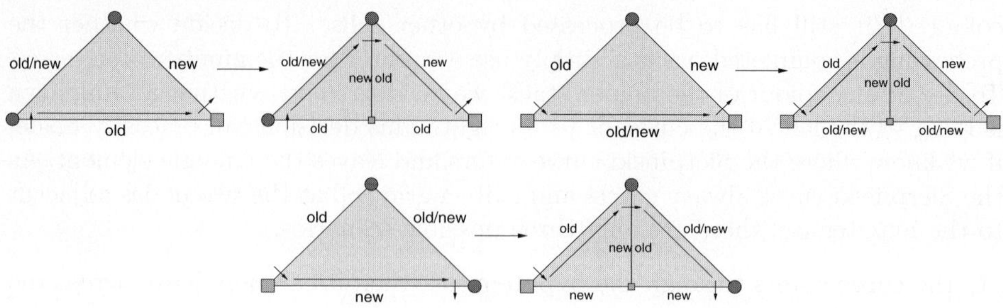

Fig. 5. Determination and recursive propagation of edge parameters

The processing of the grid can thus be managed by a set of six recursive procedures—one for each possible scenario. Each procedure implements the necessary actions in one cell of the grid tree. If the grid cell is subdivided into two child cells, the procedure will call the two respective procedures recursively. On the leaves, the operations on the discretisation level are performed.

5 Example: An Additive Multigrid Solver for Poisson's Equation on an Adaptive Grid

To add a multigrid solver to the existing scheme, we adopted an approach based on a hierarchical generating system such as introduced by Griebel [2]. The respective ansatz functions where set up according to the hierarchical basis functions introduced by Yserentant [7]. The coarse level ansatz functions – that is the coarse level grids – are integrated into our recursive discretisation tree in a way, such that every node of the tree will contain a grid cell: either one of the finest level or one of a coarser level. Thus, the tree not only provides an adaptive computational grid, but at the same time provides a hierarchy of course grids for the multigrid method. When processing the grid cells on the finest level, all parent cells (i.e. the coarse grid cells) are automatically visited, as well. Therefore, an additive multigrid method can easily be integrated into the numerical scheme. Interpolation operations are invoked before the recursive calls to child cells; restriction operations are invoked after these calls are finished. According to the choice of linear finite elements, simple linear interpolation and full-weighted restriction was used.

However, not all of the cell corners should actually carry an unknown in the multilevel sense. Consider the node opposite to the hypotenuse of a cell: in the next refinement step, the adjacent right angle will be split, and the node will be adjacent to two rectangles of half size (belonging to the next finer level). Note that the support of the respective ansatz functions will be of the same extent on these two levels. It is therefore unnecessary to generate both of these duplicate unknowns – we chose to place unknowns only at corners that are adjacent to the hypotenuse; at the opposite corners (adjacent to the right angle) values will be interpolated and no unknowns will be situated there. As a result, a corner that is adjacent to both right angles and acute angles would belong to two different levels. Such situations are therefore forbidden, and the grid generation strategy was modified to prevent these situations.

The resulting multigrid solver was tested for solving Poisson's equation on a triangular domain of unit size. A respective a priori adaptive grid, such as illustrated in figure 6, was prescribed, and the equation was discretised using linear finite elements. The additive multigrid solver was able to achieve convergence rates between 0.7 and 0.8, largely independent of the resolution and adaptive refinement of the mesh. In our test runs, we applied 40 iterations of the multigrid method, which reduced the maximum error to about 10^{-4}. Table 1 lists the run-time and memory requirements for several sample grids. It is remarkable that the algorithms requires only about approximately 19 bytes of memory per degree of freedom. To process one multigrid iteration requires approximately 1.8 seconds per one million degrees of freedom.

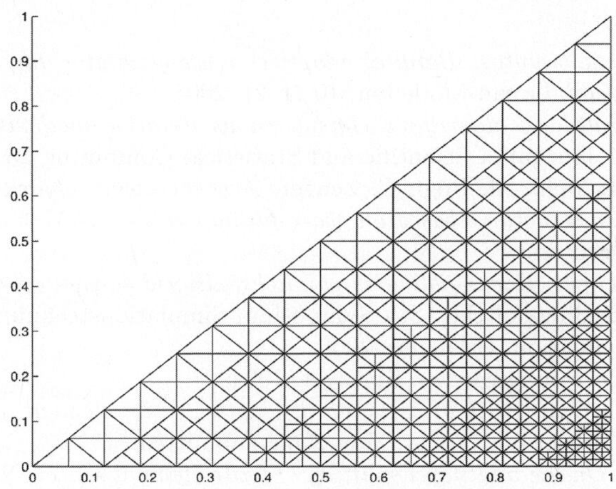

Fig. 6. Computational domain and refined grid for the solution of Poisson's equation (for this illustration, a low resolution grid was used)

Table 1. Run-time and memory requirements for several sample grids on a standard workstations (Pentium4, 3.4 GHz, 1GB main memory)

grid	depth min.	depth max.	number of cells	degrees of freedom	memory [in MB]	proc. time [per iteration]
uniform	26	26	$67.1 \cdot 10^6$	$22.4 \cdot 10^6$	397	39.7 s
adaptive	24	28	$54.3 \cdot 10^6$	$18.1 \cdot 10^6$	323	32.9 s
adaptive	22	30	$69.5 \cdot 10^6$	$23.2 \cdot 10^6$	410	41.4 s
adaptive	12	33	$82.8 \cdot 10^6$	$27.6 \cdot 10^6$	487	49,4 s
adaptive	13	33	$91.3 \cdot 10^6$	$30.4 \cdot 10^6$	538	54.7 s
adaptive	24	30	$116 \cdot 10^6$	$38.8 \cdot 10^6$	684	69.4 s

6   Conclusion

We have presented an algorithm that is able to solve Poisson's equation on an *adaptive* grid of more than 100 million cells in just a few minutes on a common PC. Thus, it has shown to be a promising approach for using full adaptivity at a memory cost and (to a certain extent) computational speed that is competitive to algorithms that are based on regular grids. This is also due to the fact that multigrid methods can be easily integrated.

Encouraged by this proof-of-concept, we are currently implemeting the algorithmic scheme for the grid generator *amatos*[1] to prove its applicability in a full-featured grid generation and simulation package. In the respective applications, the current limitation to 2D grids is still not a severe restriction. Nevertheless, we are also working on an extension of the scheme to the 3D case.

References

1. J. Behrens, et.al. *amatos: Parallel adaptive mesh generator for atmospheric and oceanic simulation.* Ocean Modelling 10 (1–2), 2005.
2. M. Griebel. *Multilevel algorithms considered as iterative methods on semidefinite systems.* SIAM Journal of Scientific and Statistical Computing 15 (3), 1994
3. F. Günther, M. Mehl, M. Pögl, C. Zenger. *A cache-aware algorithm for PDEs on hierarchical data structures based on space-filling curves.* SIAM Journal of Scientific Computing, accepted.
4. M. Mehl, C. Zenger. *Cache-oblivious parallel multigrid solvers on adaptively refined grids.* In: Proceedings of the 18th Symposium Simulationstechnique (ASIM 2005), Frontiers in Simulation, Erlangen, 2005.
5. H. Sagan. *Space Filling Curves.* Springer, 1994.
6. R.P. Stevenson. *Optimality of a standard adaptive finite element method.* Technical Report 1329, Utrecht University, May 2005, submitted.
7. H. Yserentant. *On the multilevel splitting of finite element spaces.* Numerische Mathematik 49 (3), 1986
8. G. Zumbusch. *Adaptive Parallel Multilevel Methods for Partial Differential Equations.* Habilitationsschrift, Universität Bonn, 2001.

Integrating Legacy Authorization Systems into the Grid: A Case Study Leveraging AzMan and ADAM

Weide Zhang, David Del Vecchio, Glenn Wasson, and Marty Humphrey

Department of Computer Science, University of Virginia, Charlottesville, VA USA 22904
{wz6y, dad3e, gsw2c, humphrey}@cs.virginia.edu

Abstract. While much of the Grid security community has focused on developing new authorization systems, the real challenge is often integrating legacy authorization systems with Grid software. The existing authorization system might not understand Grid authentication, might not scale to Grid-level usage, might not be able to understand the operations that are requested to be authorized, and might require an inordinate amount of "glue code" to integrate the native language of the legacy authorization system with the Grid software. In this paper, we discuss several challenges and the resulting successful mechanisms for integrating the Globus Toolkit and WSRF.NET with AzMan, a role-based authorization system that ships with Windows Server 2003. We leverage the OGSA GGF Authorization Interface and our own SAML implementation so that the enterprise can retain their existing AzMan mechanism while resulting in new, scalable mechanisms for Grid authorization.

1 Introduction

Constructing a Grid requires the integration of a potentially large number of new and existing software mechanisms and policies into a reliable, collaborative infrastructure. One of the biggest challenges in this integration is security, particularly authorization: after verifying that the person is whom they say they are (authentication), is the person allowed to perform the requested action? While the Grid community has created a number of excellent authorization systems such as CAS [1], VOMS [2], PERMIS [3], and AKENTI [4], these systems are generally assumed to be *installed* at or around the same time as the Grid software such as the Globus Toolkit [5] or WSRF.NET [6]. However, in many cases it is unrealistic to assume that the adoption of Grid technology means the abandonment of authorization mechanisms already in place. Hence the real challenge is often to integrate a *legacy* authorization system with Grid software. The legacy authorization system might be closely tied to an existing authentication system and might not be able to understand new authentication assertions/tokens. It might be designed to work with a relatively few number of users/objects and might not scale to the size of the Grid being considered. Perhaps it does not understand the requested actions and therefore cannot represent and make decisions about the proposed actions, such as "launch remote job" or "read a file via GridFTP". It is not clear how difficult it would be to get the Grid software to implement the protocol of the legacy authorization system.

This paper describes the integration challenges, approach, and lessons learned as we attempted to integrate the role-based access control (RBAC) system that is

V.N. Alexandrov et al. (Eds.): ICCS 2006, Part I, LNCS 3991, pp. 681–688, 2006.
© Springer-Verlag Berlin Heidelberg 2006

shipped with Microsoft Windows Server 2003 (Authorization Manager, or "AzMan" [7]) with the Globus Toolkit v4 and WSRF.NET as part of the University of Virginia Campus Grid [8]. A key to our solution is that this paper reports one of the first implementations of the GGF OGSA Authorization Interface [9]. Our integration not only allows an enterprise to continue using its AzMan installation as it installs Grid technology, we have found that the Grid management is further enhanced over the state of the art in improved support for dynamically modifying authorization policies, maintaining a consistent view of site-wide policies, and reducing the cost of policy management.

2 Legacy Authorization System: Overview of AzMan and ADAM

In this section, we describe AzMan, the legacy authorization system in our case study. AzMan (Section 2.1) is a general-purpose, role-based authorization architecture on Windows platforms. Section 2.2 describes ADAM, a lightweight Windows service for directory-enabled applications, which we use in combination with AzMan.

2.1 AzMan

In traditional access control mechanisms based on Access Control Lists (ACL), users are directly mapped to resource permissions using a list of authorized users for each target resource. In many situations, ACL-based systems do not scale well, in that if a new user "Fred" is introduced into the system, all of the objects which he should have access to must have their ACLs changed. Role-Based Access Control (RBAC) [10] adds a *role* layer between users and permissions. For example, assume that before Fred comes along, resources have been designed to allow certain operations according to well-defined roles such as "salesperson". Then once Fred is hired, rather than changing the ACLs on all resources Fred needs access to, he just needs to be recognized as having the "salesperson" role. Since users can typically be categorized into a number of different roles, RBAC tends to be a more scalable and flexible approach.

Conceptually, the primary purpose of AzMan is to provide a "yes" or "no" answer when asked at run-time (via a Microsoft COM API) if a particular authenticated identity is allowed to perform a particular action. AzMan also provides a graphical interface and a separate API for entering and configuring identities, roles, permissions, etc. AzMan allows for the definition of any number of access control policies, the central concepts of which are *Roles* and *Permissions*. Subjects are assigned Roles, and Roles are granted or denied Permissions for certain tasks. Roles can actually be assigned to either individual subjects or groups of security principals. Such grouping can even be computed at run-time based on an Active Directory (LDAP) lookup. In addition to dynamic groups, policies themselves can also have a dynamic element. In particular, they can reference *BizRules,* which are scripts that get executed when particular Permissions are requested, so that run-time information like "time of day" can be used to make an authorization decision.

Typically, AzMan authorization policies are grouped into named policy sets based on the application to which they apply. Note, however, that access rights are not directly associated with specific target resources (as an ACL would be for a file in the filesystem). In fact, policies are completely resource-independent; AzMan requires authorization decision requests to identify only the subject, intended task and the name of policy set. When an application that uses AzMan is initialized, it loads the authorization policy information from a policy store. AzMan provides support for storage of authorization policies locally in XML files on the AzMan server, or remotely in Active Directory (or ADAM, next section).

2.2 Active Directory Application Mode (ADAM)

Active Directory Application Mode (ADAM) is a relatively new capability in Active Directory that addresses certain deployment scenarios of directory-enabled applications [11]. In contrast to Active Directory, ADAM can be used for storing information that is not globally interesting. One example usage in an authorization decision-making context involves storing the names of policies that apply to a particular resource on a per-resource basis. ADAM is also valuable for those situations in which a particular application must store personalization data for users who are authenticated by Active Directory. Storing this personalization data in Active Directory would sometimes require schema changes to the user class in Active Directory; ADAM can be used as an alternative.

3 Leveraging AzMan and ADAM for Grid Authorization

The challenge then, is to use our legacy authorization system (AzMan and ADAM) for Grid authorization, thereby providing minimal disruption to the enterprise when attempting to deploy a Grid. In our architecture, resource information is distributed among a collection of ADAM servers based on the resource's DNS name. This allows the owner/administrator of the resources in each sub-domain to configure his or her ADAM server independently, including the authorization policy that should be applied, thereby providing the domain autonomy that is so vital to the Grid. The ADAM servers are organized hierarchically: queries to a parent ADAM server will be forwarded through to the appropriate child sub-domain.

Policy management is divided into two parts: RBAC policy management and resource-to-policy mapping. RBAC policy management includes defining roles and role permissions while resource-to-policy mapping involves defining which RBAC policies apply to a resource or resource group (the latter being the responsibility of resource owners). Out of the box, AzMan only supports assigning roles to subjects identified by Windows security tokens, which each have a unique Security Identifier (SID) [7]. In multi-organization Grids, however, X.509 Distinguished Names (DNs) are usually used to identify subjects. So for AzMan to work with Grid identities, we setup a mapping between X.509 DNs and unique custom-defined SIDs. Since this mapping could be used by several authorization servers, we made DN-to-SID lookups possible via a Web service interface. This subject-mapping Web service uses a flat XML file to store the mappings, which is fine for relatively small numbers of subjects; a more sophisticated storage mechanism (like a relational database or XML database) could easily be substituted. With this subject mapping service in place, roles in AzMan can easily be assigned to custom SIDs as desired. All domains within a

Virtual Organization (VO) share AzMan policy set names and the policy definitions they contain so that consistent authorization decisions can be made across the VO.

A typical authorization workflow is as follows (see Figure 1). A client sends a signed request message to a Web service (step 1) which defers authorization decisions to an authorization callout library. The SAML callout library contacts a SAML Authorization Web service (2), sending it the client's DN, the URL of the target service/resource, and the name of the requested operation. The SAML Authz Web service is relatively generic and relies on the AzMan Handler library (3) to provide the glue code that understands how to query AzMan for Grid authorization decisions. The first action the AzMan Handler takes is to retrieve from the Subject Mapping Service, the SID that corresponds to the subject DN (4, 5). Next, the distributed hierarchy of ADAM servers will be queried (6) to determine which RBAC policy set (7) applies to the requested resource. At this point, the AzMan engine will be invoked to make an authorization decision based on the subject's SID, the returned policy set name and the desired operation (or task). The AzMan engine will take care of retrieving the indicated set of policies from its policy store (8, 9), identifying the roles assigned to a subject SID and checking the permissions assigned to those roles. AzMan's authorization decision (10) is then relayed back through the SAML Authz Web service to the original Web service (11) which proceeds to execute the desired operation (if authorized) and return the results to the client (12).

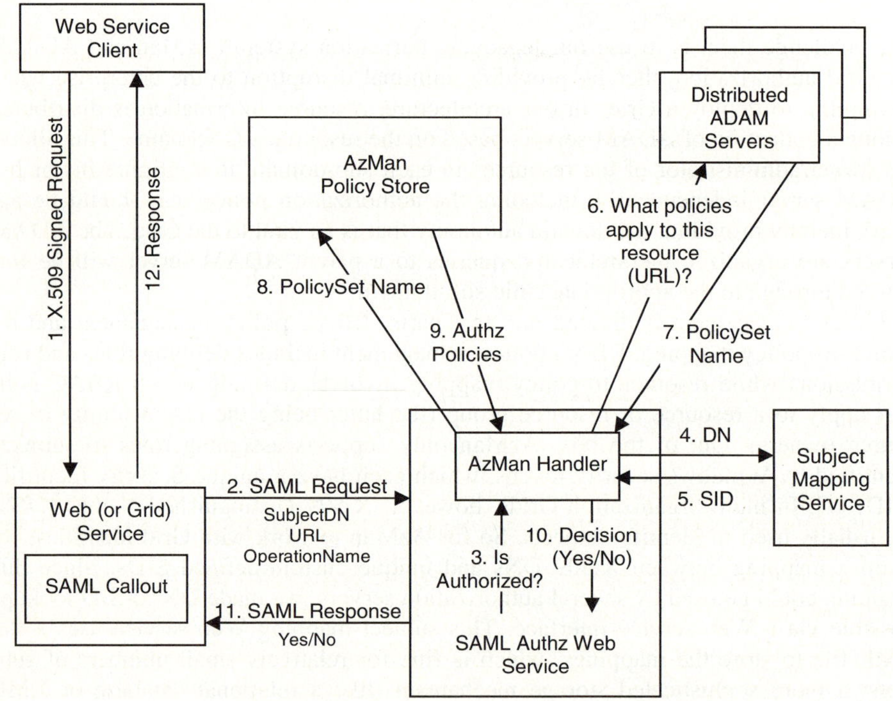

Fig. 1. Authorization Workflow using AzMan and ADAM

A key component of our approach is the OGSA Authorization Interface [9]. SAML [12] is used as a format for requesting and expressing authorization assertions to a SAML Authorization Web service. A standard XML message format for SAML requests, assertions and responses is provided. SAML defines a request/response protocol in which the request contains {subject, resource, action and supporting credentials} and the response contains either authorization assertions about the subject with respect to the resource or a yes/no decision. While, in principle, the Web service could use authorization assertions to make its own authorization decision, we have found that Web services typically can't or don't want to make such decisions. In addition, returning a yes/no answer allows the Web service owner to be separate from the entity that sets access control policy.

A Java-based SAML Callout library is distributed as part of the latest version of the Globus Toolkit as one option for enforcing access control in Grid Services. We provide a similar (and interoperable) callout library for Web services on the .NET platform. The SAML Authz Web service shown in Figure 1 is a .NET Web service that is relatively generic with regard to how it actually makes its authorization decisions. It can be configured (via a configuration file, or even at run-time) to use one or more authorization handler libraries. In this paper we've considered AzMan as our legacy authorization system, but any other authorization system could also be substituted into our architecture with appropriate authorization handler glue code. Since the SAML Authz Web service is interoperable with both .NET and Java and callout libraries, AzMan (or another authorization system) can be used to make authorization decisions for Web services running on just about any platform. Thus we accommodate the heterogeneity that is present in many organizations and multi-organization Grids.

4 Evaluation

We now assess our architecture and implementation as described in Section 3.

System Scalability: Cost of Administering Authorization Policies. Modifying authorization policy in this system requires two steps: updating the RBAC policies in the policy store, and updating the mapping of those policies to specific resources. In our system, authorization policy modification is inexpensive. The RBAC policies can be updated through AzMan's COM interface or the Microsoft Management Console (MMC), a graphical management interface. In order to update the policy that applies to a specific resource, the resource provider or authorized user changes the resource/application mapping stored in distributed ADAM servers. The worst case cost of policy modification (that affects every user, every role and every resource) in our infrastructure is U+P+R, where U is the number of users, P is the number of permissions and R is the number resources, in the VO. The vast majority of policy changes, however, only require modifications along one of these three dimensions.

Resource Scalability: Cost of Handling Dynamic Grids. In large Grid environments, many resources are not static, but instead change dynamically. In our system, resource groups and policies that govern those groups are maintained separately (in ADAM and AzMan, respectively). Resource owners/administrators can freely update information about resources in particular domains and sub-domains by changing the information stored in distributed, hierarchical ADAM servers. The resource-independent policies in the AzMan policy store will be unaffected by these resource changes with resource-to-policy mappings inspected at run-time. In contrast to this architecture, other authorization systems such as Permis [3], Akenti [4], and CAS [1] all store policies that contain resource information directly. In these systems, there is no upper bound on how many policies will be affected when a resource changes (for example, has its trust domain revoked).

Membership Scalability: Cost of Dynamic User Membership. Resources are not the only dynamic element in large-scale Grids, users change as well. Consequently, our RBAC policies stored in AzMan do not mention users directly. Instead, our strategy is to define globally unique identifiers (SIDs) that map to X.509 DNs. RBAC policies are specified in terms of roles assigned to SIDs. Changes to the set of users in a organization only requires a change to the subject DN-to-SID mapping table. Changes in role assignments are similarly localized and the policy definitions themselves are not affected by them. Some other authorization infrastructures store User Attribute Certificates or directly embed membership information into authorization policies, yielding a less scalable approach.

Resource Provider and VO Independence. Our authorization infrastructure separates the responsibility for mapping roles to permissions from the responsibility for applying policies to resources. This allows resource providers independent control over which policies apply to their resources while allowing VO administrators to define policies that (potentially) apply to all resources in the VO. This separation allows resource providers to react quickly (and separately) to change policies for their resources in the event of security breaches or other malicious activity taking place in the VO. Similarly, it allows VO administrators to effectively remove resources from the VO (by removing permissions from roles) if those resources are deemed to be acting inappropriately.

Trust Management. There are three aspects of trust that are relevant to any deployment of our system: trust of users, trust of the services involved in authorization, and trust of the authorization policies. Trust of users refers to the mechanism by which the system can trust the user and authenticate the user's identity. In our UVA Campus Grid, the UVA CA acts as an Identity Authority that services are configured to trust. Any service must therefore verify that the user's certificate is signed by the UVA CA, to establish trust. Trust of the authorization services refers to the mechanism by which the distributed ADAM servers (each of which lies in a different administrative domain) can trust the Authorization Service. In the UVA Campus Grid, the (SAML) Authorization Web service impersonates a user account to access distributed ADAM servers. This user account is set beforehand in the ADAM servers and is granted the "read" privilege so that the Authorization Service will be able to read policy names applied to particular resources. The Subject Mapping Service must also be trusted by the Authorization Service with X.509 signed messages between these two services assumed. Trust of authorization policies refers to how the Authorization Service can trust the policies that come from the AzMan service. In the UVA Campus Grid, the Authorization Service is simply configured to trust the AzMan service (and hence the policies it provides). An improvement on this would be to have each policy provider sign the policies, allowing different Authorization Services to determine if different policies come from sources they trust.

Consistent View and Flexibility of the AuthZ Policy. In our infrastructure, the authorization decision point dynamically retrieves the name of policies that apply to a resource from that resource's ADAM server. Members of a Virtual Organization (VO) will have a globally consistent view of the named policy sets and their constituent policy definitions since the AzMan policy store is accessible across the VO. In some authorization infrastructures, policies are either defined locally or signed policy certificates (in Akenti [4]) are specified for a target resource so that authorization decisions must be made locally. Without a consistent global view of policies, it becomes more difficult to make certain kinds of policy changes, for example check and modify some local policies from previously denying access to allowing a certain user access to the target resource. Also, there is also no guarantee users who possesses VO-wide administrator roles can be authorized for access across multiple administrative domains.

5 Summary

In this paper, we presented both grid authorization requirements and a new authorization infrastructure based on AzMan, authorization mechanisms included with Windows Server 2003. Our integration not only allows an enterprise to continue using its AzMan installation as it deploys Grid technology, but we have found that Grid authorization management is further enhanced over the state of the art through improved support for dynamically modifying authorization policies, maintaining a consistent view of VO policies, and reducing the cost of policy management. Future work will focus on analyzing the scalability and reliability issues the hierarchical and distributed ADAM servers introduce as well as better understanding the performance overhead of our authorization architecture.

References

[1] I. Foster, C. Kesselman, L. Pearlman, S. Tuecke, and V. Welch "The Community Authorization Service: Status and Future", *in Proceedings of Computing in High Energy Physics 03(CHEP '03),2003*

[2] R. Alfieri, R. Cecchini, V. Ciaschini, L. dell'Agnello, Á. Frohner, K. Lőrentey and F. Spataro, "From gridmap-file to VOMS: managing authorization in a Grid environment." *in Future Generation. Comput. Syst. 21, 4, 549-558(2005).*

[3] D. Chadwick, A. Otenko, "The PERMIS X.509 Role Based Privilege Management Infrastructure", in *Future Generation Comp. Syst. 19(2), 277-289(2003)*

[4] M. Thompson, S. Mudumbai, A. Essiari, W. Chin, "Authorization Policy in a PKI Environment", 2002. [Online]. Available: http://dsd.lbl.gov/security/Akenti/Papers/ NISTPKWorkshop.pdf

[5] Globus project. http://www.globus.org

[6] M. Humphrey and G. Wasson. Architectural Foundations of WSRF.NET. International Journal of Web Services Research. 2(2), pp. 83-97, April-June 2005.

[7] D. McPherson, "Role-Based Access Control for Multi-tier Applications Using Authorization Manager" [Online]. Available: http://www.netscum.dk/technet/prodtechnol/windowsserver2003/technologies/management/athmanwp.mspx

[8] M. Humphrey and G. Wasson. The University of Virginia Campus Grid: Integrating Grid Technologies with the Campus Information Infrastructure. *2005 European Grid Conference (ECG 2005)*, Amsterdam, The Netherlands, Feb 14-16, 2005.

[9] V. Welch, R. Ananthakrishnan, F. Siebenlist, D. Chadwick, S. Meder, L. Pearlman, "Use of SAML for OGSA Authorization", submitted in GGF OGSA Security Working Group, Aug 2003

[10] R. Sandhu, E. Coyne, H. Feinstein, C. Youman, "Role-Based Access Control Models", *in IEEE Computer* 29(2), 38-47 (1996)

[11] Microsoft Corporation. "Introduction to Windows Server 2003 Active Directory Application Mode". Nov 5, 2003. http://www.microsoft.com/windowsserver2003/techinfo/overview/adam.mspx

[12] Security Assertion Markup Language (SAML) OASIS Technical Committee. http://www. oasis-open.org/ committees/tc_home.php?wg_abbrev=security

Quasi-Gaussian Particle Filtering*

Yuanxin Wu, Dewen Hu, Meiping Wu, and Xiaoping Hu

Department of Automatic Control, College of Mechatronics and Automation,
National University of Defense Technology, Changsha, Hunan, P.R. China, 410073
yuanx_wu@hotmail.com

Abstract. The recently-raised Gaussian particle filtering (GPF) introduced the idea of Bayesian sampling into Gaussian filters. This note proposes to generalize the GPF by further relaxing the Gaussian restriction on the prior probability. Allowing the non-Gaussianity of the prior probability, the generalized GPF is provably superior to the original one. Numerical results show that better performance is obtained with considerably reduced computational burden.

1 Introduction

The Bayesian probabilistic inference provides an optimal solution framework for dynamic state estimation problems [1, 2]. The Bayesian solution requires propagating the full probability density function, so in general the optimal nonlinear filtering is analytically intractable. Approximations are therefore necessary, e.g., Gaussian approximation to the probability [3-9]. This class of filters is commonly called as the Gaussian filters, in which the probability of interest, e.g. the prior and posterior probabilities, are approximated by Gaussian distribution. An exception is the so-called augmented unscented Kalman filter [10] where the prior probability is encoded by the nonlinearly transformed deterministic sigma points instead of by the calculated mean and covariance from them. By so doing, the odd-order moment information is captured and propagated throughout the filtering recursion, which helps improve the estimation accuracy. This note will show that similar idea can be applied to the recently-raised Gaussian particle filtering (GPF) [7].

The GPF was developed using the idea of Bayesian sampling under the Gaussian assumption [7]. It actually extends the conventionally analytical Gaussian filters via Monte Carlo integration and the Bayesian update rule [11]. The Gaussian assumption being valid, the GPF is asymptotically optimal in the number of random samples, which means that equipped with the computational ability to handle a large number of samples the GPF is supposed to outperform any analytical Gaussian filter. The GPF also have a lower numerical complexity than particle filters [7].

This work generalizes the GPF by relaxing the assumption of the prior probability being Gaussian. Since the prior probability is allowed to be non-Gaussian, the resulting filter is named as the quasi-Gaussian particle filtering (qGPF) in the sequel. It turns out

---

* Supported in part by National Natural Science Foundation of China (60374006, 60234030 and 30370416), Distinguished Young Scholars Fund of China (60225015), and Ministry of Education of China (TRAPOYT Project).

V.N. Alexandrov et al. (Eds.): ICCS 2006, Part I, LNCS 3991, pp. 689–696, 2006.

that the qGPF outperforms the GPF in both accuracy and computational burden. The contents are organized as follows. Beginning with the general Bayesian inference, Section II derives and outlines the qGPF algorithm. Section III examines two representative examples and the conclusions are drawn in Section IV.

2 Quasi-Gaussian Particle Filtering

Consider a discrete-time nonlinear system written in the form of dynamic state space model as

$$x_k = f_{k-1}\left(x_{k-1}, w_{k-1}\right)$$
$$y_k = h_k\left(x_k, v_k\right) \tag{1}$$

where the process function $f_k : \mathbb{R}^n \times \mathbb{R}^r \to \mathbb{R}^n$ and observation function $h_k : \mathbb{R}^n \times \mathbb{R}^s \to \mathbb{R}^m$ are some known functions. The process noise $w_k \in \mathbb{R}^r$ is uncorrelated with the past and current system states; the measurement noise $v_k \in \mathbb{R}^s$ is uncorrelated with the system state and the process noise at all time instants. The probabilities of the process and measurement noises are both assumed to be known.

Denote by $y_{1:k} \triangleq \{y_1, \ldots, y_k\}$ the observations up to time instant k. The purpose of filtering is to recursively estimate the posterior probability $p(x_k \mid y_{1:k})$ conditioned on all currently available but noisy observations. The initial probability of the state is assumed to be $p(x_0 \mid y_{1:0}) \equiv p(x_0)$. The prior probability is obtained via the Chapman-Kolmogorov equation

$$p\left(x_k \mid y_{1:k-1}\right) = \int_{\mathbb{R}^n} p\left(x_k \mid x_{k-1}\right) p\left(x_{k-1} \mid y_{1:k-1}\right) dx_{k-1}. \tag{2}$$

The transition probability density $p(x_k \mid x_{k-1})$ is uniquely determined by the known process function and the process noise probability. Using the Bayesian rule, the posterior probability is given by

$$p\left(x_k \mid y_{1:k}\right) = \frac{p\left(y_k \mid x_k\right) p\left(x_k \mid y_{1:k-1}\right)}{p\left(y_k \mid y_{1:k-1}\right)} \tag{3}$$

where the normalizing constant

$$p\left(y_k \mid y_{1:k-1}\right) = \int_{\mathbb{R}^n} p\left(y_k \mid x_k\right) p\left(x_k \mid y_{1:k-1}\right) dx_k. \tag{4}$$

The likelihood probability density $p(y_k \mid x_k)$ is uniquely determined by the known observation function and the measurement noise probability. Equations (2)-(4) constitute the foundation of the optimal Bayesian probabilistic inference. Unfortunately, the exact analytic form only exists for a couple of special cases, e.g., when the system (1) is linear and Gaussian. In order to make the filtering problem tractable, approximation must be made.

Next, we start to derive the qGPF by assuming the posterior probability at time instant $k-1$ to be well approximated by a Gaussian distribution, i.e.,

$$p\left(x_{k-1} \mid y_{1:k-1}\right) \approx \mathcal{N}\left(x_{k-1}; m_{k-1}, P_{k-1}\right) \tag{5}$$

Substituting (5) and using the Monte-Carlo integration [12], the prior probability in (2) is

$$
\begin{aligned}
p\left(x_k \mid y_{1:k-1}\right) &\approx \int_{\mathbb{R}^n} p\left(x_k \mid x_{k-1}\right) \mathcal{N}\left(x_{k-1}; m_{k-1}, P_{k-1}\right) dx_{k-1} \\
&= \int_{\mathbb{R}^n} p\left(x_k \mid x_{k-1}\right) \sum_{i=1}^{M_1} \frac{1}{M_1} \delta\left(x_{k-1} - x_{k-1}^i\right) dx_{k-1} = \frac{1}{M_1} \sum_{i=1}^{M_1} p\left(x_k \mid x_{k-1}^i\right)
\end{aligned}
\tag{6}
$$

where x_{k-1}^i are random samples from the assumed posterior probability at time instant $k-1$, i.e., $\mathcal{N}\left(x_{k-1}^i; m_{k-1}, P_{k-1}\right)$, $i=1,\ldots,M_1$. The idea of the importance sampling [13, 14] is crucial to numerically implement the Bayesian rule, through which the prior probability is updated to yield the posterior probability using the information provided by the newcome observation. In view of the difficulty of drawing samples directly from the posterior probability $p\left(x_k \mid y_{1:k}\right)$, the importance sampling proposes to sample from a choice importance density $q\left(x_k \mid y_{1:k}\right)$ instead, from which random samples can be readily generated. With (6), the posterior probability in (3) is rewritten as

$$
\begin{aligned}
p\left(x_k \mid y_{1:k}\right) &\propto p\left(y_k \mid x_k\right) p\left(x_k \mid y_{1:k-1}\right) \propto \frac{p\left(y_k \mid x_k\right) \sum\limits_{i=1}^{M_1} p\left(x_k \mid x_{k-1}^i\right)}{q\left(x_k \mid y_{1:k}\right)} q\left(x_k \mid y_{1:k}\right) \\
&\propto \sum_{j=1}^{M_2} \frac{p\left(y_k \mid x_k^j\right) \sum\limits_{i=1}^{M_1} p\left(x_k^j \mid x_{k-1}^i\right)}{q\left(x_k^j \mid y_{1:k}\right)} \delta\left(x_k - x_k^j\right) \triangleq \sum_{j=1}^{M_2} \hat{w}_k^j \delta\left(x_k - x_k^j\right)
\end{aligned}
\tag{7}
$$

where x_k^j are random samples from the importance density $q\left(x_k \mid y_{1:k}\right)$ and

$$\hat{w}_k^j = \frac{p\left(y_k \mid x_k^j\right) \sum\limits_{i=1}^{M_1} p\left(x_k^j \mid x_{k-1}^i\right)}{q\left(x_k^j \mid y_{1:k}\right)}, \quad j=1,\ldots,M_2. \tag{8}$$

Considering the normalization condition, the posterior probability at time instant k is approximated by

$$p\left(x_k \mid y_{1:k}\right) = \sum_{j=1}^{M_2} w_k^j \delta\left(x_k - x_k^j\right) \tag{9}$$

where

$$w_k^j = \hat{w}_k^j \Big/ \sum_{j=1}^{M_2} \hat{w}_k^j. \tag{10}$$

Table 1. Quasi-Gaussian Particle Filtering

<table>
<tr><td>

1.   Draw samples from the posterior probability at time instant $k-1$, i.e.,
$$x_{k-1}^{i} \sim p\left(x_{k-1} \mid y_{1:k-1}\right) \approx \mathcal{N}\left(x_{k-1}; m_{k-1}, P_{k-1}\right), \quad i=1,\ldots,M_1;$$

2.   Draw samples from the important density, that is, $x_k^j \sim q\left(x_k \mid y_{1:k}\right)$, $j=1,\ldots,M_2$;

3.   Assign each sample x_k^j a weight w_k^j according to (8) and (10);

4.   Calculate the mean and covariance according to (11), then
$$p\left(x_k \mid y_{1:k}\right) \approx \mathcal{N}\left(x_k; m_k, P_k\right).$$

</td></tr>
</table>

Then approximate the posterior probability at time instant k by $\mathcal{N}\left(x_k; m_k, P_k\right)$ in which

$$m_k = \sum_{j=1}^{M_2} w_k^j x_k^j, \quad P_k = \sum_{j=1}^{M_2} w_k^j \left(x_k^j - m_k\right)\left(x_k^j - m_k\right)^{T}. \tag{11}$$

This ends the derivation and the resulting algorithm is summarized and outlined in Table I. It is clear from above that we only assume the posterior probability to be Gaussian while do not impose any restriction on the prior probability, which is the major difference from the GPF. Recall that the GPF approximates both the prior and posterior probabilities by Gaussian densities. To be more specific, the GPF approximates the discrete representation of the prior probability $p\left(x_k \mid y_{1:k-1}\right)$ by a Gaussian density, from which random samples are regenerated to be weighted by the likelihood $p\left(y_k \mid x_k\right)$, while the qGPF directly employs the discrete representation of the prior probability. This resembles the difference between the non-augmented UKF and augmented UKF [10]. By a peer-to-peer comparison between the qGPF and the GPF ([7], Table I), we see that the qGPF needs not to calculate the sample mean and covariance for the assumed Gaussian prior probability and thus has lower numerical complexity.

The following theorem says that the mean and covariance in (11) converge almost surely to the true values under the condition that the posterior probability at time instant $k-1$ is well approximated by a Gaussian distribution.

Theorem: If the posterior probability $p\left(x_{k-1} \mid y_{1:k-1}\right)$ is a Gaussian distribution, then the posterior probability expressed in (9) converges almost surely to the true posterior probability $p\left(x_k \mid y_{1:k}\right)$.

Proof: it is a very straightforward extension of Theorem 1 in [7] and thus omitted here.

It follows as a natural corollary that the mean and covariance in (11) converge almost surely to the true value. Therefore the qGPF is provably better than the GPF in accuracy because the former takes the non-Gaussianity of the prior probability into consideration. Note that the non-Gaussianity of the prior probability is not uncommon for nonlinear/non-Gaussian systems.

In theory, we could assume the posterior probability to be any other distribution, as long as the samples from the distribution were easily obtained, e.g. mixed Gaussian [15-17]. The derivation procedure and theoretical proof would be analogical to the above.

3 Numerical Results

This section examines the qGPF via the univariate nonstationary growth model and bearing only tracking, which have been extensively investigated in the literature [2, 7, 18, 19]. We also carried out the GPF for comparison. The prior probability was selected as the importance density for both filters, i.e., $q\left(x_k \mid y_{1:k}\right) = p\left(x_k \mid y_{1:k-1}\right)$.

Univariate Nonstationary Growth Model
The model is formulated as

$$x_k = f_{k-1}\left(x_{k-1},k\right) + w_{k-1}$$
$$y_k = h_k\left(x_k\right) + v_k, \quad k = 1,\ldots,N \tag{12}$$

where $f_{k-1}\left(x_{k-1},k\right) = 0.5x_{k-1} + 25\dfrac{x_{k-1}}{1+x_{k-1}^2} + 8\cos\left(1.2(k-1)\right)$, $h_k\left(x_k\right) = x_k^2/20$. The

process noise w_{k-1} and measurement noise v_k are zero-mean Gaussian with variances Q_{k-1} and R_k, respectively. In our simulation, $Q_{k-1} = 10$ and $R_k = 1$. This model has significant nonlinearity and is bimodal in nature depending on the sign of observations. The reference data were generated using $x_0 = 0.1$ and $N = 100$. The initial probability $p\left(x_0\right) \sim \mathcal{N}\left(0,1\right)$.

The mean square error (MSE) averaged across all time instants defined as $\text{MSE} = \sum_{k=1}^{N}\left(x_k - x_{k|k}\right)^2 \Big/ N$ is used to quantitatively evaluate each filter. We carried out 50 Monte Carlo runs for $M_1 = M_2 = 20, 50, 100, 200, 400$, respectively. Figure 1 shows MSEs as a function of the number of samples. The qGPF remarkably outperforms the GPF. With the same number of samples, MSE of the qGPF is less than half of that of the GPF; on the other hand, to achieve comparable performance the GPF needs at least as twice samples as the qGPF does. The average running time of the qGPF is about 20 percent less than that of the GPF.

Bearing Only Tracking
The target moves within the $s-t$ plane according to the standard second-order model

$$x_k = \Phi x_{k-1} + \Gamma w_{k-1}, \qquad k = 1,\ldots,N \tag{13}$$

where $x_k = \left[s, \dot{s}, t, \dot{t}\right]_k^T$, $w_k = \left[w_s, w_t\right]_k^T$,

$$\Phi = \begin{bmatrix} 1 & 1 & 0 & 0 \\ 0 & 1 & 0 & 0 \\ 0 & 0 & 1 & 1 \\ 0 & 0 & 0 & 1 \end{bmatrix} \text{ and } \Gamma = \begin{bmatrix} 0.5 & 0 \\ 1 & 0 \\ 0 & 0.5 \\ 0 & 1 \end{bmatrix}.$$

Here s and t denote Cartesian coordinates of the moving target. The system noise $w_k \sim \mathcal{N}(0, QI_2)$. A fixed observer at the origin of the plane takes noisy measurements of the target bearing

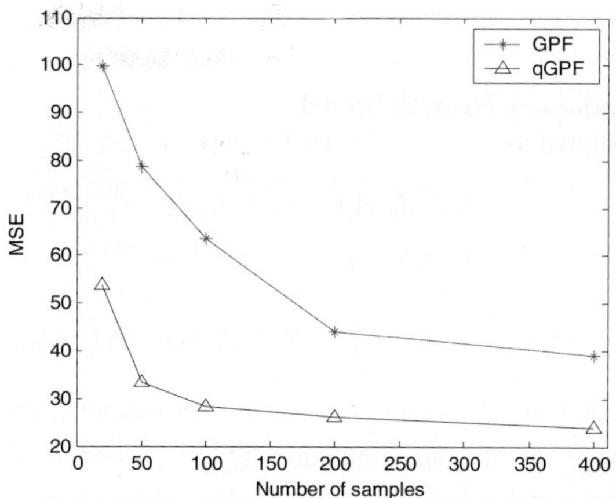

Fig. 1. MSE as a function of the number of samples for both filters

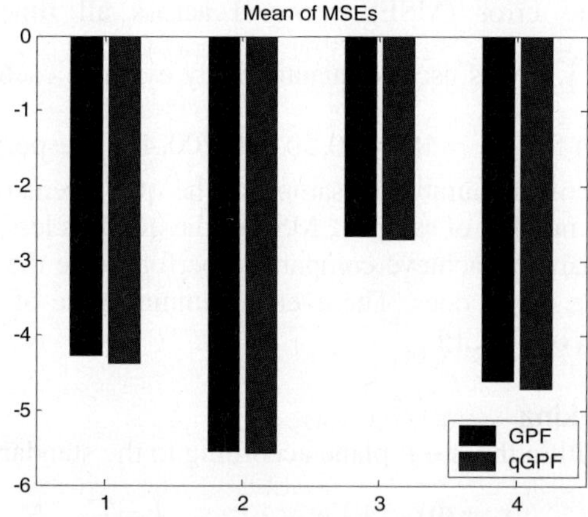

Fig. 2. Averaged MSE of all four coordinates (logarithmic in y axis)

$$y_k = \arctan\left(t_k / s_k\right) + v_k \tag{14}$$

where the measurement noise $v_k \sim \mathcal{N}\left(0, R\right)$. The reference data were generated using $Q = 0.001^2$, $R = 0.005^2$ and $N = 24$. The initial true state of the system was $x_0 = [-0.05, 0.001, 0.7, -0.055]^T$ and the initial estimate was $x_{0|0} = x_0$ with covariance $P_{0|0} = \mathrm{diag}\left(\left[0.1^2, 0.005^2, 0.1^2, 0.01^2\right]^T\right)$.

We carried out 100 random runs for $M_1 = M_2 = 1000$ and the averaged MSEs of all four coordinates are given in Fig. 2. We see that the qGPF is smaller in MSE, though marginally, than the GPF. Similar observations were obtained for various number of samples and is omitted here for brevity. In the simulation, eighteen percent of computational time was spared by using the qGPF.

4 Conclusions

This note proposes the qGPF filter that generalizes the GPF by allowing the prior probability to be non-Gaussian. It has provable superiority over the GPF. The numerical results show that the qGPF achieves (sometimes remarkably) better improvement in estimation accuracy with lower numerical complexity than the GPF. Theoretically, the posterior probability could be assumed to be any other distribution as long as it was readily sampled, e.g., mixed Gaussian. In such a case, it is promising for the qGPF to be used to construct more superior filter than the GPF-based Gaussian sum particle filter in [17].

References

[1] Y. C. Ho and R. C. K. Lee, "A Bayesian approach to problems in stochastic estimation and control," *IEEE Transactions on Automatic Control*, vol. AC-9, pp. 333-339, 1964.

[2] M. S. Arulampalam, S. Maskell, N. Gordon, and T. Clapp, "A tutorial on particle filters for online nonlinear/non-Gaussian Bayesian tracking," *IEEE Transactions on Signal Processing*, vol. 50, no. 2, pp. 174-188, 2002.

[3] K. Ito and K. Q. Xiong, "Gaussian filters for nonlinear filtering problems," *IEEE Transactions on Automatic Control*, vol. 45, no. 5, pp. 910-927, 2000.

[4] A. H. Jazwinski, *Stochastic Processing and Filtering Theory*. New York and London: Academic Press, 1970.

[5] S. J. Julier and J. K. Uhlmann, "A new extension of the Kalman filter to nonlinear systems," in *Signal Processing, Sensor Fusion, and Target Recognition VI*, vol. 3068, *Proceedings of the Society of Photo-Optical Instrumentation Engineers (SPIE)*, 1997, pp. 182-193.

[6] R. E. Kalman, "A new approach to linear filtering and prediction problems," *Transactions of the ASME, Journal of Basic Engineering*, vol. 82, pp. 34-45, 1960.

[7] J. H. Kotecha and P. A. Djuric, "Gaussian particle filtering," *IEEE Transactions on Signal Processing*, vol. 51, no. 10, pp. 2592-2601, 2003.

[8] M. Norgaard, N. K. Poulsen, and O. Ravn, "New developments in state estimation for nonlinear systems," *Automatica*, vol. 36, no. 11, pp. 1627-1638, 2000.

[9]   Y. Wu, D. Hu, M. Wu, and X. Hu, "A Numerical-Integration Perspective on Gaussian Filters," *IEEE Transactions on Signal Processing, to appear*, 2005.

[10]   Y. Wu, D. Hu, M. Wu, and X. Hu, "Unscented Kalman Filtering for Additive Noise Case: Augmented versus Non-augmented," *IEEE Signal Processing Letters*, vol. 12, no. 5, pp. 357-360, 2005.

[11]   Y. Wu, X. Hu, D. Hu, and M. Wu, "Comments on "Gaussian particle filtering"," *IEEE Transactions on Signal Processing*, vol. 53, no. 8, pp. 3350-3351, 2005.

[12]   P. J. Davis and P. Rabinowitz, *Methods of Numerical Integration*: New York, Academic Press, 1975.

[13]   A. F. M. Smith and A. E. Gelfand, "Bayesian statistics without tears: a sampling-resamping perspective," *The American Statistician*, vol. 46, no. 2, pp. 84-88, 1992.

[14]   R. Y. Rubinstein, *Simulation and the Monte Carlo Method*. New York: Wiley, 1981.

[15]   H. W. Sorenson and D. L. Alspach, "Recursive Bayesian estimation using Gaussian sums," *Automatica*, vol. 7, pp. 465-479, 1971.

[16]   D. L. Alspach and H. W. Sorenso, "Nonlinear Bayesian estimation using Gaussian sum approximation," *IEEE Transactions on Automatic Control*, vol. 17, pp. 439-448, 1972.

[17]   J. H. Kotecha and P. M. Djuric, "Gaussian sum particle filtering," *IEEE Transactions on Signal Processing*, vol. 51, no. 10, pp. 2602-2612, 2003.

[18]   G. Kitagawa, "Non-Gaussian state-space modeling of nonstationary time-series," *Journal of the American Statistical Association*, vol. 82, no. 400, pp. 1032-1063, 1987.

[19]   N. J. Gordon, D. J. Salmond, and A. F. Smith, "Novel approach to nonlinear/non-Gaussian Bayesian state estimation," *IEE Proceedings-F*, vol. 140, no. 2, pp. 107-113, 1993.

Improved Sensitivity Estimate for the H^2 Estimation Problem[*]

N.D. Christov[1], M. Najim[2], and E. Grivel[2]

[1] Technical University of Sofia, 8 Kl. Ohridski Blvd., 1000 Sofia, Bulgaria
ndchr@tu-sofia.bg
[2] ENSEIRB, Equipe Signal et Image, 1 av. du Dr A. Schweitzer,
33402 Talence, France
{najim, grivel}@tsi.u-bordeaux.fr

Abstract. The paper deals with the local sensitivity analysis of the discrete-time infinite-horizon H^2 estimation problem. An improved, nonlinear sensitivity estimate is derived which is less conservative than the existing, condition number based sensitivity estimates.

1   Introduction

In the last four decades the H^2 (Wiener-Kalman) estimators have been widely used in numerous applications in signal processing and control. However, the computational and robustness aspects of the H^2 estimation problem have not been studied in a sufficient extent and the efficient and reliable H^2 estimator design and implementation is still an open problem.

In this paper we study the local sensitivity of the discrete-time infinite-horizon H^2 estimation problem [1] relative to perturbations in the data. Using the nonlinear perturbation analysis technique proposed in [2,3] and further developed in [4,5], we derive a new, nonlinear local perturbation bound for the solution of the matrix Riccati equation that determines the sensitivity of the H^2 estimation problem. The new sensitivity estimate is a first order homogeneous function of the data perturbations and is less conservative than the existing, condition number based linear sensitivity estimates [6]-[12].

The following notations are used later on: $\mathcal{R}$ is the field of real numbers; $\mathcal{R}^{m \times n}$ is the space of $m \times n$ matrices over $\mathcal{R}$; $A^T \in \mathcal{R}^{n \times m}$ is the transpose of $A = [a_{ij}] \in \mathcal{R}^{m \times n}$; I_n is the identity $n \times n$ matrix; $\mathrm{vec}(A) \in \mathcal{R}^{mn}$ is the column-wise vector representation of the matrix $A \in \mathcal{R}^{m \times n}$; $\Pi_{n^2} \in \mathcal{R}^{n^2 \times n^2}$ is the vec-permutation matrix, so that $\mathrm{vec}(A^T) = \Pi \mathrm{vec}(A)$ for all $A \in \mathcal{R}^{n \times n}$; $\| \cdot \|_2$ and $\| \cdot \|_F$ are the spectral and Frobenius norms in $\mathcal{R}^{m \times n}$, while $\| \cdot \|$ is the induced operator norm or an unspecified matrix norm. The Kronecker product of the matrices A, B is denoted by $A \otimes B$ and the symbol $:=$ stands for "equal by definition".

---

[*] This work is supported by the European Union under Grants 15010/02Y0064/03-04 CAR/Presage No 4605 Obj. 2-2004:2 - 4.1 No 160/4605.

V.N. Alexandrov et al. (Eds.): ICCS 2006, Part I, LNCS 3991, pp. 697–703, 2006.

2   Problem Statement

Consider the linear discrete-time model

$$x_{i+1} = Fx_i + Gu_i$$

$$y_i = Hx_i + v_i, \quad -\infty < i < \infty \tag{1}$$

$$s_i = Lx_i$$

where $F \in \mathcal{R}^{n.n}$, $G \in \mathcal{R}^{n.m}$, $H \in \mathcal{R}^{p.n}$, and $L \in \mathcal{R}^{q.n}$ are known constant matrices, x_i is the state, y_i is the measured output, s_i is the desired signal, and $\{u_i\}$, $\{v_i\}$ are zero-mean white-noise processes with variance matrices $Q \geq 0$ and $R > 0$, respectively. It is assumed that the pair (F, H) is detectable and the pair $(F, GQ^{1/2})$ – stabilizable.

Given the observations $\{y_j, j \leq i\}$, the H^2 estimation problem [1] consists in finding a linear estimation strategy $\hat{s}_{i|i} = \mathcal{F}(y_0, y, y_1, \ldots, y_j)$ that minimizes the expected filtered error energy, i.e.,

$$\min_{\mathcal{F}} E \sum_{j=0}^{i} (s_j - \hat{s}_{j|j})^T (s_j - \hat{s}_{j|j}).$$

As it is well known [1], in the infinite-horizon case the H^2 estimation problem is formulated as

$$\min_{\text{causal } K(z)} \| [(L(zI - F)^{-1}G - K(z)H(zI - F)^{-1}G)Q^{1/2} \quad - K(z)R^{1/2}] \|_2$$

and its solution is given by

$$K(z) = LPH^T(R + HP_0H^T)^{-1}$$
$$+ L(I - PH^T(R + HP_0H^T)^{-1}H)(zI - F_p)^{-1}K_p$$

where $K_p = FP_0H^T R_e^{-1}$, $R_e = R + HP_0H^T$, and $P_0 \geq 0$ is the unique stabilizing ($F_p = F - K_pH$ stable) solution of the matrix Riccati equation

$$FPF^T - P + GQG^T - K_pR_eK_p^T = 0. \tag{2}$$

A state-space model for $K(z)$ can be given by

$$\hat{x}_{i+1} = F\hat{x}_i + K_p(y_i - H\hat{x}_i)$$
$$\hat{s}_{i|i} = L\hat{x}_i + LPH^T R_e^{-1}(y_i - H\hat{x}_i).$$

In the sequel we shall write equation (2) in the equivalent form

$$\bar{F}(P, D)PF^T - P + GQG^T = 0 \tag{3}$$

where $\bar{F}(P, D) = F - FPH^T R_e^{-1}(P, D)H$ and $D = (F, H)$.

Assuming that the matrices F, G, H in (1) are subject to perturbations ΔF, ΔG, ΔH, we obtain the perturbed equation

$$\bar{F}(P, D + \Delta D)P(F + \Delta F)^T - P + (G + \Delta G)Q(G + \Delta G)^T = 0 \qquad (4)$$

where $\Delta D = (\Delta F, \Delta H)$,

$$\bar{F}(P, D + \Delta D) = (F + \Delta F) - (F + \Delta F)P(H + \Delta H)^T R_e^{-1}(P, D + \Delta D)(H + \Delta H)$$

and
$$R_e(P, D + \Delta D) = R_e(P, D) + \Delta H P H^T + H P \Delta H^T + \Delta H P \Delta H^T.$$

Since the Fréchet derivative of the left-hand side of (3) in P at $P = P_0$ is invertible, the perturbed equation (4) has a unique solution $P = P_0 + \Delta P$ in the neighborhood of P_0.

Denote by $\Delta_M = \|\Delta M\|_F$ the absolute perturbation of a matrix M and let $\Delta := [\Delta_F, \Delta_G, \Delta_H]^T \in \mathcal{R}_+^3$.

The sensitivity analysis of the Riccati equation (3) consists in finding estimate for the absolute perturbation $\Delta_P := \|\Delta P\|_F$ in the solution P as a function of the absolute perturbations Δ_F, Δ_G, Δ_H in the coefficient matrices F, G, H.

A number of linear local bounds for Δ_P have been derived in [6]-[12], based on the condition numbers of (3). For instance, in [9] a perturbation bound of the type

$$\Delta_P \leq K_F \Delta_F + K_G \Delta_G + K_H \Delta_H + O(\|\Delta\|^2), \ \Delta \to 0 \qquad (5)$$

has been obtained, where K_F, K_G, K_H are the condition numbers of (3) relative to perturbations in the matrices F, G and H, respectively. However, it is possible to obtain nonlinear local sensitivity estimates for (3) that are less conservative than the condition numbers based sensitivity estimates.

The problem considered in this paper is to find a first order homogeneous local perturbation bound

$$\Delta_P \leq f(\Delta) + O(\|\Delta\|^2), \ \Delta \to 0, \qquad (6)$$

which is tighter than the condition number based perturbation bounds.

3   Main Result

Denote by $\Phi(P, D)$ the left-hand side of the Riccati equation (3) and let P_0 be the positive definite or semi-definite solution of (3). Then $\Phi(P_0, D) = 0$.

Setting $P = P_0 + \Delta P$, the perturbed equation (4) may be written as

$$\Phi(P_0 + \Delta P, D + \Delta D) = \qquad (7)$$

$$\Phi(P_0, D) + \Phi_P(\Delta P) + \Phi_F(\Delta F) + \Phi_G(\Delta G) + \Phi_H(\Delta H) + S(\Delta P, \Delta D) = 0$$

where $\Phi_P(.)$, $\Phi_F(.)$, $\Phi_G(.)$, $\Phi_H(.)$ are the Fréchet derivatives of $\Phi(P, D)$ in the corresponding matrix arguments, evaluated for $P = P_0$, and $S(\Delta P, \Delta D)$ contains the second and higher order terms in ΔP, ΔD.

The calculation of the Fréchet derivatives of $\Phi(P, D)$ gives

$$\Phi_P(Z) = \bar{F}_0 Z \bar{F}_0^T - Z \tag{8}$$

$$\Phi_F(Z) = \bar{F}_0 P_0 Z + Z^T P_0 \bar{F}_0^T$$

$$\Phi_G(Z) = GQZ + Z^T QG^T$$

$$\Phi_H(Z) = -\bar{F}_0 P_0 Z R_0^{-1} H P_0 F^T - F P_0 H^T R_0^{-1} Z^T P_0 \bar{F}_0^T$$

where $\bar{F}_0 = \bar{F}(P_0, D)$, $R_0 = R_e(P_0, D)$.

Denoting $M_P \in \mathcal{R}^{n^2 \times n^2}$, $M_F \in \mathcal{R}^{n^2 \times n^2}$, $M_G \in \mathcal{R}^{n^2 \times n^2}$, $M_H \in \mathcal{R}^{n^2 \times n^2}$ the matrix representations of the operators $\Phi_P(.)$, $\Phi_F(.)$, $\Phi_G(.)$, $\Phi_H(.)$, we have

$$M_P = \bar{F}_0 \otimes \bar{F}_0 - I_{n^2}$$

$$M_F = I_n \otimes \bar{F}_0 P_0 + (\bar{F}_0 P_0 \otimes I_n)\Pi \tag{9}$$

$$M_G = I_n \otimes GQ + (GQ \otimes I_n)\Pi$$

$$M_H = -F P_0 H^T R_0^{-1} \otimes \bar{F}_0 P_0 - (\bar{F}_0 P_0 \otimes F P_0 H^T R_0^{-1})\Pi$$

where $\Pi \in \mathcal{R}^{n^2 \times n^2}$ is the permutation matrix such that $\mathrm{vec}(M^T) = \Pi \mathrm{vec}(M)$ for each $M \in \mathcal{R}^{n \times n}$.

The operator equation (7) may be written in a vector form as

$$\mathrm{vec}(\Delta P) = N_1 \mathrm{vec}(\Delta F) + N_2 \mathrm{vec}(\Delta G) \tag{10}$$

$$+ N_3 \mathrm{vec}(\Delta H) - M_P^{-1} \mathrm{vec}(S(\Delta P, \Delta D))$$

where

$$N_1 = -M_P^{-1} M_F, \; N_2 = -M_P^{-1} M_G, \; N_3 = -M_P^{-1} M_H.$$

It is easy to show that the condition number based perturbation bound (5) is a corollary of (10). Indeed, it follows from (10)

$$\|\mathrm{vec}(\Delta P)\|_2 \leq \|N_1\|_2 \|\mathrm{vec}(\Delta F)\|_2 + \|N_2\|_2 \|\mathrm{vec}(\Delta G)\|_2$$

$$+ \|N_3\|_2 \|\mathrm{vec}(\Delta H)\|_2 + \mathrm{O}(\|\Delta\|^2)$$

and having in mind that

$$\|\mathrm{vec}(\Delta M)\|_2 = \|\Delta M\|_F = \Delta_M$$

and denoting

$$K_F = \|N_1\|_2, \; K_G = \|N_2\|_2, \; K_H = \|N_3\|_2$$

we obtain

$$\Delta_P \leq K_F \Delta_F + K_G \Delta_G + K_H \Delta_H + \mathrm{O}(\|\Delta\|^2). \tag{11}$$

Relation (10) also gives

$$\Delta_P \leq \|N\|_2 \|\Delta\|_2 + \mathrm{O}(\|\Delta\|^2) \tag{12}$$

where $N = [N_1, N_2, N_3]$.

Note that the bounds in (11) and (12) are alternative, i.e. which one is less depends on the particular value of Δ.

There is also a third bound, which is always less than or equal to the bound in (11). We have

$$\Delta_P \leq \sqrt{\Delta^\top U(N)\Delta} + \mathrm{O}(\|\Delta\|^2) \tag{13}$$

where $U(N)$ is the 3×3 matrix with elements

$$u_{ij}(N) = \|N_i^\top N_j\|_2.$$

Since $\left\|N_i^\top N_j\right\|_2 \leq \|N_i\|_2 \|N_j\|_2$ we get

$$\sqrt{\Delta^\top U(N)\Delta} \leq \|N_1\|_2 \Delta_F + \|N_2\|_2 \Delta_G + \|N_3\|_2 \Delta_H.$$

Hence we have the overall estimate

$$\Delta_P \leq f(\Delta, N) + \mathrm{O}(\|\Delta\|^2), \quad \Delta \to 0 \tag{14}$$

where

$$f(\Delta, N) = \min\{\|N\|_2 \|\Delta\|_2, \sqrt{\Delta^\top U(N)\Delta}\} \tag{15}$$

is a non-linear, first order homogeneous and piece-wise real analytic function in Δ.

4   Numerical Example

Consider a third order model of type (1) with matrices

$$F = VF^*V, \quad G = VG^*, \quad H = H^*V$$

where $V = I_3 - 2vv^T/3$, $v = [1,1,1]^T$ and

$$F^* = \mathrm{diag}(0, 0.1, 0), \quad G^* = \mathrm{diag}(2, 1, 0.1)$$

$$H^* = \mathrm{diag}(1, 0.5, 10).$$

The perturbations in the data are taken as

$$\Delta F = V \Delta F^* V, \quad \Delta G = V \Delta G^*, \quad \Delta H = \Delta H^* V$$

where

$$\Delta F^* = \begin{bmatrix} 3 & -1 & 0 \\ -1 & 2 & -9 \\ 0 & -9 & 5 \end{bmatrix} \times 10^{-i},$$

$$\Delta G^* = \begin{bmatrix} 10 & -5 & 7 \\ -5 & 1 & 3 \\ 7 & 3 & 10 \end{bmatrix} \times 10^{-i-1},$$

$$\Delta H^* = \begin{bmatrix} 1 & -1 & 2 \\ -1 & 5 & -1 \\ 2 & -1 & 10 \end{bmatrix} \times 10^{-i-1}$$

for $i = 10, 9, \ldots, 3$.

The absolute perturbations Δ_P in the solution of the Riccati equation are estimated by the linear bound (11) and the nonlinear homogeneous bound (14). The results obtained for $Q = R = I$ and different values of i are shown in Table 1. The actual relative changes in the solution are close to the quantities predicted by the local sensitivity analysis.

Table 1.

i	Δ_P	Est. (11)	Est. (14)
10	5.1×10^{-10}	9.8×10^{-9}	7.7×10^{-10}
9	5.1×10^{-9}	9.8×10^{-8}	7.7×10^{-9}
8	5.1×10^{-8}	9.8×10^{-7}	7.7×10^{-8}
7	5.1×10^{-7}	9.8×10^{-6}	7.7×10^{-7}
6	5.1×10^{-6}	9.8×10^{-5}	7.7×10^{-6}
5	5.1×10^{-5}	9.8×10^{-4}	7.7×10^{-5}
4	5.1×10^{-4}	9.8×10^{-3}	7.7×10^{-4}
3	5.1×10^{-3}	9.8×10^{-2}	7.7×10^{-3}

5   Conclusion

In this paper the local sensitivity of the discrete-time infinite-horizon H^2 estimation problem has been studied. A new, nonlinear local perturbation bound has been obtained for the solution of the Riccati equation that determines the sensitivity of the problem. The new local sensitivity estimate is a first order homogeneous function of the data perturbations and is tighter than the condition number based sensitivity estimates.

References

1. B. Hassibi, A. H. Sayed, T. Kailath: *Indefinite-Quadratic Estimation and Control: A Unified Approach to H^2 and H^∞ Theories*. SIAM, Philadelphia, 1999
2. M. M. Konstantinov, P. Hr. Petkov, N. D. Christov, D. W. Gu, V. Mehrmann: Sensitivity of Lyapunov equations. In: N. Mastorakis (Ed.): *Advances in Intelligent Systems and Computer Science*. WSES Press, N.Y., 1999, 289–292
3. M. M. Konstantinov, P. Hr. Petkov, D. W. Gu: Improved perturbation bounds for general quadratic matrix equations. *Numer. Func. Anal. and Optimiz.* **20** (1999) 717–736
4. N. D. Christov, S. Lesecq, M. M. Konstantinov, P. Hr. Petkov, A. Barraud: New perturbation bounds for Sylvester equations. In: *Proc. 39th IEEE Conf. on Decision and Control*, Sydney, 12-15 December 2000, 4233–4234
5. N. D. Christov, M. Najim, E. Grivel,D. Henry: On the local sensitivity of the discrete-time H^∞ estimation problem. In: *Proc. 15th IFAC World Congress*, Barcelona, 21-26 July 2002, paper T-Tu-A01/1091
6. M. M. Konstantinov, P. Hr. Petkov, N. D. Christov: Perturbation analysis of the continuous and discrete matrix Riccati equations. In: *Proc. 1986 American Control Conf.*, Seattle, 18-20 June 1986, vol. 1, 636-639
7. P. Gahinet, A. J. Laub: Computable bounds for the sensitivity of the algebraic Riccati equation. *SIAM J. Contr. Optim.* **28** (1990) 1461–1480
8. P. Hr. Petkov, N. D. Christov, M. M. Konstantinov: *Computational Methods for Linear Control Systems*. Prentice Hall, N.Y., 1991
9. M. M. Konstantinov, P. Hr. Petkov, N. D. Christov: Perturbation analysis of the discrete Riccati equation. *Kybernetika* **29** (1993) 18–29
10. A. R. Ghavimi, A. J. Laub: Backward error, sensitivity and refinement of computed solutions of algebraic Riccati equations. *Numer. Lin. Alg. Appl.* **2** (1995) 29–49
11. J.-G. Sun: Perturbation theory for algebraic Riccati equations. *SIAM J. Matrix Anal. Appl.* **19** (1998) 39–65
12. J.-G. Sun: Condition numbers of algebraic Riccati equations in the Frobenius norm. *Lin. Alg. Appl.* **350** (2002) 237–261

Constrained Optimization of the Stress Function for Multidimensional Scaling

Vydunas Saltenis

Institute of Mathematics and Informatics
Akademijos 4, LT-08663 Vilnius, Lithuania
Saltenis@ktl.mii.lt

Abstract. Multidimensional Scaling (MDS) requires the multimodal Stress function optimization to estimate the model parameters, i.e. the coordinates of points in a lower-dimensional space. Therefore, finding the global optimum of the Stress function is very important for applications of MDS. The main idea of this paper is replacing the difficult multimodal problem by a simpler unimodal constrained optimization problem. A coplanarity measure of points is used as a constraint while the Stress function is minimized in the original high-dimensional space. Two coplanarity measures are proposed. A simple example presented illustrates and visualizes the optimization procedure. Experimental evaluation results with various data point sets demonstrate the potential ability to simplify MDS algorithms avoiding multidimodality.

1 Introduction

Multidimensional scaling (MDS) [1, 2] is a widely used technique to visualize the dissimilarity of data points. Objects (n data points) are represented as p-dimensional vectors $Y_1, \ldots, Y_n \in R^p$ so that the Euclidean distances $d_{ij}(Y)$, ($i, j = 1, \ldots, n; i < j$) between the pairs of points correspond to the given dissimilarities δ_{ij} as closely as possible. Only representations onto a 2-dimensional space are used (p=2) as usual, since data visualization is the aim. In general, the dissimilarities δ_{ij} need not be distances between the multidimensional points.

MDS requires the multimodal Stress function optimization to estimate the model parameters, i.e. the coordinates of points (vectors Y_i) in a lower-dimensional space.

The measure of fit is usually defined by the Stress function:

$$\sigma(Y) = \sum_{\substack{i, j=1 \\ i<j}}^{n} w_{ij}(\delta_{ij} - d_{ij}(Y))^2 \ ,$$

proposed in [3]. w_{ij} are weights that may be different in various types of the Stress function. In our investigation $w_{ij} = 1$.

V.N. Alexandrov et al. (Eds.): ICCS 2006, Part I, LNCS 3991, pp. 704–711, 2006.
© Springer-Verlag Berlin Heidelberg 2006

The aim of MDS is:

$$\min_{\substack{Y \in R^{n \times d}}} \sum_{\substack{i,j=1 \\ i<j}}^{n} (\delta_{ij} - d_{ij}(Y))^2 \,. \tag{1}$$

A substantial shortcoming of MDS is the existence of local minima. The examples of proved multimodality of the Stress function are constructed (for example, [4, 5]). The number of different local minima may range from a few to several thousands. MDS algorithms that minimize Stress cannot guarantee a global minimum. In general, some advice is to use multiple random starts and select the solution with the lowest Stress value (the multiple random start method). A lot of attempts have been made to improve search procedures by a proper choice of start points, however all the strategies are computationally intensive.

2   Basic Idea

Let in our case the dissimilarities δ_{ij} in (1) be the Euclidean distances $d_{ij}(X)$ between the m-dimensional points ($m>p$) with the given coordinates $X_1,...,X_n \in R^m$ and variable vectors $Z_1,...,Z_n \in R^m$ as distinct from vectors $Y_1,...,Y_n \in R^p$ in (1) be of the same dimensionality m. Then $n \times m$ dimensional constrained minimization problem may be formulated as:

$$\min_{\substack{Z \in R^{n \times m}}} \sum_{i<j} (d_{ij}(Z) - d_{ij}(X))^2 \tag{2}$$

subject to the constraint

$$P(Z) = 0 \,. \tag{3}$$

$P(Z)$ in (3) is some nonnegative coplanarity measure of points Z. If the points in an m-dimensional (high-dimensional) space lie on a hyperplane, then the coplanarity measure must be necessarily equal to zero.

If variable coordinates Z_i are equal to given coordinates X_i in (2), then object function value is equal to zero and coplanarity measure $P(Z) > 0$. These Z_i values are a start position to constrained optimization (2) and (3) when the influence of constraint (3) is gradually increased.

The optimal coordinates Z^{opt} of the problem (2), (3) are m-dimensional and the distances between them $d_{ij}(Z^{opt})$ are the same as that between the p-dimensional optimal coordinates Y^{opt} obtained from (1):

$$d_{ij}(Z^{opt}) = d_{ij}(Y^{opt}) \,.$$

3 Coplanarity Measures

3.1 Coplanarity Measure Based on the Volumes of Tetrahedra

One of the possible coplanarity measures is based on the volumes V_{ijkl} of tetrahedra whose four vertices are multidimensional points $Z_1,\ldots,Z_n$. We use the sum of squared volumes of all possible tetrahedra as coplanarity measure:

$$P(Z) = \sum_{i=1}^{n-3} \sum_{j=i+1}^{n-2} \sum_{k=j+1}^{n-1} \sum_{l=k-1}^{n} V_{ijkl}^2 \ ,$$

where the volume V is given by the Cayley-Menger determinant [6]:

$$V_{ijkl}^2 = \frac{1}{28} \begin{vmatrix} 0 & 1 & 1 & 1 & 1 \\ 1 & 0 & d_{ij}^2 & d_{ik}^2 & d_{il}^2 \\ 1 & d_{ji}^2 & 0 & d_{jk}^2 & d_{jl}^2 \\ 1 & d_{ki}^2 & d_{kj}^2 & 0 & d_{kl}^2 \\ 1 & d_{li}^2 & d_{lj}^2 & d_{lk}^2 & 0 \end{vmatrix} .$$

For the simplicity, the notation d_{ij} is used instead of $d_{ij}(Z)$.

This coplanarity measure was used in our experimental evaluation.

3.2 Coplanarity Measure Based on Point-Plane Distances

Another possible coplanarity measure is consequent upon the coplanarity definition [7, 8]. The points $Z_1,\ldots,Z_n$ can be tested for coplanarity by finding the point-plane distances of the points Z_i, $i = 4,\ldots,n$ from the plane determined by Z_1, Z_2, Z_3 and checking if all of them are zero. If so, all the points are coplanar.

The point-plane distance from the plane determined by three points Z_1, Z_2, Z_3 may be computed [9] as follows:

$$D_i = \hat{n} \cdot (Z_k - Z_i) \ ,$$

where Z_k is any of the three points Z_1, Z_2, Z_3 and $\hat{n}$ is the unit normal

$$\hat{n} = \frac{(Z_2 - Z_1) \times (Z_3 - Z_1)}{\left|(Z_2 - Z_1) \times (Z_3 - Z_1)\right|} .$$

Then one of possible coplanarity measures may be introduced:

$$P(Z) = \sum_{i=4}^{n} D_i^2 \ .$$

The measure depends on the selection of the three points Z_1, Z_2, Z_3.

4 Constrained Optimization

The optimization problem (2), (3) was solved by the penalty function method [10]. A constrained optimization problem is transformed into a sequence of unconstrained optimization problems by modifying the objective function. In our case, we use such a sequence of unconstrained problems:

$$\min_{Z \in R^{n \times m}} \left(\sum_{i<j} (d_{ij}(Z) - d_{ij}(X))^2 + r_k P^2(Z) \right) , \ (k = 1,2,\ldots) ,$$

where k is the number of sequence, r_k is a positive penalty parameter. The problem is solved with a sequence of parameters r_k tending to ∞ :

$$r_{k+1} = \Delta r \cdot r_k . \tag{4}$$

The modified problem can be solved by the methods of unconstrained local optimization: Quasi-Newton and Conjugate gradient methods, in our case.

The Torgerson scaling technique [1, 11] may be used for recovery of coordinates of dimensionality 2 from the optimal distances $d_{ij}(Z^{opt})$. The method yields an analytical solution, requiring no iterations.

5 Simple Illustrative Example

In order to visualize and better understand the new approach and the optimization procedure, a simple unidimensional MDS illustrative example was constructed. It uses only three data points, two optimization coordinates, and ($p=1$).

Let the initial distances between three points be: $\delta_{12} = \delta_{13} = 5$; $\delta_{23} = 6$. Only two of the distances d_{13} and d_{23} will be optimized. The distance d_{12} will be fixed: $d_{12} = \delta_{12} = 5$. Then the Stress function in our case is:

$$\sigma(d_{13}, d_{23}) = (d_{13} - 5)^2 + (d_{23} - 6)^2 .$$

In our example the coplanarity measure based on the volumes of tetrahedra reduces to linearity measure L, which is based on the area of a triangle with side lengths d_{12}, d_{13}, d_{23}:

$$L(d_{13}, d_{23}) = (5 + d_{13} + d_{23})(5 + d_{13} - d_{23})(5 - d_{13} + d_{23})(-5 + d_{13} + d_{23}) .$$

L is proportional to the square of the triangle area, calculated by Heron's formula. There are three local optima of the constrained optimization problem:

1. $d_{13} = 3$; $d_{23} = 8$, with a minimal value of the Stress function $\sigma(3,8) = 8$. The constrained local optimum is marked by point A in Fig. 1.

2. $d_{13} = 2$; $d_{23} = 3$, with a minimal value of the Stress function $\sigma(2,3) = 18$. The constrained local optimum is marked by point B.

3. $d_{13} = 8$; $d_{23} = 3$, with a minimal value of the Stress function $\sigma(8,3) = 18$. The constrained local optimum is marked by point C.

The global constrained optimum is the first one.

At the beginning of optimization (point O in Fig.1) the Stress function is equal to zero, the constraint value is equal to 2304. At each step of constrained optimization, when increasing the penalty parameter r_k, the value of constraint decreases and in the last step (point A in Fig. 1) it achieves zero value. At the same time the Stress value increases and, in the last step, achieves the global optimum value 8.

A contour plot diagram demonstrates that, with slightly different data, the global optimum point may be different and, consequently, the result of constrained optimization also changes.

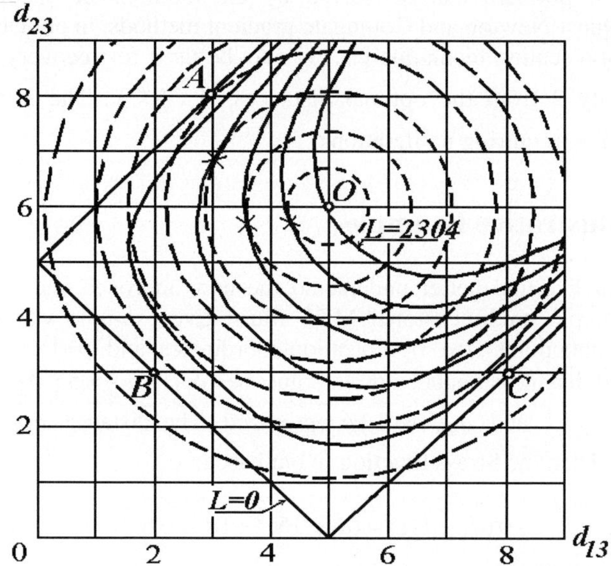

Fig. 1. The contour plots of the Stress function $\sigma(d_{13}, d_{23})$ (*dotted* contour lines) and of constraint function $L(d_{13}, d_{23})$ (*solid* contour lines). The points of constrained local minima are marked as A, B, C. The start point is denoted as O, and transitional points of constrained optimization are marked as X.

6   Experimental Evaluation

Two types of data sets were used in the experimental investigation: regular and irregular. The points of regular data sets were the vertices of a multidimensional cube of various dimensionality and the irregular ones were obtained randomly.

All the results of computational experiments with the proposed algorithm were compared with the results of global optimization obtained using usual MDS random multistart optimization of the Stress function. The number of multistart local optimizations was equal to 200-500.

Coplanarity measure based on the volumes of tetrahedra was used.

Table 1 presents the results of investigation with the regular data points of various dimensionality. The average local optimization error, the number of local minima and the probability to find a global minimum by multistart random local optimization were evaluated from the results of multistart local optimization.

The stopping rule of the constrained optimization was: $P(Z) < 10^{-7}$. The precision of local optimization was 10^{-8}.

Two local optimization methods were compared. We can see that the interval of successful values of Δr is greater for Quasi-Newton method. This method was used in the investigations of Table 2-3.

Table 1. Results of investigation with the regular data points on the vertices of a multidimensional cube of various dimensionality

	3	4
Dimensionality	3	4
Number of points	8	16
Optimal Stress value	2,854261	23,089651
Average local optimization error in %	2,42	0,69
Number of local minima	6	7
Probability of finding a global minimum by multistart random local optimization	0.68	0.74
Values of Δr from (4) (Quasi-Newton method)	1.1–10000000	1.1–700000
Values of r_1 from (4) (Quasi-Newton method)	1	1
Values of Δr from (4) (Conjugate gradient method)	1.1 – 100	1.1 – 80
Values of r_1 from (4) (Conjugate gradient method)	1	1

Table 2. Results of investigation with the random data points of dimensionality $m=4$ (number of points $n=16$)

Optimal Stress value	2,16464	1.77626
Average local optimization error in %	25.47	26.02
Number of local minima	10	8
Probability of finding a global minimum by multistart random local optimization	0.25	0.39
Δr value	100	100
r_1 value	0.01	0.01

Tables 2-3 present two examples of numerous investigations with the random data points. Table 2 presents the results with the random data points of dimensionality $m=4$ (number of points $n=16$). Table 3 presents the results with the random data points of dimensionality $m=6$ (number of points $n=20$).

In all the experiments (not only presented in Tables 1-3) the proposed constrained optimization achieved the global minimum.

Table 3. Results of investigation with the random data points of dimensionality $m=6$ (number of points $n=20$)

Optimal Stress value	8.16616	10.32240
Average local optimization error in %	22.47	9.29
Number of local minima	38	37
Probability of finding a global minimum by multistart random local optimization	0.37	0.38
Δr value	100	100
r_1 value	0.01	0.01

The proposed optimization procedure is more computation exhaustive. The number of variables is larger in comparison with the optimization by the usual approach. For example, the execution time is 5 times greater in comparison to single start of the usual approach for the data set of Table 2.

6   Conclusions

The new approach replaces the difficult multimodal optimization problem by a simpler optimization problem that uses the constrained local optimization procedure. It minimizes the Stress function in the original high-dimensional space subjected to zero planarity constraint.

This approach eliminates the problem of the initial choice of variables and the difficulties caused by the Stress function multimodality.

However, the optimization procedure is more computation exhaustive. The number of variables is larger in comparison with the optimization by the usual approach; the constrained optimization requires some steps.

We did not test any evaluations of the computational efficiency of the new approach for various data, neither did we consider possible performance improvement observations in the paper. These issues remain as a possible trend of further research.

Acknowledgements

The research was partially supported by the Lithuanian State Science and Studies Foundation, Grant No. C 03013.

References

1. Borg, L., Groenen, P.: Modern Multidimensional Scaling: Theory and Applications, Springer (1997)
2. Cox, T., Cox, M.: Multidimensional Scaling, Chapman and Hall (2001)
3. Kruskal, J.: Nonmetric Multidimensional Scaling: A Numerical Method. Psychometrica, Vol.29 (1964) 115-129
4. Trosset, M., Mathar R.: On existence of nonglobal minimizers of the STRESS Criterion for Metric Multidimensional Scaling. Proceedings of the Statistical Computing Section, American Statistical Association, Alexandria, VA, (1997) 158-162
5. Zilinskas, A., Podlipskyte, A.: On multimodality of the SSTRESS criterion for metric multidimensional scaling, Informatica, Vol. 14, No. 1, (2003) 121-130
6. Sommerville, D. M. Y.: An Introduction to the Geometry of n Dimensions. New York: Dover, (1958)
7. Abbott, P. (Ed.). In and Out: Coplanarity. Mathematica J. 9 (2004) 300-302
8. Weisstein, E. W.: Coplanar. MathWorld - A Wolfram Web Resource. http://mathworld.wolfram.com/Coplanar.html
9. Weisstein, E. W.: Point-Plane Distance. MathWorld - A Wolfram Web Resource. http://mathworld.wolfram.com/Point-PlaneDistance.html
10. Bertsekas, D. P.: Nonlinear programming. Athena Scientific (1999)
11. Torgerson, W. S.: Theory and methods of scaling. New York: Wiley (1958)

Targeted Observations for Atmospheric Chemistry and Transport Models

Adrian Sandu

Department of Computer Science, Virginia Polytechnic Institute and State
University, Blacksburg, VA 24061
`sandu@cs.vt.edu`

Abstract. The aim of this paper is to address computational aspects
of the targeted observations problem for atmospheric chemical transport
models. The fundamental question being asked is where to place the
observations such that, after data assimilation, the uncertainty in the
resulting state is minimized. Our approach is based on reducing the system along the subspace defined by the dominant singular vectors, and
computing the locations of maximal influence on the verification area.
Numerical results presented for a simulation of atmospheric pollution in
East Asia in March 2001 show that the optimal location of observations
depends on the pattern of the flow but is different for different chemical species. Targeted observations have been previously considered in
the context of numerical weather prediction. This work is, to the best of
our knowledge, the first effort to study targeted observations in the context of chemical transport modeling. The distinguishing feature of these
models is the presence of stiff chemical interactions.

Keywords: Chemical transport models, data assimilation, adjoint models, singular vectors, targeted observations.

1 Introduction

Our ability to anticipate and manage changes in atmospheric pollutant concentrations relies on an accurate representation of the chemical state of the
atmosphere. As our fundamental understanding of atmospheric chemistry advances, novel computational tools are needed to integrate observational data
and models together to provide the best, physically consistent estimate of the
evolving chemical state of the atmosphere. Such an analysis state better defines
the spatial and temporal fields of key chemical components in relation to their
sources and sinks. This information is critical in designing cost-effective emission
control strategies for improved air quality, for the interpretation of observational
data such as those obtained during intensive field campaigns, and to the execution of air-quality forecasting.

Data assimilation is the process of integrating observational data and model
predictions in order to provide an optimal analysis of the state of the system
(here, the atmosphere) and/or optimal estimates of important model parameters (e.g., the strength of the anthropogenic emissions that drive the pollutant

V.N. Alexandrov et al. (Eds.): ICCS 2006, Part I, LNCS 3991, pp. 712–719, 2006.

concentrations in the atmosphere). In a variational approach data assimilation is formulated as an optimization problem where the model parameters are estimated such that the mismatch between model predictions and observations is minimized. In atmospheric applications data assimilation is typically used to find the optimal initial state of the system; the resulting parameter estimation problems have millions of control variables.

The objective of this work is to develop techniques for optimal placement of observations in chemical transport modeling. Adaptive observations placed in well-chosen locations can reduce the initial condition uncertainties and decrease forecast errors. A number of methods were proposed to "target observations", i.e. to select areas where additional observations are expected to improve considerably the skill of a given forecast. Our proposed approach uses singular vectors to identify the most sensitive regions of the atmospheric flow and to optimally configure the observational network. The observations are placed in locations that have a maximal perturbation energy impact on the verification area at the verification time.

Singular vectors (SVs) are the directions of fastest error growth over a finite time interval [10, 14]. Buizza and Montani [1] showed that SVs can identify the most sensitive regions of the atmosphere for targeted observations. Majudmar et al.[12] compare the SV approach for observation targeting to the ensemble transform Kalman filter. Dăescu and Navon [5] discuss the adaptive observation problem in the context of 4D-Var data assimilation. Estimation of the optimal placement of adaptive observations is also discussed in [7, 11]. Leutbecher [9] derives optimal locations of observations by minimizing the variance of the assimilated field; a computationally tractable problem is obtained by projecting the covariance on the subspace of the dominant singular vectors.

The paper is organized as follows. In Section 2 we introduce the chemical transport models and the concept of singular vectors as the directions of maximal energy growth. Section 3 discusses a maximum energy impact criterion for placing the observations. Numerical results from a simulation of air pollution in East Asia are shown in Section 4. Section 5 summarizes the main findings of this work.

2   Background

2.1   3D Chemical-Transport Models

Chemical transport models solve the mass-balance equations for concentrations of trace species in order to determine the fate of pollutants in the atmosphere [16]. Let c_i be the mole-fraction concentration of chemical species i, Q_i be the rate of surface emissions, E_i be the rate of elevated emissions and f_i be the rate of chemical transformations. Further, u is the wind field vector, K the turbulent diffusivity tensor, and ρ is the air density. The evolution of c_i is described by the following equations

$$\frac{\partial c_i}{\partial t} = -u \cdot \nabla c_i + \frac{1}{\rho} \nabla \cdot (\rho K \nabla c_i) + \frac{1}{\rho} f_i(\rho c) + E_i \ , \quad t_0 \leq t \leq t_\mathrm{v} \ ,$$

$$c_i(t_0, x) = c_i^0(x) \; ,$$

$$c_i(t, x) = c_i^{\mathrm{in}}(t, x) \quad \text{for} \quad x \in \Gamma^{\mathrm{in}} \; , \quad K\frac{\partial c_i}{\partial n} = 0 \quad \text{for} \quad x \in \Gamma^{\mathrm{out}} \; , \tag{1}$$

$$K\frac{\partial c_i}{\partial n} = V_i^{\mathrm{dep}} c_i - Q_i \quad \text{for} \quad x \in \Gamma^{\mathrm{ground}} \; , \quad \text{for all} \; 1 \leq i \leq N_{\mathrm{spec}} \; .$$

Here Γ denotes the domain boundary (composed of the inflow, outflow, and ground level parts) and x is the spatial location within the domain. We will use $\mathcal{M}$ to denote the solution operator of the model (1). The state is propagated forward in time from the "initial" time t_0 to the "verification" time t_v (i.e., the final time of interest)

$$c(t_\mathrm{v}) = \mathcal{M}_{t_0 \to t_\mathrm{v}} \left(c(t_0) \right) \; . \tag{2}$$

Perturbations (small errors) evolve according to the tangent linear model (TLM)

$$\delta c(t_\mathrm{v}) = \mathbf{M}_{t_o \to t_\mathrm{v}} \, \delta c(t_0) \; , \tag{3}$$

and adjoint variables according to the adjoint model

$$\lambda(t_0) = \mathbf{M}^*_{t_\mathrm{v} \to t_o} \lambda(t_\mathrm{v}) \; . \tag{4}$$

Here M and M^* denote the solution operators of the two linearized models. A detailed description of chemical transport models, and the corresponding tangent linear and adjoint models, is given in [17].

Our main interest is to minimize the forecast uncertainty over a well defined area (the "verification domain" $\Omega_\mathrm{v} \subset \Omega$) at a well defined time (the "verification time" t_v). We define a spatial restriction operator G from the entire model domain to the verification domain:

$$G : \Omega \subset \Re^n \longrightarrow \Omega_\mathrm{v} \subset \Re^{n_\mathrm{v}} \; , \quad n_\mathrm{v} \ll n \; . \tag{5}$$

2.2   Singular Vectors

Singular vectors (SVs) determine the most rapidly growing perturbations in the atmosphere. The magnitude of the perturbation at the initial time t_0 is measured in the L^2 norm defined by a symmetric positive definite matrix E

$$\| \delta c(t_0) \|_\mathrm{E}^2 = \langle \, \delta c(t_0) \, , \; \mathrm{E} \, \delta c(t_0) \, \rangle \; . \tag{6}$$

Similarly, the perturbation magnitude at the verification time t_v is measured in a norm defined by a positive definite matrix F

$$\| \delta c(t_\mathrm{v}) \|_\mathrm{F}^2 = \langle \, \delta c(t_\mathrm{v}) \, , \; \mathrm{F} \, \delta c(t_\mathrm{v}) \, \rangle \; . \tag{7}$$

We call the norms (6) and (7) squared the "perturbation energies". The ratio between perturbation energies at t_v (over the verification domain) and at t_0 (over the entire domain) offers a measure of error growth:

$$\sigma^2 = \frac{\| \mathrm{G} \, \delta c(t_\mathrm{v}) \|_\mathrm{F}^2}{\| \delta c(t_0) \|_\mathrm{E}^2} = \frac{\langle \delta c(t_0), \, \mathbf{M}^*_{t_\mathrm{v} \to t_o} \mathrm{G}^* \, \mathrm{F} \, \mathrm{G} \, \mathbf{M}_{t_o \to t_\mathrm{v}} \, \delta c(t_0) \rangle}{\langle \delta x(t_0), \, \mathrm{E} \delta c(t_0) \rangle} \tag{8}$$

In (8) we use the fact that perturbations evolve in time according to the dynamics of the tangent linear model (3).

SVs are defined as the directions of maximal error growth, i.e. the vectors $s_k(t_0)$ that maximize the ratio σ^2 in equation (8). These directions are the solutions of the generalized eigenvalue problem

$$\mathbf{M}^*_{t_v \to t_o}\, \mathbf{G}^* \,\mathbf{F}\,\mathbf{G}\,\mathbf{M}_{t_o \to t_v}\, s_k(t_0) = \sigma_k^2\,\mathbf{E}\,s_k(t_0) \ . \tag{9}$$

The left side of (9) involves one integration with the tangent linear model followed by one integration with the adjoint model. The eigenvalue problem (9) can be solved efficiently using the software package ARPACK [8].

Using the square root of the the symmetric positive definite matrix E the generalized eigenvalue problem (9) can be reduced to a simple eigenvalue problem

$$\mathbf{E}^{-\frac{1}{2}}\,\mathbf{M}^*_{t_v \to t_o}\,\mathbf{G}^*\,\mathbf{F}\mathbf{G}\,\mathbf{M}_{t_o \to t_v}\,\mathbf{E}^{-\frac{1}{2}}\,v_k = \sigma_k^2\,v_k(t_0) \ , \quad v_k = \mathbf{E}^{\frac{1}{2}}\,s_k(t_0) \ . \tag{10}$$

Furthermore, $v_k(t_0)$ are the left singular vectors in the singular value decomposition

$$\mathbf{F}^{\frac{1}{2}}\,\mathbf{G}\,\mathbf{M}_{t_o \to t_v}\,\mathbf{E}^{-\frac{1}{2}} = U \cdot \Sigma \cdot V^T \quad \text{where} \quad \Sigma = \mathrm{diag}_k\{\sigma_k\} \ , \quad \sigma_k\,u_k = \mathbf{F}^{\frac{1}{2}}\,\mathbf{G}\,s_k(t_v) \ . \tag{11}$$

The SVs s_k are E-orthogonal at t_0 and F-orthogonal at t_v

$$\big\langle\, s_k(t_0),\, \mathbf{E}s_j(t_0)\,\big\rangle = 0 \quad \text{and} \quad \big\langle\, \mathbf{G}s_k(t_v),\, \mathbf{F}\mathbf{G}s_j(t_v)\,\big\rangle = 0 \quad \text{for} \quad j \neq k \ . \tag{12}$$

The equations (11) and (12) justify the name of "singular vectors". The singular value decomposition of the linear operator $M_{t_0 \to t_v}$, with the E scalar product at t_0 and the F scalar product at t_v, has the left singular vectors $s_k(t_0)$ and the right singular vectors $s_k(t_v)$. The singular values σ_k are the error amplification factors along each direction s_k.

The computation of singular vectors in the presence of stiff chemistry is discussed in [17], where computational challenges are reported related to the loss of symmetry due to the stiff nature of equations.

2.3 Perturbation Norms

In numerical weather prediction models variables have different physical meanings (wind velocity, temperature, air density, etc). The energy norms correspond to physical total energy, potential enstrophy, etc. Such norms provide a unified measure for the magnitude of perturbations in variables of different types.

In chemical transport models variables are concentrations of chemical species. Since all variables have the same physical meaning, and similar units, we expect that simple L^2 norms in (6) and (7) will provide a reasonable measure of the "magnitude of the perturbation". Since concentrations of different species vary by many orders of magnitude we expect that the perturbations of the more abundant species (e.g., CO) will dominate the total perturbation norms. To have a balanced account for the influence of all species it is of interest to consider the directions of maximal *relative error* growth. For practical reasons [17] it is advantageous to approximate the relative errors by the absolute errors δc_{ijk}^s scaled by "typical" concentration values w_{ijk}^s at each time instant. Therefore the choice of matrices in the norms (6) and (7) is

$$W(t) = \mathrm{diag}_{i,j,k,s}\left\{ w_{i,j,k}^s(t)\right\} , \quad E = W(t_0)^{-2} , \quad F = W(t_v)^{-2} .$$

One reason for this approximation is that the "typical" concentrations $w_{i,j,k}^s$ can be chosen to be bounded away from zero. More importantly, having the weights independent of the system state c keeps the maximization problem equivalent to a generalized eigenvalue problem.

3   Targeted Chemical Observations

We now determine those locations where perturbations have the largest energy impact over the verification area. For this, consider an initial perturbation vector δ_k equal to zero everywhere, except for one component at a given location where its value is 1. The index k spans all variables in the system, and a particular value of k identifies a single chemical component and a single location.

A certain species at a certain location is perturbed (or, equivalently, is observed, and therefore the perturbation is reduced). This vector can be written in terms of the singular vectors

$$\delta_{i,j,k}^s = \sum_m \alpha_m \, s_m\,(t_0) \, ,$$

where the expansion coefficients can be obtained by the orthogonality relations of the singular vectors

$$\alpha_m = \left\langle \delta_{i,j,k}^s , \mathrm{E}\, s_m\,(t_0)\right\rangle = (\mathrm{E}\, s_m(t_0))_{i,j,k}^s$$

The vector of perturbations at the final time is

$$\delta c(t_v) = \sum_m \alpha_m \, \mathbf{M}_{t_o \to t_v}\, s_m\,(t_0) = \sum_m \alpha_m \, \sigma_m \, s_m(t_v) .$$

Using the orthogonality of the singular vectors at the final time in the F-norm we have that the total perturbation energy is

$$\mathcal{E}_{i,j,k}^s = \left\langle \mathrm{G}\delta c(t_v) , \mathrm{FG}\delta c(t_v)\right\rangle = \sum_m \sigma_m^2 \alpha_m^2 = \sum_m \sigma_m^2 \left((\mathrm{E}\, s_m(t_0))_{i,j,k}^s\right)^2$$

A vector which has each component equal to the energy impact of the corresponding delta initial perturbation is therefore:

$$\mathcal{E} = \sum_m \sigma_m^2 \,(\mathrm{E}\, s_m(t_0))^2 = \sum_m \zeta_m^2 . \tag{13}$$

The squares of the vectors are considered in an element by element sense. Clearly this sum can be well approximated by the first several terms which correspond to the dominant singular values.

The observations should be located at those points where the energetic impact over the verification area is the largest. These points are easily identified as they are the largest entries of the E vector.

4   Numerical Results

The numerical tests use the state-of-the-art regional atmospheric chemical transport model STEM [2]. The simulation covers a region of $7200\,\mathrm{km} \times 4800\,\mathrm{km}$ in East Asia and the simulated conditions correspond to March 2001. More details about the forward model simulation conditions and comparison with observations are available in [2].

The computational grid has $\mathrm{N_x} \times \mathrm{N_y} \times \mathrm{N_z}$ nodes with $\mathrm{N_x}{=}30$, $\mathrm{N_y}{=}20$, $\mathrm{N_x}{=}18$, and a horizontal resolution of $240\,\mathrm{km} \times 240\,\mathrm{km}$. The chemical mechanism is SAPRC-99 [3] which considers the gas-phase atmospheric reactions of volatile organic and nitrogen oxides in urban and regional settings. The adjoint of the comprehensive model STEM is discussed in detailed in [16]. Both the forward and adjoint chemical models are implemented using KPP [4, 6, 15]. The forward and adjoint models are parallelized using PAQMSG [13]. ARPACK [8] was used to solve the symmetric generalized eigenvalue problems and compute the singular vectors.

We are interested in minimizing the uncertainty in the prediction of ground level ozone concentrations above Korea at 0 GMT March 4, 2001 (shaded area in Figure 1). Thus the "verification area" is Korea, and the "verification time" is

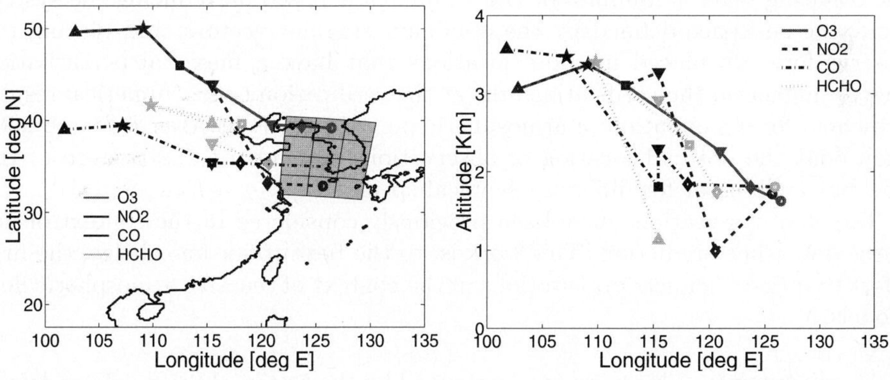

Fig. 1. Optimal placement of chemical observations using the maximum energy impact criterion. Observations of O_3, NO_2, $HCHO$, and CO are shown. The verification is ground level ozone over Korea (shaded area) at 0 GMT March 4, 2001. The observations are taken at 6h (circle), 12h (diamond), 18h (downward triangle), 24h (square), 36h (pentagon), and 48h (upward triangle) before verification time.

718   A. Sandu

0 GMT March 4 2001. Most of the uncertainty in the verification region at the final time is determined by the uncertainty along the dominant singular vectors at the initial time. The uncertainty (error) growth rates along each direction are given by the corresponding singular values.

In order to improve predictions within the verification region observations are needed in areas of maximal perturbation impact, determined using the dominant singular vectors. The optimal location of chemical observations at 6h, 12h, 18h, 24h, and 48h before the verification time is illustrated in Figure 1. As expected the optimal location of observations changes in time and drifts away from the verification area for longer intervals. Due to the different roles played by different chemical species in ozone formation, the optimal location of O_3 measurements is different than the optimal location of NO_2 or $HCHO$ observations. For example O_3 can be formed in the presence of NO_2 emissions and then transported over long distances. In contrast, the $HCHO$ lifetime is short and it can produce O_3 only locally.

5   Conclusions

The integration of observations and model predictions through data assimilation is essential for improved forecast capabilities in numerical weather prediction and in air pollution simulations. Atmospheric data assimilation is a parameter estimation problem with millions of degrees of freedom: the optimal analysis state of the atmosphere is found by minimizing the mismatch between model predictions and observations.

This paper develops a computationally tractable approach to target atmospheric chemical observations. The fundamental question being asked is where to place the observations such that, after data assimilation, the uncertainty in the resulting state is minimized. Our approach is based on reducing the system along the subspace defined by the dominant singular vectors, and placing the observations are placed in those locations that have a maximal perturbation energy impact on the verification area at the verification time. Numerical results presented for a simulation of atmospheric pollution in East Asia in March 2001 show that the optimal location of observations depends on the pattern of the flow but is different for different chemical species.

Targeted observations have been previously considered in the context of numerical weather prediction. This work is, to the best of our knowledge, the first effort to target chemical observations in the context of reactive atmospheric flow models.

Acknowledgments. This work was supported by the National Science Foundation through the awards NSF ITR AP&IM 0205198, NSF CAREER ACI–0413872, and NSF CCF–0515170, by the National Oceanic and Atmospheric Administration (NOAA) and by the Texas Environmental Research Consortium (TERC). We would like to thank Virginia Tech's Laboratory for Advanced Scientific Computing (LASCA) for the use of the Anantham cluster.

References

1. R. Buizza and A. Montani. Targeting observations using singular vectors. *Journal of the Atmospheric Sciences*, 56:2965–2985, 1999.
2. G.R. Carmichael et al. Regional-scale chemical transport modeling in support of the analysis of observations obtained during the TRACE-P experiment. *Journal of Geophysical Research*, 108:10649–10671, 2003.
3. W.P.L. Carter. Implementation of the SAPRC-99 chemical mechanism into the Models-3 framework. Technical report, United States Environmental Protection Agency, January 2000.
4. D.N. Daescu, A. Sandu, and G.R. Carmichael. Direct and adjoint sensitivity analysis of chemical kinetic systems with KPP: I – numerical validation and applications. *Atmospheric Environment*, 37:5097–5114, 2003.
5. D.N. Daescu and I.M. Navon. Adaptive observations in the context of 4d-Var data assimilation. *Meteorology and Atmospheric Physics*, 84(4):205–226, 2004.
6. V. Damian, A. Sandu, M. Damian, F. Potra, and G.R. Carmichael. The kinetic pre-processor KPP - a software environment for solving chemical kinetics. *Computers and Chemical Engineering*, 26:1567–1579, 2002.
7. R. Gelaro, R. Buizza, T.N. Palmer, and E. Klinker. Sensitivity analysis of forecast errors and the construction of optimal perturbations using singular vectors. *Journal of the Atmospheric Sciences*, 55:1012–1037, 1998.
8. R. Lehoucq, K. Maschhoff, D. Sorensen, and C. Yang. ARPACK software home page. http:// www.caam.rice.edu/ software/ ARPACK.
9. M. Leutbecher. A reduced rank estimate of forecast error variance changes due to intermittent modifications of the observing network. *Journal of the Atmospheric Sciences*, 60:729–742, 2003.
10. E.N. Lorenz. A study of the predictability of a 28 variable atmospheric model. *Tellus*, 17:321–333, 1965.
11. E.N. Lorenz and K.A. Emanuel. Optimal sites for supplementary observations: simulation with a small model. *Journal of the Atmospheric Sciences*, 55:399–414, 1998.
12. S.J. Majumdar, C.H. Bishop, R. Buizza, and R. Gelaro. A comparison of ensemble transform Kalman filter targeting guidance with ECMWF and NRL total-energy singular vector guidance. *Quarterly Journal of the Royal Meteorological Society*, 128:2527–2549, 2002.
13. P. Miehe, A. Sandu, G.R. Carmichael, Y. Tang, and D. Daescu. A communication library for the parallelization of air quality models on structured grids. *Atmospheric Environment*, 36:3917–3930, 2002.
14. F. Molteni and T.N. Palmer. Predictability and finite-time instability of the Northern winter circulation. *Quarterly Journal of the Royal Meteorological Society*, 119:269–298, 1993.
15. A. Sandu, D. Daescu, and G.R. Carmichael. Direct and adjoint sensitivity analysis of chemical kinetic systems with KPP: I – Theory and software tools. *Atmospheric Environment*, 37:5083–5096, 2003.
16. A. Sandu, D. Daescu, G.R. Carmichael, and T. Chai. Adjoint sensitivity analysis of regional air quality models. *Journal of Computational Physics*, 204:222–252, 2005.
17. W. Liao, A. Sandu, G.R. Carmichael, and T. Chai. Total Energy Singular Vector Analysis for Atmospheric Chemical Transport Models. *Monthly Weather Review*, accepted, 2005.

Model Optimization and Parameter Estimation with Nimrod/O

David Abramson[1], Tom Peachey[1], and Andrew Lewis[2]

[1] Caulfield School of Information Technology,
Monash University, Melbourne, Australia
[2] Division of Information Services, Griffith University, Brisbane, Australia

Abstract. Optimization problems where the evaluation step is computationally intensive are becoming increasingly common in both engineering design and model parameter estimation. We describe a tool, Nimrod/O, that expedites the solution of such problems by performing evaluations concurrently, utilizing a range of platforms from workstations to widely distributed parallel machines. Nimrod/O offers a range of optimization algorithms adapted to take advantage of parallel batches of evaluations. We describe a selection of case studies where Nimrod/O has been successfully applied, showing the parallelism achieved by this approach.

1 Introduction

Research in optimization concentrates on search methods; the objective function is usually only mentioned with regard to how its properties affect the validity and efficiency of the algorithm. A published algorithm will typically (see for example [20]) contain lines of the form

$$\texttt{evaluate } y = f(x_1, x_2, \ldots, x_n)$$

giving the impression that the step is minor. However for a substantial class of optimization problems the execution time of the algorithm is dominated by this evaluation step.

One such set of problems involve industrial design. Increasingly, in the design of engineering machines and structures, the prototyping stage is being replaced by computer modelling. This is normally cheaper, allows exploration of a wider range of scenarios and the possibility of optimization of the design. Consider for example a design problem in mechanical engineering, that of choosing the shape of a component that meets the functional specifications and is also optimal in the sense of giving maximal fatigue life [16]. Computation of the fatigue life involves a finite element analysis of the stress field followed by computation of perturbations produced by a range of hypothetical pre-existing cracks and calculation of the growth rate of these cracks under a given load regime.

Another class of optimization problems occurs in scientific modelling where one wishes to determine the values of underlying parameters that have given

V.N. Alexandrov et al. (Eds.): ICCS 2006, Part I, LNCS 3991, pp. 720–727, 2006.

rise to observed results. This inverse problem may be considered an optimization problem, searching through the plausible parameter space to minimize the discrepancy between the predicted and observed results. Inverse computational problems of this type are becoming common throughout many branches of science; some examples are described in a later section.

Such computational models typically take minutes or hours on a fast machine and an optimization requires that the model is executed many times. Hence it becomes attractive where possible to perform batches of these evaluations concurrently, sending the jobs to separate processors. Large clusters of processors are now commonly available for such work [1]. This paper describes a tool, Nimrod/O, that implements a variety of optimization algorithms, performing objective evaluations in concurrent batches on clusters of processors or the resources of the world computational grid.

2 The Nimrod Family of Tools

Parametric studies are explorations of the results of computational models for combinations of input parameters. Nimrod/G, [6, 4, 9], was designed to assist engineers and scientists in performing such studies using concurrent execution on a cluster of processors or the global grid. The user typically specifies a set of values for each parameter and the tasks required for a computation. Nimrod/G then generates the appropriate parameter combinations, arranges for the executions of these jobs and transfer of files to and from the cluster nodes, informing the user of progress with a graphical interface. Such an experiment may take days so failure of cluster nodes is a common problem; Nimrod/G reschedules jobs from a failed node. The number of concurrent jobs is limited only by the size of the cluster. Thus the user may achieve high concurrency without modifying the executables.

Nimrod/O [19] is a tool that provides a range of optimization algorithms and leverages Nimrod/G to perform batches of concurrent evaluations. Hence it is an efficient tool for the types of optimization problems mentioned earlier. Below we describe the operation of Nimrod/O and discuss some of the projects where it has been used. Note that Nimrod/O is not unique in offering distributed optimization. OPTIMUS Parallel [2], a commercial product, was developed at about the same time. However the focus there is narrower than that of Nimrod/O, with an emphasis on Design of Experiments and Response Surface Methods.

The Nimrod Portal [3] provides a friendly interface to the Nimrod toolset. It uses drop down menus to design and run an experiment and to select computational resources. Such resources may be added or removed as the experiment proceeds.

3 Nimrod/O

Nimrod/O requires a simple text "schedule" file that specifies the optimization task and the optimization method(s) to use. Several different algorithms and different instances of the same algorithm (varying the algorithm settings or the starting points) may be performed concurrently. But each evaluation task is funnelled through a cache to prevent duplication of jobs. If the cache cannot find the result in its database then the job is scheduled on the computational resources available. Since the Nimrod/O cache is persistent, when an aborted experiment is restarted the cache can provide results of all jobs completed earlier and hence a rapid recapitulation of the earlier work. Nimrod/O also allows separate users to share a cache so a useful database of completed jobs may be developed.

Often the objective functions produced by computational models produce multiple local optima. Further, the noise produced by discretization of the continuum may be significant and this gives a rough landscape adding further local optima as artifacts. The ability to run multiple optimizations from different starting points often reveals these multiple optima and may indicate which is global. As they are run concurrently this may be achieved without affecting the elapsed time.

The schedule file specifies the optimization in a declarative fashion. It is however simpler than standard optimization specification languages such as GAMS or AMPL as the definition of the objective function is assumed to be hidden within an executable program. An imperative section gives the commands needed to compute that objective.

An example schedule is shown in Figure 1. The first section of this specifies the parameters, their type (floats, integers or text) and ranges. Text parameter are used for categorical data; a separate optimization is performed for each combination of text values. For float parameters the "granularity" may be specified to control the rounding of values that is applied before they are sent for evaluation of the objective. The coarser granularity will improve the chances of a cache match with previous jobs, possibly at the expense of a less accurate optimum. Integer parameters are treated as floats with a granularity of 1.

```
parameter x float range from 1 to 15           method simplex
parameter y float range from 0.5 to 1.5            starts 5
parameter z float range from 0.5 to 1.0              starting points random
parameter w text select anyof "stt" "dynm"           tolerance 0.01
                                                   endstarts
constraint x >= y + 2.0^z                        endmethod
constraint {x > sin(pi*y)} or {x < 10}

                                                 method bfgs
task main                                           starts 5
  copy * node:.                                        starting points random
  node:substitute skeleton model.inp                  tolerance 0.01
  node:execute ./model.exe model.inp                   line steps 8
  copy node:obj.dat output.$jobname                 endstarts
endtask                                          endmethod
```

Fig. 1. A sample configuration file

The next section of the schedule specifies the "tasks" needed to evaluate the objective, normally to run a computational model. This includes the distribution of requisite files to the computational processors (the nodes), perhaps substitution of parameter values in input files, the execution of the evaluation programs and the return of the objective value. In this case the file `skeleton` has strings $x, $y, $z and $w replaced by their current values to form an input file `model.inp` for the computation. The executable `model.exe` performs the modelling producing a numerical result in the file `obj.dat` which is then copied back to the root node.

Finally the schedule gives the optimization method or methods to use. The example shown uses two methods, downhill simplex and BFGS. Note that several "starts" are specified, five simplex, five BFGS. This gives ten separate optimizations all of which will run concurrently if sufficient computing nodes are available.

4   Nimrod/O Algorithms

Nimrod/O is designed to solve optimization problems in engineering design and scientific modelling that typically involve a search space that is the cross product of several continuous parameters. Thus the problems are rarely of the combinatorial nature frequently encountered in operations research. Rather they require hill-climbing methods. The optimization algorithms offered by Nimrod/O reflect this requirement.

The algorithms provided fall into three categories. The first type samples the whole search space. It includes an exhaustive search with a given granularity for each numerical parameter. There is also a "subdivision search". This evaluates the space on a coarse grid then iterates, each iteration using a finer grid around the best point revealed by the previous iteration.

The second class are some traditional downhill search methods and recent variants: the direct search method of Hooke and Jeeves, the simplex method of Nelder and Mead, and some variants, the Broyden-Fletcher-Goldfarb-Shanno (BFGS) quasi-Newton method, and simulated annealing. Finally there are population based methods. EPSOC is an evolutionary programming algorithm using the ideas of self-organised criticality [13]. Nimrod/O interfaces with external genetic algorithm implementations GENEsYs and gamut.

The design of Nimrod/O allows for co-scheduling with external optimization routines. Using library functions supplied an external program can forward batches of jobs to Nimrod/O and hence take advantage of distributed computation, caching and constraint evaluation facilities.

Traditionally, search algorithm efficiency has been judged on the number of function evaluations required. In the Nimrod/O scenario execution time is dominated by function evaluations but these are processed in concurrent batches. Assuming there are sufficient computational resources to handle the largest batch then the number of batches becomes the critical factor in execution time. In implementing the optimization algorithms we have modified standard algorithms where this gives a reduced total number of batches.

For example with the traditional simplex search each iteration evaluates the objective at four points on a line. Nimrod/O offers the alternative of a line search to find the best point along that line. It also offers variants which search along several lines. Although these modifications require more evaluations, experiments suggest, [14], that batch counts are reduced and convergence expedited.

Batches of evaluations may also be augmented with jobs that may (or may not) be required at a later stage in the algorithm. This is known as "speculative computing", [10]. The Nimrod/O simulated annealing implementation can anticipate the step after next, adding tasks that may be needed then. Again this increases the total evaluations but reduces the execution time, [15]. Note that concurrent batch processing also favours the population based methods as the members of a large population may all be assessed concurrently.

5   Nimrod/O Case Studies

Nimrod/O has been successfully applied to a wide variety of optimization problems. A sample is discussed here. Data relating to the parallelism achieved in these experiments are combined at the end of this section.

Air Quality Modelling (AQM)

The model used predicted the concentration of ozone in an airshed, given concentrations of precursor chemicals together with meteorological data for the city modelled. The task [12, 5, 7] was to minimize the ozone concentration within a range of values for N and R, the concentrations of oxides of nitrogen and reactive organics respectively. Since ozone concentration is not a monotonic function of the input concentrations, the minimum does not necessarily correspond to least N and R.

Electromagnetic Modelling (EM)

The design of a test rig for mobile telephone antennas included a ferrite bead to reduce distortion of the radiation pattern [5, 7]. The finite-difference time-domain technique was used to solve Maxwell's equations for the design. The aim was to determine the dimensions and properties of the bead that minimized the losses due to testing.

Airfoil

Flow around an airfoil was modelled, [8], using the computational fluid dynamics package FLUENT. Input parameters were the thickness, camber and angle of attack of the airfoil; the aim was to maximize the ratio of lift to drag.

Quantum Chemistry (QC)

Hybrid quantum mechanics-molecular mechanics use quantum computations for small "active" regions of a molecule and classical methods for the rest. A major problem is correct coupling of the two models. This work, [21], uses the recently developed method of inserting a "pseudobond" at the junction between the models. The method was applied to an ethane molecule using a "pseudopotential" of the form $U(r) = A_1 \exp(-B_1 r^2) + A_2 \exp(-B_2 r^2)$. The task was to determine parameters A_1, A_2, B_1 and B_2 to minimize a least squares measure of the difference between the model properties and those of real ethane.

Plate Fatigue Life (PFL)
This work [16], modelled the fatigue life of plates containing an access hole, as occurs for example in stiffeners in airplane wings. This required finite element computation of the stress field and the Paris model of the growth of pre-existing cracks. The aim was to determine the hole profile under certain constraints that optimized this fatigue life.

Transformation Norm of an Integral Transform (TNIT)
The norm of the Generalized Stieltjes Transform is a long unsolved problem in mathematical analysis. A model was used to compute the ratio of the output norm to the input norm for a 2 parameter family of input functions and Nimrod/O optimized this ratio [18]. A novel aspect of this project was that multiple optimizations were performed for a parameter sweep of two further parameters; so multiple instances of Nimrod/O were launched by Nimrod/G.

5.1   Parallelism

For a single Nimrod/O optimization, if all evaluations required the same execution time then the ratio of the number of evaluations to the number of batches would give the parallelism attained. (When execution times vary then this over-estimates the parallelism as discussed in [17], since the execution time for a batch is that of the longest job.) The case studies described above used multiple optimizations, thus increasing the effective parallelism. We assume sufficient computational resources to run all optimizations concurrently. In that situation the number of batches in the longest optimization is the main determinant of the total experiment time. Figure 2 gives n, the number of optimizations, b and e, the number of batches and of evaluations for the longest optimization, B and E, the total batches and evaluations for all optimizations. Then the ratios e/b and E/B estimate the concurrency for a single optimization provided by batch evaluation. E/b estimates the concurrency for the combined optimizations provided by both batching and concurrent optimizations.

6   Conclusion

Optimization problems where evaluation of the objective function is computationally intensive are increasingly common. Nimrod/O is a tool that can expedite such problems by providing concurrent execution of batches of evaluations and concurrent multiple searches. It uses either a cluster of processors or the resources of the world computational grid; concurrency is limited only by the number of processors available. Nimrod/O offers a range of standard search algorithms and some novel ones, and is easily extensible to new algorithms. Since the total execution time is determined by the number of batches rather than the

Experiment	Method	n	b	e	B	E	e/b	E/B	E/b
AQM	BFGS	1	10	46	10	46	4.6	4.6	4.6
EM	BFGS	10	17	95	102	581	5.6	5.7	34.2
	Simplex	10	16	42	106	286	2.6	2.7	17.9
	Simplex-L	10	21	146	124	859	7.0	6.9	40.9
	Simplex-L1	10	16	144	104	843	9.0	8.1	52.7
	RSCS	10	9	54	61	371	6.0	6.1	41.2
	RSCS-L	10	17	127	116	858	7.5	7.4	50.5
	RSCS-L1	10	12	121	89	818	10.1	9.2	68.2
	EPSOC	10	20	892	200	8616	44.6	43.1	430.8
	EPSOC	10	20	1117	200	10601	55.9	53.0	530.1
Airfoil	Simplex	10	40	160	241	908	4.0	3.8	22.7
	Simplex-L	10	70	1037	463	6753	14.8	14.6	96.5
	Simplex-L1	10	28	385	183	2657	13.8	14.5	94.9
	RSCS	10	53	525	173	1717	9.9	9.9	32.4
	RSCS-L	10	71	1042	369	5385	14.7	14.6	75.8
	RSCS-L1	10	34	530	219	3301	15.6	15.1	97.1
	EPSOC	10	20	1260	200	12472	63.0	62.4	623.6
QC	BFGS	63	300	1450	3121	14650	4.8	4.7	48.8
PFL	simplex	9	26	88	132	505	3.4	3.8	19.4
TNIT	simplex	209	93	305	3048	10602	3.3	3.5	114.0

Fig. 2. Parallelism achieved by some experiments

number of evaluations modifications to some traditional search algorithms are advantageous and have been incorporated into Nimrod/O.

References

1. http://www.top500.org/lists/, accessed 3 August 2005.
2. http://www.lmsintl.com/, accessed 3 August 2005.
3. http://www.csse.monash.edu.au/~nimrod/nimrodportal/, accessed 3 August 2005.
4. Abramson D. et al. The Nimrod computational workbench: A case study in desktop metacomputing. In *Australian Computer Science Conference (ACSC 97)*, pages 17 – 26, Macquarie University, Sydney Feb 1997.
5. Abramson D. A., A. Lewis, and T. Peachey. Nimrod/O: a tool for automatic design optimisation using parallel and distributed systems. In *Proceedings of the 4th International Conference on Algorithms and Architectures for Parallel Processing (ICA3PP 2000)*, pages 497–508, Singapore, 2000. World Scientific Publishing Co.
6. Abramson D., J. Giddy, and L. Kotler. High performance parametric modeling with Nimrod/G: Killer application for the global grid? In *International Parallel and Distributed Processing Symposium (IPDPS)*, May 2000.
7. Abramson D. A., A. Lewis, and T. Peachey. Case studies in automatic design optimisation using the P-BFGS algorithm. In A Tentner, editor, *Proceedings of the High Performance Computing Symposium - HPC 2001*, pages 22 – 26, Seattle, April 2001. The International Society for Modeling and Simulation.

8. Abramson D. A., A. Lewis, T. Peachey, and C. Fletcher. An automatic design optimization tool and its application to computational fluid dynamics. In *Supercomputing*, Denver, November 2001.

9. Abramson D., R. Buuya, and J. Giddy. A computational economy for grid computing and its implementation in the Nimrod-G resource broker. *Future Generation Computer Systems*, 18(8), Oct 2002.

10. Burton F.W. Speculative computation, parallelism and functional programming. *IEEE Transactions on Computers, C*, 34:1190–1193, 1985.

11. Chong E.K.P. and S.H. Żak. *An Introduction to Optimization*. Wiley, 1996.

12. Lewis A., D. A. Abramson, and R. Simpson. Parallel non-linear optimization: towards the design of a decision support system for air quality management. In *IEEE Supercomputing 1997*, pages 1 – 13, California, 1997.

13. Lewis A., D. Abramson, and T. Peachey. An evolutionary programming algorithm for automatic engineering design. In *Parallel Processing and Applied Mathematics: 5th International Conference (PPAM 2003)*, Czestochowa, Poland, 2003.

14. Lewis A., D. A. Abramson, and T. Peachey. RSCS: A parallel simplex algorithm for the Nimrod/O optimization toolset. In *Proceedings of the Third International Symposium on Parallel and Distributed Computing (ISPDC 2004)*, pages 71–78, Cork, Ireland, 2004. IEEE Computer Society.

15. Peachey T. C., D. Abramson, and A. Lewis. Heuristics for parallel simulated annealing by speculation. Technical report, Monash University, 2001.

16. Peachey T., D. A. Abramson, A. Lewis, D. Kurniawan, and R. Jones. Optimization using Nimrod/O and its application to robust mechanical design. In *Proceedings of the 5th International Conference on Parallel Processing and Applied Mathematics [PPAM 2003]*, volume 3019 of *Lecture Notes in Computer Science*, pages 730–737. Springer-Verlag, 2003.

17. Peachey T. C., D. Abramson, and A. Lewis. *Parallel Line Search*. Springer, to appear.

18. Peachey T. C. and C. M. Enticott. Determination of the best constant in an inequality of Hardy, Littlewood and Polya. *Experimental Mathematics*, to appear.

19. Peachey T. C. *The Nimrod/O Users' Manual v2.6*. Monash University, http://www.csse.monash.edu.au/~nimrod/nimrodg/no.html, 2005.

20. Press W.H. et al. *Numerical Recipes in C*. Cambridge, second edition, 1993.

21. Sudholt W. et al. Applying grid computing to the parameter sweep of a group difference pseudopotential. In M Bubak et al., editor, *Proceedings of the 4th International Conference on Computa tional Science [ICCS 2004]*, volume 3036 of *Lecture Notes in Computer Science,*. Springer-Verlag, 2004.

The Criticality of Spare Parts Evaluating Model Using Artificial Neural Network Approach

Lin Wang[1], Yurong Zeng[2], Jinlong Zhang[1], Wei Huang[3], and Yukun Bao[1]

[1] School of Management, Huazhong University of Science and Technology,
Wuhan, 430074, China
{wanglin, jlzhang, yukunbao}@hust.edu.cn
[2] School of Computer Science and Technology, Hubei University of Economics,
Wuhan, 430205, China
zengyurong@sohu.com
[3] School of Management, Huazhong University of Science and Technology,
Wuhan, 430074, China
whuang@amss.ac.cn

Abstract. This paper presents artificial neural networks (ANNs) for the criticality evaluating of spare parts in a power plant. Two learning methods were utilized in the ANNs, namely back propagation and genetic algorithms. The reliability of the models was tested by comparing their classification ability with a hold-out sample and an external data set. The results showed that both ANN models had high predictive accuracy. The results also indicate that there was no significant difference between the two learning methods. The proposed ANNs was successful in decreasing inventories holding costs significantly by modifying the unreasonable target service level setting which is confirmed by the corresponding criticality class in the organization.

1   Introduction

Inventory control of spare parts (SPs) plays an increasingly important role in modern operations management. The trade-off is clear: on one hand a large number of spare parts ties up a large amount capital, while on the other hand too little inventory inevitably results in poor supply service level or extremely costly emergency actions [1]. There are about 5,000 SPs that are vital for safely production in a nuclear power plant in China. This plant is one of the most successful commercial power plants and the main technologies are gained from France. In addition, most of them are non-standard and purchased from France with a lead-time ranging from 8 to 20 weeks. The company is usually obliged to carry inventories consisting of over 110 millions dollars of SPs ready for maintenance.

A key distinguishing feature of the SPs inventory system is the need to specify the criticality of the items. The criticality of an item is a very important factor associated with the service level that can be defined as the probability of no shortage per replenishment from shelf. Then, we can identify the optimal control parameters according to all kinds of constrain condition. Factors such as costs of SPs, availability, storage considerations, probability of requirement of a SP, machine

V.N. Alexandrov et al. (Eds.): ICCS 2006, Part I, LNCS 3991, pp. 728−735, 2006.

downtime costs, etc., are generally considered while managing SPs inventories. Many analytical models of different inventory control systems have been discussed [2-3].However, there is no evidence that any of the works have attempted to raise the question of evaluating the criticality of SPs using systematic and well-structured procedures. Moreover, the various models described in the literature feature many assumptions that remain violated in real life. Simple and straightforward procedure such as ABC classification approach and the analytic hierarchy process (AHP) analysis have been used in practice for specifying inventory control polices and for fixing inventory review periods [4]. But the index considered, named lead time, type of spare and stock out implication, is so simple and may results in inaccuracy result. A better way to manage an inventory is thorough the development of better technique for identifying the criticality class (H, M or L) of a SP, which can also be regard as a classification problem, and management inventories from the point of view of their necessity in maintenance operation. However, the criticality of SPs needs to be evaluated and this is a difficult task which is often accomplished using subjective judgments. Therefore, identifying criticality class for a new SP in an accurate and efficient way becomes a challenge for inventory management.

Artificial neural network (ANN) is an artificial intelligence based technique, which is applicable to the classification process. The ANN can simulate a manager's utilization of perceived relationships for both quantitative and qualitative cues that provide important intermediate steps towards reaching the decision maker's final judgment. These networks have at least two potential strengths over the more traditional model-fitting techniques [5]. First, ANNs are capable of detecting and extracting nonlinear relationships and interactions among predictor variables. Second, the inferred patterns and associated estimates of the precision of the ANN do not depend on the various assumptions about the distribution of variables. The purpose of this study is to examine the classification accuracy of ANN as an aid to facilitate the decision making process of identifying the criticality of SPs. More specifically two types of learning methods, namely back propagation (BP) and genetic algorithms (GA) are used to examine the ANNs classification ability. The rest of this paper is organized as follows. Section 2 reviews the concepts of ANN. This is followed by the research methodology, and the evaluation of the classifier models. The paper concludes with summary of the findings and directions for future research.

2   Artificial Neural Networks

Like their biological counterpart, ANNs are designed to emulate the human pattern recognition function by the parallel processing of multiple inputs, and can capture the causal relationships between the independent and dependent variables in any given data set, i.e. ANNs have the ability to scan data for patterns which can be used to construct nonlinear models [6].

An ANN consists of a number of neurons, which are distributed in a number of hierarchical layers. One of the most widely implemented neural network architecture is the multilayer perceptions (MLP) model. A typical MLP used in this paper is shown in Fig. 1. This network has a three-layer, feed forward, hierarchical structure. The total number of neurons, number of neurons on each layer, as well as the number of layers determines the accuracy of the network model. The neurons in the input layer represent the attributes or stimuli in a data set. These inputs $(x_1, x_2, ..., x_n)$ initiate the activations into the network.

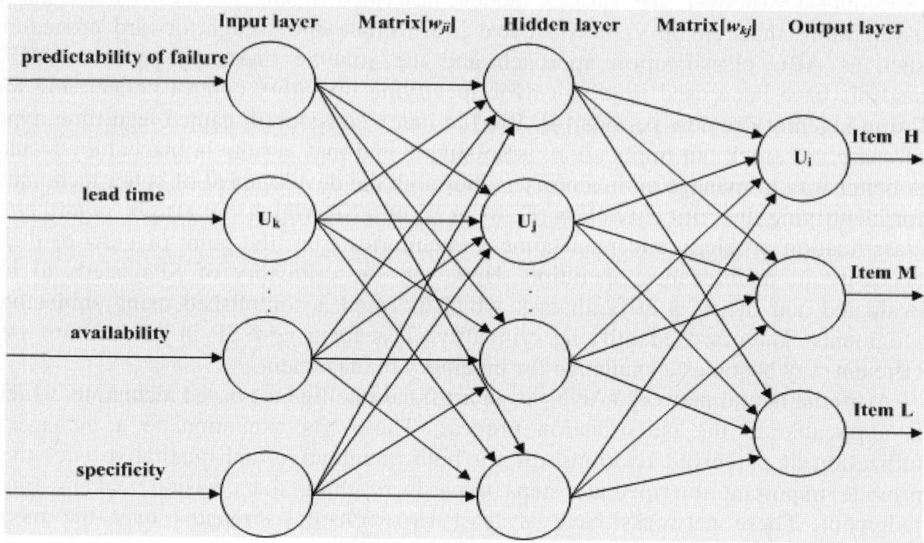

Fig. 1. Structure of the MLP ANN

As illustrated in Fig.1, these inputs are combined in the lower portion of the neuron. The upper portion of the neuron takes this sum and calculates the degree to which the sum is important using a transfer function (f); producing an individual output,

$$f \left(\sum_{i=1}^{n} w_i x_i \right)$$

where, w is weight vector $w=[w_1, w_2,...,w_n]$; and x is the input vector $x =[x_1, x_2, ...,x_n]$; for a specific neuron. The transfer function serves as a dimmer switch for turning on and off, depending on the input into the neurons. The selection of the transfer function typically depends on the nature of the output of the network. In this regard, there are a number of alternatives, including the step function, sigmoid function, hyperbolic tangent function, and linear function among others. Because the output of this study is continuous in nature and ranges from 0 to 1, this study uses the sigmoid transfer function $f(w'x) = 1/(1 + e(-w'x))$ as recommended by Zahedi [7].

Since the criticality evaluating of SPs problems are inherently non-linear in nature, it is important to create an ANN, which can approximate complex non-linear functions. This is achieved by adding hidden layers (i.e. several layers of sigmoid functions), which consist of neurons that receive inputs from the preceding cells and pass on outputs to subsequent cell layers. Although in theory a single hidden layer is sufficient to solve any function approximation problem, some problems may be easier to solve using more than a single hidden layer [8]. A commonly used learning method in ANN is the BP algorithm.

2.1 Back Propagation Algorithm Based Learning

The essence of the BP learning algorithm is to load the input/output relations within the MLP topologies so that it is trained adequately about the past to generalize the future [9]. The steps of the BP algorithm are given below:

Step 1: initialize weights to connections $[w_{jk}]$ and $[v_{ij}]$ with random weights.

Step 2: input X_i ($i=1…n$) receives an input signal and passes the signal to the hidden units Z_j.

Step 3: each of the hidden units Z_j ($i=1,…,n$) sums the weighted input signals net input to $Z_j = \theta_j + \sum_{i=1}^{n} x_i v_{ij}$ and applies the activation function to compute the output signal, where θ_k =Bias on hidden unit j.

Step 4: each of the output units Y_k ($k=1,…,n$) sums the weighted input signals net input to $Y_k = \theta_k + \sum_{j=1}^{n} z_j w_{jk}$ and applies the activation function to compute the output signal, u_k = Bias on output unit k.

Step 5: each of the output units compares its computed activation with a target to determine association error $e_k = y_k - t_k$: Based on e_k, errors at output unit Y_k is calculated as follows:

$$\delta_k = e_k \theta_k + \sum_{j=1}^{n} z_j w_{jk}$$

and sent to all neurons in the previous layers.

Step 6: when the training converges and the system error decreases below an acceptable threshold, the ANN is considered to be trained and then applied over the testing data set.

The BP algorithm however, may not provide the most efficient way to train neural networks and has in many instances resulted in inconsistent performance [10].In other words, obtaining a global solution is often dependent on the choice of starting values. An alternate approach to learning is selections, i.e. a complete behavior system is generated, by evolutionary process. Evolutionary development has been shown to be an extremely important source for generating more complexity in systems. Evolutionary development has been studied in great depth from a mathematical point of view, for instance [11].A common variant for classifying systems is called GA. This learning technique is discussed below.

2.2 Genetic Algorithm Based Learning

GA is a stochastic heuristic optimization search technique designed following the natural selection process in biological evolution, i.e. it models the nature of sexual reproduction in which the genes of two parents combine to form those of their children. When this technique is applied to problem solving, the basic premise is that an initial population of individuals representing possible solutions to a problem is created. Each of these individuals has certain characteristics that make them more or less fit as members of the population. The most fit members will have a higher probability of mating than lesser fit members, to produce progeny that have a significant chance of retaining the desirable attributes of their parents [12]. This method is very effective at finding optimal or near optimal solutions to a wide variety of problems, because it does not impose many of the limitations required by traditional methods [13-14]. It is an elegant generate-and-test strategy that can identify and exploit regularities in the environment, and converges on solutions that are globally optimal or nearly so. The GA consists of four steps, namely: initialization, reproduction, selection and convergence.

2.3 Easy-Use Subsystem

From what discussed above, we can see that this is a rather complex process. However, easy operating is an important measurement rule of any software system. So, we develop this easy-use subsystem to identify the criticality class of SPs. The main user interface of this subsystem, which is built using the Matlab software produced by Mathworks Laboratory Corporation, is shown in Fig. 2.

Fig. 2. The user interface of subsystem

3 Research Methodology

The empirical investigation was carried out using real-world data obtained from a nuclear power plant in China. A sample of 160 data sets (omitted) was used to design the network. Each data set represented a spare part and contained four types of information: predictability of failure (easy, difficult or impossible), lead time (days, ranging from 8-25 weeks), availability of spare part suppliers (easy or difficult), specificity of a spare part (standard, nor-standard). These criteria were selected based on their importance, as indicated by power plant managers through personal interviews. A five-step procedure was used to design the ANN, as discussed below.

(1) Representing spare parts in the ANN model

This study represented the identifying the criticality classes of spare parts with regard to four attributes: lead-time, predictability of failure, availability of spare part suppliers and specificity. The first criterion for judging a spare part is lead-time, which is the elapsed time between placing an order with a vendor and when that order arrives. Predictability of failure and specificity of a part is also important for criticality class. At the same time, we know that the criticality class may not be treated as "H" when the suppliers are easy to find. Inventory managers use ad hoc techniques to integrate the above criteria for criticality class classifications.

(2) Developing the neural network identifying models

Each of the networks consisted of 4 input neurons (one for each spare part characteristic), 16 hidden neurons, and 3 output neurons (namely, criticality class: H, M or L). The momentum learning parameter was chosen for its simplicity and efficiency with respect to the standard gradient. The value of the learning parameters and the threshold function were kept constant, since the purpose of this study was to evaluate the performance of the ANN topologies.

(3) Training the neural network identifying models

To assess the predictive accuracy of the ANN models the experimental sample was split into two distinct groups, namely, a training group (100 items) and a holdout data (60 items). Using the former, the network models were trained. This process is used to determine the best set of weights for the network, which allow the ANN to classify the input vectors with a satisfactory level of accuracy.

(4) Validating the ANN identifying models

After the network was trained, the holdout data (consisting of the 60 data sets) was entered into the system, and the trained ANN was used to test the selection accuracy of the network for the 60 testing data sets. This is where the predictive accuracy of the machine learning techniques compared against the criticality classes as defined by the experienced inventory managers is measured.

(5) Further validation of the classification models

To validation model's ability to classify data, another data set from another power plant was obtained; this "out of population" i.e. external data set consisting 100 spare parts was input into the predictive model only for purpose of model validation.

4   Result Analysis and Its Applications

The results of their classification for 160 randomly selected items are collected. All variables were significantly different for all the groups at $\alpha=0.01$ level, indicating that the three groups represented different populations.

In order to study the effectiveness of the ANN based classifiers, we must test the prediction accuracy by reference to expertise's judgment. As shown in Table 1, some classification errors may be inevitably and have a negative impact on replenishment decision making. So, it is necessary for managers check the classification result before using them. In practice, we can specify different target service level according to the kinds of criticality class. For example, the desire target service level that is measured by the expected shortage per replenishment cycle is specified as 99.95% for a SP with the criticality class of "H". In the same way, a certain target service level of 97% (95%) is conformed for criticality class of "M' ("L") according to the practical maintenance requirement.

Table 1. Prediction accuracy of ANNs (significant at 0.01)

	ANN (BP) (%)	ANN (GA) (%)
Overall training sample	84.5	85.8
Holding sample		
Overall classification	82.4	86.9
Item H	81.9	82.1
Item M	82.6	84.2
Item L	85.2	89.6
External sample		
Overall classification	86.0	86.3
Item H	100	100
Item M	84.2	85.2
Item L	82.8	83.9

5   Conclusions

This paper presents an ANN approach for criticality class evaluation of various SPs. Specifically, two learning methods were utilized in the ANNs, namely, BP and GA. The reliability of the models was tested by comparing their classification ability with a holdout data and an external data set. The results indicate that the ANN classifier models have a relative high predictive accuracy and acceptable. In addition, the results also indicate that there was no significant difference between the two learning methods. The use of the ANN model can prove to be a persuasive analytical tool in deciding whether the criticality class of an SP should be classified as a category H, M, or L item. By deploying this neural network based model, the unreasonable target service level setting which is confirmed by the corresponding criticality class are modified and the inventory at DaYa Bay Nuclear Power Plant in China consisting of over 100 billion dollars worth of SPs can be reduced by 6.86% while maintaining the reasonable target service level. However, although these classification models have several advantages, they also have their limitations. For example, the numbers of variables that can be input into these models are limited and many new important qualitative variables may be difficult to incorporate into the models. So, the model should not entirely replace professional judgment.

Acknowledgement

This research is partially supported by National Natural Science Foundation of China (No.70571025 and No.70401015), Chinese Postdoctoral Science Foundation (No.2005037680) and the Key Research Institute of Humanities and Social Sciences in Hubei Province-Research Center of Modern Information Management. The authors are grateful for the constructive comments of the referees.

References

1. Aronis, K.P., Magou, I., Dekker, R., Tagaras, G.: Inventory control of spare parts using a Bayesian approach: a case study. European Journal of Operations Research, 154(2004)730-739
2. Chang, P.L., Chou, Y.C., Huang, M.G.: A (r, r, Q) inventory model for spare parts involving equipment criticality. International Journal of Production Economics, 97(2005) 66-74
3. Kennedy, W.J., Patterson, J.W.: An overview of recent literature on spare parts inventories. International Journal of Production Economics, 75(2002) 201-215
4. Gajpal, P.P., Ganesh, L.S., Rajendran, C.: Criticality analysis of spare parts using the analytic hierarchy process. International Journal of Production Economics, 35(1994)293-297
5. Bishop, C.M.: Neural networks for pattern recognition. New York: Oxford University Press (1995)
6. Cai, K., Xia, J.T., Li, L.T, Gui, Z.L.: Analysis of the electrical properties of PZT by a BP artificial neural network. Computational Materials Science, 34 (2005) 166-172
7. Zahedi, F.: Intelligent systems for business: expert systems and neural networks, California: Wadsworth Publishing (1994)
8. Partovi, F.Y., Anandarajan, M.: Classifying inventory using an artificial neural network approach. Computer & Industrial Engineering, 40(2002) 389-404
9. Bansal, K., Vadhavkar, S., Gupta, A.: Neural networks based forecasting techniques for inventory control applications. Data Mining and Knowledge Discovery, 2(1998) 97-102
10. Lenard, M., Alam, P., Madey, G.: The applications of neural networks and a qualitative response model to the auditors going concern uncertainty decision. Decision Sciences, 26 (2) (1995) 209-227
11. Varetto, F.: Genetic algorithms applications in the analysis of insolvency risk. Journal of Banking and Finance, 22 (10-11) (1998) 1421-1429
12. Ferentinos, K.P.: Biological engineering applications of feed forward neural networks designed and parameterized by genetic algorithms. Neural Networks, 18(7) (2005)934-950
13. Sexton R.S., Dorsey, R.E., Johnson, J.D.: Toward global optimization of neural networks: a comparison of the genetic algorithm and back propagation. Decision Support Systems, 22 (2) (1998) 171-185
14. Kuo, R.J., Chen J.A.: A decision support system for order selection in electronic commerce based on fuzzy neural network supported by real-coded genetic algorithm. Expert Systems with Applications, 26(2) (2004)141-154

Solving Election Problem in Asynchronous Distributed Systems

SeongHoon Park

School of Electrical and Computer Engineering,
Chungbuk National Unvi. Cheongju ChungBuk 361-763, Korea
spark@chungbuk.ac.kr

Abstract. So far, the weakest failure detectors had been studied extensively for several of such fundamental problems. It is stated that Perfect Failure Detector P is the weakest failure detector to solve the Election problem with any number of faulty processes. In this paper, we introduce Modal failure detector M and show that to solve Election, M is the weakest failure detector to solve election when the number of faulty processes is less than $\lceil n/2 \rceil$. We also show that it is strictly weaker than P.

1 Introduction

The concept of (unreliable) failure detectors was introduced by Chandra and Toueg[6], and they characterized failure detectors by two properties: completeness and accuracy. Based on the properties, they defined several failure detector classes: perfect failure detectors P, weak failure detectors W, eventually weak failure detectors $\square$W and so on. In [6] and [8] they studied what is the "weakest" failure detector to solve Consensus. They showed that the weakest failure detector to solve Consensus with-any number of faulty processes is W and the one with faulty processes bounded by $\lceil n/2 \rceil$ (i.e., less than $\lceil n/2 \rceil$ faulty processes) is $\diamondsuit$W.

After the work of [8], several studies followed. For example, the weakest failure detector for stable leader election is the perfect failure detector P [7], and the one for Terminating Reliable Broadcast is also P [6].

In this paper, we first redefine the model of failure detectors and consider the weakest failure detectors to solve the stable leader election problem with the assumption that there is a majority of correct processes. We show that if f is only bounded by a value of less than $\lceil n/2 \rceil$, where n is the number of processes, the weakest failure detector to solve election is not P.

The rest of the paper is organized as follows. In Section 2 we describe our system model. In Section 3 we introduce the *Modal* Failure Detector M and show that to solve Election, M is necessary while P is not, whereas M is sufficient when a majority of the processes are correct. Finally, Section 4 summarizes the main contributions of this paper and discusses related and future work.

V.N. Alexandrov et al. (Eds.): ICCS 2006, Part I, LNCS 3991, pp. 736–743, 2006.
© Springer-Verlag Berlin Heidelberg 2006

2 Model and Definitions

Our model of asynchronous computation with failure detection is the one described in
[5]. In the following, we only discuss some informal definitions and results that are
needed for this paper.

2.1 Processes

We consider a distributed system composed of a finite set of processes $\Omega=\{1,2,..,n\}$ to
be completely connected through a set of channels. Each process has a unique id and
its priority is decided based on the id, i.e., a process with the lowest id has the highest
priority. Communication is by *message passing, asynchronous* and *reliable*. Processes
fail by crashing and the crashed process does not recover. We consider systems where
at least one process is correct (i.e. $f \langle |\Omega|$).

A failure detector is a distributed oracle which gives hints on failed processes. We
consider algorithms that use failure detectors. An algorithm defines a set of runs, and
a run of algorithm A using a failure detector D is a tuple $R = <F, H, I, S, T>$: I is an
initial configuration of A; S is an infinite sequence of events of A (made of process
histories); $T = t_0 \cdot t_1 \cdot t_2 \cdot \cdot t_k$ is a list of increasing time values indicating when each
event in S occurred where t_0 denotes a starting time; F is a failure pattern that denotes
the set $F(t)$ of processes that have crashed through any time t. A *failure pattern* is a
function F from T to 2^{Ω}. The set of correct processes in a failure pattern F is noted
correct(F) and the set of incorrect processes in a failure pattern F is noted *crashed(F)*;
H is a failure detector history, which gives each process p and at any time t, a
(possibly false) view $H(p,t)$ of the failure pattern. $H(p,t)$ denotes a set of processes,
and $q\in H(p,t)$ means that process p *suspects* process q at time t.

2.2 Failure Detector Classes, Reducibility and Transformation

Two completeness properties have been identified. *Strong Completeness*, i.e. there is
a time after which every process that crashes is permanently suspected by every
correct process, and *Weak Completeness*, i.e. there is a time after which every process
that crashes is permanently suspected by some correct process. Four accuracy
properties have been identified. *Strong Accuracy*, i.e. no process is never suspected
before it crashes. *Weak Accuracy*, i.e. some correct process is never suspected.
Eventual Strong Accuracy ($\Diamond$Strong), i.e. there is a time after which correct processes
are not suspected by any correct process; and *Eventual Weak Accuracy* ($\Diamond$Weak), i.e.
there is a time after which some correct process is never suspected by any correct
process.

The notation of *problem reduction* was first introduced in the problem complexity
theory [10], and in the formal language theory [9]. It has been also used in the
distributed computing [11,12]. We consider the following definition of problem
reduction. An algorithm A *solves* a problem B if every run of A satisfies the
specification of B. A problem B is said to be *solvable with* a class C if there is an
algorithm which solves B using any failure detector of C. A problem B_I is said to be

reducible to a problem B_2 with class C, if any algorithm that solves B_2 with C can be transformed to solve B_1 with C. If B_1 is not reducible to B_2, we say that B_1 is *harder than* B_2.

2.3 The Stable Leader Election

The *stable leader election* problem is described as follows: at any time, at most one process considers itself the leader, and at any time, if there is no leader, a leader is eventually elected. Once a process is elected to be a leader, it can't be demoted before crash. The *stable leader election* problem is specified by the following two properties:

- *Safety:* At any time, if a correct process has its *leader* set to true, then all other processes that had their *leader* set to true crashed.
- *Liveness:* At any time, there eventually exists a process that has its *leader* set to true.

3 Failure Detector to Solve Election

We define the *Modal* failure detector M, which is weaker than P. We show that, to solve Election: (1) M is necessary (for any environment), and (2) M is sufficient for any environment with a majority of correct processes. We then show that (3) P is strictly stronger than M for any environment where at least one processes can crash in a system of at least three processes.

3.1 Modal Failure Detector

Each module of failure detector M outputs a subset of the range 2^Ω. The most important property of M, denoted by *Modal Accuracy*, is that a process that was once confirmed to be correct is not suspected before crash. Let H_M be any history of such a failure detector M. Then M satisfies the following properties:

- *Strong Completeness:* There is a time after which every process that crashes is permanently suspected by every correct process.
- *Eventual Weak Accuracy:* There is a time after which some correct process is never suspected by any correct process.
- *Modal Accuracy:* Initially, every process is suspected. After that, any process that is once confirmed to be correct is not suspected before crash. More precisely:

3.2 The Necessary Condition for Election

The basic idea of our algorithm is the following. Initially, the value of FL_i and CL_i is set to Ω and Φ respectively. That means that initially every process is suspected and none is confirmed. After that each process periodically invokes election and waits until the result of election is returned. If the received leader is in FL_i, then process i removes it from FL_i and puts it into CL_i. If it is not identical with the current leader then process i puts the id of the current leader into FL_i since the leader that was once confirmed to be correct has crashed.

```
/* Algorithm executed by every process i */
1 FL_i := Ω;
2 CL_i := Φ;
3 current_leader := NULL;
4 Periodically (τ) do
5   election();
6 Upon received (leader, j) do
7 if ( j ∈ FL_i ∧ j ∉ CL_i ) then
8    FL_i := FL_i − { j };
9    CL_i := CL_i ∪ { j };
10 end-if
11 if ( current_leader ≠ j) do
12    FL_i := FL_i ∪ { current_leader };
13    current_leader := j;
14 end-if
```

Fig. 1. Emulating *M* using Election

Lemma 3.1. *The algorithm of Fig.2 uses Election to implement M.*

Proof. We show below that FL_i satisfies *Strong Completeness, Eventually Weak Accuracy* and *Modal Accuracy* properties of *M*.

• **Strong Completeness**: Once elected as a leader, the process can be demoted only if it crashes. Initially, every process is suspected by invoking FL_i :=Ωin line 2 of fig.1. Therefore, it satisfies strong completeness. After that the correct process *i* removes *j* from FL_i in line 8 of fig.1 only once at most and only if process *i* received *j* as a leader. Let assume that process *j* is elected as the leader and then crashes at time *t*, and let assume that process *i* is a correct process. Then by the *liveness* property of election, process *i* eventually receives the message (*leader, j*). Assume by contradiction that strong completeness is violated. It implies that process *i* never puts *j* into FL_i even though it has crashed. This means that process *i* invokes election in line 5, but always receive *j* as a leader in line 6 of fig.1, even though it has crashed. However, because leader process *j* crashes at time *t*, there is a time *t'* so that for every *t''> t'*, process *i* never receives process *j* as a leader by the *liveness* property of election: a contradiction. □

• **Eventually Weak Accuracy**: By contradiction, assume that eventual weak accuracy is violated. It implies that with every correct process *j*, there is a correct process *i* that suspects it. Let process *j* be elected as a leader and it doesn't crash. That is to hold, there should be a correct process *i* that never stops suspecting *j* even though *j* is elected to be the leader in the algorithm of fig.1. However, by the *liveness* property of the election algorithm of fig. 1, once correct processes *j* is elected as a leader and doesn't crash, then every correct process eventually receives the message (*leader, j*) and knows that *j* is a leader: contradiction. □

• **Modal Accuracy:** By contradiction, assume that modal accuracy is violated. By algorithm fig. 1, the predicate $j \notin FL_i(t)$ implies that at time $t'' < t$, process j is elected and removed from FL_i. The predicate $j \in FL_i(t')$ implies that at time $t' > t$, process k ($k \neq j$) is elected as a leader when j is the current leader and j is inserted to FL_i. Given that a process was once elected as a leader in stable election, the process can be demoted only if it crashes. Thus, the new leader can be returned only if the current leader crashes. That implies $j \in F(t')$. So it is a contradiction.   □

The following theorem follows directly from Lemma 3.1.

Theorem 3.1. If any failure detector D solves election, then M μ D.

3.2   The Sufficient Condition for Election

Periodically processes wait for an output from M to ensure the leader's crash. If the process receives from M the information that the current leader has crashed and at the same time the status of current leader is not false, i.e., ($current\_leader_i \neq \perp$), the process invokes *consensus* with a new candidate for leader and decides the new leader returned by *consensus*. Otherwise the process decides the current leader. We assume that every process i, either crashes, or invokes *election* in Fig.2. The new leader candidate of participant i, denoted $new\_candidate_i$, is decided by the *next* function. The current leader, denoted by $current\_leader_i$, is decided by the *consensus* function. The status of participant i whether it is a leader or not is decided by the variable, $leader_i$. That is, if the variable $leader_i$ is set true, the process i considers itself a leader.

```
function election( )
/* Algorithm executed by every process i */
1 leader_i := false;
2 current_leader_i := ⊥;
3 new_candidate_i := Next(0);
4 current_leader_i := Consensus(new_candidate_i);
5 if (current_leader_i = i ) then leader_i = true fi
6 Periodically (τ) inquiry M_i
7 Upon received H_M(i) from M_i do
8 if ((current_leader_i ∈ H_M(i))∧(current_leader_i ≠ ⊥)) then
9     new_candidate_i := Next(current_leader_i);
10    current_leader_i := ⊥;
11    current_leader_i := Consensus(new_candidate_i);
12    if (current_leader_i = i ) then leader_i := true fi
13 fi
```

Fig. 2. Transforming Consensus into Election with M

We define the **Next** function of process i in Fig.2 as follows.

$$Next(k) = \min\{ j \mid j \notin H(i, t) \wedge j \neq k \}.$$

Lemma 3.2. The algorithm of Fig.2 uses M to transform Consensus into Election.

Proof. We consider the properties of Election separately.

• *Safety*: A process that every correct process does not suspect is eventually elected as a leader by *Next* and *Consensus* functions. Let process i be the current leader elected at time t that is denoted by *current_leader* (t) = i, then clearly the process i is a correct process that the failure detector M of every correct process does not suspect at time t', $t' < t$. By *Modal Accuracy* the new leader is elected only when the current leader i has crashed. □

• *Liveness*: Consider leader i that is elected at time t in Fig.3. After that, if the leader process crashes at time t', $t' > t$, then by Strong Completeness of M, there is a time after that some correct processes suspect the current leader. There is eventually some correct process which executes line 7-11 of fig. 3. They decide a prospective leader by using the Next function and transfer it as a parameter to Consensus function. With the Validity property of Consensus, a process decides its leader only if some process has invoked consensus. By the Termination property of Consensus, every correct process eventually decides a leader that ensures the Liveness property of Election. □

We define here failure detector M. Each module of M outputs a subset of Ω. Failure detector M satisfies *Strong Completeness* and *Eventually Weak Accuracy*, together with *Modal Accuracy*. Since Consensus is solvable with *Strong Completeness* and *Eventually Weak Accuracy* for any environment with a majority of correct processes [8], then the following theorem follows from Lemma 3.2:

Theorem 3.2. M solves Election for any environment where a majority of processes are correct, $f < n/2$.

Finally, we can state the following theorem from Theorem 3.1 and Theorem 3.2.

Theorem 3.3. For any environment with $f < n/2$, M is the weakest failure detector to solve Election.

Proof : It is straightforward from Theorem 3.1 and Theorem 3.2

3.3 Modal Failure Perfection Is Not Perfection

Obviously, failure detector P can be used to emulate M for any environment, i.e., $M \mu P$. We state in the following that the converse is not true for any environment where at least one processes can crash in a system.

Theorem 3.4. $P \frown M$ for any environment where at least one process can crash in a system.

Proof. *(By contradiction).* We assume that there is an algorithm $A_{M \to P}$ that transforms M into failure detector P. Then we show the fact that P, transformed by above the algorithm, satisfies *Strong Completeness,* but it does not satisfy *Strong Accuracy:* So it is a contradiction. We denote by *output*(P) the variable used by $A_{M \to P}$ to emulate failure detector P, i.e., *output*(P)$_i$ denotes the value of that variable at a given process i. Let F_1 be the failure pattern where process 1 has initially crashed and no other process has crashed, i.e., $F_1(t_0) = \{ 1 \}$. Let H_1 be the failure detector history where all processes permanently output $\{1\}$ at t', $t' > t_0$; i.e., $\forall i \in \Omega, \exists t' \in T, t' > t_0 : H_1(i, t') = \{1\}$. Clearly, H belongs to $M(F_1)$. Since variable $output(P)$ satisfies *Strong Completeness of P* then there is a partial run of $A_{M \to P}$, $R_1 = < F_1, H_1, I, S_1, T >$ such that $\exists j \in \Omega, \exists t'' \in T, t'' \geq t' : \{1\} \subset output(P)_j$. Consider failure pattern F_2 such that correct(F_2)=Ω (F_2 is failure free) and define the failure detector history H_2 such that $\forall i \in \Omega, \forall t \in T : H_2(i, t) = \{1\}$, $t' \leq t \leq t''$ and $H_2(i, t) = \Phi$, $t > t''$. Note that $H_2 \in M(F_2)$ and $t' \leq t \leq t''$, $\forall i \in \Omega\text{-}\{1\} : H_1(i, t) = H_2(i, t)$. Consider $R_2 = < F_2, H_2, I, S_2, T >$ of $A_{M \to P}$ such that $S_1[k] = S_2[k]$, $\forall t \in T, t' \leq t \leq t''$. Let R_2 outputs a history $H_P \in P(F_2)$. Since partial runs of R_1 and R_2 for $t' \leq t \leq t''$ are identical, the resulting history H_P of process j is: $\forall t \in T, t' \leq t \leq t'': \{1\} \subset output(P)_j$. But in R_2, at t, $t' \leq t \leq t'' : 1 \in output(P)_j$ and $1 \in correct(F_2)$, which means that P violates *Strong Accuracy*: a contradiction. $\qquad\square$

4   Concluding Remarks

So far the applicability of these results to problems other than Consensus has been discussed in [6,13,14,15]. In [8], it is shown that Consensus is sometimes solvable where Election is not. In [7], it was shown that the weakest failure detector for Election is *Perfect* failure detector P, if we consider Election to be defined among every pair of processes. If we consider however Election to be defined among a set of at least three processes and at most one can crash, this paper shows that P is not necessary for Election. An interesting consequence of this result is that there exists a failure detector that is weaker than *Perfect* failure detector P to solve Election at the environment where a majority of processes are correct, $f < n/2$.

This paper introduces *Modal* failure detector M which is weaker than *Perfect* failure detector P, and shows that: (1) M is necessary to solve Election, (2) M is sufficient to solve Election, and (3) M is the weakest failure detector to solve Election when a majority of the processes are correct. A corollary of our results above is that we can construct a failure detector that is strictly weaker than P, and yet solves Election.

Is this only theoretically interesting? We believe not, as we will discuss below. Interestingly, failure detector M consists of $\Diamond S + Modal\ Accuracy$ and it helps deconstruct Election: intuitively, $\Diamond S$ conveys the pure agreement part of Election, whereas *Modal Accuracy* conveys the specific nature of detecting a leader crash. Besides better understanding the problem, this deconstruction provides some practical insights about how to adjust failure detector values in election protocols.

In terms of the practical distributed applications, we can induce some interesting results from the very structure of $\Diamond S + Modal\ Accuracy$ on the solvability of Election. In real distributed systems, failure detectors are typically approximated using time-outs. To implement the *Modal Accuracy* property, one needs to choose a large time-out value in order to reduce false leader failure suspicions. However, to implement $\Diamond S$, a time-out value that is not larger than the one for *Modal Accuracy* is needed. Therefore an election algorithm based on $\Diamond S + Modal\ Accuracy$ might reduce possibility of violating the safety condition but speed up the consensus of electing a new leader in the case of a leader crash.

References

1. G. LeLann: Distributed Systems–towards a Formal Approach. Information Processing 77, B. Gilchrist, Ed. North–Holland, 1977.
2. H. Garcia-Molina: Elections in a Distributed Computing System. IEEE Transactions on Computers, C-31 (1982) 49-59
3. H. Abu-Amara and J. Lokre: Election in Asynchronous Complete Networks with Intermittent Link Failures. IEEE Transactions on Computers, 43 (1994) 778-788
4. G. Singh: Leader Election in the Presence of Link Failures. IEEE Transactions on Parallel and Distributed Systems, 7 (1996) 231-236
5. M. Fischer, N. Lynch, and M. Paterson: Impossibility of Distributed Consensus with One Faulty Process. Journal of ACM, (32) 1985 374-382
6. T. Chandra and S.Toueg: Unreliable Failure Detectors for Reliable Distributed Systems. Journal of ACM, 43 (1996) 225-267
7. L. Sabel and K. Marzullo. Election Vs. Consensus in Asynchronous Distributed Systems. In Technical Report Cornell Univ., Oct. 1995
8. T. Chandra, V. Hadzilacos and S. Toueg: The Weakest Failure Detector for Solving Consensus. Journal of ACM, 43 (1996) 685-722
9. J. E. Hopcroft and J. D. Ullman: Introduction to Automata Theory, Languages and Computation. Addison Wesley, Reading, Mass., 1979
10. Garey M.R. and Johnson D.S: Computers and Intractability: A Guide to the Theory of NP-Completeness. Freeman W.H & Co, New York, 1979
11. Eddy Fromentin, Michel RAY and Frederic TRONEL: On Classes of Problems in Asynchronous Distributed Systems. In Proceedings of Distributed Computing Conference. IEEE, June 1999
12. Hadzilacos V. and Toueg S: Reliable Broadcast and Related Problems. Distributed Systems (Second Edition), ACM Press, New York, pp.97-145, 1993
13. R. Guerraoui: Indulgent Algorithms. In: Proceedings of the ACM Symposium on Principles of Distributed Computing, New York: ACM Press 2000
14. Schiper and A. Sandoz: Primary Partition: Virtually-Synchronous Communication harder than Consensus. In Proceedings of the 8th Workshop on Distributed Algorithms, 1994
15. R. Guerraoui and A. Schiper: Transaction model vs. Virtual Synchrony model: bridging the gap. In: K. Birman, F. Mattern and A. Schiper (eds.): Distributed Systems: From Theory to Practice. Lecture Notes in Computer Science, Vol. 938. Springer- Verlag, Berlin Heidelberg New York (1995) 121-132

A Performance Model of Fault-Tolerant Routing Algorithm in Interconnect Networks

F. Safaei[1,2], M. Fathy[2], A. Khonsari[1,3], and M. Ould-Khaoua[4]

[1] IPM School of Computer Science, Tehran, Iran
`{safaei, ak}@ipm.ir`
[2] Dept. of Computer Eng., Iran Univ. of Science and Technology, Tehran, Iran
`{f_safaei, mahfathy}@iust.ac.ir`
[3] Dept. of ECE, Univ. of Tehran, Tehran, Iran
[4] Dept. of Computing Science, Univ. of Glasgow, UK
`mohamed@dcs.gla.ac.uk`

Abstract. The Software-Based fault-tolerant routing [1] has been proposed as an efficient routing algorithm to preserve both performance and fault-tolerant demands in large–scale parallel computers and current multiprocessor systems-on-chip (Mp-SoCs). A number of different analytical models for fault-free routing algorithms have been suggested in the past literature. However, there has not been reported any similar analytical model of fault-tolerant routing in the presence of faulty components. This paper presents a new analytical model to capture the effects of failures in wormhole-switched k-ary n-cubes using Software-Based fault-tolerant routing algorithm. The validity of the model is demonstrated by comparing analytical results with those obtained through simulation experiments.

1 Introduction

Over the recent years, there exist many compute-intensive applications that require a large amount of processing power that can be merely achieved with massively parallel computers [1, 2]. The failure of components in such systems does not only reduce the machine computational power, but also deforms the interconnection network, which may consequently lead to a *disconnected* network. A network is disconnected if there exist two nodes without any fault-free path to route messages between them [2].

Direct networks have been a popular means of interconnecting mechanisms in parallel computers. The k-ary n-cube (also referred as n-dimensional, radix-k torus) is currently used as one of the most popular topologies for direct network. Moreover, the torus has been widely adopted in the last generation of practical parallel machines [2].

Network latency is one of the major factors that affect the performance communication networks. In order to minimize the network latency, *wormhole switching* (also known as wormhole routing [3]) has been widely adopted in the networks. Network throughput can be increased by organizing the flit buffers associated with each physical channel into several *virtual channels*. A virtual channel consists of a buffer, together with associated state information, capable of holding one or more flits of a message [4].

V.N. Alexandrov et al. (Eds.): ICCS 2006, Part I, LNCS 3991, pp. 744–752, 2006.

Routing is the process of transmitting data from one node to another node in a given system. Most past multicomputers have adopted deterministic routing (also widely known as *Dimension-ordered routing* [5]) where messages with the same source and destination addresses always take the same network path. Fault-tolerance is the ability of a routing algorithm to bypass faulty nodes/links in the network. The Software-Based approach [1] as an instance of fault-tolerant routing algorithm has been widely reported in literatures. The research presented in this paper use theoretical results of queuing theory to predict the message latency of the Software-Based routing scheme in the presence of faulty components.

The reminder of the paper is structured as follows. Section 2 describes some terms and backgrounds used in this paper. Section 3 presents our modeling approach while Section 4 validates the analytical model using a discrete-event simulator. Finally, Section 5 summarizes the work reported in this paper and presents possible directions for future works.

2   Preliminaries

The k-ary n-cube, where k is referred to as the radix and n as the dimension, has $N=k^n$ nodes, arranged in n dimensions, with k nodes per each dimension. Each node can be identified by an n-digit radix k address $(a_1, a_2,\ldots, a_n)$. Nodes, with address $(a_1, a_2,\ldots, a_n)$, $(b_1, b_2,\ldots, b_n)$ are connected iff there exists i, $(1 \leq i \leq n)$, such that $a_i=(b_i \pm 1) \bmod k$ and $a_i=b_j$ for $1 \leq i \leq n$; $i \neq j$. Each node consists of a Processing Element (PE) and router. A node is connected to its neighbouring nodes via the input and output channels. The *injection/ejection* channel is used by the processor to inject/eject messages to/from the network.

3   The Analytical Model

This section describes the assumptions used in the analysis, and then presents the analytical models for deterministic Software-Based routing algorithm.

3.1   Assumptions

The model is based on the following assumptions that have been widely employed in similar networks analysis studies [4-9].

- Nodes generate traffic independently of each other, and which follows a Poisson process with a mean rate of λ_g messages per cycle.
- The arrival process at a channel is approximated by an independent Poisson process.
- Message destinations are uniformly distributed across network nodes. Message length is M flits, each of which requires one-cycle to move from one router to the next.
- The local queue at the injection channel in the source node has infinite capacity. Moreover, messages are transferred through the ejection channel to the local node as soon as they arrive at their destinations.

- V (V>1) virtual channels are used per physical channel. At a given routing step, a message chooses randomly one of the available virtual channels at one of the physical channels, if available, that brings it closer to its destination.
- Each node failed with probability P_f. The probabilities of node failure in the network are equiprobable and independent of each other. Moreover, Faults are static [1-3] and distributed uniformly throughout the network such that do not disconnect it.
- Nodes are more complex than links and thus have higher failure rates [1, 2]. So, we assume only node failures.

3.2 Outline of the Analytical Model

The mean message latency is composed of the mean network latency, $\overline{L}$, which is the time to cross the network, and then the mean waiting time, $\overline{W}_s$, seen by the message in the local queue before entering the network. However, to capture the effects of the virtual channels multiplexing, the mean message latency is scaled by a factor, $\overline{V}$, representing the average degree of virtual channels multiplexing, that takes place at a given physical channel. Therefore, we can write the mean message latency as [6]

$$Mean\ message\ latency = (\overline{L} + \overline{W}_s)\overline{V} \tag{1}$$

In what follows, we will describe the calculation of $\overline{L}$, $\overline{W}_s$, and $\overline{V}$.

3.2.1 Calculation of the Mean Network Latency ($\overline{L}$)

In the Software-Based fault-tolerant routing, when a message encounters a fault, it is first re-routed in the same dimension in the opposite direction. If another fault is encountered, the message is routed in a perpendicular dimension in an attempt to route around the faulty regions [1]. At each dimension a message may encounters a node that is faulty with probability P_f or a node that is non-faulty again with probability ($1-P_f$). Thus, the average number of hops traversing each dimension, $\overline{k}_f$, is calculating by weighting average of average number of hops taken in each case. Therefore, using the method proposed in [7], $\overline{k}_f$ is calculated by

$$\overline{k}_f = \begin{cases} \left(\left(2\sum_{i=1}^{k/2-1}i + k/2\right)\Big/k\right)\cdot\left(1 + k.P_f\sum_{i=1}^{k/2}\left((-1)^i + 1\right)\Big/2i\right) & k\ is\ even \\ \left(2\sum_{i=1}^{(k-1)/2}i\Big/k\right)\cdot\left(1 + k.P_f\sum_{i=1}^{k/2}\left((-1)^i + 1\right)\Big/2i\right) & k\ is\ odd \end{cases} \tag{2}$$

Each PE generates, on average, λ_g, messages in a cycle. Resulting in a total $N\lambda_g$ newly-generated messages per cycle in the network. Since each message visits, on average, $n\overline{k}_f$ hops to cross the network, and the total number of channels in the network is $2nN$, the rate of messages received by each channel, λ_c, can be written as

$$\lambda_c = \lambda_g \overline{k}_f / 2 \tag{3}$$

Fig. 1 depicts the state transition diagram of the message flow through the network with the associated transition probability and latencies at each state. The message needs to pass r-1 intermediate nodes to reach its destination. State $L_{r,\,j}$ ($0 \le j \le r$-1) denotes that the header is at the intermediate node. Moreover, states $L_{r,\,0}$ and $L_{r,\,r}$ denotes that the header is at the source and destination nodes, respectively. Thus, the Latency seen by a message at the destination is

$$L_{r,\,n}=0$$

(4)

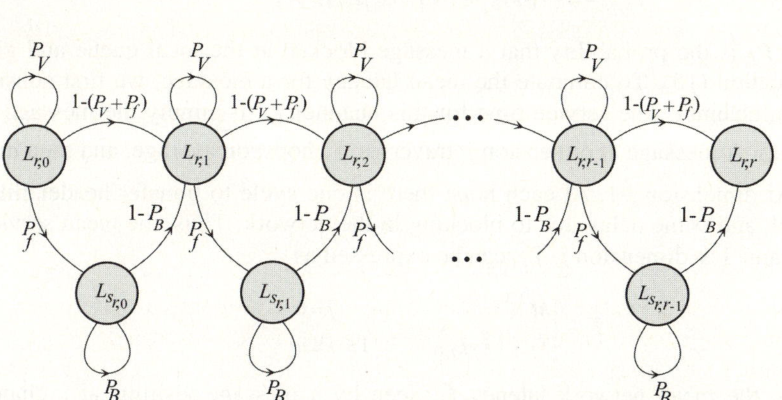

Fig. 1. The state transition diagram of the message flow in deterministic Software-Based routing

Let P_{b_j} be the probability of blocking at dimension j. A message is blocked at a given dimension j when all the virtual channels of that dimension are busy. To compute P_{b_j} two cases should be considered: (*i*) V virtual channels are busy, and (*ii*) V-1 virtual channels are busy. The number of combinations where V-1 out of V virtual channels are busy is $\binom{V}{V-1}$ of which only one combination result in all virtual channels are busy. Let P_{j_v} ($0 \leq v \leq V$) represents the probability that v virtual channels at a physical channel in dimension i are busy. Taking into account the two above cases, P_{b_j} can be calculated as

$$P_{b_j} = P_{j_v} + P_{j_{v-1}} \Big/ \binom{V}{V-1}$$

(5)

The latency $L_{r,\,j}$, seen by the message when crossing channel j, can be written as

$$L_{r,j} = [1-(P_V+P_f)](P_{b_j}\overline{W}_{c_j}+1+L_{r,j+1})+P_f(M+\varDelta+\overline{W}_s+L_{s_{r,j}})+L_{r,j}P_V$$

(6)

In the Equation (6), the first term accounts for the case when does not encounter a faulty component at channel j. The message may wait $\overline{W}_{c_j}$ with probability P_{b_j} to acquire one of the required virtual channels to advance to the next channel (i.e., j+1), where it sees latency $L_{r,\,j+1}$ to complete its journey. The second term, on the other hand, accounts for the case when the message encounters a faulty component. In this case, the message is delivered to the local queue of the current intermediate node after M cycles to account for the message transmission time. It also may suffer a delay overhead of $\varDelta$ cycles due to its re-injection at the node. The message then experiences a waiting time $\overline{W}_s$ in the local queue. Once at the head of local queue, it may experience blocking before it manages to access a virtual channel to make its next hop to continue its network journey. Therefore, the latency seen by a message at the head of local queue, $L_{s_{r,j}}$, is given by

$$L_{s_{r,j}} = (1 - P_B)(P_{b_j}\overline{W}_{C_j} + 1 + L_{r,j+1}) + P_B L_{s_{r,j}} \tag{7}$$

Where P_B is the probability that a message blocked at the local queue and given by the Equation (15). To compute the mean latency for a message, we first consider the ejection channel. The service time for this channel, $\overline{T}_0$, is simply the message length, M cycles. A message at dimension j, traverses $\overline{k}_f$ hops, on average, and then moves to the next dimension j-1. At each hope there is one cycle to transfer header flit over a channel, and some delay due to blocking in the network. Thus, the mean service time of a channel in dimension j, $\overline{T}_j$, can be expressed as

$$\overline{T}_j = \begin{cases} M & j = 0 \\ \overline{T}_{j-1} + \overline{k}_f L_{r,0} & 1 \le j \le n \end{cases} \tag{8}$$

Finally, the mean network latency, $\overline{L}$, seen by a message visiting $n\overline{k}_f$ channels to cross from source to destination is given by

$$\overline{L} = \overline{T}_n \tag{9}$$

3.2.1.1 Calculation of the Mean Waiting Time ($\overline{W}_{c_j}$). To determine the mean waiting time, $\overline{W}_{c_j}$, to acquire a virtual channel in dimension j, an M/G/1 queue is used with a mean waiting time given by [9]

$$\overline{w} = \rho_j \overline{T}_j (1 + C_{\overline{T}_j}^2) / 2(1 - \rho_j), \quad \rho_j = \lambda_c \overline{T}_j, \quad C_{\overline{T}_j}^2 = \sigma_{\overline{T}_j}^2 / \overline{T}_j^2 \tag{10}$$

Where λ_c is the traffic rate on the channel, $\overline{T}_j$ is its mean service time, and $\sigma_{\overline{T}_j}^2$ is the variance of the service time distribution. Since the minimum service time at a channel equals to the average service time of the channels in the next dimension, $\overline{T}_{j-1}$, following a suggestion proposed in [8], the variance of the service time can be approximated as

$$\sigma_{\overline{T}_j}^2 = (\overline{T}_j - \overline{T}_{j-1})^2 \tag{11}$$

Hence the mean waiting time to acquire a virtual channel in dimension j becomes

$$\overline{W}_{c_j} = \lambda_c \overline{T}_j^2 [1 + (\overline{T}_j - \overline{T}_{j-1})^2 / \overline{T}_j^2] / 2(1 - \lambda_c \overline{T}_j) \tag{12}$$

The above equations reveal that there are several inter-dependencies between the different variables of the model. For example, Equations (6) and (8) denote that $\overline{T}_j$ is a function of $\overline{W}_{c_j}$ while Equation (12) shows that $\overline{W}_{c_j}$ is a function of $\overline{T}_j$. Given that closed-form solutions to such inter-dependencies are very difficult to determine the different variables of the analytical model are calculated using iterative techniques for solving equations [9].

3.2.1.2 Calculation of the mean waiting time in the local queue ($\overline{W}_s$). The local queue
at the injection channel of a given node is
treated as an *M/G/1* queuing system, as
shown in Fig. 2. The local queue receives
two types of messages: messages that are
newly-generated by the local PE and those
that experience blocking at the input
channels of the node. Let us refer to these
two types as "newly-generated" and "transit"
messages, respectively. The local queue

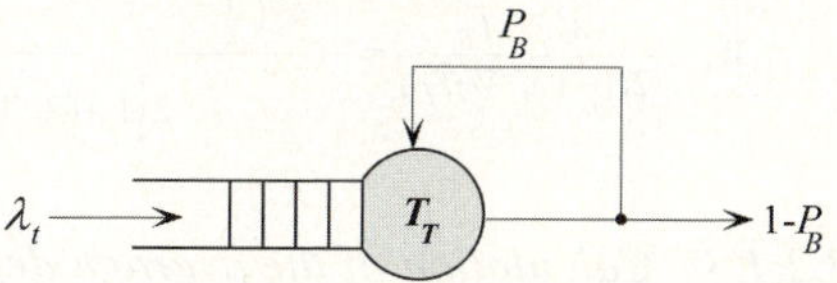

Fig. 2. The queuing model at the injection
channel of a given node

receives newly-generated messages at a mean rate of λ_g messages in a cycle. When a
transit message reaches a given node, it may suffer from blocking at the current node
with probability P_V. The rate of transit messages arriving at the local queue is given by

$$2n\,\lambda_c P_V \tag{13}$$

Therefore, the total traffic rate, λ_t , arriving at the local queue of a given node can be
written as

$$\lambda_t = \lambda_g + 2n\,\lambda_c P_V \tag{14}$$

The probability that a message blocked at the local queue, P_B, can be expressed as

$$P_B = P_V\,(\,\lambda_g + 2n\,\lambda_c)/\,\lambda_t \tag{15}$$

When a message has required ($\kappa+1$) retransmission attempts to cross the network,
it should be experienced κ transmission failures before successfully leaving the local
queue. It follows that the time interval between the start of transmission of a given
message and successfully leaving the local queue is $\overline{T}_n + \overline{W}_{c_n} + \kappa$ cycles with the
following probability

$$(1-P_B)P_B^\kappa \qquad\qquad\qquad \kappa = 0,1,2,\dots \tag{16}$$

Therefore, service time distribution in the local queue is given by

$$\Pr[T_T = \overline{T}_n + \overline{W}_{c_n} + \kappa] = (1-P_B)P_B^\kappa \qquad \kappa = 0,1,2,\dots \tag{17}$$

The first two moments of the service time in the local queue are given by

$$E[T_T] = \overline{T}_T = \sum\nolimits_{\kappa=0}^{\infty} T_T(1-P_B)P_B^\kappa = \sum\nolimits_{\kappa=0}^{\infty}(\overline{T}_n + \overline{W}_{c_n} + \kappa)(1-P_B)P_B^\kappa$$
$$= \overline{T}_n + \overline{W}_{c_n} + P_B/(1-P_B) \tag{18}$$

$$E[T_T^2] = \overline{T}_T^2 = \sum\nolimits_{\kappa=0}^{\infty} T_T^2(1-P_B)P_B^\kappa = \sum\nolimits_{\kappa=0}^{\infty}(\overline{T}_n + \overline{W}_{c_n} + \kappa)^2(1-P_B)P_B^\kappa$$
$$= (\overline{T}_n + \overline{W}_{c_n})^2 + (P_B/(1-P_B))^2 + 2(\overline{T}_n + \overline{W}_{c_n})\,P_B/(1-P_B) \tag{19}$$

A message at the local queue enters the network through any of the V virtual
channels. Using the *Pollaczek-Khinchine* [9] formula with a mean arrival rate λ_t/V and
the Equations (18) and (19), yields the mean waiting time, $\overline{W}_s$, experienced by a
message in the local queue as

$$\overline{W}_s = \frac{(\lambda_t/V)\overline{T}_T^2}{2(1-(\lambda_t/V)\overline{T}_T)} = \frac{(\lambda_t/V)\left\{(\overline{T}_n+\overline{W}_{c_n})^2 + 2(\overline{T}_n+\overline{W}_{c_n})\dfrac{P_B}{1-P_B} + \left(\dfrac{P_B}{1-P_B}\right)^2\right\}}{2\left\{1-(\lambda_t/V)(\overline{T}_n+\overline{W}_{c_n}+\dfrac{P_B}{1-P_B})\right\}} \tag{20}$$

3.2.1.3 Calculation of the average degree of virtual channels multiplexing ($\overline{V}$). The probability, P_v, that v virtual channels are busy at a given physical channel can be determined using a model proposed in [4]. In the steady state, the model yields the following probabilities

$$q_{jv} = \begin{cases} 1 & v=0 \\ q_{jv-1}\lambda_c\overline{T}_j & 0<v<V \\ q_{jv-1}\lambda_c/(1/\overline{T}_j-\lambda_c) & v=V \end{cases} \quad , \quad P_{jv} = \begin{cases} 1/\sum_{v=0}^{V} q_{jv} & v=0 \\ P_{jv-1}\lambda_c\overline{T}_j & 0<v<V \\ P_{jv-1}\lambda_c/(1/\overline{T}_j-\lambda_c) & v=V \end{cases} \tag{21}$$

Averaging over all dimensions, the average degree of multiplexing of virtual channels in the network is given by [4]

$$\overline{V} = \frac{1}{n}\sum_{j=1}^{n}\left(\sum_{v=0}^{V} v^2 P_{jv} \Big/ \sum_{v=0}^{V} v P_{jv}\right) \tag{22}$$

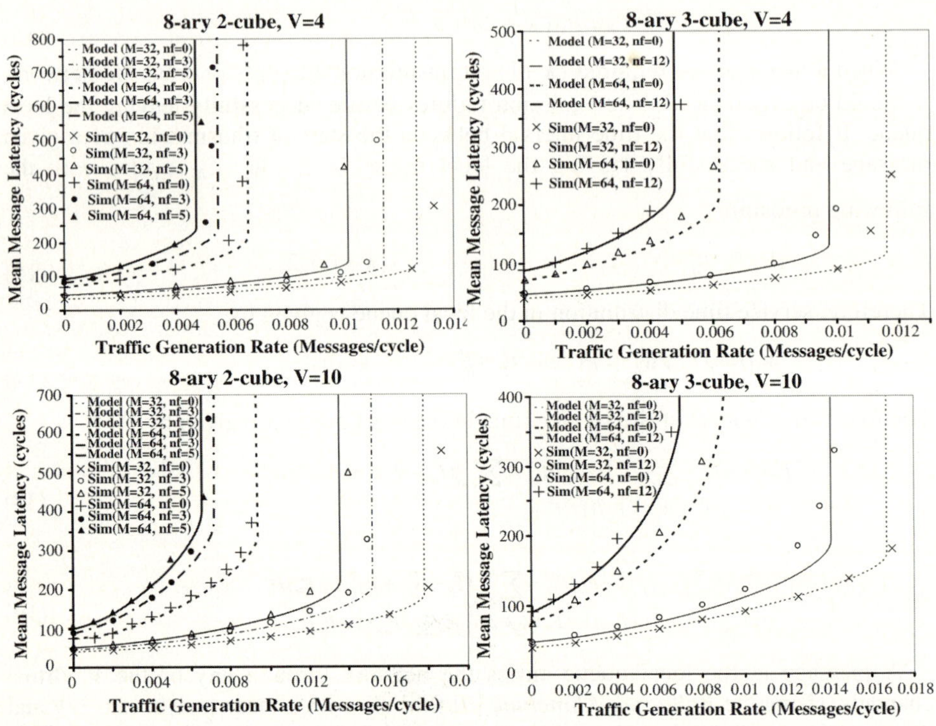

Fig. 3. The mean message latency of deterministic Software-Based routing in an 8-ary 2-cube and an 8-ary 3-cube networks with message length M=32 and 64 flits, number of random failed nodes n_f=0, 3, 5, 12 and number of virtual channels per physical channel V=4, 10.

4   Validation of the Analytical Model

Fig. 3 depicts latency results predicted by the model plotted against those provided by the discrete-event simulator for an 8-ary 2-cube and an 8-ary 3-cube using deterministic Software-Based routing, respectively. Two message lengths are used, $M=32$ and 64 flits. Random failed nodes are determined using a uniform random number generator while the delay overhead due to the message re-injection (Δ) is set to zero. Moreover, the number of virtual channels per physical channel is set to $V=4$, 10 and different number of random failed nodes is set to $n_f=0$, 3, 5 and 12. The horizontal axis in this figure represents the traffic generation rate while the vertical axis shows the mean message latency. The results reveal that the analytical model predicts the mean message latency with a good degree of accuracy in all regions. However, some discrepancies around the saturation point are apparent. This is a result of the approximations made when constructing the analytical model, e.g. the approximation used to estimate the variance of the service time distribution at a channel. This approximation greatly simplifies the model by avoiding the computation of the exact distribution of the message service time at a given channel.

5   Conclusions

A large number of fault-tolerant routing algorithms proposed in literatures for massively parallel systems, cluster-based systems, mobile systems, sensor networks, and multiprocessor systems-on-chip (MP-SoCs). The Software-Based fault-tolerant routing that has been known as an efficient routing method for reliable communication networks can route a message from source to destination, even in the presence of faulty components. In this paper, we proposed a new analytical model to compute the mean message latency of deterministic Software-Based routing in an n-dimensional torus. Simulation experiments have shown that the results predicted by the model are in good agreement with those obtained through simulations under different working conditions. Future directions may include evaluating the perfor-mance of adaptive Software-Based routing and the other well-known fault-tolerant routing algorithms using our suggested approach.

References

1. Suh, Y. J., et al.: Software-based rerouting for fault-tolerant pipelined communication, IEEE TPDS, 11 (3) (2000) 193-211.
2. Dally, W. J., Towles, B.: Principles and practices of interconnection networks, Morgan Kaufman Publishers, New York (2004).
3. Ni, L. M., McKinley, P. K.: A survey of wormhole routing techniques in direct networks, IEEE Computer, 26 (2) (1993) 62-76.
4. Dally, W. J.: Virtual channel flow control, IEEE TPDS, 3 (2) (1992) 194–205.
5. Dally, W. J., Seitz, C. L.: Deadlock-free message routing in multiprocessor interconnection networks, IEEE TC, 36 (5) (1987) 547-553.

6. Ould-Khaoua, M.: A performance model of Duato's adaptive routing algorithm in k-ary n-cubes, IEEE TC, 48 (12) (1999) 1-8.
7. Agarwal, A.: Limits on interconnection network performance, IEEE TPDS, 2 (4) (1991) 398-412.
8. Draper, J.T., Ghosh, J.: A comprehensive analytical model for wormhole routing in multicomputer systems, JPDC, 32 (2) (1994) 202-214.
9. Kleinrock, L.: Queueing Systems, Vol. 1, John Wiley, New York (1975).

Speculation Meets Checkpointing

Arkadiusz Danilecki and Michał Szychowiak

Institute of Computing Science
Poznań University of Technology
Piotrowo 3a, 60-965 Poznań, Poland
{adanilecki, mszychowiak}@cs.put.poznan.pl

Abstract. This paper describes a checkpointing mechanism destined for Distributed Shared Memory (DSM) systems with speculative prefetching. Speculation is a general technique involving prediction of the future of a computation, namely accesses to shared objects unavailable on the accessing node (*read faults*). Thanks to such predictions objects can be fetched before the actual access operation is performed, resulting, at least potentially, in considerable performance improvement. The proposed mechanism is based on independent incremental checkpointing integrated with a coherence protocol introducing little overhead. It ensures the consistency of checkpoints, allowing fast recovery from failures.

1 Introduction

Modern Distributed Shared Memory (DSM) systems reveal increasing demands of efficiency, reliability and robustness. System developers tend to deliver fast systems which would allow to efficiently parallelize distributed processes. Unfortunately, failures of some system nodes can cause loss of results of the processing and require to restart the computation from the beginning. One of major techniques used to prevent such restarts is *checkpointing*, which consists in periodically saving of the processing state (a *checkpoint*) in order to restore the saved state in case of a further failure. Then, the computation is restarted from the restored checkpoint. Only the checkpoints which represent a consistent global state of the system can be used (the state of a DSM system is usually identified with the content of the memory).

The *communication induced* (or *dependency induced*) checkpointing approach offers simple creation of consistent checkpoints, storing a new checkpoint each time a recovery dependency is created (e.g. interprocess communication), but its overhead turns out to be too prohibitive for general distributed applications. However, this approach has been successfully applied in DSM systems in strict correlation with memory coherence protocols. This correlation allows to reduce the number of actual dependencies and to significantly limit the checkpointing overhead ([2],[6]).

Speculation is a technique intended to improve the efficiency of DSM operations. The speculation methods are required to be very fast, while they do not

V.N. Alexandrov et al. (Eds.): ICCS 2006, Part I, LNCS 3991, pp. 753–760, 2006.
© Springer-Verlag Berlin Heidelberg 2006

necessary have to make correct predictions, as the cost of the mistakes is usually considered negligible. They include speculative pushes of shared objects to processing nodes before they would actually demand access [7], prefetching of the shared objects [1], or self-invalidation of shared objects [5] among other techniques.

This paper is organized as follows. Section 2 presents a formal definition of the system model and speculation operations. In Section 3 we discuss the concept of a checkpointing mechanism destined for DSM systems with speculation and propose a SpecCkpt protocol. Concluding remarks and future work are proposed in Section 4.

2   DSM System Model

A DSM system is an asynchronous distributed system composed of a finite set of sequential processes P_1, P_2, ..., P_n that can access a finite set O of shared objects. Each P_i is executed on a DSM node n_i composed of a local processor and a volatile local memory used to store shared objects accessed by P_i. Each object consists of several values (*object members*) and *object methods* which read and modify object members (here we adopt the object-oriented approach; however, our work is also applicable to variable-based or page-based shared memory). The concatenation of the values of all members of object $x \in O$ is referred to as *object value* of x. We consider here read-write objects, i.e. each method of x has been classified either as read-only (if it does not change the value of x, and, in case of nested method invocation, all invoked methods are also read-only) or read-and-modify (otherwise). Read access $r_i(x)$ to object x is issued when process P_i invokes a read-only method of object x. Write access $w_i(x)$ to object x is issued when process P_i invokes any other method of x. By $r_i(x)v$ we denote that the read operation returns value v of x, and by $w_i(x)v$ that the write operation stores value v to x.

DSM objects are replicated on distinct hosts to allow concurrent access to the same data. Concurrent processing in an asynchronous system is in general nondeterministic. A consistent state of DSM objects replicated on distinct nodes is maintained by a *coherence protocol* and depends on the assumed *consistency model*. Usually, one replica of every object is distinguished as *master replica*. The set of all replicas of a given object is referred to as *copyset*. The process holding master replica of object x is called x's *owner*. A common approach is to enable the owner an exclusive write access to the object.

The speculation introduces special part of the system, called the *predictor*, which is responsible for predicting future actions of the processes (e.g. future read and write accesses) and according reactions. Using speculation, however, an object may be fetched from its owner before the actual read access (i.e. *prefetched*), as a result of prediction. By $p_i(x)$ we will distinguish a prefetch operation on object x resulting from prediction made at process P_i.

3   Speculation and Checkpointing

3.1   Base Protocol

According to our knowledge, the impact of speculation on the checkpointing has not been investigated until now. While properly implemented speculation shall never violate the system consistency, ignoring the specific of speculation may severely damage the efficiency of checkpointing and recovery, as we will show.

We focus on prefetching techniques, but our approach should be easily adaptable to other speculation methods. In such techniques *predictor* anticipates the future read faults and prevents them by fetching respective objects in advance. The prediction may be incorrect in the sense that the process will never actually access the fetched object. Nevertheless, using speculation techniques such as the popular two level predictor MSP ([4]) turns out to increase the efficiency of most DSM applications. Since the predictor uses the underlying coherence protocol, it never violates the consistency of the memory.

Let us now consider the execution shown in Fig. 1. There is a *dependency* between processes P_1 and P_2, since P_2 fetches the value modified by P_1. To ensure the consistency in case of a subsequent failure of process P_1, the system forces P_1 to take a new checkpoint containing the previously modified object x.

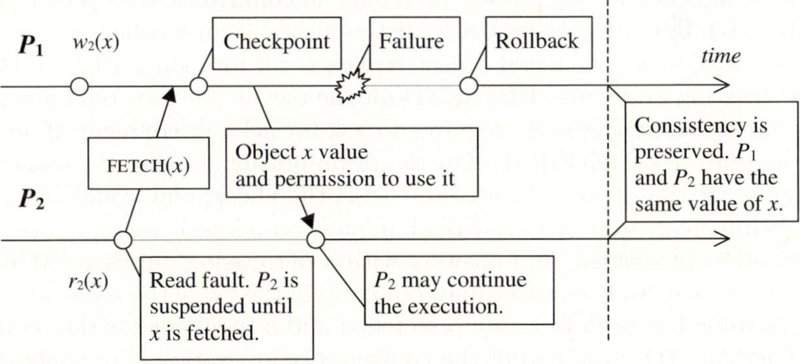

Fig. 1. Scenario without speculation. Real dependency between P_1 and P_2.

However, the situation may significantly change with speculation. In the scenario presented in Fig. 2 the predictor assumes that process P_2 will read the value modified by P_1, so it fetches the object to avoid a further read-fault. Performing that fetch, the system forces process P_1 to take a checkpoint. However, the prediction eventually turns out to be false and P_2 does not actually access x. Therefore, no real dependency was created and checkpoint was unnecessary. Unfortunately, P_1 was unable to determine that the fetch resulted from a false prediction, even if that fetch operation has been known to be speculative.

The problems presented above are summarized as follows:

– Access to objects (fetches) may result from speculation made by predictor and therefore (in case of false prediction) may not result in real dependency;

– Even when an access is marked as speculative, process has no way of determining whether true dependency between processes will ever be created, since it cannot determine whether the prediction is correct.

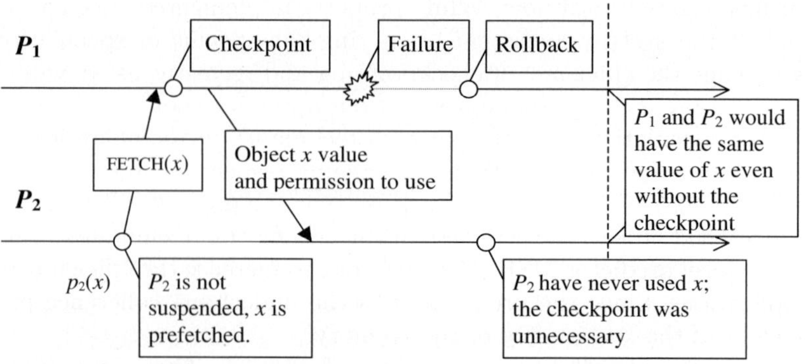

Fig. 2. Scenario with speculation. No dependency between P_1 and P_2.

A possible solution is to introduce a new replica state and decouple access requests for objects into two phases: prefetch and confirmation (Fig. 3). A speculative prefetch operation is explicitly distinguished from a coherence operation of a read access. The prefetched object replica is set into state PREFETCHED on the requesting node, and PRESEND on the owner. Further read access performed on the requesting node requires to ask for acknowledgment of accessing the object (message CONFIRM). On reception of this message the owner takes a checkpoint of the object, if necessary (e.g. the checkpoint could been taken already before reception of CONFIRM request as a result of some operations issued by other processes), and answers with a permission message (ACK).

Please note that ACK message does not contain the value the requested object (since this value has been formerly prefetched and is available for the requesting node). Therefore the overhead of the confirmation operation is in general lower than of a read-fault.

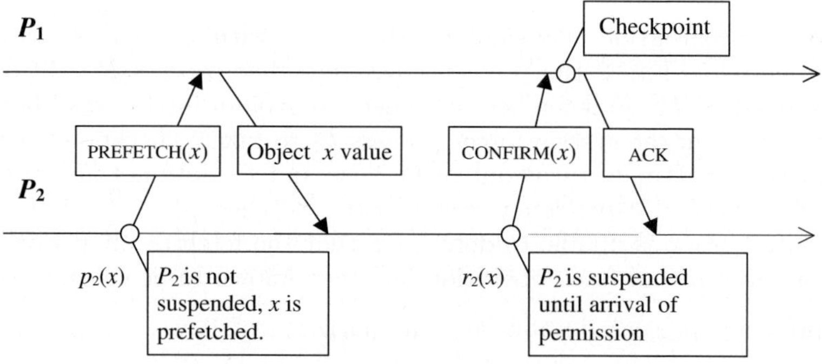

Fig. 3. Coherence decoupling

If the master replica of the considered object has been modified after a prefetch but before the corresponding confirmation it is up to the coherence protocol to decide about the acknowledgment (reading outdated values may be allowed depending on the consistency model). The coherency protocol may force the invalidation of a prefetched object before the confirmation. This invalidation will be performed exactly as for objects fetched by nonspeculative operations. Since there is no difference between those two types of operations from the point of view of the coherence, only minor modifications of coherence protocols will be necessary. The only significant difference concerns the checkpointing operations.

Our approach avoids unnecessary taking of checkpoints after a prefetch (when no real dependency is created). The checkpoint is postponed until an actual dependency is revealed on the confirmation request.

3.2   Protocol Improvement

Addressing the Protocol Efficiency. There are several possible ways to further increase the protocol efficiency. It is possible to perform a consolidated checkpoint of an entire group of objects (i.e. *burst checkpoint* [2]). This may significantly reduce the cumulative checkpointing overhead.

For instance, at the moment of further confirmation the prefetched object demanding confirmation may have already been checkpointed (during some previous burst checkpoint) and no new checkpoint will then be required. In such situation, no checkpoint overhead will be perceived by the application neither on prefetch, nor on actual read access to the prefetched object.

Another possible optimization is to send confirmations to all prefetched replicas directly after every checkpoint. The improvement of the efficiency is achieved by avoiding the need of confirmation during a further access to the replica prefetched earlier.

Addressing the Protocol Correctness. Let us consider a recovery situation presented in Fig. 4. After the value 1 of x has been checkpointed, it is modified again, to 2. Process P_2 prefetches the modified value of x from P_1. Then, P_1 fails and recovers, restoring the checkpointed value $x = 1$. Please note that the confirmation requested by P_2 cannot be granted, as it concerns a value of x that became inconsistent after the recovery.

In order to ensure the consistency, the recovered process P_1 might simply invalidate every replica prefetched from P_1 and not confirmed yet or, alternatively, refuse all confirmation requests received after the recovery. While those two solutions do prevent system from becoming inconsistent, they are far from being optimal. The first approach may unnecessarily invalidate prefetched replicas which were consistent (an unconfirmed replica may be perfectly consistent, as presented in Fig. 5), or invalidate prefetched replicas which would never be used anyway (therefore introducing unnecessary communication costs). The second approach is even worse, since it basically turns off the whole prefetching mechanism after the first failure. Optimal solution should both prevent the system from becoming inconsistent and allow the confirmation of all prefetched replicas that do not violate the consistency.

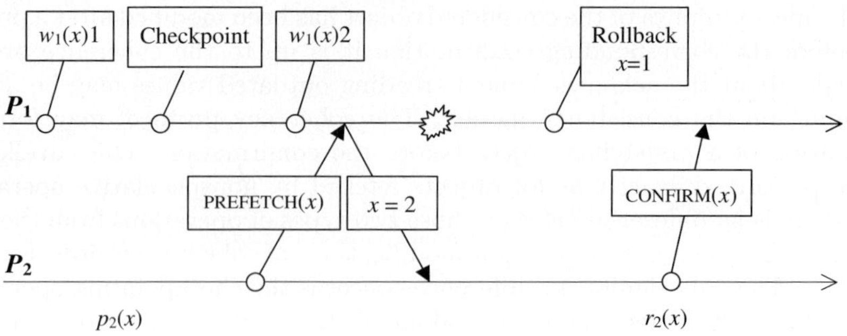

Fig. 4. Possible coherence problems with node failures

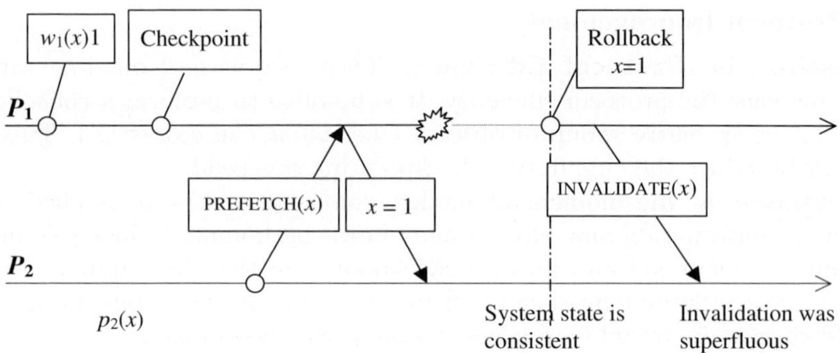

Fig. 5. Unnecessary invalidations of all prefetched replicas after the owner recovery

One intuitive and simple (and wrong, as we will soon show) solution is to use *version numbers*, increased after each meaningful change of the object (by meaningful we understand the first object modification after each checkpoint). Version numbers are stored in the checkpoints. Only replicas with version number equal to *master version number* (the version number of the master replica, restored from checkpoint after the recovery) can be confirmed, and all other confirmations would be refused.

However, this approach may result in inconsistent state after the recovery. Let us consider a simple example illustrated by Fig. 6. Owner P_1 modifies the object x, therefore increasing the version number $v(x)$ from m to $m + 1$. This version of x would be prefetched by another process P_2. Please note, that this version is not checkpointed. After the recovery, master replica of x would be rolled back to version m, and a subsequent change would again increase the version number to $m + 1$ When process P_2 would then ask for confirmation of his prefetched replica, it would appear that he has the correct version of the object, and the confirmation will be granted, possibly wrongfully.

Therefore, we investigate other possibilities, discussed in depth in [3]. Here we present one which solves all the problems described above.

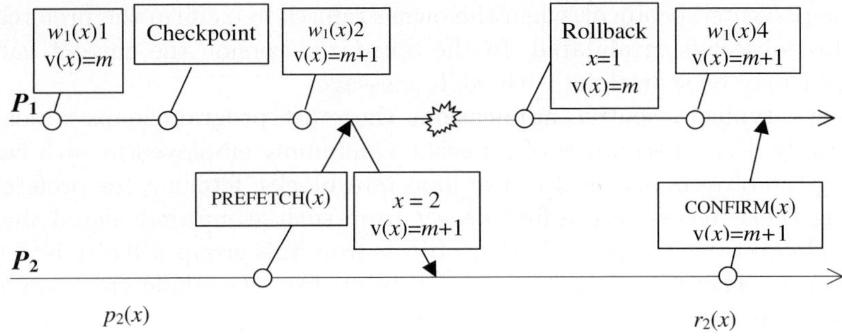

Fig. 6. Possible consistency violation in the approach with version number

The proposed checkpointing protocol, SpecCkpt, combines version numbers and the approach with invalidation of all prefetched replicas on recovery. Owners maintain a version number associated with the objects. After the recovery, the owner sends an invalidation request containing the version number restored from the checkpoint. The receiving processes invalidate the prefetched replicas only if their local version numbers are larger than the one received in the invalidation message. This approach keeps the system consistent by invalidating those and only those replicas which could violate it. The only small vice is the additional communication costs, which may be unnecessary if the invalidated prefetched replicas would never to be used anyway (i.e on mispredication).

Finally, let us present a remark about the optimization with confirmation of all prefetched replicas on every checkpoint. When using this approach, if the replica is in the prefetched state, it might be safely assumed that it is not consistent with the version of the object restored from the checkpoint. Therefore, it's enough to simply invalidate all prefetched replicas on recovery.

4   Conclusions

This paper describes an approach to checkpointing shared objects with speculation. We recognize the false dependencies and unnecessary checkpoints related to speculative operations on shared objects. We propose the operation decoupling which allows to decrease the frequency of checkpoints. Moreover, we describe additional mechanisms reducing the checkpointing overhead and enabling fast recovery. Practical verification of an implementation of SpecCkpt protocol is currently performed.

There are at least two directions in which our approach could be further studied and extended. First is to integrate the implementation of the proposed checkpointing technique with a particular coherence model. Second direction is to seek the optimizations for increasing positive effects of speculation.

Since our approach is very general, it still allows several optimizations concerning distinct characteristics of the protocol.

In the presented protocol, when the owner refuses to confirm the prefetch, the prefetched object is invalidated. In the optimized version the current value of the object may be sent along with ACK message.

In many typical scientific applications there are program loops which produce strictly defined sequence of requests. Commonly employed in such cases is grouping the objects accessed in the loop into blocks, fetching (or prefetching) them together. Access to the first object from such group may signal that the program loop started again and other objects from this group will also be fetched subsequently. Therefore, it appears useful to confirm the whole group on access to the first object.

References

1. Bianchini, R., Pinto, R., Amorim, C.L.: Data Prefetching for Software DSMs. Int. Conf. on Supercomputing, Melbourne, Australia (1998)
2. Brzeziski, J., Szychowiak, M.: Replication of Checkpoints in Recoverable DSM Systems. 21st Int. Conf. on Parallel and Distributed Computing and Networks PDCN'2003, Innsbruck, Austria (2003)
3. Danilecki, A., Szychowiak, M.: Checkpointing Speculative DSM Systems. Technical Report RA-021/05, Institute of Computing Science, Poznań University of Technology, Poznań, Poland (2005)
4. Lai, A-C., Babak Falsafi, B.: Memory Sharing Predictor: The Key to a Speculative Coherent DSM. 26th Int. Symp. on Computer Architecture (ISCA 26), Atlanta, Georgia (1999) 172–183
5. Lai, A-C., Babak Falsafi, B.: Selective, Accurate, and Timely Self-Invalidation Using Last-Touch Prediction. 27th Int. Symp. on Computer Architecture (ISCA 27), Vancouver, BC, Canada (2000) 139–148
6. Park, T., Yeom, H.Y.: A Low Overhead Logging Scheme for Fast Recovery in Distributed Shared Memory Systems. Journal of Supercomputing Vo.15. No.3. (2002) 295–320
7. Rajwar, R., Kagi, A., Goodman, J. R.: Inferential Queueing and Speculative Push. Int. Journal of Parallel Programming (IJPP) Vo. 32. No. 3 (2004) 273–284

Design and Verification for Hierarchical Power Efficiency System (HPES) Design Techniques Using Low Power CMOS Digital Logic

Taikyeong Jeong[1,*] and Jaemyoung Lee[2]

[1] Department of Electrical and Computer Engineering
University of Texas at Austin, Austin, TX 78712-1014 USA
ttjeong@mail.utexas.edu
[2] Korea Polytechnic University, Korea 429-793, ROK
lee@kpu.ac.kr

Abstract. This paper presents the design implementation of digital circuit and verification method for power efficiency systems, focused on static power consumption while the CMOS logic is in standby mode. As complexity rises, it is necessary to study the effects of system energy at the circuit level and to develop accurate fault models to ensure system dependability. Our approach to designing reliable hardware involves techniques for hierarchical power efficiency system (HPES) design and a judicious mixture of verification method is verified by this formal refinement. This design methodology is validated by the low power adder with functional verification at the chip level after satisfying the design specification. It also describes a new HPES integration method combining low power circuit for special purpose computers. The use of new circuits and their corresponding HPES design techniques leads to minimal system failure in terms of reliability, speed, low power and design complexity over a wide range of integrated circuit (IC) designs.

1 Introduction

The most important role of design and verification work is to make sure that all circuits and systems operate safely. Some designers devote countless hours to rigorously testing all integrated circuits (ICs) as part of designer's responsibility to help ensure that the system remains fail-free with minimal energy usage.

These days special purpose computers associated with energy-efficient designs are becoming more important in telecommunications, and networking systems. As one possible way of implementing energy-efficient system design, we propose the hierarchical power efficiency system (HPES) design techniques which include low power CMOS digital logic focused on stable system fault coverage.

This special purpose requirement, a low power adder design with HPES, is therefore a good example of the development of fail-free environments because it contains a well established logical prover, and uses a variety of logic.

* This work partially supported by the NASA (Grant No. NNG05GJ38G), and the corresponding author is T. Jeong.

V.N. Alexandrov et al. (Eds.): ICCS 2006, Part I, LNCS 3991, pp. 761–768, 2006.

The implementation in this paper differs from others in the following aspects. This paper proposes a new low power design method by providing a fast logical approach and low power dissipation. The outcome, such as a low power adder is introduced and a new method is derived by extending and modifying a conventional adder for the performance comparison.

To explore the design methodology for these special purpose computers, we should consider power efficiency with circuit-level implementation as well as system-level dependability. Therefore, we discuss validation of a HPES design techniques, comparing performance issues, in order to clarify the essence of design methodology in the design and verification work. By design of circuit and system validation, we show an empirical analysis of the full system reliability and emphasize the overall power efficiency.

2   Low Power CMOS Digital Circuit Design

Based on logic evaluation methods, CMOS circuits are classified into static CMOS and dynamic CMOS circuits. Static CMOS circuits have both pull-down and pull-up paths for the logic evaluation [1]. Table 1 shows the criteria of CMOS logic styles for high performance microprocessors [2].

In addition, the static power dissipation ($P[x]_{static}$) will be reduced since the threshold voltage V_t will be high when the transistors are off. So, the static power dissipation formula can be added,

$$P[x]_{static} = \frac{I_{d0n} + I_{d0p}V_{DD}}{2} \tag{1}$$

where $I_{d0x} = exp\frac{-Vt}{nV_{th}}$, n is approximately 1.5, and x can be defined as the leakage current of nMOS and pMOS. Consequently, the peak power consumption ($P[x]_{peak}$) can be summarized as follows.

$$P[x]_{peak} = i_{peak}V_{peak} = \max[p(t)] \tag{2}$$

$$P[x]_{avg} = \frac{1}{T} \int P(t)dt = \frac{V_{DD}}{T} \int i_{supply}(t)dt \tag{3}$$

We can represent the output of a circuit by a mathematical equation such as equation (3). $P[x]_{avg}$ may be a simple linear function for a linear circuit, or a complicated non-linear function for a non-linear circuit. The output of a circuit depends on the current input as well as the previous output and the values of energy storage elements such as capacitors. To build a circuit model, hspice simulation was done to gather input/output data. Then, coefficient for the linear model was determined by least mean square (LMS) error criterion, as shown in equation (2). As expected, this simple linear circuit model generates less accurate simulation result for output signal results.

Additionally, the lower the threshold voltage of a given transistor, the higher the leakage currents (I_{off}) in that transistor. Higher leakage currents may result

Table 1. Criteria of CMOS Logic Styles

Operation Structure	Remarks
Static	**Static Complementary**: CVSL(Unclocked), Complementary, Differential split-level **Static Non-Complementary**: pseudo nMOS	
Dynamic	**Dynamic Complementary**: CVSL(Clocked) **Dynamic Non-Complementary**: Domino, Zipper1, Zipper2, Nora , Latched Domino	

in higher static power dissipation in typical circuits as the threshold voltages decrease, and the leakage currents increase [3]. In one embodiment, the precharge transistor and the evaluate circuit transistors may be high-V_t transistors and may contribute to low static power dissipation since low leakage current is generated. In Figure 1, a low power CMOS digital logics are implemented.

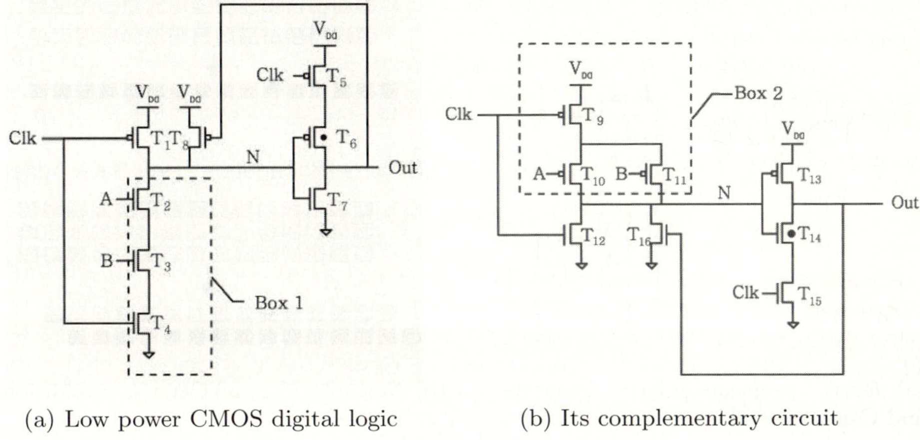

(a) Low power CMOS digital logic   (b) Its complementary circuit

Fig. 1. Schematic view of low power CMOS digital circuits

A transistor having the lower threshold voltage is referred to herein as a low-V_t transistor which is illustrated in the drawings with a dot in the center, T_6, in Figure 1(a) [4]. This circuit is meant to simultaneously control leakage currents and enhance performance could provide a boost to circuit design as V_{DD} drops below $1V$. If low power CMOS digital logic provides as much performance as it promises, this work would help provide incentive for future technology generations to have better body contacts and change the way that transistors are optimized.

Therefore, this development through the low power CMOS digital logic, is the key idea to overcome energy limitation in this special purpose computers.

As we discussed, the low power digital logic which included low leakage currents and high-threshold voltage circuits will be validated as a result of formal verification. Therefore, the design of various thereshold voltage circuit strengthens all other advantages of the circuit, such as strong logic correctness, sensitivity on noise margin, and static power dissipation.

3   Design of Low Power Adder

3.1   Low Power Dissipation in Adder Design

Figure 2 shows a one-bit full adder cell as a carry save addition which is commonly used in VLSI design. Carry propagate bit (P_i) and carry generate bit (G_i) can be defined as $X_i \oplus Y_i$ and $X_i \cdot Y_i$, respectively. Therefore, when a row of full adders is applied in parallel, 3 numbers $(X_i, Y_i$ and $C_i)$ can be reduced to 2 numbers (S_i, C_{i+1}), each of a carry bit and a save bit.

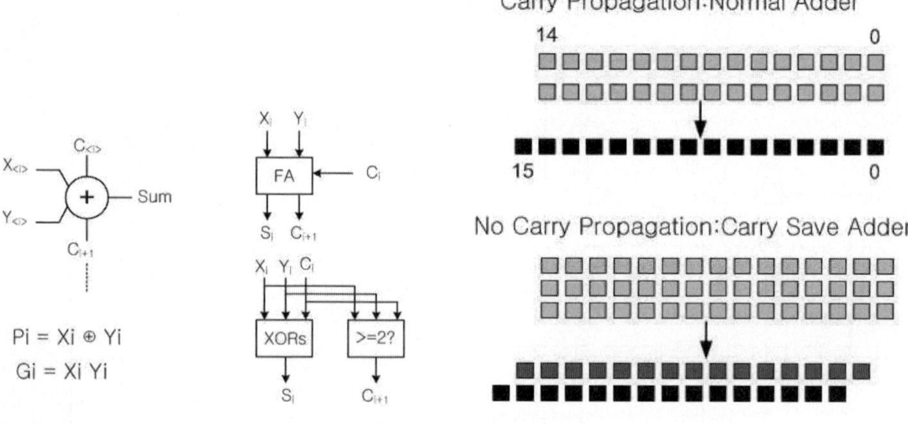

(a) Carry propagate/Carry generate and Carry save addition   (b) Normal adder and Carry save addition

Fig. 2. One-bit full adder and Propagate/Generate addition

For complex VLSI chips and systems, these 3 power reduction steps are dominant in terms of delay, power consumption and silicon area. The output, sum and carry are finally converted to one number using a fast carry-propagating adder such as a carry lookahead adder (CLA) using low power CMOS digital logic which included various thereshold voltage logic discussed in Section 2. A detail of the lookahead stage and integrated linear and non-linear logic is shown in Figure 3.

The core array occupies most of the silicon area of a large multiplier but has regularity in the design. In many cases, the core array is designed to have a fixed bit pitch in one direction since it is advantageous to have a common bit pitch for data-path operators such as multipliers, adders and register arrays. The height in the other direction has a reliability in the size depending on the need of the

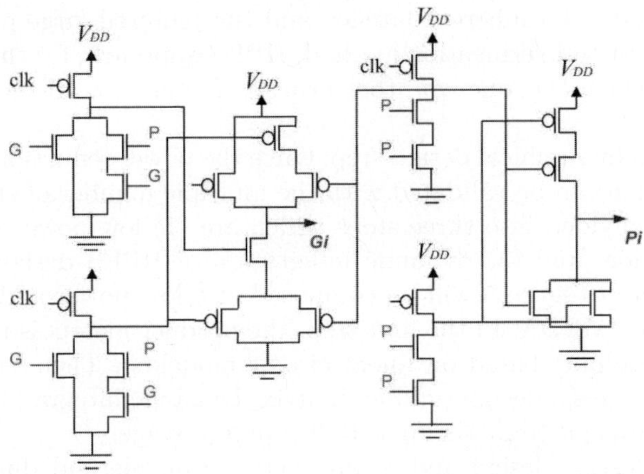

Fig. 3. Schematic view of low power adder lookahead stage

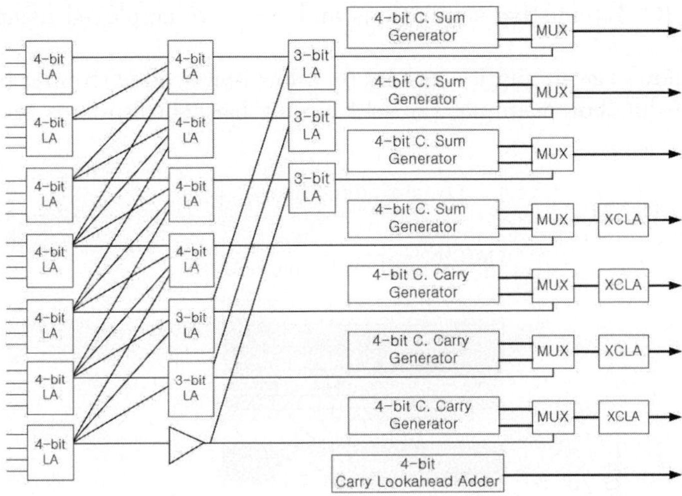

Fig. 4. Block diagram of 32-bit multiplexer core implementation based on 4-bit lookahead adder

specific operator design. Therefore, the core cells such as full adders and Booth multiplexors are stacked vertically and connected together with the same cell width. Figure 4 shows 32-bit slice constructed in this way.

3.2   Hierarchical Power Efficiency System (HPES) Design Approach

High performance design with low power adders is one of the most frequent applications encountered in microprocessor operations and signal processing applications. Due to its inherent latency and the required large power, we consider a new design and verification method, HPES approach, for this special purpose design which is one of the crucial factors to determine system performance.

We consider a hierarchical design step: Once the power reduction is generated, formal verification can be validated with the multiple number of energy savings. This process is divided into three steps which are (1) low power reduction, (2) formal verification, and (3) dynamic integration of HPES design. The overall design as shown in Figure 5 will be composed of a low power adder design and verification by LAMBDA. In the first step, the product matrix is reduced to the bottom hardware part based on linear circuit modeling. Then, static power is minimized and forms a design product matrix. Dynamic integration and bottom hardware are propagated at the final HPES design stage.

The HPES design, design and formal verification method during synthesis ensure a correct implementation and are employed to provide high coverage of other faults and 100% stuck at fault coverage. In addition, test circuitry is mathematically checked and formally proven not to interfere with the functionality of the IC. Exhaustive simulations and tests are employed using multiple simulations.

Table 2 shows the simulation results on a low power adder from a benchmark suite, ITC99 [6]. For example, the b14 circuit has 245 flip-flops and the test

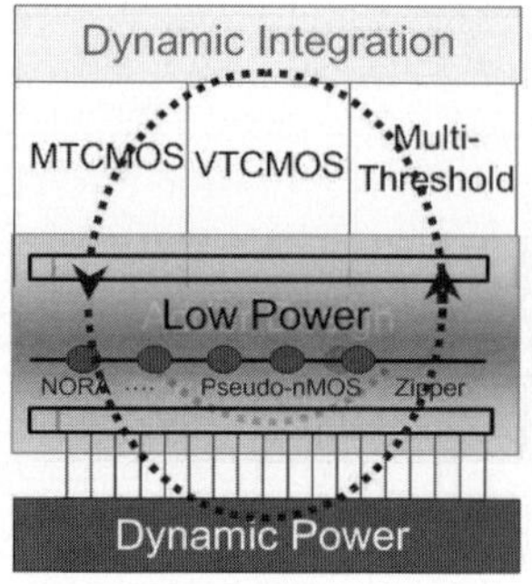

Fig. 5. Hierarchical three steps design with HPES techniques

Table 2. Simulation results of the benchmark suite on low power adder at $1.0V$ V_{DD}

Circuit	Conventional adder		Low power adder		Characteristics of Low power adder	
	no. of FFs	Fault coverage	no. of FFs	Fault coverage	high-V_t usage	Power reduction
b14	245	94.23%	295	99.65%	1,172	17.3%
b15	449	90.12%	499	92.76%	2,148	32.9%
b17	415	87.01%	465	87.11%	1,985	17.6%
$b20_1$	490	92.23%	540	94.92%	2,344	20.2%
b22	735	85.35%	785	87.53%	3,516	19.4%

results have been verified with a set of 160 stimulus vectors and the complete set of benchmark circuits. The first and second column shows a number of FFs and fault coverage from the conventional adder and low power adder, respectively. The last column represents different high-V_t that were a threshold below ground and above V_{DD}. Also, the last column implemented a power reduction rate using a low power adder for comparison. The results confirm that test length and low power adder are significantly reduced while achieving high fault coverage and energy efficiency compared to a conventional adder, with a slight increase in gate length, given that the supply voltage was scaled by at least 10% per technology generation from $0.25\mu m$ to $0.13\mu m$.

This design was verified by the hardware with functional verification at the chip-level while satisfying the design specification by formal verification tool LAMBDA. The comparison of fault coverage and power reduction rate with a benchmark circuit (i.e., $b14$) at various supply voltages range, from $1.3V$ to $0.9V$, is shown in Figure 6(a). It should be noted that the low power adder design should be determined according to the channel length of the MOS transistor and other design parameters.

In order to achieve comparable results of the chosen power adder and conventional adder, applied to the carry lookahead function, there must be a gradual descent of speed and power metrics. The conventional adder function with high fault coverage applied to the experimental setup is very similar in shape, but with low fault coverage percent. Additionally, Figure 6(b) shows comparison results of conventional and low power adder with power reduction rate.

Despite the different supply voltage scales, a very good comparison of both fault coverage and power reduction can be concluded. HPES design with low power adders is a proven design example that overcome energy limitations and will satisfy the ultimate goal of reliable system impact on minimal power usage.

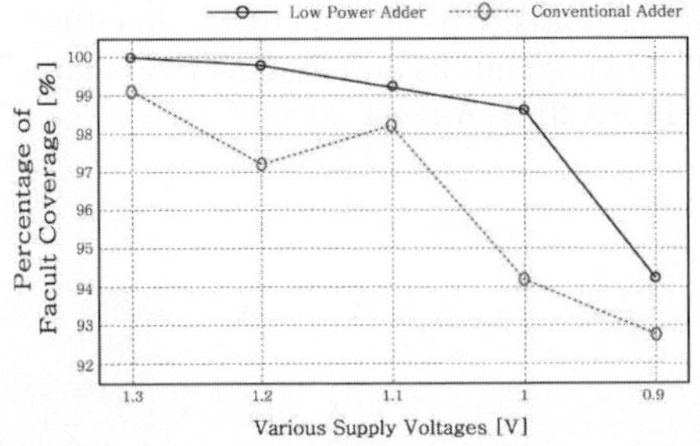

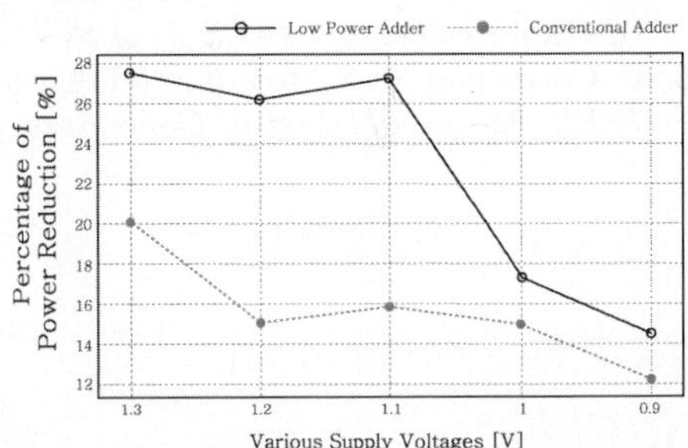

(a) Fault coverage rate at various supply voltage scales

(b) Power reduction at various supply voltage ranges

Fig. 6. Simulation results of benchmark circuit ($b14$) at various supply voltages, from $1.3V$ to $0.9V$

4   Conclusions and Future Work

The aim of the paper is to model nonlinear circuits using reliable HPES modeling options, which could be a better substitute of a vulnerable design option without losing the system performance. Moreover, the proposed design will save static power consumption and yield benefits for saving overall power dissipation. Finally, this work attempts to solve energy limitation problems using the HPES integration method that would fit within existing design specifications.

In summary, low power CMOS digital circuit power dissipation methodologies in special purpose computer architectures and its HPES design are compared with a view to both design integration and its reliability. New architectures are derived for power reduction methodologies associated with a new design of low power circuit. The better understanding of the effects of power efficiency can be used to develop accurate HPES models that can be applied into the future design technology.

References

1. R. E. Bryant, K. -T. Cheng, A. B. Kahng, K. Keutzer, W. Lamy, R. Newton, L. Pileggi, J. M. Rabaey and A. Sangiovanni-Vincetelli, "Limitations and challenges of computer-aided design tehcnology for CMOS VLSI," *Proc. of the IEEE*, vol. 89., no.3., Mar 2001
2. A. P. Chandrakasan and R. W. Brodersen, "Low power digital CMOS design," *Kluwer Academic Publishers* 1995 USA.
3. P. Pant, R. K. Roy and A Chatterjee, "Dual-threshold voltage assignement with transistor sizing for low power CMOS circuits," *IEEE Trans. on VLSI System*, vol. 9, no. 2, pp. 390-394, Apr. 2001
4. T. Jeong and A. Ambler., "Design trade-offs and power reduction techniques for high performance circuits and system," *Springer-Verlag*, ISSN: 0302-9743, 2006
5. D. Liu and C. Svensson, "Trading speed for low power by choice of supply and thershold voltages," *IEEE J. of Solid-State Circutis*, vol. 28, pp. 10-17, Jan 1993.
6. F. Corno and M. S. Reorda, "RT-level ITC 99 benchmarks and first ATPG results," *IEEE Design and Test of Computers*, pp. 44-53, July-August 2000

Dynamic Fault Tolerance in Distributed Simulation System

Min Ma, Shiyao Jin, Chaoqun Ye, and Xiaojian Liu

School of Computer Science, National University of Defense Technology,
Hunan Changsha 410073, China
maminde@163.com

Abstract. Distributed simulation system is widely used for forecasting, decision-making and scientific computing. Multi-agent and Grid have been used as platform for simulation. In order to survive from software or hardware failures and guarantee successful rate during agent migrating, system must solve the fault tolerance problem. Classic fault tolerance technology like checkpoint and redundancy can be used for distributed simulation system, but is not efficient. We present a novel fault tolerance protocol which combines the causal message logging method and prime-backup technology. The proposed protocol uses iterative backup location scheme and adaptive update interval to reduce overhead and balance the cost of fault tolerance and recovery time. The protocol has characteristics of no orphan state, and do not need the survival agents to roll-back. Most important is that the recovery scheme can tolerant concurrently failures, even the permanent failure of single node. Correctness of the protocol is proved and experiments show the protocol is efficient.

1 Introduction

Distributed simulation is popularly used in economy, education and society. The scale of simulation system develops from small to large rapidly, the platform of simulation system combines the technology of Grid and mobile agent, and furthermore, system executive time extends from hours to days even months. All these changes challenge fault tolerance in simulation system, and moreover the mobile agent system must guarantee the successful rate of migration. However there are few fault tolerance mechanisms in simulation system at present. When a simulation system fails, it will simplify restart. Some simulation systems like forecast and decision-making system have real-time requirement, the restarting method can not meet the real time requirement. It will be insignificant if the arrival results exceed the deadline.

Now, little is considered about the problem of fault tolerance in architecture of simulation system, like the popular HLA (High Level Architecture), though RTI (Run-Time Infrastructure) supplies with function of Save/Restore, the simulation entity can not guarantee the global state consistent during recovery by simple functions of Save/Restore. Multi-Agent system is a popular platform for complexity system simulation. Agents have characteristic of autonomy and intelligence, and moreover they have mobile abilities so they can migrate from node to node in network. Commonly the distributed discrete events simulation system is driven by events, so the order of interactive messages between simulation entities (usually

V.N. Alexandrov et al. (Eds.): ICCS 2006, Part I, LNCS 3991, pp. 769–776, 2006.

agent) is important. However the fault tolerance scheme in mobile agent system concerns local computing only. The requirement of simulation system must keep global state consistent and events logic causal order, so it needs to present a suitable fault tolerance protocol for simulation systems.

The paper is organized as follows. In sections 2 and 3 we state the motivation and propose a novel fault tolerance system framework. In section 4 recovery operations are discussed and the correctness is proved. Experiment and results are showed in section 5, at last a summary concludes the paper in section 6.

2 Background and Motivation

Researchers have paid more attention to fault tolerance in simulation. Most of them focus on using classic fault tolerance technology such as redundancy and checkpoint scheme. Damani and Garg introduced a fault tolerance scheme based optimistic message logging[1]. Failed simulation entity will restore checkpoint and replay messages in log, while the optimistic message logging can not guarantee all delivered messages have been logged, so orphan state may be introduced. The system must roll back relevant simulation entities to keep the global state consistent. The recovery operation will cost system much time, and moreover, this scheme can only been used in simulation system which implements optimistic time management, however local virtual time of entity can not rolled back in conservation time management. Luthi and Berchtold presented a fault tolerance framework based on active replica in HLA[2]. Every entity has several backups, the prime and backup entities compute concurrency. The output of entities will be selected by vote. This method can solve Byzantine and failure-stop error. But multi-backup costs many system resources, and it is difficult to keep consistent between prime and backup entities.

Combining the advantages of causal logging[3] and prime-backup scheme, we present a novel fault tolerance protocol in distributed simulation system. The scheme is orphan free and does not need survived entities to rollback; furthermore neither it will block the system to coordinate entities nor introduce additional messages.

3 System Model and Framework

We consider the simulation system composed of simulation nodes and interconnect network. A simulation entity can be regarded as a logic process executed in simulation nodes. An entity is called federation in HLA and agent in multi-agent system. Entities have inner states and communicate by messages. We use a graph $G(V,E)$ to illustrate the distributed topology of simulation nodes. A simulation entity locates in a node and communicates with other entities located in other nodes, an entity and the relevant entities can be presented by graph $G'(V,E)$, like Fig.1.

3.1 Prime-Backup and Heartbeat Mechanism

We use backup entity to monitor prime entity. Backup entity sends heartbeat signals to prime entity in fixed time interval, and receives updating message from the prime entity. T_{update} is set to control the updating frequency. After a time interval of T_{update}, backup entity updates once. When there is a failure happened, the backup will substitute for the prime entity, so the simulation system can continue executing.

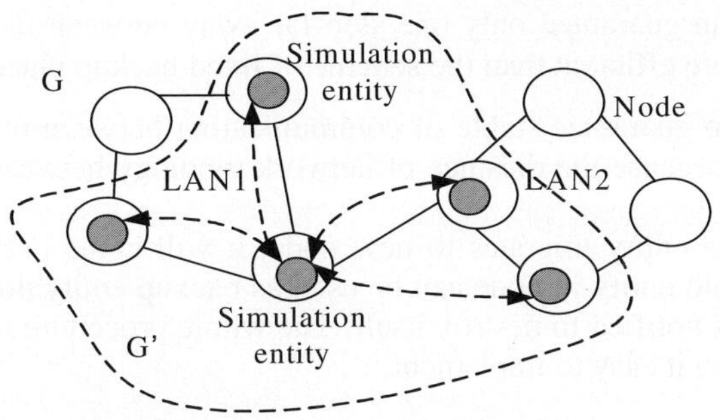

Fig. 1. Topology of simulation nodes and entities

Advantages of proposed adaptive prime-backup scheme:

1. Backup entity and prime entity are distributed in difference simulation node, so after prime entity failure, the backup will take place of prime entity quickly.
2. Though backup entities occupy some system resources, but the cost of updating is less than the cost of saving and restoring. Furthermore the checkpoint algorithm must consider the problem of checkpoint coordination and garbage recycled.
3. According output and input operation, the external environment can not be rollback. If the environment changes too much, the simulation results may be different after recovery. While using T_{update} can control the difference between system recovery before and system recovery after, so we can reduce the difference by select suitable T_{update}.

3.2   Iterative Backup Placement for Reducing Updating Cost

In order to reduce the cost of backup entity updating, we applied iterative[4] scheme to locate backup entity. Assume a simulation entity migrates between the nodes. The path is $A \Rightarrow B \Rightarrow C \Rightarrow D \Rightarrow E$ in Fig.2, when the entity arrives at the node C, the

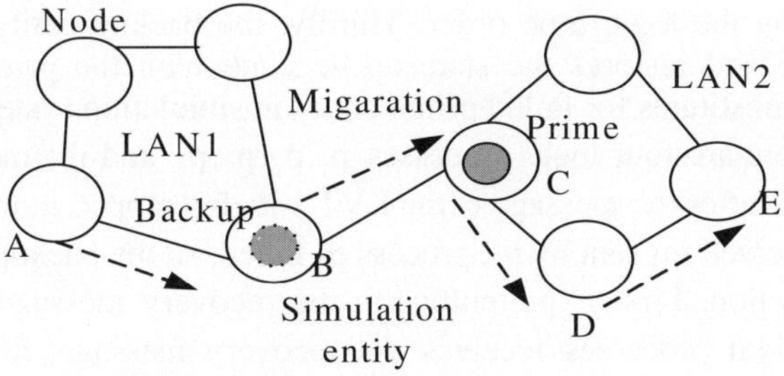

Fig. 2. Migration of simulation entity

backup entity will locate in node B, next step the prime entity migrates to node D, the backup entity will be placed in node C. That is to say the migrating path is $node_1 \Rightarrow node_2 \Rightarrow ... \Rightarrow node_n$, if prime entity is in $node_i$, then the backup entity will locate in $node_{i-1}$.

This scheme can guarantee only one step far away between the prime entity and backup, so it is more efficient than the scheme of fixed backup placement.

1. The scheme can guarantee stable of communicating between the prime entity and backup entity, because the distance of network topology between them will not be too far.
2. When the prime entity migrates to new node, it will clone itself in new node, at same time the old entity in node can be used as backup entity directly, then the old backup entity is notified to destroy itself. The whole procedure reduces backup entity migration, so it easy to implement.

4 Recovery Operation

We propose a distributed fault tolerance protocol. Every entity logs the messages received or sent within update interval T_{update} .The recovery contains several phases.

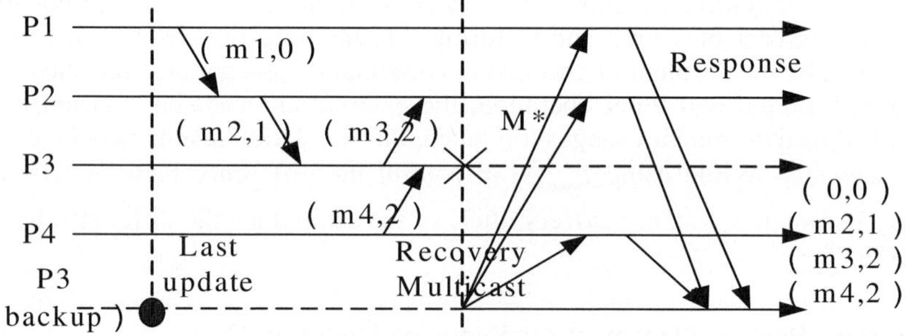

Fig. 3. Process recovery procedure operations

Firstly, the backup entity multicasts recovery announcement to relevant entities and requests them to return the concerned messages stored in history message log. Secondly, the backup entity receives the messages from relevant entities and arranges the messages by the logic time order. Thirdly, the backup entity replays the sorted message queue and restores the state up to same with the prime entity. Then the backup entity substitutes for failed prime entity in simulation system.

In fig.3, there are four logic processes p_1, p_2, p_3, p_4 and the messages m_1, m_2, m_3, m_4. The number after the message is the LVT which wrapped in message. We can see that after p_3 received m_4 sent by p_4, process p_3 failed, so the backup process of p_3 takes the recovery action. Firstly, p_3 multicasts the recovery messages m* to p_1, p_2, p_4, when the survival processes receives the recovery message, it checks the m*.lvt and wrapped process id of the sender, then it returns all the messages relevant with process id and m.lvt>m*.lvt . p_3 will receive the message set of m_2, m_3, m_4 which return by p_1, p_2, p_4. Message m_1 will not be received by p_3, because it has no relation with p_3. Secondly p_3 will arrange the received message set by the LVT order. It will be set with order $m_2 \rightarrow m_3 \rightarrow m_4$. The LVT of m_3, m_4 are both 2, so m_3, m_4 are regard as concurrency and we can place m_3 before m_4 or m_4 before m_3. The result will be same. Thirdly, p_3 replays the message $m_2 \rightarrow m_3 \rightarrow m_4$ and restores the state before failure, and then the simulation system state is consistent and continues simulating.

4.1 Correctness of Recovery Protocol

Assume the set of all processes in simulation system is $\mathbb{N}$, the set of failure processes is $\mathbb{R}$, $\mathbb{R} \subseteq \mathbb{N}$ and there is only one process fail at anytime.

Definition 1. $Depend(m)$ is set of processes whose states depend on the message m.

$$Depend(m) \overset{def}{=} \{p_j \in \mathbb{N} \mid (j \in m.dest) \vee (\exists e_j : (deliver(m) \rightarrow e_j))$$

Definition 2. $Log(m)$ is set of process which log the m in volatile memory.

$$Log(m) \overset{def}{=} (p_j \in \mathbb{N} \mid j = m.dest \vee j = m.send)$$

Definition 3. $Orphan(p)$ represent process p is an orphan process.

$$Orphan(p) \overset{def}{=} (p \in \mathbb{N}\text{-}\mathbb{R}) \wedge \exists m : (p \in Depend(m) \wedge (p_{m.send} \notin Log(m))$$

Lemma 1. If single process failed, $\forall m, \exists p \in Log(m) \wedge p \in \mathbb{N} - \mathbb{R}$

Proof: Assume that message m lost:

$$\exists m, \forall p \in \mathbb{N} - \mathbb{R} : p \notin Log(m)$$

$$\because Log(m) \overset{def}{=} (p_j \in \mathbb{N} \mid j = m.dest \vee j = m.send)$$

$$\therefore p_{m.send} \in \mathbb{R} \wedge p_{m.dest} \in \mathbb{R} \wedge (p_{m.send} \neq p_{m.dest})$$

Then there two failure processes exist. It contradicts original assumption of only one process fails; the result is correct.   □

Lemma 2. If there is only single process failure, recovery protocol guarantee no orphan process created,

$$\forall p \in \mathbb{N} : \neg Orphan(p) .$$

Proof: According to lemma 1

$$\forall m, \exists p \in Log(m) \wedge p \in \mathbb{N} - \mathbb{R}$$

There are two cases during recovery procedure:

Case 1: $p_{m.send} \notin \mathbb{R}$, then according the definition of orphan(p)

$$p \in \mathbb{N} : \neg Orphan(p)$$

Case 2: $p_{m.send} \in \mathbb{R}$, then $p_{m.send}$ will retrieve the message m from $p_{m.dest}$ in recovery operation. Then

$$p \in \mathbb{N}: \neg Orphan(p)$$

Conclude from case 1 and case 2, the result is correct. □

Lemma 3. Failure process p arranges collected messages with LVT wrapped in messages; the new message sequence $\{m'_1, m'_2, ...m'_n\}$ has the same causal order with the history message sequence $\{m_1, m_2, ...m_n\}$.

Proof: The message sequence $\{m_1, m_2, ...m_n\}$ which process p received has happen before relationship [3]:

$$m_1 \to m_2 \to ... \to m_n$$

Logic virtual time order of Messages has relation:

$$m_1.lvt \le m_2.lvt \le ... \le m_n.lvt$$

When process p failed and to be recovered, it will retrieve message set $\{m'_1, m'_2, ...m'_n\}$ from relevant processes to restore the state according to the protocol, the new message sequence has logic virtual time order

$$m'_1.lvt \le m'_2.lvt \le ... \le m'_n.lvt$$

According the protocol assumption, the message with same LVT will be regarded as concurrency events, and then we can arrange messages with the same LVT an arbitrary order, then we get

$$m'_1 \to m'_2 \to ... \to m'_n$$

According to the PWD (Piecewise Deterministic) assumption [7] we can get

$$det\,er\,min\,stic(m'_1, m'_2, ..., m'_n) = det\,er\,min\,stic(m_1, m_2, ..., m_n)$$

So the new message sequence will get same execute result with history message sequence. □

Theorem 1. The recovery protocol can guarantee the system recover to the state before failure and all processes states are global consistent.

Proof: According lemma1, failure process can retrieve entire relevant messages; according lemma2, no survived processes become orphan; according lemma3, the recovery process can arrange messages sequence to right order, then after replaying the retrieval message queue the failure process will recover state to failure before, and all processes in system are consistent. □

4.2　Recovery Protocol Extend to f Concurrently Failures

Theorem 2. If updating backup entity by history message queue, the recovery protocol can tolerate f concurrent failures.

Proof: Assume there are f ($f<N$) processes failed concurrently in system, then backup processes take recovery action, each backup process may start recovery action from different logic virtual time, but there must exist a minimal time $T_{roll\,min}$. All events and messages before the time $T_{roll\,min}$ have been logged in system, then next

timestamp $T_{roll\,min} + 1$ the state of system is deterministic by state and output messages of time $T_{roll\,min}$, so the system state can restore to $T_{roll\,min} + 1$ according to the PWD assumption, the survival processes can not push forward their local virtual time according to the synchronous mechanism of conservation time management of simulation. So failure processes can recover sequentially from $T_{roll\,min}$ to $T_{roll\,min} + 1, T_{roll\,min} + 2, \dots$, until they recover to time $T_{beforefailure}$, then the system global virtual time can be pushed forward and continue simulations.

Corollary 1. The recovery protocol can solve single node failure.

When a simulation node failed, the simulation system can recover the failed process using backup process located in other node, so a single node failure can be solved.

5   Experiment and Results

Experiment was done to test the efficiency of proposed recovery protocol in Jcass. Jcass is a multi-agent complex simulation platform developed by national university of defense technology. The Fig.4(a) shows additional time cost in percent comparing the case of system employing the fault tolerance methods with the case of system not employing. The fig.4(b) shows the additional time cost comparing executing time in fixed events and migrating failure rate with executing time in failure free. Hardware platform is four PC with Celeron 800MHz, 256M SDRAM, 10M LAN. Test was done with 200 agents and 200 times migrations. Total 1000 events executed, and events failure rate was 10% migrating failure rate was 10%. System executed about 1 hour in failure free and employing proposed method.

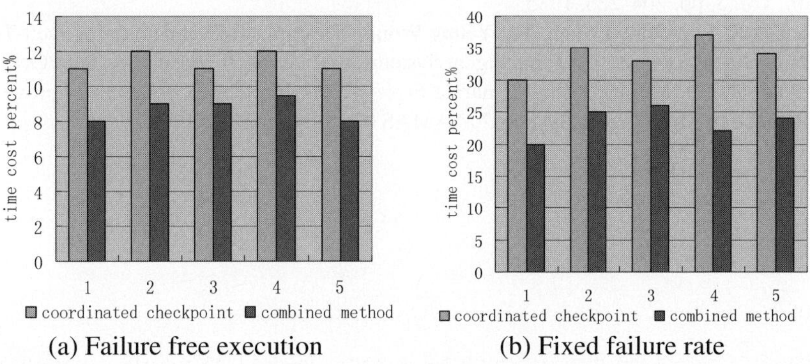

(a) Failure free execution   (b) Fixed failure rate

Fig. 4. Results of Experiment

Experiment results show that combined method has extra time cost 7% and method of checkpoint has extra time cost 11% in failure free execution. If we used fixed failure rate in execution, extra cost of combined method is about average 16%, while extra cost of checkpoint is about average 34%.

6 Conclusion

According to analysis of simulation framework and distributed recovery mechanism, we present a novel recovery scheme for simulation system. The scheme uses backup entity to monitor the prime entity and stores entity states, a recovery protocol is designed to solve the problem of system global state consistent. Correctness of recovery protocol is proved in theory. At last experiment was done to test the efficiency of recovery protocol; the experiment results show that proposed method reduces nearly half of time cost compared with checkpoint method whenever in failure free or fixed failure rate.

References

[1] Damani., "Fault -tolerant distributed simulation," presented at proceedings of the 12th workshop on parallel and distributed simulation(PADS'98), 1998.

[2] S. G. Johnnes Luthi, "F-RSS: A Flexible Framework for Fault Tolerant HLA Federations," presented at ICCS 2004.

[3] E. N. Elnozahy, D. B. Johnson, and Y. M. Wang. "A survey of rollback-recovery protocols in message-passing systems". Technical Report CMU-CS-96-181, Carnegie Mellon University, October 1996.

[4] D.Johansen, K.Marzullo, F.B.Schneider, K.Jacobsen, D.Zagorodnov."NAP: Practical Fault-Tolerance for Itinerant Computations". Technical Report TR98-1716. Department of Computer Science, Cornell University. USA. November, 1998

[5] L. Leslie, "Time, clocks, and the ordering of events in a distributed system," Commun. ACM, vol. 21, pp. 558-565, 1978.

[6] D. Agrawal, Agre,J.R., "Replicated objects in time warp simulations," presented at Proc. 1992 Winter Simulation Conference,SCS(1992), 1992.

[7] S. Rob and Y. Shaula, "Optimistic recovery in distributed systems," ACM Trans. Comput. Syst., vol. 3, pp. 204-226, 1985.

[8] Michael R. Lyu, Xinyu Chen, Tsz Yeung Wong, "Design and Evaluation of a Fault-Tolerant Mobile-Agent System," IEEE Intelligent Systems, vol. 19, no. 5, pp. 32-38, Sept/Oct, 2004.

[9] F. Alan and D. Ralph, "Using Dynamic Proxy Agent Replicate Groups to Improve Fault-Tolerance in Multi-Agent Systems", AAMAS'03, July 14-18, 2003

A Novel Supervised Information Feature Compression Algorithm

Shifei Ding[1,2] and Zhongzhi Shi[2]

[1] College of Information Science and Engineering,
Shandong Agricultural University, Taian 271018 P.R. China
[2] Key Laboratory of Intelligent Information Processing, Institute of Computing Technology,
Chinese Academy of Sciences, Beijing 100080 P.R.China
sfding@sdau.edu.cn; shizz@ics.ict.ac.cn

Abstract. In this paper, a novel supervised information feature compression algorithm is set up. Firstly, according to the information theories, we carried out analysis for the concept and its properties of the cross entropy, then put forward a kind of lately concept of symmetry cross entropy (SCE), and point out that the SCE is a kind of distance measure, which can be used to measure the difference of two random variables. Secondly, We make the SCE separability criterion of the classes for information feature compression, and design a novel algorithm for information feature compression. At last, the experimental results demonstrate that the algorithm here is valid and reliable, and provides a new research approach for feature compression, data mining and pattern recognition.

1 Introduction

Feature extraction or compression is one of the most importmant steps in pattern recognition, data mining, machine learning and so on. In order to choose a subset of the original features by reducing irrelevant and redundant, many feature selection algorithms have been studied[1,2]. The literature contains several studies on feature selection for unsupervised learning in which he objective is to search for a subset of features that best uncovers "natural" groupings (clusters) from data according to some criterion. For example, principal components analysis (PCA) is an unsupervised feature extraction method that has been successfully applied in the area of face recognition, feature extraction and feature analysis[3,4]. But the method of PCA is effective to deal with the small size and high-dimensional problems, and gets the extensive application in Eigenface and feature extraction. In high-dimensional cases, it is very difficult to compute the principal components directly[5].

Now an important question is how to deal with supervised information feature compression. In this paper, the authors are going on studying this field on the basis of these studies. Firstly, we study and discuss the cross entropy theory, and point out its shortage, then put forward a new concept of symmetry cross entropy (SCE), and prove that the SCE is a kind of distance measure. Secondly, we regard SCE as class separability criterion, and design a new algorithm of information

V.N. Alexandrov et al. (Eds.): ICCS 2006, Part I, LNCS 3991, pp. 777–780, 2006.

feature compression. At last, our experimental results indicate that the proposed algorithm here is efficient and reliable.

2 Feature Compression Algorithm

In order to set up information feature compression algorithm, we firstly discuss the following new concept of symmetry cross entropy and feature compression theorem.

2.1 Symmetry Cross Entropy

Shannon put forward the concept of information entropy for the very first time in 1948[6]. The cross entropy (CE), or the relative entropy, is used for measuring difference information between the two probability distributions. But the CE satisfies only nonnegativity, normalization and dissatisfies symmetry and triangle inequation. For this reason, we carry out the improvement, and give the following definition.

Definition 1   Suppose that $P = (p_1, p_2, \cdots, p_n)$ and $Q = (q_1, q_2, \cdots, q_n)$ are two probability vectors of discrete random variable X, and $H(P \| Q)$ and $H(Q \| P)$ are CE of P to Q and Q to P respectively. Then we define

$$D(P,Q) = H(P \| Q) + H(Q \| P) \tag{1}$$

It is called Symmetric Cross Entropy (SCE) of P and Q.

We can prove that the SCE satisfies three basic properties as follows.

1) Nonnegativity: $D(P,Q) \geq 0$, $D(P,Q) = 0 \Leftrightarrow P = Q$;

2) Symmetry: $D(P,Q) = D(Q,P)$;

3) Triangle inequation: Suppose that $W = (w_1, w_2, \cdots, w_n)$ is another probability vector, then $D(P,Q) \leq D(P,W) + D(W,Q)$.

Therefore, the SCE is a kind of distance measurement, which can be used to measure the difference of two random variables. In order to compute conveniently, we can use the function as follows in accordance with the above $D(P,Q)$, i.e.

$$H(P,Q) = \sum_{i=1}^{n} (p_i - q_i)^2 \tag{2}$$

Doing like this, we don't affect the results to select d optimal features.

2.2 Compression Theorem

Suppose that $\{X_j^{(1)}\}$ $(j = 1, 2, \cdots, N_1)$ and $\{X_j^{(2)}\}$ $(j = 1, 2, \cdots, N_2)$ are squared normalization pattern vectors which belongs to two classes. The kth feature component of $X_j^{(i)}$ is denoted by $x_{jk}^{(i)} (i = 1, 2; k = 1, 2, \cdots, n)$. The square mean of each component for every class is $\gamma_k^{(i)} = \dfrac{1}{N_i}\sum_{j=1}^{N_i}(x_{jk}^{(i)})^2$, Where $i = 1,2; j = 1,2,\cdots,n$.

Obviously $\gamma_k^{(i)} \geq 0$, and then

$$\sum_{k=1}^{n}\gamma_k^{(i)} = \sum_{k=1}^{n}\frac{1}{N_i}\sum_{j=1}^{N_i}(x_{jk}^{(i)})^2 = \sum_{j=1}^{N_i}\frac{1}{N_i}\sum_{k=1}^{n}(x_{jk}^{(i)})^2 = \sum_{j=1}^{N_i}\frac{1}{N_i} = 1 \tag{3}$$

Namely $\gamma_k^{(i)} \geq 0$ and $\sum_{k=1}^{n}\gamma_k^{(i)} = 1$. Therefore, we can comprehend $\{\gamma_k^{(i)}\}$ as the

probability distribution defined by $X_j^{(i)}$. Suppose that the (k,l) element of symmetric

matrix $G^{(i)}$ $(i = 1,2)$ is $g_{kl}^{(i)} = \dfrac{1}{N_i}\sum_{j=1}^{N_i}x_{jk}^{(i)}x_{jl}^{(i)}$. Record $\gamma^{(i)} = (\gamma_1^{(i)}, \gamma_2^{(i)}, \cdots, \gamma_n^{(i)})$, then

every components of $\gamma^{(i)}$ is element of $G^{(i)}$ $(i = 1,2)$ in diagonal line. Let

$$s = s(\gamma^{(1)}, \gamma^{(2)}) = \sum_{k=1}^{n}(\gamma_k^{(1)} - \gamma_k^{(2)})^2 \tag{4}$$

According to discussion above, we can get the following compression theorem.

Theorem 1 Suppose that $A = G^{(1)} - G^{(2)}$, and the function s defined by formula (4), then the SCE=maximum if and only if the coordinate system as s =maximum is composed of the d eigenvectors in correspondence with eigenvalues satisfied definite conditions of the matrix A .

2.3 Feature Compression Algorithm

In order to explain the effect of information feature compression, we make the eigenvalues $\lambda_i (i = 1,2,\cdots,n)$ of matrix A arrange $\lambda_1^2 \geq \lambda_2^2 \geq \cdots \geq \lambda_d^2 \geq \cdots \geq \lambda_n^2$. The

total variance sum of square is denoted by $V_n = \sum_{k=1}^{n}\lambda_k^2$, and then the variance square

ratio is defined as $V = V_d/V_n$, which can measure the degree of information compression. According to the discussion above, we can get the algorithm of information feature compression based on SCE as follows.

Step 1 Carry out square normalization processing for the two classes original data, and get data matrix $x^{(i)}(i = 1,2)$.

Step 2 Carry out centralization for obtained data matrix x , and then calculate the symmetric matrix $G^{(i)}(i = 1,2)$ and difference matrix A .

Step 3 Calculate all eigenvalues corresponding to all eigenvectors of the matrix $A = G^{(1)} - G^{(2)}$.

Step 4 Construct information compression matrix $T = (u_1, u_2, \cdots, u_d)$.

Step 5 Compress data matrix x based on $y^{(i)} = T'x^{(i)} (i = 1,2)$, and so we reach the purpose of optimal compression of information feature.

3 Conclusions

According to the definition of cross entropy, we propose a new concept of the SCE, and point out that the SEC is a kind of distance measure. Based on SCE, we set up a novel separability criterion, which can be used to measure the difference degree between the random variables, and construct a compression algorithm for information feature based on SCE.

Acknowledgements

This study is supported by the National Natural Science Foundation of China under grant No.60435010, the China Postdoctor Science Foundation under grant No.2005037439, and the Doctoral Science Foundation of Shandong Agricultural University under grant No. 23241.

References

1. Duda, R.O.,Hart, P.E. (eds.): Pattern Classification and Scene Analysis. Wiley, New York (1973)
2. Fukunaga, K. (ed.): Introduction to Statistical Pattern Recognition. Academic Press, 2nd ed., New York (1990)
3. Hand, D.J. (ed.): Discrimination and Classification. Wiley, New York (1981)
4. Nadler, M., Smith, E.P. (eds.): Pattern Recognition Engineering. Wiley, New York (1993)
5. Yang, J., Yang, J.Y.: A Generalized K-L Expansion Method That Can Deal With Small Sample Size and High-dimensional Problems. Pattern Analysis Applications 6(6) (2003) 47-54
6. Shannon, C.E.: A Mathematical Theory of Communication. Bell Syst. Tech. J. 27 (1948) 379-423

On a Family of Cheap Symmetric One-Step Methods of Order Four[*]

Gennady Yu. Kulikov and Sergey K. Shindin

School of Computational and Applied Mathematics,
University of the Witwatersrand, Private Bag 3, Wits 2050,
Johannesburg, South Africa
gkulikov@cam.wits.ac.za, sshindin@cam.wits.ac.za

Abstract. In the paper we present a new family of one-step methods. These methods are of the Runge-Kutta type. However, they have only explicit internal stages that leads to a cheap practical implementation. On the other hand, the new methods are of classical order 4 and stage order 2 or 3. They are A-stable and symmetric.

1 Introduction

When solving ordinary differential equations (ODE's) of the form

$$x'(t) = g\big(t, x(t)\big), \quad t \in [t_0, t_0 + T], \quad x(t_0) = x^0 \tag{1}$$

where $x(t) \in \mathbb{R}^n$ and $g : D \subset \mathbb{R}^{n+1} \to \mathbb{R}^n$ is a sufficiently smooth function, any one-step method reads

$$x_{k+1} = x_k + \tau_k \Phi(t_k, x_k, t_{k+1}, x_{k+1}, \tau_k), \quad k = 0, 1, \ldots, K - 1, \tag{2}$$

where $x_0 = x^0$, τ_k is a step size and the function $\Phi(t_k, x_k, t_{k+1}, x_{k+1}, \tau_k)$ is referred to as *an increment function* of method (2). One-step methods (2) possess many superior practical properties to solve ODE's (1) of different sorts. For instance, they can be A-stable and keep high order convergence rate that is not possible for multistep methods because of Dahlquist's second barrier [5]; they do not have any difficulties with a variable step size implementation (see, for example, [4], [5]); they are also of high importance when applied to Hamiltonian or reversible equations [6]. Unfortunately, one-step methods have some limitations in the sense of high execution time when solving large-scale ODE's (1). Usual representatives of one-step methods (2) are Runge-Kutta formulas, but only implicit of them are suitable for stiff problems. However, implicit Runge-Kutta methods of high order are very time-consuming because of the need to solve in general nonlinear systems of dimension ln where l is the number of stage

---

[*] The first author thanks the National Research Foundation of South Africa for partial support of his work under grant No. FA2004033000016. The second author thanks the University of the Witwatersrand, Johannesburg for a Post-Doctoral Fellowship.

V.N. Alexandrov et al. (Eds.): ICCS 2006, Part I, LNCS 3991, pp. 781–785, 2006.
© Springer-Verlag Berlin Heidelberg 2006

values at each step of the numerical integration (see [4], [5]). Hopefully, serious progress was made in this area by Bickart [1] and Butcher [2].

The aim of this paper is to present a cheap family of symmetric A-stable Runge-Kutta formulas. They are of classical order 4 and of stage order 3.

2   New Family of Symmetric One-Step Methods

Further, we suppose that ODE (1) possesses a unique solution $x(t)$ on the whole interval $[t_0, t_0 + T]$. We show how to construct the methods mentioned above.

Let us fix a subinterval $[t_k, t_{k+1}] \subset [t_0, t_0 + T]$ of the length τ_k; i.e., $\tau_k = t_{k+1} - t_k$, and consider that the exact solution of ODE (1) is known at the point t_k; i.e., $x_k = x(t_k)$. Then, on the one hand, if two additional solutions $x(t_k + c_1 \tau)$ and $x(t_k + c_2 \tau)$ evaluated at internal points of the interval $[t_k, t_{k+1}]$ are known we will be able to use a two-point quadrature formula of the form

$$x_{k+1} = x(t_k) + \tau_k b_1 g\big(t_k + c_1\tau_k, x(t_k + c_1\tau_k)\big) + \tau_k b_2 g\big(t_k + c_2\tau_k, x(t_k + c_2\tau_k)\big)$$

in order to find an approximation to the exact solution $x(t_{k+1})$.

On the other hand, if the exact solution $x(t_{k+1})$ is considered to be known we can try to approximate the values x_{k1} and x_{k2} of the exact solution evaluated at the points $t_k + c_1\tau_k$ and $t_k + c_2\tau_k$, respectively, by means of explicit formulas of the following form:

$$x_{k1} = a_{11}x(t_k) + a_{12}x(t_{k+1}) + \tau_k\Big(d_{11}g\big(t_k, x(t_k)\big) + d_{12}g\big(t_{k+1}, x(t_{k+1})\big)\Big),$$

$$x_{k2} = a_{21}x(t_k) + a_{22}x(t_{k+1}) + \tau_k\Big(d_{21}g\big(t_k, x(t_k)\big) + d_{22}g\big(t_{k+1}, x(t_{k+1})\big)\Big).$$

Thus, our task is to search the highest order one-step methods of the form

$$x_1 = a_{11}x_k + a_{12}x_{k+1} + \tau_k\big(d_{11}g(t_k, x_k) + d_{12}g(t_{k+1}, x_{k+1})\big), \tag{3a}$$

$$x_2 = a_{21}x_k + a_{22}x_{k+1} + \tau_k\big(d_{21}g(t_k, x_k) + d_{22}g(t_{k+1}, x_{k+1})\big), \tag{3b}$$

$$x_{k+1} = x_k + \tau_k\big(b_1 g(t_k + c_1\tau_k, x_1) + b_2 g(t_k + c_2\tau_k, x_2)\big) \tag{3c}$$

where a_{ij}, b_i, c_i, d_{ij}, $i, j = 1, 2$, are unknown fixed coefficients.

We first remark that the Gauss quadrature formula has the highest order 4 among all formulas of the form (3c). Therefore the coefficients b_i, c_i are determined uniquely and they are: $b_1 = b_2 = 1/2$, $c_1 = (3 - \sqrt{3})/6$, $c_2 = (3 + \sqrt{3})/6$.

Second, we require the defect (or the local error) of method (3) to be $O(\tau_k^5)$ for any ODE (1) with a sufficiently smooth right-hand side. The latter condition admits the following one-parametric family of the coefficients to provide the fourth order convergence for method (3):

$$a_{11} = \theta, \quad a_{12} = 1 - \theta, \quad a_{21} = 1 - \theta, \quad a_{22} = \theta, \tag{4a}$$

$$d_{11} = \frac{6\theta - 2 - \sqrt{3}}{12}, \quad d_{12} = \frac{6\theta - 4 - \sqrt{3}}{12}, \tag{4b}$$

$$d_{21} = \frac{4 + \sqrt{3} - 6\theta}{12}, \quad d_{22} = \frac{2 + \sqrt{3} - 6\theta}{12}. \tag{4c}$$

Below, we consider that methods (3) are based on the Gauss quadrature formula of order 4 and their coefficients satisfy conditions (4). So, it is not difficult to check that all these methods are symmetric. We refer the reader to [4] for the necessary theory. It is also quite evident that the stability functions of all the constructed methods are $R_{22}(z)$, which means the $(2,2)$-Padé approximation to the exponential function e^z (see, for example, [5]). The latter implies that our methods are A-stable. Thus, the family of methods (3) can be useful to integrate both nonstiff and stiff ODE's (including reversible problems).

3   Practical Implementation

For nonstiff ODE's, we recommend to use fixed-point iterations with both trivial predictor and nontrivial one. Particulars on implementation of this iteration in iterative Runge-Kutta methods and estimation of a sufficient number for iteration steps to provide the maximum order convergence can be found in [7].

When solving stiff ODE's we are constrained with Newton-type iterations only and have to implement the iteration in the form that does not ruin A-stability of the underlying method (see, for example, [3], [5]). From this point of view, the modified Newton iteration is a proper one. Unfortunately, the high execution cost, which is about $4n^3/3$ arithmetical operations per evaluation of the Jacobi matrix and its LU decomposition, makes it pretty unpractical for large n. Hopefully, the number of operations can be reduced with a factor of 4 by replacement of the full Jacobian of method (3) with a simplified one, as follows:

$$(I - \tau_k J/4)^2 \big(x_{k+1}^\ell - x_{k+1}^{\ell-1} \big) = -x_{k+1}^{\ell-1} + \bar{x}_k$$
$$+\tau_k \big(b_1 g(t_k + c_1\tau_k, x_{k1}^{\ell-1}) + b_2 g(t_k + c_2\tau_k, x_{k2}^{\ell-1}) \big), \tag{5}$$

where $\ell = 1, 2, \ldots, N$ and $\bar{x}_k$ is the numerical solution derived by method (3) with N Newton iteration steps per grid point; i.e., $\bar{x}_k \overset{\text{def}}{=} x_k^N$, $k = 1, 2, \ldots, K$. Here, $J \overset{\text{def}}{=} \partial_x g(t_{k+1}, x_{k+1}^0)$ be the partial derivative of the right-hand side of ODE (1) with respect to the second variable evaluated at the point (t_{k+1}, x_{k+1}^0). Note that iteration (5) implies a single LU decomposition of the matrix $I - \tau_k J/4$ and successive solutions of two linear systems with the same decomposed coefficient matrix. This feature makes methods (3) comparable to SDIRK, which are very efficient to solve stiff ODE's (see [3], [5]).

At the end, we exhibit nice practical properties of iteration (5). First, the Jacobian replacement made above does not influence the sufficient number of iteration steps with both trivial predictor and nontrivial one to provide the fourth order convergence. It follows from Theorem 3 in [8]. Second, iteration (5) is A-stable. To see this, we apply one step of iteration (5) to the Dahlquist test equation $x' = \lambda x$ where λ is a fixed complex number with a nonpositive real part and consider that $x_{k+1}^0 = x_k$. Simple calculations give the stability function of method (5) in the form $R(z) = (1 + z/4)^2/(1 - z/4)^2$, which is evidently A-stable.

4 Numerical Experiments

To test our methods, we apply the method (3) when $\theta = 1/2 + 2\sqrt{3}/9$ and with iteration (5) to the two dimensional Brusselator with diffusion and the periodic boundary conditions (see [5, p. 151–152] for full detail). We take the number of the grid points in each dimension to be equal 50. It leads to a system of ODE of dimension 5000. We solve this problem on the interval $[0, 6]$ by the method (3) and by the Gauss method of order 4 (termed also Hammer and Hollingsworth's method). Both methods are based on the same quadrature formula and they are of the same classical order. However, the stage order of our method is 3 and of the Gauss one is 2.

We use the same variable step size implementation of these two methods with modified Newton iterations and with the same step size selection mechanism based on the local error estimate evaluated by the Richardson extrapolation. We apply the modified Newton iteration in the form of algorithm (5) in method (3) and do the conventional implementation for the Gauss method. We also want to emphasize that the correct step size control requires two iteration steps per grid point in algorithm (5) and three iteration steps in the modified Newton iteration applied to the Gauss method (see, for example, [8]). The local error tolerance is chosen to be 10^{-01}.

Statistics of both integrations is presented in form of the following table and clearly displays the better performance of the method (3), at least for this test problem:

Statistics	method (3)	Gauss method
execution time (in sec.)	843.985	2456.656
number of rejected steps	6	1
number of accepted steps	44	24

References

1. Bickart, T.A.: An efficient solution process for implicit Runge-Kutta methods. SIAM J. Numer. Anal. **14** (1977) 1022–1027
2. Butcher, J.C.: On the implementation of implicit Runge-Kutta methods. BIT. **16** (1976) 237–240
3. Dekker, K., Verwer, J.G. Stability of Runge-Kutta methods for stiff nonlinear differential equations. North-Holland, Amsterdam, 1984
4. Hairer, E., Nørsett, S.P., Wanner, G.: Solving ordinary differential equations I: Nonstiff problems. Springer-Verlag, Berlin, 1993
5. Hairer, E., Wanner, G.: Solving ordinary differential equations II: Stiff and differential-algebraic problems. Springer-Verlag, Berlin, 1996
6. Hairer, E., Lubich, C., Wanner, G.: Geometric numerical integration: Structure preserving algorithms for ordinary differential equations. Springer-Verlag, Berlin, 2002

7. Kulikov, G.Yu.: On implicit extrapolation methods for ordinary differential equations, Russian J. Numer. Anal. Math. Modelling., **17** (2002) No. 1, 41–69
8. Kulikov, G.Yu., Merkulov, A.I.: Asymptotic error estimate of iterative Newton-type methods and its practical application. In: Antonio Lagana et al (eds.): Computational Science and Its Applications — ICCSA 2004. International Conference, Assisi, Italy, May 2004. Proceedings, Part III. Lecture Notes in Computer Science. **3045** (2004) 667–675

Influence of the Mutation Operator on the Solution of an Inverse Stefan Problem by Genetic Algorithms

Damian Słota

Institute of Mathematics, Silesian University of Technology,
Kaszubska 23, 44-100 Gliwice, Poland

Abstract. This paper presents the influence of choice of the mutation operator on the accuracy of a solution of a two-phase design inverse Stefan problem using genetic algorithms. In the problem to be solved, the coefficient of convective heat transfer on one boundary had to be so selected that the moving interface of the phase change (freezing front) would take the given position.

1   Introduction

In this paper we are going to find a solution of a two-phase design inverse Stefan problem [1], for which the coefficient of convective heat transfer on one boundary should be so selected that the moving interface could take the given position. The solution will consist in minimization of the functional whose value is the norm of the difference between the given interface position and the position reconstructed for the selected convective heat-transfer coefficient. For the minimization of the functional genetic algorithms were used, whereas the Stefan problem was solved by an alternating phase truncation method [4]. The paper presents the influence of choice of the mutation operator on the accuracy of the results obtained.

2   Formulation of the Problem

On the boundary of domain $D = [0, b] \times [0, t^*] \subset \mathbb{R}^2$ three components are distributed $\Gamma_0 = \{(x, 0); x \in [0, b]\}$, $\Gamma_1 = \{(0, t); t \in [0, t^*]\}$, $\Gamma_2 = \{(b, t); t \in [0, t^*]\}$, where initial and boundary conditions are given. Let D_1 (D_2) be this subset of domain D which is occupied by liquid (solid) phase, separated by the freezing front $\Gamma_g = \xi(t)$.

We will look for an approximate solution of the following problem. For given position of freezing front Γ_g, the distribution of temperature T_k in domain D_k $(k = 1, 2)$ is calculated as well as function $\alpha(t)$ on boundary Γ_2, which satisfy the following equations (for $k = 1, 2$):

$$\frac{\partial T_k}{\partial t}(x, t) = a_k \frac{\partial^2 T_k}{\partial x^2}(x, t), \qquad\qquad \text{in } D_k, \tag{1}$$

$$T_1(x, 0) = \varphi_0(x), \qquad\qquad \text{on } \Gamma_0, \tag{2}$$

V.N. Alexandrov et al. (Eds.): ICCS 2006, Part I, LNCS 3991, pp. 786–789, 2006.

$$\frac{\partial T_k}{\partial x}(x,t) = 0, \qquad\qquad \text{on } \Gamma_1, \qquad (3)$$

$$-\lambda_k \frac{\partial T_k}{\partial x}(x,t) = \alpha(t)\left(T_k(x,t) - T_\infty\right), \qquad\qquad \text{on } \Gamma_2, \qquad (4)$$

$$T_k(x,t) = T^*, \qquad\qquad \text{on } \Gamma_g, \qquad (5)$$

$$L\,\varrho_2\,\frac{d\xi}{dt} = -\lambda_1 \frac{\partial T_1(x,t)}{\partial x} + \lambda_2 \frac{\partial T_2(x,t)}{\partial x}, \qquad\qquad \text{on } \Gamma_g, \qquad (6)$$

where a_k are the thermal diffusivity in liquid phase ($k = 1$) and solid phase ($k = 2$), λ_k are the thermal conductivity, α is the coefficient of convective heat-transfer, T_∞ is the ambient temperature, L is the latent heat of fusion, ϱ_k are the mass density, and t and x refer to time and spatial location, respectively.

We will look for the $\alpha(t)$ function in the form:

$$\alpha(t) = \begin{cases} \alpha_1 & \text{for } t \le t_{\alpha_1}, \\ \alpha_2 & \text{for } t \in (t_{\alpha_1}, t_{\alpha_2}], \\ \alpha_3 & \text{for } t > t_{\alpha_2}, \end{cases} \qquad (7)$$

where $0 < t_{\alpha_1} < t_{\alpha_2} < t^*$. Let V_α^p mans a set of all functions in the form (7), where $\alpha_i \in [\alpha_i^l, \alpha_i^u]$. For the given function $\alpha(t) \in V_\alpha^p$ the problem (1)–(6) becomes a direct Stefan problem, whose solution enables finding the position of the interface $\xi(t)$ corresponding to the $\alpha(t)$ function. Using the found interface position $\xi(t)$ and the given position $\xi^*(t)$ we can build a functional which will specify the error of an approximated solution:

$$J(\alpha) = \left(\sum_{i=1}^{M}\left[\omega_i\left(\xi_i - \xi_i^*\right)^2\right]\right)^{1/2}, \qquad (8)$$

where ω_i are weight coefficients and $\xi_i^* = \xi^*(t_i)$ and $\xi_i = \xi(t_i)$ are the given and calculated points, respectively, describing the moving interface position.

3   Genetic Algorithm

For the representation of the vector of decision variables $(\alpha_1, \alpha_2, \alpha_3)$, a chromosome was used in the form of a vector of three real numbers (real number representation) [2,3]. The tournament selection and elitist model were applied in the algorithm. As the crossover operator, arithmetic crossover was applied.

The results of calculations were then compared for different mutation operators: uniform mutation ($M1$), Gaussian mutation ($M2$) and two operators of nonuniform mutation ($M3$ and $M4$) for different functions describing the uniformity of distribution. In the case of uniform mutation ($M1$) the α_i gene is transformed according to the equation:

$$\alpha_i' = \alpha_i^l + r\left(\alpha_i^u - \alpha_i^l\right), \qquad (9)$$

where r is a random number with a uniform distribution from the domain $[0,1]$, and α_i^u and α_i^l are the upper and lower limits, respectively, of variability interval of the α_i parameter, i.e. $\alpha_i \in \left[\alpha_i^l, \alpha_i^u\right]$. In the case of Gaussian mutation ($M2$) the α_i gene is transformed according to the equation:

$$\alpha_i' = \alpha_i + r(\tau), \tag{10}$$

where $r(\tau)$ is a random number with normal distribution with mean value equal to zero and variance equal to:

$$\sigma^2(\tau) = \frac{N - \tau}{N} \frac{\left(\alpha_i^u - \alpha_i^l\right)}{3}, \tag{11}$$

where τ is the current generation number, N is the maximum number of generations. In the calculations, a nonuniform mutation operator was used as well. During mutation, the α_i gene is transformed according to the equation:

$$\alpha_i' = \begin{cases} \alpha_i + \Delta(\tau, \alpha_i^u - \alpha_i), \\ \alpha_i - \Delta(\tau, \alpha_i - \alpha_i^l), \end{cases} \tag{12}$$

and a decision is taken at random which from the above formulas should be applied. Function $\Delta(\tau, x)$ was assumed in the form ($M3$ and $M4$, respectively):

$$\Delta_3(\tau, x) = x \left(1 - r^{(1 - \frac{\tau}{N})d}\right) \quad \text{or} \quad \Delta_4(\tau, x) = x\, r \left(1 - \frac{\tau}{N}\right)^d, \tag{13}$$

where r is a random number with a uniform distribution from the domain $[0, 1]$, τ is the current generation number, N is the maximum number of generations and d is a constant parameter (in the calculations, $d = 2$ was assumed).

4   Calculations

It was assumed in the calculations that: $b = 0.08$, $a_k = \lambda_k/(c_k\, \varrho_k)$ for $k = 1, 2$, $\lambda_1 = 33$, $\lambda_2 = 30$, $c_1 = 800$, $c_2 = 690$, $\varrho_1 = 7000$, $\varrho_2 = 7500$, $L = 270000$. The temperature of solidification is $T^* = 1500$, ambient temperature is $T_\infty = 50$ and initial temperature is equal $\varphi_0(x) = 1540$. The exact value of the convective heat transfer coefficient amounts to:

$$\alpha(t) = \begin{cases} 1200 & \text{dla } t \le 38, \\ 800 & \text{dla } t \in (38, 93], \\ 250 & \text{dla } t > 93. \end{cases} \tag{14}$$

For each of the mutation operators and different probability values of crossover (p_c) and mutation (p_m), calculations were carried out for ten different initial settings of a pseudorandom numbers' generator.

In the case of Gaussian mutation ($M2$), the best results were obtained for the crossover probability $p_c = 0.75$ and for mutation probability $p_m = 0.01$; the average value of the minimum found was 0.00245326, and the average value of the minimum point found $\alpha_{avg} = (1203.179, 792.496, 250.506)$.

For the remaining mutation operators, a zero value of the minimized functional was obtained. In the case of a uniform mutation $(M1)$ and a nonuniform mutation with $\Delta_3(\tau, x)$ function $(M3)$, the zero value was obtained twice, for $p_c \in \{0.7, 0.75\}$ and $p_m = 0.1$. In the case of a nonuniform mutation with $\Delta_4(\tau, x)$ function $(M4)$, this value was obtained three times, for $p_c \in \{0.7, 0.75, 0.8\}$ and $p_m = 0.1$. The results with the same value of the objective function can be subjected to further evaluation due to errors in the convective heat-transfer coefficient reconstruction. The least errors were obtained for a nonuniform mutation with $\Delta_3(\tau, x)$ function $(M3)$ and $p_m = 0.1$ and $p_c = 0.7$. The values found for the reconstructed coefficient are $\alpha_{avg} = (1200.003, 800.008, 249.999)$. A not much worse result was obtained for the same operator and crossover probability equal 0.75. In that case, the values founds were $\alpha_{avg} = (1200.01, 800.006, 249.99)$. In the remaining cases, the convective heat-transfer coefficient values were reconstructed with greater errors. Thus, in the case of the nonuniform mutation with $\Delta_4(\tau, x)$ function $(M4)$, the following values were found: $\alpha_{avg} = (1199.934, 800.072, 249.996)$ for $p_c = 0.8$, $\alpha_{avg} = (1200.037, 799.991, 249.998)$ for $p_c = 0.75$, $\alpha_{avg} = (1200.190, 799.863, 250.003)$ for $p_c = 0.7$. For the uniform mutation $(M1)$, the values determined were as follows: $\alpha_{avg} = (1200.197, 799.861, 250.001)$ for $p_c = 0.75$, $\alpha_{avg} = (1200.344, 799.720, 250.012)$ for $p_c = 0.7$.

Calculations for other values of the genetic algorithm parameters were also made, however, for none of the sets of values better results were obtained than those presented in this paper.

5   Conclusion

The paper presents the influence of choice of the mutation operator on the accuracy of a solution to the two-phase design inverse Stefan problem using genetic algorithms. The problem under consideration consisted in such selection of a convective heat transfer coefficient on one boundary that the moving interface would take the given position. Results for a uniform mutation, nonuniform mutation and Gaussian mutation have been presented. The best results were obtained for the nonuniform mutation with $\Delta_3(\tau, x)$ function and mutation probability (p_m) equal 0.1 and crossover probability (p_c) equal 0.7. The calculation results obtained show a very good approximation of the exact solution, thus corroborating the usefulness of the presented approach.

References

1. Goldman, N.L.: Inverse Stefan Problem. Kluwer, Dordrecht (1997)
2. Michalewicz, Z.: Genetic Algorithms + Data Structures = Evolution Programs. Springer-Verlag, Berlin (1996)
3. Osyczka, A.: Evolutionary Algorithms for Single and Multicriteria Design Optimization. Physica-Verlag, Heidelberg (2002)
4. Rogers, J.C.W., Berger, A.E., Ciment, M.: The Alternating Phase Truncation Method for Numerical Solution of a Stefan Problem. SIAM J. Numer. Anal. **16** (1979) 563–587

A Novel Nonlinear Neural Network Ensemble Model for Financial Time Series Forecasting

Kin Keung Lai[1,2], Lean Yu[2,3], Shouyang Wang[1,3], and Huang Wei[4]

[1] College of Business Administration, Hunan University, Changsha 410082, China
[2] Department of Management Sciences, City University of Hong Kong,
Tat Chee Avenue, Kowloon, Hong Kong
{mskklai, msyulean}@cityu.edu.hk
[3] Institute of Systems Science, Academy of Mathematics and Systems Science,
Chinese Academy of Sciences, Beijing 100080, China
{yulean, sywang}@amss.ac.cn
[4] School of Management, Huazhong University of Science and Technology,
1037 Luoyu Road, Wuhan 430074, China

Abstract. In this study, a new nonlinear neural network ensemble model is proposed for financial time series forecasting. In this model, many different neural network models are first generated. Then the principal component analysis technique is used to select the appropriate ensemble members. Finally, the support vector machine regression method is used for neural network ensemble. For further illustration, two real financial time series are used for testing.

1 Introduction

Financial market is a complex evolved dynamic market with high volatility and noise. Due to its irregularity, financial time series forecasting is regarded as a rather challenging task. For traditional statistical methods, it is extremely difficult to capture the irregularity. In the past decades, many emerging techniques, such as neural networks, were widely used in the financial time series forecasting and obtained good results.

However, neural networks are a kind of unstable learning methods, i.e., small changes in the training set and/or parameter selection can produce large changes in the predicted output. This diversity of neural networks is a naturally by-product of the randomness of the inherent data and training process, and also of the intrinsic non-identifiability of the model. For example, the results of many experiments have shown that the generalization of single neural network is not unique. That is, the neural network's results are not stable. Even for some simple problems, different structures of neural networks (e.g., different number of hidden layers, different hidden nodes and different initial conditions) result in different patterns of network generalization. In addition, even the most powerful neural network model still cannot cope well when dealing with complex data sets containing some random errors or insufficient training data. Thus, the performance for these data sets may not be as good as expected [1-2].

Recently, some experiments have been proved that neural network ensemble forecasting model is an effective approach to the development of a high performance forecasting system relative to single neural networks [3]. Meantime, some linear

V.N. Alexandrov et al. (Eds.): ICCS 2006, Part I, LNCS 3991, pp. 790–793, 2006.

ensemble methods are also presented [4-6]. Different from the previous work, this study proposes a novel nonlinear ensemble forecasting method in terms of support vector machine regression principle.

The rest of this study is organized as follows. Section 2 describes the building process of the nonlinear neural network ensemble forecasting model in detail. For further illustration, two real financial time series are used for testing in Section 3. Finally, some concluding remarks are drawn in Section 4.

2   The Building Process of the Nonlinear Ensemble Model

In this section, a triple-phase nonlinear neural network ensemble model is proposed for financial time series forecasting. First of all, many individual neural predictors are generated. Then an appropriate number of neural predictors are selected from the considerable number of candidate predictors. Finally, selected neural predictors are combined into an aggregated neural predictor in a nonlinear way.

A. Generating individual neural network predictors

With the work about bias-variance trade-off of Breiman [7], an ensemble model consisting of diverse models with much disagreement is more likely to have a good generalization. Therefore, how to generate diverse models is a crucial factor. For neural network model, there are four methods for generating diverse models.

(1) Initializing different starting weights for each neural network models.

(2) Training neural networks with different training subsets.

(3) Varying the architecture of neural network, e.g., changing the different numbers of layers or different numbers of nodes in each layer.

(4) Using different training algorithms, such as the back-propagation algorithm, radial basis function algorithm and Bayesian regression algorithms.

B. Selecting appropriate ensemble members

After training, each individual neural predictor has generated its own result. However, if there are a great number of individual members, we need to select a subset of representatives in order to improve ensemble efficiency. In this study, the principal component analysis (PCA) technique [8] is adopted to select appropriate ensemble members. Interested readers can be referred to [8] for more details.

C. Combining the selected members

Depended upon the work done in previous phases, a collection of appropriate ensemble members can be collected. The subsequent task is to combine these selected members into an aggregated predictor in an appropriate ensemble strategy. Generally, there are two ensemble strategies: linear ensemble and nonlinear ensemble.

Typically, linear ensemble strategy includes two approaches: the simple averaging [4] approach and the weighted averaging [5] approach. There are two types of weighted averaging: the mean squared error (MSE) based regression approach [6] and variance-based weighted approach [6]. The nonlinear ensemble strategy is a promising approach for determining the optimal neural ensemble predictor's weight. The literature only mentions one nonlinear approach: neural network-based nonlinear ensemble method [8]. Different from the previous work, we propose a new nonlinear ensemble method with support vector machine regression (SVMR) [9] principle.

Generally speaking, an SVMR-based nonlinear ensemble forecasting model can be viewed as a nonlinear information processing system that can be represented as:

$$\hat{y} = f(\hat{x}_1, \hat{x}_2, \cdots, \hat{x}_n) \tag{1}$$

where $(\hat{x}_1, \hat{x}_2, \cdots, \hat{x}_n)$ is the output of individual neural network predictors, $\hat{y}$ is the aggregated output, $f(\cdot)$ is nonlinear function determined by SVMR. In this sense, SVMR-based ensemble is a nonlinear ensemble method.

3 Empirical Analysis

The data set used for our experiment consists of two time series data: the S&P 500 index series, and the GBP/USD series. The data used in this study are obtained from *Datastream* (http://www.datastream. com). The entire data set covers the period from January 1, 1991 to December 31, 2002. We take daily data from January 1, 1991 to December 31, 2000 as the in-sample data sets and take the data from January 1, 2001 to December 31, 2002 as the out-of-sample data set (i.e., testing set), which are used to evaluate the good or bad performance of predictions. The root mean squared error (*RMSE*) is used the evaluation criteria over each of the two different testing sets and corresponding results are reported in Tables 1.

Table 1. A comparison of RMSE between different ensemble methods

Ensemble Method	S&P500		GBP/USD	
	RMSE	Rank	RMSE	Rank
Simple averaging	0.0115	3	0.0075	4
MSE regression	0.0159	5	0.0078	5
Variance-based weight	0.0124	4	0.0058	3
Neural network	0.0108	2	0.0044	2
SVMR	0.0098	1	0.0017	1

From Table 1, we can conclude that (1) in all the ensemble methods the SVMR-based ensemble model performs the best, followed by the neural network based ensemble method and other three linear ensemble method from a general view and (2) the nonlinear ensemble methods including neural network-based and SVMR-based method outperform all the linear ensemble methods, indicating that the nonlinear ensemble methods are more suitable for financial time series forecasting than the linear ensemble approaches due to high volatility of the financial time series. Interestedly, in the testing case of S&P 500, the simple averaging ensemble method can beat other two linear ensemble approaches. The phenomenon also reflects a basic principle, i.e., the simplest may be the best.

4   Conclusions

In this study, we propose a novel triple-phase nonlinear ensemble predictor for financial time series forecasting. The experimental results reported in this paper demonstrate the effectiveness of the proposed nonlinear ensemble approach, implying that the proposed nonlinear ensemble model can be used as a feasible approach to financial time series forecasting.

Acknowledgements

This work is partially supported by National Natural Science Foundation of China (NSFC No. 70221001); Chinese Academy of Sciences; Key Research Institute of Humanities and Social Sciences in Hubei Province-Research Center of Modern Information Management and Strategic Research Grant of City University of Hong Kong (SRG No. 7001677).

References

1. Naftaly, U, Intrator, N, Horn, D.: Optimal Ensemble Averaging of Neural Networks. Network Computation in Neural Systems 8 (1997) 283-296
2. Carney, J, Cunningham, P.: Tuning Diversity in Bagged Ensembles. International Journal of Neural Systems 10 (2000) 267-280
3. Bishop, C.M.: Neural Networks for Pattern Recognition. Oxford University Press (1995)
4. Benediktsson, J.A., Sveinsson, J.R., Ersoy, O.K., Swain, P.H.: Parallel Consensual Neural Networks. IEEE Transactions on Neural Networks 8 (1997) 54-64
5. Perrone, M.P., Cooper, L.N.: When Networks Disagree: Ensemble Methods for Hybrid Neural Networks. In Mammone, R.J. (ed.): Neural Networks for Speech and Image Processing, Chapman-Hall (1993) 126-142
6. Krogh, A., Vedelsby, J.: Neural Network Ensembles, Cross Validation, and Active Learning. In: Tesauro, G., Touretzky, D., Leen, D. (eds.): Advances in Neural Information Processing Systems. The MIT Press (1995) 231–238
7. Breiman, L. Combining Predictors. In: Sharkey, A.J.C. (ed.): Combining Artificial Neural Nets – Ensemble and Modular Multi-net Systems. Springer, Berlin (1999) 31-50
8. Yu, L., Wang, S.Y., Lai, K.K.: A Novel Nonlinear Ensemble Forecasting Model Incorporating GLAR and ANN for Foreign Exchange Rates. Computers and Operations Research 32 (2005) 2523-2541
9. Vapnik, V.: The Nature of Statistical Learning Theory. New York: Springer-Verlag (1995)

Performance Analysis of Block Jacobi Preconditioning Technique Based on Block Broyden Method*

Peng Jiang[1], Geng Yang[1], and Chunming Rong[2]

[1] School of Computer Science and Technology, P.O.Box 43, Nanjing University of Posts and Telecommunications, 210003, Nanjing, China
alice20006@hotmail.com, yangg@njupt.edu.cn
[2] Department of Electrical and Computer Engineering,
University of Stavanger, Norway
chunming.rong@uis.no

Abstract. The Block Jacobi preconditioning technique based on Block Broyden method is introduced to solve nonlinear equations. This paper theoretically analyzes the time complexity of this algorithm as well as the unpreconditioned one. Numerical experiments are used to show that Block Jacobi preconditioning method, compared with the unpreconditioned one, has faster solving speed and better performance under different dimensions and numbers of blocks.

1 Introduction

In the past few years, a number of books entirely devoted to iterative methods for nonlinear systems have appeared. The Block Broyden Algorithm was proposed and analyzed in References [1, 2]. However, the convergence speed of this algorithm is affected to some extent, for the information among the nodes is always lost. Hence, seeking for proper preconditioning methods[3] is one of the effective ways to solve this problem. Some preconditioners have been proposed and discussed in Reference [4].

This paper introduces Block Jacobi preconditioning technique based on Block Broyden method and we name this method as BJBB. It also analyzes time complexity and applies BJBB method to nonlinear systems arising from the Bratu problem.

2 Block Jacobi Method Based on Block Broyden Algorithm

2.1 General Remarks

In the following discussion, we are concerned with the problem of solving the large system of nonlinear equations as (1):

$$F(x) = 0. \tag{1}$$

---

* This work was supported in part by the Natural Science Foundation of Jiangsu Province under Grant No 05KJD520144 and the Foundation of the QingLan Project (KZ0040704006).

V.N. Alexandrov et al. (Eds.): ICCS 2006, Part I, LNCS 3991, pp. 794–797, 2006.

where $F(x) = (f_1, \cdots f_n)^T$ is a nonlinear operator from R^n to R^n, and $x^* \in R^n$ is an exact solution. Suppose that the components of x and F are divided into q blocks:

$$F = \begin{pmatrix} F_1 \\ \vdots \\ F_q \end{pmatrix} \qquad x = \begin{pmatrix} x_1 \\ \vdots \\ x_q \end{pmatrix}$$

We consider the Block Jacobi preconditioning technique based on Block Broyden method as follows:

- **Algorithm 2.1.1** BJBB Method

1. Let x^0 be an initial guess of x^*, and B^0 an initial block diagonal approximation of $J(x^0)$. Calculate $r^0 = F(x^0)$.
2. For k = 0, 1…until convergence:

   2.1 Solve $B^k s^k = -r^k$:

      2.1.1 Calculate the Block Jacobi preconditioner M:

         If the index set $S = \{1, \cdots n\}$ is partitioned as $S = U_i S_i$ with the sets S_i mutually disjoint, then the elements $m_{i,j}$ of preconditioner M is:

$$m_{i,j} = \begin{cases} a_{i,j} & \text{if and } j \text{ are in the same index subset} \\ \\ 0 & \text{otherwise} \end{cases}$$

      2.1.2 Calculate the inverse of the preconditioner.

      2.1.3 Transform the linear system as $M^{-1} B^k s^k = -M^{-1} r^k$ and solve it by Jacobi method.

   2.2 Update the solution $x^{k+1} = x^k + s^k$.

   2.3 Calculate $r^{k+1} = F(x^{k+1})$. If r^{k+1} is small enough, stop.

   2.4 Calculate $(s^k)^T s^k$ and update B^{k+1} by

$$B_i^{k+1} = B_i^k + \frac{r_i^{k+1}(s_i^k)^T}{(s^k)^T s^k} . \tag{2}$$

Then set $k = k + 1$, and go to step 2.

2.2 Time Complexity

From Reference [2], we know that the complexity of Block Jacobi method is:

$$U = \sum_{i=1}^{q} (4n_i - 1 + 2n_i^2) + L(n) + R(n) \tag{3}$$

here $L(n)$ means the complexity of solving q block linear equations $B_i^k s_i^k = -r_i^k$, and $R(n)$ is the calculative cost in step 2.3, Algorithm 2.1.1. From (3) we can know that the value of U differs in $L(n)$ for various methods.

For unpreconditioned method, we can deduce the value of $L(n)$ as follows:

$$L_n(n) = k_n \times (2\overline{n}^3 + \overline{n}^2 + \overline{n}) . \tag{4}$$

where k_n refers to the addition of number of iterations for unpreconditioned method..

For the BJBB method, we can get L(n) as follows:

$$L_p(n) = q \times (\frac{\overline{n}^4}{3} + 2\overline{n}^3 - \frac{\overline{n}^2}{3} - \overline{n}) + k_p \times (2\overline{n}^3 + \overline{n}^2 + \overline{n}) . \tag{5}$$

where k_p refers to the addition of number of iteration for BJBB method.

3 Numerical Experiments

Suppose a nonlinear partial differential equation can be written as

$$\begin{cases} -\Delta u + u_x + \lambda e^u = f, \\ u \mid_{\partial\Omega} = 0 \end{cases} (x, y) \in \Omega = [0,1] \times [0,1] . \tag{6}$$

It is known as the Bratu problem and has been used as a test problem by Yang in [2] and Jiang in [4]. In the following tests, we suppose N=110, 150 and 180, giving three grids, M1, M2 and M3, with 12100, 22500, 32400 unknowns, respectively. And we set block number q1=2000, q2=800 and q3=2500 for each grid. Table 1, 2, and 3 show the number of nonlinear iterations, which is denoted by "k" and the sum of numbers of iterations during the i-th nonlinear iteration, which is denoted by "k[i]".

Table 1. Comparison of the total number of iterations in M1, q1

	BJBB	No Preconditioner
k	3950	4499
k[500]	7106	13095
k[1500]	3533	7243

Table 2. Comparison of the total number of iterations in M2, q2

	BJBB	No Preconditioner
k	4335	4713
k[1700]	1088	5601
k[3290]	512	5020

Table 3. Comparison of the total number of iterations in M3, q3

	BJBB	No Preconditioner
k	4581	4884
k[1000]	4149	11499
k[3000]	1735	8994

To judge the performance of each method, we use data shown in Table 1 as an example. During the 1500–th iteration, the following can be known:

$$k_p = 3533,\; k_n = 7243,\; q_1 = 2000,\; \bar{n} = M_1 / q_1 = 12100 / 2000 = 6$$

According to (4), we get $L_n(n) = 3433182$ for the unpreconditioned method. According to (5), we get $L_p(n) = 3366642$ for BJBB method. Thus we find that $L_p(n) < L_n(n)$, so the performance of BJBB method is much better than the unpreconditioned one.

4 Conclusions

We have proposed Block Jacobi preconditioning technique based on Block Broyden Method for solving nonlinear systems. It shows evidently some advantages to combine Block Broyden Algorithm with preconditioners.

References

1. Yang G, Dutto L, Fortin M: Inexact block Jacobi Broyden methods for solving nonlinear systems of equations. SIAM J on Scientific Computing (1997) 1367-1392
2. Yang Geng: Analysis of parallel algorithms for solving nonlinear systems of equations. Chinese Journal of Computer (2000) 555-777 (in Chinese)
3. M. Benzi: Preconditioning Techniques for Large Linear Systems: A Survey. J Comput Phys (2002) 418-477
4. Peng Jiang, Geng Yang: Performance Analysis of Preconditioners based on Broyden Method. Applied Mathematics and Computation (Accepted for publication)

The Generic McMillan Degree:
A New Method Using Integer Matrices

E. Sagianos and N. Karcanias

Control Engineering Research Centre
School of Engineering and Mathematical Sciences
City University, Northampton Sq.
EC1V 0HB London , UK
E.Sagianos@city.ac.uk

Abstract. In this paper the problem of computing the generic McMillan degree of a *Structured Transfer Function (STF)* rational matrix is considered. The problem of determining the generic McMillan degree is tackled using genericity arguments and an optimisation procedure based on path properties of integer matrices is developed. This novel approach exploits the structure of integer matrices and leads to an efficient new algorithm for the computation of the generic value of the McMillan degree.

1 Introduction

The study of system properties based on ill-defined models is of great interest in the context of early design of large scale systems [5]. Our interest here is the study of the McMillan degree on special types of transfer functions referred to as *Structured Transfer Functions (STF)*. These are large dimension transfer functions with certain elements fixed to zero, some elements being constant, and other elements expressing the dominant dynamics of the system, which have been identified by some preliminary modelling effort. The McMillan degree indicates the complexity of the system, and can be calculated from the orders of the denominators in the Smith-McMillan form [4], [2]; this method is impractical for large dimension and uncertain models and the method suggested in [2], defining the pole polynomial as the least common multiple of the minors of all orders is used. Given an STF matrix $H(s)$, for example,

$$H(s) = \begin{bmatrix} A_1 A_2 A_3 & A_1^2 A_2 & A_2 \\ 0 & A_3^2 A_1 & A_2^3 \\ A_3 A_2 & A_1 & A_1 A_2 A_3 \end{bmatrix}, \tag{1}$$

where the A_i represent constant terms, first or second order dynamics. We can use partial fraction expansion and decompose $H(s)$ in the following manner:

$$H(s) = \underbrace{\begin{bmatrix} A_1 & A_1^2 & 0 \\ 0 & A_1 & 0 \\ 0 & A_1 & 0 \end{bmatrix}}_{H_1(s)} + \underbrace{\begin{bmatrix} A_2 & A_2 & A_2 \\ 0 & 0 & A_2^3 \\ A_2 & 0 & A_2 \end{bmatrix}}_{H_2(s)} + \underbrace{\begin{bmatrix} A_3 & 0 & 0 \\ 0 & A_3^2 & 0 \\ A_3 & 0 & A_3 \end{bmatrix}}_{H_3(s)} \tag{2}$$

V.N. Alexandrov et al. (Eds.): ICCS 2006, Part I, LNCS 3991, pp. 798–801, 2006.

where the matrices $H_i(s), i = 1, 2, 3$, are called *simple* STF matrices. The generic McMillan degree, denoted by $\delta(H)$, can be found by the sum [5]:

$$\delta(H) = \delta(H_1) + \delta(H_2) + \delta(H_3(s)) \ . \tag{3}$$

It has been shown in [5], that for simple matrices $H_i(s)$ we can define a map

$$H_i(s) \in \mathbb{R}^{m \times p}(s) \to I_i \in \mathbb{N}^{m \times p} \ ,$$

such that the entries of the integer matrix I_i, correspond to the orders of the entries in $H_i(s)$, and thus reducing the problem of determining the generic McMillan degree, into an assignment problem of integer matrices [1].

2   Algorithm for Determining the Weight of an Integer Matrix

Given a matrix $A = [a_{ij}] \in \mathbb{N}^{m \times p}, m \geq p$, we define:

Definition 1. *A $k-$length independent path $\{a_{i_1 j_1}, a_{i_2 j_2}, \ldots, a_{i_k j_k}\}$, on the matrix A, is a set of elements from the matrix such that there is no common index in the sets $\{i_1, i_2, \ldots, i_k\}$ and $\{j_1, j_2, \ldots, j_k\}$. The weight of a path is defined as the sum: $\gamma = a_{i_1 j_1} + a_{i_2 j_2} + \ldots + a_{i_k j_k}$.*

The *maximal weight* of all the independent paths of a matrix is denoted by $\gamma(A)$, and it is simply referred to as the weight of the matrix. Concerning the relationship between the generic McMillan degree of a rational matrix and the weight of the paths of an integer matrix, we have the following result [5]:

Theorem 1. The generic McMillan degree of the simple structured matrix $H_i(s)$ is equal to the maximal weight of its corresponding integer matrix I_i :

$$\delta_{gm}(H_i) = \gamma(I_i) \tag{4}$$

A matrix is called *column irreducible*, if the *highest coefficient matrix* (the Boolean matrix indicating the location of elements in a column with maximal value) has full structural rank, otherwise the matrix is called *column reducible*. The algorithm for the computation of the maximal weight uses the following result, based on the properties of Boolean matrices.

Lemma 1. *[4] If the highest coefficient matrix has full rank, then A is column irreducible and the maximal weight is given by the sum of the maximum element of each column. Otherwise A is column reducible and the sum of the maximum element of each column does not yield its maximal weight.*

The objective here is to provide a combination of p elements, which form an *independent path* (i.e. they belong in a different row and a different column of A), with the maximum *weight* $\gamma(A)$, for any irreducible or reducible matrix A.

Assume that the columns of A are ordered according to descending weight and form the table of the column contents with a resulting matrix $\mathcal{C}(A) \equiv C$, i.e.

$$
\begin{array}{cccc}
\text{col}(1) & \text{col}(2) & \cdots & \text{col}(p) \\
\hline
\delta_{11} & \delta_{12} & \cdots & \delta_{1p} \\
\delta_{21} & \delta_{22} & \cdots & \delta_{2p} \\
\vdots & \vdots & \ddots & \vdots \\
\delta_{\sigma_1(i)1} & \delta_{\sigma_2(i)2} & \cdots & \delta_{\sigma_p(i)p}
\end{array}
\tag{5}
$$

where $\delta_{1j} \geq \delta_{2j} \geq \cdots \geq \delta_{\sigma_j(i)j}, \forall j = 1, \ldots, p$ and the $\sigma_j(i), j = 1, \ldots, p$ are arrangement functions, one for each column, which take the values $\{1, \ldots, m\}$. We represent (5) by a matrix $B \in \mathbb{N}^{m \times p}$, and will refer to it as the *arrangement matrix* of A. We also define $C \in \mathbb{N}^{m \times p}$, as the *row index matrix* of A, given by:

$$
C = [\gamma_{ij}] = [\sigma_j(i)j], \quad i = 1, 2, \ldots, m \quad j = 1, 2, \ldots, p
\tag{6}
$$

where $\sigma_j(i)$ is the row coordinate of the i−th maximal element of the j−th column of A. Let $D \in \mathbb{Z}^{m \times p}$ be the *loss allocation matrix* of A, given by:

$$
D = [d_{ij}] = [b_{1j} - b_{ij}], \quad i = 1, 2, \ldots, m \quad j = 1, 2, \ldots, p
\tag{7}
$$

i.e. the amount d_{ij} is the difference of the element $a_{\gamma_{ij},j}$ from the maximum, which is $a_{\gamma_{1j},j}$. We will refer to an element of D as a *loss*. The search for the independent path with the maximum weight involves the following steps:

Algorithm

Preliminary Step: If the highest coefficient matrix has full rank, $\gamma(A) = \delta(A) = \sum_{i=1}^{p} \delta_{1i}$ and the independent path is given by the highest coefficient matrix. If A is reducible, then $\gamma(A) < \delta(A)$ and the procedure to determine $\gamma(A)$ continues as follows:

Searching Routine: In D, there will be a minimum element (*minimum loss*), denoted by m_1. Using the row index and loss allocation matrices, we can find all elements in A, for which $a_{\gamma_{1j},j} - a_{\gamma_{ij},j} = m_1$. We substitute those elements with the maximum, i.e. set $a_{\gamma_{1j},j} = a_{\gamma_{ij},j}$ for each column, and if the new highest coefficient matrix, which is a Boolean matrix, has full rank, then there definitely exists at least one independent path, which contains exactly one non-zero element from each column. We choose the independent path with the minimum loss, and denote this loss by m_{l_1}. We now have a lower bound:

$$
\sigma(A) = \delta(A) - m_{l_1} .
\tag{8}
$$

Let m_i denote a minimum element of D, after the i−th step; we have:

(i) If $m_{l_i} \geq \delta(A) - \sigma(A) = m_{l_1}$, then all independent paths, give a greater loss, we search in D for m_{i+1} :

(ii) If $m_{l_i} < m_{l_1}$, then m_{l_i} becomes the new lower bound and the search continues for m_{i+1} similarly.

The search stops when we reach a loss $m_{i+1} \geq m_{l_i}$, and the weight of A, i.e. the generic McMillan degree of the corresponding rational matrix, will be

$$\gamma(A) = \delta(A) - m_{l_i} \ . \tag{9}$$

3 Conclusions

The computation of the McMillan degree of a structured transfer function matrix has been considered using properties of column irreducibility of integer matrices, which are similar to those determining the relationship between complexity and degree of polynomial matrices [4]. The proposed algorithm avoids the general searching methods and determines the optimal solution in a small number of steps. An alternative approach for the computation of the generic McMillan degree may be developed by using standard results from a class of integer optimisation problems, more known as *optimal assignment*.

The optimal assignment problem frequently appears in Operational Research as the problem of having to assign n workers to n jobs, where we assume that the performance of the ith person for the jth job can in some sense be determined [1], [7]. Amongst the most popular methods for solving optimal assignment problems, are the *Hungarian*, and the *Bradford* [7]. Such methods provide an alternative way of looking at the problem and their performance to the case of large dimension systems has to be evaluated. The comparison of this new, algebraically based algorithm to the standard methodologies, is under investigation. There are strong indications that exploring the structural criteria based on the reduceness properties, variants of the optimal assignment algorithms may be developed which explore the sparse structure of the matrices and thus lead to algorithms with reduced complexity.

References

1. Dantzig, G.B. : Linear Programming and Extensions. Princeton University Press. (1963)
2. MacFarlane, A.G.J. , Karcanias, N. : Poles and zeros of linear multivariable systems: A survey of the algebraic, geometric and complex value theory. Int. J. of Control. **24**. (1976). 33–74
3. Reinschke, K. : Multivariable Control: A graph Theoretical Approach. Springer-Verlag. (1988)
4. Kailath, T. : Linear Systems. Prentice-Hall. (1980)
5. Karcanias, N. , Sagianos, E. , Milonides, E. : Structural Identification: The Computation of the Generic McMillan Degree. 44th IEEE Conference on Decision and Control and European Control Conference ECC 2005. Seville, Spain. (2005)
6. Vafiadis, K. : Systems and Control Problems in Early Systems Design. PhD Thesis. Control Engineering Research Centre. City University, London. (2003)
7. Mack, C. : The Bradford Method for the Assignment Problem. The New Journal of Operational Research. **1**. (1966). 17–29

State Estimation of Congested TCP Traffic Networks

Atulya Nagar, Ghulam Abbas, and Hissam Tawfik

Intelligent and Distributed Systems Laboratory, Deanery of Business and Computer Sciences,
Liverpool Hope University, Liverpool, L16 9JD. UK.
{nagara, tawfikh}@hope.ac.uk

Abstract. State Estimation is an intrinsic element of many network management systems, like Power Distribution Networks and Water Distribution Networks, where its implementation not only facilitates real-time online monitoring with better observability, but it also enables an advanced control with improved system security. This paper presents a new technique based on State Estimation to address some general shortcomings of the current Active Queue Management schemes such as RED and discusses potential issues in TCP networks in order to achieve better performance.

1 Introduction

Congestion typically refers to a situation when a TCP service either fails to fulfill a request to transfer a bulk of data, or it ends up with extensive service delays. Furthermore, data packets may also be lost in an attempt to complete the request. If the congestions are not dealt with appropriately, the packet loss rate becomes high enough, giving rise to retransmissions of lost packets and consequently cause further service delays. The Transmission control protocol (TCP) has been designed exclusively to offer a reliable service in terms of data delivery. Early implementations of TCP led to, what was known as "congestion-collapse", in which a network failed to respond altogether. This situation was soon overcome by more reliable TCP implementations [2]. However, the rapid increase in users around the globe, with a consequent increase in data requirements, has offered many threats to this reliability. These limitations are studied and remedied in this paper, by applying State Estimation mainly due to the following reasons. At first, no mathematical model is perfect and therefore may not capture all behavioral aspects of the actual physical state of the system. Numerous effects of the underlying system are deliberately left un-modeled, while the assumptions of the modeled effects are not correct under all circumstances. As such, there may be many uncertainties present in any mathematical model. Moreover, the underlying systems are driven not only by the control inputs but are often driven by disturbances or noises which cannot be modeled deterministically. The State Estimation technique proposed in this paper uses the Kalman filtering approach to try to address the general problems of RED models. We have used the discrete time model proposed in [1], as a case study in this paper, to compare the results from our State Estimator to the Simulation results of this model.

V.N. Alexandrov et al. (Eds.): ICCS 2006, Part I, LNCS 3991, pp. 802–805, 2006.

2   TCP Traffic State Estimator

The general TCP traffic flow State Estimation problem can be posed in a similar way as formulated in the Power and Water systems [6,4]. To have a more concrete description of the problem, consider the following. Let x_k be a given signal at time step k and $\mathcal{E}$ be the noise. Considering that only the sum of signal and the noise can be observed, it can be generally represented as,

$$Z = HX + \varepsilon \tag{1}$$

Where, Z is the measurement vector which is updated at each scan. X is the State vector, H is as Identity matrix ($m \times n, m \geq n$) relates state to measurement Z, and ε is the Vector of measurement errors. The error ε can arise due to a number of situations, e.g., inaccuracy of network model, measurement noise and inaccuracy of RED. The mathematical model may serve its purpose well in most cases but the assumptions of the mathematical model are not correct in all circumstances, for example, presence of a very large number of network nodes and packet-flows may affect the calculations up to a fractional level which can accumulate into a high level of measurement uncertainty when the results are used as feedback control. Moreover, an RED that operates on a router at some congested link and uses the exponentially-weighted-average-queue-length to predict packet losses and impose flow control may use wrong parameters (weights) and can consequently lead to uncertainties. A Kalman filter State Estimator is used here for this purpose. The expressions for the time and measurement updates of the Kalman filter, in order to devise a State Estimation algorithm, can be derived based on the treatment given in [1,5].

3   Results and Discussions

The derived Kalman algorithm for the queue length q, average queue length X and congestion window W is implemented in MATLAB. The following section presents comparison of the results from the Simulator [1] and the Estimator using the dumbbell and Y-shape topologies [3], the configurations of the congested connection listed in table 1.

According to the law of flow conservation [3] the flow into a congested link depends on the number of packets being injected by a sender into a link and as such, the accuracy of congestion window size is of significant importance. The congestion avoidance model [1] used in this paper, increments congestion window by *1/W* after the receipt of each acknowledgment. While this could work well for a small number of senders, it can lead to uncertainties in the presence of a large number of senders simultaneous transmitting through a link. As the acknowledgement is modeled to arrive in one round-trip time ($T_p + q/B$), which depends on the queue length q (queuing delay, q/B), the estimation assumes the round trip time to be corrupted by a small fraction (0.0005s). This fractional change is certain to occur when there is some background traffic present i.e. the congested router is also serving some other flows arriving from other nodes (note that the model assumed no background traffic). This fractional inaccuracy in the round-trip time can accumulate into large inaccuracy and consequently, the congestion window measurement becomes noisy after a few round trip times. For example, assuming 30 senders simultaneously transmitting through a queue and then calculating the round-trip time, the noise robustness becomes 0.02 packets in congestion window of each sender. The noise robustness increases at the start of the congestion soon after the first packet drop between 2nd and 3rd second. This is because the sender waits for a relatively longer round trip time during the recovery phase (note the horizontal increase of the noisy measurements). Noise robustness also increases with the increased number of senders.

Table 1. Network parameters

Variable	Description	Value
q_{min}	RED parameter	150
q_{max}	RED parameter	300
T_p	Propagation delay	0.1s
B	Bandwidth of bottleneck link	1Mbps
p_{max}	RED parameter	0.1
Weight	RED parameter	0.001

The results of estimated queue length were also compared to the results of simulated queue length [1] in the presence of certain noise. As the measurement of the queue length depends on the congestion window size of each sender transmitting through this queue $q = 1/W$, the estimator assumes five senders transmitting simultaneously, and the value of each W to be corrupted by 0.02 packets. The flow $q = \sum 1/W$ can produce the following effects on the queue size.

Table 2 lists mean queue-length of each sender observed for 10 seconds. In the presence of noise the measurement of the simulated queue length reflects wrong values. It means that, some of the buffer space at the router's queue remains unutilized due to the noisy measurements, and the packets are dropped by the router (when queue reaches its capacity) whereas, in reality, there still remain some unoccupied space. The results from the estimator match closely with the results of

Table 2. Noise Robustness

No. of Senders	Noise Robustness		
	Simulated mean Queue-length (without noise)	Simulated mean Queue-length (with noise)	Estimated mean Queue-Length (with noise)
1	69.5827	70.5673	69.5714
2	116.6457	119.3910	116.6333
3	131.3254	137.9962	131.3130
4	137.8840	150.1776	137.8696
5	141.0444	160.0377	141.0266

simulator which assumes no noise, i.e. the estimation is capable of removing the measurement noise and reflecting correct mean queue-length.

Moreover, it is also clear from the results that the measurement inaccuracy increases with the increased number of senders, consequently more packet drops will occur which in turn will lead to increased level of congestion. The accuracy of the measurements of average queue length X are of significant importance in terms of systems control, as a router using RED will drop packets as soon as X reaches q_{min}. As the measurement of the average queue depends on the actual queue length q, the noisy measurements of q can in turn affect the measurement of X resulting in early packet-drops prior to buffer filling.

4 Concluding Remarks

A State Estimation coupled with RED algorithm can provide a better control and management of the system, and security benefits.

References

1. Frommer, I., Harder, E., Hunt, B., Lance, R., Ott, E. and Yorke, J.: Two Models for the Study of Congested Internet Connections. Int. Conference on Communications and Computer Networks, (CCN 04). (2004).
2. Jacobson, V.: Modified TCP Congestion Avoidance Algorithm, Technical Report, Network Research Group LBL (1990).
3. Mathis, M. and Mahdavi, J.: Forward Acknowledgements: Refining TCP Congestion Control., in proceedings of the Int. Conf. on Applications, architectures, and protocols for computer communications, California, United States, 28-30 August (1996), Vol. 26, no. 4, p. 281-291
4. Nagar, A.K., Powell, R.S.: LFT/SDP Based Approach to the Uncertainty Analysis for State Estimation of Water Distribution Systems, IEE Journal of Control Theory and Applications, (2002) pp. 137-142, Vol. 149, issue 2.
5. Welch, G. and Bishiop, G.: An Introduction to the Kalman Filter, Technichal Report, SIGGRAPH 2001, Los Angeles Conventional Contre, (2001) 12-21.
6. Wu, F. F.: Power System State Estimation: A Survey, International Journal of EPES, Electric Power & Energy Systems, (1990), p. 80-87.

Study of Electron Transport in Composite Films Below the Percolation Threshold

Rudolf Hrach[1,2], Stanislav Novák[2], Martin Švec[1,2], and Jiří Škvor[2]

[1] Department of Electronics and Vacuum Physics,
Faculty of Mathematics and Physics, Charles University,
V Holešovičkách 2, 180 00 Prague 8, Czech Republic
`rudolf.hrach@mff.cuni.cz`
[2] Department of Physics, Faculty of Science, J. E. Purkyně University,
České mládeže 8, 400 96 Ústí nad Labem, Czech Republic
`novaks@sci.ujep.cz`

Abstract. Composite and nanocomposite films consisting of metal objects embedded in a dielectric matrix are studied by computer experiment. The electron transport through the composites is calculated in the case when the basic conductivity mechanism is the tunnel effect. It was found that the conductivity of composite film is located into tunneling clusters strongly influenced by objects arrangement in composite film.

1 Introduction

Composite and nanocomposite films represent class of promising materials with many applications in electronics, optics, catalysis and biotechnology [1]. The characteristics of such composite films depend strongly on both size and spatial distributions of embedded particles, which is influenced by used technologies – evaporation, laser deposition, plasma-assisted technologies, etc. [2], [3]. For most technologies it was observed that the resulting films contain nanometer-sized particles with excellent uniformity, at least for small filling factors.

The properties of composite films vary with the filling factor chosen. For small filling factors the structure contains individual particles completely insulated by dielectric matrix and the film has dielectric properties. With increasing filling factors the conductivity increases and tunnel current is observed below the percolation threshold. Finally, the transition into metallic state occurs.

Conventional characterization techniques of thin films fail for these new materials and novel advanced techniques are needed in order to describe the properties of composites. The goal of their morphological analysis is to characterize the forms and spatial distribution of individual objects in the film. For this purpose, the image analysis applied to micrographs of either projections or planar sections of composite films is the most suitable [4].

The investigation of electrical properties of composites theoretically leads to the nearly insurmountable difficulties, therefore a computer experiment is often used [5]. This text focuses on the study of transport properties of composite films below the percolation threshold, in order to find correlation between film structure and its electrical conductivity.

V.N. Alexandrov et al. (Eds.): ICCS 2006, Part I, LNCS 3991, pp. 806–809, 2006.

2   Model

First, several sets of simulated composite structures were prepared. The structures consist of spherical objects of fixed radii embedded in a 3D working field with characteristic dimensions from $1000 \times 1000 \times 100$ to $1000 \times 1000 \times 500$ units plus 10% margin. In the case of composite films the length unit of the model corresponds approximately to 0.1 nm. The spatial distribution of objects was determined by the static hard-sphere model. The model parameter, diffusion zone $DZ \in \langle 0, DZ_{max} \rangle$, denotes the minimum distance between edges of objects, the positions of which are generated randomly. The value of DZ sets the degree of arrangement of metal objects in the dielectric matrix; the larger the value of DZ, the more arranged structures are. In order to guaranty the sufficient precision of the results, the typical number of objects in the model varied between 1×10^3 and 1×10^4 and several structures with the same parameters were generated (about 10). Fig. 1 shows examples of prepared structures. In some cases the hard-sphere model is not satisfactory; hence, more sophisticated models corresponding to various types of experimental data were developed [6].

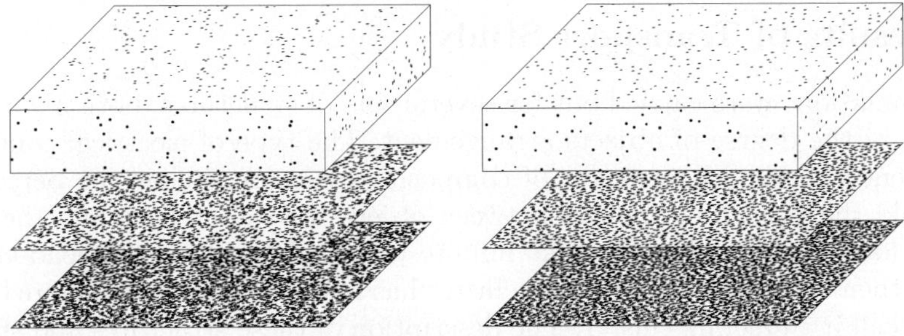

Fig. 1. Images of composite films, their sections and projections. Left – diffusion zone minimal, right – diffusion zone maximal ($DZ/DZ_{max} = 100\,\%$).

Besides the bulk (3D) composite structures, their 2D analogies also were used. These artificial structures, in contrast to 3D structures, enable better visualization of acquired results and allow easier evolution of algorithms. The 2D models of composite structures use the 1000×1000 working region with the same parameters – radii and diffusion zones – as the corresponding 3D models.

The electric current flowing through the composite film was calculated by the kinetic Monte Carlo method. The electron tunneling in a low voltage approximation was selected as a mechanism of conductivity in accordance with experimental data. The whole transport algorithm has three main parts: (i) emission of electrons from negatively biased electrode; (ii) tunneling of electrons between particular objects, causing changes in their charging; (iii) and the collection of electrons by positively biased electrode, all parts treated stochastically. The electron transport was studied as an iterative process until reaching the steady state of both potential distribution and intensities of conduction paths. Fig. 2 shows the potential distribution in 2D analogies of composite structures in the dependence on the arrangement of objects.

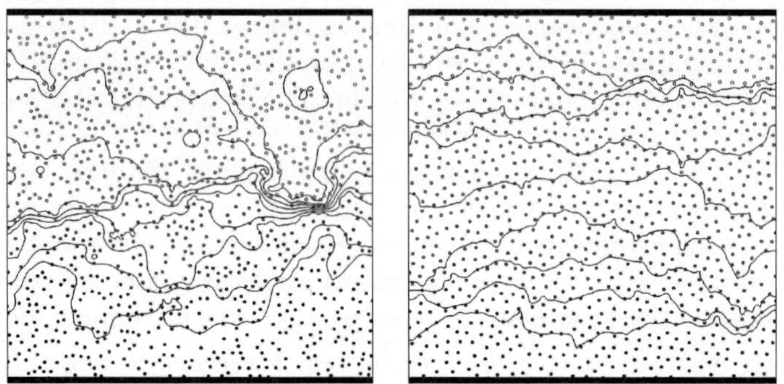

Fig. 2. Steady-state potential distribution in 2D analogies of composite films denoted by equipotentials and by colours of individual clusters. Left – $DZ/DZ_{max} = 33\%$, right – $DZ/DZ_{max} = 100\%$.

3   Results of Transport Study

The simulations were carried out for several sets of structures, both 3D and 2D, varying in the degree of objects arrangement. The type of electrical conductivity depends on the filling factor of composite structure. Below the percolation threshold the tunneling appears between objects and electrodes and the terms typical for ohmic conductivity like infinite cluster, backbone and dead-ends [7] change their meanings. The infinite cluster has now spread to much more objects and we call it tunneling cluster. The description of these structures and study of their properties is the goal of our simulations. Fig. 3 shows the tunneling clusters in structures with different degrees of ordering. The intensity of the electric current flowing through each channel is depicted by the shade of grey colour.

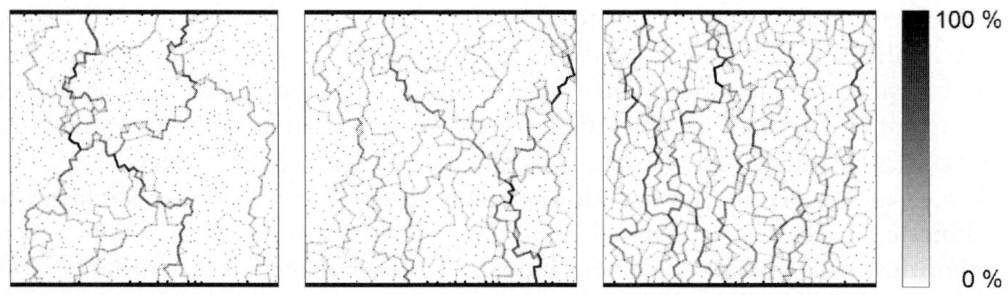

Fig. 3. Currents flowing between electrodes. From left to right – $DZ/DZ_{max} = 33, 67$ and 100 %.

During the analysis of tunneling cluster the channel with the largest intensity of electric current (so-called main conductivity channel) and its critical bond was determined. The critical bond is the bond that causes the weakest conductive connection between two objects in the main conductivity channel. Fig. 4 shows the main conductivity channels of the structures from Fig. 3; the critical bonds are also marked in the channels. Relative values of the electric current flowing through these channels are 0.151, 0.083 and 0.052 for DZ/DZ_{max} 0.33, 0.66 and 1.0, resp. (the fluctuations for various samples were about 20 %).

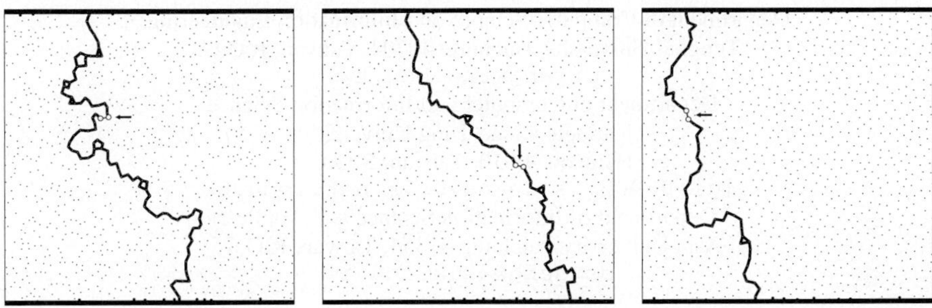

Fig. 4. Main conductivity channels and their critical bonds (indicated by arrows), i.e. the maximal resistors. From left to right – $DZ/DZ_{max} = 33$, 67 and 100 %.

4   Discussion

The electrical characteristics of composites depend on the degree of arrangement of metal objects. To describe this effect caused by the decreasing dispersion of distances between objects, one can claim: (i) the total current increases; (ii) the potential distribution in the structure becomes more uniform; (iii) the number of current paths increases; (iv) the current paths become shorter and better arranged; (v) the main conductivity channel becomes shorter and more arranged.

Acknowledgements. The work is a part of the research plan MSM0021620834 that is financed by the Ministry of Education of Czech Republic. The support of the projects 1ET400720409, GAUK-237/2005 and 1P05OC013 is acknowledged.

References

1. Naka, K., Itoh, H., Park, S.-Y., Chujo, Y. Polymer Bulletin. **52** (2004) 171
2. Faupel, J., Fuhse, C., Meschede, A., Herweg, C., Krebs, H.U., Vitta, S. Appl. Phys. A **79** (2004) 1233
3. Biederman, H., Martinu, L.: Plasma Deposition, Treatment and Etching of Polymers, Academic Press, New York (1990)
4. Novák, S., Hrach, R. Materials and Manufacturing Processes **17** (2002) 97
5. Sheng, P. Phil. Mag. B **65** (1992) 357
6. Hrach, R., Švec, M., Novák, S., Sedlák, D. Thin Solid Films **459** (2004) 174
7. Stauffer, D., Aharony, A.: Introduction to Percolation Theory, Taylor and Frencis, London (2003)

A Dynamic Partitioning Self-scheduling Scheme for Parallel Loops on Heterogeneous Clusters

Chao-Tung Yang[1,*], Wen-Chung Shih[2], and Shian-Shyong Tseng[2,3]

[1] High-Performance Computing Laboratory
Department of Computer Science and Information Engineering
Tunghai University Taichung 40704, Taiwan, R.O.C.
`ctyang@thu.edu.tw`
[2] Department of Computer and Information Science
National Chiao Tung University
Hsinchu 30010, Taiwan, R.O.C.
`{gis90805, sstseng}@cis.nctu.edu.tw`
[3] Department of Information Science and Applications
Asia University Taichung 41354, Taiwan, R.O.C.
`sstseng@asia.edu.tw`

Abstract. Loop partitioning on parallel and distributed systems has been an important problem. Furthermore, it becomes more difficult to deal with on the emerging heterogeneous PC cluster environments. In this paper, we propose a performance-based scheme, which dynamically partitions loop iterations according to the performance ratio of cluster nodes. To verify our approach, a heterogeneous cluster is built, and two kinds of application programs are implemented to be executed in this testbed. Experimental results show that our approach performs better than traditional schemes.

Keywords: Parallel loops, Self-scheduling, Cluster computing, MPI, Heterogeneous, PC cluster.

1 Introduction

As more and more inexpensive personal computers (PC) are available, clusters of PCs have become alternatives of supercomputers which many research projects cannot afford. However, it is difficult to deal with the heterogeneity in a cluster [2, 4], especially for the parallel loop scheduling problem [5].

Previous researchers [3, 6, 7] propose a two-phased self-scheduling approach, which is applicable to PC-based cluster environments. These two-phased schemes collect system configuration information, and then distribute some portion of the workload among slave nodes according to their CPU clock speed [6] or HINT measurements [7]. After that, the remaining work load is scheduled by some well-known self-scheduling scheme. Nevertheless, the performance of this approach depends on the appropriate choice of scheduling parameters. Besides, it estimates node performance only by CPU speed or HINT benchmark, which is one of the factors affecting node performance.

---

* Corresponding author.

V.N. Alexandrov et al. (Eds.): ICCS 2006, Part I, LNCS 3991, pp. 810–813, 2006.

In [3], an enhanced scheme, which dynamically adjusts scheduling parameters according to system heterogeneity, is proposed.

Previous work in [3, 7] and this paper are all inspired by [6], the α self-scheduling scheme. However, this work has different viewpoints and unique contribution. First, while [3, 6] partition α % of workload according to performance weighted by CPU clock speed in phase one, our scheme conducts the partition according to a general performance function (PF). The PF obtained by the HPL benchmark [8] can estimate performance of cluster nodes rather accurately.

Second, the scheme in [6] utilizes a fixed α value, and [3, 7] adaptively adjust the α value according to the heterogeneity of the cluster. In a word, both schemes depend on a properly chosen α value to get good performance. Nevertheless, our scheme focuses on accurate estimation of node performance, so the choice of α value is not very critical.

2 Our Approach

We propose to partition $\alpha\%$ of workload according to the performance ratio of all nodes, and the remaining workload is dispatched by some well-known self-scheduling scheme, for example, GSS [5]. Using this approach, we do not need to know the real computer performance. However, a good performance ratio is desired to estimate performance of nodes accurately.

2.1 Performance Function and Performance Ratio

We first define the Performance Function (PF) to represent the performance index of each node. In this paper, our PF for node j is defined as

$$PF_j = \frac{B_j}{\sum_{\forall node_i \in S} B_i} \tag{1}$$

where
  S is the set of all cluster nodes.
  B_i is the performance value of node i measured by the HPL benchmark [8] before each workload partitioning.

The performance ratio (PR) is defined to be the ratio of all performance functions. For instance, assume the PF of three nodes are 1/2, 1/3 and 1/4. Then, the PR is 1/2 : 1/3 : 1/4; i.e., the PR of the three nodes is 6 : 4 : 3. In other words, if there are 13 loop iterations, 6 iterations will be assigned to the first node, 4 iterations will be assigned to the second node, and 3 iterations will be assigned to the last one.

2.2 Our Algorithm

The algorithm for the master node of our approach is described as follows.

Algorithm MASTER:
```
1. Evaluate Performance Ratio by the HPL benchmark.
2. Dispatch α% of loop iterations according to the
performance ratio of nodes.
3. Master does its own computation work
4. Dispatch (100-α)% of loop iterations into the task queue
using GSS.
END MASTER
```

3 Experimental Results

We have built a heterogeneous cluster which consists of 8 PCs. The configuration of this cluster testbed is shown in Table 1.

Table 1. Hardware configuration

Host Name	CPU Type	CPU Speed	Number of CPU	RAM
hpc	Intel Xeon^{TM}	2.4GHz	2	1GB
amd1	AMD Athlon^{TM} MP	1.8GHz	2	2GB
amd1-dual1	AMD Athlon^{TM} MP	2.2Ghz	2	512MB
amd1-dual01	AMD Athlon^{TM} MP	2.0Ghz	2	2G
dna2	AMD Athlon^{TM} MP	2.0Ghz	2	2G
piii-dual1	Intel Pentium III	866MHz	2	1GB
xeon2	Intel Xeon^{TM}	3.0GHz	2	512MB
hpc2	Intel Xeon^{TM}	3.0GHz	2	1GB

We have implemented the Mandelbrot set application programs in C language, with message passing interface (MPI) directives for parallelizing code segments to be processed by multiple CPUs. In this experiment, the scheduling parameter α is set to be 50 for all two-phased schemes, except for the schemes by [7], of which α is dynamically adjustable according to cluster heterogeneity.

The Mandelbrot set is a problem involving the same computation on different data points which have different convergence rates [1]. In this experiment, execution time on the heterogeneous cluster is investigated. Figure 1(a) illustrates execution time of static scheduling, dynamic scheduling (GSS [5]) and our scheme, with input image size 64×64, 128×128 and 192×192 respectively. Experimental results show that our scheduling scheme got better performance than static and dynamic ones. In this case, our scheme for input size 192×192 got 95% and 86% performance improvement over the static one and the dynamic one respectively.

Figure 1(b) illustrates execution time of previous two-phased schemes ([6] and [7]) and our scheme, with input image size 64×64, 128×128 and 192×192 respectively. Experimental results show that our hybrid scheduling scheme got better performance than [6] and [7]. In this case, our scheme for input size 192×192 got 83% and 69% performance improvement over the static one and the dynamic one respectively.

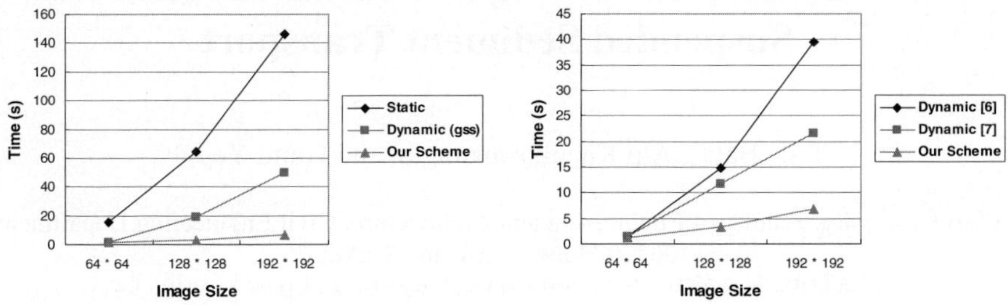

Fig. 1. Mandelbrot execution time on the heterogeneous cluster (a) Static, dynamic and our scheme; (b) Dynamic schemes [6], [7] and our scheme

4   Conclusions and Future Work

In this paper, we propose a dynamic loop partitioning scheme, and compare it with previous algorithms by experiments on the Mandelbrot application programs in our heterogeneous cluster environment. In each case, our approach can obtain performance improvement on previous schemes. In our future work, we will implement more types of application programs to verify our approach. Furthermore, we hope to find better ways of modeling the performance function, incorporating network information.

References

1. Introduction To The Mandelbrot Set, http://www.ddewey.net/mandelbrot/
2. M. Baker and R. Buyya, "Cluster Computing: The Commodity Supercomputer," International Journal of Software Practice and Experience, Vol. 29, No. 6, 2002, pp.551-575, 1999.
3. Kuan-Wei Cheng, Chao-Tung Yang, Chuan-Lin Lai, and Shun-Chyi Chang, "A Parallel Loop Self-Scheduling on Grid Computing Environments," Proceedings of the 2004 IEEE International Symposium on Parallel Architectures, Algorithms and Networks, pp. 409-414, KH, China, May 2004.
4. A. T. Chronopoulos, R. Andonie, M. Benche and D.Grosu, "A Class of Loop Self-Scheduling for Heterogeneous Clusters," Proceedings of the 2001 IEEE International Conference on Cluster Computing, pp. 282-291, 2001.
5. C. D. Polychronopoulos and D. Kuck, "Guided Self-Scheduling: a Practical Scheduling Scheme for Parallel Supercomputers," IEEE Trans. on Computers, vol. 36, no. 12, pp 1425-1439, 1987.
6. Chao-Tung Yang and Shun-Chyi Chang, "A Parallel Loop Self-Scheduling on Extremely Heterogeneous PC Clusters," Journal of Information Science and Engineering, vol. 20, no. 2, pp. 263-273, March 2004.
7. Chao-Tung Yang, Kuan-Wei Cheng, and Kuan-Ching Li, "An Efficient Parallel Loop Self-Scheduling on Grid Environments," NPC'2004 IFIP International Conference on Network and Parallel Computing, Lecture Notes in Computer Science, Springer-Verlag Heidelberg, Hai Jin, Guangrong Gao, Zhiwei Xu (Eds.), Oct. 2004.
8. http://www.netlib.org/benchmark/hpl/

3-D Numerical Modelling of Coastal Currents and Suspended Sediment Transport

Lale Balas, Alp Küçükosmanoğlu, and Umut Yegül

Gazi University, Faculty of Engineering and Architecture, Civil Engineering Department
06570 Maltepe, Ankara, Turkey
{lalebal, akucukosmanoglu, uyegul}@gazi.edu.tr

Abstract. A three dimensional hydrodynamic and suspended sediment transport model (HYDROTAM-3) has been developed and applied to Fethiye Bay. Model can simulate the transport processes due to tidal or nontidal forcing which may be barotropic or baroclinic. The Boussinesq approximation, i.e. the density differences are neglected unless the differences are multiplied by the gravity, is the only simplifying assumption in the model. The model is also capable of computing suspended sediment distributions, amount of eroded and deposited sediment. It is a composite finite difference, finite element model. At three Stations in the Bay, continuous measurements of velocity throughout the water depth and water level were taken for 27 days. Model predictions are in good agreement with the field data.

1 Introduction

The activities around enclosed or semi-enclosed coastal areas that have limited water exchange should be carefully planned and detailed researches to understand water circulations and transport processes should be performed. Since field measurements are usually costly and some times impossible due to physical inabilities, application of numerical models becomes more and more important in the simulations of coastal water bodies. Use of three-dimensional models is unavoidable in all cases where the influence of density distribution, or the vertical velocity variations can not be neglected and in the simulation of wind induced circulation [1],[2],[3]. The detailed knowledge of the water motion is very crucial for a reliable prediction of suspended sediment transport [4],[5].

An unsteady three-dimensional baroclinic circulation model (HYDROTAM-3) has been developed to simulate the transport processes in coastal water bodies[6],[7],[8]. The model consists of three components: hydrodynamic, turbulence and suspended sediment transport models. In the hydrodynamic model, full Navier-Stokes equations are solved. The Boussinesq approximation is the only simplifying assumption in the hydrodynamic model. Temperature and salinity variations are calculated by solving the three dimensional convection-diffusion equations. The two equation k-ε turbulence model is used for the turbulence modelling [6],[7]. The three dimensional conservation equation for suspended sediment where the vertical advection includes the particle settling velocity can be written as:

V.N. Alexandrov et al. (Eds.): ICCS 2006, Part I, LNCS 3991, pp. 814–817, 2006.

$$\frac{\partial C}{\partial t} + u\frac{\partial C}{\partial x} + v\frac{\partial C}{\partial y} + w\frac{\partial C}{\partial z} - w_s\frac{\partial C}{\partial z} = \frac{\partial}{\partial x}(D_x\frac{\partial C}{\partial x}) + \frac{\partial}{\partial x}(D_y\frac{\partial C}{\partial y}) + \frac{\partial}{\partial z}(D_z\frac{\partial C}{\partial z}) \tag{1}$$

where C: Suspended sediment concentration; u,v,w: Velocity components in x,y,z directions, respectively; D_x, D_y, D_z: Turbulent viscosity coefficients in x,y and z directions, respectively; w_s: Settling velocity.

Solution scheme is a composite finite element-finite difference scheme [6]. The governing equations are solved by Galerkin Weighted Residual Method in the vertical plane and by finite difference approximations in the horizontal plane, without any coordinate transformation. The water depths are divided into the same number of layers following the bottom topography.

2   Model Application to Fethiye Bay

Developed three dimensional numerical model (HYROTAM-3) has been implemented to the Bay of Fethiye. Water depths in the Bay are plotted in Fig.1. The grid system used has a square mesh size of 100x100 m. Wind characteristics are obtained from the measurements of the meteorological station in Fethiye for the period of 1980-2002. The wind analysis shows that the critical wind direction for wind speeds more than 7 m/s, is WNW-WSW direction.

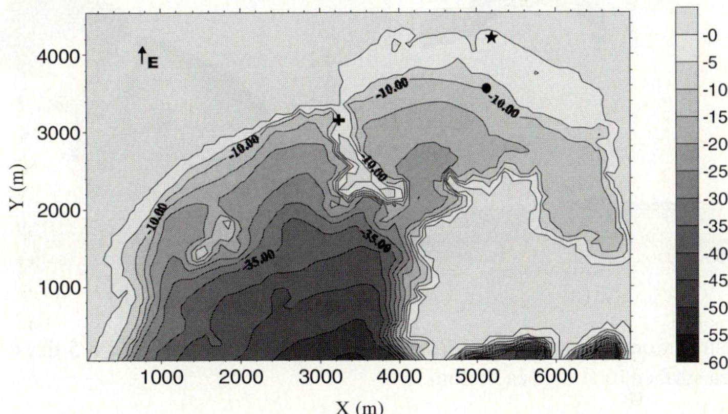

Fig. 1. Water depths (m) of Fethiye Bay where +:Station I, •:Station II,∗ : Station III

Some field measurements have been performed in the area. At Station I and at Station II shown in Fig.1, continuous velocity measurements throughout water depth, at Station III water level measurements were taken for 27 days. Simulated velocity profiles over the depth at the end of 8.5 days are compared with the measurements taken at Station I and Station II and are shown in Fig.2.

At the end of 8.5 days of simulation, for Station I, the root mean square error is 0.417 cm/s and bias is –0.095 cm/s. and for Station II, the root mean square error is 0.182 cm/s and bias is –0.027 cm/s. Comparison of results is encouraging. Model well simulates the measurement period. Predicted sediment distributions in the Bay are given in Fig.3. at the end of 8.5 days of simulation.

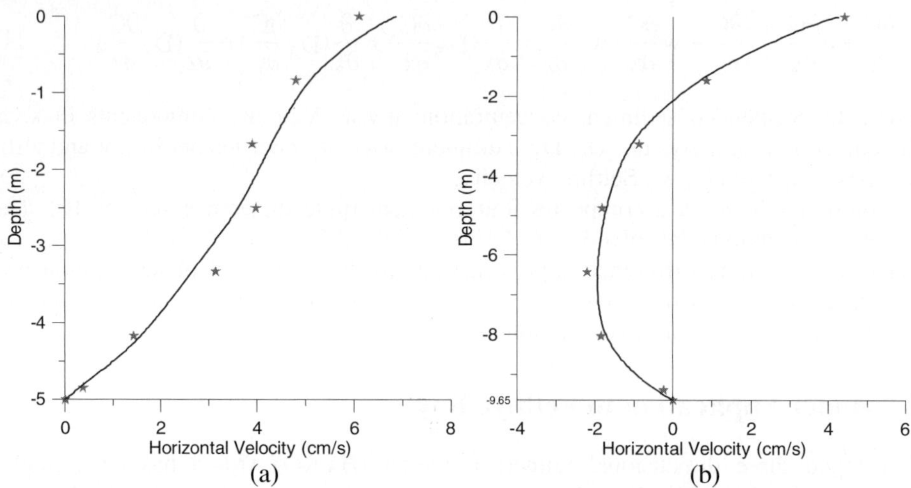

Fig. 2. Velocity profiles over the water depth at the end of 8.5 days of simulation, at Station I b) at Station II, where, solid line: simulation, dot: measurements.

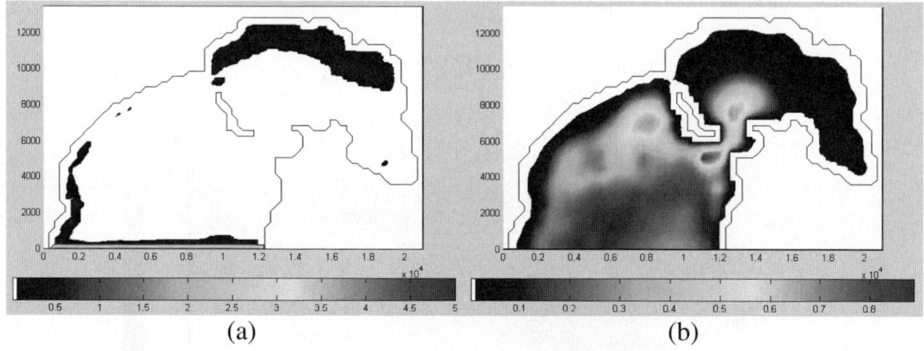

Fig. 3. Distribution of suspended sediment concentration at the end of 8.5 days of simulation a) at the sea surface b) at the sea bottom

A sensitivity study of model predictions to bottom friction coefficient is performed and C_f=0.0026 provided the best match with the measurements. In the application, measurement period has been simulated and model is forced by the recorded wind. No significant density stratification was recorded at the site. A horizontal grid spacing of $\Delta x=\Delta y=100$ m. is used. The sea bottom is treated as a rigid boundary. Model predictions are in good agreement with the measurements. During the site investigations, it has been observed that suspended sediment concentration is the highest around the island. Inner and outer parts of the Bay are rather dynamic regions where there may occur sediment deposition or resuspension depending on the blowing wind direction. Model simulations well agree to the observed regions at the site. Sediment sampling studies are still in continuation at the site.

3   Conclusions

A baroclinic three dimensional numerical model of transport processes in coastal areas has been presented. The model consists of three components; hydrodynamic, turbulence and suspended sediment transport. The developed model is able to predict suspended sediment concentration profiles, re-suspension of bottom sediments, equilibrium and deposition sites of the coastal area and determines changes in sea bed morphology quantitatively. Model has been applied to Fethiye Bay where there exist some field measurements. Model implementation to Fethiye Bay has provided realistic results and has shown the capability of model to predict complex circulation patterns.

References

1. Olsen, N. R. B.: Two-dimensional numerical modelling of flushing processes in water reservoirs. Journal of Hydraulic Research 37(1) (1999) 3-16.
2. Inoue, M., Wiseman Jr. W. J.: Transport mixing and stirring processes in a Louisiana Estuary: a model study. Estuarine Coastal and Shelf Science 50(2000) 449-466.
3. Li,C.W, Gu, J.: 3D Layered-Integrated Modelling of Mass Exchange in Semi-Enclosed Water Bodies. Journal of Hydraulic Research 39(2001) 403-411.
4. Lin, B., Falconer, R.A.: Tidal Flow and Transport Modelling Using Ultimate Quickest Scheme. Journal of Hydraulic Engineering 123(1997) 303-314.
5. Garcia-Martinez, R., Saavedra, C. I., De Power B. F., Valera E.:A two-dimensional computation model to simulate suspended sediment transport and bed changes. Journal of Hydraulic Research 37(1999) 327-344.
6. Balas, L., Özhan, E.: An implicit three dimensional numerical model to simulate transport processes in coastal water bodies. Int.Journal of Numerical Methods in Fluids 34 (2000) 307-339.
7. Balas, L., Özhan, E.: Applications of a 3-D Numerical Model to Circulations in Coastal Waters. Coastal Engineering Journal 43 (2001) 99-120.
8. Balas, L.: Simulation of Pollutant Transport in Marmaris Bay. Chineese Ocean Engineering Journal 15 (2001) 565-578.

Ontology-Driven Resource Selecting in the Grid Environments[*]

Wenju Zhang, Yin Li, Fei Liu, and Fanyuan Ma

Shanghai Jiao Tong University,
Shanghai, P.R.China, 200030
{zwj, liyin, liufei, fyma}@cs.sjtu.edu.cn

Abstract. In the Grid environments where many different implementations are available, the need for semantic matching based on a defined ontology becomes increasingly important. Especially for service or resource discovery and selection. In this paper, we propose a flexible and extensible approach for solving resource discovery and selection in the Grid Environments using ontology and semantic web and grid technologies. We have designed and prototyped an ontology-driven resource discovery and selection framework that exploits ontologies and domain knowledge base. We present results obtained when this framework is applied in the context of drug discovery grid. These results demonstrate the effectiveness of our framework.

1   Introduction

The need to discover and select entities that match specified requirements arises in many contexts in distributed systems like Peer-to-Peer networks and Grids. In such environments, many different nodes, possibly spanner across multiple organizations, need to share resource [3].

A common issue both in Peer-to-Peer and Grid is related to the fact that data and resources need to be described in a way that is understandable and usable by the community that is target user, by means of ontologies.

In this paper, we propose a flexible and extensible approach for performing Grid resource discovery and selection using an ontology-driven model and an O-Match resource rank algorithm. Unlike the traditional Grid resource matching that describe resource request properties based on symmetric flat attributes, separate ontologies are created to declaratively describe resources and job requests using an expressive ontology language. Moreover, we propose an O-Match resource rank algorithm to balance the symmetric and asymmetric matching.

The rest of this paper is organized as follows. Section 2 lists the related work. Section 3 presents ontology-driven resource discovery and selection techniques and algorithms. Section 4 presents the prototype implementation and experimental results. Finally, Section 5 gives the conclusions.

---

[*] This research work is supported in part by the the National High Technology Research and Development Program of China (863 Program), under Grant No. 2004AA104270.

V.N. Alexandrov et al. (Eds.): ICCS 2006, Part I, LNCS 3991, pp. 818–821, 2006.

2   Related Work

Related to the resource discovery and selection solution. Globus MDS and UDDI are two such examples; MDS has been widely used in the Grid community for resource discovery while UDDI has been used in the web community for business service discovery.

Different approaches for ontology-based resources matching and selecting in the grid systems also have been proposed [2]. Matchmaker [4] is a framework in order to provide a flexible strategy for the resource matching problem in the Grid. This approach is based on three ontologies: a resource ontology, a resource request ontology and a policy ontology.

3   Ontology-Driven Resource Discovery and Selection

Semantic matching is based on OWL-S [1] ontologies. The advertisements and requests refer to OWL-S concepts and the associated semantic. By using OWL-S, the matching process can perform implications on the subsumption hierarchy leading to the recognition of semantic matches despite their syntactical differences.

3.1   Ontology-Based Semantic Annotation

The meaning of services is implicitly expressed by the implementation expressed in the form of the programming language source code. The purpose of the semantic annotation is to express this intrinsic meaning explicitly and in a machine processable way. The Resource Description Framework (RDF) from W3C was designed to serve this purpose and the OWL builds on RDF to provide a way of adding domain specific vocabulary for resource description by using concepts taxonomy.

Semantic annotation of a service is developed in two stages. First, the user annotates a service with intended meaning. Next, different aspects of a service method need to be described independently as distinct resources. This stage concentrates on expressing the syntactic meaning of the service by annotating the semantics of the definition of the service method.

3.2   Ontology-Based Semantic Matching

The first step in autonomic service adaptation is to find services that are conceptually equivalent to the client's requirements. These requirements are expressed through the semantic annotation of the interface by using OWL. This ties each of the interface method to a domain concept.

A semantic matching service will need to perform two main inference operations - class and property inferencing. Each interface annotation ties the concept of a method to a ontology class.

3.3   The O-Match Algorithm

In order to enforce dynamic ontology matching, we require a flexible algorithm with the aim of facing two different requirements of the matching process. These requirements have been addressed by the O-Match algorithm for dynamic ontology matching. The aim of O-Match is to allow a dynamic choice of the kind of features to be considered in the matching process. O-Match is based on two basic functions, namely a datatype compatibility function $\mathcal{T}(dt, dt') \rightarrow \{0, 1\}$, and a property and relation closeness function $\mathcal{C}(e, e') \rightarrow [0, 1]$.

The datatype compatibility function $\mathcal{T}(dt, dt') \rightarrow \{0, 1\}$ is defined to evaluate the compatibility of data types of two properties according to a pre-defined set CR of compatibility rules. Given two datatypes dt and dt', the function returns 1 if dt and dt' are compatible according to CR, and 0 otherwise. For instance, with reference to XML Schema datatypes, examples of compatibility rules that hold between datatypes are: $xsd : integer \Leftrightarrow xsd : int, xsd : integer \Leftrightarrow xsd : float, xsd : decimal \Leftrightarrow xsd : float and xsd : short \Leftrightarrow xsd : int$.

The property and relation closeness function $\mathcal{C}(e, e') \rightarrow [0, 1]$ calculates a measure of the distance between two context elements of concepts. $\mathcal{C}(e, e')$ exploits the weights associated with context elements and returns a value in the range [0,1] proportional to the absolute value of the complement of the difference between the weights associated with the elements. For any pairs of elements e and e', the highest value is obtained when weights of e and e' coincide.

4   Prototype Implementation and Experimental Evaluation

4.1   Prototype Implementation

The ontology-based resources discovery and selection framework consists of three components: 1) resources discovery and selection engine, 2) resources database, capturing all the resources available in this domain, and 3) domain ontology knowledge base, capturing the domain model and additional knowledge about the domain.

We have developed two ontologies using OWL-S including resource ontology and domain ontology. The resource ontology provides an abstract model for describing resources, their capabilities and their relationships. The domain ontology is used during the resource selecting process. It is typically defined by the grid middleware.

4.2   Experimental Results

To verify the validity of our resources discovery and selection framework and O-Match algorithm, we conducted experiments in the context of the Drug Discovery Grid project's test bed (http://www.ddgrid.ac.cn). which comprised 8 clusters at 5 different cities. We compared the execution time and efficiency of resources discovery and selection with UDDI-based resources selection.

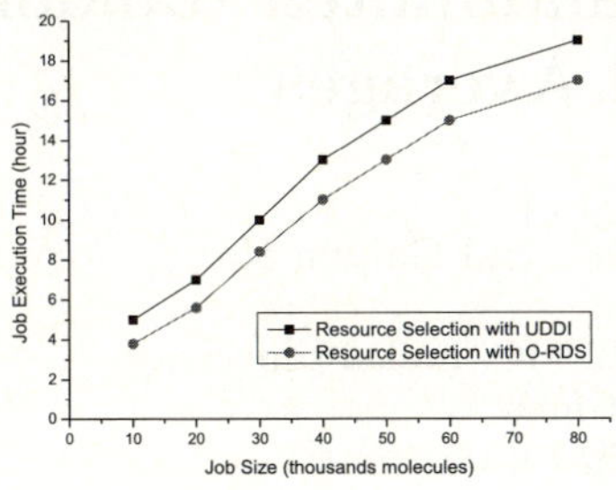

Fig. 1. The job execution time for different job sizes

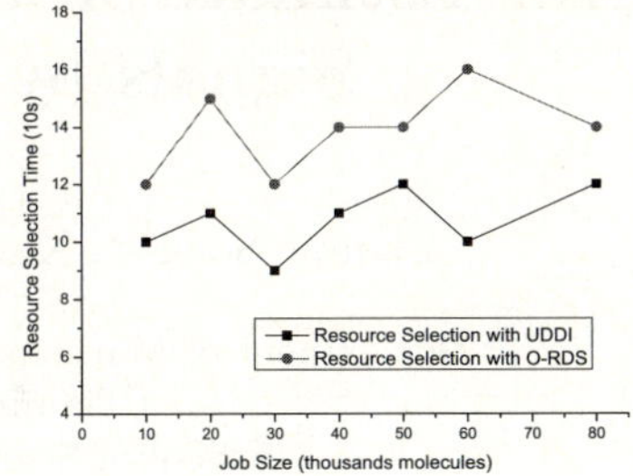

Fig. 2. The resource discovery and selection time for different job sizes

Figure 1 illustrates the job execution time for different job sizes with different resource discovery and selection algorithms. The results show that the job execution time is reduced with the ontology-based resource discovery and selection (O-RDS) algorithm. That is to say, those resources are optimal used at this scenario. Figure 2 shows that the total resource selection time increased with O-RDS algorithm. It is obviously at this dynamically resource selection environment.

5   Conclusions

We have proposed a ontology-driven resources discovery and selection framework that provides a common resources selection service for different kinds of application. This framework exploits existing ontology and semantic web and grid technologies. We have used drug discovery grid test bed to validate the design and implementation of the resource selection framework, with promising results.

References

1. Web Ontology Language (OWL). http://www.w3.org/2004/OWL/.
2. J. Kim, Y. Gil, and M. Spraragen. A Knowledge-Based Approach to Interactive Workflow Composition, Proc. ICAPS workshop Planning and Scheduling for Web and Grid Services, 2004.
3. E. Deelman et al. Mapping Abstract Complex Workflows onto Grid Environments, Journal of Grid Computing, vol. 1, 2003.
4. H. Tangmunarunkit, S. Decker, and C. Kesselman. Ontology-based Resource Matching in the Grid - The Grid Meets the SemanticWeb. In Proceedings of the 1st International Workshop on Semantics in Peer-to-Peer and Grid Computing (SemPGRID) at WWW 2003, pages 706C721, Budapest, Hungary, May 2003.

Error Estimate on Non-bandlimited Random Signals by Local Averages

Zhanjie Song[1,*], Xingwei Zhou[2], and Gaiyun He[3]

[1] Department of Mathematics and LPMC, Naikai University,
Tianjin 300071, China
School of Science, Tianjin University,
Tianjin 300072, China
songzhanjie@eyou.com, zhanjiesong@tju.edu.cn
[2] Department of Mathematics and LPMC, Naikai University,
Tianjin 300071, China
xwzhou@naikai.edu.cn
[3] School of Mechanical Engineering, Tianjin University,
Tianjin 300072, China
hegaiyun@tju.edu.cn

Abstract. We show that a non-bandlimited weak sense stationary stochastic process can be approximated by its local averages near the sampling points, and explicit error bounds are given.

It is well known that the Shannon sampling theorem plays an important role in signal processing. It states that if a function f is band-limited to $[-\Omega, \Omega]$, i.e., $f \in \mathbb{R}$ and supp $\hat{f} \subset [-\Omega, \Omega]$, where

$$\hat{f}(\omega) = \int_{-\infty}^{+\infty} f(t)e^{-it\omega}\,dt$$

is the Fourier transform of f, then f can be recovered from its sampled values at instances $k\pi/\Omega$. Specifically,

$$f(t) = \sum_{k=-\infty}^{+\infty} f\left(\frac{k\pi}{\Omega}\right) \operatorname{sinc}(\Omega t - k\pi), \tag{1}$$

where $\operatorname{sinc} t = \sin t / t$.

Since signals are often of random characters, random signals play an important role in signal processing, especially in the study of sampling theorems. But in many situations the assumption of band-limitation is not fulfilled exactly, or the correct bandwidth is unknown. For this purpose one usually uses non-bandlimited stochastic processes which are stationary in the weak sense as a model. We will give some new results on this topic in this paper.

* Supported by the Natural Science Foundation of China under Grant (60472042, 60572113) and the Liuhui Center for Applied Mathematics.

V.N. Alexandrov et al. (Eds.): ICCS 2006, Part I, LNCS 3991, pp. 822–825, 2006.

Before stating the results, let us introduce some notations. $L^p(\mathbb{R})$ is the space of all measurable functions on $\mathbb{R}$ for which $\|f\|_p < +\infty$, where

$$\|f\|_p := \left(\int_{-\infty}^{+\infty} |f(u)|^p du\right)^{1/p}, \qquad 1 \le p < \infty,$$

$$\|f\|_\infty := \operatorname*{ess\,sup}_{u \in \mathbb{R}} |f(u)|, \qquad p = \infty.$$

$B_{\Omega,p}$ is the set of all entire functions f of exponential type with type at most Ω that belong to $L^2(\mathbb{R})$ when restricted to the real line [11]. By the Paley-Wiener Theorem, a square integrable function f is band-limited to $[-\Omega, \Omega]$ if and only if $f \in B_{\Omega,2}$.

Given a probability space $(\mathcal{W}, \mathcal{A}, \mathcal{P})$ [5], a real-valued stochastic process $X(t) := X(t, \omega)$ defined on $\mathbb{R} \times \mathcal{W}$ is said to be stationary in weak sense if $E[X(t)^2] < \infty$, $\forall t \in \mathbb{R}$, and the autocorrelation function

$$R_X(t, t+\tau) := \int_{\mathcal{W}} X(t,\omega)X(t+\tau,\omega)dP(\omega)$$

is independent of $t \in \mathbb{R}$, i.e., $R_X(t, t+\tau) = R_X(\tau)$.

A weak sense stationary process $X(t)$ is said to be bandlimited to an interval $[-\Omega, \Omega]$ if R_X belongs to $B_{\Omega,p}$ for some $1 \le p \le \infty$, we note that $X(t) \in \mathcal{L}$.

Noting that any function $R_X \in B_{\Omega,p}$ is infinitely differentiable, so the process $X(t)$ belongs to Lipschitz class

$$\mathcal{Lip}_L \alpha := \{X \in \mathcal{L}; \omega(\mathcal{L}, X, \eta) \le L\eta^\alpha\} \quad (0 < \alpha \le 1),$$

where $\omega(\mathcal{L}, X, \eta) := \sup_{|h| < \eta} \|X(t+h) - X(t)\|_{\mathcal{L}}$ is the modulus of continuity in mean square, $L > 0$ is the Lipschitz constant, and $\|X(t+h) - X(t)\|_{\mathcal{L}} = \sqrt{E[(X(t+h) - X(t))^2]}$.

The convolution of two functions $f, g \in L^1$ is defined by $f * g(t) := \frac{1}{\sqrt{2\pi}} \int_{-\infty}^{\infty} f(u)g(t-u)du$. The convolution of a process $X \in \mathcal{L}$ and a function $g \in L^1$ is defined similarly by

$$X * g(t, \omega) := \frac{1}{\sqrt{2\pi}} \int_{-\infty}^{\infty} X(u, \omega)g(t-u)du.$$

It is not difficult to check that the autocorrelation function of $X * g$ is

$$R_{X*g} = R_X * g * g(\tau).$$

In 1981, Splettstösser proved the following result.

Proposition 1. ([7, Theorem 2.2]) *If the autocorrelation function of the weak sense stationary stochastic process $X(t, \omega)$ belongs to $B_{\Omega,p}$ for some $1 \le p \le 2$ and $\Omega > 0$, then*

$$\lim_{N \to \infty} E\left(\left|X(t,\omega) - \sum_{k=-N}^{N} X\left(\frac{k\pi}{\Omega}, \omega\right) \operatorname{sinc}(\Omega t - k\pi)\right|^2\right) = 0. \tag{2}$$

Proposition 2. ([7, Corollary 2.3]) *If the autocorrelation function R_X of the weak sense stationary stochastic process $X(t,\omega)$ belongs to $B_{\Omega,p}$ for some $1 \leq p \leq \infty$, where $\Omega > 0$, and satisfies*

$$|R_X(t)| = O(|t|^{-\gamma}), (|t| \to \infty) \tag{3}$$

for some $\gamma > 0$. Then the sampling expansion (2) holds.

Proposition 3. ([7, Theorem 3.1]) *If the weak sense stationary stochastic process $X(t,\omega)$ is r times differentiable (in mean square sense) for some positive integer r and with $X^{(r)} \in \mathcal{L}ip_L\alpha$ for some $\alpha \in (0,1]$, and R_X satisfies (3) for $\gamma \in (0,1]$, then for $\Omega \to \infty$*

$$\lim_{N\to\infty} E\left(\left|X(t,\omega) - \sum_{k=-N}^{N} X\left(\frac{k\pi}{\Omega},\omega\right)\operatorname{sinc}(\Omega t - k\pi)\right|^2\right) = \mathcal{O}[(\frac{\Omega}{\pi})^{-2r-2\alpha}\ln^2(\frac{\Omega}{\pi})]. \tag{4}$$

For physical reasons, e.g., the inertia of the measurement apparatus, measured sampled values obtained in practice may not be values of $f(t)$ precisely at times t_k, but only local average of $f(t)$ near t_k. Specifically, measured sampled values are

$$\langle f, u_k \rangle = \int f(t)u_k(t)dt \tag{5}$$

for some collection of averaging functions $u_k(t), k \in \mathbb{Z}$, which satisfy the following properties,

$$\operatorname{supp} u_k \subset [t_k - \frac{\sigma}{2}, t_k + \frac{\sigma}{2}], \quad u_k(t) \geq 0, \quad \text{and} \quad \int u_k(t)dt = 1. \tag{6}$$

The local averaging method in sampling was first studied by Gröchenig[4] in 1992. Butzer and Lei [2] also gave some interesting results on non-necessarily bandlimited functions in 1998. Recently Sun and Zhou [9, 10] extend some classical results on irregular sampling to local average cases. They all assume that the time intervals for averaging are symmetric. But in applications, it might not be the case. More specifically, if we want to measure the values of $f(t)$ at $k\pi/\Omega$, the measurement apparatus in fact gives a weighted average over a time interval $[k\pi/\Omega - \sigma'_k, k\pi/\Omega + \sigma''_k]$, where σ'_k, σ''_k are positive numbers. We assume that $\sigma/4 \leq \sigma'_k, \sigma''_k \leq \sigma/2$ and that the weight functions u_k are continuous, i.e,

$$\operatorname{supp} u_k \subset [t_k - \sigma'_k, t_k + \sigma''_k], \quad u_k(t) \geq 0, \quad \text{and} \quad \int u_k(t)dt = 1. \tag{7}$$

In this paper, the results on approximation of non-bandlimited weak sense stationary stochastic process by local averages near the sampling points will be given. By the property of the weak sense stationary stochastic process, the assumption (3) can be replaced by

$$R_X(t) \leq R_X(0)(1 + |t|)^{-\gamma} \quad \text{for} \quad \gamma \in (0,1]. \tag{8}$$

The following is our results, which can be proved by the Proposition of Butzer[1], Splettstösser [8], Li and Wu [5, page 291], Hausdorff-Young inequality [6, page176].

Theorem 1. *If the weak sense stationary stochastic process $X(t, \omega)$ is r times differentiable (in mean square sense) for some positive integer r, $X^{(r)} \in \mathcal{L}ip_L \alpha$ for some $\alpha \in (0, 1]$, and R_X satisfies (8). Then for $\Omega \geq \max\{\pi e^{1/(\gamma/2+r+\alpha)}, 30\pi\}$, $\delta \leq 1/\Omega$ we have*

$$\lim_{N \to \infty} E\left[\left|X(t, \omega) - \sum_{k=-N}^{N} \int_{k\pi/\Omega - \sigma'_k}^{k\pi/\Omega + \sigma''_k} u_k(t) X(t, \omega) dt \cdot \mathrm{sinc}(\Omega t - k\pi)\right|^2\right]$$

$$\leq \left(270.16 L^2 3^{2\gamma} \left(\frac{2}{\pi}\right)^{2r+2\alpha} + 153.44 R_X(0)\right) \left(1 + \frac{2(r+\alpha)}{\gamma}\right)^2 \cdot$$

$$(\frac{\Omega}{\pi})^{-2r-2\alpha} \ln^2(\frac{\Omega}{\pi}). \tag{9}$$

where $\{u_k(t)\}$ is a sequence of continuous weight functions defined by (7).

References

1. Butzer, P.L., Splettstösser, W. and Stens R. L., The sampling theorem and linear prediction in signal analysis, Jber. d. Dt. Math.-Verein., **90** (1988) 1-70.
2. Butzer, P.L., Lei, J., Errors in truncated sampling series with measured sampled values for non-necessarily bandlimited functions, Funct.Approx., **26** (1988) 25-39.
3. Ditzian, Z. and Totik, V., Moduli of smoothness, Springer-Verlag, 1987.
4. Gröchenig, K., Reconstruction algorithms in irregular sampling, Math. Comput., **59** (1992) 181-194.
5. Li, Z. and Wu, R., A course of studies on stochastic processes, High Education Press, 1987(in chinese).
6. Pinsky, M. A., Introduction to Fourier analysis and wavelets, Wadsworth Group. Brooks/Cole. 2002.
7. Splettstösser, W., sampling series approximation of continuous weak sense stationary processes, Information and Control **50** (1981) 228-241.
8. Splettstösser, W., Stens, R. L. and Wilmes, G., on the approximation of the interpolating series of G. Valiron , Funct. Approx. Comment. Math. **11** (1981) 39-56.
9. Sun, W. and Zhou, X., Reconstruction of bandlimited functions from local averages, Constr. Approx., **18** (2002) 205-222.
10. Sun, W. and Zhou, X., Reconstruction of bandlimited signals from local averages, IEEE Trans. Inform. Theory, **48** (2002) 2955-2963.
11. Zayed,A.I. and Butzer,P.L., Lagrange interpolation and sampling theorems, in "Nonuniform Sampling, Theory and Practice", Marvasti,F., Ed., Kluwer Academic, 2001, pp. 123–168.

A Fast Pseudo Stochastic Sequence Quantification Algorithm Based on Chebyshev Map and Its Application in Data Encryption

Chong Fu[1], Pei-rong Wang[1], Xi-min Ma[2], Zhe Xu[1], and Wei-yong Zhu[1]

[1] School of Information Science and Engineering, Northeastern University,
Shenyang 110004, China
fu_chong@sohu.com
[2] Computer Department, Liaoning Economic Vocational Technological Institute,
Shenyang 110036, China

Abstract. Chaos theory has been widely used in cryptography fields in recent years and the performance of the pseudo stochastic sequence quantified from chaos map has great influence on the efficiency and security of an encryption system. In this paper, an improved stochastic middle multi-bits quantification algorithm based on Chebyshev map is proposed to enhance the ability of anti reconstruction a chaos system through reverse iteration and improve the performance of the generated sequence under precision restricted condition. The balance and correlation properties of the generated sequence are analyzed. The sequence is proved to be a binary Bernoulli sequence and the distribution of the differences between the amounts of 0 and 1 is analyzed. The side lobes of auto correlation and values of cross correlation are proved to obey normal distribution $N\,(0,\,1/N)$.

1 Introduction

Chaos theory has been widely used in encryption fields nowadays due to its initial value sensitive, nonperiodic, unpredictable, and Gauss like statistical characteristics. The raw generated analog chaotic sequence could not be directly used in digital system and the binary quantification process has a great influence on system security and performance. The classic quantification method mainly has the following two disadvantages: (1) Only one bit can be generated per iteration, the computation load is very heavy. For example, to encrypt a plaintext of 1MB size, we must iterate at least $8*2^{20}$ times; (2) With the increasing of iteration, the generated raw sequence would be periodic because no processor is precision unrestricted. This will cause the quantified sequence also being periodic and making the auto and cross correlation performance worse [1]. Some scholars proposed a multi-bits quantification algorithm to reduce the computation load and extend the period of the sequence by transforming the analog value of each iteration to its binary form directly, but the security of this algorithm is relatively weak, the analog sequence can be recovered from the quantified binary sequence and the chaotic system can be reconstructed through reverse iteration [2]. In this paper, an improved stochastic middle multi-bits quantification algorithm based on Chebyshev map is proposed to strengthen the ability of anti reverse iteration and improve the performance of the generated sequence under precision restricted condition.

V.N. Alexandrov et al. (Eds.): ICCS 2006, Part I, LNCS 3991, pp. 826–829, 2006.

2 The Statistical Properties of Chebyshev Map

Chebyshev map is defined as:

$$x_{n+1} = \cos(k \cos^{-1}(x_n)), \quad x_n \in [-1,1].$$ (1)

The probability density of chaotic sequence generated by Eq.1 is [3]:

$$\rho(x) = \begin{cases} \dfrac{1}{\pi\sqrt{1-x^2}} & -1 \le x \le 1; \\ 0 & otherwise. \end{cases}$$ (2)

Let $\{x_i\}$ be chaotic sequence generated by Eq.1, we can get the following three properties based on Eq.2.

Property 1. The mean of $\{x_i\}$ is:

$$\bar{x} = \lim_{N\to\infty} \frac{1}{N} \sum_{i=0}^{N-1} x_i = \int_{-1}^{1} x\rho(x)dx = 0.$$ (3)

Property 2. The normalized auto correlation function of $\{x_i\}$ is:

$$AC(m) = \lim_{N\to\infty} \frac{1}{N} \sum_{i=0}^{N-1} (x_i - \bar{x})(x_{i+m} - \bar{x}) = \int_{-1}^{1} xf^m(x)\rho(x)dx - \bar{x}^2 = \begin{cases} 0.5 & m=0; \\ 0 & m \ne 0. \end{cases}$$ (4)

Property 3. For any two different initial values x_{01} and x_{02}, the normalized cross correlation function of the two generated sequences is:

$$CC_{12}(m) = \lim_{N\to\infty} \frac{1}{N} \sum_{i=0}^{N-1} (x_{i1} - \bar{x})(x_{(i+m)2} - \bar{x})$$

$$= \int_{-1}^{1} \int_{-1}^{1} x_1 f^m(x_2)\rho(x_1)\rho(x_2)dx_1 dx_2 - \bar{x}^2 = 0.$$ (5)

Property 2 and 3 indicates that Chebyshev map is excellent to be used to generate pseudo stochastic sequence.

3 Stochastic Middle Multi-bits Quantification Algorithm and Its Performance Analysis

The stochastic middle multi-bits quantification algorithm is as follows:

① Let $k = 4$, selecting a proper initial value x_0, getting the real value x_i from x_{i-1} by iterating Eq.1, $x_i \in [-1,1]$;

② Let $y_i = (x_i + 1)$, so $y_i \in [0,1]$;

③ Transform y_i to binary form: $(0.a_1a_2a_3...a_n)_2$, $a_i=0$ or 1, getting the multi-bits $a_1a_2a_3...a_n$, n is equal or less than the precision that the processor can provide;

④ Let $y_i' \in [0,1]$, whose decimal part is the reverse order of that of y_i, as another initial value, calculating the real value y_i' from y_{i-1}' by iterating Eq.1, $y_i' \in [0,1]$;

⑤　Let the number of bits we want to get by each iteration be l ($l < n$), $m = \lceil (y_i') \times (n - l) \rceil$, m is a stochastic integer between 0 and ($n - l$) thus;

⑥　Skip first m bits of $a_1 a_2 a_3 \ldots a_n$, getting the middle multi-bits $a_{m+1} a_{m+2} \ldots a_{m+l}$.

The rest may be deduced by analogy, concatenating all the l bits generated by each iterating together, thus getting the pseudo stochastic binary sequence with the length we needed.

Because the position m is determined by chaotic map and also stochastic, the algorithm not only has the advantages such as reducing the computation load by generating l bits per iteration and extending the period of the sequence under precision restricted condition that the original multi-bits algorithm has, but also further strengthen the ability of anti reconstruction of the chaos system. The balance and correlation performance are the two most import properties of stochastic sequence, which greatly affects the security and key space of an encryption system, and will be analyzed in the following.

Lemma 1. The pseudo stochastic sequence generated by stochastic middle multi-bits algorithm is a binary Bernoulli distributed sequence. For sequence with length N, the mean of the differences is 0 and the variance is N.

Lemma 2. For any two different sequence $\{x_{i1}\}$ and $\{x_{i2}\}$ that generated by stochastic middle multi-bits algorithm, they satisfy: When N is large enough and m is relatively small, the side lobes of auto correlation and values of cross correlation obey normal distribution $N (0, 1/N)$.

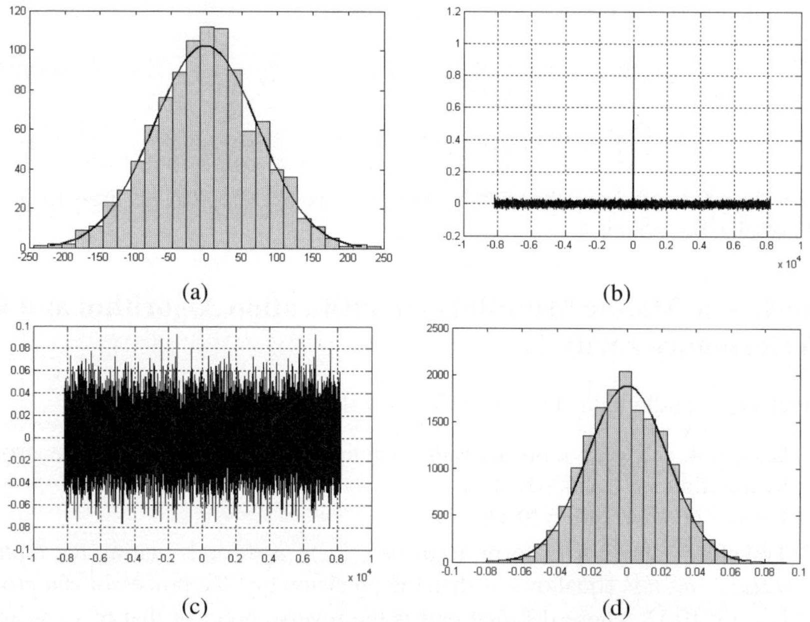

Fig. 1. The balance and auto/cross correlation performance of generated sequence. (a) balance performance, (b) auto correlation function, (c) cross correlation function, (d) distribution of cross correlation values.

The simulation result of balance property is shown in Fig. 1(a). 1000 sequences with length N=8192 are generated and the initial values are selected independently. The mean of differences is -0.9480 and the standard deviation is 92.3074, which are in good accordance with theoretical value 0 and 90.5097. The simulation results of the auto and cross correlation functions and the distribution of cross correlation values are shown in Fig. 1(b) to (d). The sequence length is 8192 and the initial values are selected as 0.60000 and 0.60001 for verify the initial value sensitive property of a chaotic system. The maximum absolute value of auto correlation side lobe is 0.0410 and the value of cross correlation is 0.0905, which are in good accordance with theoretical value 0. Above analysis indicate that the quantified sequences have excellent balance and auto/cross correlation performance.

4   Experimental Results and Conclusions

The comparison of different quantification algorithms is shown in Table 1. The plaintext to be encrypted is 1MB size and parameter l is set to 8. From Table 1 we can see, the efficiency and the sequence performance of the stochastic middle multi-bits quantification algorithm is far better than that of classic binary quantification algorithm. While compared with the multi-bits algorithm, although the encryption time is longer due to it's more complicated algorithm, but the security is high due to its powerful anti reconstruction ability. The algorithm proposed in this paper can be widely used in signal generators for generating high quality PN sequence [4], which can also be used in communication channel simulation, spread spectrum communication, etc.

Table 1. The performance comparison of sequences generated by different quantification algorithms

Algorithms	$\sqrt{\dfrac{1}{N}\sum_{m=1}^{N}[AC(m)]^2}$	$\sqrt{\dfrac{1}{N}\sum_{m=1}^{N}[CC_{12}(m)]^2}$	Encryption time (S)	Security
Binary	0.0803	0.0247	58.76	High
Multi-bits	0.0781	0.0226	31.72	Low
Stochastic multi	0.0783	0.0223	38.45	High

References

1. Heidari-Bateni, G., McGillem, C. D.: A Chaotic Direct Sequence Spread Spectrum Communication System. IEEE Transaction on Communication, Vol. 42 (1994), 1524-1527
2. Baptista, M. S.: Cryptography with Chaos. Physics Letters A, Vol. 240 (1998), 50-54
3. Geisel, T., Fairen, V.: Statistical Properties of Chaos in Chebyshev Maps. Physics Letters A, Vol. 105 (1984), 263-266
4. Lipton, J. M., Dabke, K. P.: Spread Spectrum Communications Based on Chaotic Systems. International J. Bifurcation and Chaos, Vol. 6 (1996), 2361-2374

Supporting Interactive Computational Science Applications Within the JGrid Infrastructure

Szabolcs Pota and Zoltan Juhasz

University of Veszprem, Department of Information Systems,
Egyetem u. 10., 8200, Veszprem, Hungary
{pota, juhasz}@irt.vein.hu

Abstract. Future computational grid systems must support interactive and collaborative applications to help computational scientists in using the grid effectively. This paper shortly introduces how the JGrid system, a service-oriented dynamic grid framework, supports interactive grid applications. Via providing a high level service view of grid resources JGrid enables easy access to grid services and provides an effective programming model that allows developers to create dynamic grid applications effortlessly.

1 Introduction

Due to rapid advances in grid technology and the increasing number of production grid environments, scientists now can locate suitable remote computing resources without much difficulty, but programs are still executed on the selected resource by traditional batch runtime systems. Besides depending on platform-specific details, this grid execution model does not utilize the full potential of the grid. It is still a difficult problem to connect several resources (belonging to different administrative domains) to be used at the same time for solving one particular problem. Furthermore, grid environments based on batch systems provide limited support for dynamic resource discovery and runtime exception handling.

There are grid usage scenarios where static batch execution is not adequate. Interactive applications, ranging from scientific visualization, graphics rendering through computational steering and man-in-the-loop simulations to collaborative problem solving, could also benefit to a large extent from Grid technology. Interactive computational grids should support a wide range of applications, provide effective ways for service/resource orchestration, and should integrate into the everyday user work environment.

In this paper, we describe the JGrid [1] Java/Jini based service-oriented grid framework that provides a unified, high-level service view of computational resources, and besides traditional batch execution supports interactive/collaborative grid applications.

2 The JGrid System

The goal of the JGrid project is to develop a novel service-oriented grid system that aims to create a dynamic computational service fabric in which batch and interactive

V.N. Alexandrov et al. (Eds.): ICCS 2006, Part I, LNCS 3991, pp. 830–833, 2006.

applications can discover and use computing resources on-demand, and provides a high-level, effective service-oriented programming model for developers.

The JGrid system consists of a set of extensible services that provide a complete, dynamic grid infrastructure including wide-area service discovery, security support, and core computational services (batch, compute and storage services). Services can be easily shared among multiple clients and accessed concurrently. In addition, JGrid provides uniform, seamless access to diverse grid resources.

2.1 Grid Access

Most grid systems use the web browser as their primary user interface for accessing resources, typically via grid portals. Although web technology has developed significantly, web user interfaces cannot easily provide the rich functionality of desktop environments. This is especially problematic in interactive and collaborative applications.

In JGrid services are represented by Java interfaces and accessed via service proxies implementing these interfaces. As a result, services can be accessed programmatically in a similarly seamless, protocol independent way, by simply calling methods on the proxy object. A graphical, dynamically downloadable Service UI can be also attached to these proxy objects if the service is to use by end uses. A Java service browser is also provided in JGrid for end users to access grid services and that provides rich desktop user experience and a single access point to the grid.

2.2 Key Computational Services

As the main intended use of JGrid is to solve large computational problems, we have placed great emphasis on developing computational services that support the widest range of sequential and parallel grid applications. To achieve this, we decided to create two different types of computational services. The *Batch* and *Compute Services* complement each other in providing JGrid users with a range of choices in programming languages, execution modes and inter-process communication modes.

The Batch Service [2] supports legacy applications and traditional batch execution by integrating batch runtime systems (e.g. Sun Grid Engine or Condor) in JGrid using a service-oriented interface. The integration is achieved via the standard DRMAA [3] interface, this way any batch execution environment that supports DRMAA can be integrated after re-configuration of the Batch Service.

The Compute Service supports the execution of sequential and parallel grid applications. It allows clients to execute Java programs using virtualized remote resources, which can represent single, multi-processor computers or clusters. The service, in fact, is a special Java runtime system that enables clients to execute programs in a secure and controlled way, and acts as core building block in dynamic grids that can execute interactive grid applications.

Interactivity is supported with a number of mechanisms built in the Compute Service. Compute Services use (*i*) preemptive scheduling to let submitted programs start immediately and remain responsive throughout the entire execution period. Furthermore, users can (*ii*) control (suspend, resume, cancel) program or task execution via dynamically generated proxies that connects to the remote task. The (*iii*) monitoring of applications running inside a Compute Service is also supported. Finally, users can (*iv*) spawn processes inside the remote Compute Service that creates dynamic server objects accessed via remote method invocation.

2.3 Remote Data Access

The JGrid *Storage Service* provides access to remote data in a way identical to the Java I/O model, and represents a remote file space in a platform independent manner. It enables clients to access and manipulate their own directories and files. By using proxies for file access, the protocol used to access remote files is completely hidden from the user and is interchangeable. Sharing the appropriate remote file proxies among distributed application components ensures that everybody can access the same piece of data, i.e. sharing the same data.

3 Usage Scenarios

This section shortly introduces some usage scenarios to illustrate how JGrid can be used in various computational application domains.

3.1 Long Running Numerical Applications

We used JGrid to execute a biological application performing pair-wise alignment of biological sequences based on the BioJava library [4]. The application demonstrated the reliability of JGrid in executing long running numerical applications and also showed the appropriate computational performance of a Java-based computational environment. The program was executed on one of the available Compute Services of our research group. Running tasks could access the Storage Service storing the input sequence file and write the results via its proxy as it were a local file system.

3.2 Parallel Grid Applications

We used a simple SPMD-style parallel image processing problem to illustrate parallel program execution support in JGrid. The example performs simple image processing operations, e.g. edge detection, etc. in parallel, during which nearest-neighbour communication steps are required.

Tasks of the program are spawned at runtime on dynamically discovered Compute Services. Next, the client creates the required topology by exchanging task proxies between neighbouring tasks. Once the configuration is set, the client distributes the image segments, starts the computation and displays the returned results. A successful live demonstration was performed using a client in Hong Kong and Compute Services in Hungary.

3.3 Collaborative Applications

A number of case studies were developed to demonstrate more complex ways of interactions during collaborative work. One grid application is a modified version of the C3D collaborative and distributed ray tracing application [5] developed for benchmarking Java RMI and serialization. We modified this application to be able to run on

JGrid and added the flexibility of using dynamically discovered, possibly geographically distributed services.

Another collaborative application demonstrates the use of 3D visualisation in JGrid. Using a shared grid service storing a Box World data model, multiple clients can connect to the service and view and modify the model from different locations.

4 Summary

This paper introduced the JGrid environment, a service-oriented grid system developed to run computational grid applications and some usage scenarios supporting interactive applications. In contrast to most grid environments, JGrid was designed to create dynamic, interactive and collaborative applications. Its service-oriented nature provides users with a unified view of grid resources and simplifies the addition of future services with new functionality. We are continuing work on JGrid developing full-scale, real-world computational science grid applications.

References

1. JGrid: A Jini-based Universal Service Grid, http://pds.irt.vein.hu/jgrid
2. Pota, S., Sipos, G., Juhasz, Z., Kacsuk, P.: Parallel Program Execution in the JGrid System, In Proc. 5th Austrian-Hungarian Workshop on Distributed and Parallel Systems, Springer, Kluwer International Series in Engineering and Computer Science, Vol. 777, Budapest, Hungary (2004) 13-20
3. Distributed Resource Management Application API Working Group (DRMAA-WG), http://www.drmaa.org/
4. Pocock, M., Down, T., Hubbard, T.: BioJava: open source components for bioinformatics, ACM SIGBIO Newsletter, Vol. 20, No. 2. (2000) 10-128.
5. C3D, A distributed raytracer for benchmarking Java RMI and Serialization, http://www-sop.inria.fr/sloop/C3D

Application of Virtual Ant Algorithms in the Optimization of CFRP Shear Strengthened Precracked Structures

Xin-She Yang, Janet M. Lees, and Chris T. Morley

Department of Engineering, University of Cambridge
Trumpington Street, Cambridge CB2 1PZ, UK
xy227@eng.cam.ac.uk

Abstract. Many engineering applications often involve the minimization of objective functions. The optimization becomes very difficult when the objective functions are either unknown or do not have an explicit form. This is certainly the case in the strengthening of existing precracked reinforced concrete structures using external carbon fibre reinforced polymer (CFRP) reinforcement. For a given concrete structure, the identification of the optimum strengthening system is very important and difficult, and depends on many parameters including the extent and distribution of existing cracks, loading capacity, materials and environment. The choice of these parameters essentially forms a coupled problem of finite element analysis and parameter optimization with the aim of increasing the serviceability of the structure concerned. In this paper, virtual ant algorithms combined with nonlinear FE analysis are used in the optimization of the strengthening parameters. Simulations show that the location and orientation of the CFRP reinforcement has a significant influence on the behaviour of the strengthened structure. The orientation of the reinforcement with a fixed location becomes optimal if the reinforcing material is placed perpendicular to the existing crack direction. The implication for strengthening will also be presented.

1 Introduction

Nature inspired algorithms based on swarm intelligence and the self-organized behaviour of social insects can now be used to solve many complex problems such as the travelling salesman problem and the rerouting of traffic in a busy telecom network [5]. New algorithms are often developed in the form of a hybrid combination of biology-derived algorithms and conventional methods, and this is especially true in many engineering applications. On the other hand, the development of optimum strengthening and repair strategies for existing reinforced concrete structures is a challenging task. For a given structure, the most appropriate solution will depend on many factors such as the extent and distribution of any existing cracks, the loading capacity, the type of structure, available materials and environmental considerations [2, 3, 4]. The identification of the best strengthening method is essentially a coupled optimization problem consisting

V.N. Alexandrov et al. (Eds.): ICCS 2006, Part I, LNCS 3991, pp. 834–837, 2006.

of finite element analysis and parameter optimization with the aim of increasing the serviceability of an existing structure. The optimization of parameters such as the area, spacing, location and orientation of additional CFRP reinforcing elements to strengthen a structure can be obtained by searching the parameter space and results from finite element (FE) analyses for a given set of these parameters. This paper aims to develop a simple optimization procedure to simulate different strengthening strategies for rectangular beams with dapped ends.

2   Virtual Ant Algorithms

Many problems in engineering and other disciplines involve optimizations that depend on a number of parameters, and the choice of these parameters affects the performance or objectives of the system concerned. The optimization target is often measured in terms of objective or fitness functions in qualitative models. The Virtual Ant Algorithm (VAA) starts with a troop of virtual ants, each ant randomly wonders in the phase space and in most cases; the phase space can be simply a 2-D or 3-D space. The main steps of the virtual ant algorithm are: 1) Create a initial population of virtual ants, and encode the function into virtual food; 2) Define the criterion for marking food/route with pheromone; 3) Evolution of virtual ants with time by random walking and broadcasting the best local to others if a better food location is found; 4) Evaluate the encoded concentration/locations of ants; 5) Decode the results to obtain the solution. Furthermore, in order to avoid the trapping at local maxima, a probability of $p = 0.01 \sim 0.05$ is used to perturb the position and directions of the ants. There is a tradeoff between the computational efficiency in parameter searching and the computing time of finite element analysis. For the simulations used in this paper, the number of ants is taken to be in the range of 20-60.

The optimization problem in this paper forms a coupled problem including parameter-searching and the nonlinear finite element analysis. The virtual food in our optimization is the loading capacity. For each ant and each parameter set, a nonlinear finite element analysis is carried out in the sequential manner. The nonlinear finite element model used for our simulations is the smeared crack model developed by de Borst and others [1]. The difficulty is that the explicit form of the optimization function or load capacity is not known. Standard optimization approaches usually require a known optimization function before any optimization algorithms can be applied. However, we will show that virtual ant algorithms are also applicable to the functions that are not explicitly known.

3   Simulations and Results

3.1   Strengthening of Beams with Dapped Ends

The strengthening of reinforced concrete beams with dapped ends has been studied experimentally by Taher using various types of reinforcement [6]. Consider a beam of $2200 \times 300 \times 150$ mm with dapped ends (150 mm $\times$ 150 mm), where the reinforcement bars are placed at 100 mm from the top and 50 mm from the bottom surface of the beam. Both reinforcement layers have a total area of 100 mm^2. The beam is loaded under the four-point bending conditions. The values used in the simulations for the concrete are, a Young's Modulus of elasticity E_{conc}=21000 N/mm^2, a Poisson's ratio $\nu = 0.15$, a compressive cube strength f_{cu}=53 N/mm^2, and a tensile strength f_t=6 N/mm^2. The additional CFRP strengthening reinforcement is considered as brittle elastic, but it is always within the elastic range for all the simulations. Figure 1 shows the crack stress at a load of $f = 10$ kN. We can see the stress concentration at the corners at both ends of the beam. This will lead to crack formation at these corners. If we strengthen this system with a CFRP reinforcing strap anchored near the

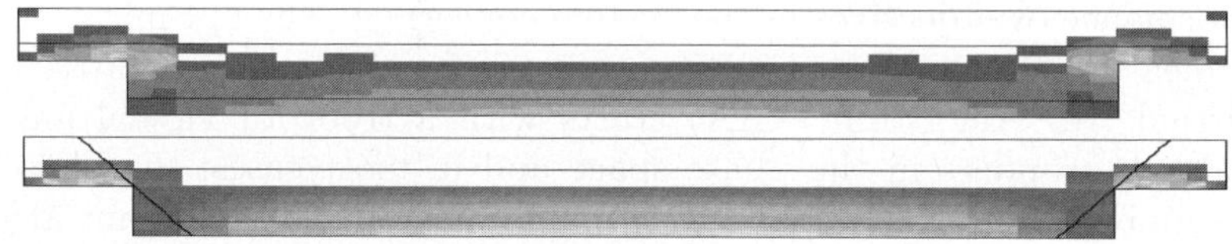

Fig. 1. Crack strain and effect of strengthening in a beam with dapped ends

top and bottom surfaces, we can study the effect of the strengthening system. The CFRP reinforcement is anchored at both ends but it is unbonded and not prestressed. The simulation of the strengthened beam gives the distribution of crack strains and crack patterns shown in the lower figure in Figure 1. We can see that the shear cracks are less extensive and have shifted away from the corners.

3.2   Optimization of Strengthening Parameters

The optimization of the strengthening parameters is carried out in the following manner. First, we use the same beam as shown in Figure 2 and vary the location (x) of the midpoint and orientation (θ) of the CFRP reinforcement, then we run the FE analysis to obtain the final load. The load is then normalised using the failure load for $x/d = 1$ and $\theta = 0$. Figure 2 shows the load variation versus the orientation and location where the dots show the updating process of the parameter set $(x/d, \theta)$ and the fitness or load function which is the normalized final load where $d = 300$ is the depth of the beam. From these figures, we can see that the final load reaches the maximum after about 60 parameter combinations, and this corresponds to the optimal angle of $\theta = 45°$. For a domain of 100 × 100 grids, the virtual ant algorithms can reach the optimal setting (after only about 60 combinations) far more quickly than a regular searching algorithm where it would be necessary to consider 10000 possible combinations. The search efficiency is therefore increased by more than two orders or 125 times. Further work will extend the current work to the strengthening and repair of various structures under various time-dependent loading conditions and history.

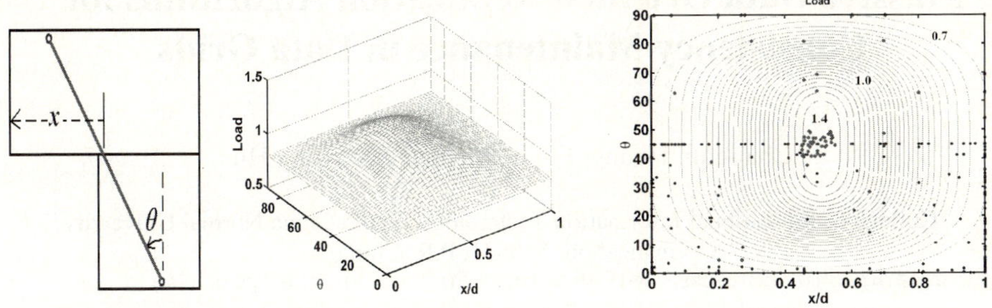

Fig. 2. Evolution of load with the location and orientation

4   Conclusions

By simulating the swarm interactions of social ants, we used the virtual ant algorithm to solve the function optimizations. An optimization problem for strengthening concrete structures is then formulated by coupling parameter searching and nonlinear finite element analysis. By using efficient parameter searching to optimize the search for the best strengthening parameters and combining it with FE simulations, we have reached the optimal load without knowing the shape of the fitness function. Simulations show that both the location and orientation of additional CFRP strengthening reinforcement elements are important. In the case of a beam with dapped ends, the optimal strengthening parameters are $x = 0.5d$ and $\theta = 45°$. As expected, this result suggests that the best option is to strengthen in the direction that is perpendicular to the direction of major cracks. Further research will consider the optimum strengthening strategy for structures with a complex geometry and time-dependent loading history.

Acknowledgement. The authors are grateful to EPSRC for their financial support (GR/S55101/01).

References

1. de Borst R.: Smeared cracking, plastity, creep and thermal loading - a unified approach, *Comp. Meth. Appl. Mech. Eng.*, **62** (1987) 89-110.
2. Lees J. M., Winistoerfer A. U. and Meier, U.: External prestressed CFRP straps for the shear enhancement of concrete, ASCE J Composites for Construction, **6** (2002) 249-256.
3. Hoult N. A. and Lees J. M.: Shear retrofitting of reinfoced concrete beams using CFRP straps, Proc. Advanced Composite Materials in Bridges and Structures, 20-23 July, Calgary, (2004).
4. Lees, J.M., Morley, C.T., Yang, X.S., Hasan Dirar, S.M.O.: Fibre reinforced polymer strengthening of pre-cracked concrete structures, Concrete, **39** (2005) 36-37
5. Solnon C.: Ants can solve constraint satisfaction problems. IEEE Trans. Evolutionary Computing, **6** (2002) 347-357.
6. Taher, S. E. M.: Strengthening of critically designed girders with dapped ends, Structures and Buildings, **158** (2005) 141-152.

Massive Data Oriented Replication Algorithms for Consistency Maintenance in Data Grids

Changqin Huang, Fuyin Xu, and Xiaoyong Hu

College of Educational Information Technology, South China Normal University,
Guangzhou, 510631, P.R. China
cqhuang@zju.edu.cn, xufy@scnu.edu.cn, huxiaoy@hotmail.com

Abstract. Based on the Grid Community (GC) for data usage, differentiated replication algorithms are proposed, by which the replica node is selected according to key replica node (*SHN*) and the other. First, The replicas are passed into *SHN*; second, if certain nodes in a GC often access the data resource, it or its nearby nodes become an optional replica node based on the storage and bandwidth. A consistency maintenance algorithm is presented to facilitate the coherence function in a differentiated manner: a pessimistic manner or an optimistic manner, and to update according to the context. The simulation results show that the differentiated mechanism can improve grid access performance.

1 Introduction

In Data Grids, the management of massive data is one of the major scientific challenges. Data replication is an important enabling technology as grid services. Replication improves availability and performance, and increase throughput. However, Data replication also brings a few hard-solved issues, such as consistency maintenance. To do well the trade-offs among performance and consistency, we present a differentiated replication approach including Optimistic replication [1].

2 The Differentiated Replication Algorithms

Because any large amount of data is produced by a single resource, we think that there is only a steady original data, named home data that resides in a certain Home Node (*HN*). Replicas exchange updates in a peer-to-peer fashion. The replicas have loosely synchronized clocks. Generally, the whole grid system consists of few Grid Communities (GC), which are perhaps a WAN or a group of nearby LANs. The GC is an important network entity, in which there exists a key node for performance as to replication, and its network connecting with the *HN* is better. The key node should store a Secondary Home Replica (*SHR*), and the node is called *SHN*, Except the *SHN*, The other nodes where the common replicas exist are not special to impact on each other's performance. The replica distribution is structured as Figure 1.

Due to the topology fact, our algorithms are to put replicas in a differentiated manner. First, pass replicas into key nodes, *SHN*. Second, if a certain node in a GC often accesses home replica or replica and its storage (or its nearby node) permission, a replica is stored into the node (or the nearby node). We assume the bandwidth of each

V.N. Alexandrov et al. (Eds.): ICCS 2006, Part I, LNCS 3991, pp. 838–841, 2006.

node is available in advance, and each node knows its nearby nodes. The *AN* denotes the number of access to home replica or replica. *T* indicates a threshold value of *AN*. If the *AN* exceeds *T*, the system should put a replica locally to improve performance. At present, the *SHN* and *T* are given in advance. *Re.NodeID* denotes the node ID, which produces a request for replica. The algorithm is described in Figure 2.

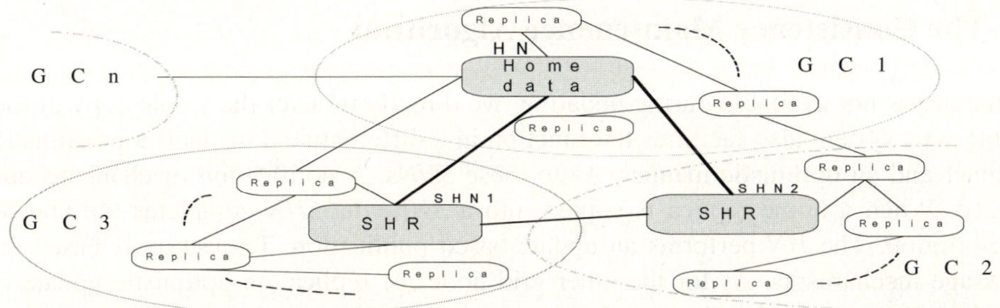

Fig. 1. The topology of replica distribution under the differentiated replication algorithms

---

```
PutReplicas (R)
//For home replica
    While find out a SHN for each grid community on the basis of statistic analyses do
            Pass a replica into the SHN  // From the HN
            Add an update and record the HN'IP and time into its Log-Replica
    For each Grid Community
        If a node's AN>the threshold value T, it is not the SHN
            If its storage capacity is adequate then
                    Pass a replica into the node  // The source is decided by the next Algorithm
                    Add an update and record the source node'IP and time into its Log-Replica
            Else find out all its nearby nodes that are running at present.
                    Sort these nodes in the bandwidth descending order and form a temp queue QUN
                    While take out a node from the QUN according to the existing order   do
                        If its storage capacity is adequate   then
                            Pass a replica into the node
                            // The source is decided by the FindHomeReplica Algorithm
                            Add an update and record the source node'IP
                            and time into its Log-Replica
                            Exit
                    Else Next
END   // Return Log-Replica
```

Fig. 2. The differentiated replication algorithm

```
FindHomeReplicas(Re)
//For each request for replica source
    If Re.NodeID is the SHN in the associated grid community then
            Return SHN (The home data is stored here)
    Else find out all nodes near Re.NodeID if these nodes are running at present.
        Sort these nodes in the bandwidth descending order and form a temp queue QUN
            While take out a node TaN from the QUN according to the existing order  do
                If a certain Log-Replica exists in the TaN then
                    Return TaN
                    Exit
                Else Next
    End  // Return TaN
```

Fig. 3. The find home replicas algorithm of the differentiated replication

The system introduces a XML log file for replica, called *Log-Replica*, which is a key basis of replica updating and update tracing. Its existence indicates the replica existence in a hosting node. Because the Log-Replica is XML format, it is common to be applied in many aspects. And X-Diff [2] is applied to produce an analysis data, and to give the information of newly updates.

3   The Consistency Maintenance Algorithm

If the diff is not available during updating, we directly transfer the whole copy of the home. The system also facilities the function in a differentiated manner: a pessimistic manner and an optimistic manner. As to these *SHN*s, A notification mechanisms enforced. When a home replica is copied into a *SHN*, the *SHN* completes the update subscription. The *HN* performs an update-based publication. The action is based on message mechanisms. As for the other grid nodes of replica, an optimistic update is invoked: the system visit the metadata to get the "time" of the home and that replica, then compare these two values to decide to take relevant actions.

MaintainConsistency(Access)
// For each update of home data
   If a home data in *HN* updates **then**
      Send an update message to all relevant *SHN*s
      For each relevant *SHN*   // update its replica (second home replica)
         If the diff is available **then**
            Copy the diff to the replica node and the replica is updated
         Else copy the whole data of the home to the replica node and the replica is updated
         Add an update and record the *HN*'IP and time into its *Log-Replica*
// For each access to home data by a replica node *ReN*
    If *ReN.Log-Replica*.lastUpdate.time=*SHN.Log-Replica*. lastUpdate.time **then**
      Directly access to the replica
      Else If *SHN.Log-Replica*.lastUpdate.time is greater than *ReN.Log-Replica*.lastUpdate.time **then**
         If the diff is available **then**
            Copy the diff to the replica and the replica is updated
         Else copy the whole data of the home to the replica and the replica is updated
         Add an update and record the source node'IP and time into its *Log-Replica*
         Directly access to the replica

Fig. 4. The consistency maintenance algorithm

4   Simulations

We adopt OptorSim [3] to simulate these applications. In our simulation, parameters are set as follows: The size of each data object is 0.5 GB and each update is 50MB. The time required to send a message from one replica to another is assumed to be negligible compared to the time between anti-entropy sessions. Experiments are described as follows: **Case 1:** The grid nodes of replica are identical to make replica and consistency maintenance; **Case 2:** Making replica and consistency maintenance applies our algorithms.

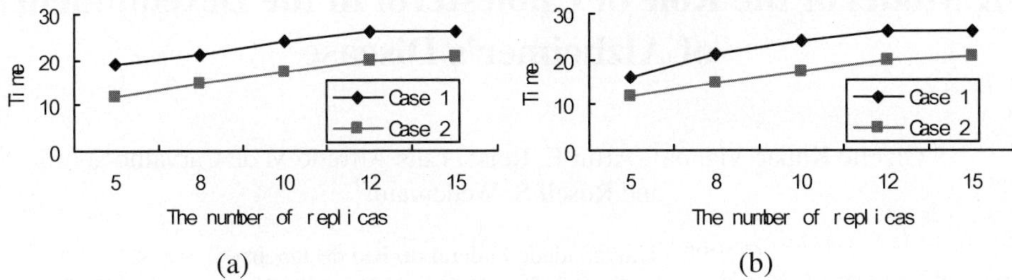

Fig. 5. (a) The time that an update has been received by all replicas for varying numbers of replicas. (b) The latency required from the time when an update is issued to the time when acknowledgments are received by each replica.

As shown in these figures, whether the first concerning time or the second, different manners for replication and updates have different performance scenarios. The Case 2 brings a better show than the Case 1, i.e. it improves data access performance that making replica and consistency maintenance applies our differentiated algorithms.

5   Conclusions

We present a suit of replication and consistency maintenance algorithms to meet the data grid requirements. A differentiated replication algorithm is proposed, by which the replica node is selected according two types: key replica node (*SHN*) and the other replica nodes. Firstly, The replicas are passed into *SHN*; Secondly, if a certain node in a GC often accesses the data resource, it or its nearby nodes become an optional replica node based on bandwidth etc. Lastly, we apply a consistency maintenance algorithm to facilitate coherence function in a two manner: a pessimistic manner, an optimistic manner. The simulation results suggest that replication algorithms are fit for consistency maintenance, and can improve grid performance of data access.

Acknowledgement. The authors wish to thank the Scientific Research Fund of Hunan Provincial Education Department(Grant No. 04A037), by which the research project has been supported.

References

1. T. W. Page, Jr., R. G. Guy, J. S. Heidemann, et al: Perspectives on optimistically replicated, peer-to-peer filing. Software --Practice and Experience, vol. 28(2): 155-180, 1998.
2. Y. Wang, D. J. DeWitt, J.Y. Cai: X-Diff: An Effective Change Detection Algorithm for XML Documents, Proc. of the 19th International Conference on Data Engineering (ICDE'03), 519-530, 2003.
3. W. H. Bell, D. G. Cameron, L. Capozza, et al: Optorsim: a Grid Simulator for Studying Dynamic Data Replication Strategies. International Journal of High Performance Computing Applications, Vol. 17(4): 403-416, 2003.

A Model of the Role of Cholesterol in the Development of Alzheimer's Disease

Gizelle Kupac Vianna[1], Artur E. Reis[1], Luís Alfredo V. de Carvalho[1], and Roseli S. Wedemann[2]

[1] COPPE - Universidade Federal do Rio de Janeiro,
Prog. Eng. de Sistemas e Computação, P.O. Box 68511, 21945-970, Rio de Janeiro, RJ, Brazil
`gkupac@gmail.com, artur_reis@yahoo.com.br, LuisAlfredo@ufrj.br`
[2] Instituto de Matemática e Estatística, Universidade do Estado do Rio de Janeiro
R. São Francisco Xavier, 524, 20550-013, Rio de Janeiro, RJ, Brazil
`roseli@ime.uerj.br`

Abstract. We present a mathematical-computational model of the development of Alzheimer's disease, based on the assumption that cholesterol plays a key role in the formation of neuropathological lesions that characterize the disease: the senile amyloid plaques and neurofibrillary tangles. The final model, conceived as a system of equations, was simulated by a computer program.

1 Introduction

We propose a mathematical-computational model, that represents the physiological alterations resulting from Alzheimer's Disease (AD) [9]. It is difficult to achieve a good understanding of this pathology, since it is generated by a combination of many complex and highly connected processes. The model thus aims at facilitating the comprehension of AD, which is associated with an adaptative biological complex system. We know of no other similar attempt to achieve such a model, in the specialized literature.

AD affects almost 7% of the population over 65 years old and up to 40% of people over 80 years of age, and it is estimated that it is responsible for almost 70% of all senile dementias. AD consists of a degenerative process, characterized by the occurrence of a series of abnormalities in the brain, selectively affecting neurons of specific regions, such as the cortex and the hippocampus. Microscopic exams reveal a great amount of two physical alterations that characterize AD and distinguish it from other diseases: extracellular senile amyloid plaques (SAP's) and intracellular neurofibrillary tangles (NT's). SAP's consist of deposits of amyloid β-peptide ($A\beta$) surrounded by distrofic axons and inflammation. NT's contain paired helical filaments, composed of hyperphosphorylated forms of Tau protein [2]. Such alterations hinder the functioning of sinapses and the feasibility of neurons, leading to neuronal loss.

Age is the most relevant risk factor for AD, as many physical alterations take place in the brain as an individual grows older. The theory of oxidative stress conceives aging as a result of the accumulation of damages to the tissues caused by free radicals [8]. Some epidemiologic studies suggest an influence displayed by high plasmatic cholesterol levels in the development of AD [5]. The disease also shows some genetic components.

V.N. Alexandrov et al. (Eds.): ICCS 2006, Part I, LNCS 3991, pp. 842–845, 2006.

2   A Computational Model and Simulations

There is uncertainty regarding the relationship between levels of cholesterol in the plasma and in the central nervous system (*CNS*), in patients with AD. This is due to the fact that plasma cholesterol and brain cholesterol are separated by the blood-brain barrier (*BBB*), that controls the exchange of substances between fluids of the brain and blood. However, there is evidence that the level of cholesterol in the frontal cortex demonstrates a significant increase when plasma cholesterol increases [1].

In [9], we have presented a mathematical model which was simulated by a computer program for the main mechanisms in AD that are influenced by the presence of cholesterol in the *CNS*, as proposed in the literature. The model consists of a set of linear equations and linear differential equations that model the concentration of substances involved in the metabolism of cholesterol, the formation of senile amyloid plaques and in the formation of neurofibrilary tangles, all of which cause neuronal loss [1, 2, 3, 4, 5, 6, 7, 8]. We consider that the level of neuronal loss will characterize AD, since it relates to the cerebral atrophy that occurs in the disease.

We used Euler's method to solve the model equations. One integration step, representing time, corresponds to 8 hours and the simulation of 100 years requires approximately 100,000 steps. A basal cholesterol diet corresponds to a daily consumption of 300mg and a high cholesterol diet to a daily consumption of 600mg. In Fig. 1, we show the production levels of SAP's and NT's from a simulation of daily ingestion of 600mg of cholesterol.

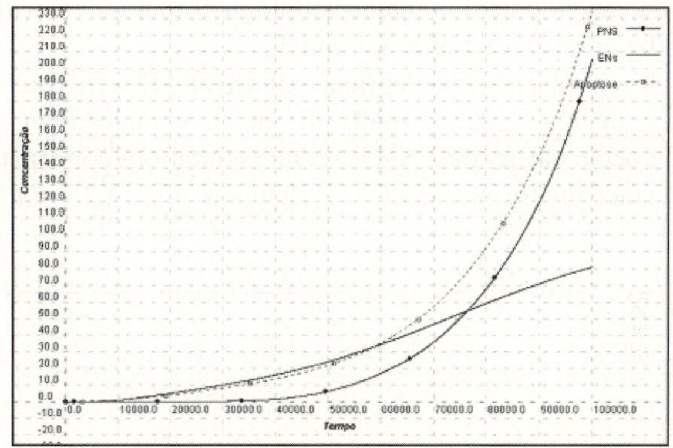

Fig. 1. $[SAP]$ (PNS) and $[NT]$ (ENs), for 600mg of cholesterol and $A\beta$ aggregation

Murray [4] hypothesized that, if the process of aggregation of $A\beta$ could be suspended, the disease would not evolve. We tested the system's behavior when the aggregation process of $A\beta$ is inhibited. The results in Fig. 2 are coherent with the hypothesis and we observe that, when the aggregation process of $A\beta$ is inhibited, suppression of the inflammatory process occurs and the formation of SAP's is inhibited. We notice a slight decrease in the formation rate of NT's. As SAP's were no longer formed,

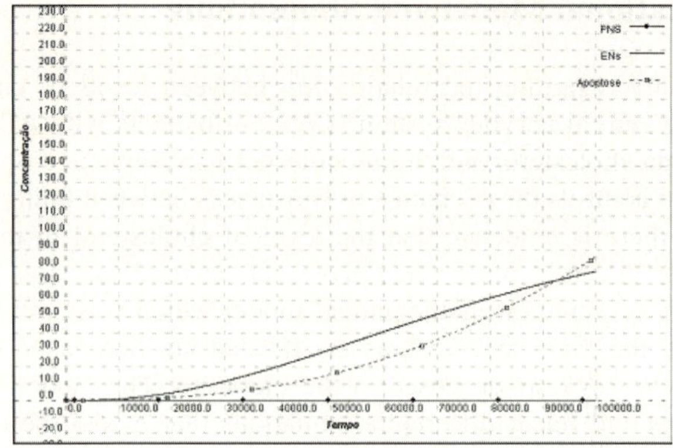

Fig. 2. $[SAP]$ and $[NT]$, for 600mg of cholesterol and inhibition of $A\beta$ aggregation

there will be no disturbance in the functioning of the *BBB* by these structures and the increase of efflux of cholesterol is avoided. Changes in the distribution of cholesterol in the neuronal membrane will occur much more slowly and the production of $A\beta$ is reduced.

There are some positive feedbacks in the system that may aggravate AD. For example, $A\beta$ and SAP's initiate an inflammatory process in the brain, which aggravates the AD development process. We inhibited the occurrence of inflammation and observed a keen reduction in the development of neuropathological alterations, with daily ingestion of 600mg, as seen by comparing Figures 1 and 3.

In Niemann-Pick's disease, NT's identical to those of AD were found, but no SAP's. Since individuals with this disease usually die much younger than AD patients, we conclude that NT's occur earlier than SAP's. Our simulations confirm this phenomenon as shown in Fig. 1, since the formation of NT's happens almost 25 years before SAP's.

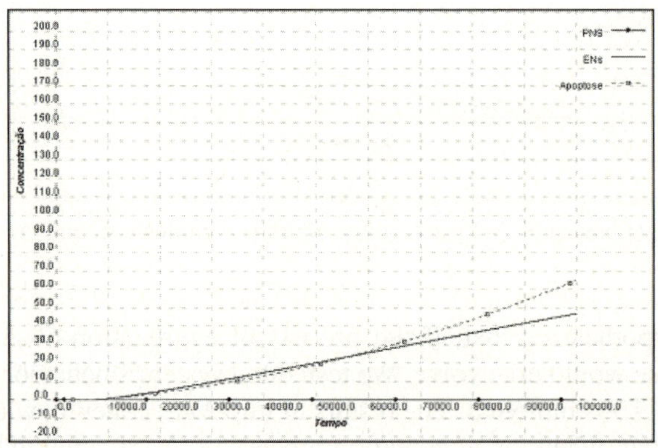

Fig. 3. Neuropathological alterations for 600mg cholesterol, without inflammation

3 Conclusions

Results from the literature and from our simulations suggest that, despite the brain's resistance to dietary lipid composition, chronic consumption of cholesterol may alter the functioning of certain cerebral proteins and even the structure of neurons. This happens because cholesterol alters the composition and certain properties of the neuronal membrane. One consequence of these alterations is the increase of $A\beta$ production and aggregation, simultaneous with a reduction of its degradation rate. These factors contribute to the increase of toxic $A\beta$ formation, leading to formation of SAP's which cause the generation of free radicals, amplifying oxidative stress OS. OS then oxidates the synaptic plasmatic membranes, creating a positive feedback over $A\beta$ production. Finally, OS and insoluble $A\beta$ trigger Tau protein phosphorylation and the neuronal structure is destabilized, generating NT's. Inflammation also plays an important role in the development of AD and its inhibition resulted in a sharp deceleration of the neurodegenerative process.

Acknowledgements. This research was developed with grants from the brazilian National Research Council (CNPq), the Rio de Janeiro State Research Foundation (FAPERJ) and the brazilian agency which funds graduate studies (CAPES).

References

1. Howland, D. et al.: Modulation of Secreted β-Amyloid Precursor Protein and Amyloid β-Peptide in Brain by Cholesterol. The J. of Biological Chemistry, **273** No. 26. (1998) 16576–16582
2. Illenberger, S. et al.: The Endogenous and Cell Cycle-Dependent Phosphorylation of Tau Protein in Living Cells: Implications for Alzheimer's Disease, Molecular Biology of the Cell, **9** (1998) 1495–1512
3. Ji, S.R., Wu, Y., Sui, S.F.: Cholesterol is an Important Factor Affecting the Membrane Insertion of β-Amyloid Peptide (Aβ-40), which May Potentially Inhibit the Fibril Formation. The J. of Biological Chem., **277** No. 8. (Feb 22, 2001) 6273–6279
4. Murray, R.K.: The Biochemical Basis of Some Neuropsychiatry Disorders. In: Murray, R.K., Granner, D.K., Mayers P.A. et al. (eds.): Harpers Biochemistry, Int. Edition, (23 ed.). Prentice-Hall (1993) 750-752.
5. Pappolla, M.A., Smith, M.A., et al.: Cholesterol, Oxidative Stress, and Alzheimer's Disease: Expanding the Horizons of Pathogenesis. In: Smith, M.A., Perry, G. (eds.): Serial Review: Causes and Consequences of Oxidative Stress in Alzheimer's Disease. Free Radical Biology and Medicine, Vol. 33, no. 2., Elsevier (2002) 173–181
6. Rissman, R.A., Poon, W.W., Jones, M.B. et al.: Caspase-cleavage of Tau is an early event in Alzheimer's disease tangle pathology. The J. of Clinical Investigation, **114** No. 1, (2004) 121–130
7. Selkoe, D.J.: Toward a Comprehensive Theory for Alzheimer's Disease. Hypothesis: Alzheimer's Disease is Caused by the Cerebral Accumulation and Cytotoxicity of Amyloid β-Protein. Annals of the New York Academy of Science, **924** (2000) 17–25
8. Wood, W.G. et al.: Brain Membrane Cholesterol Domains, Aging and Amyloid Beta-Peptides. Neurobiology of Aging, No. 23 (2002) p.685–694
9. Vianna, G.K.: Um Modelo Neurocomputacional do Papel do Colesterol no Desenvolvimento da Doença de Alzheimer. Ph.D. Dissertation, Prog. Eng. de Sistemas e Computação, Universidade Federal do Rio de Janeiro, Brazil, (2005) (in Portuguese).

Characterization of Cardiac Dynamics from Locally Topological Considerations

Victor F. Dailyudenko

Institute of Informatics Problems NAS of Belarus,
Surganov St. 6, 220012, Minsk, Belarus
selforg@newman.bas-net.by

Abstract. Evolution of cardiac activity is investigated by means of methods of nonlinear dynamics, namely the method of temporal localization on the attractor reconstructed from electrocardiogram (ECG) signal is proposed for this purpose. Convergence for the function of topological instability at changing dimensionality is proved both theoretically and numerically, independently on personal features of subjects in the latter case, that provides the opportunity to estimate the complexity of cardiac dynamics. In contrast, that instability function normalized by its average displays different kind of behaviour that somewhat differs for various persons and reflects their individual features.

1   Introduction

The computational complexity of topological algorithms being applied for cardiovascular dynamics investigation makes these algorithms rather cumbersome from standpoint of their computer time and the quantity of required experimental data N [1]. Therefore, the total time of observation and diagnosis process becomes rather long that may result in difficulties in clinical practice. So, in this paper we develop the topological method based on temporal locality approach. In comparison with the most conventional methods of nonlinear analysis based on spatial localization [1-4], the developed method allows reduction of N and computation time as well as is insensible to growing m on these characteristics.

2   The Algorithm for Exploration of Topological Instability

The method of delayed coordinates (affirmed mathematically by Takens [5]) for reconstruction of phase trajectories forming an attractor R_T^m is given by [1,2, 4-5]

$$\vec{x}_i^{(m)} = (\eta_i,\, \eta_{i+p},...,\eta_{i+(m-1)p}), \tag{1}$$

where $\eta(i\Delta t)= \eta_i$, $i =1,2,...,N$ is a time series (TS) of a kinetic variable measured with a fixed time interval Δt, $\tau = p\Delta t$ is the delay time, p is an integer. The points $\vec{x}_i^{(m)} \subset R^m$, R^m is an Euclidean phase space with a dimension $m,$ $i =1,2,..., L^{(p,m)}$,

V.N. Alexandrov et al. (Eds.): ICCS 2006, Part I, LNCS 3991, pp. 846–850, 2006.

the common quantity of the attractor points is given by $L^{(p,m)} = N - p(m-1)$. In accordance with (1), phase trajectories forming the attractor R_T^m can be represented as a superposition of p rarefied sequences $X_1, X_2, \ldots, X_p$ shifted by one sample with respect to each other, those are defined as $X_s = \{\vec{x}^{(m)}_{s+p(k-1)}\}^{L_s^{(p,m)}}_{k=1}$.

As it was recently shown, rarefying on attractor points is reasonable for numerical simulation of fractal-topological analysis [4]. Otherwise, using points that are too close together in time leads to essential underestimates of the dimension, i.e. to aggravating accuracy of the topological analysis. So, we also implement temporal rarefying of phase trajectories for creating a subset of points with decorrelated components resulting in essentially random distribution in the embedding space. It is attained by the approach that is realized in the most convenient way, namely we use only one X_s for numerical experiments. That is constructed using the sequence $\Psi_p = \{\eta_p, \eta_{2 \cdot p}, \ldots, \eta_{N_p^{(p)} \cdot p}\}$, and rarefying is determined with $p=2$. Denoting components of Ψ_p for brevity as $\Psi_p = \{\xi_1, \xi_2, \ldots, \xi_{N_p^{(p)}}\}$, we obtain that the terms of relative partition sequence $\{\mu_j^{(m)}\}$ constructed by means of segmentation of difference-quadratic TS are defined analogously [6] as follows

$$\mu_j^{(m)} = \frac{\Delta \xi_{j+m}}{\hat{\sigma}_j^{(m)}} \tag{2}$$

where $\hat{\sigma}_j^{(m)} = \sum_{i=0}^{m-1} \Delta \xi^2_{j+i}$, $\Delta \xi_j = \xi_{j+1} - \xi_j$. Similarly [6], introduce the following measure of topological instability:

$$Z_\mu(m) = \sigma(\mu_j^{(m)}), \tag{3}$$

where $\sigma(\mu_j^{(m)})$ is the mean square variance, the averaging is made over R_T^m points. For estimating the relative variance on $\{\mu_j^{(m)}\}$, introduce the following normalized instability function:

$$\tilde{Z}_\mu(m) = \frac{Z_\mu(m)}{\langle \mu_j^{(m)} \rangle} = \left(\frac{\langle (\mu_j^{(m)})^2 \rangle}{\langle \mu_j^{(m)} \rangle^2} - 1 \right)^{\frac{1}{2}}. \tag{4}$$

3   Numerical Simulations with ECG Signal

In this work the digitized ECG TS ς_i containing N=2500 points is used, $\Delta t = 2 mc$. The initial part of measured TS is shown in Fig. 1. The obtained TS is that of an adult healthy subject being under physical exercises. For decreasing linear autocorrelation effect and reduction of influence of low-frequency periodical component, we use difference TS $\eta_i = \varsigma_i - \varsigma_{i+1}$ instead of "raw" digitized ECG signal ς_i (as well as in [1]). The phase trajectories for three-dimensional attractors reconstructed from η_i by (1) at $p = 1$ and $p = 20$ respectively are shown in Fig. 2, a) and b), the latter case reveals the effect of partial decorrelation (similarly [1]).

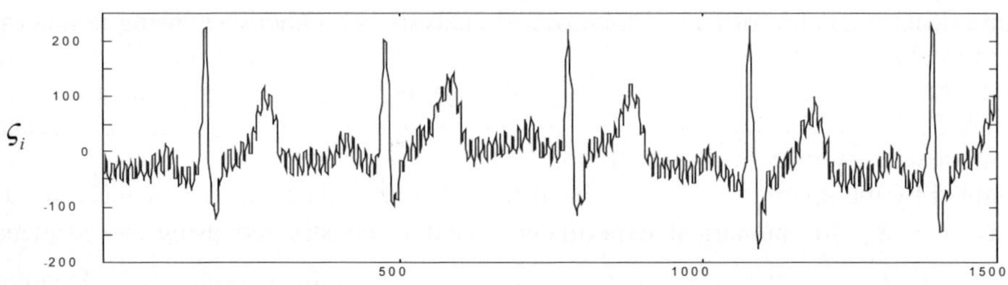

Fig. 1. The initial ECG signal taken out of the first database

For investigation of topological dependencies, we use three databases of ECG signals obtained from different groups of healthy adult subjects at the same conditions as for the ECG signal displayed in Fig. 1, the following additional normalization being used for scale unification:

$$Y(m) = \frac{Z_\mu(m)}{Z_\mu(1)}, \qquad \breve{Y}(m) = \frac{\breve{Z}_\mu(m)}{\breve{Z}_\mu(1)}. \tag{5}$$

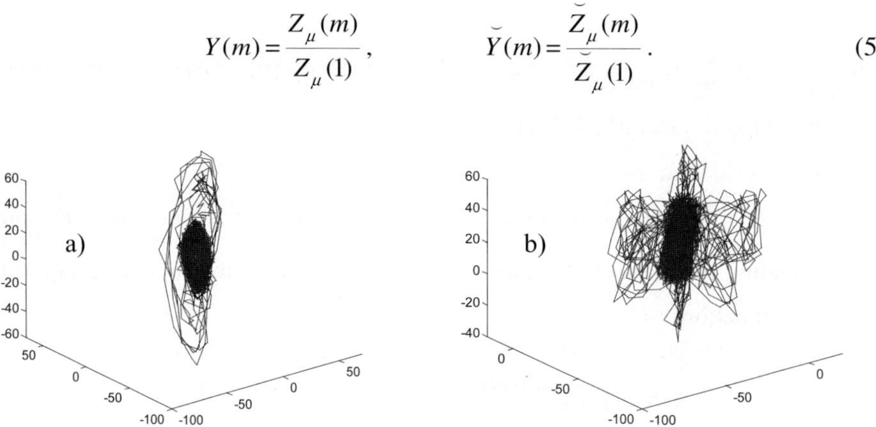

Fig. 2. Temporal evolution of cardiac activity represented through the phase trajectories reconstructed from the difference ECG TS: a) $p = 1$; b) $p = 20$

Calculated dependencies (5) are shown in Fig. 3, a) and b). These dependencies display sufficient convergence for $m \geq m_0$, i.e. m_0 is just the value of a dimension that provides preservation of topological structure of phase trajectories at enlarging dimensionality beyond m_0 and thus can be really considered as the minimal embedding dimension of the attractor. The convergence of $Y(m)$ is shown to be of the same character, independently on individual features of subjects, and one can conclude that $m_0 = 6$ is sufficient for optimal embedding of the attractor into Takens' phase space. On the other hand, the dependence $\breve{Y}(m)$ is approximately linear at $m \geq m_0$, but it differs for various persons with respect to average level of convergence and a slant angle. Evidently, it can provide some additional information concerning individual features of subjects that seems to be useful for the sake of early diagnosis.

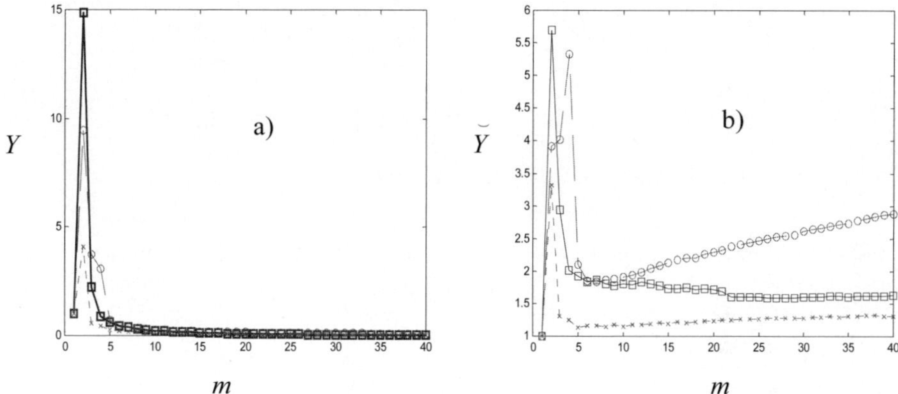

Fig. 3. The topological dependencies calculated with ECG TS from samples randomly chosen out of three different databases

4   Conclusions

The obtained results of m_0 determined on $Y(m)$ are in good coincidence with those obtained in [1] by the Grassberger - Procaccia algorithm (GPA), where slopes of the plots $\ln C(l)$ versus $\ln l$ appear to be same for $m \geq 6$. It is worth to note that in GPA N=16000 [1], while in our experiments N=2500 (moreover, the length of really used rarefied TS $N_p^{(p)} = 1250$) that means the significant reduction of data.

References

1. Bezerianos, A., Bountis, T., Papaioannou, G., Polydoropoulos, P.: Nonlinear time series analysis of electrocardiograms. Chaos. 5 (1995) 95 - 101
2. Nonnenmacher, T.F., Losa, G.A., Wribel E.R. (editors): Fractals in Biology and Medicine, Birkhauser - Verlag, (1994)
3. Wang, J., Ning, X., Chen, Y.: Multifractal analysis of electronic cardiogram taken from healthy and unhealthy adult subjects. Physica A. 323 (2003) 561 - 568

4. Albano, A.M., Passamante, A., Farrell, M.E.: Using higher-order correlations to define an embedding window. Physica D. 54 (1991) 85 - 97
5. Takens, F.: Detecting strange attractors in turbulence In: Dynamical Systems and Turbulence. Lecture Notes in Math., Springer, Berlin, 898 (1981) 366 - 381
6. Dailyudenko, V.F.: Nonlinear time series processing by means of ideal topological stabilization analysis and scaling properties investigation. In: Proc. of the SPIE's Conf. on Applications and Science of Computational Intelligence II (Apr. 1999, Orlando, Florida, USA) 3722 (1999) 108 - 119

Sliding Free Lagrangian-Eulerian Finite Element Method

Junbo Cheng, Guiping Zhao, Zupeng Jia, Yibing Chen, Junxia Cheng,
Shuanghu Wang, and Wanzhi Wen

Institute of applied physics and computational mathematics, Beijing 100088, China
cheng_junbo@iapcm.ac.cn, magpzhao@yahoo.com.cn, zpjia@iapcm.ac.cn,
chen_yibing@iapcm.ac.cn, Cheng_Junxia@iapcm.ac.cn,
Wang_Shuanghu@iapcm.ac.cn, wen_wanzhi@iapcm.ac.cn

Abstract. People usually use arbitrary Lagrangian-Eulerian method to simulate the multi-phase flowing problems, but some numerical errors may be introduced during remapping. In this paper, sliding free Lagrangian-Eulerian finite element method(SFLEFEM) is developed. In SFLEFEM compressible Eulerian equations for moving mesh are dircretized without Lagrangian step and numerical experiments prove that SFLEFEM is convergent and stable.

1   Description

In Lagrangian approach, the mesh is embedded in the fluid and moves with it, So the precise material interface can be afforded. But when dealing with complex fluid, the mesh may be distorted seriously, the computation errors will increase quickly so that the computation is stopped. For solving the problem, people usually use rezoning and remapping techniques, but some errors may be introduced. So people can't use the rezoning and remapping techniques frequently.

In Eulerian approach, the mesh of grid points is fixed and so it can be used to calculate the problem of large deformation. Because the mesh is fixed, many other techniques, such as VOF[1], Front tracking[2], Level set[3] and Phase field[4], are used to track the moving material interface. But the precision of their capturing interface is lower than that of Lagrangian approach.

Because of these shortcomings, we introduce Sliding Free Lagrangian Eulerian Finite Element Method (SFLEFEM). Our objective is to deal with multi-material physics problems with large-deformation and sliding interface. Numerical results show that SFLEFEM can simulate the multi-material problems with large-deformation and also can capture the material interface precisely.

SFLEFEM is different from Arbitrary Lagrangian-Eulerian(ALE) method. In ALE method, a Lagrangian step is first performed and the mesh is deformed according to the fluid flow. Then the improved mesh is generated based on the Lagrangian mesh and the solution is transferred from the Lagrangian mesh to the improved mesh. In SFLEFEM, a Lagrangian step is never performed. We directly calculate the fluid field from the time t^n to the time t^{n+1}. Moreover, the

V.N. Alexandrov et al. (Eds.): ICCS 2006, Part I, LNCS 3991, pp. 851–855, 2006.

852   J. Cheng et al.

grid points on the material interface or the boundary must move at the speed of Lagrangian velocity in the normal direction and can slide freely in the tangential direction. The tangential movement of boundary points can help us to generate the mesh of high quality.

In SFLEFEM, every material is calculated separately and the interaction between two neighboring materials is calculated by a new contact algorithm which is based on TENSOR contact algorithm. We will describe it in the future.

The equations of SFLEFEM are unsteady compressible Eulerian equations for moving grid points:

$$\frac{\partial \rho}{\partial t} + \triangledown \cdot (\rho \vec{D}) = -\triangledown \cdot (\rho \vec{w}), \tag{1}$$

$$\frac{\partial \rho e}{\partial t} + \triangledown \cdot (\rho e \vec{D}) = -p \triangledown \cdot (\vec{v}) - \triangledown \cdot (\rho e \vec{w}), \tag{2}$$

$$\rho(\frac{\partial v_z}{\partial t} + \vec{D} \cdot \triangledown v_z) = -\frac{\partial p}{\partial z} - \rho \vec{w} \cdot \triangledown v_z, \tag{3}$$

$$\rho(\frac{\partial v_r}{\partial t} + \vec{D} \cdot \triangledown v_z) = -\frac{\partial p}{\partial r} - \rho \vec{w} \cdot \triangledown v_r. \tag{4}$$

where $\vec{D}$ is the mesh velocity, $\vec{D} = \frac{d\vec{r}}{dt}$, $\vec{w} = \vec{v} - \vec{D}$.

Mass equation(1) is integrated on $\Omega(t)$ based upon Finite Volume method:

$$\frac{M^{n+1}-M^n}{\triangle t} = \sum_{i=1}^{4} \rho_{i,i+1} \alpha_{i,i+1} (w_r \triangle z - w_z \triangle r)_{i,i+1}, \tag{5}$$

where $\alpha_{i,i+1} = \frac{\alpha_i + \alpha_{i+1}}{2}$ ($\alpha = 1$ for the planar problem and $\alpha = r$ for the axis symmetrical problem), $(w_r)_{i,i+1} = \frac{(w_r)_i + (w_r)_{i+1}}{2}$, $(w_z)_{i,i+1} = \frac{(w_z)_i + (w_z)_{i+1}}{2}$, $\rho_{i,i+1}$ is the density along the boundary segment $(i, i+1)$. We use the second-order MUSCL[5] scheme to calculate it.

Energy equation (2) is integrated on an one arbitrary mesh $\Omega(t)$ based upon the finite volume method and the Von Neumann and Richtmyer [6] viscosity q^n is added for preventing from producing numerical oscillation, which yields

$$\frac{E^{n+1}-E^n}{\triangle t} = -(\frac{p^n + p^{n+1}}{2} + q^n)\frac{\triangle V_L^{n+1}}{\triangle t} + g_e(t). \tag{6}$$

Where $E = \int_{\Omega(t)} \rho e dV = e(t)M(t)$, $\triangle V_L^{n+1} = V(\vec{r}^n + \triangle t \vec{v}^n) - V(\vec{r}^n)$, $g_e(t) = -\int_{\Omega(t)} \triangledown \cdot (\rho e \vec{w})dV$. The discretization method for $g_e(t)$ is same as that for $-\int_{\Omega(t)} \triangledown \cdot (\rho \vec{w})dV$.

Momentum Equation(3) is discretized using finite element method, which gives

$$\frac{v_z^{n+1} - v_z^n}{\triangle t} \sum_{i=1}^{4} \frac{1}{4} M_{i,A}^{n+\frac{1}{2}} = \sum_{i=1}^{4} (p_i^{n+\frac{1}{2}} + q_i^{n+\frac{1}{2}}) \frac{r_{i,2}^{n+\frac{1}{2}} - r_{i,4}^{n+\frac{1}{2}}}{2}$$

$$- \sum_{i=1}^{4} \frac{\rho_i^{n+\frac{1}{2}}}{2} (v_{z,N}^n - \overline{v}_{z,i}^n)[(\overline{w}_r)_i(z_{i,2}^{n+\frac{1}{2}} - z_{i,4}^{n+\frac{1}{2}}) - (\overline{w}_z)_i(r_{i,2}^{n+\frac{1}{2}} - r_{i,4}^{n+\frac{1}{2}})].$$

where $(\overline{w}_r)_i = \frac{1}{4}\sum_{k=1}^{4}(w_r)_{i,k}$, $(\overline{w}_z)_i = \frac{1}{4}\sum_{k=1}^{4}(w_z)_{i,k}$, $(\overline{v}_z)_i = \frac{1}{4}\sum_{k=1}^{4}(v_z)_{i,k}$. Now the discretization of equation (3) is finished. The methods of discretization for the equation (4) are same as that for the equation (3).

2   Examples

Saltzman piston problem[7] is used to test the ability of code to simulate the shock waves that are oblique to the mesh. Exact results for the problem are that the third shock arrives at the z coordinate of 0.95 and densities in front of and behind the third shock are 10 and 20 respectively. The pure Lagrangian methods(PLM) and SFLEFEM are used to simulate the problem and the results obtained from these two methods at time t=0.925 are showed in Fig. 1.

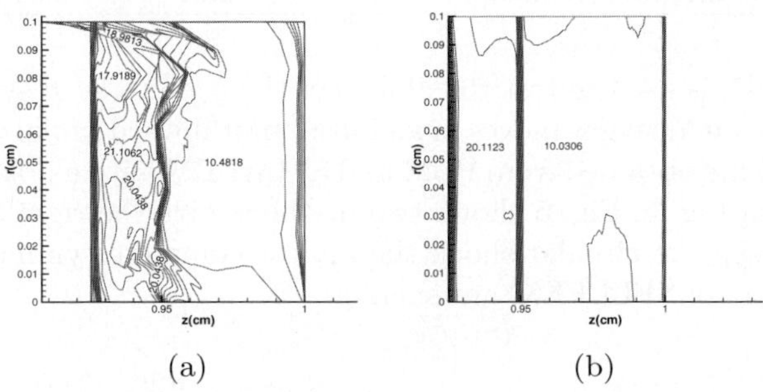

(a)   (b)

Fig. 1. (a)Density contours obtained from pure Lagrangian methods, (b)Density contours obtained from SFLEFEM

The figure of one-dimensional shock obtained from PLM is distorted seriously, especially along the top and bottom boundaries. But the density contours obtained from SFLEFEM give the right position of the shock and almost completely planar shock and the smaller density errors.

Dukowicz problem[7] is a shock refraction problem on an inclined interface and its sketch map is showed in Fig. 2. A piston moves from left to right with the constant velocity of 1.48. On the right of the piston there are two ideal gases, the interface between two gases is aligned at 30° to the horizontal.

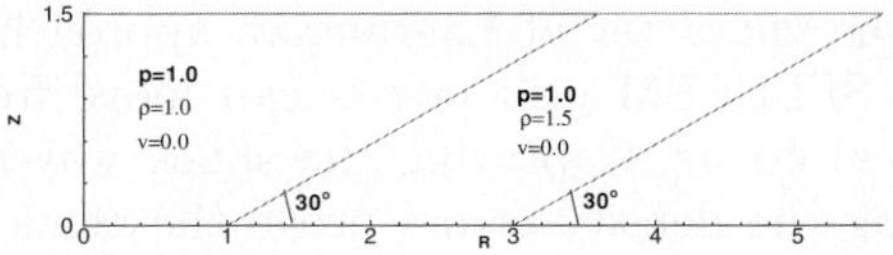

Fig. 2. Sketch map for Dukowicz problem

The initial mesh is composed of two adjacent regions. The left region is a 36×30 mesh and the right region a 40×30 uniform mesh slanted at $60°$. The density contours obtained by PLM and SFLEFEM are showed in Fig. 3 and Fig. 4 respectively at time t=1.3. They are almost same as the density contours in [7]. The main difference is where the interface arrives at on the lower boundary.

We also use the refiner mesh (72×60 zones in the left region and 80×60 zones in the right region) to calculate the problem. The interface location on the lower boundary at time t=1.3 are showed in table 1. Table 1 shows the interface location from SFLEFEM is same even if using different mesh. That shows SFLEFEM is convergent.

Table 1. Interface location on lower boundary

Algorithm mesh	PLM coarse mesh	SFLEFEM coarse mesh	PLM refine mesh	SFLEFEM refine mesh
Interface position on lower boundary	1.99	2.05	2.04	2.05

Noh problem[7] is used to test the ability of the code to keep symmetrization. In Noh problem, an ideal gas moves in an initial unit inward radial velocity. Exact solution and results obtained from PLM and SFLEFEM with a polar mesh of 50×50 are showed in Fig. 5. Fig. 5 shows two methods give the right shock location and they can keep the circular shock the almost complete symmetrization, but the density errors of SFLEFEM are smaller.

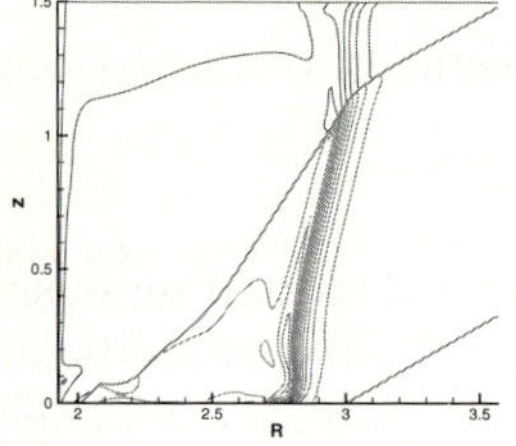

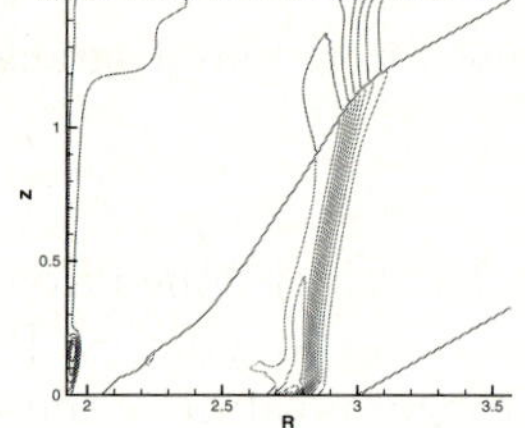

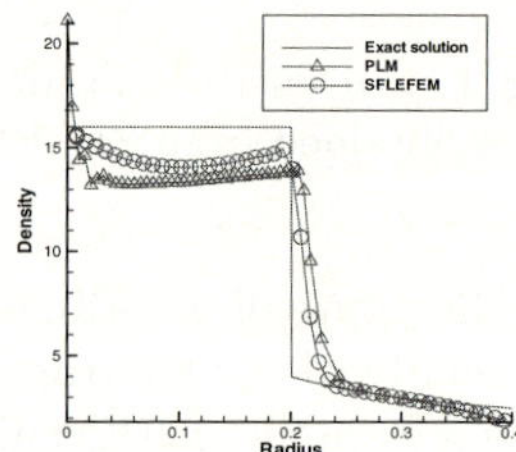

Fig. 3. Density contours from PLM

Fig. 4. Density contours from SFLEFEM

Fig. 5. Density distribution versus the radius of the center of all cells

3   Summary and Discussion

SFLEFEM has both the character of Lagrangian approach and the attribute of Euler approach. In SFLEFEM grid points can move free, which does not introduce new numerical errors. Capturing the shock waves and the interface precisely and decreasing the density errors prove the efficiency of SFLEFEM. But the more development of SFLEFEM is necessary, especially for the time discretization of momentum equations. We will use high order R-K methods to discretize the time derivatives in the momentum equations in future.

References

1. Hirt C. W. and Nichols B. D.: Volume of fluid(VOF) method for the dynamics of free boundary, J. Comput. Phys. **39** (1981) 201-225.
2. Chen I. L. and Glimm J.: Front tracking for gas dynamics, J. Comput. Phys. **62** (1986) 83-110.
3. Osher S. and Sethian J. A., Fronts propagating with curvature depend speed: Algorithm based on Hamilton-Jacobi formulation, J. Comput. Phys. **79** (1988) 12
4. Antannovskii L. K., A phase field model of capillavity.
5. David J. Benson.: An efficient, accurate, simple ALE method for nonlinear finite element programs, Comm. Pure Appl. Math. **72** (1989) 305–350
6. Von Neumann and R. D. Richtmyer.: A method for the numerical calculation of hydrodynamics shocks, J. Appl. Phys. **21**(1950)
7. J. Campbell and M. Shashkov.: A Tensor Artificial Viscosity using a Mimetic Finite Difference Algorithm, Los Alamos NM 87545, April 2000, LA-UR-00-2900

Large-Scale Simulations of a Bi-dimensional n-Ary Fragmentation Model[*]

Gonzalo Hernandez[1,2], Luis Salinas[3], and Andres Avila[4]

[1] UNAB Escuela de Ingenieria Civil, Santiago, Chile
gjho@vtr.net
[2] UChile Centro de Modelamiento Matematico, Santiago, Chile
[3] USM Departamento de Informatica, Valparaiso, Chile
[4] UFRO Departamento de Ingenieria Matematica, Temuco, Chile

Abstract. A bi-dimensional n-ary fragmentation model is numerically studied by large-scale simulations. Its main assumptions are the existence of random point flaws and a fracture mechanism based on the larger net force. For the 4-ary fragment size distribution it was obtained a power law with exponent $1.0 \leq \beta \leq 1.15$. The visualizations of the model resemble brittle material fragmentation.

1   Introduction

Fragmentation processes are complex multiphysics multiscale phenomena in Nature and Technology. Examples of fragmentation can be found in very large, medium and microscopic scale. In refs. [8, 9] there is a enumeration of some natural fragment size distributions of high energetic instantaneous breaking. This experimental evidence predicts a power-law behavior for small fragment masses, with exponent in the range [1.44, 3.54]. Several models have been proposed to explain this power-law behavior, see for instance refs. [2-7] .

In what follows the model is defined and the numerical results discussed in the case of 4-ary fragmentation.

2   Definition of the n-Ary Fragmentation Model

The hypothesis of the model are:

(a) Point flaws: The initial fragment is the unit square with $q = 100$ random point flaws that remains fixed during the fragmentation process, see ref. [1].
(b) Fracture forces: For each fragment there are fracture forces (f_x, f_y) that are applied at random positions. They correspond to uniform and independent distributed random numbers in $[0, 1]$. The fragmentation process is also self-similar, see for instance ref. [7].

[*] Supported by grants: FONDECYT 1050808, UNAB DI-07-04 and UTFSM DGIP 24.04.21.

V.N. Alexandrov et al. (Eds.): ICCS 2006, Part I, LNCS 3991, pp. 856–859, 2006.

(c) n-ary fragmentation: At each step all the fragments will be broken in n fragments independently, like in a cascade process, unless they satisfy the stopping condition (item (e)):

(c1) $\frac{n}{2}$ Fragments are obtained from the application of the forces, i.e. the fracture or cutting plane in this case is the plane perpendicular to the random direction of the larger net force.

(c2) $\frac{n}{2}$ Fragments are obtained due to the existence of flaws in the material. The $\frac{n}{4}$ cutting planes are the planes with normal perpendicular to the line defined by the point of application of the larger force and the position of one of the $\frac{n}{4}$ nearest point flaws.

(d) Mass conservation: The sum of the new fragments area will be the same of the original fragment.

(e) Random stopping: There are three situations in which the fragmentation process of a particular fragment stops:

(e1) If the fragment area is smaller than the minimal fragment size or cutoff: m_{fs}.

(e2) With probability p, see refs. [4, 5, 6].

(e3) Every fragment has a resistance $0 \leq r \leq 1$ to the breaking process. A fragment breaks only if the maximum of the net forces acting on it is greater than r. This parameter will be chosen uniform and constant.

The random stopping applies for fragments of area less or equal to the critical area a_c, introduced in order to represent the fact that greater fragments have more probability to be broken than the smaller ones.

3   Numerical Results: 4-Ary Fragmentation

The methodology for the simulations was the following:

1) The parameters p, r, a_c and m_{fs} are chosen. During all the simulations the value of q was fixed in 100.
2) The results were averaged over 5000 independent random initial conditions, characterized by the fracture forces and the point flaws distribution.
3) It was chosen 4-ary fragmentation since it is the minimal even number to produce non-binary fragmentation.
4) The fragmentation process evolves according to the rules (a) - (e) defined in section 2.

It was determined that the fragment area distribution $f(s)$ follows approximately a power law distribution. The exponent of the power-law shows an increase with respect to [5] due to the 4-ary fragmentation and the point flaws random distribution. The specific values of β are shown in table 1:

Table 1. Exponent β as a function of $p = r$

p	0.01	0.02	0.03	0.04	0.05
β	1.15	1.08	1.05	1.02	1.00

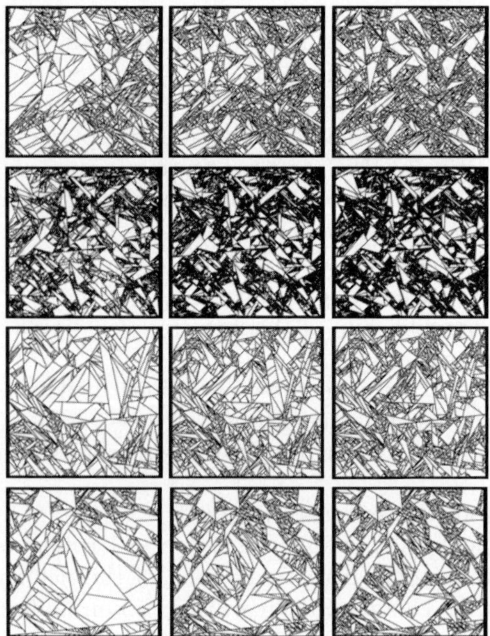

Fig. 1. Fragmentation process evolution for different set of parameters. First row: $p = r = 0.1$, $a_c = 0.01$, $m_{fs} = 0.0005$. Second row: $p = r = 0.1$, $a_c = 0.01$, $m_{fs} = 0.00005$. Third row: $p = r = 0.15$, $a_c = 0.01$, $m_{fs} = 0.0005$. Fourth row: $p = r = 0.15$, $a_c = 0.03$, $m_{fs} = 0.0005$.

In figure 1 is shown the fragmentation process evolution for different sets of parameters. The visualizations the model are very complex with patterns of fracture that resemble real fragmentation processes in brittle materials like glass. From figure 1 it can be appreciated that:

1) If m_{fs} decreases its value, the smaller fragment typical size decreases.
2) If p increases its value, the larger fragments frequency increases.
3) If a_c increases its value, the larger fragments size increases.

4   Conclusions

In this work, it was numerically studied a model for n-ary fragmentation. The main characteristic of this model is a fracture criterion based on the nearest point flaw and maximal net force. By large-scale simulations, it was determined an approximate power law behavior for the fragments area distribution. The visualizations of this dynamical system are very complex with patterns of fracture that resemble real materials fragmentation.

References

1. Abraham, F., Portrait of a Crack: Fracture Mechanics Using Parallel Molecular Dynamics, IEEE Computational Science and Engineering, Vol. 4, N. 2, pp. 66-78, 1997.
2. Aström, J. A., B. L. Holian, and J. Timonen, Universality in Fragmentation, Physical Review Letters, Volume 84, Issue 14, pp. 3061–3064, 3 April 2000.
3. Aström, J. A., R. P. Linna, J. Timonen, P. F. Møller, and L. Oddershede, Exponential and power-law mass distributions in brittle fragmentation, Physical Review E, Volume 70, Issue 2, pp. 026104-026110, August 2004.
4. Hernandez, G., Discrete Model for Fragmentation with Random Stopping, Physica A, Volume 300, Issue: 1-2, pp. 13 - 24, November 2001.
5. Hernandez, G., Two-Dimensional Model for Binary Fragmentation Process with Random System of Forces, Random Stopping and Material Resistance, Physica A, Vol. 323, Iss. 1, pp. 1 - 8, March 2003.
6. Krapivsky, P. L., E. Ben-Nai, I. Grosse, Stable Distributions in Stochastic Fragmentation, Journal of Physics A, Volume 37, Number 25, pp. 2863-2880, February 2004.
7. Krapivsky, P. L., I. Grosse and E. Ben-Nai, Scale invariance and lack of self-averaging in fragmentation, Physical Review E, Vol. 61, N. 2, pp. R993-R996, February 2000.
8. Lawn, B. R., T. R. Wilshaw, Fracture of Brittle Solids, Cambridge University Press, 1975.
9. Turcotte, D.L., Fractals and Fragmentation, Journal of Geophysical Research, Vol. 91, pp. 1921-1926, 1986.

Computationally Efficient Technique for Nonlinear Poisson-Boltzmann Equation

Sanjay Kumar Khattri

Department of Mathematics, University of Bergen, Norway
sanjay@mi.uib.no
http://www.mi.uib.no/~sanjay

Abstract. Discretization of non-linear Poisson-Boltzmann Equation equations results in a system of non-linear equations with symmetric Jacobian. The Newton algorithm is the most useful tool for solving non-linear equations. It consists of solving a series of linear system of equations (Jacobian system). In this article, we adaptively define the tolerance of the Jacobian systems. Numerical experiment shows that compared to the traditional method our approach can save a substantial amount of computational work. The presented algorithm can be easily incorporated in existing simulators.

1   Introduction

Lets consider the following non-linear elliptic problem

$$- \operatorname{div}\left(\epsilon \operatorname{grad} p\right) + f(p, x, y) = b(x, y) \quad \text{in} \quad \Omega \quad \text{and} \quad p(x, y) = p^{D} \quad \text{on} \quad \partial\Omega_D \ . \tag{1}$$

The above problem is the Poisson-Boltzmann equation arising in molecular biophysics. See the References [2, 7, 9, 10, 11, 12]. Here, Ω is a polyhedral domain in $\mathbb{R}^2$, the source function b is assumed to be in $L^2(\Omega)$ and the medium property ϵ is uniformly positive.

A Finite Volume discretization of the nonlinear elliptic equation results in a system of non-linear equations

$$\mathbf{F}(\mathbf{p}) := \mathbf{A}_1\, \mathbf{p}_h + \mathbf{A}_2(\mathbf{p}_h) - \mathbf{b}_h = 0 \ . \tag{2}$$

Here, $\mathbf{F} = [F_1(\mathbf{p}), F_2(\mathbf{p}), \cdots, F_n(\mathbf{p})]^{T}$, $\mathbf{A}_1$ is the discrete representation of the symmetric continuous operator $-\operatorname{div}\left(\epsilon \operatorname{grad}\right)$ and $\mathbf{A}_2$ is the discrete representation of the non-linear operator $f(p, x, y)$.

A Newton-Krylov method for solving the non-linear equation (2) is given by the Algorithm 1. In the Quasi-Newton method (see Algorithm 2), we are solving the Jacobian equation $(\boldsymbol{J}(\mathbf{p}_k)\, \Delta\mathbf{p} = -\mathbf{F}(\mathbf{p_k}))$ approximately. We are solving the system $\boldsymbol{J}(\mathbf{p}_k)\, \Delta\mathbf{p}_k = -\mathbf{F}(\mathbf{p}_k) + \mathbf{r}_k$ with $\|\mathbf{r}_k\|$ is chosen adaptively. The quasi-Newton iteration is given by the Algorithm 2. In the Algorithms 1 and 2, $\|\cdot\|_{L_2}$ denotes the discrete L_2 norm and $\max_{iter}$ is the maximum allowed Newton's iterations. It is interesting to note the stopping criteria in the Algorithm 2. We

V.N. Alexandrov et al. (Eds.): ICCS 2006, Part I, LNCS 3991, pp. 860–863, 2006.
© Springer-Verlag Berlin Heidelberg 2006

Algorithm 1. Newton-Krylov Algorithm

Mesh the domain;
Form the non-linear system: $\mathbf{F}(\mathbf{p})$;
Set the iteration counter: $k = 0$;
while $k \leq \max_{iter}$ *or* $\|\Delta\mathbf{p}\|_{L_2} \leq tol$ *or* $\|\mathbf{F}(\mathbf{p})\|_{L_2} \leq tol$ **do**
    Solve the discrete system : $\boldsymbol{J}(\mathbf{p}_k)\,\Delta\mathbf{p} = -\mathbf{F}(\mathbf{p_k})$ with a fixed tolerance;
    $\mathbf{p_{k+1}} = \mathbf{p_k} + \Delta\mathbf{p}$;
    k^{++};
end

Algorithm 2. Quasi-Newton-Krylov Algorithm

Mesh the domain;
Form the non-linear system: $\mathbf{F}(\mathbf{p})$;
Set the iteration counter: $k = 0$;
while $k \leq \max_{iter}$ *or* $\|\Delta\mathbf{p}\|_{L_2} \leq tol$ *or* $\|\mathbf{F}(\mathbf{p})\|_{L_2} \leq tol$ **do**
    Solve the discrete system : $\boldsymbol{J}(\mathbf{p}_k)\,\Delta\mathbf{p} = -\mathbf{F}(\mathbf{p_k})$ with a tolerance
    $1.0 \times 10^{-(k+1)}$;
    $\mathbf{p_{k+1}} = \mathbf{p_k} + \Delta\mathbf{p}$;
    k^{++};
end

---

are using three stopping criterion in the Algorithms. Apart from the maximum allowed iterations, L_2 norm of residual vector ($\|\mathbf{F}(\mathbf{p})\|_{L_2}$) and also L_2 norm of difference in scalar potential vector ($\|\Delta\mathbf{p}\|_{L_2}$) are being used as stopping criterion for the Algorithms. Generally in the literature, maximum allowed iterations and the residual vector are used as stopping criteria [9, 10, 11, and references therein]. If the Jacobian is singular than the residual vector alone cannot provide a robust stopping criteria.

2   Numerical Experiment

Let us solve (3) in the domain $\Omega = [-1, 1] \times [-1, 1]$ with $k = 1.0$ [2, 7, 8, 9]. Ω is divided into four equal sub-domains (see Figure 1) based on ϵ.

$$-\nabla \cdot (\epsilon\, \nabla p) + k \sinh(p) = f \quad \text{in} \quad \Omega \quad \text{and} \quad p(x,y) = x^3 + y^3 \qquad \text{on} \quad \partial\Omega_D \ . \tag{3}$$

For solving the linear systems, we are using ILU-preconditioned the Conjugate-Gradient (CG) method. For the Newton algorithm the tolerance of the CG method is 1.0×10^{-15}. For the quasi-Newton method the tolerance of the CG method varies with the iterations k of the Algorithm 2 as follows: $1.0 \times 10^{-(k+1)}$, $k = 0, 2, \ldots, 14$. Figures 3, 4 and 2 reports the outcome of our numerical work. The Figures 3 and 4 compares convergence of the quasi-Newton and Newton methods. The Figure 2 reports computational complexity of the quasi-Newton

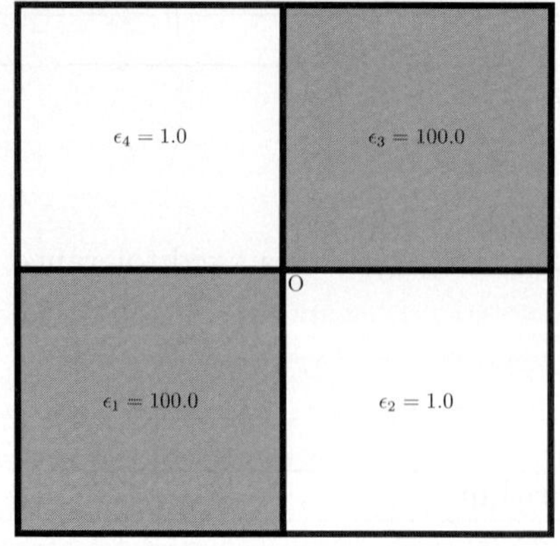

Fig. 1. Distribution of medium property ϵ in the domain $\Omega = [-1, 1] \times [-1, 1]$

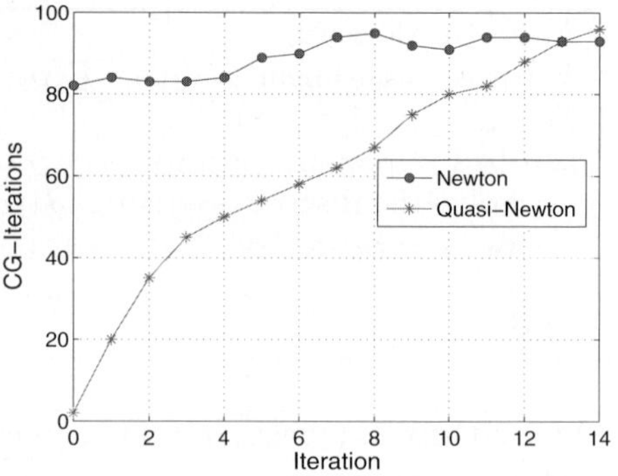

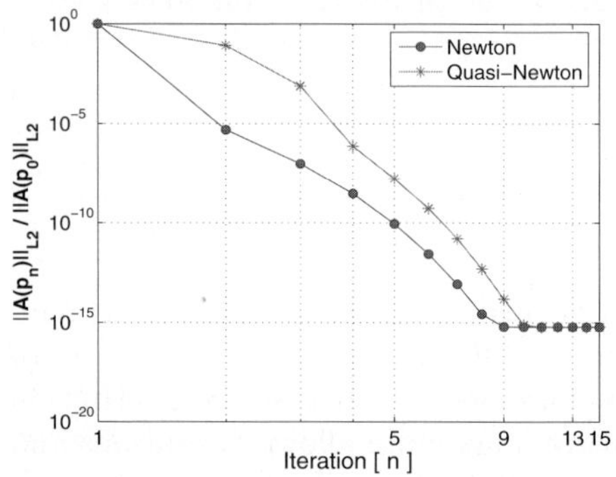

Fig. 2. Computational work required by the Quasi-Newton and Newton methods

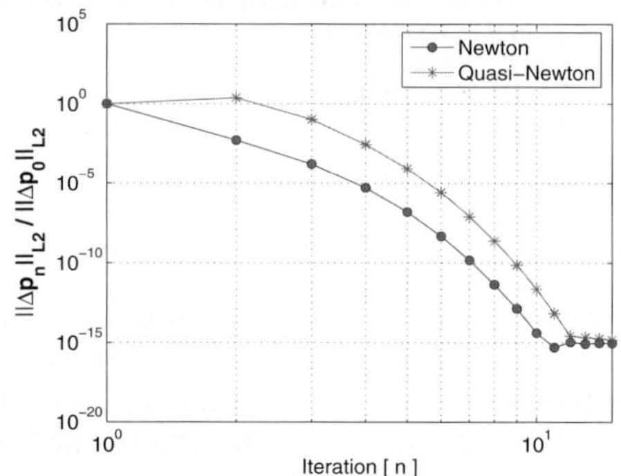

Fig. 3. Convergence of the L_2 norm of residual vector $\mathbf{A}(\mathbf{p})$

Fig. 4. Convergence of the L_2 norm of difference vector $\Delta\mathbf{p}$

and the Newton methods. It can be notice, even if initial iterations of the Newton-Krylov algorithm are solved approximately, the convergence rate of the algorithm remains unaffected. The Figure 2 shows that such an approximation saves a substantial amount of computational effort.

3   Conclusions

Quasi-Newton method for solving non-linear system of equation with symmetric Jacobian matrix is presented. Numerical work shows that the presented technique is computationally efficient compared to the traditional Newton-Krylov method. An efficient solution technique for Poisson-Boltzmann equation is of interest to the researchers in computational chemistry, bio-physics and molecular dynamics. The presented algorithm can be easily implemented in existing simulators.

References

1. Khattri, S.K.:   Analyzing Finite Volume for Single Phase Flow in Porous Media. Journal of Porous Media. Accepted for Publication, (2006).
2. Aksoyw, B.:   Adaptive Multilevel Numerical Methods with Applications in Diffusive Biomolecular Reactions. PhD Thesis, The University of California, San Diego (2001).
3. Khattri, S.K.:   Newton-Krylov Algorithm with Adaptive Error Correction for the Poisson-Boltzmann Equation. MATCH Commun. Math. Comput. Chem., **56**, (2006).
4. Chow, S.-S.:   Finite element error estimates for nonlinear elliptic equations of monotone type. Numer. Math., **54**, (1989), 373–393.
5. Eymard, R., Gallouët, T., Hilhorst, D. and Naït Slimane, Y.:   Finite volumes and nonlinear diffusion equations. RAIRO Math. Model. Numer. Anal., **32**, (1998), 747–761.
6. Lui, S.H.:   On Schwarz Alternating Methods For Non Linear Elliptic PDEs. SIAM Journal on Scientific Computing, **21**, (2000), 1506-1523.
7. Fogolari, F., Brigo, A. and Molinari, H.:   The Poisson Boltzmann equation for Biomolecular electrostatics: A Tool for Structural Biology. Journal of Molecular Recognition, John Wiley & Sons Ltd., **15**, (2002), 377–392.
8. Kuo, S.S., Altman, M.D., Bardhan, J.P., Tidor, B. and White, J.K.:   Fast Methods for Simulation of Biomolecule Electrostatics. International Conference on Computer Aided Design, (2002).
9. Host, M., Kozack, R.E., Saied, F. and Subramaniam, S.:   Treatment of Electrostatic Effects in Proteins: Multigrid-based Newton Iterative Method for Solution of the Full Nonlinear Poisson-Boltzmann Equation. Proteins: Structure, Function, and Genetics, **18**, (1994), 231–245.
10. Host, M., Kozack, R.E., Saied, F. and Subramaniam, S.:   Protein electrostatics: Rapid multigrid-based Newton algorithm for solution of the full nonlinear Poisson-Boltzmann equation. J. of Biomol. Struct. & Dyn., **11**, (1994), 1437–1445.
11. Host, M., Kozack, R.E., Saied, F. and Subramaniam, S.:   Multigrid-based Newton iterative method for solving the full Nonlinear Poisson-Boltzmann equation. Biophysical Journal, **66**, (1994), A130–A130.
12. Holst, M. and Saied, F.:   Numerical solution of the nonlinear Poisson-Boltzmann equation: Developing more robust and efficient methods. J. Comput. Chem., **16**, (1995), 337–364.
13. Baker, N., Sept, D., Holst, M. and McCammon, J.A.:   The adaptive multilevel finite element solution of the Poisson-Boltzmann equation on massively parallel computers. IBM J. Research and Development, **45**, (2001), 427–438.

Geometric Calibration for Multi-projector Tiled Display Based on Vanishing Point Theory

Yuan Guodong, Qin Kaihuai, and Hu Wei

Department of Computer Science and Technology, Tsinghua University, Beijing, China
{ygd02@mails., qkh-dcs@, huw02@mails.}tsinghua.edu.cn

Abstract. In this paper, we present a new geometric calibration method for multi-projector tiled display. Firstly, one circular feature template is attached onto a planar screen and a single un-calibrated camera observes the circular feature template and the tiled arrangement for each projector alternatively. Secondly, geometric coordinates on the planar screen for all features in each tile are computed according to the two vanishing points from the perspective template image. Thirdly, the inscribed quadrangle region in the planar screen space is calculated as the effectively region of the tiled display. Finally the geometric alignment is achieved by the correspondence between the quadrangular mesh in the projector space and the one in the display space. Our algorithm removes the limitation against the parallelism between camera image plane and planar screen plane and improves the convenience to deploy a multi-projector display system. The provided experimental results demonstrate the validity and efficiency of our new method.

1 Introduction

In recent years, many commercial multi-projector systems have been widely available, such as ImmersiveWall, PowerWall, DataWall and etc. While the high cost and technical expertise have restricted their usage to only a few large institutions and well-funded universities. However, many emerging classes of users do not have enough budgets and expertise. So it is imperative to offer new geometric calibration algorithm, which can be easily to deploy a low-cost tiled display system in existing spaces.

In this paper, a new geometric calibration algorithm for multi-projector display system is proposed. It uses an un-calibrated camera to observe a template and a set of overlapping projectors alternatively. From the template image, the three vanishing points in a 3D space are calculated. Then the coordinate of each feature on the planar screen can be reconstructed according to the vanishing point theory. From such information, the necessary image-based corrections to construct a seamless image mosaic across the projector array can be computed. By contrast with [1], the camera's orientation and its position can be laid more freely, which improves the flexibility and the configurability of multi-projector display system. Various experimental results have demonstrated the efficiency and the accurateness of the geometric calibration algorithm.

V.N. Alexandrov et al. (Eds.): ICCS 2006, Part I, LNCS 3991, pp. 864–867, 2006.
© Springer-Verlag Berlin Heidelberg 2006

2   Related Work

Geometric calibration is a fundamental step to achieve the seamless and smooth visual effect for multi-projector tiled display. As for planar screen, automatic software method for geometric calibration has been independently studied using calibrated cameras [2] [3] besides manually adjusting mean. Likai's group utilizes the camera homography to address the limitation of camera's field-of-view and to improve the scalability of multi-projector tiled system [4].To significantly reduce the infrastructure cost, R. Raskar provided a geometric calibration that could adapt to a given projector array configuration [5]. This method must compute two kinds of homographies: camera-projector homography and projector-screen homography. To simplify the computation of geometric calibration further, M. Brown et al [1] assumed that a static camera was placed in a position to observe all the output of multiple projectors, where the camera image plane is parallel to the planar screen. So it is unnecessary to calculate the above two kinds of homography. But this constraint reduces the visual range, which a camera can observe.

3   The Algorithm

First, a physical small pattern is attached to the planar screen to remove the constraint of the parallelism between the camera image plane and the planar screen,. Then a single un-calibrated camera is placed in the position where the entire output of all projectors can be captured. Finally, equally-spaced circles are projected from each of the projectors alternatively, so the camera can capture these images both from each projector and the small pattern at the same position. Figure 1 (a) ~ (e) show the template and these observed projectors.

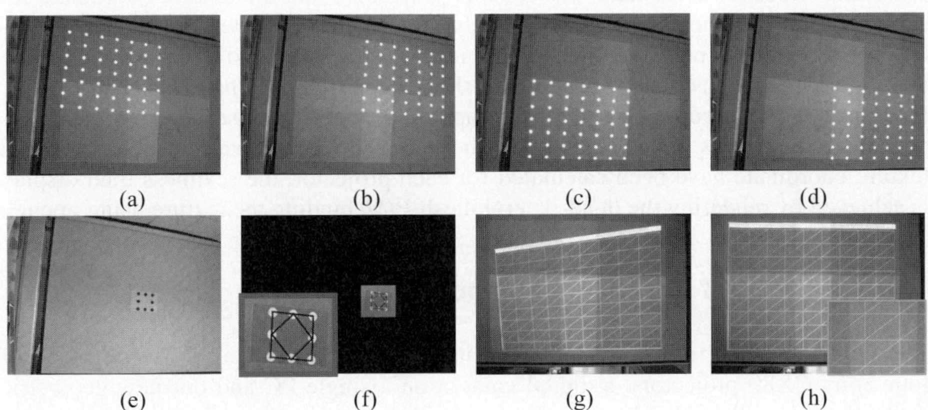

Fig. 1. (a,b,c,d) The captured images of projectors. (e) The pattern image. (f) The processed pattern image. (g) The geometric alignment result using the observed images directly. (h) The geometric calibration results by our method.

3.1 Calculating the Feature's Coordinate in the Planar Screen

From the perspective template image shown in Figure 1 (f), the two vanishing points are calculated. Then the third vanishing point is calculated according to the perspective vanishing point theory. From that, a 3D coordinate system O_XYZ is created and the plane of the projection screen is denoted as the XOZ plane. For each observed projector, the centroid of each equally-spaced feature circle is calculated to get the 3D position vector of each feature in the display space on the basis of perspective projection. Because these 3D vectors are in the same plane, they can be denoted as 2D vector.

3.2 Computing the Inscribed Display Region

Physical display region in the planar screen is the union of the output of all projectors. As mentioned in [5], finding the maximum optimal inscribed rectangle for a union of quadrilaterals is a 2D constrained linear optimization problem. Due that the problem is NP hard, the effective display region R can be calculated by using a simple case-by-case heuristic. In the planar display space, the features of each projector i can be formed to a quadrilateral mesh M_i. Then the intersection region between R and M_i can be re-meshed into a new $m \times n$ quadrilateral mesh IM_i.

3.3 Calibration Computation Between Projector Space and Display Space

For each projector i, the equally-spaced features is also formed to a quadrilateral mesh PM_i in projector space according to the same rule as in the display space. To achieve the geometric calibration for the multiple projectors, the texture and geometric coordinate for each vertex of the mesh IM_i in the projector space are needed. The details of calibration calculation are discussed as follows.

First, the minimum bounding rectangle R_i for the mesh IM_i in the display space is calculated. Second, the texture size for this projector i and the texture coordinate for each vertex by a simple 2D window transformation are calculated. Third, the one-to-one correspondence between the mesh M_i in projector space and the mesh PM_i in the display space is constructed to compute the geometric coordinate for each vertex.. Finally, the texture coordinate can be computed according to the theory of quadrilateral area coordinates. Once the size of the texture to be shown, the geometric and texture coordinate have been calculated for each projector, the seamless tiled display is achieved by rendering the quadrilateral mesh PIM_i binding the texture using opengl.

4 Experimental Results and Conclusion

The multi-projector tiled display system used in the experiments was composed of four Sony CX80 projectors, a digital camera on a single PC and our new geometric calibration algorithm. As shown in Figure 1(g), geometric alignment can be achieved by using the observed images directly. But geometric distortion occurs in the display region because of the projective distortion between the captured geometry and physical output geometry of the projectors. Figure 1(h) shows that the geometric calibration results using our algorithm.

To a planar projection screen, a new geometric calibration algorithm for multi-projector tiled display on the basis of the theories of vanishing point and quadrilateral area coordinate is presented. There are two main advantages of this algorithm. One is that it can easily achieve sub-pixel alignment accuracy. The other is that it breaks through the limitation of the parallelism between the camera image plane and the planar screen plane, thus the camera can capture much greater view-of-field in the same condition. From the experimental results, the new algorithm improves the flexibility and scalability of multi-projector tiled display system.

Acknowledgements

This work received supports from the National Natural Science Foundation of China (Grant No. 60473112), the Specialized Research Fund for the Doctoral Program of Higher Education of China (20030003053).

References

1. Michael S. Brown, W. Brent Seales: Incorporating Geometric Registration with PC-Cluster Rendering For Flexible Tiled Displays. Int. J. Image Graphics 4(4): 683-700 (2004)
2. R.Surati, Scalable Self-Calibrating Display Technology for Seamless Large-Scale Displays, PhD thesis, Dept. of Electrical Engineering and Computer Science, Massachusetts Institute of Technology, Cambridge, Mass., 1999.
3. R. Raskar et al., "Multi-Projector Displays Using Camera-Based Registration," Proc. of IEEE Visualization 1999, ACM Press, New York, Oct. 1999.
4. H. Chen, R. Sukthankar, and G. Wallace, "Scalable Alignment of Large-Format Multi-Projector Displays Using Camera Homography Trees," Proc. IEEE Visualization 2002, pp. 339-346, 2002.
5. R. Raskar, J. van Barr, and J.X. Chai. A low-cost projector mosaic with fast registration. In ACCV 2002.

Immersive Open Surgery Simulation

Ali Al-khalifah, Rachel McCrindle, and Vassil Alexandrov

School of Systems Engineering, The University of Reading, Whiteknights PO Box 225,
Reading RG6 6AY, UK
`{a.h.al-khalifah, r.j.mccrindle, v.n.alexandrov}@rdg.ac.uk`
`http://www.sse.rdg.ac.uk`

Abstract. Realistic medical simulation has great potential for augmenting or complimenting traditional medical training or surgery planning, and Virtual Reality (VR) is a key enabling technology for delivering this goal. Although, medical simulators are now widely used in medical institutions, the majority of them are still reliant on desktop monitor displays, and many are restricted in their modelling capability to minimally invasive or endoscopic surgery scenarios. Whilst useful, such models lack the realism and interaction of the operating theatre. In this paper, we describe how we are advancing the technology by simulating open surgery procedures in an Immersive Projection Display CAVE environment thereby enabling medical practitioners to interact with their virtual patients in a more realistic manner.

1 Introduction

Virtual reality (VR) is a key technology for enabling medical practitioners to visualize and interact with their models/datasets. Indeed, the potential for VR as a modelling and simulation tool for medical training or surgery planning has already been demonstrated through a number of research-based and commercial simulation systems [1, 2, 3]. However, the majority of these systems are desktop based applications and hence lack the presence, immersion, interactive and collaborative elements possible from developing VR applications within a CAVE environment. Additionally, with the exception of the US Department of Defence who are developing an immersive trauma related military medical training system [4], the majority of these applications have been restricted to modelling specific endoscopic surgical procedures.

Simulation of open surgical procedures presents a new set of challenges for researchers and developers. These challenges exist due to the complex procedures that must be modelled such as tissue deformation, suturing, incisions and bleeding of the patient, as well as the delicate movements and responses of the surgeon's hand. The nature of these procedures mean that open surgery simulations cannot be as realistically modelled on a desktop display as can an endoscopic surgical procedure.

In this paper we describe our initial work in simulating an open surgical procedure within a CAVE based environment.

V.N. Alexandrov et al. (Eds.): ICCS 2006, Part I, LNCS 3991, pp. 868–871, 2006.
© Springer-Verlag Berlin Heidelberg 2006

2   The Application

The aim of our virtual immersive application is to simulate the real-life scenario of an operating theatre with a patient lying on the operating table surrounded by surgeons and other medical staff. Through our modelling of this scenario, users of the system can surround the virtual patient and interact with it in the same manner as that of a real surgical team interacting with a real patient.

To achieve this we have initially focussed on two types of modelling procedures - surface deformations and model/object manipulations. Using a 3-D wand, the application enables the user to arbitrarily deform a pre-defined surface, the abdomen in this case, such that the virtual body can be cut open and the internal organ structures - stomach, liver, kidneys, small intestine, and the heart - exposed ready for medical intervention. Through this approach realism is added to the model as the elastic tissue of the human body can be stretched or squeezed when it come into contact with other objects or structures. This is an important characteristic for an open surgery medical simulation application.

3   Environment and Implementation

The CAVE used for the simulation is a large box shaped display device composed of 4 flat screens (three walls and a floor) approximately 10x10x10 foot in dimensions. Five to six users may step inside the CAVE space at any one time and further collaboration may also occur with other CAVE based users across a network. Stereoscopic images are back projected onto the screen by special projectors. In order for users to see the images inside the CAVE as 3D objects special liquid crystal shutter glasses are worn. Interaction with the displayed model is via a head tracking device and a 3D 6 degree of freedom mouse known as a 'wand'. In this application the wand simulates scalpel interaction.

The application is coded in the C language using the popular graphics interface OpenGL and CAVElib. OpenGL is used to model all surface and deformable objects as well as other scene aspects such as light and material properties. CAVElib is used to implement the environment interaction and for routines such as tracking, navigation, and dynamic object manipulation. The software tools were selected on the basis of their availability, ease of use and compatibility with most operating platforms. Organ models are imported into the model as individual data files, converted from various formats such as 3D-Studio (.3ds), VRML (.wrl) and SGI-Inventor (.iv) using Deep Exploration [5].

4   Results

As can be seen from Figure 1 the developed application allows us to simulate a collaborative operating environment as well as simple tissue deformation and manipulation operations and exposure of the organs pertinent to the surgical procedure being conducted (Fig 2a and 2b). Object and tool manipulations are also implemented that enable the user to interact with static non-deformable models such as surgical tools.

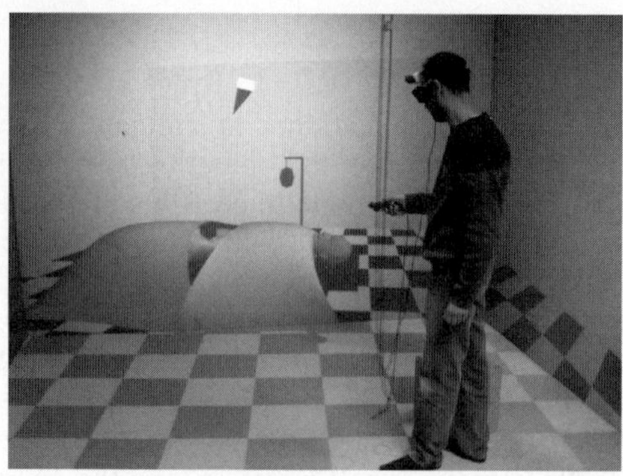

Fig. 1. Surgeon Interacting with the Virtual Patient

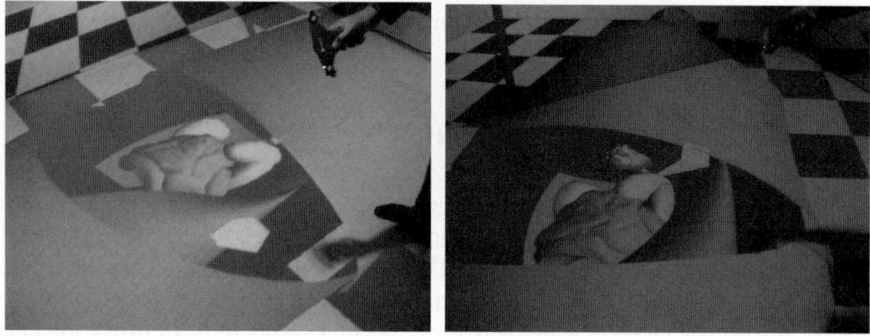

Fig. 2a and 2b. The abdominal is opened by deforming its surface and internal organs displayed ready for interventions

5 Evaluation

In order to gain feedback on our simulations and to ascertain views of practitioners on the applicability of virtual reality for medical training and surgery planning, a study, conducted via questionnaires and CAVE-based demonstrations, is underway. Initial results are encouraging as part-shown in Tables 1 and 2; based on feedback from 30 participants including medical consultants, general practitioners, researchers, surgeons, and medical educators.

Table 1. Initial feedback results in Percentages (Part 1)

Survey Question	Agree	Not Sure	Disagree
Medical practitioners will eventually benefit from the immersive simulation technology	41	28	31
Immersive displays such as the CAVE provide more powerful tools and features to the medical profession than a desktop display	47	24	29
The CAVE can be a potential medium for future medical open surgery simulations	26	52	22
I would use this technology to simulate open interventions	23	57	20

However, whilst the results show a general appreciation and acceptance of VR simulations, we have found that visual simulation alone may not be adequate to convince the medical community of the applicability of VR technology in meeting their full expectations. For example, a haptic interface will be a major enhancement to open surgery simulation adding further realism and enhancing interactions with the environment and patient.

Table 2. Initial feedback results in Percentages (Part 2)

Survey Question	High	Neutral	Low
Rate the usefulness of the CAVE in particular for medical simulations	41	12	47
Rate the usefulness of VR in general for medical simulations	53	22	25
How far would you like to see immersive CAVE technology	43	34	23

6   Conclusions and Further Work

We have successfully implemented a simple immersive application that simulates an open surgery procedure, allowing the user to open the abdominal cavity, observe deformation of elastic tissue, and to interact with internal organs. This type of surgery is hugely demanding in terms of computer power and complexity of operations involved and hence this application barely scratches the surface of such complex applications. However, it gives us a strong basis on which to develop further complex procedures and operations. The major advantage of this application is that through exploiting the CAVE technology, it supports the crucial elements associated with open surgery operations such as, immersion, interactivity, presence, collaboration and multi-user interactions. Initial feedback from practitioners involved in medical education and surgery planning has been encouraging and generally positive.

We will continue to develop the complexity of our open surgery models and simulations. Collaboration between users located at geographically different locations is being further developed. Haptic rendering inside the CAVE is another area of interest that when implemented will greatly improve the realism of open surgery simulations.

References

1. Liu A., Tendick F., Cleary K., Kaufmann C.: Survey of Surgical Simulation: Applications, Technology, & Education. Presence: Teleoperators & Virtual Env. (2003), 599-614
2. Hollands R. J., Trowbridge E. A. A virtual reality training tool for the arthroscopic treatment of knee disabilities. University of Sheffield, UK (2005)
3. Vidal F. P., Bello F., Brodlie K., John N. W., Gould D., Phillips R., Avis N. J. Principles and Applications of Medical Virtual Environments. Eurographics 2004 State of the Art reports (2004)
4. Immersion Corporation. http://immr.client.shareholder.com/ReleaseDetail.cfm?ReleaseID= 111797 (2005)
5. Right Hemisphere. http://www.righthemisphere.com/ (2005)

An XML Specification for Automatic Parallel Dynamic Programming*

Ignacio Peláez, Francisco Almeida, and Daniel González

Departamento de Estadística, I. O. y Computación
Universidad de La Laguna, c/ Astrofísico F. Sánchez s/n
38271 La Laguna, Spain
`ignacio.pelaez@gmail.com, falmeida@ull.es, dgonmor@ull.es`

Abstract. Dynamic Programming is an important problem-solving technique used for solving a wide variety of optimizations problems. Dynamic programs are commonly designed as individual applications and, software tools are usually tailored for specific classes of recurrences and methodologies. We presented in [9] a methodological proposal that allowed us to develop a generic tool [DPSKEL] that supports a wide range of Dynamic Programming formalizations for different parallel paradigms. In this paper we extent this work by including a new layer between the end users and the tool in order to reduce the development complexity. This new layer consists in a XML specification language to describe Dynamic Programming problems in an easy manner.

1   Introduction

As stated in [2] Skeletal programming has revealed as an alternative and contributes to simplify programming, to enhance portability and to improve performance. Such systems hide the parallelism to the programmer and have been characterized for been embedded entirely into a functional programming language or for integrating imperative code within a skeletal framework in a language or library. Some of these approaches can be seen at [4], [7], [3], [1], [6], [8], [5], [2]. The underlying idea of separating the specification of a problem, or an algorithm, from implementation details that are hidden to the user is present in all the proposals.

In [9] we presented a methodological proposal that allowed us to develop a generic tool (DPSKEL) that supports a wide range of Dynamic Programming formalizations for different parallel paradigms. Parallelism is supplied to the user in a transparent manner through a common sequential interface of C++ classes. The user provides the functional equation as a sequential C++ method. DPSKEL has been developed and a release for shared memory architectures has been validated using a set of tests problems representative of different classes of Dynamic Programming functional equations.

* This work has been partially supported by the EC (FEDER) and the Spanish MEC (Plan Nacional de I+D+I, TIN2005-09037-C02).

V.N. Alexandrov et al. (Eds.): ICCS 2006, Part I, LNCS 3991, pp. 872–875, 2006.

We propose to introduce a new abstraction layer between the user and the skeletons (DPSPEC from now on). This layer separates the fundamental logic behind a problem specification (the functional equation) from the specifics of the particular middleware that implements it, the instantiation using DPSKEL. This allows rapid developments and delivery of new applications. The benefits of the approach are significant to scientists:

- Reduced development time for new applications
- Improved application quality
- Increased use of parallel architectures by non expert users.
- Rapid inclusion of emerging technology benefits into their systems

We will show how this methodology can be applied without any loss of efficiency. The W3C recommendations have also been a requirement of the project. The software architecture of our transformation tools is presented in table 1. The paper has been structured as follows, we present in section 2 the XML specification for DP (DPSPEC), and in section 3 the transformation between DPSPEC and DPSKEL is stated. The paper finalizes with some concluding remarks and future lines of work.

Table 1. Software architecture of the Dynamic Programming transformation tool

Mathematics Editor (Graphical)
MathML, OpenMath, etc.
Transformer to DPSPEC
DPSPEC
Transformer to DPSKEL
DPSKEL
OpenMP, MPI, OpenMP + MPI, etc.
Parallel Architecture

2 DPSPEC a XML Specification for Dynamic Programming Problems

Although some of the XML specifications already existing (MathML, OpenMath, OMDoc, XDF, ...) could be used to specify the DP functional equation, we decided to develop our own specification adapted to DP problems. The main reasons for that decision were to reduce the elements to the minimum and to introduce some changes in the structure on specific elements appearing in classical specifications. These design issues make easier the later parsing and allow a better semantical analysis to detect data dependencies, while keeping as closer as possible to the user defined equation at the same time. The semantical analysis determines the traversing mode of the DP table. DPSPEC brings together the elements to describe piecewise defined functions, simple variables and vectors, arithmetic, logical, relational and max, min operators, and iterators. Table 2 summarizes elements available in the DPSPEC language.

Table 2. Current elements available in DPSPEC

Element	Operation
Main element	`<problem>`
Logical functions	`<and>`, `<or>`, `<not>`
Conditional	`<cond>`
Relational operators	`<lt>`, `<le>`, `<gt>`, `<ge>`, `<eq>`, `<ne>`
Binary mathematical operators	`<minus>`, `<divide>`, `<power>`
N-ary mathematical operators	`<plus>`, `<times>`, `<max>`, `<min>`
N-ary iterative operators	`<imax>`, `<imin>`
States	`<state>`
Variables	`<ci>`
Constants	`<cn>`
User defined functions	`<functiondef>`, `<function>`
User defined vectors	`<vectordef>`, `<vector>`

3 Automatic Parallelization of Dynamic Programming Problems

The automatic parallelization of Dynamic Programming problems is achieved by making explicit transformations on the DPSPEC data file holding the Dynamic Programming equation. The XML description of the formula is converted into a specific instantiation of the DPSKEL C++ skeleton to solve the problem considered. However, scientists typically represent the functional equation as mathematics expressions, using their favorite equation editor (latex, OpenOffice, etc.). Therefore, two transformations are implicitly involved in the process, the conversion of the mathematical equation to the XML specification and the transformation of the XML equation specification into the set of C++ classes. The first transformation step is currently provided automatically for many popular MathML software editors, which generate a MathML document for a given equation. DPSPEC is compatible with MathML and the conversion between a MathML document into a DPSPEC document is achieved through a XSLT preprocessing step.

The second transformation step involves a deeper analysis of the functional equation, we perform a DOM parsing of the XML functional equation to produce the proper C++ required classes of DPSKEL, the parser has been developed using the Xerces-C++ library.

4 Conclusions and Future Work

As a conclusions we can say that we have developed a XML specification language to describe Dynamic Programming problems (DPSPEC). We present a methodology that allows to generate automatically parallel applications. The methodology is based in the existence of general parallel programs (DPSKEL) that can be generated from the XML specification. The code generated is efficient

since no overhead is introduced during the transformation steps. The technique
has been validated with a wide range of cases of study. Several extensions to DP-
SPEC are in the agenda. A natural extension to the language is to support new
data types. Once a XML specification has been stated, transformations from/to
many other languages can be developed at a reasonably cost. We are also interested in the development of new transformation tools from other languages to
DPSPEC, that is the case for example OpenMath and from DPSPEC to other
languages such as WSDL to provide the interface for a web service application.

This work has been partially supported by the EC (FEDER) and the Spanish
MEC (Plan Nacional de I+D+I, TIN2005-09037-C02).

References

1. M. Aldinucci, S. Gorlatch, C. Lengauer, and S. Pelagatti. Towards parallel programming by transformation: The FAN skeleton framework. *Parallel Algorithms
and Applications*, 16(2–3):87–122, 2001.
2. Murray Cole. Bringing skeletons out of the closet: a pragmatic manifesto for skeletal
parallel programming. *Parallel Comput.*, 30(3):389–406, 2004.
3. Marco Danelutto and Massimiliano Stigliani. Skelib: Parallel programming with
skeletons in c. In *Euro-Par '00: Proceedings from the 6th International Euro-Par
Conference on Parallel Processing*, pages 1175–1184, London, UK, 2000. Springer-
Verlag.
4. John Darlington, A. J. Field, Peter G. Harrison, Paul H. J. Kelly, D. W. N. Sharp,
and Q. Wu. Parallel programming using skeleton functions. In *PARLE '93: Proceedings of the 5th International PARLE Conference on Parallel Architectures and
Languages Europe*, pages 146–160, London, UK, 1993. Springer-Verlag.
5. A. J. Dorta, J. A. González, C. Rodríguez, and F. de Sande. llc: A parallel skeletal
language. *Parallel Processing Letters*, 13(3):437–448, 2003.
6. E. Alba et al. MALLBA: A library of skeletons for combinatorial optimisation (research note). In *Proceedings of the 8th International Euro-Par Conference*, volume
2400 of *LNCS*, pages 927–932, 2002.
7. Daniel González-Morales, Francisco Almeida, F. Garcia, J. Gonzalez, Jose; L. Roda,
and Casiano Rodríguez. A skeleton for parallel dynamic programming. In *Euro-Par
'99: Proceedings of the 5th International Euro-Par Conference on Parallel Processing*, pages 877–887, London, UK, 1999. Springer-Verlag.
8. Herbert Kuchen. A skeleton library. In *Euro-Par'02: Proceedings of the 8th Euro-
Par Conference on Parallel Processing*, pages 620–629, London, UK, 2002. Springer-
Verlag.
9. Ignacio Peláez, Francisco Almeida, and Daniel González. High level parallel skeletons for dynamic programming. *Parallel Processing Letters*, To appear, 2006.

Remote Sensing Information Processing Grid Node with Loose-Coupling Parallel Structure

Ying Luo[1,3], Yong Xue[1,2,*], Yincui Hu[1], Chaolin Wu[1,3], Guoyin Cai[1], Lei Zheng[1,3], Jianping Guo[1,3], Wei Wan[1,3], and Shaobo Zhong[1]

[1] State Key Laboratory of Remote Sensing Science, Jointly Sponsored by the Institute of Remote Sensing Applications of Chinese Academy of Sciences and Beijing Normal University, Institute of Remote Sensing Applications, Chinese Academy of Sciences, P.O. Box 9718, Beijing 100101, China
[2] Department of Computing, London Metropolitan University, 166-220 Holloway Road, London N7 8DB, UK
[3] Graduate School of the Chinese Academy of Sciences, Beijing, China
jennyjordan@hotmail.com, y.xue@londonmet.ac.uk

Abstract. To use traditional algorithms and software packages on Grid system, traditional algorithms and software packages, in general, have to be modified. In this paper we focus on standards and methodologies for Grid platform within the context of the Remote Sensing Data Processing Grid Node (RSDPGN) that implements a loose-coupling parallel structure for orchestrating traditional remote sensing algorithms and software packages on the Condor platform. We have implemented 17 remote sensing applications in one system using Web service and workflow technology without any change to traditional codes. Some core algorithm codes are come from a remote sensing software package which we has neither resource codes nor APIs. Others come from the program codes accumulated by our group. The design and prototype implementation of RSDPGN are presented. The advantage and shortage of loose-coupling structure is analysed. Through a case study of land surface temperature calculation from MODIS data, we demonstrate the way to modify software packages in details. Moreover we discuss the problems and solutions based on our experience such as system architecture, the kinds of functional modules, fast data transfer, and state monitoring.

1 Introduction

The information extracted from remote sensing data plays an important role in science and society. The algorithms and software packages to extract information vary based the characteristics of sensors, spectrum, etc. Generally speaking, the process of remote sensing data is complicated and time consuming. Remote sensing software packages such as ARCGIS, ENVI, and ERDAS are expensive, and the use of them needs special skill. On the other hand, researchers might prefer to pay for only the function modules which he or she is interested in, or have a third party to do the pre-processing steps or offer some computing power on demand. Thus we need a system

---

 Corresponding author.

V.N. Alexandrov et al. (Eds.): ICCS 2006, Part I, LNCS 3991, pp. 876–879, 2006.

that provides individual function and computing power, allow user to pay for only certain functions. The emergence of Grid technology brings out a solution to this system. Grid technology is an evolution in resources sharing including hardware, data, and software. It will change the mode of software industry at both develop stage and sale stage.

The paper is organized as follows. In section 2 RSDPGN with some detail of the functions of the various components and our implementation at present will be demonstrated. Then in section 3, we will discuss the standards and methodologies for Grid platform through a case study of land surface temperature retrieval from MODIS data, and give our experiences in building RSDPGN on how to build a computing Grid node. Finally, the conclusions will be addressed in Section 4.

2 Remote Sensing Data Processing Grid Node (RSDPGN)

Remote Sensing Data Processing Grid Node (RSDPGN) is the infrastructure that manages the software and hardware resources to process remote sensing data according to users' demand. The data to be processed by our node is come from the user or a third party.

2.1 Architecture of RSDPGN

Figure1 describes the architecture of remote sensing data processing Grid node. The node has four layers:

1) Portal layer: It is an outmost user interface, which can display data and results, receive data, JOB, and algorithm, register and certificate.

2) Application layer: It is the most important layer for a processing node. The fundamental modules include reliance file transfer (RFT), security, several database (DB) such as algorithm database, remote sensing data database, user information database, and instance information database, monitor to check the status of JOB, hardware, and software, and global scheduler.

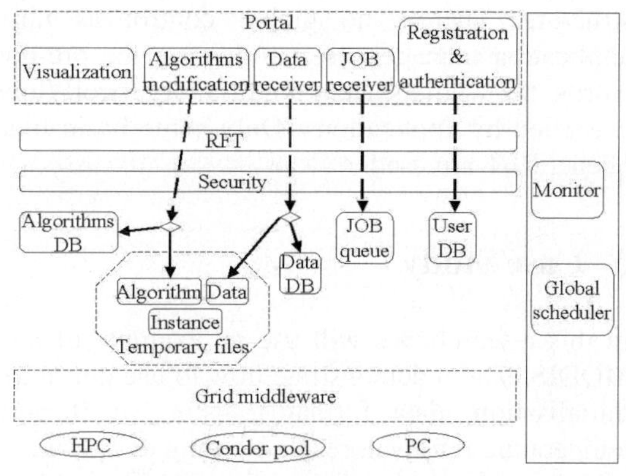

Fig. 1. Architecture of RSDPGN

3) Grid middleware layer: The communication and management to hardware resource is in this layer. We developed a toolkit named RSNkit (Remote Sensing Node toolkit) to add some functions not provided by Condor.

4) Hardware layer: The hardware entities can be PCs, HPC, or Condor pools, etc.

The node issues and registers its services to SIG manage centre periodically, responses calls of SIG, triggers services, and reports status. There an instance of an application is called a JOB. A JOB has several steps. Each step is one task that can be submitted to either the Condor pool or HPC or PC. Receiving require from SIG manager, the node will find data from user or remote data servers according to user's requirements, organize computing resource dynamically, trigger services, and monitor their running status. To decrease the total processing time of a task, the node will divide a task into several sub-tasks. The exact number and size of the sub-tasks is according to the current PC number and configure in the Condor computing pool [1] and super computer or designated by the user. Only when the task is large enough, or on the user's request, the node triggers off the super computer to do a large sub-task. The method we trigger super computer is different with that of Condor pool.

2.2 Implementation of RSDPGN

Currently, RSDPGN provides 17 remote sensing applications by MODIS data, such as aerosol optical depth retrieval, land surface temperature retrieval, soil moisture retrieval, surface reflectance retrieval, some vegetation indexes, and a series of pre-processing functions. Some core algorithm codes of remote sensing applications are from a remote sensing software package, which we has neither resource codes nor APIs. Others are program codes accumulated by our group. We implement our portal as a Web site using dynamic page technology JSP. The development work of end user software is minimum. Anyone who has a browser can use our node. In order to integrate with other Grid systems in the future, our node is implemented with Web service technology such as SOAP, XML, and WSDL. Each remote sensing application is enwrapped individually so that the system is loose coupling and parallel structure. There's no global control as normal system. Each remote sensing application manages itself following the pre-defined strategy in workflow. In other words, the whole system is not a large workflow, but many small parallel workflows classified by applications. Only some basic function modules such as monitor, JOB queue, RFT are used.

3 Case Study

In this research, we will use an example of land surface temperature retrieval from MODIS data to demonstrate how to use our node [2]. The retrieval process consists of initialisation, data format transfer, rectification, region selection, data division, temperature retrieval, result merging and result return to the users. When a user orders a temperature retrieval service via Grid portal, a JOB workflow instance is initialised once the user click the submit bottom after choosing the data to be processed. Finally the blocks of retrial result are merged, and return to the end user. The JOB status is monitored by the system monitor. The monitor of the whole node including JOB status, sub jobs' status, workflow module status, hardware status is also an important issue. Continually status query wastes much CPU cycles. Our solution is to give a speed attribute to each algorithm. So the system can estimate a probable time and use the time to decide whether to query the status.

   File transfer is one of the big issues. It is the bottleneck that increases the total processing time. There is another issue adding the complexity of the data node. The application workflow need to know not only data and metadata but also attributes such as whether it has been rectified, with what algorithm, how about the processing precision, and so on.

4 Conclusion

In this paper, we introduced our ongoing research on RSDPGN. It is a demonstration on how to integrate traditional algorithms and software packages, computing resource, and Grid in order to provide one-stop service to users. It is service-oriented. We have implemented it mainly by workflow technology. We orchestrated 17 remote sensing applications on the Grid node. We used the strategy to let each application manages itself following the pre-defined workflow. The loosely coupling structure of our node is different with common centralized control structure. It is an easy design, modifying, and extending system. Furthermore this structure is convenient to breakpoint and status query, but it adds spending of computing resource and store resource. Additionally, the rock-bottom communication, control and monitor to Condor computing pool and HPC rely on Condor and RSNkit. It is recommended to use loosely coupling structure in complex Grid system that involved in many function modules and hardware. But for the correspondingly system, it is better to use a centralized control structure.

Acknowledgement

This publication is an output from the research projects "Grid platform based aerosol fast monitoring modeling using MODIS data and middlewares development" (40471091) funded by the NSFC, China, "Dynamic Monitoring of Beijing Olympic Environment Using Remote Sensing" (2002BA904B07-2) and "863 - Remote Sensing Information Processing and Service Node" (2003AA11135110) funded by the MOST, China and "Digital Earth" (KZCX2-312) funded by CAS, China.

References

[1] Basney, J., Livny, M., and Tannenbaum, T., 1997, High Throughput Computing with Condor, *HPCU news,* Volume 1(2)
[2] Xue, Y., Cai, G. Y., Guan, Y. N., Cracknell, A. P. and Tang, J. K., 2005, Iterative Self-Consistent Approach For Earth Surface Temperature Determination. *International Journal of Remote Sensing*, Vol. 26, No. 1, 185–192.

Preliminary Through-Out Research on Parallel-Based Remote Sensing Image Processing*

Guoqing Li[1], Yan Ma[2], Jian Wang[1], and Dingsheng Liu[1]

[1] Key laboratory, Remote Sensing Satellite Ground Station, Chinese Academic of Sciences
No. 45 BeiSanHuanXi Road, P.O. Box 2434, Beijing, 100086, China
Graduate University, Chinese Academic of Sciences
No. 45 BeiSanHuanXi Road, P.O. Box 2434, Beijing, 100086, China
{gqli, yma, jwang, dsliu}@ne.rsgs.ac.cn

Abstract. As the most important and most complex Geo-Information, remote sensing data can be fast processed with cluster-based parallel computation technologies. The through-out ability of every step of such processing will affect heavily the application of remote sensing image parallel processing technologies. This paper shall discuss more detail how to set up the through-out model of remote sensing image parallel processing and presents some deeply research works on such through-out mechanism. A lot of experiments have been given to support a quantitative analysis method to adjust the performance of remote sensing parallel processing system.

1 High Through-Out Computation and Remote Sensing Image Processing

There are many research works focusing on how to improve the spatial data processing speed. HPC (high performance computation), especially based on cluster can bring the outstanding speed-up [1] [2]. On the other hand, HTC (high through-out computation) should also be paid more attention to [3].

The aim to research HTC is to build an economical and high efficiency system. High performance parallel computation can support the improvement on process speed of independent processing steps. However, the isolated effort on special step will not always cause the appearance of HTC. When we are building a HTC system, what we have to pay more attention to is the discovery and solver of bottlenecks.

Most important, remote sensing image processing has high data complexity. To design a parallel model, the matching relationship of data transfer, catching and node computing ability should be considered firstly. During certain time, HTC system can deal with most data than other system, and data flux is thought as the most important technical index for some applications.

---

* The above research is supported by the National Key Research and Development Plan of China.

V.N. Alexandrov et al. (Eds.): ICCS 2006, Part I, LNCS 3991, pp. 880–883, 2006.

2   The Through-Out Rate Model of Remote Sensing Image Processing

2.1   Process Course Model

A remote sensing image parallel process course is organized with some important steps, data loading, distribution, processing, collection and export, as shown in Fig. 1. Because the step of data loading is similar with data export, the step of data distribution is similar with data collection in the term of through-out character; such steps can be described with the steps of data loading and data distribution.

Through-out rate model

The through-out rate R is the most important index of HTC.

$$T = T_{input} + T_{dist\,ribute} + T_{process} \tag{1}$$

Where T means the full process time, T_{input} means the data loading time, $T_{dist\,ribute}$ means the data distributing time and $T_{process}$ means data processing time. If we define M is the amount of processed data, then

$$T_{input} = \frac{M}{R_{input}} \tag{2}$$

$$T_{dist\,ribute} = MAX \left\{ \frac{M_i}{R_{distributi\,on}} \right\} \quad i = 1...n \tag{3}$$

$$T_{process} = MAX \left\{ \frac{M_i}{R_{process}} \right\} \quad i = 1...n \tag{4}$$

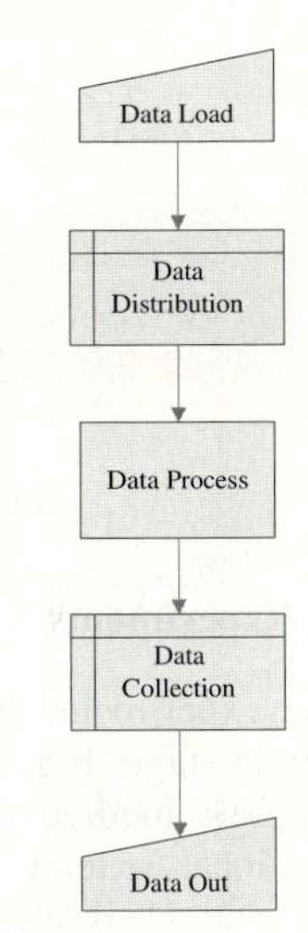

Fig. 1. process course model

If we suppose data M is averagely distributed on n nodes, then $M_i = \frac{M}{n}$. From formula (1) to (4),

$$T = \frac{2M}{R_{input}} + \frac{2M}{nR_{distributi\,on}} + \frac{M}{nR_{process}} = \frac{M}{n}\left(\frac{2n}{R_{input}} + \frac{2}{R_{distributi\,on}} + \frac{1}{R_{process}} \right) \tag{5}$$

Where R_{input} is the data loading through-out rate, and $R_{distribution}$ means the data distribution through-out rate and $R_{process}$ is the rate of data processing.

$$R = \frac{M}{T} = \frac{n}{\dfrac{2n}{R_{input}} + \dfrac{2}{R_{distributi\,on}} + \dfrac{1}{R_{process}}} = \frac{1}{\dfrac{2}{R_{input}} + \dfrac{2}{nR_{distributi\,on}} + \dfrac{1}{nR_{process}}} \tag{6}$$

If $R_{min} = MIN\{R_{input}, R_{distribution}, R_{process}\}$, $R_{max} = MAX\{R_{input}, R_{distribution}, R_{process}\}$, there are the following conclusions: (1)When Rmin ~ Rmax, the process course is in the condition of ideal through-out. In this condition, the improvement of system through-out is depended on the matching of these three through-out rates; however the limit of such improvement is not above one order of magnitude. (2)When Rmin << Rmax, the process course is in the condition of abnormal through-out. In this condition, the improvement of system through-out is depended on the tuning of the lowest course of through-out. (3)With the increasing of node number n, the through-out rate R is increasing. When n is large enough, R is limited in R_{input} It means that the increasing of computing node will bring up the improvement of system through-out capability, while the limit of such increasing is R_{input} .(4)The through-out rate R is stable in some degree, which means R has low relativity with the scale of input data.

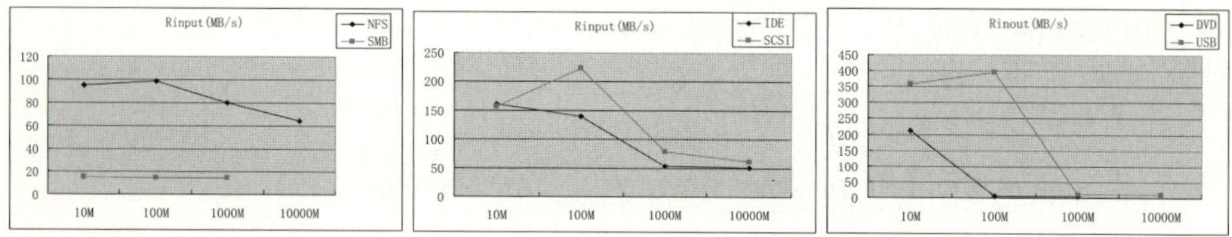

Fig. 2. The R_{input} trend of three data loading methods

2.3 Experiments

Some experiments have been taken to research the performance and character of different stage. Fig. 2 shows three methods of data loading and exporting, which are local disk method (SCSI, SATA and IDE etc.), local extend storage method (DLT, DVD and USB etc.) and network method (NFS, SMB and RCP etc.). In Fig. 3, the different scale data in cache node is distributed to the other 8 nodes in Giga LAN with methods of MPI, NFS and PIPFS [4] Fig. 4 gives the comparison of the through-put between two typical remote sensing image algorithms FFT and Wavelet within same parallel environment.

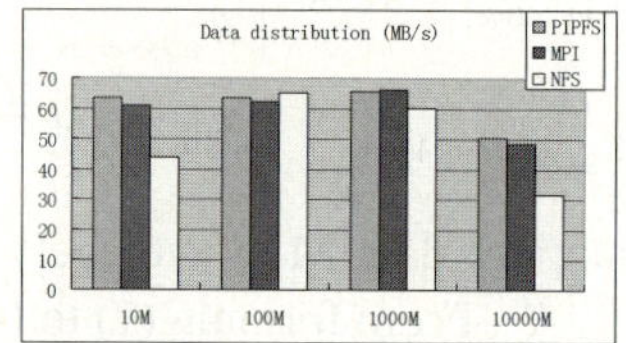

Fig. 3. the $R_{distribution}$ of three distri-bution mode

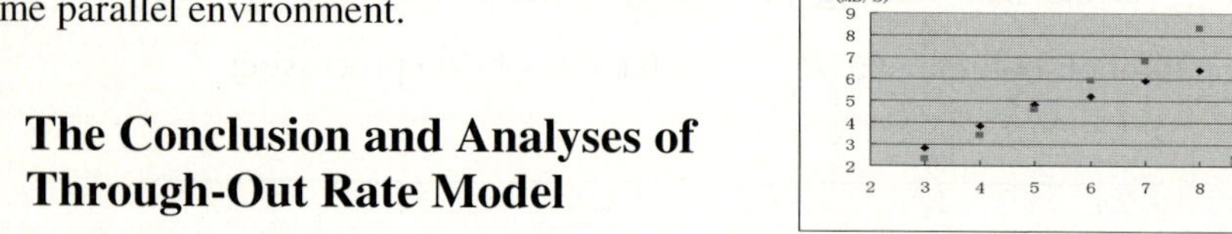

Fig. 4. $R_{process}$ trend of two typical processes

3   The Conclusion and Analyses of Through-Out Rate Model

(1) When nQ is least than the range of 40~70 MB/s, R is limited by the throughput nQ and the capability of through-out can be simply accelerated with adding computing nodes.
(2) When nQ is in the range of 40~70, R is near 40~70MB/s. The better through-out performance can be implemented by adjusting the relationship of three courses.

(3) When the nQ value is more than 70, adding computing node number will not bring the increasing of R. In such system, it is no sense to extend the scale of computer cluster.

(4) When Q is large enough, which means the process algorithm is too simple, the most cost of the process courses is spend during data loading and distributing. Only little time cost is used in the data processing. The method to improve the through-out performance of such system is depended on the accelerating of data loading and distributing, which can be accessed with advance storage technologies and combined bandwidth technology.

(5) When Q is not very large, the process algorithm is very complex and the process course will spend relative long time. The method to improve the through-out performance can be realized with the adding of computing nodes. The devoting of the improvement on storage technology will not be outstanding in such condition.

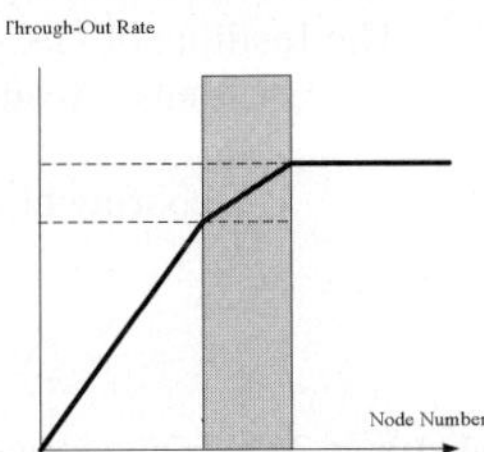

Fig. 5. The relationship between through-out rate and node number

In some degree, the above analyses are the predigesting of process courses. It will be more complex in real, for example the imbalance of load and data distribution. However, the methodology of through-out analyses is also useful for the research of such condition.

References

[1] High Performance Computing Clusters, Constellations, MPPs, and Future Directions, Jack Dongarra, Thomas Sterling, Computing in Science and Engineering, March/April 2005 (Vol. 7, No. 2) pp. 51-59

[2] Guoqing Li, Dingsheng Liu, Key Technologies Research on Building a Cluster-based Parallel Computing System for Remote Sensing, LNCS 3516, pp. 484-490

[3] IBM website:http://www-128.ibm.com/developerworks/cn/linux/cluster/hpc/part1/

[4] Keying Huang, Guoqing Li, Dingsheng Liu A parallel file system based on spatial information object, LNCS 3779, pp. 154-162

A Shortest Path Searching Method with Area Limitation Heuristics

Feng Lu[1] and Poh-Chin Lai[2]

[1] State Key Laboratory of Resources and Environmental Information System,
The Institute of Geographical Sciences and Natural Resources Research,
Chinese Academy of Sciences, Beijing 100101, P.R. China
luf@lreis.ac.cn
[2] Department of Geography, The University of Hong Kong,
Hong Kong SAR, P.R. China

Abstract. While heuristics based on geometric constructs of the networks would appear to improve performance of Dijkstra's algorithm, the fallacy of depreciated accuracy has been an obstacle to the wider application of heuristics in the search for shortest paths. The authors presented a shortest path algorithm that employs limited area heuristics guided by spatial arrangement of networks. The algorithm was shown to outperform other theoretically optimal solutions to the shortest path problem and with only little accuracy lost. More importantly, the confidence and accuracy levels were both controllable and predictable.

1 Introduction

Shortest path algorithms traditionally concerned topological or phase characteristics of a network and neglect the spatial or proximity characteristics. Such a viewpoint has resulted in a looping and radial approach to path searching which cannot prevent redundant searching even in situations when the destination nodes have been located. Some researchers have argued that the locational information (i.e. relative positions) of nodes can be employed in heuristics to inform the searching process [1][2]. Heuristics have been widely used in optimum path algorithms [3][4][5].

Most heuristics utilized local controlling strategies to guide the node search. They were efficient in locating approximately optimal one-to-one shortest paths but could not handle one-to-some or some-to-one shortest path problems.

The optimal path algorithms and heuristic strategies can be integrated to make a trade-off between efficiency and theoretical exactness, which is especially important for large-scale web applications or real time vehicular path finding. This paper highlights advantages of the integration using real-world networks of different complexity and connectedness.

2 Real-World Networks in the Study

We downloaded ten real-world networks representing road structures of varied complexity from http://www.fhwa.dot.gov/ and http://www.bts.gov/. The network set

V.N. Alexandrov et al. (Eds.): ICCS 2006, Part I, LNCS 3991, pp. 884–887, 2006.

Table 1. Characteristics of real-world networks in the study

Network set	Alabama	Georgia	Pennsylvania	New York	Texas
Number of nodes	952	2308	2640	3579	3812
Number of arcs	1482	3692	4183	5693	6340
Network set	California	Northeast USA	USA Total	Utah Detailed	Alabama Detailed
Number of nodes	5636	14009	26322	72558	154517
Number of arcs	9361	19788	44077	100533	199656

data were edited to topological correctness and then transformed into some custom-built ASCII files to store geometrical and topological information.

3   Search Heuristics Based upon Geographic Proximity

Our computation for the shortest paths used Dijkstra's algorithm implemented with a quad-heap structure. An ellipse based strategy is utilized to limit the searching within a local ellipse or its MBR generated with the source and destination nodes. Detailed description of the strategy can be found in [6].

3.1   Correlation Between Shortest Path and Euclidean Distance in Road Networks

First, 1000 sample nodes were extracted systematically from each network to reconstitute sets A and B, each containing 500 nodes. In the case of Alabama, all 952 nodes were used (Table 1). Every node in A or B would be regarded in turn as origins and destinations between which the shortest paths must be determined.

A set R of ratios $r_{ab} = p_{ab}/e_{ab}$ (e_{ab}: the Euclidean distance; p_{ab}: the shortest path distance) could be computed for each sample set. In general, we could establish a real number τ as the threshold value for the elements in R at a stated confidence level. A τ value at 95 percent would imply 95 percent confidence that the shortest path for a pair of nodes could be found within the extent built with the τ value. The τ values ranged from 1.238 (Texas) to 1.585 (Utah detailed) which appeared quite consistent across the networks.

3.2   Numerical Premise

We proposed the use of a minimum bounding rectangle (MBR) of ellipse to limit the search. Table 2 presented the average r-values against the 95 percent τ threshold of the ten networks. We would attribute a larger τ for California because of its elongated shape and Utah Detailed for its disrupted landscape. In other words, there is great likelihood (i.e. at least 95 percent of the time) that we could locate the shortest paths (SP) between any node pairs within the ellipse MBR. Our attempt to compare success-rates of the elliptical versus MBR search limits showed that the MBR is superior to the ellipse; the former had fewer SPs beyond bound.

Table 2. τ values and comparison between elliptical and MBR limitation areas

Network set	Average ratio $r_{ab} = p_{ab}/e_{ab}$	Threshold τ (at 95 % confidence)	Number of SP beyond the ellipse	Number of SP beyond the MBR	Real confidence for ellipse search (%)	Real confidence for MBR search (%)
Alabama	1.201	1.379	1588	840	99.30	99.63
Georgia	1.162	1.286	1013	619	99.59	99.75
Pennsylvania	1.179	1.393	761	453	99.70	99.82
New York	1.191	1.375	750	406	99.70	99.84
Texas	1.128	1.238	1220	369	99.51	99.85
California	1.170	1.426	1163	802	99.53	99.68
Northeast USA	1.166	1.334	438	250	99.82	99.90
USA	1.139	1.267	226	109	99.91	99.96
Utah Detailed	1.308	1.585	717	405	99.71	99.84
Alabama Detailed	1.209	1.334	822	549	99.67	99.78
Average	1.185	1.362	-	-	99.64	99.81

The last two columns in Table 2 recorded that more than 99% of the shortest paths could be identified within the limited areas. The higher confidence levels meant that the limited search was extremely effective because solutions to shortest paths could be found 99% of the time in one round of search. In other words, only a very small percentage of the shortest paths (an average of 0.36% for ellipse and 0.19% for MBR) would be found beyond the areas.

Table 3. Accuracy and efficiency of ellipse MBR limitation for solving shortest paths

Network set	Possibility of SPs beyond the MBR but near-SPs found (%)	Difference of near-SPs found and real SPs (%)	Time saving with ellipse MBR limitation (%)
Alabama	0.070	13.60	29.25
Georgia	0.108	6.94	34.41
Pennsylvania	0.069	9.10	16.44
New York	0.021	8.06	24.18
Texas	0.060	3.29	37.64
California	0.207	17.74	15.35
Northeast USA	0.077	7.46	32.83
USA	0.070	6.28	33.51
Utah Detailed	0.030	1.50	23.74
Alabama Detailed	0.026	2.56	35.73
Average	0.074	7.65	26.68

3.3 Establishing Accuracy and Computational Efficiency

Our experiment showed that very few shortest paths (i.e. an average of 0.074%) would be mis-calculated with the ellipse MBR limitation (i.e. an average of 7.65% longer than the actual shortest paths). The results indicated that the MBR method would not cause excessive loss in the computational accuracy.

Another experiment was conducted to verify the computational efficiency of the presented method for resolving the one-to-one shortest paths with the 1000 sample nodes (all 952 nodes for Alabama). It showed that an average 26.68% saving in computational time was realized with only minimal loss in accuracy (i.e. 0.074% of the optimum paths would be 7.65% longer than the actual shortest paths on average). The results were listed in Table 3.

4 Conclusion

The theoretically optimal solution and heuristic strategies could be integrated to offer a reasonable trade-off between efficiency and accuracy. The ellipse MBR method we put forth allows users to solve the shortest paths for a stated confidence and in less time. Our experiment showed that more than 99.8 percent of shortest paths could be found within the proposed MBR along with 26.68 percent time saving on average, with an average of only 0.074 percent of the optimal paths exceeding 7.65 percent of the lengths of the actual shortest paths.

References

1. Miller H.J. and Shaw S.L. Geographic Information Systems for Transportation: Principles and Applications. New York: Oxford University Press (2001).
2. Zhao Y.L. Vehicle Location and Navigation Systems. Boston: Artech House Publishers (1997).
3. Fisher P. F. A primer of geographic search using artificial intelligence, Computers and Geosciences, 16(1990) 753-776.
4. Car A. and Frank A. General principles of hierarchical spatial reasoning-the case of wayfinding, Proceedings of the 6th International Symposium on Spatial Data Handling, (1994) 646-664.
5. Holzer M., Schulz F., and Willhalm T., Combining speed-up techniques for shortest-path computations, Lecture Notes in Computer Science, 3059((2004) 269-284.
6. Feng Lu, Yanning Guan, An optimum vehicular path solution with multi-heuristics, Lecture Notes in Computer Science, 3039(2004) :964-971.

FPGA-Based Hyperspectral Data Compression Using Spectral Unmixing and the Pixel Purity Index Algorithm

David Valencia and Antonio Plaza

Department of Computer Science, University of Extremadura
Avda. de la Universidad s/n, E-10071 Caceres, Spain
{davaleco, aplaza}@unex.es

Abstract. Hyperspectral data compression is expected to play a crucial role in remote sensing applications. Most available approaches have largely overlooked the impact of mixed pixels and subpixel targets, which can be accurately modeled and uncovered by resorting to the wealth of spectral information provided by hyperspectral image data. In this paper, we develop an FPGA-based data compression technique based on the concept of spectral unmixing. It has been implemented on a Xilinx Virtex-II FPGA formed by several millions of gates, and with high computational power and compact size, which make this reconfigurable device very appealing for onboard, real-time data processing.

1 Introduction

Our focus in this work is to design a hyperspectral data compression technique able to reduce significantly the large volume of information contained in hyperspectral data while, at the same time, being able to retain information that is crucial to deal with mixed pixels and subpixel targets. A mixed pixel is a mixture of two or more different substances present in the same pixel[1]. A subpixel target is a mixed pixel with size smaller than the available pixel size (spatial resolution). So, it is embedded in a single pixel and its existence can only be verified by using the wealth of spectral information provided by hyperspectral sensors. In this case, spectral information can greatly help to effectively characterize the substances within the mixed pixel via spectral unmixing techniques. However, a major drawback of spectral-based data compression methods for hyperspectral imaging is their computational complexity[2]. The possibility of real-time, onboard data compression is a highly desirable feature to overcome the problem of transmitting a sheer volume of high-dimensional data to Earth control stations via downlink connections. In this work, we develop and FPGA-based compression algorithm based on spectral unmixing concepts. Section 2 develops the lossy compression algorithm. Section 3 maps the algorithm in hardware using systolic array design. Section 4 evaluates the algorithm in terms of both mixed-pixel and subpixel characterization accuracy, using a real image data set collected by the NASA Jet Propulsion Laboratory's Airborne Visible Infra-Red Imaging Spectrometer (AVIRIS). Performance data in a Xilinx Virtex-II FPGA are also given. Section 5 concludes with some remarks.

V.N. Alexandrov et al. (Eds.): ICCS 2006, Part I, LNCS 3991, pp. 888–891, 2006.

2   Hyperspectral Data Compression Algorithm

The first step of the algorithm consists of extracting endmembers from the input data. A well-known approach to accomplish this goal is the PPI algorithm[3], which proceeds by generating a large number of N-dimensional (N-D) random unit vectors called "skewers" through the dataset. Every data point is projected onto each skewer, and the data points that correspond to extrema in the direction of a skewer are identified and placed on a list. The number of times a given pixel is placed on the list is tallied, and the t pixels with the highest tallies are considered the final endmembers.

Using the set of t endmembers above, an inversion model is required to estimate the fractional abundances of each of the endmembers at the mixed pixels. Here, we use a commonly adopted technique as the second step of our compression algorithm, i.e., the linear spectral unmixing (LSU)[1] technique. Suppose that there are t endmembers $\{\mathbf{e}_i\}_{i=1}^{t}$ in a hyperspectral image scene, and let $\mathbf{f}$ be a mixed pixel vector. LSU assumes that the spectral signature of $\mathbf{f}$ can be represented by a linear mixture of $\mathbf{e}_1, \mathbf{e}_2, \ldots, \mathbf{e}_t$ with appropriate abundance fractions specified by an abundance vector $\mathbf{a} = \{\, a_1, a_2, \ldots, a_t \,\}$. Then, we can model the spectral signature of an image pixel $\mathbf{f}$ by a linear regression form $\mathbf{f} = \mathbf{e}_1 \cdot a_1 + \mathbf{e}_2 \cdot a_2 + \cdots + \mathbf{e}_t \cdot a_t$. Two constraints are usually imposed on this model to produce adequate solutions[1], i.e., abundance sum-to-one constraint: $\sum_{i=1}^{t} a_i = 1$, and abundance non-negativity constraint: $a_i \geq 0$ for $1 \leq i \leq t$.

Taking advantage of PPI and LSU, the idea of the proposed data compression algorithm is to represent a hyperspectral image cube by a set of fractional abundance images[4]. More precisely, for each N-D pixel vector $\mathbf{f}$, its associated LSU-derived abundance vector $\mathbf{a}$ of t dimensions is used as a fingerprint of $\mathbf{f}$ with regards to t endmembers obtained by the PPI algorithm. The implementation of the PPI/LSU algorithm can be summarized as follows:

1.   Use the PPI to generate a set of t endmembers $\{\mathbf{e}_i\}_{i=1}^{t}$.
2.   Use the LSU algorithm to estimate the corresponding endmember abundance fractions $\mathbf{a} = \{\, a_1, a_2, \ldots, a_t \,\}$ and approximate each pixel vector as $\mathbf{f} = \mathbf{e}_1 \cdot a_1 + \mathbf{e}_2 \cdot a_2 + \cdots + \mathbf{e}_t \cdot a_t$. Note that this is a reconstruction of $\mathbf{f}$.
3.   Construct t fractional abundance images, one for each PPI-derived endmember.
4.   Apply lossless predictive coding to reduce spatial redundancy within each of the t fractional abundance images, using Huffman coding to encode predictive errors.

3   FPGA-Based Hardware Implementation

Fig. 1 shows our proposed systolic array design for FPGA implementation of the PPI algorithm. The nodes labeled as "dot" in Fig. 1 perform the individual products for the skewer projections, while the nodes labeled as "max" and "min" respectively compute the maxima and minima projections after the dot product calculations have been completed (asterisks in Fig. 1 represent delays). In Fig. 1, $s_j^{(i)}$ denotes the j-th value of the i-th skewer vector, with $i \in \{1, \ldots, k\}$, and $j \in \{1, \ldots, b\}$, where b is the

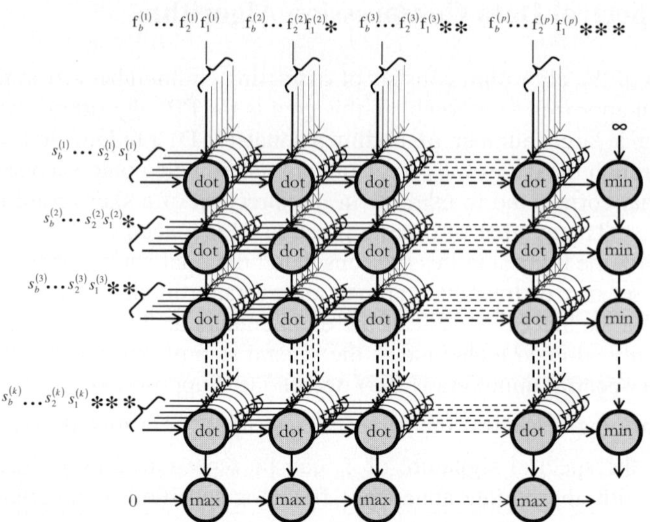

Fig. 1. Systolic array design for the proposed FPGA implementation of the PPI algorithm

number of bands. Similarly, $f_j^{(m)}$ denotes the reflectance value of the j-th band of the m-th pixel, with $m \in \{1,...,p\}$ and p is the total number of pixels in the scene.

We are still working towards the implementation of the LSU algorithm in hardware, so timing results in this work will be related with the optimization introduced by PPI-based FPGA design. The synthesis was performed using Handel-C, a hardware design and prototyping language. The implementation was compiled by using the DK3.1 software package. We also used other tools such as Xilinx ISE 6.1i to carry out automatic place and route (PAR), and to adapt the implementation to the Virtex-II FPGA used in experiments.

4   Experimental Results

The algorithm was applied to a real hyperspectral scene collected by an AVIRIS flight over the Cuprite mining district in Nevada, which consists of 614x512 pixels and 224 bands (available online from http://aviris.jpl.nasa.gov). Table 1 reports the spectral angle-based similarity scores[1] among U.S. Geological Survey reference signatures (see http://speclab.cr.usgs.gov) and the PPI-extracted endmembers from the resulting images after data compression (the lowest the scores, the highest the similarity). As the compression ratio was increased, the quality of extracted endmembers was decreased. We also included results by two standard methods: the wavelet-based JPEG2000 multicomponent[5] and the set partitioning in hierarchical trees (SPIHT)[6]. Results in Table 1 show that such 3-D techniques, which enjoy success in classical image compression, did not find equal success in hyperspectral image compression.

Table 1. Spectral similarity scores among ground-truth spectra and the endmembers extracted and from several reconstructed versions of the original image after compression

Mineral	Original image	PPI/LSU			JPEG2000			SPIHT		
		20:1	40:1	80:1	20:1	40:1	80:1	20:1	40:1	80:1
Alunite	0.063	0.069	0.078	0.085	0.112	0.123	0.133	0.106	0.119	0.129
Buddingt.	0.042	0.053	0.061	0.068	0.105	0.131	0.142	0.102	0.125	0.127
Calcite	0.055	0.057	0.063	0.074	0.102	0.128	0.139	0.097	0.122	0.134
Kaolinite	0.054	0.059	0.062	0.071	0.114	0.140	0.151	0.110	0.134	0.146
Muscovite	0.067	0.074	0.082	0.089	0.123	0.145	0.167	0.116	0.139	0.152

Finally, Table 2 shows the resource utilization by our systolic array-based implementation of PPI/LSU on the Xilinx Virtex-II XC2V6000-6 FPGA, which compressed the full AVIRIS scene in only 39 seconds approaching (near) real time performance.

Table 2. Summary of resource utilization for the FPGA-based implementation.

Number of gates	Number of slices	Percentage of total	Maximum operation frequency (MHz)
526944	12418	36%	18.032

5 Conclusions

On-board compression of hyperspectral imagery has been a long-awaited goal by the remote sensing community. This paper investigated the importance of mixed pixels in hyperspectral data compression and further proposed an innovative, application-oriented data compression technique which is based on the pixel purity index (PPI) algorithm and linear spectral unmixing (LSU). Experimental results demonstrate that the algorithm provides very high compression ratios with no apparent spectral signature degradation. A systolic array-based FPGA implementation of the algorithm on a Xilinx Virtex-II XC2V6000-6 FPGA was also provided.

References

1. Chang, C.-I: Hyperspectral Imaging: Detection and Classification. Kluwer, New York (2003)
2. Motta, G., Storer, J. (eds.): Hyperspectral Data Compression. Springer-Verlag, Berlin Heidelberg New York (2005)
3. Chang, C.-I, Plaza, A.: A Fast Iterative Implementation of the Pixel Purity Index Algorithm. IEEE Geoscience and Remote Sensing Letters, 3 (2006) 63–67
4. Du, J., Chang, C.-I: Linear Mixture Analysis-Based Compression for Hyperspectral Image Analysis. IEEE Trans. Geoscience and Remote Sensing, 42 (2004) 875–891
5. Taubman, D.S., Marcellin, M.W.: JPEG2000: Image Compression Fundamentals, Standard and Practice. Kluwer, Boston (2002)
6. Said, A., Pearlman, W.A.: A New, Fast, and Efficient Image Codec Based on Set Partitioning in Hierarchical Trees. IEEE Trans. Circuits and Systems, 6 (1996) 243–350

Advertisement-Aided Search in a P2P Context Distribution System

Irene Sygkouna, Miltiades Anagnostou, and Efstathios Sykas

Computer Networks Laboratory, School of Electrical and Computer Engineering, National Technical University of Athens (NTUA), Greece
`{isygk, miltos, sykas}@telecom.ntua.gr`

Abstract. We study a P2P proactive search mechanism based on the dissemination of advertisements for the new sources. The system design goal of limiting the state maintained by each peer and ensuring search efficiency is the driving reason for exploiting the hierarchical network model of the small-world idea. The results testify the theoretical bounds on search time and provide a view on the search time in relation to the directory capacity requirements of the peers.

1 Introduction

Context-aware services (CASs) need a flow of information from and about their environment in order to be able to adapt to it. Context producers, consumers and brokers are employed, with the last ones acting as mediatory players between producers and consumers. A consumer is a CAS that addresses context requests during its operation to the nearest broker. Producers are all the context sources. The need to support CASs that are highly robust and can scale well with the number of nodes and information sources points to Peer-to-Peer (P2P) architecture. The system faces the challenge to ensure efficient and scalable distribution of context information. Toward this direction we employ the *advertisement dissemination* mechanism, which implies that once a new source becomes available, the respective broker propagates an advertisement of the source to the remaining brokers. A broker may store a received advertisement in a local directory. Assuming infinite directory sizes, each peer broker stores each advertisement and thus maintains knowledge of all the sources available. However, as the system scales to a high number of sources, the length of the directories may become unbounded. Thus, our intention is to limit the directory capacity requirements of each peer and at the same time ensure low path-length for any search request. The idea is to make the peers store advertisements selectively, according to the philosophy of small-world model. A requested object is then located in a bounded number of steps with a greedy search algorithm. We consider a set of brokers forming an overlay network. Each peer broker maintains the *Local Sources Directory* (LSD), with entries to the local sources, and the *Remote Sources Directory* (RSD), which caches directory entries for sources maintained by other peers. We assume that each source is described by a structure, the *context object*, that consists of a name and a set of attributes determined by (*key, value*) pairs. Search algorithms from literature are either request broadcast-based or advertisement-based. In the former case, the related mechanisms do not scale well with the network size since a search request is broadcast throughout the network,

V.N. Alexandrov et al. (Eds.): ICCS 2006, Part I, LNCS 3991, pp. 892–895, 2006.

but improvements, such as *Random Walk*, have been proposed. Mechanisms of the latter case have been also proposed [1, 2] but have not addressed scalability problems.

2 Search Mechanism with Advertisements

Inspired by the *hierarchical network model* [3] of the small-world phenomenon we propose an advertisement dissemination mechanism, which proceeds in a way to make the information maintained in the RSDs reflect the links of the graph generated from the hierarchical model. The sources maintained by the overlay are classified according to a hierarchy, based on the names of the respective context objects. Thus, each source corresponds to a specific leaf of a complete b-ary context tree. Note that we allow more than one source to reside at a specific leaf since multiple sources may produce context objects of the same context name but different attributes.

Once a source enters the overlay through a node, an advertisement is disseminated to the rest of the peer nodes according to a flooding algorithm. A node stores a received advertisement if there is still room in its RSD and no other entry for a source belonging to the same leaf with the new source exists. Once the RSD becomes full, an *elementary distribution* of the entries is performed by corresponding each of the relevant remote sources to a local source in the following way: each remote source in turn is assigned to the nearest, in terms of tree distance, local source provided that this is feasible (does not exceed the threshold determined by the balanced distribution of the entries among the local sources). This way, a list of "concentrated" sources L_{S_i} is created for each local source S_i, $i \in \{1, 2, \ldots, LSD\_SIZE\}$. Note that the distance between any two sources X_i, X_j, denoted by $dist(X_i, X_j)$, is measured by the height of the least common ancestor of the leaves hosting X_i and X_j, respectively, in the tree. Note also that this procedure needs to be repeated every time a new source becomes available though a peer. When an advertisement of a new source is received by a node that would cause its RSD to exceed its size, the replacement scheme will take place. Assuming that the advertisement refers to a source X, a replacement can take place if the RSD does not contain any entry for a source that belongs to the same leaf with X. In particular, the following steps take place:

1. Among the local sources, the one that is closest to X in terms of tree distance, say S_X, is selected, namely: $dist(S_X, X) = \min_{S_i \in LSD} dist(S_i, X)$.

2. From the respective list of concentrated sources L_{S_X}, a source that is furthest from S_X in terms of tree distance, say S_{X_MAX}, is selected: $dist(S_{X\_MAX}, S_X) = \max_{X_i \in L_{S_X}} dist(X_i, S_X)$. S_{X_MAX} will then compete with X for membership in the list. Based on the intuition from the model, S_X should have a connection to S_{X_MAX} with probability proportional to $p_1 = b^{-dist(S_X, S_{X\_MAXt})}$, and a connection to X with probability proportional to $p_2 = b^{-dist(S_X, X)}$.

According to the normalizing probabilities, X will therefore replace S_{X_MAX} with probability equal to $\dfrac{p_2}{p_1 + p_2}$.

Note that the advertisement of every new source that belongs to the same leaf with a local source is definitely stored, provided that the storage capacity is not exceeded.

The following greedy search algorithm is then applied: Once a peer receives a request and a matching source is not found in the LSD, it then searches its RSD. If a matching source is found, the request is forwarded to the peer pointed by the RSD. Otherwise, it looks up the nearest source (in terms of tree distance) in the RSD to the one requested, whose home-broker has not received the same request message from the current node in the past, in order to avoid cycles. It then forwards the request to the corresponding peer. The procedure is repeated until a matching source is located or the TTL is exceeded. Note that a step forward may incur a divergence from the target source if it cannot find in the RSD a source that is at least in the same distance from the target with the one selected during the previous step. Thus, two versions of the search algorithm are tested: the *Advertisement Flooding-Forward* (AF-F) makes only forward steps, while the *Advertisement Flooding-Backwards* (AF-B) makes a move backwards if it does not want to continue from its current node.

3   Simulation Results

We study and quantify the potential gain from the proposed mechanism with an event-based multithreaded simulation. As a reference for comparison we use *Random*, which applies for preprocessing a random replacement scheme at the RSDs, and for searching the same greedy algorithm. A number of sources join the overlay sequentially. After all the peer nodes have provided the same number of sources, a set of requests are initiated. We assume that the request popularity follows *Zipf* distribution [4] (a=0.6). The performance of the mechanisms is well captured by the *request hit ratio* (the ratio of the number of successful requests to the total number of requests) and the *average path-length per request* (the ratio of the number of hops incurred across all requests to the total number of requests). If a request fails, it is deemed to have incurred TTL number of hops. Based on the hierarchical network model, the RSD_SIZE, not counting the entries for the sources that do belong to the same leaf with a local source, is computed as $RSD\_SIZE = LSD\_SIZE \cdot \left(c \cdot log_b^{\,2}\, n\right)$, with n denoting the total number of leaves in the complete b-ary context tree and c a constant. In effect, we will vary the RSD_size by varying the parameter c. Note that we set the LSD_SIZE to the size of the LSD just prior to executing the requests. We consider a b-ary tree with b=4 and height equal to 5, for a total number of leaves equal to 1024. We also consider 200 peers. Setting the TTL to 20 and the LSD_SIZE initially to 10 (Fig. 1a) and then to 30 (Fig. 1b), we intend to explore how the performance metrics vary with respect to the RSD_SIZE. Parameter c will vary from 0.2 to 1 with a step equal to 0.2, changing the percentage of the RSD_SIZE to the total number of sources to 2.5%, 5%, 7.5%, 10%, and 12.5%, respectively. The metrics are calculated over 100 requests. The figure shows that the proposed mechanism clearly outperforms the Random, achieving high hit ratio and testifying the theoretical bound on search time ($O(log_b n)$). Obviously, as the value of c increases, the performance improvement, compared to the Random, becomes less evident. Moreover, as the LSD_SIZE increases this performance divergence decreases more quickly with an increase in c. Thus, the value of the proposed algorithms is mainly identified under limited storage capacity of the RSD, since in this case the small-world connections are clearly shaped. Finally, the AF-F provides slightly better results compared to the AF-B.

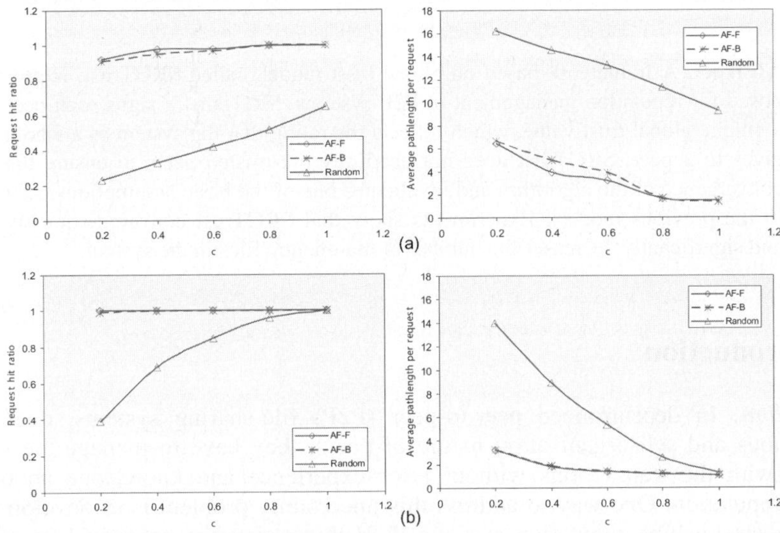

Fig. 1. Performance metrics after (a) 2000 and (b) 6000 sources have joined the system

References

1. A. Ranganathan and R.H. Campbell, "Advertising in a Pervasive Environment", Proceedings of the 2nd ACM International Workshop on Mobile Commerce, Atlanta, Georgia, pp.10-14, September 2002.
2. T. Finin, O. Ratsimor, A. Joshi and Y. Yesha, "eNcentive: A Framework for Intelligent Marketing in Mobile Peer-To-Peer Environments", Proceedings of the 5th International Conference on Electronic Commerce (ICEC'03), New York, NY, USA, pp.87-94, October 2003.
3. J. Kleinberg, "Small-world Phenomena and the Dynamics of Information", Advances in Neural Information Processing Systems (NIPS) 14, 2001.
4. L. Breslau, P. Cao, L. Fan, G. Phillips, and S. Shenker, "Web Caching and Zipf-like Distributions: Evidence and Implications", INFOCOM (1), pp.126–134, 1999.

A Reputation Management Framework Based on Global Trust Model for P2P Systems[*]

Jingtao Li, Xueping Wang, Yongqiang Chen, and Gendu Zhang

School of Information Science & Engineering, Fudan University, Shanghai, 200433, China

Abstract. A framework based on global trust model, called SRGTrust, is proposed for reputation management in P2P systems. SRGTrust assigns each peer a unique global trust value, which reflects the rating that the system as a whole gives to a peer. SRGTrust does not need any pre-trusted peers to ensure the convergence of the algorithm and invalidates one of the basic assumptions used in the previous models. Experiments show that SRGTrust converges quickly and significantly decreases the number of inauthentic files in the system.

1   Introduction

Motivation. In decentralized peer-to-peer (P2P) file-sharing systems, due to the anonymous and self-organization nature of peers, they have to manage the risk involved with the transactions without prior experience and knowledge about each other's reputation. One way to address this uncertainty problem is to develop strategies for establishing reputation systems [2,3] that can assist peers in assessing the level of trust they should place on the quality of resources they are receiving.

Challenge. The very core of the reputation mechanism in a P2P system is to build a distributed reputation management system that is efficient, scalable, and secure in both trust computation and trust data storage and dissemination. The main challenge of building such a reputation mechanism is how to effectively cope with various malicious collectives of peers who know one another and attempt to collectively subvert the system by providing fake or misleading ratings about other peers [2,3].

Contribution. We present a framework, called **SRGTrust** (Global **T**rust model based on **S**imilarity-weighted **R**ecommendations), for reputation management in P2P systems, which includes a global trust model for quantifying the trustworthiness (section 2), and a decentralized implementation of such model over a structured P2P overlay network (section 3). Previous global trust models such as eigentrust [3] are based on the assumption that the peers with high trust value will give the honest recommendation. We argue that this assumption may not hold in all cases. In SRGTrust, each peer i is assigned a unique global trust value that based on similarity-weighted recommendations of the peers in the network who has interacted with peer i; Finally, A series of simulation-based experiments show that the SRGTrust algorithm is robust and efficient to cope with various malicious collectives (section 4).

---

[*] This work is supported by the National Natural Science Foundation of China (No. 60373021).

V.N. Alexandrov et al. (Eds.): ICCS 2006, Part I, LNCS 3991, pp. 896–899, 2006.
© Springer-Verlag Berlin Heidelberg 2006

2   The Trust Model

Assume n denotes the number of the peers in a system. Each peer i rates another peer j from which he tries to download files by rating each download as either positive or negative, depending on whether i was able to download an authentic file from j or not. The sum of the ratings of all of i's interactions with j is called a **local trust value** S_{ij}. $S_{ij}=G_{ij}-F_{ij}$ wherein G_{ij} denotes the number of positive downloads and F_{ij} denotes the number of negative downloads.

A **normalized local trust value** L_{ij} is defined as follows:

$$L_{ij} = \frac{\max (S_{ij},0)}{\sum_{j} \max (S_{ij},0)} \; . \tag{1}$$

L_{ij} is a real value between 0 and 1, and $\sum_{j} L_{ij} = 1$. If $\sum_{j} \max(S_{ij},0)=0$, let $L_{ij} = {T_i}/{n}$. $L_{ii}=0$

for every peer i; otherwise, i can assigns arbitrarily high local trust values to itself.

One critical step in our model is to compute the similarity between rating opinions of peers and then to weight the peers' recommendations by that value. The similarity between them is measured by computing the cosine of the angle between these two vectors [1]. **The similarity** between peers i and j, denoted by C_{ij}, is given by

$$C_{ij} = \frac{B_i * B_j}{\|B_i\| \cdot \|B_j\|} \; , \tag{2}$$

where " * " denotes the dot-product of the two vectors. Here B_i denotes the **rating opinion vector** of peer i, defined as $B_i= [B_{i1}, B_{i2}, ..., B_{in}]$ where $B_{ik}(k=1, ...,n)$ is the rating opinion of peer i on peer k. B_{ik} is defined as follows:

If $i{\neq}k$, and $G_{ik}+F_{ik}= 0$, then $B_{ik}= 0$;

$$\text{if } i{\neq}k, \text{ and } G_{ik}+F_{ik} > 0, \text{ then } B_{ik} = \begin{cases} G_{ik}/(G_{ik}+F_{ik}) & G_{ik} \geq F_{ik} \\ -F_{ik}/(G_{ik}+F_{ik}) & G_{ik} < F_{ik} \end{cases} \; . \tag{3}$$

For each peer i, we let $B_{ii} =1+ \varepsilon$ where ε is an arbitrarily small positive constant.

A **global trust value vector**, $T = [T_1, T_2, ... , T_n]^T$, is given by

$$T_i = \sum_{k \in U_i}(L_{ki} \cdot C_{ki} \cdot T_k) \; , \tag{4}$$

where T_i is the **global trust value** of peer i. U_i denotes the set of the peers who have interacted with peer i. The global trust value of peer i is the sum of the weighted recommendations of the peers that have interacted with i in a single iteration. After several iterations of asking friends of friends by using the above formula, peer i will have a view of the entire network. The credibility of the recommendations of a peer is different from that of the peer itself. So, different from eigentrust [3], we weight the recommendations of peer k by the similarity between peers k and i.

3 Distributed Implementation

Trust Data Management. There is no central database in P2P environments. Trust data that are needed to compute the trust measure for peers are stored across the network in a distributed manner. Each peer has a trust value manager that is responsible for feedback submission and trust evaluation, a small database that stores a portion of the global trust data, and a data locator for placement and location of trust data. In our implementation, the DHT algorithm, Chord [4], is used to assign trust value managers to peers. The detailed methods will not be given here due to space constraint.

The SRGTrust Algorithm is executed to compute a global trust value vector in a dynamic and decentralized fashion at each peer.

Primitives. Submit (ID_i, (ID_j, ID_k), Value1, Value2): Submitting *Value1* and *Value2* to the trust value managers of peer *i*. We use the *Value1* and *Value2* to refer to the normalized local trust value that peer *j* place in *k* and rating opinion vector of peer *j*, the meaning will be clear from context; *Query (ID_j, T_j, L_{ji}, B_j)*: Querying the global trust value, recommendation to peer *i*, and rating opinion vector of peer *j*.

Algorithm 1. SRGAgorithm1, peer *i* as an ordinary peer.
Update_SubmitTrustdata( )   // submits its rating (G_{ij}, F_{ij}) to peer *j* after a transaction
{   If (a good transaction) $G_{ij} \leftarrow G_{ij}+1$;
      else $F_{ij} \leftarrow F_{ij}+1$;
      Submit(ID_i, (ID_i, ID_j), G_{ij}, F_{ij});   }
      // submits G_{ij} and F_{ij} to M_i, and triggers the UpdateLocaltrust( ) in M_i

Algorithm 2. SRGAgorithm2, peer *i* as a trust value manager of peer *u*.
UpdateLocaltrust( )   // updates L_{uv} and B_{uv} after receiving a Submit primitive from *u*
{   Verify the consistency of G_{uv} and F_{uv};
      Compute S_{uv}, L_{uv}, B_{uv};
      Submit (ID_v, (ID_u, ID_v), L_{uv}, null);   }   // submits L_{uv} to M_v
CalcGlobaltrust( )   // computes the global trust value of peer *u*
{   for (every $j \in U_u$ ($j \neq u$))
      {   *Query (ID_j, T_j, L_{ju}, B_j)*;

$$C_{ju} \leftarrow \frac{B_j * B_u}{\|B_j\| \cdot \|B_u\|} \; ;$$ // computes the similarity between peers *u* and *j*

$T_u \leftarrow T_u + L_{ju} \cdot C_{ju} \cdot T_j$;   }
      return T_u;   }

4 Experiments

Simulation Setup. As a test bed for the experiments, we use the Query Cycle Simulator [5] which simulates a P2P file-sharing network. Trust values in this network are used to bias download sources. We shall demonstrate the algorithms' performance under two typical threat models. **Threat Model IM:** Individual malicious peers, called IM peers, always provide an inauthentic file when selected as a download source and they always set their satisfaction values to $S_{ij}= F_{ij} -G_{ij}$; **Threat Model CM:**

Malicious peers, called CM peers, always provide an inauthentic file when selected as a download source and each CM peer i let $L_{ij}=1/\|CMs\|$ for any $j \in CMs$ where CMs denotes the set of all CM peers, and $\|CM\|$ denotes the total of all CM peers.

***Results*. Proportion of Authentic Downloads (PAD)**, the ratio of the number of authentic downloads to the number of all downloads viewed by all good peers when we use three algorithms separately, are showed in Fig. 1. The PAD is higher than 80% even if IM, or CM peers make up a half of all peers in the network when our scheme is activated.

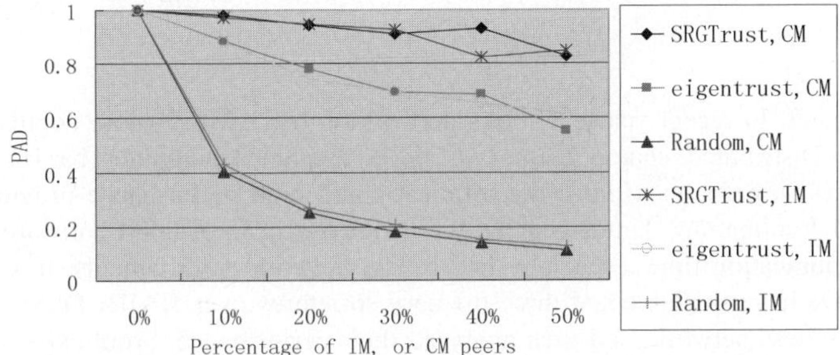

Fig. 1. PAD against different percentage of malicious peers. "Random" means a P2P system where no reputation system is implemented. We set up a network consisting of 500 peers. The initial distribution of the global trust values is a uniform probability distribution over all n peers, that is $T_i=1/n$ ($i=1, \ldots, n$). After issuing a query, a peer waits for incoming responses from the peers that have the file he is looking for, selects a download source whose global trust value is the highest among those peers, and downloads the file. The last two steps are repeated until the peer receives a complete, authentic copy of the file.

References

1. Sarwar, B., Karypis, G., Konstan, J., Riedl, J.: Item-Based collaborative filtering recommendation algorithms. Proceedings of the 10th International World Wide Web Conference. ACM Press. (2001) 285–295
2. Xiong, L., Liu, L.: PeerTrust: Supporting Reputation-Based Trust for Peer-to-Peer Electronic Communities. IEEE TRANSACTIONS ON KNOWLEDGE AND DATA ENGINEERING 16(7). IEEE Press. 843–857
3. Kamvar, S.D., Schlosser, M.T.: The eigentrust algorithm for reputation management in P2P networks. Proceedings of the 12th international conference on World Wide Web. ACM Press. (2003) 640–651
4. Stoica, I., Morris, R., Karger, D.: Chord: A scalable peer-to-peer lookup service for Internet applications. Technical Report. MIT. (2002). http://pdos.csail.mit. edu/chord/
5. Schlosser, M.T., Condie, T.E., Kamvar, S.D.: Simulating a File-Sharing P2P Network. Proceedings of First Workshop on Semantics in P2P and Grid Computing. (2003) 69–80

A Comparative Study of Memory Structures for DSM Systems on Wireless Environments[*]

Hsiao-Hsi Wang, Kuang-Jui Wang, Chien-Lung Chou,
Ssu-Hsuan Lu, and Kuan-Ching Li

Parallel and Distributed Processing Center
Dept. of Computer Science and Information Management
Providence University Shalu, Taichung 43301 Taiwan
{hhwang, g9371014, g9371001, g9234024, kuancli}@pu.edu.tw

Abstract. In recent years, wireless networking has become more popular than ever. Distributed shared memory (DSM) combines computer hardware resources in order to achieve the efficiency and high performance provided by parallel computing. Unfortunately, the major overhead of DSM software is the communication time, especially in wireless network environments. In this paper, we have implemented three memory structures over JIAJIA DSM system on wireless network, and then analyzed their performance. From experimental results, we could find out the relation between communication time and memory layouts. In addition, we have also discovered relationship between characteristics of application programs and memory structures. Experimental results of five well-known benchmark applications show that a suitable memory layout can effectively reduce communication overhead in wireless network. We have analyzed advantages and disadvantages of these memory structures, to improve future designs of wireless DSM systems.

1   Introduction

Distributed Shared Memory (DSM) software system is an excellent technique and easy alternative for concurrent computing, since computers can exchange data via high speed network to achieve higher performance. In original DSM system architectures, the common shared memory is made by combining all computing nodes' local memories. The home pages are located on all nodes, so nodes often need to access data in remote nodes. This induces remote access latencies, especially when using wireless networking technologies. Due to a number of limitations present in wireless computing environments, such as narrow bandwidth in wireless communication, unstable connectivity and data synchronization in mobile terminals are present.

S. Yokoyama et al. proposed a memory management architecture named Memory Management Architecture for Mobile Computing Environment (MMM) [1]. From initial concepts present in MMM, we have some ideas. We built a wireless DSM system over JIAJIA DSM software [2] and compare three different layouts of home

[*] This research is supported by National Science Council (NSC), Taiwan, under grant no. NSC 94-2745-E-126-002-URD.

V.N. Alexandrov et al. (Eds.): ICCS 2006, Part I, LNCS 3991, pp. 900–903, 2006.

pages, which are Hybrid Memory Structure (HMS), Centralized Memory Structure (CMS) and Distributed Memory Structure (DMS). We expect to find out main drawbacks when using wireless network and to improve these problems. In addition, we also expect to find out some relation between the characteristics of application programs and memory structures, such as the bandwidth, packet losing rate, and disconnection.

The remaining of this paper is organized as follows. In section 2, we compare the difference among three memory layouts. In section 3, we bring up experimental performance evaluation and finally, some conclusions and future works in section 4.

2 Memory Structures

In this section, we will describe three memory structures and compare differences between them, as show in Figure 1.

(A). Distributed Memory Structure (DMS): In JIAJIA, the pages are initially distributed among all nodes in the original memory layout. If any of nodes occurs page faults and needs to get pages from remote nodes, nodes will communicate with each other. If there are many nodes that need to send data at the same time, it maybe occur competition for network bandwidth or data retransmission. Due to the fact that bandwidth of wireless network is limited, it may not be suitable for wireless DSM system.

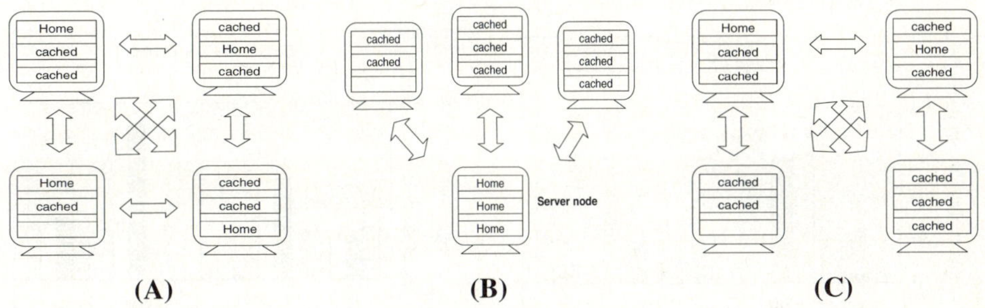

Fig. 1. (A). Distributed Memory Structure. (B). Centralized Memory Structure. (C). Hybrid Memory Structure.

(B). Centralized Memory Structure (CMS): We changed the original memory layout in JIAJIA software system, according to the design of MMM. If some node needs to access data, it will request to server. The server node maintains the consistency of the common memory areas and provides pages for client. Other nodes only have cache pages. In other words, when the page fault occurs on server node it only takes pages from itself, as also the times of communication will reduce, since the communication is simpler than DMS.

(C). Hybrid Memory Structure (HMS): In Hybrid Memory Structure (HMS), half of nodes have home pages and caches and the other nodes just have caches. In order to reduce communication overhead on server node, but still can reduce the times of communication, so we have an idea that located home pages on half of nodes. It can have better load balancing between communication and the percentage of data access hits.

3 Performance Evaluation

We have built a wireless DSM system platform for our experiments, and the hardware computing platform we used for our investigation is constructed using 4 PCs, each containing one Intel P4 3GHz CPU, 256 MB DDR memory, ASUS USB wireless network adapter 54MB/s and Fedora Core 3 OS with kernel version 2.6.9. We evaluate the performance of DMS, CMS, and HMS structures running five different parallel applications: IS from NAS [3], LU from SPLASH2 [4], Merge, SOR, and TSP. Table 1 shows the problem size and characteristics of these applications.

Table 1. Characteristics of Benchmark Applications

Application	Size	Memory (MB)	Barriers	Locks
IS	2^{24}, 2^{10}, 10	0.32	32	40
LU	1024*1024, 32	10.28	68	0
Merge	200*7500	7.36	5	0
SOR	512, 256, 50	10.24	101	0
TSP	19 (cities)	0.99	2	687

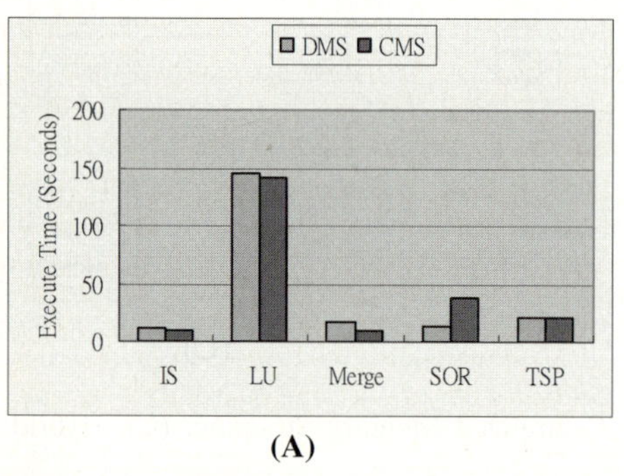

(A)

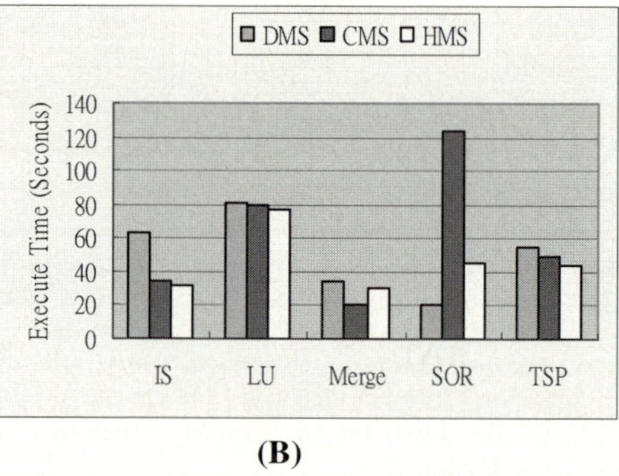

(B)

Fig. 2. Execution Time of Parallel Applications over Wireless DSM Platform. (A). Using 2 Wireless Computing Nodes (B). Using 4 Wireless Computing Nodes.

Figure 2(A) shows the execution time of original JIAJIA and CMS on two wireless nodes. We can see that some applications like IS and Merge on CMS can have better performance than that on DMS. If page faults happen, it will need to transmit the data among every node on DMS, since the bandwidth is limited in the wireless environment. If there are too many nodes transmitting data at the same time, it will cause bandwidth competition, data loss and data retransmission. Therefore, the performance will degrade rapidly, and this phenomenon can be seen obviously under four nodes DSM environment.

Additionally, we find that the characteristics of the application programs are important factors, such as Merge and SOR. In SOR, the original structure initially distributes red and block arrays across all nodes, but CMS allocates all shared data in one node. CMS opposites to the characteristic of SOR itself and causes the hits rate dropping and the communication time raising a lot. The characteristic of Merge is completely opposite to SOR. The best performance of Merge is in CMS and the worst performance of SOR is also in CMS, as in Figure 2(B).

4   Conclusions and Future Work

According to experimental results, we believe that is very sensitive the relation between the home location and performance. It is essential to locate the home pages correctly and the CMS can reduce the cost of communication. Under the wireless network environment, it is not reliable as in wired network, as it has narrow bandwidth. As large amount of data need to be transmitted, packets easy to collide as also to lose in wireless network. If we can improve the problems in this respect, it will improve global DSM system efficiency.

In addition, there exist close relation between the characteristics of applications and the percentage of access hit rates. It may cause performance come down fast if the location of home pages and the access pattern of application program are not suitable. The percentage of hits and increase efficiency can be improved if we can offer a suitable structure for home pages layout.

As future work, we will continue to probe the DSM software system in wireless environment and find out a number of main factors for improvement. We are looking for more effective methods to reduce the amount of message passing, to increase the percentage of hit rates for reducing the burden of the network, and to improve the efficiency of the wireless DSM environment.

References

[1] S. Yokoyama, T. Mizuno, and T. Watanabe, "A Proposal of a Memory Management Architecture for Mobile Computing Environments," in the Proceedings of Database and Expert Systems Applications, 2000, pp. 28-32.

[2] W. Hu, W. Shi, and Z. Tang, "Reducing System Overheads in Home-Based Software DSMs," in the Proceedings of 13th International and 10th Symposium on Parallel and Distributed Processing, 1999, pp. 167-173.

[3] D. Bailey, J. Barton, T. Lasinski, and H. Simon, "The NAS Parallel Benchmarks," Technical Report 103863, NASA, 1993.

[4] S. Woo, M. Ohara, E. Torrie, J. Singh, and A. Gupta, "The SPLASH-2 Programs: Characterization and Methodological Considerations," in the Proceedings of 22nd Annual International Symposium on Computer Architecture, 1995, pp.24-36.

Coordinated Exception Handling in J2EE Applications

Paweł L. Kaczmarek, Bogdan Krefft, and Henryk Krawczyk

Faculty of Electronics, Telecommunications and Informatics
Gdańsk University of Technology, Poland
pawel.kaczmarek@eti.pg.gda.pl, bogdan.krefft@bluemedia.pl,
henryk.krawczyk@eti.pg.gda.pl

Abstract. In the paper, we present a method of exception handling in J2EE applications. It is proposed to create a dedicated component that is responsible for handling two types of exceptions: those concerning more than one object and those occurring commonly in an environment. The component, referred to as Remote Exception Handler, is an extension of the middleware layer of a computer system, which enables its use without modifications of application source code. Concerning highly distributed architectures, many cooperating Remote Exception Handlers placed on different nodes are created. The solution has been implemented in practice in JBoss Application Server as an additional service of the platform.

1 Introduction

Exception handling (*eh*) is commonly used to increase software fault tolerance and consequently software dependability. An exception is a special event, usually a result of an error, that causes a change in the control flow and requires special handling. *Eh* in J2EE is based on the sequential Java language, although extensions have been designed. RemoteException is thrown if a remote object cannot fulfill a service or if there is a problem in communication with a server.

1.1 Related Work on Exception Handling in Distributed Environments

Distributed programming platforms supply a range of extensions that support fault tolerance and application management. The J2EE environment is equipped with Java Management Extension (JMX) [1] as a means of management and monitoring of applications, devices, services and JVM. Container Managed *EH* [2] (CMEH) is proposed as a server-side fault tolerance mechanism in J2EE application servers. CMEH framework is based on intercepting calls of component methods by an interceptor augmented with *eh* features. A system of recovering from exceptional situations by microreboots has been proposed in the JAGR system [3]. The H2O system creates a reliable platform of distributed computing, including self-organizing applications [4]. *Eh* in distributed environments is also addressed in [5] [6] and [7] for different platforms.

V.N. Alexandrov et al. (Eds.): ICCS 2006, Part I, LNCS 3991, pp. 904–907, 2006.
© Springer-Verlag Berlin Heidelberg 2006

Our solution differs from the ones described above as we propose specialized components that can cooperatively handle exceptions. We focus on the J2EE environment. We present the theoretical concept and describe the implementation of our solution in following sections. The source code of the implementation is available at author's web page (see Sect. 3).

2   Local and Remote Exception Handling

Eh in distributed systems may require the knowledge about system state or cooperation between a group of objects. Additionally, there are exceptions that occur commonly in an environment.

We propose a dedicated Remote Exception Handler (REH) that supplies predefined handling functions. REH extends the middleware layer by adding a service for handling two kinds of exceptions: those occurring commonly in the environment and those concerning more than a single object. REH operates in three main fields: performing handling actions, suggesting repair actions to objects and gathering information about system state.

Application programmer is freed from implementing handling functions, which reduces the required work and enables faster application development. As an example consider that a group of objects throw repeatedly runtime exceptions (e.g. a corrupted data base access pol). REH detects the situation and eliminates the cause (creates a new pol). The automated repair applies to those exceptions that are not part of the business specification (e.g. an input value exceeds a bound).

2.1   Integration of REH with a Distributed System

REH is integrated with an existing distributed system in three alternative ways:

- with interceptor cooperation - an exception is caught by an interceptor and redirected to REH, REH performs a handling function, it returns a normal result or throws an exception if the handling fails,
- direct cooperation with an object or service - if an object detects an anomaly, it invokes an REH method to repair the situation,
- self-activation - REH performs independently a handling function.

The methods of integration require some modifications in the middleware layer as explicit invocations of REH methods need to be added. It is also possible to extend REH's functionality with application specific handling functions. A detailed description of REH implementation and use is presented in Sect. 3.

2.2   Multi-remote Exception Handler Architecture

Many REH components are proposed for highly distributed architectures to overcome reliability and performance problems. The multi-REH architecture, however, requires additional solutions to enable inter-REH communication and synchronization. On REH creation, the component sends notification to other

*REH*s to register its existence. Each component maintains a list of active *REH*s together with their references. They communicate with each other during execution to interchange information about system and application state. It is assumed that inter-*REH* communication is reliable for active (visible) *REH* components. Only those components participate in a recovery. Fig. 1 shows the architecture of multi-*REH* cooperation.

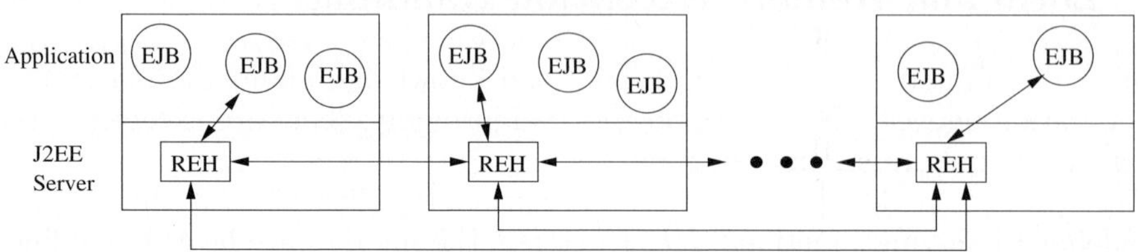

Fig. 1. Multi-*REH* architecture

Typically, *REH* performs repair actions independently, however, it can communicate with another *REH* to request the handling of an exception. The second handler attempts the handling and returns the result. In the current implementation, each *REH* sends a request to another arbitrary *REH*. If the request fails in the second handler, an exception is returned to the caller.

3   Exception Controllers in JBoss Application Server

The *REH* architecture has been implemented in JBoss Application Server [8]. The JBoss architecture consists of a micro-kernel and components implementing individual services. JBoss is an open-source server that can be easily extended.

The *REH* implementation in JBoss consists of two subcomponents as new JBoss services: Local Exception Controller (*LEC*) and Remote Exception Controller (*REC*). *LEC* implements basic exception handling scenarios that concern a single instance of the server. *REC* implements inter-*REH* cooperation, registration and communication between different instances of JBoss servers.

The creation and maintenance of *REH*s requires similar resources to other services in JBoss. Sending a request to a *REH* component is comparable to a remote invocation. Therefore, the performance degradation of a system using *REH* depends on the number of thrown exceptions.

The source code of the implementation and its configuration files are available at author's web page. The page contains also an exemplary application:

www.eti.pg.gda.pl/~pkacz/j2ee_exception.html

3.1   Required Configuration

The implemented extensions need to be integrated with JBoss, which requires configuration changes. Most changes concern the JBoss environment, however, minor modifications must be made in application deployment descriptors. The following configuration items are involved: deployment descriptors of *LEC* and *REC*, the configuration of remote servers cooperating with the local *REC*, the configuration of the EJB container. The first two points concern configuration changes made in JBoss independently of any application. The last point concerns configuration changes required in an application run on the platform: bean's container must be specified and container definition must be copied.

4   The Use of Exception Controllers

The implemented extension has been tested on an exemplary application - Duke's Bank. Supposing an EJB method throws a non-business exception (see Sect. 2), examples of handling functions are: (i) simple reinvocation - the interceptor catches the exception and requests *LEC* to reinvoke the method, (ii) ordering a reinvocation on a remote server, (iii) running garbage collection - if the exception is OutOfMemory, *LEC* enforces garbage collection to free unused memory, the request is repeated, (iv) invalidating a bean - it is assumed that bean's invocations can degrade the system, further invocations are rejected.

An explicit invocation of a *LEC* method from a regular object is also possible, which requires the same operations as a typical remote invocation: looking up the JNDI and creating object's reference.

Exception handling in distributed systems can be supported with the use of Remote Exception Handler, which extends the range of handled exceptions. After the extensions are made to the middleware layer of a system, *REH* can be used by all applications running in the environment, which seems a significant advantage of the solution. The work was supported in part by KBN under the grant number 4T11C 00525.

References

1. Armstrong, E., Ball, J., Bodoff, S.: The J2EE 1.4 Tutorial. Sun Microsystems Inc. (2004)
2. Simons, K., Stafford, J.: Container-Managed Exception Handling Framework. Department of CS, Tufts University, Medford, MA, USA (2004)
3. Candea, G., Kiciman, E., Zhang, S., Keyani, P.: JAGR: An Autonomous Self-Recovering Application Server. In: 5th Int. Wrkshp on Active Middleware Services. (2003)
4. Kurzyniec, D., Wrzosek, T., Drzewiecki, D., Sunderam, V.: Towards Self-Organizing Distributed Computing Frameworks: The H2O Approach. Parallel Processing Letters (2003)
5. Feitelson, D.G.: Exception Propagation in the ParPar System. Technical report, Inst. of Comp. Science, The Hebrew Univ. of Jerusalem (1998)
6. Romanovsky, A.: Practical Exception Handling and Resolution in Concurrent Programs. Comput. Lang. Vol. 23 (1997)
7. Kaczmarek, P.L., Krawczyk, H.: Remote Exception Handling for PVM Processes. LNCS 2840 Conf. X Conf. EuroPVM/MPI (2003)
8. JBoss Inc.: JBoss Admin Development Guide. (2004)

Efficient Unilateral Authentication Mechanism for MIPv6

Yoon-Su Jeong[1], Bong-Keun Lee[2], Keon-Myung Lee[1], and Sang-Ho Lee[1]

[1] Department of Computer Science, Chungbuk National University, Chungju, Chungbuk, Korea
bukmunro@netsec.cbnu.ac.kr, {kmlee, shlee}@chungbuk.ac.kr
[2] Department of Multimedia Computer, Busan Kyungsang College, Chungbuk, Korea
rbk@bsks.ac.kr

Abstract. We present a unilateral authentication protocol for protecting IPv6 networks against abuse of mobile IPv6 primitives. The proposed protocol imposes minimal computational requirements on mobile nodes, uses as few messages as possible. It is also easy to implement, economic to deploy and lightweight in use. We formally verifies the correctness of the protocol using the finite-state analysis tool mur ϕ, which has been used previously to analyze hardware designs and security properties of several protocols.

1 Introduction

Recently mobility technology is regarded as indispensable feature in internet area according to widely use of mobile nodes such as notebook computer, internet accessible mobile phone. Mobile IPv6 is a key protocol that supports mobility in IPv6 network. But Mobile IPv6 specification does not specify how to gain access to network when mobile node is away from its home.

The most difficult adoption issue for any form of authenticated Mobile IPv6 is reliance on authentication infrastructure. Recent Diameter specification deals with IPv6 mobility support [1]. However, The [1] does not state any specific protocol to be used between mobile node and AAA client. This paper proposes efficient user authentication to provide a minimum level of unilateral authentication of binding update. We assume that MIPv6 is used to authenticate the user and Diameter conveys MIPv6 authentication message between foreign AAA(AAAF) and home AAA(AAAH). In this paper, diameter is responsible for supporting IP mobility features such as home agent assignment and key distribution which is needed to securely send BU message to its home agent.

2 Related Works

Brandner, Mankin and Schiller [4] proposed a framework called Purpose Built Keys. An advantage of this framework is that it does not require any security infrastructure. However, purpose-built keys provide authentication if and only

V.N. Alexandrov et al. (Eds.): ICCS 2006, Part I, LNCS 3991, pp. 908–911, 2006.

if the initial hash of the public key is received correctly by correspondent node (CN). This might not be the case. An attacker could intercept the hash and send the hash of a different key (which it owns) to CN. Subsequently, it can pretend to be mobile node (MN) without CN being any wiser. The authors of the draft were, of course, aware of this weakness.

Le and Faccin[2] proposed two protocols for authenticating binding updates. The first assumes that both the MN and the CN share security associations with two AAA servers. The second protocol proposed in [2] involves an unauthenticated Diffie-Hellman key exchange between MN and CN. The resulting key is subsequently used to authenticate binding updates. The authors recognize that this protocol is vulnerable to a man-in-the-middle(MITM) attack but state that "due to the properties of IP" such an attack will always be detected.

3   Proposed Protocol

In this section, our main goal is to institute a mechanism for setting up security associations between MN-CN pairs that will allow binding updates to be authenticated.

3.1   Assumption

We consider the following design assumptions:

- Each MN(Mobile Node) has a pre-established bidirectional security association with its HA(Home Agent) using which they can authenticate each other.
- HA's are capable of authenticating each other using public key cryptography.
- Identity verification should not rely on the existence of a global PKI.
- Minimize the number of messages and bytes sent between the participating entities.
- Consider the computational capabilities of the MN's and CN's.
- Resist to DoS attacks that CN does not need to create state before the third message.

3.2   Protocol Description

Our protocol considers (a) an initialization phase in which MN and CN set up an authentication key and (b) an update phase in which MN sends an authenticated binding update to CN using the key obtained from phase (a).

Initialization Phase. The initialization phase for the protocols are the elliptic curve parameters that are common to both entities and consist of an elliptic curve E defined over a finite field F_q, generating element G(a point of the elliptic curve) $\in E(F_q)$, n is order of G in $E(F_q)$, and h is cofactor of n,i.e.,h= $\sharp\, E(F_q)/n$.

We will assume that static public keys are exchanged via certificates. $Cert_{HA_{MN}}$ denotes HA_{MN}'s public-key certificate, containing her static public key Y_a,

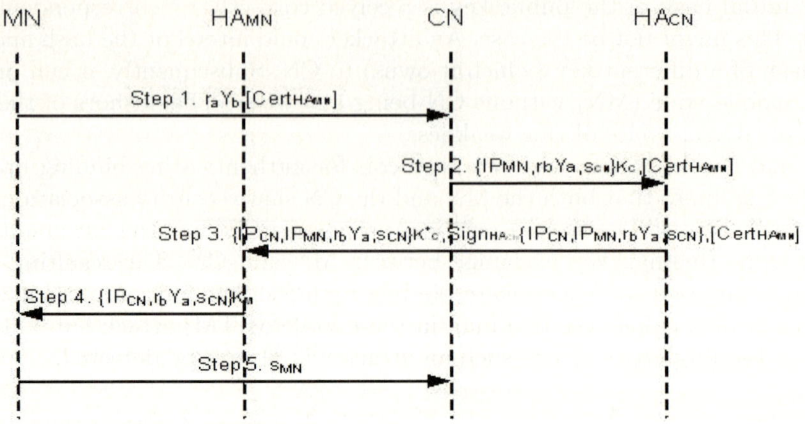

Fig. 1. Registration protocol

ephemeral key(random number) r_a and a certifying authority HA_{CN}'s signature. The security of proposed protocol in this paper is based on the Diffie-Hellman problem in elliptic curve group(ECDHP). The protocol is shown in Figure 1 as follows.

- Step 1: A MN send a message($r_a Y_b$ and $Cert_{HA_{MN}}$) to CN when MN already knows a group and generator that is acceptable to CN.
- Step 2: Upon receiving message 1, CN generates $r_b Y_a$, computes $K_D==$ h$((r_b/x_b)M_a + x_b Y_a)=h(r_a r_b + x_a x_b)$G and S_{CN} and sends a message to HA_{CN} encrypted with their shared key, K_C.
- Step 3: HA_{CN} forwards to HA_{MN} the session parameters that it recovers by decrypting message2. It is necessary to include CN's IP address since the same MN might be executing this protocol parallely with multiple correspondent nodes.
- Step 4: HA_{MN} forwards the session parameters to MN in message 4, encrypted with K_M, their shared key.
- Step 5: MN encrypts the session parameters with the newly computed elliptic curve secret based on the ECDLP.

In this protocol, CN authenticates herself to MN through the chain of trust CN $\rightarrow HA_{CN} \rightarrow HA_{MN} \rightarrow$ MN. MN could also use the same trust chain in the other direction to authenticate herself to CN.

Update Phase. Once MN and CN have set up a shared secret, K_D, MN can easily send an authenticated binding update(BU) by executing the following Step 1 protocol.

$$\text{Identity, BU, h}(K_D, \text{Identity, BU})$$

Here, h(...) is a keyed cryptographic hash function [3].

4 Analysis of the Protocol

The use of elliptic curve key exchange ensures that the protocol provides perfect forward secrecy(PFS)[5]. If the perfect forward secrecy is not a requirement, an alternative would be to replace the exchange of certificate by an exchange of nonces(n_{MN} and n_{CN}). n_{MN} could be sent in the clear in Step 1, while n_{CN} is sent encrypted in Step 2, 3, 4. The shared secret could be derived from a hash of these nonces, e.g., HMAC(n_{MN},n_{CN}).

A key exchange protocol requires each participant to use some fresh information in every run of the protocol. Any adversary which possesses a certificate could intercept Step 1 and then send Step 3 without HA_{MN} being any wiser.

The basic 5-Step protocol is susceptible to denial of service attacks. A useful property of our protocol is that since home agents do not create state, memory denial of service attacks are not possible on the home agents. Preventing computation denial of service attacks on the home agents reduces to the problem of detecting, without performing expensive computations, whether a message has been replayed. Preventing denial of service attacks on HA_{MN} without adding extra message, appears to be more difficult. Adding a sequence number to message 3 and including it in the signature alleviates the problem somewhat.

5 Conclusions

In response to the requirement that all location information about a mobile node in IPv6 should be authenticated, we proposed a protocol for authenticating binding updates. The other requirements are taken into accounts as well: the computational load on the nodes and on the routers is minimized, by eliminating expensive encryption operations and keeping the number of steps at minimum. Therefore, we believe that our protocol addresses the main issues in Mobile IPv6 authentication and makes best use of whatever infrastructure is available.

References

[1] A. J. Menezes, P. C. van Oorschot, S. A. Vanstone. *Handbook of Applied Cryptography*. CRC Press, (1996).

[2] F. Le, S. M. Faccin. *Dynamic Diffie Helman based Key Distribution for Mobile IPv6*. Internet Draft, (April 2001).

[3] S. Bradner, A. Mankin, J. I. Schiller. *A framework for purpose Built Keys(PBK)*. Internet Draft. (February 2001).

[4] H. Krawczyk, M. Bellare, R. Canetti. *HMAC:Keyed-Hashing for Mesage Authentication*. RFC 2104. (February 1997).

[5] D. Dill. *The Mur ϕ Verification System*. In Proc. 8th International Conference on Computer Aided Verification, pages 390-393. (1996).

Optimal Constant Weight Codes

Igor Gashkov

Karlstad University, Department of Engineering Sciences, Physics and Mathematics 65188
Karlstad Sweden
Igor.Gachkov@kau.se

Abstract. A new class of binary constant weight codes is presented. We establish new lower bound and exact values on A(n, 2k, k + 1), in particular, A(30, 12, 7) = 9, A(48, 16, 9) = 11, A(51,16, 9) = 12, A(58, 18, 10) = 12.

An (n, d, w) constant weight binary code is a code of length n, code distance d in which all code words have the same number of "ones" . The number of "ones" is w. We will denote the maximal possible size (a number of code words) of an (n, d, w) constant weight code by $A(n, d, w)$.

The most important and interesting problem is finding the largest possible size $A(n, d, w)$ of an (n, d, w) constant weight code (hereafter called optimal code). The results of code searching used to be put in tables of optimal codes. The first lower bound appeared in 1977 in the book of MacWilliams and Sloane ([1], pp.684-691). A table of binary constant weight codes of length $n \leq 28$ with explicit constructions for most of the 600 codes was presented in the encyclopedic work of E E. Brouwer, J. B. Shearer, N. J .A .Sloane [2]. Today Neil J. A. Sloane presents his table of constant weight codes [3] online and performs continual updates.

Let us consider an $(n, 2k, k + 1)$ constant weight code with length n, code distance $2k$ and weight of all $k + 1$. Johnson's upper bound (se, [2], p. 525) in this case is

$$A(n,2k,k+1) \leq \frac{kn}{(k+1)^2 - n} \quad \text{(if denominator is positive)} \tag{1}$$

Theorem. $A(n = \dfrac{k^2 + 3k + 2}{2}, 2k, k + 1) = k + 2$ holds for all k.

The research was support by The Royal Swedish academy of Sciences.

Proof. The code is constructed from representation $\dfrac{k^2 + 3k + 2}{2}$ element set

$M = \{1,2,3,..., \dfrac{k^2 + 3k + 2}{2}\}$ as union of $k + 2$ subsets M_i with $k + 1$ elements

V.N. Alexandrov et al. (Eds.): ICCS 2006, Part I, LNCS 3991, pp. 912–915, 2006.
© Springer-Verlag Berlin Heidelberg 2006

in every subset and $\left|M_i \cap M_j\right| = 1$ for $i \neq j$. We now give explicit construction of the code.

For $k = 2$ and $M = \{1,2,3,4,5,6\}$ we can find 4 subsets $M_1 = \{1,2,4\}, M_2 = \{1,3,5\}, M_3 = \{2,3,6\}, M_4 = \{4,5,6\}$ where all vectors of constant weight code $(6,4,3)$ are:

$$\begin{aligned}
v_1^{k=2} &= (1,1,0,1,0,0) \\
v_2^{k=2} &= (1,0,1,0,1,0) \\
v_3^{k=2} &= (0,1,1,0,0,1) \\
v_4^{k=2} &= (0,0,0,1,1,1)
\end{aligned}$$

(2)

For $k = 3$ we obtain all code words of the code $(10,6,4)$ based on code words $v_1^{k=2}, v_2^{k=2}, v_3^{k=2}, v_4^{k=2}$

$$\begin{aligned}
v_1^{k=3} &= (v_1^{k=2},1,0,0,0) = (1,1,0,1,0,0,1,0,0,0) \\
v_2^{k=3} &= (v_2^{k=2},0,1,0,0) = (1,0,1,0,1,0,0,1,0,0) \\
v_3^{k=3} &= (v_3^{k=2},0,0,1,0) = (0,1,1,0,0,1,0,0,1,0) \\
v_4^{k=3} &= (v_4^{k=2},0,0,0,1) = (0,0,0,1,1,1,0,0,0,1) \\
v_5^{k=3} &= (0,0,0,0,0,0,1,1,1,1)
\end{aligned}$$

(3)

For $k = j$, we obtain all code words using the same construction the upper bound follows from (1)

Corollary 1. $A\left(n = \dfrac{k^2 l^2}{2} + \dfrac{kl^2}{2} + k + 1, 2lk, lk + 1\right) \geq kl + l$ for all k and l.

Proof. Using the same method as in the theorem we can construct the $n = \dfrac{k^2 l^2}{2} + \dfrac{kl^2}{2} + k + l$ element set as union of $kl + l$ subsets with $kl + 1$ elements in every subset. The second Johnson bound (se, [5], p. 525) yields the upper bound as $A(n,d,w) \leq kl + l$ for $l = 3,4,5$.

Example 1. For $l = 3$ and $k = 2$, we have a nontrivial optimal code with parameters $(30,12,7)$. The optimal code consists of 9 vectors of weight 7. In this case the set $M = \{1,2,...,30\}$. We have following subsets and code words respectively.

$$M_1 = \{1,2,3,4,5,6,19\} \leftrightarrow v_1 = (1,1,1,1,1,1,0,0,0,0,0,0,0,0,0,0,0,0,1,0,0,0,0,0,0,0,0,0,0,0)$$
$$M_2 = \{7,8,9,10,11,12,19\} \leftrightarrow v_2 = (0,0,0,0,0,0,1,1,1,1,1,1,0,0,0,0,0,0,1,0,0,0,0,0,0,0,0,0,0,0)$$
$$M_3 = \{13,14,15,16,17,18,19\} \leftrightarrow v_3 = (0,0,0,0,0,0,0,0,0,0,0,0,1,1,1,1,1,1,1,0,0,0,0,0,0,0,0,0,0,0)$$
$$M_4 = \{1,7,13,20,21,22,29\} \leftrightarrow v_4 = (1,0,0,0,0,0,1,0,0,0,0,0,1,0,0,0,0,0,0,1,1,1,0,0,0,0,0,0,1,0)$$
$$M_5 = \{2,8,14,23,24,25,29\} \leftrightarrow v_5 = (0,1,0,0,0,0,0,1,0,0,0,0,0,1,0,0,0,0,0,0,0,0,1,1,1,0,0,0,1,0)$$
$$M_6 = \{3,9,15,26,27,28,29\} \leftrightarrow v_6 = (0,0,1,0,0,0,0,0,1,0,0,0,0,0,1,0,0,0,0,0,0,0,0,0,0,1,1,1,1,0)$$
$$M_7 = \{4,10,16,20,23,26,30\} \leftrightarrow v_7 = (0,0,0,1,0,0,0,0,0,1,0,0,0,0,0,1,0,0,0,1,0,0,1,0,0,1,0,0,0,1)$$
$$M_8 = \{5,11,17,21,24,27,30\} \leftrightarrow v_8 = (0,0,0,0,1,0,0,0,0,0,1,0,0,0,0,0,1,0,0,0,1,0,0,1,0,0,1,0,0,1)$$
$$M_9 = \{6,12,18,22,25,28,30\} \leftrightarrow v_9 = (0,0,0,0,0,1,0,0,0,0,0,1,0,0,0,0,0,1,0,0,0,1,0,0,1,0,0,1,0,1)$$

$$(4)$$

We can find that $A(30,12,7) = 9$, $A(58,18,10) = 12$ and $A(51,16,9) = 12$.

Corollary 2.

$$A(n = \frac{k^2 l^2}{2} + \frac{3kl^2}{2} + l^2 - kl - l + k + 1, 2l(k+1) - 2, l(k+1)) \geq kl + 2l - 1$$

for all k and l

Example 2. In the case $l = 3$ we obtain optimal codes and particular for $k = 2$ we have an optimal code $A(48,16,9) = 11$. The constant weight code with parameters $(48,16,9)$ we obtain using construction in Example 1. The set $M = \{1,2,...,48\}$ and subsets TM_i

$$TM_i = \{M_i, 30+2i-1, 30+2i\}, for \qquad i = 1,...,9$$
$$TM_{10} = \{31,33,35,37,39,41,43,45,47\}$$
$$TM_{11} = \{32,34,36,38,40,42,44,46,48\}$$

$$(5)$$

M_i is subset from Example 1.

All code words and parameters of the code can be found using MATHEMATICA

```
In[3] := TM=
{{1,2,3,4,5,6,19,31,32},{7,8,9,10,11,12,19,33,34},
{13,14,15,16,17,18,19,35,36},{1,7,13,20,21,22,29,37,38},
{2,8,14,23,24,25,29,39,40},{3,9,15,26,27,28,29,41,42},
{4,10,16,20,23,26,30,43,44},{5,11,17,21,24,27,30,45,46},
{6,12,18,22,25,28,30,47,48},{31,33,35,37,39,41,43,45,47},
{32,34,36,38,40,42,44,46,48}}
```

```
MM={};Do[If[Count[K[[i]],1]==9,AppendTo[MM,K[[i]]]],{i,
1,Length[K]}]
Union[Flatten[Table[Count[Mod[MM[[i]]+MM[[j]],2],1],{i,
1,Length[MM]},{j,i+1,Length[MM]}],1]]
```

```
(* We calculate all code words and code distance*)
```

```
Out[3]=
```

```
⎛1 1 1 1 1 1 0 0 0 0 0 0 0 0 0 0 0 0 1 0 0 0 0 0 0 0 0 0 0 0 0 1 1 0 0 0 0 0 0 0 0 0 0 0 0 0 0 0⎞
⎜0 0 0 0 0 0 1 1 1 1 1 1 0 0 0 0 0 0 1 0 0 0 0 0 0 0 0 0 0 0 0 0 0 1 1 0 0 0 0 0 0 0 0 0 0 0 0 0⎟
⎜0 0 0 0 0 0 0 0 0 0 0 0 1 1 1 1 1 1 1 0 0 0 0 0 0 0 0 0 0 0 0 0 0 0 0 1 1 0 0 0 0 0 0 0 0 0 0 0⎟
⎜1 0 0 0 0 0 1 0 0 0 0 0 1 0 0 0 0 0 0 1 1 1 0 0 0 0 0 0 1 0 0 0 0 0 0 0 1 1 0 0 0 0 0 0 0 0 0 0⎟
⎜0 1 0 0 0 0 0 1 0 0 0 0 0 1 0 0 0 0 0 0 0 1 1 1 0 0 0 1 0 0 0 0 0 0 0 0 0 0 1 1 0 0 0 0 0 0 0 0⎟
⎜0 0 1 0 0 0 0 0 1 0 0 0 0 0 1 0 0 0 0 0 0 0 0 1 1 1 1 0 0 0 0 0 0 0 0 0 0 0 0 0 1 1 0 0 0 0 0 0⎟
⎜0 0 0 1 0 0 0 0 0 1 0 0 0 0 0 1 0 0 0 1 0 0 1 0 0 1 0 0 1 0 0 0 0 0 0 0 0 0 0 0 0 0 1 1 0 0 0 0⎟
⎜0 0 0 0 1 0 0 0 0 0 1 0 0 0 0 0 1 0 0 0 1 0 0 1 0 0 1 0 0 1 0 0 0 0 0 0 0 0 0 0 0 0 0 0 1 1 0 0⎟
⎜0 0 0 0 0 0 0 0 0 0 0 0 0 0 0 0 0 0 0 0 0 0 0 0 0 0 0 0 0 0 1 0 1 0 1 0 1 0 1 0 1 0 1 0 1 0 1 0⎟
⎜0 0 0 0 0 0 0 0 0 0 0 0 0 0 0 0 0 0 0 0 0 0 0 0 0 0 0 0 0 0 0 1 0 1 0 1 0 1 0 1 0 1 0 1 0 1 0 1⎟
⎝0 0 0 0 0 1 0 0 0 0 0 1 0 0 0 0 0 1 0 0 0 1 0 0 1 0 0 1 0 0 1 0 0 0 0 0 0 0 0 0 0 0 0 0 0 0 1 1⎠
```

16

Matrix in out [3] gives all code vectors for optimal constant weight code with parameter (48, 16, 9) and we calculated what code distance between all pair code vectors is 16.

Conclusion

We establish new lower bound and exact values on A(n, 2k, k + 1), in particular, A(30, 12, 7) = 9, A(48, 16, 9) = 11, A(51,16, 9) = 12, A(58, 18, 10) = 12.

References

1. MacWilliams, F. J., and Sloane, N. J. A. (1977) The Theory of Error - Correcting Codes.North - Holland, Amsterdam.
2. E.E. Brouwer, J.B. Shearer, N.J.A .Sloane - A new table of Constant weight codes. IEEE Transactions of information theory, v 36, No 6 (1990)
3. Neil J. A. Sloane: Home Page http://www.research.att.com/~njas/codes/Andw/,Table of Constant Weight Binary Codes, online version

Measure on Time Scales with Mathematica

Ünal Ufuktepe and Ahmet Yantır

Izmir Institute of Technology and Yasar University
Izmir, Turkey
unalufuktepe@iyte.edu.tr, ahmet.yantir@yasar.edu.tr

Abstract. In this paper we study the Lebesgue Δ-measure on time scales. We refer to [3, 4] for the main notions and facts from the general measure and Lebesgue Δ integral theory. The objective of this paper is to show how the main concepts of *Mathematica* can be applied to fundamentals of Lebesgue Δ- and Lebesgue ∇- measure on an arbitrary time scale and also on a discrete time scale whose rule is given by the reader. As the time scale theory is investigated in two parts, by means of σ and ρ operators, we named the measures on time scales by the set function **DMeasure** and **NMeasure** respectively for arbitrary time scales.

1 Introduction

Time Scales works can be found in [1, 2,5]. The software *Mathematica* is one of the most powerful tool in discrete and continuous analysis. Computational works on time scale calculus are collected in **TimeScale** package [6]. As probability theory is established on continuous and discrete analysis, we generalize **TimeScale** package in order to calculate the measure on time scales as a first step of probability theory and statistics.

In this paper first we give an introduction to the time scales, and we present the connection of the Lebesgue measure with the Lebesgue Δ-measure on an arbitrary bounded time scale T such that $\min T = a$ and $\max T = b$. We set out many basic concepts from measure theory to the Δ-measurable space. Finally we use *Mathematica* to illustrate the Lebesgue Δ- and ∇- measures to set out the differences between Lebesgue measure and Lebesgue Δ (∇)- measure.

2 Δ- and ∇- Measure on Time Scale

Let T be a time scale, $a < b$ be points in T, and $[a, b)$ be half closed bounded interval in T, σ and ρ be the forward and backward jump operators respectively on T. Let

$$\Im_1 = \{[a', b') \textstyle\bigcap T : a', b' \in T, a' \leq b'\}$$

be the family of all left closed and right open intervals of T. Then $\Im_1$ is a semiring. Here $[a', a') = \emptyset$. $m_1 : \Im_1 \to [0, \infty]$ is a set function which assigns to each interval its length: $m_1([a', b')) = b' - a'$. So if $\{I_n\}$ is a sequence of disjoint intervals in $\Im_1$ then $m_1(\bigcup I_n) = \sum m_1(I_n)$.

V.N. Alexandrov et al. (Eds.): ICCS 2006, Part I, LNCS 3991, pp. 916–919, 2006.

Let $E \subset \mathrm{T}$. By Carathéodory extension, outer measure of E is

$$m_1^*(E) = \inf_{E \subset \bigcup_n I_n} \sum m_1(I_n)$$

where $I_n \in \mathfrak{I}_1$. If there is no such covering of E, then $m_1^*(E) = \infty$.

Definition 1. *A set $E \subset \mathrm{T}$ is said to be Δ- measurable if for each set A*

$$m_1^*(A) = m_1^*(A \cap E) + m_1^*(A \bigcap E^c)$$

where $E^c = \mathrm{T} - E$. If E is Δ-measurable then E^c is also Δ-measurable. Clearly $\emptyset$ and T are Δ- measurable.

Let $\mathfrak{M}(m_1^*) = \{E \subset \mathrm{T} : E \text{ is } \Delta \text{ measurable}\}$ be a family of Δ- measurable sets.

Corollary 1. $\mathfrak{M}(m_1^*)$ *is a σ algebra.*

Definition 2. *The restriction of m_1^* to $\mathfrak{M}(m_1^*)$ is called Lebesgue Δ- measure and denoted by μ_Δ.*

So $m_1^*(E) = \mu_\Delta(E)$ if $E \in \mathfrak{M}(m_1^*)$.

Similarly, if we take $\mathfrak{F}_2 = \{(a', b'] : a', b' \in \mathrm{T}, a' \leq b'\}$ where $(a', a']$ is understood as an empty set then $m_2 : \mathfrak{F}_2 \to [0, \infty]$ such that $m_2((a', b']) = b' - a'$ is a countably additive measure. Then $\mathfrak{M}(m_2^*)$ is the set of ∇- measurable sets and μ_∇ is Lebesgue ∇- measure on T.

Proposition 1. *Let $\{E_n\}$ be an infinite decreasing sequence of Δ- measurable sets, that is, a sequence $E_1 \supset E_2 \supset \cdots \supset E_n \supset \cdots$, $E_i \in \mathfrak{F}_1$ for each i, $\cap E_i \in \mathfrak{F}_1$ and $m_1^*(E_1) < \infty$. Then*

$$m_1^*\left(\bigcap_{n=1}^{\infty} E_i\right) = \lim_{n \to \infty} m_1^*(E_n).$$

3 ∇ and Δ Measures with Mathematica

Theorem 1. *For each $t_0 \in \mathrm{T} - \{\min \mathrm{T}\}$ the ∇- measure of the single point set $\{t_0\}$ is given by $\mu_\nabla(\{t_0\}) = t_0 - \rho(t_0)$.*

Proof. **Case 1.** Let t_0 be left scattered. Then $\{t_0\} = (\rho(t_0), t_0] \in \mathfrak{F}_2$. So $\{t_0\}$ is ∇ measurable and $\mu_\nabla(\{t_0\}) = t_0 - \rho(t_0)$.

Case 2. Let t_0 be left dense. Then there exists an increasing sequence $\{t_k\}$ of points of T such that $t_k \leq t_0$ and $t_k \uparrow t_0$. Since $\{t_0\} = \bigcap_{k=1}^{\infty} (t_k, t_0] \in \mathfrak{F}_2$. Therefore $\{t_0\}$ is ∇ measurable. By continuity

$$\mu_\nabla(\{t_0\}) = \mu_\nabla\left(\bigcap_{k=1}^{\infty} (t_k, t_0]\right) = \lim_{n \to \infty} \mu_\nabla((t_n, t_n]) = \lim_{n \to \infty} t_0 - t_n = 0$$

which is the desired result since t_0 is left dense.

Theorem 2. *If $a, b \in \mathrm{T}$ and $a \leq b$ then $1)\mu_\nabla((a,b]) = b - a$, $2)\mu_\nabla((a,b)) = \rho(b) - a$, 3) If $a, b \in \mathrm{T} - \min \mathrm{T}$ then $\mu_\nabla([a,b)) = \rho(b) - \rho(a)$ and $\mu_\nabla([a,b]) = b - \rho(a)$.*

Proof. From the definition $\mu_\nabla((a,b]) = b - a$.

$$\mu_\nabla((a,b]) = \mu_\nabla((a,b)\bigcup\{b\}) = \mu_\nabla((a,b)) + \mu_\nabla(\{b\}) = \mu_\nabla((a,b)) + b - \rho(b)$$

$$b - a = \mu_\nabla((a,b)) + b - \rho(b)$$

So $\mu_\nabla((a,b)) = \rho(b) - a$.

iii) Let $a, b \in \mathrm{T} - \min \mathrm{T}$.

$$\mu_\nabla([a,b)) = \mu_\nabla(\{a\}\bigcup(a,b)) = \mu_\nabla(\{a\}) + \mu_\nabla((a,b)) = a - \rho(a) + \rho(b) - a = \rho(b) - \rho(a)$$

$$\mu_\nabla([a,b]) = \mu_\nabla([a,b)\bigcup\{b\}) = \mu_\nabla([a,b)) + \mu_\nabla(\{b\}) = \rho(b) - \rho(a) + b - \rho(b) = b - \rho(a)$$

Theorem 3. *For each $t_0 \in \mathrm{T} - \{\max \mathrm{T}\}$ the single point set $\{t_0\}$ is Δ- measurable and its Δ- measure is given by $\mu_\Delta(\{t_0\}) = \sigma(t_0) - t_0$.*

Proof. **Case 1.** Let t_0 be right scattered. Then $\{t_0\} = [t_0, \sigma(t_0)) \in \mathfrak{F}_1$. So $\{t_0\}$ is Δ- measurable and $\mu_\Delta(\{t_0\}) = \sigma(t_0) - t_0$.
Case 2. Let t_0 be right dense. Then there exists a decreasing sequence $\{t_k\}$ of points of T such that $t_0 \leq t_k$ and $t_k \downarrow t_0$. Since $\{t_0\} = \bigcap_{k=1}^{\infty} [t_0, t_k) \in \mathfrak{F}_1$. Therefore $\{t_0\}$ is Δ- measurable. By proposition 1

$$\mu_\Delta(\{t_0\}) = \mu_\Delta(\bigcap_{k=1}^{\infty} [t_0, t_k)) = \lim_{n \to \infty} \mu_\Delta([t_0, t_n)) = \lim_{n \to \infty} t_n - t_0 = 0$$

which is the desired result since t_0 is right dense.

Every kind of interval can be obtained from an interval of the form $[a,b)$ by adding or subtracting the end points a and b. Then each interval of T is Δ-measurable.

Theorem 4. *If $a, b \in \mathrm{T}$ and $a \leq b$ then $1)\mu_\Delta([a,b)) = b - a$, $2)\mu_\Delta((a,b)) = b - \sigma(a)$, 3) If $a, b \in \mathrm{T} - \max \mathrm{T}$ then $\mu_\Delta((a,b]) = \sigma(b) - \sigma(a)$ and $\mu_\Delta([a,b]) = \sigma(b) - a$.*

To illustrate these properties with mathematica, our **TimeScale** package must be loaded.
In[1]:= << TimeScale'
Let the time scale is as follows
In[2]:= $T = \{5 <= x <= 7 || x == 15/2 || 9 <= x <= 11 || x == 12 || x == 18\}$
We must define set function with respect to the interval or a single point
In[3]:= ClosedSet={ closed,a,b,closed }; OpenSet={ open,a,b,open };

RSemiClosedSet={ open,a,b,closed };
   LSemiClosedSet = { open,a,b,closed };
Spoint={ closed,a,closed };
DMeasure[ClosedSet[a_,b_]]:=sigma[b]-a;
   NMeasure[OpenSet[a_,b_]]:=b-sigma[a];
DMeasure[LSemiClosedSet[a_,b_]]:=b-a;
DMeasure[RSemiClosedSet[a_,b_]]:=sigma[b]-sigma[a];
DMeasure[Spoint[a_]]:=sigma[a]-a;
Now, we would like to find the measures of $\{5\}$ and $(5, 11]$ **In[4]:= DMeasure[Spoint[7]]**
Out[4]:= $\frac{1}{2}$
In[5]:= DMeasure[RSemiClosedSet[5,11]]
Out[6]:=7
Mathematica applications of ∇-measure also can be done as Δ-measure. The sigma operator must be replaced by the r operator also.

4   Conclusion

In this work we worked on Lebesgue Δ-measure and Lebesgue ∇-measure on time scales with *Mathematica*. We investigated each of these two measures in two parts, arbitrary time scales and discrete time scales. To do these we improved the **TimeScale** package. In the future, we will work on generalizing the probability theory on Time Scales with *Mathematica*.

Acknowledgement

This work is supported by The Scientific & Technological Research Council of Turkey.

References

1. Bohner,M.& Peterson,A., Dynamic Equations on Time Scales, Birkhäuser Boston, 2001.
2. Bohner,M.& Peterson,A., Advances in Dynamic Equations on Time Scales, Birkhäuser Boston, 2004.
3. Guseinov, G.S., Integration on time scales, J.Math. Anal.Appl. 285, 1, 107-127, (2003).
4. Guseinov, G.S. & Kaymakalan, B., Basics of Riemann delta and bale integration on time scales, J.Difference Equ. Appl. 8, 11, 1001-1017, (2002).
5. Hilger, S.: Analysis on measure chains-a unified approach to continuous and discrete calculus, 1990, Results Math. 18, 18-56.
6. Yantir, A.& Ufuktepe Ü., Mathematica Aplications on Time Scales for Calculus, 2005, Lecture Notes in Computer Science, 3482, 529-537.

Mackendrick: A Maple Package Oriented to Symbolic Computational Epidemiology

Juan Ospina[1] and Doracelly Hincapie[2]

[1] EAFIT University
School of Sciences and Humanities
Department of Basic Sciences
Division of Physical Engineering
Logic and Computation Group
Medellin, Colombia
judoan@epm.net.co
[2] University of Antioquia
National School of Public Health
Department of Basic Sciences
Group of Epidemiology
Medellin, Colombia
doracely@guajiros.udea.edu.co

Abstract. A Maple Package named *Mackendrick* is presented. Such package is oriented to symbolic computational epidemiology.

1 Introduction

We present here, the maple package *Mackendrick* which we have constructed for the solution of certain problems in symbolic computational epidemiology. Our package does not incorporate any kind of element of artificial intelligence, but for some of the problems that we solved, will be very funny to have some computer algebra system with artificial intelligence. The problems that we can solve here are linear problems but such problems only can be solved using computer algebra, due the involved calculations are very tedious and long as to be implemented by hand using pen and paper only.

Our emblematic problems are situations of spatial propagation of directly transmitted diseases when boundary conditions are involved at the form of endemic boundaries from where the disease spreads towards the interior of the habitat. More over, we consider here the extra complication that arises from the inclusion of the effects of heterogeneity of contact between individuals.

A fundamental epidemiological magnitude is the well know basic reproductive rate, denoted R_0. The principal function of the our package *Mackendrick* is the computation of the explicit analytical form of the R_0 for certain spatial models of disease diffusion with heterogeneity effects. We need here, computer algebra, because that it is required is a symbolic expression for R_0 and not a number or a graphic. Due, our package is constructed under maple platform, then our package has numerical and graphical computational power too.

V.N. Alexandrov et al. (Eds.): ICCS 2006, Part I, LNCS 3991, pp. 920–923, 2006.

The package is loaded with

```
restart:
with(Mackendrick);
```

and the notification is

```
[dielou, difumemoestra, memo, memoyf, mysol,prosize,sir,veneco,
venecomemo];
```

which is the list of procedures that are contained within *Mackendrick*.
In the following sections, the commands of *Mackendrick* are presented.

2   The Command *mysol*

For example, the procedure *mysol* solves the following problem:

$$\frac{\partial}{\partial t} u\left(r,t\right) - \frac{\delta_1 \left(\frac{\partial}{\partial r} u\left(r,t\right) + r \frac{\partial^2}{\partial r^2} u\left(r,t\right)\right)}{r}$$

$$-\delta_2 \int_0^t \frac{M_0\left(t-\tau\right)\left(\frac{\partial}{\partial r} u\left(r,\tau\right) + r \frac{\partial^2}{\partial r^2} u\left(r,\tau\right)\right)}{r} d\tau$$

$$- \left(\beta_1 S_0 - \gamma_1\right) u\left(r,t\right) - \beta_2 S_0 \int_0^t u\left(r,\tau\right) M_1\left(t-\tau\right) d\tau +$$

$$\gamma_2 \int_0^t u\left(r,\tau\right) M_2\left(t-\tau\right) d\tau = 0 \qquad (1)$$

with the boundary condition

$$u(a,t) = \mu_b e^{-\eta t}. \qquad (2)$$

The procedure *mysol* needs as inputs the specifical forms of the functions $M_0(t)$, $M_1(t)$ and $M_2(t)$. Here we present two cases.

2.1   Without Memory

For example when
$$M_0(t) = 0,\, M_1(t) = 0,\, M_2(t) = 0, \qquad (3)$$

and with the instructions

```
M0:=0:M1:=0:M2:=0:
mysol(M0,M1,M2);
```

Mackendrick produces the following solution [1]

$$u\left(r,t\right)=\frac{\mu_b e^{-\eta t} J_0\left(\sqrt{\lambda\left(-\eta\right)}r\right)}{J_0\left(\sqrt{\lambda\left(-\eta\right)}a\right)}+\sum_{i=1}^{1}\sum_{n=1}^{\infty}\frac{-2\,e^{S_{i,n}t}\mu_b J_0\left(\frac{\alpha_n r}{a}\right)\alpha_n}{\left(S_{i,n}+\eta\right)\left(J_1\left(\alpha_n\right)\right)a^2\left(\frac{d}{dS_{i,n}}\lambda\left(S_{i,n}\right)\right)}.$$

$$\tag{4}$$

The corresponding basic reproductive rate is given by

$$R_0=\frac{\beta_1 S_0 a^2}{\gamma_1 a^2+\alpha_n{}^2\delta_1}.\tag{5}$$

The function $\lambda(s)$ at (4) has the form

$$\lambda\left(s\right)=-\frac{s-\beta_1 S_0+\gamma_1}{\delta_1},\tag{6}$$

and the parameters denoted $S_{i,n}$ at (4) are the solutions of the equation on s

$$-\frac{s-\beta_1 S_0+\gamma_1}{\delta_1}=\frac{\alpha_n{}^2}{a^2},\tag{7}$$

where α_n are the zeroes of the Bessel function $J_0(x)$ [2].

2.2 With Exponential Memory

$$M_0(t)=e^{-\epsilon_0 t},\,M_1(t)=e^{-\epsilon_1 t},\,M_2(t)=e^{-\epsilon_2 t},\tag{8}$$

and with instructions

```
M0:=exp(-epsilon[0]*t):M1:=exp(-epsilon[1]*t):M2:=exp(-epsilon[2]*t):
mysol(M0,M1,M2);
```

Mackendrick produces the following solution

$$u\left(r,t\right)=\frac{\mu_b e^{-\eta t} J_0\left(\sqrt{\lambda\left(-\eta\right)}r\right)}{J_0\left(\sqrt{\lambda\left(-\eta\right)}a\right)}+\sum_{i=1}^{4}\sum_{n=1}^{\infty}\frac{-2\,e^{S_{i,n}t}\mu_b J_0\left(\frac{\alpha_n r}{a}\right)\alpha_n}{\left(S_{i,n}+\eta\right)\left(J_1\left(\alpha_n\right)\right)a^2\left(\frac{d}{dS_{i,n}}\lambda\left(S_{i,n}\right)\right)}.$$

$$\tag{9}$$

The corresponding basic reproductive rate is given by

$$R_0=\frac{S_0 a^2\epsilon_2\epsilon_0\left(\beta_1\epsilon_1+\beta_2\right)}{\epsilon_1\left(\alpha_n{}^2\delta_1\epsilon_0\epsilon_2+\alpha_n{}^2\delta_2\epsilon_2+\gamma_2 a^2\epsilon_0+\gamma_1 a^2\epsilon_2\epsilon_0\right)}.\tag{10}$$

The function $\lambda(s)$ at (9) has the form

$$\lambda\left(s\right)=\left(s-\beta_1 S_0+\gamma_1-\frac{\beta_2 S_0}{s+\epsilon_1}+\frac{\gamma_2}{s+\epsilon_2}\right)\left(-\delta_1-\frac{\delta_2}{s+\epsilon_0}\right)^{-1},\tag{11}$$

and the parameters denoted $S_{i,n}$ at (9) are the solutions of the equation on s

$$\sqrt{\left(s-\beta_1 S_0+\gamma_1-\frac{\beta_2 S_0}{s+\epsilon_1}+\frac{\gamma_2}{s+\epsilon_2}\right)\left(-\delta_1-\frac{\delta_2}{s+\epsilon_0}\right)^{-1}}=\frac{\alpha_n}{a}.\tag{12}$$

3   The Command *veneco*

The procedure *veneco* solves the following problem

$$
\frac{d}{dt}X_i\left(t\right) - \frac{\beta\, n X_i\left(t\right)}{k} + \gamma\, X_i\left(t\right) - \frac{2\beta\, n\left(\sum_{j=1}^{k}\nu\, X_j\left(t\right) - \nu\, X_i\left(t\right)\right)}{k} = 0. \quad (13)
$$

with the instruction

```
veneco(n);
```

our Mackendrick gives the following form of the basic reproductive rate [3]

$$
R_{0,k} = \frac{\beta\, n\left(1 + 2\nu\, k - 2\nu\right)}{\gamma\, k} \tag{14}
$$

4   Conclusions

We believe that the Maple package *Mackendrick* can be useful within the domain of symbolic computational epidemiology. Our *Mackendrick* can solve certain complex spatial epidemic models. The method of solution that *Mackendrick* incorporates is the Laplace transform technique with the application of the Bromwich integral and residue theorem for the realization of the inverse Laplace transform [4]. Also, *Mackendrick* involves certain theorem of Linear Algebra, which is presented in [3]. This theorem must be introduced *ad hoc* but it is possible that with the future development of artificial intelligence, such theorem can be proved directly by the computer algebra system that is the background of *Mackendrick* . We hope that at the future our package can be extended and applied to more numerous and complex problems in mathematical epidemiology.

References

1. Hincapie, D., Ospina, J.: Basic reproductive rate of a spatial epidemic model using computer algebra software. In Valafar, F., Valafar, H., eds.: Proceedings of the 2005 International Conference on Mathematics and Engineering Techniques in Medicine and Biological Sciences. (2005)
2. Bowman, F.: Introduction to Bessel Functions. Dover Publications Inc., New York (1958)
3. Rodriguez, D.J., Torres-Sorando, L.: Models of infectious diseases in spatially heterogeneous environments. Bulletin of Mathematical Biology **63** (2001) 547–571
4. Apostol, T.M.: Mathematical Analysis. Addison-Wesley Publishing Company, Reading, Massachusetts (1988)

The Effect of the Theorem Prover
in Cognitive Science

Tadashi Takahashi[1] and Hidetsune Kobayashi[2]

[1] Dept. of Mathematics and Informatics, Faculty of Human Development, Kobe
University, 3-11, Tsurukabuto, Nada-ku, Kobe 657-8501, Japan
[2] Dept. of Mathematics, College of Science and Technology, Nihon University, 8-14,
Kanda-Surugadai 1-chome, Chiyoda-ku, Tokyo 101-8308, Japan

Abstract. Humans use strategies to solve problems. Strategies are used
as knowledge to plan solutions and decide procedures. A computer algebra system with a theorem prover is being developed. We must consider
the theorem prover from not only the perspective of its effect on cognitive
science, but also from the perspective of mathematical studies.

1 Introduction

According to the three-level human behavior model of Rasmussen, automatic
human actions can be classified into the three levels of skill-, rule- and knowledge-
based actions([6]). A skill-based action is a response that occurs in less than 1
second ([4]). A chain of skill-based actions is a rule-based action. Thinking about
how to solve a problem is a knowledge-based action.

Skill-based actions are performed smoothly without intentional control. Rule-
based actions require a great deal of repetitive practice in order to be transferred
to the skill-based level. First, the external conditions must be recognized, then
the rules for composing the act are combined with the conditions required to
carry out the behavior. Knowledge-based actions require the recognition of external conditions, the interpretation of these conditions, the construction of a
psychological model for considering solutions, planning, and finally, the use of
the other two behavior levels to carry out the action. This is a process model in
which mastery of behavior requiring thought is internalized to the point where
it can be carried out unconsciously. Mistakes can be explained as omitted steps,
or for example, as pushing the wrong nearby button in smoothly carried out
skill-based actions. In the case of knowledge-based actions, illusion can lead to
error. In the present study, this process was analyzed using Rasmussen's three-
level human behavior model in order to identify what functions are essential
to facilitating smooth action and learning. Behavior used to learn about problems and how to solve them is classified in detail according to the three-level
model. Humans act by classifying issues and their relationships by consciously
combining them. Humans control themselves by constantly observing, thinking
about, evaluating, and integrating their behavior in order to achieve accuracy,
continuity, consistency, and normality ([3]). Classified factors can be separated
into the same three levels as the general actions.

V.N. Alexandrov et al. (Eds.): ICCS 2006, Part I, LNCS 3991, pp. 924–927, 2006.

2   Strategy

Strategies are used as knowledge to plan solutions and decide procedures. When these procedures, in general or for the most part, obtain the correct answer, the procedure is called a heuristic; however, such heuristics do not always result in a correct solution.

Strategies are used even when human beings solve mathematical problems. Recognition knowledge and experience are used as "doing it like this is effective in this case". The ability to rapidly reference knowledge is required for strategies based on experience. Furthermore, the recognition of thoughts and feelings controls. The famous book by the mathematician Polya, "How to solve it"([5]), showed the processes of mathematical problem solving; however, one can not learn how to use heuristics in problem solving just by reading a book. In researching problem solving, there are two contrasting concepts. The first emphasizes insight, flash, and senses, while the second emphasizes experiential knowledge. The former concept employs a strong tendency to perceive that strategies of thought are learned through the experience of problem solving. In other words, it is assumed that an intuitive feelings and specific technical abilities can be acquired. In the latter concept, it is assumed that problem solving ability arises from the accumulation of rules inherent to the domain provided by an individual problem. Such differences depend on the problem's nature, domain, and level, and the type of person involved in the learning process. In addition, it is difficult to establish clear boundary lines between these two concepts. In problem solving, experiential knowledge plays a large role. Heuristics are general ideas or algorithms (a procedure providing the correct solution), and are widely used. Heuristics are equal to "the logic of a thought". Examples of extremely general strategies are "try to draw a figure if you come across a difficult problem", and "search for similar problems that you have experience with". There are also concrete strategies we are familiar with, such as "A problem requiring the comparison of quantities requires two differences, and a transform formula" and "try to make clauses that differ next to each other for number sum sequence problems" ([1]).

3   Theorem Prover

As a representative of a theorem prover, the Isabelle/HOL system was used. Research on formalizing abstract algebra in Isabele/HOL is based on work by Hidetsune Kobayashi. This study focuses on researching mathematics, and in particular, on training researchers in the technics of proving ([2]). In the area of mechanical theorem proving, Kobayashi gave a decision procedure for what he called abstract algebra, based on algebraic method. It is really surprise to prove many abstract algebra theorems whose traditional proofs need enormous amounts of human intelligence. One of the key observations of Kobayashi is that theorems in abstract algebra can be relatively easily dealt with by a lot of lemmata, completely from former methods. The power of the method can be shown by experiments on computers in which many abstract algebra theorems were proved. The success of Kobayashi's method stimulated researchers to apply the connection of lemmata images. This research on formalizing abstract algebra in Isabelle/HOL is being conducted in order to develop a computer algebra system that supports mathematical study focused on abstract algebra. The system combines methods of automated theorem proving and also integrates programming in a natural way.

This method are of interest to researchers both in artificial intelligence (AI) and in algebraic modeling because they have been used in the design of programs that, in effect, can prove or disprove conjectured relationships between, or theorems about, abstract algebraic objects.

It is interesting to note that theorems have been verified by this method. In a limited sense, this "theorem prover" is capable of "reasoning" about algebraic conjectures, an area often considered to be solely the domain of human intelligence.

This research aims at extending current computer systems using facilities for supporting mathematical proving. The system consists of a general higher-order predicate logic prover and a collection of special provers. The individual provers imitate the proof style of human mathematicians and produce human-readable proofs in natural language presented in nested cells. The long-term goal of this research is to produce a complete system, which supports mathematicians. On the meta-level, we can write explicit programs for reasoning tactics using Isabelle/HOL.

4   Knowledge Base in Cognitive Science

When researchers use the theorem prover for the acquisition of knowledge or skills, we must consider a "tool" to be a "symbol device". A symbol device exists between the researchers and the research subject. Operation activity occurs between a symbol device and the researching subject. In cognitive science, two difficulties exist, one in the interaction between the researcher and the symbol device, and one in the interaction between the symbol device and the research subject. Therefore, we must overcome these difficulties in order to effectively utilize the theorem prover in cognitive science. Moreover, we must assess the benefits of considering the integration of the theorem prover from the perspective of the relationship between mathematical knowledge and mathematical concepts. When theorem provers are used in mathematical studies, researchers achieve a result through their efforts. Then, the researchers must investigate whether conceptual problems exist or whether they simply do not appreciate how the theorem prover works. By using a theorem prover effectively, researchers become aware of numerous mathematical ideas. This is made possible by incorporating the results of research in cognitive science. In carrying out a seven-phase model of human action, "the formation of a series of intentions or actions" must be performed smoothly. The effective use of a theorem prover in cognitive science is influenced by the contents of mathematical thought, and research and understanding of mathematics can further influence general idea formation. The theorem prover influences the "perception - interpretation - evaluation" phases of evaluation. The foundations of this model were studied by Rasmussen as the three-level control model of individuals actions ([7]). We can use the theorem prover as a material object that is available for the assessment of human activity. The use of the theorem prover can establish automatic and routine procedures. Controlling this automation is essential, especially in research on though processes. There are three methods for creating a theorem proof (by hand, by mind, and with a computer). A researcher's point of view of cognitive science considers the relationship between the brain and mind as the relationship between hardware and software in a computer. According to this point of view, the science of the mind is a special science, the science of thought.

5   Conclusion

In the three-level model of human behavior, operations and strategies can be identified and considered in relation to human thought processes in order to facilitate error-free problem solving. In consideration of surface features and conditions, similar problems can be recognized and suitable problem-solving methods can be identified. In addition, it was found that contents of the subconscious can be raised to the knowledge-based action level in order to support the expression process and the achievement of efficient functioning.

The technology of theorem prover automated reasoning. The ultimate goal of mathematics is technology. To do mathematics is gaining knowledge and solving problems by reasoning. Theorem prover is a powerful tool for researching mathematics. Researchers should appreciate the possibility of sharing cognitive level with such technology.

References

1. Ichikawa S.: Psychology of Learning and Education, Iwanami Shoten, 91 (1995) (in Japanese).
2. Kobayashi H, Suzuki H and Ono Y.: Formalization of Henzels Lemma, 18th International Conference, TPHOLs, Oxford, UK, Emerging Trends Proceedings Oxford Research Report (2005).
3. Kozuya T. (ed.): Memory and Knowledge (Cognitive Psychology Lecture 2), University of Tokyo Press, 17 (1978) (in Japanese).
4. Polson P. G. and Kieras D. E.: A Quantitative Model of the Learning and Performance of Text Editing Knowledge, Proceedings of ACM CHI'85 Conference on Human Factors in Computing Systems. (1985).
5. Polya G.: How to solve it. Doubleday (1957).
6. Rasmussen J.: Recognition engineering of interface, Keigakushuppan (1990) (in Japanese).
7. Tamura H.: Human interface. Ohm-sha (1998) (in Japanese).

Designing Next-Generation Training and Testing Environment for Expression Manipulation

Rein Prank, Marina Issakova, Dmitri Lepp, and Vahur Vaiksaar

University of Tartu, Institute of Computer Science, Liivi Str 2, 50409 Tartu, Estonia
{prank, marinai, dmitri, vax}@ut.ee

Abstract. T-algebra is a project for creating an interactive learning environment for basic school algebra. Our main didactical principle has been that all the necessary decisions and calculations at each solution step should be made by the student, and the program should be able to understand the mistakes. This paper describes the design of our Action-Object-Input dialogue and different input modes as an instrument to communicate three natural attributes of the step: choice of conversion rule, operands and result.

1 Introduction

When the student solves an expression manipulation task, he should at each step

1. Choose a transformation rule corresponding to a certain operation.
2. Select the operands (certain parts of expressions or equations) for this rule.
3. Replace them with the result of the operation.

For proper learning of this difficult area an environment should be available where all the necessary decisions and calculations at each solution step would be made by the student, and the environment would be able to understand the mistakes.

Traditional classroom cannot be such environment because the teachers are not able to discover and correct the mistakes in time. Existing software does not address the spectrum of arising problems in whole complex. On the one hand, some programs require from the student only execution of the first and (partially) the second substep. Using computer algebra systems, the student selects at best only the transformation rule and a part of the expression; the transformation itself is made by the computer. In addition, computer algebra systems have only a small number of very powerful operations (Simplify, Solve, ...) that enable to get the answer of a task in one step but not to build traditional step-by-step solution. The latter deficiency has been overcome in rule-based learning environments [1, 6] that have much more detailed rules. But they also perform the selected transformations mainly automatically.

Some other programs (for instance APLUSIX [4]) use paper-and-pencil like dialog design where a transformation step consists purely of entering the next line. However, such input does not provide the program with information about the decisions made by the student at stages 1 and 2. As a consequence, practically the only error that can be diagnosed is the non-equivalence with the previous line.

In the University of Tartu our first environment for step-wise expression manipulation (in Propositional Logic) was implemented about fifteen years ago [5]. The step

V.N. Alexandrov et al. (Eds.): ICCS 2006, Part I, LNCS 3991, pp. 928–931, 2006.

dialogue was built using Object-Action scheme. The student had to mark some subformula and enter the result of conversion (there was also a working mode with conversion rules in the menu). The program was able to verify separately selection of operand and performed conversion.

Using our long positive experience, we try now to integrate the above-described rule-oriented and input-oriented approaches. In 2004 we received funding from Estonian School Computerisation Foundation 'Tiger Leap' and launched a project, called T-algebra, for expression manipulation tasks in four areas: calculation of the values of numerical expressions; operations with fractions; solving of linear equations, inequalities and linear equation systems; simplification and factoring of polynomials.

2   General Design of Solution Steps – Action-Object-Input Scheme

Working with T-algebra, each step consists of three substeps:

1. Selection of the operation from the menu,
2. Marking the operand(s) in expression,
3. Entering the result of the operation.

The following figure demonstrates third substep after selection of rule *Collect like terms* and marking of appropriate operands.

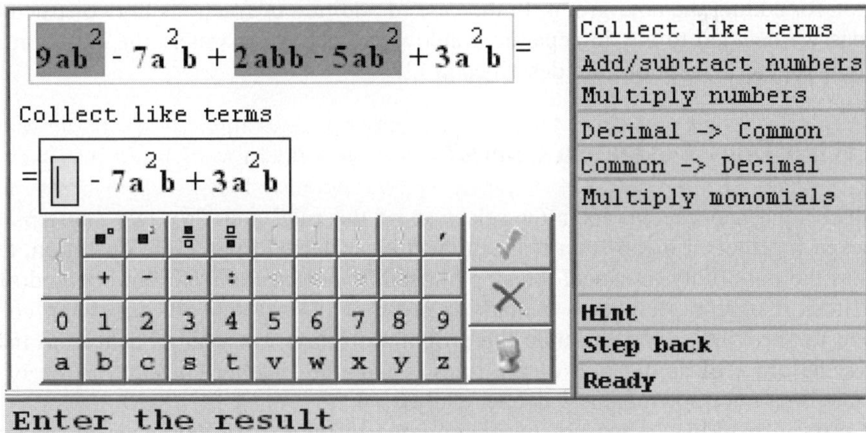

Fig. 1. Entering the result in free input mode

How much does the amount of input grow in comparison with simple entering of the result? Our step dialogue requires from the student two preceding actions. First of them is only one click on the mouse. At the second substep the student has to mark one or more parts of the expression. As a compensation, the program is able to copy the rest of the expression automatically to the next line. If the dialogue would consist only of the third substep of our step, the student should enter the passive parts of the expression from the keyboard or mark and copy them one by one. This means that addition of the second substep does not change the amount of the work.

For third substep the program has three input modes: *free, structured* and *partial* mode (Figures 1 and 2 demonstrate corresponding input areas).

In free input mode the program generates one input box inside of the expression in the next line. The student should enter in the box one expression replacing all marked parts from the previous line.

In structured input mode the information about the actual rule and operands is used to create the pattern of the result automatically, using different input boxes for signs, coefficients, variables and exponents.

Partial input mode is a modification of structured input where the program fills some boxes automatically and leaves to the student only the boxes for the components that are crucial for particular rule (signs, coefficients, exponents).

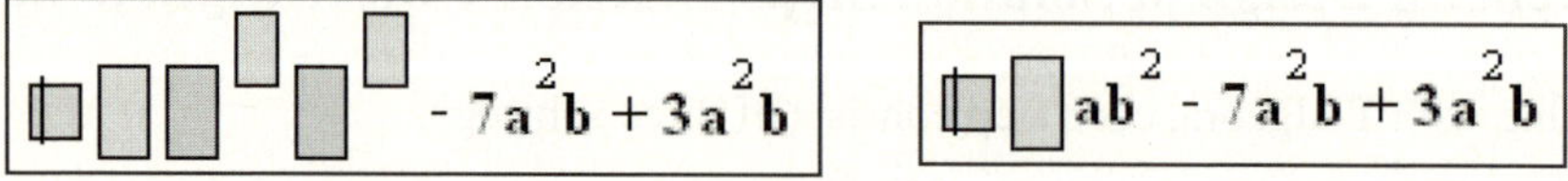

Fig. 2. Input areas for $6ab^2$ in structured and partial mode

There are a few transformations that need some addition to our general input scheme. For example, in the case of multiplication of polynomials the program creates only one monomial-structured group of boxes and our virtual keyboard contains a button for adding the next group. In the case of addition of fractions in two structured modes we decided to ask in separate window and check separately the common denominator, etc. Exceptions are described in [3].

3   What Does T-Algebra Check?

T-algebra has no checkpoint after selection of the rule, and first two substeps are checked together. It means that if the student has selected impossible operation, then he has the possibility to cancel the step himself. If he selects irrelevant operand(s) as well then T-algebra diagnoses a marking error. If the selected rule does not correspond to "official" algorithm then the program displays a warning because in many cases the order of application of the rules is insignificant but current version of T-algebra does not implement any deeper analyse.

Errors in marking (when not misclicks) tend to refer to serious gaps in student's understanding of the whole task, priority order of operations or of selected rule. T-algebra checks the following items:

1. Are the marked parts syntactically correct subexpressions?
2. Do the marked terms have the form required for operands of selected rule? (The like terms for combining should be monomials etc.)
3. Do the marked terms satisfy compatibility requirements for selected rule? (For example, are they really like terms?)
4. Do the marked terms satisfy location requirements (belong to the same sum or product etc)?

The input of third substep is checked in the following order.

1. Is the entered expression syntactically correct?
2. Has the input the structure required by the result of application of the selected rule? For example, the result of multiplication of monomial with polynomial should be the sum of monomials and the result of combining like monomials should be one monomial.
3. Is the entered expression equivalent to the marked (i.e., replaced) part and is the result of substitution equivalent to the previous line?
4. For some rules the program checks yet whether the operation was really performed. For example, 6/12 cannot be reduced to 60/120 or 5/10 or 6/12. Some details of error checking in equation solving tasks are described in [2].

4 Conclusion

Our main interest has been such a dialogue from where the program could understand the solution steps of the student and give intelligible feedback for weaker students who tend to make many mistakes. In our first version of the program we are not able to realize all the potential of our interface and to implement very detailed diagnosis. We start from giving some level of feedback and add further details if the experiments demonstrate that the pupils really need more explanations. But we have seen that implementation of the items listed in Section 3 was quite easy and the error messages that refer to the rules and operands were helpful for the pupils.

References

1. Beeson, M.: Design Principles of Mathpert: Software to support education in algebra and calculus. In: Kajler, N. (ed.): Computer-Human Interaction in Symbolic Computation. Springer-Verlag (1998) 89-115
2. Issakova, M.: Possible Mistakes During Linear Equation Solving on Paper and in T-Algebra Environment. In: Proceedings of the 7th International Conference on Technology in Mathematics Teaching, Bristol, UK, Vol. 1 (2005) 250-258
3. Lepp, D.: (2005) Extended Solution Step Dialogue in Problem Solving Environment T-Algebra. In: Proceedings of the 7th International Conference on Technology in Mathematics Teaching, Bristol, UK, Vol. 1 (2005), 267-274
4. Nicaud, J., Bouhineau, D., & Chaachoua, H.: Mixing microworld and cas features in building computer systems that help students learn algebra. International Journal of Computers for Mathematical Learning, 5(2), (2004) 169-211
5. Prank, R.: Using Computerised Exercises on Mathematical Logic. Informatik-Fachberichte, Vol. 292, Springer-Verlag (1991) 34-38
6. Ravaglia, R., Alper, T., Rozenfeld, M., Suppes, P.: Succesful pedagogical applications of symbolic computation. In: Kajler, N. (ed.) Computer-Human Interaction in Symbolic Computation, Springer-Verlag (1998) 61-88

Automatic Node Configuration Protocol Using Modified CGA in Hierarchical MANETs

Hyewon K. Lee

School of Computing, Soongsil University, Seoul, Korea
kerenlee@nate.com

Abstract. The CGA is designed to prevent address spoofing and stealing and to provide digital signature to users without any help from security infrastructures, but fake key generation and address collision appear in flat-tiered network. To solve these problems, CGA defines security parameter (SEC), which is set to high value when high security is required. Although CGA with high SEC makes attackers be difficult to find fake key, it brings an alarming increase in processing time to generate CGA. On the contrary, the probability to find a fake key is high if low SEC is applied to CGA. In this paper, MCGA applicable to as well public networks as ad-hoc networks is proposed. Address collision problems are settled by employing hierarchy. Using MCGA, no previous setup is required before communication, and automatic node configuration is feasible.

1 Introduction

Mobile ad-hoc network (MANET) is a multi-hop wireless network without any prepared base station. It is capable of building a mobile network automatically without any help from DHCP servers or routers. Routing protocols to find shortest or optimistic route have been proposed, but these assume that nodes have been pre-configured. MANETconf [1], node configuration protocol [2] and prophet address allocation [3] have been proposed; however, these do not consider how to generate address.

CGA is designed to solve address spoofing and stealing attacks in IPv6. CGA offers digital signature without any help from CA, which is proper to ad-hoc nodes that have low processing power and memory capacity. However, fake key generation and address collision appear in flat-tiered network due to 64-bit-taken operation. To solve it, CGA defines SEC and allows a node to generate address only when the specific condition is satisfied. When SEC is set to high value, it becomes difficult for attackers to find fake key, but processing time to generate CGA increases incredibly.

2 Proposed Modified CGA (MCGA)

MCGA is composed of 64-bit subnet prefix learned from network and 64-bit interface identifier generated by individual node. For interface identifier, a random number or NIC address may be used. For subnet prefix, local-scoped prefix, FE80::/64 is used. When a node moves and gets different kinds of IEEE 802.11 beacon message, then it may accept new subnet prefix. Address format is similar to CGA except the SEC field.

V.N. Alexandrov et al. (Eds.): ICCS 2006, Part I, LNCS 3991, pp. 932–935, 2006.

MCGA generation process is detailed as follows: build key pair using RSA algorithm. Generate random number for MODIFIER, and set collision count to 0. Concatenate MODIFIER, collision count and user's public key. Put the concatenation into MD5. Take the first 64 bits from 128 bits key value generated by MD5, and set interface identifier to them. Set the u and g bits of interface identifier to 0. Concatenate subnet prefix and interface identifier, and put the concatenation into MCGA. If DAD is done successfully, allocate the MCGA to interface. If the check goes wrong and collision count is equal to 3, build new MODIFIER. Else, add 1 to collision count and build new concatenation.

Once a node enters into network, it generates MCGA and will request duplication check to the nearest agent, which lookups its resource table and gives appropriate answer to the requester. If requested address is not registered, the agent will give positive answer to the requester, and vice versa [2]. If no duplication is found, the agent will give final answer to the requester. When there are n nodes in a network, the probability that at least 1 fail occurs in n generations can be expressed as (1). In hierarchical network, a node in logically higher position holds information about all address resources in network, so duplication check between two nodes in the different logical positions seems to be enough. In this paper, opti-DAD [5] is employed for DAD. A node is able to initiate communication with others using on-pending address, which reduces delay due to long DAD process. Even though duplication ratio for address generation is very low, unallocated address may go to on-pending state concurrently by different nodes, and priority from arbitrary contention algorithm can be used.

$$1 - \frac{2^{64}P_n}{(2^{64})^n} \tag{1}$$

If there are a hash function ($h(\ )$) and two different inputs (m_1, m_2,), $h(m_1) \neq h(m_2)$ is true. Picking specific part from hash's output, though, provokes collision. When we think of CGA, 2^{96} cases are mapped to one 64-bit identifier, mathematically. The probability for collision is proved by 'birthday problem,' and it becomes 0.63. Attackers are easily able to build fake key pair corresponding to origin key pairs by brute-force. Once fake key pair is found, an origin node encounters with address spoofing or stealing attack. For example, $User_A$ builds its key pair and generates its CGA_A. If $User_B$ finds a fake key pair which yields CGA_A and begins to send a message to $User_C$. Once $User_C$ receives the message, it will identify $User_B$ as proper owner for CGA_A. Unless $User_A$'s key is disclose, $User_B$ cannot mimic $User_A$'s signature nor can it decrypt any message from $User_A$, but nodes can be induced into wrong communication.

When MCGA is applied to hierarchical ad-hoc network, an agent holds information about all address resources in network. If a stranger sends any message with different parameters for registered address, intermediate agent will notice it and drop the message. Let's go back to above example. If $User_A$ and $User_C$ locate at the same MUnit, $User_C$ notices that the message from $User_B$ is strange and drop it. If $User_A$ and $User_C$ locate at different regions, any message from $User_B$ will be dropped by any intermediate agent between $User_B$ and $User_C$.

3 Performance Evaluation

Processing time for CGA is the sum of requisition time of proper MODIFIER, generation time of interface identifier and delay due to duplication check, as (2). Processing time for MCGA is the sum of generation time of random number for MODIFIER, generation time of interface identifier and delay due to duplication check, as (3).

$$L_{CGA} = \left\lfloor \frac{m}{2} + 1 \right\rfloor l_{MOD} + (m+1)(l_{SHA} + l_{DAD}) \tag{2}$$

$$L_{MCGA} = \left\lfloor \frac{m}{2} + 1 \right\rfloor l_{RV} + (m+1)(l_{MD5} + l_{DAD}) \tag{3}$$

$$2l_d(d + \frac{8s}{b}) \le l_{DAD} \le 2l_d(d + r + \frac{8s}{b}) \tag{4}$$

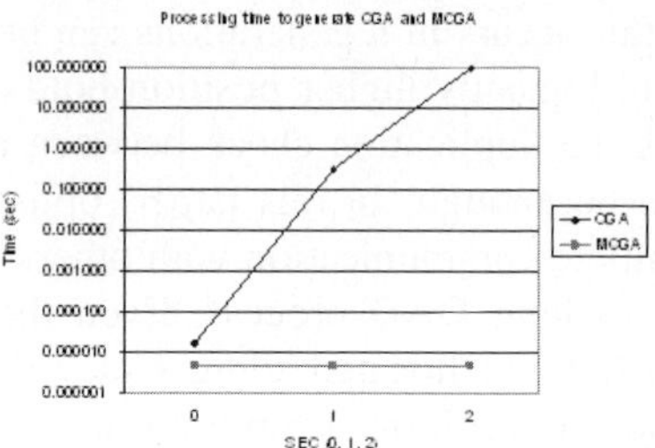

Fig. 1. Processing time for CGA and MCGA, respectively

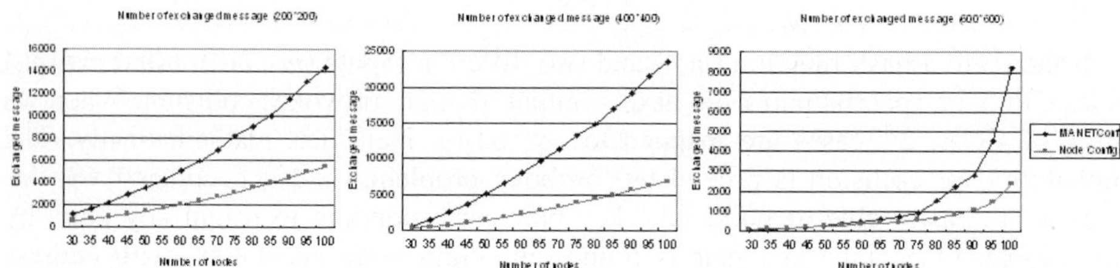

Fig. 2. The number of exchanged message between nodes

From (1), the number of duplication during address generation is assumed as 0. Generation time of MODIFIER has higher value than random number ($l_{MOD} \gg l_{RV}$). It is known that to process SHA takes more time than MD5 ($l_{SHA} \gg l_{MD5}$). From convergence time from [6], DAD time (l_{DAD}) can be expressed as (4).

System for simulations has following resources; CPU Pentium 4.3GHz and Memory 1GB. For operating system, Linux is employed, especially Kernel 2.4. Fig. 1 shows variation on generation time of 3000 CGAs and 3000 MCGAs. SEC is set to 0, 1 and 2, respectively. The average execution time for MCGA is 4.77 µs while the average execution time for CGA when SEC is set to 0 is 15.57 µs. The unit, µs, is very small, but address generation by ad-hoc nodes will need more time. For example, 400 MHz ad-hoc node will perform the process 10 times slower than the above system. No address duplication is occurred in both CGA and MCGA generations. Propagation delay is not considered in this simulation. Fig. 1 clearly proves that processing time for CGA increases dramatically when SEC increases, and processing time is strongly affected by SEC. When SEC is set to 3, it requires more than 200 hours. CGA with larger than 3 seems to be inappropriate as well public network as ad-hoc network.

The next simulation starts with initial network configuration and finishes when the number of nodes becomes 100. Ad hoc nodes distributed in two dimensional region of size 200×200, 400×400 and 600×600m^2, respectively. The network is randomly generated with the constraint that the graph be fully connected. Each node randomly moves with 1~2 m/s to random directions, and it is equipped with a radio transceiver, which is capable of transmitting a signal from 80 m. Processing delay for transmitting a message is randomly chosen between 5 ms and 10 ms. Propagation delay is 500 ms. MCGA with link-local prefix is used for address. Each node is supplied by a battery with enough power to at least make it able to carry out a complete operation. As shown in Fig. 2, the number of exchanged packets for node configuration increases as the number of nodes increases. MANETConf requires more packets. The result from this simulation obviously shows that node configuration protocol offers scalability to large network. Network topology and size give an effect on the number of packets.

4 Conclusions

In this paper, MCGA which is proper to hierarchical ad-hoc network is proposed. The MCGA has shorter processing time than CGA and offers digital signature with no additional overheads. To solve fake key and collision problems, we adopt hierarchical network structure. The MCGA is applicable to as well public networks as ad-hoc network. Simulations show that processing time for MCGA is reduced down 3.3 times and 68,000 times, compared to CGA with SEC 0 and SEC 1, respectively, Further, CGA with SEC 3 is inappropriate for both ad-hoc and public networks.

References

1. Nesargi, S. and Prakash, R., "MANETconf: Configuration of Hosts in a Mobile ad Hoc Network," INFOCOM, IEEE, 2002
2. Lee, H. and Mun, Y., "Node Configuration Protocol based on Hierarchical Network Architecture for Mobile Ad-Hoc networks," LNCS 3090, 2004
3. Zhou, H., Ni, L. and Mutka, M., "Prophet Address Allocation for Large Scale MANET," INFOCOM, IEEE, 2003
4. Aura, T., "Cryptographically Generated Address," RFC 3972, IETF, 2005
5. Moore, N., "Optimistic DAD for IPv6," work in progress, IETF, 2004
6. Kulik, J., Heinzelman, W. and Balakrishnann, H., "Negotiation-Based Protocols for Disseminationg Information in Wireless Sensor Networks," 2002

Route Optimization in NEMO Environment with Limited Prefix Delegation Mechanism*

Jungwook Song, Sunyoung Han**, and Kiyong Park

College of Information and Communication
Department of Computer Science and Engineering
Konkuk University
1 Hwayang, Gwangjin, Seoul 143-701, Korea
{swoogi, syhan, kypark}@cclab.konkuk.ac.kr

Abstract. The 4G network will must be ALL-IP network and IPv6 is destiny. Already IPv6 has extended as Mobile IPv6 for host mobility and Mobile IPv6 is extending for whole network mobility. There are many possible mobile networks, and they would be nested as they change locations. Most serious problem in a nested mobile network is the complexity of routing path of packets, and the complexity grows as the level of nesting increases. In this paper, we propose 'limited prefix delegation mechanism', which delivers packets through the optimized path even though mobile networks are nested. We show the effectiveness of our mechanism by solved problems.

1 Introduction

As information and communication technologies are rapidly progressed, there are increasing sorts of informative devices that can access wireless network while they move their locations. There are many researches on wireless and mobile network. When IPv6 is widely deployed, almost electronic devices will be connected to the Internet. These two key cores the 'Mobile Access Network' and 'IPv6', are necessary for the 'Ubiquitous Computing'.

If IPv6 prevails and mobile access network is established, it would be a common case that tens or hundreds of mobile nodes change their locations at the same time. Because existing Mobile IPv6 has been designed to support host mobility only[1], they do not smoothly support the current movement of many hosts, i.e. movement of network that consisted with two or more hosts or PANs(Personal Area Networks) or network in moving vehicles. To support movement of whole network, IETF nemo WG is making study an extension of Mobile IPv6 and already has the protocol standard that called the 'NEMO Basic Support Protocol'[2].

---

* This research was supported by the MIC(Ministry of Information and Communication), Korea, under the ITRC(Information Technology Research Center) support program supervised by the IITA(Institute of Information Technology Assessment).

** Corresponding author.

V.N. Alexandrov et al. (Eds.): ICCS 2006, Part I, LNCS 3991, pp. 936–939, 2006.

In general situation, however, any networks and hosts can change their location simultaneously, and they can be nested. Direct application of [2] to the nested mobile network causes complicated routing path from MNN(Mobile Network Node) to the CN(Correspondent Node). Therefore, it is important to optimize the routing path of packets in nested mobile network. The route optimization problem is a part of main issues that working progress[3, 4]. The problems of nested mobile network are described in [3].

In this paper, we try to solve the route optimization problem in nested mobile network using the 'limited prefix delegation mechanism'. Instead of by opening bi-directional tunnel between the MR(Mobile Router) and its HA(Home Agent), the packet routing path can be optimized by opening tunnel directly from the MR to the CN. And the CN sends packet with the routing header. Our mechanism opens one or two tunnels, and we can optimize the routing path in both directions.

2   Limited Prefix Delegation Mechanism

The 'limited prefix delegation mechanism' we propose in this paper differs from [2]. Our mechanism sends all packets directly through tunnel between the MR and the CN. The packet from MNN to the CN is tunneled on MR, and the CN sends the packet to the MNN with the routing header. To achieve this, we add new RA(Router Advertisement) option on MR. And we also modified Binding Update process and tunneling process on MR.

2.1   Addition of New Router Advertisement Option

To all MRs in nested mobile network open just one tunnel to the CN directly, delegating access router's prefix is required. The MR that attached access router directly, it is simple that delegating prefix. But, the nested MRs cannot recognize access router's prefix. So, Upper link MR must notify access router's prefix to nested MRs. To achieve this, we add new RA option called 'delegated prefix option' as shown in Fig. 1.

If the MR does not receive this option, MR delegating prefix from uplink router's(access router's) original RA message, and install new care-of address on

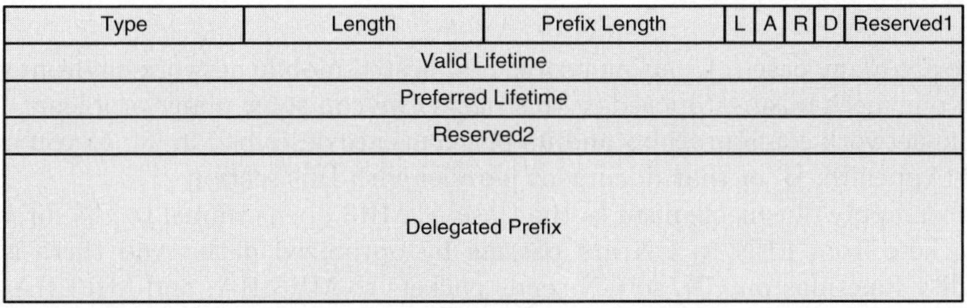

Type	Length	Prefix Length	L	A	R	D	Reserved1
Valid Lifetime							
Preferred Lifetime							
Reserved2							
Delegated Prefix							

Fig. 1. Delegated Prefix Option

its egress interface. And the MR adds delegated prefix option to its RA message. The new RA message consists of prefix option that has prefix of home link, and delegated prefix option that has delegated prefix from access router.

If the MR receives delegated prefix option, MR delegating prefix from this delegated prefix option, and then same as before case. So, all nested MR can have care-of address that is subnet address of visited access router on top of nested mobile network. And, all other MNNs silently ignore delegated prefix option from RA message, thus there are no changes on MNNs.

2.2   Modification of Tunneling on Mobile Router

In our mechanism, the MR sends packet through direct tunnel to the CN instead of sending through the tunnel to its HA. If outgoing packets have source address with mobile network prefix(from home link prefix), the MR should open direct tunnel to the CN. In this scheme, the MRs do not care about the types of MNN, whether it is plain host or not: it applies the same scheme on all packets passing by that have source address with mobile network prefix.

2.3   Modification of Binding Update Process

When a MR detects its movement, it performs Binding Update process. In the same way, the MR sends Binding Update to the CN when a MNN sends packets to a CN or a CN sends packets to a MNN. The CN that received Binding Update can send packets to the MR directly. The MR maintains the 'CN table' to keep the information if Binding Update was sent to the CN or not. The MR must initialize this table when it moves to another location.

2.4   Processing Routing Header on Mobile Router

If the CN was enabled Mobile IPv6 and received Binding Update from the MR, the CN sends packet to the MR's care-of address with the routing header type 2, that has the MNN's original address. When the MR receives this packet, it examines that packet has the routing header, and if exists, the MR forwards packet to the MNN that is original destination.

3   Solved Problems

There are many cases of routing problem in nested mobile network environment. With our mechanism proposed in this paper, we can solve many problems. The mobile network configurations and its problems are described in [3], so you must refer 'Appendix B' of that document parallel with this section.

When apply our mechanism to the Case A, MR3 opens tunnel to CN directly, so packets from LFN to CN are passing by optimized path. And there is no mobility functions on CN, so CN sends packets to MR3_HA, and MR3_HA can open tunnel to MR3 directly. We can achieve near-optimized path. When apply our mechanism to the Case B, MR3 opens tunnel to CN directly, so packets from LFN to CN are passing by optimized path. And CN has mobility function; CN sends packets to MR3 directly with the routing header. Thus, we can achieve optimized path in both directions. The Case C, after Mobile IPv6 route optimization, the rest are same processes as Case A.

The Case D is similar to Case A. When apply our mechanism in this- case, MR3 try to opening tunnel to CN directly, and MR5_HA receives and forwards it via bi-directional tunnel between MR5_HA to MR5. When MR5 receives packet from MR3, MR5 send Binding Update to MR3. So, MR3 sends packet directly to MR5 with the routing header. The same process occurs on reverse direction. Thus, we can achieve optimized path in both directions with only two tunnels. The Case E is similar to Case B and The Case F is similar to Case C. After Mobile IPv6 route optimization, the rest are same processes as Case D.

The Case G is similar to Case A. Same process occurs as Case D. And the Case H, same process occurs as Case E. The Case I, as for Case C and Case F, Mobile IPv6 Route Optimization cat not performed. But, MR3 and MR5 perform Route Optimization like Case D.

The Case J, no special function is necessary for optimization of their communication. The Case K is similar to Case H. Two nodes may initiate Mobile IPv6 route optimization. Same as Case J, no special function is necessary. The Case L, Mobile IPv6 Route Optimization cannot be preformed. With our mechanism, similar to Case D, MR3 and MR5 send Binding Update to each other. Thus, we can achieve optimized path in both directions.

4   Concluding Remarks

In this paper, we proposed the 'limited prefix delegation mechanism' which optimizes the routing path in nested mobile network environment. From previous section, almost problems in [3] could be solved by our proposed mechanism.

The 'limited prefix delegation mechanism' opens just one or two tunnels, and it establishes direct path bypassing MRs' HAs. To achieve the optimized path, we add new 'delegated prefix option' on the MR's RA message, modify tunneling target on the MR, modify Binding Update process on the MR. We just add or modify small functions on the MR only. So, we can get much better results with little changes.

References

1. D. Johnson, C. Perkins, J. Arkko: Mobility Support in IPv6: RFC3775, IETF, 2004
2. V. Devarapalli, R. Wakikawa, A. Retrescu, P. Thubert: Network Mobility (NEMO) Basic Support Prototol: RFC3963, IETF, 2005
3. C. Ng, P. Thubert, M. Watari, F. Zhao: Network Mobility Route Optimization Problem Statement: IETF nemo WG Draft, 2005
4. C. Ng, F. Zhao, M. Watari, P. Thubert: Network Mobility Route Optimization Solution Space Analysis: IETF nemo WG Draft, 2005

A Target Tracking Method to Reduce the Energy Consumption in Wireless Sensor Networks

Hyunsook Kim and Kijun Han[*]

Department of Computer Engineering,
Kyungpook National University, Daegu, Korea
hskim@netopia.knu.ac.kr, kjhan@knu.ac.kr

Abstract. Reducing the energy consumption is one of the most critical issues in wireless sensor networks and an accurate prediction is especially required in tracking. It is desirable that only the nodes surrounding the mobile target should be responsible for observing the target to save the energy consumption and extend the network lifetime as well. In this paper, we propose an efficient tracking method based on prediction through the error correction, which can minimize the number of participating sensor nodes for target tracking. We show that our tracking method performs well in terms of saving energy regardless of mobility pattern of the mobile target.

1 Introduction

Object tracking is emerging as one of the new attractive applications in large-scale wireless sensor networks such as wild animal habit monitoring and intruder surveillance in military regions. Tracking of the mobile targets has lots of open problems to be solved including target detection, localization, data gathering, and prediction.

In the localization problem, excessive sensors may join in detection and tracking for only a few targets. And, if all nodes have to always wake up to detect a mobile target, there are a lot of waste of resources such as battery power and channel utilization. So, if each node uses timely its energy to execute tasks, the network lifetime may be extended as a whole. This raises the necessity for prediction of the moving path of the mobile target to maintain the number of participating nodes in tracking as small as possible.

Many tracking protocols in large-scale sensor networks have been proposed to solve the problems concerned with tracking of the mobile targets from various angles [1, 2, 3, 4].

In this paper, we present a tracking method by predicting the location of the mobile target in 2-dimensional WSN, based on linear estimation.

The rest of this paper is organized as follows. Our proposed tracking method is presented in Section 2. Next, in Section 3, we present some simulation results. Finally, Section 4 concludes the paper.

---

[*] Correspondent author.

V.N. Alexandrov et al. (Eds.): ICCS 2006, Part I, LNCS 3991, pp. 940–943, 2006.

2   Proposed Tracking Method

2.1   Approximate Prediction Step

First, since we assume that the mobile target does not change its direction or speed so abruptly in the sensing field, the location of the mobile object at the time instant of (n+1) is approximately predicted by estimating the velocity when the mobile target will move during the time interval [n, n+1]. Fig.1 shows the concept of tracking of the mobile target.

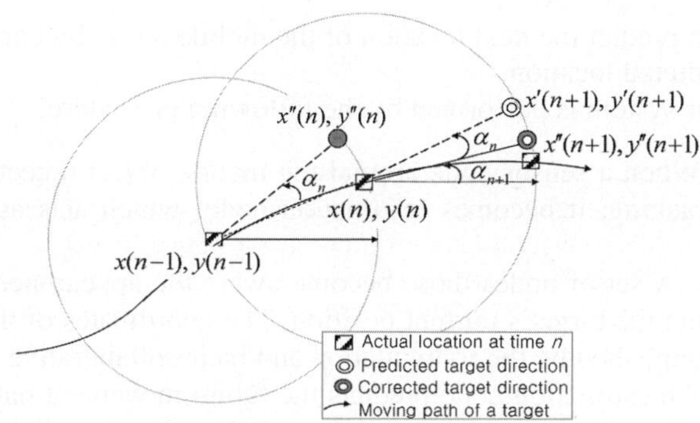

Fig. 1. Tracking of the mobile target in WSN

Given the current location $(x(n), y(n))$, the first predicted location of the mobile object at the time instant of (n+1), denoted by $(x'(n+1), y'(n+1))$ is

$$x'(n+1) = x(n) + \tilde{v}_x(n+1), \quad y'(n+1) = y(n) + \tilde{v}_y(n+1) \tag{1}$$

where $\tilde{v}_x(n+1)$ and $\tilde{v}_y(n+1)$ represent the future speed estimates of the mobile object in the direction of x and y, respectively.

These speed estimates based on the previous history are given by

$$\tilde{v}_x(n+1) = \frac{\sum_{i=n-h+1}^{n} \tilde{v}_x(i)}{h}, \quad \tilde{v}_y(n+1) = \frac{\sum_{i=n-h+1}^{n} \tilde{v}_y(i)}{h}, \tag{2}$$

where h is a predefined number of the past history based on which we predict the next moving factor. Hence, the future speed is a moving average of acceleration of an object. That is, the present movement of a mobile object means a reflection of the patterns of the moving history.

The estimate obtained in this step makes it possible to exactly predict the future location of the mobile object that moves linearly. However, the estimate is no longer effective when the mobile object moves in non-linear fashion since only velocity information in used to predict the future location. So, we need a correction mechanism to get a more exact estimate.

2.2 Correction Step

Let us express the prediction error by the angle between the actual location and the previously predicted, denoted by α. Then we have

$$\cos \alpha_n = \frac{x''(n)-x(n-1)}{\sqrt{[x''(n)-x(n-1)]^2+[y''(n)-y(n-1)]^2}} - \frac{x(n)-x(n-1)}{\sqrt{[x(n)-x(n-1)]^2+[y(n)-y(n-1)]^2}} \tag{3}$$

Then we can predict the new moving direction of the mobile target by

$$\cos \alpha_{n+1} = \frac{x'(n+1)-x(n)}{\sqrt{[x'(n+1)-x(n)]^2+[y'(n+1)-y(n)]^2}} - \cos \alpha_n \tag{4}$$

Finally, we can predict the next location of the mobile target by correcting an angle from the first predicted location.

Tracking in our system is performed by the following procedure.

1. Discovery: When a sensor node around the mobile object detects the target and initializes tracking, it becomes 'estimation node' which acts as a master node temporarily.
2. Localization: A set of nodes those become aware the appearances of the mobile target compute the target's current position. The coordinates of the mobile target may be accomplished by the triangulation and their collaborative works.
3. Estimation: An estimation node predicts the future movement path of the mobile target, and transmits message about the approaching location to its neighbor nodes. The prediction is carried out by two steps: approximate a prediction and correction step that is explained above. The moving factors of a mobile target, such as direction and velocity, can be obtained by sensor nodes through collecting moving patterns of the tracked target.
4. Communication: As the mobile target moves, each node may hand off an initial estimate of the target location to the next node in turn. At that time, each node changes its duty cycle along the movement of the target.

3 Simulation

We carry out experiments to measure the missing rate and wasted energy.

To model the movement behavior of the mobile target, we use the Random Way Point model (RWP) and Gauss-Markov mobility model. Energy consumption used for simulation is based on some numeric parameters obtained in [2]. Our prediction method is compared with the *Least Squares Minimization* (LSQ) to evaluate the performance of accuracy. LSQ is a common method used for error reduction in estimation and prediction methods.

As shown in Fig. 2 (a), our scheme offers a smaller missing rate than LSQ regardless of mobility model used in simulation. In RWP, since the moving pattern of the mobile target is random, the prediction error increases. As the sensing range becomes larger, the missing rate decreases as well. Due to the inaccurate prediction of location, some nodes miss the target because the real location of the target is out of the sensing range.

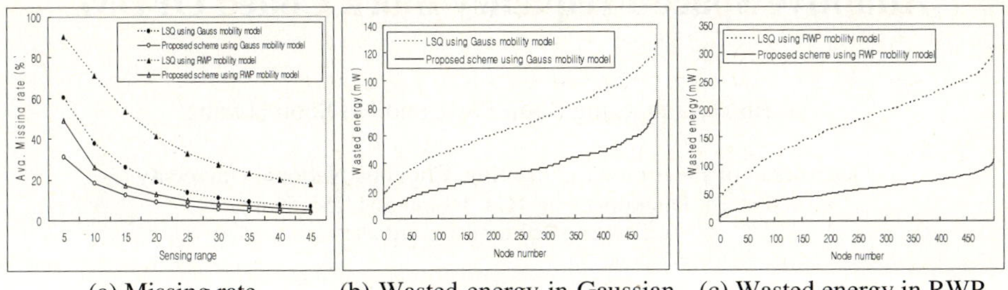

(a) Missing rate   (b) Wasted energy in Gaussian Markov mobility   (c) Wasted energy in RWP

Fig. 2. Wasted energy

As shown Fig. 2(b) and 2(c), energy consumption is greatly influenced by an accuracy of prediction. As described earlier, we can extend the network lifetime by avoiding such unnecessary energy consumption at nodes that do not need to join in tracking. This figure indicates that our scheme can decrease the number of participating nodes and thus reduce the energy consumption too.

4 Conclusion

Power conservation and an accurate prediction are important issues for object tacking in wireless sensor networks. In this paper, we propose an efficient tracking method using a moving average estimator to decide the future location of the mobile target. And we improve the accuracy of prediction through the error correction. Simulations results show that our estimation method performs accurately, which contributes to saving energy and thus extending the network lifetime as well regardless of mobility pattern of the mobile target by reducing the number of participating nodes in tracking.

Acknowledgement

This research is supported by Program for the Training of Graduate Students for Regional Innovation.

References

1. C. Gui and P. Mohapatra. : Power conservation and quality of surveillance in target tracking sensor networks. *In Proceeding of the ACM MobiCom*, Philadelphia, PA, September (2004)
2. Y. Xu, J. Winter, and W.-C. Lee. : Prediction-based strategies for energy saving in object tracking sensor networks. *In Proceedings of the Fifth IEEE International Conference on Mobile Data Management (MDM)*, USA, January (2004)
3. Y.Xu and W.C.Lee. : On Localized Prediction for Power Efficient Object Tracking in Sensor Networks. *In 1st International Workshop on Mobile Distributed Computing*, May (2003)
4. H. Yang and B. Sikdar. : A Protocol for Tracking Mobile Targets using Sensor Networks. *In IEEE International Workshop on Sensor Network Protocols and Applications*, May (2003)

Adaptive Space-Frequency Block Coded OFDM

Tae Jin Hwang, Sang Soon Park, and Ho Seon Hwang

Department of Electronic Engineering, Chonbuk National University
664-14, Duckjin-Dong 1Ga, Jeonju 561-756, Korea
tjhwang@chonbuk.ac.kr

Abstract. This paper discusses an adaptive modulation technique combined with space-frequency block coded OFDM(SFBC OFDM) over frequency selective channels and evaluates the performance in terms of the outdated channel state information(CSI) in mobile environments. This paper employs the Alamouti's diversity scheme in multiple input multiple output OFDM (MIMO OFDM) and an adaptive modulation with enhanced performance. Through the various simulations, the performance of SFBC OFDM employing adaptive modulation is compared with the performance of fixed modulation. Also, in adaptive modulation scheme, the effects of the outdated CSI under mobile environments are shown

1   Introduction

In order to improve the performance of OFDM system in frequency selective and multipath fading environments, this paper presents an adaptive bit allocation combined with SFBC OFDM. The perfect CSI ensures a desired efficiency/ performance of adaptive modulation scheme. In MIMO OFDM system, by making use of SVD the MIMO channel on each subcarrier is decomposed into parallel non-interfering single input single output(SISO) channels. But, a SFBC OFDM system with Alamouti's diversity scheme in [1] does not require the SVD for the CSI. Assuming the availability of the perfect CSI at the transmitter, the performance gains of adaptive modulation have been demonstrated. This paper examines the impact on performance of an adaptive OFDM system, which combined with SFBC scheme, due to the outdated CSI in mobile fading channel.

2   Adaptive Space Frequency Block Coded OFDM

This paper considers a MIMO OFDM system employing the Alamouti's diversity scheme in Fig 1. To begin with, let $\mathbf{H}_{ij}(n)$ be the following diagonal matrix whose diagonal elements are the frequency responses of the channel impulse responses h_{ij} between the i-th transmit antenna and the j-th receive antenna during the n-th time slot

$$\mathbf{H}_{ij}(n) = \text{diag}[H_{ij}(n,0) \cdots H_{ij}(n,N-1)], \quad i=1,2, j=1,2 \tag{1}$$

V.N. Alexandrov et al. (Eds.): ICCS 2006, Part I, LNCS 3991, pp. 944–947, 2006.

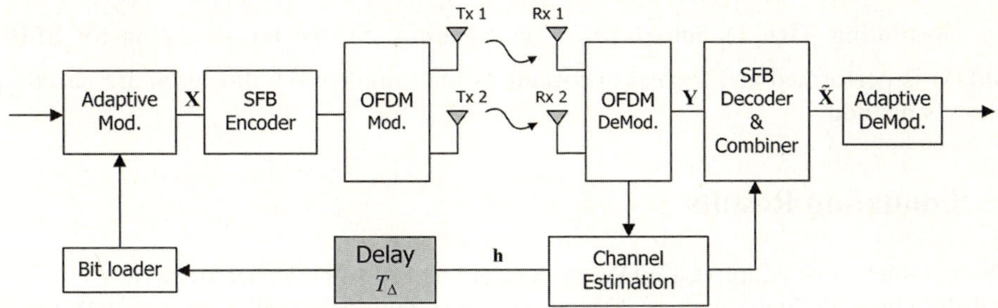

Fig. 1. Block diagram of an adaptive SFBC OFDM

The data symbol vector $\mathbf{X}(n)$ is coded into two vectors $\mathbf{X}_1(n)$ and $\mathbf{X}_2(n)$ by the space-frequency encoder block as

$$\mathbf{X}_1(n) = [X(n,0) \quad -X^*(n,1) \quad \cdots \quad X(n,N-2) \quad -X^*(n,N-1)]^T$$
$$\mathbf{X}_2(n) = [X(n,1) \quad X^*(n,0) \quad \cdots \quad X(n,N-1) \quad X^*(n,N-2)]^T. \tag{2}$$

Let $\mathbf{Y}_j(n)$ be the n-th received OFDM symbol from the j-th receive antenna. The receive symbol vector can be represented by the even and odd component vectors as follows

$$\mathbf{Y}_{j,e}(n) = \mathbf{H}_{1j,e}(n)\mathbf{X}_{1,e}(n) + \mathbf{H}_{2j,e}(n)\mathbf{X}_{2,e}(n) + \mathbf{W}_{j,e}(n),$$
$$\mathbf{Y}_{j,o}(n) = \mathbf{H}_{1j,o}(n)\mathbf{X}_{1,o}(n) + \mathbf{H}_{2j,o}(n)\mathbf{X}_{2,o}(n) + \mathbf{W}_{j,o}(n), \qquad j = 1,2. \tag{3}$$

Assuming the frequency responses between adjacent subcarriers are approximately constant, the combined signals can be rewritten by [2]

$$\tilde{\mathbf{X}}_e(n) = \left(\left|\mathbf{H}_{11,e}(n)\right|^2 + \left|\mathbf{H}_{12,e}(n)\right|^2 + \left|\mathbf{H}_{21,e}(n)\right|^2 + \left|\mathbf{H}_{22,e}(n)\right|^2 \right) \mathbf{X}_e(n) + \tilde{\mathbf{W}}_e(n)$$
$$\tilde{\mathbf{X}}_o(n) = \left(\left|\mathbf{H}_{11,o}(n)\right|^2 + \left|\mathbf{H}_{12,o}(n)\right|^2 + \left|\mathbf{H}_{21,o}(n)\right|^2 + \left|\mathbf{H}_{22,o}(n)\right|^2 \right) \mathbf{X}_o(n) + \tilde{\mathbf{W}}_o(n) \tag{4}$$

In this paper, we consider the bit allocation scheme in [3]. Assume M-QAM is employed for each subcarrier, $b(n,k)$ bits per symbol are sent for the k-th subcarrier in the n-th OFDM symbol. According to [4][5], given the channel frequency response $H(n,k)$, the instantaneous bit error rate(BER) can be approximated by

$$P_e(n,k) = 0.2 \left\{ -\frac{1.6\dfrac{E_s}{N_0}\left|H(n,k)\right|^2}{2^{b(n,k)} - 1} \right\} \tag{5}$$

Let us consider the bit allocation in SFBC OFDM system. From the equation (4), the decoupled CSI for bit allocation is as follows

$$\left|\mathbf{H}(n,k)\right|^2 = \left|H_{11}(n,k)\right|^2 + \left|H_{12}(n,k)\right|^2 + \left|H_{21}(n,k)\right|^2 + \left|H_{22}(n,k)\right|^2 \tag{6}$$

By substituting $\left|\mathbf{H}(n,k)\right|^2$ into $\left|H(n,k)\right|^2$ in equation (5), the bit allocation for SFBC OFDM is performed and the next procedures for complete bit allocation are based on Chow's method.

3 Simulation Results

The parameters of adaptive OFDM system are as follows. Carrier frequency is 2GHz and the channels bandwidth is 20MHz which is divided equally among 2048 tones. The channel is based on COST 207 for a hilly terrain area [6] and the SISO channels associated with different couples of transmit/receive antennas are statistically equivalent and independent. The RMS delay spread is $5\mu s$. A total of 4096 information bits transmitted in each OFDM frame The velocity of mobile station is 60km/h. We allocate 0, 2, 4, or 6 bits to each subcarrier. To compare with adaptive OFDM, the conventional OFDM scheme, called as an uniform OFDM, is uniformly modulated by 16-QAM. For the simulation according to feedback delay T_Δ , the minimum feedback delay is $81\mu s$ and the maximum delay is $810\mu s$.

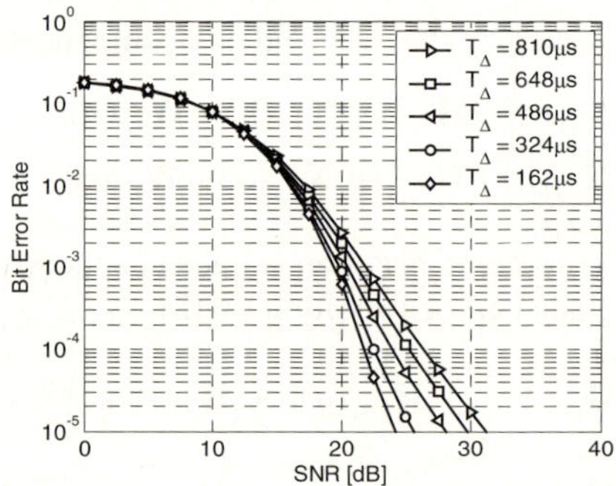

Fig. 2. BER curves of adaptive SFBC OFDM with 2Tx/1Rx antenna scheme according to feedback delay

Let us consider the question of how the performance of adaptive OFDM system with diversity scheme appears in mobile fading environments. Fig. 2 indicates that the feedback delay has an effect on the performance of adaptive SFBC OFDM with 2Tx/1Rx antenna scheme. As expected, the BER is gradually degraded as the feedback delay increases. This performance degradation is due to unavailability of CSI at the transmission time. Let us examine that an adaptive OFDM system employing space-frequency block coding scheme is one of ways how to overcome the effects of feedback delay. Fig. 3 indicates the simulation result when the feedback delay is $324\mu s$ and $810\mu s$, respectively. An adaptive OFDM system with 1Tx-1Rx scheme needs the additional power more than about 7 dB. It is shown that there is no merit of adaptive 1Tx-1Rx OFDM when feedback delay is long. On the other hand, in case of 2Tx-1Rx scheme, adaptive SFBC OFDM systems need the additional power of about 1dB and

3dB, respectively. We can see that adaptive OFDM system employing a diversity scheme mitigates the effect of feedback delay. From this result, the performance degradation of adaptive OFDM due to the outdated CSI can be mitigated by diversity technique and we can refer the adaptive 2Tx-2Rx SFBC OFDM as an excellent system which have scarcely power loss in spite of severe feedback delay.

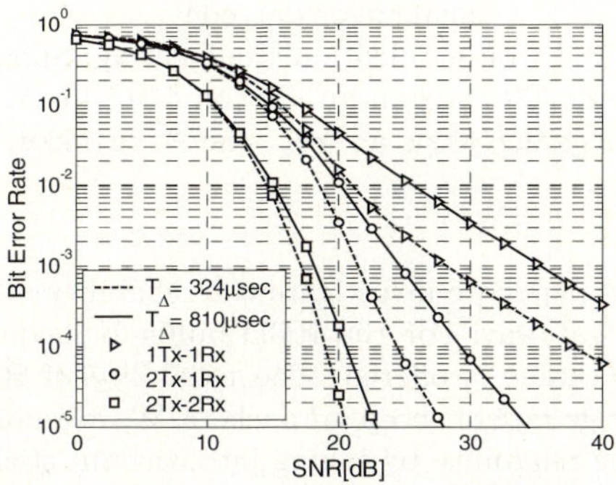

Fig. 3. BER curves according to diversity scheme in case that feedback delays are 324μs and 810μs respectively

4 Conclusions

In this paper combining an adaptive bit allocation scheme with SFBC OFDM system has been discussed. It has been illustrated that the CSI from SVD of MIMO channel is identical to the CSI from SFBC OFDM. From the various simulations, the performance of adaptive SFBC OFDM has been evaluated. Particularly, the BER performances according to the feedback delay have been indicated in detail. In the results, it is very interesting that the diversity schemes mitigate the effect of long feedback delay for adaptive OFDM. Most of all, adaptive SFBC OFDM with 2Tx-2Rx antenna scheme has made an excellent performance in spite of a severe feedback delay.

References

1. S. M. Alamouti, "A simple transmit diversity technique for wireless communications," *IEEE J, Select. Areas Comm.*, vol. 16, pp. 1451-1458, Oct. 1998.
2. H. Bolcskei and A. Paulraj, "Space-frequency coded broadband OFDM systems," *in Proc. Of Wireless Comm. Networking Conf.*, pp. 1-6. Sept. 2000.
3. P. S. Chow, J. M. Cioffi and J. A. C. Bingham, "A practical discrete multi-ton transceiver allocation algorithm for data transmission over spectrally shaped channels," *IEEE Trans. on Comm*, vol. 43, pp. 773-775, Apr 1995.
4. S. T. Chung and A. J. Goldsmith, "Degrees of freedom in adaptive modulation: a unified view," *IEEE trans. On Comm.*, vol. 49, pp. 1561-1571, sep. 2001
5. S. Ye, R. S. Blum and L. L. Cimini, "Adaptive modulation for variable-rate OFDM systems with imperfect channel information," *Proc, VTC 2002*, pp. 767-771.
6. M. Patzold, *Mobile Fading Channels*, Wiley, 2002.

On Modelling Reliability in RED Gateways*

Vladimir V. Shakhov[1], Jahwan Koo[2], and Hyunseung Choo[2,**]

[1] Institute of Computational Mathematics and Mathematical Geophysics of SB RAS,
Prospect Akademika Lavrentjeva, 6, Novosibirsk 630090, Russia
shakhov@skku.edu
[2] School of Information and Communication Engineering, Sungkyunkwan University
Chunchun-dong 300, Jangan-gu, Suwon 440-746, South Korea
jhkoo@songgang.skku.ac.kr, choo@ece.skku.ac.kr

Abstract. In this paper we investigate the reliability of Random Early Detection (RED) gateway. For the RED buffer behavior a new model based on Markov chains is offered. The reliability of RED gateway is defined by an average rate of accepted packets. We examine the impact of RED tuning on the reliability by taking into account stick-slip nature of traffic intensity. The goal of the proposed model is to improve RED buffer management by using a technique of traffic intensity change detection.

1 Introduction

Congestion occurs on a communication link whenever the amount of traffic injected on that link exceeds its capacity. This excessive traffic causes queueing delays of packets based on buffer fill up, and in extreme cases packets are lost due to buffer overflow. To avoid such a extreme situation effectively, IETF has recommended Random Early Detection (RED) as the default queue management scheme for the next generation Internet gateways [1].

The RED mechanism is first described in [2]. The basic idea of RED is that a gateway employing RED detects congestion earlier by computing the average queue length and drops randomly the packets in buffer if the computed length remains between minimum and maximum thresholds configured manually by network administrators. Although RED has some merits points, the selection of optimal values on parameters is still an open issue according to network and traffic situation. Moreover, it is shown that a static RED cannot provide better results than tail drop in general [3]. Thus, a dynamic RED mechanism is the recent focus of investigations. We present our vision of dynamic RED implementation and RED schema quality in this paper.

The remainder of the paper is organized as follows. Section 2 provides the basic notation and proposed reliability modelling approach. For the RED buffer

---

* This research was supported by the MIC(Ministry of Information and Communication), Korea, under the ITRC(Information Technology Research Center) support program supervised by the IITA(Institute of Information Technology Assessment), IITA-2005-(C1090-0501-0019).

** Corresponding author.

V.N. Alexandrov et al. (Eds.): ICCS 2006, Part I, LNCS 3991, pp. 948–951, 2006.
© Springer-Verlag Berlin Heidelberg 2006

behavior a new model based on Markov chains is offered. The reliability of RED gateway is defined by an average rate of accepted packets. In Section 3 we examine the impact of RED tuning on the reliability by taking into account stick-slip nature of traffic intensity. The reason of RED gateway management modification by a point-of-change detection technique is discussed. Section 4 is a brief conclusion.

2   RED Gateway Reliability Modelling

Let us consider a RED scheme in detail. RED utilizes two thresholds, min threshold h and max threshold H, and a exponentially-weighted moving average (EWMA) formula to estimate the average queue length, $Q_{avg}(t) = (1 - W_q)Q_{avg}(t-1) + W_qQ$, where $Q_{avg}(t)$ is average queue length at time t, Q is instantaneous queue length at time t and W_q is a weight parameter, $0 \leq W_q \leq 1$. A gateway implementing RED accepts all packets until the queue reaches h, after which it drops a packet with a probability as follows $\pi(Q_{avg}) = max_p(Q_{avg} - h)/(H-h)$, where max_p is the maximal packet drop probability. When the queue length reaches H, all packets are dropped with a probability of one.

Let us note that a low traffic load has a little relevance to the reliability problem. The focus will be on the case of high traffic intensity. Therefore, it is reasonable to take $W_q = 1$. Actually, if the offered load is excessive, then the probability of buffer overflow is high. Hence, we have to use an aggressive strategy of packet rejection. Assume that packet arrivals form a Poisson process with rate λ. The processing times of the packets in gateway are independent exponentially distributed random variables with mean $1/\mu$. These are generally accepted in the literature. Packets are processed in their order of arrivals. Probability of Q equals $n, n = 0 \ldots H - h$ we denote by p_n.

The state diagram is given by Figure 1 where the number Q is the state index. As the state diagram shows, when Q exceeds the min threshold h, a part of incoming traffic should be dropped. Thus, the buffer enters state $h + i$ with probability $\alpha_{i-1}, i = 2...H - h$. Let us remark $\alpha_i = 1 - \pi(h + i)$. Writing down and solving the steady-state balance equations, we get

$$p_i = \rho^i p_0, i = 1 \ldots h + 1,$$

$$p_{h+1+i} = \rho^{h+1+i}\alpha_1\alpha_2 \ldots \alpha_i p_0, i = 1 \ldots H - h + 1,$$

where $\rho = \lambda/\mu$ and $p_0 = ((\rho^{h+2} - 1)/(\rho - 1) + \sum_{i=1}^{H-h-1} \rho^{h+1+i}\alpha_1 \ldots \alpha_i)^{-1}$.

Let us now define a reliability of RED gateway. The rate of packets dropping before threshold H depends on Q. Then Average Drop Rate is calculated by

$$ADR = \sum_{i=h+1}^{H} \pi(i) * p_i.$$

Here $\pi(H) = 1$, that means total-lot blocking of packets.

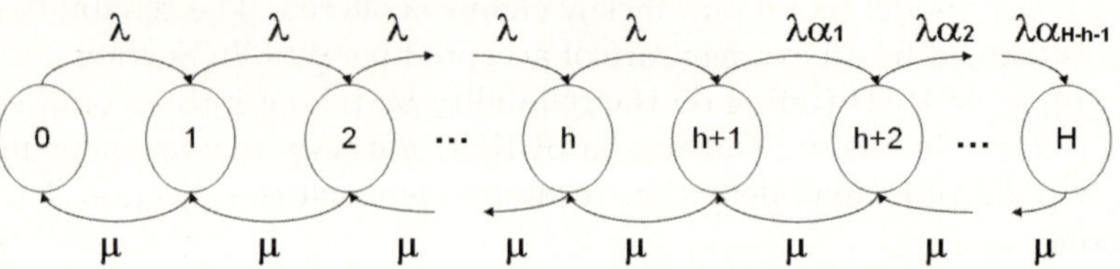

Fig. 1. Markov chain for RED gateway buffer

Actually, we have estimated an unreliability of the gateway. The packets are normally (but not always) assumed to be rejected with rate ADR. It is reasonable *to define* that a *reliability R* of gateway equals $1 - ADR$. It is a metric of successful packet acceptance or quality of RED schema. As we can see the considered reliability drastically depends on traffic intensity. Thus, the choice of appropriate RED parameters depends on the arrival rate.

3　Improving the Reliability of RED Gateway

It is often convenient to assume that the external packet traffic entering a RED gateway can be modelled by stationary stochastic process that has a constant packets arrival rate. This approximates a situation where the arrival rate changes slowly with time and constitutes what we refer to as the quasistatic assumption. When there are jumps of offered load intensity, this assumption is violated. In this case a behavior of packet arrivals remains quasistatic property on separate time durations. Using results of previous section, the optimal RED parameters can be calculated for concrete packets arrival rate. But produced RED tuning can be the worst for other packets arrival rate. In other words, if probability of gateway buffer overflow is very small, then it is not reasonable to worsen reliability (packets acceptance) by RED using. Drop Tails approach is more preferable. On the other hand, if incoming flow has got a hard rate (for example, because of DDoS attack) then protection policy should be aggressive.

Please refer to Figure 2. Calculations of reliability are made under h=2, H=128. The figure shows that the RED gateway reliability degrades along with traffic load increase. It was expected. But we can see the pattern of reliability behavior is also changed. If $\rho = 1.2$ then preferable $max_p = 0.6$. But optimal max_p value is differ under other ρ. If a change of traffic intensity is detected then use of optimal parameters max_p gives a reliability improvement from 0.5 percent (case of $\rho = 1.1$) to 3.8 percent (case of $\rho = 1$).

Thus, reliability of the RED gateway can be improved if RED parameters are recalculated for differ packets arrival rates. For this purpose, a mechanism of incoming flow intensity detection can be applied. The estimation of a moment of random process parameter changing is known as point-of-change problem (discard problem). Appropriated algorithm for discard detection under Poisson process had been proposed in [4].

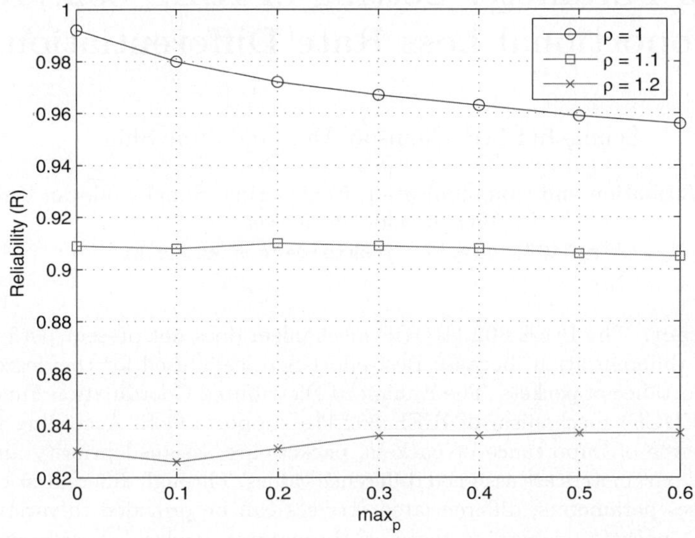

Fig. 2. Reliability behavior for RED parameter max_p

4   Conclusion

On of the key network technologies is RED gateway management. For these reasons, it is important to consider the nature of RED parameters, and the manner in which it depends on the reliability of the RED gateway. In this paper the model of RED buffer behavior is offered. Its use requires simplifying assumptions but the proposed model provides a basic for adequate reliability estimations. We have shown that the stick-slip nature of traffic intensity has an essential impact on the reliability of RED gateway. Thus, dynamical tuning of RED parameters has an advantage. By this reason we offer to include point-of-change detection technique in RED gateway management.

References

1. B. Braden, *et al.*, "Recommendations on queue management and congestion avoidance in the Internet," RFC 2309, IETF, April 1998.
2. S. Floyd and V. Jacobson, "Random early detection gateways for TCP congestion avoidance," IEEE/ACM Transactions on Networking vol. 1, no. 4, pp. 397-413, August 1993.
3. M. May, C. Diot, B. Lyles, and J. Bolot, "Reasons not to deploy RED," Seventh International Workshop on 31 May-4 June 1999, IWQoS '99, pp. 260 - 262, 1999.
4. V. Shakhov, H. Choo, Y.-C. Bang, "Discord model for detecting unexpected demands in mobile networks," Future Generation Comp. Syst. 20(2), pp. 181-188, 2004.

Control Parameter Setting of IEEE 802.11e for Proportional Loss Rate Differentiation

Seung-Jun Lee, Chunsoo Ahn, and Jitae Shin

School of Information and Communication Engineering, Sungkyunkwan University,
Suwon, 440-746, Korea
{lsj6467, navy12, jtshin}@ece.skku.ac.kr

Abstract. The IEEE 802.11 DCF mechanism does not present performance differentiation, because Best-effort-Service is used for the degree of importance of packets. The Enhanced Distributed Coordination Function (EDCF) mechanism of IEEE 802.11e supports QoS. According to the degree of importance of packets, packets are assigned priority and control parameters are assigned difference values. Through differentiation of theses parameters, differentiated services can be provided to various priority packets (classes) in terms of throughput, packet loss rate, and delay. In this paper, parameters of the IEEE 802.11e EDCF mechanism for Proportional Loss Rate Differentiation Service(PLDS) between adjacent priority classes are investigated through mathematical analysis and network simulation.

1   Introduction

The DCF mechanism only provides a best-effort service, even though the importance of packets or type of packets (real-time or non real-time) exists [1].IEEE 802.11e Enhanced DCF (EDCF) supports QoS. The MAC method of the EDCF mechanism is similar to the DCF, with the exception that EDCF applies a different value to control parameters, according to the packet type, to support QoS [3]. The Markov chain model is used frequently for mathematical analysis of the IEEE 802.11 mechanism [2]. However, Markov chain analysis in [2] had two problems, ignoring the cases of dropping packets and the frozen slot time. Ref. [4] uses a modified Markov chain model for mathematical analysis of the IEEE 802.11e EDCF mechanism. In the case of [4],this includes two cases that are not considered in [2]. In this paper, based on [4] and [5], we will find out which control parameters such as the CW_{min}/CW_{max}, retry limit, AIFS, and so on., in IEEE 802.11e EDCF mechanism have the most effect on loss rate. In addition, in order to obtain a proportional and differentiated performance between adjacent priority classes, a method of setting the most dominant parameter in QoS, is proposed.

2   Mathematical Model of Packet Loss Rate

In order to support QoS, IEEE 802.11e EDCF classifies the packets and the packet map into four access categories (ACs), according to priority. AC

V.N. Alexandrov et al. (Eds.): ICCS 2006, Part I, LNCS 3991, pp. 952–955, 2006.

denotes AC[i] (i=0,1,2,3) for distinguishing priority. A smaller value of i represents higher priority. Each AC[i] is assigned different values of parameters (CW$_{\min}$[i]/CW$_{\max}$[i], AIFS[i], etc). A default value of IEEE 802.11e EDCF parameters is defined in [3].

The backoff stage increases by one, in the case where a station does not transmit a packet. If collision occurs continuously, the backoff stage continues to increase. As presented in Fig. 1 of ref. [5], the backoff stage increases until the maximum retry limit, i.e., L. At the maximum retry limit, if the packet collision occurs with the backoff stage, and it is the same as the value of maximum retry limit, the packet will be dropped. The packet-loss-rate is defined by the following equation.

Let $P_{i,\text{loss}}$(i=0, , N-1) denote the packet-loss probability for the priority i class.

$$P_{i,\text{loss}} = p_i^{L_i+1} \tag{1}$$

As p_i of Eq. (1) is the probability that a transmitted packet collides, refer to Eq. (9) of ref. [5].

3   Analysis of Proportional Differentiation Service in Loss Rate

In Proportional Loss Rate Differentiation Service(PLDS), the goal is to have the packet loss rate ratio between the adjacent i-th priority class and $i+1$-th priority class, at a certain value (K_L) such as Eq. (2).

$$\frac{P_{i+1,\text{loss}}}{P_{i,\text{loss}}} = \frac{p_{i+1}^{L_{i+1}+1}}{p_i^{L_i+1}} \triangleq K_L \tag{2}$$

The desired loss rate ratio is assigned a K_L. When taking the logarithm to Eq. (2),

$$\log K_L = (L_{i+1} + 1) \log p_{i+1} - (L_i + 1) \log p_i \tag{3}$$

Arrange Eq. (3) in terms of L_{i+1}. Then $\log p_i$ and $\log p_{i+1}$ are considered as nearly same value for our approximation. Therefore,

$$L_{i+1} \cong L_i + \frac{\log K_L}{\log p_{i+1}} = L_i + \frac{\log K_L}{a\,(n) \log n} \tag{4}$$

Where n and $a(n)$ are the number of nodes and a related constant value to compensate the relationship between $\log p_{i+1}$ and $\log n$, respectively. From Eq. (4), guidance of PLDS between adjacent classes can be derived. When the value of L_i is provided, the K_L for the desired ratio between adjacent priority classes is assigned, and the collision probability p_{i+1} calculated according to the station number (n) is obtained, L_{i+1} can be found out, and should be set for PLDS.

4   Numerical and Simulation Results

In this section, the validation of PLDS through numerical and simulation results is provided. For basic parameter setting, the IEEE 802.11 FHSS system parameters of ref. [2] are referred to.

In both numerical and simulation by the NS-2 simulator, priority classes are divided into classes 0, 1, and 2 (class 0 represents the highest priority). The number of each priority class station is assumed to be identical. The numerical and simulation results are performed at the number of each priority class station, from 4 to 15. That is, the number of total stations (or nodes) varies from a minimum value of 12 to a maximum value of 45 nodes.

In order to present validation of PLDS, except for retry limit (L_i), which most influencing parameter for PLDS, it is assumed that other control parameters remains as the same cross classes. The CW_{min} of all classes is 16, maximum backoff stage of all class is 2, and initial value of L_0, i.e., maximum retry limit of the highest priority class 0, is fixed at 4. When the node number is counted and the ratio (K_L=2) for PLDS is provided, then L_1, i.e., 3, and L_2, i.e., 2, can be calculated through Eq. (2). However the calculated value through Eq. (2) is not the exact difference with the number 1. The result is actually a decimal smaller than the number 1. However, the value of the maximum retry limit should have an exact positive number. Therefore, the difference of the retry limit between adjacent priority classes is approximately applied as 1. Fig. 1(a) presents numerical results for the loss rate of classes and Fig. 1(b) presents the simulation results for loss rate of classes. When Fig. 1(a) is compared with Fig. 1(b), little difference exists. This is due to NS-2 random execution. In addition, Fig. 1(c) (regarding analysis) and Fig. 1(d)(regarding simulation) confirms whether the desired ratio is obtained through PLDS.

When the number of the station (or node) is minimal, i.e., the nodes for each class consist of approximately 2~6 numbers, PLDS deviates considerably from the desired ratio. This affects the absolute value of $\log p$.

5   Conclusion

In this paper, mathematical analysis of the loss rate for the proposed PLDS is derived, in order to provide proportional and differentiated services among different priority classes in the IEEE 802.11e mechanism. In addition, validation of PLDS is proven through NS-2 simulation results.

As a result, the desired loss-rate according to the type of packets is provided in the case where data is transmitted. The maximum retry limits are very sensitive to packet loss, and are applied differently to each packet, depending on the type of transmitted packets. Therefore, users can directly control the condition of a network. If the packet should not be lost, a larger maximum retry limit is applied to recover packet loss. In addition, if the packet is less affected by dropping, the packet takes maximum retry limit that is smaller than that of sensitive packet. Through this technique, the complexity of a network can be reduced.

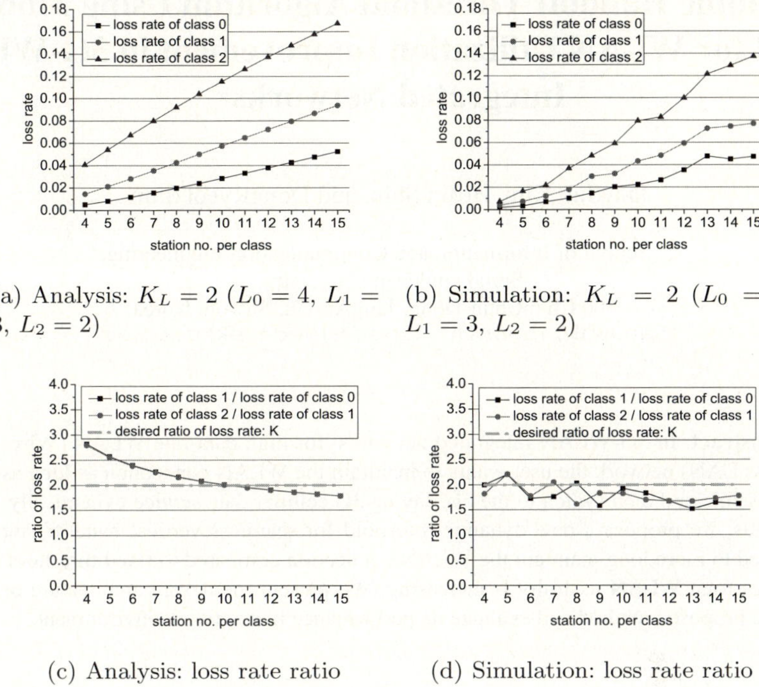

(a) Analysis: $K_L = 2$ ($L_0 = 4$, $L_1 = 3$, $L_2 = 2$)

(b) Simulation: $K_L = 2$ ($L_0 = 4$, $L_1 = 3$, $L_2 = 2$)

(c) Analysis: loss rate ratio

(d) Simulation: loss rate ratio

Fig. 1. Analysis and simulation results

Acknowledgements

This research was supported by the Ministry of Information and Communication (MIC), Korea, under the Information Technology Research Center (ITRC) support program supervised by the Institute of Information Technology Assessment (IITA) (IITA-2005-(C1090-0502-0027)).

References

1. IEEE Standard for Wireless LAN Medium Access Control(MAC) and Physical Layer(PHY) Specifications, P802.11, Nov. 1997.
2. G. Bianchi, "Performance Analysis of the IEEE 802.11 Distributed Coordination Function," IEEE Journal on Selected Areas in Communications, vol. 18, no. 3, pp. 535-547, Mar. 2000.
3. IEEE 802.11 WG, Draft Supplement to Part 11: Wireless Medium Access Control (MAC) and physical layer (PHY) specifications: Medium Access Control (MAC) Enhancements for Quality of Service (QoS), IEEE 802.11e/D2.0, Nov. 2001.
4. Yang Xiao, "Performance Analysis of IEEE 802.11e EDCF under Saturation Condition,"IEEE Communications Society, 2004.
5. Seung-Jun Lee, Chunsoo Ahn, and Jitae Shin, "Control Parameter Setting of IEEE 802.11e for Proportional Throughput," ICOIN 2006.

Dynamic Handoff Threshold Algorithm Using Mobile Speed for WLAN Utilization Improvement in 3G-WLAN Integrated Networks*

JangSub Kim, HoJin Shin, and DongRyeol Shin

School of Information and Communication Engineering,
Sungkyunkwan University,
300 ChunChun-Dong, JangAn-Gu, Suwon, Korea
{jangsub, hjshin, drshin}@ece.skku.ac.kr

Abstract. In 3G-WLAN integrated networks, for high data-rate WLAN (Wireless LAN) network the user wants to maintain the WLAN connection as long as possible and then switch to the overlaying 3G cellular data service dynamically. Thus, we propose a new dynamic threshold for seamless vertical handoff, are used to more long maintain the WLAN connection compared to fixed threshold, thus total WLAN usability is increasing. We present the design architecture of the proposed method and evaluate its performance in a network environment.

1 Introduction

3rd Generation cellular and WLANs will complement each other to provide ubiquitous high-speed wireless Internet connectivity to mobile users. Therefore, it is important to consider dual mode users roaming in between 3G cellular and WLANs. In order to provide a convenient access of both technologies in different environments, interworking [1] of the two networks are regarded as a very important work.

In this paper, we propose a new mechanism for obtaining link layer indication is dynamically transported to the upper layer (network layer). The dynamic threshold as function of mobile node speed be used to information of triggers for low latency MIPv4 [2] and fast MIPv6 [3]. To extend the WLAN service time we will find the optimal value for dynamic threshold to relate with RSS (Received Signal Strength) and mobile speed. The fixed threshold value which is not considered user speed cannot be fully used to maintain WLAN service as long as possible. But the proposed dynamic threshold which is adapted user speed has more advantage than general mechanism. We show improvement of the utilization of during WLAN service.

In Section 2, problems are formulated, and core part of algorithmic for dynamic threshold. Simulations are performed in Section 3 to validate the proposed approach. Finally, the summary of the result are presented in the conclusion section.

---

* This research is supported by the ubiquitous Autonomic Computing and Network Project, the Ministry of Information and Communication (MIC) 21st Century Frontier R&D Program in Korea.

V.N. Alexandrov et al. (Eds.): ICCS 2006, Part I, LNCS 3991, pp. 956–959, 2006.

2 Handoffs Optimization

This section presents the vertical handoff triggers analyses. In case of fixed threshold value to trigger, the handoff from 3G cellular network to WLAN carried out irrespective of mobile node speed. When a mobile node moves from one network to another, if the preparation time of fast handoff is larger than WLAN sojourn time related to mobile node speed, the handoff failed and occurred the packet loss. If the mobile node speed is too slow in case of fixed threshold value, instead, handoffs are triggered too late and thus WLAN service time is reduced. Thus in this paper, we have considering dynamic threshold as function of mobile node speed. In this case the mobile node speed is too slow, threshold is alter to small value, instead, handoffs are triggered too early and thus WLAN service time is increasing and total utilization is increasing.

To find the optimal dynamic threshold values (T_{Hd} and T_{Ld}) we set up a test to relate RSS and user speed. Generally, the channel propagation model used for RSS is given by [4]. For analysis simplicity, we are assumed the RSS on WLAN link shown as Fig. 1.

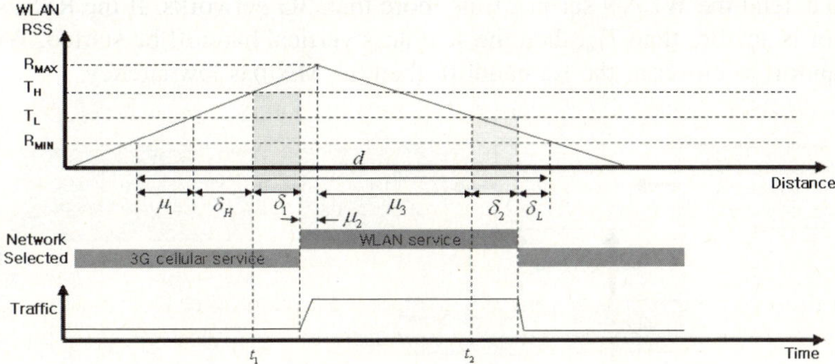

Fig. 1. A vertical handoff from 3G cellular to WLAN

We use the following variables to determine the vertical handoff analysis. Where R is received signal strength, d is diameter of WLAN coverage, δ_1 and δ_2 are handoff prepare time from 3G to WLAN and from WLAN to 3G, respectively. Specifically, it shows the WLAN signal observed by the client over time. At t_1, when the signal strength exceeds the threshold, T_H, the client will attempt to use the WLAN airlink. Similarly at time t_2, when the signal strength drops below the threshold, T_L, the client will revert to the 3G airlink. Two thresholds, T_H and T_L, are used to avoid unnecessary handoffs that can result in poor connection. In this paper, two threshold is using to instead of dwell timer.

We omit the detailed derivation of dynamic threshold values for lack of space, thus we show the resulting and proposed perspective of that. The user speed is related with the slope of the RSS. We have to find the dynamic threshold satisfied the following equation.

$$T_{Hd} = \frac{R(t+1)-R(t)}{T(t+1)-T(t)} \cdot \delta_H \;,\; T_{Ld} = \frac{R(t+1)-R(t)}{T(t+1)-T(t)} \cdot (\delta_2 + \delta_L) \tag{1}$$

The procedure is now concerned with the T_{Hd} and T_{Ld} in which can be written as a function of velocity (V), and hence finding the value of T_{Hd} and T_{Ld}. If shadowing fading is existed, the slope of the RSS (dBm per unit second or dBm per unit meter) be calculated to use smoothing method, following by

$$\nabla R_{t+1} = \alpha \nabla R_{t+1} + (1 - \alpha) \nabla R_t \qquad (2)$$

Thus, as we are considering the previous quantity, we avoid to abrupt change quantity in the WLAN environments. α is the memory factor.

The proposed perspective of dynamic threshold calculation procedure is shown in Fig. 2. To calculate the dynamic threshold, firstly, a MN has to scan the AP signal. If the RSS is larger than R_{MIM}, the MN has to measure a change quantity of the RSS per unit time. Because the change quantity of that is proportion to the mobile velocity, we can be estimated the mobile velocity. At this time, if the GPS or ToA information is available, the mobile velocity is correction for the higher prediction accuracy. Using (2), we calculate the dynamic threshold. This dynamic threshold is sensitivity to the mobile velocity. According to the MN velocity, the dynamic threshold becomes altered to extend the WLAN service time more than 3G networks. If the RSS is larger than T_{Hd} or is smaller than T_{Ld}, then the seamless vertical handoff be started. We make the L3 handoff as closer as the L2 handoff, then the MN has low latency.

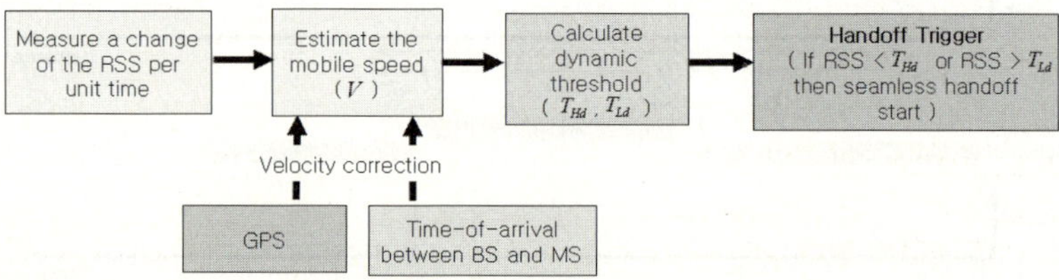

Fig. 2. The procedure of the dynamic threshold calculation

3   Numerical Analysis

The proposed procedure is tested with a number of numerical examples for the overlaid structure. The WLAN constitutes the lower layer of the two-layer hierarchy. The WLAN are overlaid by a large 3G networks, which forms the upper cell layer. In our system, mobile node are traversing the coverage are of the WLAN and 3G network. The diameter of the WLAN is assumed to be $d = 100m$. The downward (from 3G to WLAN) vertical handoff preparation time and the upward (from WLAN to 3G) vertical handoff preparation time are assumed to $\delta_1 = 500ms$ and $\delta_2 = 500ms$, respectively. And we are assumed to $\delta_H = 1\sec$ and $\delta_L = 1\sec$. From Fig. 1, the WLAN service time is given by

$$WLAN_{service\ time} = \mu_2 + \mu_3 + \delta_2$$

Fig. 3 illustrate the WLAN service time as user speed. As the figure indicated, the WLAN service time of the proposed method is longer than the fixed threshold below the user speed with 30 m/sec. Thus, the dynamic threshold to low speed user have add benefit that the WLAN service time is a more long than fixed threshold. In the mobile velocity at 1m/sec, the WLAN service time is extended to 1.7 times, compared to fixed threshold. Thus, we have to apply the dynamic threshold in the mobile velocity at below 30m/sec. Our simulation is easily extended to the real received signal model for longer the WLAN service time.

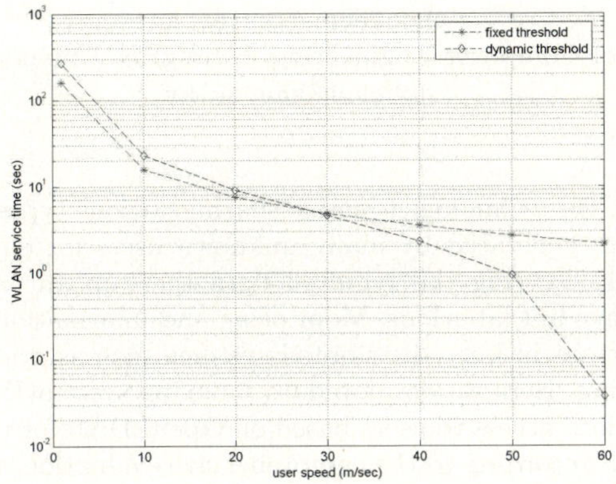

Fig. 3. WLAN service time as user speed

4   Conclusion

We have proposed a dynamic threshold for slow moving user and simply analysis the dynamic threshold in order to improve the utilization of WLAN more than 3G cellular network. The analysis results show the dependency of the WLAN utilization improvement upon the dynamic threshold, T_{Hd} and T_{Ld}. The dynamic threshold has shown to be an important system parameter that the system providers should determine to produce better utilization improvement.

References

1. K. Ahmavaara, H. Haverinen, and R. Pichna, "Interworking architecture between 3GPP and WLAN systems," IEEE Commun. Mag., vol.41, no.11, pp.74-81, Nov. 2003.
2. Mobile IP Working Group, "Low Latency Handoffs in Mobile IPv4," draft-ietf-mobileip-lowlatency-handoffs-v4-00.txt, Feb. 2001..
3. G. Tsirtsis et al., "Fast Handovers for Mobile IPv6," draft-ietf-mobileip-fast-mipv6-00.txt, Feb. 2001.
4. Amir Majlesi, Babak H. Khalaj, "An Adaptive Fuzzy Logic Based Handoff Algorithm for Interworking between WLANs and Mobile Networks," IEEE PIMRC 2002.

Efficient Data Indexing System Based on OpenLDAP in Data Grid*

Hongseok Lee[1], Sung-Gon Mun[1], Eui-Nam Huh[2], and Hyunseung Choo[1,**]

[1] School of Information and Communication Engineering,
Sungkyunkwan University, Korea
`choo@ece.skku.ac.kr`
[2] Dept. of Computer Engineering, KyungHee University, Korea
`johnhuh@khu.ac.kr`

Abstract. Grid technologies enable sharing various types of large-scale data resources generated by many unknown users for day by day jobs done at work. Finding the required data there in an easy manner is really necessary but laborious. Monitoring and information service(MIS) is very important in huge distributed systems such as grid and effective in data probing. Here we develop data indexing system(DIS) to provide the efficient data access to users based on OpenLDAP for the distributed environment. According to the comprehensive evaluation, DIS shows the better performance in terms of response time and scalability for the large number of users. It is also expected that the proposed system can be applied to globus system.

1 Introduction

Proportional to the increasing number of scientific disciplines, large data collections are emerging as important community resources. In domains as diverse as global climate change, high energy physics, and computational genomics, the volume of interesting data is currently measured in terabytes and will soon total petabytes. This combination of dataset size, geographic distribution of users and resources, and computationally intensive analysis results in complex and stringent performance demands that cannot be satisfied by existing data management infrastructure. A large scientific collaboration may generate many queries, each involving access to gigabytes or terabytes of data. The efficient and reliable execution of these queries may require careful management of terabyte caches, gigabit data transfer over wide area networks, scheduling of data transfers and supercomputer computation[1].

At present, few studies have been published that quantitatively evaluate the performance of the current monitoring and information services in distributed

---

* This research was supported by the Ministry of Information and Communication, Korea under the Information Technology Research Center support program supervised by the Institute of Information Technology Assessment, IITA-2005-(C1090-0501-0019).

** Corresponding author.

V.N. Alexandrov et al. (Eds.): ICCS 2006, Part I, LNCS 3991, pp. 960–964, 2006.
© Springer-Verlag Berlin Heidelberg 2006

systems. The Relational Grid Monitoring Architecture(R-GMA)[2] monitoring system is an implementation of the Grid Monitoring Architecture(GMA). It is based on the relational data model and Java Servlet technologies. Its main use is the notification of events that is, a user can subscribe to a flow of data with specific properties directly from a data source. Hawkeye[3] is a tool developed by the Condor group and designed to automate problem detection, for example to identify high CPU load, high network traffic, or resource failure within a distributed system.

In this paper, DIS based on Open source implementation of the Lightweight Directory Access Protocol(OpenLDAP) which is included in Monitoring and Discovery Service(MDS2)[4], is developed in Globus environment. The Globus Toolkit(GT)[5] has been developed since the late 1990s to support the development of distributed computing applications and infrastructures. MDS2 is the grid information service used in GT. It uses an extensible framework for managing static and dynamic information regarding the status of a computational grid and all its components: networks, computing nodes, storage systems, instruments, and so on. MDS2 is built on top of OpenLDAP[6]. That is the predominant Internet directory access protocol and hence is also used in Public Key Infrastructure. It stores certificates and provides efficient access methods by harnessing storage means with communication mechanisms.

2   Proposed DIS

2.1   DIS Architecture

The topology consists of four Linux machines with hostnames monet{156, 157, 158, and 164}.skku.ac.kr set up on a 100Mbps LAN. Fig. 1 presents DIS architecture with MDS2. Host monet164 serves Grid Index Information Service(GIIS) and remaining monet{156, 157, and 158} nodes only serving Grid Resource Information Service(GRIS) include DIS. Each GRIS reads the information regarding the science data, and generates the Lightweight Directory Interchange Format(LDIF) data objects. Then, it registers LDIF to GIIS. Therefore, GIIS maintains entire information such as the index of data names and sizes of each host that operates on the system. Hence, the authorized user accesses to GIIS and searches for information that the user wants. DIS interoperates with MDS2 on the Globus Toolkit, therefore the architecture that supports many GIISs can be expanded to prevent the single point of failure.

2.2   Development Procedure for DIS

In order to develop DIS, the first step is to define schema, which will be represented in the Directory Information Tree(DIT). We must follow the OpenLDAP policy schema, Object Identifier(OID), and its naming convention, because a directory service based on OpenLDAP is used. One objectclass is

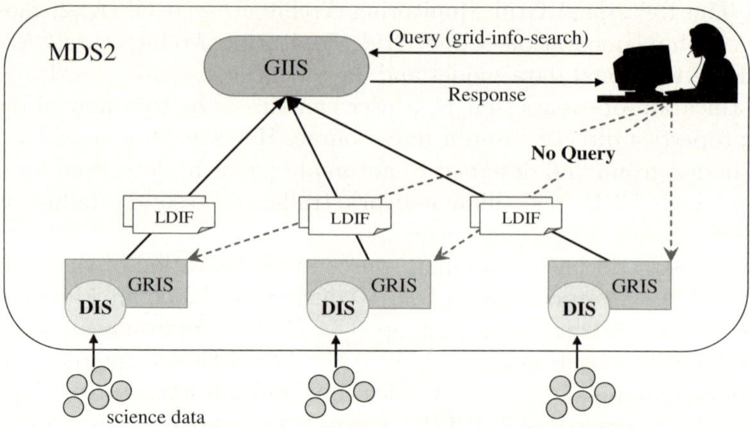

Fig. 1. DIS architecture with MDS2

defined with six attributes, in order for metadata of data to represent the information
in DIS.

The second step is to create Data Information Provider Module(DIPM) which extracts the specific information from the data. It is a program that represents the data of every host maintained in Virtual Organization(VO). In other words, DIPM reads and represents information of the specific directory in the host. Now only the name and size of data for the information are defined. In this research the aim is to expand DIPM according to the characteristic of science data stored in the host. DIPM operates by reading the name and size of each data, also the count and size of entire data. It is the expansion module of the slapd server using the OpenLDAP generic modules API, operating on the cache memory in the slapd server, and GRIS backend. DIPM is called by the function of *fork()* and *exec()* in GRIS backend and returns LDIF data objects based on the previously defined schema. In other words, it receives single data that consist of *add* and *delete* commands, such as the configuration file by an input. It creates the LDIF data objects, the specific information according to the LDAP schema, and then transmits this to the GRIS backend as an output. As a result, DIPM is a core module in DIS.

The last step is that enables DIS by the modification of the environment configuration. In order to interoperate with MDS2, DIS is identified by MDS2. Some information must be modified to the *grid-info-resource-ldif.conf* file that is a configuration file of MDS2. For interoperation of the proposed system with MDS2, the required information includes a Distinguish Name(DN), objectclass, information of DIPM, arguments, cache time, time limit, size limit, and so on. The relatively short interval time of the cache memory, which is 1,800 seconds, is defined because of the characteristics of the science data that has considerable information movement.

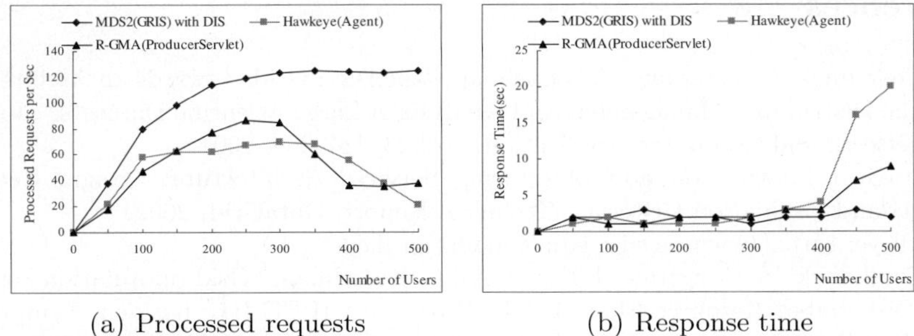

(a) Processed requests   (b) Response time

Fig. 2. Performance comparison with information servers

2.3   Performance Evaluation

The goal of the experiment is to compare the scalability of the existing information systems with the proposed DIS in the grid environment. The experiment is conducted on three different systems. The R-GMA and Hawkeye are selected for the comparison with DIS included in MDS2. The components on these systems are classified because of the heterogeneity in terms of system architecture. At this point, GRIS with DIS in MDS2 is compared to ProducerServlet in R-GMA and Agent in Hawkeye. In this environment, the number of requests processed per second and response time on each server are measured. The number of concurrent users is considered up to 500 users by running individual user processes. The request is the average number of requests processed by a service component per second and the response time is the average amount of time required for a service component to handle a request from a user. The values reported in each experiment are the average over all values recorded over 10 minutes time span. In the Figs. 2(a) and 2(b), the request on GRIS with DIS has a linearly increasing trend and stabilized for the requests processed per second, while both Hawkeye and R-GMA show quite unstable trends and rather slightly fluctuate. In addition, the response time on GRIS is superior to other systems. Especially, the proposed system demonstrates 4 times and 10 times better than R-GMA and Hawkeye, respectively, in terms of response time.

3   Conclusion

In this paper, DIS for a data management in distributed environment is developed. It consists of Data Information Provider Module, and creates metadata that indices information for the data. Then it registers metadata regarding the specific information to the previous MDS2. Through the interoperation to MDS2, it keeps compatibility for the user of Globus Toolkit. In our future work, the DIS will be expanded to include the search engine for meteorological data. The system will also be included in the gridsphere for web service. This will allow users to access to the system using a web browser, and search for the required data. Therefore, it will provide the better convenience for users.

References

1. A. Chervenak, I. Foster and C. Kesselman, "the Data Grid: Towards an Architecture for the Distributed Management and Analysis of Large Scientific Datasets," Journal of Network and Computer Applications, vol.23, 187-200, 2001.
2. "DataGrid Information and Monitoring Services Architecture: Design, Requirements and Evaluation Criteria," Technical Report, DataGrid, 2002.
3. Hawkeye: http://www.cs.wisc.edu/condor/hawkeye
4. K. Czajkowski, S. Fitzgerald, I. Foster and C. Kesselman, "Grid Information Services for Distributed Resource Sharing," In Proc. 10th IEEE International Symposium on High Performance Distributed Computing(HPDC-10), IEEE Press, 2001.
5. Globus Forum: http://www.globus.org/
6. W. Yeong, T. Howes and S. Kille, "Lightweight Directory Access Protocol," IETF RFC 1777, March 1995.

A Home-Network Service System Based on User's Situation Information in Ubiquitous Environment*

Yongyun Cho, Joohyun Han, Jaeyoung Choi, and Chae-Woo Yoo

School of Computing, Soongsil University,
1-1 Sangdo-dong, Dongjak-gu, Seoul 156–743, Korea
{yycho, jhhan}@ss.ssu.ac.kr,
{choi, cwyoo}@comp.ssu.ac.kr

Abstract. In this paper, we propose a home-network service system that can support home services appropriate to user's situation information in ubiquitous computing environments. The suggested system uses a uWDL workflow service scenario describing a user's situation information as service execution constraints to support context-aware home services. The suggested system consists of a context handler and a context mapper. The context handler represents contexts described in a uWDL document as a context subtree, which expresses not only context data but also relation information among services into the fields of its node. The context mapper uses a context comparison algorithm for context comparison between context subtrees and user's situation information. The algorithm can distinguish contexts that have the same values but different types with user's contexts and selects a context that has all together values and types entirely equal to those of user's contexts.

1 Introduction

For a smart home, execution of all the home services must be dependent on user's situation contexts, which are dynamically generated in ubiquitous environments [1]. uWDL (ubiquitous Workflow Definition Language) is a workflow language based on a structural context model which expresses context information as transition constraints of workflow services [2]. Through a uWDL workflow service scenario, an user can describe what services must be executed according to situation information. For execution of context-aware services, we need a method that can recognize contexts in a scenario and select a service correspondent with situation information.

In this paper, we present a uWDL-based home-network system that represents contexts described in a uWDL workflow service scenario document as rule-based context subtrees, and derives service transition according to user's state information in ubiquitous environments.

---

* This work was supported by Korea Research Foundation Grant (KRF-2004-005-D00172).

V.N. Alexandrov et al. (Eds.): ICCS 2006, Part I, LNCS 3991, pp. 965–968, 2006.

2   Related Work

Context in a ubiquitous environment means any information that can be used to characterize the situation of an entity [3]. In ubiquitous environments, context can be expressed with a RDF-based triplet form [2]. RDF(Resource Description Framework) [4] is a language to describe resource's meta-data and it expresses a resource as a triplet of {subject, predicate, objective}. The existing workflow languages, such as BPEL4WS [5], WSFL [6], and XLANG [7], do not include any element to describe context information in ubiquitous computing environments as transition conditions of services.

uWDL [2] can describe context information as transition conditions of services through the <context> element consisting of the knowledge-based triplet - subject, verb, and object. The uWDL reflects the advantages of current workflow languages such as BPEL4WS, WSFL, and XLANG, and also contains rule-based expressions to interface with the DAML+OIL ontology language [8].

3   A uWDL-Based Home-Network Service System

3.1   A System Architecture

Figure 1 shows the architecture of the suggested uWDL scenario-based smart-home system, which is aware of user's situation information in ubiquitous computing environments.

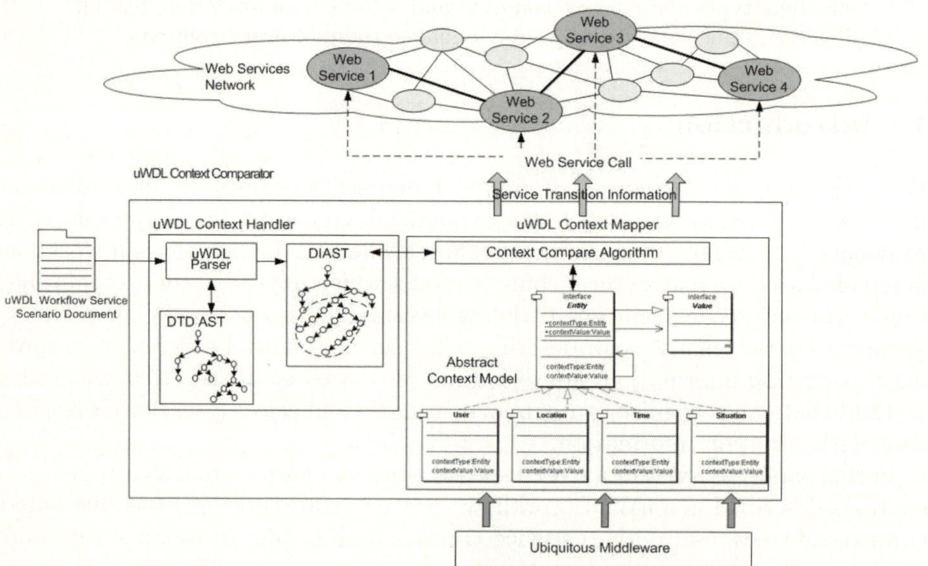

Fig. 1. A smart-home system's architecture based in a uWDL workflow service scenario

As shown in Figure 1, the suggested system supports context-aware home services using a uWDL service scenario, in which an user's situation information is described as a services execution condition. A uWDL context mapper in Figure 1 uses a context subtree for context comparison with user's situation information. Through the context comparison, the uWDL context mapper searches home service networks for a home service appropriate to a user's situation information generated from ubiquitous computing environments. If it finds a service corresponding to a specific user's situation information, the uWDL context mapper calls the service to offer context-aware home service to users.

3.2 A uWDL Context Handler and a uWDL Context Mapper

As a result of a parsing, the uWDL handler makes a DIAST (Document Instance Abstract Syntax Tree) that represents the structure information of a uWDL document as a tree data structure. At this time, a context described as RDF-based triplet entity in a uWDL scenario is constructed as a subtree of the parse tree. Contexts that the context mapper uses for the comparison are described in a triplet based on RDF. Context information from the sensor network can be embodied as a triplet consisting of subject, verb and object according to the structural context model based in RDF. The context mapper extracts context types and values of the entity objectified from sensors. It then compares the context types and values of the objectified entity with those of the DIAST's subtree elements related to the entity. In the comparison, if the context types and values in the entity coincide with the counterpart in the DIAST's subtree, the context mapper drives the service workflow. For that, we define a context embodied with a structural context model from the sensor network as OC and a context described in a uWDL scenario as UC. OC means a context objectified with the structural context model, and UC means a context described in a uWDL scenario. Also, OCS and UCS that mean each set of OC and UC.

4   Experiments and Results

For an experiment, we have developed an uWDL scenario for home-network services through PDA. The example scenario is as follows: John has a plan to go back his home at 8:00 PM, take a warm bath, and then watch a recorded TV program, which he wants to see after a bath. When John arrives to his apartment, an RFID sensor above the apartment door transmits John's basic context information (such as name, notebook's IP address) to the uWDL home server. If the conditions, such as user location, situation, and current time, are satisfied with contexts described in the uWDL workflow service scenario, then the server will prepare warm water. When he sits the sofa in the living room after he finishes a bath, the service engine will turn on the power of TV in the living room and play the TV program that was recorded already. For the experiment, we use a Pentium IV 3.0 GHZ computer with 1GB memory based in Windows XP OS as a uWDL home service engine and a PDA with 512MB memory based in Windows CE for the experiment. Figure 2 shows the result.

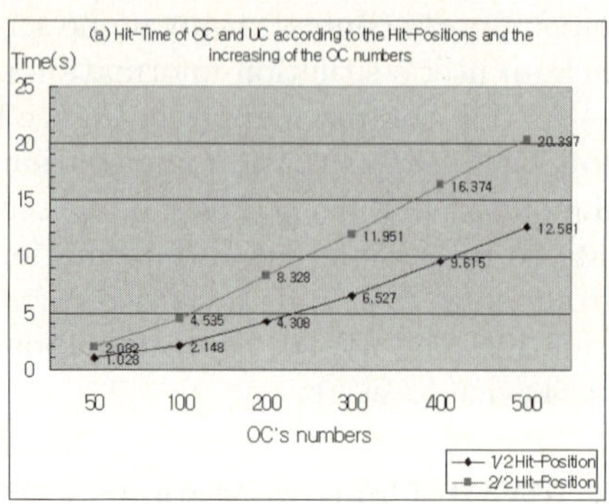
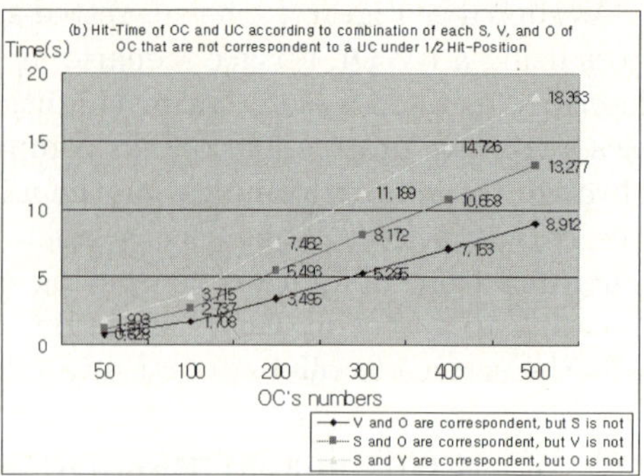

Fig. 2. A result of hit-time of OC and UC according to hit-position and conditions of OC's *s*, *v*, and *o*

5 Conclusion

In this paper, we presented a home-network service system that can recognize a uWDL workflow service scenario document, and can drive home services according to a user's situation information. Through experiments, we defined contexts described in a uWDL scenario as OC and contexts objectified from ubiquitous computing environments. We showed an experiment in which the uWDL mapper compared contexts of UCSs and OCSs through the context comparison algorithm, and measured hit-times and service transition accuracy to verify the efficiency of the algorithm. Through the results, we found that the hit-times were reasonable in spite of increment in the OCs amounts. Therefore, this uWDL context comparator will contribute greatly to the development of the context-aware application programs in ubiquitous computing environments.

References

1. W. Keith Edwards and Rebecca E. Grinter, Ubicomp 2001, LNCS 2201, pp. 256-272, 2001.
2. Joohyun Han, Yongyun Cho, Jaeyoung Choi: Context-Aware Workflow Language based on Web Services for Ubiquitous Computing, ICCSA 2005, LNCS 3481, pp. 1008-1017, (2005)
3. Anind k. Dey: Understanding and Using Context, Personal and Ubiquitous Computing, Vol 5, Issue 1, pp.69-78 (2001)
4. W3C: RDF/XML Syntax Specification, W3C Recommendation (2004)
5. Tony Andrews, Francisco Curbera, Yaron Goland: Business Process Execution Language for Web Services, BEA Systems, Microsoft Corp., IBM Corp., Version 1.1 (2003)
6. Frank Leymann: Web Services Flow Language (WSFL 1.0). IBM (2001)
7. Satish Thatte: XLANG Web Services for Business Process Design, Microsoft Corp. (2001)
8. R. Scott Cost, Tim Finin: ITtalks: A Case Study in the Semantic Web and DAML+OIL, University of Maryland, Baltimore County, IEEE (2002) 1094-7167

Simple-Adaptive Link State Update Algorithm for QoS Routing[*]

Seung-Hyuk Choi[1], Myoung-Hee Jung[1], Min Young Chung[1,**],
Mijeong Yang[2], Taeil Kim[2], and Jaehyung Park[3]

[1] School of Information and Communication Engineering, Sungkyunkwan University
300, Chunchun-dong, Jangan-gu, Suwon, Kyunggi-do, 440-746, Korea
{zealion, aodgl, mychung}@ece.skku.ac.kr
[2] Broadband Converged Network Division, ETRI
161, Gajeong-dong, Yuseong-gu, Daejeon, 305-700, Korea
{mjyang, tikim}@etri.re.kr
[3] Department of Computer Engineering, Chonnam National University
300, Yongbong-dong, Buk-ku, Gwangju, 500-757, Korea
hyeoung@chonnam.ac.kr

Abstract. To guarantee Quality of Service, routers should automatically determine routing paths in order to satisfy service requirements efficiently, based on link state information as well as network topology. Link State Database (LSDB) in routers should be well managed in order to reflect the current state of all links effectively. However, there is a trade-off between the exact reflection of the current link status and its update cost. In this paper, a simple-adaptive LSU algorithm to adaptively control the generation of link state update messages is proposed and its performance is compared with those of four existing algorithms by intensive simulations.

1 Introduction

In general, in order to guarantee QoS, routers determine routing paths by considering link state information as well as network topology. Therefore, it is important that routers know link state information to calculate the routing paths, i.e., information in Link State Database (LSDB) resided in all routers should be well managed. To reflect the link status in the LSDB, routers transmit LSU messages to their neighbors. If routers generate Link State Update (LSU) messages inordinately, the router performance is reduced due to the processing of LSU messages [1]. However, in the case that link state information is not updated appropriately, route setup requests may be rejected even though routes exist. This problem with QoS routing therefore can be characterized by the trade-off

[*] This study was partially supported by the Smart(Ubiquitous) Seoul Project titled "An Intelligent Convergence System of Urban Information for Smart(Ubiquitous) Cities" which was operated by the University of Seoul and funded by City of Seoul, Korea(South).

[**] Corresponding author.

V.N. Alexandrov et al. (Eds.): ICCS 2006, Part I, LNCS 3991, pp. 969–972, 2006.

between the accuracy of link state information and the overhead incurred by exchanging this information [2].

Many algorithms exist for determining the instant of transmitting LSU messages, i.e., Period Based (PB), Threshold Based (TB), Equal Class Based (ECB), Unequal Class Based (UCB) LSU algorithm [3], [4], Dynamic Threshold Based (DTB) [5], and Second-moment Based (SB) [3] LSU algorithms. The existing LSU algorithm uses fixed value(s) in order to decide transmission of LSU messages, therefore they cannot effectively perform under different network topologies and traffic conditions. erfectly to network conditions.

2   The Proposed Algorithm

Since the existing algorithms use fixed value(s) in order to determine the transmission instant of LSU messages, they may perform ineffectively depending on varying network topology and traffic conditions. To overcome this disadvantage, a Simple Adaptive (SA) LSU algorithm is proposed, where parameter values adaptively change as the link state changes.

In the proposed algorithm, in order to determine whether to transmit LSU messages, routers observe available bandwidth and the number of serving connections on a link. For every request for a connection setup (or release), the available bandwidth increases (or decreases) by the corresponding request bandwidth, and the number of connections served increases (or decreases) by one. If LSU messages are flooded, all receiving routers update their LSDB as information contained in LSU messages.

Let B_n be the available bandwidth value stored in the LSDB. In addition, let $\widetilde{B}(t)$ and $\widetilde{N}(t)$ be the current available bandwidth on links and the number of serving connections on a link, respectively, immediately prior to making the decision to transmit LSU messages. Routers transmit LSU messages to their neighbors if the following condition is satisfied.

$$|B_n - \widetilde{B}(t)| \ \geq \ \frac{\widetilde{B}(t)}{\widetilde{N}(t)} \tag{1}$$

$$\Longleftrightarrow \ \frac{|\widetilde{U}(t) - U_n|}{1} \ \geq \ \frac{C}{\widetilde{N}(t)} - \frac{\widetilde{U}(t)}{\widetilde{N}(t)}, \tag{2}$$

where $\widetilde{U}(t)$ denotes unavailable bandwidth immediately prior to making a decision to transmit LSU messages, U_n indicates unavailable bandwidth stored in the LSDB, and C expresses the total capacity of one link.

Consequently, in the SA LSU algorithm, if the variation in used bandwidth per service is equal to or larger than the mean available bandwidth per connection, expected at the instant of the next link state update, LSU messages are transmitted. Unlike the other algorithms, since the SA LSU algorithm only uses the control parameters about link state, it is able to effectively update the LSDB.

3 Performance Evaluation

In order to evaluate the performance of the proposed algorithm, an MCI topology consisting of 18 nodes and 30 bidirectional T3 links (45 Mbps) is considered [3]. Each connection request is defined as $(s,\ d,\ b_{req})$, where s, d, and b_{req} denote source node, destination node, and request bandwidth, respectively. For each connection, s and d are different and randomly selected from 18 nodes. In addition, b_{req} is uniformly chosen in (3Mbps, 7Mbps), and it assumes that connection requests arrive as a Poisson process with the mean arrival rate λ and holding duration of connections is determined by the exponential distribution with the average service rate μ. Therefore, total offered traffic load is equal to λ/μ.

As performance measures, the number of updates per unit time and blocking probability are defined. The number of updates per unit time is defined that the total number of LSU messages transmitted is divided by the total simulation time. The blocking probability of connection requests is defined as

$$P_{block} = \frac{N_{block}}{N_{total\ request}} \tag{3}$$

where $N_{total\ request}$ is the total number of connection requests and N_{block} is the number of blocked connection requests.

In general, performance of the existing algorithms relies heavily on the used values of control parameters. For each of them, the specified value(s) are determined as the pseudo-optimal value(s) of parameter(s) if the blocking probability is close to that of the basic LSU algorithm in which LSU messages are generated whenever the available bandwidth varies. For the MCI network and $\mu = 1/sec.$, the pseudo-optimal values of control parameters considered for the the existing three algorithms are $th = 0.2$ for TB, $R_0 = 0.1$ and $\Delta th = 0.5$ for DTB, and $th = 0.03$ for SB LSU algorithm. These pseudo-optimal values are used in order to evaluate the performance of the three existing algorithms.

For $\mu = 1/sec.$, the blocking probabilities of the four existing algorithms, and the proposed algorithm is presented in Fig. 1. The blocking probability of the basic LSU algorithm provides the least margin of those, because this algorithm makes an update of the status-changed information immediately when routers sense link status change. The blocking probability of the connection requests increases in proportion to the traffic load. Each existing LSU algorithm with the pseudo-optimal value(s) demonstrates a blocking probability which is analogous to that of the basic LSU algorithm. In addition, the blocking probability in the proposed LSU algorithm is similar to that of the basic LSU algorithm.

Fig. 2 presents the LSU rates varying traffic load. In the three existing LSU algorithms, the parameter value(s) are fixed.

On the other hand, in the SA LSU algorithm, the values used to determine LSU message transmissions are calculated considering network traffic conditions. Therefore, routers effectively transmit LSU messages and make the smaller number of updates than the existing LSU algorithms.

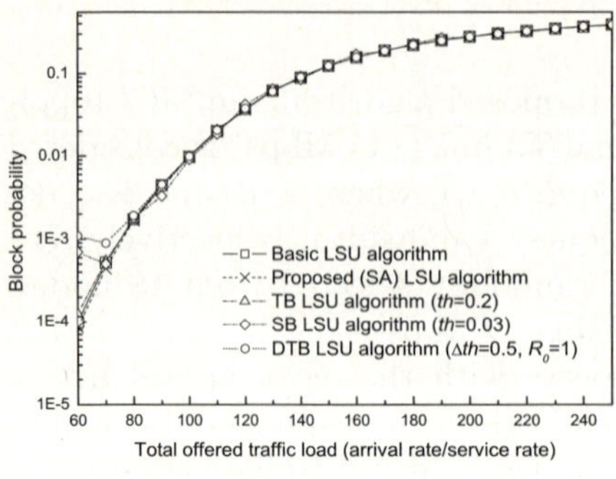

Fig. 1. Blocking probability   **Fig. 2.** LSU rate

4   Conclusions

In this paper, the SA LSU algorithm is proposed and its performance is evaluated by simulations. Existing LSU algorithms decide to transmit LSU messages by fixed parameter value(s). Therefore, the associated performance may degrade with varying network conditions. However, since the SA LSU algorithm decides whether to transmit or not LSU messages by parameters which are calculated from traffic conditions, the associated performance dose not depend on traffic conditions. In addition, in the proposed LSU algorithm, while the LSU messages are transmitted less than existing LSU algorithms, the blocking probability of connection requests is similar to that of existing LSU algorithms. Additionally, because the SA LSU algorithm requires simple computations, it may be easily implemented in QoS routers.

References

1. Apostolopoulos, G., Guerin, R., Kamat, s.: Implementation and Performance Measurements of QoS Routing Extensions to OSPF. IEEE INFOCOM 1999, Vol. 2., New York (1999) 680-688
2. Jia, Y., Nikolaidis, I., Gburzynski, P.: Multiple Path Routing in Networks with Inaccurate Link State Information. ICC 2001, Vol. 8., IEEE, Helsinki (2001) 2583-2587
3. Zhao, M., Zhu, H., Li, V.O.K., Ma, Z.: A Stability-Based Link State Updating Mechanism for QoS Routing. ICC 2005, Vol. 1., IEEE, Seoul (2005) 33-37
4. Ma, Z., Zhang, P., Kantola, R.: Influence of Link State Updating on the Performance and Cost of QoS Routing in an Intranet. 2001 IEEE Workshop on High Performance Switching and Routing, Dallas (2001) 375-379
5. Ariza, A., Casilari, E., Sandoval, F.: QoS routing with adaptive updating of link states. Electronics Letters, Vol. 37., IEEE, (2001) 604-606

Throughput Analysis and Enhancement for CSMA Based Wireless Networks

Younggoo Kwon

Konkuk University, 1 Hwayang-dong, Kwangjin-gu, Seoul, 143-701, Korea
ygkwon@konkuk.ac.kr

Abstract. We studied the performance of contention based medium access control (MAC) protocols. We used a novel technique for estimating the throughput, and other parameters of interest, of such protocols. In this paper, a new assumption for the theoretical throughput limit in the distributed CSMA based MAC algorithm is introduced. Through the performance analysis and simulation studies, the proposed algorithm shows significant performance improvements in CSMA based wireless networks.

1 Introduction

In many performance analysis papers for the binary exponential backoff based CSMA algorithms, the performance analysis starts from the assumption that all stations have the same average contention window range in steady state [1]-[3]. The same average contention window range for all active stations can be understood that all stations have the same average probability of packet transmission in steady state. The performance of many binary exponential backoff based MAC algorithms, including the IEEE 802.11 MAC algorithm, can be explained well by using the assumption that all stations have the same average contention window range in steady state. Furthermore, the optimum value which will minimize the wasting overheads during the contention procedure can be derived for a given number of active stations [1]-[3]. However, there are still the wasting overheads come from the inherent limitation of the assumption that all stations have the same contention window range for packet transmission in steady state.

2 Enhanced CSMA

Under high traffic load and under some ergodicity assumption, we can obtain the simplified expression for the throughput:

$$\rho = \frac{\bar{m}}{E[N_c](E[B_c] \cdot t_s + \bar{m} + DIFS) + (E[B_c] \cdot t_s + \bar{m} + SIFS + ACK + DIFS)} \tag{1}$$

where $E[N_c]$ is the average number of collisions, $E[B_c]$ is the average number of idle slots.

V.N. Alexandrov et al. (Eds.): ICCS 2006, Part I, LNCS 3991, pp. 973–976, 2006.

From this result, we can see that the theoretical throughput limit would be the following: a successful packet transmission must be followed by another successful packet transmission without any overheads, in which case, $E[N_c] = 0, E[B_c] = 0$.

If we could develop a contention-based MAC algorithm, which assigns a backoff timer 0 to the station in transmission while assigns all other stations' backoff timers to ∞ for each contention cycle, then we could achieve the perfect scheduling, leading to the theoretical throughput limit. Unfortunately, such a contention-based MAC algorithm does not exist in practice. However, this does provide us the basic idea how to improve the throughput performance in the MAC algorithm design. We can use the operational characteristics of the perfect scheduling to design more efficient contention-based MAC algorithm. One way to do so is to design a MAC protocol to approximate the behavior of perfect scheduling. To achieve the similar operational characteristics of perfect scheduling, the proposed MAC algorithm provides the following design factors: Large contention window range & small idle slots, long-term fairness, backoff timer realignment.

3   Performance Evaluations

We consider two kinds of contention window sizes, one for the whole contention procedure including deferring conditions, $E[CW]$, and the other for the case of transmitting a packet, $E[CW]_{PkSend}$. The relation between the average contention window size for each contention procedure $E[CW]$ and the probability of a successful packet transmission for one station $p_{suc,1}$ is given by the following equation.

$$p_{suc,1} = \sum_{i=1}^{E[CW]} \frac{1}{E[CW]} \cdot (\frac{E[CW] - i}{E[CW]})^{M-1} \qquad (2)$$

Furthermore, the summation of the probability of collision and the probability of deferring for one station is given by the following equation: $1 - p_{suc,1} = p_{col,1} + p_{defer,1}$.

The contention window size for each contention procedure is increased by the increasing factor (IF) when a station experiences either a collision or a deferred situation, and goes to the minimum value with a successful packet transmission. Therefore, the average contention window size for each contention period is

$$E[CW] = p_{suc,1} \times minCW + (1 - p_{suc,1}) \times E[CW] \times IF \qquad (3)$$

If we use the above equations (2) and (3), we can use an iterative process to obtain the average contention window size for each contention procedure $E[CW]$ and the probability, $p_{suc,1}$, of a successful packet transmission for one station. If the number of stations in the network is M, the total probability of successful packet transmission for the whole network is $p_{suc,total} = p_{suc,1} \cdot M$ and the total average probability of collision for the whole network is $p_{col,total} = 1 - p_{suc,total}$.

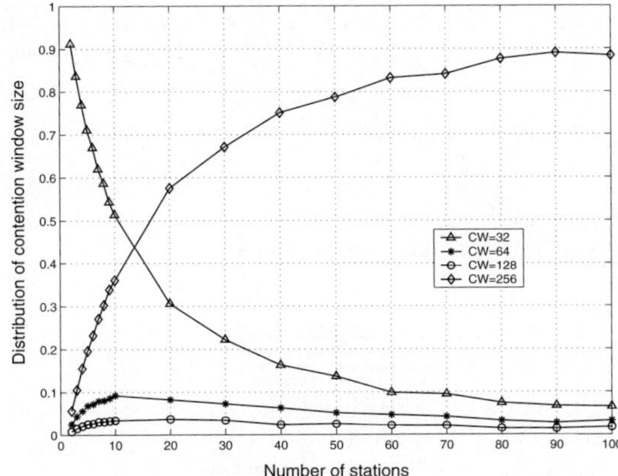

Fig. 1. Distribution for Contention Window Size

Now, we can calculate the average number of collisions for a successful packet transmission

$$E[N_c] = \frac{p_{col,total}}{p_{suc,total}} \tag{4}$$

To calculate the average idle backoff slot number $E[IdleSlot]/t_{slot}$, we need the probability of sending a packet at each contention window size $E[CW]_{PkSend}$. In Figure 1, an example of the distribution of contention window sizes for sending a packet in steady state is shown for the $minCW = 32$, $maxCW = 256$, $IF = 2$ case. In the 10 station case, 51% of stations have contention window size of sending a packet at $CW = 32$, 9% of stations have $CW = 64$, 3% of stations have $CW = 128$, and 37% of stations have $CW = 256$. As the number of stations increase, we can see the operational characteristics of the proposed algorithm follow those of the perfect scheduling. The average contention window size of sending a packet is

$$E[CW]_{PkSend} = 32 \times p_{PkSend,32} + 64 \times p_{PkSend,64} + 128 \times p_{PkSend,128}$$
$$+256 \times p_{PkSend,256} \tag{5}$$

where $p_{PkSend,i}$ is the probability that a packet is transmitted with a contention window size i.

The average number of idle backoff slots is given by the following equation

$$E[IdleSlot]/t_{slot} = \sum_{i=1}^{E[CW]_{PkSend}-1} i \times \frac{(E[CW]_{PkSend} - i)^{M-1}}{(E[CW]_{PkSend})^M} \tag{6}$$

In Figure 2, the throughput results of the proposed algorithm, the improved DCF (under the assumption that all stations have the same contention window range) and the IEEE802.11 MAC are shown for 10 and 50 contending stations, respectively. We can see that the throughput of the proposed MAC is significantly

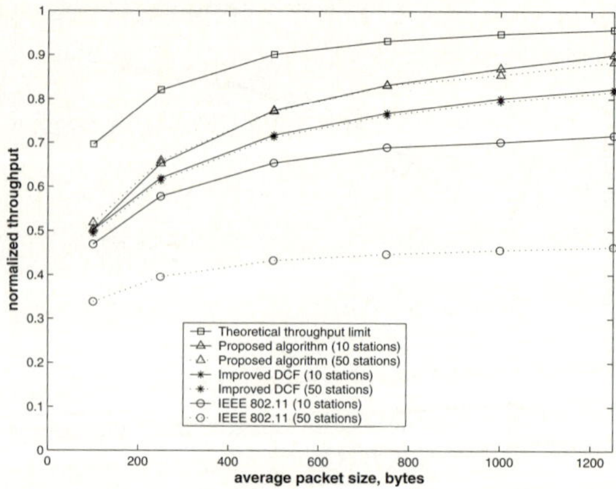

Fig. 2. Throughput results

improved compared with other MAC algorithms and close to the theoretical limit of perfect scheduling as the average packet size is increased.

4   Conclusions

The general assumption that all stations have the same contention window range in steady state results in sub-optimal solutions for performance analysis of distributed contention-based MAC algorithms because of its inherent limitations. We present a new assumption for the theoretical throughput limit from observing the operational characteristics of the perfect scheduling CSMA algorithm. The proposed algorithm based on the new assumption significantly improves the throughput performance and still provides easy implementation property in wireless networks. Extensive performance analysis and simulation studies for various performance factors have demonstrated that the proposed algorithm reduces the wasting overheads come from each contention procedure.

References

1. Bharghavan, V.: MACAW: A Media Access Protocol for Wireless LAN's. SIG-COMM'94, London, England, Aug. (1994) 212-225
2. Bianchi, G.: Performance Analysis of the IEEE802.11 Distributed Coordination Function. IEEE Journal on Selected Areas in Commun. **18** (2000) 535-547
3. Cali, F., Conti, M., Gregori, E.: Dynamin Tuning of the IEEE 802.11 Protocol to Achieve a Theoretical Throughput Limit. IEEE/ACM Trans. on Networking **8** (2000) 785-799
4. Aad, I., Castelluccia, C.: Differentiation mechanisms for IEEE 802.11. IEEE INFOCOM Anchorage, AK, USA (2001)

Efficient Password-Authenticated Key Exchange for Three-Party Secure Against Undetectable On-Line Dictionary Attacks[*]

Jeong Ok Kwon[1,2], Kouichi Sakurai[1], and Dong Hoon Lee[2]

[1] Department of Computer Science and Communication Engineering,
Kyushu University, Japan
pitapat@cist.korea.ac.kr, sakurai@csce.kyushu-u.ac.jp
[2] Center for Information Security Technologies (CIST), Korea University, Korea
donghlee@korea.ac.kr

Abstract. A password-authenticated key exchange (PAKE) protocol in the three-party setting allows two users communicating over a public network to agree on a common session key by the help of a server. In the setting the users do not share a password between themselves, but only with the server. In this paper, we explore the possibility of designing a round-efficient three-party PAKE protocol with a method to protect against undetectable on-line dictionary attacks without using the random oracle. The protocol matches the most efficient three-party PAKE protocol secure against undetectable on-line dictionary attacks among those found in the literature while providing the same level of security. Finally, we indentify the relations between detectable on-line and undetectable on-line dictionary attacks by providing counter-examples to support the observed relations[1].

Keywords: Cryptography, password-authenticated key exchange, dictionary attacks, round complexity, mobile network security.

1 Protocols for PAKE in the Three-Party Setting

We now present our two protocols $3\mathcal{PAKE}1$ and $3\mathcal{PAKE}1$ for PAKE in the three-party setting. $3\mathcal{PAKE}1$ is designated to protect passwords from detectable on-line and off-line dictionary attacks and is explicit authentication. $3\mathcal{PAKE}2$ is designated to enhance $3\mathcal{PAKE}1$ against undetectable on-line dictionary attacks in addition to detectable on-line and off-line dictionary attacks, yet still requires four rounds and is implicit authentication. In this paper, we assume the parties can transmit messages simultaneously.

---

[*] This work was done while the first author visits in Kyushu Univ. and was supported by the Ministry of Information & Communications, Korea, under the Information Technology Research Center (ITRC) Support Program.
[1] This is a very partial version of the full paper available at http://cist.korea.ac.kr/pitapat/JeongOkKwonTriS.ps

V.N. Alexandrov et al. (Eds.): ICCS 2006, Part I, LNCS 3991, pp. 977–980, 2006.

$3\mathcal{P}\mathcal{A}\mathcal{K}\mathcal{E}1$

Public information. Two primes p, q such that $p = 2q + 1$, where p is a safe prime such that the decision Diffie-Hellman problem is hard to solve in $\mathbb{G}$. A finite cyclic group $\mathbb{G}$ has order q, and g_1, g_2 are generators of $\mathbb{G}$ both having order q, where g_1 and g_2 must be generated so that their discrete logarithmic relation cannot be known. A hash function H from $\{0,1\}^*$ to $\{0,1\}^l$. A message authentication code (MAC), $\mathsf{Mac} = (\mathsf{Mac.gen}, \mathsf{Mac.ver})$. Given a random key k, $\mathsf{Mac.gen}$ computes a tag τ for a message M; we write this as $\tau = \mathsf{Mac.gen}_k(M)$. $\mathsf{Mac.ver}$ verifies the message-tag pair using the (shared) key, and returns 1 if the tag is valid or 0 otherwise. F is a pseudo random function family.

Initialization. Each user U_i for $i \in \{1, 2\}$ obtains pw_i at the start of the protocol using a password generation algorithm $\mathcal{PG}(1^k)$ which on input a security parameter 1^k outputs a password pw_i uniformly distributed in a password space **password**. Then U_i sends $v_{i,1} = g_1^{H(U_i\|S\|pw_i)} mod\ p$ and $v_{i,2} = g_2^{H(U_i\|S\|pw_i)} mod\ p$ which are verifiers of the password to the server S over a secure channel. Upon receiving the verifiers, S stores them in a password file with an entry for U_i. The indices are cyclic, i.e., U_{n+1} is U_1, where $n = 2$.

Round 1. An initiator U_i for i = 1 broadcasts (U_i, U_{i+1}, S).

Round 2. Each user U_i chooses a random number $x_i \in \mathbb{Z}_q^*$, computes $X_{i,S} = g_1^{x_i} \cdot v_{i,2}\ mod\ p$, and sends $(U_i, X_{i,S})$ to the server.

Round 3. Upon receiving $(U_i, X_{i,S})$, S selects a random number $s \in \mathbb{Z}_q^*$ and computes $X_{S,i} = (X_{i+1,S}/v_{i+1,2})^s \cdot v_{i,2}\ mod\ p$, and sends $(S, X_{S,i})$ to each user U_i for $i = \{1, 2\}$.

Round 4. Upon receiving $(S, X_{S,i})$, each user U_i computes $k_i = (X_{S,i}/v_{i,2})^{x_i}\ mod\ p$ and $\tau_i = \mathsf{Mac.gen}_{k_i}(U_i\|U_{i+1}\|S)$ and sends (U_i, τ_i) to U_{i+1}.

Key computation. Upon receiving (U_i, τ_i), each user U_{i+1} computes $\mathsf{Mac.ver}_{k_{i+1}}(\tau_i)$. Each user U_{i+1} halts if $\mathsf{Mac.ver}$ returns 0 or computes the session key $sk_{i+1} = F_{k_{i+1}}(U_1\|S\|U_2)$ otherwise.

Remark. We have observed that the relations between detectable and undetectable on-line dictionary attacks. This is, the security against detectable on-line dictionary attacks does not necessarily imply the security undetectable on-line dictionary attacks and vice versa. To support the observed relations we consider counter-examples presented in this paper. For the first relation, consider our first protocol $3\mathcal{P}\mathcal{A}\mathcal{K}\mathcal{E}1$. $3\mathcal{P}\mathcal{A}\mathcal{K}\mathcal{E}1$ is secure against detectable on-line dictionary attacks but it is insecure against undetectable on-line dictionary attacks. For the second relation, consider our second protocol $3\mathcal{P}\mathcal{A}\mathcal{K}\mathcal{E}2$. We can modify $3\mathcal{P}\mathcal{A}\mathcal{K}\mathcal{E}2$ to secure against undetectable on-line dictionary attacks but to insecure against detectable on-line dictionary attacks. If the MAC verification between each user and the server is removed in Round 4 of $3\mathcal{P}\mathcal{A}\mathcal{K}\mathcal{E}2$, the protocol does not secure against on-line dictionary attacks anymore because the users cannot detect and log the failed guesses. Therefore we must independently consider the both attacks and provide methods to resistant to both attacks when design secure protocol for 3-party PAKE.

$3\mathcal{PAKE}2$

Same as $3\mathcal{PAKE}1$ except the following points.

Round 2. Each user U_i chooses a random number $x_i \in \mathbb{Z}_q^*$, computes $X_{i,S} = g_1^{x_i} \cdot v_{i,2} \bmod p$, and sends $(U_i, X_{i,S})$ to the server S. For $i = \{1, 2\}$, S chooses random numbers $y_i \in \mathbb{Z}_q^*$, computes $X_{S,i} = g_1^{y_i} \cdot v_{i,2} \bmod p$, and sends $(S, X_{S,i})$.

Round 3. Upon receiving $(S, X_{S,i})$, each user U_i computes $k_{i,S} = (X_{S,i}/v_{i,2})^{x_i} \bmod p$ and $\tau_{i,S} = \mathsf{Mac.gen}_{k_{i,S}}(U_i\|S\|X_{i,S}\|X_{S,i})$ and sends $(U_i, \tau_{i,S})$ to S.

Round 4. Upon receiving $(U_i, \tau_{i,S})$, S computes $\mathsf{Mac.ver}_{k_{S,i}}(\tau_{i,S})$ where $k_{S,i} = (X_{i,S}/v_{i,2})^{y_i} \bmod p$ for $i = \{1, 2\}$. S halts if at least one of $\mathsf{Mac.ver}_{k_{S,i}}$ returns 0 or proceeds to the next process otherwise. S selects a random number $s \in \mathbb{Z}_q^*$, and computes $Y_{S,i} = (g_1^{x_{i+1}})^s$ and $\tau_{S,i} = \mathsf{Mac.gen}_{k_{S,i}}(U_i\|U_{i+1}\|Y_{S,i})$ and sends $(S, Y_{S,i}, \tau_{S,i})$ to each user U_i.

Key computation. Upon receiving $(S, Y_{S,i}, \tau_{S,i})$, each user U_i computes $\mathsf{Mac.ver}_{k_i}(\tau_{S,i})$. Each user U_i halts if $\mathsf{Mac.ver}$ returns 0 or computes the session key $sk_{i+1} = F_{k_{i+1}}(U_1\|S\|U_2)$ otherwise, where $k_i = (Y_{S,i})^{x_i} \bmod p$.

SECURITY OF $3\mathcal{PAKE}2$.

1. $3\mathcal{PAKE}2$ is secure against *detectable on-line dictionary attacks* since the users are able to detect a failed guess by the MAC verification in Round 4. $3\mathcal{PAKE}2$ is resistant to *undetectable on-line dictionary attacks* since the server is able to depart honest requests from malicious attempts. If the MAC verification in Round 3 is failed, the server will notice that whose password to being a target of undetectable on-line dictionary attacks and it be at a crisis. After a small amount of failed guesses the server reacts and informs the target user to stop any further use of the password and to change the password into a new one. $3\mathcal{PAKE}2$ is secure against *off-line dictionary attacks* via the difficulty of the decision DDH problem.

2. Key secrecy means that no computationally bounded adversary should learn anything about session keys shared between two honest users. Especially in the 3-party key exchange model existing a server, more desirable security notion is that the participating server should not learn anything about session keys shared between two honest users, additionally. $3\mathcal{PAKE}2$ provides strong *key secrecy* since passwords are unguessed from dictionary attacks and in order to learn some significant information on session keys, the adversary $\mathcal{A}$ including $\mathcal{S}$ has to solve the DDH problem or break the MAC which is selectively unforgeable against adaptively chosen message attacks.

3. Forward secrecy means that even if long-term secret keys of one or more users are compromised, the secrecy of previous session keys established by honest users without any interference by the adversary is not affected. $3\mathcal{PAKE}2$ provides *forward secrecy* under the DDH assumption. That is, even if the

passwords of users are compromised, any adversary cannot learn any significant information on previously established session keys under the DDH assumption.

4. A protocol is secure against known key attack (or Denning-Sacco attack) if the following conditions hold. First, even with the session key from an eavesdropped session an adversary cannot gain the ability to impersonate the legitimate user directly. Second, an adversary cannot gain the ability to performing off-line dictionary attacks on the users' passwords from using the compromised session keys which are successfully established between honest entities. Third, a legitimate user $\mathcal{U}_i$ knowing pw_i does not learn any information about the partner U_{i+1}'s password pw_{i+1} from the session key established with U_{i+1}. In $3\mathcal{PAKE}2$, the session keys are computationally independent from each other. Thus, a compromised old session key is not helpful for $\mathcal{A}$ in attempting known key attack. To break the second condition $\mathcal{A}$ has to solve the DDH problem. The security on the third condition is also provided under the DDH assumption. That is, to successfully mount an off-line dictionary attack from the session key established with U_{i+1}, $\mathcal{U}_i$ knowing pw_i has to solve the DDH problem, in which one's password pw_i is not helpful in attempting the off-line dictionary attack against pw_{i+1}.

2   Concluding Remarks

In this paper, we have proposed a three-party PAKE protocol secure against detectable/undetectable on-line and off-line dictionary attacks without making use of the random oracle model. However, we only provided informal security analysis. Proving the security of our protocol in the standard model is the subject of ongoing work.

A Publish-Subscribe Middleware for Real-Time Wireless Sensor Networks

Mohsen Sharifi, Majid Alkaee Taleghan, and Amirhosein Taherkordi

Computer Engineering Department, Iran University of Science and Technology
`msharifi@iust.ac.ir,{alkaee, taherkordi}@comp.iust.ac.ir`

Abstract. The specific characteristics of Wireless Sensor Networks (WSNs) have changed Quality of Service (QoS) support in these networks to a challenging task. In this paper, we propose a dispatcher as part of a message routing component in publish/subscribe WSNs. Some works consider link-based solutions to support real-time parameters in WSNs. These works do not take into account the dynamic behavior of WSNs with probable damaged nodes and links. The use of dispatcher can reduce the average message delay, whether the message has high priority or low priority. The dispatcher uses a scheduler to support real-time parameters, such as delay, and selects messages from two separate queues, namely, QoS queue and non-QoS queue. Simulation results show that our approach really reduces the average delay and increases the delivery rate for both QoS messages and non-QoS messages.

1 Introduction

Due to real-time requirements, high degree of faults, noise, and non-determinism caused by the uncontrolled aspects of environment [8], a fundamental issue in realizing a WSN is how to route applicative information, i.e., messages controlling the operation of various sensors and the data gathered by them. In WSNs' context, few works consider QoS parameters and especially real-time parameters, such as, delay in their solutions [2].

Most of works [3, 6] consider the sink node as an interface between a WSN and user applications, so that there is only one sink node in the WSN, which has higher capabilities and does not have the constraints that the other nodes have. They assume the links between nodes do not change frequently and the networks have a specific organization. A recent work [2] considers the sink as a *role* that can be assigned to every one of nodes, and not necessarily only the node which has higher resources.

Some protocols with QoS support at network layer, require much energy, are too complex, and consider only one QoS factor, i.e. either fault tolerance or real-time [1]. A more recent solution [4] considers both of these two QoS factors, but it does not support the data-centric style of communication and energy-awareness.

Mires [7] is a message-oriented middleware and needs a pre-configuration phase to detect publishers and so it cannot operate in dynamic environments with changing behavior. It does not consider QoS parameters in its design and implementation too.

The semi-probabilistic approach [2] has proposed an approach to route events based on a semi-probabilistic broadcast schema. It shows a semi probabilistic way against flooding approach to transmit the WSNs' messages. It is one of the

V.N. Alexandrov et al. (Eds.): ICCS 2006, Part I, LNCS 3991, pp. 981–984, 2006.

(probabilistic) approaches proposed in [5] for mobile ad-hoc networks. The semi-probabilistic approach has not taken into account real-time properties in its work and uses probability when a node cannot find any subscription for sending event. Therefore, this approach has its own overheads too.

In this paper, we propose an approach to consider the delay time from publishing time (start time) until when they are received by subscribers (end time). The approach uses the concepts from the semi-probabilistic approach. It does not need the configuration phase of nodes that some approaches like cluster-based approaches require. QoS parameters are injected in interests, and subscribers send their interests to network. Our approach uses *QoS-queue* and *non-QoS-queue* to buffer QoS and non-QoS messages, respectively, in order to reduce the delay times of messages and to manage these messages as quickly as possible. This means that QoS messages are scheduled according to their real-time constraints.

The remainder of paper is organized as follows. Section 2 describes our approach. Section 3 is dedicated to simulative comparison of our approach with the work reported in semi-probabilistic approach. Section 4 concludes the paper.

2 Our Approach

According to what has been proposed in the semi-probabilistic approach, frequent changes in WSNs do not allow considering the links concept between nodes. In publish-subscribe based systems, both subscribers and publishers can be QoS-aware and use QoS parameters to distinguish QoS messages. Subscribers can add a QoS parameter (e.g. delay) to interests, and publishers can add priority to detected events when they detect or sense an object.

There are two steps in our mechanism; the first step is the dissemination of interests, and the second step is the dispatching of events. After an interest is specified, one copy of the interest is stored in *local-subscription* storage and the interest is published to the network. Neighboring nodes, which are in subscriber's radio range, receive the interest. When a node receives the interest, it stores the interest in its *non-local-subscription* storage.

We use a value called *subscription horizon* which denotes the range that the interests are broadcasted and is the hop count; the same value is considered in the semi-probabilistic approach. In the event propagation step, publishers publish events according to their available sensors and operation specification of events. Every node has a QoS messages queue and a non-QoS messages queue.

The remaining part of our mechanism is the dispatching of messages based on their priorities and delays. These two parameters help the scheduler to send messages. Events are removed from queues based on the priority of their subjects. They are checked against local-subscription and non-local-subscription. If the event matches the local-subscription, a subscription node receives it. If the event matches to the non-local-subscription, it is re-broadcasted to other nodes. Otherwise, it uses *event propagation threshold* and if it matches, it re-broadcasts the event.

Since the re-broadcast of message can produce duplicate messages in the network, we use a message identifier which consists of publisher identifier and the time the event has been published; the publisher identifier is a local identifier in vicinity.When a message is forwarded and removed from a node' queues, it may be possible that it receives the same message again. Thus, we keep the publisher identifier and the last event time for that received publisher event to prevent re-broadcasting of older messages that are looped back to this node.

3   Simulation and Evaluation

To evaluate how well our mechanism supports real-time QoS in WSNs, the JiST simulator (http://jist.ece.cornell.edu/) was used. The simulation filed was 120 meters x 120 meters and consisted of 40 nodes with radio range 40 meters, and simulation was run for 40 seconds. We considered two criticality levels based on subscribers' interests: *High* and *Low,* and used 2 subscribers and 4 publishers. The publisher average rate and the average message transmission time were nearly 2 events per second and 1 second, respectively. The propagation threshold and subscription horizon in the two approaches were 0.5 and 2, respectively.

The semi-probabilistic approach does not consider QoS properties when messages are forwarded. But, in our approach, we consider QoS properties and the QoS scheduler uses these properties to dispatch messages. The scheduler operates according to *first delay* selection criteria. The message delay is measured by deducting the arrival time from the generation time of the message. Fig. 1 shows the delivery and the average delays of messages in both approaches.

The results show that our approach really reduces the average delay (i.e. the time between an event is generated until it is received by a subscriber) and increases the delivery rate (i.e., the ratio between the expected and actual receipt time of events) for both QoS messages and non-QoS messages.

Message waiting time in queue can change the message overheads and consequently the delay. In broadcast-based communication, one message may be in network for a long time and increase the overheads thereof. We have used a message identifier to remove message duplication in queues.

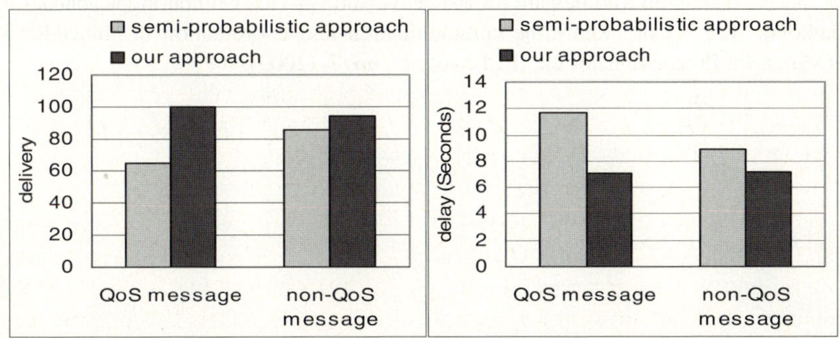

Fig. 1. The delivery rate and average delay of QoS messages and non-QoS messages

4   Conclusion and Future Works

QoS consideration in large-scale distributed middleware systems can raise the applicability of these systems in a variety of domains with favorable results. Unfortunately, few works have been reported which consider real-time deadlines in the context of WSNs. We considered this quality parameter and the goal was to reduce delays in message transmissions between nodes of WSNs.

We used two queues and a method to add messages to QoS queue and non-QoS queue. We evaluated our scheduler with first delay deadline and showed the average delay and delivery. Evaluation results showed lower average delay and higher delivery rates for QoS messages and non-QoS messages.

We intend to carry out more simulations in future and include more QoS parameters into published messages in order to reduce the duplicate message propagations in the network.

References

[1]   Chen, D., Varshney, P.K.: QoS support in wireless sensor networks: a survey. In: Proceedings of the International Conference on Wireless Networks (ICWN), Las Vegas, USA, (2004).

[2]   Costa, P., Picco, G.P., Rossetto, S.: Publish-subscribe on sensor networks: a semi-probabilistic approach. In: Proceedings of the 2nd IEEE International Conference on Mobile Ad Hoc and Sensor Systems (MASS), Washington DC, USA, (2005).

[3]   Delicato, F.C., et al.: A service approach for architecting application independent wireless sensor networks. In: Cluster Computing, Springer Science, (2005), 211-221.

[4]   Felemban, E., et al.: Probabilistic QoS guarantee in reliability and timeliness domains in wireless sensor networks. In: Proceedings of IEEE INFOCOM, vol.4, (2005), 2646-2657.

[5]   Ni, S., et al.: The broadcast storm problem in mobile ad hoc networks. In: Proceedings of ACM Mobicom, (1999).

[6]   Sharifi, M., Taleghan, M.A., Taherkordi, A.: A middleware layer mechanism for QoS support in wireless sensor networks. In: 5th International Conference on Networking (ICN), Mauritius, (2006).

[7]   Souto, E., et al.: A message-oriented middleware for sensor networks. In: Proceedings of the 2nd Workshop on Middleware for Pervasive and Ad-Hoc Computing, Canada, (2004).

[8]   Stankovic, J.A., et al.: Real-time communication and coordination in embedded sensor networks. In: Proceedings of the IEEE, vol. 91, no. 7, (2003).

Performance Evaluation of a Handover Scheme for Fast Moving Objects in Hierarchical Mobile Networks

In-Hye Shin[*], Gyung-Leen Park[**], and Junghoon Lee

Department of Computer Science and Statistics,
Cheju National University, Jeju, Korea
{ihshin76, glpark, jhlee}@cheju.ac.kr

Abstract. Reducing the handover latency has been one of the most critical research issues in Mobile IPv6. The research includes the fast handover, the hierarchical handover, and variations of them. This paper proposes a variation of the hierarchical handover and develops analytical models to compare the performance of the proposed scheme with that used in the hierarchical mobile networks. The performance evaluation shows that the proposed scheme is very effective for applications like Telematics where mobile nodes move fast. The paper also gives readers the threshold values with which they can select an optimal handover scheme for the given applications.

1 Introduction

The mobility support in Mobile IPv6 [1] is one of the most important research issues to provide many mobile users with the seamless Internet services. In the Internet environments, when a Mobile Node (MN) moves and attaches itself to another network, it needs to obtain a new IP address. Mobile IPv6 describes how the MN maintains connectivity to the Internet when it changes its Access Router (AR) into another. The process is called handover. During this process, there is a time period when the MN is unable to send or receive the IPv6 packets due to link switching delay and IP protocol operations. The time period is called handover latency. Reducing the handover latency has been a critical research issue to support the seamless service for mobile users. There have been many standardization works such as the hierarchical Mobile IPv6 mobility management (HMIPv6) [2], [3], the fast handover for Mobile IPv6 (FMIPv6) [3], [4], and the simultaneous bindings for fast handover [5]. The paper focuses on the handover in the hierarchical mobile networks. We propose a variation, which can be adapted to applications like Telematices where MNs move fast.

The rest of the paper is organized as follows. Section 2 proposes a new handover scheme. Section 3 presents the analytical models and shows the result of the performance evaluation. Finally, Section 4 concludes the paper.

---

[*] This research was supported by the MIC(Ministry of Information and Communication), Korea, under the ITRC support program supervised by the IITA (IITA-2005-C1090-0502-0009.

[**] Corresponding author.

V.N. Alexandrov et al. (Eds.): ICCS 2006, Part I, LNCS 3991, pp. 985–988, 2006.

2   The Proposed Approach

Figure 1 describes the basic operation of the proposed handover scheme. The main idea of the proposed scheme is that it carries out the Mobility Anchor Point (MAP) binding with an MN's Home Address instead of its Regional Care-of Address (RCoA). When an MN enters a new MAP domain as shown in Figure 1, it discovers a new MAP domain as it receives the MAP Option advertised by the MAP. The MAP informs the visiting MNs of its presence by sending the Router Advertisement (RA) message including its prefix information [7]. And then MN needs to configure only on-Link Care-of Address (LCoA), by appending its interface identifier to the prefix sent by the RA message [6], without configuring a new RCoA when it moves into the MAP domain. The MN initializes the MAP registration with its LCoA and its Home Address. The LCoA is used as the source address of the Binding Update (BU) message as done in the HMIPv6. The Home Address is included in the Home Address Option. The MAP will bind the LCoA to the Home Address instead of the RCoA. Thus, the proposed scheme does not have to perform the Duplicate Address Detection (DAD) process for the RCoA any more. Note the fact that the MN's Home address replaces its RCoA. The MAP may include an On-link Care-Of address Test (OCOT) Option in the Binding Acknowledgement (BA). The OCOT Option is in detail specified in [2]. The last process is identical as done in the HMIPv6.

After the MAP registeration, the MN must register its MAP's IP address, included in the MAP Option, with its HA or CNs by sending a BU message that specifies the binding of the MAP's IP address and its home address. The home address is put in the Home Address Option and the MAP's IP address can be involved in the source address field or the Alternate Care-of Address Option. The IP address of the MAP is used instead of the RCoA.

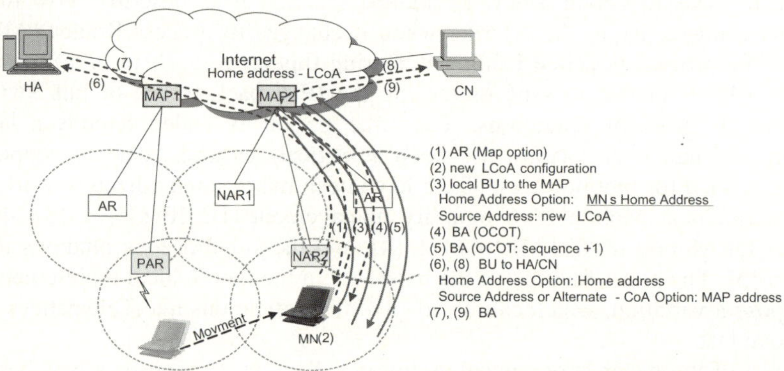

Fig. 1. The Handover Process Supporting Macro Mobility in the Proposed Scheme

The proposed scheme deletes the overhead of configuring the RCoA and the corresponding DAD procedure while adding the overhead of the home address in the data packet. Thus, the proposed scheme will be appropriate for the application like Telematics where a mobile user moves very fast across MAP domains.

3 Performance Evaluation

Table 1 depicts the notations used in the analytical models. The default MTU size for IPv6 packets on an Ethernet is 1500 bytes [8]. The delay for the DAD is set to the MAX_RTX_SOLICITION_DELAY (1 second), the maximum transmission delay of the NS message [7]. The total of the packet size includes the IPv6 Basic Header and some of optional IPv6 Extension Headers. The other IPv6 Extension Headers are omitted due to the variety of its size and the optional preference.

Table 1. The Notations Used in the Model

Notation	Description	Value
T_{sim}	The Total Simulation Time (min)	15
V_{MN}	The MN's Speed (km/min)	1
S_{MAP}	The Size of MAP (km)	1~16
N_{MAP}	The total number of Map Update : $\lfloor (T_{sim} \times V_{MN})/S_{MAP} \rfloor$	
p	The Frequency of Packet Transmission (1/min)	1~30
N_{pkt}	The Total number of Packet Transmission	
V_{Net}	The Network Transfer Speed (byte per second) : $\{50(Kbps) \times 1024\}/8 = 6400$	6400
D_{Net}	The Network Delay per MTU (s) : $p \times MTU/V_{Net}$	
MTU	The Size of MTU (byte)	1500
$Data$	The Size of Transfer Data (byte)	
IPH	The Size of IP Header (byte) [8]	40
$IPEH$	The Size of IP Extension Header (byte) [8]	
$Auth$	The Size of Authentication Header (byte) [9]	20
$DestOp$	The Size of Destination Options Header (byte) [8]	20
$Frag$	The Size of Fragment Header (byte) [8]	8
D_{DAD}	The Delay for DAD (s)	1
$HMIPv6$	The Delay using the Hierarchical Mobile IPv6 during S_{MAP} (s)	
$proposed$	The Delay using the proposed scheme during S_{MAP} (s)	

The following HMIPv6 and proposed represent the additional delay for the hierarchical handover and that for the proposed scheme, respectively.

$$HMIPv6 = N_{MAP} \times D_{DAD} + \left\lceil \frac{Data}{MTU - \{IPH + IPEH(Auth + Frag)\}} \right\rceil \times T_{sim} \times D_{Net} \qquad (1)$$

$$proposed = \left\lceil \frac{Data}{MTU - \{IPH + IPEH(Auth + Frag + DestOp)\}} \right\rceil \times T_{sim} \times D_{Net} \qquad (2)$$

Figure 2-(a) shows that the larger the size of MAP is, the smaller the difference between two schemes will become. Figure 2-(b) shows that the delay of the proposed scheme is less than that of HMIPv6 at the same rate. The difference results from the DAD cost and is small in this case. Figure 2-(c) and Figure 2-(d) provide the threshold value in the specific point, respectively.

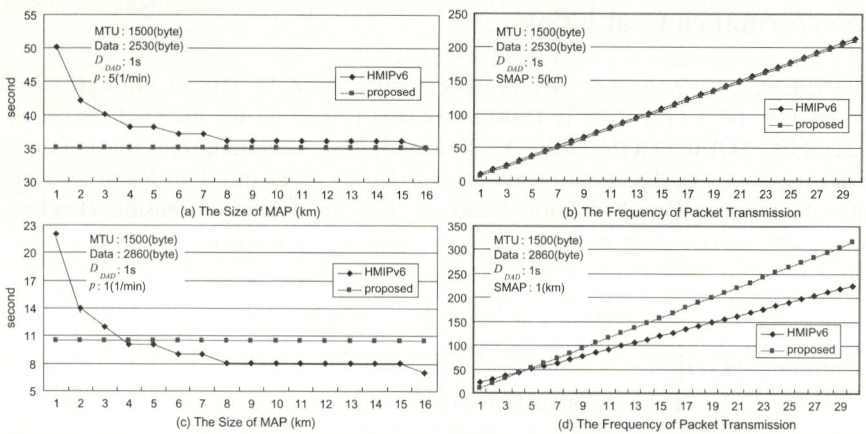

Fig. 2. The Delay according to the Size of the MAP and the Frequency of Packet Transmission

4 Conclusion

The paper proposes a variation of the handover scheme used in the HMIPv6. The paper also develops the analytical models to compare the performance of the proposed scheme and that used in the HMIPv6. The result shows that the proposed scheme is very effective regardless of the frequency of the packet transmission if it enables the packet to transfer by a MTU and the MN moves fast. In particular, the proposed scheme can be adapted to Telematics services where an user drives fast and sometimes receives the Internet services using the terminal in the car.

References

1. Johnson, D., Perkins, C., Arkko, J.: Mobility Support in IPv6. IETF RFC 3775 (2004)
2. Soliman, H., Castelluccia, C., Malki, K., Bellier, L.: Hierarchical Mobile IPv6 Mobility Management (HMIPv6). Internet Draft, IETF, draft-ietf-mobileip-hmipv6-08.text (2003)
3. Jung, H., Koh, S.: Fast Handover Support in Hierarchical Mobile IPv6. IEEE Conference Proceeding, Vol. 2. (2004) 551-554
4. Koodli, R.: Fast Handovers for Mobile IPv6. IETF RFC 4068 (2005)
5. Hsieh et al., R.: S-MIP: A Seamless Handoff Architecture for Mobile IP. in Proc. INFOCOM 2003 (2003)
6. Thomson S., Narten, T.: IPv6 Stateless Address Configuration. IETF RFC 2462 (1998)
7. Narton, T., Nordmark, E., Simpson, W.: Neighbor Discovery for IP Version 6 (IPv6). IETF RFC 2461 (1998)
8. Deering, S., Hinden, R.: Internet Protocol, Version 6 Specification. IETF RFC 2460 (1998)
9. Kent, S., Atkinson,R.: IP Authentication Header. IETF RFC 2402 (1998).

Longest Path First WDM Multicast Protection for Maximum Degree of Sharing*

Hyun Gi Ahn, Tae-Jin Lee**, Min Young Chung, and Hyunseung Choo

Lambda Networking Center
School of Information and Communication Engineering
Sungkyunkwan University
440-746, Suwon, Korea
Tel.: +82-31-290-7145
{puppybit, tjlee, mychung, choo}@ece.skku.ac.kr

Abstract. In this paper, we investigate efficient approaches and algorithms for protecting multicast sessions against any single link failure while establishing multicast sessions in WDM mesh networks. Since a single failure may affect whole nodes in a multicast group and causes severe service disruption and a lot of traffic loss, protecting critical multicast sessions against link failure such as fiber cut becomes important in WDM optical networks. One of the most efficient algorithms is optimal path pair-shared disjoint paths (OPP-SDP). In this algorithm every source-destination (SD) pair has the optimal path pair (working and protection path) between the source and destination node. Since degree of sharing among the paths is essential to reduce the total cost and blocking probability, we propose the longest path first-shared disjoint paths (LPF-SDP) algorithm which decides the priority of selection among SD pairs in a resource-saving manner. Our LPF-SDP is shown to outperform over OPP-SDP in terms of degree of sharing and blocking probability.

1 Introduction

The growth of wavelength division multiplexing (WDM) technology has opened the gate for bandwidth-intensive applications [1]. In addition, the Internet services expand and multicast applications such as video conference, interactive distance learning, and a large-scale online games become more popular [2]-[6]. In high-speed WDM networks it becomes more and more important to protect multicast sessions against various types of failures. Therefore we propose longest path first-shared disjoint paths (LPF-SDP) which determines the order of selecting SD pairs appropriately. Our LPF-SDP algorithm is shown to achieve higher performance compared to the OPP-SDP algorithm. This paper is organized as follows. We propose multicast protection algorithms based on longest path first in Section 2. Performance evaluation of the proposed algorithms is presented in Section 3. Finally, we conclude in Section 4.

---

* This work was supported in parts by Brain Korea 21 and the Ministry of Information and Communication, Korea.

** Corresponding author.

V.N. Alexandrov et al. (Eds.): ICCS 2006, Part I, LNCS 3991, pp. 989–992, 2006.

2 Proposed Multicast Protection Algorithms

In an attempt to improve performance we propose an LPF-SDP heuristic algorithm which determines the priority of selecting SD pairs in a cost-saving manner. In our proposed LPF-SDP, for every destination node of a multicast session, we find an optimal path pair between a source and a destination node successively, and update the link cost along the already-found optimal path pair to zero. Since the cost of already-found optimal path pairs is updated to zero, the probability of sharing the links increases, if the route is the longest one, resulting in reduction of total cost. In this sense we propose the concept of longest path first order when the first SD pair is selected. If the SD pair with the longest path is established as the first path pair, forthcoming working and protection pairs can have more chance to share more links on the longest path. This mechanism decreases the total cost for establishing multicast trees. After the first path pair

```
Input : G = (V, E), S = {s, d_1, ..., d_k}
Output : P(s, d_i)={P_w(s,d_i),P_p(s,d_i)} : OPP between s and d_i, i = 1, ..., k
         P_w(s,d_i) : working path for SD pair between s and d_i
         P_p(s,d_i) : protection path for SD pair between s and d_i
01: Algorithm SPF-SDP(G, S)
02:    P(s, d_i) = FIND_MAX_COST_OPP(G, S)
03:    Update link cost = 0 for all edges of P(s, d_i)
04:    S = S - {d_i}
05:    While (S ≠ {s})
06:       P(s, d_j) = FIND_MIN_COST_OPP(G, S)
07:       Update link cost = 0 for all edges of P(s, d_j)
08:       S = S - {d_j}
09:    Merge OPPs from P(s, d_1) to P(s, d_k) to make multicast trees T_w and T_p
```

Fig. 1. Proposed SPF-SDP algorithm

```
Input : G = (V, E), S = {s, d_1, ..., d_k}
Output : P(s, d_i)={P_w(s,d_i),P_p(s,d_i)} : OPP between s and d_i, i = 1, ..., k
         P_w(s,d_i) : working path for SD pair between s and d_i
         P_p(s,d_i) : protection path for SD pair between s and d_i
01: Algorithm LPF-SDP(G, S)
02:    P(s, d_i) = FIND_MAX_COST_OPP(G, S)
03:    Update link cost = 0 for all edges of P(s, d_i)
04:    S = S - {d_i}
05:    While (S ≠ {s})
06:       P(s, d_j) = FIND_MAX_COST_OPP(G, S)
07:       Update link cost = 0 for all edges of P(s, d_j)
08:       S = S - {d_j}
09:    Merge OPPs from P(s, d_1) to P(s, d_k) to make multicast trees T_w and T_p
```

Fig. 2. Proposed LPF-SDP algorithm

is established, forthcoming working and protection path pairs can be established randomly or sequentially. We call this algorithm The First Longest Path Once-Shared Disjoint Paths (FLPO-SDP). Fig. 1 and Fig. 2 summarize the proposed SPF-SDP and LPF-SDP algorithm.

3   Performance Evaluation

We evaluate and compare the performance of LPF-SDP with that of OPP-SDP, FLPO-SDP, and SPF-SDP in terms of the total network cost on the sample network with 24nodes and 43links. In our simulation, a multicast session of size k is assumed to be established and protected. We repeat the experiment for 10,000 different multicast sessions of the same size k of a multicast group. The size of a multicast group k varies from 1 to 23 (unicast to broadcast).

Fig. 3 shows the average total cost versus session size k as the session size increases in the sample network with 24 nodes and 43 links. Since the number of links in the optimal path pairs increases as the session size increases, the average total cost grows in general as the session size increases. We notice that the LPF-SDP outperforms over the other schemes because this scheme results in the best degree of sharing. The performance of LPF-SDP is about 6% higher compared to the OPP-SDP algorithm.

Fig. 4(a) shows average total cost versus p_e as p_e increases in random networks with 24 nodes. Since the number of links in random networks increases as p_e increases, the average cost decreases in general due to increased connectivity. We also note that LPF-SDP outperforms among the schemes because the LPF-SDP scheme inherits more degree of sharing. The performance of LPF-SDP is about 5.4% higher compared to the OPP-SDP algorithm. And Fig. 4(b) shows the average cost versus session size k as the session size increases in random networks with 24 nodes. Since the number of links in the optimal path pairs increases as the session size increases, the average cost increases as well.

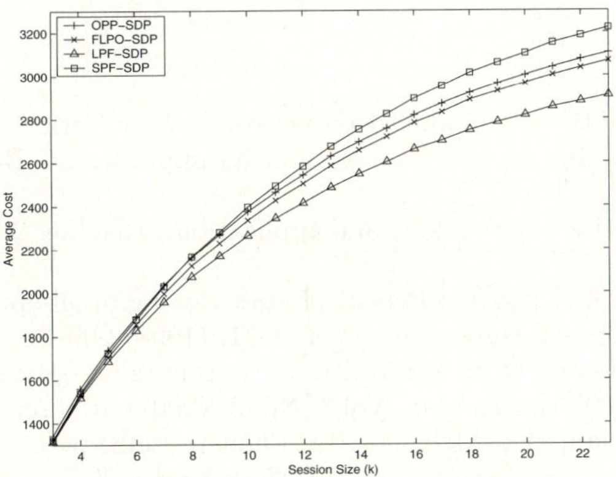

Fig. 3. Average cost versus session size (k) in the sample network

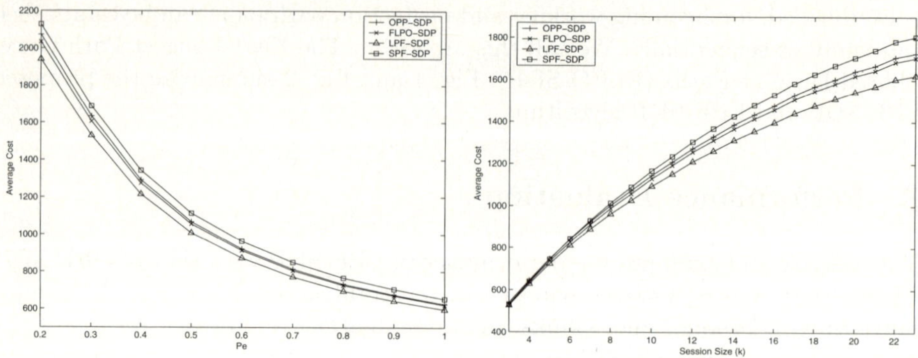

(a) Average cost versus p_e in random networks ($k = 19$)

(b) Average cost versus session size (k) in random networks ($p_e = 0.3$)

Fig. 4. Average total cost of establishing multicast session on random networks

4   Conclusion

In this paper, we have proposed the LPF-SDP algorithm to efficiently solve the multicast protection problem in WDM optical networks. The optimal solution by ILP has very high complexity and requires large computational cost, and OPP-SDP randomly determines the priority of selecting SD pairs. Our proposed LPF-SDP is shown to be an efficient algorithm to determine the priority of selection among SD pairs appropriately. The simulation results both for static multicast sessions and for dynamic ones in the sample network and random networks show that our LPF-SDP yields the least total network cost and the blocking performance than OPP-SDP. The proposed LPF-SDP multicast protection scheme holds the property of resource-saving while providing survivability for high-speed multicast applications in optical WDM networks.

References

1. Mukherjee, B.: Optical communication networks, New York: McGraw Hill. (1997)
2. Paul, S.: Multicasting on the Internet and its applications, Boston, MA: Kluwer. (1998)
3. Miller, C.K.: Multicast networking and applications, Reading, MA: Addison-Wesley. (1999)
4. Malli, R., Zhang, X., Qiao, C.: Benefit of multicasting in all-optical networks, Proc. SPIE Conf. All-Optical Networking, Vol. 2531. (1998) 209–220
5. Sun, Y., Gu, J., Tsang, D. H. K.: Multicast routing in all-optical wavelength routed networks, Optical Networks Mag. Vol.5. No. 3. (2001) 101–109
6. Znati, T., Alrabiah, T., Melhem, R.: Point-to-multi-point path establishment schemes to support multicasting in WDM networks, Proc. the 3rd IFIPWorking Conf. Optical Network Design Modeling, Paris in France. (1999) 456–466

Multi-scale CAFE Modelling for Hot Deformation of Aluminium Alloys

M.F. Abbod[1], I.C. Howard[3], D.A. Linkens[2], and M. Mahfouf[2]

IMMPETUS
Institute for Microstructural and Mechanical Process Engineering, The University of Sheffield
[1] School of Engineering and Design, Brunel University, Uxbridge UB8 3PH, UK
[2] Department of Automatic Control and Systems Engineering,
[3] Department of Mechanical Engineering
University of Sheffield, Sheffield S1 3JD, UK
Maysam.Abbod@brunel.ac.uk

Abstract. The multi-Scale CAFE modelling system utilises Cellular Automata, Finite Elements and a Hybrid Modelling technique which combines neuro-fuzzy models and physical equations to simulate hot deformation of Al-1%Mg aluminium alloys using the commercial finite element software package ABAQUS^{TM}. This paper addresses the issue of capturing microstructural details and providing macro linkage by simulating two phenomena. The first defines a suitable length scale such that numerical models are sufficient in detail and are appropriate in terms of computational time. The second is the feasibility using Cellular Automata (CA) as an additional technique that can be used in conjunction with a conventional Finite Elements (FE) representation to model material heterogeneity and related properties. This is done by identifying an abstract scale in between the micro and macro scales, termed the "mesoscale" to obtain a multi-scale CAFE modelling technique that utilises the CA technique to represent initial and evolving microstructural features at an appropriate length obtained using an overlying FE mesh.

1   Introduction

During hot deformation of aluminium alloys, rolling will change particularly the deformation texture and recrystallisation texture. Since aluminium and its alloys have high stacking-fault energy, hot deformation will not provide mechanical twining. Incorporating microstructural details in thermomechanical processing models is well recognised by many researchers [4], [6].

Finite elements (FE) method can be used with numerical formulations that describe the behaviour of different material models to elicit the response of a structure to strain. The material microstructure model can be expressed mathematically by a implicit semi-empirical methodology which is based on the use of trigonometric functions that relate a change in equivalent strain to a change of equivalent stress. FE models based on this approach [3] are successful in predicting the macro behaviour of metal flow stress and load, but fail to model the evolution of the underlying microstructure. In addition, constants used in these formulations should be employed with caution outside the conditions beyond which they have been tested.

V.N. Alexandrov et al. (Eds.): ICCS 2006, Part I, LNCS 3991, pp. 993–996, 2006.

The microstructural evolution and its effects during the rolling process can be modelled using physically-based models. These relationships relate the evolution of the microstructal variable to a change in the equivalent strain [5]. Conventional FE models based on this approach use a statistically-averaged value to represent the evolution of an internal variable that is linked to each finite element [2]. The local variations, either present initially or evolving during the process itself, can be monitored individually.

An alternative method is to model the microstructure explicitly. This methodology requires the reduction of the size of each finite element to the size of a single grain or even lower which computationally extensive. This paper introduces a novel framework which couples hybrid modelling of thermomechanical processing [7], the mathematical tool of Cellular Automata within the FE code to form a Cellular Automata-based Finite Element (CAFE) model. Results for applications of the augmented CAFE model as applicable to hot deformation are presented.

2   The Multi-scale Microstructure CAFE Model

The CAFE model [1] is a framework to capture different strata of initial and evolving microstructures during deformation through a materials-mechanics formulism. CAFE is based on three steps: 1) approximation of relevant microstructure, 2) distribution of macro variables on the meso domain and 3) averaging meso-variables to the macro domain.

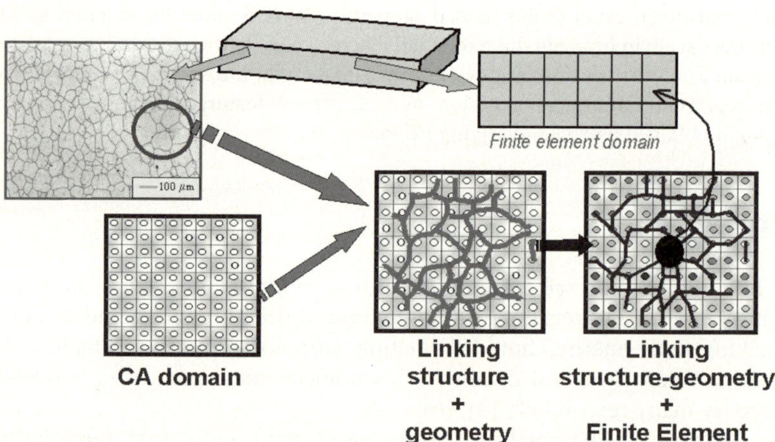

Fig. 1. Linking Structure and Geometry to incorporate microstructural details into geometric space of the Finite Element domain

CAFE is based on the structure shown in Fig. 1 which illustrates the methodology of linking the microstructure to the Finite Element domain using CA. The material slab under investigation is discretised into 18 rectangular finite elements. Each finite element cell simulates the microstructure of aluminium alloys and comprises equiaxed grains of the order of 80~100 microns. The spatial information is mapped onto the CA domain, the size of which is determined by the size of each CA cell (marked by open circles). Finally, the CA domain is linked to the integration point (black circle in figure) of each finite element. The next step in the CAFE methodology is to transfer the independent macro level variables into the meso level CA variables. These macro variables are supplied by the overlying FE either at its nodes, centroids or Gaussian integration points.

To start the simulation of a microstructure, grain nuclei are randomly scattered over the domain. A second-level array of CA cells of $0.1\sim5\mu$m size represent these nuclei. All nuclei can be assumed to be present at the start of the simulation as is the case with site-saturated nucleation or they can continue to appear during the simulation as a function of the deformation variables. The present work employs the first approach and uses a random function to generate and locate these nuclei on the domain. All first-level CA cells except those occupied by the nuclei have a "zero" orientation that represents the matrix within which the grain can grow. The presence or absence of nuclei is an internal variable for the first-level CA cell. Other internal variables are the spatial location of the CA cell and the orientation of the microstructure it will represent. This form of linking a microstructural entity to a spatial geometric entity allows embedding different structure-geometry relationships in a multi-level CA.

In the initial simulation validation stage, a Plane Strain Compression (PSC) model was simulated. Both the specimen and tool are modelled using 4-noded quadrilateral elements. All the stock elements are initialised with a starting temperature of 400°C and the tool elements are at 390°C. Friction at the tool-stock interface is modelled using the Amonton-Coulomb law with a constant coefficient of friction of 0.1. The heat transfer coefficient at the interface is 80 kW/m²K. The instantaneous tool velocity changes with the instantaneous specimen height and is controlled to achieve a constant nominal strain-rate of 3 s^{-1}. The stock is thickness-reduced by 41% during the deformation. The simulation begins with the introduction of a representative microstructure into each finite element. Results for the von-Mises stress and the strain are shown in Fig. 2 which gives the advantages of showing the microstructure of the material as well as the grains structure.

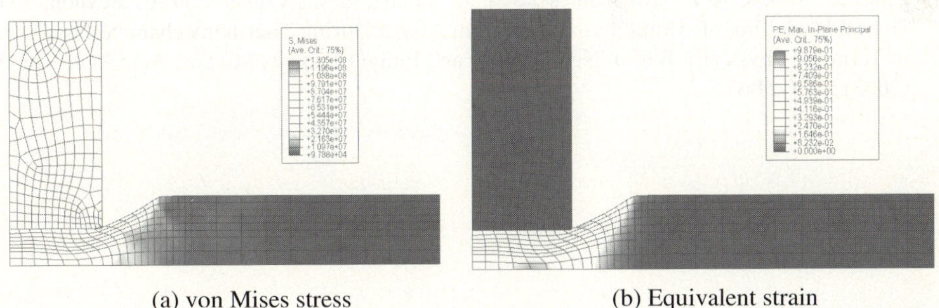

(a) von Mises stress   (b) Equivalent strain

Fig. 2. The multi-scale CAFE simulation for PSC test of commercial purity aluminium alloy, T$_{strip}$=400°C, T$_{tool}$=390°C

3 Conclusions

This paper presents a generic method for modelling the microstructure within the continuum formulation of finite element structural modelling. The Multi-Scale CAFE framework was applied to model the behaviour of the materials during hot deformation using a hybrid modelling technique. The model has simulated the microstructure of the material during hot deformation. The material behaviour was developed from the physically-based relations that relate the stress evolution to the total dislocation density using the CAFE approach. Recrystallisation behaviour was modelled using a critical dislocation density and subgrain size at a CA level.

Acknowledgements

The authors gratefully acknowledge the UK EPSRC (Engineering and Physical Sciences Research Council) for their financial support under grant number GR/R70514/01.

References

1. Beynon, J.H., Das, S., Howard, I.C., Palmiere, E.J., Chterenlikht, A.: The Combination of Cellular Automata and Finite Elements for the Study of Fracture: The CAFE Model of Fracture. 14th European Conference on Fracture - ECF 14, Cracow, Poland (2002).
2. Busso, E.P.: A Continuum Theory for Dynamic Recrystallisation with Icrostructure-Related Length Scales. International Journal of Plasticity, Vol. 14, No. 4-5 (1998) 319-353.
3. Dutta, K.: Finite Element Modelling of Hot Rolling. PhD Thesis. The University of Sheffield (1996).
4. Jonas, J.J., Sellars, C.M., McG Tegart, W.J.: Strength and Structure Under Hot Working Conditions. Metallurgical Reviews, Vol. 130 (1969) 1–24.
5. Marthinsen, K., Nes, E.: A General Model for Metal Plasticity. Material Science and Engineering. Vol. A234-236 (1997) 1095-1098.
6. Sellars, C.M., Whiteman, J.A.: Recrystallisation and Grain Growth in Hot Rolling. Metal Science, (1979) 187-194.
7. Zhu, Q., Abbod, M.F., Talamantes-Silva, J., Sellars, C.M., Linkens, D.A., Beynon, J.H.: Hybrid Modelling of Aluminium-Magnesium Alloys During Thermomechanical Processing in Terms of Physically-Based, Neuro-Fuzzy and Finite Elements Models. Acta Mat, Vol. 51 (2003) 5051–5062.

Construction of Small World Networks Based on K-Means Clustering Analysis

Jianyu Li[1], Rui Lv[1], Zhanxin Yang[1], Shuzhong Yang[2],
Hongwei Mo[3], and Xianglin Huang[1]

[1] School of Computer Science and Software,
Communication University of China, Beijing, China
`{lijianyu, lvrui, yangzx, huangxianglin }@cuc.edu.cn`
[2] School of Computer and Information Technology, Beijing
Jiaotong University, 100044 Beijing, China
`yangshuzhong@163.com`
[3] School of Automation, Harbin Engineering University, 150001 Harbin, China

Abstract. In this paper we present a new method to create small world networks based on K-means clustering analysis. Because of the close relationship between the small world networks and the data with clustering characteristics, the resulting networks based on K-means method have many properties of small world networks including small average distance, right skewed degree distribution, and the clustering effect. Moreover the constructing process also has shown some behaviors including networks formation and evolution of small world networks.

1   Introduction

In recent years, the discovery of small-world, scale-free and community properties of many natural and artificial complex networks has stimulated a great deal of interest in studying the underlying organizing principles of various complex networks, which has led to dramatic advances in this emerging and active field of research [1, 2, 3, 4, 5].

Considering the features of complex networks, a novel idea about how to construct the networks is to find the relationship between networks and data with clustering features. Unlike the model of Watts and Strogatz [1], we find an alternate route to generate the complex networks, especially to construct the small-world networks based on k-mean cluster analysis.

Many data sets from nature and society have the clustering characteristics. Like most data sets, most real networks also have the clustering property, especially in social networks [1]. In social networks, there are usually some groups of nodes (also called communities or modules), where nodes in each group are more likely connected with each other than to the rest of the networks [6]. One question could be given: "Could clustering analysis be used to construct relationships between them? Could networks be generated using data cluster analysis? From the above analysis, cluster analysis method (K-means method) could serve as a bridge exploring the relationship between them, and construct the networks based on the given data.

V.N. Alexandrov et al. (Eds.): ICCS 2006, Part I, LNCS 3991, pp. 997–1000, 2006.

2   The New Method Based on KM Algorithm

Procedure of our method based on k-means algorithm is given as below:

1. Place k points into the space represented by the objects that are being clustered. These points represent initial group centroids;
2. Assign each point to the group that has the closest centroid;
3. When all points have been assigned, recalculate the positions of the k centroids. Find the points nearest to the centroids and connect the nearest points obtained at the ith and the $(i+1)th$ step in the same cluster. The measurement used here is Euclidean distance;
4. Repeat Steps 2 and 3 until the centroids no longer move. This produces a separation of all the points into groups from which the metric to be minimized can be calculated;
5. Repeat Steps 1, 2, 3 and 4 until all the possible and different initial C_n^k cases are carried out.

Remark 1. There exist many factors, which affect the clustering analysis results such as number of clusters, the algorithm (k-means or fuzzy k-means method may be used), the data distribution, and the measurement of the data. Since the resulting complex networks are generated by the K-means method, the structure and topology of the networks will be affected by these factors.

3   Properties of the Resulting Networks

The properties of the clustering path
In order to analyze the networks in detail, all the clustering paths are to be divided into two classes. One is called local clustering path whose vertices belong to the same cluster, the other is called global clustering path whose vertices come from different clusters, see Fig.1. Generally the local paths are relatively short and won't fluctuate remarkably, at the end it terminates near the same cluster centroid. The global paths are long and they will go through the points which belong to other clusters, but later it will become stable and stop near one cluster centroid. From above analysis, we can conclude the paths show clustering properties. In any cases, because of the feature of K-means method, all paths avoid the long range connections and the global paths play the role of shortcuts to connect different clusters.

Statistics properties of the networks
In this part we will investigate properties of the resulting networks. We quantify the structural properties of these networks by their characteristic path length L, clustering coefficient C, and degree distribution $P(k)$. The characteristic path length, L, is the path length averaged over all pairs of nodes. The path length $d(i,j)$ is the number of edges in the shortest path between nodes i and j. The clustering coefficient is a measure of the cliqueness of the local neighborhoods. For a node with k neighbors, then at most $k(k-1)/2$ edges can exist among

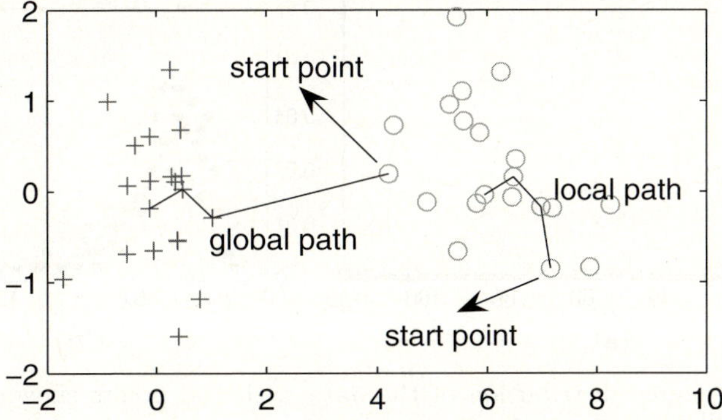

Fig. 1. The resulting clustering paths. Here the data consist of two classes of points which are represented by the plus (+) and circle (o) sign.

them. The clustering of a node is the fraction of these allowable edges that occur. The clustering coefficient, C, is the average clustering over all the nodes in the graph. The degree of a vertex in a network is the number of edges incident on (i.e., connected to) that vertex. We define $P(k)$ to be the fraction of vertices in the network that have degree k.

Example 1

The data (x, y) was generated from two bivariate independent Gaussian distributions (X_1, X_2), $X_i \sim N(\mu_i, \sigma_i)$ and (Y_1, Y_2), $Y_i \sim N(\mu_i, \sigma_i), i = 1, 2$ with different means $\mu_1 = 0, \mu_2 = 3$, and same variances $\sigma_1 = \sigma_2 = 1$. Half of the data was generated from (X_1, X_2), (Y_1, Y_2) respectively.Hence the number of clusters is two.

The average path length L of the resulting networks is close to that of a random graph with the same size and average degree, 2.4 compared with 3.07, but its clustering coefficient C is more than 20 times higher than a random graph, 0.4005 vs. 0.0175. From the analysis, the resulting networks show small world feature.

In Fig.2(a), we show the networks' degree distributions. The degree distributions of the networks are often highly skewed and differs from the Poisson distribution. It's at the border between Poisson and power law distribution.

Example 2

The Data (x_1, x_2) was generated from independent bivariate Gaussian distributions (X_1, X_2), $X_i \sim N(\mu, \sigma)$ with different means $\mu = 5$ and variances $\sigma = 1$, and independent bivariate standard Cauchy distributions (Y_1, Y_2), the probability density function of is

$$\frac{1}{\pi \gamma [1 + (\frac{x - x_0}{\gamma})^2]}$$

where $x_0 = 0, \gamma = 1$. The data has 400 points, and 200 points was obtained from $(X_1, X_2), (Y_1, Y_2)$ respectively.

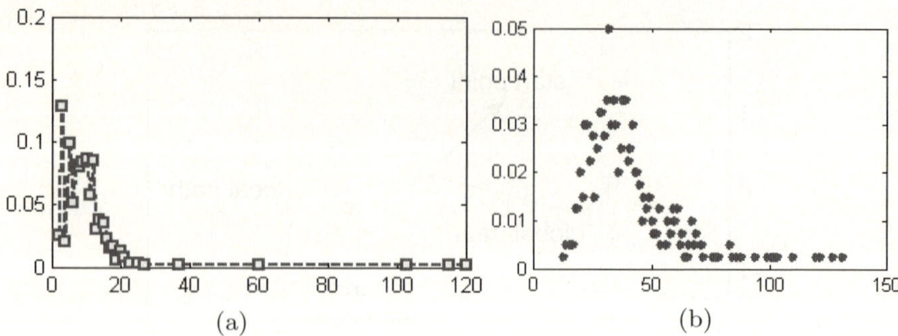

(a)   (b)

Fig. 2. (a)The degree distribution of the data with 600 points shown in example 1, (b)The degree distribution of the data with 600 points shown in example 2

This resulting graph has a small world structure, with: $C = 0.5004 >> C_{rand} = 0.04$, and $L = 2.9270$ and $L_{rand} = 3.3219$ are very close.

The degree distribution in Fig.2(b) showed Poisson distribution feature with a long tail.

4   Conclusion and Research in the Future

In this paper we presented the method based on k-means clustering analysis to create the networks. Although it is easy to understand the method, the idea is novel. We use the clustering thought which may be the nature of the universe and construct the relationships between data and networks. The experiments show that the generating networks has many properties which distinguishes itself from other existing complex networks. The generating networks can describe many phenomena of complex networks such as small world effect, and the evolution behavior. Moreover, a good and proper explanation for the formation of the small world networks topology is presented due to the given data's clustering feature. Finally, in future work, the factors that influence the formation and topology of the resulting networks will be discussed.

References

1. D. J. Watts et al: Collective dynamics of 'small-world' networks. Nature 393, 440-442 (1998).
2. S. N. Dorogovtsev et al: Evolution of networks. Advances in Physics 51, 1079-1187.
3. M. E. J. Newman: The structure and function of complex networks. SIAM Review 45, 167-256 (2003).
4. A.L. Barabási and R. Albert: Emergence of scaling in random networks. Science , 286, 509-512 (1999).
5. A. R. Barabási, A.L. and H. Jeong: Scale-free characteristics of random networks: The topology of the World Wide Web. Physica A 281, 69-77 (2000).
6. M. E. J. Newman and M. Girvan: Finding and evaluating community structure in networks. Phys. Rev. E 69:026113 (2004).

Spatiotemporal Data Mining with Cellular Automata

Karl Fu and Yang Cai

Carnegie Mellon University, Visual Intelligence Studios,
Cylab, CIC-2218, 4720 Forbes Ave.
Pittsburgh, PA 15213 USA
ycai@cmu.edu

Abstract. In this paper, we describe a cellular automata model for predicting biological spatiotemporal dynamics in an imagery data flow. The Bayesian probability-based algorithm is used to estimate the algal formation in a two-dimensional space. The dynamics of the cellular artificial life is described with diffusion, transport, collision and deformation. We tested the model with the historical data, including parameters, such as time, position and temperature.

1 Introduction

Spatiotemporal data are widely used in remote sensing community. The imagery data usually contain multi-spectrum information such as visible, infrared, UV, or micro-wave signals. Scientists need to find out the physical properties behind the data flow and predict the future trends such as global warming, flood, harmful algae blooms or river plumes. However, in many cases, the inverse physics process is complicated. Heuristics has to be used for approximate estimations.

Cellular Automata (CA) has been used for modeling artificial lives [2-13], as well as, at the large scale, urban and economic dynamics [5]. Furthermore, CA has been employed in simulating and verifying various types of physical interactions such as the expansion patterns of different lattice gases, surface tension of liquids and diffu-sion-limited aggregation [13]. Although CA's rules are simple and local, it can simu-late complex spatiotemporal interactions that are challenging to other computational methods. In this paper, we present a two-dimensional cellular artificial life model for predicting the spatiotemporal dynamics in an imagery flow. The model consists of a set of parameters and rules that control spatiotemporal dynamics in form of a shape, such as diffusion, transport, collision. Similar to the conventional cellular automata, artificial lives in this model interact with their neighbors *individually*. However, they form a coherent shape as a *lump*.

2 Formation of the Artificial Life

Given historical spatiotemporal data of the location of the alga, it is possible to model the formation of the artificial life in a cellular grid. In this model, we consider the location of the algae, time and physical parameters, such as temperature and wind.

V.N. Alexandrov et al. (Eds.): ICCS 2006, Part I, LNCS 3991, pp. 1001 – 1004, 2006.

For historical images I_j where $j \in \{1,2,3,...,i\}$, $I_j(x, y)$ is a logical '1' if the pixel is occupied by algae, and a logical '0' if it's not occupied by an alga where $x \in \{1,2,3,...,a\}$, $y \in \{1,2,3,...,b\}$. The total number of images given is i, the horizontal dimensions of the image is denoted a, and the vertical dimensions of the image is denoted b. To predict the location of the algal, evidence such as the time of the year, or the surface temperature are given as condition. Let e be the condition given (i.e. temperature and/or time) and I_j^e be the predicted image given e as condition:

$$P(I_j(x, y) \mid e) = P(e \mid I_j(x, y)) * P(I_j(x, y)) / P(e)$$

$$I_j^e(x, y) = (P(I_j(x, y) \mid e) \geq \alpha ,$$

where α is a threshold for when $I_j^e(x, y)$ is inhabited by algae.

The probabilities under those conditions/evidences, such as time and temperature, can be calculated. The Figure 1 shows examples of the estimated target locations.

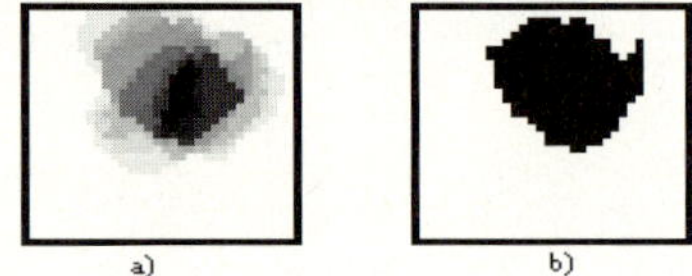
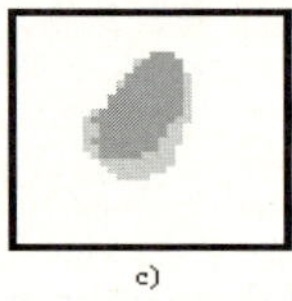
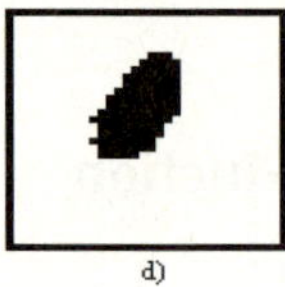

Fig. 1. a) Probability of location being inhabited by algae using a single input temperature =15°C. b) predicted image of using 15°C as input $\alpha = .3$, c) probability of location being inhabited by algae using a single input time = January, d) predicted image of using January as input.

Prediction using one input becomes much more inaccurate during months with cold temperature because during certain time of the year no alga is detected when the temperature falls below 15°C. In Figure 1, a) and c) are poor probability models because of lack of evidences for the Bayesian model to make an accurate prediction; this is indicated by the large number of cells in each image with very low probability (light colored cells).

3 Dynamics of the Artificial Life

The growth diffusion function changes the object shape by a predefined structuring element. The dilation or erosion morphology is based on the assumption of a glass-like surface where no friction exists. They can only deform the outline of a shape *uniformly* based on the population. It doesn't reflect the intrinsic or external non-linear or random factors that contribute to the shape process. To simulate the non-uniform shape growth, we investigate the percolation clustering [18-19]. Figure 2 shows an example of the percolation process started from a given shape.

Algae's growths and movements are often affected by environmental factors such as external forces: winds and currents, and the presence of obstacles. A collision causes deformation. The level of a deformation depends on the elasticity of an object. For a

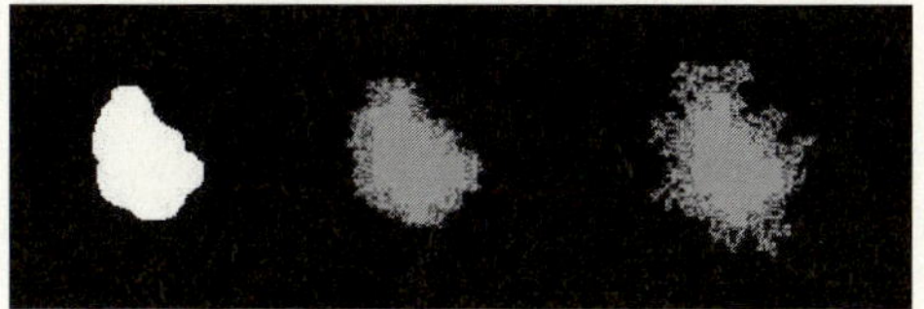

Fig. 2. Diffusion with resistance

rigid object, it will stop at the collision contact point and keep the same shape. We classify objects as rigid, elastic and flexible. Rigid objects don't deform after the impact. Flexible objects, on the other hand, completely deform along the outline of the collided hard objects. The two extreme cases are rather easy to describe. Assume we have an elastic object and a rigid object, we use the following rules to model the deformation: 1) Detect the collision points; 2) Move a portion of the collided cells to 'sideways', the ratio is proportion to the elasticity of the object. The bigger elasticity, the more cells move sideways. 3) Continue to deform until the motion stops.

4 Results

We verify the algorithm with a case study of predicting the movement of harmful algal bloom (HAB) of *Karena Brevis*. In this case, the HABs are 'lumps', rather than individual cells. The input data are simulated satellite images and cell counts. The predicted results from the model are compared to the actual images under the same conditions, such as time and temperature. The error of the 30 trials is within 5.40%. Without using both variables, the error is 8.24%. While using both variables, the error is only 4.68%. With increasing amount of evidences, the prediction accuracy increases substantially.

5 Conclusions

In this paper, we use the cellular artificial life to anticipate complex spatiotemporal phenomena of algal blooms, including formation, diffusion, collision, transport, and deformation. By combining the statistical model with cellular automata, we found the new approach yields decent accuracy to predict the algal location and the shape dynamics. However, the predictability of the hybrid model depends on the historical data and its fitness to its complex environment.

Acknowledgement

This study is supported by NASA ESTO grant AIST-QRS-04-3031. We are indebted to our collaborators in NOAA, Dr. Richard Stumpf and Mr. Timothy Wynne for their expertise in oceanography and databases. The authors appreciate the comments and suggestions from Dr. Karen Meo, Dr. Kai-Dee Chu, Dr. Steven Smith, Dr. Gene Carl Feldman and Dr. James Acker from NASA. Also, many thanks to Professor Christos Faloulus, graduate student Daniel Chung of Carnegie Mellon University for their input.

References

[1] Lewin, K. Field Theory in Social Science. Harper and Row, 1951

[2] von Neumann, John, 1966, *The Theory of Self-reproducing Automata*, A. Burks, ed., Univ. of Illinois Press, Urbana, IL.

[3] Gardtner, M. 1970. Mathematical game. The fantastic combinations of John Conroy's new solitaire game "life". Scientific American vol. 223, pp. 120–120.

[4] Margolus, N. and Toffoli, T. Cellular Automata Machines: A New Environment for Modeling. The MIT Press, 1987.

[5] Yeh, A. G. O. and Li, X. *Urban Simulation Using Neural Networks and Cellular Automata for Land Use Planning*. Symposium on Geospatial Theory, Processing and Applications, Ottawa 2002.

[6] Wolfram, S. Statistical Mechanics of Cellular Automata. *Rev. Mod. Phys.* 55, 601-644, 1983.

[7] Wolfram, S. Universality and Complexity in Cellular Automata. *Physica D* 10, 1-35, 1984.

[8] Wolfram, S. Twenty Problems in the Theory of Cellular Automata. *Physica Scripta* T9, 170-183, 1985.

[9] Wolfram, S. *A New Kind of Science*. Champaign, IL: Wolfram Media, pp. 60-70 and 886, 2002.

[10] Wolfram, S. Cellular automata as models for complexity, *Nature* 311:419 (1984).

[11] Wolfram, S. Statistical mechanics of cellular automata, *Rev. Mod. Phys.* 55:601 (1983).

[12] Vichniac, G. Simulating physics with cellular automata, *Physica* 10D:96 (1984).

[13] Hardy, J., O. de Pazzis, and Y. Pomeau, Molecular dynamics of a classical lattice gas: transport properties and time correlation functions, *Phys. Rev.* A13:1949 (1976).

[14] Leyton, M. Symmetry, Causality, Mind, MIT Press, 1999

[15] Sonika, M. etc. Image Processing, Analysis, and Machine Vision, PWS Publishing, 1999

[16] Zhou, S., Chellappa, R. and Moghaddam, B. Visual tracking and recognition using apprearance-adaptive models in particle filters. IEEE Trans. On Image Processing, Vol. 13, No. 11, pp. 1491-1506, 2004

[17] Bandini, S. and Pavesi, G. Controlled generation of two-dimensional patterns based on stochastic cellular automata, Future Generation Computer Systems, vol. 18 (2002) 973-981

[18] Stauffer, D., Aharony, A. Introduction to percolation Theory, Taylor & Francis, London, 1992

[19] Essam, J.W. Percolation theory, *Rep. Prog. Phys.* 43:833 (1980)

MicroCASim: An Automata Network Simulator Applied to the Competition Between Microparasites and Host Immune Response

Luciana B. Furlan[1,2], Henrique F. Gagliardi[2,3], Fabrício A.B. da Silva[3], Ivan T. Pisa[1], and Domingos Alves[2]

[1] Departamento de Informática em Saúde (DIS), UNIFESP, São Paulo, SP, Brazil
lubenzoni@yahoo.com.br, ivanpisa@dis.epm.br
[2] Laboratório de Computação Científica Aplicada à Saúde Coletiva (LCCASC),
UNISANTOS, Santos, SP, Brazil
henrique.gagliardi@gmail.com, quiron@unisantos.br
[3] Programa de Mestrado em Informática, UNISANTOS, Santos, SP, Brazil
fabricio@unisantos.br

Abstract. In this work we had developed an alternative model framework to study the concept of immunological memory by proposing as a simplified model for the interaction dynamics between a population of pathogens and the immunological system. The model is based on a probabilistic cellular automata which allows relatively easy inclusion of some of the real immunological systems in a transparently biological manner. The resulting virtual laboratory is called Microscopic Cellular Automata Simulator (*MicroCASim*), a software whose basic idea is to create a tool to assist the visualization of the interaction dynamics of these entities *in silico*.

1 Introduction

The antigen-specific mechanisms of lymphocytes (T cells and B cells) are the most advanced and most precise mechanism of host defense. These cells are also responsible for the development of immunological memory, a hallmark of the adaptive immune response. This memory, which stores its encounters with invaders, enables humans to rapidly clear, or even prevent altogether, infection by pathogens which they have been previously infected [1]. Whether such immunity is maintained through constant exposure to infection via long-lived clones of lymphocytes that are able to recognize specific antigens and maintain antibody production in the absence of repeated exposure, or via the persistence of the microparasite at low levels of abundance within the host, remains unclear stil [1].

In the complex area of the interacting immune system, when there is potentially insufficient information available to construct a detailed model, cellular automata can be used [1-3]. There have been few models using the concept of cellular automata, in which the body is depicted as a grid [3,4]. We address this issue by describing in this work the implementation of Microscopic Cellular Automata Simulator (MicroCASim), a software to simulate the competition between the population of *effectors* cells (T and B lymphocytes) and the population of pathogens, based on a generalized

V.N. Alexandrov et al. (Eds.): ICCS 2006, Part I, LNCS 3991, pp. 1005–1008, 2006.

probabilistic cellular automata [4]. The remainder of the paper is organized as follows. In the next Section we present the set of cellular automata rules that govern the interaction of the populations of effectors cells and pathogens. The MicroCASim itself is presented in the Section 3. Finally, some concluding remarks are stated in Section 4.

2   The Alternative Probabilistic Cellular Automata Model

Consider a discrete dynamical system where a population of N entities is distributed on the sites of a bi-dimensional square lattice (the virtual lymph node) $M = m_{ij}$ (where i and j may vary from (1,L) for L, $N = L$ x L). Each individual site is assigned to receive three specific attributes: (1) a spatial address or lattice position (i,j); (2) a set of possible *occupation states* where each site is either empty or occupied by a T cell, a B cell and a pathogen (Fig. 1), and finally (3) an period τ_i, specifying the number of units of time an entity of type i can die.

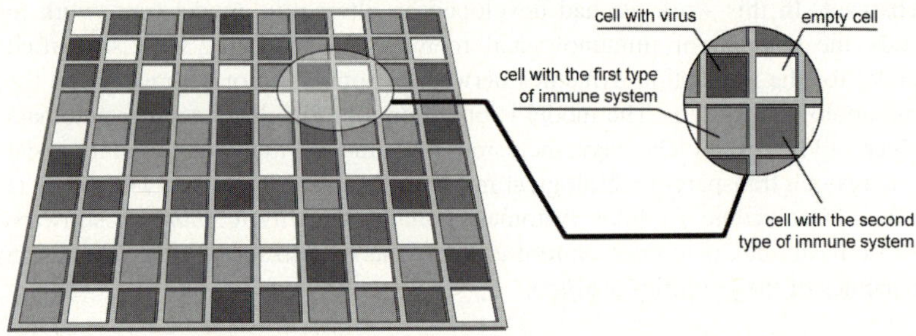

Fig. 1. Schematic representation of the virtual lymph grid

The dynamics of the system is modeled by three main features: the mobility of its inner elements, the competition between its elements and its reproduction along the time. Mobility is modeled as a diffusion process using the Tofoli-Margolus scheme [5]. We use two diffusion steps within each simulation time step that are defined as follows: the lattice is divided in blocks of 2x2 cells and each block has a p_{RE} probability to rotate $90°$ and p_T to translate.

The competition is modeled by a predator-pray mechanism of interaction which the effectors cells can kill each virus with an estimated probability when a local contact is established with the pathogen on its neighborhood. In equation 1, we can see the definition of the probability where a virus occupying a site being eliminated due the presence of n_1 effectors cells from type 1 and n_2 from type 2 in its neighborhood:

$$Pc = 1 - (1 - \lambda_1)^{n_1} * (1 - \lambda_2)^{n_2} \tag{1}$$

which λ_i is the probability of a type i immune system cell to kill a virus cell and $1 - \lambda_i$ being the probability of this type of cell to do not kill a virus cell when rounded by n_i cells of type i. Therefore, the defined probability in equation 1 is the probability that both immune system cells to eliminate the virus. Schematically, this mechanism works as follows:

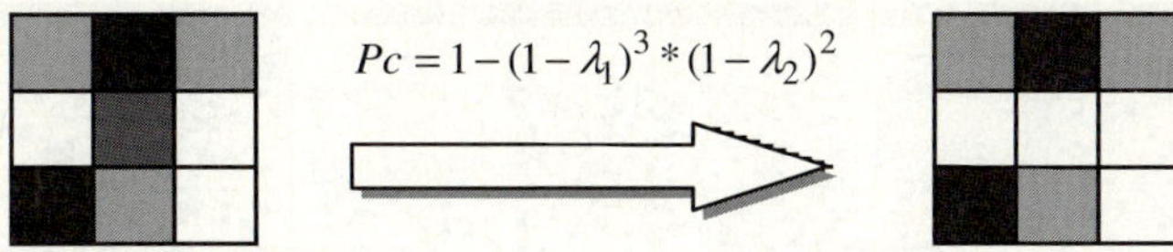

$$Pc = 1 - (1 - \lambda_1)^3 * (1 - \lambda_2)^2$$

Fig. 2. In the figure we observe that *central cell* changed from a *vírus* to an *empty cell*. The *empty cell* is represented by *white spaces*. Here we have $n_1=3$ and $n_2=2$.

For effectors cells, first we ensure that a constant amount of cells will enter the system after a particular instant of time τ that the first virus has entered in the system and replicated itself. Here, the effectors cells of type i depend on the number of viruses entered the system. The same kind of dependency also occurs for the effectors cells of the type j that depends on the number of effectors cells of type i to enter on the system. After that, this procedure is modified to allow that entrance of effectors cells is proportional to the quantity of virus in the system, i.e.:

$$p_{Ci} = \frac{\rho_1 \left(N_v + N_{Cj}\right)}{N} \tag{2}$$

whose ρ_i is a parameter related to the probability of the immune system recognize the virus (eq. 2). The replication can be divided into two phases: virus and immune system replication, which replicates it selves by its own way. Every element of the system has its own time life.

3   The *MicroCASim* Software System

The MicroCASim software system is a simulation environment for the study and analysis of these models with their own probabilistic rules. The simulation was implemented using object oriented methodology and the C++ programming language was chosen for this. The system architecture is composed of four distinct modules: specification, simulation, visualization and analysis. In the specification module the user can configure all the parameters of each available model through a setup window. At each simulation update, the replication, competition and mobility mechanisms of elements are executed, respectively and the resulting data is sent to the visualization module to display the simulation process to the user (Fig. 3).

The visualization module offers also controls to adjust the velocity, simulation animation settings, and a graphic window to visualize simulation current status. Particularly, the Fig. 3 displays a very interesting simulation scenario. We can observe the model with action of a regulatory network with two kinds of immune system cells interacting and a replicating antigen. Its worth to note that there is a stationary behavior due to the coexistence between the B and T cells, which produces in the system an immunological memory effect.

Thus, if the same virus infects again the system, its response will be faster than the previous one, eliminating the menace in a short time. Furthermore, through the analysis module we are able to see the average progression of the infection and the immunological response by generating series of stochastic realizations. Hence, it is possible to sweep all the space of parameters of the present models.

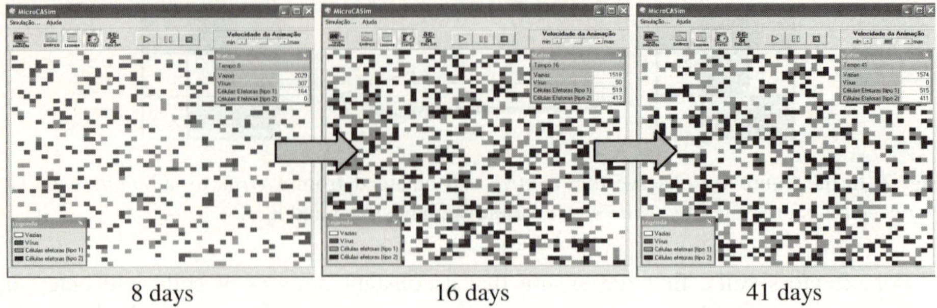

8 days   16 days   41 days

Fig. 3. A model simulation: the model with 3 types of cells

4   Conclusions

The most important feature of the model presented here is the explicitness of individual contact process and mixing. This model is a good platform where to start further expansions like incorporation of various types of interacting immunological entities. Furthermore, owing to its extreme simplicity, this formulation may be useful, in the sense of having the value of an approximation, to tackle problems in immunology.

Acknowledgements

This work was supported by (FAPESP: Proc. 02/03564-8).

References

1. Anderson, R.M., May, R.M.: Infectious Diseases of Humans. Oxford, U.K.: Oxford University Press (1991).
2. Seiden, P.E., Celada, F.: A model for simulating cognate recognition and response in the immune system. J. Theor. Biol., Vol. 158. (1992) 329-357.
3. Celada, F., Seiden, P.E.: A computer model of cellular interaction in the immune system. Immunology Today, Vol. 13. (1992) 56-62.
4. Alves, D., Hass, V., Caliri, A.: The predictive power of R_0 in an epidemic probabilistic model. Journal of Biological Physics, Vol.29. 1 (2003) 63.
5. Tofoli, T., Margolus, N.: Cellular Automata Machines: A New Environment for Modeling. MIT Press (1987).

Simulated Annealing: A Monte Carlo Method for GPS Surveying

Stefka Fidanova

IPP – BAS, Acad. G. Bonchev str. bl.25A, 1113 Sofia, Bulgaria
stefka@parallel.bas.bg

Abstract. This paper describes simulated annealing technique,which is a Monte Carlo method, to analyze and improve the efficiency of the design of Global Positioning System (GPS) surveying networks. The paper proposes various local search procedures which can be coupled with the simulated annealing technique.

1   Introduction

The GPS is a satellite-based navigation system that permit users to determine their three-dimensional position. Sequencing play a crucial role in GPS management. In the last decades simulated annealing is considered as a heuristic method that uses a Monte Carlo global minimization technique for minimizing multi-variance functions [2]. The concept is based on the manner in which liquids freeze or metal recrystallize in the process of annealing. When related to the positioning the process of designing a GPS surveying network can be described as follows. A number of receivers r are placed at stations n and take simultaneous measurements from satellites. The receivers are then moved to other stations for further measurements. The problem is to search for a best order in which to consecutively observe these sessions to have the best schedule at minimum cost (time) i.e.: $Minimize: \quad C(V) = \sum_{p \in R} C(S_p)$, where: $C(V)$ is the total cost of a feasible schedule V; S_p is the rout of the receiver p in a schedule; R is the set of receivers.

To maximize the benefit to use simulated annealing, different Local Search (LS) procedures have been coupled with this technique to improve the performance and explore the search space more effectively. The rest of this paper is organized as follows. Section 2 outlines the simulated annealing method applied to GPS surveying networks. The search strategy of the local search is explained in Section 3 and followed by different local search procedures. Several case studies and the obtained numerical results are reported in Section 4.

2   Simulated Annealing

Simulated annealing is a flexible and robust technique which derived from physical science. This method has proved to be a local search method and can be

V.N. Alexandrov et al. (Eds.): ICCS 2006, Part I, LNCS 3991, pp. 1009–1012, 2006.

successfully applied to the majority of real-life problems [1, 2, 3]. The algorithm starts by generating an initial solution and by initializing the so-called temperature parameter T. The temperature is decreased during the search process, thus at the beginning of the search the probability of accepting uphill moves is high and it gradually decreases. Initial solution is a random solution. The structure of the simulated annealing algorithm is shown below:

Step1.
$$\begin{cases} \text{Initializing the cooling parameters:} \\ \text{Set the initial starting value of the temperature parameter, } T > 0; \\ \text{Set the temperature length, } L; \\ \text{Set the cooling ratio, } F; \\ \text{Set the number of iterations, } K = 0 \end{cases}$$

Step2.
$$\begin{cases} \text{Select a neighbor } V' \text{ of } V \text{ where } V' \in I(V) \\ \text{Let } C(V') = \text{the cost of the schedule } V' \\ \text{Compute the move value } \Delta = C(V') - C(V) \end{cases}$$

Step3.
$$\begin{cases} \text{If } \Delta \leq 0 \text{ accept } V' \text{ as a new solution and set } V = V' \\ \text{ELSE} \\ \text{IF } e^{\Delta/T} > \Theta \text{ set } V = V', \\ \text{where } \Theta \text{ is a uniform random number } 0 < \Theta < 1 \\ \text{OTHERWISE retain the current solution } V. \end{cases}$$

Step4.
$$\begin{cases} \text{Updating the annealing parameters using the} \\ \text{cooling schedule } T_{k+1} = F * T_k \ \ k = \{1, 2, \ldots\} \end{cases}$$

Step 5. IF the stopping criteria is met THEN

Step6.
$$\begin{cases} \text{Show the output} \\ \text{Declare the best solution} \\ \text{OTHERWISE Go to step 4.} \end{cases}$$

3   Local Search Strategies

LS procedure perturb given solution to generate different neighborhoods using a move generation mechanism [6]. A move generation is a transition from a schedule V to another one $V' \in I(V)$ in one step (iteration). In Saleh [4], a local search procedure that satisfies the GPS requirements has been developed. This procedure is based on the sequential session-interchange. The potential pair-swaps are examined in the order $(1, 2), (1, 3), \ldots, (1, n), (2, 3), (2, 4), \ldots, (n-1, n)$, n is the number of sessions. The solution is represented by graph. The nodes correspond to the sessions and the edges correspond to the observation order. For comparison reason this sequential local search procedure is called procedure (a). In this paper two new local search procedures have been developed. In the first, see Figure 1, A and B are chains of nodes (part of the solution) while $\{0, i\}$ and $\{i+1, n\}$ are, the first and the last nodes of the chains A and B respectively. In this procedure only one exchange of new edge $[n-1, 0]$ has been performed each iteration and this can be done by selecting one edge $[i, i+1]$ to be removed. For comparison reason this local search procedure is called (b1).

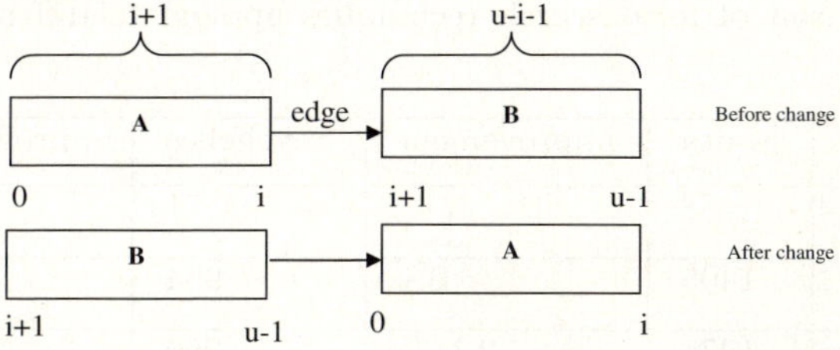

Fig. 1. Local search structure (b1) with 1-exchange of edges

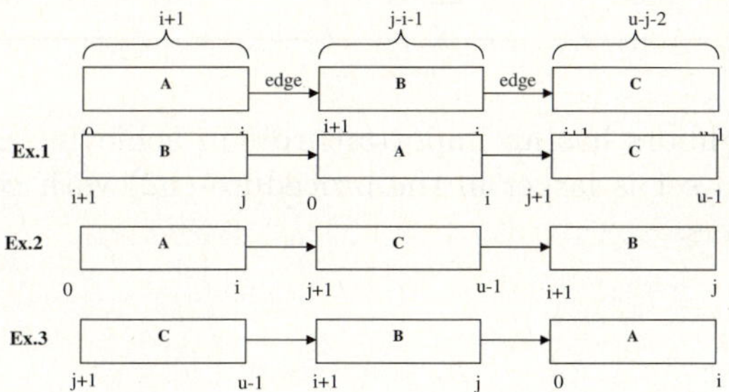

Fig. 2. Local search structure (b2) with 2-exchange of edges

In the second procedure, as shown in Figure 2, A, B and C are chains of nodes (part of the solution), where $\{0, i\}$, $\{i + 1, j\}$ and $\{j + 1, n - 1\}$ are the first and the last nodes of the chains A, B and C respectively. In the case of two edge exchange, there are several possibilities to build a new solution where the edges $[i, i + 1]$ and $[j, j + 1]$ are selected to be removed in one iteration. For comparison reason this local search procedure is called (b2).

4   Experimental Results

This section reports on the computational experience of the simulated annealing coupled with local search procedures using real GPS networks. The first network is a GPS network of Malta. The initial schedule with a cost of 1405 minutes was composed of 38 sessions observed. The second network is a GPS network of Seychelles. The initial schedule with cost of 994 minutes was composed of 71 sessions observed.

Where: C_{INT} - The cost of the initial solution; C_a - Schedule using local search procedure (a); C_{b1} - Schedule using local search procedure (b1); C_{b2} Schedule using local search procedure (b2).

We compare simulated annealing algorithm coupled with Saleh and Dare [4, 5] LS procedure, with developed in this paper LS procedures (b1) and (b2). The analysis of the results in Table 1 shows the advantage of the local search procedure (b2). We use the following parameters: initial temperature 8, final temperature 3, cooling parameter 0.9 for all tests. In the simulated annealing the

Table 1. Comparison of local search techniques applied to different types of GPS networks

Data	Malta	improvement	Seychelles	improvement
n	38		71	
C_{INT}	1405	0%	994	0%
C_a	1375	2.14%	969	2.52%
C_{b1}	1285	8.54%	994	0%
C_{b2}	1105	21.34%	707	28.87%

selected set of neighbors has an important role in achieving good results. The size of this selected set is larger in the procedure (b2) with comparison to the other two procedures.

5   Conclusion

In this paper two local search procedures have been developed and compared with local search procedure from [4, 5], procedure (a). The comparison of the performance of the simulated annealing of these procedures applying to different GPS networks is reported. The obtained results are encouraging and the ability of the developed techniques to generate rapidly high-quality solutions for observing GPS networks can be seen. The problem is important because it arises in mobile phone communications too.

Acknowledgments. The author is supported by European community grant BIS 21++.

References

1. Dowsland K., *Variants of Simulated Annealing for Practical Problem Solving* In Rayward-Smith V. ed., Applications of Modern Heuristic Methods, Henley-on-Thames: Alfred Walter Ltd. in association with UNICOM, (1995);
2. Kirkpatrick S., Gelatt C.D., Vecchi P.M.:*Optimization by Simulated Annealing*, Science 220, (1983),671–680;
3. Rene V.V.: *Applied Simulated Annealing*, Berlin, Springer, (1993);
4. Saleh H. A., Dare P.: *Effective Heuristics for the GPS Surveying Network of Malta: Simulated Annealing and Tabu Search Techniques*, J. of Heuristics, Vol 7(6), (2001), pp. 533-549;
5. Saleh H. A., Dare P.: *Heuristic Methods for Designing Global Positioning System Surveying Network in the Republic of Seychelles*, The Arabian Journal for Science and Engineering 26, (2002), pp. 73-93;
6. Schaffer A. A., Yannakakis M.: *Simple Local Search Problems that are Hard to Solve*, Society for Industrial Applied Mathematics Journal on Computing, Vol 20, (1991), pp. 56-87.

Novel Congestion Control Scheme in Next-Generation Optical Networks*

LaeYoung Kim, SuKyoung Lee, and JooSeok Song

Dept. of Computer Science, Yonsei University, Seoul, Korea
{leon, sklee, jssong}@cs.yonsei.ac.kr

Abstract. In this paper, to improve the burst loss performance, we actively avoid contentions by proposing a novel congestion control scheme that operates based on the highest (called peak load) of the loads of all links over the path between each pair of ingress and egress nodes in an Optical Burst Switching (OBS) network.

1   Introduction

The most important design goal in Optical Burst Switching (OBS) [1] networks is to reduce burst loss. The authors of [2] and [3] propose congestion control schemes pointing out that without controlling the offered load, burst loss rate will eventually go up to a very large number. In [2], a TCP congestion control scheme for OBS networks is studied. In this scheme, as in a normal TCP connection, the ingress node determines its burst sending rate based on the received ACK packet from the egress node. In [3], each ingress node adjusts its burst sending rate continually according to the information about burst loss received from all core nodes. However, these congestion control schemes have a problem that they may be triggered inappropriately because most of burst loss randomly occurs due to the bufferless nature of OBS networks although the network is not congested. In this paper, we propose a novel congestion control scheme that operates based on the highest (called peak load) of the loads of all links over the path between each pair of ingress and egress nodes (referred as flow) in an OBS network.

2   Congestion Control Scheme Based on Peak Load

Note that in OBS networks, when only one outgoing link over the path for a burst is congested even though the other outgoing links over the path are not congested, the burst will be dropped with a high probability in an OBS network without any contention resolution scheme. Thus, we propose a congestion control scheme based on peak load over the path for each flow.

In the proposed scheme, each egress node sends a LOAD message containing *Peak-Load* field set to 0 to all reachable ingress nodes over the backward path

---

* This research was supported by the MIC(Ministry of Information and Communication), Korea, under the ITRC(Information Technology Research Center) support program supervised by the IITA(Institute of Information Technology Assessment).

V.N. Alexandrov et al. (Eds.): ICCS 2006, Part I, LNCS 3991, pp. 1013–1017, 2006.

every T_L units of time. The LOAD message is sent on a control channel like a control packet. We assume that T_L for each flow is in proportion to the total number of hops of the flow. When the core node receives the LOAD message which is originated by egress node, E destined to ingress node, I from the link (k, j), it compares the value of *Peak-Load* field in the received LOAD message and calculated load on link (j, k) for the flow (I, E). If the calculated load is greater than the value of *Peak-Load* field, the core node copies its calculated load to the *Peak-Load* field in the LOAD message and forwards the message to the next node towards the destination of the message, I. Otherwise, the core node forwards the LOAD message to the next node without any change. In this way, each ingress node can receive the LOAD message including the highest load on the path to each reachable egress node. Each core node, j maintains the following information to calculate the load of outgoing link (j, k) for each flow which shares the link:

$T_{(I,E)}$: Time interval between two consecutive LOAD messages for flow (I, E)

$B_{(I,E)}$: Duration of all incoming bursts which want to use the outgoing link (j, k) during the interval, $T_{(I,E)}$

Whenever the core node receives the LOAD message for the flow (I, E), it calculates the load on the link (j, k) for the flow (I, E) as $\frac{B_{(I,E)}}{T_{(I,E)}}$. Each ingress node, I maintains the following information for every flow to egress node, E:

$R_{(I,E)}$: Current burst sending rate for flow (I, E)

$NAK_{(I,E)}$: Number of NAK messages received from the congested node (i.e., core node where a burst is dropped) on path (I, E) during the interval between two consecutive LOAD messages

---

Table 1 shows the procedure of congestion controller in our scheme. As shown in this table, if ingress node, I receives a LOAD message originated by egress node, E, congestion controller at the ingress node compares the value of *Peak-Load* field in the LOAD message with $LOAD_{TH}$ which is a threshold for tolerable level of congestion at high load on each link. If the *Peak-Load* is greater than $LOAD_{TH}$, the congestion controller decreases the burst sending rate. Otherwise, the congestion controller checks whether the value of $NAK_{(I,E)}$ is zero or not. If $NAK_{(I,E)}$ is zero, the congestion controller increases the burst sending rate. Finally, $NAK_{(I,E)}$ is set to 0 as line 6 in Table 1. The proposed congestion control scheme adopts the well-known Additive Increase/Multiplicative Decrease (AIMD) algorithm and the detailed algorithm for burst sending rate adjustment is as follows.

Table 1. The procedure of congestion controller at ingress node, I

```
1   If (Receive LOAD message from egress node, E) {
2       /* If Peak-Load in the LOAD message is greater than LOAD_TH */
3       If (Peak-Load > LOAD_TH) decrease the burst sending rate;
4       Else /* Check whether NAK_(I,E) is zero or not */
5           If (NAK_(I,E) == 0) increase the burst sending rate;
6       NAK_(I,E) = 0; /* Reset NAK_(I,E) */ }
```

$$\textbf{Increase}: R_{(I,E)} \leftarrow min(R_{(I,E)} + (\alpha \times H), MAX_{RATE}) \tag{1}$$

where α is the increase constant whose unit is a burst and H is the total number of hops between the ingress node, I and the egress node, E. We assume that each ingress node can send bursts up to MAX_{RATE} to achieve fairness among all the flows.

$$\textbf{Decrease}: R_{(I,E)} \leftarrow R_{(I,E)} \times (1 - \beta) \tag{2}$$

where β is the decrease constant. As shown in Table 1, if the peak load is greater than $LOAD_{TH}$, the burst sending rate will be proactively decreased to avoid contentions regardless of burst loss. Although the ingress node has received one or more NAKs during the interval between two consecutive LOAD messages, the burst sending rate will not be decreased if the peak load is not greater than $LOAD_{TH}$. As a result, our scheme tries to avoid an unnecessary decrease of the sending rate, that has been a problem in the existing schemes [2, 3] because they decrease the sending rate only according to burst loss. If the peak load is not greater than $LOAD_{TH}$ and the ingress node has not experienced any burst loss on the path (I, E) during the interval between two consecutive LOAD messages, the burst sending rate will be increased by an amount proportional to the total number of hops on the path as in Eq. 1. Because T_L for each flow is in proportion to the total number of hops of the flow as mentioned above, the congestion controller for a flow with larger hop count is activated less frequently than that for flows with smaller hop count. Thus, for a flow with larger hop count, the congestion controller increases its sending rate more than for flows with smaller hop count to achieve fairness among all the flows.

3   Performance Evaluation

In this section, we evaluate the performance of the proposed Peak Load-based Congestion Control scheme (PL-CC) comparing with existing OBS protocol using TCP Congestion Control (TCP-CC) [2] and conventional OBS without Congestion Control (no-CC). The simulations are performed using the ns-2 simulator. In our simulation, we use the 14-node NSFNet with 21 links where the link capacity is 10 Gbps and the number of flows is 72. Bursts are generated from all the ingress nodes. It is assumed that burst arrivals follow the Poisson process and their lengths are negative exponentially distributed with mean 1 Mbits. We set T_L to $30 \times H$ msec for each flow. T_L is about three times the average Round Trip Time (RTT) of each flow since the average delay per link in the simulation topology is about 5 msec. We set α to 3 (bursts) while setting β to 0.2. From the simulation results obtained by varying the $LOAD_{TH}$ and the MAX_{RATE}, we have learned that the proper values for $LOAD_{TH}$ and for MAX_{RATE} are 0.05 and 150 Mbps, respectively, for our simulation topology.

Fig. 1 (a) plots the burst loss rate versus traffic load for no-CC, TCP-CC and the proposed PL-CC. For TCP-CC, the increase constant is set to 1 while the decrease constant is set to 0.5 as in TCP. We see that the burst loss rate of PL-CC is much lower than TCP-CC as well as no-CC at all loads. From the

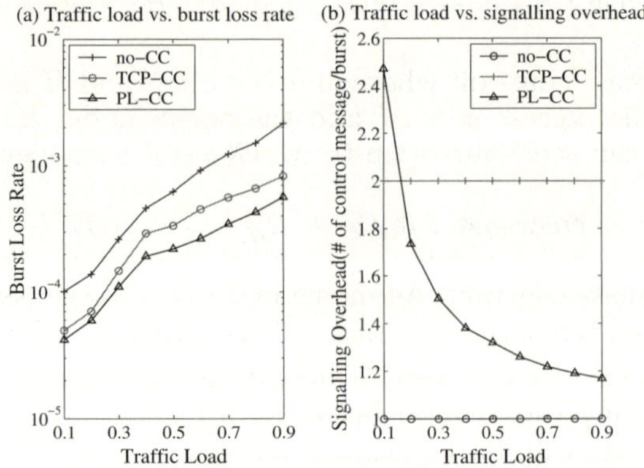

Fig. 1. (a) Traffic load vs. burst loss rate (b) Traffic load vs. signalling overhead

results, the burst loss performance improvement of the proposed PL-CC over TCP-CC ranges from 15.36% at the load of 0.1 to 31.46% at the load of 0.9.

To investigate signalling overhead due to congestion control for our scheme, we run simulation tests for all the three schemes. Fig. 1 (b) plots the count of control messages per burst versus traffic load. At first, we see that for no-CC, a control message per burst is sent regardless of the traffic load because control message contains only control packet for no-CC. For TCP-CC, count of control message per burst reaches nearly 2.0 regardless of the traffic load because whenever the egress node receives a burst, it sends an ACK to the ingress node. For PL-CC, control message contains LOAD and NAK messages as well as control packet. In this scheme, the LOAD message is periodically generated. Thus, we observe from Fig. 1 (b) that for our scheme, the count of control message per burst decreases significantly as the traffic load increases. We also see that for our scheme, additional control messages is generated significantly less compared to TCP-CC at most ranges of the traffic load.

Finally, we investigate throughput fairness for our scheme by using the well-known Jain fairness index. For our scheme, the Jain fairness index decreases as the traffic load increases. Specifically, the Jain fairness index is about 0.998 and 0.956 when the traffic load is 0.1 and 0.9, respectively. However, these results prove that our scheme can maintain acceptable throughput fairness among all the flows at all loads.

4   Conclusions

Simulation results indicated that the proposed congestion control scheme significantly reduces the burst loss rate while maintaining a reasonable fairness in OBS networks. Simulation results also showed that our scheme can balance between the performance gain in terms of the burst loss rate and the operation cost in terms of the signalling overhead.

References

1. Qiao, C., Yoo, M.: Optical Burst Switching (OBS) - A New Paradigm for an Optical Internet, Journal of High Speed Networks, Vol.8, No.1 (January 1999) 69–84.
2. Wang, S.Y.: Using TCP Congestion Control to Improve the Performances of Optical Burst Switched Networks, IEEE ICC '03, Vol.2 (May 2003) 1438–1442.
3. Maach, A., Bochman, G.V., Mouftah, H.: Congestion Control and Contention Elimination in Optical Burst Switching, Telecommunication Systems, Vol.27, Issue 2–4 (October–December 2004) 115–131.

A New Fairness Guaranteeing Scheme in Optical Burst Switched Networks

In-Yong Hwang, SeoungYoung Lee, and Hong-Shik Park

Information and Communications University
103-6 Munji-Dong, Yuseong-gu, Daejeon, Korea
{iyhwang, seoungyoung, hspark}@icu.ac.kr

Abstract. Optical burst switching (OBS) paradigm has been proposed to achieve the packet switching level throughput in the all optical domains. To meet the recent OBS commercialization requirement, we study the edge/core node combined (ECNC) OBS networks where the core node performs the edge node function as well. In this type of networks, the fairness problem occurs because the remaining offset-time decides the priority under the JET-based OBS. To solve this problem, we propose a fairness guaranteeing scheme by grouping bursts based on the destination and the remaining offset-time and assigning them to each wavelength group. Also, we propose a new burst generating and scheduling scheme for the ingress node and analyze the networks throughput.

Keywords: OBS, fairness, QoS, JET, wavelength grouping, offset-time, mesh networks.

1 Introduction

The emergence of wavelength division multiplexing (WDM) technology is considered as a solution to satisfy the tremendously increasing demands of transmission bandwidth driven by the growth of IP-based data traffic. At the same time, the necessity to make the next-generation optical Internet architecture is augmented, which can transport IP packets directly over the optical layer without opto-electro-optic (O/E/O) conversions, like optical packet switching (OPS). Although OPS which can achieve higher utilization is attractive, there are practical limitations such as optical buffer and all optical processing to implement OPS by using today's technology. Presently, optical burst switching (OBS) technology is under study as a solution for the optical Internet backbone in the near future since OBS technology can cut-through data messages without O/E/O conversion and guarantee the Class of Service (CoS) without any buffering.

In the JET-based OBS, several schemes have been proposed to provide CoS differentiation by assigning different offset-times [1]. To differentiate CoS, the source node assigns different offset-time according to the traffic classes. However, in a mesh network, the length of offset-time will be different due to the remaining number of hops even though data burst (DB) have the same class. Different offset-time among the same class invokes fairness problems in OBS networks.

V.N. Alexandrov et al. (Eds.): ICCS 2006, Part I, LNCS 3991, pp. 1018–1021, 2006.

Recently, researches about OBS commercialization have been arisen for the real network, i.e. mesh-type network. To meet this trend, we design the edge/core node combined (ECNC) OBS networks, where the OBS nodes can generate data burst, perform the wavelength grouping and assign Transit Data Bursts (TDB) that are bursts from previous nodes to wavelength groups to solve fairness problems. Then, we propose a burst scheduling scheme, which does not interrupt the TDB and analyze the network throughput. The analytic results show that the network throughput increases when the offset-time of transit bursts is sufficiently large. The remainder of the paper is structured as follows. Section 2 presents the fairness problem and fairness guaranteeing scheme in OBS networks. Section 3 provides the throughput analysis and results of ECNC networks, and the conclusion follows in Section 4.

2 Fairness Guaranteeing Scheme

In the mesh network, every node has the ability of edge and core nodes. To transmit data in OBS networks, the source node assembles packets to make a DB [2] and the DB follows a Burst Control Packet (BCP) after the some offset-time [1]. The offset-time is determined by the number of hops between the source and destination and QoS of the bursts. Therefore, the offset-times of BCPs deciding a priority are not the same in OBS nodes even though the DB has the same CoS, which may incur the fairness problem. To solve this fairness problem, we design the OBS node in Figure 1.

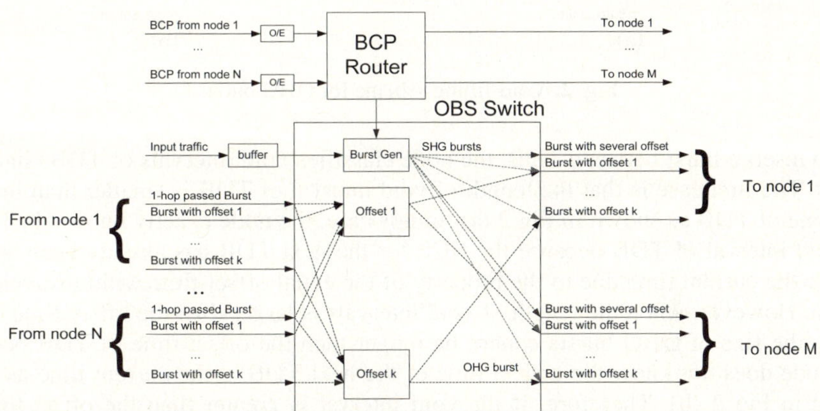

Fig. 1. The Architecture of ECNC OBS node for mesh networks

Entry bursts to the network are assorted to the one-hop-going (OHG) bursts and several-hop-going (SHG) bursts. The SHG bursts are transmitted to the next hop through the exclusive wavelength group. Thus, SHG bursts do not influence the performance of TDB. In case of OHG bursts, it can be inserted between TDBs due to the ability of knowing the void interval among TDBs. One-hop passed (OHP) bursts, which traveled only one-hop with exclusive wavelengths with the several offset-times, are classified according to the offset-time and QoS and then compete with TDBs under the equal condition. After competing and bursts grouping process, the bursts are assigned to the corresponding wavelength group. Thus, the QoS level of bursts can be kept the same because the offset-times within the same wavelength is equal.

3   Throughput Analysis

We consider the network throughput as the channel utilization of the wavelength groups and assume traffic load is equally distributed among wavelength group. There are several researches to utilize the void interval between bursts [3-4]. If the information contained in the BCP is used proprerly, the network utilization can be dramatically improved. Ideally, OHG bursts can utilize the remaining bandwidth of output channels if OHG bursts were fully inserted in the void intervals between TDB. But the available bandwidth cannot be utilized to the maximum because there is a limitation to predict available void intervals among TDBs as explained later.

The void filling scheme is illustrated in Figure 2. We can estimate the available void intervals among TDBs exactly with information the size of OHG bursts contained in the BCP.

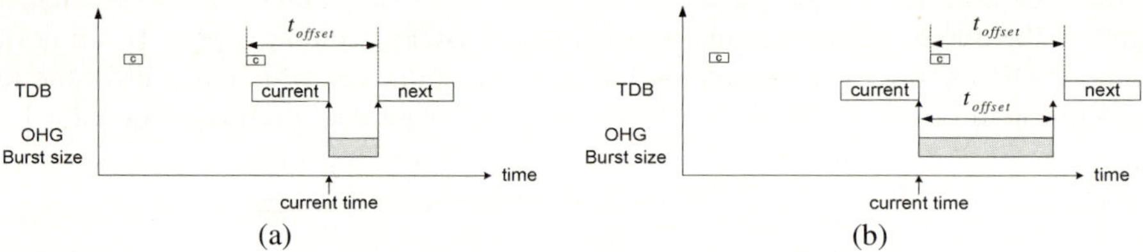

Fig. 2. Void-filling scheme for OHG bursts

To insert a burst between TDB, the node classifies void intervals of TDBs into two cases. The first case is that the length of void interval of TDB is smaller than the offset-time of TDB as shown in Fig 2 (a). In this case, the node exactly knows the length of void interval of TDB because the BCP for the next TDB has already been arrived before the current time due to the property of the equal offset-time within wavelength group. However, when the length of void intervals is larger than the offset-time of the TDB, the size of OHG bursts cannot be longer than the offset-time of TDB because the node does not know the arrival time of the next TDB at the current time as illustrated in Fig 2 (b). Therefore, if the void interval is greater than the offset-time of TDB or the BCP does not arrive until the end of current DB, the burst length of the OHG bursts should be limited to the size of offset-time. This is the reason why we do not utilize void intervals to the maximum. To verify the performance of the proposed burst scheduling scheme, we compare the network throughput through the simulation. Also, to investigate the relationship between the network throughput and the offset-time of TDBs, we change the offset-time of TDB.

The simulation results are shown in Fig 3, where the number 2, 4, 6, 8 and 10 in the legend mean the ratio of the offset-time to the mean burst length of TDBs. If the ratio is large, the offset-time increases accordingly. As shown in the simulation results, the network throughput improves when the offset-time increases as expected in Fig 2.

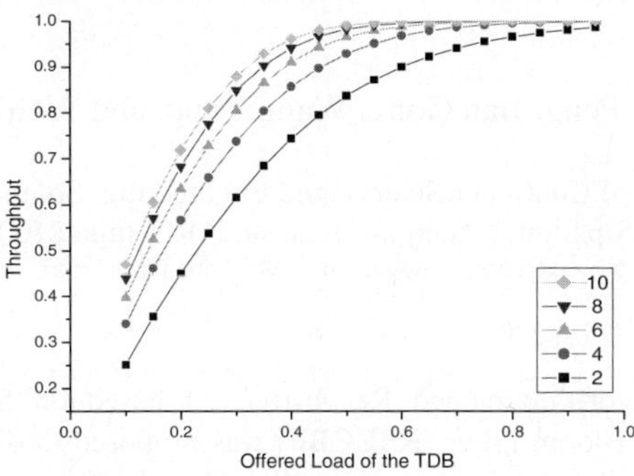

Fig. 3. Network throughput vs. offset-time of TDBs

4   Conclusion

We designed ECNC OBS node and proposed the fairness guaranteeing scheme, and the void filling scheme for ECNC OBS networks. The simulation results show that the network throughput is improved when the ratio of the offset-time to the size of TDBs is large. This means that the burst size should be minimized and the offset-time should be large sufficiently to maximize the throughput. The mathematical analysis is needed to analyze the accurate network throughput as a further work.

Acknowledgements

This work was supported in part by the MIC and IITA through the ITRC support program and by the KOSEF through its ERC program.

References

1. M. Yoo and C. Qiao.: A New Optical Burst Switching Protocol for Supporting Quality of Service. *Proc. SPIE All Optical Comm. Syst.: Architecture, Control Network Issues*, Vol. 3531, 1998, pp.395-405.
2. Ge, A., Callegati, F., and Tamil, L.S.: On Optical Burst Switching and Self-similar Traffic. IEEE Communications Letters, Volume 4, Issue 3, March 2000, pp. 98 – 100.
3. S. Y. Lee., I. Y. Hwang. and H. S. Park.: The Influence of Burst Length in Optical Burst Switching System with Completely Isolated Classes. *IEEE, ICACT 2005*, Vol. 1, 2005, pp.303-306.
4. Barakat N. and Sargent E.H., The Influence of Low-class Traffic Load on High-class Performance and Isolation in Optical Burst Switching, *IEEE ICC* 2004, Vol 3, 2004, pp. 1554-1558.

Disclosing the Element Distribution of Bloom Filter

Yanbing Peng, Jian Gong, Wang Yang, and Weijiang Liu

Department of Computer Science and Engineering, Southeast University
Sipailou 2, Nanjing, Jiangsu, P.R. China 210096
{ybpeng, jgong, wyang, wjliu}@njnet.edu.cn

Abstract. An algorithm named Reconstruction based on Semantically Enhanced Counting Bloom Filter (**RSECBF**) was proposed to disclose the distribution of original element from semantically enhanced Counting Bloom Filter's hash space. The algorithm deploys **DBSM,** which directly selects some bits from original string as the reversible hash function. The overlapping of hash bit strings in this paper brings the ability to confirm the homogenetic hash strings. The efficiency of **RSECBF** in the principal component analysis was verified by the DDoS detection in a published trace.

1 Introduction

Bloom Filter was created by Bloom in 1970 [1] to compress the searching space of string and cut down the error rate by multi-hash functions. The Bloom Filter became a popular tool in networking studies. In the studies on network [2], the Bloom Filter was implemented as an ingenious classifier. The newest paper on reversible sketches appeared in IMC 2004 [3]. The sketcher needs uniform distributed hash functions derived from the non-uniform distributed original strings, which makes it anfractuous.

The key issue of the reconstruction from different multi-hash strings to original string is to confirm that these strings are homogenetic. The Directly Bit String Mapping (**DBSM**) hash functions select some bits directly from the original string. The overlapped bit string between different hash functions confirms their homogenetic property, which means that they are from the same original string.

2 Properties of Semantically Enhanced Counting Bloom Filter

We found out that if the hash function is treated as aggregating rules in partition the original strings, the hash function can be any distributed. We call Counting Bloom Filter as semantically enhanced one when its hash functions select some bits directly from the original string. For example, a hexadecimal string, *0xabcd*, when using directly selected hash functions, the higher two octets is *0xab*, and the lower two octets is *0xcd*. The two hash functions, as a part of the original string, have semantical implication. These semantically enhanced hash functions are reversible and easy to be calculated, which are called as Directly Bit String Mapping (**DBSM**).

For **DBSM** in the space-independent Counting Bloom Filter:

Property 1: *One original string is mapped into the hash space once and only once.*

V.N. Alexandrov et al. (Eds.): ICCS 2006, Part I, LNCS 3991, pp. 1022–1025, 2006.

Property 1 can be induced from the work principle of Bloom Filter. It suggests that, when an original string x_i appears C_{xi} times in set S, its hash strings will appears in each hash space at least C_{xi} times.

The ordinary length of hash string is a trade-off between space efficiency and accuracy. The more the amount of the hash functions are, the more the calculations are needed, but the longer the hash string is, the larger the space overhead is. The general length of the hash string is in the range of 8~24 bits, and a 16 bits' hash string is a suitable choice for most of the applications.

When the original string is unclear, it is a key point to confirm which hash strings are homogenetic for the different hash strings. The overlapped bits among different **DBSM** hash strings deployed in this paper really benefit the homogenetic judgement. The hash values from two hash functions may be homogenetic if their overlapped bits have the same value. The hash strings can be combined with each other to a longer one by the patchworks, i.e., the overlapped string. When the splice is failed, the shorter string would be the distribution of the original strings.

For 32-bit original string such as IP address, the highest 16-bit string is selected as hash function H_h, the middle 16 bits as H_m, and the lowest 16 bits string as H_l. The overlapping relationship can be described by **Property 2**:

Property 2: *The sum of those counters in H_h with the same value overlapped string is equal to that in H_m. So is it between H_m and H_l.*

Property 2 can be induced from **Property 1** and the overlapping relationship between two hash functions. If a string, e.g. IP address a.b.c.d appears C times in set S, the short string a, b, c, d, which also expresses their location in the original string, must appears at least C times in respective hash space. When a.b.c.d is active, the short string in respective hash space would be active too. When a.b in H_h actively appears C times, b.c in H_m as an active homogenetic candidate of a.b in H_h, would appear at least C times in H_m, or the string a.b.0.0/16 should be an active aggregation.

3 Original String Distribution Discovery

If the hash strings' value is Pareto distributed in the Counting Bloom Filter's hash space, the principal component analysis in hash space will disclose the remarkable changes on the aggregation by *Top N*. This method is very suitable for those large scaled abnormality detections which change IP address greatly in distribution. For example, Eddie Kohler *et al.* [4] found that the active IP addresses are Pareto distributed, so the distributions of hash function IP_h, IP_m, IP_l are Pareto distributed too. The scanning behavior aggregates its source at the launchers' IP address, but the DDoS behavior focuses its aggregation character on the victims' destination IP address, which can be disclosed easily by **Property 2** and the principal component analysis in the Counting Bloom Filter's hash space.

The next paragraph proposes an algorithm, which is called **R**econstruction with **S**emantic **E**nhanced **C**ounting **B**loom **F**ilter (**RSECBF**), to reconstruct the original string, or disclose the aggregation character of the original strings.

> Get the Top N for H_p and H_q by their counters
> Get the most active hash string $x.y$ and $y.z$ in H_p and H_q
> Analyze the principal component according to Property 2, for each y
>   if there is no $y.z$ responding to $x.y$, it suggests that $x.y$ is an active prefix;
>   else if y.z is the principal component, $x.y.z$ is an active original element;
>   subtract x.y.*/x.y.z in their respective counter;
> Repeat the analysis till the active hash string is exhausted

The access to hash space string is k times per original string; ordinarily k and N in the *Top N* are very small. The primary calculation of **RSECBF** is the overhead in the sorting of the *Top N* which can be carried out by some quick methods in paper [5]. The space size m is 2^{16} for 16-bits string; so the spatial complexity is $O(m + N)$, and the calculation complexity is $O((m + NlogN)* k)$.

4 Actual Effect of RSECBF

The trace set published by Caida.org [6] was analyzed and the function of the distribution discovery was verified. Since the packets in this trace are known as backscatter packets which contains only responses from the attacked victim that went back to other IP addresses. It provided us a perfect chance to disclose the abnormal behavior and reconstruct their distribution of IP addresses and ports.

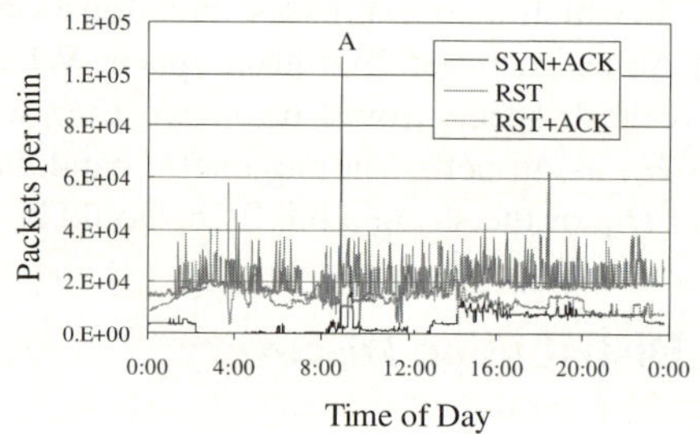

Fig. 1. The SYN+ACK, RST, RST+ACK packets' waves at Backscatter DDoS Trace [6]

The Backscatter's SYN+ACK/RST packet is a response packet for the spoofed SYN packet or other TCP packet (except RST packet) in DDoS attack, so its amount per minute can be used as an index to discover burst of abnormal behavior. Peak A in Figure 1 shows the burst of SYN+ACK/RST at 8:52 am, and then IP addresses and ports of related packets in Peak A are dug by **RSECBF** algorithm.

Table 1 shows that the two most active source IP hosts are 65.61.216.31 and 61.152.96.19 as well as the most active port is 80. The destination IP addresses are aggregated at 0.166.0.0/16 subnet at random in Table 3. The results suggest that 65.61.216.31 should be the victim, a web host.

The results imply that **RSECBF** has high efficiency in reconstructing the distribution characteristic of IP from their hash space when abnormal behavior occurs.

Table 1. Top N in SYN+ACK's Counting Bloom Filter hash space, 107005 packets

DBSM	H_h		H_m		H_l		*PORT*	
	Hash	Hits	Hash	Hits	Hash	Hits	Hash	Hits
Source IP and source port	**65.61**	**104585**	**61.216**	**104585**	**216.31**	**104585**	**80**	**106974**
	61.152	2214	152.96	2214	96.19	2214	21	24
	219.139	143	139.24	143	240.176	143	14364	5
Destination IP and Port	**0.166**	**104598**	166.33	1130	0.44	10	1336	189
	0.16	19	166.28	1078	2.237	10	1234	185

5   Conclusion and Future Works

The improvements of Bloom Filter in this paper extend the hash function to non-uniform distributed one with semantically implication. The reversible hash function, Directly Bit String Mapping (**DBSM**), makes the distribution discovery of original element easily. The overlapped bit string between different **DBSM** hash functions merges the homogenetic hash string into the original string by a simplified algorithm, which is called **RSECBF** for the Pareto distributed hash functions. The high efficiency of **RSECBF** in DDoS attacking detection is verified in a published trace.

Acknowledgement

This paper is supported by the National Basic Research Program of China (973 Program) 2003CB314804; the National High Technology Research and Development Program of China (2005AA103001)

References

1. Burton, H.B.: Space/time trade-offs in hash coding with allowable errors. Communications of the ACM, 13(7) , (1970):422–426
2. Andrei, B., Michael, M.: Network Applications of Bloom Filters: A Survey. Internet Math., no. 4, (2003)485–509
3. Robert, S., Ashish, G., Elliot, P., Yan, C.: Reversible Sketches for Efficient and Accurate Change Detection over Network Data Streams. IMC 2004, Taormina, Sicily, Italy, 207-212.
4. Eddie, K., Jinyang, L., Vern, P., Scott, S.: Observed Structure of Addresses in IP Traffic. Internet Measurement Workshop (2002).
5. Fabrizio, A., Clara, P: Outlier Mining in Large High-Dimensional Data Sets. IEEE Transactions on Knowledge and Data Engineering, vol. 17, no. 2, (2005) 203-215
6. http://www.caida.org/data/passive/backscatter_tocs_dataset.xml

An Efficient Key-Update Scheme for Wireless Sensor Networks

Chien-Lung Wang[1], Gwoboa Horng[1], Yu-Sheng Chen[1], and Tzung-Pei Hong[2]

[1] Department of Computer Science, National Chung-Hsing University
Taichung 40227, Taiwan, R.O.C.
{phd9004, gbhorng, s9356047}@cs.nchu.edu.tw
[2] Department of Electrical Engineering, National University of Kaohsiung
Kaohsiung 811, Taiwan, R.O.C.
tphong@nuk.edu.tw

Abstract. A novel key-update scheme is proposed for wireless sensor networks. The center server in a wireless sensor network first broadcasts a series of randomly generated code slices to sensor nodes. Upon receiving all the code slices, the sensor nodes find their neighboring coordinators to generate a permutation of slice numbers and send this permutation back to the center server. The center server and the sensor nodes can thus assemble a common program based on the permutation to derive their common key. Subsequent key-updates can then be done by this program based on the previous keys. The proposed scheme is simple, efficient, and is secure if the sensor nodes cannot be compromised within a short bound of time.

1 Introduction

Sensor networks are a kind of ad-hoc networks [3] and widely used in real applications. In a sensor network, each sensor node is deployed in a different location and is in charge of perceiving local information and reporting to the center server. A sensor node is usually limited by its computing power, memory, and battery power. These constraints make public-key algorithms infeasible for sensor nodes. In the past two decades, a lot of researches about security protocols [1] were proposed, including several key pre-distribution schemes [2][4][5].

In real applications, sensor nodes are usually deployed in a large number in order to cover a sufficiently large area. For instance, a military aircraft may scatter a lot of tiny sensor nodes over a certain terrain to gather information. Tens of thousands of sensor nodes may be required in this case. Since the amount of sensor nodes used in an application is usually large, it is better for them to be provided as cheaply as possible. If all sensor nodes used in an application are the same, they can be manufactured uniformly and the production costs can thus be reduced. Specifically, "uniform" means that each sensor node is equipped with the same hardware, software, and initial settings. It may be criticized that each sensor node with the same keys are dangerous. This thus causes a trade-off issue between security and production cost.

An efficient key-update scheme is necessary if all sensor nodes are initially equipped with the same keys. The keys should be updated immediately after the

V.N. Alexandrov et al. (Eds.): ICCS 2006, Part I, LNCS 3991, pp. 1026–1029, 2006.

deployment of sensor nodes since a compromise of the initial key of a sensor node may crash the entire sensor network. Besides, a good key-update scheme should be prompt and efficient. In this paper, a key-update scheme is thus proposed for wireless sensor networks.

2   The Proposed Key-Update Scheme

The proposed key-update scheme is divided into two parts: server part and sensor part. In this paper, all sensor nodes are assumed the same except for their IDs. The center server and all sensor nodes initially share a key K_c. This key is only used to initialize encryption keys and should be annihilated as soon as the deployment is done. To prevent the catastrophic consequence of compromising a sensor node, a short time bound t is set for the key-update phase. Each sensor node will begin a timer immediately after its physical deployment. If a sensor node does not finish its key-update phase within time t, it should sacrifice himself. That is, it will annihilate its initial key and stop its functionalities for keeping the security of the whole sensor network. The initial key can thus be protected in this way. The proposed scheme needs to estimate the time required to compromise a sensor node and chooses t as small as possible. The proposed key-update scheme executed respectively in the center server and sensor nodes is described below.

3   Execution on the Center Server - Broadcasting Code Slices

The center server first prepares some operators to generate random code slices. These operators are pairwise non-commutative. A code slice is composed of an operand and an operator. An operand can be any integer and an operator can be one of the following operators: *addition*, *multiplication*, *division, exponentiation*, *logarithm*, *shift*, etc. For example, +2, *4, ^3 are possible code slices and six possible combinations can be derived from the three code slices without repetition. Six possible programs can thus be assembled in the example. In general, if there are m code slices, then $m!$ possible programs can be obtained. Finally, the center server broadcasts these code slices to the wireless sensor network.

4   Execution on the Sensor Nodes – Coordinating and Assembling

In a sensor network, a sensor node can only communicate directly with its neighbors within a short range. A header is usually chosen from a subnet of sensor nodes as a relay. In the proposed scheme, the one with the most neighbors is elected as the header. In the dark, frogs cry to locate and identify other frogs. Similarly, each sensor node can send out its own ID to notify its neighbors. At the first stage, each sensor node sends out its ID and counts the number of its neighbors. At the second stage, each sensor node announces this number. Based on this information, headers can be elected and located by their neighbors (see Fig. 1). This method is called an echo algorithm.

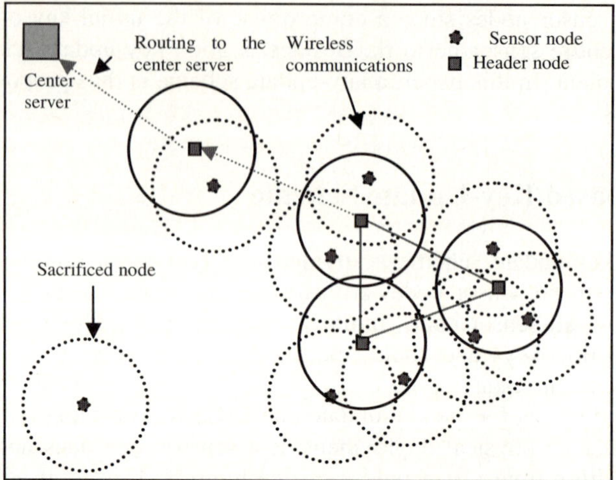

Fig. 1. System architecture: a center server and randomly distributed sensor nodes in a wireless sensor network. A solid circle represents the subnet of a certain header elected by its neighbor nodes. When a node needs to send a message to the center server, the message is first sent to the header of its subnet, then the header routes the message to the center server.

After headers are elected from sensor nodes, the next step is to form a secret program. Let the subnet of a header be defined as the network formed from the header itself and all its neighbors. Thus, there are many subnets in a wireless sensor network. Some subnets may overlap.

Each header will generate a random permutation of m objects, denoted by a string r_m, upon receiving code slices P_1 to P_m from the center server. It then deliveries r_m to its subnet and routes a path to the center server. With the permutation r_m, the code slices can then be assembled together to get a common program P. r_m should also be encrypted with the key K_c before it is sent out to prevent eavesdropping.

After an agreement of the program P has been made between the center server and the subnet of a certain header, subsequent key-updates can be done by executing the program P on the previous key of the subnet, i.e. $k_j = P(k_{j-1})$. Initially, the key of all subnets is $k_c = k_0$. But after the key-update procedure, each subnet should have its own key agreed with the center server.

5 Security Analysis

In the proposed scheme, an attacker may tape all the code slices $P_1, ..., P_m$ and try to recover a secret program P of a certain subnet by rearranging the slices. But lacking the permutation information r_m, the attacker can only guess possible ones. Since each permutation of the code slices can result in an executable program, the probability for a guessed program to equal the program P is only $1/m!$. This is why the m code slices are required to be pairwise non-commutative; otherwise, the probability may be larger than $1/m!$. The security of the proposed protocol thus depends on the size of m. m may be increased for better security at the cost of program-assembling time. For instance, $m!>2^{64}$ for a choice of $m=21$, and $m!>2^{128}$ for a choice of $m=35$.

Suppose that the average number of neighbor nodes is k. Each node in a subnet then needs $2*(1+k)$ operations. Each header node needs an extra encryption. Each node can thus easily assemble the program P from the m code slices. The choice for the size of m provides the flexibility between security and efficiency.

6   Conclusion

A novel key-update scheme has been proposed for wireless sensor networks. The proposed scheme is simple, efficient and secure if the sensor nodes are assumed not to be compromised within a short time bound t. It is thus feasible for the proposed scheme to be applied to a resource-constrained wireless sensor network. In the future, we will attempt to improve the proposed scheme with other constraint considerations.

Acknowledgement

This research was supported by the National Science Council of the Republic of China under contract NSC- 94- 2213- E- 005- 028.

References

1. Perrig, A., Szewczyk, R., Wen, V., Culler, D., Tygar, J.D.: SPINS: Security Protocols for Sensor Networks. The Seventh ACM Annual International Conference on Mobile Computing and Networking (2001) 189–199
2. Price, A., Kosaka, K. and Chatterjee, S.: A Key Pre-distribution Scheme for Wireless Sensor Networks. Wireless Telecommunications Symposium (2005) 253–260
3. Stajano, E., Anderson, R.: The Resurrecting Duckling: Security Issues in Ad-Hoc Wireless Networks. The Seventh International Workshop on Security Protocols (1999)
4. Ramkumar, M., Memon, N., Simha, R.: Pre-Loaded Key Based Multicast and Broadcast Authentication in Mobile Ad-Hoc Networks. Globecom, San Fransisco, CA, (2003)
5. Ramkumar, M., Memon, N.: On the Security of Random Key Predistribution Schemes. The Fifth Annual IEEE Information Assurance Workshop, NY (2004)

Estimating Average Flow Delay in AQM Router

Ming Liu, Wen-hua Dou, and Rui Xiao

Computer College, National University of Defense Technology
Hunan, 410073, P.R. China
liutomorrow@hotmail.com

Abstract. Delay measurement plays an important role in network QOS control. Much work has been done about the measurement of end-to-end delay, this paper describes a mechanism estimating flow Number and average delay in AQM router in the Internet. This estimate is obtained without collecting or analyzing state information on individual flows.

1 Introduction

The estimation of connections and delay is useful for network QOS control and resource management. Much work has been done about the measurement of end-to-end delay and several models were proposed[1]. Variously, this paper describes a mechanism estimating flow Number and average delay in an AQM[2] router in the Internet.

A linearized TCP/AQM dynamic model was developed[3]. The forward-path transfer function of the plant $P(s) = P_{tcp}(s) \times P_{queue}(s) \times e^{-sR_0}$ was given by

$$P_{tcp}(s) = (\frac{R_0 C^2}{2N^2})/(s + \frac{2N}{R_0^2 C}) \, , \quad P_{queue}(s) = (\frac{N}{R_0})/(s + \frac{1}{R_0})$$

where N is load factor (number of TCP connections), R_0 is round trip time, C is the link capacity. The actuating signal produced by AQM controller also reflects the information of flows that traverse the router.

2 Estimating Average Flow Delay in an AQM Router

2.1 Estimating the Flow Number

By TCP/AQM model, we need estimate the flow number firstly. Several approaches have been proposed previously to estimate the number of active flows competing for bandwidth[4,5,6]. Stabilized RED (SRED) compares the arrival packet with recently arrived packets in a 'Zombie List' and estimates the number of flows based on the hit probability. Hash Rate Estimation (HRE) uses a similar method while bases on the comparison of the incoming packet with the randomly selected backlogged packets in the queue. Flow Random Early Detection (FRED) estimates the number of flows that have packets queuing in the buffer. In FPIP, routers maintain a state table of long

V.N. Alexandrov et al. (Eds.): ICCS 2006, Part I, LNCS 3991, pp. 1030–1033, 2006.

flows to record the arrival time (*prevtime*) of the packet arrived lately from each long flow.

The approaches described above are effective in specific environments, we use the reference of the method "hit and miss" here. The main idea is to compare, whenever a packet arrives at some buffer, the arriving packet with a randomly chosen packet that recently preceded it into the buffer.

Considering the following situation, there are n flows passing through the router; r_i and p_i respectively denote the flow rate and the buffer occupancy of flow i. Intuitively, r_{hit} is the rate that the flow ID of incoming packet matches that of one randomly selected packet in the queue and r_{miss} denotes the rate that two predescribed packet are not of the same flow. The active flow number N_{act}, are shown as:

$$N_{act} = \frac{r_{hit} + r_{miss}}{r_{hit}} = \frac{1}{r_{hit}} , \qquad r_{hit} = \sum_{i=1}^{n} r_i p_i / C \; ; \quad r_{miss} = \sum_{i=1}^{n} r_i (1 - p_i) / C$$

We calculate an estimate r_{hit} for the hit frequency around the time of the arrival of packet. (Let Hit(t) = 0 if no hit and Hit(t) = 1 if hit):

$$r_{hit}(k) = (1 - \alpha) r_{hit}(k - 1) + \alpha Hit(k)$$

with $0 < \alpha < 1$, here $\alpha = 2.5 \times 10^{-4}$.

We use simple network topology with a single bottleneck link between r1 and r2 as depicted in Figure 1. C is 3750pkt/s which corresponds to 15Mbps with an average packet size of 500B. Connections are established between s_i and d_i. The propagation delay ranges uniformly between 40ms and 220ms, our target queue length is 100 packets. r1 runs PI and supports ECN marking, while the other router runs Drop Tail, buffer size is 800 packets.

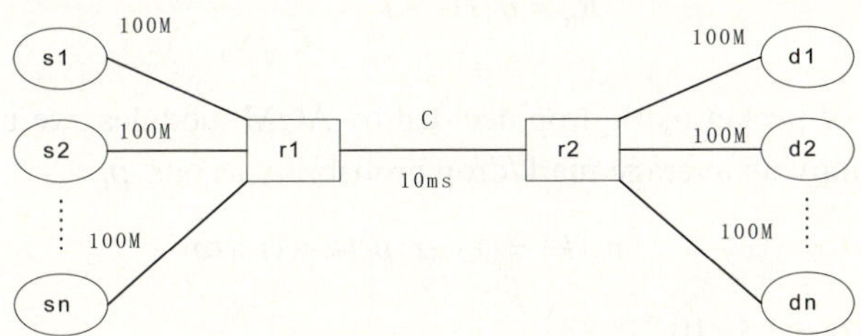

Fig. 1. Simple network topology

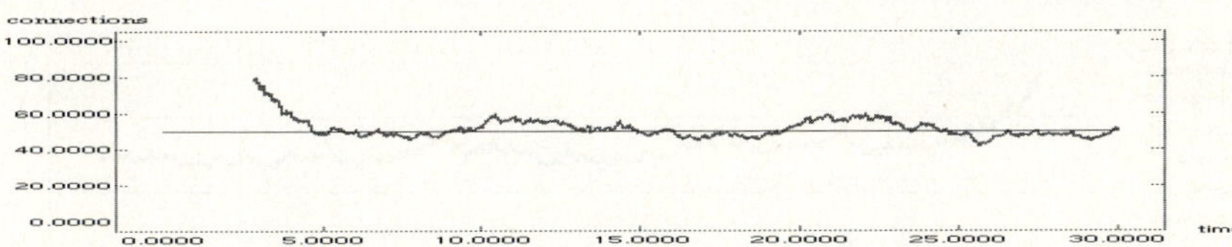

Fig. 2. the estimated flow number N_{act} in experiment 1

In experiment 1, the number of FTP flows is 50 at the beginning, the total simulation lasted for 30s. As shown in the figure, the estimated value N_{act} can converge to the real flow number.

In experiment 2, the number of FTP flows is 50 at the beginning, 50 FTP flows join the link 20 seconds later, 50 FTP flows join the link when t=40s, 50 FTP flows leave the link 30 seconds later, the total simulation lasted for 100s. As shown in the experiment result, the estimated value N_{act} can converge to the real flow number, and change with the real flow number n.

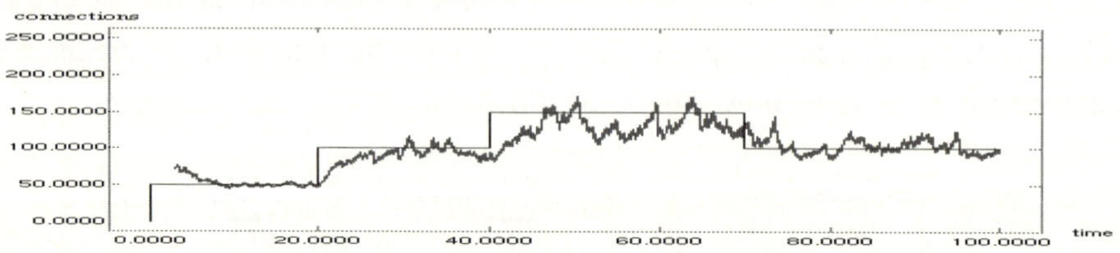

Fig. 3. the estimated flow number N_{act} in experiment 2

2.2 Estimating the Average Flow Delay

Using linearization techniques near equilibrium point of non-linear differential equation [3], (W_0, q_0, p_0) is the equilibrium point which meet

$$\left.\frac{dW(t)}{d(t)}\right|_{(W_0,q_0,p_0)} = \frac{1}{R_0} - \frac{W_0^2}{2R_0}p_0 = 0, \qquad \left.\frac{dq(t)}{d(t)}\right|_{(W_0,q_0,p_0)} = \frac{N}{R_0}W_0 - C = 0$$

W_0, q_0, p_0 can be calculated from: $W_0^2 p_0 = 2, W_0 = R_0 C / N$, so

$$R_0 = q_0 / C + T_p = \frac{\sqrt{2}N}{C\sqrt{p_0}}$$

p is probability of packet mark/drop decided by AQM modules, we use the exponentially weighted moving average mark/drop probability as our p_0:

$$p_0(k) = (1-\alpha)p_0(k-1) + \alpha p$$

$0 < \alpha < 1$, here $\alpha = 2.5 \times 10^{-5}$

We still use the network topology depicted in figure 1. In experiment 3, the propagation delay between the end and the router is 5ms, other settings are same as

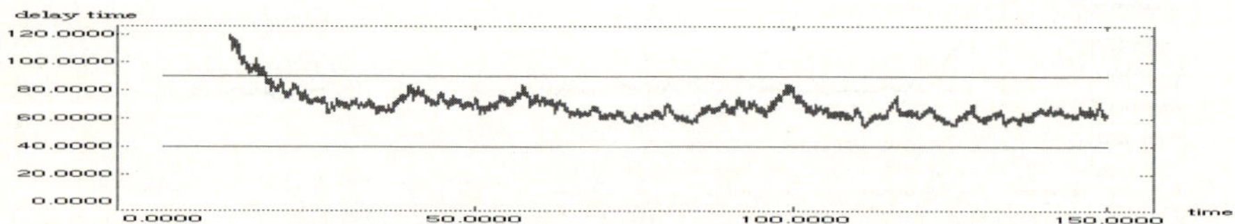

Fig. 4. The estimated delay time in experiment 3

experiment 1, so the propagation delay is 40ms, and the queueing delay is during 0~50ms.The number of FTP flows is 50 at the beginning, the total simulation lasted for 150s. As shown in the figure, the estimated delay time is within the range of real delay time.

In experiment 4, the propagation delay between the end and the router is 10ms, other settings are same as experiment 3, so the propagation delay is 60ms, and the queueing delay is during 0~50ms. As shown in the figure, the estimated delay time is within the range of real delay time.

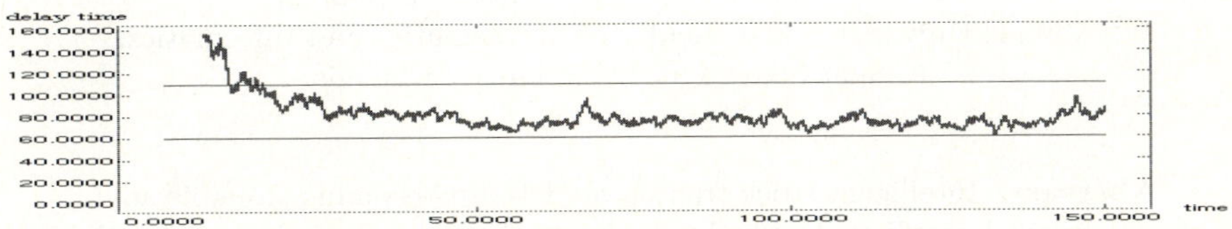

Fig. 5. The estimated delay time in experiment 4

3   Conclusion

This paper describes a mechanism estimating flow Number and average delay in a router in the Internet. This estimate is obtained without collecting or analyzing state information on individual flows.

There are also some limits with the estimating mechanism. It will take a period of time for the estimation to converge to the real value, this depends on the AQM scheme deployed in the router. Sometimes the estimated values are smaller than reality: if all flows have the same rate, the value N_{act} equals the real value n, otherwise, the estimated number will be smaller; the estimated average delay will be smaller than reality when nonresponsive flows exist. Suitable adjustment of the estimation value need further attention.

References

1. G. Almes, S. Kalidindi, M. Zekauskas. A One-way Delay Metric for IPPM, RFC 2679, September 1999
2. B. Braden, D. Clark, J. Crowcroft, B. etc, Recommendations on Queue Management and Congestion Avoidance in the Internet, RFC2309, April 1998.
3. Vishal Misra, Wei-Bo Gong, and Don Towsley, Fluid-based Analysis of a Network of AQM Routers Supporting TCP Flows with an Application to RED[A], in Proceedings of ACM SIGCOMM 2000[C], Stockholm, Sweden, 2000.
4. Li J.S.and Leu M.S. Network Fair Bandwidth Share Using Hash Rate Estimation. Networks, 40(3):125-141. 2002.
5. Lin D and Morris R. Dynamics of Random Early Detection. In Proceedings of the ACM SIGCOMM 1997, pp.127-137. Cannes, France. 1997.
6. Ott T. J., Lakshman T.V. and Wong L.H. SRED: Stabilized RED.In Proceedings of the IEEE INFOCOM 1999, pp.1346-1355. New York, USA. 1999.

Stock Trading System: Framework for Development and Evaluation of Stock Trading Strategies

Jovita Nenortaitė[1] and Alminas Čivilis[2]

[1] Vilnius University, Kaunas Faculty of Humanities, Department of Computer Science, Muitines 8, 44280 Kaunas, Lithuania
[2] Vilnius University, The Faculty of Mathematics and Informatics, Naugarduko 24, 03225 Vilnius, Lithuania

Abstract. Intelligent stock trading models are becoming valuable advisors in stock markets. While developing such models a big importance is given to its evaluation and comparison with other working models. The paper introduces trading system for stock markets, which is adesigned as a framework for development and evaluation of intelligent decision–making models.

1 Introduction

There is a number of existing stock trading systems, which allow users to make technical and fundamental analysis of stock markets' changes, to analyze the historical fluctuations of stocks or options prices etc. Systems like WinnerStock-Picks [8], NasTradingSystem [3], TradingForProfits [6] generate trading signals and allow the users to make an analysis of behavior of stock market through the application of technical analysis. Trading system UltraTradingSystem [7] for generating of trading signals is using a balanced approach of technical analysis and news research. All these systems miss an intelligent mechanism for decision–making and do not allow to integrate other analytical tools and to make comparison of trading results, which were achieved by using different trading models and strategies. Many commercial software packages for technical analysis offer both a comprehensive programming language, and a simulation mode, where the performance can be computed. However, most available products do not take this very seriously, and real trading simulation with a multi–stock portfolio is seldom possible [2].

The realization of the proposed real time trading system was developed using MATLAB. The system allows to download real time data, develop different intelligent trading models, compare trading results and make a detailed analysis of the trading results.

The paper is organized as follows: first sections describes the system and introduces its architecture, the second section is focused on the analysis of trading system applications for the analysis of different strategies. This section presents two trading strategies and discusses the analysis possibilities of the results. Finally the conclusions and future work is given.

V.N. Alexandrov et al. (Eds.): ICCS 2006, Part I, LNCS 3991, pp. 1034–1037, 2006.

2   Stock Trading System

The development of trading system was instigated by several reasons:

- *The need of framework for evaluation of stock trading algorithms and strategies.*
- *The need of real data management for the analysis of trading algorithms.*
- *The need to increase speed of computations.*

The architecture of the developed trading system is presented in Fig. 1.

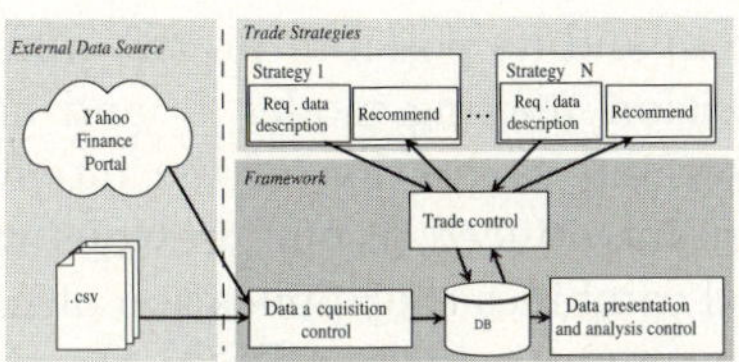

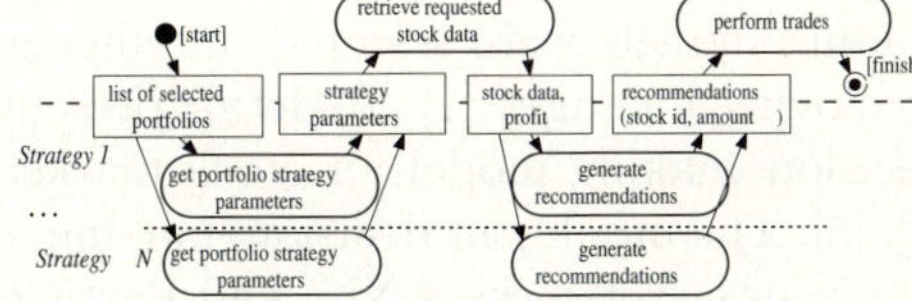

Fig. 1. Trading System Architecture   **Fig. 2.** Trading Scenario

Architecturally the system is divided into three parts: external data source and two internal parts — framework and trading strategies. External data source is only used to retrieve real time or historical stock data. The most important part of the system is the framework. The main tasks of the framework are: to download data from external data source; to preprocess downloaded data and store it in the database; to prepare data according to settings of the trading strategies; to perform trades for the created portfolios; to provide graphical user interface for the above mentioned controls; to provide means for data analysis. Based on a created portfolio and stock returns, data presentation and analysis control allows to do technical analysis of the results. One of the main requirements for the developed framework was to provide common interfaces for trading strategies and in this way to ensure easy integration of new trading strategies. Each trading strategy has to be implemented according to the given template. From an perspective of object oriented paradigm, each trading strategy is an object with two main methods. One provides information about required data quantity for given strategy and the other performs trades.

After the trade action is started the system scenario could be described in the following steps(see Fig. 2): (1) user is asked to enter interval dates for sequential historical trades or trades are started for the current day; (2) system for all selected portfolios determines trading strategies and asks each strategy to issue date requests; (3) system checks for data in the database (at this moment system can inform the user about missing data and can ask to retrieve data from external data source); (4) strategy performs calculations and returns recommendations; (5) after the recommendations are retrieved from strategies the framework performs trades and updates the database.

For the possibility to tune the analyzed strategy, framework allows for each strategy to change strategy settings. This is made through strategy options window, which is called from the main window using 'strategy options' button. This feature allows to use several portfolios with the same strategy but with different strategy parameters, as all parameters are referenced to strategy and portfolio. This allows to analyze the importance of different strategy parameters and results.

3 Trading System Application for Analysis of Different Strategies

For the analysis of the proposed trading system and its possibilities two different trading models were selected: intelligent decision–making model [5] and model of moving averages [1]. A detail presentation and evaluation of the intelligent decision–making model for stock markets are introduced in our previous works [4], [5]. The intelligent decision–making model combines the application of Artificial Neural Networks(ANN) and Particle Swarm Optimization(PSO) algorithm. The second model is a well known moving averages method.

There were created two portfolios with identical coincidental parameters. For the analysis of the models there were taken 130 stocks from SP500 index group and time period from 01-Jan-2000 to 01-Mar-2000. The main window of introduced trading system is presented in Fig. 3.

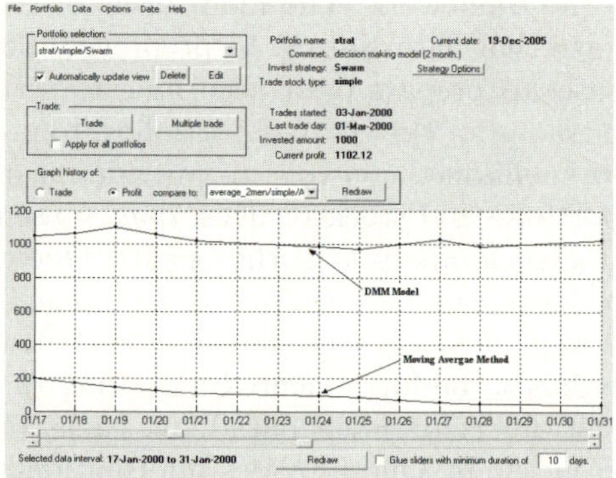

Fig. 3. Trading Systems Main Window

Figure 3 presents trading results of two portfolios. The descriptions of the selected portfolio are shown on the right side of the screen. Each portfolio is described by its name, investment strategy, trade stock type (stocks or indexes), trade start and last dates, and investment amount. The user can select what strategy options he/she wants to use. For intelligent decision–making model user can select the periodicity of trading, how many stocks should be recommended for the investment, size of the sliding window, number of deltas and the size of population (nubmer of ANNs). For the analysis of the results there is possibility to draw the graphs which allow to make detailed analysis on training of ANNs; selection of stocks or ANN which have shown the best performance. for the method of moving average the user is able to select the periodicity of trading, number of recommended stocks and size of period which is used for moving average calculation.

Trading results of each portfolio can be compared to the results of any other selected portfolio. For example, comparison shown in the Fig. 3, shows that the trading using intelligent decision–making model has much better performance for the selected time period. It is important to mention, that for any implemented trading strategy there is a possibility to create many portfolios with different strategy parameters. In that case, the presented stock trading system makes the evaluation of the trading strategies easy and fast.

4 Conclusions and Future Works

Because of the limited space available, this paper is merely an introduction to the architecture and framework provided by stock trading system. It was decided to emphasize on presentation of the general idea. The paper gives an illustration and explains the architecture and scenario of stock trading system. The main task of this paper was to show that the presented stock trading system is a powerful tool for development and tuning of intelligent decision–making models. The benefits of trading system were shown through the implementation of two models. It was shown, that new trading models can be easily integrated, extended, and evaluated with different parameters for the identical data sets.

References

1. LeBaron B.: Do Moving Average Trading Rule Results Imply Nonlinearities in Foreign Exchange Markets?. Social Science Research, 1992, 1–43.
2. Hellstrom Th.: ASTA - a Tool for Development of Stock Prediction Algorithms. Theory of Stochastic Processe, **5(21)**, 1999, 22-32.
3. NASTradingSystem (Swing Trading System). http://www.nastradingsystem.com/ (Accessed 15th of December 2005)
4. Nenortaite J., Simutis R.: Adapting Particle Swarm Optimization to Stock Markets. Intelligent Systems Design and Applications. 5th International Conference on Intelligent Systems Design and Application (IEEE), 2005, 520–525.
5. Nenortaite J., Simutis R.: Stocks' Trading System Based on the Particle Swarm Optimization Algorithm. Lecture Notes in Computer Science, **3039**, 2004, 843-850.
6. Trading for Profits. http://www.tradingforprofits.com/ (Accessed 15th of December 2005)
7. UltraTradingSystem.com http://www.ultratradingsystem.com/ (Accessed 15th of December 2005)
8. WinnerStockPicks.com (Daily Trading System) http://www.winnerstockpicks.com/ (Accessed 15th of December 2005)

A Quantum Hydrodynamic Simulation of Strained Nanoscale VLSI Device

Shih-Ching Lo[1] and Shao-Ming Yu[2]

[1] National Center for High-Performance Computing, Computational Engineering Division
No. 7, R&D 6th Rd., Science-Based Industrial Park, Hsinchu, 300, Taiwan
`sclo@nchc.org.tw`
[2] National Chiao-Tung University, Department of Computer and Information Science
No. 1001, Ta Hsueh Road, Hsinchu, 300, Taiwan
`smyu.cis91g@nctu.edu.tw`

Abstract. Strained silicon field effect transistor (FET) has been known for enhancing carrier mobility. The stained Si channel thickness, the $Si_{1-x}Ge_x$ composition fraction and the $Si_{1-x}Ge_x$ layer thickness are three crucial parameters for designing strained Si/SiGe MOSFET. Mobility enhancement and device reliability may be unnecessarily conservative. In this paper, numerical investigation of drain current, gate leakage and threshold voltage for strained Si/SiGe MOSFET are simulated under different device profiles. According to our results, the optimal combination of parameters are as follows: stained Si channel thickness is 7 nm, Ge content is 20%, and the $Si_{1-x}Ge_x$ layer thickness should be chosen between 20~50 nm.

1 Introduction

The introduction of strained Si and SiGe in CMOS technology is a means of improving the performance of Metal-Oxide-Semiconductor Field Effect Transistors (MOSFETs) in the deep submicron era [1-2]. A general approach to introduce biaxial tensile strain is using a virtual substrate of SiGe [1-2]. The underlying SiGe layer serves as an anchor to constrain the lattice of the strained silicon on top. Therefore, the electron mobility, and hence nMOSFET drive current performance, is enhanced. The stained Si channel thickness (T_{Si}), the $Si_{1-x}Ge_x$ composition fraction (x) and the $Si_{1-x}Ge_x$ layer thickness (T_{SG}) are three crucial parameters for designing strained Si/SiGe MOSFET. In this study, computer-aided design (CAD) approach is used to optimize the structure of strained Si/SiGe device. Drain current, gate leakage and threshold voltage are simulated and discussed.

2 Quantum Transport Models

The density-gradient (DG) model is considered to couple with the classical transport models. The hydrodynamic model (HD) [3] provides a very good compromise of velocity overshoot and the impact ionization generation rates. For the sake of saving computing time, DD model is used while drain bias (V_D) is low (< 0.1). For high-drain bias, hydrodynamic model is considered. The DD model is given as

V.N. Alexandrov et al. (Eds.): ICCS 2006, Part I, LNCS 3991, pp. 1038–1042, 2006.

$$\nabla \varepsilon \cdot \nabla \phi = -q(p - n + N_D - N_A), \tag{1}$$

$$q\, \partial n/\partial t - \nabla \cdot \mathbf{J}_n = -qR, \tag{2}$$

$$q\, \partial p/\partial t + \nabla \cdot \mathbf{J}_p = -qR, \tag{3}$$

where $\mathbf{J}_n = -qn\mu_n \nabla \phi_n$ and $\mathbf{J}_p = -qp\mu_p \nabla \phi_p$ are the electron and hole current densities. $\phi_n = -\nabla\phi - \nabla n(kT/\mu_n)$ and $\phi_p = -\nabla\phi + \nabla p(kT/\mu_n)$. R is the generation-recombination term. In the hydrodynamic model, the carrier temperatures T_n and T_p are not assumed to be equal to lattice temperature T_L, together with DD model, up to three additional equations can be solved to find the temperatures, which are

$$\partial W_n/\partial t + \nabla \cdot \mathbf{S}_n = \mathbf{J}_n \cdot \nabla E_C + dW_n/dt\big|_{coll}, \tag{4}$$

$$\partial W_p/\partial t + \nabla \cdot \mathbf{S}_p = \mathbf{J}_p \cdot \nabla E_V + dW_p/dt\big|_{coll}, \tag{5}$$

$$\partial W_L/\partial t + \nabla \cdot \mathbf{S}_L = dW_L/dt\big|_{coll}, \tag{6}$$

where $\mathbf{J}_n = \mu_n\big(n\nabla E_C + k_B T_n \nabla n + f_n^{td} k_B n \nabla T_n - 1.5nk_B T_n \nabla \ln m_e\big)$ and $\mathbf{J}_p = \mu_p\big(p\nabla E_V - k_B T_p \nabla p - f_p^{td} k_B p \nabla T_p - 1.5 p k_B T_p \nabla \ln m_h\big)$ are current densities. $\mathbf{S}_L$, $\mathbf{S}_n$ and $\mathbf{S}_p$ are energy fluxes and $dW_n/dt\big|_{coll}$, $dW_p/dt\big|_{coll}$ and $dW_L/dt\big|_{coll}$ are the collision terms.

According to DG method, an additional potential Λ is introduced into the classical density formula, which reads: $n = N_C \exp((E_F - E_C - \Lambda)/k_B T)$. In this study, Λ is given as $\Lambda = \gamma \hbar^2 \beta \big[\nabla^2(\phi + \Lambda) - \beta(\nabla\phi + \nabla\Lambda)^2/2\big]/12m$ [4]. The quantum transport systematic equations are discretized by the box discretization [5] and solved. After the drain current (I_D) is obtained, the gate leakage (I_G) and threshold voltage (V_{TH}) are determined.

3 Simulated Results and Discussion

In the numerical studies, a 40 nm strained Si/SiGe nMOSFETs is simulated. The device profile and simulated scenario are given in Fig. 1. According to previous experimental studies [1-2], the effects of interface trap are also considered in the numerical simulation.

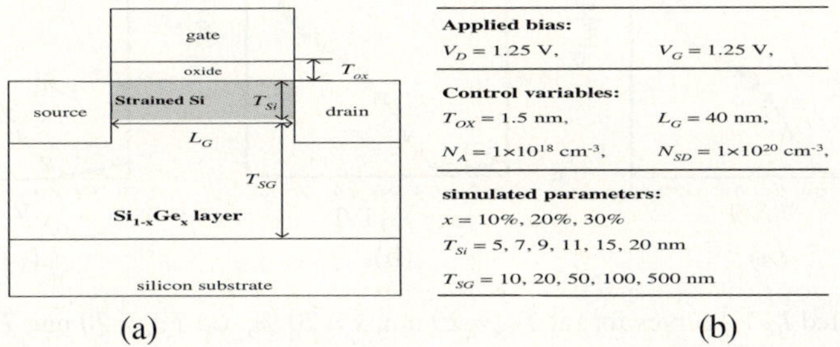

(a)   (b)

Fig. 1. (a) Simulated strained Si/SiGe nMOSFET and (b) scenario of simulation

Figures 2 and 3 illustrate part of the results to show the dependence of I_D and I_G on T_{SG}, T_{Si} and x, respectively. Fig. 2(a) demonstrates increasing device performance with increasing Ge content in the SiGe layer due to higher strain in the Si channel. However, a higher Ge content also induced higher gate leakage, which is caused by mismatch of lattice and diffusion of Ge. I_G is shown in Fig. 3(a). Fig. 2(b) and Fig. 3(b) illustrate I_D and I_G under different T_{Si} with $x = 20\%$ and $T_{SG} = 20$ nm. A thinner T_{Si} performs a larger I_D. The reason is that a thin strained Si channel can prevent the stress relaxation and present a better performance. The lattice mismatch proportionally decreases with increasing T_{Si}. In another word, the relaxation of tensile stress in the strained Si channel could be suppressed by decreasing T_{Si}, $i.e.$, the mobility enhancement would be larger in the thin T_{Si} device than in the thick T_{Si} device. The observation in the nanoscale device is different to long-channel devices. Unfortunately, a thin T_{Si} may induce large interface trap caused by Ge diffusion. In this study, I_D and I_G of $T_{Si} = 5$ nm is the largest. I_G of $T_{Si} = 7, 9, 11$ and 15 do not show much difference. Fig. 2(c) and 3(c) demonstrate I_D and I_G dependence on T_{SG}. A thicker T_{SG}

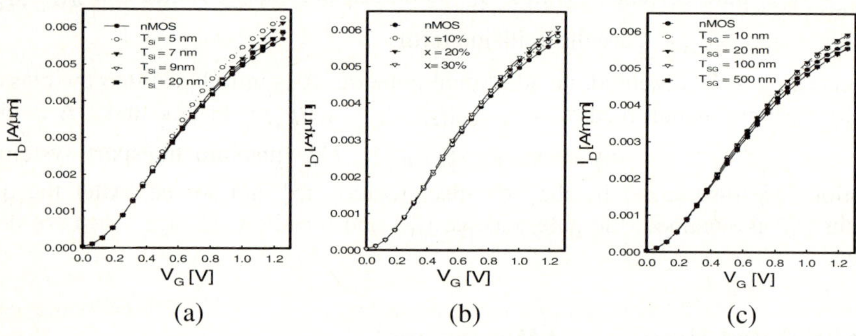

(a)   (b)   (c)

Fig. 2. Simulated I_D-V_G curves for (a) $T_{SG} = 20$ nm, $x = 20$ %, (a) $T_{SG} = 20$ nm, $T_{Si} = 9$ nm and (c) $T_{Si} = 9$ nm, $x = 20$ %

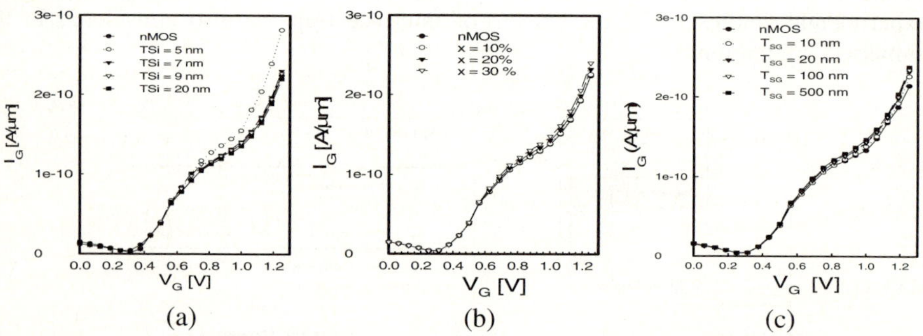

(a)   (b)   (c)

Fig. 3. Simulated I_G-V_G curves for (a) $T_{SG} = 20$ nm, $x = 20$ %, (a) $T_{SG} = 20$ nm, $T_{Si} = 9$ nm and (c) $T_{Si} = 9$ nm, $x = 20$ %

provides a larger stress, but a larger number of lattice mismatches is also induced. Moreover, a thicker T_{SG} may increase Ge diffusion to MOS interface, which contributes to an increased interface state.

For further discussion, Fig. 4 illustrates the V_{TH} shift (V), I_D enhancement (%) and $\Delta I_G/I_G$ (%) of the whole simulation scenario. According to the figure, if $T_{Si} \gtrsim 14$ nm, V_{TH} shift may be larger than 0.01 V. As $T_{Si} \lesssim 14$ nm, V_{TH} shift is small enough to be neglected. From Fig. 4 (b), the largest I_D enhancement is achieved by $T_{Si} = 5$ nm, $T_{SG} = 20$ nm and Si$_{70\%}$Ge$_{30\%}$. However, the I_G induced by the interfacial state is too large to be accepted. $T_{Si} = 5$ nm presents a sudden increase of interfacial states. $\Delta I_G/I_G$ is given in Fig. 4(c). $T_{Si} = 7$ nm is chosen. Although $x = 30\%$ may have a better improvement of drain current, it also have a serious problem of Ge diffusion. Therefore, $x = 20\%$ is suggested. Since a thin SiGe layer cannot provide enough stress to improve drain current, the best case occurs between $T_{SG} = 20 \sim 50$ nm.

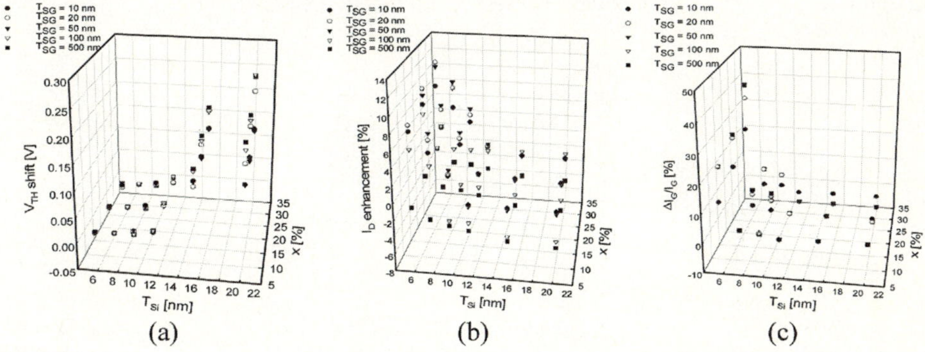

Fig. 4. (a) V_{TH} shift (V), (b) I_D enhancement (%), and (c) $\Delta I_G/I_G$ (%) for the simulated scenario

4 Conclusions

In this study, optimal profile of strained Si/SiGe device is suggested by numerical simulation. Hydrodynamic model is employed and the effect of interface state is considered in the simulation. Considering the improvement of performance and reliability, we suggested that the optimal stained Si channel thickness is 7 nm, Ge content is 20%, and the Si$_{1-x}$Ge$_x$ layer thickness is between 20~50 nm. The optimal design may obtain an 8~11% improvement of performance and maintain the same level of gate leakage for a 40 nm nMOSFET.

Acknowledgement

The authors would like to thank Professor Y. Li of Department of Communication Engineering, National Chiao Tung University, Taiwan for useful discussion.

References

1. Kwa, K. S. K., Chattopadhyay, S., Olsen, S. H., Driscoll, L. S., and O;Neill, A. G.: Optimization of Channel Thickness in Strained Si/SiGe MOSFETs, in Proc. ESSDERC (2003) 501-504
2. Olsen, S. H., O;Neill, A. G., Driscoll, L. S., Chattopadhyay, S., Kwa, K. S. K., Waite, A. M., Tang, Y. T., Evans, A. G. R., and Zhang, J.:Optimization of Alloy Composition for High-Performance Strained-Si-SiGe N-Channel MOSFETs, IEEE Trans. Elec. Dev. 51 (2004) 1156-1163
3. Bløtekjær, K.: Transport Equations for Electrons in Two-Valley Semiconductors, IEEE Trans. Elec. Dev. ED-17 (1970) 38-47
4. Wettstein, A., Schenk, A. and Fichtner, W.: Quantum Device-Simulation with the Density-Gradient Model on Unstructured Grids, IEEE Trans. Electron Devices 48 (2001) 279-284
5. Bürgler, J. F., Bank, R. E., Fichtner, W., and R. K. Smith,: A New Discretization Scheme for the Semiconductor Current Continuity Equations, IEEE Trans. CAD 8 (1989) 479-489.

Implementing Predictable Scheduling in RTSJ-Based Java Processor

Zhilei Chai[1,2], Wenbo Xu[1], Shiliang Tu[2], and Zhanglong Chen[2]

[1] Center of Intelligent and High Performance Computing, School of Information Engineering,
Southern Yangtze University
214122 Wuxi, China
[2] Department of Computer Science and Engineering, Fudan University
200433 Shanghai, China
zlchai@fudan.edu.cn

Abstract. Due to the preeminent work of the RTSJ, Java is increasingly expected to become the leading programming language in embedded real-time systems. To provide an efficient real-time Java platform, a Real-Time Java Processor (HRTEJ) based on the RTSJ was designed. This Java Processor efficiently implements the scheduling mechanism proposed in the RTSJ, and offers a simpler programming model through meliorating the scoped memory. Special hardwares are provided in the processor to guarantee the Worst Case Execution Time (WCET) of scheduling. In this paper, the scheduling implementation of this Java Processor is discussed, and its WCET is analyzed as well.

1 Introduction

Currently, to provide an efficient Java platform suitable for real-time applications, many different implementations are proposed. These implementations can be generally classified as **Interpreter** (such as RJVM [1] and Mackinac [2]), **Ahead-of-Time Compiler** (Anders Nilsson et al [3]) and **Java Processor** (such as aJile-80/100 [4] and JOP [5]). Comparing with other implementing techniques, Java Processor is preferably being used in embedded systems because of its advantages in execution efficiency, memory footprint and power consumption.

The scheduling predictability is a basic requirement for real-time systems. The Real-Time Specification for Java (RTSJ) [6] makes some major improvements to Java's thread scheduling. Many of the current real-time Java platforms implement scheduling based on the RTSJ, such as Mackinac, RJVM and aJile etc, most of which allow threads to be allocated in heap memory. JOP implements a new and simpler real-time profile other than the RTSJ.

In this paper, the scheduling implementation in our RTSJ-based real-time Java Processor is introduced. None of the thread in this Processor is allocated in heap and the interference of Garbage Collector is totally avoided. Comparing with JOP, the profile of the HRTEJ Processor is more close to the RTSJ and dynamic thread creation and termination are supported.

V.N. Alexandrov et al. (Eds.): ICCS 2006, Part I, LNCS 3991, pp. 1043–1046, 2006.

2   Scheduling Implementation in the HRTEJ Processor

Based on the optimization method proposed in our previous work [7], to guarantee the real-time performance of the HRTEJ Processor, standard Java class files was processed by the *CConverter* (the program we designed to preprocess the Java class file) before being downloaded into the processor's memory. During this phase, all the process interfering predictability such as Class loading, verifying and resolution are handled. Some other optimizing operations are processed simultaneity. All the Classes needed in the application are loaded and linked before execution. The memory layout produced by the *CConverter* is displayed as a binary sequence. All of the strings, static fields and other data can be accessed by their addresses directly.

2.1   Thread Management Mechanism in the HRTEJ Processor

There are some thread related registers in the HRTEJ Processor to facilitate the predictability of the scheduling as follows:

Run_T, Ready_T, Block_T and Dead_T: *n*-bit (n is the width of the data path) registers to record the queues of threads which are running, ready, blocked and dead. A thread can be put into a queue by setting corresponding bit of that register to '1' according to its priority.

ThisThread: recording the object reference of current running thread.

Wait_Base: The base address of the static fields WaitObject0~n-1 in figure 1.

STK_base0~n-1: the stack base address of each thread.

LTMAddr0~n-1: the *LTMomory* base address of each thread.

The HRTEJ Processor can support *n* threads at most with unique priority from 0 to n-1 (0 is the highest priority).These threads can be created and terminated dynamically.

Creating a new thread just as creating a general object, but the object reference of this thread should be put into the corresponding static field '*Thread0~n-1*' according to its priority. The *Scheduler* terminates a thread by moving the corresponding '1' from other queues to the *Dead_T*. The *Scheduler* always chooses the thread corresponding to the leftmost '1' in *Ready_T* to dispatch and execute.

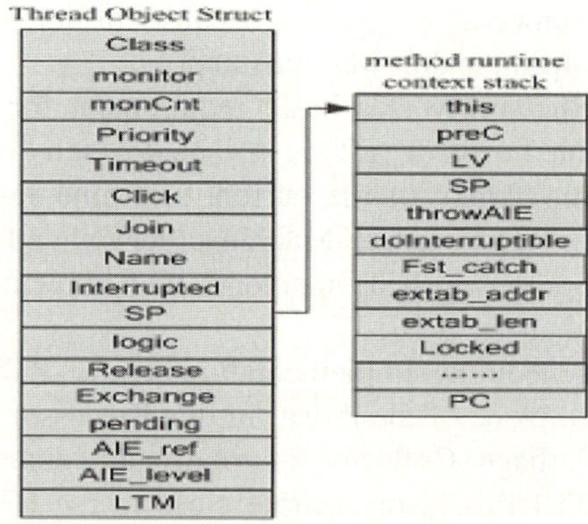

Fig. 1. Tread Object Structure of the HRTEJ Processor

When scheduling occurs, the context of the thread being preempted is saved at the top of its stack. The thread object and its corresponding context are shown in fig. 1.

Wait Method Implementation: When a thread calls the wait method and blocks itself, it records the reference of the waited object in corresponding static field *WaitObject0~n-1*. This static field will be checked when notifying a thread. *WaitObject0~n-1* is used to record the locked object waited by each thread.

Join Method Implementation: Using the instance field '*join*' to record the object reference of the thread to wake up when current thread is finished.

Priority Inheritance Implementation: If a thread wants to enter a synchronization block which another lower priority thread is in, then the Priority Inheritance must be taken. In the HRTEJ Processor, a simple method to implement the Priority Inheritance is adopted. Two threads sharing the same synchronization object exchange their priority, and record the old priority in the *Exchange* field. When the thread exits the synchronization area, it takes the original priority back again.

As discussed above, special hardwares are used in the HRTEJ Processor to ensure the predictability of WCET. The clock cycles of thread scheduling, dispatching, and other thread related mechanisms are all predictable in the HRTEJ Processor. The implementations of other scheduling related mechanisms described in [7] and [8], will not be discussed anymore.

3   Evaluation and Discussion

The HRTEJ differs from JOP in supporting dynamic thread creation and termination, ATC, nested scoped memory, and dynamic allocation of shared objects, which provides a more flexible programming model. Table 1 shows the comparison of some bytecodes execution cycles between the JOP and the HRTEJ. It is displayed that the average execution cycles of the HRTEJ is smaller than that of JOP.

Table 1. Clock Cycles of Bytecode Execution Time

	HRTEJ(min)	HRTEJ(max)	JOP
iload iadd	3	3	2
iinc	8	8	11
ldc	7	8	10
if_icmplt	8	10	6
getfield	6	7	25
getstatic	7	8	17
iaload	3	3	30
invoke	42	43	128
invoke static	39	39	101
dup	2	2	x
new	10	12	x
iconst_x	2	2	x
astore_x / aload_x	3	5	x
return	20	20	x
goto	3	5	x

Estimating the WCET of tasks is essential for designing and verifying real-time systems. As a rule, static analysis is a necessary method for hard real-time systems. Hence, the WCET of an application (*Demo.java*) is statically analyzed in this paper to demonstrate the real-time performance of the HRTEJ Processor.

The bytecodes compiled from *Demo.java* can be mainly partitioned into 3 parts (one part a thread). In each part, the WCET of the general bytecode is known according to table 1. For the finite loop in thread t0, its WCET can be calculated as *100*WCET (general code + LTMemory + start() + Scheduling)*. The LTMemory operation is predictable as mentioned in [8], and the WCET of the scheduling and method start() is also predictable. So, the real-time performance of the whole application can be guaranteed.

Furthermore, the maximal allocation of the LTMemory space in this application is *S(t0)+S(t1)* instead of *S(t0)+100*S(t1)*.   S(t) denotes the space of thread t. Another advantage of this processor is that Java programmers just need creating and entering a LTMemory space to use instead of denoting the memory size.

4   Conclusions

The multithreading mechanism is vital for real-time systems to handle the concurrent events in the real world. The RTSJ defines more accurate semantics for the predictable scheduling. It makes Java become popular in embedded real-time systems. In this paper, the RTSJ based scheduling mechanism implemented in our Java Processor is introduced. With special architectural supporting, all the WCET of the thread related mechanisms are predictable. Because heap memory is not used, this Processor is suitable for hard real-time applications.

References

1. http://www.cs.york.ac.uk/rts/
2. G. Bollella, B. Delsart, R. Guider, C. Lizzi, and F. Parain, "Mackinac: making HotSpot/spl trade/ real-time," presented at Eighth IEEE International Symposium on Object-Oriented Real-Time Distributed Computing, ISORC 2005, 45 – 54.
3. A. Nilsson and S. G. Robertz, "On real-time performance of ahead-of-time compiled Java," presented at Eighth IEEE International Symposium on Object-Oriented Real-Time Distributed Computing, ISORC 2005, 372 – 381.
4. http://www.ajile.com/
5. M. Schoeberl, "JOP: A Java Optimized Processor for Embedded Real-Time Systems", http://www.jopdesign.com/thesis/thesis.pdf, 2005
6. G. Bollela, J. Gosling, B. Brosgol, P. Dibble, S. Furr, D. Hardin and M. Trunbull, "The Real-Time Specification for Java", Addison Wesley, 1st edition,2000.
7. Z. L. Chai, Z. Q. Tang, L. M. Wang, and S. L. Tu, "An Effective Instruction Optimization Method for Embedded Real-Time Java Processor," 2005 International Conference Parallel Processing Workshops, Oslo, Norway, pp. 225-231, 2005.
8. Z. L. Chai, Z. L. Chen, and S. L. Tu, "Framework of Scoped Memory in RTSJ-Compliant Java Processor", Mini-Micro Systems, accepted.

The Improvement of NetSolve System

Haiying Cheng[1], Wu Zhang[1], and Wenchao Dai[2]

[1] School of Computer Engineering and Science, Shanghai University,
Shanghai 200072, PRC
`sunnychy@163.com, zhang@staff.shu.edu.cn`
[2] Dept. of Computer Science and Technology, East China Normal University,
Shanghai 200062, PRC
`wenchaodai@163.com`

Abstract. NetSolve is a kind of grid middleware used for high performance computing. In this article, the architecture and operational principle of NetSolve are first analyzed, the limitations of the Netsolve system are pointed out, and then Web Service Server and Server Proxy are put forward as the improvement of NetSolve System to solve these limitations.

1 Introduction

Thanks to advance in hardware, networking infrastructure and algorithms, computing intensive problems in many areas can now be successfully attacked. Various mechanisms have been developed to perform computations across diverse platforms. But until recently, there are still some difficulties in using high performance computer and grid technology on a large scale.

NetSolve, underway at the University of Tennessee and at the Oak Ridge National Laboratory, is a project that makes use of distributed computational resources connected by computer networks to efficiently solve complex scientific problems. This article is meant to probe into the implementation of NetSolve and describe its two extensions based on the super Cluster ZQ2000 of Shanghai University.

2 NetSolve

NetSolve is a client-server system that enables users to solve complex scientific problems remotely. The system searches for computational resources on the network, chooses the best one available, and solves a problem using retry for fault tolerance, and returns the answers to the user. It has three components:

• **Agent:** The agent maintains a database of NetSolve servers along with their capabilities and dynamic usage statistics for use in scheduling decisions. It attempts to find the server that will service the request, balance the load amongst its servers, and keep track of failed servers.

V.N. Alexandrov et al. (Eds.): ICCS 2006, Part I, LNCS 3991, pp. 1047 – 1050, 2006.

• **Server:** The NetSolve server is a daemon process that awaits client requests. It can run on single workstations, clusters of workstations, SMPs, or MPPs. One key component of the server is the ability to wrap software library routines into NetSolve services by using an Interface Definition Language (IDL) .

• **Client:** The NetSolve client uses application programming interfaces (APIs) to make a request to the NetSolve system, with the specific details required with the request.

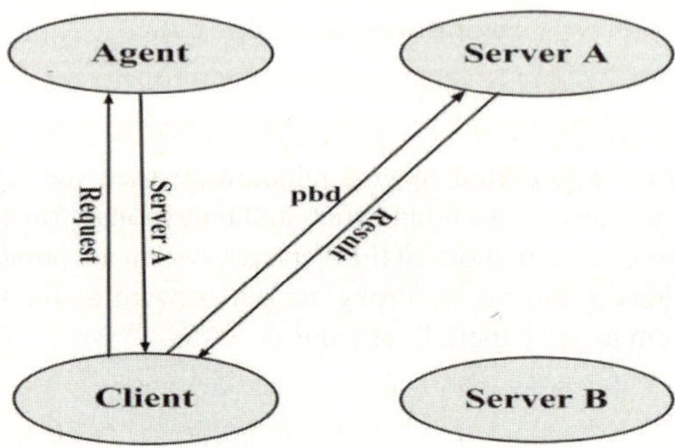

Fig. 1. NetSolve Architecture

3   Limitations of the Netsolve system

However, NetSolve system doesn't have a universal interface, its interface protocol can only be applied in the clients of NetSolve system. Therefore, other systems can't communicate with NetSolve directly. Besides, it can't provide a friendly and all-purpose interface for people who want to use the high performance computing resources. Another limitation is: unless clients, agent, servers are located in the same local network, real IP address is needed for them, as they communicate with each other via IP address. But for security and limitation of IP resources concern, many clusters, like ZQ2000 high performance cluster computer of Shanghai University, doesn't have real IP address for each node of the cluster. Therefore, NetSolve clients from outside the local network can't access any server on the node of ZQ2000.

To solve these limitations, we have two ways, the first is to make use of the Web Services technology to add a Web Services server into the NetSolve system, and the second is to add a server proxy into the NetSolve system.

4   Web Services Server

Web Services may use HTTP protocol, which can offer a friendly interface to facilitate calling services and make real IP address no longer a necessity. After adding a Web Services server into NetSolve, the work flow of the system changes as follows:

(1) Developing tools that supporting Web Services can be used to query the service which exist in Web Services server. Service and its interface described in WSDL format are returned to users from the server. After that, clients can call the service, just like calling a common local process. (2) Clients send simple object access protocol requests to Web Services server. On receiving the requests, the Web Services server converts it into a service execution request to NetSolve system. NetSolve system can then select the best server to execute the corresponding service. (3) NetSolve system returns the results of a service to Web Services server after it is completed. Clients can then get simple object access protocol responses described in XML format from the Web Services server.

Web Services server turns services provided by NetSolve system into its services, which is needed by Web Services call interface. The Web Services server and NetSolve agent are installed on the same computer. All services provided by NetSolve servers are registered in NetSolve agent.

5 Server Proxy

The common structure of super-computer is that they have one or several pre-servers which possess real IP address and the rest fake IP address (local network IP address). The pre-servers have double IP address like Gateway. If we run a Server Proxy on pre-server or on Gateway, it can set up a bridge between the client and the interior NetSolve servers.

Server proxy must run on the pre-server or gateway of super-computer. It can ensure the success of connecting between client and server. First, client can connect to server proxy, because pre-server has real IP. Meanwhile, pre-server and interior server are in the same local area network (LAN), the pre-server can also connect to the interior servers. Another problem is that all the data must pass through pre-server, which may cause a bottleneck. Start up several server proxies in different pre-server, and each server proxy manage a few of the interior servers can solve this problem. Moreover, in order to improve the efficiency, we can set up a connection pool between the server proxy and the interior servers.

Agent manages server proxy. It has the right to kill server proxy. Server proxy registers to Agent at first time. If server proxy dies, Agent will delete the server proxy's information of the relevant interior servers.

6 Conclusions

These two kinds of improvement of NetSolve system have been successfully applied in the ZQ2000 high performance cluster computer of Shanghai University. There are many ways in which NetSolve should be further extended, we should constantly improve our system to catch up with the fast development of numerical computing.

References

1. J. Dongarra, Network-enabled Solvers: A Step Toward Grid-based Computing, http://www.siam.org/siamnews/12-01/solvers.htm
2. D. Arnold, Sudesh Agrawal, S. Blackford, and J. Dongarra. Users' Guide to NetSolve V1.4. Technical report, Computer Science Dept., University of Tennessee, May 2001. http://icl.cs.utk.edu/netsolve.
3. D.C. Arnold, D.Bachmann and J.Dongarra. Request Sequencing: Optimizing Communication for the Grid. [J].In Euro-Par 2000-Parallel Processing, August 2000.
4. H. Casanova and J. Dongarra, Applying NetSolve's network enabled server, IEEE Comput. Sci. & Eng., 5:3 (1998), 57-66.

Grid Information Service Based on Network Hops*

Xia Xie, Hai Jin, Jin Huang, and Qin Zhang

Cluster and Grid Computing Lab
Huazhong University of Science and Technology, Wuhan, 430074, China
hjin@hust.edu.cn

Abstract. Grid information service influences outcome of applications on grid platforms directly. In this paper, network coordinate is introduced in grid information service mechanism to locate each grid node. With the hop count generation algorithm, network hops between user and resource providers can be forecasted, and results can be submitted to grid information service, which offers a list of resource providers with network hops increasing so that scheduler can work more efficiently. Performance proves that it is suitable for time-sensitive applications or applications with special restriction of network hops.

1 Introduction

Grid system can offer several key services and one of them is *Grid Information Service* (GIS) [4]. *Relational Grid Monitoring Architecture* (R-GMA) offers a global view of the information [1]. Globus Toolkit's *Monitoring and Discovery System* (MDS) defines and implements mechanisms for service and resource discovery and monitoring in distributed environments. The latest version MDS4 [6] is defined in the new WS-Resource Framework and WS-Notification specifications.

In this paper, GIS based on network hops is proposed. By using the network hops, the network distance between the user and resource providers can be calculate. A list of candidate resource providers with hop count in increasing order can be returned automatically. User can choose the resource provider with minimum network hops.

2 Design Principle

We predict the hop count between two nodes according to their network coordinates. Some notations are defined as follows. L_{ij} stands for actual hop counts between node i and node j. X_i is the network coordinates of node i. E is the system squared-error. L is an aggregate of nodes and the sum of nodes is Listlength(L). d is a constant defined according to user's requirement.

$$E = \sum_i \sum_j (L_{ij} - \|x_i - x_j\|)^2 \tag{1}$$

---

* This paper is supported by National Science Foundation of China under grant 60125208 and 90412010, ChinaGrid project of Ministry of Education of China.

V.N. Alexandrov et al. (Eds.): ICCS 2006, Part I, LNCS 3991, pp. 1051–1054, 2006.

$\left\| x_i - x_j \right\|$ is the distance of network coordinates from node i to j in appointed coordinate space. Hop count algorithm calculates the unit-length vector from node i to j by minimizing Eq.1. In order to reduce the network traffic, we change the policy that each node running hop count algorithm only communicates with part of the other nodes. These parts of the other nodes make up of an aggregate L and the nodes in L must not less than $d+1$.

Hop count algorithm generates a unit-length vector in randomly chosen direction and detaches two nodes at a same location so that network coordinate of each node can be quickly convergent. The node running the hop counts algorithm keeps on updating network coordinates periodically. If the network topology changes, node will update automatically. With the network coordinates of node i and j, the third information service node can forecast the network hops between these two nodes though node i does not communicate or measure the actual hop count with node j.

For hierarchical information service, node can get its network coordinates by running the network hop count algorithm. New user node or new resource provider node gets the IP address, network coordinates and estimated error of other nodes through the information providers and composes of its own L. For information service in the P2P style, each node runs the hop count algorithm to get its own network coordinates. New node gets the IP address, network coordinates and estimated error of other nodes through an arbitrary node to make up of L.

Index service is embedded with an optimizer of network hop count. Function of the optimizer is to calculate the hop count between user and candidate resource provider through their network coordinates and sort the hop count in increasing order. Index service returns a list of candidate resource providers to users but not ensures the availability of each resource provider. Index service also does not include policy information and resource provider need not to publish its own policy. For user who does not select the attached service, user can negotiate with each result item one by one. Only after authenticated can the resource provider deal with the submitted task.

3 Experiments and Analysis

Two performance metrics are used here: average response time and throughput of information discovery. Average response time of information discovery is the time from information discovery requirement being sent out to the time the results return. It is an average value for multiple simulations. We call the general information service without optimizer "random information service".

We use a grid simulation, JFreeSim [2]. JFreeSim is a grid simulation tool based on multiple tasks, multiple schedulers and multiple resources model. As a modular and extensible simulation tool, JFreeSim realizes many entity modeling and communication mechanisms between all entities, and makes system simulation according with the characteristics of the grid environment.

We use the Inet [3] topology generator to create 8,000-node wide-area AS-level network with a variable number of 500 client nodes with 4 client nodes per stub. ModelNet [5] is used to emulate wide-area bandwidth and latencies. Transit-transit and transit-stub links are 155 Mb/s and client-stub links are 2 Mb/s. Latencies are based on the Inet topology. The 500 client nodes are simulated by 11 PCs running

JFreeSim. Hop count algorithm is implemented in each information entity, grid user entity or resource entity generated by JFreeSim. Each PC has 1.6 GHz Pentium IV processor and 512 MB RAM with Redhat 9.0 operating system. All PCs are connected using Gigabit Ethernet. One of 11 PCs simulates information service and information service uses MDS4. The number of resource providers is 50, 100, 200, and 400, respectively. Other 10 PCs are used to simulate resource providers and users. The number of simulated grid user is 80, 160, 240, 320, 400, and 480, respectively.

Fig. 1 describes the average response time of information discovery for random information service and GIS based on network hops. For the later, it is also divided into two types: one is resource discovery with the range of hops, which is less than 10 and submitted by users; another is resource discovery without giving the range of network

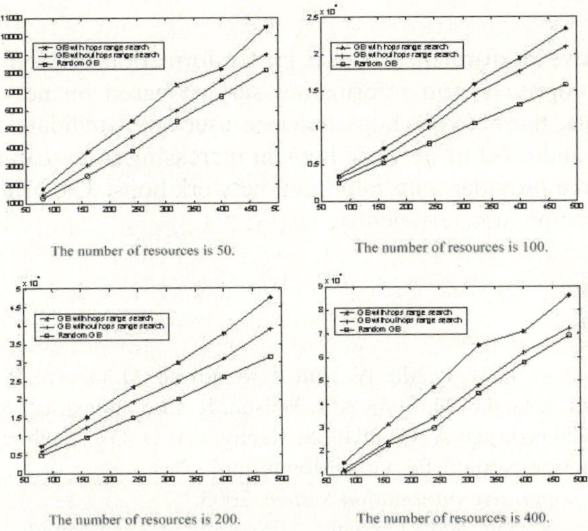

Fig. 1. Average response time for information discovery

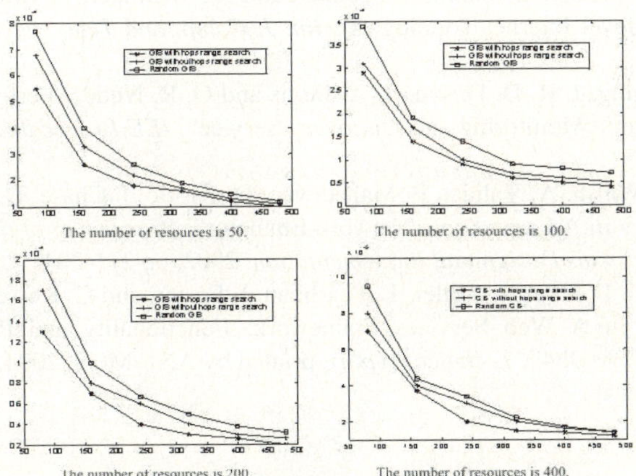

Fig. 2. Throughput for information discovery

hops. Comparing these figures, we can draw conclusions as follows: 1) GIS based on network hops has longer average response time; 2) information discovery with the range of hops has longer average response time than information discovery without giving the range of network hops; 3) more information provided by resource providers, more average response time.

Fig. 2 describe throughput of information discovery. We find: 1) throughput of information discovery in GIS based on network hops is smaller than random information service; 2) more resource providers, less throughput; 3) information discovery with the range of hops has smaller throughput than information discovery without giving the range of network hops.

4 Conclusions

In order to improve performance of the grid information service in time-sensitive application, we propose a grid information service based on network hops. Using network coordinate, the network hops between user and candidate resource provider can be predicted, and a list of network hops in increasing order can be returned. User can use the resource provider with minimum network hops. The grid resource sharing and cooperation can be more efficiently.

References

1. A. W. Cooke, A. J. G. Gray, L. Ma, W. Nutt, J. Magowan, M. Oevers, P. Taylor, R. Byrom, L. Field, S. Hicks, J. Leake, M. Soni, A. J. Wilson, R. Cordenonsi, L. Cornwall, A. Djaoui, S. Fisher, N. Podhorszki, B. A. Coghlan, S. Kenny, and D. O'Callaghan, "R-GMA: An Information Integration System for Grid Monitoring", *Proceeding of the 10th International Conference on Cooperative Information System*, 2003.
2. H. Jin, J. Huang, X. Xie, and Q. Zhang, "JFreeSim: A Grid Simulation Tool Based on MTMSMR Model", *Proceedings of 6th International Workshop on Advanced Parallel Processing Technologies*, Hong Kong, China, 2005.
3. H. Chang, R. Govindan, S. Jamin, S. Shenker, and W. Willinger, "Towards Capturing Representative AS-level Internet Topologies", *Int. J. Comp. and Tele. Net.* 44 (2004), pp.737-755.
4. H. N. L. C. Keung, J. R. D. Dyson, S. A. Jarvis and G. R. Nudd, "Performance Evaluation of a Grid Resource Monitoring and Discovery Service", *IEE Proceeding on Software*, 150, 2003, pp.243–251.
5. K. Yocum, K. Walsh, A. Vahdat, P. Mahadevan, D. Kostic, J. Chase, D. Becker, "Scalability and Accuracy in A Large-Scale Network Emulator", *Proceeding of the Fifth Symposium on Operating Systems Design and Implementation*, 2002, pp.271-284.
6. J. M. Schopf, M. D'Arcy, N. Miller, L. Pearlman, I. Foster, and C. Kesselman, "Monitoring and Discovery in A Web Services Framework: Functionality and Performance of the Globus Toolkit's MDS4", *Technical report*, printed by ANL/MCS, 2004, pp.1248-1260.

Security and Performance in Network Component Architecture*

Dongjin Yu[1,2], Ying Li[3], Yi Zhou [3], and Zhaohui Wu[3]

[1] Zhejiang Gongshang University, 310035 Hangzhou, China
`yudongjin@hz.cn`
[2] Zhejiang Institute of Computing Technology, 310006 Hangzhou, China
[3] College of Computer Science, Zhejiang University, 310027 Hangzhou, China
`{cnliying, zhouyi, wzh}@zju.edu.cn`

Abstract. In the Internet computing environment, clients usually invoke remote components to accomplish computation, which leads to many security problems. Traditional network component architecture model focuses on business logic, and takes little consideration for security problems. This paper proposes Secure Network Component Architecture Model (SNCAM). It classifies clients according to their credibility levels on component-side. Next, it presents the main idea of this paper, which is the security domain. To reduce the overhead introduced by the security mechanism, a method called security agent is also proposed.

1 Introduction

In the Internet computing environment, it is very pervasive that clients may not implement all the services they need. When clients need certain services, they can invoke remote components to get them, leading to the network component architecture model [7]. In such a model, local programs invoke remote components to accomplish the tasks, which may be required to transmit a great amount of data between clients and remote components. If the data is transmitted on the network without encryption, malicious listeners may get and change it. Traditional network component architecture model [6] focuses on business logic, and takes little consideration for security problems when business logic is invoked. All the parameters and returned results are transmitted on the network as plain text, causing security hazard.

The Common Component Architecture (CCA) [2] provides the means to manage the complexity of large-scale scientific software systems and to move toward a "plug and play" environment for high-performance computing. In the design of CCA, the considered requirements are: Performance, Portability, Flexibility and Integration. CCA gives concerns to performance issues, but does not address the requirements of security, which is very important in distributed component computing environment.

There are two measures to ensure secure communication between computers. One is data-encryption, which is further divided into an algorithm and an encryption key.

---

* The work is supported in part by Natural Science Foundation of Zhejiang Province of China with grant No. Z104550, Zhejiang Provincial Science and Technology Department Key Projects with grant No. 2005C21024, of China.

V.N. Alexandrov et al. (Eds.): ICCS 2006, Part I, LNCS 3991, pp. 1055 – 1058, 2006.

The negotiation of encryption key is always made using public-key technology, such as that of the IETF "SPKI"(Simple Public-Key Infrastructure).

The second measure is authentication or access-control. Many research works [9] address this problem, but few combine the authentication process with the data-encryption process for an integrated security framework. A language called Ponder [4] defines a declarative, object-oriented language for specifying policies for the security and management of distributed systems. Ponder focuses on specifying authentication rules, but lacks the security while data is being transmitted after the authentication.

Another issue is performance. Most of the existing security architectures do not address the problem of performance reduction [3]. Actually, we can do many performance improvements under the security framework.

In response to this situation, we propose the Secure Network Component Architecture Model (SNCAM) to ensure secure communication of clients and remote components in both ways: data-encryption and access-control. We also take into consideration for performance reduction owing to security mechanisms.

SNCAM adds security mechanisms to the traditional model. It forces components and clients to negotiate encryption algorithms and exchange an encryption key in every session for future communication. It also forces the authentication of clients to ensure that clients have the right to access the services provided by components. The participants of SNCAM are clients and security domains. A security domain consists of many business components and one security component. The relationship between these participants is: 1. a client invokes a business component; 2. the business component invokes security service provided by the security component, and the result indicates whether it should give the client the opportunity to access the business service; 3. if the client is granted the right to access the business service, the business component may also invoke other business services provided by components either in the domain or outside the domain.

2   Client Classification and Security Domains

SSH [1] (Secure SHell) is a protocol for secure remote login and other secure network services over an insecure network. It divides the process of establishing a secure connection into five phases: version exchange, algorithm negotiation, key exchange, authentication and session. The version exchange phase is to negotiate the protocol and software versions that the client and the server are both compatible to. In the session phase, the communication between up-level applications of the client and the server has already begun. These two phases are irrelevant to security mechanisms. Therefore, we can simplify the 5-phase SSH protocol into 3 phases: algorithm negotiation, key exchange and authentication. These three phases are loosely coupled: algorithm negotiation is to decide which algorithms to use during encryption and authentication; key exchange is to decide the key that is used to do encryption and decryption; authentication is to verify the client's access right. To accomplish the above three tasks, we can design three separate pieces of programs, which are also called as *security procedures*.

From the viewpoint of a server-side component, different clients may have different credibility. We classify these clients into 4 classes according to their credibility in the ascending order, resulting in four credibility levels: A-SEC, B-SEC, C-SEC, and D-SEC. For example, clients whose credibility levels are B-SEC are more credible than clients of A-SEC. When a client passes a security procedure, its credibility level goes up. In SSH, the exchanged key is temporarily valid. When a certain period of time expires, the key becomes invalid. The client and the server then have to re-exchange the key (rekey). The clients' credibility levels can go up and down as well. For example, a client whose credibility level is C-SEC or D-SEC can migrate back to B-SEC.

The Implementation of security procedures is the implementation and organization of the program that accomplishes the tasks of algorithm negotiation, key exchange, and authentication. Security procedures, which reside on component servers, read information about a client's credibility level from storage and may change it.

In many situations, the implementations of security procedures are similar. Hence it is reasonable to combine components that have the same implementation of security procedures into a domain, and configure a security component that implements these security procedures. The security component controls and authorizes clients to access services provided by components in that domain. In this way, other components in that domain need not implement security procedures. Instead, they can use the security service provided by the security component. All the components in one security domain can be classified into two categories: business and security. The business components accomplish business logic and the security component is responsible for controlling access.

3   Security Agent

A Security Agent method provides an express way for direct communication between distrusted clients and components. The discussion below regards clients as components, and we call components that provide services as object components.

The direct trust relationship between components is a 2-tuple $DT(C_i, C_j)$, which means that component C_i and C_j can communicate directly with each other without any security procedures. The interface trust relationship between components is a 2-tuple $IT(C_i, C_j)$, which means that C_i and C_j cannot communicate directly with each other without any security procedures. But they can avoid that by communicating with the third party components.

When there is no direct trust relationship but interface trust relationship between client components and object components, it is certain that we can find a sequence of components which begins in a client component and ends in an object component, where every two adjacent elements are directly trusted. This sequence is named as the security agent sequence. Every element in it is named as the security agent.

4   Experiments and Results

We take the size of a message as a parameter of a task, and set up two environments, one using security agents, and the other not. We set the size of message in turn as: 4K, 16K, 64K, 128K, and 512K. The clients' credibility levels are fixed to B-SEC. The obtained data is shown in Table 1, with the format of "without agent/with agent":

Table 1. Performance of security agents

message size	client 1	client 2	client 3
4K	7.4/2.9	8.1/3.6	10.4/4.1
16K	11.6/10.5	14.1/14.3	15.3/14.8
64K	34.3/43.7	47.6/51.9	50.9/59.1
128K	56.2/85.3	68.3/102.4	88.2/120.7
512K	196.3/304.8	237.9/395.6	279.8/463.7

The above data is the time of transmitting messages of the corresponding size from a client to a server for 40 times, measured in seconds. In Table 1 we can see that for light-weight tasks, using security agents is better. But for heavy-weight tasks, it is better to carry out security procedures and then communicate with the server directly.

References

1. T. Ylönen, T. Kivinen, M. Saarinen, T. Rinne, and S. Lehtinen. "SSH Protocol Architecture", Internet Engineering Task Force, work in progress, (2002).
2. D.E. Bernholdt, W.R. Elwasif, James A. Kohl, and Thomas G. W. Epperly. A component architecture for high-performance computing. In Proceedings of the Workshop on Performance Optimization via High-Level Languages (POHLL-02), New York, NY, June 2002.
3. R.L. Rivest, B.Lampson. SDSI - A Simple Distributed Security Infrastructure. Working document from http://theory.lcs.mit.edu/cis/sdsi.html
4. N. Damianou, N. Dulay, E. Lupu, M. Sloman, Ponder: A Language for specifying Management and Security Policies for Distributed Systems, Imperial College Research Report DoC2001, January,2001.
5. Lacoste, G. (1997) SEMPER: A Security Framework for the Global Electronic Marketplace. IIR London.
6. Li Yang, Zhou Yi, Wu ZhaoHui. Abstract Software Architecture Model based on Network Component. Journal of Zhejiang University, Engineering Science, 2004, 38(11):1402-1407.
7. A. Thomas, Enterprise JavaBeans: Server Component Model for Java, White Paper, Dec. 1997, http://www.javasoft.com/products/ejb/.
8. M. S. Sloman, J. Magee, "An Architecture for Managing Distributed Systems", Proceedings of the Fourth IEEE InternationalWorkshop on Future Trends of Distributed Computing Systems, pp. 40-46, IEEE Computer Society Press, September, 1993.
9. Lupu, E. C. and M. S. Sloman (1997b). Towards a Role Based Framework for Distributed Systems Management. Journal of Network and Systems Management 5(1): 5-30.

Design and Implementation of a Resource Management System Using On-Demand Software Streaming on Distributed Computing Environment*

Jongbae Moon[1], Sangkeon Lee[1], Jaeyoung Choi[1], Myungho Kim[1], and Jysoo Lee[2]

[1] School of Computing, Soongsil University, Seoul, 156-743, Korea
{comdoct, seventy9}@ss.ssu.ac.kr, {choi, kmh}@ssu.ac.kr
[2] Korea Institute of Science and Technology Information
jysoo@kisti.re.kr

Abstract. As distributed computing has become a large-scale environment such as grid computing, software resource management is rising as the key issue. In this paper, we propose a new resource management system, which manages software resources effectively, and its prototype implementation. This system uses an existing on-demand software streaming technology to manage software resources. This system greatly reduces traditional software deployment costs and associated installation problems. In this system, an added node can also execute a requested job without installation of applications.

1 Introduction

As computing technology advances, computing systems, which are geographically dispersed, are interconnected through networks and cooperate to achieve intended tasks. Moreover, not only system architecture but also software is more complicated. Such computing systems, which are called distributed systems, are composed of various types of resources and provide high throughput computing and high reliability. As distributed computing has evolved from a simple cluster to complicated grid computing, providing a reliable and efficient distributed computing environment largely depends on the effective management of these resources [1].

Resource management has always been a critical aspect, even in centralized computing environments and operating systems [2, 3]. Managing resources is much simpler in centralized systems than in distributed systems, since the resources are confined to a single location, and in general the operating system has full control of them. In distributed systems, however, these resources are scattered throughout distributed computing environment and no single entity has full control of these resources. Thus, the resource management in distributed computing environments is more difficult.

Various types of resources, which include both hardware resources and software resources, exist in distributed computing systems. Many researches focus on hardware resource management [4, 5, 6, 7]. For example, some researches focus on monitoring

---

* This work was supported by the National e-Science Project, funded by Korean Ministry of Science and Technology (MOST).

V.N. Alexandrov et al. (Eds.): ICCS 2006, Part I, LNCS 3991, pp. 1059–1062, 2006.

work nodes, selecting available work nodes, allocating jobs to available work nodes, and balancing loads between work nodes. In this paper, we only focus on software resource management, because installing various kinds of software in many work nodes, keeping software up to date, monitoring software status, and performing routine maintenance require more efforts.

Recently, on-demand software streaming has been used to make managing PC Labs easy [8]. On-demand software streaming technology streams an application code itself to a client computer, which then executes it natively. On-demand software streaming provides the benefits of server-based computing. Server-based computing offers potential of reducing the total cost of computational services through reduced system management cost and better utilization of shared hardware resources [9]. Currently, only a few companies including [10, 11, 12, 13, 14] are known to possess software streaming technology. On the other hand, software streaming is already used to manage software resources in many offices and school PC Labs, and its demand is increasing.

However, distributed systems do not support controlling streamed software resource. In this paper, we propose a new resource management system that can effectively manage software resources including streamed software in a distributed system. The proposed system adapts software streaming technology to a distributed system. We implemented a prototype system including a job broker and a job management module. The proposed system greatly reduces traditional software deployment costs and associated installation problems.

2   Proposed Resource Management System

We have implemented a prototype system by using Z!Stream [14] that was developed for Linux by SoftOnNet, Inc., and was implemented in C and PHP language. Also we used MSF (Meta Scheduling Framework)[15] system to build workflow. In this paper, we implemented a broker service and a job management module with Java language to support multiple platforms.

2.1   System Architecture

The architecture of the proposed system is shown in Fig. 1. The system consists of several modules and a Z!Stream package. A job broker collects work node status information and requests jobs to those work nodes using a workflow. The job broker is composed of several modules, which are a workflow interpreter, a streaming map, work node monitor, and a job submission module. A job management module resides at a work node with Z!Stream client and executes requested jobs. The job management module consists of a status detector module, software detect module, and a job execution module. If the requested application does not exist, the job execution module requests the application to Z!Stream server. After Z!Stream server streams the application code to a work node, then the work node executes it natively.

2.2   Implementation of the Job Broker

The job broker consists of a streaming map, a workflow interpreter, a work node monitor, and a job submission module. The job broker keeps the streaming map,

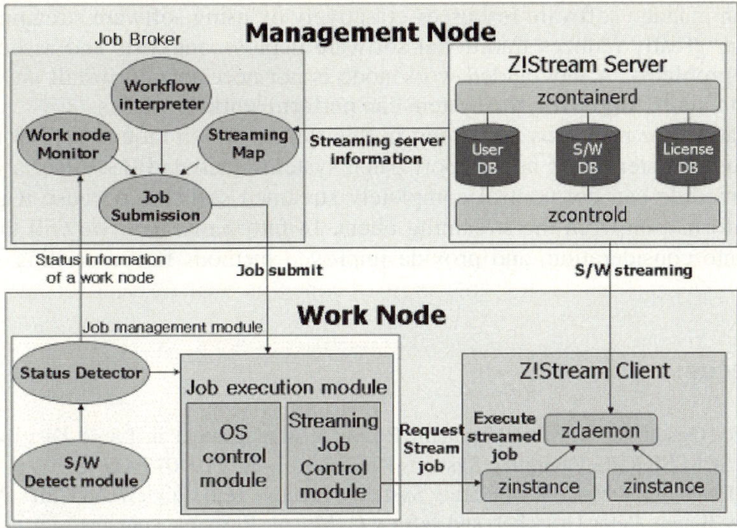

Fig. 1. Architecture of the proposed system

which is a list of streamed software that a work node can stream and run. The work node monitor collects status information of work nodes. The status information includes running job's status and system load information. The job status is one of the following: WAIT, RUNNING, and DONE. The workflow interpreter builds a workflow according to user requests, and selects work nodes to request jobs. The requested jobs may consist of one job step, but most of them are generally batch jobs, which can consist of several job steps. To execute batch jobs, workflow can be used. The workflow interpreter selects appropriate nodes to execute workflow jobs.

2.3 Implementation of the Job Management Module

The job management module consists of a status detector, a software detector, and a job execution module. The status detector collects conventionally-installed software list and process information, which the job execution module is performing. Then, the status detector forwards the list and information to a job broker. The job execution module uses a job queue. The OS control module controls conventionally-installed applications by calling system calls. The streaming job control module controls streamed jobs by calling interface, which the streaming client provides.

3 Conclusions and Future Works

In this paper, we proposed a new resource management system which manages software resources effectively. The proposed system adapts on-demand software streaming technology to existing distributed systems. We implemented a broker and a job management module. The broker receives user requests and builds workflow if necessary, and then selects work nodes and requests jobs to work nodes. The job management module runs requested jobs and control the jobs including streamed jobs. This system can manage software resources effectively by using software streaming. Also, this system greatly reduces traditional software deployment costs and associated installation problems. A new added work node is not necessary to install any applications to run a job. Moreover, the system can perform workflow tasks.

There are some problems remaining in this system which require further research. The proposed system does not support batch systems such as PBS. Also, a job management module can not control completely streamed software because it gets only limited information from the streaming client. In future research, we will take batch systems into consideration, and provide improved methods for controlling streamed software.

References

1. Andrzej Goscinski, Mirion Brearman: Resource Management in Large Distributed Systems, ACM SIGOPS Operating Systems Review Vol. 24, (1990) 7–25
2. A. S. Tanenbaum: Modern Operating Systems, Prentice Hall, Englewood Cliffs, NJ (1992)
3. Gaurav Banga, Peter Druschel, and Jeffrey C. Mogul: Resource containers: A new facility for resource management in server system, Proceedings of the 3rd USENIX Symposium on Operating Systems Design and Implementation (ODSI), New Orleans, LA (1999)
4. D.L. Eager, E.D. Lazowska and J. Zahorjan: Adaptive Load Sharing in Homogeneous Distributed Systems, IEEE Trans. on Software Eng. Vol. SE-12, No. 5, May.
5. I.S. Fogg: Distributed Scheduling Algorithms: A Survey, Technical Report No. 81, University of Queensland, Australia.
6. A. Goscinski: Distributed Operating Systems. The Logical Design, Addison-Wesley (1989)
7. Y.-T. Wang and R.J.T. Morris: Load Sharing in Distributed Systems, IEEE Trans. On Computers, Vol. C-34. No. 3, March.
8. AppStream Inc.: Solving the unique PC management challenges IT faces in K-12 schools. 2300 Geng Road, Suite 100, Palo Alto, CA 94303.
9. Fei Li, Jason Nieh: Optimal Linear Interpolation Coding for Server-based Computing, Proceedings of the IEEE International Conference on Communications (2002)
10. AppStream: http://www.appstream.com/
11. Softricty, Inc.: http://www.softricity.com/
12. Stream Theory: http://www.streamtheory.com/
13. Exent Technologies: http://www.exent.com/
14. SoftOnNet, Inc.: http://www.softonnet.com/
15. Seogchan Hwang, Jaeyoung Choi: MSF: A Workflow Service Infrastructure for Computational Grid Environments, Lecture Notes in Computer Science, Vol. 3292. Springer-Verlag, Berlin Heidelberg New York (2004) 445–448

Pipelining Network Storage I/O*

Lingfang Zeng**, Dan Feng**, and Fang Wang

Key Laboratory of Data Storage System, Ministry of Education
School of Computer, Huazhong University of Science and Technology, Wuhan, China
dfeng@hust.edu.cn, zenglingfang@tom.com

Abstract. In this paper, with introducing pipelining technology, network storage (e.g. network-attached storage, storage area network) and segmenting the reasonable pipelining stage are studied, which may have significant direction for enhancing performance of network storage system. Some experiments about pipelining I/O are implemented in the Heter-RAID. These results show that I/O pipelining scheduling can take advantage of pipelining features to achieve high performance I/O over network storage system.

1 Introduction

In computer architecture, pipelining is an implementation technique whereby multiple instructions are overlapped in execution; it takes advantage of parallelism that exists among the actions needed to execute an instruction.

Reference [1], [2] and [3] discuss the parallel I/O process policy. Reference [2] proposes a strategy for implementing parallel-I/O interfaces portably and efficiently. And they define an abstract-device interface for parallel I/O, called ADIO. Any parallel-I/O API can be implemented on multiple file systems by implementing the API portably on top of ADIO, and implementing only ADIO on different file systems.

Computer storage system has been the bottleneck in all computer systems. Parallel storage is one of the efficient ways to solve this I/O bottleneck problem. In our previous work [4], we provide the key technology to construct a parallel storage system is the implementation of the low I/O level parallel operations. To construct the high performance parallel storage system, the parallelism of storage devices is studied in detail: the research on the parallel operations of multiple storage devices within the multiple strings, the study of multithread I/O scheduling among the storage devices on a string as well as the related characteristic, parallel I/O scheduling and the analysis of disk array system based on the fact to reduce the computational complexity of the code and the cost of the parallel I/O scheduling. Reference [5] [6] [7] analyze some stage in cluster application environment. However, they are all limited in storage device and seldom study the pipelining storage in the whole network environment.

---

* This paper is supported at Huazhong University of Science and Technology by the National Basic Research Program of China (973 Program) under Grant No. 2004CB318201, National Science Foundation of China No.60273074, No.60303032, Huo Yingdong Education Foundation No.91068.

** Corresponding authors.

V.N. Alexandrov et al. (Eds.): ICCS 2006, Part I, LNCS 3991, pp. 1063–1066, 2006.
© Springer-Verlag Berlin Heidelberg 2006

2   Network Storage I/O Pipelining

2.1   I/O Pipelining Strategy and Practice in Network Attached Storage (NAS)

Heter-RAID [9] is a network attached storage system which adopts virtual SCSI commands encapsulated by operation type (read/write), start sector, sector number and other information to execute I/O operation. There are multiple virtual SCSI commands in the Heter-RAID command queue within system resources. Pipelining producer/consumer policy divides the I/O cycle of virtual SCSI command into different stages and uses buffer technology to smooth work speed of different function components, the policy overlaps disk I/O and CPU computation to improve system performance.

According to overlap degree of I/O scheduling processes, pipeline operation of network attached storage system can be divided into two methods, one is fixed pipeline scheduling, and the other is flexible pipeline scheduling. Multi processes will execute by fixed scheduling sequence in fixed pipeline scheduling. Otherwise, flexible pipeline scheduling judges the completing sequence and overlaps multi processes freely. In the following Section 3, tests prove that the above two kinds of pipeline scheduling can improve bandwidth utilization when a lot of clients access Heter-RAID.

2.2   I/O Pipelining Strategy in Storage Area Network (SAN)

Figure 1 shows that the object-based storage [8] architecture is built from some components: compute nodes, metadata server (MS), storage nodes, high speed network and high speed networks etc. The MS provides some information necessary to directly access data in those storage nodes. Those compute nodes first contact with the MS and get the information about data. Then those compute nodes send a request to some storage node for wanted data. Obviously, in SAN, those storage nodes, such as object storage nodes, mirroring RAID and archive tape library, which facilitate to copy or move data from one storage node to the other.

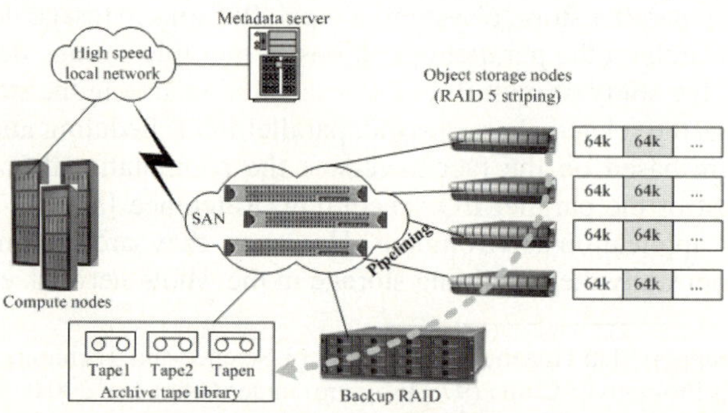

Fig. 1. Object storage system I/O pipelining strategy in SAN

3 Experiment Results

Two kinds of experiments are proved in order to show the advantage of I/O pipelining strategy. One is performed in single storage node which is FC RAID configured 0-level. The other is tested in above mentioned NAS environment.

Intel Iometer for windows is adopted as test tool to test sequential access RAID in order to compare the write-only/read-only performance under non-pipelining policy and pipelining policy, respectively. Sequential write-only and sequential read-only are also presented changing with different *Outstanding* I/O (e.g. 1, 2, 4, 8 and 16) and different transfer request size (e.g. 512bytes, 2kbytes, 8kbytes, 32kbytes, 64kbytes and 128kbytes). The number of **Outstanding** I/Os represents IOMeter loads. Tweaking the number of simultaneous *Outstanding* requests affects the aggregate load the tested RAID is under. Test results are shown in Figure 2 and 3.

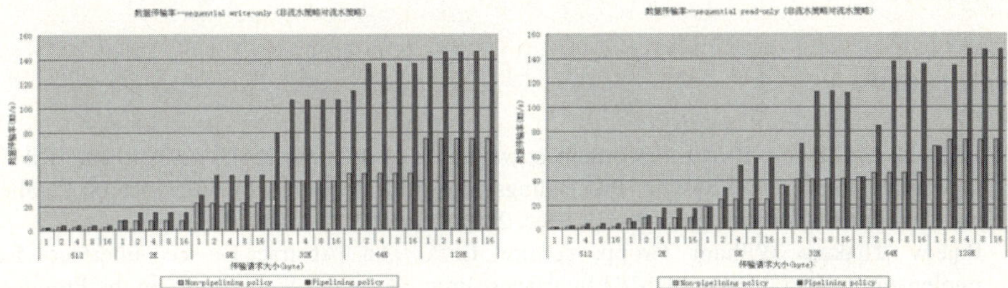

Fig. 2. Data transfer rate (sequential write-only/read-only) under non-pipelining policy/pipelining policy

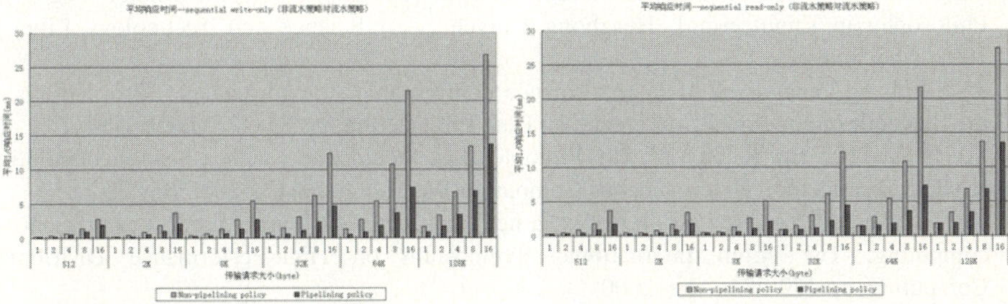

Fig. 3. Average I/O response time (sequential write-only/read-only) under non-pipelining and pipelining policy

For the sequential write-only and sequential read-only, the data transfer rate under non-pipelining policy and pipelining policy are shown in Figure 2. For the sequential write-only and the sequential read-only, when the pipelining policy is adopted, the data transfer rate doubles those under non-pipelining policy, except for 512byte transfer data request in Figure 2 (there is not very noticeable) and some *Outstanding* 1 in Figure 2. It is because a depth of 1 is an extremely linear load which, combined with a 100% random test and is not representative of any real kind of access. Also, in Figure 3, the average response time has the similar effect under the circumstances.

Moreover, from Figure 2, the data transfer rate and the average response time all have obviously improved performance when the test changes from *Outstanding* 1 to *Outstanding* 2. But these have not had any effect changing from *Outstanding* 2 to *Outstanding* 16, the other way round, the average response time increasing.

4   Conclusions

The advantage of pipelining over parallelism is specialization (this is to say, different stage finishes different part of a task). The task can be partitioned and assigned to objects according to the specialized stages, and those stages can be kept busy by performing only their stage on repetitive tasks. In this paper, network storage I/O is discussed under pipelining technology. The experiment results show that pipelining solution improves the network storage I/O latency and enhances the storage system throughout.

References

1. J. Hsieh, C. Stanton, R. Ali. Performance evaluation of software RAID vs. hardware RAID for Parallel Virtual File System. Proceedings of Ninth International Conference on Parallel and Distributed Systems, pp:307–313. Dec. 2002.
2. Rajeev Thakur, William Gropp, Ewing Lusk. An Abstract-Device Interface for Implementing Portable Parallel-I/O Interfaces. Proc. of the 6th Symposium on the Frontiers of Massively Parallel Computation, pp: 180-187, October 1996.
3. ROMIO: A High-Performance, Portable MPI-IO Implementation. Website, 2005. http://www-unix.mcs.anl.gov/romio/
4. Feng Dan. Research on Parallel Storage System. [Dissertation for the Doctor Degree of Philosophy in Engineering]. Huazhong University of Science and Technology Library. 1997. (In Chinese)
5. A. Biliris, E. Panagos. A high performance configurable storage manager. Proceedings of the Eleventh International Conference on Data Engineering, pp:35–43. March 1995.
6. K. Hwang, Hai Jin, R. Ho, W. Ro. Reliable cluster computing with a new checkpointing RAID-x architecture. Heterogeneous Computing Workshop, pp:171-184. May 2000.
7. Kai Hwang, Hai Jin, Roy Ho. RAID-x: a new distributed disk array for I/O-centric cluster computing. The Ninth International Symposium on High-Performance Distributed Computing, pp:279-286. Aug. 2000.
8. M. Mesnier, G.R. Ganger, and E. Riedel. Object-based storage. Communications Magazine, IEEE, Vol 41, Issue: 8, pp. 84–90, Aug. 2003.
9. Fang Wang, Jiang-Ling Zhang, Dan Feng, Tao Wu, Ke Zhou. Adaptive control in Heter-RAID system. International Conference on Machine Learning and Cybernetics, pp.842-845, 4-5 Nov. 2002.

Modeling Efficient XOR-Based Hash Functions for Cache Memories[*]

Sung-Jin Cho[1], Un-Sook Choi[2], Yoon-Hee Hwang[3], and Han-Doo Kim[4]

[1] Division of Mathematical Sciences,
Pukyong National University, Busan 608-737, Korea
sjcho@pknu.ac.kr
[2] Department of Multimedia Engineering, Tongmyoung University
Busan 608-711, Korea
choies@pknu.ac.kr
[3] Department of Information Security, Graduate School,
Pukyong National University, Busan 608-737, Korea
yhhwang@pknu.ac.kr
[4] School of Computer Aided Science, Inje University
Gimhae 621-749, Korea
mathkhd@inje.ac.kr

Abstract. In this paper, we design new XOR-based hash functions, which compute each set index bit as XOR of a subset of the bits in the address. These are conflict-free hash functions which are different types according to m is even or odd.

1 Introduction

Hash functions are used in processors to augment the bandwidth of an interleaved multibank memory or to enhance the utilization of a prediction table or a cache [1]. Bank conflicts can severely reduce the bandwidth of an interleaved multibank memory and conflict misses increase the miss rate of a cache. Therefore it is important that a hash function has to succeed in spreading the most frequently occurring patterns over all indices. Vandierendonck et al.[2] constructed XOR-based hash functions which provided conflict-free mapping for a number of patterns for multibank memories and skewed-associative caches. Their functions map $2m$ bits to $m(= 2k)$ bits which are conflict free. But they didn't construct hash functions which are conflict free when m is odd. Also they constructed two XOR-based hash functions for skewed-associative caches. But the degree of interbank dispersion between two hash functions is less than $2m$, which is the maximum degree between them. So they changed the basis of one hash function to overcome this problem.

In this paper, we design new XOR-based hash functions, which compute each set index bit as XOR of a subset of the bits in the address by using the concepts of rank and null space ([3], [4]). These are conflict-free hash functions which are

[*] This work was supported by KOSEF:R01-2006-000-10260-0.

V.N. Alexandrov et al. (Eds.): ICCS 2006, Part I, LNCS 3991, pp. 1067–1070, 2006.

different types according to m is even or odd. To apply the constructed hash functions to the skewed-associative cache, we show that the degree of interbank dispersion between two hash functions is maximal.

2   Modeling Efficient XOR-Based Hash Functions

A hash function is a function from $\{0,\cdots,2^n-1\}(:=\mathbf{A})$ of n-bit addresses to $\{0,\cdots,2^m-1\}(:=\mathbf{S})$ of m-bit indices $(m < n)$. An n-bit address $\mathbf{a}$ is represented by a bit vector $a_1,\cdots,a_n$. A hash function mapping n bits to m bits is represented as a binary matrix H with n rows and m columns. Since H is surjective, the image of $\mathbf{A}$ is $\mathbf{S}$. Therefore $rank(H) = dim(\mathbf{S}) = m$, where $dim(\mathbf{S})$ is the dimension of $\mathbf{S}$. The bit on the i-th row and the j-th column is 1 when address bit a_i is an input to the XOR-gate computing the j-th set index bit. Consequently, the computation of the set index $\mathbf{s}$ can be expressed as the vector-matrix multiplication over $GF(2)$, where addition is computed as XOR and multiplication is computed as logical AND, denoted by $\mathbf{s} = \mathbf{a}H$.

Since matrices are considered as linear transformations, they can be characterized by their null spaces. The null space of a matrix H is the set of all addresses which map to index $\mathbf{0}$, namely $N(H) = \{\mathbf{x} : \mathbf{x}H = \mathbf{0}\}$.

In this section we model efficient XOR-based hash functions when the number of address bits is $2m$, where m is even or odd. We propose the new model of XOR-based hash functions for two cases. We give XOR-based hash functions by the following.

Definition 2.1. We define XOR-based hash functions of four types as the following:

(i) $m = 2k - 1$ $(k \in \mathbf{N})$, where $\mathbf{N}$ is the set of all positive integers.
$H_1 = (T_1, I_{2k-1})^t$, $H_3 = (T_3, I_{2k-1})^t$, where I_{2k-1} is the identity matrix,

$$T_1 = (t_{ij})_{(2k-1)\times(2k-1)}, \quad t_{ij} = \begin{cases} 1, & \text{if } 1 \leq i \leq k, 1 \leq j \leq k - i + 1, \\ 1, & \text{if } 1 \leq i \leq 2k - 1, j = 2k - i, \\ 0, & \text{otherwise .} \end{cases}$$

$$T_3 = (t_{ij})_{(2k-1)\times(2k-1)}, \quad t_{ij} = \begin{cases} 1, & \text{if } i = j = 1, \\ 1, & \text{if } 1 \leq i \leq 2k - 1, j = 2k - i, \\ 1, & \text{if } k \leq i \leq 2k - 1, 3k - i - 1 \leq j \leq 2k - 1, \\ 0, & \text{otherwise .} \end{cases}$$

(ii) $m = 2k$ $(k \in \mathbf{N})$

$H_2 = (T_2, I_{2k})^t$, $H_4 = (T_4, I_{2k})^t$, where I_{2k} is the identity matrix,

$$T_2 = (t_{ij})_{(2k)\times(2k)}, \quad t_{ij} = \begin{cases} 1, & \text{if } 1 \leq i \leq k, i \leq j \leq k, \\ 1, & \text{if } 1 \leq i \leq 2k - 1, j = 2k - i, \\ 0, & \text{otherwise .} \end{cases}$$

$$T_4 = (t_{ij})_{(2k)\times(2k)}, \quad t_{ij} = \begin{cases} 1, & \text{if } k + 1 \leq i \leq 2k, k + 1 \leq j \leq i, \\ 1, & \text{if } 1 \leq i \leq 2k - 1, j = 2k - i, \\ 0, & \text{otherwise .} \end{cases}$$

Theorem 2.2. (i) All $m \times m$ matrices $T_i's$ in Definition 2.1 are nonsingular, where $i = 1, 2, 3, 4$.

(ii) Each $T_i \oplus I$ is nonsingular, where T_i is in Definition 2.1.

Theorem 2.3. Let $H = (T, I_m)^t$be the hash function in Definition 2.1, where $T = (\mathbf{t_1}, \mathbf{t_2}, \cdots, \mathbf{t_m})^t$ and $I_m = (\mathbf{e_1}, \mathbf{e_2}, \cdots, \mathbf{e_m})^t$. And let

$$M_{ij} = (\mathbf{t_{m+1-i}}, \cdots, \mathbf{t_{m-1}}, \mathbf{t_m}, \mathbf{e_{m+1-j}}, \cdots, \mathbf{e_m})^t$$

where $i + j = m$. Here A^t is the transpose of A. Then $rank(M_{ij}) = m$.

By Theorems 2.2 and 2.3, the modeled XOR-based hash functions such as the functions defined in Definition 2.1 map the patterns(rows, columns, (anti)diagonal, and rectangles) without conflicting.

3   Conflict-Free Mapping in 2-Way Skewed-Associative Caches

In this section, we construct XOR-based hash functions for a 2-way skewed-associative cache of the same size that maps the same patterns conflict-free.

Definition 3.1. We define the degree of interbank dispersion(DID) between two XOR-based hash functions H_1, H_3 by $2m \times m$

$$DID(H_1, H_3) = rank[H_1\ H_3]$$

		Value of H_1						
		0	1	2	3	4	5	6	7
Value of H_3	0	0:0	5:2	4:5	1:7	7:1	2:3	3:4	6:6
	1	5:3	0:1	1:6	4:4	2:2	7:0	6:7	3:5
	2	4:7	1:5	0:2	5:0	3:6	6:4	7:3	2:1
	3	1:4	4:6	5:1	0:3	6:5	3:7	2:0	7:2
	4	7:5	2:7	3:0	6:2	0:4	5:6	4:1	1:3
	5	2:6	7:4	6:3	3:1	5:7	0:5	1:2	4:0
	6	3:2	6:0	7:7	2:5	4:3	1:1	0:6	5:4
	7	6:1	3:3	2:4	7:6	1:0	4:2	5:5	0:7

Fig. 1. Illustration of H_1 and H_3 and interbank dispersion

The hash functions of a 2-way skewed-associative cache should be designed such that the DID is maximal for every pair of hash functions. Vandierendonck et al.[2] defined the DID between two hash functions H_1 and H_3 by using the concepts of supremum(infimum) of two functions and the concept of the dimension of column space. But we defined the DID by using only the concept of the rank of the augmented matrix $[H_1\ H_3]$. Every address in main memory is mapped to a set in bank 1 by H_1 and to a set in bank 2 by H_3. These functions are illustrated in a two-dimensional plot(Fig. 1). Each axis is labeled with the possible set indices for that bank. Every address is displayed in the grid in a position that corresponds to its set indices in each bank. For two vectors x and y in the same row, $xH_3 = yH_3$ but $xH_1 \neq yH_1$. Similarly, for two vectors u and v in the same column, $uH_1 = vH_1$ but $uH_3 \neq vH_3$. Therefore this skewed-associative cache can avoid conflict by H_1 and H_3. Fig. 1 shows that the DID of H_1 and H_3 is maximal. Also the element $(5 : 3)$ represents $(101\ 011)$.

The following theorem characterize the maximality of the DID of H_1 and H_3.

Theorem 3.3. The degree of interbank dispersion between H_1 and H_3 is maximal if and only if $N([H_1\ H_3]) = \{\mathbf{0}\}$.

Lemma 3.4. Let T_1 and T_3 be matrices in Definition 3.1, where $m = 2k-1 (k \geq 3)$. Then $rank[T_1 \oplus T_3] = m$.

The following theorem shows that the DID between the proposed hash functions H_1 and H_3 is maximal. We can see that these functions map $2m$-bit address to m-bit set index without conflict for a 2-way skewed-associative cache.

Theorem 3.5. Let $H_1 = (T_1, I_m)^t$ and $H_3 = (T_3, I_m)^t$ for T_1 and T_3 are matrices in Definition 2.1, where $m = 2k - 1 (k \geq 3)$. Then $rank[H_1\ H_3] = 2m$.

Corollary 3.6. Let $H_1 = (T_1, I_m)^t$ and $H_3 = (T_3, I_m)^t$ for T_1 and T_3 are matrices in Definition 2.1, where $m = 2k - 1 (k \geq 3)$. Then $N([H_1\ H_3]) = \{\mathbf{0}\}$.

We can show that Theorem 3.5 and Corollary 3.6 hold for H_2 and H_4, where $m = 2k$.

4   Conclusion

In this paper we designed new XOR-based hash functions by using the concepts of rank and null space. We constructed conflict-free hash functions which are different types according to m is even or odd. We showed that the DID between the proposed hash functions is maximal. By the result we showed that these functions map without conflict for a 2-way skewed-associative cache.

References

1. A. Seznec, A New Case for Skewed-Associativity, Technical Report PI-1114, IRISH (1997)
2. Hans Vandierendonck and Koen De Bosschere, XOR-Based Hash Functions, *IEEE Trans. Computers* **54** (2005) 800-812
3. S.J. Cho, U.S. Choi, Y.H. Hwang, Y.S. Pyo, H.D. Kim and S.H. Heo, Computing Phase Shifts of Maximum-Length 90/150 Cellular Automata Sequences, *Lecture Notes in Computer Science* **3305** (2004) 31-39
4. S.J. Cho, U.S. Choi and H.D. Kim, Behavior of Complemented CA whose Complement Vector is Acyclic in a Linear TPMACA, *Mathematical and Computer Modelling* **36** (2002) 979-986

Maintaining Gaussian Mixture Models of Data Streams Under Block Evolution

J.P. Patist, W. Kowalczyk, and E. Marchiori

Free University of Amsterdam, Department of Computer Science,
Amsterdam, The Netherlands
`{jpp, wojtek, elena}@cs.vu.nl`

Abstract. A new method for maintaining a Gaussian mixture model of a data
stream that arrives in blocks is presented. The method constructs local Gaussian
mixtures for each block of data and iteratively merges pairs of closest components. Time and space complexity analysis of the presented approach demonstrates that it is 1-2 orders of magnitude more efficient than the standard *EM*
algorithm, both in terms of required memory and runtime.

1 Introduction

The emergence of new applications involving massive data sets such as customer click
streams, telephone records, or electronic transactions, stimulated development of new
algorithms for analysing massive streams of data, [2, 4].

In this paper we address the issue of maintenance of a Gaussian mixture model of a
data stream under block evolution, see [5], where the modeled data set is updated periodically through insertion and deletion of sets of blocks of records. More specifically,
we consider block evolution with a restricted window consisting of a fixed number of
the most recently collected blocks of data. The window is updated one block at a time
by inserting a new block and deleting the oldest one. Gaussian mixture models may be
viewed as an attractive form of data clustering.

Recently, several algorithms have been proposed for clustering data streams, see e.g.,
[1], [6], or [8]. In our approach, we apply the classical *EM* algorithm, [3], to generate
local mixture models for each block of data and a greedy merge procedure to combine
these local models into a global one. This leads to a dramatic reduction of the required
storage and runtime by 1-2 orders of magnitude.

2 Maintenance of Gaussian Mixture Models

A Gaussian mixture model with k components is a probability distribution on $\mathcal{R}^d$
that is given by a convex combination $p(x) = \sum_{s=1}^{k} \alpha_s p(x|s)$ of k Gaussian density
functions:

$$p(x|s) = (2\pi)^{-d/2}|\Sigma_s|^{-1/2}\exp(-(x - \mu_s)^\top \Sigma_s^{-1}(x - \mu_s)/2), \ \ s = 1, 2, \ldots, k,$$

each of them being specified by its mean vector μ_s and the covariance matrix Σ_s.

V.N. Alexandrov et al. (Eds.): ICCS 2006, Part I, LNCS 3991, pp. 1071–1074, 2006.

Given a set $\{x_1, \ldots, x_n\}$ of points from $\mathcal{R}^d$, the learning task is to estimate the parameter vector $\theta = \{\alpha_s, \mu_s, \Sigma_s\}_{s=1}^k$ that maximizes the log-likelihood function $L(\theta) = \sum_{i=1}^n \log p(x_i; \theta)$. Maximization of the data log-likelihood $L(\theta)$ is usually achieved by running the *Expectation Maximization (EM)* algorithm [3]. For fixed values of d and k, the time and space complexity of this algorithm is linear in n.

Let us suppose that data points arrive in blocks of equal size and that we are interested in maintaining a mixture model for the most recent b blocks of data. An obvious, but expensive solution to this problem would be to re-run the *EM* algorithm after arrival of each block of data. Unfortunately, for huge data sets this method could be too slow.

In our approach, b local mixtures are maintained, one for each block. Mixtures are stored as lists of components. When a new block of data arrives, all components from the oldest block are removed and the *EM* procedure is applied to the latest block to find a local mixture model for this block. Finally, all bk local components are combined with help of a greedy merge procedure to form a global model with k components.

Two Gaussian components $(\mu_1, \Sigma_1, \alpha_1)$, $(\mu_2, \Sigma_2, \alpha_2)$ are merged into one component (μ, Σ, α) using the following formulas:

$$\mu = \frac{\alpha_1\mu_1 + \alpha_2\mu_2}{\alpha_1 + \alpha_2}, \Sigma = \frac{\alpha_1\Sigma_1 + \alpha_2\Sigma_2 + \alpha_1\mu_1\mu_1^T + \alpha_2\mu_2\mu_2^T}{\alpha_1 + \alpha_2} - \mu\mu^T, \alpha = \alpha_1 + \alpha_2.$$

The greedy merge procedure systematically searches for two closest components and merges them with help of the above formulas until there are exactly k components left.

The distance between components is measured with help of the Hotelling T^2 statistic, [7], which is used for testing whether the sample mean μ is equal to a given vector μ_0:

$$H^2(\mu, \alpha, \Sigma, \mu_0) = \alpha(\mu - \mu_0)^T \Sigma^{-1}(\mu - \mu_0).$$

The Hotelling distance between components C_1 and C_2 is then defined as follows:

$$dist_H(C_1, C_2) = (H^2(\mu_1, \alpha_1, \Sigma_1, \mu_2) + H^2(\mu_2, \alpha_2, \Sigma_2, \mu_1))/2.$$

Let us note that for a fixed value of d the merging process requires $O((bk)^2)$ steps and for large values of b and k it may be prohibitively slow. Therefore, we propose a more efficient method, called *k-means Greedy*, which is a combination of the k-means clustering and the greedy search. The combined algorithm reduces the initial number of components from bk to l, where $bk >> l > k$, with help the k-means clustering procedure and then the greedy search is used to reduce the number of components from l to k. A further speed-up can be achieved by using the Euclidean distance measure applied to $\mu's$ in the "k-means phase", rather than the expensive Hotelling statistic.

3   Space and Time Complexity Analysis

The main advantage of the proposed model construction and maintenance technique is the reduction of the required memory. Instead of storing the last n points, we store only the last block of data with up to n/b points and b local models. Now we will analyze the relation between the values of n (window size), d (data dimensionality), k (the number of components; the same for both local and the global model), b (the number of blocks), and the memory reduction rate. We measure memory size in terms of the number of stored values.

When modeling data with multi-dimensional Gaussian distributions it is common to consider two models: a general one, with no restrictions on the covariance matrix Σ (other than being symmetric an positive definite), or a restricted one, where the covariance matrix is supposed to be diagonal. Let us first consider the case with full covariance matrices. Every component involves $1 + d + d(d+1)/2$ parameters (1 for prior, d for μ, and $d(d+1)/2$ for Σ), thus the total number of values that have to be stored, $M(b)$, is $bk(1 + d + d(d+1)/2) + dn/b$.

In order to find the optimal value of b that minimizes the above expression let us notice that the function $f(x) = \alpha x + \beta/x$, where $\alpha, \beta, x > 0$, reaches the minimum $2\sqrt{\alpha\beta}$ for $x = \sqrt{\beta/\alpha}$. Therefore, the optimal number of blocks, b_{opt}, is given by $b_{opt} = \sqrt{nd/k(1 + d + d(d+1)/2)}$, thus $M(b_{opt}) = 2\sqrt{ndk(1 + d + d(d+1)/2)}$. Hence, the optimal memory reduction rate, $R(b_{opt})$, satisfies:

$$R(b_{opt}) = nd/\sqrt{2ndk(1 + d + d(d+1))}.$$

In the case of diagonal covariance matrices similar reasoning gives:

$$R(b_{opt}) = nd/2\sqrt{ndk(1 + 2d)}.$$

To get a better insight into the relation between the compression rate and other parameters we produced two plots for a fixed $n = 50.000$, $d = 10$, and k ranging between 1 and 10, and $b = 1, 2 \ldots, 250$, see Figure 1.

Memory reduction rate is largest for small values of k: for $k = 2$ this rate is about 30-80 while for $k = 10$ it drops to 13-25, depending on d and the model type.

Finally, let us notice that the optimal number of blocks can be interpreted as the time speed-up factor. Indeed, for fixed values of d and k, the time complexity of the *EM*

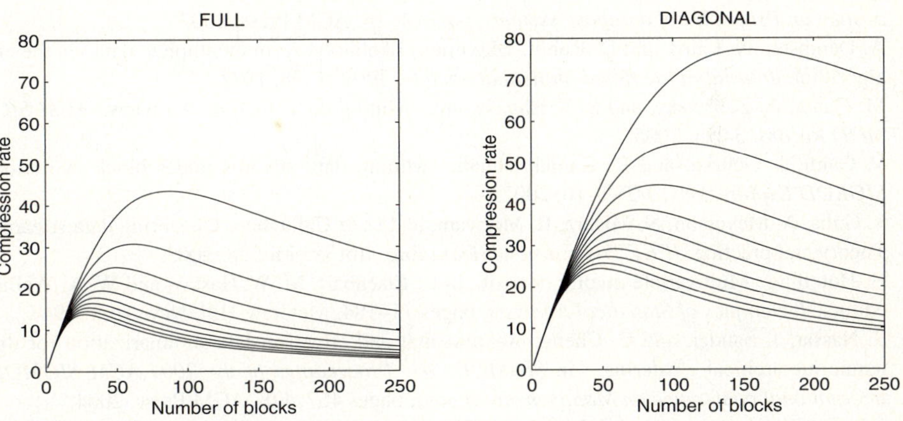

Fig. 1. The compression rate as a function of the number of blocks and the number of mixture components, $d = 10$. Each curve corresponds to another value of $k = 1, 2, \ldots, 10$. Lower curves correspond to larger values of k. The window size $n = 50.000$.

algorithm is linear in n. Therefore, the run time of this algorithm on a block of data of size n/b is b times smaller than when applied to the whole data set of n points. The additional time that is needed for updating the global model depends on the merging strategy and is $O((bk)^2)$ in case of greedy search, and $O(bk)$ when a combination of k-means and greedy search is applied (provided the value of parameter l is small enough, i.e., $l < \sqrt{bk}$). Taking into account that both k and b are relatively small compared to n, the influence of this factor on the overall run-time of the algorithm may be neglected. In practice, in case of relatively small (but realistic) values of k and d the speed-up factor ranges between 30-150 times.

4   Conclusions and Future Work

We presented a local approach for maintaining a Gaussian mixture model over a data stream under block evolution with restricted window. The proposed approach is 1-2 orders of magnitude more efficient than the standard *EM* algorithm, both in terms of required memory and runtime.

In our future research we would like to address three issues: 1) impact of our heuristic approach on the accuracy of models (some initial results are already reported in [9]), 2) dynamic identification of the optimal number of components (so far we assumed this number was known in advance and fixed), and 3) incremental modeling of mixtures of non-Gaussian distributions, e.g., mixtures of multinomials.

References

1. C. C. Aggarwal, J. Han, J. Wang, and P. Yu. A framework for clustering evolving data streams. In *VLDB*, pages 81–92, 2003.
2. B. Babcock, S. Babu, M. Datar, R. Motwani, and J. Widom. Models and issues in data stream systems. In *PODS '02: Proceedings of the twenty-first ACM SIGMOD-SIGACT-SIGART symposium on Principles of database systems*, pages 1–16. ACM Press, 2002.
3. A. Dempster, N. Laird, and D. Rubin. Maximum likelihood from incomplete data via the em algorithm. *Journal of the Royal Statistical Society*, 39(B):1–38, 1977.
4. M. Gaber, A. Zaslavsky, and S. Krishnaswamy. Mining data streams: A review. *ACM SIGMOD Record*, 34(1), 2005.
5. V. Ganti, J. Gehrke, and R. Ramakrishnan. Mining data streams under block evolution. *SIGKDD Explorations*, 3(2):1–10, 2002.
6. S. Guha, A. Meyerson, N. Mishra, R. Motwani, and L. O'Callaghan. Clustering data streams: Theory and practice. *IEEE Trans. Knowl. Data Eng.*, 15(3):515–528, 2003.
7. H. Hotelling. Multivariate quality control. In C. Eisenhart, M. W. Hastay, and W. A. Wallis, editors, *Techniques of Statistical Analysis*, pages 11–184. McGraw-Hill, New York, 1947.
8. S. Nassar, J. Sander, and C. Cheng. Incremental and effective data summarization for dynamic hierarchical clustering. In *SIGMOD '04: Proceedings of the 2004 ACM SIGMOD international conference on Management of data*, pages 467–478. ACM Press, 2004.
9. J. Patist, W. Kowalczyk, and E. Marchiori. Efficient Maintenance of Gaussian Mixture Models for Data Streams. *http://www.cs.vu.nl/~jpp/GaussianMixtures.pdf*, Technical Report, Vrije Universiteit Amsterdam, 2005.

An Adaptive Data Retrieval Scheme for Reducing Energy Consumption in Mirrored Video Servers

Minseok Song

School of Computer Science and Engineering,
Inha University, Korea
`mssong@inha.ac.kr`

Abstract. We propose a new energy-aware data retrieval scheme (EDR) for mirrored video servers. We start by analytically determining the retrieval period that balances bandwidth and buffer size. We then propose a data retrieval scheme in which the period can be dynamically changed to reflect server utilization, with the aim of giving disks more chance to enter low-power mode. Simulation results show that it saves up to 36% energy compared with conventional video server operation.

1 Introduction

Today's data centers typically consume a lot of power. Large power consumption is a serious economic problem to service providers [5]. For example, a medium-sized 30,000 ft^2 data centers requires 15 MW which costs \$ 13 million per year [5]. Cooling systems for higher heat densities are prohibitively expensive and running them makes up a significant proportion of the power cost of running a data center [2, 3]. In addition, it has been shown that 15 °C above ambient can double the failure rate of a disk drive [2].

To reduce power consumption in servers, modern disks have multiple power modes [2, 5]: in active mode the platters are spinning and the head is seeking, reading or writing data, in idle mode the disk spins at full speed but there is no disk request, and in low-power mode the disks stops spinning completely and consumes much less energy than active or idle mode. Most of the schemes for energy conservation in data centers involve switching disks to low-power modes whenever that is possible without adversely affecting performance [2, 5]. These schemes primarily aim to extend the period during which a disk is in low-power mode. Because returning from low-power to active mode requires spinning up the disk, the energy saved by putting the disk into low-power mode needs to be greater than the energy needed to spin it up again. But this concept is hardly relevant to video servers due to the long duration of video streams.

To address this, we propose an energy-aware data retrieval (EDR) scheme for mirrored video servers. Based on an analytical model for the optimal data retrieval size, we propose a data placement scheme in which an appropriate block size can be dynamically selected and a data retrieval scheme which adaptively retrieves data from the primary and backup copies to permit disks to go to low-power mode as often as possible.

The rest of this paper is organized as follows. We explain video server model in Section 2. We then propose a new data retrieval scheme in Section 3. We validate the proposed scheme through simulations in Section 4 and conclude the paper in Section 5.

V.N. Alexandrov et al. (Eds.): ICCS 2006, Part I, LNCS 3991, pp. 1075–1078, 2006.

2   Video Server Model

We use round-based scheduling for video data retrieval: time is divided into equal-sized periods, called rounds, and each client is served once in each round. We partition a video object into blocks and distribute them over multiple disks. The size of the striping unit denoted as the maximum amount of contiguous data stored on a single disk is the amount of data retrieved during a round so that only one seek overhead is incurred in each data retrieval [1]. Data retrieved during the period R of a round are grouped together and constitute a segment. Then each segment is stored in a round-robin fashion across disks.

For fault-tolerance, we use a replication technique where the original data is duplicated on separate disks. We refer to the original data as the primary copy (PMC) and call the duplicated data the backup copy (BCC). The server reads data from the PMC when all disks are operational; but when a disk fails, it uses the BCC for data retrieval. Among various replication schemes, chained declustering (CD) shows the best performance [4]. Let us assume that a disk array consists of D homogeneous disks. In the CD scheme, the primary copy on disk i has a backup copy on disk $(i+1)$ mod D. We place the backup copy on one disk as in the CD scheme.

3   Energy-Aware Data Retrieval

The round length plays an important role in determining the requirements for server resources (i.e. disk bandwidth and buffer) [1, 4]. We now describe the optimal round length that balances the requirements for disk bandwidth and buffer. Let us assume that each video V_i has data rate of dr_i (in bits/sec), $(i = 1, ..., N)$. Let p_i be the access probability of video V_i, where $\sum_{i=1}^{N} p_i = 1$. Let B be the total buffer size. We use SCAN scheduling. Since double buffering is used for SCAN scheduling [1], the buffer utilization BU for c clients is expected as follows:

$$BU = \sum_{i=1}^{N} \frac{2R \times p_i \times dr_i}{B} c. \tag{1}$$

The bandwidth utilization for a disk is usually defined as the ratio of total service time to R [1]. We use a typical seek time model in which a constant seeking overhead (seek time + rotational delay) of T_s is required for one read of contiguous data [1]. Let tr be the data transfer rate of the disk. Since there are D disks in the server, the disk bandwidth utilization DU for c clients can be estimated as follows:

$$DU = \frac{1}{D} \sum_{i=1}^{N} p_i \Big(\frac{T_s + R \times \frac{dr_i}{tr}}{R} \Big) c. \tag{2}$$

If $DU = BU$, the server is able to balance the use of buffer and disk bandwidth. From Equations (1) and (2), we obtain the optimal round length OR as follows:

$$OR = \frac{\frac{B}{tr} + B\sqrt{\frac{1}{tr^2} + \frac{8 \times D \times T_s}{B \times \sum_{i=1}^{N} p_i \, dr_i}}}{4 \times D}. \tag{3}$$

To allow disks to go to low-power mode, EDR permits dynamic data retrieval from either the PMC or the BCC. Let CDU_i be the bandwidth utilization of disk i. We define two states: if $\forall i\ CDU_i \leq 0.5$, then the server is in *energy reduction (ER)* state; otherwise, the server is in *normal* state. In the normal state, data is retrieved from the PMC on every disk; while in the ER state, data is retrieved only from odd-numbered disks. Since the data on disk i is replicated as a backup copy on disk $(i + 1) \bmod D$, the disk loads incurred in reading the PMC of disks $2i$ are shifted to the BCC of disks $(2i + 1) \bmod D$ ($i = 1, ..., \lfloor \frac{1}{2}D \rfloor$). The disks $2i$ are able to go to low-power mode because they are not accessed. Even though the disk loads of disks $2i$ are shifted to disks $(2i+1) \bmod D$, the utilization of disk $(2i + 1) \bmod D$ does not exceed 1 because $\forall i$, $CDU_i \leq 0.5$.

If $\forall i$, $CDU_i \leq 0.5$, even-numbered disks go to low-power mode, and so reducing the disk bandwidth utilization below 0.5 is important for energy reduction. We observe from Equation (2) that extending the round length reduces disk bandwidth utilization; but this increases the buffer overhead, by Equation (1). EDR has *common*, *major* and *minor* cycles, with lengths of $3OR$, OR and $\frac{3}{2}OR$, respectively. From Equation (3), we observe that a cycle of length OR leads to the balanced use of disk bandwidth and buffers. Thus, choosing the major cycle results in the balanced use of disk bandwidth and buffers. But by selecting the minor cycle for the round length we reduce the energy consumption because a long round is advantageous in terms of disk throughput.

To extend the time spent in low-power mode, the server adaptively adjusts the round length. Assuming that all data is retrieved from the PMC, EDR calculates ADU, the maximum disk bandwidth utilization among all disks when the major cycle is selected, and IDU, the maximum bandwidth utilization when the minor cycle is chosen. Additionally, let ABU and IBU be the buffer utilizations when major and minor cycles are selected, respectively. Use of the major cycle may produce the normal state, in circumstances under which the minor cycle would permit the system to go to ER state. As a consequence, if possible, EDR tries to select the minor cycle as the round length for energy reduction. But the server needs to check the buffer condition (i.e. that $IBU \leq 1$) because if that constraint is not met, using the minor cycle may lead to buffer overflow. If $IBU \leq 1$ and $IDU \leq 0.5$, EDR selects the minor cycle as the round length; otherwise, the major cycle is chosen. Because the minor cycle is only chosen if $IDU \leq 0.5$, we can easily see that *the server is in ER state if the minor cycle is chosen*.

The change of round length should not cause additional seek overhead. To meet this constraint, we split a data segment into 6 sub-segments, where the size of each sub-segment corresponds to the data retrieved during a round of length $\frac{1}{2}OR$. The 6 sub-segments are stored contiguously and constitute a segment. In normal state, two contiguous subsegments are read during OR, but in ER state, the server retrieves three contiguous subsegments during $\frac{3}{2}OR$.

4   Experimental Results

To evaluate the effectiveness of our scheme, we performed simulations. Our server has 16 disks, each of which employs an IBM Ultrastar36Z15 disk whose parameters are shown in [5]. We choose 12 ms for T_s and 1 GB for B. The arrival of client requests is assumed to follow a Poisson distribution. We also assume that all videos are 90 minutes

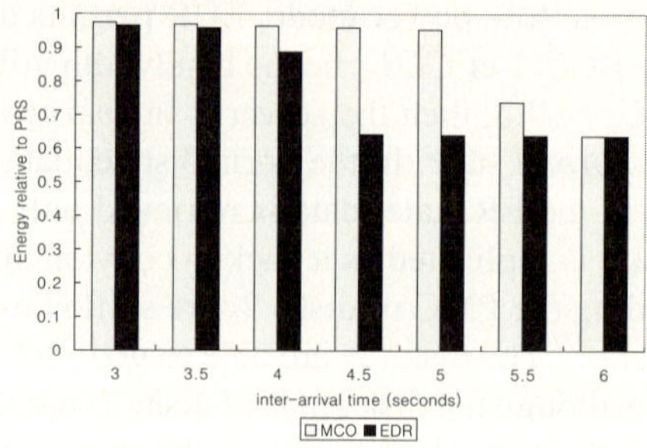

Fig. 1. Energy consumption relative to PRS for various inter-arrival times

long and have the same bit-rate of 1MB/sec. We compare EDR with two other methods; the PRS scheme that does not allow data retrieval from the BCC, and the MCO scheme that permits data retrieval from the BCC but does not allow adaptive cycle adjustment. We assess how the energy consumption depends on the inter-arrival time of requests over 24 hours. Fig. 1 shows the energy consumption of MCO and EDR schemes relative to PRS. The EDR scheme exhibits the best performance under all workloads, saving between 4% to 36% over the conventional PRS scheme. Compared with MCO, EDR reduces energy consumption by up to 33%. This is because by adaptively increasing the round length from the major to the minor cycle, EDR reduces the disk bandwidth utilization, which results in more opportunities to stay in low-power mode.

5   Conclusions

We have proposed a new energy-aware data retrieval scheme for mirrored video servers. An analytical model shows how to balance the use of disk bandwidth and buffer and based on this, we have proposed a data retrieval scheme which adaptively retrieves data from the primary and backup copies to give the server more chances to operate in low-power mode. Experimental results show that our scheme enables the server to achieve appreciable energy savings under a range of workloads.

References

1. E. Chang. *Storage and Retrieval of Compressed Video.* PhD thesis, University of California at Berkeley, 1996.
2. S. Gurumurthi. *Power Management of Enterprise Storage Systems.* PhD thesis, Pennsylvania State University, 2005.
3. E. Pinheiro and R. Bianchini. Energy conservation techniques for disk-array-based servers. In *Proceedings of ACM/IEEE Conference on Supercomputing*, pages 88–95, June 2004.
4. M. Song and H. Shin. Replica striping for multi-resolution video servers. In *Proceedings of IDMS/PROMS (LNCS 2515)*, pages 300–312, November 2002.
5. Q. Zhu and Y. Zhou. Power aware storage cache management. *IEEE Transactions on Computers*, 54(5):587–602, 2005.

Author Index

Lecture Notes in Computer Science

For information about Vols. 1–3906

please contact your bookseller or Springer

Vol. 4011: Y. Sure, J. Domingue (Eds.), The Semantic Web: Research and Applications. XVIII, 723 pages. 2006.

Vol. 4007: C. Àlvarez, M. Serna (Eds.), Experimental and Efficient Algorithms. XI, 329 pages. 2006.

Vol. 4004: S. Vaudenay (Ed.), Advances in Cryptology - EUROCRYPT 2006. XIV, 613 pages. 2006.

Vol. 4003: Y. Koucheryavy, J. Harju, V.B. Iversen (Eds.), Next Generation Teletraffic and Wired/Wireless Advanced Networking. XVI, 582 pages. 2006.

Vol. 3998: T. Calamoneri, I. Finocchi, G.F. Italiano (Eds.), Algorithms and Complexity. XII, 394 pages. 2006.

Vol. 3994: V.N. Alexandrov, G.D. van Albada, P.M.A. Sloot, J. Dongarra (Eds.), Computational Science – ICCS 2006, Part IV. XXXV, 1096 pages. 2006.

Vol. 3993: V.N. Alexandrov, G.D. van Albada, P.M.A. Sloot, J. Dongarra (Eds.), Computational Science – ICCS 2006, Part III. XXXVI, 1136 pages. 2006.

Vol. 3992: V.N. Alexandrov, G.D. van Albada, P.M.A. Sloot, J. Dongarra (Eds.), Computational Science – ICCS 2006, Part II. XXXV, 1122 pages. 2006.

Vol. 3991: V.N. Alexandrov, G.D. van Albada, P.M.A. Sloot, J. Dongarra (Eds.), Computational Science – ICCS 2006, Part I. LXXXI, 1096 pages. 2006.

Vol. 3990: J. C. Beck, B.M. Smith (Eds.), Integration of AI and OR Techniques in Constraint Programming for Combinatorial Optimization Problems. X, 301 pages. 2006.

Vol. 3987: M. Hazas, J. Krumm, T. Strang (Eds.), Location- and Context-Awareness. X, 289 pages. 2006.

Vol. 3986: K. Stølen, W.H. Winsborough, F. Martinelli, F. Massacci (Eds.), Trust Management. XIV, 474 pages. 2006.

Vol. 3984: M. Gavrilova, O. Gervasi, V. Kumar, C.J. K. Tan, D. Taniar, A. Laganà, Y. Mun, H. Choo (Eds.), Computational Science and Its Applications - ICCSA 2006, Part V. XXV, 1045 pages. 2006.

Vol. 3983: M. Gavrilova, O. Gervasi, V. Kumar, C.J. K. Tan, D. Taniar, A. Laganà, Y. Mun, H. Choo (Eds.), Computational Science and Its Applications - ICCSA 2006, Part IV. XXVI, 1191 pages. 2006.

Vol. 3982: M. Gavrilova, O. Gervasi, V. Kumar, C.J. K. Tan, D. Taniar, A. Laganà, Y. Mun, H. Choo (Eds.), Computational Science and Its Applications - ICCSA 2006, Part III. XXV, 1243 pages. 2006.

Vol. 3981: M. Gavrilova, O. Gervasi, V. Kumar, C.J. K. Tan, D. Taniar, A. Laganà, Y. Mun, H. Choo (Eds.), Computational Science and Its Applications - ICCSA 2006, Part II. XXVI, 1255 pages. 2006.

Vol. 3980: M. Gavrilova, O. Gervasi, V. Kumar, C.J. K. Tan, D. Taniar, A. Laganà, Y. Mun, H. Choo (Eds.), Computational Science and Its Applications - ICCSA 2006, Part I. LXXV, 1199 pages. 2006.

Vol. 3979: T.S. Huang, N. Sebe, M.S. Lew, V. Pavlović, M. Kölsch, A. Galata, B. Kisačanin (Eds.), Computer Vision in Human-Computer Interaction. XII, 121 pages. 2006.

Vol. 3978: B. Hnich, M. Carlsson, F. Fages, F. Rossi (Eds.), Recent Advances in Constraints. VIII, 179 pages. 2006. (Sublibrary LNAI).

Vol. 3976: F. Boavida, T. Plagemann, B. Stiller, C. Westphal, E. Monteiro (Eds.), Networking 2006. Networking Technologies, Services, and Protocols; Performance of Computer and Communication Networks; Mobile and Wireless Communications Systems. XXVI, 1276 pages. 2006.

Vol. 3975: S. Mehrotra, D.D. Zeng, H. Chen, B. Thuraisingham, F.-Y. Wang (Eds.), Intelligence and Security Informatics. XXII, 772 pages. 2006.

Vol. 3973: J. Wang, Z. Yi, J.M. Zurada, B.-L. Lu, H. Yin (Eds.), Advances in Neural Networks - ISNN 2006, Part III. XXIX, 1402 pages. 2006.

Vol. 3972: J. Wang, Z. Yi, J.M. Zurada, B.-L. Lu, H. Yin (Eds.), Advances in Neural Networks - ISNN 2006, Part II. XXVII, 1444 pages. 2006.

Vol. 3971: J. Wang, Z. Yi, J.M. Zurada, B.-L. Lu, H. Yin (Eds.), Advances in Neural Networks - ISNN 2006, Part I. LXVII, 1442 pages. 2006.

Vol. 3970: T. Braun, G. Carle, S. Fahmy, Y. Koucheryavy (Eds.), Wired/Wireless Internet Communications. XIV, 350 pages. 2006.

Vol. 3968: K.P. Fishkin, B. Schiele, P. Nixon, A. Quigley (Eds.), Pervasive Computing. XV, 402 pages. 2006.

Vol. 3967: D. Grigoriev, J. Harrison, E.A. Hirsch (Eds.), Computer Science – Theory and Applications. XVI, 684 pages. 2006.

Vol. 3966: Q. Wang, D. Pfahl, D.M. Raffo, P. Wernick (Eds.), Software Process Change. XIV, 356 pages. 2006.

Vol. 3965: M. Bernardo, A. Cimatti (Eds.), Formal Methods for Hardware Verification. VII, 243 pages. 2006.

Vol. 3964: M. Ü. Uyar, A.Y. Duale, M.A. Fecko (Eds.), Testing of Communicating Systems. XI, 373 pages. 2006.

Vol. 3962: W. IJsselsteijn, Y. de Kort, C. Midden, B. Eggen, E. van den Hoven (Eds.), Persuasive Technology. XII, 216 pages. 2006.

Vol. 3960: R. Vieira, P. Quaresma, M.d.G.V. Nunes, N.J. Mamede, C. Oliveira, M.C. Dias (Eds.), Computational Processing of the Portuguese Language. XII, 274 pages. 2006. (Sublibrary LNAI).

Vol. 3959: J.-Y. Cai, S. B. Cooper, A. Li (Eds.), Theory and Applications of Models of Computation. XV, 794 pages. 2006.

Vol. 3958: M. Yung, Y. Dodis, A. Kiayias, T. Malkin (Eds.), Public Key Cryptography - PKC 2006. XIV, 543 pages. 2006.

Vol. 3956: G. Barthe, B. Gregoire, M. Huisman, J.-L. Lanet (Eds.), Construction and Analysis of Safe, Secure, and Interoperable Smart Devices. IX, 175 pages. 2006.

Vol. 3955: G. Antoniou, G. Potamias, C. Spyropoulos, D. Plexousakis (Eds.), Advances in Artificial Intelligence. XVII, 611 pages. 2006. (Sublibrary LNAI).

Vol. 3954: A. Leonardis, H. Bischof, A. Pinz (Eds.), Computer Vision – ECCV 2006, Part IV. XVII, 613 pages. 2006.

Vol. 3953: A. Leonardis, H. Bischof, A. Pinz (Eds.), Computer Vision – ECCV 2006, Part III. XVII, 649 pages. 2006.

Vol. 3952: A. Leonardis, H. Bischof, A. Pinz (Eds.), Computer Vision – ECCV 2006, Part II. XVII, 661 pages. 2006.

Vol. 3951: A. Leonardis, H. Bischof, A. Pinz (Eds.), Computer Vision – ECCV 2006, Part I. XXXV, 639 pages. 2006.

Vol. 3950: J.P. Müller, F. Zambonelli (Eds.), Agent-Oriented Software Engineering VI. XVI, 249 pages. 2006.

Vol. 3947: Y.-C. Chung, J.E. Moreira (Eds.), Advances in Grid and Pervasive Computing. XXI, 667 pages. 2006.

Vol. 3946: T.R. Roth-Berghofer, S. Schulz, D.B. Leake (Eds.), Modeling and Retrieval of Context. XI, 149 pages. 2006. (Sublibrary LNAI).

Vol. 3945: M. Hagiya, P. Wadler (Eds.), Functional and Logic Programming. X, 295 pages. 2006.

Vol. 3944: J. Quiñonero-Candela, I. Dagan, B. Magnini, F. d'Alché-Buc (Eds.), Machine Learning Challenges. XIII, 462 pages. 2006. (Sublibrary LNAI).

Vol. 3943: N. Guelfi, A. Savidis (Eds.), Rapid Integration of Software Engineering Techniques. X, 289 pages. 2006.

Vol. 3942: Z. Pan, R. Aylett, H. Diener, X. Jin, S. Göbel, L. Li (Eds.), Technologies for E-Learning and Digital Entertainment. XXV, 1396 pages. 2006.

Vol. 3941: S.W. Gilroy, M.D. Harrison (Eds.), Interactive Systems. XI, 267 pages. 2006.

Vol. 3940: C. Saunders, M. Grobelnik, S. Gunn, J. Shawe-Taylor (Eds.), Subspace, Latent Structure and Feature Selection. X, 209 pages. 2006.

Vol. 3939: C. Priami, L. Cardelli, S. Emmott (Eds.), Transactions on Computational Systems Biology IV. VII, 141 pages. 2006. (Sublibrary LNBI).

Vol. 3936: M. Lalmas, A. MacFarlane, S. Rüger, A. Tombros, T. Tsikrika, A. Yavlinsky (Eds.), Advances in Information Retrieval. XIX, 584 pages. 2006.

Vol. 3935: D. Won, S. Kim (Eds.), Information Security and Cryptology - ICISC 2005. XIV, 458 pages. 2006.

Vol. 3934: J.A. Clark, R.F. Paige, F.A. C. Polack, P.J. Brooke (Eds.), Security in Pervasive Computing. X, 243 pages. 2006.

Vol. 3933: F. Bonchi, J.-F. Boulicaut (Eds.), Knowledge Discovery in Inductive Databases. VIII, 251 pages. 2006.

Vol. 3931: B. Apolloni, M. Marinaro, G. Nicosia, R. Tagliaferri (Eds.), Neural Nets. XIII, 370 pages. 2006.

Vol. 3930: D.S. Yeung, Z.-Q. Liu, X.-Z. Wang, H. Yan (Eds.), Advances in Machine Learning and Cybernetics. XXI, 1110 pages. 2006. (Sublibrary LNAI).

Vol. 3929: W. MacCaull, M. Winter, I. Düntsch (Eds.), Relational Methods in Computer Science. VIII, 263 pages. 2006.

Vol. 3928: J. Domingo-Ferrer, J. Posegga, D. Schreckling (Eds.), Smart Card Research and Advanced Applications. XI, 359 pages. 2006.

Vol. 3927: J. Hespanha, A. Tiwari (Eds.), Hybrid Systems: Computation and Control. XII, 584 pages. 2006.

Vol. 3925: A. Valmari (Ed.), Model Checking Software. X, 307 pages. 2006.

Vol. 3924: P. Sestoft (Ed.), Programming Languages and Systems. XII, 343 pages. 2006.

Vol. 3923: A. Mycroft, A. Zeller (Eds.), Compiler Construction. XIII, 277 pages. 2006.

Vol. 3922: L. Baresi, R. Heckel (Eds.), Fundamental Approaches to Software Engineering. XIII, 427 pages. 2006.

Vol. 3921: L. Aceto, A. Ingólfsdóttir (Eds.), Foundations of Software Science and Computation Structures. XV, 447 pages. 2006.

Vol. 3920: H. Hermanns, J. Palsberg (Eds.), Tools and Algorithms for the Construction and Analysis of Systems. XIV, 506 pages. 2006.

Vol. 3918: W.K. Ng, M. Kitsuregawa, J. Li, K. Chang (Eds.), Advances in Knowledge Discovery and Data Mining. XXIV, 879 pages. 2006. (Sublibrary LNAI).

Vol. 3917: H. Chen, F.-Y. Wang, C.C. Yang, D. Zeng, M. Chau, K. Chang (Eds.), Intelligence and Security Informatics. XII, 186 pages. 2006.

Vol. 3916: J. Li, Q. Yang, A.-H. Tan (Eds.), Data Mining for Biomedical Applications. VIII, 155 pages. 2006. (Sublibrary LNBI).

Vol. 3915: R. Nayak, M.J. Zaki (Eds.), Knowledge Discovery from XML Documents. VIII, 105 pages. 2006.

Vol. 3914: A. Garcia, R. Choren, C. Lucena, P. Giorgini, T. Holvoet, A. Romanovsky (Eds.), Software Engineering for Multi-Agent Systems IV. XIV, 255 pages. 2006.

Vol. 3911: R. Wyrzykowski, J. Dongarra, N. Meyer, J. Waśniewski (Eds.), Parallel Processing and Applied Mathematics. XXIII, 1126 pages. 2006.

Vol. 3910: S.A. Brueckner, G.D.M. Serugendo, D. Hales, F. Zambonelli (Eds.), Engineering Self-Organising Systems. XII, 245 pages. 2006. (Sublibrary LNAI).

Vol. 3909: A. Apostolico, C. Guerra, S. Istrail, P. Pevzner, M. Waterman (Eds.), Research in Computational Molecular Biology. XVII, 612 pages. 2006. (Sublibrary LNBI).

Vol. 3908: A. Bui, M. Bui, T. Böhme, H. Unger (Eds.), Innovative Internet Community Systems. VIII, 207 pages. 2006.

Vol. 3907: F. Rothlauf, J. Branke, S. Cagnoni, E. Costa, C. Cotta, R. Drechsler, E. Lutton, P. Machado, J.H. Moore, J. Romero, G.D. Smith, G. Squillero, H. Takagi (Eds.), Applications of Evolutionary Computing. XXIV, 813 pages. 2006.